Sixth Edition

Heating and Cooling ESSENTIALS

Don Crawshaw

Jerry Killinger

LaDonna Killinger

Gregory D. Smith

Publisher

The Goodheart-Willcox Company, Inc.

Tinley Park, IL

www.g-w.com

Library of Congress Control Number: 2025936154

ISBN 979-8-89737-160-0

1 2 3 4 5 6 7 8 9 – 27 – 30 29 28 27 26 25

The Goodheart-Willcox Company, Inc. Brand Disclaimer: Brand names, company names, and illustrations for products and services included in this text are provided for educational purposes only and do not represent or imply endorsement or recommendation by the author or the publisher.

The Goodheart-Willcox Company, Inc. Safety Notice: The reader is expressly advised to carefully read, understand, and apply all safety precautions and warnings described in this book or that might also be indicated in undertaking the activities and exercises described herein to minimize risk of personal injury or injury to others. Common sense and good judgment should also be exercised and applied to help avoid all potential hazards. The reader should always refer to the appropriate manufacturer's technical information, directions, and recommendations; then proceed with care to follow specific equipment operating instructions. The reader should understand these notices and cautions are not exhaustive.

The publisher makes no warranty or representation whatsoever, either expressed or implied, including but not limited to equipment, procedures, and applications described or referred to herein, their quality, performance, merchantability, or fitness for a particular purpose. The publisher assumes no responsibility for any changes, errors, or omissions in this book. The publisher specifically disclaims any liability whatsoever, including any direct, indirect, incidental, consequential, special, or exemplary damages resulting, in whole or in part, from the reader's use or reliance upon the information, instructions, procedures, warnings, cautions, applications, or other matter contained in this book. The publisher assumes no responsibility for the activities of the reader.

The Goodheart-Willcox Company, Inc. Internet Disclaimer: The Internet resources and listings in this Goodheart-Willcox Publisher product are provided solely as a convenience to you. These resources and listings were reviewed at the time of publication to provide you with accurate, safe, and appropriate information. Goodheart-Willcox Publisher has no control over the referenced websites and, due to the dynamic nature of the Internet, is not responsible or liable for the content, products, or performance of links to other websites or resources. Goodheart-Willcox Publisher makes no representation, either expressed or implied, regarding the content of these websites, and such references do not constitute an endorsement or recommendation of the information or content presented. It is your responsibility to take all protective measures to guard against inappropriate content, viruses, or other destructive elements.

Image Credits. Front cover (Left to right): Fieldpiece; narai chal/Shutterstock.com

Preface

Heating and Cooling Essentials is a thorough, easily understood, and up-to-date textbook for those entering the refrigeration, heating, and air-conditioning field, or HVACR. The text is a practical blend of theory, on-the-job skills-building procedures, and customer service soft-skill training. It is designed for use by first-year students in HVACR programs at career centers, technical schools, adult continuing education classes, or community colleges. This text may also be used in apprenticeship programs and can serve as an excellent reference for working technicians.

Heating and Cooling Essentials is written from a service technician's point of view, but the service experience is appropriate for the introductory HVACR student. The ability to combine on-the-job skills with theory is a must to become a successful service technician. The content of this textbook provides a solid foundation in the fundamentals and integrates practical service procedures with technical theory. The book is a complete and accurate guide to performing essential service procedures on many types of systems.

An important aspect of *Heating and Cooling Essentials* is its emphasis on career information and workplace skills. Chapter 1 covers an overview of the HVACR industry, explains why HVACR is a good career option, and discusses different career paths in the HVACR industry. Also discussed are some of the physical, cognitive, communication, and technical skills needed to be successful in the HVACR industry.

A textbook is not a storybook that is read only once. It is designed to be read over and over again. Just because you master a chapter today does not mean that you will remember everything about that chapter later in your training or even when you enter the workforce. Refresh your knowledge by rereading chapters already covered even after completing your education.

This textbook will provide you with a solid foundation for further study of more specialized aspects of the HVACR field. Remember that to be successful means not only mastering your hands-on-skills but also understanding operational theory and having great customer soft skills.

About the Authors

Donald (Don) L. Crawshaw is the former HVAC Program Coordinator at Salt Lake Community College in Salt Lake City, UT. Prior to his position at Salt Lake Community College, Mr. Crawshaw served as Department Chairman and Instructor of the HVAC program at Pikes Peak Community College (PPCC) in Colorado Springs, CO. Mr. Crawshaw has over 30 years of experience in the HVACR field, having worked in the areas of installation; service and energy management for residential, commercial, and supermarket systems; and has achieved CMHE (Certified Master HVACR Educator) certification from HVAC Excellence.

Gregory (Greg) Dean Smith Jr. has worked in the HVACR industry since 2002 but returned to school to earn his Florida state-certified HVAC contractor's license in 2015. After earning his certification, he ran a successful HVAC company until being offered a position at Locklin Technical College in Milton, FL, where he taught secondary and postsecondary HVAC students. Since starting at Locklin, Mr. Smith has earned awards for CTE Rookie of the Year, Teacher of the Year, the Zach Butler award, and was named one of the HVAC Excellence Top 25 HVAC instructors in 2025. Mr. Smith is working towards his Certified Master HVACR Educator certificate, already earning subject matter expert certifications in Electrical Theory, Air-Conditioning, and Teaching Methodologies.

Mr. Smith currently serves as the HVACR Department head for Locklin Technical College where he supervises the postsecondary HVAC program as well as the HVAC apprenticeship program. He also continues to teach adult postsecondary HVAC courses.

Jerry Killinger and **LaDonna Killinger** established this text with excellent content and organization. Mr. Killinger had extensive experience installing and servicing air conditioning, refrigeration, and heating systems. He worked with industrial, commercial, and residential systems ranging from heat pumps to boilers to cooling towers to refrigerated food cases to solid-state environmental control equipment. Jerry spent over 20 years teaching traditional classes as well as various seminars and workshops on refrigeration, air conditioning, and electricity for industrial maintenance personnel at facilities of major manufacturers across the United States.

Reviewers

The author and publisher wish to thank the following industry and teaching professionals for their valuable input into the development of *Heating and Cooling Essentials*.

Ezequiel Gonzalez
Eastern Iowa Community College
Davenport, IA

Patrick Heeb
Long Beach City College
Long Beach, CA

William Highley
Columbus State Community College
Columbus, OH

Cary Hogg
Central Georgia Technical College
Robins, GA

Mark Kohnen
Kaskaskia College
Centralia, IL

Gary Larson
Ozarks Technical Community College
Springfield, MO

Barbara MacQueen
Vancouver Island University–Cowichan Campus
Duncan, BC, Canada

Dr. Christopher Molnar
Porter and Chester Institute
Stratford, CT

Benjamin Nusz
Mid-State Technical College–Wisconsin Rapids Campus
Wisconsin Rapids, WI

Lawrence P. Penar
Oakton Community College
Des Plaines, IL

R. Jeff Polk
Central New Mexico Community College
Albuquerque, NM

Robert Recko
Quinsigamond Community College
Worcester, MA

T. Jason Taylor
Cleveland State Community College
Cleveland, TN

Chad Vlietstra
Prairie State College
Chicago Heights, IL

Freddie Williams
Lanier Technical College
Gainesville, GA

Acknowledgments

The author and publisher would like to thank the following companies, organizations, and individuals for their contribution of resource material, images, or other support in the development of *Heating and Cooling Essentials.*

Aeroquip Corp.
Alco
Alfa Laval Inc.
Allied Signal
American Electric Co.
American Tool Companies, Inc.
Amprobe
Amtrol
Bacharach, Inc.
Baltimore Aircoil Company
Beckett
Big Blu
Bitzer
Bohn Heat Transfer
Browning
Calumet Refining Co.
Carlin Combustion Technology, Inc.
Carrier Corporation
ClimateMaster
CooperTools
Copeland Corp.
Cutler-Hammer
Danfoss
Delavan
DiversiTech Corporation
Dunham-Bush
DuPont
Elf Atochem
Emerson Climate Technologies
The Energy Conservatory
ESCO Group
Fasco
Fastenal
Fieldpiece
Fluke
Frigidaire
Fuller Co.
Fujitsu General America, Inc.
Goodman Manufacturing Company, L.P.
Grasslin Controls Corporation
Halstead
Hartford Compressors, Inc.
H.B. Fuller Co.
Henry Valve Co.
Hilmor
Home Heating Systems Inc.
Honeywell
Hussmann Corporation
Imperial Eastman
ITW Ramset
JB Industries
Johnson Controls, Inc.
Johnstone Supply
John Walker
J. W. Harris Co., Inc.
Klein Tools, Inc.
Kramer
LaRoche Chemicals
Lau
Lennox International

LoLo Building Products
Lucas-Milhaupt
Mac Tools, Inc.
Malco Products, Inc.
Manchester Tank
The Marley Cooling Tower Company
Mastercool Inc.
Maytag Corporation
McQuay International
Metal Products
Milwaukee Electric Tool Corp.
Mitsubishi Electric, HVAC Advanced Products Division
Mobil Oil Corporation
MSA – The Safety Company
Mueller Industries, Inc.
National Refrigerants
Nibco
NuTone, Inc.
Paragon Electric Co.
Penn
Pikes Peak Community College
Prest-O-Lite
Ranco
RB&W Corporation
RCD Corporation
Reed Manufacturing Co.
Refrigeration Technologies
Rheem Manufacturing Co.
The Ridge Tool Company
Ritchie Engineering – YELLOW JACKET Products Division
Robinair Mfg. Corp.
Sealed Unit Parts Co., Inc.
Simpson Electric Co.
Smith Equipment, Division of Tescom Corp.
Spectronics Corporation
Sporlan Division–Parker Hannifin Corporation
Standard Refrigeration Company
Star Expansion Company
State of Ohio
Suntec Industries, Inc.
Tecumseh
Thermal Engineering
Thermodyne
TIF Instruments, Inc.
Trane
Triumph Twist Drill Co.
United Refrigeration
Uniweld
Vaco Products
Van Steenburgh Engineering Laboratories, Inc.
Vermont American
Vilter Mfg. Co.
Virginia KMP
Watsco
Westermeyer Industries, Inc.
Westwood Products, Inc.
White-Rodgers Division, Emerson Climate Technologies
Worthington Enterprises
York
Zero Zone, Inc.

Goodheart-Willcox HVACR Advisory Board

Goodheart-Willcox would like to thank the past and current members of the Goodheart-Willcox HVACR Advisory Board who have generously shared their insight, thoughts, and expertise in the development of our HVACR products. Their commitment to HVACR education inspires us. The following are the HVACR Advisory Board members at the time of publication of this edition of *Heating and Cooling Essentials*:

Jerry Alderman, Ivy Tech Community College, IN
David Blais, Ivy Tech Community College, IN
Dan Bracciano, Goodheart-Willcox author, FCA LLC, MI
Bill Brown, Brownson Technical, CA
David Feutz, Grand Rapids Community College, MI
Steven Gutsch, Chippewa Valley Technical College, WI
Samuel Heath, Southern Technical College, FL
Mark Kohnen, Kaskaskia College, IL
Fred Kosier, Chippewa Valley Technical College, WI
John Labriola, West Virginia Northern Community College, WV
Stephen Manbeck, Pennsylvania College of Technology
Robert McNabb, Northshore Technical Community College, LA
William Ramsey, W.R. Bristow Manufacturers' Agents, Inc, GA
Richard Ruscigno, Universal Technical Institute, NJ
Gregory Smith, Radford M. Locklin Technical College, FL
Travis Southerland, Southern Tech, OK
Don Steeby, Grand Rapids Community College, MI

New to This Edition

The sixth edition of *Heating and Cooling Essentials* has been updated to reflect key changes in the HVACR industry, improve the instructional design, consolidate the book's focus on fundamental theory and skills, and enhance chapter review questions and lab activities in the Lab Workbook.

- New Chapter 3, *Safety*, covers essential safety guidelines and industry hazards relevant to introductory HVACR technicians.
- Updated Chapter 15, *Refrigerants*, includes a thorough introduction of refrigerants. This chapter was updated to integrate zeotropic refrigerants and A2L refrigerants.

- New Procedure features throughout the text highlight step-by-step instruction for key processes performed by HVACR technicians.
- All learning outcomes, summary items, and end-of-chapter review questions are aligned with the corresponding section in the chapter, allowing for instructor flexibility in presenting and assigning textbook content in the classroom.
- A heavily-updated Lab Workbook now includes chapter review questions for assessing comprehension of chapter content and over 70 lab activities for hands-on practice of key procedures and skills required for entry-level HVACR technicians.
- Updated digital assets include HVACR videos and animations with related assessments and a new HVACR Virtual Toolbox.

Credentialing Partners and Support

Goodheart-Willcox appreciates the value of industry credentials, certifications, and accreditation. We are pleased to partner with leading organizations to support students and programs in achieving credentials. Integrating industry-recognized credentialing into a career and technical education (CTE) program provides many benefits for the student and for the institution. By achieving third-party certificates, students gain confidence, have proof of a measurable level of knowledge and skills, and earn a valuable achievement to include in their résumés. For educators and administrators, industry-recognized credentials and accreditation validate learning, enhance the credibility of programs, and provide valuable data to measure student performance and help guide continuous program improvement.

Curriculum Correlations

Heating and Cooling Essentials aligns with curriculum standards for HVAC Excellence and American Council of Construction Education (ACCE) accreditation.

HVAC Excellence is a not-for-profit organization that serves the HVACR industry with the goal of supporting and improving HVACR education and training. HVAC Excellence provides many services to HVACR education and training, including awarding program accreditation, professional certifications, and instructor credentials.

The American Council of Construction Education (ACCE) is a not-for-profit organization with the goal of promoting and improving construction education. ACCE's HVACR Industry Certification Program (formerly PAHRA) incorporates industry-validated curriculum standards developed by the Air-Conditioning, Heating, and Refrigeration Institute (AHRI).

To see how *Heating and Cooling Essentials* correlates to HVAC Excellence and ACCE standards, visit the Correlations tab at www.g-w.com/heating-cooling-essentials-2027. For more information on HVAC Excellence and ACCE, please visit their websites at www.hvacexcellence.org and www.acce-hq.org.

Features of the Textbook

The instructional design of this textbook includes student-focused learning tools to help you succeed. This visual guide highlights these features.

Chapter Opening Materials

Each chapter opener contains a list of learning outcomes, a list of technical terms, and an introduction to the chapter. **Learning Outcomes** clearly identify the knowledge and skills to be gained when the chapter is completed. These outcomes are labeled and correlated to related assessment questions to facilitate reporting and remediation. **Technical Terms** list the key words to be learned in the chapter. **Introduction** provides an overview and preview of the chapter content.

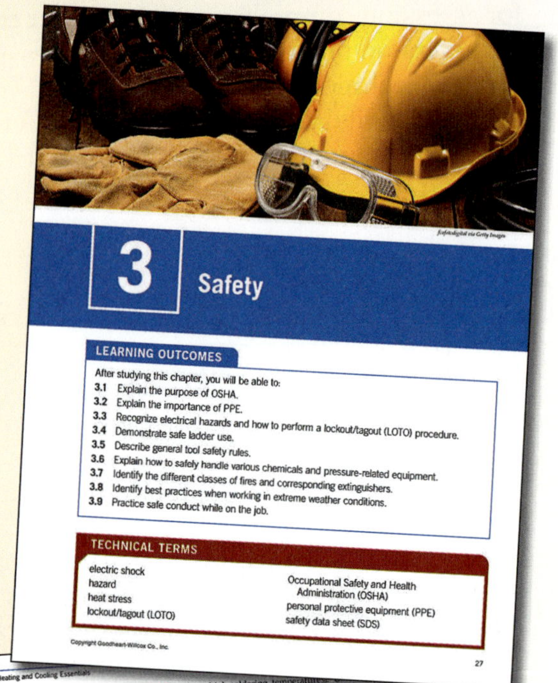
3 Safety

LEARNING OUTCOMES

After studying this chapter, you will be able to:

3.1 Explain the purpose of OSHA.
3.2 Explain the importance of PPE.
3.3 Recognize electrical hazards and how to perform a lockout/tagout (LOTO) procedure.
3.4 Demonstrate safe ladder use.
3.5 Describe general tool safety rules.
3.6 Explain how to safely handle various chemicals and pressure-related equipment.
3.7 Identify the different classes of fires and corresponding extinguishers.
3.8 Identify best practices when working in extreme weather conditions.
3.9 Practice safe conduct while on the job.

TECHNICAL TERMS

electric shock
hazard
heat stress
lockout/tagout (LOTO)
Occupational Safety and Health Administration (OSHA)
personal protective equipment (PPE)
safety data sheet (SDS)

Copyright Goodheart-Willcox Co., Inc.

27

Additional Features

Additional features are used throughout the body of each chapter to further learning and knowledge. **Safety Notes** alert you to potentially dangerous materials and practices. **Cautions** alert you to practices that could potentially damage equipment or instruments. **Procedures** are highlighted throughout the textbook to provide step-by-step instruction for key processes performed by service technicians. **Pro Tips** provide advice and guidance that are especially applicable to on-the-job situations.

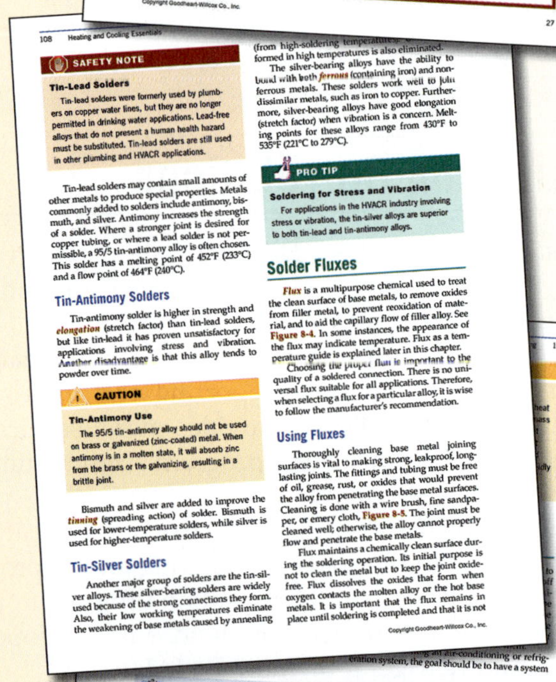
108 Heating and Cooling Essentials

SAFETY NOTE

Tin-Lead Solders

Tin-lead solders were formerly used by plumbers on copper water lines, but they are no longer permitted in drinking water applications. Lead-free alloys that do not present a human health hazard must be substituted. Tin-lead solders are still used in other plumbing and HVACR applications.

Tin-lead solders may contain small amounts of other metals to produce special properties. Metals commonly added to solders include antimony, bismuth, and silver. Antimony increases the strength of a solder. Where a stronger joint is desired for copper tubing, or where a lead solder is not permissible, a 95/5 tin-antimony alloy is often chosen. This solder has a melting point of 452°F (233°C) and a flow point of 464°F (240°C).

Tin-Antimony Solders

Tin-antimony solder is higher in strength and *elongation* (stretch factor) than tin-lead solders, but like tin-lead it has proven unsatisfactory for applications involving stress and vibration. Another disadvantage is that this alloy tends to powder over time.

CAUTION

Tin-Antimony Use

The 95/5 tin-antimony alloy should not be used on brass or galvanized (zinc-coated) metal. When antimony is in a molten state, it will absorb zinc from the brass or the galvanizing, resulting in a brittle joint.

Bismuth and silver are added to improve the *tinning* (spreading action) of solder. Bismuth is used for lower-temperature solders, while silver is used for higher-temperature solders.

Tin-Silver Solders

Another major group of solders are the tin-silver alloys. These silver-bearing solders are widely used because of the strong connections they form. Also, their low working temperatures eliminate the weakening of base metals caused by annealing (from high-soldering temperatures) ... formed in high temperatures is also eliminated.

The silver-bearing alloys have the ability to bond with both *ferrous* (containing iron) and nonferrous metals. These solders work well to join dissimilar metals, such as iron to copper. Furthermore, silver-bearing alloys have good elongation (stretch factor) when vibration is a concern. Melting points for these alloys range from 430°F to 535°F (221°C to 279°C).

PRO TIP

Soldering for Stress and Vibration

For applications in the HVACR industry involving stress or vibration, the tin-silver alloys are superior to both tin-lead and tin-antimony alloys.

Solder Fluxes

Flux is a multipurpose chemical used to treat the clean surface of base metals, to remove oxides from filler metal, to prevent reoxidation of material, and to aid the capillary flow of filler alloy. See Figure 8-4. In some instances, the appearance of the flux may indicate temperature. Flux as a temperature guide is explained later in this chapter.

Choosing the proper flux is important to the quality of a soldered connection. There is no universal flux suitable for all applications. Therefore, when selecting a flux for a particular alloy, it is wise to follow the manufacturer's recommendation.

Using Fluxes

Thoroughly cleaning base metal joining surfaces is vital to making strong, leakproof, long-lasting joints. The fittings and tubing must be free of oil, grease, rust, or oxides that would prevent the alloy from penetrating the base metal surfaces. Cleaning is done with a wire brush, fine sandpaper, or emery cloth, Figure 8-5. The joint must be cleaned well; otherwise, the alloy cannot properly flow and penetrate the base metals.

Flux maintains a chemically clean surface during the soldering operation. Its initial purpose is not to clean the metal but to keep the joint oxide-free. Flux dissolves the oxides that form when oxygen contacts the molten alloy or the hot base metals. It is important that the flux remains in place until soldering is completed and that it is not

Copyright Goodheart-Willcox Co., Inc.

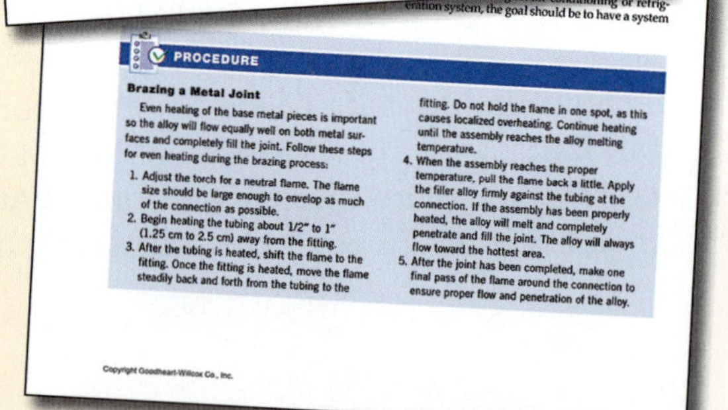
137

... air-conditioning or refrigeration system, the goal should be to have a system

PROCEDURE

Brazing a Metal Joint

Even heating of the base metal pieces is important so the alloy will flow equally well on both metal surfaces and completely fill the joint. Follow these steps for even heating during the brazing process:

1. Adjust the torch for a neutral flame. The flame size should be large enough to envelop as much of the connection as possible.
2. Begin heating the tubing about 1/2" to 1" (1.25 cm to 2.5 cm) away from the fitting.
3. After the tubing is heated, shift the flame to the fitting. Once the fitting is heated, move the flame steadily back and forth from the tubing to the fitting. Do not hold the flame in one spot, as this causes localized overheating. Continue heating until the assembly reaches the alloy melting temperature.
4. When the assembly reaches the proper temperature, pull the flame back a little. Apply the filler alloy firmly against the tubing at the connection. If the assembly has been properly heated, the alloy will melt and completely penetrate and fill the joint. The alloy will always flow toward the hottest area.
5. After the joint has been completed, make one final pass of the flame around the connection to ensure proper flow and penetration of the alloy.

Copyright Goodheart-Willcox Co., Inc.

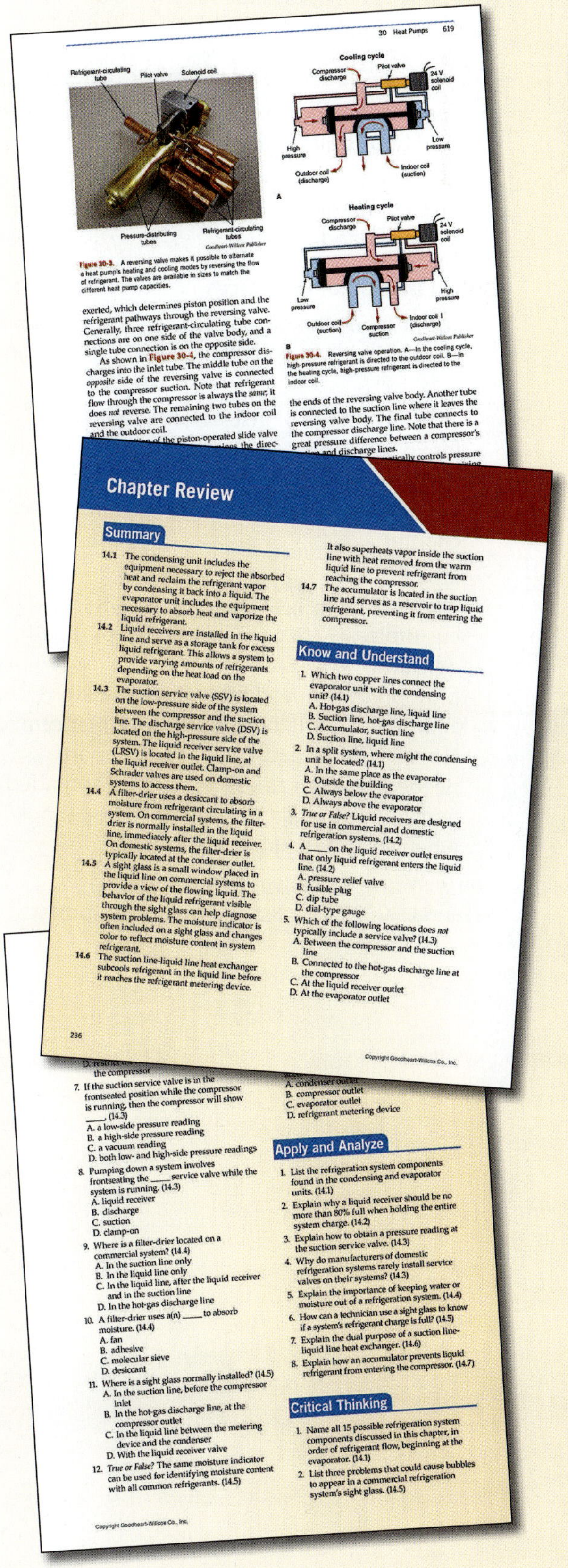

Illustrations

Updated **illustrations** are designed to clearly and simply communicate the specific topic. Photographic images have been updated to show the latest tools and equipment.

End-of-Chapter Content

End-of-chapter material provides an opportunity for review and application of concepts. A concise **Summary** provides an additional review tool and reinforces key learning outcomes. This helps you focus on important concepts presented in the text. **Know and Understand** questions enable you to demonstrate knowledge, identification, and comprehension of chapter material. **Apply and Analyze** questions extend learning and develop your abilities to use learned material in new situations and to break down material into its component parts. **Critical Thinking** questions develop higher order thinking and problem-solving abilities.

TOOLS FOR STUDENT AND INSTRUCTOR SUCCESS

Student Tools

Student Text

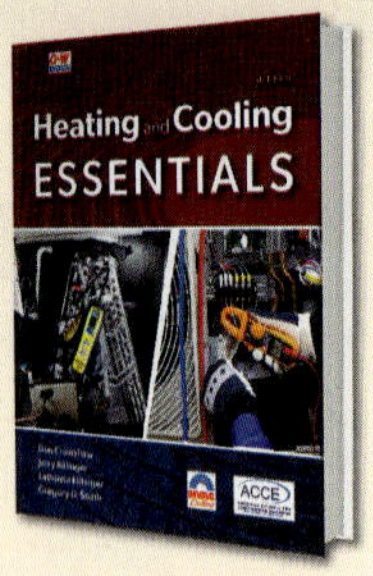

Available in print and digital, *Heating and Cooling Essentials* is an introductory text that focuses on theory, skills, and procedures used by professional HVACR service technicians.

Lab Workbook

- Minds-on practice includes chapter review questions. Hands-on practice includes lab activities designed to prepare technicians for procedures that will be performed in the field.
- Print and digital options are available.

G-W Digital Companion

- For digital users, e-flash cards, vocabulary exercises, and end-of-chapter Review Questions allow interaction with content to create opportunities to increase achievement.

EduHub LMS-Ready Content

- Students will enjoy intuitive digital resources in your school's preferred platform.
- The digital platform encourages application and discovery.
- HVACR videos, animations, and Follow-the-Heat™ animations provide students with visualizations of fundamental concepts essential for understanding in the HVACR trade.
- HVACR Troubleshooting Activities prepare students to be critical-thinking technicians who diagnose HVACR system issues.
- HVACR Virtual Toolbox provides students with e-flash cards, activities, and assessments to help students identify key tools and instruments used in the HVACR industry.

Instructor Tools

LMS Integration

Integrate EduHub LMS-Ready Content into your preferred platform for a seamless user experience for both you and your students.

A Complete Teaching and Learning Package

- The Digital Textbook and Online Lab Workbook build student knowledge.
- G-W Assessment gives students easy access to assignments, tests and quizzes, instant feedback, and grades.

Online Instructor Resources

- Tap into time-saving preparation tools: answer keys, editable lesson plans, and other teaching aids.
- Take advantage of Instructor's Presentations for PowerPoint® with fully customizable, richly illustrated slides that help you teach and visually reinforce the key concepts from each chapter.
- Easily create, customize, edit, and print prebuilt questions and tests using G-W Assessment, our cloud-based platform. No more downloads, uploads, or imports needed! Student assessments are autograded and flow automatically into your gradebook when you integrate with your LMS.

G-W Assessment
Assess learning • Gain insights • Build skills

See www.g-w.com/heating-cooling-essentials-2027 for a list of all available formats.

G-W EduHub®

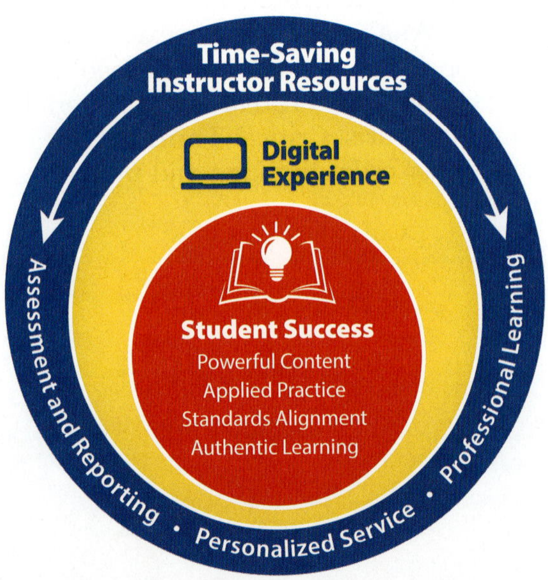

Brief Contents

Contents

Feature Contents

PROCEDURES

PeopleImages.com – Yuri A/Shutterstock.com

1 Careers in Heating, Ventilation, Air Conditioning, and Refrigeration

LEARNING OUTCOMES

After studying this chapter, you will be able to:

1.1 Recognize the employment outlook of the HVACR industry and some of the different HVACR career options.

1.2 List the physical, mental, communication, and technical skills needed to be successful in HVACR.

1.3 Explain how to begin a career in the HVACR industry and how to complete the job application process by preparing a résumé, cover letter, and for a job interview.

TECHNICAL TERMS

building engineer
commercial refrigeration technician
controls technician
counter sales representative
energy auditor
heating, ventilation, air conditioning, and refrigeration (HVACR)
installer
ladder diagram
mechanical skill
pictorial diagram
reference
résumé
service technician
sheet metal fabricator
technical training instructor

Introduction

Heating, ventilation, air conditioning, and refrigeration (HVACR) is an industry that affects the lives of everyone. From heating and cooling our homes and businesses to the refrigerators that keep food and medical products at proper temperatures, HVACR is all around us. For many years, people have used fire and ice to keep things warm and cool. In the 1800s, scientists began to discover how to use machinery and fluids like alcohol and ammonia to provide cooling. Throughout the 1900s, the process of refrigeration and air conditioning really took off. Some highlights of that time are as follows:

- 1902: Willis Carrier invents a way to condition the air for a New York publishing company. His invention was intended to help the printing process by controlling temperature and humidity. Some call Willis Carrier the creator of modern air conditioning.
- 1931: The invention of the window air conditioner by H.H. Schultz and J.Q. Sherman. However, these were initially extremely expensive.
- 1939: Packard Motor Car Company develops the first air-conditioned automobile.
- 1950s: Residential window air conditioners became affordable for the public to purchase.
- 1970s: Central air-conditioning systems began to replace window units.

Because of these achievements, we now enjoy our heated and cooled homes and the convenience of many HVACR appliances. But there are many other ways that the HVACR industry affects our lives. The following are some of those other ways that we do not often consider:

- Medical laboratories use refrigeration for blood, medicine, and research.
- Commercial refrigeration and ice machines are used in hotels, restaurants, and grocery stores.
- Water fountains provide chilled drinking water.
- Funeral parlors use refrigerated body boxes.
- Airplanes have climate-controlled cabins, making it possible for us to fly long distances at high altitude in comfort.
- Freight trucks and other shipping vessels with cooling units for the transportation of perishables, **Figure 1-1**.
- Vending machines provide readily consumable chilled or heated food products.

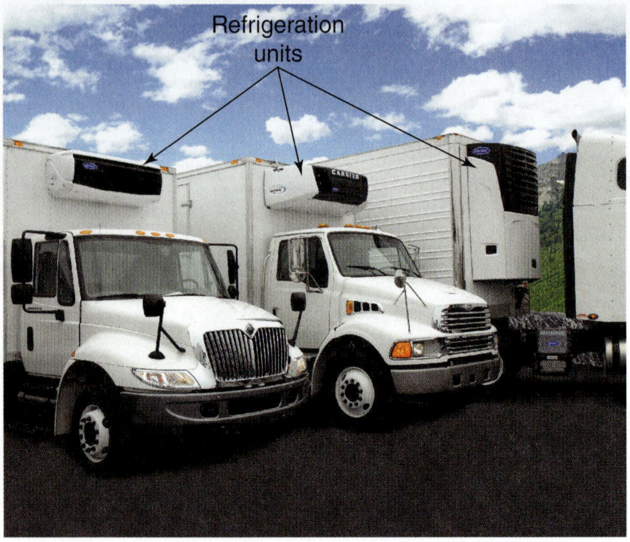

Carrier Corporation, Subsidiary of United Technologies Corp

Figure 1-1. Refrigeration units on trucks allow products to be preserved during transportation.

- Life-saving heating and cooling measures used in space travel.

It is hard to imagine a life without heating, air conditioning, or refrigeration equipment providing us with comfort and safety. In this chapter, many HVACR career paths are discussed, along with required skills needed to pursue a successful career in HVACR.

1.1 HVACR as a Career

HVACR is a vast industry with many career options to explore. This section will discuss the employment outlook in the HVACR industry, highlighting some of the diverse career paths available.

Employment Outlook

The industry demand for HVACR technicians and installers is rising. According to the US Bureau of Labor Statistics, employment of HVACR technicians and installers is expected to grow 9% from 2023 to 2033. Furthermore, many job openings are projected due to members of the HVACR industry that are retiring and will need to be replaced . With expanding businesses and a shrinking workforce, HVACR workers are in increasing demand.

Additionally, locations everywhere will always need professionally trained HVACR service technicians. Homes and businesses will always need heating and cooling; supermarkets and convenience stores will always need refrigeration. Heating, cooling, and refrigeration service can never be imported—meaning that there will always be a need for local HVACR technicians to install equipment and to keep it running properly. For more information on the occupational outlook of the HVACR trade, visit the US Bureau of Labor Statistics website.

HVACR Careers

It is a common misconception that the only two career options in the HVACR industry are the installer and the service technician. However, these are only two of many careers available to someone wanting to enter the HVACR workforce. There is a demand for qualified people in many distinct areas of the industry.

Nevertheless, the HVACR service technician and installer are two entry-level positions in which many HVACR workers choose to start their careers. HVACR residential ***service technicians*** regularly repair and maintain heating and cooling systems in single-family homes, **Figure 1-2**. In warm weather, they work on the air-conditioning system. In wintery weather, they work on the heating system. HVACR ***installers*** are responsible for installing HVAC systems, repairing or replacing faulty components, and performing maintenance tasks.

Counter sales representatives work behind the counter of a supply house and need to be knowledgeable about various parts and manufacturers, **Figure 1-3**. They can be a great resource to service technicians in the field.

Neil Lockhart/Shutterstock.com

Figure 1-2. A service technician taking pressure measurements with a manifold gauge on a residential condensing unit.

Due to the increasing concern of energy usage and conservation, the need is arising for ***energy auditors***. Their job is to inspect buildings to find where improvements can be made to make the building more energy efficient. Energy auditors often perform a blower door test and a thermographic scan to look for areas of energy leakage, **Figure 1-4**. They should be knowledgeable about building construction to properly perform an audit.

Image courtesy of Johnstone Supply

Figure 1-3. A counter sales representative at a distributor counter helping service technicians.

Photo courtesy of FLIR Systems, Inc.

Figure 1-4. Energy auditor using a thermal imaging camera while performing a blower door test. This test will help the auditor locate any air leaks in the home.

Technical training instructors provide students and technicians with the basic set of skills to work on a variety of systems, **Figure 1-5**. To be a complete service technician not only requires hands-on skills, but also a knowledge and understanding of electrical theory and the refrigeration cycle as well. These are best learned in a classroom training environment. Instructors and trainers can work for technical colleges, trade unions, or equipment manufacturers.

Sheet metal fabricators shape and manipulate sheet metal into a variety of useful products, **Figure 1-6**. They work on sheet metal with hand tools and machines both in the shop and out in the field.

Image courtesy of Pikes Peak Community College

Figure 1-5. A technical training instructor showing students how to troubleshoot a wiring control problem.

Image courtesy of Metal Products

Figure 1-6. Sheet metal fabricator bending a ductwork fitting. Sheet metal is commonly used for ductwork in a building's ventilation system.

Commercial refrigeration technicians are needed in both the supermarket and the food service field, **Figure 1-7**. Keeping our perishable food supplies preserved requires service technicians skilled in the supermarket field of commercial refrigeration. Their responsibility is to keep supermarket refrigeration equipment up and running properly. Commercial refrigeration technicians also work on refrigeration systems such as water chillers, ice machines, and walk-in coolers or freezers. Commercial refrigeration technicians may specialize in working on one particular type of system or many different systems.

With the increasing development of more sophisticated commercial automation controls, ***controls technicians*** have the responsibility of installing, programming, and servicing commercial HVAC building and supermarket equipment with automated control systems, **Figure 1-8**.

Another prominent position in the HVACR industry is the ***building engineer***. Building engineers are responsible for working in-house to maintain the HVAC equipment in commercial buildings like colleges, apartments, offices, hospitals, and factories. They usually work for the building owner or manager, **Figure 1-9**.

There are vast opportunities for growth in the HVACR industry, with many leadership positions available. Directors of maintenance and operations, supervisors, project managers, or lead workers are just a few of the leadership positions

Goodheart-Willcox Publisher

Figure 1-7. Commercial refrigeration technician working on a supermarket system.

Goodheart-Willcox Publisher

Figure 1-8. Controls technician working on a building automation system.

Goodheart-Willcox Publisher

Figure 1-9. Building engineer inspecting a chiller cooling system for a community college campus.

responsible for managing the work of other HVACR technicians. Leaders usually start out as service technicians and work their way up to these positions.

An industry as wide and diverse as HVACR has numerous specialty positions and others in supporting areas of the business. **Figure 1-10** lists some of the diverse types of jobs and work areas available. The HVACR industry is so varied that whatever your niche in life is, you can find a career path.

1.2 Job Skills

A variety of skills are necessary to work in HVACR. These skills can be broken down into the following:

- Physical skills
- Mental skills
- Communication skills
- Technical skills

Physical Skills

To work in the HVACR field, certain physical abilities and skills are required. Technicians should be able to lift large and sometimes heavy objects. An individual should be highly mobile and able to climb up stairs, ladders, and scaffolding. HVACR work may also require crawling and maneuvering into and out of tight spots, such as cramped mechanical closets, crawl spaces, and attics.

Careers in the HVACR Industry

Blueprint Creation/Reading	Government Regulation / Inspector	Purchasing
Consulting	Green Technology Specialist	Safety Director
Controls Technician	HVAC Equipment Dealer	Scheduler
Customer Service	IAQ Inspector	Sheet Metal Fabrication
Design Engineer	Installer	Stocker
Dispatcher/Coordinator	Manager	Supply Chain Specialist
Distributor	Manufacturer	Tech Support Specialist
Documentation Specialist	Marketing	Technical Training Instructor
Energy Auditor	Product Manager	Trade Organization Administration
Equipment Sales	Product Support	Truck and Shop Maintenance
Estimator	Product Test Engineer	Vending Machines
Field Supervisor	Programming	Vendor Management
Field Technical Specialist	Project Manager	

Goodheart-Willcox Publisher

Figure 1-10. Various jobs and occupational areas of HVACR work.

Technicians need to have endurance and stamina to work in hot, humid conditions during the warm season and in frigid, windy conditions during the cold season. Work may be indoors, outdoors, in the rain, in the snow, under the hot sun, or wherever the job requires.

Mental Skills

While physical skills are of vital importance to HVACR work, of equal importance are mental skills. A worker that cannot follow directions cannot do the job properly. HVACR technicians need to be able to read and understand manufacturer instructions. They also need to be able to take measurements and perform mathematical calculations. When diagnosing unfamiliar systems, technicians must be able to follow flowcharts or troubleshooting guides. They must be able to use their minds to critically think through a problem, using their troubleshooting skills so they are not wasting valuable time examining items that do not need to be examined.

Math and Measurements

When installing ductwork and various devices, technicians use math and geometry when sizing and cutting materials. They use rulers and tape measures for sizing parts and prior to drilling and cutting holes. A technician should be able to compare measurements, recognize if there are any inaccuracies, determine the reasons for the inaccuracies, and figure out the best solution.

When performing building heat load loss/gain calculations, a technician takes measurements of wall, window, door, ceiling, and floor dimensions to determine area. These values are then used in formulas for calculations. Formulas are an important part of the HVACR field, and a technician must be able to use these formulas. Some formulas deal with compression ratios, gas laws, Ohm's law, and heat transfer, **Figure 1-11**.

Technicians may have to calculate costs for job estimates and jobs completed, such as for installing a new furnace, replacing a compressor, or filling out a work order for a service call. A work order includes length of labor time, material costs,

Commonly Used HVAC Formulas and Conversions

Formulas	Conversions
Ohm's Law: E = I × R or I = E/R or R = E/I	1 Btu = the amount of heat needed to raise 1lb of water 1°F
Power Law (in watts): P = I × E	1 Ton of Refrigeration = 12,000 Btu/hr
CFM = Btu/hr (sensible) / (1.08 × Temperature)	28″ of WC = 1 psi
U Factor: 1 / R = U	°F = °C × 9/5 + 32
Sensible Heat: Btu/hr = 1.08 × CFM × DT	1 hp = 33,000 ft-lb. /1 minute
Boyle's Law: $P_1 \times V_1 = P_2 \times V_1$	1 hp = 746 Watts
Charles' Law: $P_1 / T_1 = P_2 / T_2$	1 watt = 3.413 Btu
General Gas Law: $(P_1 \times V_1) / T_1 = (P_2 \times V_2) / T_2$	°C = (°F – 32) × 5/9
Dalton's Law: $P_1 + P_2 + P_3 \ldots = P_T$	
Work = Force × Distance	
rpm of motor: (60 Hz × 120) / (# of poles)	
Total Heat: Btu/hr = 4.5 × CFM × Δh	
CFM = Area (sq. ft.) × Velocity (ft./min.)	
Round Duct Area = $(\pi)r^2$	
Net Oil Pressure = Oil Pump Pressure – Suction (Crankcase) Pressure	
Compression Ratio = Discharge Pressure Absolute / Suction Pressure Absolute	
Total Duct Pressure = Static Pressure + Velocity	
Rectangular Duct Area = L × W	

Goodheart-Willcox Publisher

Figure 1-11. A list of commonly used formulas and conversions in HVACR work.

hardware/accessory prices, and any other associated fees, **Figure 1-12**.

While the use of computers, tablets, and apps have reduced the need for technicians to do math in their head or on paper, technicians still need to know and understand common formulas to be able to do these calculations manually. The time may come when an electronic device is unavailable, and they will need to be able to perform the calculations themselves.

Prints and Schematics

An important mental skill is reading various prints and schematics. HVACR technicians and installers need to have a good understanding of construction prints. This skill is necessary when installing equipment, ductwork, refrigerant lines, and piping in buildings. Installing refrigerant lines and piping is one thing. Installing it *properly* and *in the right location* is another. There are other trades

LOGO [Company Name / Logo]

WORK ORDER

W.O. #: **[123456]**
W.O. Date :

[Street Address]
[City, ST ZIP]
Phone: [000-000-0000]
Fax: [000-000-0000]
[Web Address]

Requested By: [Customer Name]
Customer ID: [abc1]
Department:

JOB	BILL TO	SHIP TO (if different)
[Enter description of work]	[Name] [Company Name] [Street Address] [City, ST ZIP] [Phone]	[Name] [Company Name] [Street Address] [City, ST ZIP] [Phone]

QTY	DESCRIPTION	TAXED	UNIT PRICE	LINE TOTAL
2	Part XYZ	x	150.00	300.00
5	Hourly Labor for ABC (5 hours)		50.00	250.00
				-
				-
				-
				-
				-
				-
				-
				-
				-
				-
				-
				-

SUBTOTAL	$	550.00
TAXABLE		300.00
TAX RATE		6.875%
TAX	$	20.63
S & H	$	-
OTHER	$	-
TOTAL	$	**570.63**

Other Comments or Special Instructions
1. Total payment due 30 days after completion of work
2. Please refer to the W.O. # in all your correspondence
3. Please send correspondence regarding this work order to:
   [Name, Phone #, Email]

Make checks payable to
[Enter Company Name]

I agree that all work has been performed to my satisfaction.

Completed Date: ____________

Signature: ____________ Date: ____________

Goodheart-Willcox Publisher

Figure 1-12. A work order template used to calculate the bill total for a customer after a service call.

like electricians, plumbers, fire safety system installers, and others that also have equipment to install. Everyone needs to be able to read the print to make sure they are installed in the proper location. Prints and schematics serve as the directions from the architect, engineers, and construction contractors to the trades workers on how to properly construct a safe and functional building.

Electrical and electronic schematics come in a variety of forms. A ***pictorial diagram*** is an electrical diagram that shows how components are wired together and the approximate physical location of components within aI unit, **Figure 1-13**. A ***ladder diagram*** is an electrical diagram that shows the logic of system operation. This diagram shows the order in which devices turn on and what action causes another action to happen, **Figure 1-14**.

Symbols are used on wiring diagrams to help identify the different components that are in the equipment. These symbols help to simplify the wiring diagram and make it easier to read. Technicians need to be able to recognize the different symbols for components.

Troubleshooting

HVACR technicians must be problem solvers, being able to think through problems to produce answers. While physical abilities and technical skills are necessary, a technician that does not cultivate a strong mental awareness and a calculating mind will not become a good troubleshooter. A technician must not only be able to take pressure and voltage measurements but must also be able

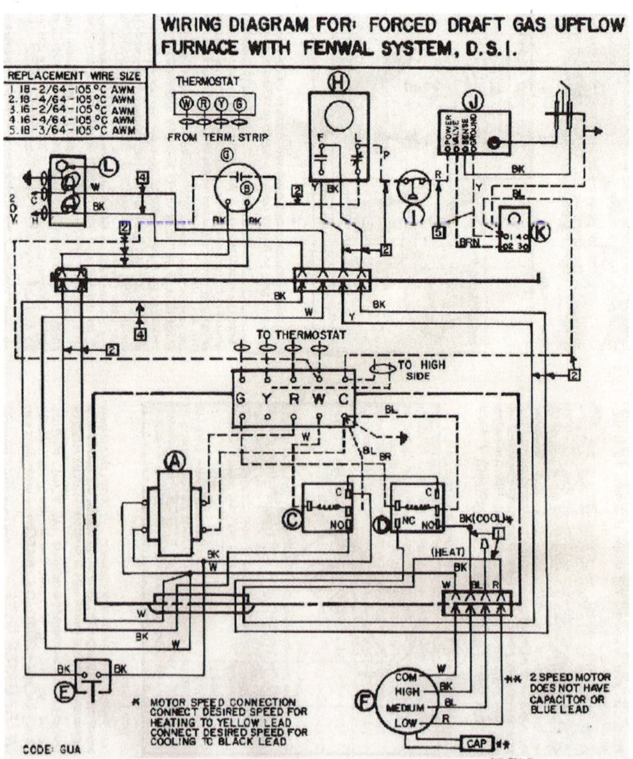

Goodheart-Willcox Publisher

Figure 1-13. Pictorial diagram of a heating and cooling unit.

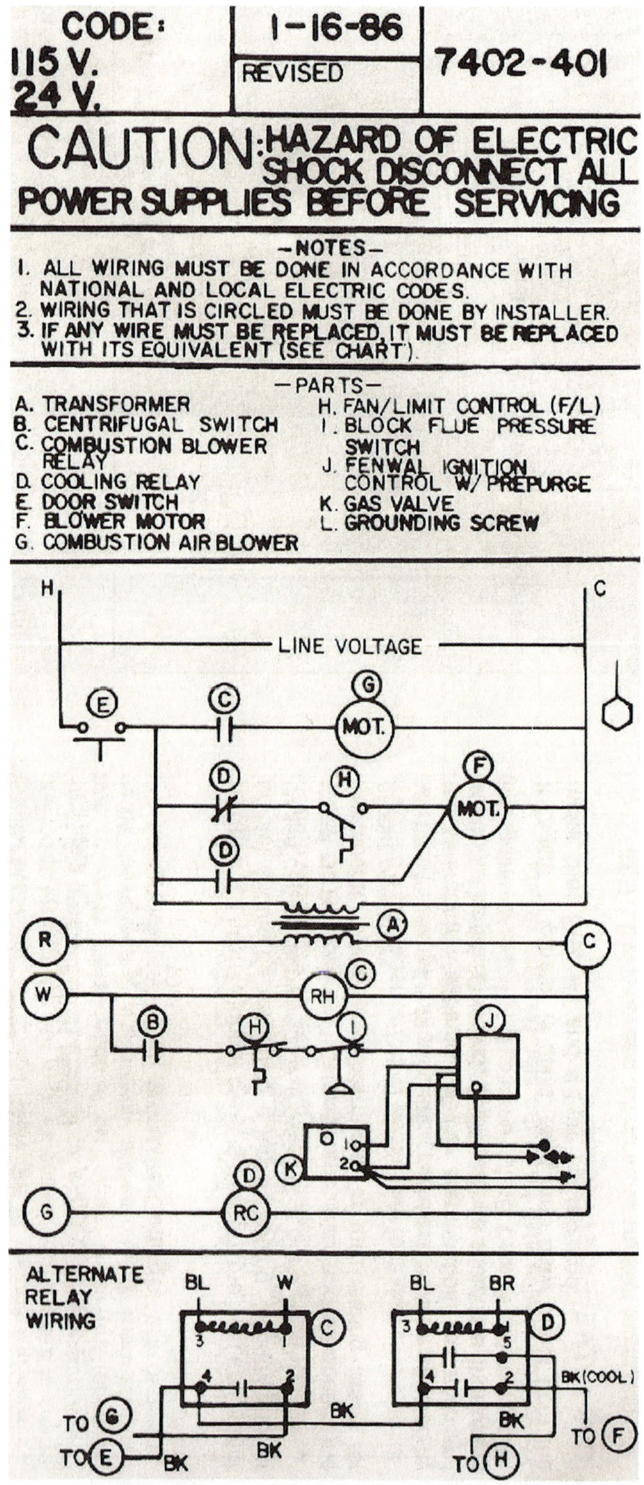

Goodheart-Willcox Publisher

Figure 1-14. Ladder diagram of the same heating and cooling unit shown in Figure 1-13.

interpret the measurements and apply them to ensure proper equipment operation.

Understanding how a system should be functioning can help a technician narrow down a problem. Additionally, manufacturers often provide troubleshooting charts that can help technicians know where to start when there is a system fault. Troubleshooting is a skill that improves with acquired knowledge and experience. Some technicians may have to focus their time practicing a specific area of troubleshooting, like diagnosing electrical problems for instance.

Focus

One of the most important mental skills that technicians must cultivate is the ability to focus. HVACR work may take place in a variety of settings that may be loud, hot, cold, humid, dirty, elevated, in poor lighting, or in cramped spaces. Technicians will need to overlook distractions to focus on the task at hand. This will take practice and concentration, but it is necessary and beneficial to being a good and safe technician.

Attitude and State of Mind

How technicians approach their work can have compounding effects. These can be positive or negative. Achieving the right mindset can make the difference between a good day and a difficult day. Technicians cannot control the weather, the customer, or the service conditions; they can only control their own attitude.

HVACR work can be demanding. Service calls may come in at 5:00 pm, 5:00 am, on a weekend, or holiday. The boss will have to send someone on these jobs. Getting upset or angry about these times will not help matters. In most cases, it will make matters worse. Calmly accept the responsibility and do the job well.

Technicians must be self-motivated. When making a service call, a technician will often be alone and must make the important decisions necessary to get a customer's equipment back up and running. At the same time, technicians must be aware that there are times when they must ask for guidance. Through experience and advice from senior technicians, this insight will begin to develop.

Technicians must realize the gravity and importance of their work. Hundreds to thousands of dollars of equipment and business can be lost due to a system failure in a commercial or even residential system. Timely and proper work is crucial and can help customers avoid financial hardship.

To succeed in HVACR, technicians need to be able to work outside their comfort zone. This means not being afraid to work on a new or unfamiliar piece of equipment. By gaining a thorough grasp of basic system operation and the principles of thermodynamics and electricity, a technician can apply past work experience and tackle new challenges.

Communication Skills

Effective communication skills are required with HVACR work. While much of the work may involve working on HVACR systems, technicians may interact with a variety of people throughout the course of a workday. They may receive their service calls from a dispatcher. Then they will talk with customers. They may need to get parts at the supply house and later report to their service manager or supervisor. In all these cases and others, technicians must be in command of their communication skills, **Figure 1-15**.

Technicians may face a variety of difficult situations—technicians may have to interact with a customer that is angry at them or their company. The supply house may not have the part they need to finish their service call, or they may have to fill in for another technician who called in sick. In all these examples, a technician should actively listen

Goodheart-Willcox Publisher

Figure 1-15. HVACR technicians must have effective communication skills. A technician is shown here explaining the bill to the customer.

to the problem and explain how they can offer a solution. Keeping the focus on problem solving is a good example of effective communication.

Good communication skills are necessary for nearly any career that involves working with people. Keep in mind that the customer sees the service technician as a representative of the service company. The technician is the only face a customer may ever know from the company. That is a great responsibility. See Chapter 2, *Customer Relations,* for more information on working with customers.

Technical Skills

A unique aspect of the HVACR trade is the multiple technical skill sets that are required. Such skills are required to accurately install, maintain, and troubleshoot a variety of systems.

Mechanical Skills

Technicians should be mechanically inclined. Having ***mechanical skill*** means being able to work well with one's hands. This will include using hand and power tools to assemble and disassemble units. A technician must have the knowledge and skills to use these tools and others. Some of the specific types of tools used in HVACR are covered in Chapter 4, *Hand Tools and Instruments.*

Using Instruments

Along with hand tools, service technicians use many distinct types of testing instruments. Technicians taking the time to learn how to use these instruments will make their troubleshooting tasks easier. Some of these instruments are used to measure system pressure, like a refrigerant manifold gauge. Some instruments are used to measure temperature, like digital thermometers. Being able to measure system pressures and temperatures allows a service technician to correctly diagnose equipment problems, **Figure 1-16**.

Goodheart-Willcox Publisher

Figure 1-16. This technician is using two instruments at once. The manifold gauge is being used to monitor system pressures, while the clamp-on ammeter is measuring electrical current during system operation.

Some of these instruments are also used to make sure that HVAC equipment is operating correctly and at peak efficiency. Combustion analyzers are used to check furnace combustion flue gases. Refrigerant leak detectors are used to check for system leaks. Thermographic cameras are used for finding heat loss locations in buildings.

Technicians should be willing to learn to use instruments that may operate differently based on a manufacturer's preference. Not all instruments are the same. Technicians should read instructions based on different models to fully understand an instrument's operation and controls.

Piping and Tubing Skills

HVACR technicians work with copper, black iron, and plastic tubing and fittings on a regular basis; and run refrigeration lines, water lines, and drain lines. They must know how to properly cut, bend, and shape several types of piping and tubing. They must develop skills in how to join piping and tubing. Some of these skills include soldering, brazing, flaring, and using compression and clamp-on fittings.

The technician may also be required to work with steel gas lines for natural gas and propane for heating equipment. The technician will need to know how to cut, thread, and join steel pipes and fittings. Being able to make leakproof joints is necessary when installing pipe and fittings, **Figure 1-17**.

Goodheart-Willcox Publisher

Figure 1-17. Even if a system is leakproof when installed, years of operation can result in leaks that must be detected and fixed using piping and tubing skills. This technician is pointing to a leaky joint.

Sheet Metal Skills

Service technicians as well as system installers need to be able to work with sheet metal. Whether it is installing a new HVAC system in new construction or adding a component to an existing ductwork system, having sheet metal skills is important. Being able to use sheet metal hand tools, such as snips, folding tongs, hand notchers, crimpers, hammers, and other tools, can make an installation or retrofit much easier. Most of the time, ductwork and transitions are built and knocked together in the shop, but being able to manipulate and modify ductwork and transitions in the field is a valuable skill.

Electrical Skills

Electrical skills involve installing and hooking up electrical wiring and troubleshooting electrical problems. Service technicians must be able to use and read an electrical multimeter and understand its reading. Sometimes troubleshooting involves working on equipment with the power still turned on. Technicians are able to lower the risks of working on live electrical equipment if they know how to do so safely. For instance, turning the power off to a condensing unit and verifying with a multimeter that no electricity is flowing through the circuit is a safe practice when servicing the unit, **Figure 1-18**. Electrical safety will be discussed in greater detail in Chapter 3, *Safety*.

Depending on local codes, the technician might have to run electrical conduit for power lines and control wiring. Even if local codes require an electrician to run power lines, it may be the job of the technician to make the final equipment connections for both high power and control power.

1.3 Starting a Successful Career in the HVACR Industry

Success in the HVACR industry begins with obtaining a strong working knowledge of the fundamental concepts of heating and cooling systems. Being an active participant in the education process involves learning in and outside of the classroom. This involves acquiring necessary industry training and professional job experience.

Goodheart-Willcox Publisher

Figure 1-18. The technician has removed a device from an electrical circuit for testing.

Treat Education as Employment

The best way to succeed in HVACR education is to treat the time spent training in the classroom as employment. Be as serious about education as a full-time job. This means showing up to class on time, being engaged in the time spent in the classroom, and doing the work assigned. It also means asking for clarification if a concept or theory is unclear. If a technician does not know what the boss expects, how can the work be done properly? Communication is a two-way street, and both parties are responsible for understanding one another. When in doubt, ask. Always maintain calm and polite speech to foster mutual respect with all participants involved. In short, it is imperative to be a professional.

To work long-term in HVACR, technicians need to have the desire to continue their education, even at the cost of their own money or time. Manufacturers continue to improve their equipment, incorporate modern designs, and add more controls. Systems continue to get more complicated, and technicians need to keep up with the changes. To be successful, always look for ways to improve technical skills and knowledge.

Online Resources

The internet is a versatile tool that can help to enhance a student's HVACR education. A web search of any HVACR topic can be extremely useful. There are numerous online videos on many HVACR topics. Equipment manufacturer websites often contain a wealth of information, service manuals, parts lists, and even training videos, **Figure 1-19**.

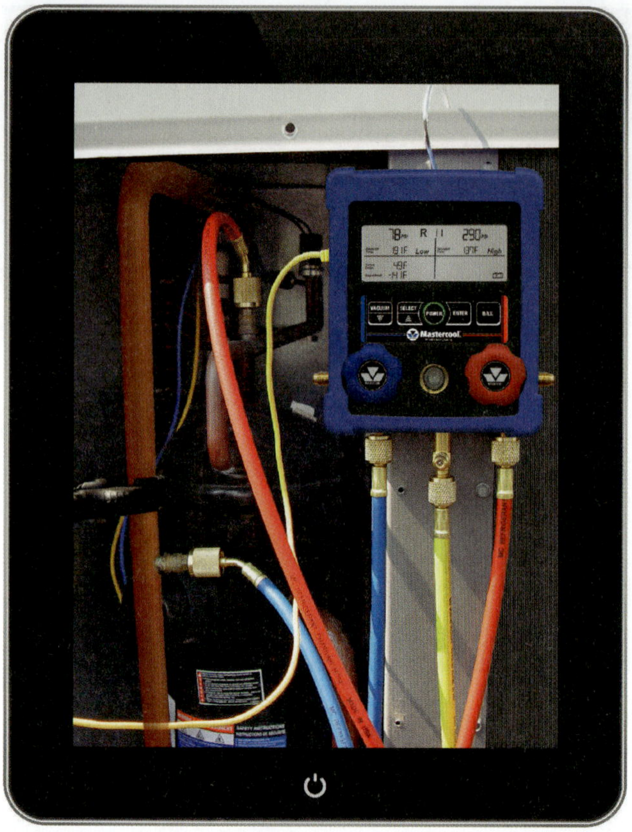

Mastercool Inc.

Figure 1-19. Mobile devices can be used to access useful online information while on-site at a job.

Obtaining Field Experience

One of the best ways to enhance the education process is to get out into the HVACR world as soon as possible. Classroom training will make more sense after putting it into practice in the field. The following sections describe three ways to accomplish this.

Work Study

Postsecondary HVACR programs frequently hire work studies or lab assistants. While this does not get one out into the field, it does allow more time around the lab equipment and other HVACR instructors.

Internships

If an HVACR program has an internship program, make use of it. An internship allows the student the opportunity to see firsthand what goes on in an HVACR company without needing to commit to a full-time position with the company. Enroll in multiple internships if possible, picking companies that have different areas of emphasis, such as residential heating and cooling, light commercial heating and cooling, food service, or supermarket refrigeration. This will provide a better feel for what areas in which to seek employment.

Early Employment

Find part-time employment with an HVACR company that will hire a student that is still in school. This will enhance HVACR education, providing real-world experience while still learning in the classroom.

Applying for a Job

Once you are ready to apply for a position, the first step is to get to research the service companies in your area. Consider the options available, and once you find a company you would like to work for, start preparing for the application and interview process. This section provides guidance on how to complete important documents needed to apply for a job and prepare for a job interview.

Résumé

A ***résumé*** is a brief outline of your education, work experience, and other qualifications for work. A well-written résumé can help you get an interview. An example appears in **Figure 1-20**. Make sure that your résumé is precise and without errors. Always bring a hard copy of your résumé to an interview, be sure to print it on neutral-colored paper (white, gray, or cream colors).

Some employers request an electronic résumé that is sent via email or attached with an online application. You can also post an electronic résumé to a number of online job-search sites. Employers may search for key terms among résumés of prospective candidates that match their descriptions of an ideal job candidate.

Along with the résumé, you need to develop a list of references. A ***reference*** is an individual who will provide important information about you to a prospective employer. A reference can be a teacher, school official, previous employer, or any other adult outside your family who knows you well.

You will need at least three references. Always get permission from each person to use their name as a reference before actually doing so. Your list of references, along with their titles, phone numbers, and addresses, should be kept private. Share this list only with an employer who has interviewed you and asks for your references.

You can also have your references write letters of recommendation for you. These give an employer a more in-depth look at your skills. Choose people who know you well. Make sure you choose references who are good writers, since they will be representing you. Ask as many people as possible. Then you can choose the best letters to submit to employers.

Cover Letter

The cover letter is often the first contact you have with a potential employer. It can make a lasting impression. It should be neat and follow a standard form for business letters. If providing a physical copy, make sure the paper is ivory, white, or a neutral color and free of smudges and mistakes. Use a standard font to give the letter a professional look. Be sure to check spelling and punctuation. Have several people read the letter and offer advice for improving it. You should include your résumé with your cover letter.

A sample cover letter appears in **Figure 1-21**. The letter should be brief and to the point. It should include the following items:

- Title of the job you seek
- Where you heard about the job
- Your strengths, skills, and abilities for the job
- Reasons you should be considered for the job
- When you are available to begin work
- Request for an interview

Job Application Forms

A prospective employer may ask you to complete a job application form before having an interview. The job application form highlights the information the employer needs to know about you, your education, and your prior work experience. Employers often use these forms to screen applicants for the skills needed on the job. You might complete a form in a personnel or employment office.

The appearance of the application form can give an employer the first opinion about you. Fill out the form accurately, completely, and neatly. How well you accomplish that can determine whether you get the job. When asked about salary, write *open* or *negotiable*. This means you are willing to consider offers.

Michael J. Garcia

134 Lincoln Street
Wilton, CA 93232

(212) 555-1234
mjgarcia22@e-mail.com

Career Objective
To obtain an entry-level position in the HVACR field.

Work Experience
Builder's Supply, Holloton, CA — September 2025–present
Retail Sales Associate
- Provide assistance to customers in hardware, HVACR, and electrical departments.
- Review department inventory.
- Maintain neat and orderly displays.

Builder's Supply, Holloton, CA — March 2024–September 2025
Receiver/Stocker
- Unload, store, and stock merchandise.
- Operated fork trucks and hand trucks.

Tony's Grocery Store, Wilton, CA — May 2023–March 2024
Cashier
- Operated checkout scanner and bagged groceries.
- Answered customer questions.
- General stocking and cleaning throughout store.

Education
Associate degree in Heating, Ventilation, Air Conditioning, and Refrigeration — May 2024
Oceanside Community College
- GPA: 3.22/4.0
- Coursework included electrical wiring, commercial refrigeration, and residential heating and cooling systems.
- Participated in SkillsUSA chapter.

Community Service
Habitat for Humanity, volunteer, summers of 2022, 2023, 2024
Wilton Food Bank, volunteer, 2022–present

References
Available upon request.

Goodheart-Willcox Publisher

Figure 1-20. A résumé should be neat, easy to read, and include relevant work experience and education.

Michael J. Garcia
134 Lincoln Street
Wilton, CA 93232
(212) 555-1234
mjgarcia22@e-mail.com

September 24, 2025

Mr. James Williamson
Williamson Heating and Cooling, LLC
4392 East 134th Street
Wilton, CA 93232

Dear Mr. Williamson:

The HVACR service technician position you advertised on the Career Finder website is exactly the type of job I am seeking. After reviewing the job description and requirements, it was clear that my experience, skills, and interests are a perfect match for this opportunity.

While obtaining my associate's degree in Heating, Ventilation, Air Conditioning, and Refrigeration from Oceanside Community College, I gained both the theoretical knowledge and the hands-on skills required for this position. In my current position as a retail sales associate at Builder's Supply, I have developed strong customer service skills and gained a better understanding of HVACR components and systems. I am eager to apply the skills I have learned and to continue gaining new skills.

Please find my résumé enclosed with this letter. I would greatly appreciate an opportunity to interview for this position. Please contact me at your convenience by phone or email to schedule an interview. I look forward to hearing from you.

Sincerely,

Michael J. Garcia

Michael J. Garcia

Goodheart-Willcox Publisher

Figure 1-21. A cover letter should be succinct and focus on skills or experience that may not be reflected on your résumé. The letter should conclude with a request for an interview.

Be sure to send or give the form to the correct person. The name of the correct person often appears on the form. Tips for completing the job application appear in **Figure 1-22**.

Online Application Forms

Many employers now request electronic applications, either through their company website or independent job-search website. When filling out an online application, it is extremely important to include key terms for which the employer may search. This will help you stand out from the many other applications the employer will receive.

When preparing your application, be sure to save it in the appropriate format. If a preferred format is not given, it is best to save the application in a file format specified by the application or in PDF file format if no format is given. This will enable the employer to find specific search terms in your document. Be sure to complete all the fields of the application. Many job-search sites have sample forms on which you can practice before attempting a real application.

Tips for Completing a Job Application

- ✓ Follow the instructions for filling out the form. If you are asked to print, type, or use black ink, be sure to do so. Be as neat as possible.
- ✓ Complete every question in the form. If some questions do not apply to you, draw a dash or write or "NA" (for not applicable) so the employer knows you did not overlook it.
- ✓ You may wish to omit your Social Security Number and write "will provide if hired."
- ✓ You can write "open" or "negotiable" for any question regarding salary requirements.
- ✓ For each former job, there may be a question asking your reason for leaving the job. Avoid writing any negative comments about yourself or a former employer.

Goodheart-Willcox Publisher

Figure 1-22. While many companies have applicants fill out applications online, these tips are important to keep in mind in case a company does require you to fill out a physical application.

Job Interview

The interview gives you the opportunity to learn more about a company and to convince the employer that you are the best person for the position. The employer wants to know if you have the skills needed for the job. Adequate preparation is essential for making a lasting, positive impression. Here are some ways to prepare for the interview:

- Research the employer and the job. Know the mission of the employer and specifics about the job. Also, try to learn what the company looks for when hiring new employees.
- Be prepared to answer questions. Go over the list in **Figure 1-23** and prepare answers for each.
- List the questions you want answered. For example, do you want to know if there is on-the-job training? Or opportunities for advancement?
- List the materials you plan to take. This seems simple enough. However, if you wait to grab items at the last minute, you will likely forget something important.
- Decide what to wear. Dress appropriately, usually one step above what is worn by your future coworkers. For instance, business casual clothing is acceptable for future HVACR technicians who will do manual labor and may wear a company uniform. Regardless, always appear neat and clean.
- Practice the interview. Have a friend or family member interview you in front of a mirror until you are happy with your responses.
- Know where to go for the interview. Verify the address of the interview location by checking the site beforehand, if possible. Plan to arrive ready for the interview at least 10 minutes early.

Good preparation will make you feel more confident and comfortable during the interview. Be polite, friendly, and cheerful. Use a firm handshake. Maintain eye contact at all times. Answer all questions carefully and as completely as you can. Be honest about your abilities. Avoid chewing gum and fidgeting. Also be aware of questions you legally do not have to answer, such as those related to age, marital status, religion, or family background.

A prospective employer may ask you to take employee tests. Some employers administer tests to job candidates to measure their knowledge or skill level under stress. Since all employers support a drug-free workplace, most will likely require you to take a drug test if hired.

After the interview, send a follow-up email to the employer within 24 hours, thanking them for the interview. If you get a job offer, respond to it quickly. If you do not receive an offer after several interviews, evaluate your interview techniques and seek ways to improve them.

Common Interview Questions and Responses

Question	Response
What can you tell me about yourself?	Briefly summarize your abilities as they relate to the job qualifications or your career goals. Do not provide a general life history.
Why do you want to work for this company?	Tell what you know about the company. Explain how your abilities match the company's needs.
Why do you think you would like this kind of work?	Relate the job requirements to your successful past experiences.
What are some of the projects you worked on in school?	Briefly summarize a project or coursework relevant to the job qualifications.
What other jobs have you had?	Focus on jobs with skills that relate to the jobs you are seeking.
Why did you leave your last job?	Be honest. However, avoid saying anything negative about your previous employer.
Have you ever been fired from a job? If so, why?	Answer honestly. If you have been fired, share what you learned from the experience. Avoid trying to blame others.
What are your major strengths and weaknesses?	Select a strength that relates to the job qualifications. Be honest when selecting a weakness but give an example of how you've worked to improve on it.
Have you ever had a conflict with a coworker? How did you handle it?	Briefly describe the situation and how you handled it. Avoid placing all the blame on the other person. Explain what you learned from the experience.
What do you expect to be paid?	If possible, determine the salary range before the interview. Say that you are willing to discuss the salary or state a range you feel comfortable with.
What are your future plans?	Describe how the need to learn and grow is important to you. Confine your answer to the company with which you are interviewing.

Goodheart-Willcox Publisher

Figure 1-23. Common interview questions and guidelines for preparing strong responses. A good practice is to share your responses with family or friends and ask for feedback.

Chapter Review

Summary

1.1 The HVACR industry is growing in demand with expanding businesses and a shrinking workforce. Many career options are available including HVACR service technicians, installers, technical training instructors, building engineers, and various leadership positions.

1.2 HVACR work requires good physical, mental, communication, and technical skills. A unique aspect of the HVACR trade is the multiple technical skills required such as being able to work well with one's hands.

1.3 A career in the HVACR industry begins by investing in education and obtaining relevant work experience. Once ready to apply for a position, research companies and apply for a position at a company you would like to work for. Completing a professional résumé, cover letter, and job application can help you gain an interview. Good preparation will make you feel more confident and comfortable during the interview.

Know and Understand

1. *True or False?* The industry demand for HVACR technicians and installers is expected to grow. (1.1)
2. *True or False?* The only entry-level positions in the HVACR industry are service technicians and installers. (1.1)
3. A(n) ____inspects buildings to find where improvements can be made to make the building more energy efficient. (1.1)
   A. parts clerk
   B. energy auditor
   C. technical training instructor
   D. controls technician
4. The ability to use a ruler and take measurements is an example of a ____ skill. (1.2)
   A. physical
   B. technical
   C. mental
   D. mechanical
5. Having ____skill means being able to work well with one's hands. (1.2)
   A. physical
   B. technical
   C. mental
   D. mechanical
6. Sheet metal is commonly used in HVAC work for ____. (1.2)
   A. ductwork
   B. furnaces
   C. piping and tubing
   D. electrical wiring
7. *True or False?* Running electrical conduit for power lines and control wiring should never be done by an HVACR technician. (1.2)
8. A(n) ____allows a student to obtain field experience without committing to a full-time position with a company. (1.3)
   A. work study
   B. internship
   C. apprenticeship
   D. HVACR program
9. A ____is a brief outline of your education, work experience, and other qualifications for work. (1.3)
   A. reference
   B. cover letter
   C. résumé
   D. portfolio
10. It is good practice to send an employer a follow-up email after an interview within ____hours. (1.3)
    A. 2
    B. 24
    C. 48
    D. 72

Apply and Analyze

1. List three advantages to choosing a career in HVACR. (1.1)
2. How can a technician work to improve their troubleshooting skills? (1.2)
3. What are three piping and tubing skills required in HVACR work? (1.2)
4. What are three instruments that technicians may use to gather data? (1.2)
5. Why is it important to spend time preparing for a job interview? (1.3)

Critical Thinking

1. Explain why physical skills are necessary in HVACR work. (1.2)
2. What is the difference between a pictorial diagram and a ladder diagram? How are these schematics used in HVACR work? (1.2)
3. What information should be included on a cover letter? (1.3)

SrdjanPav via Getty Images

2 Customer Relations

LEARNING OUTCOMES

After studying this chapter, you will be able to:

2.1 Explain the importance of having good customer relations and how to demonstrate professionalism when making first impressions and dealing with difficult customers.

2.2 Demonstrate how to begin and end a service call.

TECHNICAL TERMS

callback
constructive criticism
customer interview
customer relations
difference of opinion
negative criticism
people skills
professionalism

Introduction

HVACR service technicians not only need to have good mechanical service skills but also good people skills. This chapter is intended to help you practice the skills needed to interact with customers and other people. Being successful in a service business requires having strong ***people skills***, or the ability to work well with people. The service technician must be able to think quickly and apply ways to work effectively with people, **Figure 2-1**.

2.1 Customer Relations

Customer relations is the way a business interacts with the public, particularly customers. Having strong people skills is a good foundation for establishing excellent customer relations; the two go hand-in-hand. Displaying clear and honest communication, maintaining a professional appearance, and being respectful are a few examples of good people skills to exercise when working with customers. These skills can help a business thrive, as customers tend to keep coming back to a business that employs technicians with strong people skills. These skills can also help technicians navigate difficult situations, such as an angry customer. Customers encounter problems just like everyone else, and they sometimes become upset. Regardless of the situation, the goal should be to stay professional and try to solve the issue at hand.

Goodheart-Willcox Publisher

Figure 2-1. In school or in training classes, learning how to work with people is just as important as learning how to work with heating, cooling, or refrigeration systems. Good people skills are beneficial whether you work as a technician employed by a company or operate your own business.

Some experts estimate that, for businesses, about 15% of financial success is due to technical skills and 85% is due to people skills. Employers are eager to hire those who have good people skills. These specific skills are easy to learn, but practice must be put forth to master them and make them a habit. This requires time, persistence, and daily application.

Making a Good First Impression

First impressions are important. This principle is especially relevant to customer service. What the customer sees before they speak to you can set the stage for failure or success. First impressions may be made based on your physical appearance, mannerisms, and the condition of your service vehicle. The customer will continue to evaluate you during and after speaking with you. Good people skills are important to showcase when interacting with customers, especially when meeting them for the first time.

Physical Appearance and Hygiene

Your physical appearance, hygiene, and manner of dress convey a lasting message about your work ethic, **Figure 2-2**. A sloppy appearance is seen by customers as an indication of sloppy work habits and poor character traits. There is no excuse for having an unkept appearance.

Clean, professional work attire is essential for any customer-facing job. HVACR service technicians frequently wear a pair of blue jeans and an appropriate work shirt. Work attire may vary depending on the company you work for, so be sure to always follow company policy. Maintaining a professional appearance not only can help make a good first impression, but it also shows respect for yourself, your company, and the customer.

Service Vehicle

How and where you park your service vehicle says a lot about the level of respect you have for your customer. Unless you have equipment to unload, the service vehicle should be parked out in the street, along the curb (not on the curb or in the driveway of the customer). If you need to bring

fstop123 via Getty Images

Figure 2-2. As a field representative of your company, remember that your performance, attitude, and appearance will speak louder than words. Demonstrating professional appearance and a desire to solve the customer's problem will help you succeed in the HVACR field.

heavy equipment inside, you should check with the customer first to see if you can park closer to the home. Also, having a vehicle that is dirty on the inside or outside can indicate a disorganized or careless work ethic to the customer. Similar to your tools and equipment, keep the service vehicle clean and properly maintained.

Introducing Yourself to a Customer

In business, a handshake is customary when introducing yourself to a customer. When performed properly, a handshake will make a good impression, **Figure 2-3**. When introducing yourself, maintain eye contact and be succinct. State your purpose for being there and be certain you learn the customer's name and remember it. Repeat the name by saying, "Glad to meet you, John," or "It's a pleasure to meet you, Mr. Smith." This repetition not only helps you remember the name, but it also displays respect for the other person.

fstop123 via Getty Images

Figure 2-3. A customer's first impression of you determines how they will respond. A neat, clean uniform, friendly smile, and firm handshake will go a long way toward making a good impression on the customer.

Dealing with Difficult Customers

Service technicians may encounter customers that are angry or upset. When this happens, it is important to remain calm and maintain professionalism. ***Professionalism*** is the act of being respectful, courteous, and carrying oneself in a businesslike manner. The following sections describe some ways to be professional when dealing with difficult customers.

Ask Questions

If a customer is angry or upset, keep the focus on how you can solve the problem. Ask questions and allow the other person to answer. If you disagree, *do not interrupt*. Listen patiently and encourage the other person to express their viewpoint fully. Asking questions for clarification presents customers with an opportunity to fully express their ideas and opinions. It also gives the service technician the chance to listen and show due respect to the customer. Having the customer's full perspective can help you efficiently resolve the issue.

Be Diplomatic

Never permit yourself to become a participant in an argument, especially with a customer. An argument is easily avoided by listening attentively and remaining calm. You may encounter someone with a ***difference of opinion*** than your own. A difference of opinion is when two people have different ideas about a topic or issue. Hearing out the other person and keeping opinions to yourself can help avoid an argument. Empathizing with the customer's frustrations can be a good way to de-escalate an upset customer. No matter what a customer may say or do, always remain friendly, courteous, and considerate.

Handling Criticism

Customers may offer criticism to you about your service or your company. To avoid escalating the situation, listen carefully to their feedback and decide what kind of criticism you are receiving. ***Negative criticism*** is unfair criticism or a harsh judgment. If you are receiving negative criticism, remain professional and try to earn the customer's confidence in your work. ***Constructive criticism***, on the other hand, is feedback given with the goal of helping the other person learn or improve something. For example, a customer may repeatedly point out that they cannot hear you when you are talking. Do not take this criticism personally; listen to the feedback and take it as a potential opportunity for improvement. This type of criticism will help you grow as a technician and as a professional.

The Ten Commandments of Good Service

The "Ten Commandments of Good Service" have been around for a long time, but the principles they illustrate always remain current:

1. Customers are not dependent on us. We are dependent on them.
2. Customers are not an interruption of our work. They are the purpose of it.
3. Customers do us a favor when they call. We are not doing them a favor by serving them.
4. Customers are a part of our business, not outsiders.
5. Customers are not cold statistics. They are real people with the same feelings and emotions we have.
6. Customers are not people to argue with or match wits against.
7. Customers are people who bring us their needs. It is our job to fill those needs.
8. Customers deserve the most courteous and attentive treatment we can give them.
9. Customers are the people who make it possible to pay your salary.
10. Customers are the lifeblood of any business.

The Ten Commandments of Good Service imply one golden rule: *The customer is the most important person in any business.* Without customers, you would not have a job. The customer is aware of this business principle and has every right to expect good service and fair treatment. If the customer does not receive good service and fair treatment, they will go elsewhere.

Remember that service technicians are field representatives of their company. Good mechanical skills and good people skills will reflect favorably on the company. You must remember that you are selling *service*, and that performance speaks louder than words.

2.2 Service Calls

Service calls require constant use of good people skills. It is up to the service technician to communicate effectively with customers. Customers may become apprehensive when something is wrong with their heating and cooling equipment. They may have questions like, "Will it catch fire or explode?" "Will it be out of service for days?" "How expensive will the repair be?" Customers usually do not have the technical knowledge a technician does, so it is up to the service technician to address any questions or concerns they may have, **Figure 2-4**.

Beginning the Service Call

On arrival, introduce yourself to the customer, show proper identification, and conduct a customer interview. A ***customer interview*** involves asking the customer questions about the problems

SDI Productions via Getty Images

Figure 2-4. When interacting with customers on a service call, be sure to listen carefully as they describe the problem. Without technical knowledge, customers may not use industry terms. A good practice is to repeat the problem back to them to ensure you interpreted their description correctly.

they are experiencing with their HVAC system. Effective troubleshooting requires a clear understanding of the problem. A thorough customer interview will help you get to the root of the problem quicker, which can save time and prevent a future callback. A ***callback*** is a return visit necessary to correct a problem that was not resolved during the initial service call.

If your service call is a callback, follow the same procedure as you would for any other service call. If the customer is angry or upset, follow the guidelines discussed earlier in this chapter. If your initial service calls result in multiple callbacks, then these are good opportunities to focus on improving your troubleshooting methods.

Customer Safety and Security

An HVACR equipment malfunction may cause a sense of danger and fear in customers. All service calls are important for this reason. A "basic" service call (for example, an air conditioner is "making a funny noise") is just as important as a call to perform a major overhaul. A customer does not always know if the issue is an emergency or not.

If you notice a customer is afraid, *never* make light of this fear. You can ease a customer's concerns by conducting a thorough customer interview and fixing the issue. Once you have made the repair, explain how you went about troubleshooting the issue. Educating the customer can help ease any future concerns they may have when dealing with an equipment malfunction.

Customer Complaints

It is always best practice to make the customer feel like they are heard, especially if they have complaints. As discussed earlier, be sure to *listen carefully* and do *not* interrupt them. Analyze what is being said and determine exactly why the customer is upset. Try to honestly see things from the customer's point of view.

After hearing the customer's complaints, concentrate on finding solutions. Ask the customer what can be done to make things right. Most customers want far less than you think. A replacement or repair, a minor adjustment to a bill, or an apology will often resolve the problem. Let the customer know you are genuinely interested in offering help. Contact your service manager if the customer's request is beyond your authority to fulfill.

Always be receptive to customer complaints. Every business produces some unhappy customers. Situations can be turned around if customers have an opportunity to voice their complaints, know they are being heard, and have their problem solved quickly and effectively.

Closing the Service Call

At the end of the service call, it is necessary to assure the customer that you have solved the problem. Explain the work you performed in terms the customer can understand. If you are unable to correct the problem, contact another technician or supervisor for help. Sometimes the repair requires the help of another tradesperson, such as an electrician or a plumber. In any case, explain to the customer clearly what must be done to resolve the problem. Offer suggestions so that the customer can select a course of action.

Chapter Review

Summary

2.1 Having good people skills is a strong foundation for establishing excellent customer relations. A technician with good customer relations can help a business thrive.

2.1 Wearing professional work attire and keeping the service vehicle clean are essential to making a good first impression with a customer. Introducing yourself in a polite, professional manner is good practice when first interacting with a customer. Service technicians may encounter customers that are angry or upset. When this happens, it is important to remain calm and maintain professionalism. Ask questions, listen carefully, and focus on solving their problem.

2.1 The Ten Commandments of Good Service imply one golden rule: *The customer is the most important person in any business.*

2.2 Begin a service call with a customer interview. Listen carefully, ask questions, and ease their fears or concerns. Focus on solving the problem. At the end of the service call, explain the work you performed in terms the customer can understand.

Know and Understand

1. Experts estimate that businesses with financial success can attribute that success to ___% technical skills and ___% people skills. (2.1)
   A. 85, 15
   B. 15, 85
   C. 50, 50
   D. 75, 25
2. *True or False?* Clean and professional work attire can help a technician make a good first impression with customers. (2.1)
3. *True or False?* It is *not* the responsibility of the service technician to keep a service vehicle clean and properly maintained. (2.1)
4. A(n) ___ is when two people have different ideas about a topic or issue. (2.1)
   A. argument
   B. negative criticism
   C. difference of opinion
   D. constructive criticism
5. According to the Ten Commandments of Good Service, customers are ___. (2.1)
   A. statistics
   B. dependent on businesses
   C. the least important person in any business
   D. the most important person in any business
6. Service technicians are ___ of their company. (2.2)
   A. owners
   B. representatives
   C. experts
   D. customers
7. Effective troubleshooting requires ___. (2.2)
   A. a clear understanding of the problem
   B. years of experience
   C. the newest HVAC service equipment
   D. industry certifications
8. A return visit necessary to correct a problem that was *not* resolved during the initial service call is called a ___. (2.2)
   A. customer interview
   B. reverse call
   C. system malfunction
   D. callback
9. *True or False?* Customers usually do *not* have the technical knowledge to know whether an issue is an emergency or not. (2.2)
10. If you cannot solve a problem for a customer, then ___. (2.2)
    A. make the cheapest repair you can and do not charge them for it
    B. explain to the customer clearly what must be done to resolve it
    C. give up and leave
    D. redo the customer interview

Apply and Analyze

1. Why is customer relations important to a business? (2.1)
2. Describe three ways to make a good first impression with customers. (2.1)
3. What is the difference between negative criticism and constructive criticism? (2.1)
4. Explain why "basic" service calls are just as important as calls that require a lot of work. (2.2)

Critical Thinking

1. On a no heat service call, you discover that the previous technician who worked on the furnace left a wire disconnected. How do you report this to the customer without negatively criticizing the previous technician? (2.1)
2. On a service call, you learn that a burst water pipe in the wall is the cause of the owner's dehumidifier constantly running. Due to the nature of the repair, you decide it is best for a trained plumber to complete the repair. How do you communicate this issue to the customer? (2.2)

fcafotodigital via Getty Images

3 Safety

LEARNING OUTCOMES

After studying this chapter, you will be able to:

3.1 Explain the purpose of OSHA.
3.2 Explain the importance of PPE.
3.3 Recognize electrical hazards and how to perform a lockout/tagout (LOTO) procedure.
3.4 Demonstrate safe ladder use.
3.5 Describe general tool safety rules.
3.6 Explain how to safely handle various chemicals and pressure-related equipment.
3.7 Identify the different classes of fires and corresponding extinguishers.
3.8 Identify best practices when working in extreme weather conditions.
3.9 Practice safe conduct while on the job.

TECHNICAL TERMS

electric shock
hazard
heat stress
lockout/tagout (LOTO)
Occupational Safety and Health Administration (OSHA)
personal protective equipment (PPE)
safety data sheet (SDS)

Introduction

A properly trained service technician puts safety first. Being aware of potential hazards can help keep yourself and others safe. Following proper procedures, planning ahead, and wearing protective equipment can also ensure safety when working with a potential hazard. A ***hazard*** is a potentially harmful situation or substance. This chapter identifies some of the hazards associated with the HVACR industry and how to properly handle working with these hazards.

3.1 Occupational Safety and Health Administration (OSHA)

In 1970, the US Congress created the ***Occupational Safety and Health Administration (OSHA)*** as part of the Occupational Safety and Health Act of 1970. OSHA is a government organization responsible for worker safety and health. By creating and enforcing workplace safety standards, OSHA carries out its mission to provide workers with safe working conditions. The OSHA Job Safety and Health Poster must be posted in all workplaces and serves as a reminder to all employers and workers of their right to work in a safe environment, **Figure 3-1**.

OSHA offers OSHA 10 and OSHA 30 certifications for workers to learn safe working practices. OSHA 10 is a 10-hour class intended to provide workers with awareness of common job-related safety and health hazards. OSHA 30 is a 30-hour safety course that provides a greater variety of safety subjects and in-depth, industry-specific training. This course is intended for supervisors and workers with safety and health responsibilities. Some HVACR companies require new employees to have both the OSHA 10 and OSHA 30 certifications before hiring.

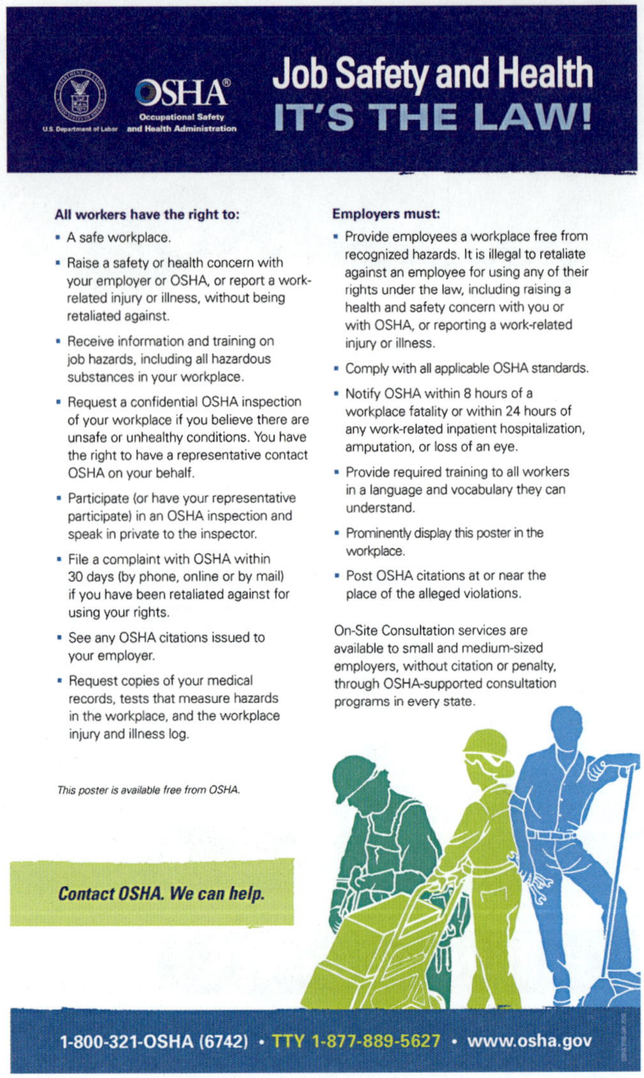

OSHA

Figure 3-1. OSHA publishes and enforces industry standards to help improve workplace safety. Employers are responsible for following these standards.

3.2 Personal Protective Equipment (PPE)

An HVACR service technician must always wear the appropriate personal protective equipment when on the job. According to OSHA, ***personal protective equipment (PPE)*** is equipment worn to minimize exposure to hazards that can cause serious workplace injuries and or illnesses. In some cases, OSHA requires employers to supply PPE when it is needed to ensure the safety of workers. PPE for a service technician includes items such as gloves, safety glasses, steel-toed boots, earplugs or muffs, hard hats, respirators, and appropriate clothing. A service technician should also be familiar with their company's rules and regulations on PPE. OSHA requires employers to train workers required to use PPE to know:

- When it is necessary
- Which kind is necessary
- How to properly put it on, adjust, wear, and take it off
- The limitations of the equipment
- Proper care, maintenance, useful life, and disposal of the equipment

Due to the variety of working conditions service technicians may encounter, it is important for technicians to be prepared with various types of PPE in their service vehicle, **Figure 3-2**.

3.3 Electrical Safety

Extreme care must be taken when working around electrical equipment. There is a risk of electric shock if proper procedures are not followed. ***Electric shock*** occurs when a discharge of electrical current passes through the body. A shock occurs when a person becomes part of an electrical circuit. Direct physical contact is not required to receive a shock; high-voltage sources can pass electrical current through the air or other objects surrounding the source. Wiring thermostats, taking electrical measurements on devices, changing an electrical component, and working with power tools are just a few electrical tasks service technicians may perform.

MSA Safety, Inc.

Figure 3-2. A properly prepared HVACR service technician should know when and how to use various types of PPE. Testing and maintenance of PPE should be done periodically to ensure properly functioning equipment.

SAFETY NOTE

Electrical Safety

The following lists some safety rules to keep in mind when working around electrical equipment:

- Only use hand and power tools that are electrically insulated. Never use power tools with damaged extension cords or missing ground prongs.
- The human body is a conductor of electricity. To avoid becoming part of a circuit, avoid touching electrical terminals and connections with any uninsulated part of your body. When performing electrical work, always wear insulated safety boots and gloves. Avoid wearing jewelry while working on electrical circuits.
- Water is also a conductor of electricity. Be mindful of your environment. Do not stand in wet or damp areas when working on live electrical equipment.
- Ensure power to equipment is off when operating on it. Note that a circuit may still carry current in the wiring or components even when the power is off. Use electrical testing equipment to verify that no electric current is flowing through a part of the system you need to work on.
- Use proper lockout/tagout procedures while working on a system.

Lockout/Tagout (LOTO)

A ***lockout/tagout (LOTO)*** is a safety procedure used to isolate or shut off sources of electricity and communicate that someone is working on the equipment, **Figure 3-3**. This prevents accidental startup of a machine or system while maintenance or repair work is being done. The energy source (usually a circuit breaker) is deactivated and locked, and a tag is placed on it to alert others that the system is being worked on. Once work is complete, the lock and tag are removed and the breaker can be turned back on. For more information on electrical safety, see Chapter 22, *What is Electricity?*

SocoXbreed/Shutterstock.com

Figure 3-3. A lock and tag are placed on this electrical control box to let other workers know that work is being done on the equipment.

3.4 Ladder Safety

HVACR technicians may need to use a ladder to reach equipment. Proper safety procedures should be followed when working on a ladder to prevent falls and potential injury to others.

SAFETY NOTE

Stepladder Safety

When using a stepladder, always be aware of the area around you and follow these safety procedures:

- Inspect the ladder for damage. Never use a damaged ladder, and if you discover a ladder is damaged, report it to a supervisor.
- Most stepladders have four legs; some older models have three legs. When using a four-leg stepladder, make sure the lock spreaders are pushed down and locked in place. Make sure you have all four feet firmly on the floor or ground.
- Ladders are designed for only one user at a time.

(Continued)

SAFETY NOTE

Stepladder Safety *(continued)*

- Stepladders come in many sizes. Make sure you use a ladder with enough height to reach the work area. Never stand on the top two rungs; you could easily lose your balance or cause the ladder to become top-heavy resulting in it falling over while you are on top of it.
- When working on a ladder, keep your body centered within the ladder rungs and have the work area in front of you. Do not lean off the sides to reach your work area. Always maintain three points of contact with the ladder when working on one.
- If you need to move your ladder, get off and move it. Do not try to move the ladder while you are still on it.
- Do not use a ladder on slippery surfaces.
- Do not climb a stepladder that is leaning against a wall.
- Use the appropriate ladder for the job. Only use a fiberglass ladder when working around electricity, **Figure 3-4**.

SAFETY NOTE

Extension Ladder Safety

When using an extension ladder, follow these safety procedures:

- Inspect the ladder for damage. Never use a damaged ladder, and if you discover a ladder is damaged, report it to a supervisor.
- When setting up an extension ladder, make sure to properly set up the ladder to avoid it slipping out from under you. The top of the ladder needs to be a minimum of 3′ above the height you are climbing. The horizontal distance of the ladder from the wall to the base of the ladder should be 1/4 of the length of the ladder, **Figure 3-5**.
- If you have another person available, have them stand at the bottom of the ladder to help hold it while you climb. Just like when using a stepladder, make sure you have three points of contact with the ladder as you are climbing.
- When using a fiberglass ladder, be sure to climb the ladder with your hands on the rungs, not the rails.
- If you need to carry tools and equipment up the ladder, use a rope or hoist.
- Use the appropriate ladder for the job. Use only a fiberglass ladder when working around electricity.

Werner Co.

Figure 3-4. Metal ladders conduct electricity, so they should be avoided when working around electricity. Fiberglass is a poor conductor of electricity, so fiberglass ladders are safer to use around electrical applications.

Goodheart-Willcox Publisher

Figure 3-5. Proper extension ladder placement.

3.5 Tool Safety

HVACR service technicians work with a variety of tools and instruments. To ensure safety to the technician, customer, and equipment, technicians need to have the proper tools and know how to use them properly. Using a tool for a purpose other than its intended use is a hazard. Read manufacturer instructions or ask a supervisor how to use a tool if you are unsure.

Tools should be periodically inspected for damage or wear. Repair tools if they can be fixed; if not, replace them immediately. Tools with blades used to perform cutting work can be a hazard if the blade is dull because more effort is required from the user to make a cut. This can lead to slipping or kickback of the tool, which can damage equipment or injure the user. Tools commonly used by service technicians and related tool safety guidelines will be covered in more detail in Chapter 4, *Hand Tools and Instruments.*

3.6 Pressure and Chemical Safety

Service technicians work with a variety of chemicals such as refrigerants, cleaning agents, and fuel sources. Many of these chemicals are contained in pressurized cylinders or containers that can cause serious injury if improperly handled. Service technicians must be familiar with how to safely handle these materials.

Refrigerant Safety

In the HVACR trade, refrigerants are used in refrigeration and air-conditioning systems. Proper use and handling of refrigerants are governed by Section 608 of the Clean Air Act. The Clean Air Act will be discussed in more detail in Chapter 16, *Refrigerant Recovery and Recycling.* Refrigerants must be handled with great care. Liquid refrigerant can burn skin on contact. Always wear gloves and safety glasses when working with refrigerants. If you are dealing with a major refrigerant leak in a system, do not try to stop the leak even if you have gloves on. Your gloves can freeze to your hands. Additionally, refrigerant can displace oxygen and make it hard to breathe. Some refrigerants contain chlorine and if exposed to high heat (like an open flame), they become toxic and release a deadly gas called *phosgene.* Make sure that you follow all ventilation standards when working around equipment that may contain refrigerant. Always wear a respirator when working on equipment in tight spaces, **Figure 3-6**.

Some refrigerants are highly flammable. Refrigerants are stored in high-pressure cylinders. These cylinders should be handled with care and stored out of sunlight and away from heat. If a refrigerant cylinder gets too hot, this increases the internal pressure which can lead to a burst. More information on refrigerant safety is covered in Chapter 16, *Refrigerant Recovery and Recycling.*

High-Pressure Cylinder Safety

Aside from refrigerant cylinders, service technicians work with other high-pressure cylinders and tanks—such as oxygen and acetylene soldering and brazing torches, nitrogen, and carbon dioxide. Cylinders must be transported standing up with the valve cover caps on (if applicable) during transportation. If the valve on top of the tank gets broken off, the tank can become a dangerous projectile.

MSA Safety, Inc.

Figure 3-6. Many different cartridges are available for respirators. When working in attics, crawlspaces, or basements, be sure to have a respirator on hand with appropriate cartridges.

Safety Data Sheets (SDSs)

According to OSHA's Hazard Communication Standard (HCS), chemical manufacturers, distributors, or importers must provide ***safety data sheets (SDSs)*** (formerly *material safety data sheets*) for every hazardous chemical to communicate safe handling information on these hazards. Before working with any hazardous chemical, a technician should know where to locate the SDS for that chemical. The information contained in the SDS is broken down into the following 16 sections:

- Section 1: Identification
- Section 2: Hazard(s) identification
- Section 3: Composition/information on ingredients
- Section 4: First-aid measures
- Section 5: Fire-fighting measures
- Section 6: Accidental release measures
- Section 7: Handling and storage
- Section 8: Exposure controls/personal protection
- Section 9: Physical and chemical properties
- Section 10: Stability and reactivity
- Section 11: Toxicological information
- Section 12: Ecological information
- Section 13: Disposal considerations
- Section 14: Transport information
- Section 15: Regulatory information
- Section 16: Other information

3.7 Fire Safety

Service technicians will occasionally have to use a torch to repair equipment. When working with fire, pay close attention to the area around the work zone. Be aware of any flammable materials nearby and where to locate a fire extinguisher anytime you are using a torch. Always wear gloves and safety glasses when working with torches. More on working with torches will be covered in Chapter 8, *Soldering* and Chapter 9, *Brazing*.

Fire Extinguishers

Knowing how to properly use fire extinguishers can save lives and property. When working with torches or around flammable materials, always know where the nearest fire extinguisher is located. Be sure to have the correct extinguisher for the application. Fire extinguishers contain different extinguishing agents such as water, carbon dioxide, dry chemical or wet chemical, depending on the kind of fire the extinguisher is intended for. There are five primary types of fire extinguishers, as described in **Figure 3-7**:

- Class A. For use with ordinary materials like cloth, wood, and paper.
- Class B. For use with combustible and flammable liquids like grease, gasoline, oil, and oil-based paints.
- Class C. For use with electrical equipment like appliances, tools or other equipment that is plugged into an outlet. Class C fire extinguishers use an agent that does not conduct electricity.
- Class D. For use with flammable metals.
- Class K. For use with vegetable oils, animal oils, and fats in cooking appliances.

The **PASS** acronym is helpful to remember the correct way to use a fire extinguisher:

1. **P**ull the pin. Hold the extinguisher with the nozzle pointing away from you and release the locking mechanism.
2. **A**im low. Point the extinguisher at the base of the fire.
3. **S**queeze the lever slowly and evenly.
4. **S**weep the nozzle from side to side.

3.8 Weather Safety

HVACR service technicians may work outdoors on rooftops, or in confined spaces where they may be exposed to extreme weather conditions. Workers who are exposed to extreme heat or work in hot environments may be at risk of ***heat stress***. Heat stress can result in heat stroke, exhaustion, cramps, or rashes. Heat can also increase the risk of injuries in workers as it may result in sweaty palms, fogged-up safety glasses, and dizziness. Burns may also occur because of accidental contact with hot surfaces or steam.

Workers at greater risk of heat stress include those who are 65 years of age or older, are overweight, have heart disease or high blood pressure, or take medications that may be affected by extreme heat. Prevention of heat stress in workers is important. Dressing in light, breathable

Fire Classifications

Class	Description	Requires	New Symbol	Old Symbol
A	Ordinary Combustibles (Materials such as wood, paper, and textiles)	Cooling/quenching		A
B	Flammable Liquids (Liquids such as grease, gasoline, oils, and paints)	Blanketing or smothering		B
C	Electrical Equipment (Wiring, computers, switches and any other energized electrical equipment)	A nonconducting agent		C
D	Combustible Metals (Flammable metals such as magnesium and lithium)	Blanketing or smothering		D
K	Kitchen Fires (Grease, fat, and oil fires in commercial kitchens)	Blanketing or smothering		K

Fire Extinguishers

Type	Description	Typically Approved for Use On	Not for Use On
Pressurized Water	Water under pressure	A	B C D K
Carbon Dioxide (CO_2)	Carbon dioxide (CO_2) gas under pressure	B C	A D K
Foam	Aqueous film-forming foam (AFFF) or film-forming fluoroprotein (FFFP)	A B	C D K
Dry Chemical, Multipurpose Type	Typically contains ammonium phosphate	A B C	D K
Dry Chemical, BC Type	May contain sodium bicarbonate or potassium bicarbonate	B C	A D K
Dry Powder	May contain sodium chloride, sodium carbonate, copper, or graphite	D	A B C K
Wet Chemical	May contain potassium acetate, potassium citrate, or potassium carbonate	K	B C D

Goodheart-Willcox Publisher

Figure 3-7. Fire classifications and corresponding fire extinguishers for use with each class of fire.

clothing; working in the shade if possible; and staying hydrated are good measures when working in extreme heat.

Workers may also work in extremely cold weather conditions. When working outside in colder weather, workers should wear adequate layers of clothing to prevent frostbite. Boots should be worn when working in wet conditions. Employers should provide training to workers to understand the safety measures their company has in place when working in extreme weather.

3.9 Safe Conduct on the Job

Demonstrating safe conduct means putting safety first while on the job. It is an HVACR service technician's responsibility to practice safe conduct when working with hazardous materials, in hazardous conditions, and around other people. Putting safety first is also the best way to prevent accidents and injuries to yourself or others on the job.

A service technician must do a lot of driving to travel from one customer's home to another. When driving, a technician should drive undistracted—which means paying attention to the road ahead, not texting, and not driving while impaired. The US Department of Labor reported that drug and alcohol use in the workplace causes 65% of on-the-job accidents. Drugs and alcohol on the job do not mix, driving or otherwise. If you must make or take a phone call while driving, do so only with a hands-free set up, or wait until you can pull over. Do not eat while driving. The company you work for may have their own rules when it comes to driving a company vehicle, so it is best to follow these policies.

Working while distracted or goofing off can also lead to accident or injury. While on the job, remain focused on the task at hand and wait until an appropriate time to engage in personal conversations with coworkers or customers. Additionally, you may encounter a situation where you do not know how to perform a task or cannot perform it alone. Recognizing when you need help to complete a task is a great way to practice safe conduct.

Chapter Review

Summary

3.1 OSHA is a government organization responsible for worker safety and health. They enforce workplace safety standards and provide worker training.

3.2 Service technicians are responsible for knowing how and when to use personal protective equipment (PPE) to minimize exposure to hazards.

3.3 There is a risk of electric shock when working around electrical equipment. Exercise caution by following proper procedures and ensuring no electric current is flowing through equipment while working on it. A lockout/tagout (LOTO) procedure involves the use of a lock on an electrical source and tag to communicate to other workers that someone is working on the equipment.

3.4 Proper safety procedures should be followed when working on a ladder. Inspect a ladder for damage before use, set it on firm ground, and avoid using metal ladders when working around electricity.

3.5 Service technicians need to have the proper tools and know how to use them properly. Using a tool for a purpose other than its intended use is a hazard.

3.6 Handle refrigerants and high-pressure cylinders with care. Improperly handled cylinders can burst. Safety data sheets (SDSs) contain safe handling information about hazardous chemicals. Manufacturers, distributors, and importers are required by OSHA to provide SDSs for their chemical products.

3.7 Service technicians are required to know the five different classes of fires (A, B, C, D, and K) and the corresponding types of extinguishers to use with those fires.

3.8 Service technicians may work in extreme weather conditions. To avoid heat stress or frostbite, workers should wear appropriate clothing.

3.9 Demonstrating safe conduct on the job means putting safety first. Driving responsibly, staying focused, and asking for help are great ways to practice safe conduct.

Know and Understand

1. *True or False?* OSHA only enforces workplace safety standards and does not create standards. (3.1)
2. Which of the following describes a service technician's responsibility regarding PPE? (3.2)
   A. Supply all their own PPE when dangerous situations are discovered.
   B. Be familiar with OSHA's and their company's rules and regulations on PPE.
   C. Wear safety glasses at all times.
   D. Seek out PPE training on their own time.
3. *True or False?* An electric shock only occurs when physically coming in contact with an electrical source. (3.3)
4. A ____ is a safety procedure used to prevent accidental startup of a system while it is being worked on. (3.3)
   A. lockout
   B. PASS method
   C. lockout/tagout (LOTO)
   D. short circuit
5. Stepladders are designed to hold how many people at one time? (3.4)
   A. One
   B. Two
   C. Three
   D. Four
6. Always maintain ____ point(s) of contact when climbing a ladder. (3.4)
   A. One
   B. Two
   C. Three
   D. Four

7. Extension ladders should extend at least ____′ above the height you are climbing. (3.4)
   A. 1
   B. 2
   C. 3
   D. 4
8. *True or False?* Using a tool for a purpose other than its intended use is considered a hazard. (3.5)
9. If refrigerant that contains chlorine is mixed with carbon monoxide, it may release ____. (3.6)
   A. phosgene
   B. oxygen
   C. nitrogen
   D. carbon dioxide
10. To find the ingredients of a chemical, check section ____ of the chemical's SDS. (3.6)
    A. 1
    B. 3
    C. 9
    D. 12
11. A refrigerant fire should be extinguished with a Class ____ extinguisher. (3.7)
    A. A
    B. B
    C. C
    D. D
12. *True or False?* Heat stress only affects workers who are 65 years of age or older. (3.8)
13. According to the US Department of Labor, ____% of on-the-job accidents are caused by drug and alcohol use. (3.9)
    A. 2
    B. 25
    C. 50
    D. 65

Apply and Analyze

1. List two requirements OSHA enforces employers to follow regarding PPE use. (3.2)
2. Describe how to properly perform a lockout/tagout when working on an HVACR system. (3.3)
3. Why is a dull tool considered hazardous? (3.5)
4. What is the purpose of a safety data sheet (SDS)? (3.6)

Critical Thinking

1. A service technician is working on a commercial rooftop unit in extreme heat. What safety precautions should they take? (3.7)
2. What should a service technician do if they do not know how to complete a task? (3.9)

FuatKose via Getty Images

4 Hand Tools and Instruments

LEARNING OUTCOMES

After studying this chapter, you will be able to:

4.1 Identify the various types of wrenches and pliers used by service technicians.

4.2 Describe how to select the correct screwdriver and hammer for the job.

4.3 Explain how to use a hacksaw to cut metal objects, drill bits and their uses, and the four common types of snips used in HVACR work.

4.4 Explain how to safely use punches and cold chisels.

4.5 Identify various instruments used in HVACR and their applications.

TECHNICAL TERMS

arbor
cheater
digital multimeter (DMM)
dressed
flutes
hex-head
manifold gauge
manometer
mushroomed
points
psychrometer
serrated
shank
socket
struck tool
swing space
torque
universal joint

Introduction

The job of the refrigeration, heating, and air-conditioning technician consists of performing mechanical operations and using common tools and materials. Tool skills are valuable to the technician and are acquired through experience and practice. In this chapter, tools and instruments commonly used by HVACR service technicians are discussed. Successful technicians select quality tools, care for them properly, and skillfully use them. Misuse of tools can result in delays, callbacks, and injuries.

4.1 Gripping and Clamping Tools

In most HVAC installation and service work, materials need to be gripped or held in place. Wrenches and pliers are commonly used to grip, turn, clamp, or remove various types of fasteners and materials on HVAC systems.

Wrenches

Wrenches are the most widely used type of hand tool. They are used to hold or turn nuts, bolts, cap screws, and other fasteners and can be used to clamp parts together. Applying too much ***torque*** (turning force) can strip bolt threads or break off bolt heads. Quality wrenches are designed to keep leverage and load in safe balance.

CAUTION

Wrench Leverage

Never use a pipe extension or other type of "***cheater***" object to increase the leverage of a wrench.

Always select a wrench with an opening that exactly fits the nut. Wrench size is determined by measuring across the jaw opening. The wrench size is stamped on the side of the tool.

Open-End Wrenches

An open-end wrench has an open jaw on each end, **Figure 4-1**. Each end is a different size, and the jaw opening is set at an angle. The angle permits using the wrench when there is only a small amount of space to turn the nut or bolt. This space is referred to as ***swing space***. The wrench can be turned over to obtain a new grip on the nut.

CAUTION

Open-End Wrenches

A slipping open-end wrench could round nut edges. Open-end wrenches work best on *loose* nuts and bolts.

Box-End Wrenches

A box-end wrench is closed on both ends to surround the nut or bolt head. See **Figure 4-2**. Each end has an opening of a different size. Opening sizes are stamped on the wrench. Box wrenches are available in 6-point or 12-point types. ***Points*** are the "teeth" that grip the edges of the nut or the bolt head. The 6-point wrench is the stronger of the two since it has more grip area. A box-end wrench will not slip easily, so it should be used on bolt heads or nuts that are tight or partially rounded.

Combination Wrenches

Combination wrenches combine the features of open-end and box-end wrenches. They have one open end and one box end, **Figure 4-3**. Both

Goodheart-Willcox Publisher

Figure 4-1. The openings are different sizes at each end of an open-end wrench.

Goodheart-Willcox Publisher

Figure 4-2. Box-end wrenches surround the nut or bolt head, making them much less likely to slip than open-end wrenches.

Goodheart-Willcox Publisher

Figure 4-3. The open end and box end of a combination wrench are usually the same size.

ends are usually the same size. The combination wrench is designed for several tasks. The open end is useful where little swing space is available. The box end provides the grip area needed for final tightening (or loosening of tight bolts).

Refrigeration Service Wrenches

Refrigeration service wrenches are designed to easily turn refrigeration service valve stems. The wrench in **Figure 4-4** is a ratchet-type with square openings to fit most valve stems. A quick-flip reversing lever permits instant reversal for turning service valve stems in either direction. This wrench is a must in every technician's tool kit.

Flare-Nut Wrenches

The flare-nut wrench is a special type of box wrench with an *opening* in the box, **Figure 4-5**. The opening permits the wrench to first slip over tubing and then slide onto a flare nut. Each end of the flare-nut wrench has a different-sized opening.

Adjustable Wrenches

An adjustable wrench has a movable lower jaw, **Figure 4-6**. Like an open-end wrench, this wrench tends to slip. When using an adjustable wrench, make sure it is *tightly* adjusted to the nut. Pull only in the direction that will put force on the *fixed* side of the jaw. See **Figure 4-7**.

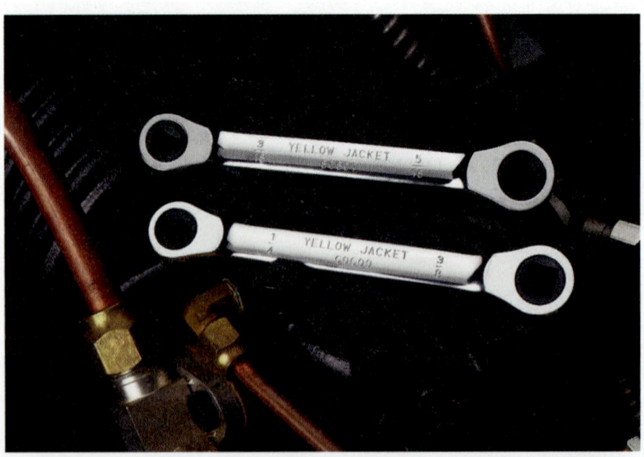

Ritchie Engineering – YELLOW JACKET Products Division

Figure 4-4. The ratchet feature of the refrigeration service wrench makes it easy to turn a service valve stem, even in tight quarters. This wrench is a necessity for every technician.

Goodheart-Willcox Publisher

Figure 4-5. The opening in the box end of the flare-nut wrench allows it to slip over tubing and fit on a flare nut tightly.

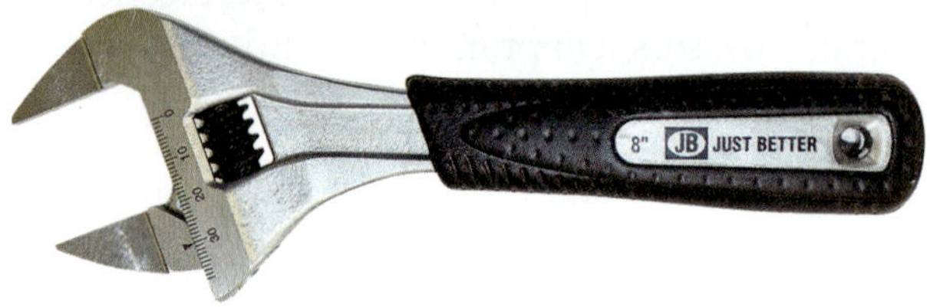

JB Industries

Figure 4-6. An adjustable wrench tends to slip, unless the movable jaw is tightened firmly on the nut or bolt head.

Allen Wrenches

An Allen wrench, also known as a hex key, is a hardened steel shaft with a special hexagonal (6-sided) shape. It is used most often to turn setscrews. Allen wrench sets are available in "jackknife" style, as long individual wrenches, or with T-handles for applying increased torque. See **Figure 4-8**. Long wrenches allow easy access to setscrews used to anchor fan blades and pulleys to a shaft. To prevent damage to the wrench, be sure it is fully inserted in the screw before turning.

Socket Wrenches

A ***socket*** is an individual, cylinder-shaped box-end tool, used with a handle to perform the same turning tasks as other wrenches. One end of

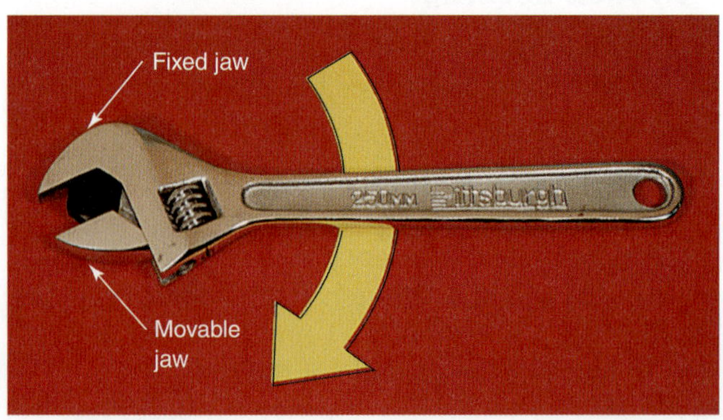

John Walker

Figure 4-7. When using an adjustable wrench, pull only in the direction shown, so that force is exerted on the fixed jaw.

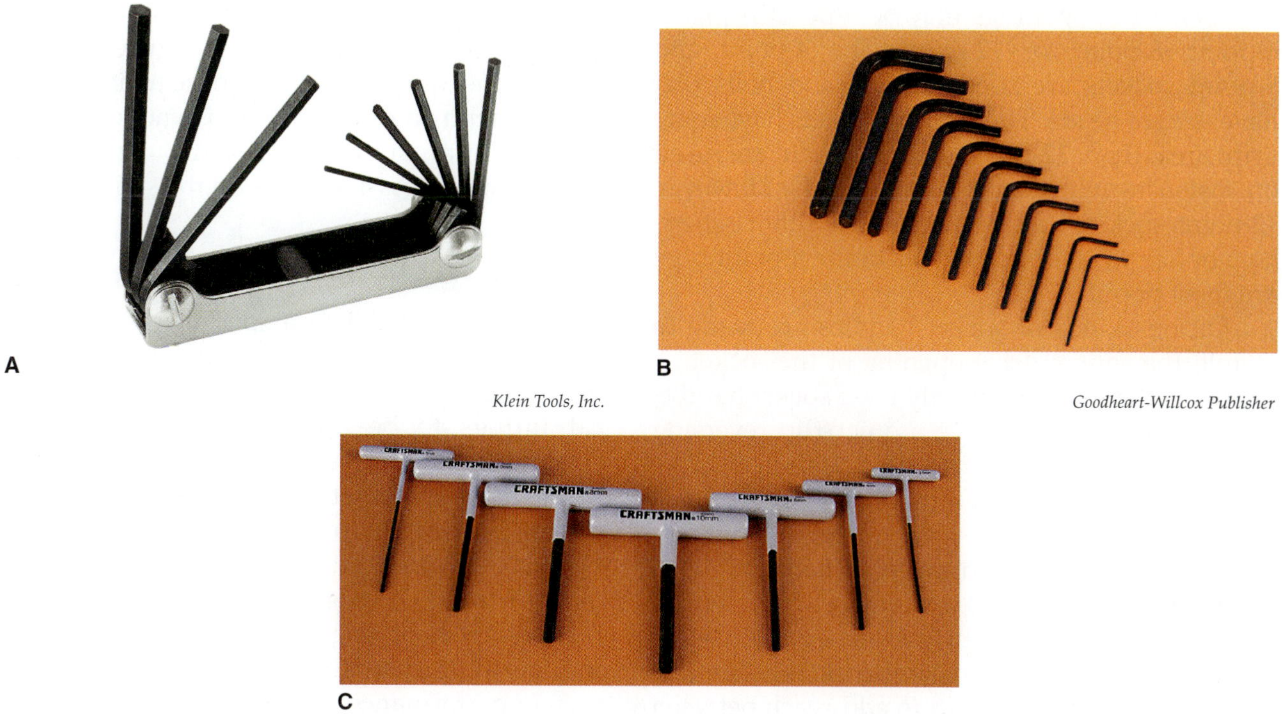

Klein Tools, Inc.

Goodheart-Willcox Publisher

Goodheart-Willcox Publisher

Figure 4-8. Allen wrenches, or hex wrenches, fit hexagonal recesses in setscrews and other fasteners. A—"Jackknife"-style key set. B—Individual wrenches. C—T-handle wrenches.

the socket fits over the bolt head or nut. The other end has a square opening for attaching the drive. Sockets are available in multiple sizes and are often sold in sets like the one shown in **Figure 4-9**.

As shown in **Figure 4-10**, the square drive opening of a socket may be 1/4″, 3/8″, 1/2″, or 3/4″. The drive size is a measure of the socket's capacity to withstand torque. Small drives are used for small fasteners; large drives are used for large fasteners.

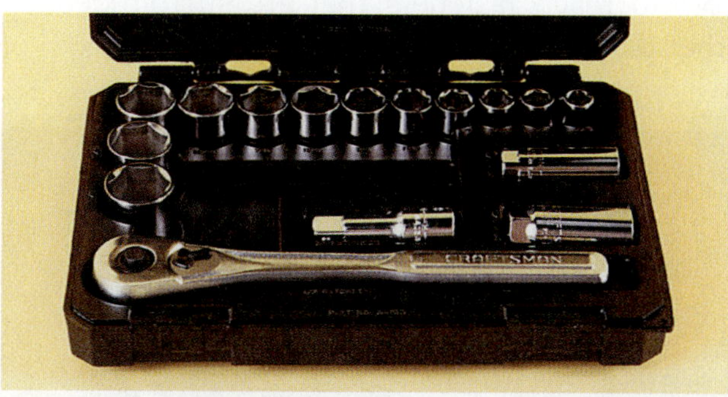

Goodheart-Willcox Publisher

Figure 4-9. A typical socket wrench set.

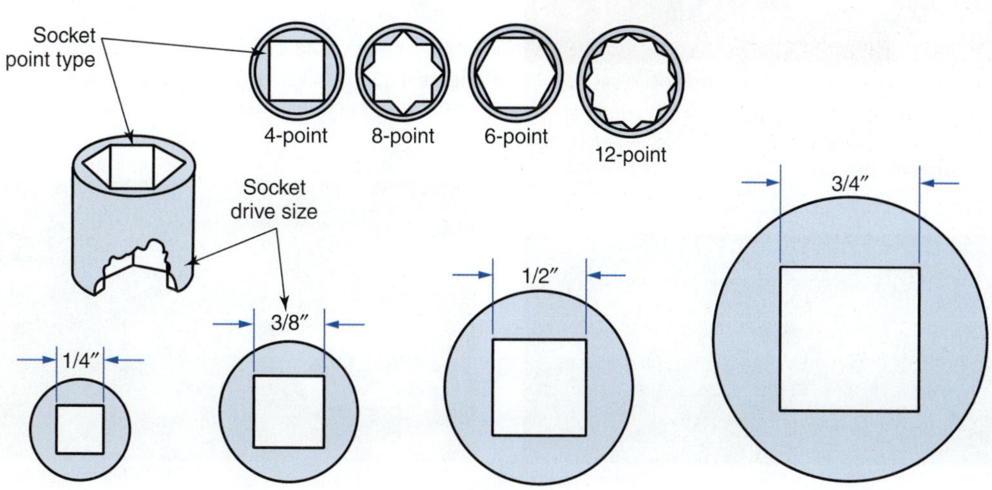

Goodheart-Willcox Publisher

Figure 4-10. Sockets are available in several drive sizes and point configurations.

Socket nut openings may be 4-point (square), 6-point, 8-point, or 12-point. The 4-point and 8-point sockets are used on nuts and bolts that have square heads. The 6-point sockets have more grip area than 12-point sockets; both are used on fasteners with hex (6-sided) heads. However, 12-point sockets can be used with either *square-head* or ***hex-head*** bolts. Sockets of the 6-point type are most popular.

Various kinds of socket handles are made to fit into the square drive opening of the socket. A ratchet is the most commonly used socket handle, **Figure 4-11**. Ratchets have a quick-flip reversing lever for instant reversal of socket rotation, but like any tool, careful use is advised as excessive force can damage the ratchet.

A breaker bar is the strongest socket handle. It is used when breaking loose extremely tight nuts and bolts. After the bolt or nut is loosened, the ratchet handle is used to finish removing the fastener.

Extension bars are used to add reach between the socket and its handle, **Figure 4-12**. They are available in different lengths and can be combined for even greater reach. Extension bars make it possible to reach fasteners that are surrounded by other parts.

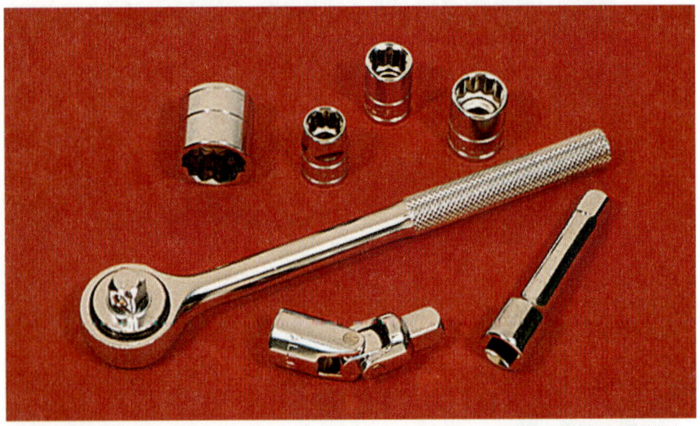

Goodheart-Willcox Publisher

Figure 4-11. Ratchet handles allow tightening or loosening of fasteners when only limited swing space is available. A reversing lever allows easy change of direction.

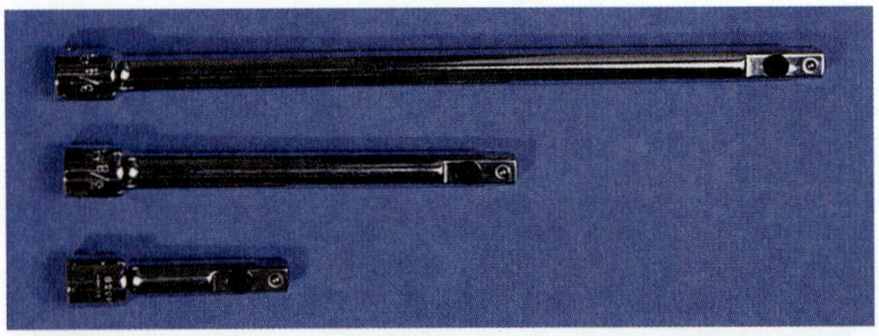

Goodheart-Willcox Publisher

Figure 4-12. Extension bars are used, singly or in combination, to allow tightening or removal of fasteners that otherwise could not be reached.

A ***universal joint*** is a type of swivel that permits reaching around objects, **Figure 4-13**. It is used between the socket and drive handle and can be used with an extension bar. Do not use the universal joint at an extremely sharp angle. A sharp angle places excessive stress on the swivel.

Torque Wrenches

A torque wrench is a special wrench consisting of a handle with a pressure gauge attached to it that allows nuts, bolts, and other mechanical fittings to be tightened to specific torque specifications. Torque wrenches are available in mechanical and digital styles, **Figure 4-14**. Being able to set fittings to manufacturer torque specifications ensures proper equipment safety and performance. Overtightening fittings can create issues such as damaged threads or systems forming leaks which will then decrease efficiency in system performance.

Using Wrenches Safely

When using wrenches, follow these safety rules to prevent injury to yourself or damage to the tool:

- Never use a wrench to do the job of another tool.

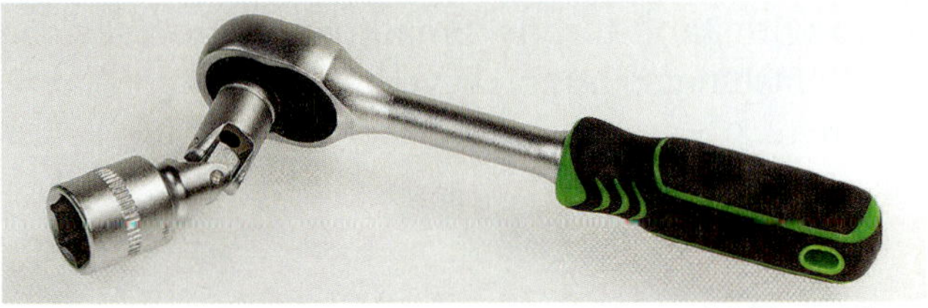

anmbph/Shutterstock.com

Figure 4-13. Universal joints make it possible to reach off-center or hard-to-reach fasteners.

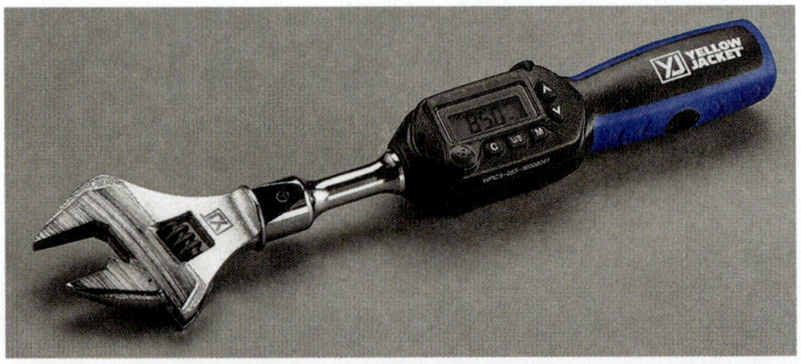

Ritchie Engineering – YELLOW JACKET Products

Figure 4-14. Digital torque wrenches are available that can give real-time readings on a digital display of the amount of torque being applied.

- Never use a wrench opening that is too large for the fastener.
- Never push a wrench beyond its capacity. Quality wrenches are designed and sized to keep leverage and intended load (torque) in safe balance. The safest wrench is a box or socket type because it is less likely to slip off the bolt head.
- Never expose a wrench to excessive heat.
- Never *push* on a wrench unless necessary. Always *pull* on a wrench to protect your knuckles. If you must push, use the open palm of your hand, with fingers slightly curled.
- Never cock or tilt an open-end wrench.
- Never depend on plastic-dipped handles for protection against electrical shock. These handles provide comfort and a firm grip, *not* shock protection.
- Repairing box-end, open-end, or combination wrenches is not recommended. Such wrenches should be discarded and replaced if they have bent handles, rounded or damaged box points, or jaws that are spread, nicked, or battered.
- Ratchets and adjustable wrenches often can be repaired by replacing the damaged parts. However, an adjustable wrench with a bent handle or a fixed jaw that is spread or damaged should be discarded and replaced. Also discard and replace bent socket wrench handles and extensions, and cracked or battered sockets.

SAFETY NOTE

Repair or Replace

Worn or broken tools are neither safe nor efficient. They can be dangerous to the user and those working nearby. Repair or replace worn tools as required. The cost of a new tool is small when compared to the cost of an injury or of time wasted by substandard performance.

Pliers

Pliers are used to grip, cut, crimp, hold, or bend various materials. Each type does its job better than another type. Choosing the correct type of pliers improves efficiency.

Lineman's Pliers

Lineman's pliers are heavy-duty pliers available in various sizes, **Figure 4-15**. These pliers have a strong side-cutting feature for cutting large wires and should be in every technician's tool kit.

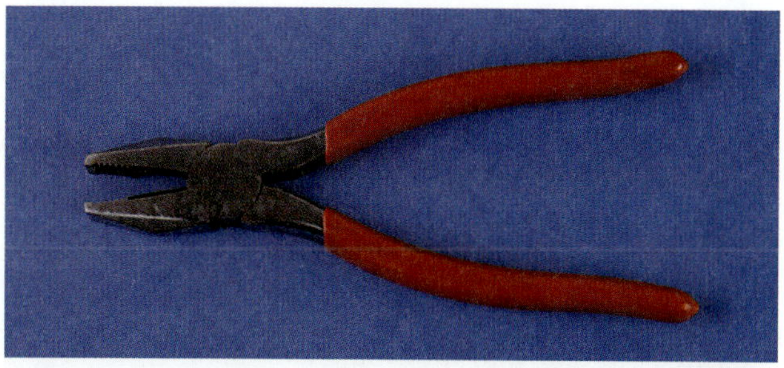

Goodheart-Willcox Publisher

Figure 4-15. Lineman's, or side-cutting, pliers are a basic item of the technician's toolbox. Plastic-covered handles, like those shown on these pliers, provide a better grip on the tool. However, plastic handles do *not* protect against electrical shock.

Long-Nose Pliers

Long-nose pliers are available in three nose designs: needle, round, and flat (sometimes called "duck bill"), **Figure 4-16**. They allow the technician to reach into tight places and perform work that would be awkward or difficult with any other tool. They are available with or without side cutters.

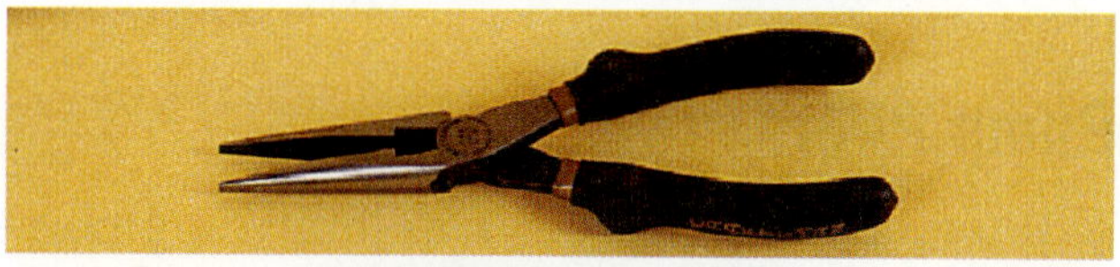

Needle-nose pliers

Goodheart-Willcox Publisher

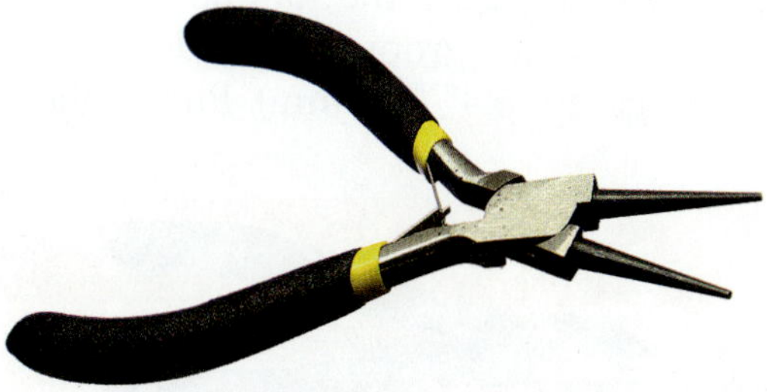

Round-nose pliers

JV Korotkova/Shutterstock.com

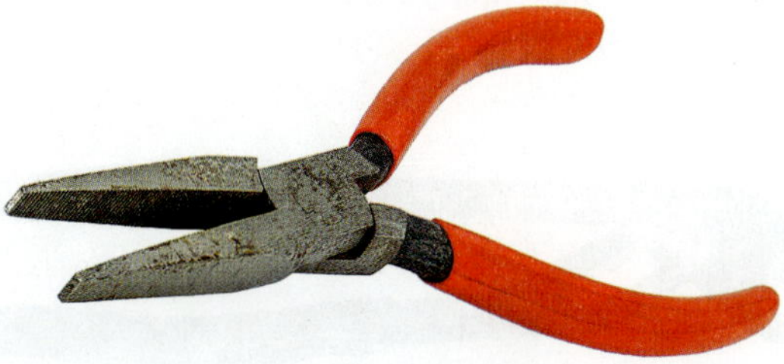

Flat-nose pliers

Silo/Shutterstock.com

Figure 4-16. Long-nose pliers are available in needle-, round-, and flat-nose types.

Most long-nose pliers are designed for electrical use but are useful in other ways.

CAUTION

Long-Nose Pliers Damage

To avoid damage to long-nose pliers, never *pry* or try to bend *stiff wire* with the pliers' tips.

Avoid exposing these pliers to excessive heat; the jaw tips will bend outward and render the tool useless. The long-nose pliers recommended for refrigeration, heating, and air-conditioning work are the heavy-duty 8″ (203 mm) size, with cutter, in the round pattern.

Diagonal Cutting Pliers

Diagonal cutters are designed for cutting electrical wires, cotter pins, nails, and other types of fasteners. See **Figure 4-17**. When cutting wire, always cut at a right angle to avoid damage to the pliers. Do not rock the pliers from side to side. The 8″ (203 mm), heavy-duty type is best for refrigeration, heating, and air-conditioning work.

Pump Pliers

Pump pliers, also known as utility or groove-joint pliers, are used by all types of mechanics, **Figure 4-18**. The jaws are positioned and locked into place by engaging the tongue in the proper groove. A series of grooves provides a range of jaw openings up to 4″ (102 mm). Pump pliers will not slip even under heavy pressure. They will grip round, square, flat, and hexagonal objects and can apply limited torque without damaging the equipment.

Goodheart-Willcox Publisher

Figure 4-17. Diagonal cutting pliers are used for heavier cutting tasks than other types of pliers.

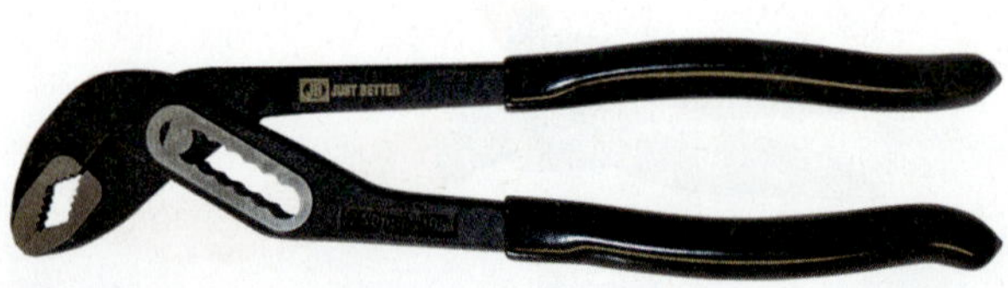

Goodheart-Willcox Publisher

Figure 4-18. Pump pliers provide a strong, nonslip grip on either round or square objects.

Slip-Joint Pliers

Slip-joint pliers are an old but still widely used style of pliers, **Figure 4-19**. The slip joint provides two jaw positions, and the pliers may have a shear-type wire cutter. These pliers are designed for gripping, turning, and bending.

Adjustable Pliers

There are different types of adjustable pliers available. At least one type permits one-handed adjustment of jaw size by merely squeezing the handles. See **Figure 4-20**. Some types of adjustable pliers may not have a locking feature.

Locking Pliers

Locking pliers are available with straight or curved jaws, in a variety of sizes, **Figure 4-21**. These pliers use a compound leverage system to lock the jaws for holding various shapes and sizes of work. It is a combination tool that functions as pliers, a wrench, portable vise, or clamp. Because of their clamping power, locking pliers can free

Goodheart-Willcox Publisher

Figure 4-19. Slip-joint pliers are an "all-purpose" tool for gripping and holding objects.

Goodheart-Willcox Publisher

Figure 4-20. This type of adjustable pliers offers one-handed operation. Jaw size adjustment is done automatically when the handles are squeezed.

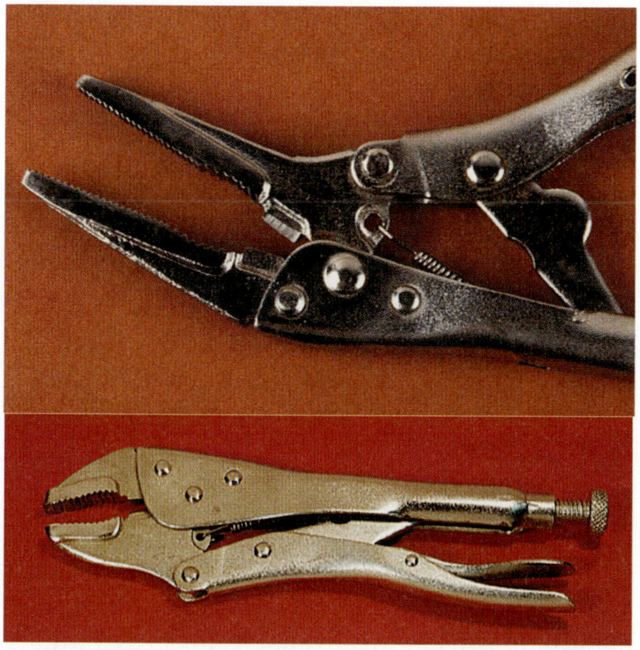

Goodheart-Willcox Publisher

Figure 4-21. Locking pliers are versatile and can be used to hold objects, freeing both hands for other tasks. They are available with several types of jaws.

both hands for doing other work (such as brazing, sawing, or drilling). Locking pliers can sometimes be used to unscrew fasteners that have stripped or rounded heads. Since they will mar surfaces, however, locking pliers and any style of pliers should never be used on good nuts and bolts. Attempts to repair this tool are not recommended.

Using Pliers Safely

When using pliers, follow these safety rules to prevent injury to yourself or damage to the tool:

- Never use pliers to do the job of another tool.
- Never push pliers beyond their capacity. Bending stiff wire with light pliers or the tip of needle-nose pliers can spring or break the tool.
- Never expose pliers to excessive heat. Direct flame on metal can draw the temper (hardness) and ruin the tool. Cutting pliers are especially vulnerable to high heat.
- Never cut hardened wire with ordinary pliers. Also, never rock pliers from side to side when cutting wire, nor bend the wire back and forth against the cutting knives. These practices can dull or nick cutting edges.
- Never cut any wire or metal unless your eyes and your fellow workers' eyes are protected. Safety goggles are an absolute must when working with pliers.
- Never depend on plastic-dipped handles to insulate you from electricity. They are intended for comfort and improved grip, not protection against electric shock.
- While repairing pliers is not recommended, regular maintenance is important. Dull cutting knives may be sharpened, and the hinge should receive a drop of oil occasionally. Pliers that are cracked, broken, sprung, or have nicked cutting knives should be discarded and replaced.

4.2 Fastening and Driving Tools

Many different fasteners are used in HVAC systems to secure components in place. Screwdrivers and hammers are commonly used to install or remove fasteners depending on the application.

Screwdrivers

Screwdrivers are used to drive or withdraw threaded fasteners, such as wood screws, machine screws, and self-tapping screws. Screwdrivers are available in a wide variety of shapes and sizes, Figure 4-22. Choosing the right screwdriver is vital for fast and efficient driving and removal of screws. Screwdriver tips are designed to tolerate a limited amount of strain or torque. A screwdriver that is too long, too short, or does not fit the screw properly can waste time and cause damage to the fastener. See Figure 4-23.

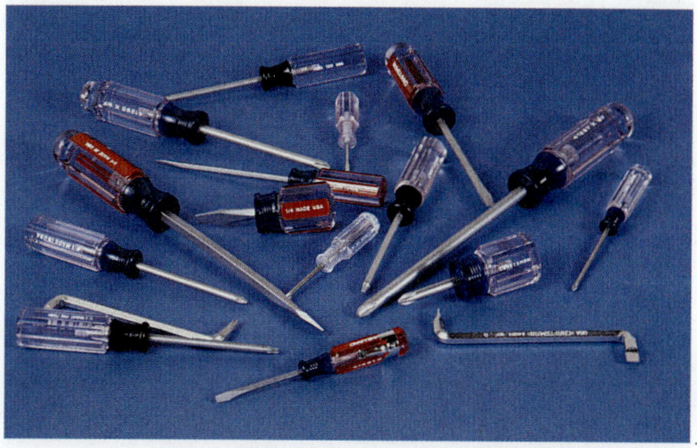

Goodheart-Willcox Publisher

Figure 4-22. Screwdrivers are available in a wide variety of sizes, shank lengths, and tip shapes.

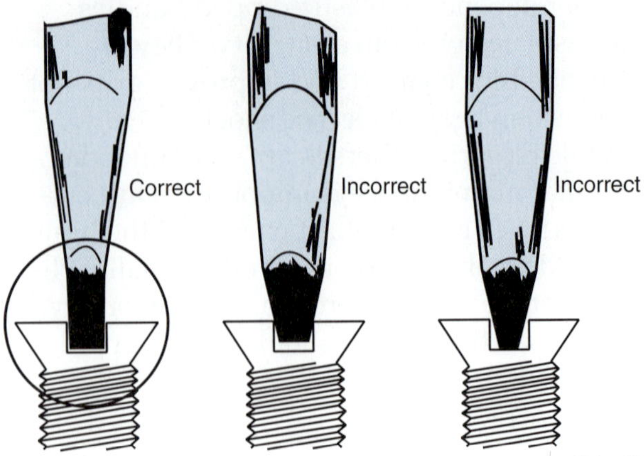

State of Ohio

Figure 4-23. The tip of the screwdriver must fit the slot properly to avoid damaging the slot or slipping and causing injury.

Special Screwdrivers

Reversible screwdrivers have a different tip at each end of the ***shank*** (shaft). The driver handle is removable and fits at either end of the shank. The most common reversible screwdriver has a standard (straight-blade) tip at one end and a Phillips tip at the other.

Interchangeable bit screwdrivers allow different types and sizes of bits to be mounted to suit the task at hand, **Figure 4-24**. Straight-blade, Phillips, Torx®, square socket, and nut-driver tips are available. Some models provide storage in the handle for several bits.

The screw-holding screwdriver is indispensable for working in close quarters or electrical boxes. Screw-holding drivers are used only to get the screw firmly started. After starting, a conventional driver is used for further driving. There are two types of screw-holding screwdrivers: one that holds by means of pressure against the sides of the screw slot and one that has spring-steel fingers to grip the screw head. The use of a screw-holding screwdriver with spring-steel fingers is shown in **Figure 4-25**.

Screwdrivers for Torx®, Allen, Pozidriv®, and other special screwhead patterns are available in many sizes and lengths. **Figure 4-26** shows these special patterns.

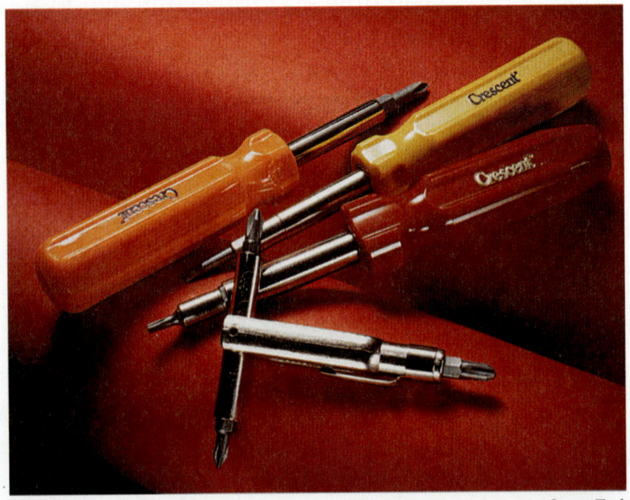

CooperTools

Figure 4-24. Reversible and interchangeable bit screwdrivers are versatile, since they allow easy selection of the tip needed for a job.

Nut Drivers

A nut driver is a special tool with a 6-point socket at the tip of a screwdriver-type handle, **Figure 4-27**. Nut drivers are useful for removing and replacing small hex head (6-sided) screws and nuts. Nut drivers are available in sizes ranging from 3/16″ to 5/8″. Handles are usually color-coded for quick selection of size. Hollow shaft nut drivers are considered best for general use.

Goodheart-Willcox Publisher

Figure 4-25. In one type of screw-holding screwdriver, spring-steel fingers grip the head of the screw and hold it firmly against the blade.

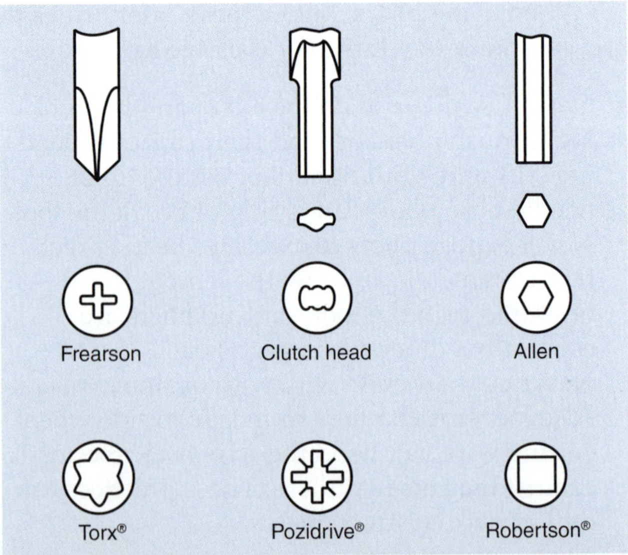

Vaco Products

Figure 4-26. Besides standard and Phillips screwhead patterns, there are several special patterns like those shown here. The Torx® pattern is widely used in the automotive industry, for example.

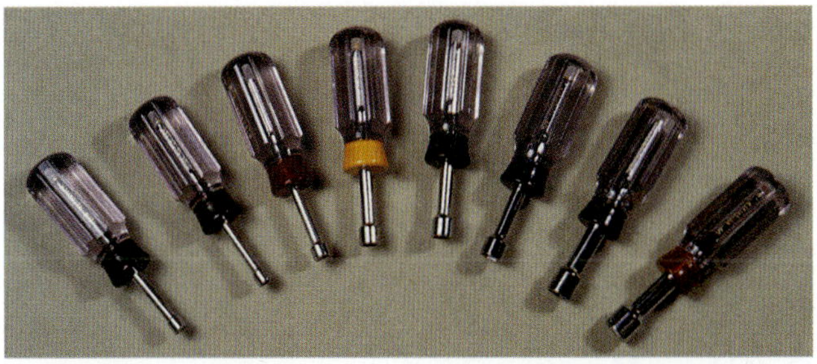

Goodheart-Willcox Publisher

Figure 4-27. Nut drivers, like these hollow-shaft types, usually have handles that are color-coded for easy identification.

Using Screwdrivers Safely

When using screwdrivers, follow these safety rules to prevent from injuring yourself and damaging the tool:

- Always fit the tool to the work. The size of the screw and the type of drive opening determine which driver to use.
- Never use a driver to do the job of another tool.
- Never push a driver beyond its capacity. For heavy work that requires a wrench to help do the turning, use a square-shank driver.
- Never use a driver at an angle to the screw. The driver could slip and damage the screw head or cause injury.
- Never expose a driver to excessive heat.
- Never depend on a driver's handle or covered blade for insulation from electricity.
- Do not attempt to repair most types of drivers. Drivers with cracked handles, bent or twisted shafts, or worn tips should be discarded and replaced. The tip of a straight-blade driver, however, can be ***dressed*** (squared) on a bench grinder. Be careful to avoid letting the tip get hot. This will draw the temper causing the tip to become soft and easily damaged.

Hammers

Hammers are used for various driving (striking) operations. It is important to use the correct hammer and to use it properly. For example, a nail hammer should not be used to strike a chisel or other hardened tool. Likewise, a soft-faced hammer should not be used to drive a nail; nor should a sledge-type hammer be used to drive a brad.

Nail Hammers

Nail hammers are generally made in two patterns: curved claw and straight claw, **Figure 4-28**. They are available in various head weights for different types of nailing: 16 oz (450 g) and 20 oz (570 g) hammers are for general use; 22 oz (625 g) and 28 oz (795 g) hammers are used for heavy-duty framing work. Framing hammers usually have longer handles than those for general use. Handles are made of wood, fiberglass, or steel.

CAUTION

Suitable Hammer

Never drive hardened-steel cut nails or masonry nails with a nail hammer. These nails will shatter under the force of an indirect or glancing blow. Instead, use an engineer's (sledge-type) hammer.

Ball Pein Hammers

Ball pein hammers are one of the most frequently used hammers, **Figure 4-29**. They are designed for striking chisels and punches and for riveting, shaping, and straightening metal. Ball pein hammers are available with head weights ranging from 2 oz to 48 oz (55 g to 1360 g), depending on the task. When striking a tool like a chisel or punch, the hammer face should be 3/8″ (9.5 mm) larger than the struck tool.

Curved-claw hammer

CooperTools

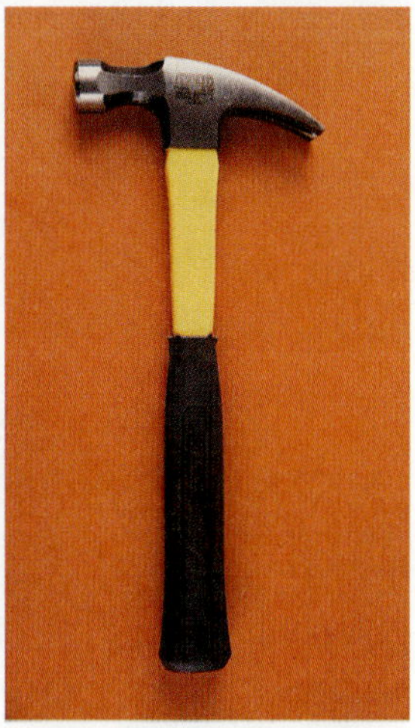

Straight-claw hammer

Goodheart-Willcox Publisher

Figure 4-28. Nail hammers are available in a variety of weights and in either curved-claw or straight-claw patterns.

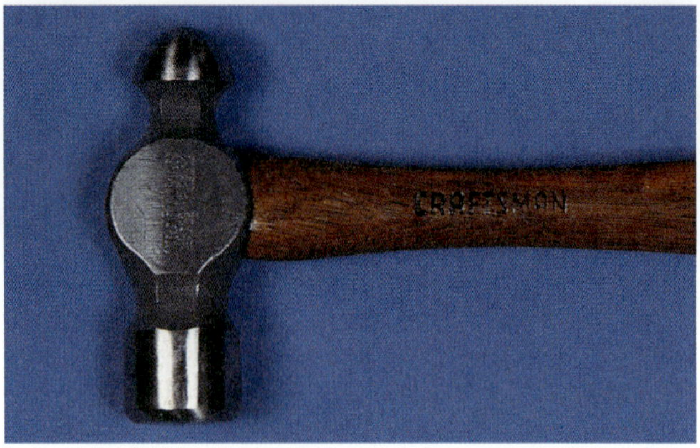

Goodheart-Willcox Publisher

Figure 4-29. The ball pein hammer is the proper tool to use when striking cold chisels or punches.

Soft-Faced Hammers

The heads of soft-faced hammers or mallets are made of wood, rawhide, rubber, plastic, copper, or brass, **Figure 4-30**. They are used where steel hammers would mar or damage the workpiece.

> **CAUTION**
>
> **Driving Nails and Screws**
>
> Never use a soft-faced hammer to strike sharp metal objects or to drive nails or screws.

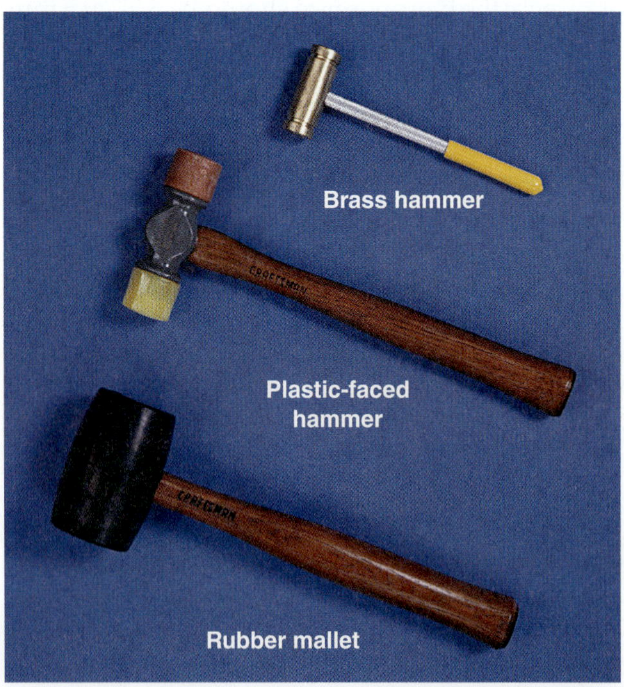

Goodheart-Willcox Publisher

Figure 4-30. The heads of soft-faced hammers or mallets are made from such materials as rawhide, rubber, wood, brass, or plastic.

Engineer's Hammer

The engineer's hammer, or sledge-type hammer, is designed for heavy-duty striking, **Figure 4-31**. Weights are typically 2 1/2–4 lb (1.1–1.8 kg). This type of hammer is commonly used for striking spikes, cold chisels, and hardened nails.

Setting Hammers

Setting hammers are designed for working on sheet metal, **Figure 4-32**. They have square corners and straight sides to form sharp corners and to close seams. The corners and sides of the setting hammer are easily chipped if incorrectly used.

Using Hammers Safely

When using hammers, follow these safety rules to prevent injury to yourself or damage to the tool:

- Always align the face of the hammer squarely with the surface being struck. Avoid glancing blows.
- Never use a nail hammer to strike a chisel or other hardened objects, such as masonry nails. The hammer face may chip and cause injury.
- Always wear safety goggles when using a hammer to drive nails or strike objects.

Goodheart-Willcox Publisher

Figure 4-31. The engineer's hammer, or sledge, is used for such tasks as driving hardened nails.

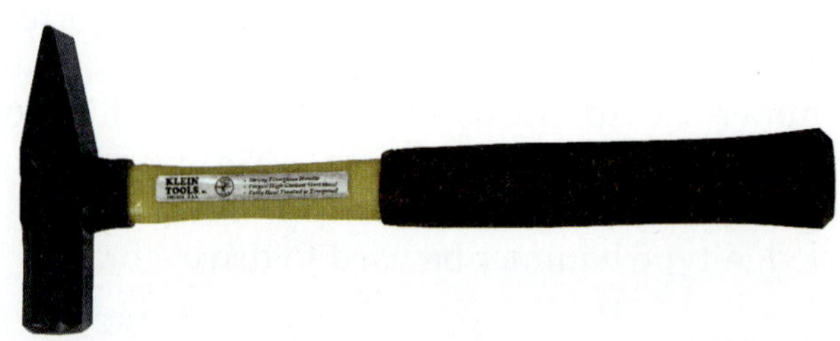

Klein Tools, Inc.

Figure 4-32. Setting hammers have square heads that make them ideal for forming seams and corners in sheet metal.

- Never strike one hammer with another hammer.
- Never use a hammer that has a loose or damaged handle.
- Never use the side or "cheek" of a hammer for striking any object, except in the case of the setting hammer.
- Never grind or redress the face of a hammer. If the hammer head is damaged, discard and replace it.
- Discard hammers that have damaged heads, as they are dangerous to use. Damage is indicated by dents, chips, excessive wear, or a broken claw. Damaged wood or fiberglass handles can usually be replaced.

4.3 Cutting Tools

Service technicians may have to cut into building materials or system components depending on the application. Hacksaws and drill bits are useful in HVAC work due to their flexibility in cutting various types of building materials.

Hacksaws

Hacksaws are used to cut all types of metal objects, **Figure 4-33**. Hacksaw blades are selected according to the number of teeth per inch: 18 (coarse), 24 (medium), and 32 (fine). The blade should be fastened tightly in the frame, with the blade teeth pointed away from the handle. At least two teeth should contact the metal being cut, or the saw will catch and bind.

You can use one or two hands on the saw to make cuts. Press down lightly on the forward stroke and release pressure on the backstroke. Avoid rapid strokes, else the blade will overheat and become dull. Use 50 to 60 strokes per minute.

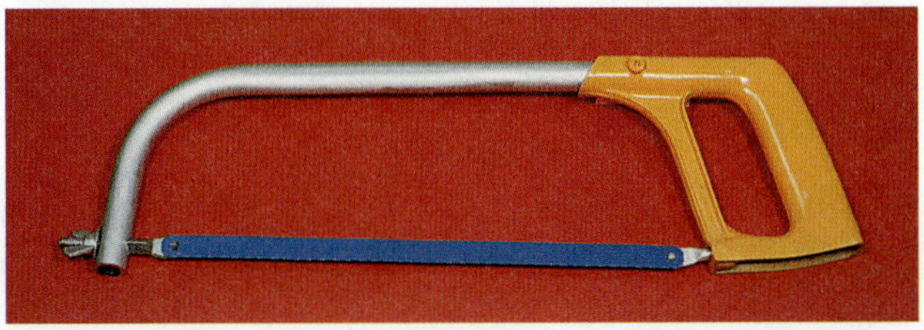

Goodheart-Willcox Publisher

Figure 4-33. The hacksaw is an important tool for cutting metal. Blade teeth should face away from the handle, so the saw cuts on the forward stroke.

Small or "mini" hacksaws are used for working in confined areas or for flush cutting, **Figure 4-34**. These small hacksaws are indispensable to the technician.

Drill Bits

Various types of drill bits are used for boring and cutting holes. Each is designed for use on a specific type of material: drill bits designed for wood cannot be used on metal or concrete; bits for metal can be used on wood but not concrete; bits for concrete cannot be used on metal or wood.

Twist Drill Bits

Twist drill bits are used to cut small-diameter holes in metal or wood. Bits made of high-speed steel are most common and are sometimes carbide-tipped. A twist drill bit has three principal parts: a *shank* that is clamped into the drill chuck, a *body* with two spiral grooves called ***flutes***, and a cone-shaped cutting end called the *point*. See **Figure 4-35**. The flutes act as channels for the escape of metal chips from the hole being drilled. Twist drills can be purchased separately or in sets, with drill sizes ranging from 1/16″ to 1/2″. See **Figure 4-36**.

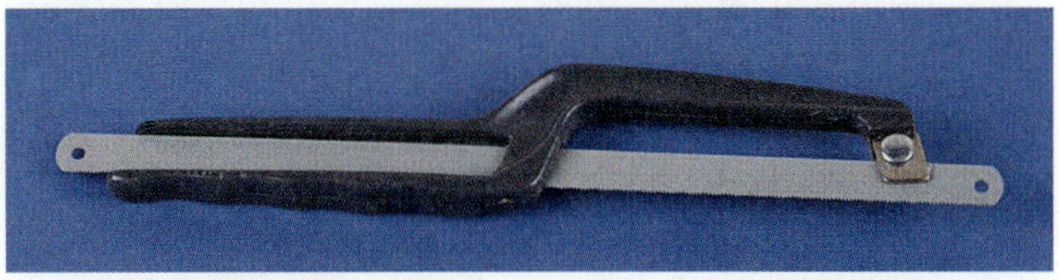

A

B

Vermont American

Figure 4-34. For flush-cutting metal or working in close quarters, the mini-hacksaw is the ideal tool. A—A frame-type mini-hacksaw. B—A model with a contour grip.

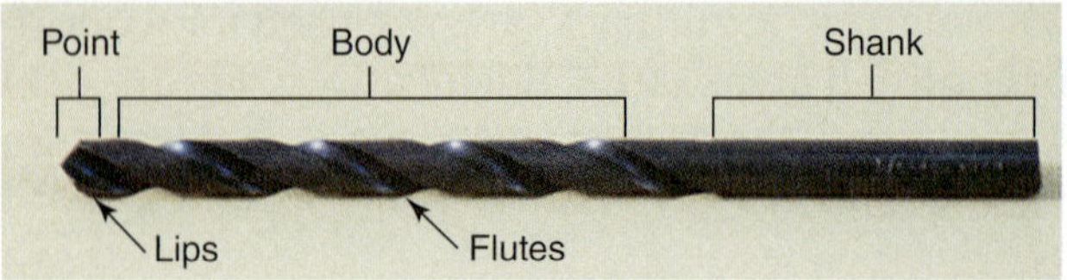

Goodheart-Willcox Publisher

Figure 4-35. Parts of a typical twist drill bit.

Goodheart-Willcox Publisher

Figure 4-36. Twist drill bits are available individually or in sets. Most often, a technician starts with a set, then replaces bits individually as necessary.

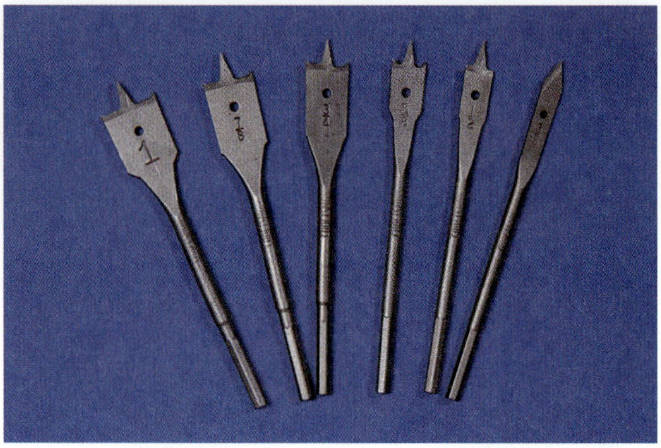

Goodheart-Willcox Publisher

Figure 4-37. Spade-type bits are used to bore medium to large holes in wood and similar nonmetal materials. Sizes are stamped on each bit.

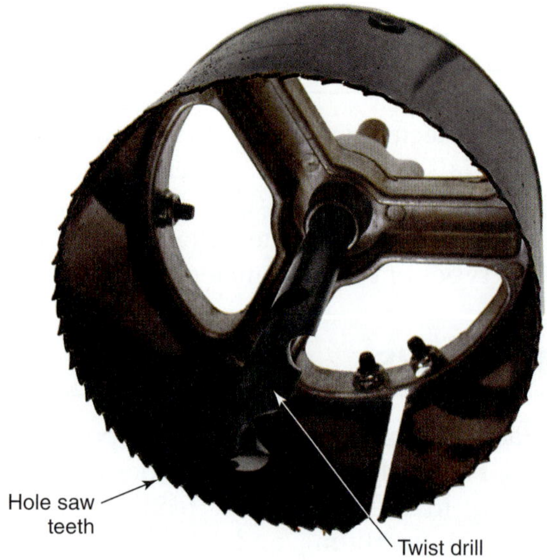

Malco Products, Inc.

Figure 4-38. Hole saws are attached to an arbor that includes a twist drill for boring a pilot hole.

Wood Bits

The *spade-type wood bit* is designed to cut straight or angled holes in wood, plastic laminate, plaster, foam insulation, and similar materials. Like twist drills, they are available as individual bits or in sets, in sizes ranging from 1/4″ to 1 1/2″ (6 mm to 38 mm). As shown in **Figure 4-37**, sizes are stamped on each bit. Spade-type bits can be resharpened with a file or grinder. Extension bars are available for extended reach.

Auger-type wood bits are used for drilling large holes and are available in sizes up to 2 1/2″ (63.5 mm). The pilot screw tip draws the bit into the wood. Cutting blades can be sharpened with a file.

Hole Saws

High-speed hole saws are designed for cutting steel, aluminum, copper, brass, sheet metal, stainless steel, wood, or plastics. See **Figure 4-38**. The high-speed cutting edges are welded to steel bodies to produce clean, accurate holes. Shatter-resistant construction makes them safe and durable. Hole saw sets are available and include sizes ranging from 9/16″ (14 mm) to as much as 5″ (127 mm), **Figure 4-39**. Individual hole saws are quickly attached to an ***arbor*** (spindle or axle) that has a twist drill for starting the pilot hole.

Masonry Bits

Masonry bits are used for drilling brick, stone, concrete, and ceramic materials, **Figure 4-40**. Such holes are often needed to allow use of concrete anchors or similar fasteners. The special carbide

Malco Products, Inc.

Figure 4-39. Hole-saw sets include one or more arbors and a range of saw sizes for holes of different diameters.

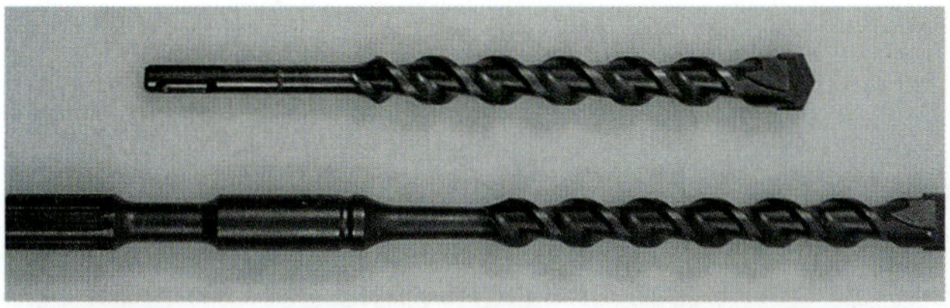

American Tool Companies, Inc.

Figure 4-40. Carbide-tipped masonry drills make fast work of holes in concrete, brick, and similar materials. Note the reduced shank on the drill at top. It allows use of the drill in a chuck with an opening smaller than the drill size.

tip resists dulling. Flutes remove dust from the hole so that the bit cuts at maximum efficiency without clogging. Bits may be resharpened on a bench grinder using a silicon carbide grinding wheel. Sizes range from 1/8″ to 1 1/2″ (3 mm to 38 mm).

Star drills are designed for hand-drilling holes in masonry, **Figure 4-41**. The bit is held in place with one hand while struck with an engineer's hammer held in the other hand. The bit must be rotated slightly after each blow. The four cutting edges steadily chip into the material with each blow.

SAFETY NOTE

Masonry Bit Usage

Always wear safety goggles when using any form of masonry bit. *Never* use a star drill with a struck face that is chipped or ***mushroomed*** (flattened and spread out from being struck).

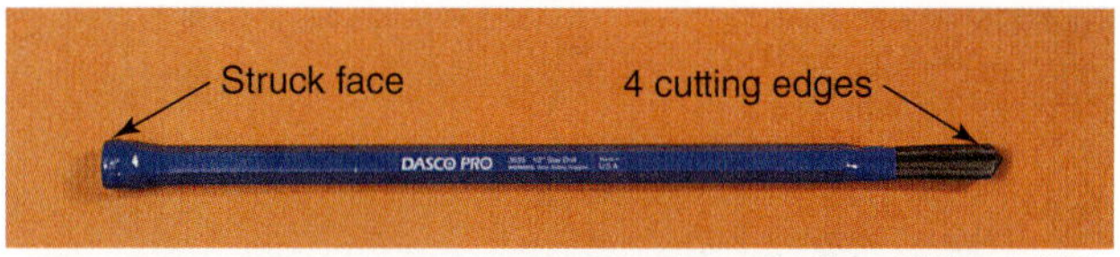

Goodheart-Willcox Publisher

Figure 4-41. The star drill is struck with a sledge-type hammer to make holes in concrete or masonry materials.

Step Bits

Step bits are special purpose drill bits that allow thin metal to be drilled out to wide holes, **Figure 4-42**. These bits are often used during installations to drill holes in units, disconnect boxes, Unistrut channels, and other metals or plastics where holes need to be enlarged or made wider. The bits are cone shaped so that as the bit cuts one hole, it can step up to the next size easily.

Snips

Snips are used to cut sheet metal, screens, gaskets, and straps. There are four basic types in common use: straight-pattern, combination, duck-bill, and aviation. Snips are made for right-handed use, although they can be used in either hand.

Straight Snips

Straight-pattern snips are used for making straight-line cuts, **Figure 4-43**. Curved cuts can be made if the curve is not extreme. Sizes vary from 7″ to 16″ (178 mm to 406 mm) in overall length. The most popular size is 10″ (254 mm).

Combination and Duckbill Snips

Combination snips and duckbill snips can be used for straight-line cutting as well as cutting curves in either direction, **Figure 4-44**. When

Vinokurov Alexandr/Shutterstock.com

Figure 4-42. Step bits are used to drill wide holes in metal or plastic. Note the spiraled cone shape that allows stepping up from one size to the next as the bit is inserted deeper.

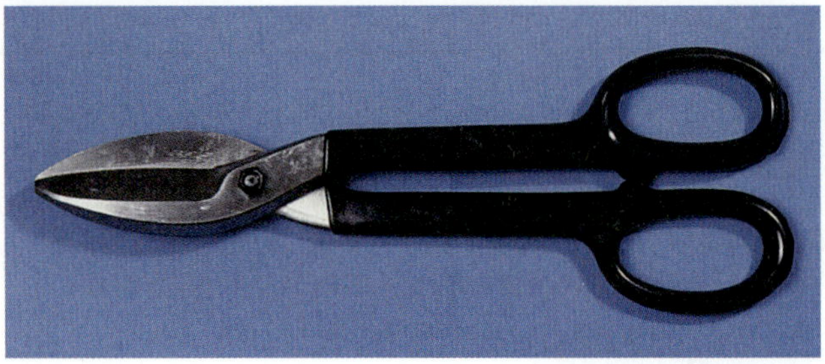

Goodheart-Willcox Publisher

Figure 4-43. Straight-pattern snips are used primarily for straight cuts but can make gentle curves if necessary.

Goodheart-Willcox Publisher

Figure 4-44. Duckbill snips are designed for making curved cuts in sheet metal or similar materials.

used for straight-line cutting, both types tend to bend the metal, so they require more effort than a straight-pattern snip. The duckbill snip will cut smooth curves in sheet metal in either direction, **Figure 4-45**.

Aviation Snips

Aviation snips are popular in all fields of metalwork. Their compound lever action requires less effort to cut metal than other types of snips, **Figure 4-46**. The blades are made of alloy steel,

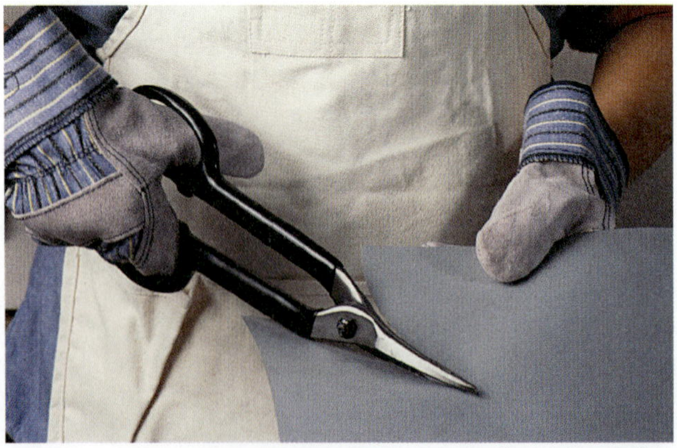

Goodheart-Willcox Publisher

Figure 4-45. Smoothly curved cuts in either direction can be made with duckbill snips.

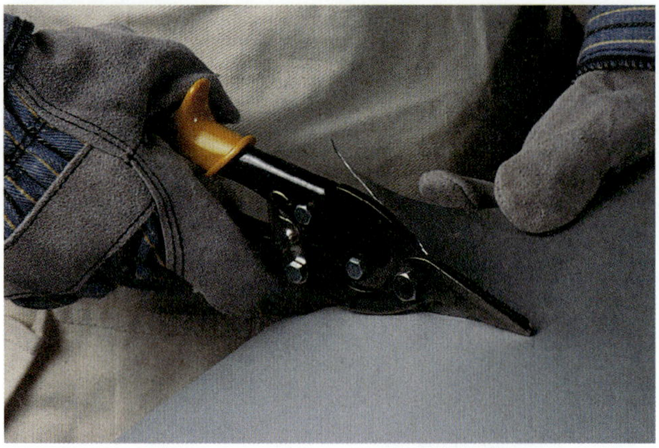

Goodheart-Willcox Publisher

Figure 4-46. Aviation snips use compound lever action to make cutting easier. They are available in models designed for straight cuts, left-hand curves, or right-hand curves.

so they are also harder and tougher than regular snips. The cutting blades are ***serrated*** (toothed or ridged) to prevent metal from slipping out when pressure is applied, **Figure 4-47**. The standard size for the professional aviation snip is 10″ (254 mm).

Aviation snips are available in three types of cut patterns. The straight pattern is for cutting straight lines. The left-hand pattern is for cutting curves to the left. The right-hand pattern is for cutting curves to the right.

Using Snips Safely

When using snips, follow these safety rules to prevent injury to yourself or damage to the tool:

- Always wear safety goggles when using snips.
- Be careful of the sharp cutting edges on snip blades.
- Always wear gloves to protect against cuts from sharp metal edges.

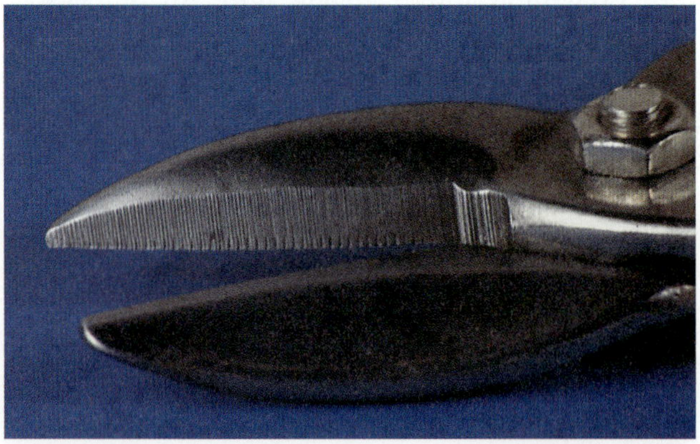

Goodheart-Willcox Publisher

Figure 4-47. Aviation snips are made from tough alloy steel and have serrated jaws to keep metal from slipping.

- Use snips only for cutting soft metal or other soft materials.
- Use only hand pressure for cutting. Never hammer on the snip or use your foot to put extra pressure on the cutting edge.
- Perform periodic maintenance to keep snips efficient and safe. Occasionally oil the pivot bolt and protect the cutting edges from damage. Sharpen the edges as necessary. Never try to sharpen an *aviation* snip, however. This removes the serrations that grip the metal.

4.4 Struck Tools

Punches and cold chisels are referred to as ***struck tools***, because they are struck with a hammer. They are made of special alloy steel that is heat-treated and drawn to provide maximum resistance to impact, **Figure 4-48**. They are designed to direct the force of each blow toward the center, or body, of the tool. Tools must be struck squarely. Blows that are off-center can damage the chisel or material or cause injury.

SAFETY NOTE

Punch and Chisel Eye Protection

Always wear eye protection when using chisels or punches.

Punches

A center punch and prick punch are similar, but the prick punch has a sharper point. Either one can be used to mark parts before disassembly. This ensures proper alignment when the parts are reassembled. A punch is also used to make an indentation in metal before drilling a hole. The indentation prevents the drill bit from wandering (moving off center).

A pin punch has a straight shank for driving pins or rivets out of a hole. A similar tool, the drift punch, is often used for aligning holes in different sections of material.

SAFETY NOTE

Damaged Punch Point

Never use a punch with a dull, chipped, or deformed point. Such a damaged punch could slip and cause injury.

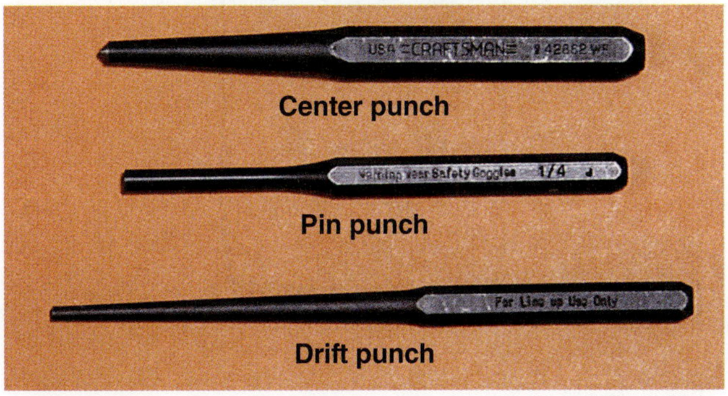

Goodheart-Willcox Publisher

Figure 4-48. Punches are usually struck with a ball pein hammer.

Cold Chisels

Cold chisels are used to cut off damaged or badly rusted nuts, bolts, rivet heads, or other fasteners, **Figure 4-49**. The angle and thickness of the cutting edge are designed to give maximum cut

Goodheart-Willcox Publisher

Figure 4-49. A cold chisel can be used to cut through bolts, pins, or other metal fasteners or parts.

and durability. A *dull* cold chisel can slip and cause injury. Sharpen the hardened cutting edge as needed with a hand file or a whetstone (special sharpening stone).

SAFETY NOTE

Mushroomed Chisel

Never use a chisel that has a mushroomed head (striking face) since chips of metal could fly off and cause injury. Use a grinder to dress the chisel (remove the mushroomed metal and form a smooth chamfer), **Figure 4-50**.

Using Punches and Cold Chisels Safely

When using punches and cold chisels, follow these safety rules to prevent injury to yourself or damage to the tools:

- Always wear eye protection when working with chisels and punches.
- Always strike a punch or chisel squarely. An off-center blow could cause damage or injury.
- Never use a cold chisel to cut or split materials such as stone or concrete.
- Never use a dull chisel or a punch that has a chipped or deformed point.
- Never use a chisel or punch with a mushroomed head. Grind off the mushroom to form a chamfer. Avoid overheating when grinding.

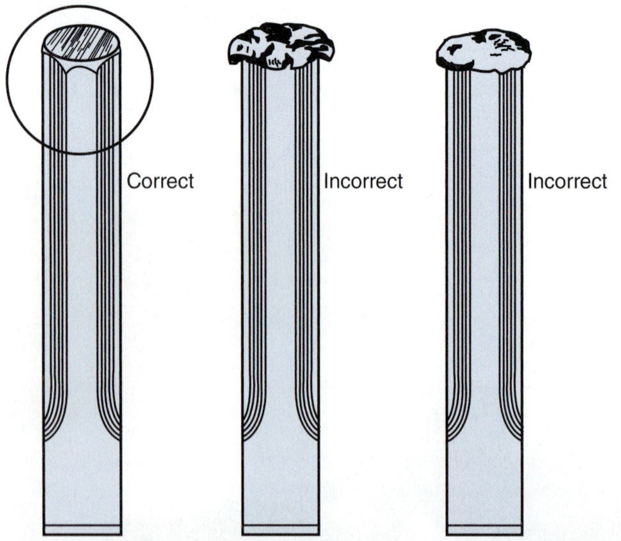

State of Ohio

Figure 4-50. A chisel or punch with a mushroomed head is dangerous because bits of metal can fly off when the tool is struck. A properly dressed chisel or punch is ground to a smooth chamfer as shown.

4.5 Instruments

Data-collecting instruments are valuable tools for HVACR technicians. The following section is an overview of some common instruments used by service technicians. These instruments are covered in greater detail in later chapters.

Digital Multimeters (DMMs)

A ***digital multimeter (DMM)*** is one of the most important instruments an HVACR technician will use due to its ability to accurately measure many different electrical variables of an HVACR system. Digital multimeters are offered by numerous manufacturers, and all have similar functions that include being able to read voltage, resistance, amperage, capacitance, temperature, and more depending on the model chosen, **Figure 4-51**. These electrical readings can help a technician diagnose and troubleshoot electrical problems in an HVACR system.

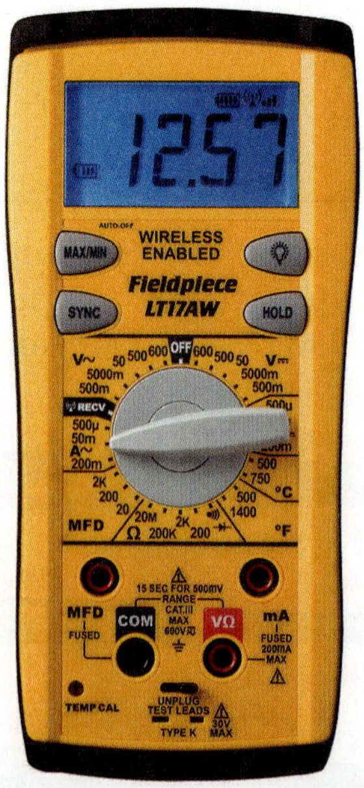

Fieldpiece

Figure 4-51. A digital multimeter (DMM) is a versatile tool for HVACR technicians. This DMM shown is capable of measuring voltage, resistance, capacitance, amperage, frequency, and temperature.

Temperature-Sensing Instruments

Temperature-sensing instruments are valuable tools used to accurately measure temperatures and diagnose equipment issues. These temperature-sensing instruments range from simple analog pocket thermometers to more advanced instruments like digital psychrometers.

Pocket Thermometers

Pocket thermometers are instruments that can either come in analog or digital types. Both types generally have a temperature-sensing stem that can be placed into an air stream or held in hand to take ambient readings, **Figure 4-52**.

K-Type Thermocouples

K-type thermocouples are temperature-sensing probes that are made of two dissimilar metals. They can generate a small ranging voltage that corresponds with varying temperatures. These temperature-sensing accessories must be used with meters that offer the correct plug in for them. Meters range from dedicated temperature meters to digital multimeters to digital manifold gauges, **Figure 4-53**.

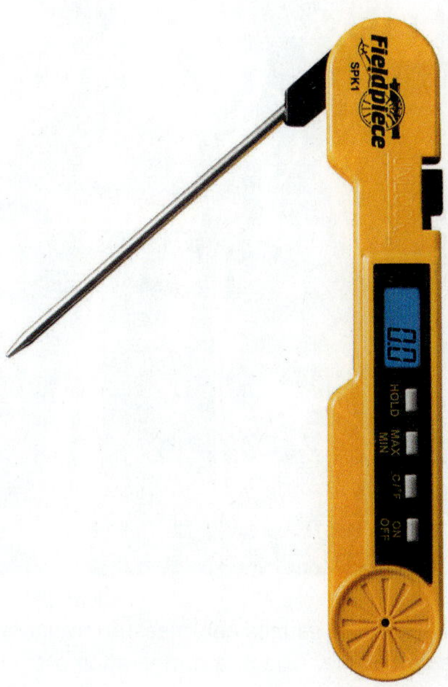

Fieldpiece

Figure 4-52. A digital pocket thermometer can be used to take ambient temperature readings.

Teodor Costachioiu/Shutterstock.com

Figure 4-53. K-type thermocouples can sense a wide temperature range, making it useful in many different HVACR applications.

Psychrometers

Psychrometers are instruments that allow technicians to take measurements related to air. Psychrometers come in analog and digital types. Analog psychrometers can read dry bulb (not affected by moisture) and wet bulb (affected by moisture) air temperatures. Digital psychrometers can read dry and wet bulb temperatures as well as the relative humidity and dew point of air. Analog sling psychrometers are handheld and must be spun to gather data while digital psychrometers are able to be placed directly into a stream of air to gather data, **Figure 4-54**.

Manometer

Manometers are instruments used to measure air and gas pressure. In HVACR applications, this instrument is helpful in measuring the pressure of a fan pushing and pulling air through the system and ductwork, **Figure 4-55**. Being able to gather data on the pressure in a system is an important part of understanding and diagnosing HVACR equipment operation. Systems are dependent on having proper static pressure to provide correct airflow. Manometers are also used to measure gas pressure in gas-fired furnaces.

There are two types of manometers: analog and digital. Both types measure low pressure in inches of water column (in. WC). While there are various types of analog manometers, the industry standard is increasingly turning to digital manometers due to their ability to provide quick, accurate readings and are often easier to use.

Fieldpiece

Figure 4-54. A digital psychrometer can be placed directly into ductwork to gather information about air, including relative humidity and dew point.

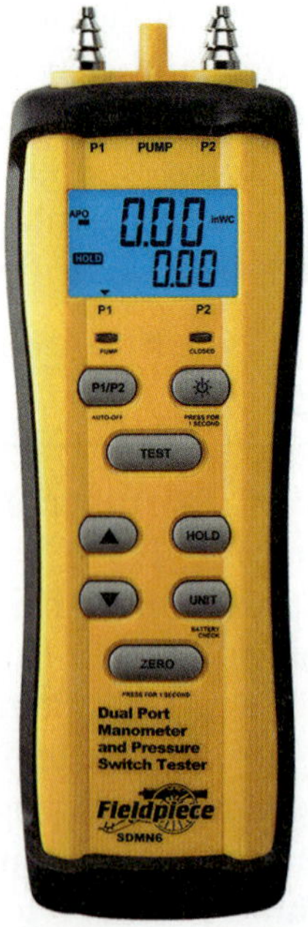

Fieldpiece

Figure 4-55. A digital manometer can provide rapid, accurate air pressure measurements (in in. WC). These instruments are useful in checking pressure switches.

Manifold Gauges

Manifold gauges are another important instrument that every HVACR technician should have and be adept in using. Manifold gauges inform a technician of what the refrigerant of a system is doing inside. Manifold gauges read pressure on two sides of a system at the same time. The two sides are often referred to as the high side and low side of the system. The pressure readings will help a technician understand, diagnose, and decide on a proper repair when needed on the refrigeration circuit of the system, **Figure 4-56**.

SAFETY NOTE

Manifold Gauges

Care must always be taken when using manifold gauges due to high risk of refrigerant burns. Gloves are essential when connecting these gauges to an HVACR system.

InkkStudios via Getty Images

Figure 4-56. Manifold gauges can measure refrigerant pressure on the high side and low side of a refrigeration system. The correct manifold gauge must be chosen based on the refrigerant used in the system due to the distinct operating pressures of refrigerants.

Chapter Review

Summary

4.1 Wrenches are used to hold or turn nuts, bolts, cap screws, and other fasteners and can be used to clamp parts together. Open-end, box-end, combination, refrigeration service, and socket wrenches are some common types. Pliers are used to grip, cut, crimp, hold, or bend various materials. Choosing the correct type of pliers improves efficiency.

4.2 Screwdrivers are used to drive or withdraw threaded fasteners. Select the correct size and shape screwdriver for the application. Specialty screwdrivers and nut drivers are available for various fasteners. Hammers are used for various driving operations. Always align the face of the hammer squarely with the surface being struck.

4.3 Hacksaws are used to cut all types of metal objects. Blades are selected according to the number of teeth per inch. Various types of drill bits are used for boring and cutting holes. Each is designed for use on a specific type of material. There are four basic types of snips commonly used in HVAC work: straight-pattern, combination, duckbill, and aviation.

4.4 Punches and chisels are referred to as struck tools, because they are struck with a hammer to mark or remove material from a workpiece.

4.5 HVACR instruments are used to accurately gather internal data about equipment operation. Instruments—such as digital multimeters, temperature-sensing instruments, and manometers—can collect information that may not be seen, heard, or felt. The data may include information on electrical performance, temperature, air and refrigerant pressures, or properties of air.

Know and Understand

1. When using a wrench, avoid using too much ____ to avoid stripping bolt threads. (4.1)
   A. swing space
   B. torque
   C. upward force
   D. downward force
2. A(n) ____ wrench has a hexagonal shape and is most often used to turn setscrews. (4.1)
   A. adjustable
   B. combination
   C. Allen
   D. socket
3. *True or False?* It is preferred to pull on a wrench, rather than push. (4.1)
4. ____ pliers allow a technician to reach into tight places and perform work that would be awkward or difficult with any other tool. (4.1)
   A. Lineman's
   B. Long-nose
   C. Diagonal
   D. Slip-joint
5. *True or False?* Pliers with plastic-dipped handles offer insulated protection from electric shock. (4.1)
6. The size and the ____ determine which screwdriver to use. (4.2)
   A. type of drive opening
   B. tip material
   C. torque rating
   D. length
7. Steel-cut nails and masonry nails should be driven with a(n) ____ hammer. (4.2)
   A. nail
   B. ball pein
   C. engineer's
   D. soft-faced
8. *True or False?* Always attempt to grind or redress damaged hammer heads before discarding them. (4.2)

9. When cutting metal with a hacksaw, ensure at least ____ teeth are always in contact with the metal. (4.3)
   A. two
   B. three
   C. four
   D. six
10. A twist drill bit has three principal parts: a shank that is clamped into the drill chuck, a body with two spiral grooves called *flutes*, and a cone-shaped cutting end called the ____. (4.3)
    A. bit
    B. point
    C. blade
    D. arbor
11. A drill bit with a face that is flattened and spread out from being struck is referred to as ____. (4.3)
    A. chipped
    B. stripped
    C. mushroomed
    D. pancaked
12. Which of the following snips has a serrated blade? (4.3)
    A. Straight-pattern
    B. Combination
    C. Duckbill
    D. Aviation
13. Periodic maintenance of snips involves occasionally applying ____ to the pivot bolt. (4.3)
    A. oil
    B. water
    C. heat
    D. special lubricant
14. A ____ punch is used for aligning holes in different sections of material. (4.4)
    A. center
    B. prick
    C. drift
    D. pin
15. Cold chisels can be sharpened using a hand file or ____. (4.4)
    A. grinder
    B. sandpaper
    C. another chisel
    D. whetstone
16. *True or False?* A cold chisel should never be used to split materials. (4.4)
17. When looking to obtain an amperage reading on an electrical circuit, the best instrument to use would be a ____. (4.5)
    A. manometer
    B. psychrometer
    C. thermocouple
    D. digital multimeter
18. Which instrument is best used to determine relative humidity? (4.5)
    A. Manometer
    B. Psychrometer
    C. Voltmeter
    D. Manifold gauge
19. Manifold gauges are mainly used to read an HVACR system's ____. (4.5)
    A. airflow
    B. total energy efficiency
    C. pressure
    D. temperature changes

Apply and Analyze

1. Describe the various functions of locking pliers. (4.1)
2. Explain proper maintenance for a straight-blade driver with a dulled tip. (4.2)
3. Describe how to properly cut metal with a hacksaw. (4.3)
4. Why should aviation snips never be sharpened? (4.3)
5. Explain how punches are used. (4.4)
6. Explain the benefits of knowing how to properly use a digital multimeter and when a technician may use one. (4.5)

Critical Thinking

1. Explain when tools may need to be repaired or replaced. (4.1)
2. Explain how to use snips safely. (4.3)

popovaphoto via Getty Images

5 Fasteners

LEARNING OUTCOMES

After studying this chapter, you will be able to:

5.1 Describe various thread-related terms, standard thread forms, and how to properly use taps and dies.
5.2 Identify screw types and sizes.
5.3 Identify and correctly use bolts, nuts, and washers.
5.4 Explain the use of various anchors.
5.5 Recognize when to use setscrews, blind rivets, and threaded rods.

TECHNICAL TERMS

American National Acme Thread
American National Standard Thread
blind hole
blind rivet
bolt
bolt extractor
Brown and Sharpe Worm Thread
cap screw
crest
die
flutes
International Thread
machine screw
major diameter
mandrel
masonry anchor
minor diameter
Molly bolt
nut
root
screw
screw thread
setscrew
shank
tap
thread angle
threaded rod
threads per inch (TPI)
toggle bolt
Unified Thread
washer

Introduction

Screws, bolts, nuts, washers, and anchors are among the fasteners that should be familiar to the entry-level technician. These fasteners and others commonly used in HVACR work are discussed in this chapter.

5.1 Threads

To use fasteners properly, it is helpful to be familiar with the following thread information:

- Thread terminology
- Standard thread forms and thread measuring methods
- Proper tap-drill size for a screw thread

Thread Terminology

A ***screw thread*** is a helical ridge of uniform section formed inside a hole (such as a nut) or on the outside of a fastener (such as a screw or bolt). Threads cut on an outside surface are referred to as "external;" those cut on an inside surface are called "internal."

The following terms describe the parts of a screw thread and are shown in **Figure 5-1**:

- ***Major diameter*** is the widest measurement from the outside edges of the threads.
- ***Minor diameter*** is the smallest measurement from the inside edges of the threads.
- ***Root*** is the bottom area of two adjoining threads.
- ***Crest*** is the top edge of two adjoining threads.
- ***Threads per inch (TPI)*** are determined by counting the number of crests (or roots) per inch of threaded section.
- ***Thread angle*** is the V-shaped angle of the threads. The sides of the threads normally form a 60° angle.

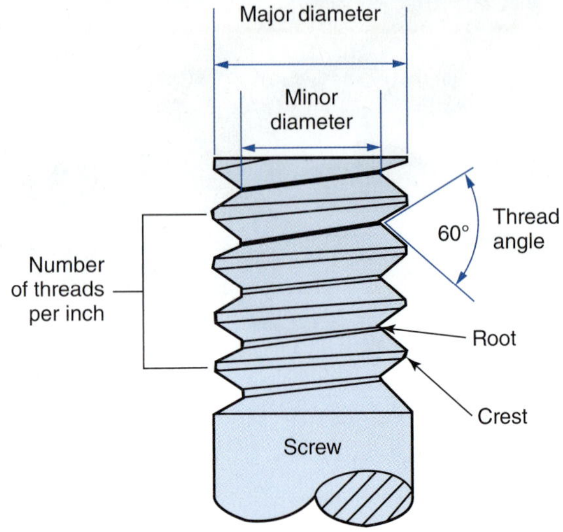

Goodheart-Willcox Publisher

Figure 5-1. Parts of the screw thread. The number of threads per inch (TPI) determines whether a thread is coarse or fine. A lower TPI indicates a coarse thread; a higher TPI indicates a fine thread.

Standard Thread Forms

Of the standard thread forms, those most widely used in North American industry today are the American National Standard Thread and Unified Thread forms. These two forms are popular because they can be mass-produced easily and economically. The thread form in widespread use in most parts of the world is the International Thread, also known as the ISO Metric Thread Series.

American National Standard

American National Standard Thread has three major series of threads: National Fine (abbreviated NF), National Coarse (NC), and National Pipe (NP).

NF has a greater number of threads per inch than NC, and the threads are not as deep. Both NC and NF threads maintain constant major and minor diameters.

NP threads, however, are tapered (sloped) 1/16" in diameter for every inch of threaded length. This taper causes the threads to bind together, making an increasingly tight seal as a pipe fitting is tightened. Pipe threads are self-sealing due to compression of the V-threads as the fittings are connected.

> **CAUTION**
>
> **Matched Threads for Fasteners**
>
> All fastener components must share the same style of threads and the same number of threads per inch. For example, an NF nut cannot be screwed onto an NC bolt, **Figure 5-2**. Attempts to force mismatched threads to mate can severely damage threads and render the fastener useless. Damaged threads can sometimes be dressed by rethreading them with a tap or die of the proper size.

NC threads are widely used to make bolts, nuts, and screws where quick assembly is desired in materials such as cast iron, steel, bronze, brass, aluminum, and magnesium. The NF thread series is also used in the manufacture of nuts, bolts, and screws. NF threads are usually specified if the thread engagement is short, or the internal threads are cut into a thin-walled material. NP threads are used in the assembly of steel pipe and cast-iron fittings for water lines and natural gas lines. Some heavy-duty electrical conduits also use pipe threads.

Unified Screw Threads

The American National screw thread form was the standard in the United States for many years. However, some industry leaders advocated a thread system that would permit interchangeability of screw threads with other countries, especially those following British standards.

British Standard Whitworth Thread was not interchangeable with American National Thread. American National had a thread angle of 60°, with roots and crests slightly flattened. The Whitworth thread had a thread angle of 55° with rounded roots and crests.

During World War II, England and Canada agreed to change the Whitworth thread angle to 60°, but to retain the rounded crests and roots. This new thread form, called ***Unified Thread***, was interchangeable with American National threads. Since that time, American National Fine has been interchangeable with Unified National Fine (UNF), as have been National Coarse and Unified National Coarse (UNC).

Unified Thread is now the standard thread, but the American National coarse and fine thread series will continue to be seen on gauges, tools, and prints for several years. For this reason, it is necessary to be familiar with both systems.

Goodheart-Willcox Publisher

Figure 5-2. Threads of distinct types and styles cannot be used together. The 24 TPI on the 1/2" nut will not mate properly with the 27 TPI on the bolt.

Other Thread Forms

The standard form used in most parts of the world is ***International Thread***, also known as the *International Standards Organization (ISO) Metric Thread*. The ISO metric thread has a 60° angle with slightly flattened roots and crests. Use of this thread form is growing in North America. The main difference between International Thread and Unified Thread is the depth of the threads. The two thread forms are not interchangeable.

American National Acme Thread has a 29° angle and is used for feed screws, jacks, and vises. ***Brown and Sharpe Worm Thread*** also has a 29° angle, but the depth of the root is greater than that of the Acme thread. This thread is used to mesh with worm gears and transmit motion. A self-locking feature makes it usable for winches and steering mechanisms.

Cutting Internal Threads

A ***tap*** is a tool used to cut threads inside a hole. Taps are made of special hardened steel and threaded for about half their length. The threaded grooves are called ***flutes***, and the edge of each thread forms a cutting edge, **Figure 5-3**.

Taps are available in three types: taper, plug, and bottoming, **Figure 5-4**. On the *taper* tap, the first six threads are tapered for easy starting. On the *plug* tap (the most common), only three or four threads are tapered. The *bottoming* tap is not tapered; it is used to cut threads to the bottom

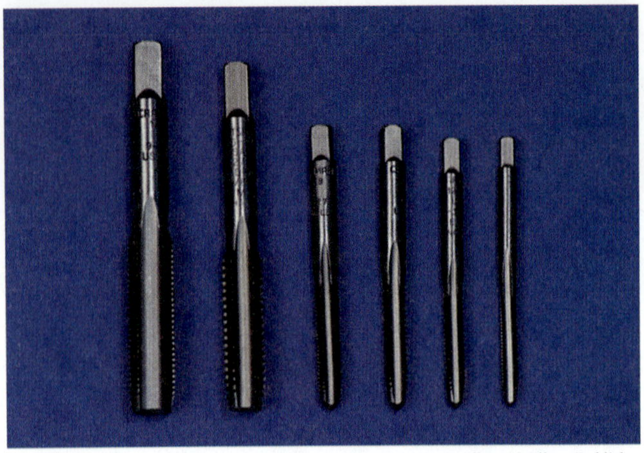

Goodheart-Willcox Publisher

Figure 5-3. Taps are made in many sizes and thread types.

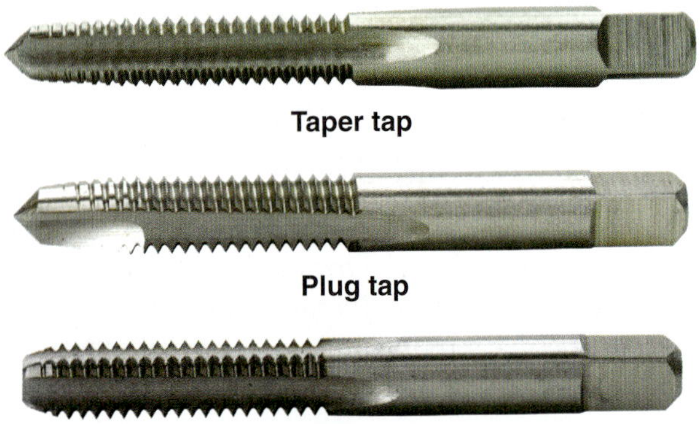

Images courtesy of Fastenal®

Figure 5-4. A taper tap is designed for easy starting. A plug tap is the most common. A bottoming tap is used to cut threads to the bottom of a hole.

of a hole. The size is stamped on the unthreaded portion of the tap, called the ***shank***. The size is determined by the outside diameter and the number of threads per inch.

Before using a tap, a pilot hole must be drilled into the workpiece. The size of the pilot hole is critical. The hole must be smaller than the tap's outside diameter to allow for cutting of full threads inside the hole. The proper pilot hole size (drill size) can be determined by consulting tables depending on the thread type, like those in **Figures 5-5**.

Once the pilot hole has been drilled, the tap is inserted and properly aligned with the work surface. If the tap is large, use a wrench to turn it. Small taps are usually turned with a special tap wrench, **Figure 5-6**. Use slight downward pressure to start the cut but release the pressure once the tap threads begin to bite. Use a lightweight oil to lubricate the tap during the cutting process. Reverse the tap every two or three turns to clear metal shavings from the threads. The cutting process should

National Coarse Threads

Tap Size	Threads per Inch	Drill Size
#5	40	#38
#6	32	#36
#8	32	#29
#10	24	#25
#12	24	#16
1/4″	20	#7
5/16″	18	F
3/8″	16	5/16″
7/16″	14	U
1/2″	13	27/64″
9/16″	12	31/64″
5/8″	11	17/32″
¾″	10	21/32″
7/8″	9	49/64″
1″	8	7/8″

National Fine Threads

Tap Size	Threads per Inch	Drill Size
#5	44	#37
#6	40	#33
#8	36	#29
#10	32	#21
#12	28	#14
¼″	28	#3
5/16″	24	I
3/8″	24	Q
7/16″	20	25/64″
1/2″	20	29/64″
9/16″	18	33/64″
5/8″	18	37/64″
3/4″	16	11/16″
7/8″	14	13/16″
1″	14	15/16″

National Pipe Threads

Tap Size	Threads per Inch	Drill Size
1/8″	27	11/32″
1/4″	18	7/16″
3/8″	18	19/32″
1/2″	14	23/32″
3/4″	14	15/16″
1″	11 1/2	1 5/32″
1 1/4″	11 1/2	1 1/2″
1 1/2″	11 1/2	1 23/32″
2″	11 1/2	2 3/16″

Goodheart-Willcox Publisher

Figure 5-5. Tap and drill sizes for National Coarse (NC), National Fine (NF), and National Pipe (NP) threads.

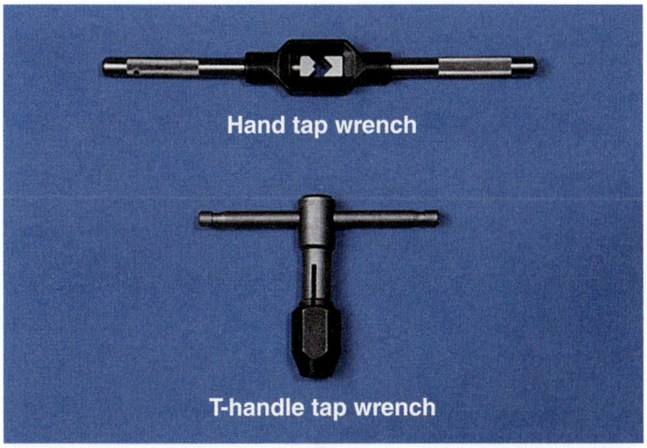

Goodheart-Willcox Publisher

Figure 5-6. A hand tap wrench provides greater leverage when using large taps. A T-handle tap wrench is used for small taps.

continue until the tap turns smoothly with little pressure.

A ***blind hole*** is a pilot hole that does not penetrate completely through a component. To thread a blind hole, use a taper tap until it touches the bottom of the component hole. Then change to a plug tap. Finally, complete the thread with a bottoming tap.

Cutting External Threads

A ***die*** is a tool used to cut threads around the outside of a piece of metal, such as a bolt, rod, or pipe. A threading die may be round, square, or hexagonal. See **Figure 5-7**. The thread cutters may be built-in or replaceable. The leading edges of the cutters are ground away slightly to make starting easier.

Dies are available in all sizes from small to large and are often sold in sets with matching taps, **Figure 5-8**. The die selected depends upon the kind of thread desired (NF, NC, or NP) and the diameter of the material being threaded.

To use a die, first clamp the workpiece in a vise. Fasten the die into a die stock or holder, **Figure 5-9**. Then, place the die squarely over the end of the work. Apply downward pressure while turning the die slowly to the right until the threads bite. Use lightweight oil on the die during the cutting process and stop applying the downward pressure once the threads are started. The die should be backed off after each half-turn to release the metal chips.

ZHMURCHAK/Shutterstock.com

Figure 5-7. Various sizes of round- and hexagonal-type threading dies. Square dies are also available.

5.2 Screws

Screws fasten parts securely while, at the same time, allowing the parts to be readily removed. Screws are selected according to the job and the material to be fastened. Some screws are intended for use in wood, others in thin metals, and still others in thick metals. The type of threads used will determine the holding power.

It is often necessary to drill a pilot hole before inserting a screw. The pilot hole must be smaller in diameter than the actual screw so the screw threads can grip the material.

Goodheart-Willcox Publisher

Figure 5-8. A large tap and die set.

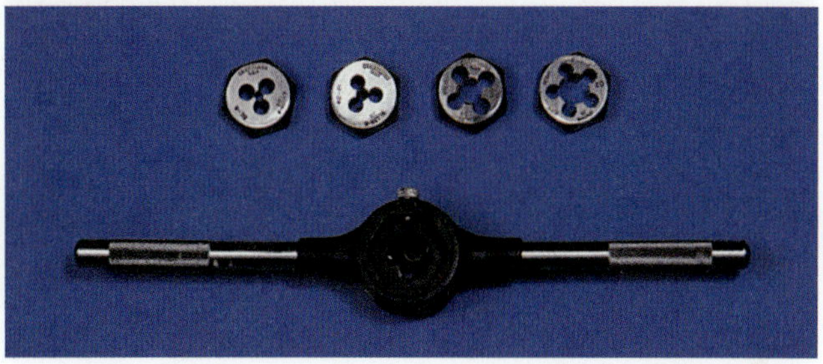

Goodheart-Willcox Publisher

Figure 5-9. A hexagonal-type die stock holder.

Screw Sizes

Screw sizes are measured by both length and diameter. The length is designated in inches (or fractions of an inch), but the diameter is indicated by a gauge number. Screws are available in gauges ranging from 0 (about 1/16″) to 24 (about 3/8″) and are available from 1/4″ to 6″ in length. The most common screw gauge numbers are 2 through 16. For example, a 6×1 screw would be gauge number 6 in diameter and 1″ long. The gauge number is always listed first, and the length is second.

Screw Threads

Different types of threads are used for fastening different types of materials, **Figure 5-10**. For example, the threads on a wood screw are different from those on a sheet metal screw. The screw should always match the job, based on the material being fastened.

Screw Head Styles

Screws are available in many head styles, **Figure 5-11**. Screw heads are selected for holding ability and appearance, depending on the application. Head styles are often available in distinct types of drives. The drive determines what type of screwdriver must be used, **Figure 5-12**.

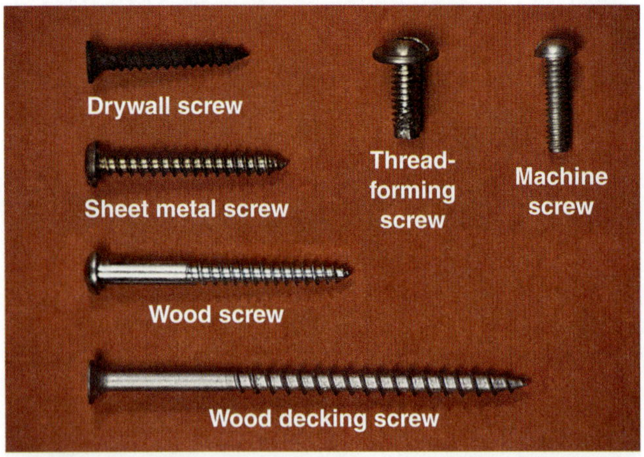

Goodheart-Willcox Publisher

Figure 5-10. Types of screws.

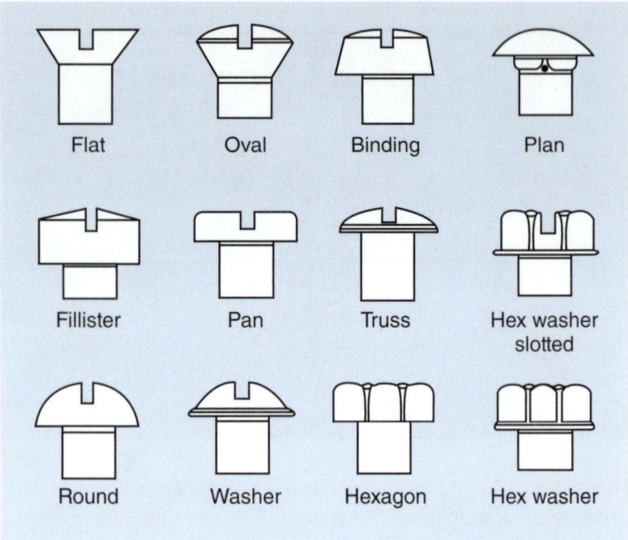

Goodheart-Willcox Publisher

Figure 5-11. Many different screw head styles are available.

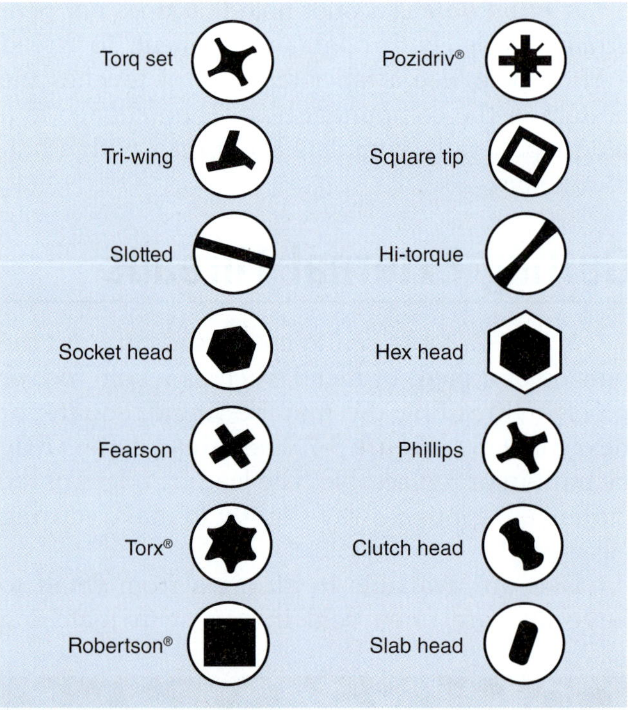

Goodheart-Willcox Publisher

Figure 5-12. Styles of screw head drives.

Self-Drilling Screws

Self-drilling screws are designed to drill and thread holes in one quick and easy step. A pilot hole is not required, **Figure 5-13**. These screws are used for anchoring sheet metal and metal ductwork. They have hex-type heads, slotted or unslotted, and are available in two-point types, **Figure 5-14**.

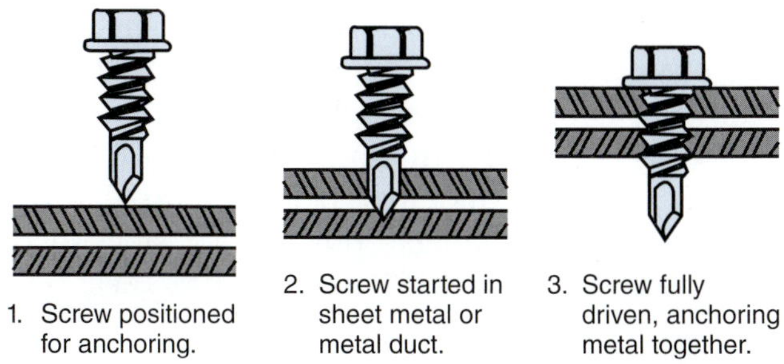

Goodheart-Willcox Publisher

Figure 5-13. Self-drilling sheet metal screws do not require a pilot hole. Shown here is a self-drilling screw being driven into metal material.

Malco Products, Inc.

Figure 5-14. Two-point types are available in self-drilling sheet metal screws.

Quick installation of self-drilling screws requires a special hex chuck driver for use with an electric drill. The driver head is magnetized to hold the screw while it is being driven. Hex chuck drivers are available in 1/4″, 5/16″, and 3/8″ sizes.

Machine Screws

Machine screws fall into a unique category, filling the void between a screw and a bolt. As shown in **Figure 5-15**, they are small bolts with screw-type heads.

A machine screw is designated first by its diameter number, second by its number of threads per inch, and finally by its length. For example: 1/4-20 × 1 indicates a machine screw that is 1/4″ in diameter, has 20 threads per inch (NC), and is 1″ long. To fit this screw, a 1/4-20 machine nut would be required.

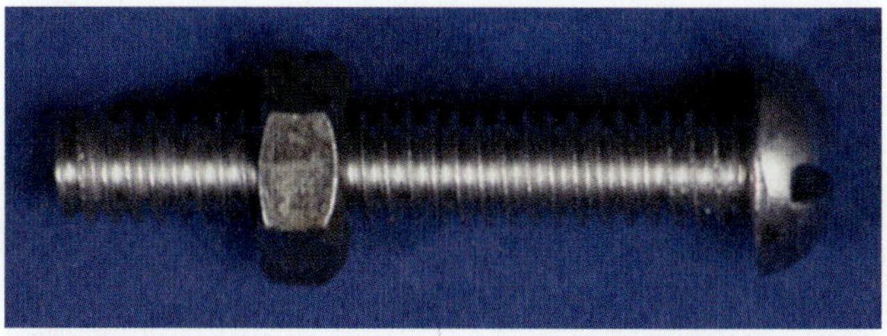

Goodheart-Willcox Publisher

Figure 5-15. Machine screws are small bolts. The screw and nut must be the same diameter and have matching threads.

Machine screws are available in eleven diameters. Eight are in numbered sizes (like screws) and three are in fractional sizes (like bolts). As the diameter gets larger, the numbers get bigger. A table of common machine screw sizes and threads is shown in **Figure 5-16**. Machine screws are available in a variety of thread forms, head shapes, and drives, **Figure 5-17**.

5.3 Bolts, Nuts, and Washers

Bolts, nuts, and washers are used in conjunction with each other for maximum securing of system parts. Various types of bolts are available so the correct type of nuts and washers should be selected according to the type of bolt used.

Bolts

A ***bolt*** is a fastener that is used with a nut and is tightened by turning the nut. Bolts are normally used for fastening heavy metal parts together. **Figure 5-18** shows a variety of bolts. Bolts are made in both square and hex head forms and have either square or hex nuts.

Machine Screw Sizes and Threads

Machine Screw Number	Diameter (″)	National Fine (TPI)	National Coarse (TPI)
#2	0.086	64	56
#3	0.099	56	48
#4	0.112	48	40
#5	0.125	44	40
#6	0.138	40	32
#8	0.164	36	32
#10	0.190	32	24
#12	0.216	28	24
1/4	0.250	28	20
5/16	0.3125	24	18
3/8	0.375	24	16

Goodheart-Willcox Publisher

Figure 5-16. Table of common machine screw sizes and threads.

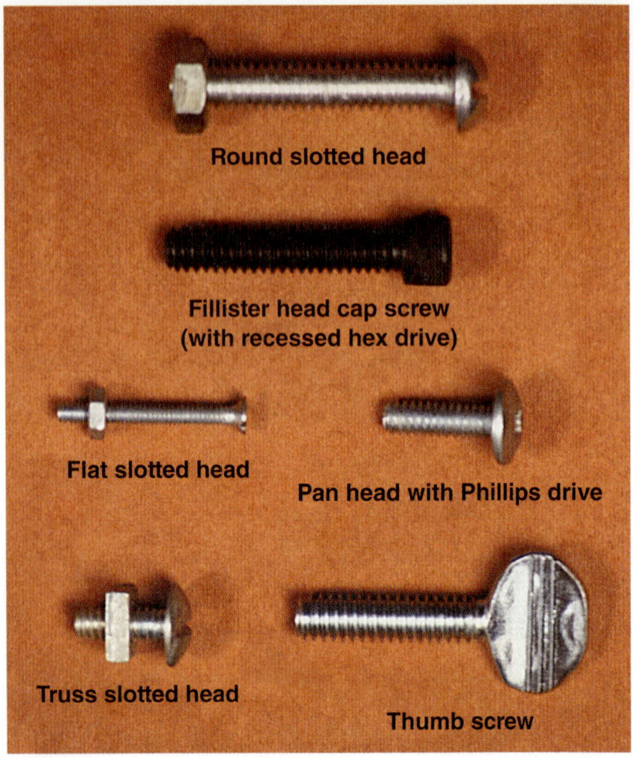

Goodheart-Willcox Publisher

Figure 5-17. Machine screws are available in a variety of head shapes and drives.

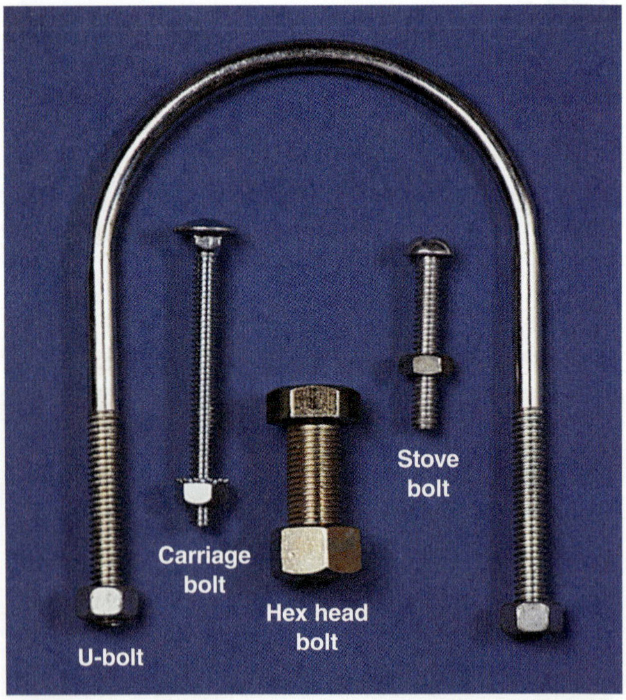

Goodheart-Willcox Publisher

Figure 5-18. Bolts are available in various types.

Machine bolts (and threaded rods) are available in all common thread forms, with coarse threads being most common. Bolt sizes are designated the same as machine screws: diameter first, then number of threads per inch, and then length. For example: 1/2-13 × 3 indicates a bolt 1/2″ in diameter, 13 threads per inch (NC), and 3″ long.

Bolt sizes ordinarily range from 1/4-20 to 1/2-13, and their lengths range from 3/8″ to 6″. Much larger sizes are available but are not stocked by most hardware stores.

Carriage Bolts

Carriage bolts are usually used to anchor wood to metal. The head of the carriage bolt includes a countersunk square shoulder along with the round head. This square shoulder keeps the bolt from turning when the nut is tightened. Carriage bolts are also used when the head is not accessible to a wrench.

Stove Bolts

Stove bolts are made with slotted round, flat, or oval heads. They were originally used in stove construction but are now sold as general utility bolts. Stove bolts are similar to machine screws, except the smallest size is 1/4″ in diameter.

Bolt Extractors

Occasionally, a bolt or screw may be overtightened, causing the head to break off or the fastener to snap off inside a hole. In either case, the bolt or screw must be removed and replaced with a new one. Removing a broken bolt or screw is not hard, but it is time-consuming and requires the use of a ***bolt extractor***. This tool is typically paired with a socket wrench or drill to grip broken fasteners.

To remove a broken bolt or screw, first drill a hole in the end of the broken bolt, as shown in **Figure 5-19**. Insert a bolt extractor of the proper size in the hole and then use a wrench to turn the extractor counterclockwise. This causes the extractor to wedge (bite) into the hole and unscrew the bolt.

Nuts

Nuts are available in a wide variety of types and sizes for use with different types of bolts and machine screws, **Figure 5-20**. Different nuts serve specialized purposes: acorn nuts for a finished appearance, wing nuts for finger tightening, and slotted or castle nuts for locking with a cotter pin

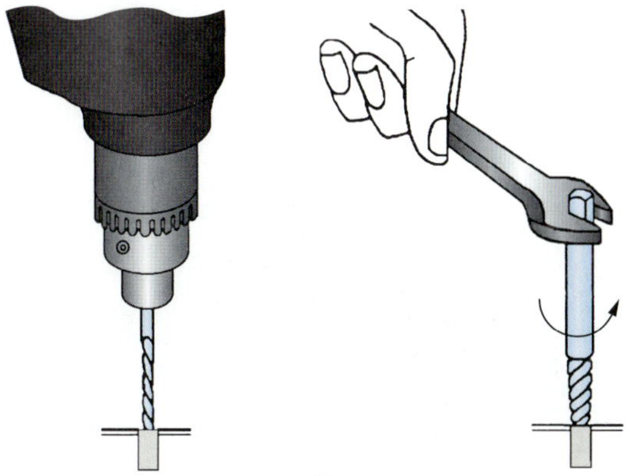

Goodheart-Willcox Publisher

Figure 5-19. Removing a broken bolt with a bolt extractor. A—Drilling a hole in a broken bolt. B—An extractor is inserted into the drilled hole and turned counterclockwise with a wrench to "back out" the broken fastener.

Goodheart-Willcox Publisher

Figure 5-20. Many types of nuts are available.

or wire. Nuts are available in various finishes and degrees of hardness.

Washers

Washers are used to extend the gripping area of a fastener or to prevent a nut from loosening. See **Figure 5-21**. *Flat* washers are used to extend the surface area of a bolt head or nut. This prevents the bolt head or nut from being drawn into the surface of the material being fastened. *Lock* washers prevent a nut from loosening on a bolt. Lock washers are available in four styles: spring, helical, internal tooth, and external tooth.

5.4 Anchors

Distinct types of anchors are used to fasten materials firmly to various surfaces. The choice of anchor depends upon the surface and the load (weight of the object or material). Plastic screw

Goodheart-Willcox Publisher

Figure 5-21. Washers are either the flat or lock type.

anchors, expanding metal fasteners, and toggle bolts are used for solid or hollow walls. Masonry anchors are used for solid surfaces, such as concrete or brick. See **Figure 5-22**.

Plastic Screw Anchors

Plastic screw anchors are light-duty anchors. They are used to fasten light loads to solid walls, floors, or other surfaces. As shown in **Figure 5-23**, the anchor is expanded against the sides of a hole by the screw, providing the grip needed to hold the load.

Hollow Wall Anchors

Special expanding metal fasteners, often called ***Molly bolts***, are used to anchor light objects to

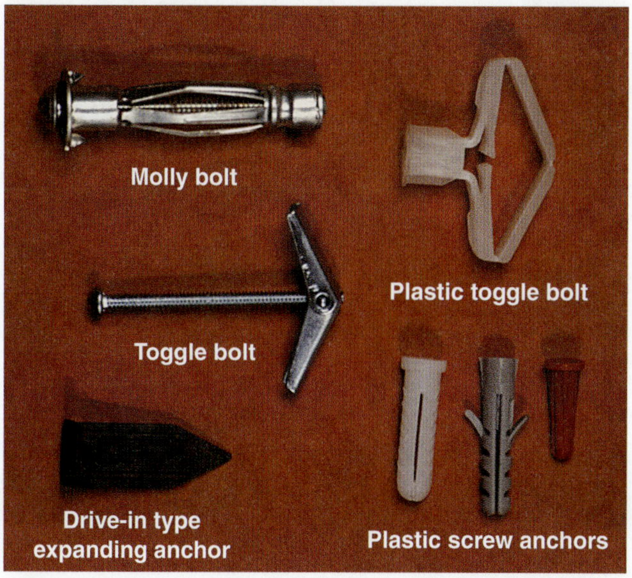

Goodheart-Willcox Publisher

Figure 5-22. Several types of anchors.

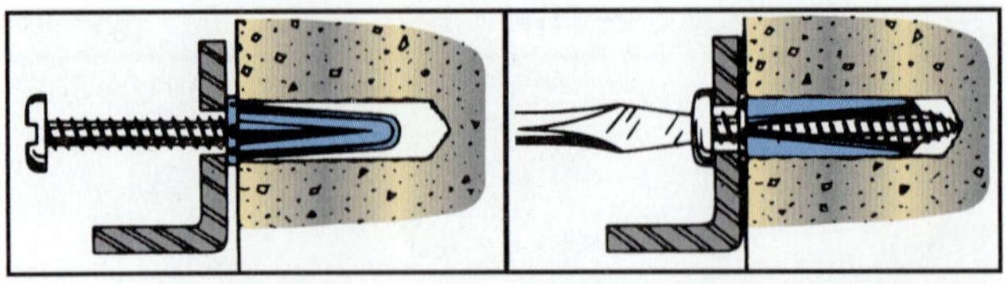

Star Expansion Company

Figure 5-23. A plastic screw anchor is inserted into a drilled hole. It is expanded and wedged against the sides of the hole as the screw is tightened.

hollow walls. As shown in **Figure 5-24**, the fastener is inserted into a hole drilled through the wall and, as the screw is turned, the legs are pushed outward to grip the inside wall surface. A variation is a drive-in-type expanding anchor, which does not require drilling a hole. The flat metal anchor is hammered into place, then expanded by a screw. See **Figure 5-25**.

Toggle Bolts

A ***toggle bolt*** is used to fasten heavy objects to a hollow wall. A hole large enough to pass the spring-loaded wings must be drilled into the wall. The wings will then spread out and bear against the inside of the wall, allowing the bolt to be screwed in, **Figure 5-26**.

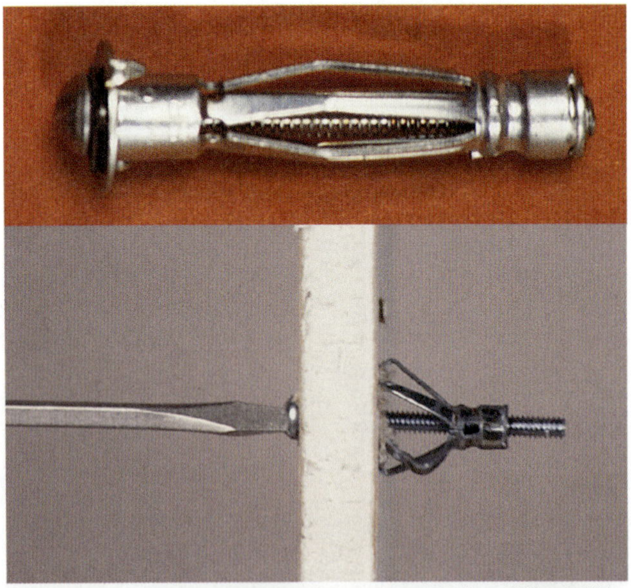

Goodheart-Willcox Publisher

Figure 5-24. A hollow wall anchor (top) is inserted into a drilled hole, then expanded by tightening the screw. The legs are permanently expanded inside the hollow wall (bottom), allowing the screw to be removed and replaced, as necessary.

Goodheart-Willcox Publisher

Figure 5-25. The screw expands the legs of a drive-in-type anchor after it is hammered into a hollow wall.

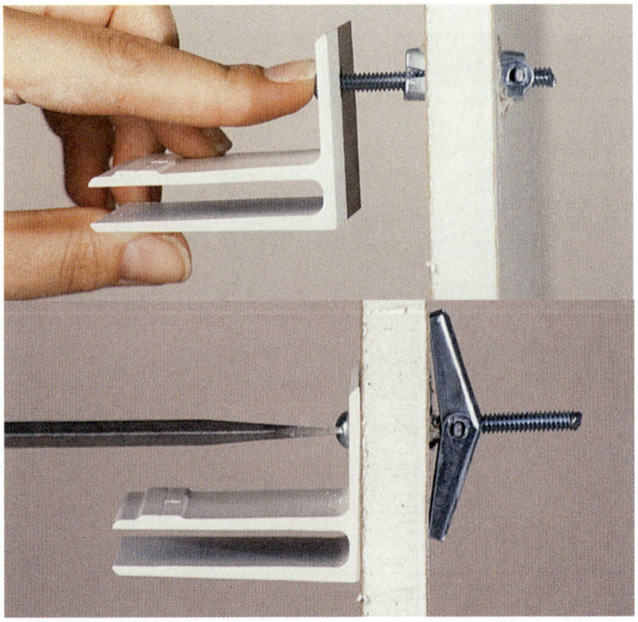

Goodheart-Willcox Publisher

Figure 5-26. A toggle bolt has two spring-loaded wings. After the toggle bolt is inserted through a hole into the hollow center of the wall (top), the wings are forced open by spring pressure. The wings are drawn against the inner wall surface by tightening the screw (bottom).

Masonry Anchors

Masonry anchors are available in many forms and styles. They are often used to fasten brackets to a masonry surface for such tasks as supporting lengths of conduit or copper tubing. See **Figure 5-27**.

Masonry anchors may be light-, medium-, or heavy-duty. Each type is available in different sizes for added strength and holding power.

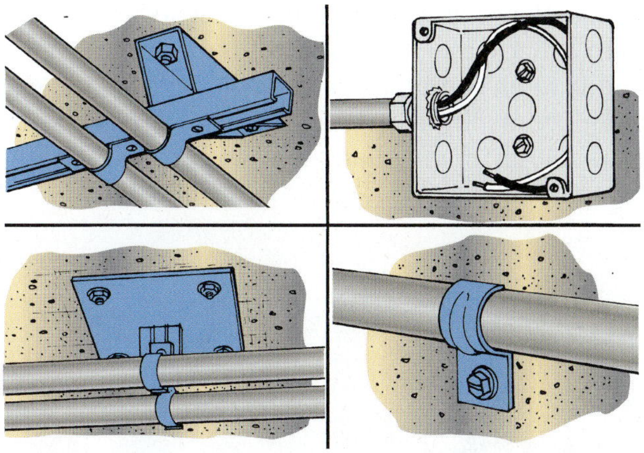

ITW Ramset

Figure 5-27. Masonry anchors are used to fasten brackets, junction boxes, and conduit clamps to masonry walls.

In general, anchors are installed by drilling a hole the size of the anchor into concrete or other material. When a bolt is inserted and tightened, the anchor expands and grips the inside of the hole, **Figure 5-28**. Sometimes, a special tool is required to "set" the anchor in the hole before inserting the bolt.

5.5 Other Fasteners

Besides screws and bolts, there are several other types of fasteners used in special situations or on specific jobs.

Setscrews

Setscrews are used to anchor pulleys and fan blades to motor shafts. As shown in **Figure 5-29**, setscrews do not have heads because they are usually recessed inside a threaded hole. Setscrews are normally driven by an Allen (hex) wrench. Setscrews that *do* have heads are called ***cap screws***.

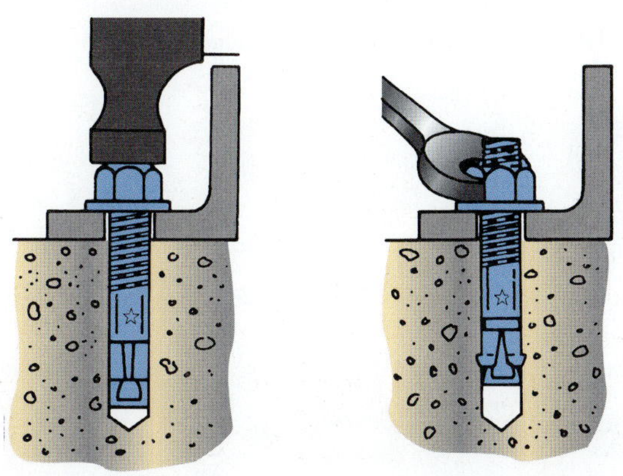

Star Expansion Company

Figure 5-28. Wedge anchors are driven into a hole drilled in masonry, then expanded by tightening the bolt.

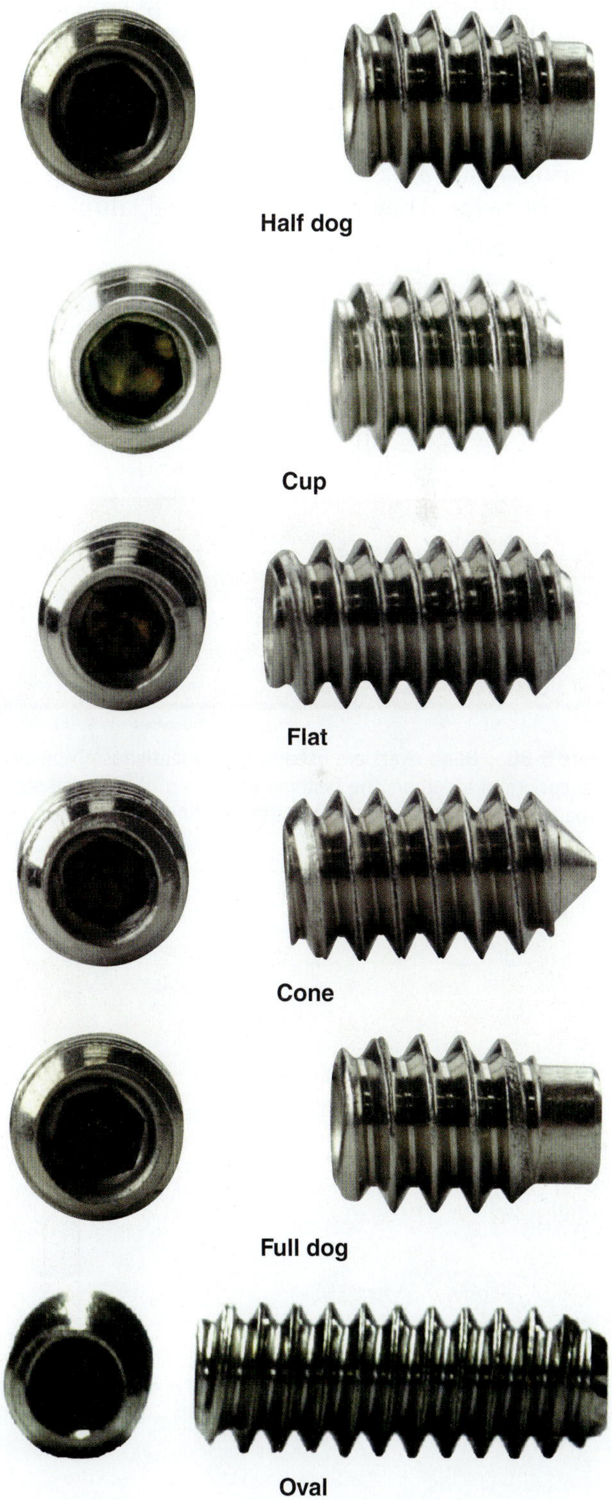

Images courtesy of Fastenal®

Figure 5-29. Various setscrews with different point styles.

Blind Rivets

A ***blind rivet*** (or pin rivet) is used to join two pieces of sheet metal for a strong connection that will not loosen under vibration. The blind rivet consists of a rivet and a nail-like pin called a ***mandrel***, **Figure 5-30**. The mandrel is designed to break off at the crimp after setting the rivet body with a hand tool, **Figure 5-31**. No special skill is required to set these rivets.

Blind rivets are available in five diameters, ranging from 3/32″ to 1/4″, with 1/8″ being the most popular. They are made of steel, aluminum, stainless steel, and copper.

Threaded Rod

A ***threaded rod*** is a metal rod that is threaded its entire length. This fastener is used in many applications because it can be cut to any length with a hacksaw. It is used to make pipe hangers and is often combined with steel angle to suspend components from a ceiling. Threaded rods are available in many diameters with coarse or fine threads.

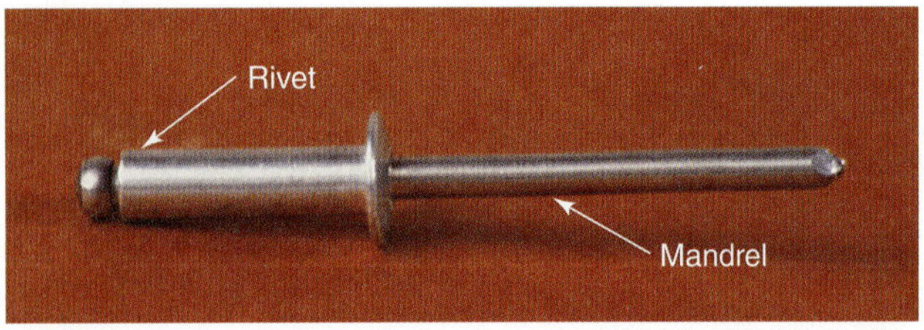

Goodheart-Willcox Publisher

Figure 5-30. Blind rivets are used to permanently fasten pieces of sheet metal together. The fastener consists of the rivet itself and a mandrel that is snapped off after forming a rivet head.

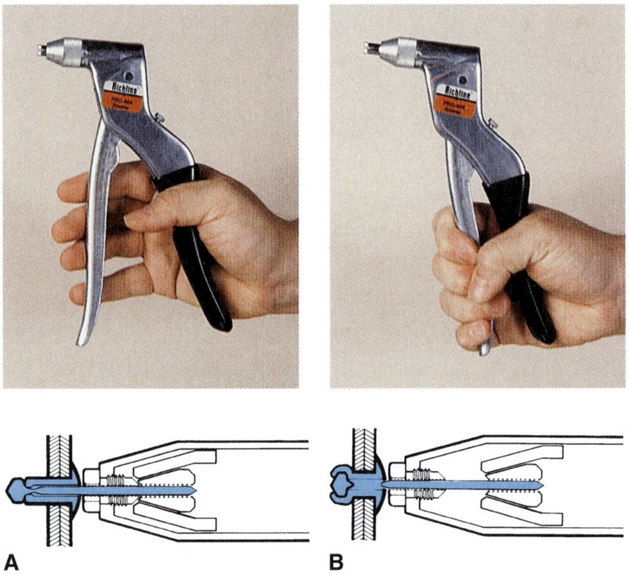

Malco Products, Inc.

Figure 5-31. Installation of a blind rivet. A—The mandrel of the blind rivet is gripped by the tool, and the rivet is inserted through the material to be fastened. B—When the tool handles are squeezed together, the mandrel is pulled out, deforming the rivet material to form a head. The mandrel is then snapped off, leaving the rivet in place.

Chapter Review

Summary

5.1 To use fasteners properly, it is helpful to be familiar with the parts of a screw thread: major and minor diameter, root, crest, threads per inch (TPI), and the thread angle. American National Standard Thread and Unified Thread are the standard thread forms most widely used in North America. A tap is used to cut threads inside a piece of metal. A die is used to cut threads around the outside of a piece of metal.

5.2 Screws are selected according to the job and the material to be fastened. Screws are available in gauges ranging from 0 (about 1/16″) to 24 (about 3/8″) and are available from 1/4″ to 6″ in length. Machine screws are designated first by its diameter number, second by its number of threads per inch, and finally by its length.

5.3 Bolts are used in conjunction with nuts and washers to fasten heavy metal parts together. Various types are available for different applications.

5.4 The choice of anchor depends upon the surface and the load (weight of the object or material). Plastic screw anchors, expanding metal fasteners, and toggle bolts are used for solid or hollow walls. Masonry anchors are used for solid surfaces, such as concrete or brick.

5.5 Setscrews are used to anchor pulleys and fan blades to motor shafts. A blind rivet (or pin rivet) is used to join two pieces of sheet metal for a strong connection that will not loosen under vibration. A threaded rod is used in many applications because it can be cut to any length with a hacksaw.

Know and Understand

1. The _____ of a screw is the bottom area of two adjoining threads. (5.1)
   A. minor diameter
   B. crest
   C. root
   D. thread angle
2. The standard thread forms most widely used in North American industry today are the _____ and _____ forms. (5.1)
   A. American National Standard Thread, Unified Thread
   B. American National Standard Thread, International Thread
   C. Unified Thread, International Thread
   D. American National Acme Thread, American National Standard Thread
3. The _____ tap is used for starting an internal thread. (5.1)
   A. shank part of the
   B. taper
   C. plug
   D. bottoming
4. If a coarse external thread is desired, a die that matches the _____ thread series should be selected. (5.1)
   A. NF
   B. NP
   C. NC
   D. UNF
5. A 2 × 6 screw has a gauge number of _____. (5.2)
   A. 0
   B. 2
   C. 6
   D. 12
6. *True or False?* Wood screws and sheet metal screws have the same threads. (5.2)
7. The head of a _____ bolt includes a countersunk square shoulder along with the round head. (5.3)
   A. stove
   B. toggle
   C. carriage
   D. Molly

8. To prevent a nut from loosening on a bolt, _____ are used. (5.3)
   A. flat washers
   B. lock washers
   C. anchors
   D. blind rivets
9. A _____ does not require a predrilled hole to install and is used to hold light objects to hollow walls. (5.4)
   A. Molly bolt
   B. drive-in type expanding anchor
   C. toggle bolt
   D. masonry anchor
10. *True or False?* Masonry anchors can be used to support lengths of conduit or copper tubing. (5.4)
11. Setscrews are normally driven by _____. (5.5)
   A. screwdrivers
   B. power drills
   C. Allen wrenches
   D. socket wrenches
12. A _____ is used to join two pieces of sheet metal for a strong connection that will not loosen under vibration. (5.5)
   A. blind rivet
   B. setscrew
   C. toggle bolt
   D. threaded rod
13. The threaded rod is cut to length with a _____. (5.5)
   A. pair of adjustable pliers
   B. handsaw
   C. hacksaw
   D. pair of aviation snips

Apply and Analyze

1. How are threads per inch on a screw determined? (5.1)
2. What's the difference between a tap and a die? (5.1)
3. Describe when to use a self-drilling screw. (5.2)
4. List the length, diameter, and number of threads per inch of a machine bolt designated as 1/4-20 × 4. (5.3)
5. Explain how to properly install a toggle bolt anchor. (5.4)
6. Explain when setscrews, blind rivets, and threaded rods are typically used. (5.5)

Critical Thinking

1. Explain how screw sizes are determined. (5.2)
2. How can a technician remove a fastener that has broken off into a workpiece? (5.3)

Mueller Brass Co.

6 Working with Copper Tubing

LEARNING OUTCOMES

After studying this chapter, you will be able to:

6.1 Identify types and sizes of copper tubing and how tubing differs from pipe.

6.2 Describe how to properly cut, ream, deburr, and bend copper tubing.

6.3 Describe how to connect copper tubing using swaged connections, wrought fittings, and mechanical connections.

TECHNICAL TERMS

ACR tubing	flare plug	pipe	tee
annealed	flare tee	psi	telescoping
chamfer	flare union	P-trap	tubing
compression fitting	flaring	ratchet tubing bender	tubing cutter
coupling	flaring block	reamed	tubing expander
elbow	hard-drawn copper	reducing coupling	Type K
flare bonnet	hydraulic swage	soft copper tubing	Type L
flare cap nut	inside diameter (ID)	spinner swage	Type M
flare elbow	lever-type tubing bender	spring bender	union
flare fitting	offsets	street elbow	work-hardening
flare nut	outside diameter (OD)	swaging	wrought fitting

Introduction

Copper tubing used for air conditioning and refrigeration work is called ***ACR tubing***. ACR stands for Air Conditioning and Refrigeration. ACR tubing differs from copper tubing used for general plumbing work. When ACR tubing is manufactured, the inside of the tubing is dehydrated to remove all moisture. The tubing is then charged (filled) with low-pressure nitrogen gas and sealed with a cap at each end to keep the tubing safe from contamination by oxygen and moisture in the air. If oxygen atoms were to combine with copper atoms (a process called *oxidation*), a layer of copper oxide would form inside the tubing. The caps also keep out dirt and other foreign matter that could contaminate a refrigeration system. Caps or plugs should be replaced after cutting a length of tubing.

Plumbing copper tubing is cheaper than ACR copper tubing, but plumbing copper is not used in refrigeration systems because it is not protected against contamination. Plumbing copper also differs from ACR copper in size. Plumbing copper is measured by its nominal ***inside diameter (ID)***, while ACR copper is measured by its ***outside diameter (OD)***. See **Figure 6-1**.

6.1 Copper Tubing

Copper tubing is available in soft and hard-drawn types. ***Soft copper tubing*** has been ***annealed*** (softened by heating it to a bright cherry red color and permitted to cool) so it can be bent easily. ***Hard-drawn copper tubing*** is rigid and cannot be bent easily unless it is first annealed. Both soft copper and hard copper are further classified according to the thickness of the tubing wall. Three thicknesses are available: K, L, and M.

- ***Type K*** has the thickest wall. It is used in applications where abuse or corrosion might occur. Type K is usually hard-drawn tubing and is used in commercial refrigeration systems.
- ***Type L*** is the most common tubing. It has a medium-thick wall and is used for residential and commercial applications. Most ACR tubing, both soft and hard, is L thickness. When ACR copper tubing is ordered, Type L is automatically provided unless another thickness is specified.
- ***Type M*** is a thin-walled tubing. It is not used in refrigeration systems because it does not meet safety code requirements. Some manufacturers use Type M copper tubing to construct water-carrying coils. Plumbers may use it for small drain lines and for other noncritical applications.

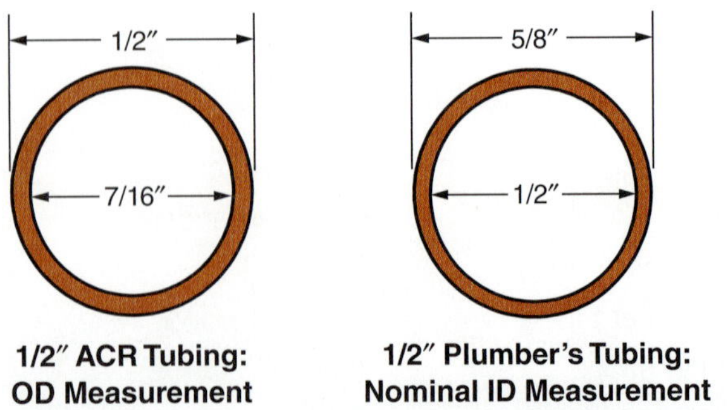

Goodheart-Willcox Publisher

Figure 6-1. ACR tubing is sized by actual outside diameter. Plumber's tubing is measured by nominal inside diameter (the OD usually will be 1/8″ larger than its ID).

Soft Copper Tubing

Type L soft copper tubing is available from 1/8″ to 1 1/8″ OD and is usually sold in 50-foot coils, **Figure 6-2**. As noted earlier, these coils are dehydrated and sealed at the factory.

Because soft copper is easily bent, it must be supported with clamps or brackets every four to six feet. Soft copper tubing tends to harden as a result of vibration, oxidation, and bending. This is called ***work-hardening***. Work-hardened copper will crack at stress points, especially when it is flared or formed at the tubing ends.

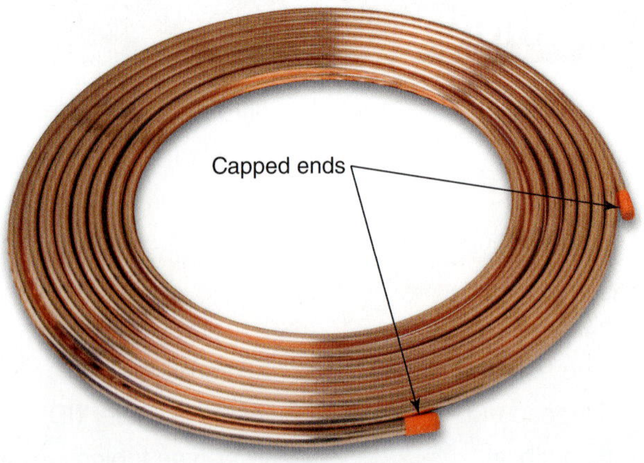

Mueller Brass Co.

Figure 6-2. Soft copper ACR tubing is supplied in 50-foot coils. To prevent contamination, it is dehydrated, filled with nitrogen gas, and capped at both ends.

Unrolling Soft Copper Coils

Soft copper coils should be handled with care because the tubing is easily damaged. Kinks, bends, flat spots, or dents will make the material unfit for use. Bending and kinking can be avoided by unrolling the coil properly. As shown in **Figure 6-3**, unrolling is done by supporting the coil upright with one hand and holding the free end of the tubing stationary on a flat surface with the other hand (or a foot). The coil is then rolled in a straight line to the desired length. Do *not* unroll an excessive amount because it is difficult to recoil the tubing without bends or kinks. After cutting off the desired length, replace the cap or plug on the end of the coil to prevent contamination.

Hard-Drawn Copper Tubing

Hard-drawn copper tubing is hard and rigid and either Type L or Type K thickness. It comes in standard 20-foot lengths that are dehydrated, charged with nitrogen, and sealed with rubber plugs at each end, **Figure 6-4**. Hard-drawn copper tubing cannot be bent easily, so soldered or brazed fittings are used when making connections or changing directions. Because of its rigidity, hard-drawn copper tubing requires fewer supports or brackets than soft copper tubing, and assembly is quicker. It is available in sizes ranging from 3/8″ OD to over 6″ OD.

Goodheart-Willcox Publisher

Figure 6-3. Uncoil soft copper tubing carefully to avoid kinking. Hold one end flat on the floor or other surface and unroll as much as needed.

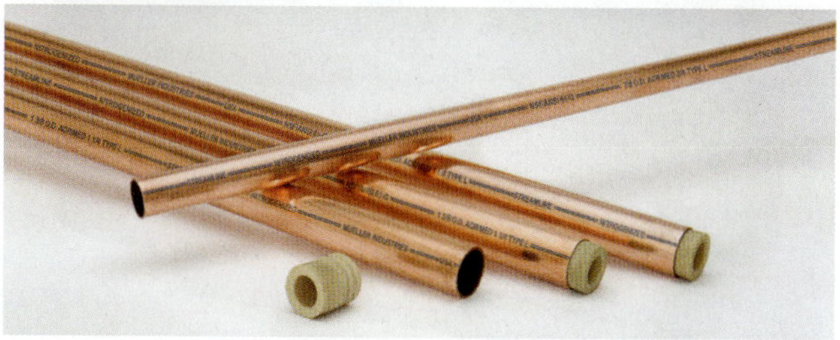

Mueller Brass Co.

Figure 6-4. Rubber plugs are used to seal the ends of lengths of hard-drawn copper tubing after it is dehydrated and filled with nitrogen gas.

Tubing vs. Pipe

In the early days of refrigeration and air conditioning, plumbers performed the installation and service work. These early systems used ammonia and other corrosive chemicals as refrigerants. Iron and steel pipe had to be used to avoid the chemical reaction that occurs when these chemicals are used with copper tubing and fittings. The development of noncorrosive refrigerants has made possible the use of copper tubing in almost all refrigeration and air-conditioning systems today.

The distinction between tubing and pipe is primarily wall thickness and the resulting joining method. ***Tubing*** is considered *thin-walled* material (regardless of whether it is Type K, L, or M) compared to steel and plastic pipe. The term "tubing" is generally applied to materials such as copper and aluminum that are joined by means other than threads. ***Pipe*** is the term used to describe *thick-walled* materials, such as steel or plastic. Threads are cut into the pipe wall, allowing lengths of pipe to be joined with threaded fittings that screw into place.

Fitting different sizes of tubing inside one another is known as ***telescoping***. This procedure is sometimes used to perform emergency repairs when the proper fitting is not available. Telescoping can be performed with ACR tubing because of its OD sizing system and wall thickness. As shown in **Figure 6-5**, ACR 1/4″ copper tubing will fit snugly inside 3/8″ copper tubing, 3/8″ tubing will fit snugly inside 1/2″ tubing, 1/2″ will fit snugly inside 5/8″, and so on.

CAUTION

Telescoping Copper

When telescoping copper, care must be taken as to not reduce or expand too much. Improper telescoping can result in a restriction or metering effect in the copper tubing.

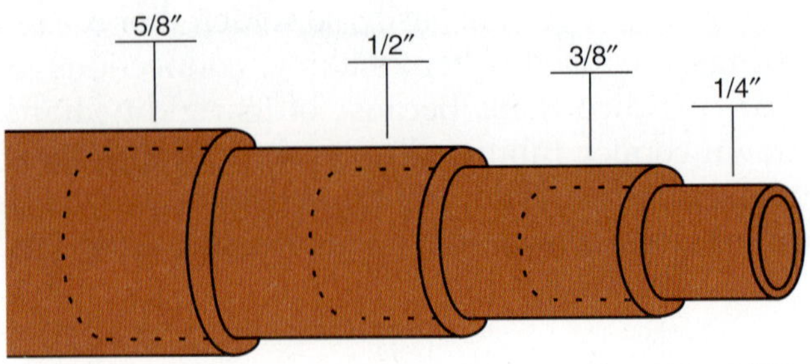

Goodheart-Willcox Publisher

Figure 6-5. Accurate OD sizing allows different diameters of ACR tubing to telescope inside each other.

6.2 Working with Copper Tubing

The ability to properly perform such operations as cutting, bending, and joining copper tubing is a basic requirement for success as an HVACR technician. Careful attention to the correct use of tools and development of good work habits will result in trouble-free installations and satisfied customers.

Cutting Tubing

Cutting copper tubing is a simple task, but it must be performed properly. Care must be exercised not to damage the ends being cut. The most common and accurate way to cut copper tubing is to use a tubing cutter, **Figure 6-6**.

Goodheart-Willcox Publisher

Figure 6-6. Tubing cutters produce an accurate, clean 90° cut, resulting in strong, well-made joints.

Tubing cutters make an accurate 90° cut on either hard or soft copper tubing. They are available in several sizes for use on different tubing size ranges or for special purposes, **Figure 6-7**. The tubing is positioned on the cutter rollers with the cutter blade accurately located at the cut point. Rotating the cutter knob will raise or lower the cutter blade. The blade is pressed firmly against the metal, and the tool is rotated around the copper tubing. After each rotation, the knob is tightened to force the blade lower into the cut. Several rotations are needed to properly complete the cut.

CAUTION

Flat Spots

Avoid excessive blade pressure, which will flatten the tubing and may cause a severe burr on the inside of the tubing ends. See **Figure 6-8**.

Routinely examine the cutter blade. Keep it sharp, as a dull blade will not cut a single clean groove. Instead, it will "track," making rows of shallow cuts on the outside of the tubing.

Tubing cutters should be used only on copper or aluminum tubing. Using the cutter on steel tubing or electrical conduit will quickly dull the blade, making a replacement necessary.

A hacksaw is the second (and less desirable) method of cutting copper tubing. This method has two main disadvantages: it is difficult to obtain a precise 90° cut, and the sawing process creates tiny metal chips that can contaminate the inside of the

JB Industries

Figure 6-7. This tiny, special-purpose tubing cutter can be used to cut tubing up to 5/8″ OD where little clearance is available.

Goodheart-Willcox Publisher

Figure 6-8. Excessive pressure from the tubing cutter blade may cause flat spots and severe burrs on the tubing. Tubing at left was cut with correct pressure; tubing at right, with excessive pressure. Note the flat spot that resulted.

tubing. To obtain the smoothest possible cut, use a saw blade with at least 32 teeth per inch. Be careful to remove all metal chips from the tubing.

Reaming and Deburring

After the copper tubing is cut, the tube ends must be ***reamed***, removing burrs and scraping the ends to produce a flat surface. This procedure can be performed with a pointed reamer blade built into the tubing cutter, but the procedure can be performed more easily and neatly with a handheld reamer. Use care to prevent copper chips or burrs from entering the tubing. Hold the tubing upside down or at an angle during the reaming process so the chips will fall to the floor. See **Figure 6-9**.

It is important to properly dress the cut tubing end. Burrs or ridges on the inside of the hole will cause problems in assembly, as will thin or uneven edges. The thickness of the tubing *at the cut* should match the thickness of the rest of the tubing, **Figure 6-10**. A little time and effort spent dressing the cut will prove worthwhile.

Bending Copper Tubing

Several types of tubing benders are available for making accurate bends in tubing without causing flats, kinks, or dents. Flats and kinks are eliminated, not for the sake of appearance, but because they would restrict the flow of liquid or gas through the tubing.

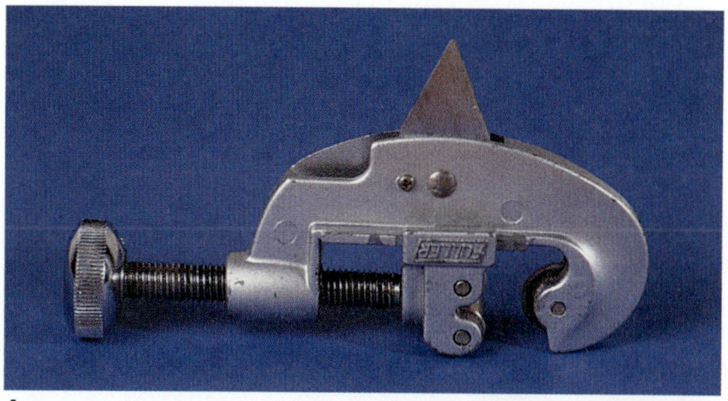

A Goodheart-Willcox Publisher

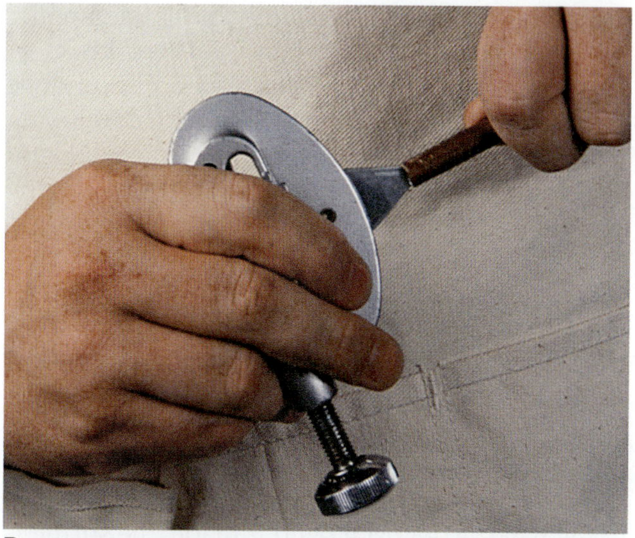

B Goodheart-Willcox Publisher

C JB Industries

Figure 6-9. Reaming tools and procedure. A—Reamer blade built into a tubing cutter. B—Hold the tubing with the open end down while reaming to prevent contamination of the tubing with metal chips from the burrs. C—Handheld tubing reamer with replaceable blades.

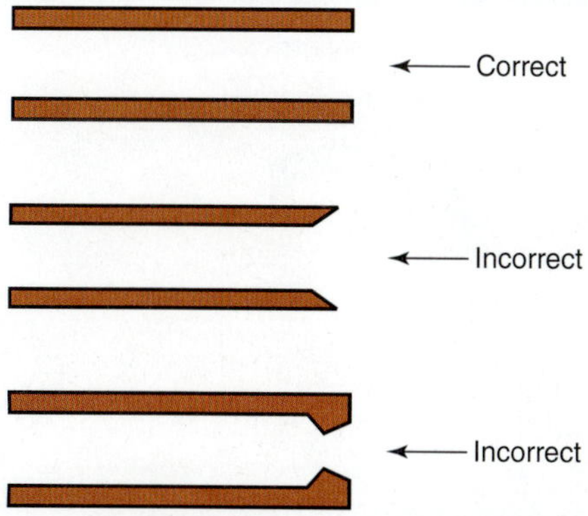

Goodheart-Willcox Publisher

Figure 6-10. Correctly and incorrectly dressed cut tubing. A—Tubing is correctly reamed, with proper wall thickness at cut end. B—Tubing is excessively reamed; wall is too thin. C—Tubing is not reamed sufficiently; wall is too thick.

In a refrigeration system, the tubing must carry liquid or vapor from one component to another. The tubing is carefully sized to carry a specific flow, so the technician must be careful not to create unintended restrictions, such as flats and kinks. Restrictions to flow reduce overall system pressure, resulting in a decrease of efficiency.

Hand Bending

Soft copper greater than 1/4″ in diameter can sometimes be bent successfully by hand. Smaller tubing is easily bent by hand without kinks or flats, unless the bend is sharp. Bending is done carefully, a little at a time, to avoid developing severe flats or kinks that must be removed. Hand bending requires skill and should be done only by experienced technicians.

Spring Benders

Spring benders provide an efficient, low-cost method to bend soft copper tubing, **Figure 6-11**. Spring benders are available in a variety of sizes to fit tubing from 1/4″ OD to 3/4″ OD. These springs slip over the tubing to completely cover the area of the bend. After each bend is made, the spring is slid along the tubing to the next section to be bent.

When a bend is sharp, spring benders tend to bind, or stick, to the tubing. This tendency to bind can be overcome by bending the tubing a little farther than needed, then bending it back to relieve pressure on the spring. Push, *do not pull*, on the spring to remove it from the tubing. Pulling can permanently separate the spring coils, making the bender unfit for further use. If the bend is still too tight to slide the spring off the tubing, simply twist the spring to "unscrew" it.

Little effort is needed to accomplish proper bends in smaller tubing with a spring bender. Larger tubing, however, requires more physical force to bend, so spring benders are used on larger tubing primarily to accomplish slight bends or curves. Although the spring bender is designed for use on the outside of tubing, a smaller spring is sometimes inserted into the tubing to make a bend at the tubing end.

Lever-Type Benders

Lever-type tubing benders are easy to operate and are calibrated to allow accurate short-radius bends up to 180°, **Figure 6-12**. Some benders are designed to fit only one size of tubing, while others can be used with a range of sizes. Lever benders can be used on soft copper, aluminum, steel, stainless steel, and Types K and L hard copper tubing.

For the technician, making accurate bends in copper tubing is almost a daily task, especially on installation jobs. Bending is much faster than installing a fitting and is less likely to leak than fittings. Making accurate 45° and 90° bends requires some practice. Lever benders are designed to make this task easier and more accurate.

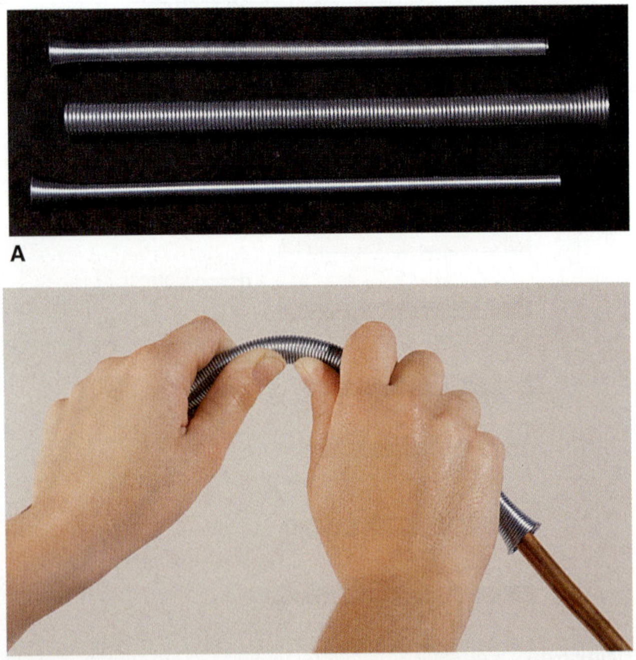

A

B

Goodheart-Willcox Publisher

Figure 6-11. Spring benders. A—Used mostly with small-diameter, soft copper tubing, spring benders are available in several sizes. B—Using a bending spring on soft copper tubing is a two-handed operation.

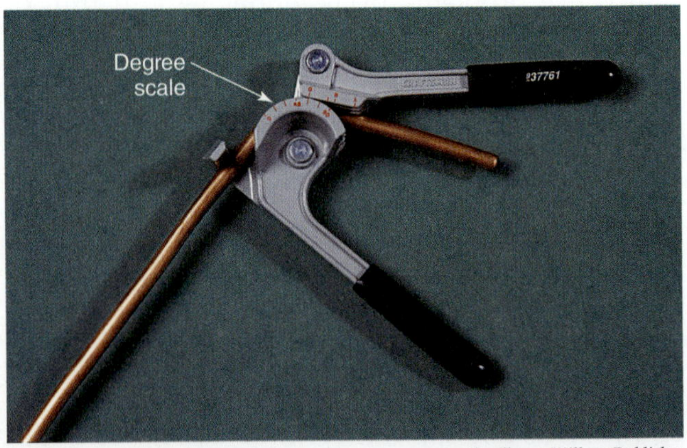

Goodheart-Willcox Publisher

Figure 6-12. Lever-type tubing benders can make short-radius bends of up to 180° without flattening the tubing. Note the scale used determines bend degrees.

Ratchet Tubing Benders

Ratchet tubing benders are an easy way to bend tubing, **Figure 6-13**. They often come in a kit that will bend tubing from 1/4″ to 7/8″. Kits also often include a tubing cutter and reamer. This type of bender is useful when having to make a bend on tubing in a tight location where it would be hard to use other types of benders. Ratchet tubing benders can also be used on aluminum and light wall hydraulic tubing.

6.3 Connecting Copper Tubing

The walls of copper tubing are too thin to make strong threaded connections, so other means are employed to join lengths of tubing. Two approaches are used:

- Permanent connections using soldering or brazing on swaged copper tubing or wrought fittings
- Mechanical connections using threaded fittings that can be easily disconnected to make repairs or to replace defective parts

The ability to work with copper tubing and the various fittings used to make connections is a requirement for anyone entering the HVACR field.

Swaging Copper Tubing

Swaging involves enlarging the diameter of one end of a length of soft copper tubing so the end of another length can be slipped into it. See **Figure 6-14**. The connection is then soldered or brazed to make a strong, leakproof joint. Swaging is the preferred method of joining soft tubing since the process requires little time, and only one brazed joint is needed to complete the connection (compared to two joints for a fitting).

Swaging Methods

Swaged connections can be made using various swaging tools. Common swaging methods include the punch-type and the screw-type. Both types of joints require the use of special hand tools.

Punch-Type Swage

The tubing is clamped into a special tool called a *flaring block* with just enough tubing protruding through the block to accomplish the enlarging process, **Figure 6-15**. The depth of the finished swage should be equal to the original tubing diameter. For example, 1/4″ tubing is swaged 1/4″

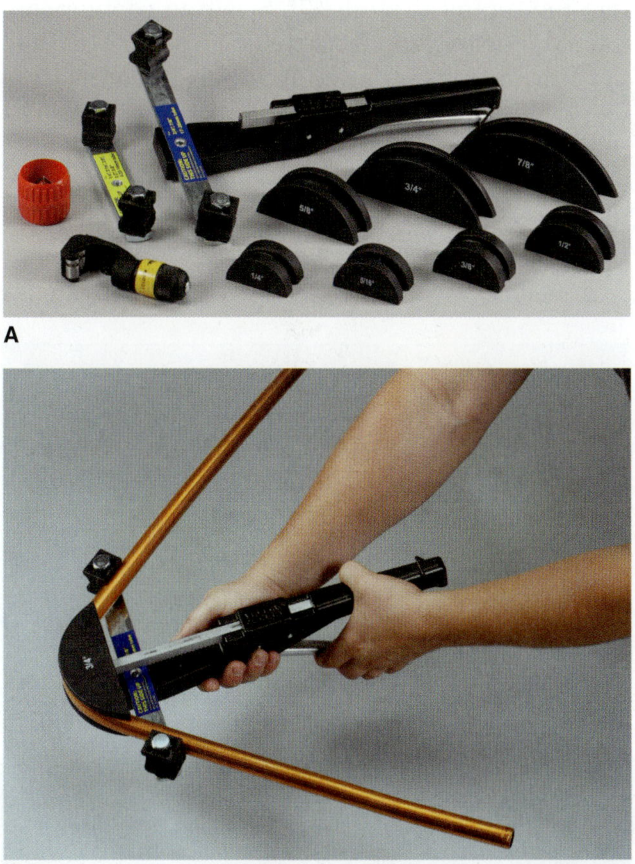
A

B

Ritchie Engineering – YELLOW JACKET Products Division

Figure 6-13. Ratchet benders. A—Ratchet bender kit for bending 1/4″ to 7/8″ tubing. B—Ratchet bender bending 3/4″ tubing.

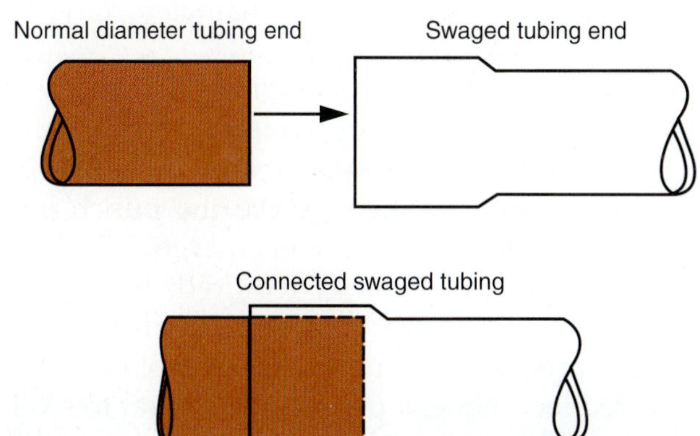

Goodheart-Willcox Publisher

Figure 6-14. Swaging joins tubing by enlarging the diameter of one piece so that it forms a socket for another piece. The joint is then soldered or brazed for strength and leak proofing.

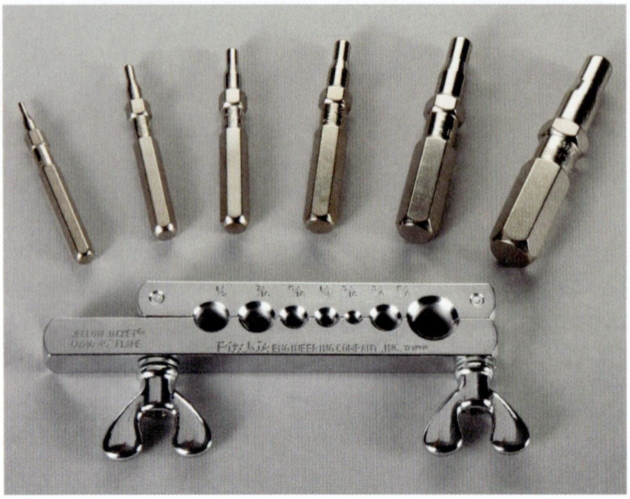

A Ritchie Engineering – YELLOW JACKET Products Division

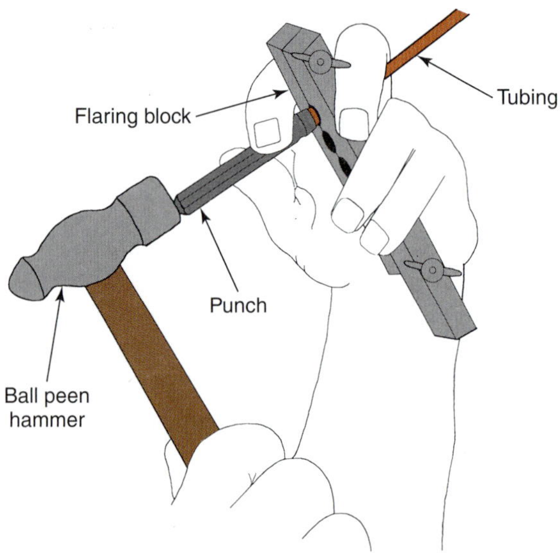

B Goodheart-Willcox Publisher

Figure 6-15. A—A punch-type swaging tool set. B—A punch-type swaging tool uses a flaring block to hold the tubing. A swaging punch is driven into the tubing end by striking with a ball peen hammer. This stretches the opening of the tubing end.

deep, and 1/2″ tubing is swaged 1/2″ deep. Swaging punches are available in diameters ranging from 3/16″ to 7/8″.

Swaging is done by clamping the tubing in the flaring block, then positioning the proper size swaging punch at the tubing end. A ball peen hammer is used to slowly drive the punch into the tubing end. The punch will enlarge the tubing end to the precise diameter required. The punch should be held firmly and straight during the swaging process. Striking several easy blows is the preferred technique; it permits the copper to swell slowly, resulting in a straighter and more accurate swage. Determining how hard to strike the punch to obtain the desired swage will come with a little practice.

SAFETY NOTE

Driving a Punch-Type Swage

Using less force when driving a punch-type swage not only helps with producing accurate swages, but is also easier on the knuckles if a blow is missed.

When tubing is inserted into the swage, the fit should be tight and straight. Such a fit is best sealed with proper soldering and brazing. Loose or crooked connections are difficult to solder or braze, resulting in a weak joint. Chapter 8, *Soldering* and Chapter 9, *Brazing* cover additional procedures for connecting tubing.

Screw-Type Swage

With this method, the tubing is clamped into the flaring block with the correct amount of tubing protruding. As shown in **Figure 6-16**, a screw-type yoke is fitted over the tubing end and twisted to lock it onto the flaring block. By turning the handle on the yoke, the swaging adapter is slowly forced into the tubing to stretch it. Swaging adapters are interchangeable and range in OD from 1/8″ to 3/4″.

The screw-type swaging tool is often easier to use in close quarters than the punch-type tool. The screw-type tool precisely aligns the adapter with the tubing, producing nearly perfect swages every time.

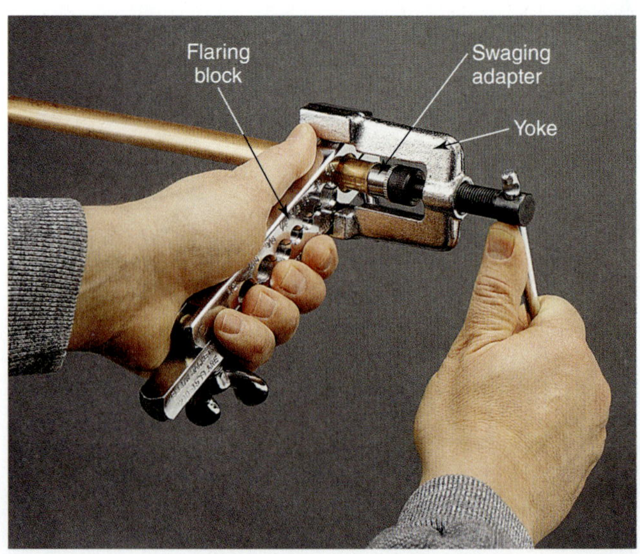

Ritchie Engineering – YELLOW JACKET Products Division

Figure 6-16. The screw-type swaging tool consists of a yoke, a flaring block, and swaging adapter for different tube sizes.

Tubing Expander Swaging

A ***tubing expander*** is a type of swaging tool that uses a lever-actuated expanding mechanism, **Figure 6-17**. Select the proper expander head for the tubing size. Connect the expander head to the end of the lever. Insert the expander head into the end of the tubing and squeeze the handles together. This expands the end of the tubing to prepare a swaged joint. These tools can also be used to re-round tubing and fittings that have been bent out of shape. They come in kits that can make swaging connections from 3/8″ to 1 5/8″ in soft copper tubing. They are also useful in making swaging joints in small lengths of tubing, which can be difficult to do with other types of swaging tools.

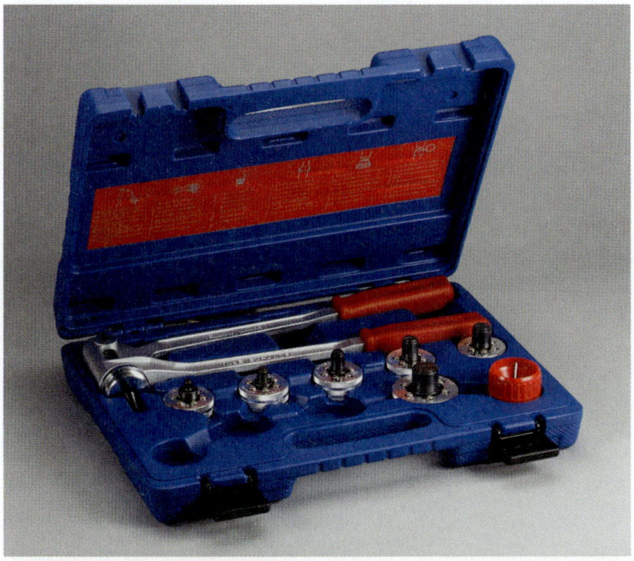

A

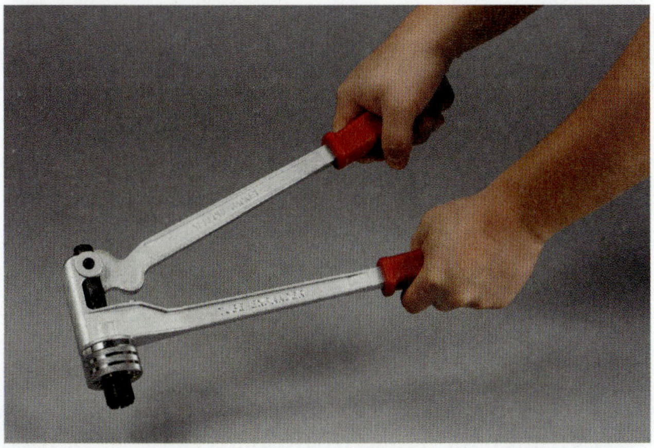

B

Ritchie Engineering – YELLOW JACKET Products Division

Figure 6-17. Tubing expander. A—Tubing expander kit for tubing sizes 1/4″ to 7/8″. B—Tubing expander being used to make a swaged expansion on the end of copper tubing.

Hydraulic Swage

A ***hydraulic swage*** is similar to the tubing expander swage in that an appropriate expander head can be selected to match the tubing size. The hydraulic swage, however, differs from the tubing expander as this swage utilizes hydraulic pressure to expand the tubing. Its convenient design allows it to be operated with one hand. Hydraulic swage tools are generally more compact which allows them to be used in tighter areas as well. Hydraulic swage kits are available with sizes comparable to other swaging tools, **Figure 6-18**.

Spinner Swage

A ***spinner swage*** is designed to be inserted into a drill and uses high speed to create friction that heats up the copper. The copper can then be easily expanded as the swaging tool slides into the copper. Spinner swages are designed with stops to keep the swage from going too deep, which can help the technician gauge the proper depth of the swage.

Using Wrought Fittings

Wrought fittings are sometimes called *sweat* fittings because of the *sweat soldering* or brazing method used to join them to either hard-drawn or soft copper tubing. These fittings are usually made of copper, but sometimes are brass. They do *not* have threads. They are sized according to the copper tubing used and are readily available in sizes ranging from 1/4″ to 6″.

Wrought fittings are used where the connection is permanent, such as long runs of either soft or hard copper tubing. These fittings are *required* when using hard-drawn copper tubing. The brazed connection is stronger than the actual

hilmor

Figure 6-18. This compact hydraulic swage tool can swage tubing from 5/16″ to 1 5/8″.

copper tubing, but the connection *must* be properly brazed and leakproof.

Wrought fittings are designed to fit tightly over copper tubing for easy soldering or brazing. Avoid dropping or roughly handling fittings since a dent in a fitting can make it unsuitable for use.

Types of Fittings

Fittings are available in many different types and sizes, so almost any connection is possible, **Figure 6-19**. The names and shapes of fittings are common to many trade areas (plumbing, electrical, HVACR), but the sizes, threads, and types of material often differ.

Couplings

The fitting used to connect two lengths of copper tubing is the ***coupling***, **Figure 6-20**. Couplings are available in all sizes and usually have openings of the same diameter on each end. A 5/8″ coupling, for example, is used to connect two sections of 5/8″ hard copper tubing. These fittings also can be obtained as ***reducing couplings*** by specifying the sizes on each end. For example, 7/8″ × 5/8″ signifies a reducing coupling with one end that is 7/8″ and another that is 5/8″.

Elbows

Available in either 45° or 90° angles, ***elbows*** are tubing couplings that bend to change direction. They usually have female fittings on each end, **Figure 6-21**. Elbows are used to change the direction of a tubing run. A *short-radius* elbow bends quite sharply, while a *long-radius* elbow makes a more sweeping turn. The long-radius type is preferred because it offers less restriction to the flow of refrigerant. Like couplings, elbows are sized according to the OD of the tubing on which they are used. A 1/2″ 90° elbow, for example, would have 1/2″ female openings on each end to accept 1/2″ copper tubing.

Goodheart-Willcox Publisher

Figure 6-20. Couplings are used to connect two pieces of tubing in a straight line. Most couplings have openings of the same diameter at both ends. A reducing coupling, like the one on the right, has openings of different sizes, allowing a larger-diameter piece of tubing to be connected to tubing with a smaller diameter.

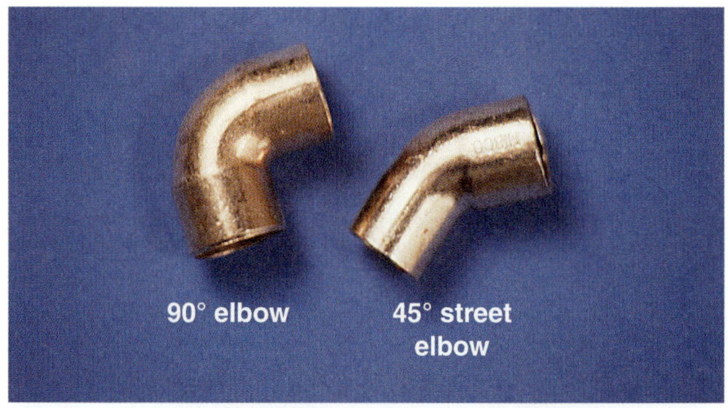

Goodheart-Willcox Publisher

Figure 6-21. Elbows are used to make 45° or 90° changes in direction. Standard elbows have female fittings at both ends. Street elbows have a female fitting at one end and a male fitting at the other.

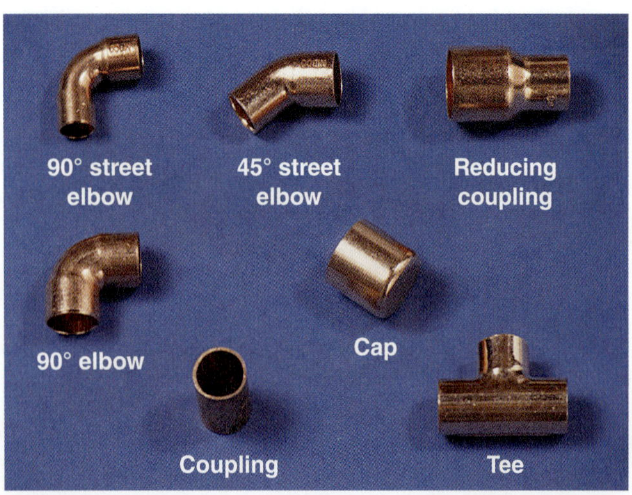

Goodheart-Willcox Publisher

Figure 6-19. Typical fittings used to make permanent copper tubing connections by sweat soldering or brazing.

A ***street elbow*** has a female end and a male end. Both openings are normally sized for the same size tubing. The male end of the street elbow is the same size as the tubing, while the female end is sized to fit the outer diameter. They are used frequently in HVACR work to make ***offsets*** (changes in direction that extend only a short distance). See **Figure 6-22**. Street elbows reduce the number of brazed connections needed, compared to using regular elbows and short lengths of tubing.

Tees

Fittings called ***tees*** are used to connect a branch circuit to an existing line of copper tubing. A tee numbering system guarantees the correct size, type, and placement of an opening. As shown in

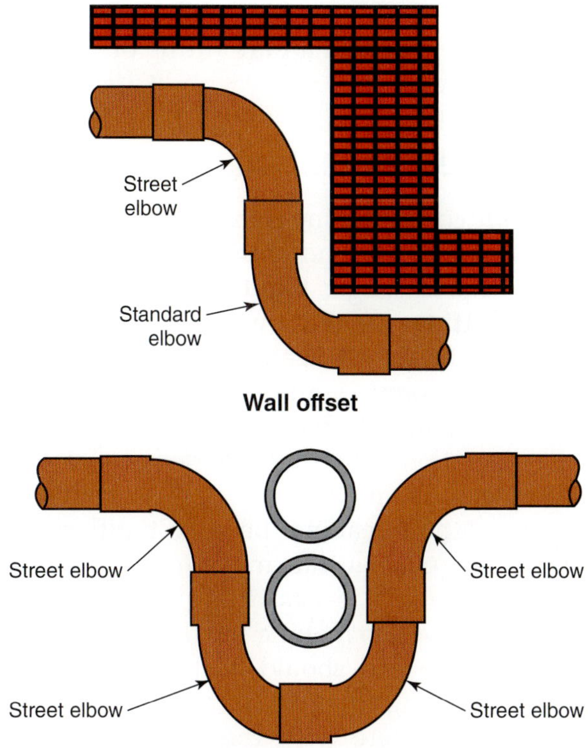

Goodheart-Willcox Publisher

Figure 6-22. Street elbows are used to make offsets (changes in direction) similar to those shown. Using street elbows for offsets reduces the number of brazed connections needed.

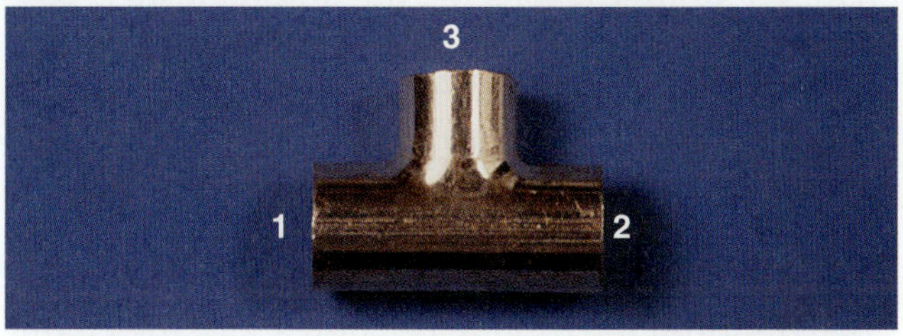

Goodheart-Willcox Publisher

Figure 6-23. A standard numbering system is used for the openings in a tee. The straight-through connections are numbered 1 and 2, and the branch is assigned number 3.

Figure 6-23, the straight-through connections are numbered 1 and 2, and the branch is 3. With this numbering system, it is possible to order a tee with different sizes at each opening. For example, a wrought tee sized at 7/8″ × 5/8″ × 1/4″ would have the following measurements:

- Straight-through end: 5/8″
- Straight-through end: 7/8″
- Branch: 1/4″

Other Fittings

Wrought fittings are available to satisfy almost any need. Special fittings, such as ***P-traps*** (named for the shape into which the tubing is bent), return bends, unions, and caps can save much time and trouble when installing HVACR equipment.

Unions

The ***union*** combines aspects of mechanical and permanent-type fittings. It has three parts: two shoulders and a nut for pulling the shoulders together, **Figure 6-24**. The shoulders of the union are brazed to the ends of the joined copper tubing. The nut pulls on one shoulder while screwing onto the other shoulder, providing a mechanical connection. This results in a leakproof joint that can be disconnected. A union is used in such applications as a drain line that may need to be disconnected for cleaning.

Using Mechanical Connections

Mechanical tubing connections can be divided into two types: compression fittings and flare fittings. Compression-type connections are never used on refrigeration systems since they cannot withstand the high pressure and vibration associated with such systems.

Flared connections are common on all types of refrigeration systems because they can withstand high pressure and some vibration. Failure of a flare connection can usually be traced to abuse or improper installation.

Compression-Type Connections

Compression fittings are commonly used in heating applications that involve only low

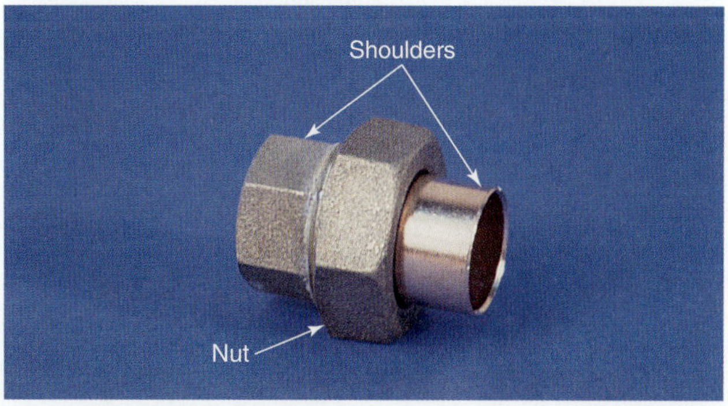

Goodheart-Willcox Publisher

Figure 6-24. A union combines permanent connections with a mechanical system that allows the union to be disconnected. The locking nut pulls the shoulders tightly together, making it possible to "break" the connection for cleaning or repair work.

pressure. They also are widely used for connecting gasoline, natural gas, propane, water, and air lines, where excessive pressure and vibration are not involved. Compression fittings are simple, efficient, and easy to assemble. See **Figure 6-25**.

Other than making sure tubing ends are cut square and burrs are removed, no special preparation is needed when using compression fittings. Simply slip the nut and compression ring over the tubing, then insert the tubing into the fitting until it rests against a shoulder. Slide the compression ring into position and tighten the nut.

Tightening the nut causes the ring to be compressed between mating surfaces, providing a leakproof connection. When the connection is disassembled, the ring will remain attached to the tubing and cannot be removed. The ring is often called a *sleeve* or *ferrule*.

Compression fittings are *never* used on refrigeration systems because of vibration and high pressure. However, some residential air-conditioning systems use special compression fittings to join tubing from the indoor unit to the outdoor unit. These field connections are simple, inexpensive, and quick.

Fittings are sized according to the outside diameter of the tubing, ranging from 1/8″ to 7/8″. Nuts and compression rings needed for installation are provided with all compression fittings, such as unions, elbows, and tees. Many adapters are available to connect tubing with a compression fitting on one side and a flare fitting or pipe connection on the other.

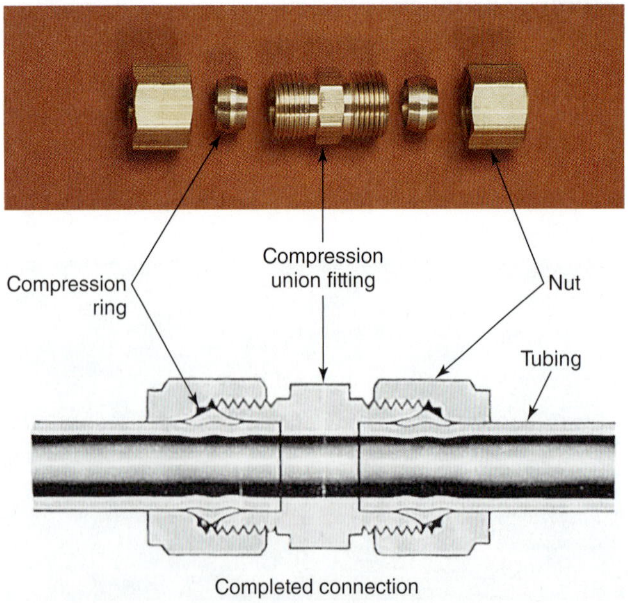

Figure 6-25. Compression connections are made by compressing (squeezing) a sleeve or ring between a fitting and a nut. Compression fittings are used in low-pressure residential heating applications.

Flare-Type Connections

Flaring copper tubing is the process of expanding or spreading the end of the tube into a funnel shape with a 45° angle, **Figure 6-26**. All refrigeration flare fittings are made with a 45° angle so the tubing will fit snugly against the fitting. A flare nut is used to compress the flare against the fitting to obtain a tight, leakproof, metal-to-metal contact.

Refrigeration tubing connections must withstand at least 500 ***psi*** (pounds per square inch) of pressure without leaking. Because the flare connection is a mechanical, metal-to-metal contact without gaskets, it is vital that proper attention and care be given to making the flares.

The tubing end should be properly reamed before attempting to make flares. Burrs or rough edges will interfere with the smooth metal-to-metal contact and permit leakage.

The tubing is clamped in a flaring block with its end protruding slightly above the ***chamfer*** (beveled edge) on the block's top side. A screw-type yoke with a special flaring adapter is then clamped onto the block and automatically centered above the tube. See **Figure 6-27**. Turning the screw will force the cone-shaped adapter into the tubing end, spreading it until it forms a 45° angle against the chamfer, **Figure 6-28**.

Flare Defects

Extending the tubing too high above the chamfer will result in a flare that is too wide. This

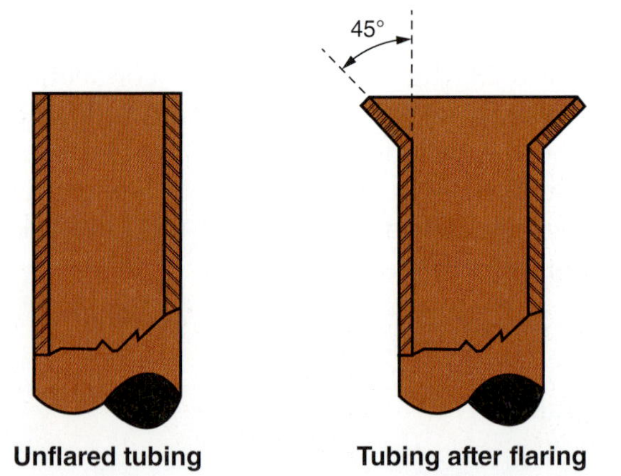

Figure 6-26. When copper tubing is flared, the tubing end is expanded to a funnel shape. The 45° angle matches the angle used on flare fittings.

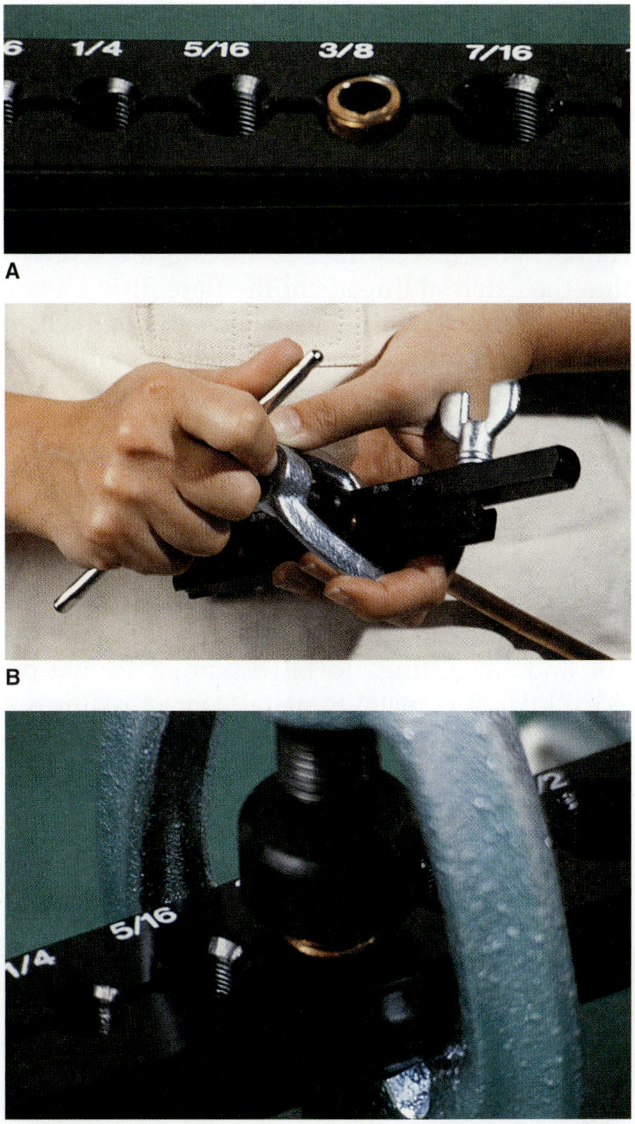

Goodheart-Willcox Publisher

Figure 6-27. Forming a flare. A—The tubing is clamped in the flaring block. B—The yoke is slipped over the block and twisted into place. C—The adapter is screwed into the tubing to form the flare.

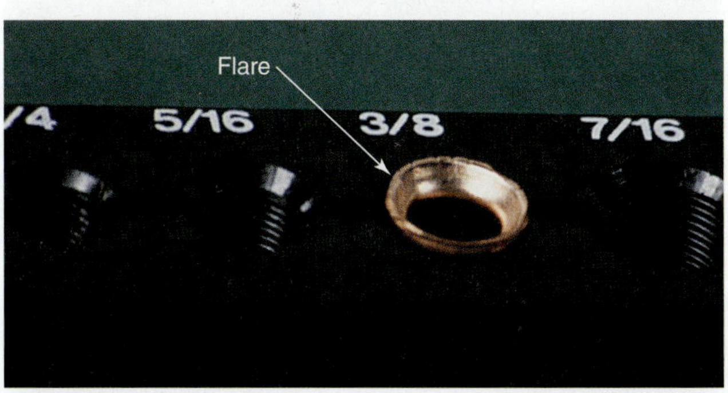

Goodheart-Willcox Publisher

Figure 6-28. The flaring tool's cone-shaped adapter stretches the end of the tubing and forces it against the chamfer, forming a 45° flare.

prevents the nut from sliding over the flare. If the tubing is too low in the chamfer, the result is a small flare that can pull free from the flare nut. A properly made flare will almost fill the bottom of the flare nut without binding or rubbing the threads. **Figure 6-29** shows a correctly made flare and some that are incorrectly made.

> **CAUTION**
>
> **Flare Breakage**
>
> Overtightening the yoke screw while forming the flare will create a thin area where the adapter meets the chamfer of the block. This thin area weakens the flare and can cause it to break off when the nut is tightened.

When preparing a flare connection, *always* remember to place the flare nut on the tubing *before* making the flare. Many technicians have at one time or another made a perfect flare and then discovered that the flare nut was not on the tubing. To resolve such a problem, first cut off the flare,

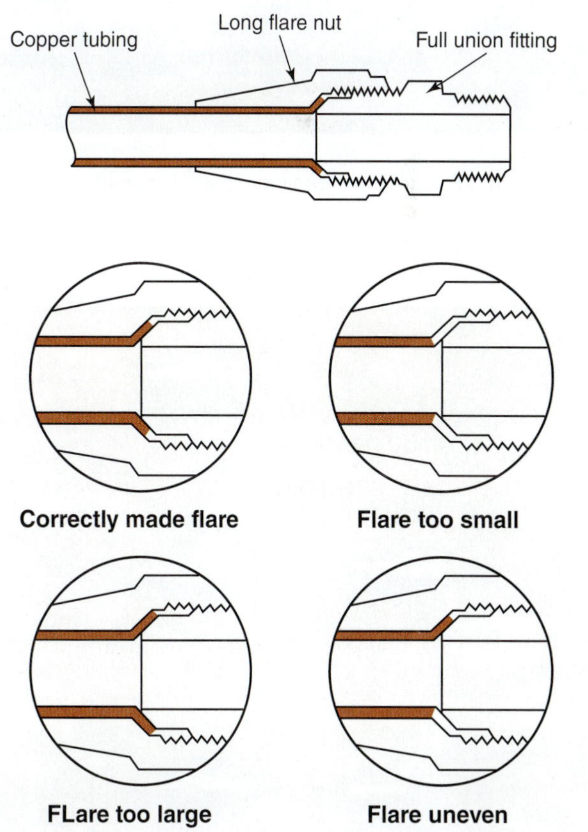

Goodheart-Willcox Publisher

Figure 6-29. A correctly made flare is vital to achieving a leakproof connection.

install the nut on the tubing, and make a new flare. Most tubing cutters have a special groove located on the rollers to ensure only the flare is removed, **Figure 6-30**.

Flare Fittings

Flare fittings have threads for a mechanical connection and a beveled contact area for the flared end of a tube. These fittings are intended for use on soft copper tubing. They are usually drop-forged brass and are accurately machined to form a 45° flare face (area where tubing joins the fitting). See **Figure 6-31**.

Flare fittings are sized according to the copper tubing size and are commonly available from 3/16″ to 3/4″. The threads for all flare fittings are Society of Automotive Engineers (SAE) National Fine. This means they cannot be connected to other threads, such as pipe threads, National Coarse, or ISO Metric.

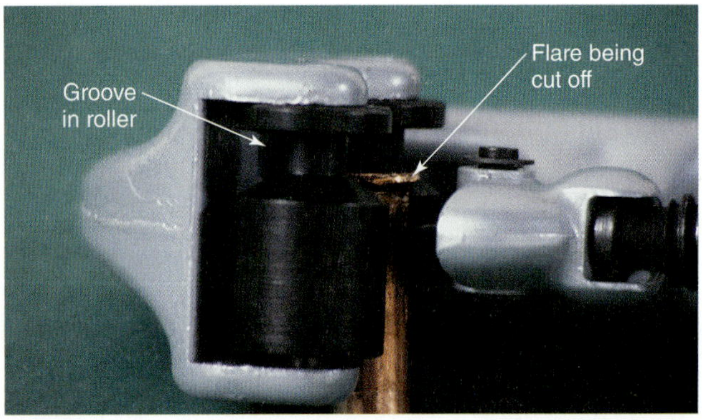

Goodheart-Willcox Publisher

Figure 6-30. A groove in the roller of the tubing cutter allows just the flare to be cut off so the tubing is not shortened any more than necessary.

Flare Nuts

A ***flare nut*** is a piece of hardware that holds flared tubing against a flare fitting. It secures the connection when its internal threads are screwed into the external threads of the flare fitting. Flare nuts are sized according to the hole through which the tubing is inserted. Since the hole fits only over the tubing, any flats or kinks in the tubing will interfere with the fit. Threads are located on the inside of the nut, and the bottom of the nut contains a perfect 45° flare to exactly match the flare on the tubing end. The tubing flare should almost fill the area at the bottom of the nut.

When the nut is screwed onto a fitting, the flare on the tubing is compressed between the nut and the fitting. To be leakproof to 500 psi (3500 kPa), this metal-to-metal contact requires a perfect match without burrs or ridges to interfere with the seal.

Flare nuts can be obtained in either short or long style, **Figure 6-32**. The short flare nut is the most commonly used in high-pressure refrigeration systems. The long style is used mostly in low-pressure systems like water and natural gas.

Sometimes, a *reducing flare nut* is used to attach tubing to a fitting of another size. This eliminates the use of additional fittings and reduces the number of possible leaks.

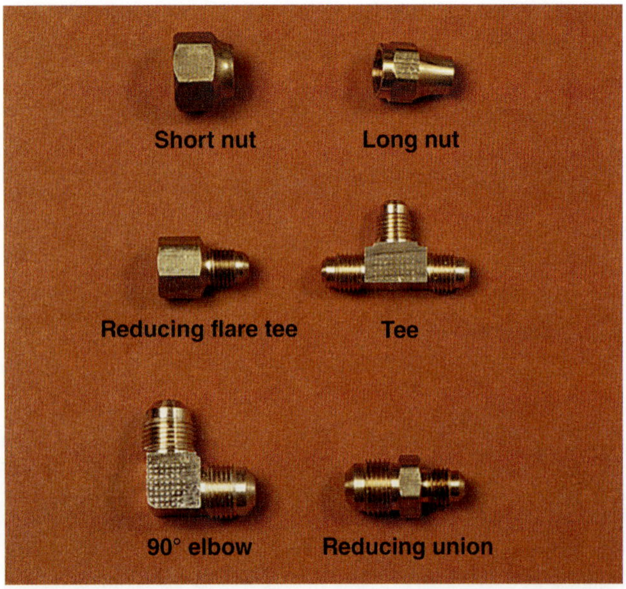

Goodheart-Willcox Publisher

Figure 6-31. Flare fittings are normally machined from brass.

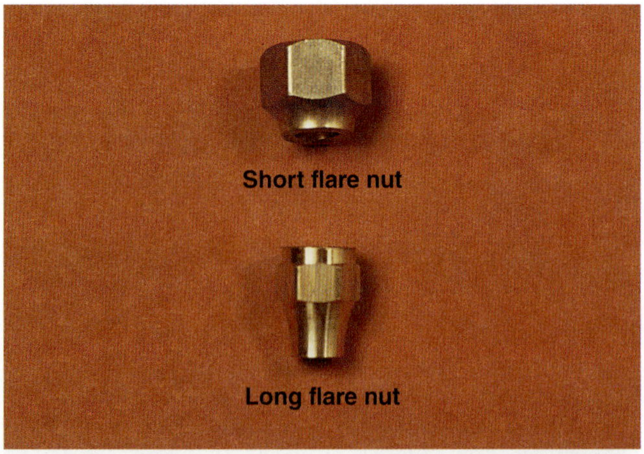

Goodheart-Willcox Publisher

Figure 6-32. Flare nuts are made in both short and long styles. Short flare nuts are built heavier and need to be used on high-pressure refrigeration systems. Long flare nuts are used on lower pressure systems.

Flare Unions

A full ***flare union*** is a fitting used to connect two flare nuts of the same size. It is called a *full union* because each end of the fitting is the same size. A 3/8″ full union would have a 3/8″ male flare on each end. A 3/8″ full union is a quick way of saying, "3/8″ MFT × 3/8″ MFT union" (MFT = Male Flare Thread).

It is possible to obtain a union that has a male fitting on one end and a female fitting on the other. Such a union might be described as 3/8″ MFT × 3/8″ FFT (FFT = Female Flare Thread).

A reducing union performs the same function as a full union, except the sizes of the flare nuts are different. It is used to reduce from one size tubing to another, such as 1/2″ MFT × 3/8″ MFT.

Flare Elbows

The ***flare elbow*** serves the same function as the flare union while providing an accurate bend of either 45° or 90°. Flare elbows (often called "ells") can be obtained in any tubing size, as conventional or reducing types, and with male or female fittings. For example: 90° flare elbow, 1/4″ MFT × 3/8″ MFT (or FFT).

Flare Tees

The ***flare tee*** is a flare fitting with three connections. This fitting makes it possible to connect a branch to an existing line of copper tubing. The tee numbering system guarantees the correct size, type, and placement of the opening. With the numbering system, it is possible to order a tee with openings of different sizes. For example: flare tee, 1/2″ MFT × 3/8″ MFT × 1/4″ MFT. Position 1 is 1/2″ MFT, position 2 is reduced to 3/8″ MFT, and position 3 (the branch) is 1/4″ MFT. Of course, any of the positions on the tee can be ordered with female threads.

Flare Plugs

Available for all tubing sizes, the ***flare plug*** is used to seal a flare nut or similar female-threaded opening. This seal can be temporary or permanent but is usually temporary until proper repairs can be made.

Flare Cap Nuts

A female nut used to seal off a male-threaded fitting is called a ***flare cap nut***. This seal can be permanent or temporary. The flare cap nut is used extensively on service valves and other such devices. The cap nut prevents dirt and other foreign matter from entering the system through an access fitting.

Flare Bonnets

The ***flare bonnet*** is made of copper and is used to convert an ordinary flare nut into a cap nut. Such devices come in handy when a cap nut is not available. The bonnet is placed inside a flare nut, which can be used to cap a male-threaded fitting.

Chapter Review

Summary

6.1 Copper tubing is available in soft and hard-drawn types. Soft copper tubing can be easily bent; hard-drawn copper tubing is rigid and cannot be bent unless first annealed. Both are available in three thicknesses: K, L, and M. Tubing is considered thin-walled material and applies to materials such as copper and aluminum that are joined by means other than threads. Pipe is considered thick-wall material, such as steel or plastic, and is joined with threaded fittings.

6.2 Copper tubing can be cut using a tubing cutter or hacksaw. Using a tubing cutter is the most common and accurate way to make a 90° cut on copper tubing. After tubing is cut, the tube ends must be reamed and deburred to create a clean and flat surface. Copper tubing can be bent by hand or by using spring benders, lever-type tubing benders, or ratchet tubing benders.

6.3 Swaging involves enlarging the diameter of one end of a length of soft copper tubing so the end of another length can be slipped into it. Wrought fittings are installed using sweat soldering or brazing to form a permanent connection. Mechanical tubing connections are divided into two types: compression fittings and flare fittings. Flared connections are common on all types of refrigeration systems, whereas compression fittings are only used on low pressure heating applications.

Know and Understand

1. Most ACR tubing, both soft and hard, is _____ thickness. (6.1)
   A. K
   B. L
   C. M
   D. 1/2″
2. Soft copper tubing that is _____ will crack at stress points, especially when it is flared or formed at the tubing ends. (6.1)
   A. annealed
   B. work-hardened
   C. brazed
   D. Type K thickness
3. The distinction between tubing and pipe is primarily _____ and _____. (6.1)
   A. application, joining method
   B. length, material used
   C. wall thickness, joining method
   D. sizing convention, working temperatures
4. *True or False?* Tubing cutters can only be used to cut soft copper tubing. (6.2)
5. After copper tubing is cut, the tubing should be dressed so its thickness _____ matches the thickness of the rest of the tubing. (6.2)
   A. inside the tubing
   B. outside the tubing
   C. at the cut
   D. on the opposite side of the cut
6. Tubing that is in a difficult to reach location can be bent using a _____ bender. (6.2)
   A. spring
   B. lever-type tubing
   C. ratchet tubing
   D. hand
7. Enlarging the diameter of one end of a length of soft copper tubing so the end of another length can be slipped into it is referred to as _____. (6.3)
   A. reaming
   B. deburring
   C. flaring
   D. swaging
8. A _____ is useful in making swaging joints in very small lengths of tubing, which can be difficult to do with other types of swaging tools. (6.3)
   A. flaring block
   B. screw-type swage
   C. punch-type swage
   D. tubing expander

9. Which of the following statements does *not* describe wrought fittings? (6.3)
   A. They are joined with sweat soldering or brazing.
   B. They are used where the connection is permanent.
   C. They are threaded.
   D. They are sized according to the copper tubing used.
10. A(n) _____ is a fitting used to change the direction of a tubing run. (6.3)
    A. coupling
    B. elbow
    C. tee
    D. union
11. *True or False?* Compression fittings cannot withstand high pressure, so they are rarely used on refrigeration systems. (6.3)
12. The angle on all refrigeration flare fittings is _____°. (6.3)
    A. 30
    B. 45
    C. 60
    D. 90
13. A flare _____ is used to connect a branch to an existing line of copper tubing. (6.3)
    A. elbow
    B. union
    C. tee
    D. bonnet

Apply and Analyze

1. What difference in the processing of soft copper tubing makes it easily bent as compared to hard-drawn copper tubing? (6.1)
2. Why is it important to remove burrs from the cut ends of copper tubing? (6.2)
3. Explain how to use a spring bender to bend soft copper tubing. (6.2)
4. What are two advantages of swaging copper tubing over joining tubing with a fitting? (6.3)
5. Explain the difference between a 90° elbow and a 90° street elbow. (6.3)

Critical Thinking

1. Explain the difference between tubing and pipe. (6.1)
2. Why are soldering and brazing or using mechanical connections the preferred methods of joining copper tubing instead of creating threaded copper tubing? (6.3)

Mueller Industries, Inc.

7 Working with Pipe

LEARNING OUTCOMES

After studying this chapter, you will be able to:

7.1 Identify characteristics of steel pipe, tools and fittings commonly used, and how to repair leaks in steel pipe.

7.2 Explain the various types and sizes of rigid plastic pipe, pipe preparation tasks, solvent-welding process, and how to repair leaks in plastic pipe.

7.3 Identify alternative methods of making pipe connections.

TECHNICAL TERMS

ABS
adapter
chain-type pipe wrench
close nipple
corrosion
CPVC
crimp-on pipe fitting
galvanized
jaw-type pipe wrench
malleable fitting
nipple
nominal size
pipe compound
pipe cutter
pipe die
pipe fitting
pipe reamer
pipe thread
pipe vise
push-on pipe fitting
PVC
shoulder nipple
solvent cement
solvent-welding
strap-type pipe wrench
tapered

Introduction

Steel or plastic pipe, depending upon the application, may be required for HVACR installation and repair jobs. Steel pipe has traditionally been used for water lines, drains, and natural gas supply lines to furnaces. However, plastic pipe is rapidly replacing steel in many applications involving water flow, drainage, and heating systems. Steel pipe is required for R-717 (ammonia) systems. Plastic pipe is often used to connect certain accessories to a system. Such accessories include water cooling towers, water-cooled condensers, and condensate drains. Local and state building codes specify the acceptable uses of each type of pipe. Technicians must develop and perfect the skills needed to work with steel and plastic pipe. Improper pipe installation can produce leaks in a system, which can be costly and dangerous.

7.1 Steel Pipe

Steel pipe is available in unfinished (black) or ***galvanized*** (zinc-coated) forms, **Figure 7-1**. Galvanized pipe is usually specified for water and drain lines because it is less likely to rust than unfinished pipe. Generally, black pipe is used for natural gas applications. However, local codes or installation rules will govern specific applications.

Steel pipe is available in a standard length of 21 feet. There are three main grades or strengths: standard (Schedule 40), extra strong (Schedule 80), and double extra strong. Standard grade (Schedule 40) is acceptable for most plumbing applications.

The International Fuel Gas Code (IFGC) states that gas lines using steel pipe must be at least Schedule 10. Local codes may require a stronger grade of pipe. Always check local codes before installation.

Mueller Brass Co.

Figure 7-1. Galvanized and black steel pipe.

Measuring Pipe

Steel and plastic pipe are measured by their inside diameter (ID). ID sizing relates to the pipe bore, which is called ***nominal size***. As discussed in Chapter 6, *Working with Copper Tubing*, this method is different from the method used to size ACR copper tubing. ACR tubing is measured by its outside diameter (OD). See **Figure 7-2**.

The ID, or nominal size, is not precise because wall thicknesses of pipe vary slightly. Nominal size is, however, considered acceptable for determining sizes for pipe and pipe fittings. Sizes given for all plumbing pipe and fittings, as well as for plumber's copper tubing, are *nominal sizes*. **Figure 7-3** compares nominal size, ID, and OD of steel pipe.

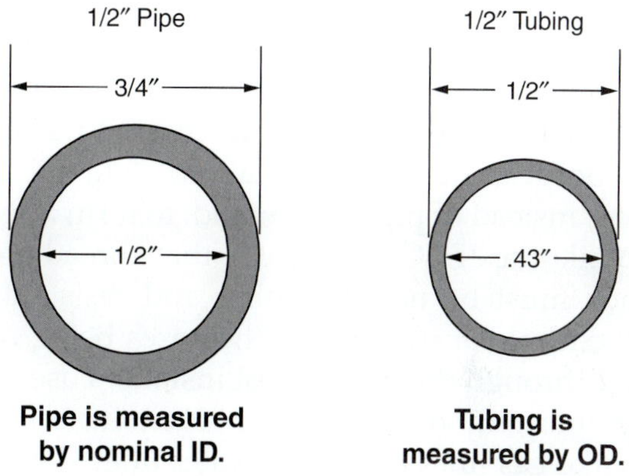

Goodheart-Willcox Publisher

Figure 7-2. Pipe measurements and ACR tubing measurements differ. Pipe is measured by its nominal inside diameter (ID), while ACR tubing is measured by its more precise outside diameter (OD). Note that the wall thickness of pipe is much greater than that of tubing.

Pipe Sizes (Inches)

Nominal Size	Inside Diameter	Outside Diameter
1/8″	0.269	0.405
1/4″	0.364	0.540
3/8″	0.493	0.675
1/2″	0.662	0.840
3/4″	0.824	1.050
1″	1.049	1.315
1 1/4″	1.380	1.660
1 1/2″	1.610	1.900

Goodheart-Willcox Publisher

Figure 7-3. Nominal sizes, ID, and OD of steel pipe commonly used in plumbing work.

Nominal size copper tubing is used by plumbers for water lines and drains. If a plumber asked to borrow 1/2″ copper tubing from an HVACR technician, the ACR tubing provided would be 5/8″, since it would have the desired ID of 1/2″. Check tubing and pipe usage before ordering and installing.

Pipe Threads

Pipe walls are quite thick as compared to tubing walls. This makes it possible to cut threads in pipe walls to connect fittings. ***Pipe thread*** is different from other types of threads in that they are ***tapered*** (sloped) 1/16″ for every inch of length, **Figure 7-4**.

Pipe threads are specially formed, tapered V-threads made in a conical spiral (like the threads on a wood screw). The taper causes the threads to bind together and make a seal as pipe and fitting are tightened. Approximately 1/2″ of the threads is usable before the fitting begins to tighten on the taper. The remaining threads are mostly incomplete. A special ***pipe compound*** (usually referred to by plumbers as *pipe dope*) is normally brushed onto the threads before assembly to ensure a strong, leakproof seal. In some applications, Teflon™ tape is used, instead of pipe compound, to form a seal.

While the IFGC only states that thread compounds must be nonhardening and resistant to the chemical constituents of the gases being conducted through the pipes, most installers use pipe compound only on permanent pipe joint connections like gas lines and main water lines. Teflon™ tape is used mostly for semipermanent joints like water lines under a kitchen or bathroom sink. Local codes may require different requirements.

Pipe threads differ from National Coarse (NC) threads and National Fine (NF) threads in several ways. Both NC and NF are *nontapered*, machine-type threads and are based on the outside diameter of the fastener. Pipe thread sizes are based on the nominal size of the pipe.

Tools Used with Steel Pipe

Cutting and threading steel pipe requires the use of tools specially designed for these tasks. Tools for working with steel pipe are described in the following paragraphs.

Pipe Vise

A ***pipe vise*** is a tool used to securely hold steel pipe in position for cutting and threading. There are two basic types: the yoke vise and the chain vise. See **Figure 7-5**. Some vises mount on a bench, while others are portable and free standing on tripod-type legs. Some vises are equipped with an electric motor to rotate the pipe in either direction for easy, fast cutting and threading.

Pipe Cutter

Steel pipe is cut using a ***pipe cutter***. This tool clamps around the outside of the pipe and uses a wheel blade to rotate and cut around the pipe. It works in the same manner as a copper tubing cutter. However, a pipe cutter is larger than a tubing cutter. It is designed for heavy-duty use, with a special blade for cutting steel pipe. Pipe cutters are available in different sizes, each able to cut several different diameters of steel pipe, **Figure 7-6**.

When cutting steel pipe, allow about 1/2″ of extra pipe length at each end for threading. This is necessary because about 1/2″ of thread will disappear into each connecting fitting. Steel pipe can

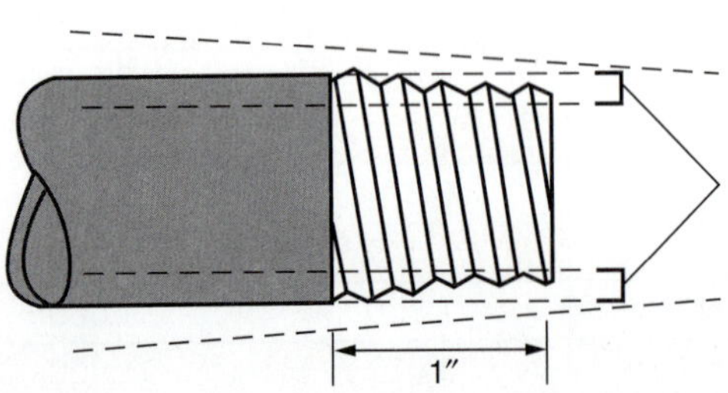

Goodheart-Willcox Publisher

Figure 7-4. Pipe thread tapers 1/16″ for every inch of length. Tapering helps form a tight seal between the fitting and the pipe.

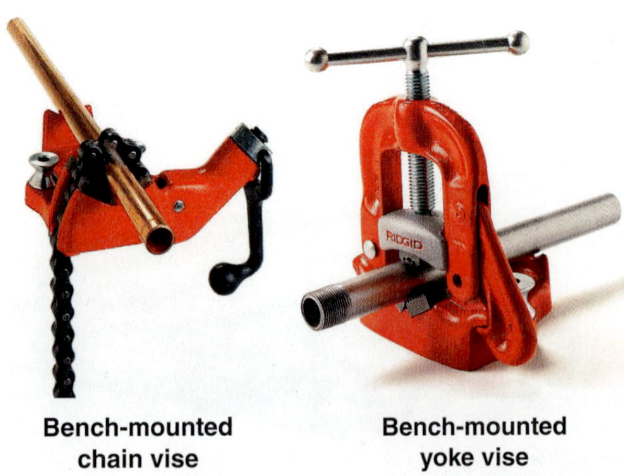

The Ridge Tool Company

Figure 7-5. The two types of pipe vises.

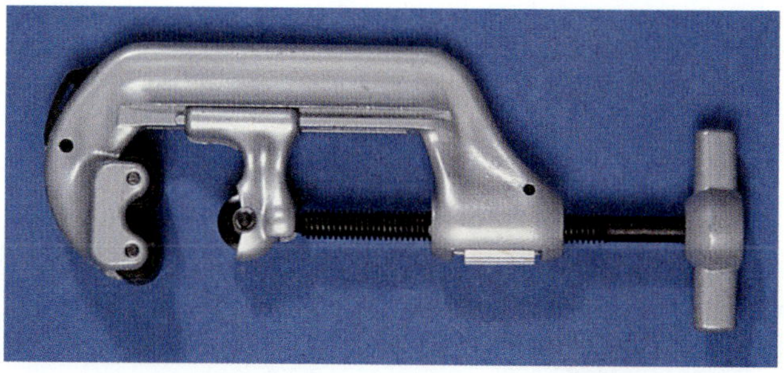

Goodheart-Willcox Publisher

Figure 7-6. A pipe cutter used for steel pipe is larger than a tubing cutter used for copper tubing. Different types are available for use with various ranges of pipe sizes.

also be cut with a hacksaw, but the work must be done carefully to obtain a straight cut.

Pipe Reamer

Regardless of the cutting method used, the cut ends of pipe must be reamed to remove the burr on the inside of the hole. This is done with a ***pipe reamer*** designed for use on steel pipe, **Figure 7-7**.

To ream a pipe, begin by securing the pipe horizontally in a pipe vise. Do *not* attempt to use a pipe wrench to hold the pipe while reaming. A wrench could slip, causing damage to the pipe or injury to the technician. Select the proper reamer size to match the size of the pipe being reamed. Insert the taper reamer into the end of a pipe that has a burr. While applying a small amount of forward pressure, turn the handle of the reamer in a clockwise direction. When the burr has been removed leaving a smooth inside on the pipe end, the pipe has been properly reamed. Depending on how bad the burr is, this might take two or three turns of the handle. This reaming process can be used for both black and galvanized pipe. The burr can also be removed using a file, but this practice takes more time and does not produce as smooth of an end.

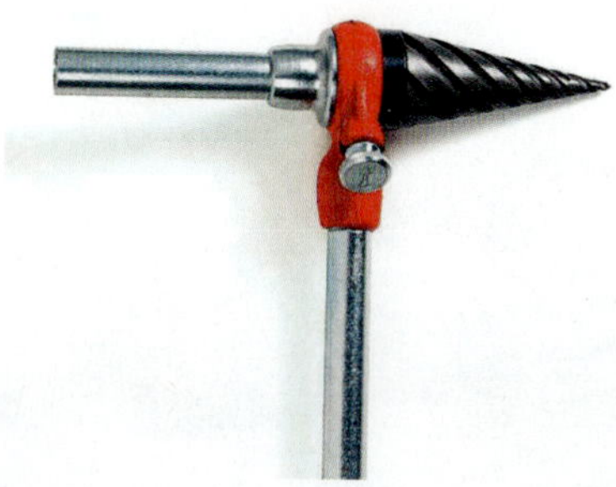

The Ridge Tool Company

Figure 7-7. This spiral pipe reamer will deburr pipe in nominal sizes ranging from 1/8″ to 2″. A ratchet handle permits use in tight quarters.

Pipe Wrench

Pipe wrenches are designed for holding and turning steel pipe and fittings, **Figure 7-8**. They range in size from 3/8″ to 8″ and are available in several types. The regular ***jaw-type pipe wrench*** or the ***chain-type pipe wrench*** may be used when tooth marks from the wrench jaws are not objectionable. To avoid tooth marks, a ***strap-type pipe wrench*** is used.

When using the jaw-type wrench it is important to maintain a gap between the back of the hook jaw and the pipe, **Figure 7-9**. This concentrates pressure at two points and produces maximum gripping action and rotating force. Always *pull* in the correct direction on the wrench handle. This will cause the jaw teeth to bite into the pipe. Pulling in the wrong direction will cause the wrench to lose its grip and

Jaw-type pipe wrench

JB Industries

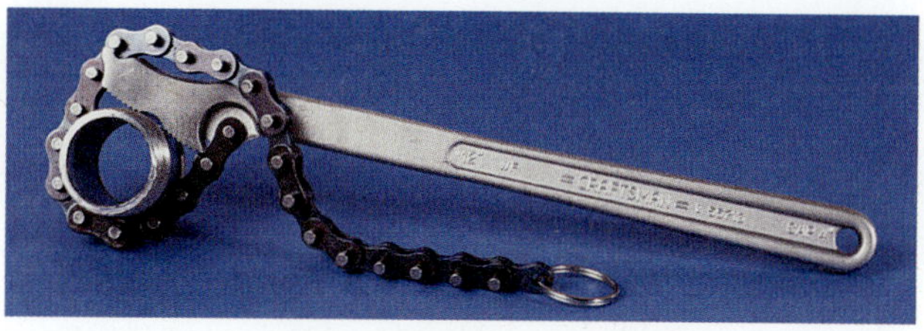

Chain-type pipe wrench

Goodheart-Willcox Publisher

Strap-type pipe wrench

Reed Manufacturing Co.

Figure 7-8. Basic types of pipe wrenches. The jaw-type pipe wrench has sharp teeth to provide a firm grip on pipe and fittings. A chain-type wrench is often used with larger pipe. In situations where tooth marks on the pipe or fitting would be objectionable, a pipe wrench with a fabric strap, instead of a chain, may be used.

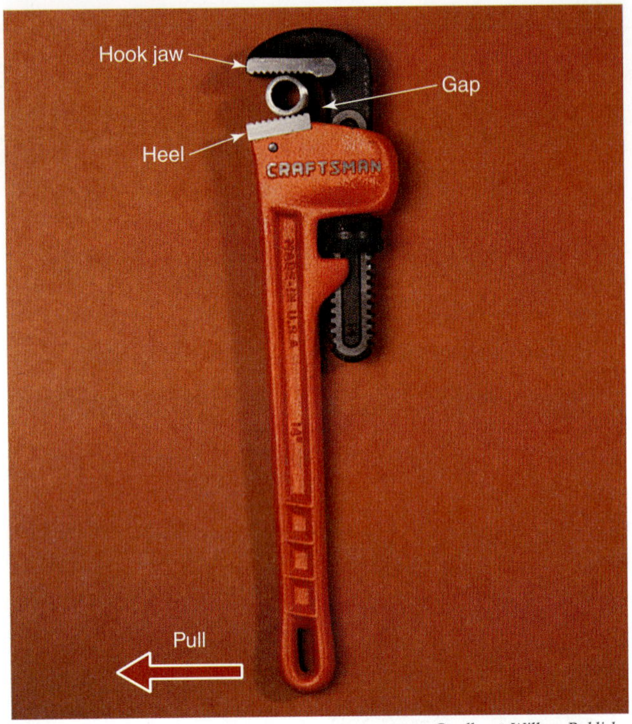

Goodheart-Willcox Publisher

Figure 7-9. A jaw-type pipe wrench should be used properly. By pulling in the direction shown, the teeth will grip the pipe tightly. Pulling in the opposite direction will cause the grip to loosen and the wrench to slip. Leaving a slight gap between the pipe and the back of the jaw will concentrate the force at two points for a better grip.

slip around the pipe. Pipe wrenches are normally used as a pair, one for holding the pipe and the other for turning the fitting.

Pipe Die

A ***pipe die*** is a special tool used to cut threads on the outside of steel pipe. As shown in **Figure 7-10**, the die is an individual tool that fits into a special handle used to turn the die on the end of the pipe. Each die is accurately sized to fit a particular size of steel pipe. When changing pipe size, the die must also be changed. The ratchet-type handle shown is reversible, allowing the die to be unscrewed from the pipe once the threads are cut. A special thread cutting oil is used to keep the die from getting too hot, which could damage the cutting edges.

Fittings for Steel Pipe

Pipe fittings are components used to join sections of pipe. These fittings are made of annealed cast iron. Annealed cast iron fittings are called ***malleable fittings***. They can withstand more bending, pounding, and internal pressure than ordinary cast iron.

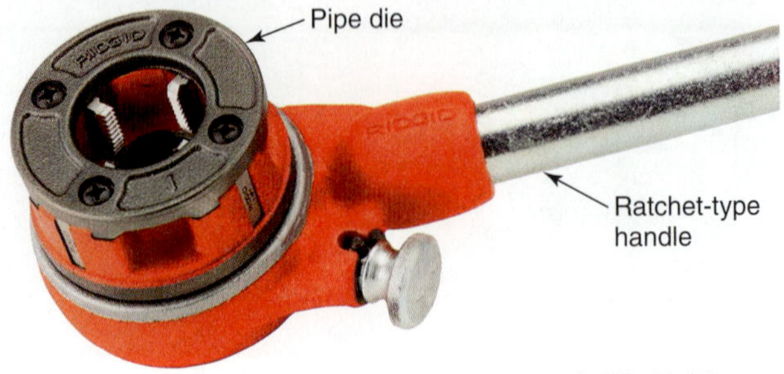

The Ridge Tool Company

Figure 7-10. A pipe die is used to cut threads on the outside of a pipe. A different die must be used for each pipe size. The ratchet-type handle shown can be reversed to "back off" the die after the threads have been cut.

Pipe fittings are similar to those for copper tubing described in Chapter 6, *Working with Copper Tubing*. The major difference is that steel pipe fittings are threaded rather than smooth like brazed tubing fittings. As previously mentioned, pipe fittings are always sized according to the nominal size (ID) of the pipe being used. This can be confusing for someone accustomed to dealing with ACR tubing that is sized by OD, but a little practice and size comparison will quickly eliminate any confusion. Types of fittings used with steel pipe are described in the following paragraphs.

Elbows

Pipe elbows, called "ells," are necessary when changing the direction of a piping run. They are readily available in all sizes and in 90° or 45° angles, **Figure 7-11**. Pipe elbows are also available as "street ells," which means one end of the elbow is female and the other end is male.

Couplings

A coupling is a short fitting with female threads at each end, used for connecting two lengths of pipe in a straight line, **Figure 7-12**. Reducing

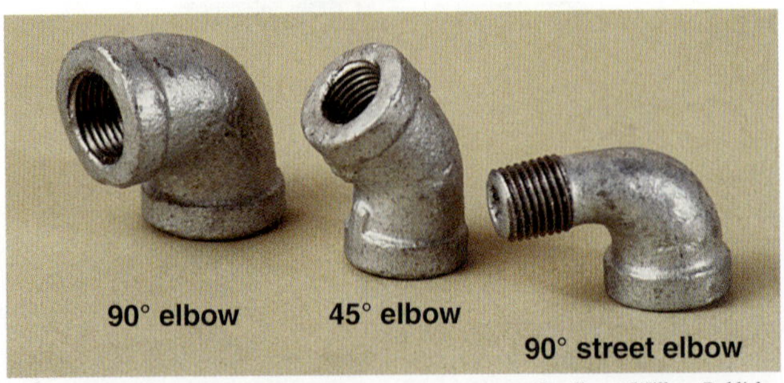

Goodheart-Willcox Publisher

Figure 7-11. Elbows are available in all nominal pipe sizes.

Goodheart-Willcox Publisher

Figure 7-12. Couplings connect pipe lengths in a straight line.

couplings are used to connect two different sizes of pipe in a straight line.

Tees

Tees are used to create a branch from an existing line, **Figure 7-13**. Pipe tees have numbered openings just like those used on copper tubing. Numbers 1 and 2 are straight through, and number 3 is the branch.

Pipe tees can be obtained in almost any combination of sizes, all with female threaded openings. A 1/2″ tee has three 1/2″ openings. Each size for a reducing tee must be specified according to the numbering system. For example: a 1/2″ × 3/4″ × 1/4″ tee has a branch sized at 1/4″.

Reducing Bushings

A reducing bushing has male threads on the outside and female threads on the inside, **Figure 7-14**. It is used to reduce the size of a female opening. For example, a water valve with 3/4″ female openings can be used on 1/2″ pipe by installing a reducing bushing (3/4″ × 1/2″) on each side of the valve. Reducing bushings are available in a wide variety of sizes.

Goodheart-Willcox Publisher

Figure 7-13. A tee permits the connection of a branch to a main pipe. The numbering system for tee openings is universal. Openings 1 and 2 are the main line. Opening 3 is the branch line.

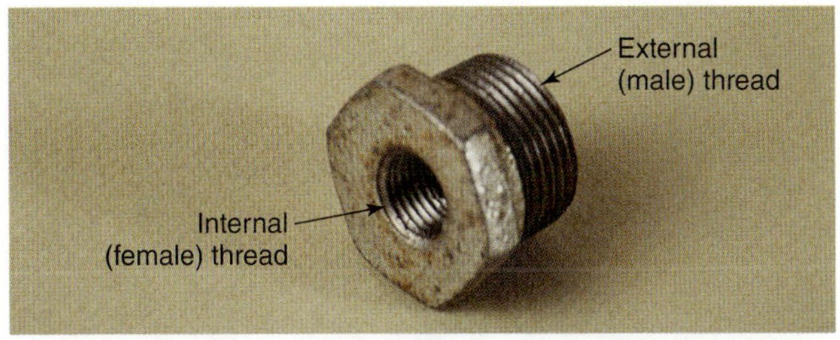

Goodheart-Willcox Publisher

Figure 7-14. A reducing bushing threads into a female opening (such as on an elbow) to allow a smaller pipe to be connected.

Caps and Plugs

Caps and plugs are used to seal openings. Caps have female threads and are used to close the end of a pipe. Plugs have male threads and are used to close openings in fittings. All nominal sizes are readily available.

Unions

A union is a special fitting required to complete a run of pipe, **Figure 7-15**. Pipe threads are all right-hand threads, making it impossible to assemble or disassemble the last length of threaded pipe with just a coupling or other fitting. Trying to use a coupling or other fitting would tighten on one pipe end and untighten on the other pipe end. Therefore, a union must be used to complete the final connection. Pipe unions provide a means of connecting two runs of pipe while making it possible to disconnect the piping when repairs are required. A union should always be installed when the possibility exists that the pipe may need to be disconnected.

Nipples

Nipples are short lengths of pipe that are threaded on each end and used to connect fittings that are close together, **Figure 7-16**. Nipples can be

Goodheart-Willcox Publisher

Figure 7-15. Since all pipe threads are right-handed, a run of pipe cannot be completed unless a union is installed. The union also allows pipes to be disconnected for repairs.

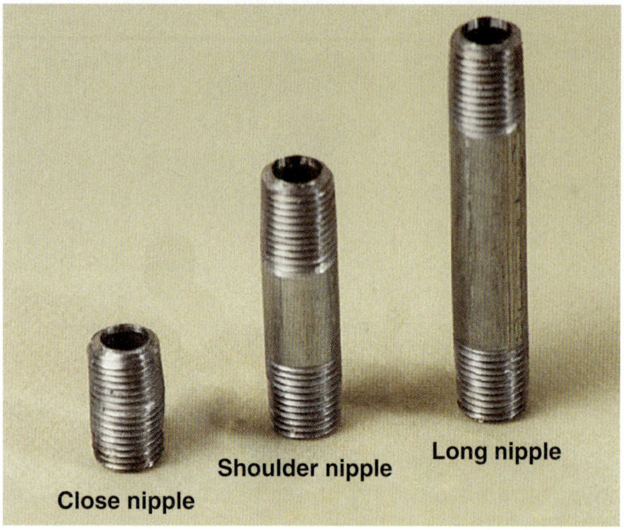

Goodheart-Willcox Publisher

Figure 7-16. Pipe nipples are short pieces threaded at both ends. They are useful in connecting fittings that are close together. The long nipple is used for connecting fittings up to 12″ apart.

purchased in any pipe size, in lengths up to 12″. Length is measured from end to end, including threads.

A short nipple is called a ***close nipple*** or *all-thread nipple* because it is threaded along its entire length. A ***shoulder nipple*** has a short section of unthreaded pipe. These nipples are convenient when making short connections.

Repairing Steel Pipe

Leaks in steel pipe are usually caused by loose connections or by ***corrosion*** (rusting) of the pipe. Loose connections can be tightened but rusted or otherwise damaged sections must be replaced. Rusted and leaking connections at a fitting are caused by removal of the galvanized coating during the threading procedure. The defective fitting or section usually must be removed and replaced. Pipe repair is shown in **Figure 7-17**.

Steel pipe used as a drain line sometimes can be repaired by removing the damaged section and using an adapter to replace the steel pipe with rigid plastic pipe. Check local building codes before using this repair procedure since not all codes permit the use of plastic pipe.

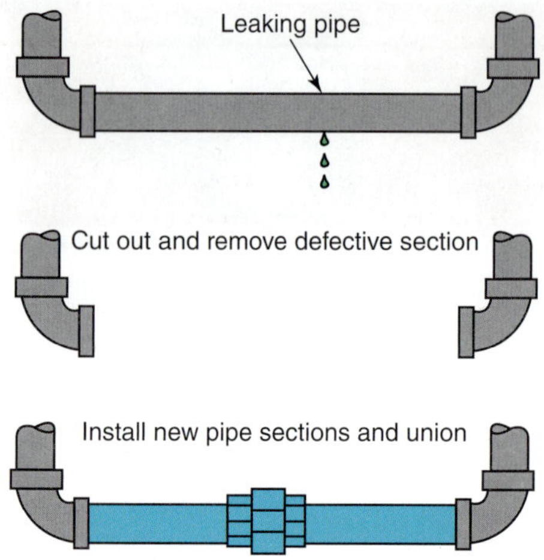

Goodheart-Willcox Publisher

Figure 7-17. Steel pipe can be repaired by cutting out the defective section and replacing it with two new lengths of pipe and a union.

7.2 Rigid Plastic Pipe

Plastic pipe was once considered a "do-it-yourself" material, used chiefly for farm water systems and drains. Now, rigid plastic piping is widely used. It is employed in chemical and food processing, natural gas distribution and supply, shipboard installations, laboratory and industrial waste disposal, municipal water treatment, industrial and residential plumbing, and a host of other applications.

Despite the growing popularity of this material, it is sometimes difficult to obtain reliable, up-to-date, and comprehensive information. When in doubt, consult local building codes before installing plastic pipe.

Advantages of Plastic Pipe

Plastic pipe has become an acceptable and inexpensive means of transferring many types of liquids and is replacing steel pipe in numerous applications. Plastic pipe has outstanding resistance to nearly all acids, caustics, salt solutions, and other corrosive liquids. It does not corrode, rust, scale, or pit on inside or outside surfaces. It does not rot and resists the growth of bacteria, algae, and fungi that could cause offensive odors or create serious sanitation problems.

Since its inner wall is smooth, plastic pipe provides maximum flow rates and abrasion resistance. Moreover, it exhibits minimal buildup of sludge and slime.

Plastic pipe is tough and strong. It has tensile and burst strengths sufficient to handle the operating pressures that fall within the temperature range of the plastic used to make it. However, as the temperature increases, the burst point becomes lower.

Rigid plastic pipe is easier to handle, join, and install than metal piping and is less expensive. The low cost of materials and the savings in labor account for its popularity.

Types of Plastic Pipe

There are three types of rigid plastic pipe:

- ***ABS*** (acrylonitrile-butadiene-styrene) plastic pipe and fittings are black and used only for drain, waste, and vent piping, **Figure 7-18**. There are two grades of ABS, Schedule 40 and Service. Building codes generally require Schedule 40.
- ***PVC*** (polyvinyl chloride) plastic pipe is white and is used for cold water supply, drain lines, and some venting. PVC pipe is rated at a temperature of 73°F (23°C) and a pressure of 100 psi (700 kPa). PVC pipe and fittings are available in diameters from 3/8″ to 16″, and in two grades, Schedule 40 and Schedule 80.
- ***CPVC*** (chlorinated polyvinyl chloride) plastic pipe is tan and can be used for hot and cold water lines or drains, provided local building codes permit. CPVC is available in two grades, Schedule 40 and Schedule 80. It is rated to take a water pressure of 100 psi (700 kPa) at a temperature of 180°F (82°C).

Pressure-rated plastic pipe is designed for water supply but can be used for drains. Pressure-rated pipe provides a complete system with piping of uniform strength. Plastic pipe is *never* used within a refrigeration system itself but often replaces the steel pipe normally used for drains or support systems.

Plastic Pipe Sizes

Rigid plastic pipe and fittings are manufactured to conform to nominal pipe sizing of steel pipe, with matching outside diameters. The inside (flow) diameter of plastic pipe changes slightly with the schedule number. Both Schedule 40 and Schedule 80 plastic pipes have the same outside diameter, but Schedule 80 pipe has a thicker wall. The thicker wall of Schedule 80 plastic pipe allows it to be threaded like steel pipe. However, it can also be solvent-welded like Schedule 40 material. Schedule 80 pipe is a heavier-duty material than Schedule 40; the difference can be compared to that between Type L and Type K copper tubing.

Schedule 40 PVC and Schedule 40 CPVC are the most commonly used types of plastic pipe. Both are available in 20′ lengths. Schedule 40 pipe and fittings are joined by using solvent cement. Do *not* attempt to thread Schedule 40 plastic pipe because its walls are too thin.

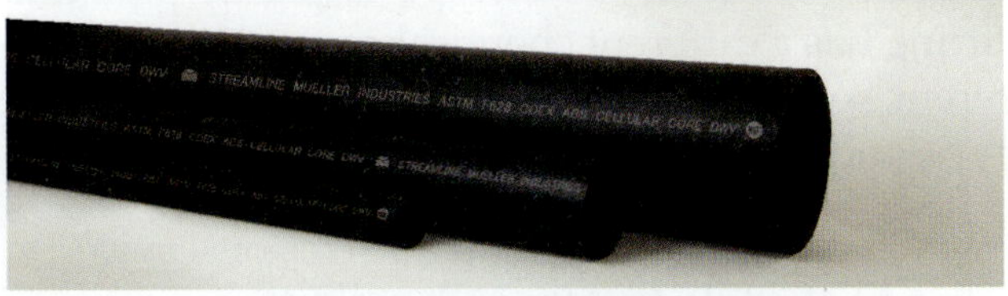

Mueller Brass Co.

Figure 7-18. ABS pipe in different sizes.

Pipe Joining Preparation

A major difference between plastic pipe and steel pipe is the method used for joining. Steel pipe is joined *mechanically* using threaded fittings; plastic pipe is usually joined *chemically* by a process called ***solvent-welding***. The solvent-welding assembly process is much faster and easier, resulting in labor savings.

Proper preparation of rigid plastic pipe for joining is important to achieving a satisfactory installation. Pipe must be cut square, burrs removed, and ends beveled to ensure joints will be properly solvent-welded and leakproof.

Cutting

Plastic pipe can be easily cut with almost any type of saw, but for best results, use a fine-tooth blade. For properly square end cuts, use a miter box as shown in **Figure 7-19**. There are two main types of plastic pipe cutters. One is much like the rolling cutters with a blade wheel used on copper tubing. Such pipe cutters may be used when the cutting wheel is designed for use on plastic pipe. Another type of plastic pipe cutter is similar in appearance to pruning shears, **Figure 7-20**.

Deburring and Beveling

All burrs, chips, and other loose material must be removed from the inside and outside surfaces of a pipe before joining is attempted. Use a knife, deburring tool, or a half-round, coarse file to remove all burrs.

All pipe ends should be beveled to approximately the dimensions shown in **Figure 7-21**.

Goodheart-Willcox Publisher

Figure 7-19. For clean, square-cut plastic pipe ends, a miter box is recommended. A fine-tooth saw blade will give the best results.

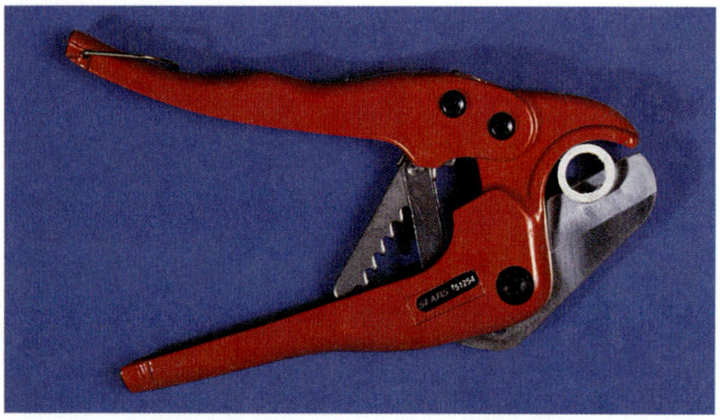

Goodheart-Willcox Publisher

Figure 7-20. This type of plastic pipe cutter, similar to pruning shears, has a ratchet feature that allows quick, clean, one-handed cuts for plastic pipe up to 1 1/2″ OD.

Beveling makes it easier to socket the pipe in the fitting. It also minimizes the chance of wiping solvent cement from inside the fitting when the pipe is inserted.

Cleaning and Priming

Surfaces to be solvent-welded must be clean and dry. Use a clean cloth to wipe away all loose dirt and moisture from the inside and outside surfaces of the pipe end. Also wipe inside the socket of the fitting.

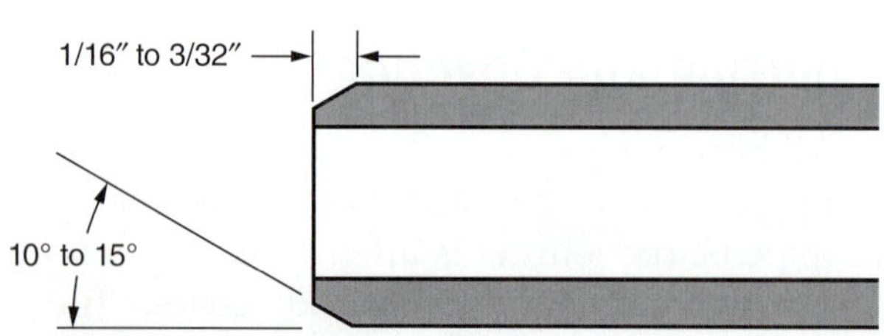

Nibco

Figure 7-21. For easier assembly of fittings to pipe, the cut end of the pipe should be beveled as shown.

The function of the primer is to penetrate and soften the bonding surfaces of plastic pipe and fittings. Primer is a high-strength product that penetrates rapidly. It is effective on the hard-finished, high-gloss products now being produced. Apply primer to the pipe end with a dauber or a natural-bristle paint brush for a length approximately half the pipe diameter.

SAFETY NOTE

Primer Skin Contact

It is best to apply primer using a tool, such as a dauber or brush. Using a rag to apply primer is *not* recommended since repeated skin contact may cause irritation or blistering. Protect yourself by using the proper application tools.

Next, apply primer freely to the socket of the fitting. Keep the surface wet and the applicator in motion for 5 to 15 seconds. Redip the applicator as necessary. In cold weather, allow more time for proper penetration before applying the cement to the joint.

Solvent-Welding

Solvent-welding (commonly referred to as *cementing*) is a simple operation, but each detail of the procedure is important and must be followed closely. Over 90% of plastic pipe joint failures are due to poor cementing techniques or outright carelessness. It is important to follow instructions carefully and resist shortcuts. When properly joined, PVC, ABS, and CPVC pipe and fittings produce pressure-tight joints, whether those joints are made in the shop or the field. Skill and knowledge are required to obtain joints of consistently good quality.

Solvent Cement

Solvent cement is normally purchased in one-pint metal containers with an application dauber attached to the lid. Each type of plastic pipe and fitting has a different chemical composition. A different solvent cement is required for each plastic type and should be used only with that material:

- ABS cement is used with all sizes of service and Schedule 40 ABS pipe and fittings.

- CPVC is used with all sizes of Schedule 40 and Schedule 80 CPVC pipe and fittings.
- PVC *light-duty* is used with Schedule 40 pipe and fittings 6″ or smaller in diameter. PVC *heavy-duty* is used with Schedule 80 pipe and fittings 6″ or smaller in diameter. PVC *extra heavy-duty* is used with all PVC pipe and fittings 6″ or larger in diameter.
- As previously mentioned, primer is applied before solvent-welding to clean and soften the bonding surfaces of rigid plastic pipe and fittings.

When not in use, solvent cement containers should always be covered to prevent evaporation. Do not use thinner in solvent cement. If cement is lumpy and shows signs of thickening, it should be discarded.

Special care must be exercised if two different types of plastic pipe are to be joined. Local codes usually forbid such combinations unless a mechanical joint or special adapter is used.

SAFETY NOTE

Solvent Cements and Primers

Solvent cements and primers are extremely flammable and emit dangerous vapors. They are harmful if swallowed, and fumes can cause eye irritation. Repeated or prolonged skin contact causes skin irritation. Keep away from heat, sparks, and open flames. Use only with adequate ventilation. Avoid contact with eyes, skin, and clothing. Avoid prolonged breathing of vapor. Close containers after each use.

First Aid: In case of *skin* contact, flush with water. For *eyes*, flush with water for at least 15 minutes and seek medical attention. Thoroughly wash contaminated clothing before reuse. If *swallowed*, do *not* induce vomiting. Call a physician immediately.

Making Plastic Pipe Joints

When properly solvent-welded, plastic pipe joints will be mechanically strong and leakproof. The solvent cement used to make the joints softens the two mating surfaces, allowing them to fuse. When the solvent evaporates, the pipe and fitting will have formed one piece.

Applying Cement

After cleaning and preparing the joint with primer, apply a generous coating of cement around the outside surface of the pipe to a width slightly more than the socket depth of the fitting, **Figure 7-22**. Use the dauber supplied with the cement or a natural bristle brush (the solvent in the cement could dissolve synthetic bristles). Do not worry about excess cement on the pipe. The excess cement will be forced out as a bead when the pipe is inserted into the fitting.

Next, apply an even layer of cement around the inside socket of the fitting, **Figure 7-23**. Avoid applying too much cement. Excess cement inside the socket could be pushed into the fitting and form an unwanted restriction to flow through the system.

Joining Pipe and Fitting

Immediately after applying cement to both pieces, insert the pipe to the full socket depth of the fitting. While inserting the pipe, rotate the pipe

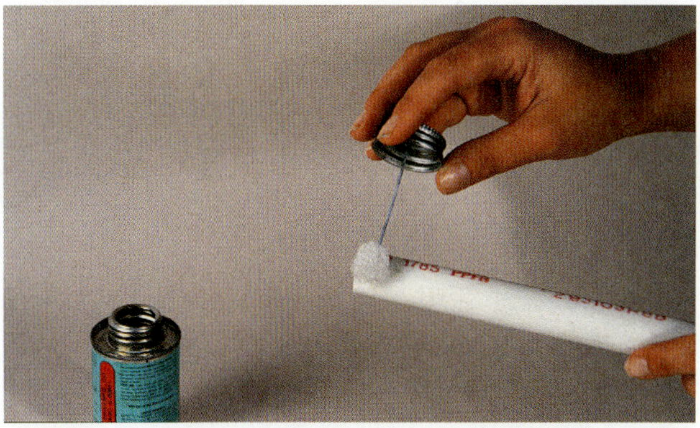

Goodheart-Willcox Publisher

Figure 7-22. Apply solvent cement to the end of the pipe for a distance slightly greater than the depth of the fitting's socket.

Goodheart-Willcox Publisher

Figure 7-23. A full, even coat of solvent cement should be applied inside the fitting's socket. Avoid excessive application, which could later restrict fluid flow through the fitting.

(or the fitting) about one-quarter turn to ensure complete and even distribution of the cement, **Figure 7-24**. Hold the joint firmly together for a minimum of 20–30 seconds. This will prevent the pipe from moving or "backing out" of the socket until the cement has set sufficiently. Keep the newly cemented joint immobile for about two minutes after joining.

PRO TIP

Solvent-Welding in Proper Conditions

Do *not* attempt to solvent-weld under the following conditions:

- In rain (surfaces must be clean and dry)
- In temperatures below 40°F (4°C)
- In direct sunlight at temperatures above 90°F (32°C)

Check primer, solvent cement, and pipe manufacturer data for product performance in varying conditions.

Do not discard empty primer or solvent cans, brushes, or daubers near plastic pipe. Concentrated fumes or drippings from such material could soften the pipe and cause it to fail.

Fittings for Plastic Pipe

Plastic pipe fittings are readily available in all types and sizes and use the same names as those applied to fittings for steel pipe. See **Figure 7-25**. Do not mix different types of plastic pipe and fittings: use ABS pipe with ABS fittings, PVC pipe with PVC fittings, Schedule 40 pipe with Schedule 40 fittings, and so on. Dissimilar types of plastics will not bond together properly, allowing leaks to develop.

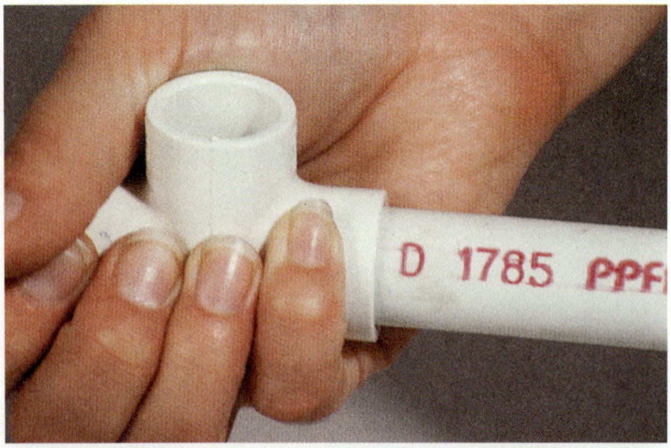

Goodheart-Willcox Publisher

Figure 7-24. Twist the joint one-quarter turn as it is being assembled and hold it in position until the cement begins to set. Correctly made solvent-welded joints are leakproof and mechanically strong.

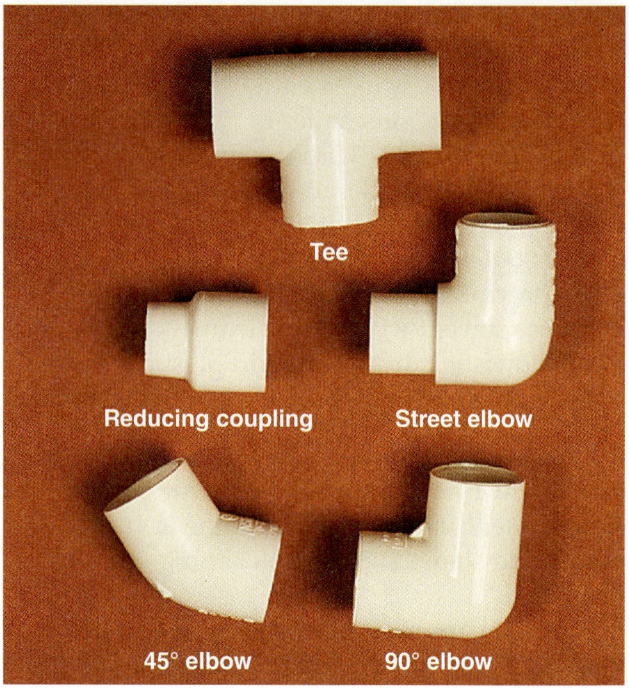

Goodheart-Willcox Publisher

Figure 7-25. Fittings for rigid plastic pipe.

Repairing Plastic Pipe

Leaks in plastic pipe usually are repaired by cutting out the defective section of pipe and replacing it. Often, this can be done by simply inserting a coupling. An example of such a repair is shown in **Figure 7-26**. Be certain to use the same type of plastic pipe and the correct solvent cement.

7.3 Alternate Methods of Pipe Connections

While refrigeration tubing will continue to use brazed copper lines, contractors are always looking for new and improved ways to connect water, drain, and hydronic heating lines that save time but still provide tight and leakproof connections without needing to solder or glue these connections. The connectors mentioned in this section are designed to be used on copper, CPVC, PEX, and other types of plastic tubing and pipe, depending

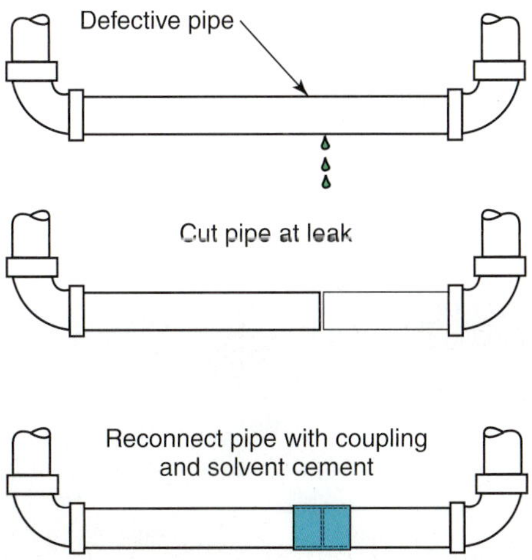

Goodheart-Willcox Publisher

Figure 7-26. Repairing a leak in plastic pipe involves cutting out the defect and installing a coupling or short section of new pipe.

on the manufacturer. These types of connections usually work on tubing from 3/8″ to 1″. Some of these alternate methods are push-on connectors, crimp-on pipe connectors, and adapters.

Push-On Pipe Connections

Push-on pipe fittings are designed to be installed without any type of installation tool, **Figure 7-27**. Just cut the end of the pipe or tubing and insert it into the fitting for a quick, leakproof connection. These connectors can be used for potable water, hydronic hot water heating systems, and condensate drain lines. While these fittings may cost more than sweat or flare fitting, the cost savings will be found in the reduced labor required for installation. Push-on type of fittings can be removed and reused using a special removal tool.

Push-on pipe fittings and related parts

Mueller Brass Co.

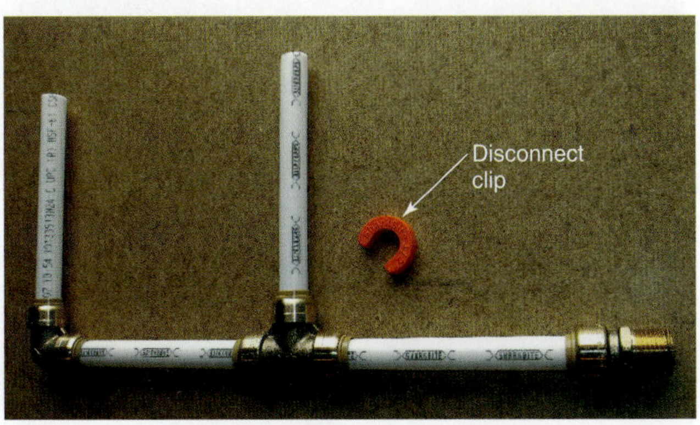

Installed push-on pipe fittings

Goodheart-Willcox Publisher

Close-up of push-on pipe joint

Goodheart-Willcox Publisher

Figure 7-27. Push-on pipe fittings that can be connected without special tools.

Crimp-On Pipe Connections

Crimp-on pipe fittings use a special installation tool to crimp the fitting around the pipe, **Figure 7-28**. After inserting the pipe into the fitting, use the crimping tool to squeeze the fitting around the pipe to make a rigid and leakproof seal. These fittings can also be used on potable water, hydronic hot water heating systems, and condensate drain lines. Depending on the type used, a crimping ring may need to be installed on the end of the pipe before inserting it into the fitting. Some types of fittings have the compression ring built onto the fitting. If the fitting needs to be removed, often it will have to be cut out. Depending on its design, once such a fitting is used and cut out, it may not be reused.

Adapters

Sometimes, it is necessary to connect one type of pipe or tubing to another (for example, steel pipe to PVC or copper tubing to steel pipe). Special fittings, called ***adapters***, are available for making such connections. See **Figure 7-29**. These adapters can be elbows, tees, or couplings in various combinations of male or female fittings. Almost any combination is possible in adapters.

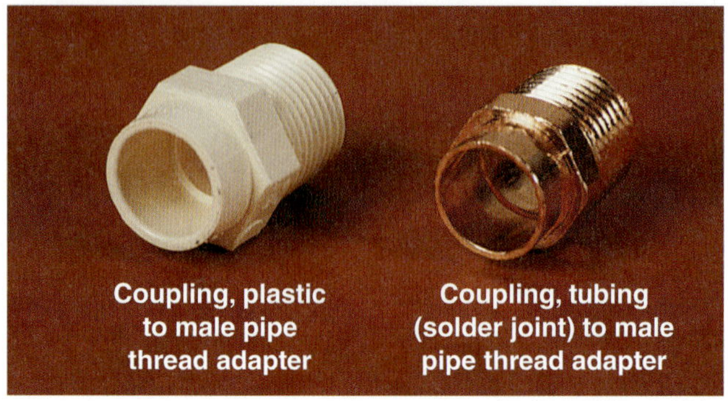

Goodheart-Willcox Publisher

Figure 7-29. Adapters used with plastic pipe.

Crimp-on pipe fittings

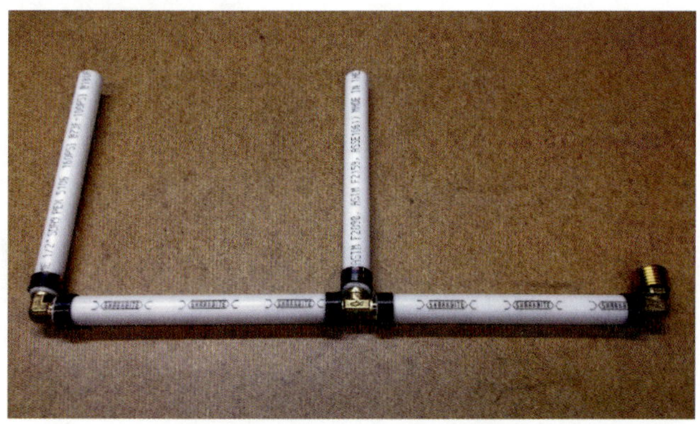

Installed crimp-on fittings

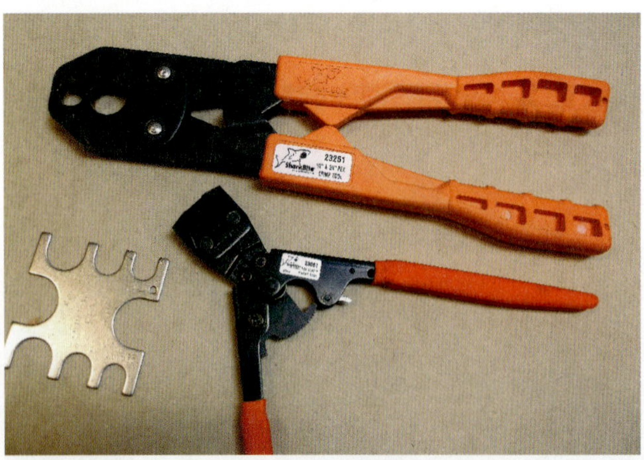

Crimp-on tools

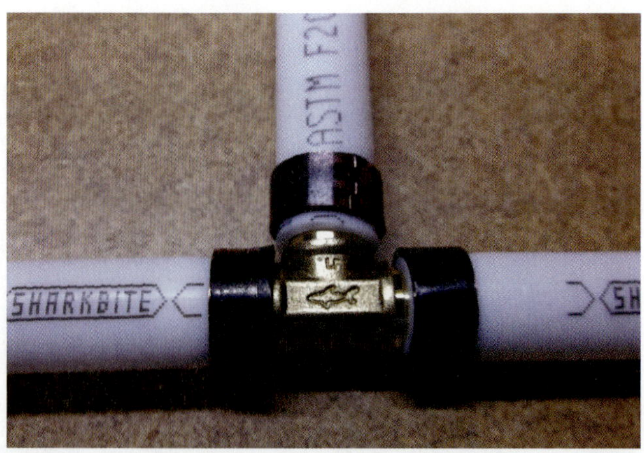

Close-up on a crimp-on pipe joint

Goodheart-Willcox Publisher

Figure 7-28. Crimp-on pipe fittings are connected with special tools.

Chapter Review

Summary

7.1 Steel pipe is available in black or galvanized forms. Steel pipe is measured by its inside diameter (nominal size). Steel pipe is thick, allowing it to be threaded. Steel pipe can be cut with pipe cutters, reamed with pipe reamers, and gripped with special pipe wrenches. A pipe die is used to cut external threads on steel pipe.

7.1 Pipe fittings are made of annealed cast iron, making them malleable. Pipe fittings are similar to copper tubing fittings except they are sized according to the corresponding nominal size of the pipe. Loose pipe connections can be fixed by tightening the fittings. Rusted or otherwise damaged pipe sections must be removed and replaced.

7.2 ABS, PVC, and CPVC are the three types of rigid plastic pipe used in HVACR and related plumbing work. Rigid plastic pipe and fittings are sized according to their nominal size, like steel pipe. Before joining, rigid plastic pipe must be cut square, burrs removed, ends beveled, and surfaces clean and primed. A different solvent cement is required depending on the plastic pipe used. Cement is applied to the prepped joint and fitting, then the fitting is inserted into the pipe and held together until the cement has set.

7.2 Leaks in plastic pipe are repaired by cutting out the defective section of pipe and replacing it. Often, this is done using a coupling.

7.3 Push-on and crimp-on pipe connectors are two alternate methods of connecting plastic tubing and pipe. Special installation tools are required with crimp-on fittings, but not with push-on fittings. Adapters are used to connect one type of pipe or tubing to another.

Know and Understand

1. Nominal pipe sizes are measured by _____ diameter. (7.1)
   A. outside
   B. inside
   C. total
   D. half their
2. Pipe threads are different from other types of threads in that they are _____. (7.1)
   A. self-sealing
   B. pre-threaded
   C. tapered
   D. nontapered
3. A tool used to cut external threads on steel pipe is called a pipe _____. (7.1)
   A. cutter
   B. vise
   C. die
   D. reamer
4. *True or False?* Pipe fittings are considered malleable since they are made of annealed cast iron. (7.1)
5. A(n) _____ is a short fitting with female threads at each end, used for connecting two lengths of pipe in a straight line. (7.1)
   A. elbow
   B. tee
   C. reducing bushing
   D. coupling
6. Pipe _____ provide a means of connecting two runs of pipe while making it possible to disconnect the piping when repairs are required. (7.1)
   A. unions
   B. nipples
   C. elbows
   D. caps and plugs
7. Rusted and leaking connections at a fitting are caused by removal of the _____ during the threading procedure. (7.1)
   A. burrs
   B. pipe compound
   C. galvanized coating
   D. Teflon tape

8. PVC pipe is available in _____ grade(s). (7.2)
   A. Schedule 40 and Service
   B. Schedule 40 and Schedule 80
   C. Schedule 10
   D. Schedule 40
9. *True or False?* Plastic pipe and fitting sizes are manufactured to conform to actual size, with matching inside diameters. (7.2)
10. The function of the primer on plastic pipe is to _____. (7.2)
    A. clean the surface of the pipe
    B. remove burrs, chips, and other loose materials
    C. penetrate and soften the bonding surface
    D. color the pipe to the desired finish
11. Excess solvent cement inside the socket of a fitting _____. (7.2)
    A. is not an issue because the excess cement will be forced out as a bead when the pipe is inserted
    B. should be avoided because the excess cement could create an unwanted restriction
    C. is desired because some cement is lost in the joining process
    D. does not affect the bond
12. Leaks in plastic pipe can be repaired by removing the defective section of pipe and simply inserting a new pipe with a(n) _____. (7.2)
    A. reducing bushing
    B. coupling
    C. adapter
    D. extra layer of solvent cement
13. Push-on and crimp-on pipe connectors work on tubing from _____″ to _____″. (7.3)
    A. 1/4, 3/8
    B. 1/4, 1/2
    C. 3/8, 1
    D. 1, 2
14. To connect one type of pipe to another, such as copper to steel or steel to PVC, use special fittings called _____. (7.3)
    A. couplings
    B. adapters
    C. reducing bushings
    D. unions

Apply and Analyze

1. What does *nominal size* mean as it relates to steel and plastic pipe? (7.1)
2. Describe how to use a pipe reamer to ream the cut ends of a steel pipe. (7.1)
3. Explain how to fix a leak in steel pipe caused by a rusted or damaged pipe. (7.1)
4. Why is excess cement not a concern on the pipe but should be avoided on the fitting? (7.2)
5. Describe two alternate types of pipe connectors. (7.3)

Critical Thinking

1. What are some of the differences between pipe and copper tubing used in HVACR work? (7.1)
2. What characteristics of steel pipe and plastic pipe might account for the difference in their joining methods? (7.2)

Flaviya/Shutterstock.com

8 Soldering

LEARNING OUTCOMES

After studying this chapter, you will be able to:

8.1 Select proper alloys, fluxes, and fuel gases for soldering.

8.2 Describe how to safely use an air-acetylene torch.

8.3 Describe how to properly solder joints and take apart joints for various materials.

8.4 Compare the use of air-acetylene torch tips for brazing with those used for soldering.

TECHNICAL TERMS

acetone
acetylene
alloy
balling
brazing
capillary action
elongation
epoxy
ferrous
flow point
flux
heat sink
liquidus
melting point
plastic range
pressure regulator
solder
soldering
solidus
tare weight
tinning
working pressure

Introduction

The ability to make strong, neat, leakproof solder connections is a fundamental skill for every HVACR technician. Good, soldered connections must be made quickly and correctly *on the first try.* Too much heat or not enough heat will result in a poor connection. Trying to correct a badly made connection is more difficult than starting over. Paying attention to details and repeatedly practicing can help improve soldering techniques.

8.1 Soldering Overview

Soldering is the process of joining two pieces of base metal using a filler metal. Often the filler metal is an ***alloy*** (a compound of two or more metals). This alloy filler metal is referred to as ***solder***. This alloy usually consists of tin and another metal with a low-melting point, such as lead, antimony, or silver. A soldered connection is excellent for conducting heat or electricity and for making joints leakproof. However, a soldered joint is not as mechanically strong as the base metals it brings together.

As mentioned, soldering joins two base metal pieces. The base metals may be the same or different types of metal. Heat is applied to the base metals, though *not* enough heat to melt either piece. Once enough heat has been applied to the base metals, filler metal alloy is applied to the joint seam. The heat of the base metals melts the alloy.

Soldering is an adhesion process that works much like gluing two pieces of wood together. In soldering, the alloy acts as the "glue," and heat melts the alloy, which causes it to adhere to the surface of the base metals, **Figure 8-1**.

PRO TIP

Melting Heat

The heat from the torch should *not* be used to melt the filler metal directly. Rather, heat is applied to the base metals, and the heat of the base metals is used to melt the filler metal. Do *not* use a torch to directly melt the filler metal. This could result in an inadequate solder joint.

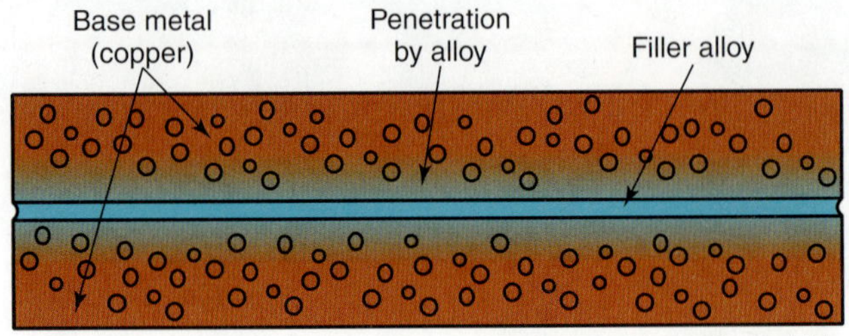

Goodheart-Willcox Publisher

Figure 8-1. The filler metal alloy melts and adheres to the surfaces of the base metal pieces, entwining (wrapping around) the molecules of the base metal to form a strong bond.

When a soldering alloy reaches its ***flow point***—the point where melted solder flows as a liquid—the molten alloy is drawn by capillary action into the narrow space between the base metals. ***Capillary action*** refers to the process where an alloy is drawn in by indirect heat to fill a gap between pieces of base metals. The molecules of the alloy penetrate and join with the molecules of the base metal (without melting the base metal). For maximum joint strength, the gap between base metals should be as small as possible, and the layer of alloy between the base metal pieces should be thin.

PRO TIP

Soldering and Brazing Terms

Soldering joins metals at less than 840°F (448°C). Joining metals at temperatures above 840°F (448°C) is called ***brazing***. The American Welding Society (AWS) established this distinction between soldering and brazing. In the HVACR industry, sometimes the meanings of soldering and brazing are used interchangeably. Therefore, it is important to know that *soft soldering* refers to joining that is done at temperatures below 840°F (450°C), and *hard soldering* refers to the high-temperature process that AWS calls *brazing*.

Two factors determine whether to solder or braze and what type of alloy to use. They are the following:

- Physical properties of the metals being joined.
- Demands that will be placed on the connection after it has been formed.

In refrigeration and air-conditioning work, soldering is used to install copper water lines and drains. Water lines are not under high pressure nor are they subject to vibration. On the other hand, copper tubing used for refrigeration systems should be brazed to withstand high pressure and vibration.

Soldering Alloys

Filler metals (soldering alloys) can be obtained in a variety of forms: foil, tape, sheet, wire, flux-cored or solid bar, preform, or paste. One-pound spools of solder in wire form are the most popular for HVACR work, **Figure 8-2**. Depending on the application, different tin-lead or tin-silver alloys are used.

© J. W. Harris Co., Inc.

Figure 8-2. Lead-free solders like this tin-silver type must be used for joints in any tubing that will carry drinking water. Tin-lead and tin-antimony solders are also available for various applications.

The point at which solder begins to melt is its melting point or ***solidus***. The point at which solder begins to flow as a liquid is its flow point or ***liquidus***. The range of temperatures between solidus and liquidus is called a solder's ***plastic range***. **Figure 8-3** shows the composition and plastic range of several common solders. A wide plastic range provides easier control of the alloy during the soldering process.

A numbering system indicates the percentage of each metal in the solder. For example, in the tin-lead group, 50/50 indicates 50% tin and 50% lead, and 60/40 indicates 60% tin and 40% lead.

Tin-Lead Solders

Tin-lead solders comprise the largest group of filler metals used in soldering. They are used for joining most metals and have good corrosion resistance. Various tin-lead alloys are produced to meet the differing needs of soldering tin, copper, brass, bronze, sheet iron, and sheet steel.

CAUTION

Leak Prevention

Tin-lead solders are *not recommended* for high-stress or vibration areas. Stress and vibration will cause the solder to crack, permitting the joint to leak. Soldering is best used on condensate drain lines and similar nonstress applications.

The 50/50 solder has a ***melting point*** of 360°F (182°C), and a *flow point* of 420°F (215°C). The 60/40 solder is slightly stronger; it melts at 360°F (182°C) and reaches its flow point at 375°F (190°C).

Soldering Alloys

Solder	Composition Percentage	Melting Point (Solidus) in °F	Flow Point (Liquidus) in °F
50/50	50% tin / 50% lead	360	420
40/60	40% tin / 60% lead	360	460
60/40	60% tin / 40% lead	360	375
95/5	95% tin / 5% antimony	452	464
silver solder 96/4	96% tin / 4% silver	430	430
silver solder 94/6	94% tin / 6% silver	430	535

Goodheart-Willcox Publisher

Figure 8-3. The plastic range of a solder is the span from its melting point (solidus) to its flow point (liquidus), as shown in the right-hand column of this table of common soldering alloys.

SAFETY NOTE

Tin-Lead Solders

Tin-lead solders were formerly used by plumbers on copper water lines, but they are no longer permitted in drinking water applications. Lead-free alloys that do not present a human health hazard must be substituted. Tin-lead solders are still used in other plumbing and HVACR applications.

Tin-lead solders may contain small amounts of other metals to produce special properties. Metals commonly added to solders include antimony, bismuth, and silver. Antimony increases the strength of a solder. Where a stronger joint is desired for copper tubing, or where a lead solder is not permissible, a 95/5 tin-antimony alloy is often chosen. This solder has a melting point of 452°F (233°C) and a flow point of 464°F (240°C).

Tin-Antimony Solders

Tin-antimony solder is higher in strength and ***elongation*** (stretch factor) than tin-lead solders, but like tin-lead it has proven unsatisfactory for applications involving stress and vibration. Another disadvantage is that this alloy tends to powder over time.

CAUTION

Tin-Antimony Use

The 95/5 tin-antimony alloy should not be used on brass or galvanized (zinc-coated) metal. When antimony is in a molten state, it will absorb zinc from the brass or the galvanizing, resulting in a brittle joint.

Bismuth and silver are added to improve the ***tinning*** (spreading action) of solder. Bismuth is used for lower-temperature solders, while silver is used for higher-temperature solders.

Tin-Silver Solders

Another major group of solders are the tin-silver alloys. These silver-bearing solders are widely used because of the strong connections they form. Also, their low working temperatures eliminate the weakening of base metals caused by annealing (from high-soldering temperatures). Oxide scale formed in high temperatures is also eliminated.

The silver-bearing alloys have the ability to bond with both ***ferrous*** (containing iron) and nonferrous metals. These solders work well to join dissimilar metals, such as iron to copper. Furthermore, silver-bearing alloys have good elongation (stretch factor) when vibration is a concern. Melting points for these alloys range from 430°F to 535°F (221°C to 279°C).

PRO TIP

Soldering for Stress and Vibration

For applications in the HVACR industry involving stress or vibration, the tin-silver alloys are superior to both tin-lead and tin-antimony alloys.

Solder Fluxes

Flux is a multipurpose chemical used to treat the clean surface of base metals, to remove oxides from filler metal, to prevent reoxidation of material, and to aid the capillary flow of filler alloy. See **Figure 8-4**. In some instances, the appearance of the flux may indicate temperature. Flux as a temperature guide is explained later in this chapter.

Choosing the proper flux is important to the quality of a soldered connection. There is no universal flux suitable for all applications. Therefore, when selecting a flux for a particular alloy, it is wise to follow the manufacturer's recommendation.

Using Fluxes

Thoroughly cleaning base metal joining surfaces is vital to making strong, leakproof, long-lasting joints. The fittings and tubing must be free of oil, grease, rust, or oxides that would prevent the alloy from penetrating the base metal surfaces. Cleaning is done with a wire brush, fine sandpaper, or emery cloth, **Figure 8-5**. The joint must be cleaned well; otherwise, the alloy cannot properly flow and penetrate the base metals.

Flux maintains a chemically clean surface during the soldering operation. Its initial purpose is not to clean the metal but to keep the joint oxide-free. Flux dissolves the oxides that form when oxygen contacts the molten alloy or the hot base metals. It is important that the flux remains in place until soldering is completed and that it is not

© J. W. Harris Co., Inc.

Figure 8-4. This liquid soldering flux removes oxidation from copper tubing. It prevents new oxidation from occurring during the soldering process.

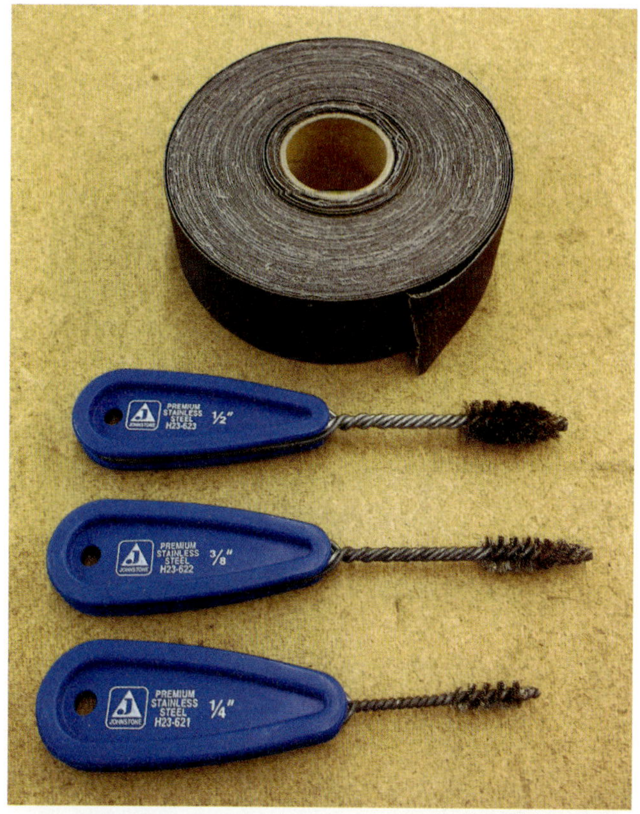

Goodheart-Willcox Publisher

Figure 8-5. Wire brushes, sandpaper, or emery cloth are used to clean tubing and fittings before they are soldered. The brushes are made specifically for common fitting sizes and should be used only for fittings of the specified size.

blown away or vaporized. Keep soldering time to a minimum. More time means more heat, and more heat means more oxide formation.

As noted, flux serves several functions, all of them necessary to provide a sound connection. Along with removing surface oxides and acting as a shield against new oxide formation, flux increases the "flow" properties of the alloy, yielding to the alloy as it melts into the joint.

The choice of flux in any soldering process is determined by the metals being joined and the filler alloy being used. Since there are many solders and fluxes, follow the manufacturer's recommendations for the best flux for the situation.

CAUTION

Avoid Excess Flux

Special care must be exercised when using fluxes to solder copper tubing or brass fittings. Fluxes are harmful contaminants inside copper tubing. Avoid excess flux. Apply only the thinnest possible film.

The flux best suited to the tin-lead, tin-antimony, and tin-silver solders is the noncorrosive type in paste form. It consists of zinc chloride (which may contain a trace of ammonium chloride) mixed into a petroleum jelly. When this flux is heated, it loses moisture, then melts and partly decomposes to form hydrochloric acid. The acid dissolves the oxides, floats the oxide particles ahead of the molten solder, and promotes the capillary action of the solder joint.

CAUTION

Flux Application

Flux is applied only to the portion of the tubing that is to be inserted into the fitting. Do *not* apply flux to the fitting socket.

Paste-type flux should *not* be applied to the tip of the cut end of the tubing. It should remain about 1/8″ (3 mm) from the end, as shown in **Figure 8-6**. The tubing is then inserted into the fitting and rotated to spread the flux evenly. This procedure

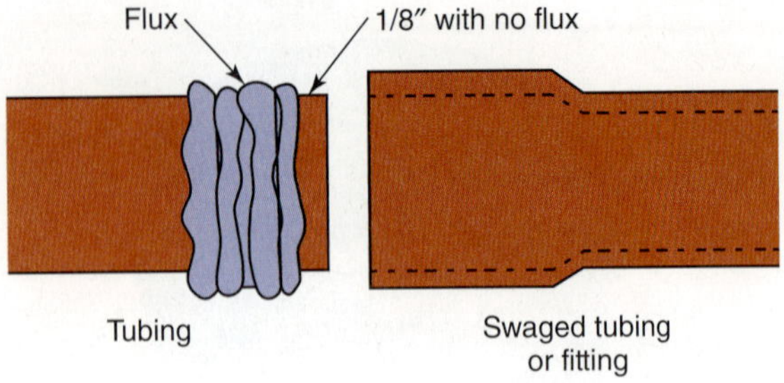

Goodheart-Willcox Publisher

Figure 8-6. Flux should be applied to the tubing, not the fitting. As shown, approximately 1/8″ of tubing at the cut end should *not* be fluxed. This will help avoid contaminating the system with soldering flux.

prevents excess flux from entering the fitting and contaminating the refrigeration system.

CAUTION

Acid Flux Use

Although it is commonly used in other types of soldering, an acid flux composed of 71% zinc chloride and 29% ammonium chloride should *never* be used in refrigeration work. Using such a highly corrosive flux on a refrigeration system will damage the system.

When soldering electrical connections, a noncorrosive rosin flux must be used. Rosin flux is produced from pine tar. Sometimes activators are added to rosin flux to improve its spreading action.

Fuel Gases in Soldering

The melting of the solder alloy (filler metal) and the heating of the base metals is normally done with a fuel gas torch. In these torches, fuel gas is mixed with atmospheric air (21% oxygen) to produce a soldering flame. Flame temperature is determined by the type of fuel gas used. Acetylene produces the highest temperature, followed by *MPS*, or methylacetylene-propadiene stabilized gas—often referred to by its trade names, *MAP* or *MAPP*. Propane, butane, natural gas, and manufactured gas produce lower temperatures respectively.

8.2 The Air-Acetylene Torch

The air-acetylene torch combines fuel gas from a cylinder with combustion air drawn from the atmosphere, **Figure 8-7**. Because atmospheric air contains only 21% oxygen, the air-acetylene torch produces a low-temperature flame. The low heat levels used in soldering reduce stress on the base metals and minimize the formation of oxides inside the tubing. Reduced stress and minimal oxide formation make the air-acetylene torch the ideal tool for soft soldering.

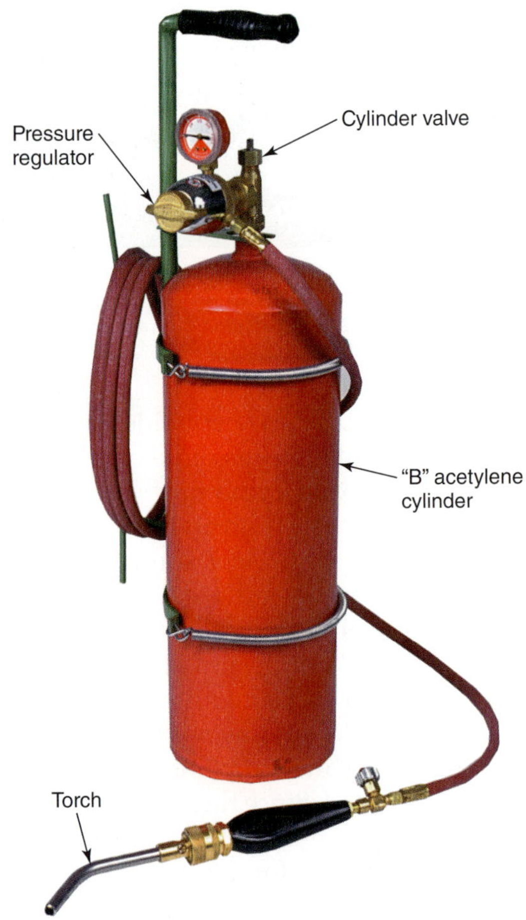

Uniweld

Figure 8-7. A typical air-acetylene torch outfit for soldering. Combustion air combines with fuel gas at the tip of the air-acetylene torch.

Acetylene

Acetylene is a colorless, gaseous hydrocarbon formed by the chemical reaction of water and calcium carbide. It is two parts carbon and two parts hydrogen, so its chemical formula is C_2H_2. It is used chiefly in soldering, brazing, and welding work. When burned with oxygen, acetylene will produce the highest flame temperatures obtainable (approximately 5660°F or 3126°C).

Acetylene Cylinders

Acetylene cylinders, also called *tanks*, are available in various sizes, **Figure 8-8**. The smallest cylinder (size MC) contains 10 ft^3 of acetylene. *MC* stands for *Motor Car*. These cylinders were originally used to fuel acetylene-burning cars and motorcycle headlights. The next largest cylinder (size B) contains 40 ft^3 of acetylene.

MC and B size cylinders are most frequently used by service technicians, but larger cylinders are occasionally used on large installation jobs. Each acetylene cylinder is equipped with a valve that has a threaded outlet for installing a pressure regulator. The size of the pressure regulator must be compatible with the size of the cylinder.

All acetylene cylinders are manufactured to Department of Transportation (DOT) requirements. Cylinders are thoroughly tested by the DOT Bureau of Explosives before being used commercially. A fusible plug located in the bottom of each cylinder will melt at approximately 212°F (100°C) to relieve dangerous pressure buildup.

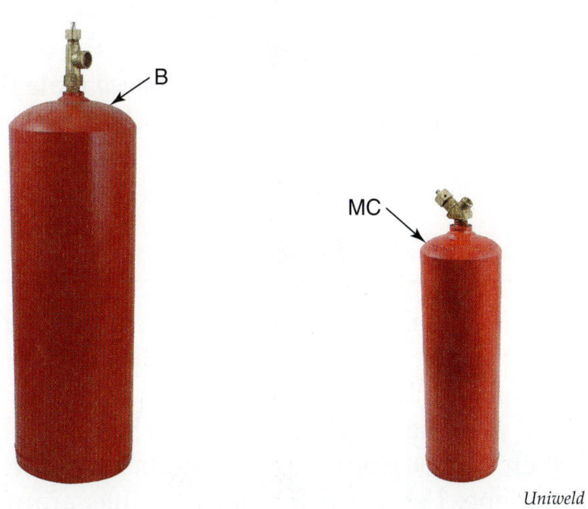

Figure 8-8. Common acetylene cylinder sizes are *B* and *MC*. The B cylinder holds 40 ft^3 of acetylene; the MC cylinder holds 10 ft^3. All acetylene cylinders must be tested and approved for use.

SAFETY NOTE

Acetylene Cylinders

Acetylene can be unstable, unsafe, and explosive at working pressures above 15 pounds per square inch (psi) or 103 kilopascals (kPa). However, the gas can be safely stored in cylinders at a higher pressure by using acetone.

An acetylene cylinder is filled with a porous substance that absorbs ***acetone***, a liquid solvent. Next, a precisely measured amount of acetone is introduced into the cylinder and absorbed by the porous substance. Then, the cylinder containing the porous material and acetone is weighed. The weight is stamped on the cylinder. This is the ***tare weight***, which is the weight of the container before it is filled with acetylene. Finally, acetylene is pumped into the cylinder where it is absorbed by the acetone. As pressure is relieved during use, the acetylene gas is released to flow out of the torch. Each time the cylinder is refilled, the tare weight should be checked and acetone added as needed.

PRO TIP

Acetylene Cylinder Position

Acetylene cylinders should be used in an upright position to avoid loss of acetone and poor flame quality.

Safety Precautions with Acetylene

Safety-consciousness is a mark of good craftsmanship and is guided by common sense. Observe the following precautions for using acetylene safely:

- Never permit fuel gas to escape and accumulate. Acetylene concentrations ranging from 3% to 90% in air or oxygen are explosive.
- Never store acetylene or other fuel cylinders in a closed, unventilated area.
- Do not expose fuel gas tanks to high temperatures or sources of ignition.
- When setting up an air-acetylene torch, check all fittings for leakage by spraying them with a soapy water solution. Leaks will blow bubbles. A soapy solution can be made by mixing concentrated dishwashing detergent and water

in a 1:1 ratio. A squirting dispenser is handy for applying the solution. Do not use the torch until all fitting leaks have been eliminated. Never use a flame to detect leaks.

- Never use a cylinder that is leaking acetylene. If gas leaks around the valve stem when the valve is opened, close the valve and tighten the packing nut. This will compress the packing around the valve stem and should stop the leak. If the leak does not stop, close the cylinder valve, tag the cylinder a "leaker," and return it to your supplier. Keep the "leaker" away from all sources of ignition.
- Never open the acetylene cylinder valve more than one-half turn.
- Always leave the valve key or wrench on the valve stem while a cylinder is in use, so the acetylene can be turned off quickly in case of emergency. Never use pliers or a wrench on the cylinder valves. Use only the proper valve key or valve key wrench. Always close the cylinder valve and regulator when work is completed.
- Handle all fuel cylinders with care. Do not drop or roll them.
- Never allow the torch flame or an electric arc to contact the cylinder.

Pressure Regulators

An acetylene cylinder is pressurized to 250 psi (1725 kPa) at 70°F (21°C). A ***pressure regulator*** is used to reduce cylinder pressure to working pressure, **Figure 8-9**. The ***working pressure*** (flow pressure) will vary according to the type of torch tip being used, but 5 psi (34 kPa) is considered normal.

SAFETY NOTE

Acetylene Working Pressure

Never allow acetylene working pressure to exceed 15 psi (103 kPa). Always keep working pressure below this level for safe torch use.

Pressure regulators are available with either one or two gauges. The single-gauge regulator shows working pressure traveling through the hose to the torch tip. This working pressure is adjustable by means of a knob or handle located on the front of the pressure regulator. On two-gauge models, the second gauge shows the level of acetylene in the cylinder (full, 3/4, 1/2, 1/4, or empty).

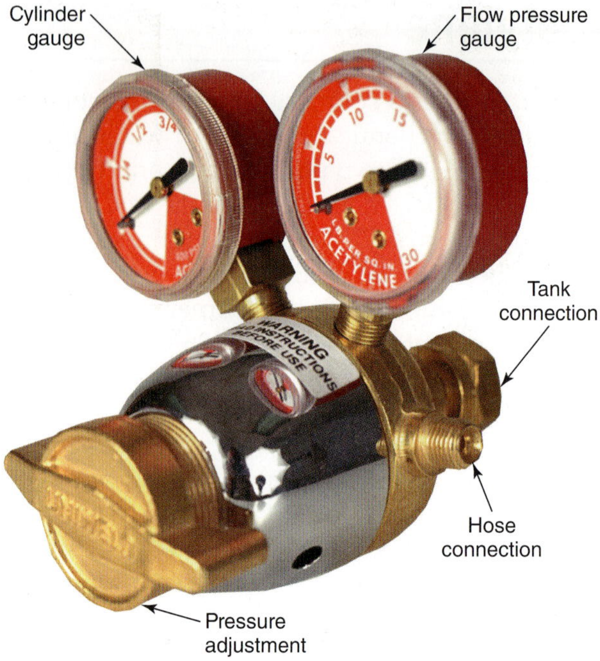

Uniweld

Figure 8-9. A two-gauge acetylene pressure regulator. The cylinder gauge shows the acetylene level, while the flow pressure gauge provides a reading of the working pressure supplied to the torch. Pressure is adjusted by rotating the handle on the front of the regulator.

The pressure adjustment on a regulator is designed to open and close in directions opposite those of ordinary valves. Turning the adjustment knob *counterclockwise* will *close* the pressure regulator, while a *clockwise* rotation will *open* the regulator.

Acetylene regulators and gauges are color-coded red and have special cylinder connections to eliminate confusion with other types of regulators. The outlet hose connection on the acetylene regulator is a special *left-hand threaded* male fitting. The hose fitting can be either small (size "A") or large (size "B").

Acetylene Hose

The female hose fitting screws onto the male fitting of the regulator outlet. All acetylene connections are left-hand threads to prevent the user from mistakenly connecting the hose to oxygen, which may be used at a much higher pressure than acetylene. Left-hand threaded fittings are readily recognized because of the notches, or grooves, cut around the middle of the outside edges of the nut. **Figure 8-10** shows check valves for oxygen and

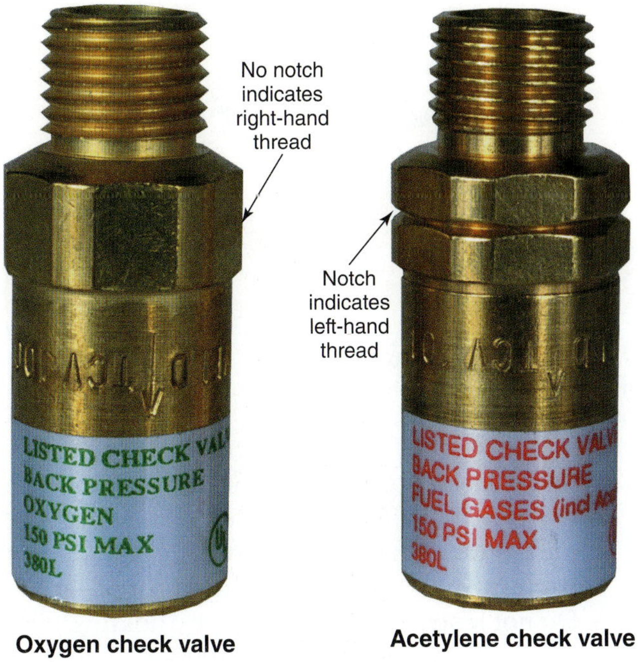

Uniweld

Figure 8-10. Check valves are safety devices used to prevent the reverse flow of gases. The threads of the connectors for oxygen and acetylene are different to prevent accidentally interchanging them. The acetylene check valve (right) has a left-hand thread and is identified by a groove or notch around the nut. The oxygen valve (left) has right-hand threads and no notch or groove.

acetylene hoses. Notice the notch on the acetylene check valve.

Acetylene hose is a special grade designated by the letters *RM*. The hose is red and is measured by the inside diameter (ID). It is available in 3′, 6′, 12 1/2′, or 25′ lengths. The light- and medium-duty hoses are 3/16″ ID; the heavy-duty hose is 1/4″ ID.

The hose connector is available in two sizes. Most technicians prefer size A, a light-duty (also called *aircraft*) hose. It has a 3/8-24 female nut with a left-hand thread. A size B hose is for medium- or heavy-duty use. It has a 9/16-18 female nut with a left-hand thread. Adapters are available for connecting the two fitting sizes.

Air-Acetylene Torch Handles

One end of the torch handle has a threaded connection for an acetylene hose; the other end of the torch handle has a threaded connection for a torch tip, **Figure 8-11**. The handle also has a finger-operated valve to control and adjust gas flow. The valve is manually opened and closed with ease. It should not be overtightened or valve seat damage could occur.

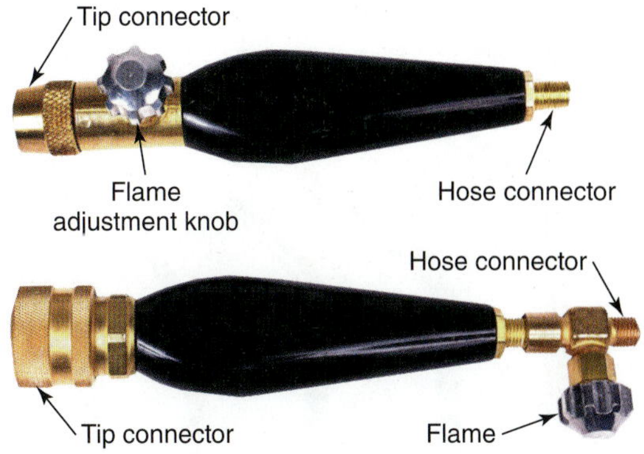

Uniweld

Figure 8-11. Air-acetylene torch handles have connections for the acetylene hose at one end and for the torch tip at the other. As shown, a finger-operated knob for flame adjustment may be at the hose end of the handle or at the tip end.

Air-Acetylene Torch Tips

Different sizes and styles of torch tips are available for different applications, **Figure 8-12**. Tips can be easily interchanged by loosening a locknut on the torch handle and unscrewing the tip. No tools are needed.

SAFETY NOTE

Changing Torch Tips

Exercise care when changing torch tips. Torch tips become quite hot when a flame is produced. If a torch tip proves to be too small for a given application, be sure to wear thick safety gloves or allow it to cool after use before handling it for replacement. Severe burns may occur if skin contact is made with a hot torch tip.

Lighting the Torch

Before lighting a torch, be sure to have all setup and arrangements complete. Inspect the joint to be made. Determine the proper torch tip size based on the type and size of material being joined. Then, determine the proper filler metal and flux to use.

The procedure for lighting an air-acetylene torch is simple, but certain steps must be followed to ensure safety:

1. Inspect the equipment for damage and make sure all valves and regulators have been turned off.
2. Open the cylinder valve slowly, using the proper valve key wrench. Open

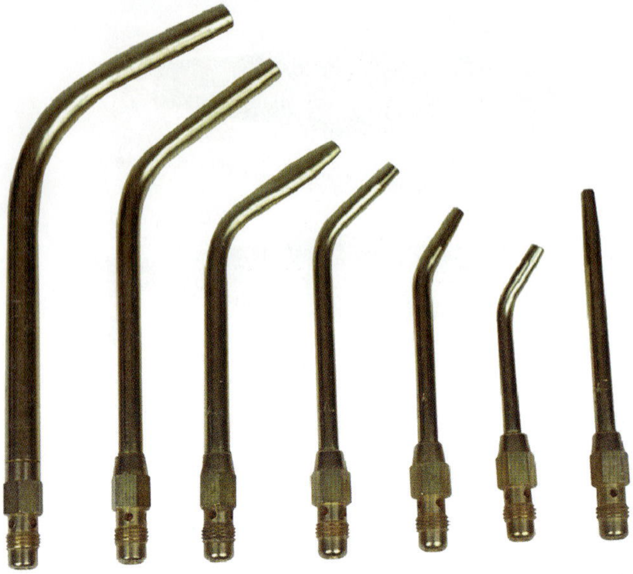

Uniweld

Figure 8-12. Air-acetylene torch tips are available in a variety of sizes for different applications.

the valve one-quarter to one-half turn counterclockwise. Leave the wrench on the valve stem so the cylinder can be turned off quickly. If the pressure regulator is the two-gauge type, the contents gauge will immediately register the level of acetylene in the cylinder.

3. Open the torch handle valve one-half turn. Slowly turn the pressure-regulator adjusting screw clockwise until the delivery gauge indicates 5 psi (34 kPa) flow pressure. Vent the escaping gas safely.
4. Turn off the torch handle valve using fingertip pressure only.
5. Point the tip of the torch away from any object or person. The torch is now adjusted to the correct working pressure and ready to be lit.
6. Slightly "crack" open the torch handle valve. Use a flint striker to ignite the acetylene gas coming out of the torch tip. Hold the striker near the end of the torch tip, but do not block the gas flow, **Figure 8-13**.

SAFETY NOTE

Flint Strikers

A flint striker is the only safe device to use when lighting any torch. *Never* use a match or cigarette lighter.

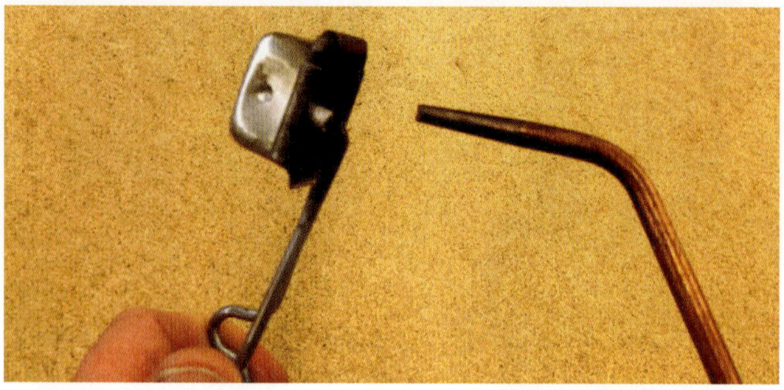

Goodheart-Willcox Publisher

Figure 8-13. An air-acetylene torch should never be lit with a match or cigarette lighter. For safety, always use a flint striker, also known as a spark lighter.

7. Open the handle valve slowly until the desired flame size is obtained. The hottest part of the flame should be about 1/8″ (3 mm) from the tip of the inner cone, **Figure 8-14**. Do not restrict the inner cone by holding the flame too close to the workpiece. Such a practice will interfere with proper mixing and combustion of the fuel-air mixture and result in poor flame quality.
8. Once the workpiece is heated to a temperature that will melt the solder, back the flame away just far enough to maintain the proper flow temperature of the solder.

The maximum temperature obtainable with an air-acetylene torch system is 3600°F (1982°C), regardless of the tip size. Large tips will provide more heat output, but the temperature of the flame will remain the same. The workpiece will seldom, if ever, reach this temperature, since solders will usually melt in the range of 400°F to 500°F (204°C to 260°C). Avoid overheating the base metals.

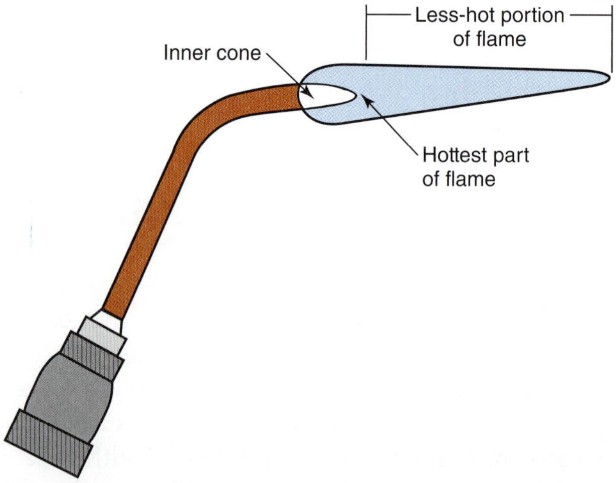

Goodheart-Willcox Publisher

Figure 8-14. The flame can be adjusted with a knob on the handle. The hottest area of the flame is just in front of the inner cone's tip.

Shutting Off the Torch

The torch can be shut down temporarily by merely closing the handle valve to turn off the flame. It can then be laid aside until it is needed again. However, if the unit is not going to be used for a long time, it should be completely shut down as follows:

1. Close the valve on the torch handle, using fingertip pressure only. This will kill the flame.
2. Tightly close the cylinder valve.
3. Reopen the torch handle valve to bleed off (release) any acetylene trapped between the hose and regulator. Bleeding should continue until both gauges read zero.
4. Turn the adjusting screw on the regulator all the way out (counterclockwise) to close the regulator. This will help to avoid gas loss if the cylinder or torch valves leak.
5. Rewind the hose and place the unit in safe storage.

8.3 Soldering Techniques

It is important to apply heat properly in any soldering operation. Several heat sources are available, such as soldering irons, soldering guns, and a variety of torches, but this text addresses only the air-acetylene torch and open-flame technique.

Torch selection is determined by the type and size of the soldering job. Large tips providing high heat, or even an oxyacetylene torch, may be used by skilled operators. Less-skilled operators should use the air-acetylene torch, with its soft flame and lower temperatures, to avoid overheating the base metals. A variety of tips may be used with an air-acetylene torch.

PRO TIP

Heat and Torch Tip Relationship

More heat is obtained with larger torch tips. The smaller the torch tip, the lower the heat output. The larger the torch tip, the higher the heat output. Selecting the proper size tip (or flame) requires experience. In general, small tubing requires a small flame, and large tubing requires a large flame. See **Figure 8-15**.

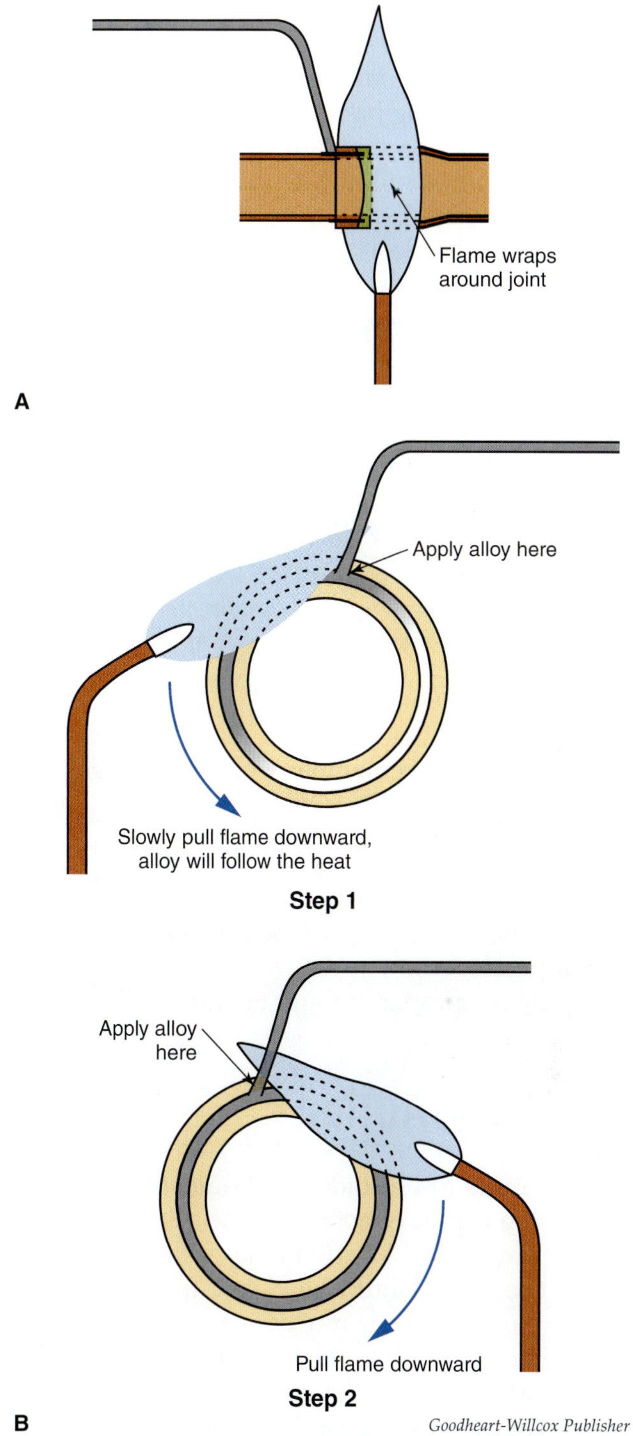

Goodheart-Willcox Publisher

Figure 8-15. Techniques for soldering small and large tubing. A—For small tubing, use a direct flame around the bottom of the joint. The alloy will flow around the tubing toward the heat. B—For large tubing, work the flame slowly around the tubing as shown, applying small amounts of filler alloy.

Dirty tips or faulty adjustment may produce a sooty flame that deposits carbon on the workpiece. This carbon deposit will prevent the alloy from flowing properly and result in a poor joint.

For consistently good soldering, keep in mind that the alloy will flow properly only when the joining pieces are heated equally and evenly to the solder's flow point. *Do not apply the flame directly to the alloy.* Use the flame to heat the base metals to the solder's melting point. Then, back off the flame just far enough to maintain the proper temperature until melted alloy fills the joint. Apply only enough alloy to form a bright shiny bead at the point where the pieces meet.

PRO TIP

Filler Metal Balling

When solder is applied to a heated joint, it should flow smoothly. If it forms a globule or ball on the work, the base metal temperature may be too low, or the joint surfaces are dirty. ***Balling*** can also occur if the torch flame was directed onto the solder, melting a drop from the end. It is difficult to get these globules of solder to flow properly.

Key points to remember when soldering are to:

- Make the joint fit.
- Clean the joint thoroughly.
- Apply the proper flux.
- Heat the joint evenly to the correct temperature.
- Apply the proper amount of alloy filler metal.

Capillary Action

All joints to be soldered should fit tightly for maximum capillary action by the alloy. To obtain maximum capillary action (and greater strength), the entire joint should be close-fitting and well-supported to prevent movement. When the filler alloy enters the joint, the alloy molecules have a greater attraction to the base metal than to each other. The alloy is attracted to each wall of the base metal and "walks" itself into the fitting to completely fill the joint.

This capillary attraction makes it possible to solder a *vertical up (uphill) joint* because the alloy will be drawn up into the joint, **Figure 8-16**. When soldering an uphill joint, heat the tubing first, then apply heat to the fitting. It is important to heat both pieces evenly. If the tubing is overheated, the alloy may run downward, rather than be drawn upward into the joint. Remember, alloy flows *toward* the heat.

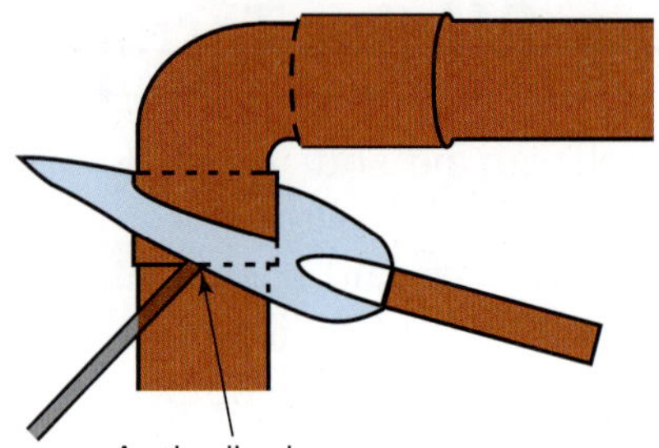

Goodheart-Willcox Publisher

Figure 8-16. Solder will be drawn up into a vertical joint by capillary action. The alloy will travel toward the heat.

Soldering Copper Tubing

An HVACR service technician may frequently have to solder copper tubing.

PROCEDURE

Soldering Copper Tubing

Follow this procedure to make mechanically sound and leakproof connections with copper tubing and fittings:

1. Cut tubing ends square and remove all inside and outside burrs with a reamer. See Chapter 6, *Working with Copper Tubing* for details on deburring and reaming copper tubing.
2. Clean the copper tubing and fitting thoroughly down to bare metal before making the soldered joint. Joining surfaces must be clean to make a sound solder joint.
3. Apply flux to cleaned tubing before soldering. Flux should be applied sparingly and kept away from the tubing end by about 1/8″ (3 mm).
4. Always move the torch in short arcs; do not let it remain in one spot too long. Heat the tubing first by applying the flame next to the fitting. Heating the tubing first will conduct heat inside the fitting. Work the flame around the tubing and fitting to bring the temperature up *equally* in both pieces.

(Continued)

PROCEDURE

Soldering Copper Tubing *(continued)*

5. Use the flux as a temperature guide. Apply heat until the flux stops bubbling and becomes fluid and clear. Apply heat to the fitting as well to ensure the flux inside the fitting is clear. A clear flux indicates the alloy flow temperature is being reached.
6. Back the flame away slightly to reduce the heat intensity. Sweep the flame back and forth over the assembly to maintain uniform heat in both parts.
7. Apply the alloy. The heat of the base metals, not the flame, should melt the alloy. Feed the alloy into the joint between the tube and fitting. It will quickly melt and completely fill the joint area.
8. Remove the flame when the alloy begins flowing into the joint. Do not continue feeding alloy after the joint is filled.

Soldering Aluminum

Aluminum is now used for many applications where copper or steel were once standard. Aluminum is an excellent metal for conducting heat and is widely used in the manufacture of evaporators for domestic refrigerators. However, soldering aluminum presents special problems not normally found with other metals.

When soldering aluminum, attention must be given to such details as surface preparation, solder composition, temperature, and the application of heat. Soldering pure aluminum and the many aluminum alloys requires special techniques and materials. For example, the aluminum used in the manufacture of evaporators is thin, so heat applied during the soldering process must be precisely controlled. Aluminum and its alloys are weak when hot, a condition called *hot-shortness*. The metal does not change color as a warning that it is approaching its melting point. Instead, it will suddenly collapse. For this reason, the metal must be properly supported during the soldering process.

The corrosion-resistance of aluminum results from oxides that form quickly on the metal's surface when it is exposed to air. To make a sound soldered connection, the oxides must be removed. Additional oxides are formed by the application of heat. Oxides have a higher melting point than aluminum. A special flux is used that will combine chemically with the oxides to form a scum, or *slag*. The slag rises to the surface of the molten solder, where it will not interfere with the soldering process.

The solder must be melted before it can flow into the joint. The melting point of the solder will determine the minimum temperature needed to make the connection. The solders used for aluminum have a higher melting point than those used for copper or steel, so aluminum must be heated to 100°F–200°F (55°C–110°C) higher than copper or steel. The increased temperature is the major reason that soldering aluminum and aluminum alloys is difficult.

The solders used for aluminum can be placed into four groups: zinc-based, zinc-cadmium–based, tin-zinc–based, and tin-lead–based solders. See **Figure 8-17**. These solders may contain quantities of other metals to obtain certain properties or qualities.

Soldering aluminum requires skill and proper equipment. These requirements can usually be met under carefully controlled manufacturing processes. Such control is not readily available in the field where service and repair work must be performed. The aluminum used in the manufacture of evaporators on domestic refrigerators presents a typical problem. The metal is thin, which promotes good heat transfer for the cooling process. However, because of the thin metal, these evaporators are easily punctured by sharp instruments. A puncture will result in loss of refrigerant, thus stopping the refrigeration process. In the field, soldering such a hole is almost impossible. Repairing the damage would be time-consuming and expensive. The customer is usually advised to purchase a new refrigerator in this situation.

Aluminum Soldering Alloys

Solder	Melting Point
Zinc-based	700°F to 820°F (370°C to 440°C)
Zinc-cadmium	510°F to 750°F (265°C to 400°C)
Tin-zinc	550°F (290°C) and higher
Tin-lead	450°F (230°C) and higher

Goodheart-Willcox Publisher

Figure 8-17. Alloys used for soldering aluminum have melting points considerably higher than those used to solder copper or steel.

Repairs to a hole in an aluminum evaporator *can* be performed but not by soldering. ***Epoxy*** is a compound that chemically hardens and can be used as a filler, leak stopper, or an adhesive. Epoxies are readily available for making such repairs in aluminum and have proven to be highly successful. See **Figure 8-18**. An epoxy kit contains all necessary materials and instructions for proper use.

The same epoxy compound can be used to repair leaks in joints between aluminum tubing and copper tubing. The original aluminum-to-copper connection is made at the factory, using welding techniques that are not available in the field. Sometimes leaks will develop at this special connection. Epoxy compounds provide a satisfactory and reliable means for repairing joints.

Taking Apart a Soldered Joint

A leak or other unfavorable conditions are reasons for redoing a soldered joint. Before this, the soldered joint must be taken apart. Review previous instructions regarding torch use before taking apart a soldered joint.

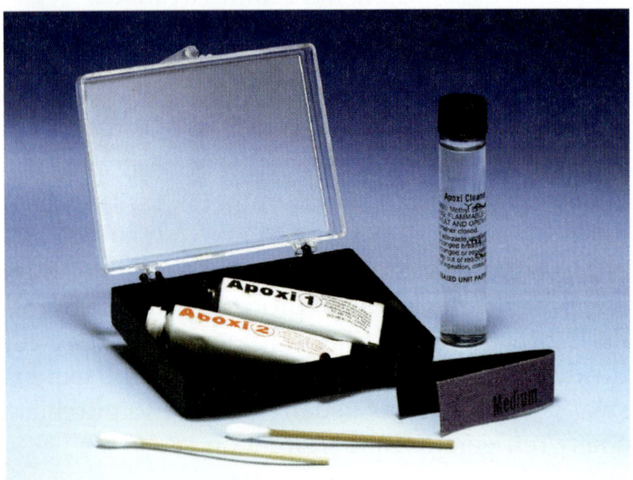

Sealed Unit Parts Co., Inc.

Figure 8-18. Repairs to aluminum that cannot be made in the field with soldering can often be accomplished with epoxy compounds. Epoxies harden quickly and will withstand pressure and the effects of refrigerants.

PROCEDURE

Taking Apart a Soldered Joint

Follow this procedure to take apart a soldered joint:

1. Clean the joint that needs to be disconnected.
2. Apply flux to the joint.
3. Heat the tubing *and* fitting evenly.
4. When the alloy becomes fluid throughout the joint, the tubing should be easily removed from the joint. Do not twist or apply force to the tubing to remove it. Twisting or force will cause severe damage to the hot fitting and tubing.

To resolder a disconnected joint, first clean the tubing end and the inside of the fitting. Then, apply flux and solder the joint as previously described.

SAFETY NOTE

Redoing a Soldered Joint

Never apply heat to tubing that is *under pressure*. Applying heat increases pressure. The molten alloy and fitting may be blown violently apart. Be sure to release all pressure from the tubing before heating a soldered connection.

8.4 Air-Acetylene Brazing

Special tips have been designed for use on the regular air-acetylene soft-soldering torch to increase the working temperature of the flame, making it possible to perform some brazing operations. Some of these special tips are constructed with a gas diffuser inside. The diffuser slows the gas flow and causes the flame to start burning and expanding inside the tip. This creates a high-velocity gas stream that completes the burning process outside the tip, providing a high-heat working flame. It is important to keep a full gas flow and flame to avoid overheating the tip.

Torch Tips for Brazing

A major difference when using these special tips is the work (flow) pressure of the acetylene. High-velocity tips require 15 psi (103 kPa) work pressure from the acetylene pressure regulator. See **Figure 8-19**. When using these high-velocity tips, the regulator is usually turned all the way in (fully clockwise) for maximum pressure. The tips are designed so that the flow of acetylene draws atmospheric air into the base of the tip, resulting in more oxygen mixing with the acetylene. Another design mixes the gases near the tip end, using a cupped tip to stabilize the high-velocity flame.

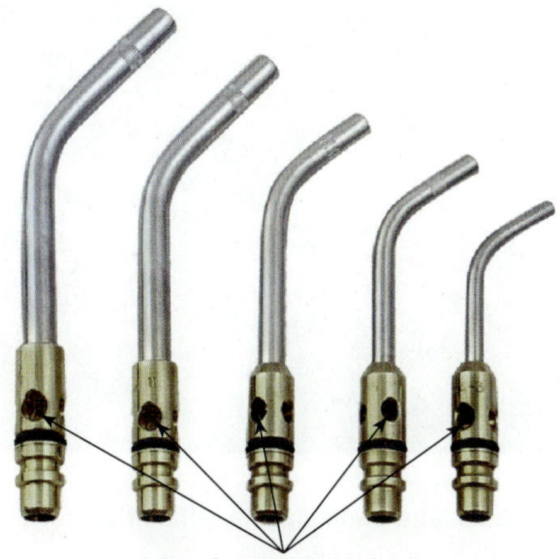

High-velocity air-acetylene torch tips

Air-acetylene outfit with high-velocity tip *Uniweld*

Figure 8-19. High-velocity tips for air-acetylene torches allow the use of higher temperatures for brazing work. Inlets draw in atmospheric air for more oxygen resulting in better combustion.

These tips are available from various manufacturers under different trade names. They are designed for use on a regular air-acetylene handle. Various sizes of tips are available to accommodate different flame sizes.

Because all high-velocity tips must be operated at 15 psi (103 kPa) work pressure, the resulting high-heat flame is usually nonadjustable. Some manufacturer's tips are fully adjustable at the torch handle, giving the operator more flame control. Settings may include idle, soft (medium), and high-intensity (full) operation.

Special Tip Advantages

Special tips offer several advantages. The flame is pencil-shaped (thin), short, and concentrated for close work. The short flame allows soldering or brazing of joints within an inch of walls or woodwork without damaging or discoloring them. It is an all-weather flame, as the high velocity allows it to be used in wind, rain, and subzero temperatures.

The tips develop an efficient high-heat-transfer zone at the tip of the inner cone of the flame. This forms a concentrated heat flow that wraps around the workpiece to heat it evenly and quickly. The flame temperatures in this high-heat zone range from 1800°F (982°C) for propane, to over 2400°F (1315°C) for MAPP, to over 2600°F (1427°C) for acetylene. This gives the user a choice of tip sizes and gases to meet the heating requirements, **Figure 8-20**.

Special Tip Disadvantages

The major disadvantage of these special tips is that they cannot be used for brazing a connection located near a heat sink. A ***heat sink*** is any heavy metal device, such as a valve or compressor, that tends to draw heat away from the brazing area, **Figure 8-21**. With these tips, even a moderate-sized heat sink will steal the heat rapidly enough to greatly prolong the brazing process. Also, the supply of acetylene is quickly exhausted because of the high-flow rates.

It is recommended that all brazing connections be performed with an oxyacetylene torch or high-heat air-gas tips, reserving soft flame air-acetylene torch tips for soft-soldering procedures. The use of air-gas tips, when possible, saves oxygen for applications requiring higher temperatures, such as cutting, welding, and brazing near heat sinks.

Uniweld

Figure 8-20. Specialty torch tips are helpful for maintaining equal distribution of heat or when working in a difficult spot. Note that these torch tips are more commonly used for brazing with oxyacetylene rather than with air-acetylene.

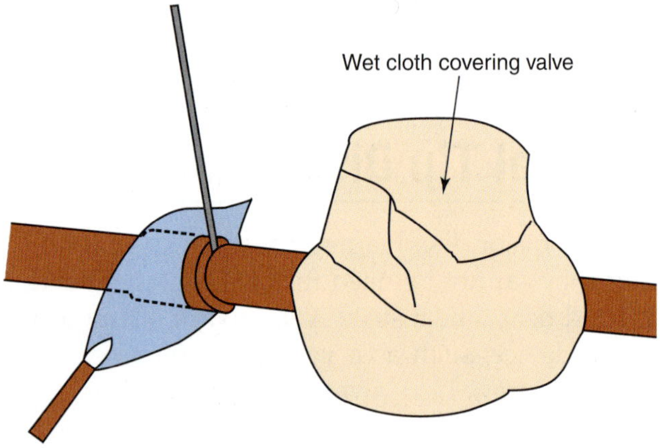

Aeroquip

Figure 8-21. A heavy metal component, such as a valve, will act as a heat sink and draw heat away from the area being brazed. This heat could damage the component. To prevent damage, the valve or other component is sometimes wrapped in a wet cloth.

Tubing sizes appropriate for several high-velocity heat tips, when soldering or brazing, are listed in **Figure 8-22**. Notice the difference in the tubing sizes when changing from soldering to brazing with the same size tip. This table does not consider the presence of a possible heat sink.

Tips vs Tubing Size

Tip No.	Soft Soldering, Copper Tubing Size	Brazing, Copper Tubing Size
#3	1/4″ to 1 1/2″	1/4″ to 7/8″
#4	1″ to 2″	5/8″ to 1 1/8″
#5	1 1/2″ to 3″	7/8″ to 1 5/8″
#6	2″ to 4″	1 1/8″ to 2 1/8″

Goodheart-Willcox Publisher

Figure 8-22. Tubing sizes that can be soldered or brazed with several sizes of high-velocity heat tips.

Chapter Review

Summary

8.1 When deciding what type of alloy to use for soldering or brazing, the physical properties of the metals being joined and the demands that will be placed on the connection must be considered. The choice of flux in any soldering process is determined by the metals being joined and the filler alloy being used. Fuel gas torches are used to melt solder alloy. Acetylene and MAPP gas are commonly used to produce higher flame temperatures. Propane, natural gas, and manufactured gas produce lower flame temperatures.

8.2 The air-acetylene torch combines fuel gas from a cylinder with combustion air drawn from the atmosphere. Safely handling the acetylene cylinder and operating the torch with proper procedures are essential when soldering with an air-acetylene outfit.

8.3 When soldering, make sure the joint fits, clean the joint thoroughly, apply the proper flux, heat the joint evenly to the correct temperature, and apply the proper amount of alloy filler metal. Additionally, consider the base metals being soldered and adjust accordingly. Aluminum, for instance, requires special attention to surface preparation, solder composition, temperature, and heat application as compared to other metals. To take apart a soldered joint, clean the joint, apply flux to it, heat the tubing and fitting evenly, and remove the joint once the alloy becomes consistently fluid.

8.4 Special torch tips are available for brazing applications, allowing high-velocity flames to be used in close working conditions. One disadvantage of special tips is they cannot be used for brazing near a heat sink.

Know and Understand

1. Solder usually consists of _____ and another metal with a low melting point. (8.1)
   A. silver
   B. lead
   C. tin
   D. aluminum
2. When an alloy is drawn into a gap by indirectly applying heat, this process is referred to as _____. (8.1)
   A. flow point
   B. capillary action
   C. melting
   D. elongation
3. A wide _____ provides easier control of an alloy during the soldering process. (8.1)
   A. torch flame
   B. elongation
   C. plastic range
   D. tinning action
4. *True or False?* Tin-silver alloys are a better choice than tin-lead or tin-antimony alloys for refrigeration work due to their higher strength and elongation. (8.1)
5. Before applying flux to a joint that is to be soldered, thoroughly clean the surface of oil, grease, rust, and oxides by using a(n) _____. (8.1)
   A. soap and water cleaning solution
   B. clear flux
   C. emery cloth
   D. acid flux
6. In soldering applications, _____ gas produces the highest temperature. (8.1)
   A. MAPP
   B. acetylene
   C. propane
   D. natural
7. All acetylene cylinders are manufactured to _____ requirements. (8.2)
   A. Department of Energy (DOE)
   B. Environmental Protection Agency (EPA)
   C. American Welding Society (AWS)
   D. Department of Transportation (DOT)

8. The fusible plug on acetylene cylinders will melt at _____°F to relieve pressure buildup. (8.2)
   A. 70
   B. 212
   C. 400
   D. 3600
9. On two-gauge pressure regulators, the second gauge shows the _____. (8.2)
   A. working pressure
   B. level of acetylene in the cylinder
   C. speed of the acetylene to the torch tip
   D. temperature of the flame at the torch tip
10. The hottest area of the air-acetylene flame should be about _____" from the tip of the inner cone. (8.2)
    A. 1/8
    B. 1/4
    C. 1/2
    D. 1
11. *True or False?* The larger the torch tip, the lower the heat output; the smaller the torch tip, the higher the heat output. (8.3)
12. When soldering an uphill joint, _____. (8.3)
    A. heat the fitting first, then the tubing
    B. heat the tubing first, then the fitting
    C. heat the alloy first, then the fitting
    D. heat the fitting first, then the alloy
13. Repairing a hole in an aluminum evaporator can be done by _____. (8.3)
    A. soldering
    B. brazing
    C. applying epoxy
    D. replacing the aluminum piece
14. When taking apart a soldered joint, be sure to heat _____. (8.3)
    A. only the tubing
    B. only the fitting
    C. the tubing and fitting evenly
    D. the fitting then the tubing
15. A _____ inside the torch tip creates the high-heat flame for air-acetylene brazing applications. (8.4)
    A. smaller tip
    B. second fuel gas source
    C. gas diffuser
    D. heat sink

Apply and Analyze

1. Which two factors determine what type of alloy to use and whether to solder or braze a joint? (8.1)
2. List three purposes of using flux prior to soldering. (8.1)
3. How are acetylene hose connections distinguished from oxygen hose connections? (8.2)
4. Explain how to properly light an air-acetylene torch. (8.2)
5. What are two ways that the process of soldering aluminum differs from other metals? (8.3)
6. What is a major difference in special air-acetylene tips used for brazing compared to those used for soldering? (8.4)

Critical Thinking

1. Explain how capillary action aids in soldering joints. (8.1)
2. Why should heat never be directly applied to the solder alloy? (8.3)

Goodheart-Willcox Publisher

9 Brazing

LEARNING OUTCOMES

After studying this chapter, you will be able to:

9.1 Explain why brazing is the preferred method for joining heating, air-conditioning, and refrigeration equipment and the appropriate brazing alloys and fluxes to use.

9.2 Identify uses of the oxyacetylene torch; related hazards; and the proper equipment and operating conditions when preparing, lighting, and shutting down the torch.

9.3 Connect and operate oxyacetylene equipment for brazing metal with three types of brazing connections and troubleshoot issues that can occur with these connections.

TECHNICAL TERMS

backfire
brazing
brazing alloy
cadmium-bearing alloy
carburizing flame
downhill joint
flashback
hard solder
horizontal joint
inert
multiflame tip
neutral flame
oxidizing flame
oxyacetylene torch
oxygen regulator
uphill joint

Introduction

Brazing and soldering are similar metal-joining processes. The same skills are required for each process, and proper use of the flame is critical to each. Brazing and soldering differ in the temperatures and alloys used to join metals. By definition, ***brazing*** joins metals above 840°F (448°C). However, in practice, it is usually accomplished at about 1200°F (648°C).

Like soldering, brazing is performed by heating the base metals to a specified temperature *above* the melting point of the filler metal alloy but *below* the melting point of the base metals (for example, copper melts at 1981°F or 1083°C), **Figure 9-1**. The melted filler alloy is distributed between the close-fitting base metal pieces by capillary attraction.

© *J. W. Harris Co., Inc.*

Figure 9-1. Filler alloys for use in brazing come in a variety of shapes, sizes, and compositions.

9.1 Brazing Overview

Brazing is the principal method of assembling heating, air conditioning, and refrigeration equipment. A typical large system may contain hundreds of brazed joints. Brazing permits the joining of similar and dissimilar metals, thin and thick sections, and metals with different melting points. A brazed joint is:

- Stronger than the base metals that it joins
- Tolerant of vibration without cracking
- Leakproof under high pressures
- Noncorrosive
- Excellent at transferring heat

All of these attributes make brazing the preferred method for joining tubing on refrigeration systems. In fact, many local, state, and federal codes *require* brazing on all refrigeration systems.

Some local codes also require that licensed technicians make the brazed connections. A license is usually obtained by paying a fee and submitting test samples of properly brazed connections. Leakproof connections are essential to ensure a properly functioning refrigeration system.

Brazing Alloys

Brazing alloys are filler metals designed for filling brazing joints. They may be generically referred to as *filler metals*. Brazing alloys usually contain a certain amount of silver, which has led to the term *silver soldering*. However, that term is incorrect. Silver-bearing solder is an alloy of tin and silver with a melting point of 430°F (221°C). This *soldering* alloy should not be confused with the silver-bearing *brazing* alloys that have melting temperatures ranging from 1200°F to 1500°F (593°C to 816°C). These brazing alloys are sometimes called ***hard solders***, a term that is commonly used but not universally accepted.

Brazing alloys are available in a variety of forms, **Figure 9-2**. The preferred form is a 20″ (50.8 cm) length of 1/8″ (3 mm) flat rod. The rod is packaged in 1 lb (0.45 kg) or 5 lb (2.27 kg) tubes. Larger containers are also available.

As noted in Chapter 8, *Soldering*, the term *melting point* (*solidus*) refers to the temperature at which an alloy starts to melt, while *flow point* (*liquidus*) is the temperature at which the alloy becomes a liquid. The difference between the solidus and liquidus temperatures is called the *plastic range*. When compared to soldering alloys, brazing alloys have a relatively wide plastic range.

Since brazing filler metal is an alloy composed of different metals with different melting points, the filler metals used in brazing may separate during the heating process. The metal with the lowest melting temperature in the alloy may flow into the joints, leaving the higher melting-point metals behind. This causes a change in the color and strength of the alloy, which makes it difficult to get the remaining alloy to flow. Heating the base metals quickly prevents separation of the alloy and minimizes oxidation.

Flat brazing rods

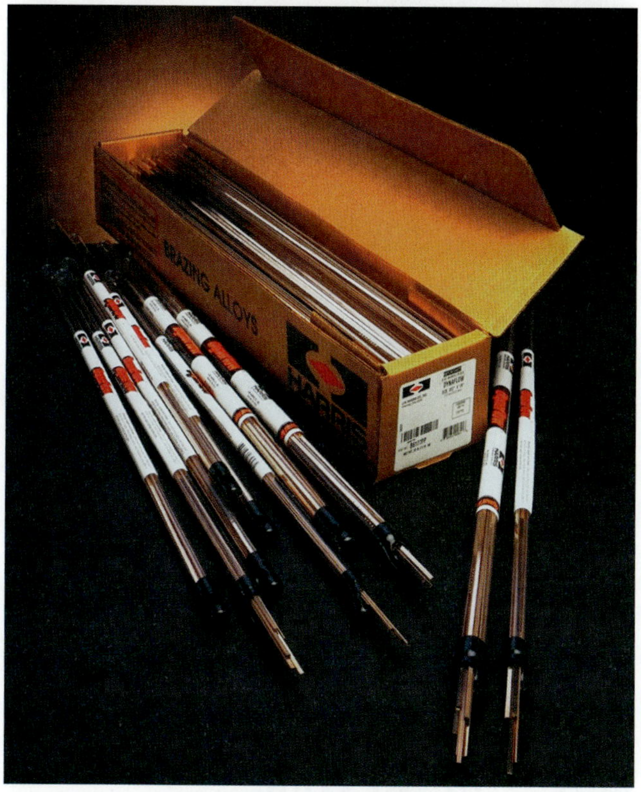

Tube brazing rods

© J. W. Harris Co., Inc.

Figure 9-2. Flat 1/8″ (3 mm) brazing rod (top) is the form preferred by many technicians. Brazing rod is also packaged in 1 lb (0.45 kg) tubes (bottom).

Separation of filler metals is a feature that can be advantageous when brazing a loose connection. The higher-melting-point alloys will span the gap. However, the connection will not be as strong as one made with an alloy whose metals did not separate. Silver-brazing alloys generally melt at approximately 1200°F (593°C) and flow at about 1500°F (816°C), which gives a plastic range of about 300°F (165°C).

Alloy Compositions

Brazing alloys are composed of varying combinations of copper (Cu), silver (Ag), phosphorus (P), and zinc (Zn). The composition of a given brazing alloy determines its use, strength, and cost. Manufacturers have their own numbering system for identification of a particular alloy. However, the American Welding Society (AWS) uses a special numbering system to identify brazing alloys by their composition, regardless of manufacturer. The AWS number consists of the letter *B* to indicate a *brazing alloy*, followed by the chemical abbreviations of the metals used in the alloy, and ending with an identifying number. See **Figure 9-3**.

Copper-Phosphorus Alloy (AWS BCuP-2)

This is a low-cost alloy that is about 93% copper and 7% phosphorus. It is sometimes called *phos-copper* and eliminates the need for a flux when brazing copper to copper; the phosphorus acts as a deoxidizing agent and flux. When joining copper tubing to a brass or bronze fitting, however, a flux is needed.

CAUTION

Copper to Steel Joints

Do *not* use phos-copper alloy for joining copper to steel. This combination would result in a brittle joint.

When using phos-copper for connecting copper to copper, the surfaces of the connection should be cleaned to bare metal.

Copper-Phosphorus-Silver Alloy (AWS BCuP-5)

Many alloys of silver have been developed to provide different melting points, flow properties, strengths, and colors. Silver-brazing is one of the best methods for making connections in refrigeration components that are leakproof and able to withstand severe conditions.

An alloy of 80% copper, 5% phosphorous, and 15% silver (AWS BCuP-5) has been the industry leader for many years. This alloy allows the creation of highly ductile and strong connections when brazing copper to copper, without using a flux. In addition, with the use of a flux, this alloy can braze copper to brass, bronze, or steel.

Chemical Composition of Brazing Alloys

AWS Number	Composition Percentage	Melting Point (in °F)	Flow Point (in °F)
BCuP-2	93% copper / 7% phosphorus	1310	1485
BCuP-3	89% copper / 6% phosphorus / 5% silver	1190	1485
BCuP-4	87% copper / 7% phosphorus / 6% silver	1190	1480
BCuP-5	80% copper / 5% phosphorus / 15% silver	1190	1500
BAg-5	31% copper / 46% silver / 23% zinc	1250	1370
BAg-7	22% copper / 56% silver / 17% zinc / 5% tin	1145	1200
BAg-20	38% copper / 30% silver / 32% zinc	1250	1410

Goodheart-Willcox Publisher

Figure 9-3. The American Welding Society's identification system for brazing alloys assigns a specific letter/number combination to each alloy. This table shows the chemical compositions and the melting and flow (solidus and liquidus) points for some common brazing alloys.

Copper-Phosphorus-Silver Alloy (AWS BCuP-4)

Silver-bearing alloys are expensive, and the higher the silver content, the more expensive the alloy. Alloys with silver content lower than 15% have proven economical and have nearly the same characteristics as the 15% alloys. An example is AWS BCuP-4, which contains 6% silver. This alloy was developed to perform nearly as well as the higher silver-content alloys. It is widely used in the HVACR industry because of its economy.

Alloy Additives

The metals zinc and cadmium are used as additives in brazing alloys that are 35% or more silver. These additives have superior capillary action properties and form excellent brazes on all metals except aluminum and magnesium. They also lower the melting and flow points of the alloys. Despite these benefits, ***cadmium-bearing alloys*** should be avoided whenever possible.

SAFETY NOTE

Cadmium Fumes

Cadmium is a toxic metal when molten. It emits highly poisonous cadmium oxide fumes that can cause illness or death. Adequate ventilation must be provided when using cadmium-bearing alloys.

Cadmium-free alloys that meet all brazing requirements are available, **Figure 9-4**. HVACR technicians should adopt cadmium-free brazing alloys wherever possible to protect their health. The danger of cadmium poisoning is real, and the risk is unnecessary. The nontoxic alloys eliminate the danger of cadmium oxide fumes and can be used on both ferrous and nonferrous metals.

© J. W. Harris Co., Inc.

Figure 9-4. Cadmium-free brazing alloys eliminate the danger of poisonous cadmium oxide fumes that are generated when using cadmium-bearing filler alloy.

Fluxes for Brazing

Common fluxes for silver brazing are mixtures of boric acid, borates, fluorides, and fluoborates. See **Figure 9-5**. These chemical powders form a paste when water or alcohol is added. Their containers usually have an application brush. If the paste becomes too dry, it may be thinned with water and stirred. The flux must be kept clean to avoid contaminating the joint.

By watching the behavior of the flux, a technician can determine the temperature at different

© J. W. Harris Co., Inc.

Figure 9-5. Fluxes for brazing with silver-bearing alloys are packaged in containers of various sizes. Most packages have a built-in applicator brush.

stages of the brazing process. Observe the flux while heating and take note of these visual cues:

- At 212°F (100°C), the flux will dry out as the water boils off.
- When the temperature reaches about 600°F (316°C), the flux will start to bubble and turn milky white.
- At about 1100°F (593°C), the flux will turn into a clear liquid. At this point, apply the filler metal alloy.

When brazing, the heating process should be performed as evenly and quickly as possible to reduce oxide formation. Brazing that is performed quickly reduces stress on the base metals. **Figure 9-6** lists visual signals to the various stages in the heating process.

CAUTION

Avoid Excess Flux

Special care must be exercised when using fluxes to braze copper tubing or brass fittings in a refrigeration system. Fluxes are harmful contaminants inside the tubing of a refrigeration system. Avoid excess flux. Apply only the amount necessary.

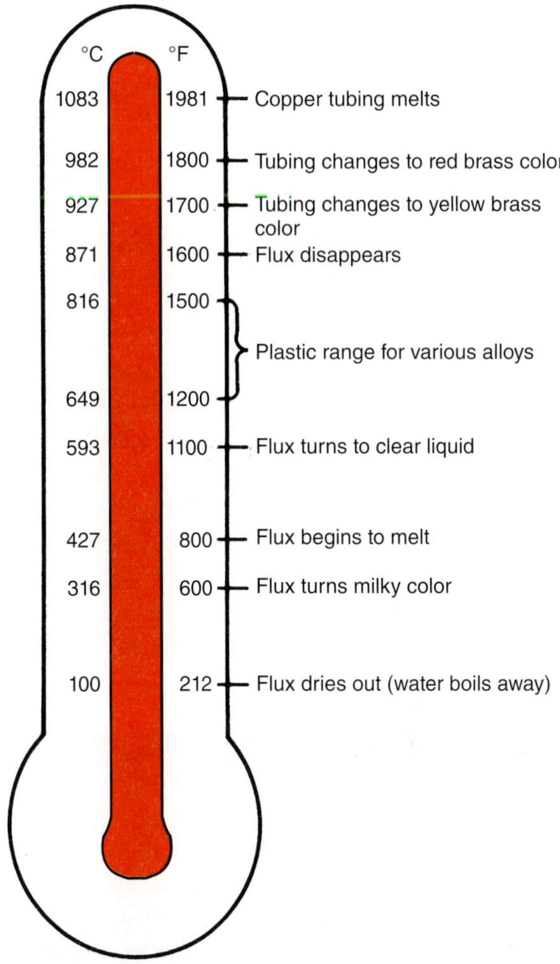

Goodheart-Willcox Publisher

Figure 9-6. Visual signals to important temperatures involved in the brazing process.

9.2 The Oxyacetylene Torch

Heating the base metals for brazing is normally done with a torch. Flame temperature is determined by the type of fuel gas used. Oxygen mixed with any fuel gas will give the highest attainable flame temperature.

The most versatile heat source for brazing is an oxygen-acetylene torch, often referred to as an ***oxyacetylene torch***. It provides a heat range that handles all types of joining work, regardless of the materials.

Using an oxyacetylene torch properly and safely and achieving good results is not difficult. Unsafe conditions or poor results are usually due to an improperly adjusted flame and incorrect pressure. The technician should make every effort to acquire good habits working with this tool and maintain these habits with every use. The key is practice, practice, practice.

Where the Oxyacetylene Process Is Used

The oxyacetylene process is used in many fields where metal work is involved. When acetylene is burned with oxygen, it produces a flame so hot that it can melt, fuse, or cut any metal.

In the hands of a skilled operator, the oxyacetylene torch makes it possible to quickly solder, braze, weld, or cut all metals. The intensely hot flame concentrates considerable heat in a small area, so the brazing process can be performed quickly.

SAFETY NOTE

Oxygen and the Flame

The presence of oxygen rapidly increases the rate of burning for almost any ignited material, especially oil and grease. Also, a concrete surface on which material is being cut will chip explosively when overheated by an oxyacetylene flame. Flying concrete chips can cause injury, so be aware of concrete materials when using an oxyacetylene torch.

Dressing for Safety

An oxyacetylene torch flame can reach up to 6300°F (3482°C). The workpiece can reach almost 3000°F (1649°C). The process can produce flying sparks, molten metal, slag, fumes, and intense light rays, all of which can harm the torch operator or persons nearby.

SAFETY NOTE

Protective Clothing

For your safety, take a practical "head-to-toe" approach to protective clothing worn when brazing. Wear hair and head coverage and welding and cutting goggles with safety-tempered dark lenses (shade 5 is standard). Wear thick gloves and proper shoes. Do not wear any flammable articles or clothing that has been exposed to flammables (oil, grease, wax, or solvent). Sparks and molten material will find their way to unprotected areas, so protect yourself thoroughly before starting.

Gas Cylinders and Related Equipment

Oxygen and acetylene cylinders are sized according to the amount of gas in cubic feet the cylinder will hold. The cylinder sizes are designated by a letter/number combination. The table in **Figure 9-7** matches oxygen and acetylene cylinders by size.

HVACR service technicians seldom use the large fuel gas cylinders common in welding operations. Technicians typically prefer smaller cylinders that can be easily transported to rooftops, basements, and other locations where HVAC equipment is installed. See **Figure 9-8**. The small oxyacetylene units perform the same duty as the large units, except the operating pressures are different, and the supply of oxygen and acetylene does not last as long in the small units. To overcome the supply problem, technicians stock one or two extra cylinders for additional capacity. Carrying stands and two-wheel carts are available to hold the small cylinders.

Cylinder Sizes for Matched Systems

Acetylene	Oxygen Equivalent
10 cu ft (MC)	20 cu ft (AA or R-Oxy)
40 cu ft (B)	40 cu ft (A)
60 cu ft (#2)	60 cu ft (J) or 80 cu ft (JJ)

Goodheart-Willcox Publisher

Figure 9-7. The different cylinder sizes that can be matched for use in an oxyacetylene brazing and cutting outfit.

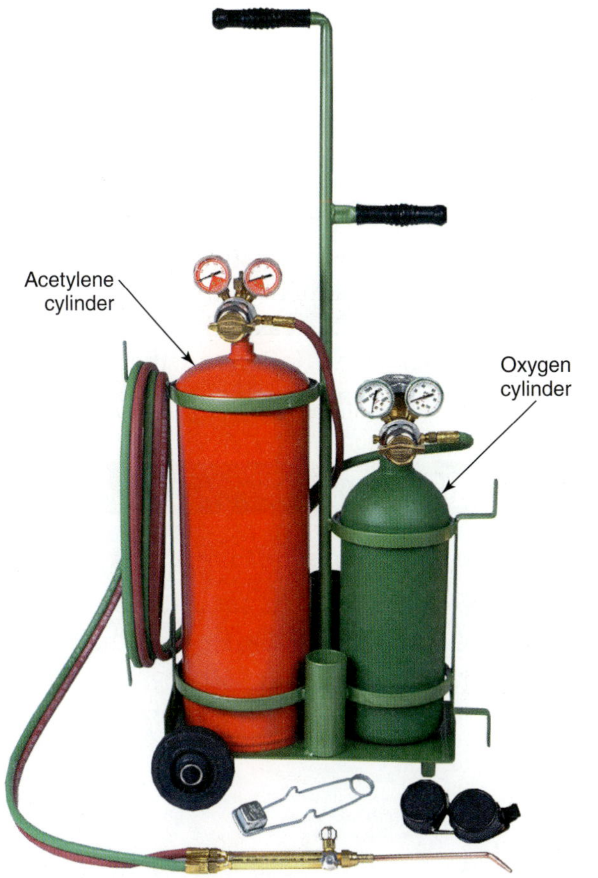

Uniweld

Figure 9-8. A portable oxyacetylene outfit with a size B (40 ft^3) acetylene cylinder and a size R (20 ft^3) oxygen cylinder.

An oxyacetylene outfit that uses small cylinders, regulators, hoses, and torches is called a light-duty, or *aircraft-type*, unit. This type is preferred by many service technicians because of its portability. All the components are matched and have 3/8″ (size "A") fittings at the regulator, hoses, and torch handle.

The small oxyacetylene outfits also require small brazing and cutting tips. These, in turn, require small operating (work or flow) pressures. Follow the manufacturer's recommendations for the type of equipment available.

Cylinder Pressures

Oxygen cylinders are charged to 2200 psi at 70°F (15 169 kPa at 21°C), but the actual pressure in the cylinder will vary with the temperature. All gases expand when heated and contract when cooled. If the temperature exceeds 70°F (21°C), the pressure in a full cylinder will rise above 2200 psi. A safety relief device is built into oxygen cylinder valves to protect against explosion from extremely high pressures. Such pressures might occur if the cylinders were exposed to fire.

SAFETY NOTE

Oxygen Cylinder Storage

Oxygen cylinders should *never* be used or stored where they might become overheated.

Oxygen Cylinder Safety

An oxygen cylinder valve is designed to operate at high pressures. Large oxygen cylinders have an iron cap screw down over the valve to protect it during shipment or handling. This cap should always be in place when the cylinder is not in use.

The cylinder valve has a *double seat* that prevents leakage around the valve stem when the valve is fully opened for operation, Figure 9-9. To open and close the valve, hand pressure is sufficient. A wrench should not be used.

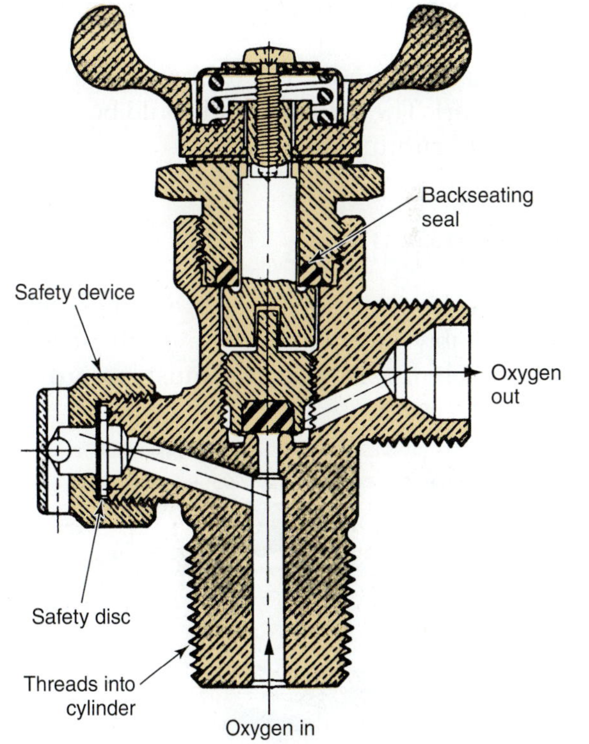

Goodheart-Willcox Publisher

Figure 9-9. An oxygen cylinder valve has a double-seating feature (backseating seal) that prevents leakage around the stem when the valve is fully opened. Note the safety disc; it will rupture if oxygen pressure becomes dangerously high.

Observe the following precautions when using or handling oxygen cylinders:

- Carefully secure cylinders in place to keep them from falling over.
- Never store cylinders and equipment in unventilated and confined spaces, in closed vehicles, or near any source of heat or ignition.
- Close cylinder valves securely when not in service or when empty.
- Avoid dangerous pressure imbalance and cylinder contamination by never allowing any cylinder (especially oxygen) to become completely empty. Stop using a cylinder by the time pressure reaches a level between 25 psig and 50 psig.
- For maximum protection against contamination, install external reverse-flow check valves on the regulator or torch handle, **Figure 9-10**.
- Never allow oil or grease to come in contact with oxygen cylinder valves. Oxygen reacts violently with oil and grease.
- Do not use a cylinder that has a leaking valve. Carefully move the cylinder outdoors and then notify your gas supplier.
- Do not expose cylinders to torch flames or electric arcs.
- Never use a cylinder, full or empty, as a roller or support. The cylinder wall could be damaged and rupture or explode.

Oxygen Regulators

The working (or flow) pressure at the torch tip is considerably lower than the available cylinder pressure. Obviously, a device must be provided to reduce high cylinder pressure to working pressure. Also, to maintain a steady and uniform flame, the pressure of the gases reaching the torch tip must not vary (despite steadily decreasing cylinder pressure). An ***oxygen regulator*** performs both these important functions, **Figure 9-11**.

The regulator is adjustable to permit any flow pressure, and it will maintain that pressure without further adjustment until the cylinder approaches empty. It is easy to recognize a nearly empty cylinder by variations in the flame or by checking the pressure gauge.

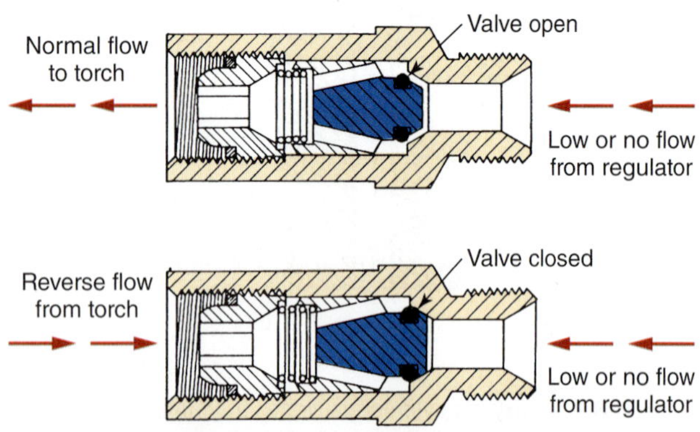

Uniweld

Figure 9-10. Check valves are designed to prevent contamination of fuel gas cylinders by allowing flow in only one direction. Reverse flow causes the check valve to close.

PRO TIP

Replace Cylinders before Empty

Always replace a cylinder *before* it becomes empty and unsafe. Stop using a cylinder by the time cylinder pressure reaches 25–50 psig. Low pressure can result in cylinder contamination.

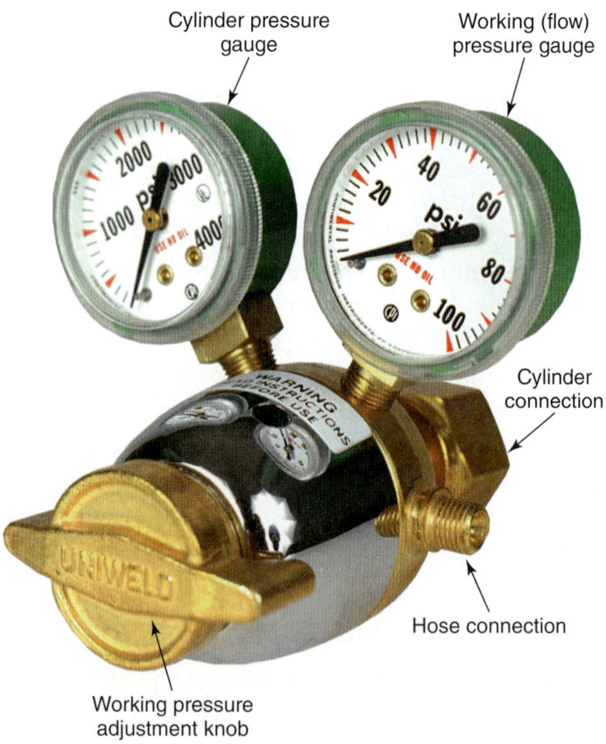

Uniweld

Figure 9-11. An oxygen regulator reduces the high-pressure flow of gas from the cylinder to a much lower-pressure flow through the hose and torch. It has separate gauges for the cylinder pressure and the working (flow) pressure. Adjustment of the working pressure is done with a knob on the front of the regulator.

An oxygen regulator is attached directly to the oxygen cylinder valve and contains an outlet for connecting the hose leading to the torch handle. Most regulators have two gauges. The high-pressure gauge indicates the pressure in the cylinder. The lower pressure gauge indicates the work (flow) pressure being supplied to the torch handle.

A pressure-adjusting screw on the front of the regulator sets flow (working) pressure. When this screw is turned completely to the left (counterclockwise), the valve is closed and no gas can pass through the regulator. As the screw is turned to the right (clockwise), the regulator opens and permits gas to flow to the torch handle. Gas flow is immediately indicated on the flow gauge. Working pressure is changed by turning the adjusting screw until the desired flow is registered on the gauge.

PRO TIP

Regulator Controls

Remember that gas regulators work the opposite of most manual valves. Clockwise (right) opens a regulator valve. Counterclockwise (left) closes it.

Adjusting Flow Pressure

Pressure cannot be adjusted properly unless the gas is actually flowing. The torch handle valve must be open to permit the gas to be released into the atmosphere. Adjust the flow pressure as if the torch were lit and operating. After the pressure is adjusted, turn off the torch handle valve to stop the flow.

There must be separate regulators for oxygen and acetylene. The two regulators are *not* interchangeable. The regulators and gauges are color-coded: green for oxygen and red for acetylene. They also have special cylinder connections to prevent an acetylene regulator from being connected to an oxygen cylinder or vice versa. Always return both regulator screws to the *off* ("backed out" or counterclockwise) position when shutting down the outfit.

Regulator Safety

The following procedures for regulators should be followed each time the equipment is set up and used:

- Use a regulator only with the gas for which it is intended. Oxygen regulators must be used only for oxygen service.
- Keep regulators and cylinder connections free of dirt, dust, grease, and oil. Do not use a regulator if it is damaged or has oil or grease on it.
- Do not *crack* (slightly open) fuel gas or oxygen cylinders near a flame or any source of ignition. Make sure you are in a well-ventilated area and stand clear of the valve outlet.
- Make sure the threads engage properly when connecting the regulator to the cylinder valve. Tighten it with a wrench, but do not use excessive force.
- To shut off the regulator, turn the pressure-adjusting screw counterclockwise until tension is fully released. The regulator should always be shut off when not in use to avoid gas loss if the cylinder or torch valves leak.
- Never stand directly in front of or behind regulators when opening a cylinder valve. Open the valve slowly. Open it fully only after the high-pressure gauge indicator stops moving. Open acetylene valves one-quarter (1/4) to one-half (1/2) turn. Open oxygen valves fully to seal the double-seated valve.

Oxyacetylene Hoses

Oxyacetylene hoses combine specially reinforced rubber and fabric layers to make them strong and flexible. Only hoses specifically made for welding and cutting should be used. Two lengths of hose, one for oxygen and one for acetylene, connect the gas regulators to the torch handle. To prevent confusion, the hoses are color-coded: green for oxygen and red for acetylene. To further guard against interchanging the hoses, the oxygen fittings (both male and female) have right-hand threads, while the acetylene fittings (male and female) have left-hand threads and a notch cut into the female connector. It is not possible to screw a left-hand nut onto a right-hand fitting.

Twin hoses are designed for easy handling, **Figure 9-12**. They are furnished in lengths of 12.5″, 25″, or 50″. The most popular is the 12.5″ length with an inside diameter of 3/16″.

Torch Handle

Nuts on the oxyacetylene hoses screw onto matching threaded fittings on the base of the

Uniweld

Figure 9-12. Twin color-coded hoses (red for acetylene, green for oxygen) are convenient and less likely to tangle than individual hoses.

torch handle, **Figure 9-13**. These hose connections should be firmly tightened with a wrench. The torch handle also contains two finger-operated valves to adjust the flow of oxygen and of acetylene to the gas mixer in the welding tip or cutting attachment.

Oxyacetylene Tips

Various welding tips are available to fit the torch handle, **Figure 9-14**. Tips are selected according to the type or size of flame desired. Tip sizes range from #000 (smallest) to #5 (largest). Sizes #0, #2, and #4 are commonly used, with #4 being the most preferred. The tip is screwed onto the torch handle only by hand, using firm pressure. Do not use a wrench. The tip is equipped with special rubber O-rings to provide a leakproof seal when attached to the torch handle. Overtightening this connection could damage the O-rings.

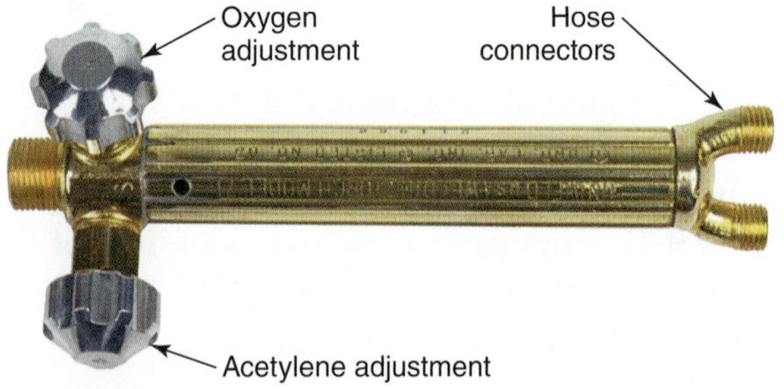

Uniweld

Figure 9-13. The torch handle has separate adjustment knobs for acetylene and oxygen flow. To prevent mismatching of hoses, oxygen connectors have a right-hand thread, and acetylene connectors have a left-hand thread.

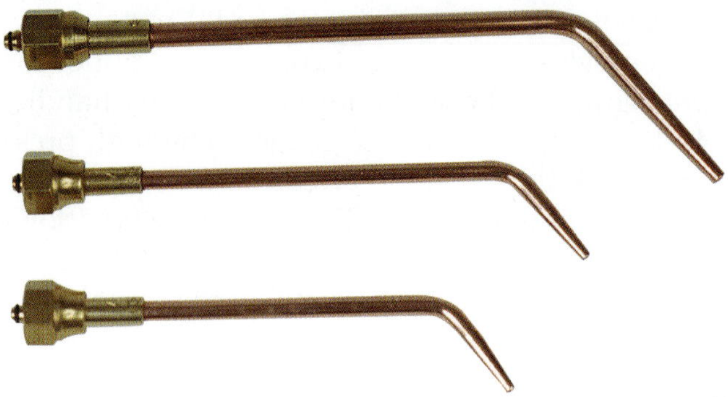

Uniweld

Figure 9-14. Welding tips come in different sizes. The tip is hand-screwed onto the torch handle.

Welding tips produce a small, concentrated, intensely hot flame. These tips are used to fuse steel or, by using a filler welding rod, to gas weld. Welding tips are frequently used for brazing copper to copper or copper to steel tubing, but care must be exercised to prevent overheating the metal.

Copper melts at 1981°F (1083°C), so it is easy to melt copper tubing. The flame must be moved, or *backed away*, from the metal to prevent overheating and making a hole in the tubing. The hottest part of the flame is at the tip of the inner cone, as shown in **Figure 9-15**. The temperature of the metal can be controlled by rotating the flame or by pulling the inner cone away from the workpiece.

The proper tip size depends entirely upon the flame size desired. Many factors are involved in determining the proper tip size for a given application. Factors include the type of metal, thickness of the metal, operation (brazing or welding), and skill of the operator. Trial and error, or some practice with the different tips, will reveal the tip size that produces the desired flame. In general, thick metals or large copper tubing require large tips; thin metals or small copper tubing require small tips. The smallest tips are suitable for delicate work and are used by jewelers.

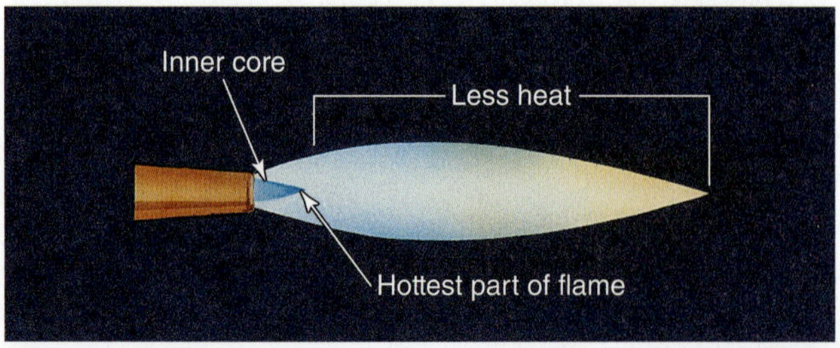

Lucas-Milhaupt

Figure 9-15. The hottest part of a neutral oxyacetylene flame is just ahead of the inner cone. Heating of the metal can be controlled by moving the inner cone closer or farther away.

Multiflame Tips

Many technicians prefer to braze with a ***multiflame tip***, sometimes called a *rosebud tip*, which produces several small flames in one tip, **Figure 9-16**. The flames tend to wrap around the tubing. This feature is useful because the flames heat the metal to the desired temperature quickly without overheating it.

The flames produced by such a tip are not large or long but do provide intense heat to the workpiece. Such flames permit a skilled operator to rapidly braze on any tubing size. The multiflame tip is the only tip used for *all* types of brazing jobs. The small multiflame tip (#15) will successfully braze 1/8″ to 3″ (3 mm to 76 mm) copper tubing with relative ease. The same tip will braze larger tubing, but heating will take longer. The next tip size (#30) is recommended for best results in such cases.

Multiflame tips and single tips have some similarities and differences. The amount of gas required to operate the multiflame tip is slightly greater than the amount for a single tip, but the difference is not significant. The flow pressures required for the multiflame tip are the same as those for a single tip. The holes in the multiflame tip are small. In fact, it requires several of these small flames to equal the diameter of the single tip flame. However, the multiflame tip can wrap around the tubing and distribute heat more evenly to the workpiece. This important feature speeds the brazing process and helps prevent hot spots.

Multiflame tip

Uniweld

Multiflame tip application

Goodheart-Willcox Publisher

Figure 9-16. The multiflame tip helps complete brazing jobs quickly since the flame tends to wrap around the fitting for rapid, even heating.

Specialty Tips

Several specialty torch tips are available for specific purposes, **Figure 9-17**. A pincer tip directs two flames on opposite sides of a joint. A hook tip provides equally distributed heat around a joint using multiple flames. Hook tips are available in different sizes and with different numbers of orifices for more or fewer flames. A reversing valve torch tip is shaped like a two-pronged fork. It has six orifices that direct a flame on each side of the three tubing connections that are side by side on a reversing valve. These may be called *reversing valve tuning forks*, since they resemble musical tuning forks.

Backfire and Flashback

Brazing tips are designed to operate at the pressure recommended by the manufacturer. Most torch systems are set up to operate with acetylene pressure of 5–7 psi (34–48 kPa) and an oxygen pressure of 10–12 psi (69.0–82.7 kPa). Then the handle valve can be adjusted to conform to the desired type of flame. If the flame produced by this procedure is undesirable, the unit should be shut down and the tip size changed. Do *not* change flow pressures at the regulator.

Improper operation of the torch tip may cause the flame to extinguish with a loud cracking sound. This is called a ***backfire***. A backfire may be caused by touching the brazing tip against the workpiece, but the most common cause is flow pressures that are too low.

Flashback is a condition in which the flame burns back inside the tip. Sometimes the flame may extend back through the hoses and to the regulators. This condition is revealed by a shrill hissing or squealing sound. When flashback occurs, the unit should immediately be shut down and allowed to cool before relighting. Flashback is an indication that something is *very wrong*. This condition can

Two-flame pincer tip

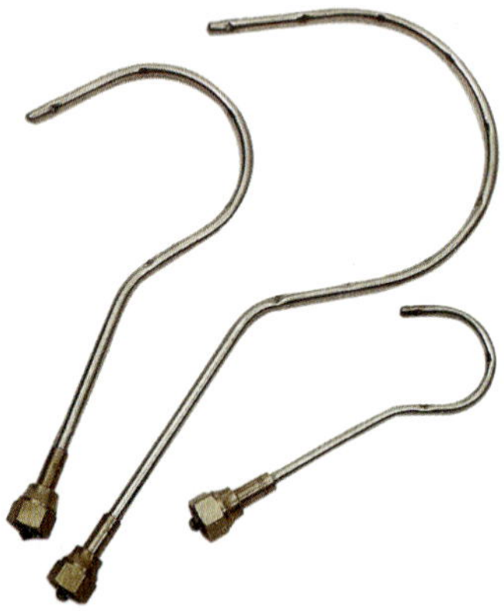

Multiflame hook tips

Reversing valve (tuning fork) tip

Uniweld

Figure 9-17. Three types of specialty brazing tips.

be caused by a clogged orifice, but it is more commonly caused by incorrect oxygen and acetylene pressures.

Backfire and flashback are conditions usually brought about by the operator attempting to soften the flame by choking the flow pressures, either at the handle valves or the regulators. A skilled operator *can* produce a softer flame by adjusting the handle valves of an oxyacetylene setup, but this procedure has its limits. Such an operator can produce a flame that is suitable for soft soldering, but a less-skilled person should resort to an *air-acetylene* torch for a softer flame.

The Oxyacetylene Flame

There are three types of oxyacetylene flames, **Figure 9-18**. The different flames are obtained by controlling the amount of oxygen and acetylene supplied to the torch tip:

- A ***neutral flame*** results from supplying equal amounts of oxygen and acetylene.
- An ***oxidizing flame*** results from supplying an excess amount of oxygen.
- A ***carburizing flame*** (also called a *reducing* flame) results from supplying excess acetylene.

An oxidizing flame (excess oxygen) will burn the base metal and produce oxides that interfere with the flow of alloy. A carburizing flame (excess acetylene) deposits carbon on the base metals, interfering with the brazing process. The *neutral* flame, which heats the base metal and neither carburizes nor oxidizes, is the correct flame to use when brazing.

A neutral flame results from using an equal mixture of oxygen and acetylene (1:1). In this union of two gases, the oxygen burns up the carbon and hydrogen in the acetylene so the flame releases only heat and harmless gases.

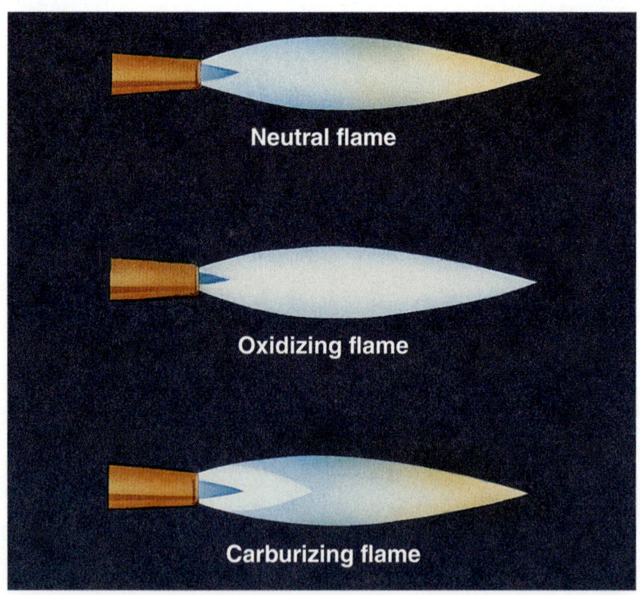

Lucas-Milhaupt

Figure 9-18. The three types of oxyacetylene flame.

The *neutral* flame has two distinct sections:

- A slightly rounded, blue-white inner cone
- An outer flame (The acetylene feather has disappeared into the inner cone.)

The *oxidizing* flame has two distinct sections:

- A sharp, pale-blue inner cone
- A shorter, more ragged outer flame than shown by a neutral flame

The *carburizing* flame has three distinct sections:

- A blue inner cone
- A yellow-white "acetylene feather"
- An outer flame

The difference between the neutral and oxidizing flames is slight. Not much visual distinction exists between the inner cone of each flame.

PRO TIP

Brazing Flames

Care should be taken to remove *only* the acetylene feather when adjusting for a neutral flame. It is better to err on the acetylene side (a slightly carburizing or reducing flame) than on the oxygen side (an oxidizing flame).

Although the neutral flame is said to result from supplying equal amounts of the two gases, a *perfectly neutral flame* requires 2 1/2 parts oxygen to 1 part acetylene. The regulators are adjusted to provide the 1:1 ratio in the inner cone of the flame, and *atmospheric air* provides the other 1 1/2 parts oxygen needed to complete the combustion process.

Preparing to Use an Oxyacetylene Torch

The procedure for getting started with an oxyacetylene torch consists of opening the cylinder valves and adjusting the pressure regulators for the tip size being used. The specific steps in this procedure are to:

1. Inspect the equipment for any damage. Make certain all valves and regulators are turned off.
2. Open the acetylene cylinder valve slowly, turning it *counterclockwise* with the proper wrench. Open the valve only one-quarter to one-half turn. Leave the wrench on the valve stem so the cylinder can be turned off quickly in case of an emergency.
3. Open the acetylene valve on the torch handle one full turn. Slowly rotate the adjusting screw on the acetylene regulator *clockwise* until the flow pressure gauge indicates 5 psi (34 kPa) of pressure. Close the acetylene torch handle valve, using only finger-tight pressure.
4. Open the oxygen cylinder valve slowly, turning it *counterclockwise* until the regulator's cylinder (high-pressure) gauge reaches its maximum setting. When opening the oxygen cylinder valve, always stand to one side of the regulator in case of a malfunction. The cylinder valve must be opened slowly to avoid damage to the regulator in case of a sudden surge of about 2200 psi (15,168 kPa) of pressure. After the high-pressure gauge reaches its maximum reading, continue to turn the oxygen cylinder valve until it is fully open. This will backseat the valve and prevent leakage around the stem.
5. Open the oxygen valve on the torch handle one full turn. Next, turn the oxygen regulator's adjusting screw *clockwise* until the flow pressure gauge indicates a pressure of 10 psi (69 kPa). Close the torch handle valve, using only finger-tight pressure (too much pressure on these sensitive valves can damage the ball seat).

PRO TIP

Pressure Regulator Settings

Always check manufacturer recommendations for the correct pressure regulator settings based on the specific torch tip being used.

Lighting the Torch

These are the proper steps for lighting an oxyacetylene torch and adjusting it to a neutral flame:

1. Once proper flow pressures have been established, crack (open slightly) the acetylene torch handle valve. Use a flint striker (also called a *spark lighter*) to ignite the acetylene gas coming out of the torch tip. Hold the flint striker near the end of the tip, but do not cover the end. Note that a flint striker is the only safe device to use when lighting a torch. Never use matches or a cigarette lighter.

SAFETY NOTE

Tip Direction When Lighting a Torch

When lighting a torch, always be sure the torch tip is pointed away from any ignition source, object, or person that could be harmed by the flame.

2. Slowly open the acetylene handle valve. When sufficient acetylene is flowing, the flame will stop releasing soot.
3. Slowly open the oxygen valve on the torch handle. As the oxygen is fed into the flame, the inner cone will develop, and the acetylene feather will appear. As the amount of oxygen is increased, the acetylene feather will draw back into the inner cone. When the acetylene feather disappears, the inner cone will lose its blurred edge and become round and smooth; this is a *neutral flame*. If more oxygen is added, a sharp pale blue inner cone will appear, indicating an *oxidizing flame*. A neutral or slight carburizing flame is preferred for brazing.

Shutting Down the Torch

If the torch is to be shut down for a short time, simply cut off the gas flow by closing the torch handle valves. Then, lay the torch aside (on a nonflammable surface) until it is needed again.

If the unit is not going to be used for a long time, the system should be shut down completely following these steps:

1. Close the hand valves on the torch handle in the proper order. Turn off the oxygen first to avoid a backfire into the mixer. Then, quickly shut off the acetylene valve to avoid forming soot. Note that if the acetylene is shut off first, the flame can burn back to the oxygen supply. Acetylene, however, cannot burn back without oxygen. Shutting off the oxygen first will also prevent formation of soot inside the mixer and torch that might plug the passages.
2. Tightly close both cylinder valves.
3. Reopen each valve on the torch handle separately to bleed the gases from the hoses and regulators. Continue until each regulator's working or flow pressure gauge reads zero.
4. Close the adjusting screw on each regulator by turning it fully *counterclockwise*. This will help avoid gas loss if cylinder or torch valves leak.
5. Close both valves on the torch handle. Do not overtighten; they should be only finger-tight.
6. Rewind the hose and place the unit in safe storage.

9.3 Making Brazed Connections

Like soldering, brazing is a four-step process:

1. Thoroughly clean the base metal pieces.
2. Apply flux when appropriate.
3. Heat the base metal pieces evenly to the melting point of the filler metal alloy.
4. Apply filler metal alloy to the joint.

Cleaning the Base Metal

Dirty metal surfaces will prevent the alloy from traveling into the joint and penetrating the metal. Brazed connections depend upon capillary action to draw the alloy into the joint. Dirt on metal surfaces prevents capillary action and keeps the alloy from flowing or adhering properly. It is much like applying paint to a dirty wall. The paint does not flow properly, and when the dirt falls off, so does the paint.

The cleaning technique for brazing is the same as for soldering. Use sandpaper, emery cloth, a wire brush, or other abrasive to remove dirt and oxidation. The cleaned metal surface should shine.

Using Flux

In brazing, as in soldering, flux is a chemical compound applied to joint surfaces to prevent oxidation of the metal surfaces as they are heated. Flux is normally supplied in paste form and is applied with a small brush just before brazing. It melts and becomes active during the heating process, absorbing oxide and floating it away from the flowing alloy.

Flux is seldom used in refrigeration and air-conditioning system brazing that joins *copper-to-copper.* The flux is eliminated by using a filler metal alloy that contains phosphorus, which acts as a fluxing agent. However, when brazing dissimilar metals, such as copper to brass or copper to steel, use of the phosphorus-bearing alloy will result in a brittle joint. For this reason, a different alloy is used, and a flux *is* required.

CAUTION

Brazing Fluxes

Brazing fluxes cause metal salts to form. Methyl alcohol (used as an antifreeze) can react with aluminum and cause corrosion. Moreover, poor brazing practice can introduce flux (acid) into the system.

CAUTION

Thick Material Brazing

When heating assemblies for brazing, extra heat may be needed for sections that have heavier mass or greater thickness. The heavy section will heat more slowly. Also, different metals conduct heat at different rates. Copper, for example, is a good thermal conductor; it carries heat away more rapidly than steel. Never heat base metals to the point where they begin to melt!

Flowing Nitrogen When Brazing

After brazing a copper joint and allowing it to cool down, small black flakes can be rubbed off the copper pipe. This is oxidation. If there is oxidation on the outside of the joint, it will also be on the inside. Every brazed joint can develop these oxidation flakes that could plug components in the system. A liquid line filter-drier will catch some of these flakes, but it may not catch all of them.

When installing an air-conditioning or refrigeration system, the goal should be to have a system

PROCEDURE

Brazing a Metal Joint

Even heating of the base metal pieces is important so the alloy will flow equally well on both metal surfaces and completely fill the joint. Follow these steps for even heating during the brazing process:

1. Adjust the torch for a neutral flame. The flame size should be large enough to envelop as much of the connection as possible.
2. Begin heating the tubing about 1/2″ to 1″ (1.25 cm to 2.5 cm) away from the fitting.
3. After the tubing is heated, shift the flame to the fitting. Once the fitting is heated, move the flame steadily back and forth from the tubing to the fitting. Do not hold the flame in one spot, as this causes localized overheating. Continue heating until the assembly reaches the alloy melting temperature.
4. When the assembly reaches the proper temperature, pull the flame back a little. Apply the filler alloy firmly against the tubing at the connection. If the assembly has been properly heated, the alloy will melt and completely penetrate and fill the joint. The alloy will always flow toward the hottest area.
5. After the joint has been completed, make one final pass of the flame around the connection to ensure proper flow and penetration of the alloy.

that is tight, dry and *clean* on the inside. To help eliminate oxidation flakes from accumulating, nitrogen needs to flow through the tubing as it is brazed. Nitrogen is an ***inert*** (chemically inactive) gas and is nonflammable. These qualities allow nitrogen flowing through the tubing to displace the oxygen and any other gases in the tubing so that oxidation cannot form. Only a small amount needs to flow through the tubing as it is brazed. If pressure is too high, problems will develop while brazing the joints. Acceptable flow pressure ranges from 3 psi to 5 psi, **Figure 9-19**.

The nitrogen is vented (permitted to escape) after passing the brazing area. Venting eliminates the pressure buildup that would otherwise result from the heat of the brazing torch acting on the confined gas. An escape or vent *must* be provided. If pressurized gas is allowed to build up in the tubing, it will blow melted brazing alloy out of the connection.

CAUTION

Nitrogen Gas Pressure

Nitrogen is a compressed gas supplied in cylinders under pressures as high as 2350 psig (16 215 kPa). Since these pressures are much higher than the 2–3 psig (14–21 kPa) needed, a pressure-regulating valve must always be used on nitrogen gas cylinders, **Figure 9-20**. A refrigeration hose (or manifold gauge) is used to transfer the nitrogen gas from the outlet of the flow pressure regulator to the system tubing.

SAFETY NOTE

Pressurizing a System

Never pressurize a system with oxygen. Oxygen will react violently with oil in the compressor crankcase and cause an explosion.

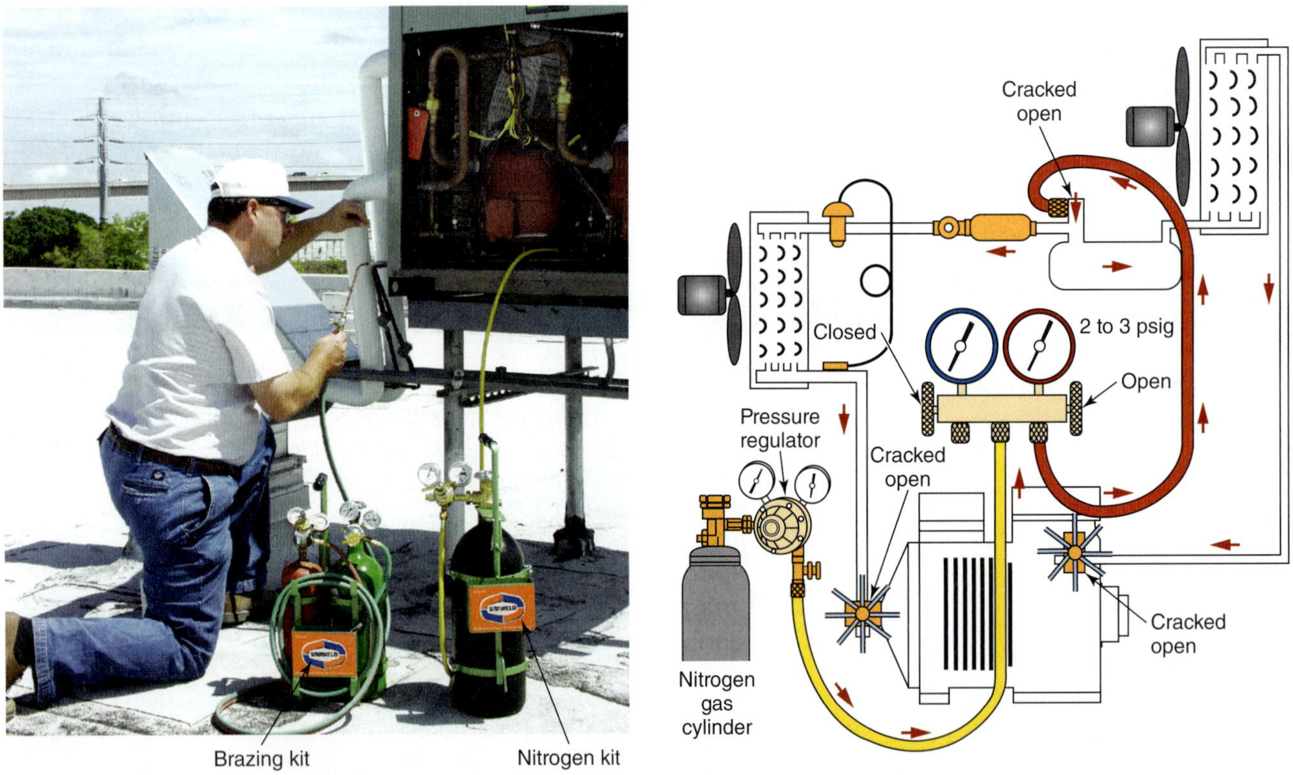

Uniweld *Goodheart-Willcox Publisher*

Figure 9-19. This technician is brazing a joint with low-pressure nitrogen flowing through the tubing to prevent oxidation from occurring within the tubing. As shown in the system diagram, the suction and discharge service valves can be cracked open to vent the gas and prevent pressure buildup inside the system.

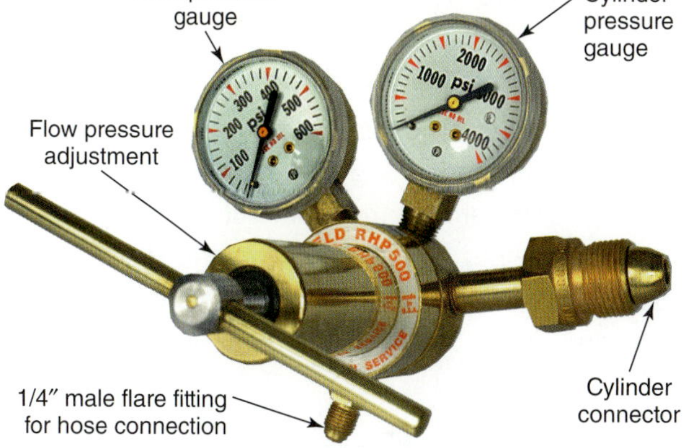

Uniweld

Figure 9-20. A medium-duty nitrogen regulator capable of regulating output gas flow pressures from 0 psig to 600 psig. It can be fitted on cylinders filled with gases up to 4000 psig.

Types of Joints

There are three general types of brazing connections, identified by the direction the filler alloy flows: downhill joints, uphill joints, and horizontal joints. Recommended procedures for making each of these connections follow.

Downhill Joint

In a ***downhill joint*** (also known as a *vertical down joint*), the fitting is below the point where the alloy is applied, **Figure 9-21**. Bring the entire joint area to brazing temperature quickly and evenly. Heat the tubing first and then the fitting. When brazing temperature is reached, apply a little extra heat to the fitting as the alloy is melting. Since alloy always flows toward the heat, this will help it to penetrate the fitting.

Goodheart-Willcox Publisher

Figure 9-21. In a downhill joint, the alloy is applied above the fitting and is drawn downward into the joint.

Uphill Joint

In an ***uphill joint*** (also known as a *vertical up joint*), the fitting is above the point where the alloy is applied, **Figure 9-22**. Start by heating the tubing, then transfer the heat to the fitting. Sweep the flame back and forth from the fitting to the tubing, all around the joint area. When brazing temperature is reached, keep the flame on the fitting while applying alloy to the connection. This heating pattern will draw the alloy upward into the joint. Do not overheat the tubing since this will cause alloy to run downward, rather than be drawn upward into the joint.

Horizontal Joint

In a ***horizontal joint***, the tubing and the fitting are on the same level. Heat both the tubing and the fitting quickly and evenly, beginning with the tubing. When brazing temperature is reached, apply the alloy first to the bottom of the joint, as shown in **Figure 9-23**. Then, add alloy along the

Goodheart-Willcox Publisher

Figure 9-22. Applying heat to the fitting after the entire joint area has been heated will help draw the alloy upward into the uphill joint.

Goodheart-Willcox Publisher

Figure 9-23. On a horizontal joint, alloy is applied at the bottom of the joint. This is so the joint will be filled by capillary action and not rely on gravity, which may not fill the gap properly.

two sides of the joint. Lastly, add alloy at the top of the joint. With proper heating, capillary action will draw the alloy into the fitting and completely around the tubing.

CAUTION

Capillary Action vs. Gravity

Do *not* rely solely on gravity to fill the seam of a horizontal joint. Starting at the top could result in alloy flowing around the seam but not filling the joint entirely. Begin applying filler metal at the bottom and work upward to ensure that the seam is filled and sealed completely.

Troubleshooting Brazed Connections

Brazing is a skill that requires practice to develop and maintain. Sometimes, the process fails to produce satisfactory results. Refer to the following brazing problems and related causes when troubleshooting brazing issues.

Problem: The alloy melts and forms a bead but does not flow into the joint.
Causes:

- The tubing is hot, but the inside is not up to brazing temperature. Review the heating procedure. Alloy always travels to the heat.
- Flux was destroyed by excessive heat. Use less heat or apply a heavier coating of flux.

Problem: The alloy balls up and does not flow into the joint.
Causes:

- The base metal is not up to brazing temperature, and the alloy was melted by the flame, not the heat of the base metals.
- The joint was overheated and destroyed the flux.
- Base metal pieces were not cleaned.

Problem: The alloy flows away from, rather than into, the joint.
Causes:

- The fitting is not heated to brazing temperature.
- The flame is directed away from the fitting.

Overcoming Heat Sink Problems

The oxyacetylene process is especially useful when brazing tubing connections to a valve, compressor, or other heavy metal object. Typically, during brazing, a heavy metal object will act as a heat sink, absorbing applied heat and drawing it away from the brazing area. The heat loss prevents the connection from achieving brazing temperature quickly, as desired. Worse, the excessive heat may damage the object that acts as a heat sink.

The oxyacetylene torch makes it possible to overcome heat sink problems. The intense flame heats the brazing area rapidly, allowing the operation to be completed quickly, before the heat sink can absorb much heat.

When the brazing operation involves a heat sink that could be damaged, precautions must be taken. Common practice is to wrap the heat sink with damp rags, **Figure 9-24**. Commercial paste products are also available to apply to the heat sink before brazing. Both methods will adequately protect the heat sink from damage if brazing is not prolonged.

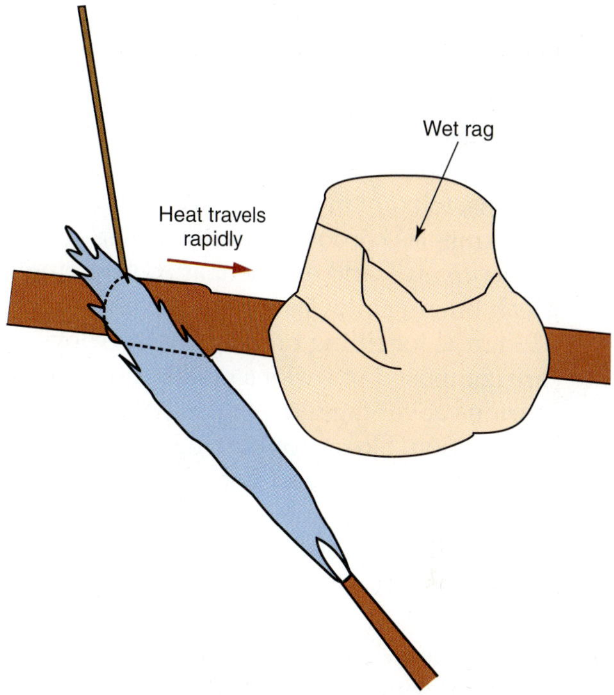

Goodheart-Willcox Publisher

Figure 9-24. If a valve or other component that acts as a heat sink might be damaged by the heat of a brazing operation, precautions must be taken. Wrapping the heat sink in wet rags is one method for keeping the temperature down.

Chapter Review

Summary

9.1 Brazing permits the joining of similar and dissimilar metals and creates joints that are stronger than the base metals it joins, tolerant of vibration without cracking, leakproof under high pressures, noncorrosive, and excellent at transferring heat. Brazing alloys are filler metals designed for filling brazing joints. Alloys are composed of various metals, offering various advantages depending on the application. Fluxes for brazing are mixtures of boric acid, borates, fluorides, and fluoborates. A technician can determine the temperature at different stages of the brazing process by watching the behavior of the flux.

9.2 The oxyacetylene torch uses oxygen and acetylene to produce a heat range that handles all types of joining work, regardless of the materials. It is commonly used to solder, braze, weld, or cut metals. An oxyacetylene torch flame can reach extremely high temperatures, and the process can produce flying sparks, molten metal, slag, fumes, and intense light rays. The technician must take a "head-to-toe" approach to the protective clothing worn when brazing.

9.2 Oxygen and acetylene cylinders are sized according to the amount of gas in cubic feet the cylinder will hold. Torch tips are selected according to the type or size of flame desired. Multiflame tips are preferred due to their ability to heat metal quickly without overheating it.

9.2 A backfire is when a flame is extinguished with a loud cracking sound. Flashback is a condition in which the flame burns back inside the tip. Both indicate improper operation and should be corrected. A neutral flame results from supplying equal amounts of oxygen and acetylene. An oxidizing flame results from supplying an excess amount of oxygen. A carburizing flame results from supplying excess acetylene.

9.2 Before starting an oxyacetylene torch, cylinder valves and pressure regulators must be properly opened and adjusted for the tip size being used. Then, the torch is lit by operating the acetylene and oxygen valves to achieve a neutral flame. The torch is shut down by closing the valves in the proper order.

9.3 Like soldering, brazing is a four-step process: First, thoroughly clean the base metal pieces. Second, apply flux when appropriate. Third, heat the base metal pieces evenly to the melting point of the filler metal alloy. Last, apply filler metal alloy to the joint.

9.3 There are three general types of brazing connections, identified by the direction the filler alloy flows: downhill joints, uphill joints, and horizontal joints. Paying close attention to the behavior of the filler alloy or any part of the torch operation process will point to any brazing issue encountered.

Know and Understand

1. *True or False?* Brazing allows the joining of similar and dissimilar metals. (9.1)
2. All brazing alloys usually contain a certain amount of _____. (9.1)
   A. aluminum
   B. silver
   C. copper
   D. zinc
3. The copper-phosphorus brazing alloy (AWS BCuP-2) is used for connecting copper to _____ without the need for a flux. (9.1)
   A. steel
   B. copper
   C. brass
   D. bronze

4. Despite their ability to form excellent brazed joints, _____ alloys should be avoided due to the toxic fumes they emit when brazed. (9.1)
   A. cadmium-bearing
   B. AWS BCuP-2
   C. AWS BCuP-4
   D. AWS BCuP-5
5. When brazing temperatures reach about 1100°F (593°C), flux will _____. (9.1)
   A. dry out as the water boils off
   B. start to bubble and turn milky white
   C. turn into a clear liquid
   D. evaporate slowly
6. *True or False?* An oxyacetylene torch can solder, braze, and weld metals but cannot cut through metals. (9.2)
7. An oxyacetylene torch flame can reach temperatures as high as _____°F. (9.2)
   A. 600
   B. 3000
   C. 6000
   D. 6300
8. Oxygen and acetylene cylinders are sized according to the _____. (9.2)
   A. size of the torch tip used
   B. amount of heat (in Btu) required for brazing
   C. amount of gas in cubic feet the cylinder will hold
   D. amount of pressure required
9. Generally, the proper torch tip size depends entirely upon the _____. (9.2)
   A. type of fuel gas used
   B. amount of cylinder pressure
   C. flame size desired
   D. type of joint desired
10. The _____ tip is the only tip used for all types of brazing jobs. (9.2)
    A. #0
    B. #2
    C. multiflame
    D. reversing valve torch
11. The most common cause of a(n) _____ is using flow pressures that are too low. (9.2)
    A. flashback
    B. backfire
    C. excess amount of slag
    D. excess amount of flying sparks
12. The type of flame that should normally be used when brazing with an oxyacetylene torch is the _____ flame. (9.2)
    A. carburizing
    B. oxidizing
    C. neutral
    D. reducing
13. *True or False?* The acetylene cylinder valve is opened by turning it clockwise. (9.2)
14. When shutting down an oxyacetylene torch, the _____ valve should be turned off first. (9.2)
    A. acetylene torch handle
    B. oxygen torch handle
    C. acetylene cylinder
    D. oxygen cylinder
15. Flux is seldom used in refrigeration and air-conditioning system brazing that joins _____. (9.3)
    A. copper to steel
    B. copper to tin
    C. copper to copper
    D. copper to aluminum
16. In a(n) _____ joint, the fitting is above the point where the alloy is applied. (9.3)
    A. downhill
    B. uphill
    C. horizontal
    D. vertical down
17. A common practice used to prevent heat loss when brazing around a potential heat sink is to _____. (9.3)
    A. apply less heat to the joint
    B. use a smaller torch tip
    C. applying additional flux to the joint
    D. wrap the heat sink with damp rags

Apply and Analyze

1. How can a technician use flux to determine the temperature at different stages of the brazing process? (9.1)
2. How can backfire and flashback conditions be prevented when brazing? (9.2)
3. What visual cues indicate a neutral flame is achieved with an oxyacetylene torch? (9.2)
4. What are the flow pressure regulator settings for acetylene and oxygen when using an oxyacetylene torch for brazing? (9.2)

5. Why is flux seldom used in certain refrigeration and air-conditioning system brazing applications? (9.3)
6. What may be the cause(s) of a brazing alloy that balls up and does not flow into a joint? (9.3)

Critical Thinking

1. Give three reasons brazing is the preferred procedure for joining tubing on refrigeration and air-conditioning systems. (9.1)
2. Explain why you should stop using an oxygen cylinder when pressure reaches a level between 25 psig and 50 psig. (9.2)
3. When brazing, you notice that the filler alloy is flowing away from, rather than into, the joint. How might you solve this issue? (9.3)

sturti via Getty Images

10 Mathematics for Technicians

LEARNING OUTCOMES

After studying this chapter, you will be able to:

10.1 Read the place values of numbers.

10.2 Recognize how to read whole numbers, fractions, and decimals, and how to convert fractions to decimals.

10.3 Recognize US Customary and SI units and make linear, two-, and three-dimensional measurements.

10.4 Convert Fahrenheit temperature readings to Celsius and vice versa.

10.5 Demonstrate how to rearrange formulas for problem-solving.

TECHNICAL TERMS

area of a circle
Celsius
cubic measure
decimal number
decimal point
denominator
diameter
digit
dimensions
exponent
factor
Fahrenheit
formula
fraction
height
improper fraction
linear measure
lowest terms
mixed decimal number
mixed number
number
numerator
place value
proper fraction
quotient
radius
reducing
remainder
rough measurement
square measure
square unit
surface area
volume
whole number

Introduction

Math skills are essential in the refrigeration, heating, and air-conditioning trade. Making accurate measurements, properly sizing parts, installing equipment, and completing work orders and time sheets are daily duties requiring computational skills. Apprentices and advanced technicians alike depend on their math abilities for successful job performance and advancement.

Can you calculate a time sheet to the quarter-hour? How many cfm (cubic feet per minute) of air is the furnace producing? These are two examples of the sort of mathematical operations that a technician might be required to perform on the job. This chapter will help refresh the math skills needed in preparation for a position as an HVACR service technician.

10.1 Reading Place Values

A ***number*** is a figure or word that denotes quantity. The numbers 0 to 9 are called ***digits***. They are used alone, or in combination, to indicate "how many." When digits are used in combination with each other, each one is given a special position called a ***place value***. Although the digits change, the positions never vary. Working with numbers is much easier when the place value is clearly understood.

Place value is shown in **Figure 10-1**. The number shown, 7,654,321 reads, *seven million, six hundred fifty-four thousand, three hundred twenty-one*. The smallest number, positioned farthest to the right, is a single digit from 0 to 9 (in this case, the digit is 1). This position is called *ones*. Other numbers are added to the left of the ones position. The next position has a value of *ten*, and the next to the left of that has a value of *one hundred*. Thus, the number 321 is made up of 3 hundreds (300), 2 tens (20), and a one (1), or 300 + 20 + 1 = 321. To make reading and writing large numbers easier, groups of three digits are separated by commas, starting from the ones' place.

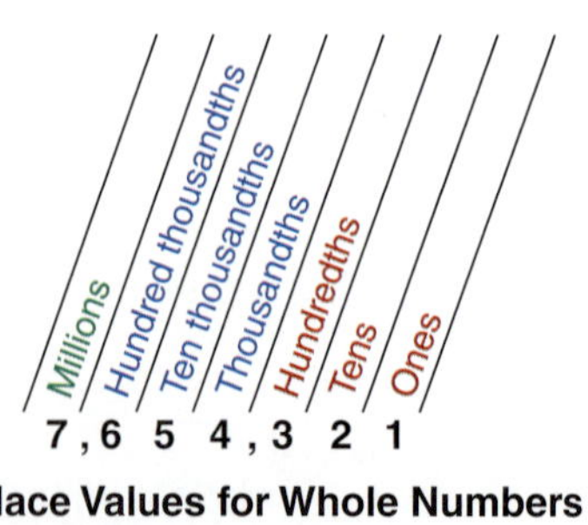

Goodheart-Willcox Publisher

Figure 10-1. Place value is important in reading whole numbers. The positions of the digits, from right to left, determine their place values.

10.2 Whole Numbers, Fractions, and Decimal Numbers

Numbers can take different forms, including the forms of whole numbers, proper fractions, improper fractions, mixed numbers, and decimals, all which will be discussed in this section. A ***whole number*** is any positive number that contains no fractional parts. It is a complete unit. The numbers 0, 1, 2, 3, 4, etc. are whole numbers.

Fractions

Fractions are a means of indicating parts of a whole number. For instance, a 1/4-inch mark on a ruler or tape measure designates a fraction (or part) of a *whole* inch. A payroll time sheet may be rounded off to the nearest 15 minutes or 1/4 of a *whole* hour.

Fractions are either written with one number above another number, or with the two numbers separated by a diagonal line, as shown in **Figure 10-2**. The top number on a fraction, the ***numerator***, indicates the number of parts available. The bottom number, the ***denominator***, indicates the number of parts required to make a whole.

A fraction is read by stating the numerator first, then the denominator. The fraction 5/16 is read, *five-sixteenths*. The numerator **(5)** indicates

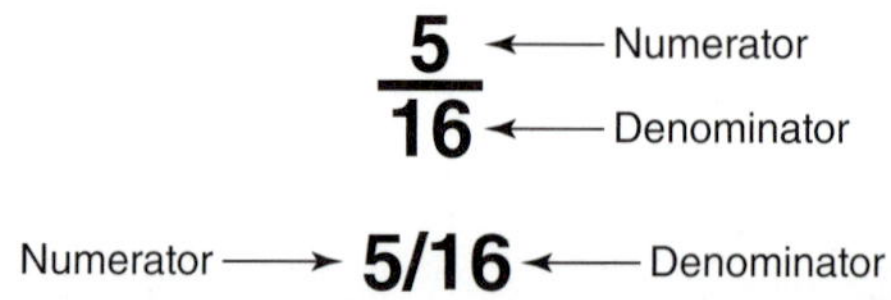

Goodheart-Willcox Publisher

Figure 10-2. Two ways to write a fraction: one number above another or side-by-side separated by a diagonal line. The number on top or to the left is called the numerator. The number on the bottom or to the right is the denominator.

there are 5 parts available; the denominator **(16)** indicates 16 parts are required to make one whole.

Proper Fractions

When the numerator of a fraction is *smaller* than the bottom number, the fraction is termed *proper*. For example, the fractions 1/4, 3/8, 1/2, and 7/16 are examples of ***proper fractions***.

Improper Fractions

In an ***improper fraction***, the numerator is the same or larger than the denominator. Examples are 3/3, 9/8, 7/6, 19/16, 5/4, 8/8. These fractions are called *improper* because the top number is *equal to* or *greater than* the amount required to make a whole. When the numerator and denominator are the same, the fraction is *equivalent* (equal) to the whole number 1. Thus, 4/4, 8/8, and 16/16 are all equivalent to 1. For ease of understanding, improper fractions should be converted to whole or mixed numbers.

Mixed Numbers

In a ***mixed number***, a whole number and a fraction are used together. Examples are 3 1/2, 6 5/8, 4 2/3, and 8 9/16. Mixed numbers are read by inserting the word *and* between the whole number and the proper fraction. Thus, 6 5/8 is read, *six and five-eighths.*

To convert an improper fraction (such as 7/4) to a mixed number, *divide* the numerator by the denominator. Dividing the numerator **(7)** by the denominator **(4)** yields a ***quotient*** (division result) of **1** and a ***remainder*** (the number left over) of **3**. The quotient becomes the whole number, and the remainder becomes the numerator. It is placed above the original denominator. Thus, the improper fraction 7/4 can be converted to the mixed number 1 3/4.

Reducing Fractions

Reducing a fraction to its lowest terms means changing the numerator and the denominator to the smallest possible numbers without changing the *value* of the fraction. It is done by dividing the numerator and denominator by a whole number that divides evenly into both. For example, 5/10 is equal to 1/2 because both numbers can be evenly divided by **5**. The fraction 1/2 cannot be reduced any further; it is in its ***lowest terms***. Reducing fractions to their lowest terms makes them easier to understand and use.

In the same way, 25/100 can be reduced to 5/20 by dividing both numbers by **5**. This fraction could be further reduced to 1/4 by again dividing both numbers by **5**. The fraction 1/4 is the fraction 25/100 reduced to lowest terms.

Figure 10-3 illustrates how fractions can be reduced. Notice that as slices are removed, the fractions of what is missing and what remains can be further reduced to their lowest terms.

Decimal Numbers

Decimal numbers are simply another way of expressing fractions (or parts) of numbers. The number of parts is always located to the right of a decimal point. Our money system makes use of decimals. For example, five dollars and fifty cents is written $5.50. The whole number **5** occupies the ones' position, while **50** represents 50 parts of a whole dollar. Since a dollar can be broken down to 100 parts (cents or pennies), the decimal number 0.50 stands for *half* that amount.

Pizza cut into six pieces (6/6 equals 1)

Pizza with one piece missing (1/6 missing, 5/6 remains)

Pizza with two pieces missing (2/6 missing, 4/6 remains)

Pizza with three pieces missing (3/6 missing, 3/6 or 1/2 remains)

Goodheart-Willcox Publisher

Figure 10-3. Slices of a pizza provide examples of reducing fractions to lowest terms.

Reading Decimal Numbers

The decimal number 0.50 should be read, *fifty one-hundredths* because of place value. The digits appearing to the right of a decimal point are given specific values, just as they are to the left for whole numbers. The place value of the digit farthest to the right determines how many parts are required to make a whole. It is *not* the number in that digit, but the *place value* of the digit that determines what will make a whole.

The number shown in **Figure 10-4** indicates a quantity divided into many parts because the digit farthest to the right **(4)** is in the *ten thousandths* position. This decimal number reads *one thousand two hundred thirty-four ten thousandths*. When reading decimal numbers, the position of the digit farthest to the right indicates how many parts are needed to equal one whole (in this case, 10,000). The actual digits indicate how many parts are available (1234).

Three digits to the right of a decimal indicate that 1000 parts are required to equal one whole. For example, 0.187 is one hundred eighty-seven *one thousandths*. In the same way, 0.75 is seventy-five *one hundredths*, and 0.2 is two *tenths*. Zeros are used as place holders for unoccupied positions: twenty-five *ten thousandths* = 0.0025.

Mixed Decimal Numbers

Decimal numbers with digits before *and* after the decimal point, such as 16.75, are referred to as ***mixed decimal numbers***. The digits to the left of the decimal point give the number of wholes. The digits to the right of the decimal point give the fractional parts. The mixed decimal number 16.75 is read, *sixteen and seventy-five one hundredths*. The word *and* is used to indicate the ***decimal point***. In industry, the word *point* is often used instead and the place value is not read. For example, the number 3.485 is read, *three point four eight five*.

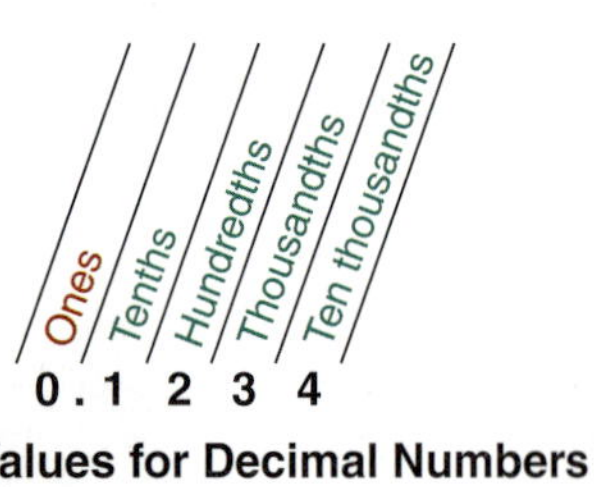

Goodheart-Willcox Publisher

Figure 10-4. Decimal numbers are read from left to right. The position of a digit to the right of the decimal point determines its place value.

Expressing Fractions of Time in Decimals

A practical application of both fractions and decimal numbers is a payroll time sheet for keeping track of hours or fractions of hours worked.

For simplicity, many companies "round off" time to the nearest quarter hour. For example, 10 minutes would be rounded to 1/4 hour (15 minutes). Twenty minutes also would be rounded to 1/4 hour. The fractions 1/4, 1/2, and 3/4 can also be expressed in decimal terms, as 0.25, 0.50, and 0.75. Converting fractions to decimals makes it easier to calculate a worker's pay simply because decimals are easier to multiply than fractions. If a technician's pay rate is $20 per hour, and the technician works 25 hours and 45 minutes, it is much easier to multiply $20 by 25.75 than by 25 45/60 (or, in simplest terms, 25 3/4).

If a person works more than 40 hours per week (or 8 hours per day in some states), they usually qualify for overtime pay for the extra hours. Overtime pay is usually 1.5 (one and a half) times the regular hourly rate. Pay for work on holidays may be double the time or 2 times the regular hourly rate. For example, if a technician's regular pay is $20 per hour, the overtime rate would be $30 per hour, and the double-time rate would be $40 per hour.

The following example shows time worked by a technician earning $20 per hour:

Monday:	9.5 hours
Tuesday:	10.25 hours
Wednesday:	8.0 hours
Thursday:	11.75 hours
Friday:	11.25 hours
Saturday:	9.0 hours
Total hours worked during week:	59.75

40 regular hours at $20 = $800
19.75 overtime hours at $30 = $592.50
Total pay amount for week = $1392.50

10.3 Measuring Dimensions

The application of fractions and decimal numbers to heating and cooling work becomes apparent when the technician must measure dimensions. ***Dimensions*** are measurements of length, width, and depth. Sometimes, the terms *height* or *thickness* are used in place of *depth*. New technicians are often assigned to installation projects where their skills in making accurate measurements are quickly tested.

Measurements are made in either the US Customary or International System of Units (abbreviated SI) systems. The US Customary system uses basic units of inch, foot, and yard dimensions; SI uses the millimeter, centimeter, and meter. Most nations of the world, other than the United States, use the SI system. The units of the two systems are compared as follows.

US Customary		SI
1 inch	=	2.54 centimeters (25.4 millimeters or 0.0254 meter)
1 foot	=	30.48 centimeters (304.8 millimeters or 0.3048 meter)
1 yard	=	91.44 centimeters (914.4 millimeters or 0.9144 meter)

Linear (Length) Measure

A measurement in one dimension is a ***linear measure***, or the distance from one point to another. Length, width, and depth are all linear measures. So are the radius, diameter, and circumference of a circle, and measurements around the perimeter of any other flat shape. A typical use of the length dimension would be measuring a piece of electrical conduit to fit a specific installation. Tools used to measure length include rulers, steel tape measures, yardsticks or metersticks, chains (for land measurement in *rods* or *meters*), and odometers (for measurements in miles or kilometers).

In refrigeration, heating, and air-conditioning work, length is normally measured in compound units of feet, inches, and fractions of an inch (for example, 6 feet, 2 1/4 inches) or in meters, centimeters, or millimeters. In SI, compound units are not used. Instead, dimensions are given in decimal fractions (for example, 2 meters, 30 centimeters is written as 2.3 meters).

Abbreviations for the units of length are commonly used: *ft* for feet, *in* for inches, *m* for meter, *cm* for centimeter, and *mm* for millimeter. On architectural drawings using US Customary measurements, symbols may be used: (′) indicates feet and (″) indicates inches. For example, a drawing might show a length measurement as:

|←—— 4′ 7 3/16″ ——→|

This measurement in compound units reads, *four feet, seven and three-sixteenths inches*. The inches are mixed numbers; therefore, the word *and* is used to separate whole inches from fractional inches.

The equivalent measurement in SI units would be shown as:

|←—— 140.18 cm ——→|

This measurement is read, *one hundred forty point one eight centimeters*.

Using Measuring Tools

Folding rules and measuring tapes are frequently used to obtain accurate measurements. Such measuring devices are illustrated in **Figure 10-5**. The ability to read measuring tools accurately is vital to successful job performance.

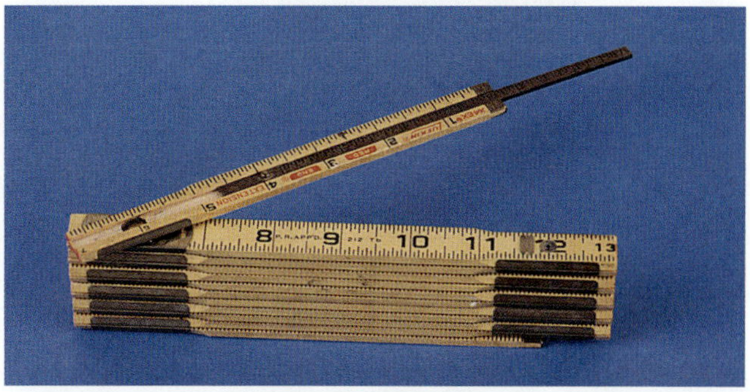

Wood folding rule

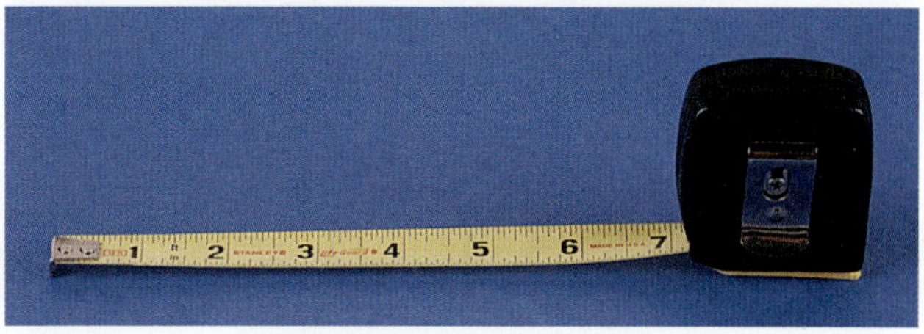

Steel measuring tape

Goodheart-Willcox Publisher

Figure 10-5. Common measuring devices.

Incorrect measurements waste time and materials, resulting in poor craftsmanship and cost overruns.

Folding rules are usually 6′ (approximately 2 m) long. Tape measures are available in a variety of lengths, ranging from 6′ to 100′. Some have both US Customary and SI divisions. The blades of these tape measures are usually steel, **Figure 10-6**. In the 50′ and 100′ (15 m and 30 m) lengths, nonmetallic blades reinforced with fiberglass are available. Also available are tapes equipped with a battery and small lightbulb for reading measurements in dimly lit areas, such as attics and basements.

Reading a US Customary Ruler

Each foot on a standard ruler or tape is calibrated to contain exactly 12 inches, numbered from 1 to 12, as shown in **Figure 10-7**. Measurements in inches can sometimes be expressed as fractions of a foot. A measurement of 6 inches, for example, may be read as 1/2 foot. Since 1 foot contains 12 inches, or 12/12, 6 inches equal 6/12, which can be reduced to its lowest terms, 1/2. In the same way, 4 inches equal 4/12, which reduces to 1/3 foot, and 9 inches equal 9/12, which reduces to 3/4 foot.

Sixteenths

On standard rulers and tapes, an inch is divided into sixteen equal parts, called *sixteenths*. Therefore, 1 inch equals 16/16, as shown in **Figure 10-8**. Because sixteenths are tiny segments, noticeably short lines are used on the ruler to indicate these divisions.

Eighths

Two tiny sixteenth divisions equal 2/16, which reduced to lowest terms is 1/8. As shown in **Figure 10-9**, the mark for *eighths* is slightly longer than the one for sixteenths.

Fourths

Two eighths (or four sixteenths) equal one-fourth of an inch. The longer mark used to divide the inch into *fourths* is shown in **Figure 10-10**.

JB Industries

Figure 10-6. Measuring tapes usually have steel blades for durability, but long tapes may have nonmetallic blades. This tape has a magnetic end hook and measures up to 30′.

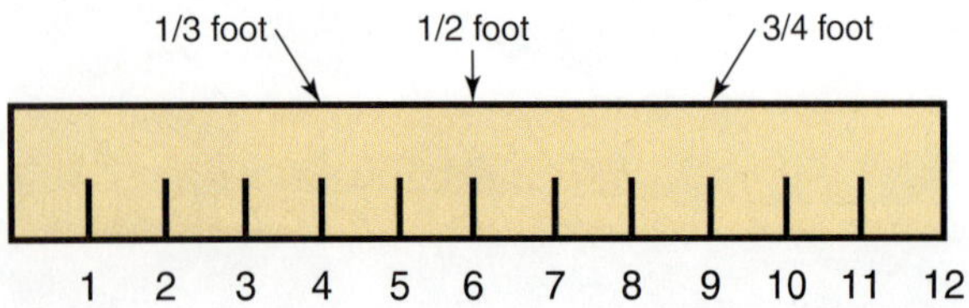

Goodheart-Willcox Publisher

Figure 10-7. In the US Customary system, each foot of a ruler or tape is divided into 12 equal inches. Measurements can also be expressed as fractions of a foot, as shown.

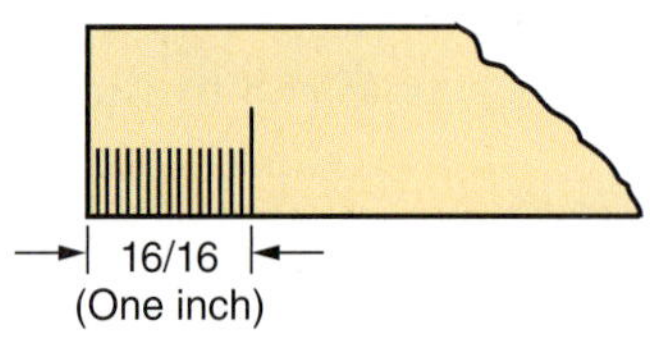

Goodheart-Willcox Publisher

Figure 10-8. Sixteenths are the smallest division of an inch shown on most rulers and measuring tapes.

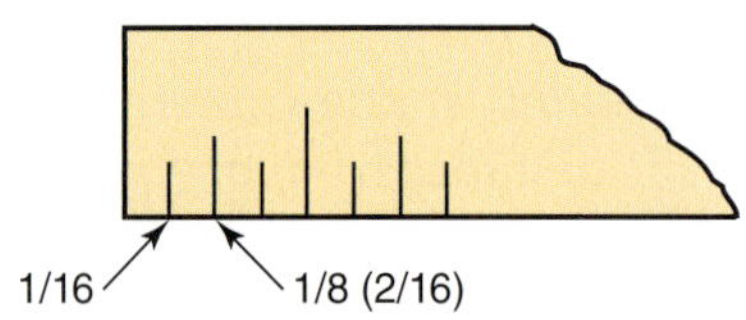

Goodheart-Willcox Publisher

Figure 10-9. One-eighth inch is equivalent to two-sixteenths. Eighth marks on measuring devices are slightly longer or darker than sixteenth marks.

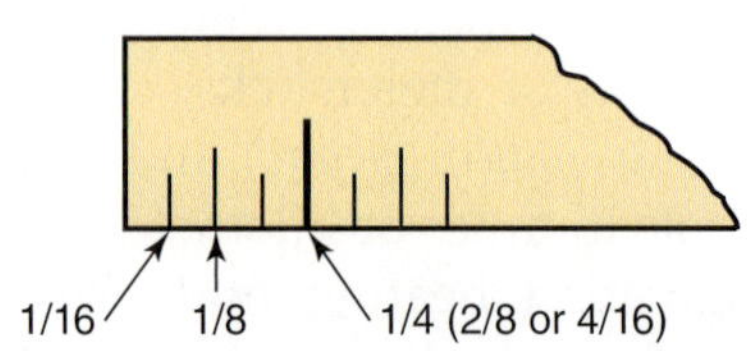

Goodheart-Willcox Publisher

Figure 10-10. One-fourth inch is equivalent to two-eighths or four-sixteenths. The fourth marks are longer and darker than the eighth and sixteenth marks.

Halves

The *half-inch* mark is longer than the marks for the fourths, eighths, or sixteenths divisions, as shown in **Figure 10-11**.

Reading an SI Ruler

Metric rulers usually show only two divisions: centimeters and millimeters. Measuring tapes normally indicate meter divisions as well.

The SI system is based on units of 10. The smallest unit on the ruler is the *millimeter.* As shown in **Figure 10-12**, ten millimeters make up one *centimeter,* which is set off with a longer line. (Approximately 21/2 or 2.54 centimeters equal one inch.) Ten centimeters, in turn, make up a *decimeter.* This unit is not widely used and is seldom even marked on rulers or tapes. Ten decimeters make up one *meter.* One hundred centimeters (*centi* means hundred) or 1000 millimeters (*milli* means thousand) are also equal to one meter.

Square (Area) Measure

A measurement of two dimensions, length and width, is known as ***square measure***. A two-dimensional measurement gives ***surface area***. To find the area of a rectangular surface, multiply the first dimension (length) by the second dimension (width). See **Figure 10-13**. To find the area of a triangle, multiply the length by the width, then divide by 2. Finding the area of a circle is covered later in this chapter.

Surface area measurements aid in placing equipment and determining the amount of paint or tile needed to cover floors, walls, and ceilings.

The surface area of any shape is expressed in ***square units***: square inches (in^2) or square feet (ft^2), square centimeters (cm^2) or square meters (m^2), and so on. The units used for land area, acres and hectares, are inherently square measures, so there

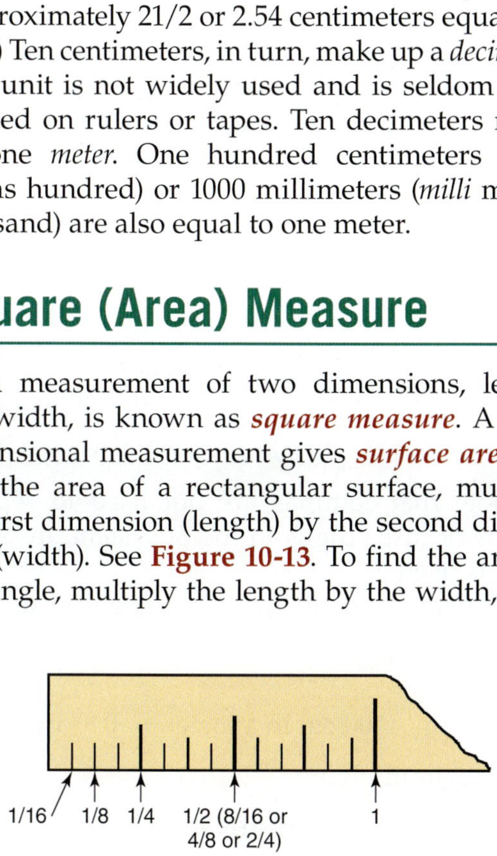

Goodheart-Willcox Publisher

Figure 10-11. Half-inch marks are longer and darker than fourth-inch marks, but shorter and usually less dark than marks indicating each full inch. A number is usually placed at each inch mark as well.

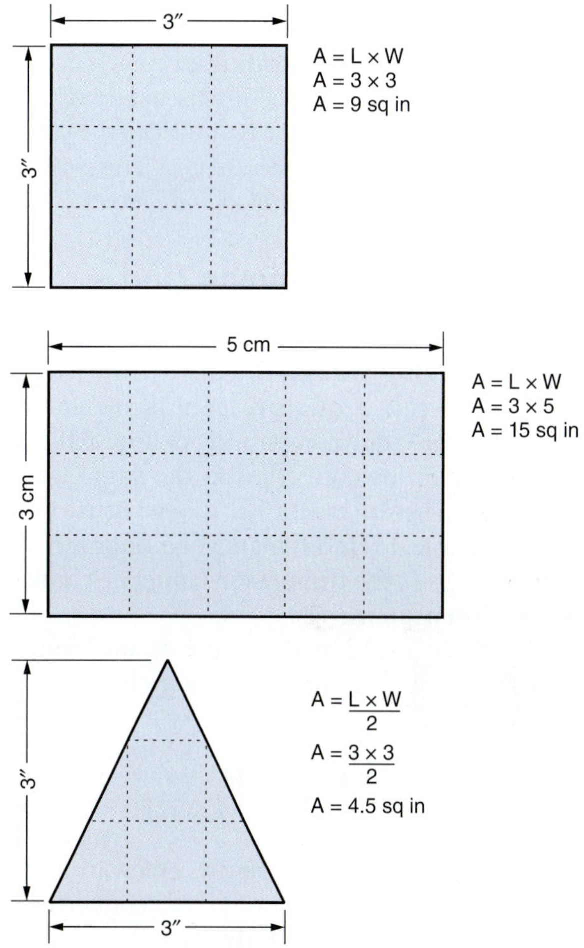

Goodheart-Willcox Publisher

Figure 10-13. To find surface area, multiply length by width. To find the area of a triangle, multiply the length by the width, then divide by 2. The result is expressed in square units. In this figure, two areas are square inches, the other in square centimeters.

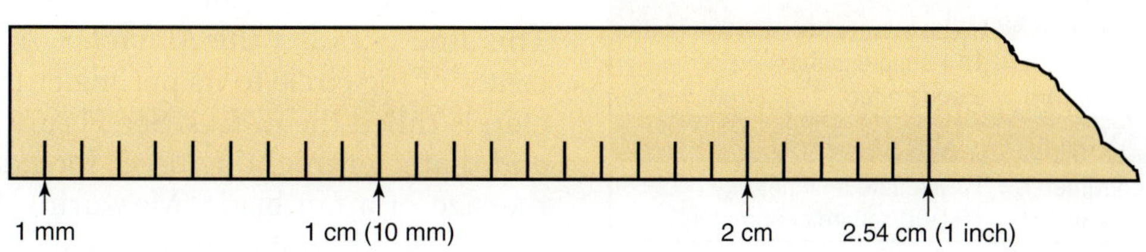

Goodheart-Willcox Publisher

Figure 10-12. On a meter stick or metric ruler, each centimeter is divided into 10 millimeters. Note that approximately 2.5 centimeters (more precisely, 2.54 cm) is equal to 1 inch.

is no need to use the word *square* when referring to them.

Reading Exponents or Power Numbers

An ***exponent***, or power, is an abbreviation used to indicate a multiplication process. The exponent is a numeral placed above and to the right of a base number. For example, in the number 10^2, the numeral 2 is an exponent or power that tells how many times the base number 10 (also called a ***factor***) is used for multiplying. Thus, 10^2 means 10 × 10. The factor (10) was used twice. In math terms, a number that is multiplied by itself (such as 10 × 10) is said to be *squared*. Thus, 10^2 can be read *10 squared* or *10 to the second power.*

Converting to a Common Unit

A mixture of feet and inches cannot be used when multiplying measurements to find surface area. Neither can a mixture of millimeters and centimeters, or centimeters and meters. All measurements must be converted to the same unit (a common unit) before multiplying. See **Figure 10-14**.

For example, to find the area of a floor measuring 10′-6″ × 12′-4″, the dimensions must be changed to a common unit, inches.

Since there are 12 inches in a foot, multiply the number of feet by 12, then add the inches:

$$10'\text{-}6'' = (10' \times 12'' = 120'' + 6'' = 126'')$$
$$12'\text{-}4'' = (12' \times 12'' = 144'' + 4'' = 148'')$$
$$126'' \times 148'' = 18{,}648 \text{ in}^2$$

Since large areas are usually given in square feet, the square inches should be converted to square feet. This is done by dividing by 144. Each square foot contains 12″ by 12″, or 144 sq in.

$$18{,}648 \text{ in}^2 \div 144 \text{ in}^2 = 129.5 \text{ ft}^2$$

Area Measurement

US Customary
1 square foot = 144 square inches
1 square yard = 9 square feet
SI
1 square centimeter = 100 square millimeters
1 square meter = 10 000 square centimeters

Goodheart-Willcox Publisher

Figure 10-14. Common measuring units for area in both US Customary and SI systems.

The same problem can be solved by converting all measurements to feet and the decimal equivalents of fractions of feet:

$$10'\text{-}6'' = 10.5'$$
$$12'\text{-}4'' = 12.3333'$$

Thus:

$$10.5' \times 12.3333' = 129.4996 \text{ ft}^2 \text{ (rounded to 129.5 for convenience)}$$

Converting *metric* dimensions to a common unit is simpler since all metric units are multiples of 10. For example, a contractor needs to find the area of a countertop that is 85 cm wide and 2 m long. First, the measurements are put into a common unit. Either convert the 2 m dimension to centimeters or convert the 85 cm dimension to a decimal fraction of a meter:

$$2 \text{ m} \times 100 \text{ cm} = 200 \text{ cm}$$

Then, multiply length by width:

$$85 \times 200 = 17\,000 \text{ square centimeters (or written as } 17{,}000 \text{ cm}^2\text{).}$$

To convert this figure to *square meters*, divide by 10 000 (100 cm × 100 cm = 10 000 cm^2 or 1 m^2):

$$17\,000 \text{ cm}^2 \div 10\,000 \text{ cm}^2 = 1.7 \text{ m}^2$$

Note that commas are not used to separate groups of three digits in the SI system. Instead, a space is used.

The same result can be reached, in just one step, by converting 85 cm to its equivalent decimal fraction:

$$85 \text{ cm} \times 100 \text{ cm in a meter} = 0.85 \text{ m}$$

Then multiplying:

$$0.85 \text{ m} \times 2 \text{ m} = 1.7 \text{ m}^2$$

Area of a Circle

Measuring and cutting circles are tasks that occur often in heating and cooling work. A circle is measured by a straight line across its widest part. This line is called the ***diameter***. A line from the center of the circle to its perimeter (half the diameter) is called the ***radius***. See **Figure 10-15**. Radius and diameter might be used, for example, to find the size of a fan blade. Measuring from the center of the hub to the outside edge of one blade, as shown in **Figure 10-16**, will give the radius. Then, just multiply by 2 to find the diameter.

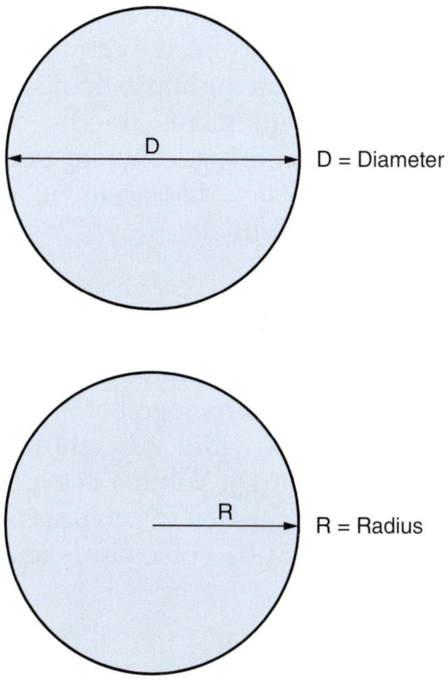

Goodheart-Willcox Publisher

Figure 10-15. The diameter is the distance across the widest part of a circle. The radius is half the diameter.

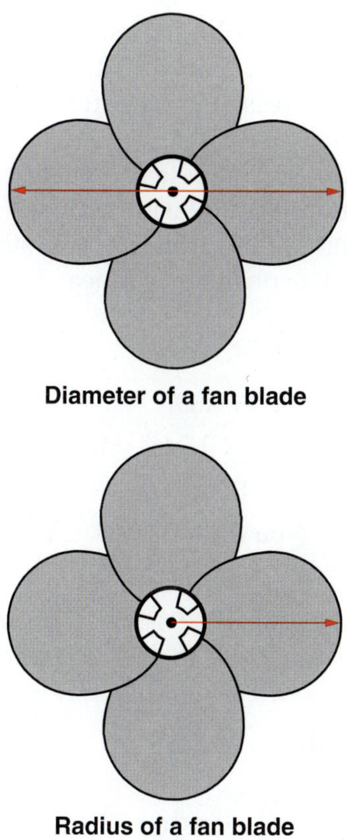

Goodheart-Willcox Publisher

Figure 10-16. Finding the diameter and the radius of a fan blade.

Finding the ***area of a circle*** is sometimes necessary. The most common method or formula for finding the area of a circle is:

$$A = \pi r^2$$

where

A = area
π (pi) = 3.1416
r = radius

Reading Formulas

A ***formula*** is a mathematical expression used as a standard to determine a value. Many formulas, like the one for finding the area of a circle, do not show a multiplication sign (×); instead, the × is understood. However, functions other than multiplication are indicated by symbols. The formula $A = \pi r^2$ means:

Area of a circle = 3.1416 (the value of π, pronounced "pi") × radius × radius

Special rules govern the order (first, second, and so on) in which operations within mathematical formulas are performed:

1. Sections enclosed within parentheses ( ) are performed first. If there are parentheses within parentheses, the operations are performed from the innermost set outward.
2. Base numbers must be multiplied to the specified power.
3. Multiplication and division are performed in the order given, reading left to right.
4. Addition and subtraction are performed in the order given, reading left to right.

Memorizing the acronym **PEMDAS**, which stands for the phrase "**P**lease **E**xcuse **M**y **D**ear **A**unt **S**ally," provides a simple aid for remembering the order in which functions are performed. The first letter of each word provides the clue:

- **P**arentheses
- **E**xponents
- **M**ultiply
- **D**ivide
- **A**dd
- **S**ubtract

Now, using the formula $A = \pi r^2$, find the area of a circle whose radius is 3″ as follows:

$$A = \pi r^2$$
$$A = 3.1416 \times 3^2$$
$$A = 3.1416 \times (3 \times 3)$$
$$A = 3.1416 \times 9$$
$$A = 28.2744 \text{ in}^2$$

Now, using the formula $A = \pi r^2$, find the area of a circle whose diameter is 9″ as follows:

$$A = \pi r^2$$

First, find radius:

$$r = 9 \div 2$$
$$r = 4.5$$

Then, plug into the formula:

$$A = 3.1416 \times 4.5^2$$
$$A = 3.1416 \times (4.5)^2$$
$$A = 3.1416 \times 20.25$$
$$A = 63.62 \text{ in}^2$$

Cubic (Volume) Measure

Cubic measure, using the three dimensions of length, width, and ***height*** (or *depth*), makes it possible to determine ***volume***. The difference between area and volume is the addition of a third measurement, height. See **Figure 10-17**.

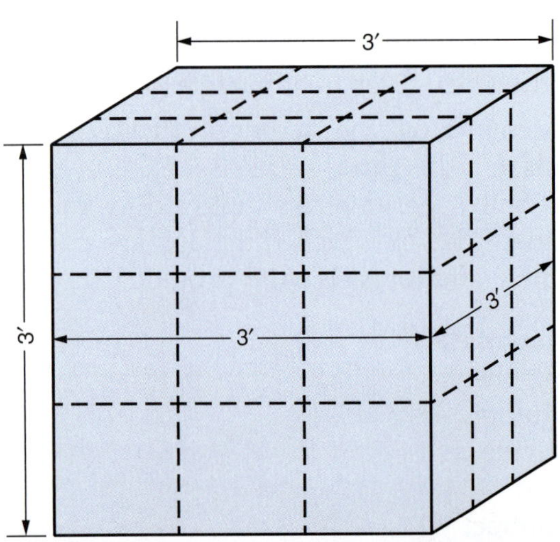

Goodheart-Willcox Publisher

Figure 10-17. By measuring length, width, and height, you can compute the volume of an object, such as this cube. In this example, 3′ × 3′ × 3′ = 27 cu ft. The volume could also be expressed as 3^3 or as "three cubed."

Volume measurements are necessary, for example, when figuring out the number of cubic feet (ft^3) space in a room or house being air conditioned. Volume measurements are also necessary for such tasks as figuring concrete needed for slabs, determining piston displacement, and finding the capacity of a cylinder.

Volume of a Rectangle

To find the volume of a rectangular solid, multiply the three dimensions together. The formula is: volume = length × width × height (or depth), or V = L × W × H (or D). Volume is expressed in cubic terms: cubic inches (cu in), cubic centimeters (cu cm), cubic feet (cu ft), cubic yards (cu yds), or cubic meters (cu m).

1 cu in = 16.387 cu cm
1 cu ft = 1728 cu in = 28 316.846 cu cm
1 cu yd = 27 cu ft = 0.7645 cu m

A perfect cube would have identical measurements in all three dimensions. For example, a cube measuring 12″ in all three dimensions would have a volume of 1728 cu in (12 × 12 × 12) = 1728). This cube can be expressed as 12^3, and is read, *12 cubed* or *12 to the third power*.

For example, a concrete pad needs to be poured before an outdoor condensing unit can be installed for a residential air-conditioning system. What is the volume of a concrete pad measuring 4′ long, 3′ wide, and 6″ deep?

The formula is:

$$V = L \times W \times D$$
$$V = 4 \times 3 \times 0.5 \; (6'' = 0.5')$$
$$V = 12 \times 0.5$$
$$V = 6 \text{ cu ft (or } 6 \text{ ft}^3)$$

Knowing that an 80-pound bag of concrete mix will make 2/3 cu ft (0.66 cu ft) of concrete, how many bags are needed to pour the pad in the previous calculation?

$$6 \div 0.66 = 9 \text{ bags}$$

Volume of a Cylinder

Finding the volume of a cylinder also involves three dimensions. Since a cylinder is round, the area of a circle is found first, and then it is multiplied by the height dimension. The formula for finding the volume of a cylinder is: $V = \pi r^2 H$. For example, to find the volume of a cylinder that is 3″ in diameter and 5″ tall, as shown in **Figure 10-18**:

$$V = \pi r^2 H$$

Remember to apply **PEMDAS**:

$$V = 3.1416 \times (1.5 \times 1.5) \times 5 \text{ (diameter = 3″, so radius = 1.5″)}$$

$$V = 3.1416 \times 2.25 \times 5$$

$$V = 3.1416 \times 11.25$$

$$V = 35.343 \text{ cu in (or } 35.343 \text{ in}^3)$$

Using cylinder volume measurements is useful in HVACR work. For example, it is possible to determine the amount of vapor a compressor can compress (compressors are rated in cubic feet per minute, or cfm). Using the cylinder volume from the previous calculation, assume you are calculating the cfm for a two-piston compressor, operating at 1800 revolutions per minute (rpm). Since this compressor has two pistons, the cylinder volume must be doubled:

$$35.343 \times 2 = 70.686 \text{ cu in (or } 70.686 \text{ in}^3)$$

Next, multiply the volume by the rpm. This will result in a cubic *inches* per minute measurement:

$$70.686 \text{ in}^3 \times 1800 \text{ rpm} = 127{,}234.8 \text{ in}^3 \text{ per minute}$$

Finally, convert in^3 per minute to ft^3 per minute (since $1728 \text{ in}^3 = 1 \text{ ft}^3$, divide by 1728):

$$127{,}234.8 \text{ in}^3 \div 1728 \text{ in}^3 = 73.63 \text{ cfm}$$

10.4 Temperature Scales

While the ***Fahrenheit*** (°F) temperature scale is most common in the US, the ***Celsius*** (°C) scale is used in almost all other countries. Technical literature may use either scale or both. While the Fahrenheit scale is used in refrigeration, heating, and air-conditioning work in the US, a technician should also be able to use the Celsius scale. Fahrenheit temperatures can be converted to Celsius, or vice versa. See **Figure 10-19** for scale comparisons.

Converting Celsius to Fahrenheit

There are two methods for converting Fahrenheit to Celsius or Celsius to Fahrenheit. One method uses fractions; the other uses decimals. Either method will yield the correct answer, but the decimal method is easier to use.

$$\text{Fahrenheit} = (1.8 \times \text{Celsius}) + 32$$

To convert a temperature of 40°C to Fahrenheit:

$$°F = (1.8 \times 40) + 32$$

$$°F = 72 + 32$$

$$°F = 104$$

Converting Fahrenheit to Celsius

Using the decimal method to convert Fahrenheit to Celsius is similar to the previous method described. When converting to Celsius, however, the number 32 is subtracted from the Fahrenheit value at the beginning, then divided. When converting to Fahrenheit, the number 32 is added after multiplying.

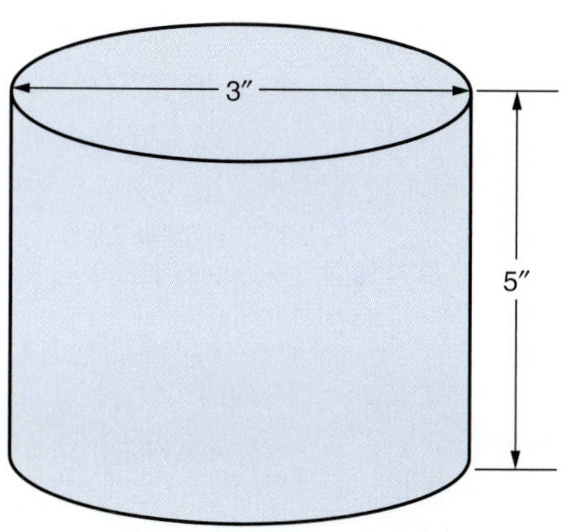

Goodheart-Willcox Publisher

Figure 10-18. To find the volume of a cylinder, first find the area of a circle; then multiply by the height dimension.

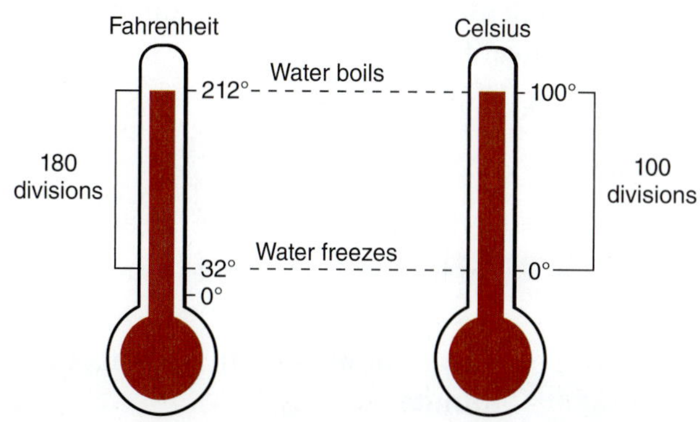

Goodheart-Willcox Publisher

Figure 10-19. Fahrenheit and Celsius temperature scales compared. Note that the Fahrenheit scale has nearly twice as many divisions as the Celsius scale between freezing and boiling points.

$$\text{Celsius} = (\text{Fahrenheit} - 32) \div 1.8$$

To convert 75°F to Celsius:

$$°C = (75 - 32) \div 1.8$$
$$°C = 43 \div 1.8$$
$$°C = 23.9 \text{ (or rounded up to 24)}$$

It is worth noting that the Fahrenheit and Celsius scales read the same at –40°. This can be proved by using these formulas.

10.5 Rearranging Formulas

Along with the formulas already talked about in this chapter, there are some other formulas that are used in the HVAC field. These are formulas that will assist a technician in many areas of the HVACR industry. The following formulas will be discussed in later chapters:

- Ohm's Law
- Watt's Law
- Boyle's Law
- Charles' Law
- Dalton's Law
- Gay-Lussac's Law
- Combined Gas Law
- Specific heat formula
- Sensible and latent heat formulas
- Total heat formula
- Heat transfer formula
- Cfm (Cubic feet per minute) formulas

It is important to remember that formulas can be written or rearranged in different ways depending on the calculation needed. For example, take the sensible heat formula:

$$Q_S = 1.08 \times cfm \times \text{TD}$$

where

Q_S = Sensible heat value of the air

1.08 = An industry accepted value for calculating sensible heat

cfm = The amount of air in cubic feet per minute

TD = The temperature difference between two temperatures

The way this formula is currently written, it is used to find the amount of sensible heat in an air sample (Q_S), because the amount of air that was moving is known (cfm) and how much heat it had picked up (TD). This formula can be changed to determine how much air is moving through a furnace. This requires the formula to be rearranged to solve this other value.

To find the cfm, it needs to be on one side of the equal sign by itself. In math, if a function is performed to both sides of a formula, the formula's value does not change. So, by dividing both sides of the formula by "1.08 × TD," the formula can now be rewritten:

$$Q_S = 1.08 \times \text{cfm} \times \text{TD}$$
$$\frac{Q_S}{1.08 \times \text{TD}} = \frac{1.08 \times \text{cfm} \times \text{TD}}{1.08 \times \text{TD}}$$
$$\frac{Q_S}{1.08 \times \text{TD}} = \frac{\cancel{1.08} \times \text{cfm} \times \cancel{\text{TD}}}{\cancel{1.08} \times \cancel{\text{TD}}}$$
$$\frac{Q_S}{1.08 \times \text{TD}} = \text{cfm}$$

Now, the amount of air moving through a furnace (cfm) can be determined. Q_S would equal the output heat rating of the furnace, and 1.08 is still the value used. TD would be the difference between the temperature of the supply air and the return air of the furnace.

Any formula can be rewritten or rearranged by performing a function or functions to both sides until the value that is needed is on one side of the equal sign by itself. This process will be discussed as these formulas are introduced in other chapters.

Chapter Review

Summary

10.1 The numbers 0 to 9 are called digits. They are used alone, or in combination, to indicate "how many." When digits are used in combination with each other, each one is given a special position called a *place value*.

10.2 A whole number is any positive number that contains no fractional parts. Fractions are a means of indicating parts of a whole number, indicated by a numerator and denominator. Decimal numbers express fractions (or parts) of numbers. Converting fractions to decimals makes it easy to perform certain calculations, like worker's pay on a time sheet for example.

10.3 Measurements are made in either the US Customary or International System of Units (SI). The US Customary uses basic units of inch, foot, and yard dimensions; SI uses millimeter, centimeter, and meter. A measurement in one dimension is a linear measure, or the distance from one point to another. Measuring tools, like rulers and tapes, are used to make linear measurements. Square measure is a measurement of two dimensions, which gives an object's surface area. Cubic measure is a measurement of three dimensions, which gives an object's volume. All of these measurements are applicable to HVACR work.

10.4 A technician should be familiar with both the Fahrenheit (°F) scale and the Celsius (°C) scale and how to convert between the two. Fahrenheit is most common in the US, while the Celsius scale is used in almost all other countries.

10.5 Formulas can be rearranged based on the variables available to find a missing variable. Performing a function or functions to both sides of the equal sign can be done until the value that is needed is one side of the equal sign by itself.

Know and Understand

1. The number 3720 is read _____. (10.1)
   A. *three seven, two zero*
   B. *three hundred seventy-two zero*
   C. *three thousand, seven hundred twenty*
   D. *three million, seven thousand, two hundred*
2. *True or False?* A whole number contains no fractional parts. (10.2)
3. The bottom number of a fraction is the _____. (10.2)
   A. commutator
   B. denominator
   C. designator
   D. numerator
4. When the numerator is larger than the denominator, then the number is a(n) _____. (10.2)
   A. proper fraction
   B. improper fraction
   C. mixed number
   D. reducing fraction
5. The improper fraction 5/4 can be converted to _____. (10.2)
   A. 3/8
   B. 5/8
   C. 1 1/4
   D. 1 1/2
6. The fraction 12/120 reduced to its lowest terms is _____. (10.2)
   A. 1/10
   B. 3/30
   C. 4/40
   D. 6/60
7. The decimal number 3.1416 is read *in industry* as _____. (10.2)
   A. *three and one four one six*
   B. *three and one thousand four hundred sixteen ten thousandths*
   C. *three point one four one six*
   D. *three point one thousand four hundred sixteen ten thousandths*

8. A technician works a 45-hour work week, with an overtime pay of 1.5 times the regular hourly rate. The hourly rate is $27 per hour. What is the total amount of pay the technician earned this week? (10.2)
   A. $1080
   B. $1215
   C. $1282.50
   D. $1822.50
9. There are _____ centimeters in 1″. (10.3)
   A. 2.54
   B. 25.4
   C. 30.48
   D. 304.8
10. When reading a US Customary ruler, how many sixteenths are in 1″? (10.3)
    A. 4
    B. 6
    C. 8
    D. 16
11. *True or False?* Rough measurements are often more accurate than measurements made with a measuring tool. (10.3)
12. What is the area of a triangle with a length of 6 ft and a width of 8 ft? (10.3)
    A. 10 ft^2
    B. 12 ft^2
    C. 24 ft^2
    D. 48 ft^2
13. Half the distance across the widest part of a circle is called the _____. (10.3)
    A. radius
    B. diameter
    C. area
    D. circumference
14. What is the volume of a concrete pad measuring 6′ long, 3′ wide, and 8″ deep? (10.3)
    A. 6 ft^3
    B. 9 ft^3
    C. 12 ft^3
    D. 144 ft^3
15. Convert 32°F to Celsius. (10.4)
    A. 90°C
    B. 18°C
    C. 1°C
    D. 0°C
16. Any formula can be rewritten or rearranged by performing a function or functions to _____. (10.5)
    A. the left side of the equal sign
    B. the right side of the equal sign
    C. both sides of the equal sign
    D. only the constant variable(s)

Apply and Analyze

1. How would you spell out the number 6.435? (10.2)
2. Convert 12 feet to centimeters. (10.3)
3. Calculate 18′-2″ × 12′-6″ and convert the answer to decimal format. Round to the nearest hundredths place. (10.3)
4. Calculate the area of a fan blade's span that measures 24″ in diameter. (10.3)
5. What happens to water at 100°C? Converting to Fahrenheit may help identify the answer. (10.4)

Critical Thinking

1. Convert the number 25/16 to a proper fraction and a decimal. (10.2)
2. Calculate the cfm (cubic feet per minute) of a four-piston compressor, operating at 1800 rpm (revolutions per minute). A single piston measures 4″ in diameter and is 7″ tall. (10.3)

Alexander Konradi/Shutterstock.com

11 Basic Thermodynamic Principles

LEARNING OUTCOMES

After studying this chapter, you will be able to:

11.1 Recognize the first two laws of thermodynamics and the three methods of heat transfer.

11.2 Explain how a substance undergoes a change of physical state while considering sensible heat, specific heat, latent heat, saturated conditions, superheat, and subcooling.

11.3 Explain the ton of refrigeration effect.

TECHNICAL TERMS

- absolute zero
- atom
- boiling point
- British thermal unit (Btu)
- change of state
- cold
- conduction
- convection
- First Law of Thermodynamics
- forced convection
- Ice Melting Equivalent (IME)
- insulator
- kilojoule (kJ)
- latent heat
- latent heat of condensation
- latent heat of freezing
- latent heat of melting
- latent heat of sublimation
- latent heat of vaporization
- molecule
- natural convection
- physical states
- radiation
- refrigeration
- saturated condition
- saturation point
- Second Law of Thermodynamics
- sensible heat
- specific heat
- subcooled
- sublimate
- superheated vapor
- thermodynamics
- ton of refrigeration effect

Introduction

Refrigeration is often thought of as a cold or cooling process, where cold is added to a space. Refrigeration does *not* add cold, it *removes heat*. ***Refrigeration*** is the process of removing unwanted heat and carrying it away from the space.

The term ***cold*** describes a lack of heat, the condition produced by the *removal of heat*. Heat is measured in British thermal units (Btus). Heating and cooling systems are sized according to the amount of Btu needed to heat or cool a space. For example, a refrigerator produces *cold* by removing a certain amount of heat (in Btu) from inside the cabinet, then releasing that heat to the air outside the cabinet. Heat can leak into a refrigerator through its walls and insulation. Warm air enters whenever the door is opened, or when warm foods are placed inside the refrigerator. To accomplish refrigeration, the heat must be removed faster than it enters. The refrigeration system simply transfers heat from inside the cabinet to the air outside of it, just as water would be bailed from a leaking boat and thrown back into a lake, **Figure 11-1**.

11.1 Thermodynamics and Heat Transfer

Thermodynamics is the science that deals with the mechanical action of heat. There are several basic thermodynamic principles called the *laws of thermodynamics*. Two of these laws are important to the study of refrigeration and air conditioning. Since the HVACR technician constantly works with the controlled movement of heat, an understanding of basic thermodynamic principles is vital.

First Law of Thermodynamics

The ***First Law of Thermodynamics*** states that energy cannot be created or destroyed, but energy can be converted from one form to another. This law applies to heat and other forms of energy, such as electrical, mechanical, light, chemical, and atomic.

Heat can be generated by converting another form of energy into heat energy. For example, electrical energy is converted to heat energy by such devices as a toaster, electric range, hair dryer, water heater, or space heater. **Figure 11-2** illustrates an example of energy conversion.

Second Law of Thermodynamics

The ***Second Law of Thermodynamics*** states that heat always travels from hot to cold. See **Figure 11-3**. Heat energy is always in motion. For example, a metal spoon placed in hot coffee will absorb heat from the coffee, causing the coffee to begin to cool. How fast heat travels depends on the temperature difference between the two objects and the material the objects are made of. The greater the temperature difference and the

Popova Valeriya/Shutterstock.com

Figure 11-1. A refrigeration system is similar to bailing out a boat. When bailing, you move water from inside the boat to outside. In a refrigeration system, you move heat from inside a house, refrigerator, or other space to the outside.

Alexander Konradi/Shutterstock.com

Figure 11-2. An electric range stove is a device that converts electrical energy to heat energy.

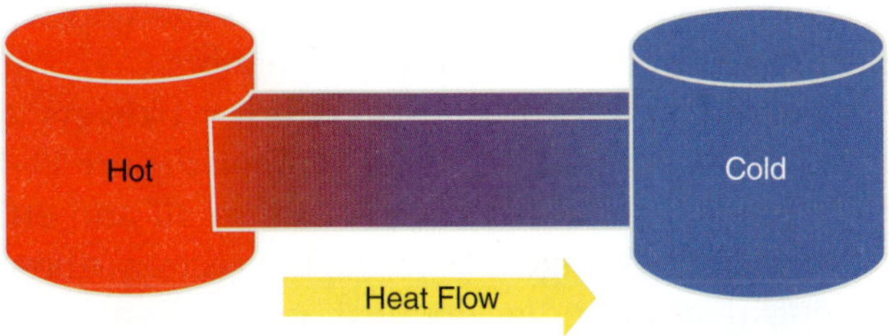

OSweetNature/Shutterstock.com

Figure 11-3. Heat always travels from hot to cold. The greater the temperature difference, the faster the rate of heat transfer.

better the heat transfer material, the faster heat will travel.

How Heat Travels

Since refrigeration deals with the movement of heat, it is necessary to know *how* heat travels. There are three basic methods by which heat is transferred from one substance to another: radiation, conduction, and convection. Sometimes, a combination of these methods is used.

Radiation

Radiation is the transfer of heat by waves such as light waves or radio waves. The sun's energy is transferred by means of radiant heat waves, which travel through space in a straight path. Radiant heat waves are absorbed by objects, not the air they pass through. Just like moving from the shade into direct sunlight on a hot, sunny day, the effects of these heat waves can be felt. The difference is felt because the body immediately absorbs the radiant heat waves.

Dark-colored materials tend to absorb radiant heat waves. Light-colored materials tend to reflect them. Clothing manufacturers apply this principle when designing garments for different seasons or climates.

An automobile parked in the hot sun with the windows closed will absorb radiant heat waves, causing the inside of the automobile to become hot, **Figure 11-4**.

aappp/Shutterstock.com

Figure 11-4. Radiant heat causes the interior of a closed car to be quickly heated by the sun on a hot, sunny day. A light-colored windshield screen reflects the radiant heat waves, reducing the heat in the car.

Conduction

Conduction is the flow of heat through a substance from one end to the other. An iron skillet provides a good example of conduction. If the skillet is placed on a hot fire, heat will travel to the handle until it becomes just as hot as the skillet body, **Figure 11-5**.

Heat flow by conduction can take place between two substances or objects if they are touching each other. Conduction is improved by the amount of physical contact involved. Most metals will conduct heat well. Copper and aluminum are excellent conductors of heat, hence why these two metals are used extensively in refrigeration systems.

Substances that are poor heat conductors are called ***insulators***. Examples of insulators are cork, wood, fiberglass, mineral wool, and polyurethane foams. Insulators cannot totally stop heat flow but can slow it significantly. Another example of this

Samuel V.Rocheleau/Shutterstock.com

Figure 11-5. If an iron skillet is placed over a fire, the heat will travel from the skillet body to the handle by means of conduction.

is fiberglass insulation used on homes. The insulation does not entirely stop heat flow, evidenced by the need for the air conditioning on in the summer or the heat on in the winter, but prevents a home from being completely exposed to the full impact of outside weather.

Convection

Convection is the movement of heat by means of a carrier, such as air or water. In a forced-air heating system, air is heated by a furnace and discharged into the living areas to warm them. The air is then returned to the furnace to be heated again.

Movement of convection currents can be either *natural* or *forced*. ***Natural convection*** involves slow-moving currents of air in which lighter, warm air rises and heavier, cold air falls. This method was once used for furnaces and refrigerators.

Forced convection uses fans or blowers to increase the amount of air movement, **Figure 11-6**. This procedure permits the use of small and efficient heat exchangers. Forced convection allows a central air handler to heat or cool an entire building.

11.2 States of Matter and Heat

All substances are made up of particles known as ***atoms***. An atom is the smallest physical particle of any substance that can exist by itself and retain

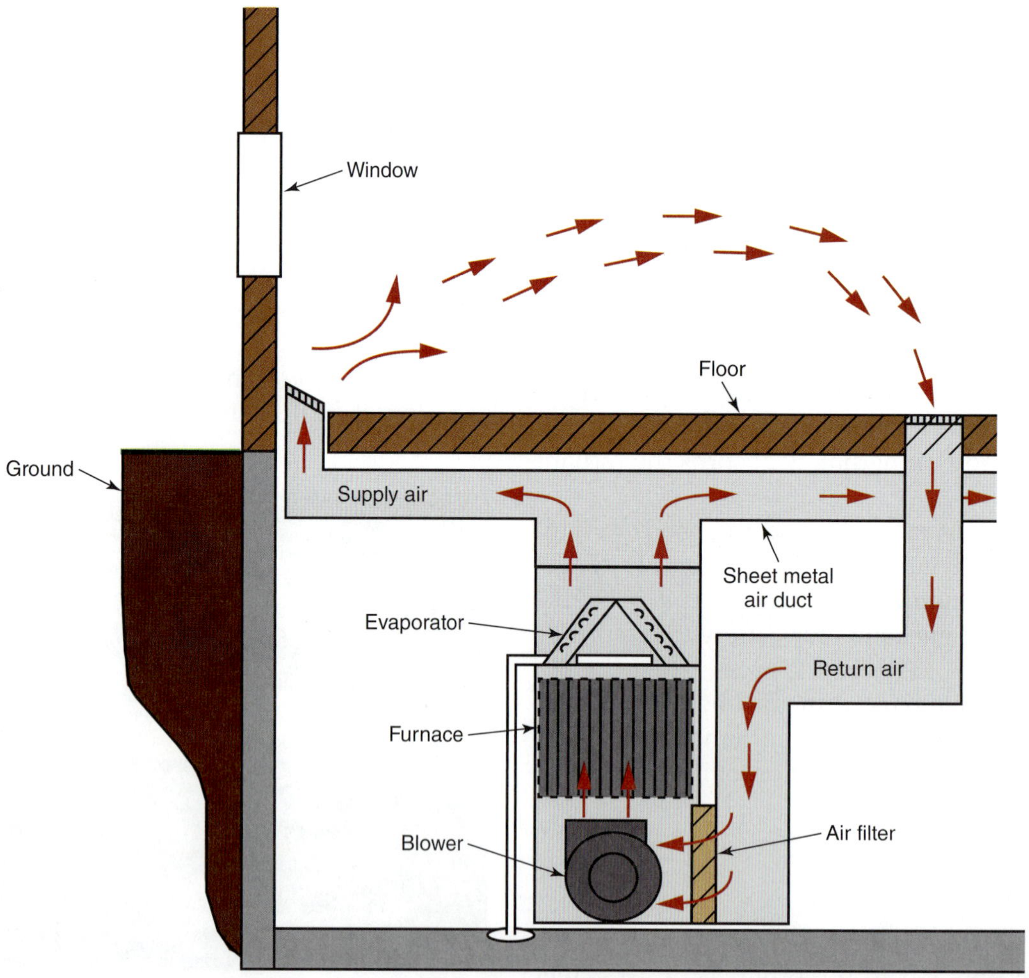

Figure 11-6. Forced convection transfers heat more efficiently than does natural convection. A modern central air-conditioning system uses forced convection to remove heat and distribute the cooled air throughout the living area of a house.

its chemical properties. A group of atoms that are bonded together are called ***molecules***. Molecules represent the smallest unit in a chemical compound that can take part in a chemical reaction. For example, a water molecule (H_2O) consists of two atoms of hydrogen and one atom of oxygen.

Regardless of the substance they make up, atoms are always in rapid motion. As the temperature of a substance increases, the motion of the atoms and molecules also increases, **Figure 11-7**. As the temperature drops, the molecules slow down. Even in a solid, such as ice or steel, the molecules are moving unless all heat is removed from the substance. If all heat is removed (at ***absolute zero***, –460°F or –273°C), molecular motion will stop completely.

Atoms are like building blocks arranged in certain patterns to form different substances. In a solid, molecules still move or vibrate, but distances are limited. The force holding the molecules in place is strong in a solid. Much heat energy is required to overcome this force and permit the molecules to move and form a new pattern.

When heat flows from a warmer to a cooler substance, the faster-moving molecules give up some of their energy to the slower-moving ones. As a result, the faster molecules slow down, and the slower ones move a little faster.

Physical States

Most substances can exist in three ***physical states***: solid, liquid, or gas. See **Figure 11-8**. As heat is added or removed, the substance may go through a change of state. Water is the substance commonly used to illustrate this process because it can exist as a solid (ice) below 32°F (0°C), a liquid (water) between 32°F and 212°F (0°C and 100°C), and a gas (steam) above 212°F (100°C). These

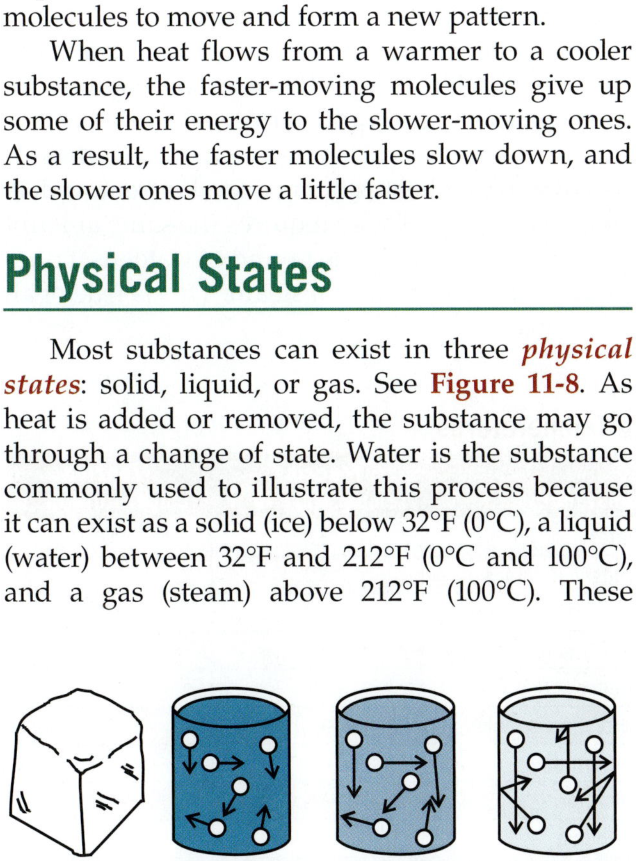

Goodheart-Willcox Publisher

Figure 11-7. Molecules in a substance speed up or slow down with the rise or fall of temperature. Molecular motion is slowest in a solid, faster in a liquid, and fastest in a gas.

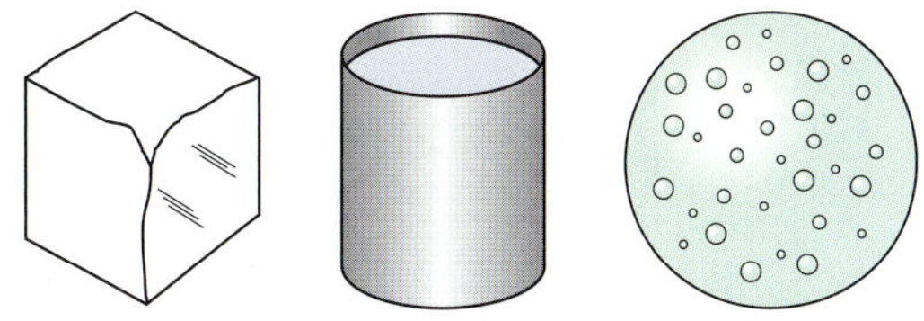

Goodheart-Willcox Publisher

Figure 11-8. Most substances can exist in any of the three states of matter: solid, liquid, or gas.

temperatures are based at sea level (different elevations affect these temperatures).

When a substance is changing from one physical state to another, the temperature level remains *constant* (does not change) until all the molecules are rearranged to the new pattern. For example, water can be a liquid at 212°F(100°C), and a gas at 212° F(100°C). This principle is difficult to understand—the question is often asked, "If the temperature does not rise or fall during a change of state, where does the heat energy go?" The energy is being used to shift the molecules into a different pattern.

A liquid allows more movement than a solid, but the movement of the molecules is still limited to a specific pattern. When changing from a liquid to a gas, which is referred to as a substance's ***boiling point***, a large amount of heat energy (970 Btu/lb for water) is needed to shift the molecules and permit them to move in total freedom in all directions as a gas.

Change of State

A ***change of state*** occurs when the temperature and speed of the moving molecules reach a certain level. At this precise level, the molecules will rearrange themselves into a different pattern. The shift in pattern causes the substance to change from a solid to a liquid or from a liquid to a gas.

This rearrangement of molecules is best illustrated by adding heat to ice that has a temperature of 32°F (0°C). See **Figure 11-9**. The additional heat will *not* raise the temperature of the ice, but the heat will provide the energy needed to cause the molecules to shift and rearrange themselves into a liquid (water) pattern. More heat does *not* always mean higher temperature.

Once the ice becomes water, additional heat will cause the water molecules to increase their speed of motion. After the change of state, additional heat will raise the temperature. As more heat is added, temperature and speed will continue to increase

Goodheart-Willcox Publisher

Figure 11-9. A change of state occurs at a certain temperature for a given substance. For water, the change from solid to liquid comes at 32°F (0°C), and the change from liquid to gas at 212°F (100°C).

until the water is heated to its boiling point (212°F or 100°C). Additional heat at this temperature will provide the energy necessary to cause the molecules to shift their pattern again, becoming steam.

Of course, the reverse of this is also true. By *removing* heat from a substance, its temperature will decrease, until it reaches a change of state temperature. Then removing heat will cause it to change its state with temperature remaining steady during the process. The molecules will slow down, shifting their pattern to a new physical state (gas to liquid, liquid to solid).

Temperature and Change of State

Because the molecules are different for each substance, the temperature at which a change of state will occur is *different* for each substance. For example, water boils at 212°F (100°C), while ethyl alcohol boils at 173°F (78°C).

Because they are common and familiar to everyone, the three states of water (ice, liquid water, and water vapor or steam) are used to explain the basic principles of temperature and change of state. The principles are applied to substances that change state at very high or very low temperatures, as well. Take a substance like liquid ammonia that *boils* at –28°F (–33°C) and *freezes* at –107°F (–77°C). See **Figure 11-10** for a table of boiling and freezing temperatures for various substances.

Heat energy must be removed to cause a gas to condense back to a liquid. Likewise, heat must be removed to cause a liquid to freeze into a solid. The same amount of heat energy is involved, regardless of whether heat is being added or removed. A change of physical state requires the same amount of heat energy to boil one pound of water as it does to condense one pound of steam. This is true of all substances.

Boiling and Freezing Temperatures

Substance	Boiling Temperature in °F	Boiling Temperature in °C	Substance	Freezing Temperature in °F	Freezing Temperature in °C
Water	212	100	Water	32	0
Ethyl alcohol	173	78	Fruits/vegetables	30	–1
Chloroform	143	62	Seafood	28	–2
Butane	31	1	Beef and pork	28	–2
Ammonia	–28	–33	Poultry	27	–3
Propane	–43	–42	Carbon tetracholoride	–9	–23
Carbon dioxide	–109	–78	Linseed oil	–11	–24
Acetylene	–118	–83	Chloroform	–81	–63
Oxygen	–287	–177	Ammonia	–107	–77
Nitrogen	–321	–196	Acetone	–139	–95
Hydrogen	–423	–253	Ether	–177	–116
Helium	–452	–269	Ethyl alcohol	–179	–117

Goodheart-Willcox Publisher

Figure 11-10. Each substance has a specific temperature at which it changes state (freezes or boils). The table lists temperatures for some common foods and chemical substances.

Sensible Heat

Sensible heat is heat that causes a change in temperature but not a change of state. When a substance is heated and the temperature rises because of the added heat, the added heat is sensible heat. Likewise, if heat is removed and the temperature of the substance falls, the heat being removed is sensible heat. This is heat that can be measured with a thermometer or a thermostat in a home, for example.

British Thermal Unit

The ***British thermal unit (Btu)*** is the basic unit used to measure the *quantity* of heat. A thermometer measures only the temperature of a substance. It cannot measure the amount of heat required to reach a certain temperature.

Many years ago, scientists realized a standard method to measure quantities of heat was needed. Since water was such a common substance, they placed a quantity of water that weighed exactly one pound into a container. Next, they measured the precise amount of heat required to raise the temperature of the water one degree Fahrenheit. See **Figure 11-11**. The standard they established was accepted by all scientists and is still used today:

1 Btu = the amount of heat that will raise the temperature of 1 lb of water by 1°F

If more than one pound of water is involved, it will require one Btu for each pound of water to raise temperature 1°F. Likewise, if the temperature is to be raised more than one degree, it will require one Btu for each degree of temperature change. For example:

How many Btu are required to raise the temperature of 10 pounds of water from 72°F to 82°F?

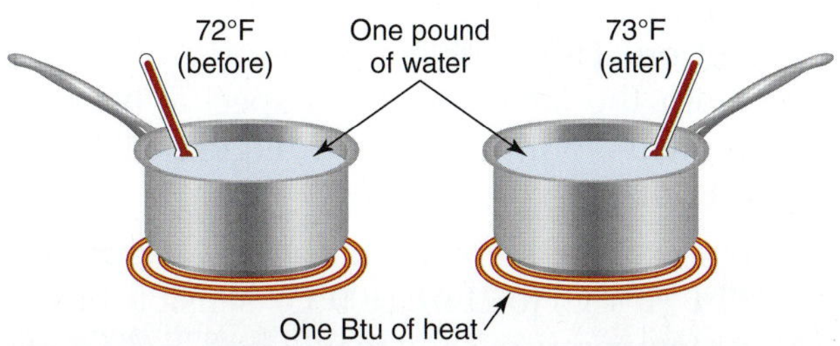

Figure 11-11. The Btu (British thermal unit) is the standard used to measure quantities of heat. The Btu is the amount of heat needed to raise the temperature of 1 lb of water by 1°F.

Btu = wt × tc (weight × temperature change)

Btu = 10 lb × 10°F

Btu = 100

If a substance is cooled by removing heat, the amount of heat removed is figured the same way. If heat is removed from water, the weight is multiplied by the temperature difference to obtain the number of Btu removed in cooling the water. For example:

How much heat must be removed to cool 50 pounds of water from 75°F to 65°F?

Btu = wt × tc

Btu = 50 lb × 10°F

Btu = 500

Kilojoule

In the SI system, the unit used to measure quantities of heat is the joule (J). It is, however, a tiny unit. For practical use in refrigeration work, the ***kilojoule (kJ)*** is used. A kilojoule is 1000 J. To raise the temperature of one kilogram (kg) of water by one degree Celsius (1°C), the amount of heat required is 4.187 kJ.

Latent Heat

Latent heat is heat energy that causes a change of state but no temperature change. During a change of state, the temperature will remain constant until the process of change is completed. Latent heat is sometimes called *hidden* heat because it cannot be measured with a thermometer. The heat energy is used to rearrange a substance's molecules into a different pattern, and the temperature cannot change until all the molecules have been rearranged.

Each substance has a temperature/pressure point at which a change of state will occur. For example, under normal atmospheric pressure, ice will melt at 32°F (0°C), or water will freeze at 32°F (0°C). Therefore, at 32°F (0°C), it is possible to have all ice, all water, or a combination of water and ice. See **Figure 11-12**. The temperature will remain the same until the change of state is completed. It requires exactly 144 Btu of heat to melt 1 lb of ice. The same amount of heat must be removed to freeze 1 lb of water. This is called the *latent heat of melting* or the *latent heat of freezing*.

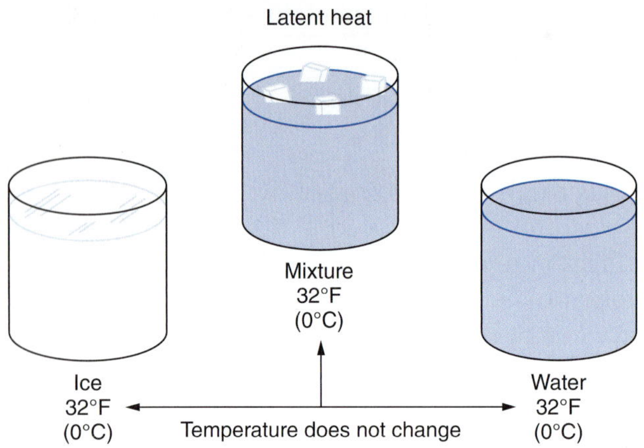

Goodheart-Willcox Publisher

Figure 11-12. Latent heat causes a change of state without increasing the temperature of the substance. The temperature will remain constant until the change of state is complete. As shown, while ice is changing to liquid water, the temperature remains constant at 32°F (0°C).

When water is heated, the temperature will stop rising at 212°F (100°C) because the water begins to boil into steam. During this change of state, the temperature will remain at 212°F (100°C) until all the water has turned into steam. It requires 970 Btu to change 1 lb of water into steam (latent heat of vaporization). If the steam is trapped in a closed container, and 970 Btu are removed by cooling the steam, the steam will condense back to water (latent heat of condensation).

In short, the four latent heats just described are as follows:

- ***Latent heat of melting***—the amount of heat added to change a solid to a liquid without changing the substance's temperature.
- ***Latent heat of freezing*** (or *fusion*)—the amount of heat removed to change a liquid to a solid without changing the substance's temperature.
- ***Latent heat of vaporization*** (or *evaporation*)—the amount of heat added to change a liquid to a vapor without changing the substance's temperature.
- ***Latent heat of condensation***—the amount of heat removed to change a gas (or vapor) to a liquid without changing the substance's temperature.

Figure 11-13 shows latent heat of vaporization/condensation values for some typical refrigerants. Water is included as a reference value. Latent heat of vaporization/condensation forms the basis of all refrigeration systems. Refrigerants are the vital working fluids in any refrigeration or air-conditioning system. They absorb heat as they change from a liquid to a gas (evaporate). The gas then travels to a place where the heat is removed, causing the gas to return to a liquid state (condense). The refrigerant is then ready for reuse (another cycle).

Latent Heat of Common Refrigerants

Refrigerant	Latent Heat of Vaporization or Condensation (Btu/lb)
Water	970.4
R-32	362.1
R-123	73.2
R-1234yf	170.7
R-134a	93.4
R-290 (Propane)	183.1
R-404A	86.4
R-407C	107.2
R-410A	117.4
R-502	68.96
R-507	84.35
R-600a (Isobutane)	343.68
R-717 (Ammonia)	565.0

Goodheart-Willcox Publisher

Figure 11-13. Latent heat of vaporization/condensation values for several common refrigerants. Water is provided as a reference value but is also considered a refrigerant. Since the refrigerants are used only in the liquid and gaseous states, no values are provided for the latent heat of freezing or the latent heat of melting.

The specific heat formula (*Btu = weight [wt] × specific heat [sp ht] × change in temperature [tc]*) can be applied to discover Btu quantities in the heating or cooling process dealing with sensible heat. When dealing with latent heat, use this formula, *Btu = wt × L ht*, where *L ht* equals the latent heat of a substance. The following problem illustrates different heat values, and the steps required to allow for sensible and latent heats.

Problem: How much heat must be removed to convert 10 lb of steam at 220°F (104°C) into ice at 12°F (–11°C)?

Solution: This problem is solved in five easy steps, with the latent heat and specific heat values obtained from the tables in **Figure 11-13** and **Figure 11-14**.

1. Lower the temperature of the steam from 220°F (104°C) to 212°F (100°C) (sensible heat). The temperature change will be 8°F (4°C). We must stop at 212°F (100°C), the point where the latent heat of condensation comes into effect.

Specific Heats

Substance	Specific Heat (Btu/lb)
Acetone	0.514
Alcohol	0.680
Aluminum	0.24
Ammonia	1.090
Beef	2.345
Benzene	0.412
Bread	1.993
Butter	1.373
Cheese	2.077
Chicken	3.316
Chloroform	2.340
Copper	0.095
Fish	3.550
Ice	0.487
Iron	1.373
Oranges	3.751
Stainless steel	0.12
Steam	0.460
Steel	0.116
Water	1.000

Goodheart-Willcox Publisher

Figure 11-14. Specific heats for several foods and other substances.

Btu = wt × sp ht × tc

Btu = 10 lb × .46 × 8°F

Btu = 36.8

2. Condense the steam into water (latent heat).

   Btu = wt × L ht

   Btu = 10 lb × 970.4

   Btu = 9704

3. Lower the temperature of the water from 212°F (100°C) to 32°F (0°C) (sensible heat). The temperature change will be 180°F (100°C). We must stop at 32°F (0°C), the point where the latent heat of freezing comes into effect.

   Btu = wt × sp ht × tc

   Btu = 10 lb × 1 × 180°F

   Btu = 1800

4. Freeze the water into ice (latent heat). Recall that the latent heat of freezing for water is 144 Btu.

   Btu = wt × L ht

   Btu = 10 × 144

   Btu = 1440

5. Lower the temperature of the ice from 32°F (0°C) to 12°F (–11°C) (sensible heat). The temperature change will be 20°F (11°C).

   Btu = wt × sp ht × tc

   Btu = 10 × .487 × 20°F

   Btu = 4.87 × 20

   Btu = 97.4

Add the answers to obtain the total amount of Btu removed to change 10 lb of steam into ice at 12°F (–11°C).

Lowered steam temperature to 212°F (100°C) = 36.8 (sensible heat)
Condensed steam into water = 9704 (latent heat)
Lowered water temperature to 32°F or 0°C = 1800 (sensible heat)
Changed the water into ice = 1440 (latent heat)
Lowered the temperature of ice to 12°F or –11°C = 97.4 (sensible heat)
Total = 13,078.2 Btu

Imagine the steam, water, and ice are enclosed in a copper container that weighs 15 lb. The copper must be included in the problem since its temperature also must be lowered. The copper will remain a solid at the temperatures involved, so its temperature can be lowered directly from 220°F (104°C) to 12°F (–11°C). Therefore, its temperature change will be 208°F (115°C).

Btu = wt × sp ht × tc

Btu = 15 × .095 × 208°F

Btu = 1.425 × 208

Btu = 296.4

To calculate the total Btus of the change of state of steam to ice and the temperature change of both the copper container and the water, add both Btu totals.

Btu of steam to ice and temperature change = 13,078.2 Btu
Btu of copper container temperature change = 296.4 Btu
Total = 13,374.6 Btu (new total of heat to be removed)

Latent Heat of Sublimation

Latent heat of sublimation describes the process in which a substance changes directly from a solid to a vapor, without passing through the liquid phase. A common example of this process is solid carbon dioxide (CO_2) or "dry ice." This substance will bypass the liquid state and ***sublimate*** (go directly from a solid to a gas).

Another example of sublimation is a tray of ice cubes left in the freezer for an extended period. The tray may have been full of ice to begin with but will sublimate if not used. When returning from vacation, for example, the ice cube tray may be half full or almost empty.

To review, the five latent heats and their related state changes include:

- Sublimation = solid to vapor
- Melting = solid to liquid
- Freezing (fusion) = liquid to solid
- Vaporization (evaporation) = liquid to vapor
- Condensation = vapor to liquid

Specific Heat

Specific heat is the amount of heat required to raise the temperature of 1 lb of a substance 1°F. This definition is almost the same as that for a Btu, except it considers *all* substances, not just water. The specific heat of water is 1 Btu, but the amount of heat required (in Btu/lb) to cause a temperature change in other substances will vary with each substance. Each substance requires different amounts of Btu per pound. Scientists have already calculated the specific heats for most substances. These figures are readily available in technical manuals or other references in a library. **Figure 11-14** lists specific heats for some common substances.

Determining the number of Btu needed to raise or lower the temperature of a substance involves three factors:

- The number of pounds of the substance involved
- The number of degrees the substance's temperature is to be raised
- The specific heat for the substance

To find the number of Btu required, multiply the *weight (wt)* times the *specific heat per pound (sp ht)* times the desired *temperature change (tc)*.

The formula, *Btu = wt × sp ht × tc*, makes it possible to quickly figure quantities of heat. For example:

How many Btu must be removed to cool 50 lb of water from 75°F to 55°F?

$$\text{Btu} = \text{wt} \times \text{sp ht} \times \text{tc}$$
$$\text{Btu} = 50 \text{ lb} \times 1 \times 20°\text{F}$$
$$\text{Btu} = 50 \times 20$$
$$\text{Btu} = 1000$$

How many Btu must be added to 150 lb of ice to change the temperature from –20°F to 30°F? The table in **Figure 11-14** shows the specific heat of ice to be 0.487.

$$\text{Btu} = \text{wt} \times \text{sp ht} \times \text{tc}$$
$$\text{Btu} = 150 \text{ lb} \times 0.487 \times 50°\text{F}$$
$$\text{Btu} = 73.05 \times 50$$
$$\text{Btu} = 3652.5$$

How many Btu must be added to 100 lb of steam to change the temperature from 225°F to 245°F? The table in **Figure 11-14** shows the specific heat of steam to be 0.460.

$$\text{Btu} = \text{wt} \times \text{sp ht} \times \text{tc}$$
$$\text{Btu} = 100 \text{ lb} \times .46 \times 20°\text{F}$$
$$\text{Btu} = 46 \times 20$$
$$\text{Btu} = 920$$

Saturated Condition

Saturated condition is a term that refers to the boiling/condensing point of a substance. When the temperature/pressure combination of a substance is such that the substance can *change its state*, the condition is called the ***saturation point***. The boiling/condensing point of a substance is dictated by a specific combination of temperature and pressure.

For water at atmospheric pressure, saturated condition is achieved at 212°F (100°C), **Figure 11-15**. At this temperature/pressure, it is possible to have all water, a mixture of water and steam, or all steam. Whenever a mixture of vapor and liquid exists, it is *saturated* because it is in the process of changing state.

Superheated Vapor

The term ***superheated vapor*** describes a gas that has been heated to a temperature *above* its boiling point. The air is composed of superheated vapors, since the gases that make up air (nitrogen, oxygen, and carbon dioxide) have boiling points well below zero at atmospheric pressure.

Figure 11-16 shows steam in a superheated condition. Since water vaporizes (becomes steam)

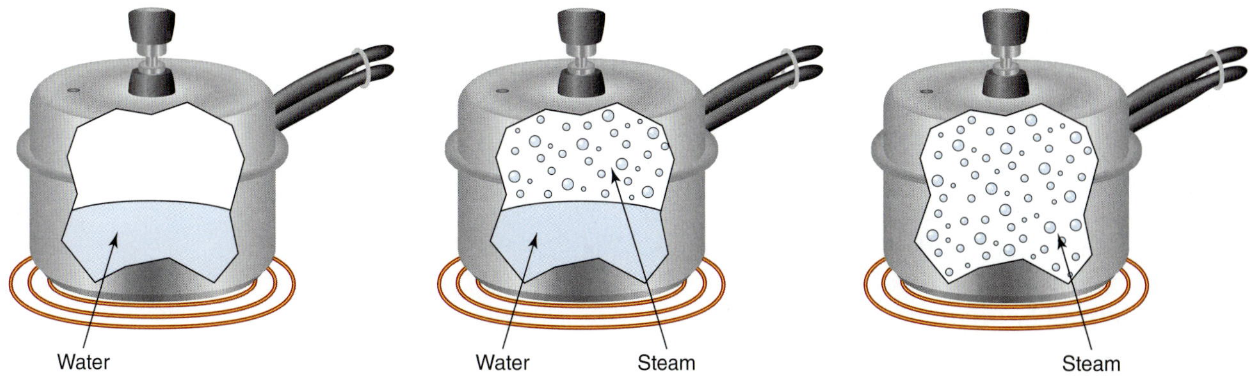

Goodheart-Willcox Publisher

Figure 11-15. The saturation point of a substance is a specific temperature at a specific pressure. For water, it is 212°F (100°C) at atmospheric pressure. Under saturated conditions, the substance can be water, steam, or a mixture of water and steam, all at the same temperature/pressure.

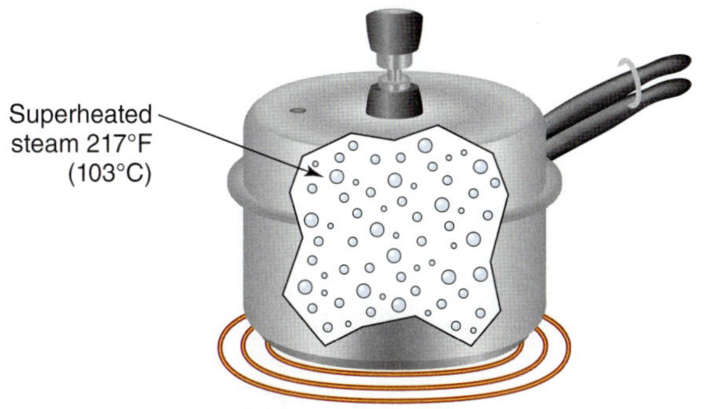

Goodheart-Willcox Publisher

Figure 11-16. When a substance is superheated, sensible heat is added after the substance changes its state. The steam (water vapor) shown has been superheated 5°F (3°C) to a temperature of 217°F (103°C).

at 212°F (100°C), any heat energy added to the vapor will raise its temperature (sensible heat), making it a superheated vapor. The specific heat of steam is 0.46, so it requires 0.46 Btu/lb to superheat steam to any temperature above its saturation (boiling) point. Steam at 217°F (103°C) would be superheated by 5°F (217 – 212 = 5) or 3°C.

PRO TIP

Vapors and Gases

The terms *gas* and *vapor* are often used interchangeably to describe a substance in the gaseous state. In a strictly scientific sense, a gas refers to a substance in only the gas phase and a vapor refers to a substance that is a mixture of two phases. In practical, everyday applications in the HVACR field, however, this distinction is not used. This book follows that trade practice.

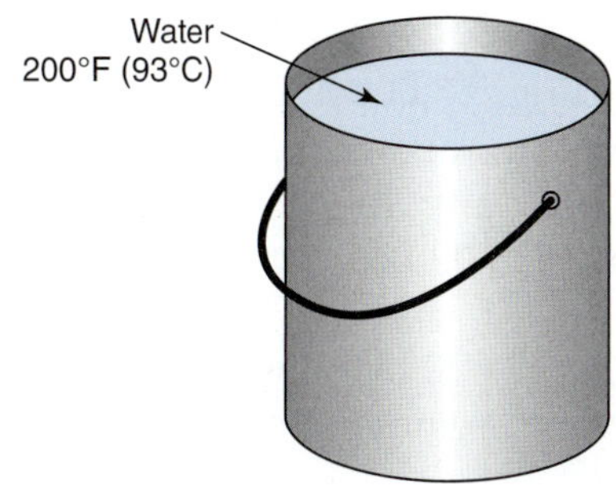

Goodheart-Willcox Publisher

Figure 11-17. A substance that has a temperature below its saturation point is said to be subcooled. This container of liquid has been subcooled to 200°F (93°C), which is 12°F (7°C) below its saturation point 212°F (100°C).

Subcooled Liquid

A liquid at a temperature *below* its saturation point can be described as ***subcooled***. An example of subcooling is shown in **Figure 11-17**. Water at 200°F (93°C) is subcooled 12°F (7°C) since the saturation point of water is 212°F (100°C).

PRO TIP

Superheat and Subcooling

Superheat and *subcooling* are important measurements for a technician to understand and calculate. This knowledge will help in understanding what is happening to refrigerant inside of the refrigeration system and how it relates to proper system capacity and function. These concepts will be discussed in more detail in later chapters.

11.3 Ton of Refrigeration Effect

Ice (water in a solid state) was the forerunner of all refrigeration systems and is still used today to compare the refrigerating effect of various systems. The unit of measurement is called a *ton* and refers to the ***Ice Melting Equivalent (IME)***—the amount of heat absorbed in melting one ton (2000 lb) of ice at 32°F (0°C) in exactly 24 hours. See **Figure 11-18**.

As discussed earlier, 1 lb of ice will absorb 144 Btu when it melts. Therefore, 288,000 Btu are needed to melt one ton of ice (2000 lb × 144 Btu). Any refrigeration system capable of removing 288,000 Btu in 24 hours is, therefore, called a *one-ton system*.

By dividing 288,000 by 24, it can be determined how many Btu are being removed each hour (12,000). Thus, any refrigeration system capable of removing 12,000 Btu per hour is also a one-ton system.

Dividing 12,000 by 60 will identify how many Btu are being removed each minute (200). An absorption rate of 200 Btu per minute is also equivalent to one ***ton of refrigeration effect***.

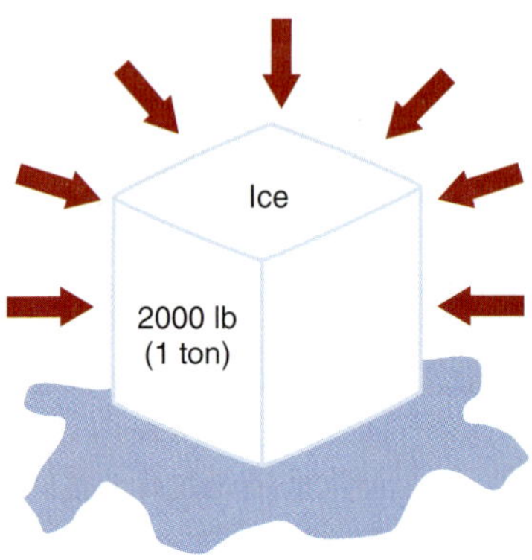

Goodheart-Willcox Publisher

Figure 11-18. A ton of refrigeration effect is equal to the removal of 288,000 Btu of heat by a block of ice in a 24-hour period (12,000 Btu/hr or 200 Btu/min).

Thus, there are three ways of expressing a Btu removal rate equal to a one-ton system. One of these methods usually appears on the data plate of the system:

- 288,000 Btu/24 hrs
- 12,000 Btu/hr
- 200 Btu/min

An air-conditioning system rated at three tons would be capable of removing 36,000 Btu/hr (600 Btu/min). A half-ton window unit would be capable of removing 6000 Btu/hr (100 Btu/min). Most systems are rated on a per-hour basis, but sometimes the data plate rating can be in Btu/min, **Figure 11-19**.

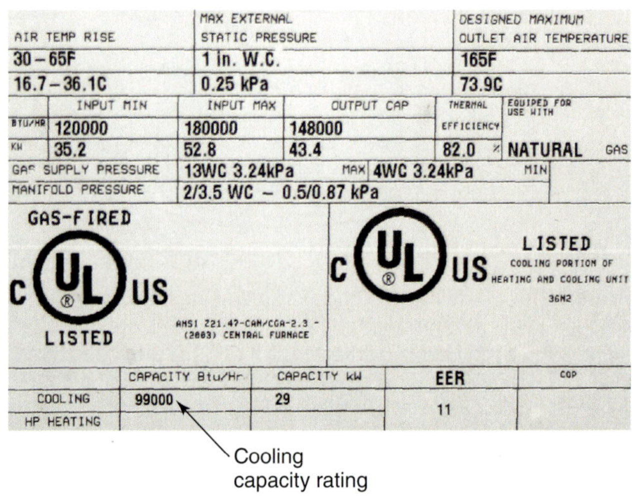

Goodheart-Willcox Publisher

Figure 11-19. This commercial heating and cooling rooftop unit has a cooling capacity rated at 99,000 Btu per hour. Multiply this value by 24 hours to get 2,376,000 Btu per day. Divide that by 288,000 Btu (1 ton) to get 8.25 tons of refrigeration. This system provides a lot of cooling.

Chapter Review

Summary

11.1 The First Law of Thermodynamics states that energy cannot be created or destroyed, but energy can be converted from one form to another. The Second Law of Thermodynamics states that heat always travels from hot to cold. There are three basic methods by which heat is transferred from one substance to another: radiation, conduction, and convection.

11.2 Most substances can exist in three physical states: solid, liquid, or gas. As heat is added or removed, the substance may go through a change of state, which involves the rearranging of the substance at the molecular level to produce this change.

11.2 Sensible heat is heat that causes a change in temperature but not a change of state. Latent heat is heat energy that causes a change of state but no temperature change. Specific heat is the amount of heat required to raise the temperature of 1 lb of a substance 1°F.

11.2 Latent heat of melting refers to the amount of heat added to change a solid to a liquid without changing the substance's temperature. Latent heat of freezing (or fusion) refers to the amount of heat removed to change a liquid to a solid without changing the substance's temperature. Latent heat of vaporization (or evaporation) refers to the amount of heat added to change a liquid to a vapor without changing the substance's temperature. Latent heat of condensation refers to the amount of heat removed to change a gas (or vapor) to a liquid without changing the substance's temperature. Latent heat of sublimation describes the process in which a substance changes directly from a solid to a vapor, without passing through the liquid phase.

11.2 Saturated condition is a term that refers to the boiling/condensing point of a substance. Superheated vapor describes a gas that has been heated to a temperature above its boiling point. Subcooled liquid refers to a liquid at a temperature below its saturation point.

11.3 Ice is used to measure the refrigerating effect of various systems using Ice Melting Equivalent (IME) as the unit of measure. An absorption rate of 200 Btu per minute is equal to one ton of refrigeration effect.

Know and Understand

1. *True or False?* The Second Law of Thermodynamics states that heat always travels from cold to hot. (11.1)
2. The sun's heat waves are an example of heat moving by _____. (11.1)
   A. radiation
   B. convection
   C. conduction
   D. insulation
3. In a forced-air heating system, _____ is used to move heated air through a system. (11.1)
   A. conduction
   B. natural convection
   C. forced convection
   D. radiation
4. Atoms are always _____. (11.2)
   A. still
   B. in rapid motion
   C. changing state
   D. removing heat
5. When a substance is changing from one physical state to another, the temperature level _____. (11.2)
   A. is increasing
   B. is decreasing
   C. remains constant
   D. is not a factor considered
6. *True or False?* A change of physical state requires the same amount of heat energy to boil one pound of water as it does to condense one pound of steam. (11.2)

7. Sensible heat is heat that _____. (11.2)
   A. causes a change of state but not a change in temperature
   B. causes a change in temperature but not a change of state
   C. cannot be measured with a thermometer
   D. raises the temperature of 1 lb of a substance 1°F
8. *True or False?* At 32°F (0°C), a container of water can be all water, all ice, or a combination of water and ice. (11.2)
9. The amount of heat removed to change a liquid to a solid without changing the substance's temperature is referred to as latent heat of _____. (11.2)
   A. melting
   B. freezing
   C. vaporization
   D. sublimation
10. Calculate the amount of latent heat required to condense 25 lb of steam into water. (11.2)
   A. 3600 Btu
   B. 24,248.5 Btu
   C. 24,260 Btu
   D. 24,271.5 Btu
11. The process in which a substance changes directly from a solid to a vapor, without passing through the liquid phase is referred to as latent heat of _____. (11.2)
   A. condensation
   B. vaporization
   C. melting
   D. sublimation
12. How many Btu must be added to heat 5 lb of water from 75°F to 212°F? (11.2)
   A. 315.1 Btu
   B. 685 Btu
   C. 664,724 Btu
   D. 665,409 Btu
13. Whenever a substance is composed of a mixture of vapor and liquid, the substance is referred to as _____ because it is in the process of changing state. (11.2)
   A. sublimated
   B. vaporized
   C. saturated
   D. superheated
14. Steam at 215°F is considered _____. (11.2)
   A. subcooled by 3°F
   B. subcooled by –3°F
   C. superheated by 3°F
   D. superheated by 0°F
15. When a liquid is at a temperature lower than its saturation point, it is referred to as _____. (11.2)
   A. subcooled
   B. superheated
   C. saturated
   D. fused
16. Which of the following is *not* equivalent to one ton of refrigeration effect? (11.3)
   A. 288,000 Btu/24 hrs
   B. 12,000 Btu/hr
   C. 200 Btu/min
   D. 16.67 Btu/sec

Apply and Analyze

1. How does the color of a material affect radiant heat flow? (11.1)
2. Explain what happens on a molecular level to a substance when it undergoes a change of physical state. (11.2)
3. List three factors that need to be considered when determining the number of Btu needed to raise or lower the temperature of a substance. (11.2)
4. If 970 Btu are removed from trapped steam, what will happen to the steam? (11.2)
5. Give an example of a substance that goes through sublimation. (11.2)
6. What does the Ice Melting Equivalent (IME) unit of measurement refer to and how is it used in HVACR systems? (11.3)

Critical Thinking

1. Describe how an HVACR system might use a combination of heat transfer methods to move heat. (11.1)
2. How many Btu need to be added to change 5 lb of ice at 30°F to steam at 215°F? Refer to **Figure 11-13** and **Figure 11-14** to obtain specific heat and latent heat values. (11.2)
3. How many tons of refrigeration effect can be obtained from a window air-conditioning unit rated at 18,000 Btu/hr? (11.3)

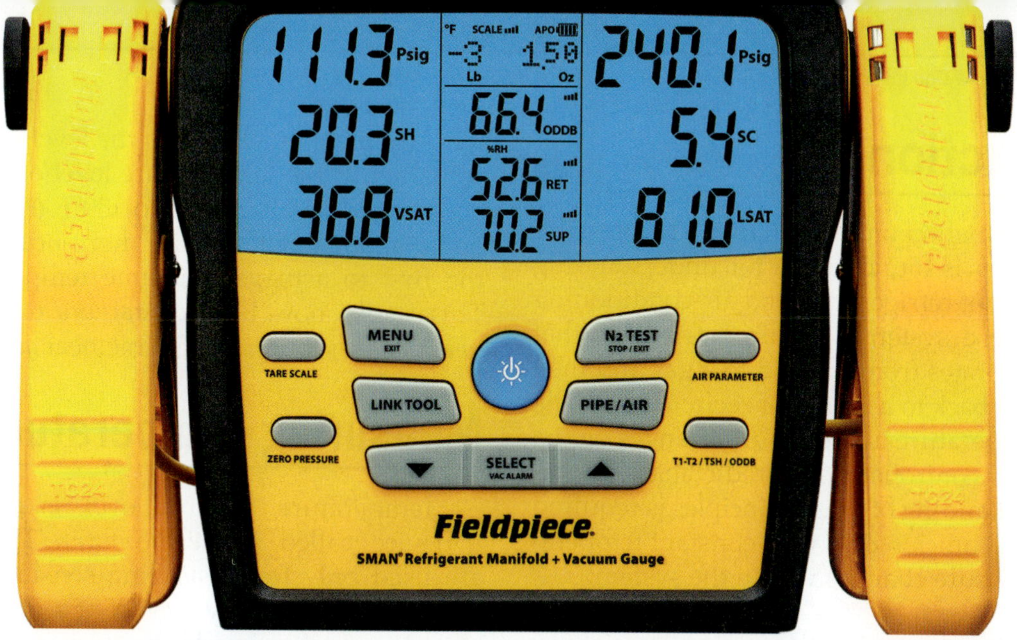

Fieldpiece

12 Temperature and Pressure

LEARNING OUTCOMES

After studying this chapter, you will be able to:

12.1 Describe temperature, the four types of temperature scales, and ambient temperature.

12.2 Explain how atmospheric pressure, vacuum, absolute pressure, and gauge pressure are measured.

12.3 Apply Boyle's Law, Charles' Law, Combined Gas Law, and Dalton's Law to air-conditioning and refrigeration systems.

12.4 Explain how pressure-temperature charts are used in HVACR.

TECHNICAL TERMS

absolute pressure	centigrade	Kelvin scale	psig
absolute zero	Charles' Law	manifold gauge	pull a vacuum
ambient temperature	compound gauge	mercury barometer	Rankine scale
atmospheric pressure	Dalton's Law	partial vacuum	recalibration
boiling point	Fahrenheit scale	perfect vacuum	temperature
Boyle's Law	gauge pressure	pointer flutter	thermometer
Celsius scale	in. Hg	psia	variable

Introduction

The information on temperature and pressure in this chapter is foundational for understanding the operation of refrigeration and air-conditioning systems. These systems contain refrigerant that readily evaporates from a liquid to a gas and then is condensed back to a liquid for recirculation. The effect of temperature and pressure changes on the refrigerant is controlled and predictable. A thorough understanding of these principles is required to help the technician deal with constant temperature and pressure changes within the system.

12.1 Temperature

Temperature does not reveal the *quantity* of heat in a substance. It merely indicates the *intensity*, or heat level (how hot or cold a substance is). In the molecular theory of heat, temperature indicates the speed of motion of a single molecule. Temperature specifies how hot a substance is but does *not* indicate how many Btu were needed to reach that temperature level.

For example, consider two containers of water over identical heat sources. In the first container, there is one gallon of water. In the second container, there are five gallons. See **Figure 12-1**. If the same *amount* of heat is applied to each container, the temperature level of the water in the first container will rise more quickly than the level in the second container. If the temperature level of the water in both containers is raised to 180°F (82°C), the second container will require more Btu of heat than the first to achieve the same temperature level. This shows how both containers have the same heat *intensity*, but *not* the same heat *quantity*.

Measuring Temperature

Temperature, or heat level, is measured by a device called a ***thermometer***. Thermometers are used only for measuring sensible heat. Bulb thermometers usually consist of a glass tube calibrated in *degrees*. Inside the tube is a liquid, such as mercury or colored alcohol, that will expand or contract at a known rate with changes in temperature. Expansion or contraction will cause the liquid to rise or fall within the tube, where its level can be compared to the degree marks. Most HVAC service technicians now use digital thermometers, **Figure 12-2**. These are more durable, accurate, and easier to read. Also, some models allow the user to measure two different temperatures at once.

In the US, temperature is usually measured in degrees Fahrenheit, but in other parts of the world the Celsius scale is used.

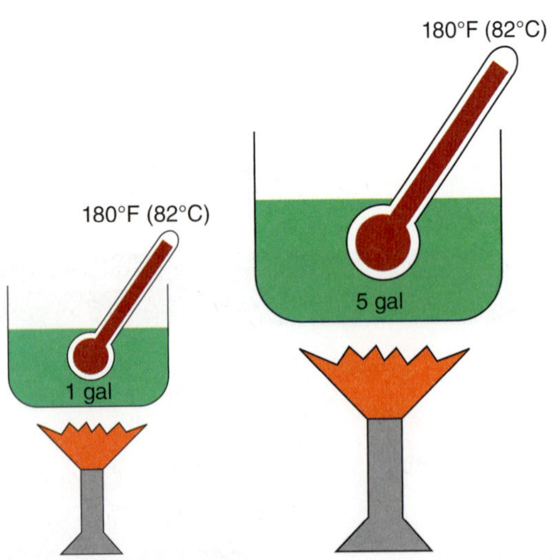

Goodheart-Willcox Publisher

Figure 12-1. To reach the same temperature, the larger quantity of water (right) would require more Btu of heat than the smaller quantity of water (left).

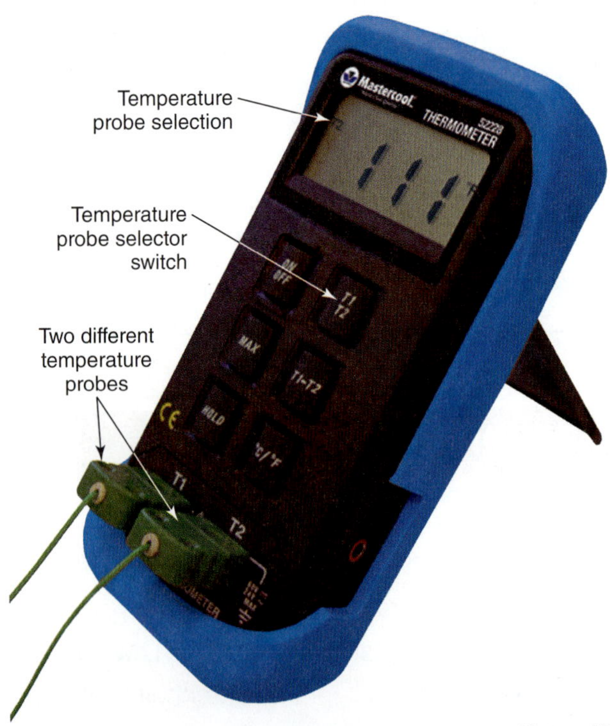

Mastercool Inc.

Figure 12-2. Digital thermometers are used by most HVACR service technicians to take the required temperature readings on a system.

Fahrenheit Scale

In 1714, the German scientist Gabriel Fahrenheit (1686–1736) became the first person to use mercury as the liquid in a thermometer. Mercury allowed more accurate measurement of temperature changes than other liquids previously used. To calibrate his thermometer, Fahrenheit chose the freezing point of a mixture of salt and water as the *zero* point. Ordinary water freezes at 32° on the ***Fahrenheit scale*** and boils at 212°. As shown in **Figure 12-3**, there are exactly 180 divisions (*degrees*) between the freezing and boiling points of water (212 – 32 = 180). Any temperature that occurs below zero is expressed as a negative or "minus." For example, a temperature of 10 degrees below zero is written as *–10°F.*

Absolute Zero

When Fahrenheit devised his scale, he set the zero point at the temperature where saltwater freezes since this was the lowest temperature he could achieve in the laboratory. However, the lowest possible temperature is far below 0°F.

Scientists have established that heat is always present until a temperature of –460°F (–273°C) is achieved. This is the lowest possible temperature because *all* heat has been removed. It is, therefore, called ***absolute zero***. By comparison with absolute zero, the coldest weather we might experience is quite warm. Atmospheric air of 0°F actually contains many Btus of heat.

Celsius Scale

The Swedish astronomer Anders Celsius (1701–1744) believed the Fahrenheit scale was impractical for laboratory use because its 180 divisions between the freezing and boiling points of water required troublesome mathematical calculations. To overcome this problem, Celsius developed his own scale in 1742. He called it the ***centigrade*** ("hundred steps") scale because it had exactly 100 divisions between the freezing (0°C) and boiling (100°C) points of water, **Figure 12-4**. This new scale was quickly adopted by other scientists. Today, it is used by most countries of the world. Since 1948, the centigrade scale officially has been known as the ***Celsius scale***.

Rankine and Kelvin Scales

For some types of scientific work, the use of *absolute temperature scales*—scales with no negative numbers—were a great convenience. In the mid-1800s, two Scottish scientists, William Rankine and William Thomson (Lord Kelvin), developed two different approaches to the absolute temperature scale.

In the ***Rankine scale***, Fahrenheit divisions are used. Absolute zero is 0°R, the freezing point

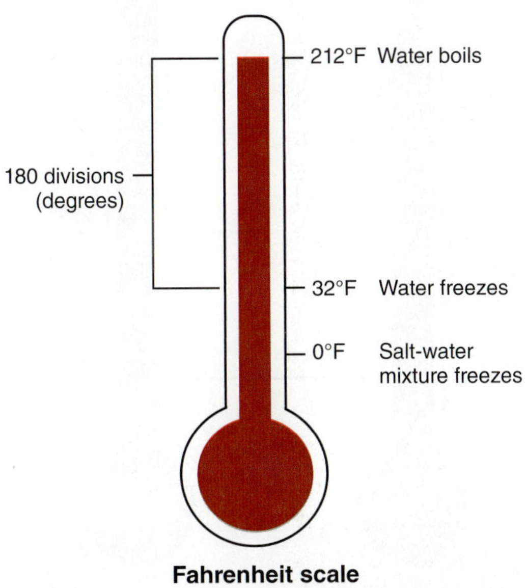

Goodheart-Willcox Publisher

Figure 12-3. The Fahrenheit temperature scale has 180 divisions between the freezing and boiling points of water. Zero is set at the point where a salt-water mixture freezes.

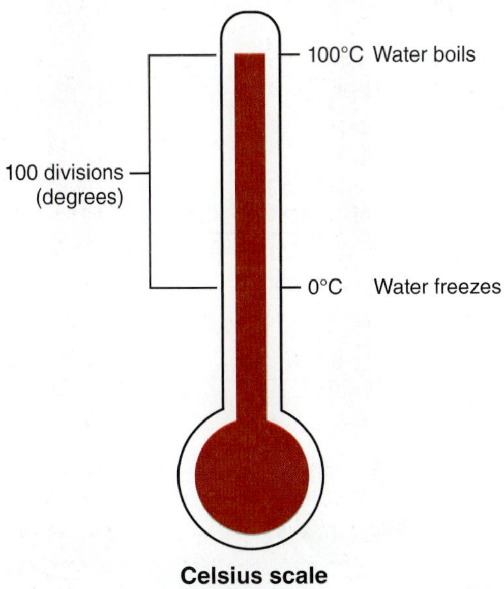

Goodheart-Willcox Publisher

Figure 12-4. The Celsius temperature scale has 100 divisions between the freezing and boiling points of water. The scale is easier to use than the Fahrenheit scale, especially in scientific work.

of water is 492°R, and the boiling point of water is 672°R. The Rankine scale is also known as the *Fahrenheit Absolute* (F_A) scale.

Kelvin's approach was identical, but he preferred the Celsius scale. Absolute zero is 0K, water freezes at 273K, and water boils at 373K. The ***Kelvin scale***, also called the *Celsius Absolute* (C_A) scale, is widely used by scientists. **Figure 12-5** shows a side-by-side comparison of the four temperature scales. Note that Kelvin temperature values are written with just the number and then the letter *K*. Kelvin values do not include the degree symbol (°), as is used in the other temperature scales.

Ambient Temperature

The term ***ambient temperature*** is used frequently in the refrigeration field. It simply refers to the temperature surrounding an object, typically a motor or condenser. Ambient temperatures can have a great influence on the operating conditions of a system or part of a system. For example, a motor may be rated to give maximum efficiency at any temperature that does *not* exceed 72°F (40°C) above *ambient*. The word *ambient* makes communication easier when referring to the temperature conditions surrounding an object.

Many times, a refrigeration system will have components located outside the building that are connected to other components located inside the building. The ambient temperature of the inside components is controlled, and thus remains relatively constant. However, the components located outside are exposed to all types of weather conditions, **Figure 12-6**. These ambient temperature conditions can be rather extreme compared to those experienced by indoor components.

12.2 Pressure

Generally, pressure can be understood as the force applied to an object or substance in a specified area. Similar to temperature, pressure can be measured in various ways. Understanding how to measure pressure, how these measurements relate to temperature, and how pressure overall affects HVACR systems is essential when servicing these systems.

Atmospheric Pressure

Our planet is surrounded by a gaseous atmosphere that consists of about 78% nitrogen and 21%

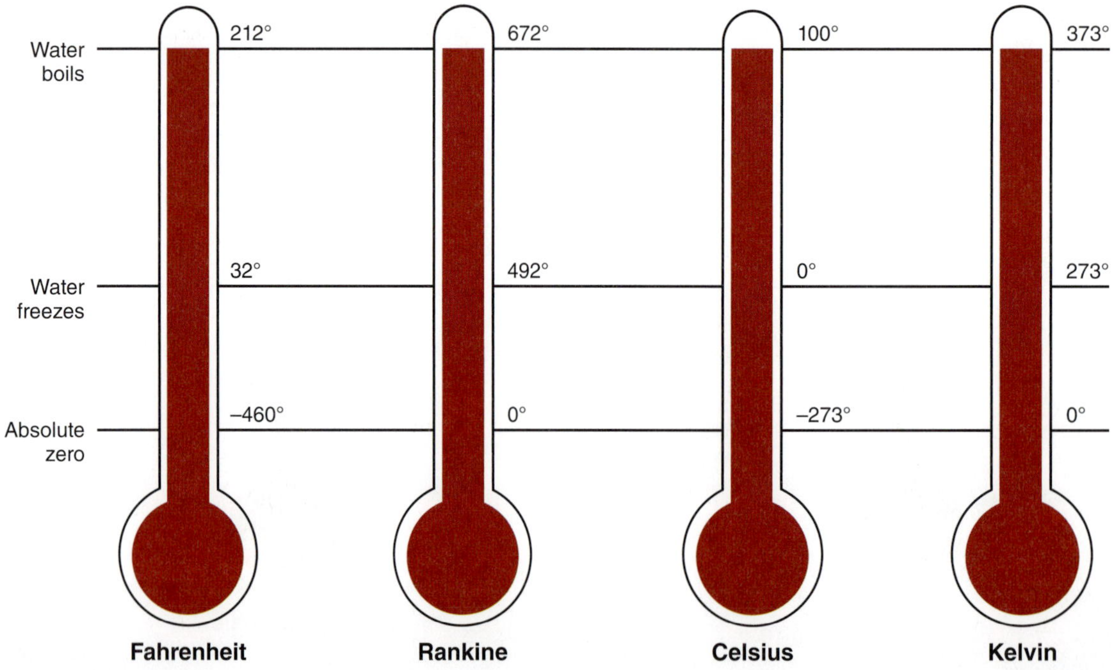

Goodheart-Willcox Publisher

Figure 12-5. A comparison of the four temperature scales. Both the Rankine and Kelvin scales use absolute zero as their zero point. The Rankine scale is based on Fahrenheit degree divisions. Kelvin is based on Celsius divisions.

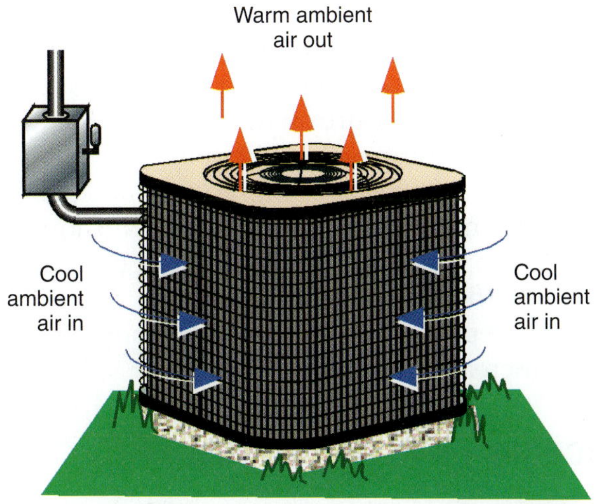

Goodheart-Willcox Publisher

Figure 12-6. Ambient temperature refers to the temperature surrounding an object. Cool ambient air is drawn through the outdoor condenser of an air-conditioning system. As the moving air absorbs heat from the condenser coils, it becomes warm ambient air.

oxygen. The remaining 1% is composed of other gases. The atmosphere extends approximately 50 miles above the earth's surface and is held in place by the force of gravity. The gases that make up the atmosphere are composed of tiny molecules traveling in all directions. Even though they are small, these molecules are not weightless. The weight of the molecules exerts a downward force, known as ***atmospheric pressure***, upon the earth's surface. This pressure is greatest at sea level and is measured at 14.7 psi (101.3 kPa). At higher altitudes, such as a mountaintop, the atmospheric layer is much thinner and exerts less pressure. See **Figure 12-7**.

Because the atmosphere is composed of a mixture of gases, pressure is exerted in all directions. At sea level, anyone or anything exposed to the atmosphere will have 14.7 pounds of pressure exerted against every square inch of surface area.

Atmospheric Pressure and Boiling Point

Atmospheric pressure has a direct influence on the ***boiling point*** of water and other liquids. If the amount of pressure applied on a liquid is changed, the boiling point will also change. At sea level, where atmospheric pressure is 14.7 psi (101.3 kPa), water boils at 212°F (100°C). At the summit of a mountain 15,000 feet high, however, atmospheric pressure is only 8.3 psi (57.3 kPa), so water boils at 184°F (84°C).

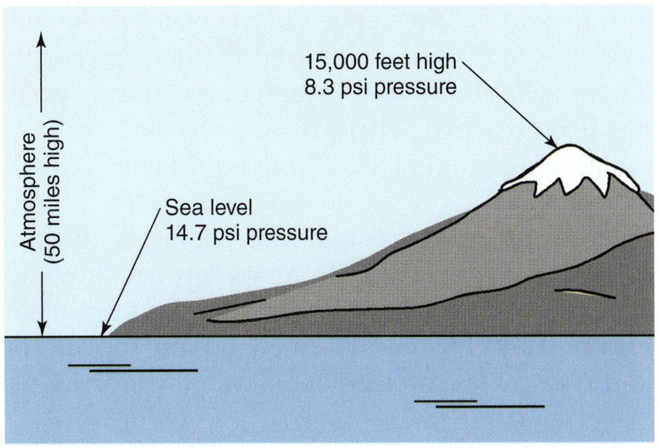

Goodheart-Willcox Publisher

Figure 12-7. Atmospheric pressure (14.7 psi at sea level) is a result of the weight of gas molecules in the 50-mile-high thick layer of air surrounding the earth. The higher the altitude, the lower the pressure because the blanket of air is thinner (and thus weighs less).

When a liquid substance, such as water, reaches its boiling point (saturation point), added latent heat causes a change of state. The temperature of the substance remains constant until the change is completed. To illustrate this process, consider an open pan of water over a flame on a kitchen range, **Figure 12-8**. In this example, the water is at atmospheric pressure, so it will boil when a temperature of 212°F (100°C) is reached. The water cannot get hotter, *regardless of the amount of heat applied*. It will only boil at a faster rate. The temperature will remain constant until the change of state is completed, and all the liquid has changed to a gas.

Atmospheric pressure creates surface tension on top of the water. The tension opposes the water molecules trying to break through the surface and become a gas. As the temperature of the water increases, the speed of the water molecules

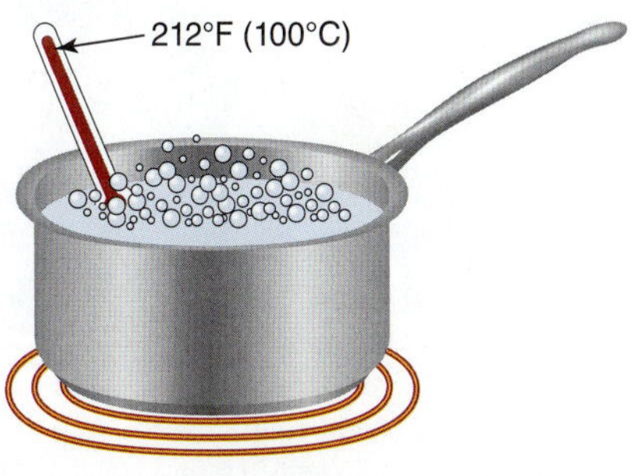

Goodheart-Willcox Publisher

Figure 12-8. At normal atmospheric pressure (14.7 psi), water boils at 212°F (100°C). The water will remain at that temperature until its state is completely changed (turns to steam). Adding more heat will merely make the water boil faster.

also increases. If the speed of the water molecules increases to the point where surface tension can no longer keep them from escaping, the boiling point has been reached. There is now a mass escape of molecules changing state from liquid to gas (water to steam).

The pressure cooker was devised to permit water to be heated to temperatures above its normal boiling point without changing state, **Figure 12-9**. Placing a sealed lid on top of the kettle traps the molecules of water vapor in the small space between the lid and the surface of the boiling water. As space becomes crowded with escaped water molecules, the pressure increases. This increased pressure on the surface of the water prevents the escape of additional molecules. In effect, the water stops boiling. To continue the process of changing the water's state from liquid to gas, the temperature of the water must be raised to overcome the increased surface pressure. Thus, as the surface pressure is increased, the boiling point is also increased.

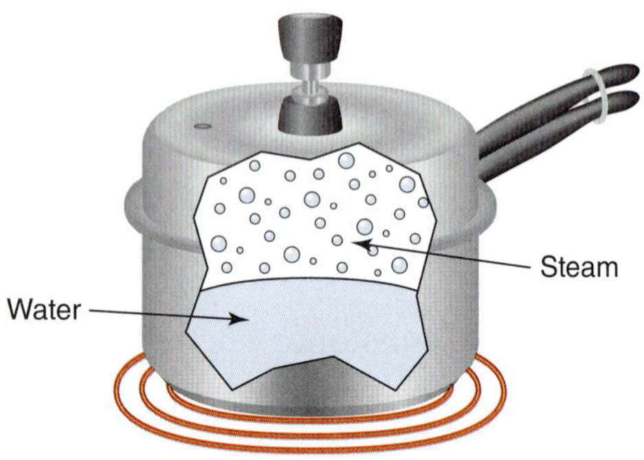

Pressure Above Atmospheric	Boiling Point
5 psi	228°F (109°C)
10 psi	244°F (118°C)
15 psi	250°F (121°C)
35 psi	281°F (138°C)
85 psi	328°F (164°C)

Goodheart-Willcox Publisher

Figure 12-9. By enclosing a liquid and increasing pressure (as in a pressure cooker), the boiling point of the liquid can be raised. Examples of boiling point increases are shown.

Vacuum

While an increase in surface pressure results in an increase in the boiling point, the *reverse* is also true: reducing surface pressure reduces the boiling point. Any pressure less than atmospheric is called a ***partial vacuum***. When *all* atmospheric pressure is removed, it is called a ***perfect vacuum***. If a perfect vacuum was achievable, water would boil at –90°F (–67.8°C) without a fire or other external heat source, **Figure 12-10**.

HVACR service technicians typically work with refrigerant pressures that are above and below atmospheric pressure. That is why a good understanding of vacuum principles is a must.

Refrigeration systems are designed to operate with only refrigerant and oil circulating inside them. However, air and moisture do enter the system when the unit is assembled or serviced. Air is a mixture of gases and water vapor. Air is noncondensable in a refrigeration system, meaning they do not change phase or contribute to cooling.

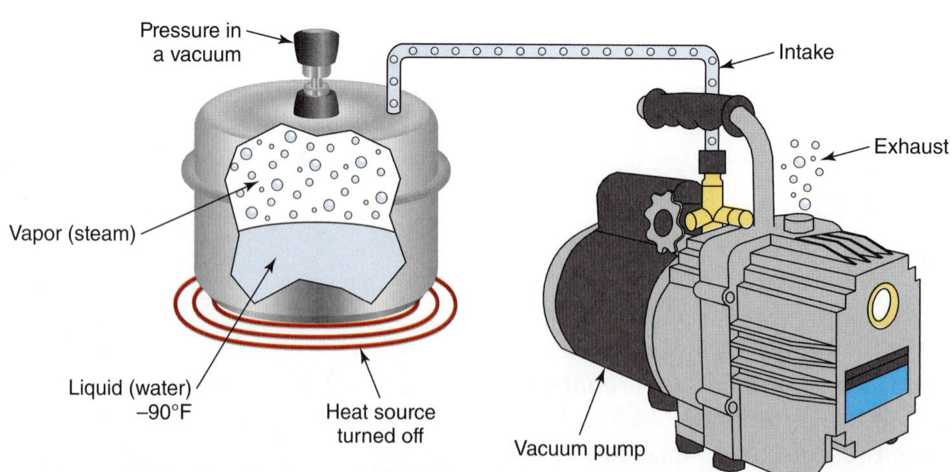

Goodheart-Willcox Publisher

Figure 12-10. Reducing atmospheric pressure on a liquid (creating a partial or perfect vacuum) lowers the temperature at which it will boil. In this example, a vacuum pump is used to remove all atmospheric pressure from the water in the closed container. In such a perfect vacuum, water will boil at –90°F (–68°C) without an external heat source.

Noncondensables in a refrigeration system cause high-discharge pressure (according to Dalton's Law of Gases, which is discussed later in this chapter). Moisture or water vapor in a refrigeration system forms hydrochloric and hydrofluoric acid, which can cause compressor burnout. Proper servicing removes noncondensable gases and moisture from the system prior to charging the system with a single, pure refrigerant. Also, proper maintenance procedures prevent air from infiltrating a charged system.

To ***pull a vacuum*** means to lower the pressure in a refrigeration system below atmospheric pressure. The procedure is performed with a vacuum pump such as the one depicted in **Figure 12-10**. The vacuum pump is a suction mechanism that removes any gases and reduces the pressure inside the system to a low vacuum. The vacuum pump removes moisture by lowering the boiling point of water and then removing the resulting vapor.

Measuring Vacuum

There are several methods used to measure vacuum, depending on the purpose and accuracy required. One instrument used to measure atmospheric pressure is the ***mercury barometer***. As shown in **Figure 12-11**, this device consists of a hollow glass tube about 34″ high, sealed on one end and open at the other. The glass tube is filled with mercury, turned upside down, and the open end placed in a dish half-filled with mercury.

The level of mercury in the glass tube will correspond to the atmospheric pressure being exerted on the mercury in the open dish. The normal atmospheric pressure of 14.7 psi (101.3 kPa) exerted on the exposed mercury in the dish will force a 29.92″ column of mercury up into the tube. From this, it can be determined that every pound of atmospheric pressure is equal to 2.035 inches of mercury in the column.

$$\text{in. Hg} \div \text{psi} = \text{Inches of mercury per pound of pressure}$$

$$29.92 \div 14.7 = 2.035$$

Mercury was once referred to as *quicksilver* or liquid silver. In Latin, the term *liquid silver* is *hydrargyrum*. The abbreviation of *hydrargyrum* is *Hg*, which became the chemical symbol for mercury. To indicate inches of mercury in vacuum measurements, the abbreviation ***in. Hg*** is used. For example, a partial vacuum of 25″ of mercury would be written *25 in. Hg*.

Microns

Deep vacuums are required before charging refrigeration systems. Deep vacuums are measured in microns for extreme accuracy. Atmospheric pressure equals 759,999 microns and a deep vacuum is about 500 microns. Pulling a deep vacuum requires a vacuum pump and electronic micron vacuum gauge, **Figure 12-12**. A technician's vacuum pump should be capable of pulling a vacuum of 500 microns.

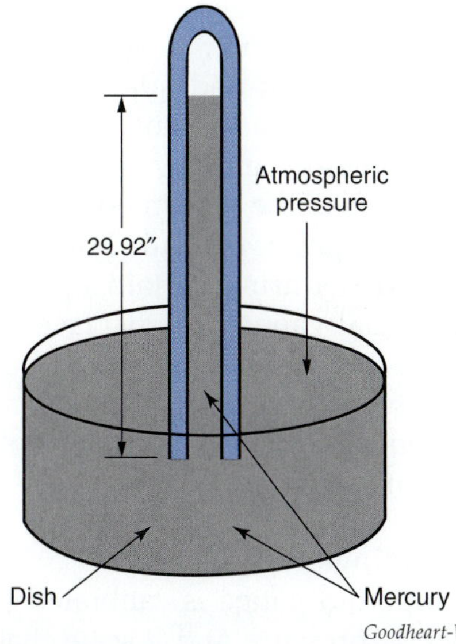

Goodheart-Willcox Publisher

Figure 12-11. A mercury barometer. Atmospheric pressure on the mercury in the open dish supports a column of mercury inside the tube. By marking the tube in increments of whole and fractional inches, pressure readings can be made. Normal atmospheric pressure will support a 29.92″ column of mercury in such a device. Vacuum is expressed in "inches of mercury" (in. Hg).

Ritchie Engineering – YELLOW JACKET Products Division

Figure 12-12. A vacuum pump and vacuum micron gauge are used to pull a deep vacuum on a refrigeration or air-conditioning system.

Millimeters of Mercury (mm Hg)

Vacuum is sometimes expressed in millimeters of mercury (mm Hg). The atmosphere will support a column of mercury 760 mm (29.92″) high. Millimeters of mercury (mm Hg) is considered an *absolute* measurement and is seldom used by service personnel. To achieve a perfect vacuum in a refrigeration system, for example, the pressure inside the system would have to be reduced to 0 psi, 29.92 in. Hg, 0 microns, or 0 mm Hg. This represents a perfect vacuum, which has never been achieved.

Absolute Pressure

Absolute pressure is any pressure above a perfect vacuum and is expressed in terms of *pounds per square inch absolute*, or ***psia***. Absolute pressure measurements allow both vacuums and pressures above atmospheric pressure to be expressed in the same units. Note that a reading of 10 psia would be a partial vacuum (which could also be expressed as 20 in. Hg). However, a reading of 20 psia would be well above the atmospheric pressure of 14.7 psia.

Gauge Pressure

Gauge pressure is expressed in *pounds per square inch gauge*, or ***psig***. It is different from absolute pressure (psia) in that it ignores vacuum. Readings in psig measure only pressures *above* atmospheric pressure. Gauges are calibrated to read 0 psi at atmospheric pressure (14.7 psia). A reading in psig can easily be converted to the corresponding absolute value (psia) by adding atmospheric pressure. For example: 20 psig + 14.7 = 34.7 psia. A simple way to remember the difference is that psig is pressure *above* atmospheric, and psia *includes* atmospheric pressure.

High-Pressure Gauge

All refrigeration systems are divided into two sides according to internal pressures. One side of the system operates at high pressure and the other side at low pressure. A high-pressure gauge makes readings on the high-pressure side of a system, **Figure 12-13**. The outermost scale on the gauge shown displays readings from 0 to 800 psig. The innermost scales on the gauge represent the corresponding saturation temperature in degrees Fahrenheit (°F) at specific pressures.

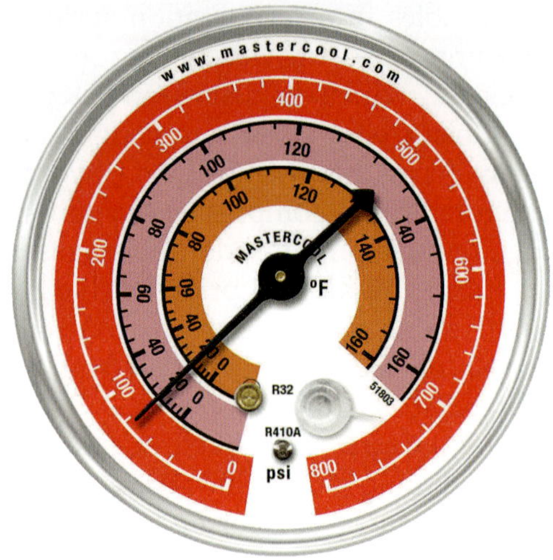

Mastercool, Inc.

Figure 12-13. A high-pressure gauge, used only for reading pressures above atmospheric. The zero point on this gauge actually represents a pressure of 14.7 psi (normal atmospheric pressure). To avoid confusion, pressure readings made with such a gauge are identified as "pounds per square inch gauge" (psig). Absolute readings are 14.7 psi higher and are identified as "psia."

The two refrigerants represented in **Figure 12-13** are R-32 and R-410A, and they range from 0°F to 160°F. These temperatures will vary depending on the refrigerant they represent. Readings are in psig because the zero point is calibrated at atmospheric pressure. A high-pressure gauge is usually color-coded red.

Compound Gauge

In refrigeration work, a service technician must frequently check and adjust the operating pressures of the system. As noted above, the system has high-pressure and low-pressure sides. The low-pressure side of the system normally operates at or above atmospheric pressure but may slip into the vacuum area during certain operating conditions. A special gauge that displays two scales makes it possible to read pressures both above and below atmospheric. Such gauges are called ***compound gauges*** and are usually color-coded blue. See **Figure 12-14**.

Pressure Readings

The compound gauge is calibrated with zero at atmospheric pressure and a scale that is accurate up to a certain pressure. In **Figure 12-14**, the high pressure is 300 psig, though this level may be higher for different compound gauges for different refrigerants. A retard area from 300 psig to 500 psig is provided to protect the gauge against

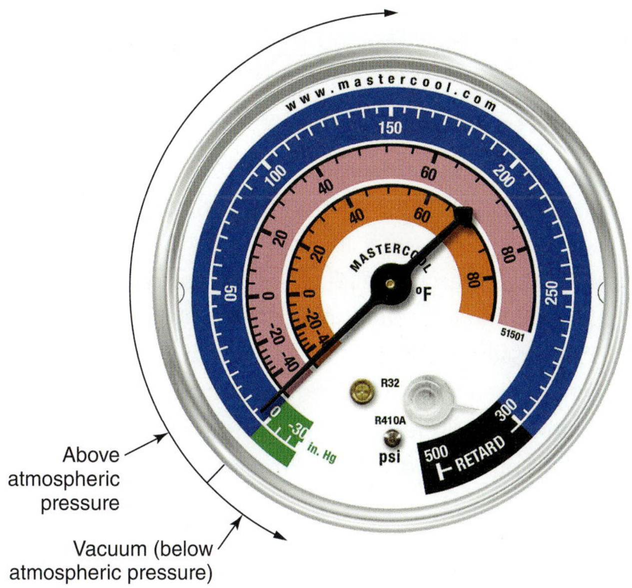

Mastercool, Inc.

Figure 12-14. A compound gauge makes possible both pressure and vacuum readings. The zero point is atmospheric pressure; pressure readings are above zero and vacuum readings are below. Vacuum values are the opposite of the scale used with the mercury barometer: normal atmospheric pressure is zero, and a perfect vacuum is 30 in. Hg.

overpressure but is not accurate. Pressure readings on the compound gauge are expressed as pounds per square inch gauge (psig).

Vacuum Readings

The other scale on the compound gauge is used for vacuum readings. This scale is calibrated downward in inches of mercury (0 in. Hg to 30 in. Hg). This numbering system is the reverse of the one used on the mercury barometer to measure atmospheric pressure changes. On the compound gauge, 30 in. Hg indicates a perfect vacuum. When pulling a vacuum on a system, do not rely on this gauge. Only a micron gauge should be used to get a proper vacuum in the system.

Rounding Off Readings

For practical purposes, most technicians "round off" atmospheric pressure from 14.7 psia to 15 psia and perfect vacuum from 29.92 in. Hg to 30 in. Hg. To see how these rounded units relate to one another, divide the higher value by the lower.

$$30 \text{ in. Hg} \div 15 \text{ psia} = 2$$

Then in rounded terms, every pound of atmospheric pressure equals two inches of mercury. This is called a 2:1 ratio (2 in. Hg = 1 psia).

As described earlier, when the needle is in the pressure scale (above atmospheric), the reading is expressed in pounds (psig). When the compound gauge needle is in the vacuum scale, the reading is expressed in inches of mercury (in. Hg). This distinction is important. Always indicate the units when expressing a measured value.

Manifold Gauge

A ***manifold gauge*** is a pressure-checking device that is equipped with compound and high-pressure gauges, control valves, and hose connectors, Figure 12-15. It is used to determine the operating characteristics inside a system. It also has control valves and connectors for hoses to the service valves. Service valves are discussed in Chapter 14, *Other Refrigeration System Components.* The manifold gauge provides a convenient way of checking both high-side and low-side pressures which, in turn, help the technician determine needed repairs. The manifold gauge is also used while performing the repairs.

The manifold gauge has a compound gauge (blue) and a high-pressure gauge (red) for a wide range of readings, Figure 12-16. Just as with other pressure gauges, the gauges on the manifold gauge are designated for use in pressure-temperature ranges for specific refrigerants.

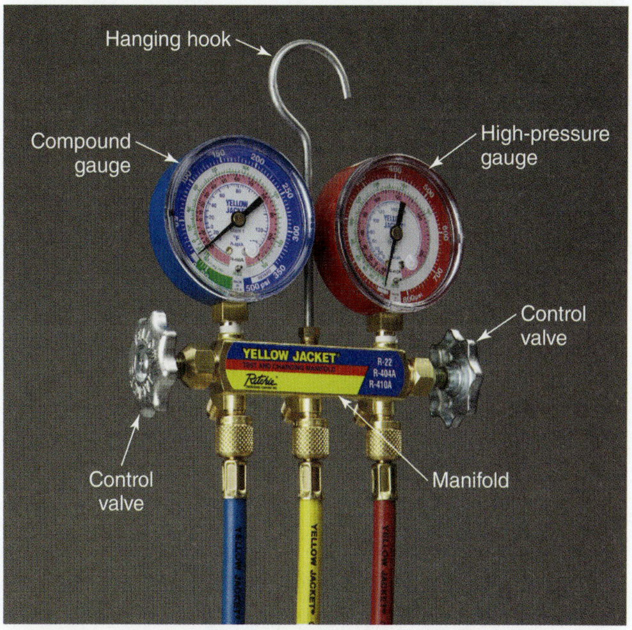

Ritchie Engineering – YELLOW JACKET Products Division

Figure 12-15. The manifold gauge is the technician's basic tool for determining system pressures. They are equipped with two gauges, two control valves, and three connection hoses.

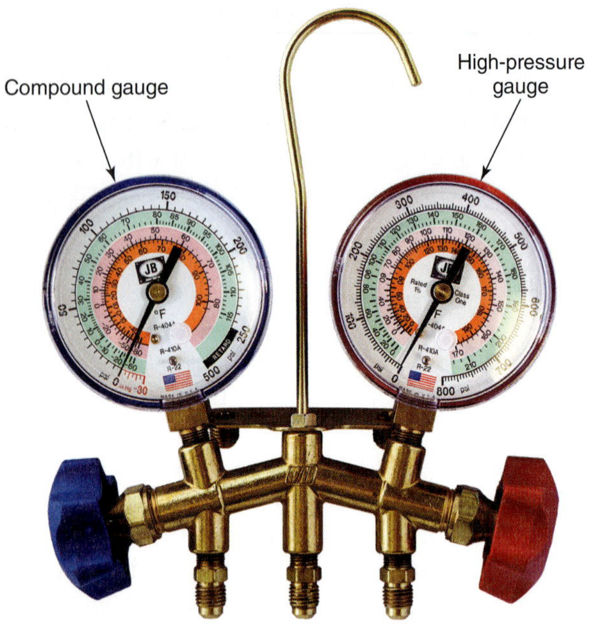

Figure 12-16. The compound gauge is used for both pressure and vacuum readings, while the high-pressure gauge is used only for pressure readings.

PRO TIP

Gauges and Different Refrigerants

Keep in mind that just because a certain refrigerant is listed on a gauge, this does not mean that this gauge cannot be used with other refrigerants. In other words, in most cases, a pressure gauge is a pressure gauge. It can measure pressure on most systems, regardless of the refrigerant. What is important is the pressure reading. When measuring pressure using a gauge that does *not* have a saturation temperature scale for the refrigerant in the system, consult a pressure-temperature (P/T) card for that refrigerant's saturation temperature.

CAUTION

Gauges, Specific Refrigerants, and High-Pressure Limits

Exercise discretion and caution when measuring pressures using gauges that do *not* have a refrigerant's corresponding temperature scale. Confirm the refrigerant's pressure does not exceed your gauge max pressure. Also, note that these gauges should never be used on CO_2 or ammonia systems. There are special gauges needed for those systems as well.

PRO TIP

Pointer Flutter

The rapid compression strokes of the compressor pistons sometimes create pressure pulsations that cause a gauge pointer to swing above and below the actual pressure reading. Called ***pointer flutter***, this condition does not indicate problems with the compressor or any other type of defect. The correct pressure reading is obtained at the center of the flutter. Special pressure gauges are available with a built-in pulsation dampener to prevent pointer flutter.

Note that refrigeration gauges only reveal *saturation* temperature for a given pressure. The gauges do *not* reveal superheat (temperature above saturation) or subcooling (temperature below saturation). Because the gauges only read saturation, a thermometer is used to determine if the actual temperature is above or below the saturation point (superheat or subcooling). Superheat and subcooling are discussed in more detail in Chapter 13, *Basic Refrigeration System Cycle and System Components.*

Recalibrating Gauges

Refrigeration gauges are sensitive instruments, initially calibrated to produce accurate readings. While these gauges can withstand some abuse, they should be handled with care. It is not unusual for the gauges to require ***recalibration*** in the field because of use and handling.

To gain access to the recalibration screw, the clear crystal face on the gauge must be unscrewed, **Figure 12-17**. The recalibration screw is located on the gauge face, just below the needle hub. Remove the hose on the manifold body immediately beneath the gauge to expose the inlet port to atmospheric pressure. Use a suitable screwdriver to slowly turn the recalibration screw until the pointer lines up with 0 psig (atmospheric pressure).

Manifold Gauge Body

The compound gauge and the high-pressure gauge are connected to their hoses by special bypass passages through the manifold body. A hose connection is located directly beneath each gauge. The passage between each gauge and its connecting hose is *never closed*, **Figure 12-18**. The two hand valves located at each end of the manifold control access to the center hose only. These

Goodheart-Willcox Publisher

Figure 12-17. Unscrew the clear cover of the gauge to reach the recalibration screw on the dial face. For accurate readings, gauges must be recalibrated, or zeroed, periodically.

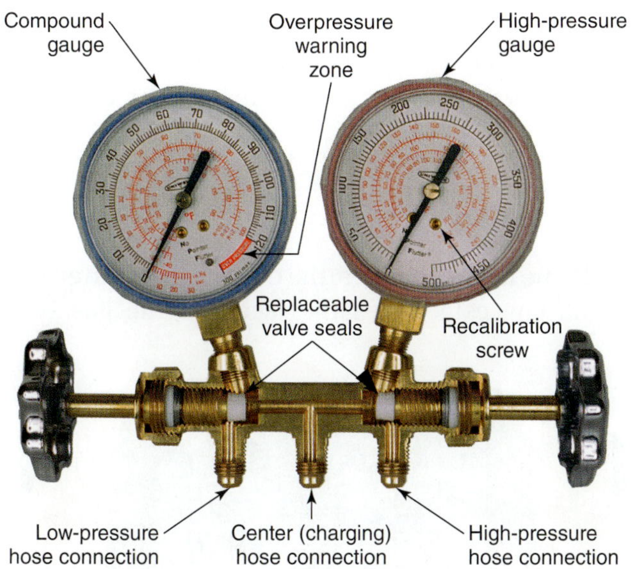

Uniweld

Figure 12-18. Cross section of a manifold gauge body. The hand valves are used only for access to the center hose.

valves do *not* control the bypass passages to the gauges.

Refrigeration Hoses

Refrigeration hoses are designed to withstand working pressures of 500 psig to 750 psig (3450 kPa to 5175 kPa). The hoses are normally color-coded: *blue* for low pressure, *red* for high pressure, and *yellow* for the center connection. The hoses are also designed to remain flexible under most temperature conditions for easy handling. Each hose has a straight connector at one end and an angled (45°) connector at the other, **Figure 12-19**. The straight ends connect to the 1/4″ flare fittings on the manifold body. The angled ends easily connect to the proper service valve by finger-tightening. Angled ends contain a valve-core depressor for connection to Schrader (core-type) valves. Due to the Clean Air Act, the end of the hoses that connect to the service valves should have shut off valves attached to the ends to help reduce the loss of refrigerant from the hoses. These can be built into the hoses or can be added later. See **Figure 12-20**.

The center hose is for adding refrigerant to a system (*charging*), removing refrigerant from a system (*recovery*), or emptying a system with a vacuum pump (*evacuating*). These processes are discussed in Chapter 16, *Refrigerant Recovery and Recycling*.

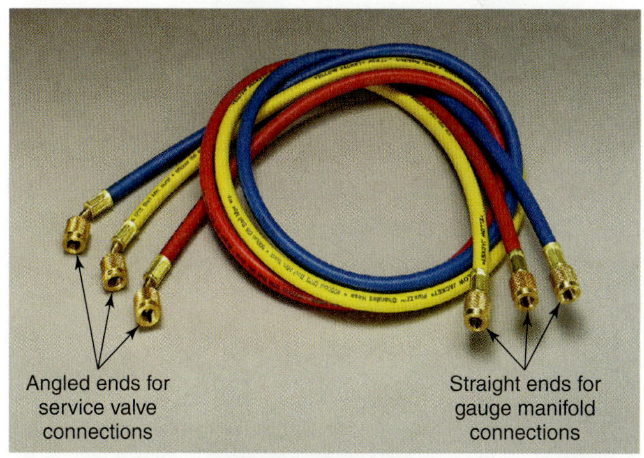

Refrigeration hoses

Ritchie Engineering – YELLOW JACKET Products Division

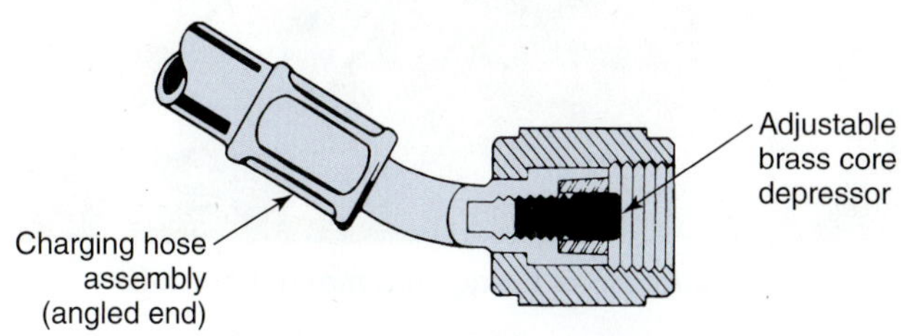

Angled end hose connector with valve core depressor

Goodheart-Willcox Publisher

Figure 12-19. A refrigeration hose is designed to withstand high pressures and temperature extremes. The straight connector attaches to the manifold body, and the angled connector attaches to the service valve or Schrader valve.

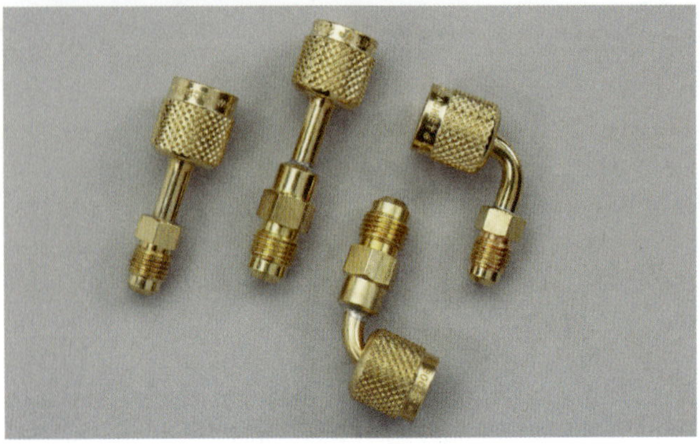

Low loss fittings

Ritchie Engineering – YELLOW Jacket Products Division

Built-on low loss fittings with built-in shutoffs

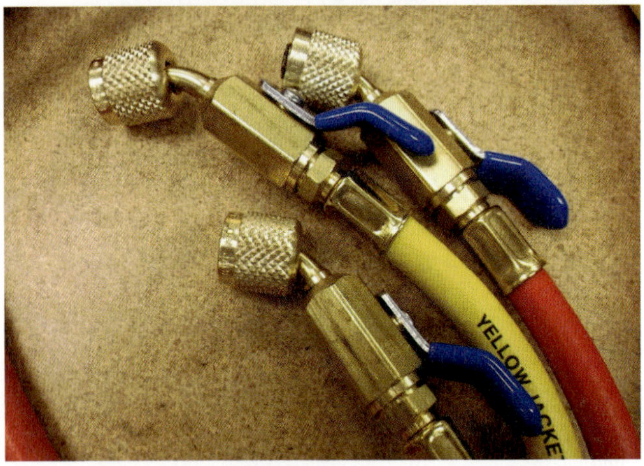

Built-on low loss fitting with manual shutoffs

Images courtesy of Johnstone Supply

Figure 12-20. Built-on or aftermarket add-on ends for refrigerant hoses.

Digital Gauge

Pressure measurements need to be carefully taken for proper diagnostics and equipment function. Digital gauges offer a convenient solution for obtaining accurate pressure readings of both high-side and low-side pressures by combining these gauges all in one housing, **Figure 12-21**. Digital gauges offer a library of different refrigerants and their corresponding temperatures. They also display other useful information such as precalculated superheat and subcooling measurements. Most digital gauges also offer a built-in micron gauge along with temperature clamps and one button calibration for ease of use. Many manifold gauges are available with digital gauges.

12.3 Gas Laws

Refrigerants are the vital working fluids in any system. They are the vehicles that transport heat from one location to another. All refrigeration systems are constantly changing liquid to gas and gas back to liquid for another cycle. To read pressure gauges and diagnose system problems accurately, a technician must fully understand the laws defining the behavior of gases. Gas laws involve three factors:

- Temperature
- Pressure
- Volume

If one factor changes, the other two are affected. A gas law establishes the fact that the gas will

Fieldpiece

Figure 12-21. This digital manifold gauge contains a library of different refrigerants to select from when taking pressure readings. It also contains a built-in vacuum gauge.

always behave according to the rule, with no exceptions. These laws are so exact that the behavior of the temperature-pressure-volume relationship can be predicted and controlled.

Boyle's Law

Robert Boyle (1627–1691) was an English scientist who was among the first to practice chemistry as a true science. He is best known for the gas law that bears his name. ***Boyle's Law*** explains the relationship between the pressure and the volume of a gas if the temperature does *not* change (remains constant). Boyle's Law states:

The pressure of a gas varies inversely (opposite) as the volume, provided the temperature remains constant. In other words, if temperature stays steady, pressure will increase as volume decreases. Also, if temperature stays steady, pressure will decrease as volume increases.

Figure 12-22 illustrates Boyle's Law. If 2 cubic feet (cu ft) of gas at 50 psia are compressed to 1 cu ft (volume reduced by half), the pressure will double to 100 psia, so long as the temperature does *not* change.

Of course, the opposite is true as well. If the volume is doubled, the pressure is halved. Thus, when either the pressure or volume is changed, the corresponding pressure or volume is changed in the opposite direction (in exact proportion), as long as temperature remains constant.

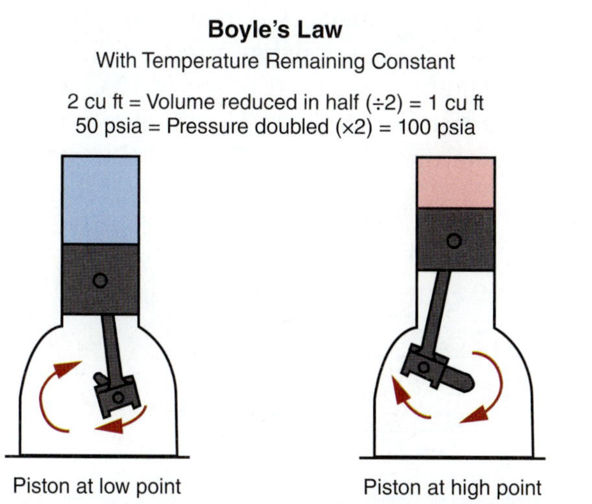

Goodheart-Willcox Publisher

Figure 12-22. Boyle's Law of Gases describes the inverse relationship of pressure and volume when temperature is constant. If volume is reduced by half, pressure will double. The reverse is true as well. If volume is doubled, pressure will be halved.

Gas laws can be proven mathematically. When working with gas law formulas, remember *absolute* pressures and temperatures must always be used. The formula for Boyle's Law is:

$$P_1 \times V_1 = P_2 \times V_2$$

where

$$P_1 = \text{initial pressure}$$
$$V_1 = \text{initial volume}$$
$$P_2 = \text{final pressure}$$
$$V_2 = \text{final volume}$$

Notice that temperature does not appear in Boyle's formula because the temperature remains constant. Only the pressure and volume change. The initial conditions are on the left of the equal sign, and the final conditions are on the right. The equal sign means *the same as*. In other words, when multiplying the initial conditions, the answer will be the same as when multiplying the final conditions. For example, in the equation $10 \times 3 = 5 \times 6$, both sides equal 30.

In practice, the problem will always provide three numbers, or *values*, but the unknown fourth value must be discovered by division.

$$10 \times 3 = 5 \times ?$$
$$30 = 5 \times ?$$
$$30 \div 5 = \frac{5 \times ?}{5}$$
$$6 = \frac{\cancel{5} \times ?}{\cancel{5}}$$
$$6 = ?$$

Try an actual Boyle's Law problem. What is the final volume if 6 cu ft of gas at 35 psig is compressed to 85 psig, providing the temperature stays constant? Remember to use *absolute* pressures by adding 15 psi to each pressure value, so 35 psig becomes 50 psia, and 85 psig becomes 100 psia.

$$P_1 \times V_1 = P_2 \times V_2$$
$$50 \text{ psia} \times 6 \text{ cu ft} = 100 \text{ psia} \times ?$$
$$300 \text{ psia cu ft} = 100 \text{ psia} \times ?$$
$$\frac{300 \text{ psia cu ft}}{100 \text{ psia}} = \frac{100 \text{ psia} \times ?}{100 \text{ psia}}$$
$$\frac{300 \cancel{\text{psia}} \text{ cu ft}}{100 \cancel{\text{psia}}} = \frac{\cancel{100 \text{ psia}} \times ?}{\cancel{100 \text{ psia}}}$$
$$3 \text{ cu ft} = ?$$

In this problem, the pressure doubled: from 50 psia to 100 psia. Therefore, the volume should be reduced by half, from 6 cu ft to 3 cu ft.

Charles' Law

As with every great invention or discovery, someone will improve upon it or carry the experiment a step further than the originator. In 1787, French scientist Jacques Charles (1746–1823) expanded upon the work done by Boyle a century earlier. He was aware that temperature would not always stay constant, so he performed various experiments to determine how gases would behave when temperature became a ***variable*** (a factor that changes).

In these experiments, Charles maintained one constant factor (either pressure or volume). He was mainly concerned with what happened to the pressure or volume when the temperature changed. He did indeed prove that gases behave consistently with temperature changes. ***Charles' Law*** states:

"At a constant pressure, the volume of a confined gas varies directly as the absolute temperature. At a constant volume, the absolute pressure varies directly as the absolute temperature."

It is actually two laws expressed in one sentence. Charles' Law *first* deals with constant pressure, *then* with constant volume. We will consider the two laws separately.

Constant Pressure

The formula for Charles' Law is:

$$V_1 \times T_2 = V_2 \times T_1$$

where

V_1 = initial volume
T_1 = initial temperature
V_2 = final volume
T_2 = final temperature

Notice that T_2 and T_1 do not follow the form of Boyle's equation, with "initial on the left and final on the right" of the equal sign. Pressure and volume do follow this formula, but temperature does *not*. Otherwise, the equation works just like Boyle's Law.

Always remember to use absolute temperatures whenever applying gas laws. Fahrenheit is easily converted to Fahrenheit Absolute (Rankine) by adding 460°. Absolute zero is 460° lower than regular Fahrenheit zero (0°F). For example, 30°F + 460° = 490°F_A (490°R).

Celsius is just as easily converted to Celsius Absolute (Kelvin) by adding 273° to the Celsius temperature. Absolute zero is 273° lower than regular Celsius zero. For example, 10°C + 273° = 283°C_A (283K).

Charles proved that, with a constant pressure, a temperature increase will result in a volume increase (in exact proportion). In other words, temperature and volume will change together. If one goes up, the other goes up. If one goes down, the other goes down.

With constant pressure:

$$\text{Temperature} \uparrow = \text{Volume} \uparrow$$
$$\text{Temperature} \downarrow = \text{Volume} \downarrow$$

Example 1: A volume of 5 cu ft of gas at 40°F is raised to 140°F at constant pressure. What is the final volume? Remember use *absolute* temperature, so add 460° to the 40°F to get 500°R, and add 460° to 140°F to get 600°R.

$$V_1 \times T_2 = V_2 \times T_1$$
$$5 \text{ cu ft} \times 600°\text{R} = ? \times 500°\text{R}$$
$$3000°\text{R cu ft} = ? \times 500°\text{R}$$
$$\frac{3000°\text{R cu ft}}{500°\text{R}} = \frac{? \times 500°\text{R}}{500°\text{R}}$$
$$\frac{3000°\cancel{\text{R}} \text{ cu ft}}{500°\cancel{\text{R}}} = \frac{? \times \cancel{500°\text{R}}}{\cancel{500°\text{R}}}$$
$$6 \text{ cu ft} = ?$$

where

$$T_1 = 40°\text{F} + 460° = 500°\text{F}_A$$
$$T_2 = 140°\text{F} + 460° = 600°\text{F}_A$$

The temperature went up by 100°, from 500°R to 600°R, while the volume went from 5 cu ft to 6 cu ft. The volume and temperature increased in exact proportion, just as Charles' Law states.

Example 2: At a constant pressure, what is the new temperature of 2 cu ft of gas at 12°C when the volume is increased to 4 cu ft? Remember to use *absolute* temperature by adding 273° to 12°C to get 285K.

$$V_1 \times T_2 = V_2 \times T_1$$
$$2 \text{ cu ft} \times ? = 4 \text{ cu ft} \times 285\text{K}$$
$$2 \text{ cu ft} \times ? = 1140\text{K cu ft}$$
$$\frac{2 \text{ cu ft} \times ?}{2 \text{ cu ft}} = \frac{1140\text{K cu ft}}{2 \text{ cu ft}}$$
$$\frac{\cancel{2 \text{ cu ft}} \times ?}{\cancel{2 \text{ cu ft}}} = \frac{1140\text{K} \cancel{\text{cu ft}}}{2 \cancel{\text{cu ft}}}$$
$$? = 570\text{K}$$

In this example, the volume was doubled (from 2 cu ft to 4 cu ft), so the temperature should also double, from 285K to 570K. See **Figure 12-23**.

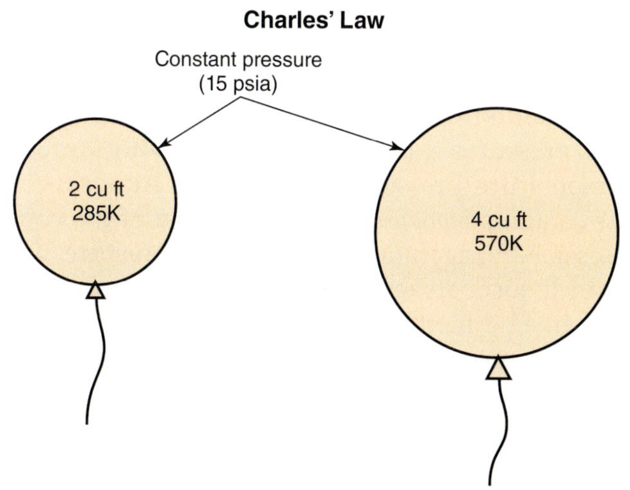

Goodheart-Willcox Publisher

Figure 12-23. According to Charles' Law of Gases, volume and temperature change in exact proportion to each other if pressure is constant. In this example, doubling the volume of the gas also doubles the temperature.

Constant Volume

This part of Charles' Law is important for technicians because refrigeration systems are completely sealed and, therefore, maintain a *constant volume*. The pressures and temperatures in these systems are easily controlled because the volume remains constant. Charles' Law proves that when the pressure goes up, the temperature also goes up. Likewise, when the temperature goes down, the pressure also goes down in exact proportion. This law illustrates a major principle in refrigeration: if the pressure is controlled, the temperature will also be controlled, and vice versa.

With constant volume:

$$\text{Temperature} \uparrow = \text{Pressure} \uparrow$$
$$\text{Temperature} \downarrow = \text{Pressure} \downarrow$$

When solving Charles' Law problems, remember to convert gauge pressure and temperature readings to absolute pressure and absolute temperature.

Example 1: What is the final pressure of a confined gas at 40°F and 35 psig, if the temperature is raised to 90°F? Firstly, convert temperature and pressure to absolute scales.

$$T_1 = 40°\text{F} + 460° = 500°\text{R}$$
$$T_2 = 90°\text{F} + 460° = 550°\text{R}$$
$$P_1 = 35 \text{ psig} + 15 \text{ psi} = 50 \text{ psia}$$

Now follow the Charles' Law formula for constant volume.

$$P_1 \times T_2 = P_2 \times T_1$$
$$50 \text{ psia} \times 550°\text{R} = ? \times 500°\text{R}$$
$$27{,}500°\text{R psia} = ? \times 500°\text{R}$$
$$\frac{27{,}500°\text{R psia}}{500°\text{R}} = \frac{? \times 500°\text{R}}{500°\text{R}}$$
$$\frac{27{,}500°\text{R psia}}{500°\text{R}} = \frac{? \times 500°\text{R}}{500°\text{R}}$$
$$55 \text{ psia} = ?$$

In this example, the temperature was raised from 500°R to 550°R. Therefore, the pressure should also have increased in exact proportion, from 50 psia to 55 psia.

Example 2: What is the final temperature of a confined gas at 20 psig and 4°C, if the pressure is changed to 55 psig? Firstly, convert temperature and pressure to absolute scales.

$$P_1 = 20 \text{ psig} + 15 \text{ psi} = 35 \text{ psia}$$
$$P_2 = 55 \text{ psig} + 15 \text{ psi} = 70 \text{ psia}$$
$$T_1 = 4°\text{C} + 273° = 277\text{K}$$

Now follow the Charles' Law formula for constant volume.

$$P_1 \times T_2 = P_2 \times T_1$$
$$35 \text{ psia} \times ? = 70 \text{ psia} \times 277\text{K}$$
$$35 \text{ psia} \times ? = 19{,}390\text{K psia}$$
$$\frac{35 \text{ psia} \times ?}{35 \text{ psia}} = \frac{19{,}390\text{K psia}}{35 \text{ psia}}$$
$$\frac{\cancel{35 \text{ psia}} \times ?}{\cancel{35 \text{ psia}}} = \frac{19{,}390\text{K } \cancel{\text{psia}}}{35 \ \cancel{\text{psia}}}$$
$$? = 19{,}390\text{K} \div 35$$
$$? = 554\text{K}$$

In this example, the pressure was doubled (from 35 psia to 70 psia), so the temperature should also double (277K to 554K). Converting these temperatures back to Celsius (4°C and 281°C) shows how a large increase in pressure (doubling) can have a drastic effect on temperature in a constant volume.

Combined Gas Law

In some areas of engineering, situations can develop in which there is no constant. All three factors (temperature, pressure, and volume) will change. A "combined gas law" was formulated to handle this situation. While it requires higher mathematics than are used in everyday refrigeration work, it is valuable for making scientific calculations and understanding engineering problems involving the performance of different gases. The formula combines Boyle's Law and Charles' Law:

$$\frac{P_1 \times V_1}{T_1} = \frac{P_2 \times V_2}{T_2}$$

Dalton's Law of Partial Pressures

From 1803 to 1810, the English schoolteacher John Dalton (1766–1844) conducted a series of experiments that greatly added to the existing knowledge of gases and atoms. Dalton experimented with a *mixture* of gases rather than a single, pure gas. ***Dalton's Law*** established that in a mixture of gases, each gas behaves as if it occupies the space alone. To obtain the total pressure of a confined mixture of gases, the pressure for each gas in the mixture must be added. For example, if refrigerant R-134a, oxygen, and nitrogen are occupying the same space, add the pressure for each gas to obtain the total pressure. See **Figure 12-24**. The formula is:

$$\text{Total pressure} = P_1 + P_2 + P_3$$

where

P_1 = pressure of first gas
P_2 = pressure of second gas
P_3 = pressure of third gas

This system of partial pressures is the basis of operation for old absorption-type refrigeration systems. Such systems contain at least two gases, such as ammonia and hydrogen. Today, however, the majority of refrigeration systems are compression-type, designed to use a single, pure gas. If another gas (such as atmospheric air) enters the system, then Charles' Law is no longer sufficient to explain system behavior. Dalton's Law takes effect.

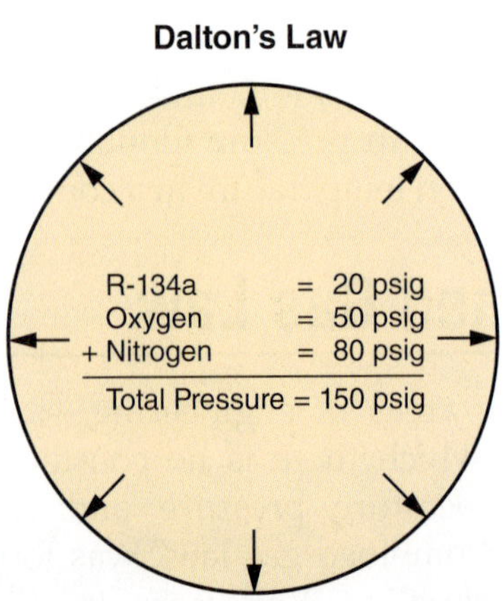

Goodheart-Willcox Publisher

Figure 12-24. Dalton's Law of Partial Pressures specifies that in mixed-gas situations, the pressures of all the gases are added to obtain the total pressure.

Major problems will develop when another gas is permitted to enter a compression-type system. The pressures will not correspond with familiar temperature/pressure relationships. Atmospheric air contains *moisture*, a violent enemy of the compression refrigeration system. The moisture will quickly form an acid and destroy the motor windings in a hermetic compressor.

Poor service procedures are the primary cause of air entering a system. Various methods can be used to rid a system of air and moisture, but severe cases require major repairs. These procedures are not difficult but are time-consuming and expensive. The problem can usually be prevented with proper servicing.

12.4 Pressure-Temperature Charts

Thanks to the principles discovered years ago by Jacques Charles, pressure-temperature (P/T) charts have been compiled to illustrate the pressure-temperature relationship of refrigerants. A chart like the one shown in **Figure 12-25** is a working tool for all HVACR service technicians and engineers. Pocket-size versions of these charts are usually available free from local refrigeration parts suppliers. Having these charts on-hand can save valuable time as these pressure-temperature measurements are often needed when servicing a system. Knowing the corresponding temperature or pressure that a refrigerant should measure in at can help confirm proper operation of a system. A well-prepared service technician should always carry access to P/T charts while they are on the job.

Whenever a pure gas is confined within a constant volume (refrigeration system), the pressure-temperature (P/T) chart will reveal the *saturation temperature and pressure* for the refrigerants listed. For example, if refrigerant R-134a is at saturation at 10°F, the pressure will be 12 psig.

PRO TIP

Saturation Charts

P/T charts are sometimes referred to as *saturation charts* due to their inclusion of a substance's saturation point (boiling/condensing point).

Pressure-Temperature (Saturation) Chart

°F	°C	R-22	R-134a	R-290	R-410A	R-404A Bubble Liquid	R-404A Dew Vapor	R-407C Bubble Liquid	R-407C Dew Vapor	R-1234yf	R-600a
−40	−40	0.5	14.8	1.4	11.6	4.9	4.3	2.7	4.6	11.5	21.6
−35	−37	2.6	12.5	3.4	14.9	7.5	6.8	5.1	0.9	8.9	20.4
−30	−34	4.9	9.9	5.7	18.5	10.3	9.6	7.7	1.6	6	18.9
−25	−32	7.4	6.9	8.1	22.5	13.4	12.7	10.6	3.9	2.8	17.4
−20	−29	10.1	3.7	10.7	26.9	16.8	16	13.7	6.5	0.4	15.6
−15	−26	13.2	0.6	13.6	31.7	20.5	19.7	17.2	9.3	2.3	13.7
−10	−23	16.5	1.9	16.7	36.8	24.6	23.6	20.9	12.3	4.4	11.5
−5	−21	20.1	4	20.1	42.5	28.9	27.9	25	15.7	6.7	9.2
0	−18	24	6.5	23.7	48.6	33.7	32.6	29.5	19.4	9.1	6.6
5	−15	28.2	9.1	27.6	55.2	38.8	37.7	34.3	23.5	12	3.8
10	−12	32.8	11.9	31.8	62.3	44.3	43.1	39.5	27.9	14.9	0.7
15	−9	37.7	15	36.3	70	50.2	49	45.2	32.7	18.1	1.3
20	−7	43	18.4	41.1	78.3	56.6	55.3	51.2	37.9	21.6	3.1
25	−4	48.8	22.1	46.3	87.3	63.4	62.2	57.7	43.5	25.4	5
30	−1	54.9	26.1	51.8	96.8	70.7	69.3	64.7	49.6	29.4	7.1
35	2	61.5	30.4	57.7	107	78.6	77.1	72.2	56.1	33.8	9.4
40	4	68.5	35	63.9	118	86.9	85.4	80.2	63.2	38.4	11.8
45	7	76	40.1	70.6	130	96.8	94.2	88.8	70.7	43.4	14.4
50	10	84	45.5	77.6	142	105	104	97.9	78.8	48.8	17.2
55	13	92.6	51.3	85.1	155	115	114	108	87.5	54.5	20.2
60	16	102	57.5	93	170	126	124	118	96.8	60.6	23.5
65	18	111	64.1	101.4	185	137	136	129	107	67	26.9
70	21	121	71.2	110.2	201	159	147	141	117	73.9	30.6
75	24	132	78.8	119.5	217	162	160	153	129	81.3	34.5
80	27	144	86.8	129.3	235	175	173	166	141	89	38.6
85	30	156	95.4	139.7	254	190	188	180	153	97.2	43
90	32	168	104	150.5	274	205	202	194	167	106	47.7
95	35	182	114	161.9	295	220	218	209	181	115	52.7
100	38	196	124	173.9	317	237	235	226	196	125	57.9
105	41	211	135	186.5	340	254	252	242	212	135	63.4
110	43	226	147	199.6	365	273	270	260	229	146	69.3
115	46	243	159	213.4	391	292	290	279	247	157	75.4
120	49	260	171	227.8	418	312	310	299	266	169	81.9
125	52	278	185	242.9	446	333	331	319	286	182	88.7
130	54	297	199	258.7	476	356	354	341	307	195	95.8
135	57	317	214	275.1	507	379	377	363	329	209	103
140	60	337	229	292.3	539	404	402	387	352	223	111
145	63	359	246	310.2	573	430	428	412	377	239	120
150	66	382	263	328.9	608	457	455	438	403	255	128

Goodheart-Willcox Publisher

Figure 12-25. Pressure-temperature charts show pressure-temperature relationships for various refrigerants. These quick reference charts are easy to use. For example, at 40°F, the pressure for refrigerant R-22 is 68.5 psig. For refrigerant R-134a at the same temperature, the pressure is only 35 psig. Some refrigerants, like R-404A and R-407C in this chart, have different temperatures for liquid and vapor saturation.

Chapter Review

Summary

12.1 Temperature is a measure of the heat intensity or heat level of a substance. Temperature can be measured in degrees Fahrenheit, degrees Celsius, degrees Rankine, and in Kelvin. Ambient temperature refers to the temperature surrounding an object. This is typically referred to when working on outdoor components, such as a motor or condenser, to factor how surrounding temperature can influence the refrigeration system.

12.2 Atmospheric pressure refers to the weight of molecules in the air that exert a downward force. Atmospheric pressure is measured at 14.7 psi (101.3 kPa) at sea level. Any pressure less than atmospheric is referred to as a partial vacuum. When all atmospheric pressure is removed, it is called a perfect vacuum. Vacuum is measured with a mercury barometer, in inches of mercury (in. Hg).

12.2 Absolute pressure is any pressure above a perfect vacuum and is expressed in pounds per square inch absolute (psia). Gauge pressure measures pressures above atmospheric pressure and is expressed in pounds per square inch gauge (psig).

12.2 A high-pressure gauge takes psig readings on the high-pressure side of a refrigeration system and compound gauges take psig and in. Hg readings on the low-pressure side of a refrigeration system. A manifold gauge combines both high-pressure and compound gauges and can be connected to both the high side and low side to take system pressure readings.

12.3 Boyle's Law explains the relationship between the pressure and the volume of a gas if the temperature does not change (remains constant). Charles' Law states that at a constant pressure, the volume of a confined gas varies directly as the absolute temperature. At a constant volume, the absolute pressure varies directly as the absolute temperature. The Combined Gas Law accounts for situations where there is no constant among the three variables: temperature, pressure, and volume. Dalton's Law states that in a mixture of gases, each gas behaves as it occupies the space alone.

12.4 Pressure-temperature charts show the pressure-temperature relationships of various refrigerants. These charts can save valuable time as these pressure-temperature measurements are often needed when servicing a system.

Know and Understand

1. Temperature indicates the _____ of heat, *not* the quantity of heat. (12.1)
   A. scope
   B. speed
   C. intensity
   D. energy in Btu
2. On the Fahrenheit scale, there are _____ divisions between the freezing and boiling points of water. (12.1)
   A. 32
   B. 100
   C. 180
   D. 212
3. On the Celsius scale, there are _____ divisions between the freezing and boiling points of water. (12.1)
   A. 32
   B. 100
   C. 180
   D. 212
4. The Rankine and Kelvin scales are called *absolute temperature scales* because they _____. (12.1)
   A. only deal in temperatures above atmospheric pressure
   B. record temperature to the thousandths decimal place
   C. do not have negative numbers
   D. record temperature based on water's freezing and boiling points

5. *True or False?* Ambient temperature refers to the temperature conditions surrounding an object. (12.1)
6. Because the atmosphere is composed of a mixture of gases, pressure is exerted _____. (12.2)
   A. at a downward force
   B. at an upward force
   C. in all directions
   D. depending on the dominant gas in the atmosphere
7. When all atmospheric pressure is removed, it is called _____. (12.2)
   A. absolute pressure
   B. absolute zero
   C. partial vacuum
   D. perfect vacuum
8. *True or False?* Vacuum is measured in inches of water column. (12.2)
9. An absolute pressure reading of 8 psia could also be expressed as _____ in. Hg. (12.2)
   A. 8
   B. 12
   C. 16
   D. 20
10. A gauge pressure reading of 25 psig would have an absolute value of _____ psia. (12.2)
    A. 10.3
    B. 14.7
    C. 39.7
    D. 40
11. A compound gauge makes it possible to read pressures _____. (12.2)
    A. above atmospheric pressure
    B. below atmospheric pressure
    C. both above and below atmospheric pressure
    D. on both the high-pressure side and low-pressure side of a system
12. A vacuum reading of 20 in. Hg is equal to _____ psia. (12.2)
    A. 0.5
    B. 1
    C. 1.33
    D. 1.5
13. Which of the following describes the correct connection of a manifold gauge's components to a system when preparing to take pressure readings? (12.2)
    A. The blue hose is connected to the high-pressure side, the red hose is connected to the low-pressure side, and the yellow hose is connected to the center.
    B. The yellow hose is connected to the high-pressure side, the blue hose is connected to the low-pressure side, and the red hose is connected to the center.
    C. The red hose is connected to the high-pressure side, the blue hose is connected to the low-pressure side, and the yellow hose is connected to the center.
    D. The red hose is connected to the high-pressure side, the yellow hose is connected to the low-pressure side, and the blue hose is connected to the center.
14. Boyle's Law says that the pressure and volume relationship of a gas will _____, as long as the temperature stays constant. (12.3)
    A. vary inversely
    B. vary directly
    C. not be affected
    D. affect temperature
15. One of the formulas for Charles' Law is used for constant _____, the other for constant _____. (12.3)
    A. temperature, pressure
    B. temperature, volume
    C. pressure, volume
    D. pressure, saturation
16. If air enters a compression-type system, then _____ takes effect. (12.3)
    A. Boyle's Law
    B. Charles' Law
    C. Combined Gas Law
    D. Dalton's Law
17. *True or False?* A pressure-temperature chart includes the saturation temperature and pressure for a refrigerant confined within a constant volume. (12.4)

Apply and Analyze

1. Explain how a thermometer measures temperature. (12.1)
2. Convert 78°F to the Kelvin scale. (12.1)
3. Explain how atmospheric pressure affects the boiling point of a liquid. (12.2)
4. What's the difference between absolute pressure and gauge pressure? (12.2)
5. Describe how to recalibrate refrigeration gauges. (12.2)
6. Using Boyle's Law, calculate the final volume if 10 cu ft of gas at 25 psig is compressed to 70 psig, provided the temperature stays constant. (12.3)
7. Using Charles' Law, calculate the final volume if 18 cu ft of gas at 52°F is raised to 160°F at constant pressure. (12.3)
8. What does Dalton's Law of Partial Pressures refer to? (12.3)
9. Using **Figure 12-25**, what is the pressure of R-600a if it is at a saturation temperature of 30°F (–1°C)? (12.4)

Critical Thinking

1. What is ambient temperature and how is ambient temperature applicable to HVACR work? (12.1)
2. Explain why a perfect vacuum is not achievable in a refrigeration system. (12.2)
3. Using Boyle's Law, what is the final gauge pressure if 50 cu ft of gas at 35 psig is compressed to 25 cu ft? (12.3)

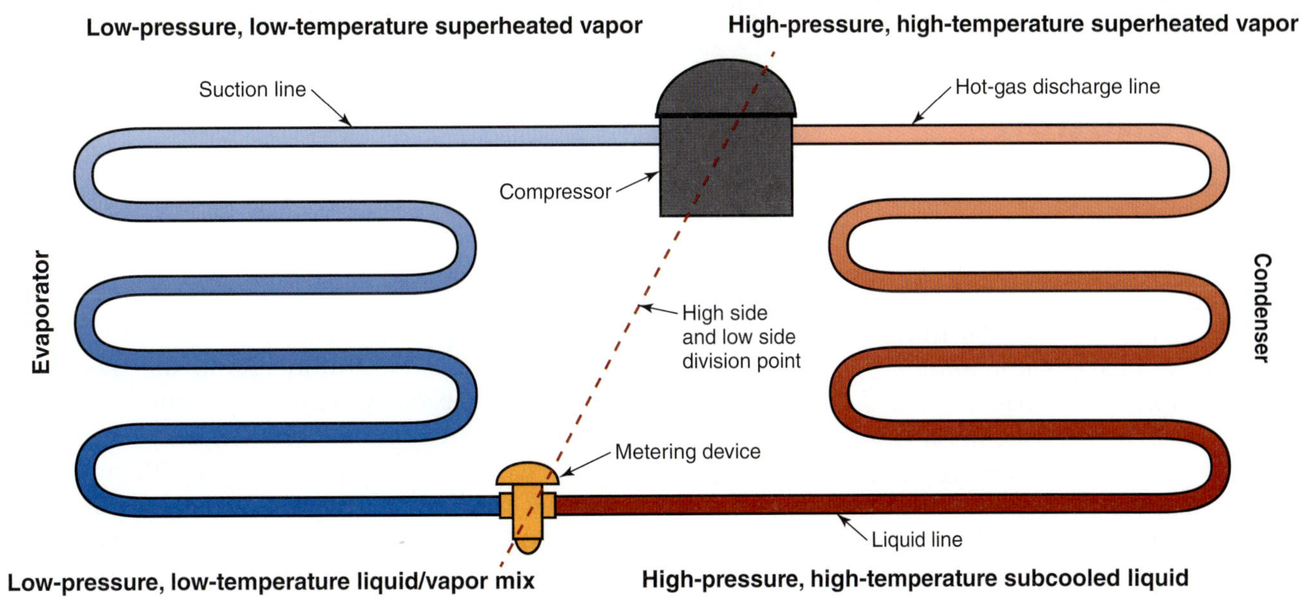

Goodheart-Willcox Publisher

13 Basic Refrigeration Cycle and System Components

LEARNING OUTCOMES

After studying this chapter, you will be able to:

13.1 Explain the purpose and operation of each component in the refrigeration system.

13.2 Identify variations in types and installation details for evaporators, suction lines, compressors, condensers, and the water-regulating valve.

TECHNICAL TERMS

- accumulator
- basic refrigeration cycle
- compressor
- condensation
- condenser
- cut-in
- dissipated
- evaporation
- evaporator
- forced convection
- high-pressure side
- hot-gas discharge line
- liquid line
- low-pressure side
- off cycle
- on cycle
- pressure drop
- reciprocating
- refrigerant metering device
- series connection
- subcooled
- suction line
- superheat
- temperature difference (td)
- vertical riser

Introduction

All refrigeration and air-conditioning systems are designed to contain and control a circulating refrigerant. This chapter describes and illustrates the ***basic refrigeration cycle***, in which the circulating refrigerant absorbs unwanted heat at one location and carries it to another location where it will be released. Refrigerants are the vital working fluids in any refrigeration system. They absorb heat by ***evaporation*** (changing from a liquid to a gas). Then, they release the heat through ***condensation*** (changing from a gas back to a liquid). In a properly operating system, the refrigerant will not break down, wear out, or become lost.

The compression system used in refrigeration is quite simple if the principles explained in Chapters 11 and 12 are understood. A thorough understanding of the basic refrigeration system is important to the success of a technician since "complicated" systems are simply basic systems with one or more components added to make them more efficient, versatile, or serviceable.

The refrigeration system components serve only to control the circulating refrigerant. Learning the basic system is not difficult. Committing it to memory will greatly assist in understanding later areas of study.

PRO TIP

Refrigeration Systems

Keep in mind that the basic refrigeration cycle and components discussed in this chapter also apply to air-conditioning systems. *Refrigeration system* is a phrase often used broadly to refer to any system that removes heat, not just refrigerators. These distinctions are important to note depending on the context.

13.1 Refrigeration System Components

The basic refrigeration system consists of four basic components, **Figure 13-1**. These components are arranged so the refrigerant travels in a continuous cycle through refrigerant tubing (suction line, hot-gas discharge line, and liquid line). The four basic components include the following:

- Evaporator
- Compressor
- Condenser
- Refrigerant metering device

It is necessary to understand the theory and operation of each system component, because the malfunction of one component can significantly influence the operation of others. Being able to narrow a problem down to a particular component is important. Understanding the condition of the refrigerant inside each component is important as well.

Evaporator

The ***evaporator*** is a heat-exchanging device located inside the area where heat is to be removed, **Figure 13-2**. The evaporator performs the main purpose of the system—*refrigeration*. It absorbs heat from where it is unwanted (such as a room or the interior of a refrigerator). Heat always travels from hot to cold, so warm air passing through the fins of the evaporator gives up heat to the cooler evaporator. Liquid refrigerant entering the evaporator leaves it as vapor.

The heat travels through the fins to the copper tubing of the evaporator, **Figure 13-3**. Then, the heat is absorbed by liquid refrigerant traveling through the copper tubing. Heat absorption (latent heat) causes the liquid to boil (evaporate). Liquid refrigerant inside the evaporator boils at a low temperature. A refrigerant with a low boiling point is selected because a temperature difference must exist between the air and the refrigerant for heat to flow.

Superheat

Saturated conditions (liquid at its boiling point) exist inside the evaporator tubing. A refrigerant metering device ensures that *all* liquid boils off in the evaporator, and only superheated vapor leaves the evaporator outlet. The normal amount of ***superheat*** (temperature above saturation) at the evaporator outlet will vary depending on manufacturer and the type of metering device being used. A system operating at the correct superheat will provide a maximum cooling effect and prevent liquid from leaving the evaporator. For

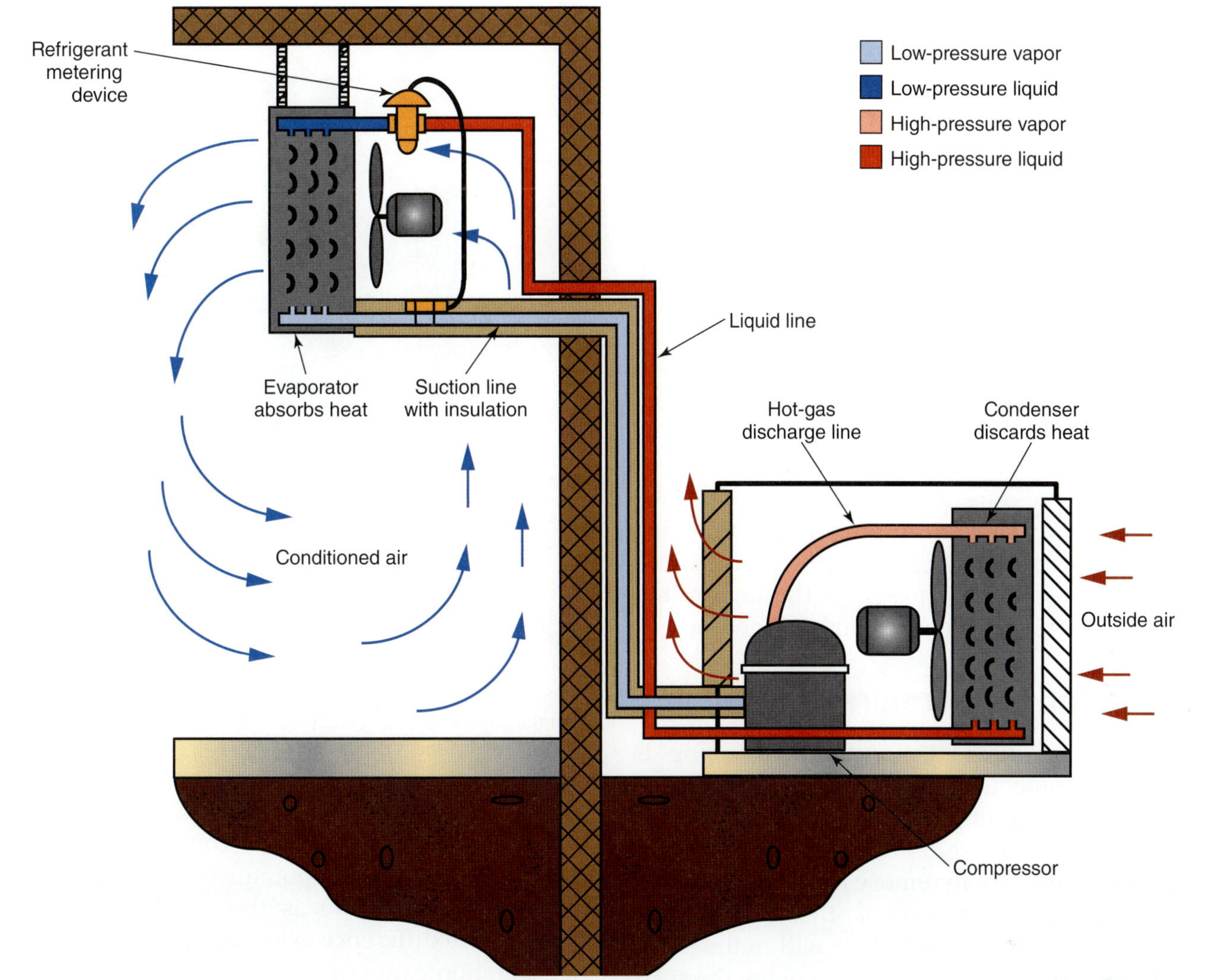

Goodheart-Willcox Publisher

Figure 13-1. The basic refrigeration system consists of four basic components. The evaporator unit is mounted inside where heat is absorbed, and the condensing unit is located outside for disposing heat to the atmosphere. The suction and liquid lines serve to connect the two units. The hot-gas discharge line discharges high-pressure, high-temperature gas from the compressor to the condenser.

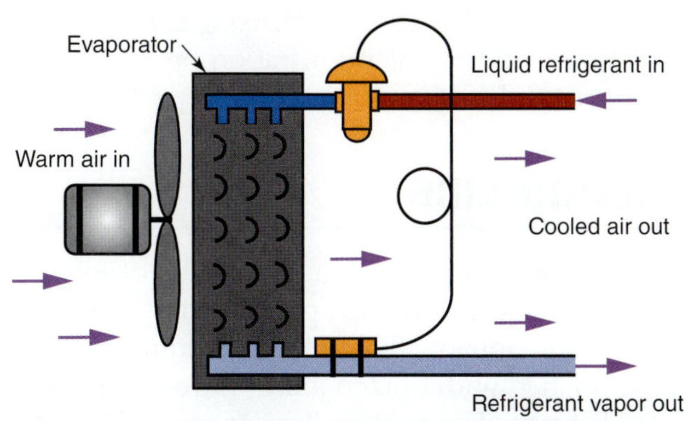

Goodheart-Willcox Publisher

Figure 13-2. As warm air passes through the fins of the evaporator, the air gives up heat. The heat is conveyed by the fins to the copper tubing and, in turn, to the liquid refrigerant. The added heat causes the refrigerant to boil and evaporate (change to a gas).

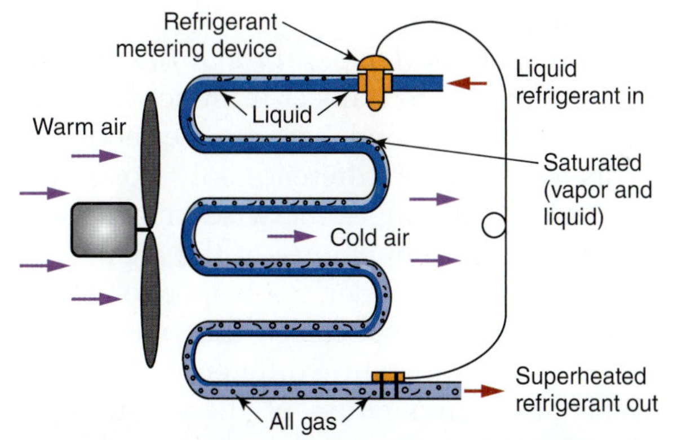

Goodheart-Willcox Publisher

Figure 13-3. As the liquid refrigerant inside the evaporator tubing absorbs heat from the air, it reaches a saturated condition and evaporates to a gaseous state.

efficient operation of the system, the amount of superheat at various points must be determined and controlled.

For example, a pressure reading at the suction line for R-134a may be 12 psig (83 kPa), which corresponds to 10°F (–12°C) saturation temperature. However, if a thermometer is clamped to the suction line at the inlet to the suction line (at the service valve), the reading would be about 25°F (–4°C). The difference in temperature readings amounts to 15°F (25 – 10 = 15) or 8°C. Therefore, the refrigerant vapor contains 15°F (8°C) of superheat as it enters the suction service valve.

Common refrigerants have boiling points ranging from –21°F to –50°F (–29°C to –45°C) at atmospheric pressure. Any liquid refrigerant boiling at this low temperature-pressure combination will indeed create the temperature difference needed to cause heat to travel from the air, through the evaporator, and into the liquid refrigerant.

Evaporator Temperature Difference

Refrigerant inside the evaporator tubing is in a low-temperature, low-pressure, saturated condition. The boiling point of the refrigerant is deliberately controlled to provide the temperature difference necessary to remove heat.

The precise function of the refrigeration system is to create a temperature within the evaporator that is about 20°F (11°C) colder than air passing through the evaporator. See **Figure 13-4**. The refrigerant is coldest, about 10°F (5.5°C) colder than the evaporator. The evaporator, in turn, is about 10°F (5.5°C) colder than the air flowing past its fins. The fins increase the surface area of the tubing and aid in the transfer of heat from the air to the tubing.

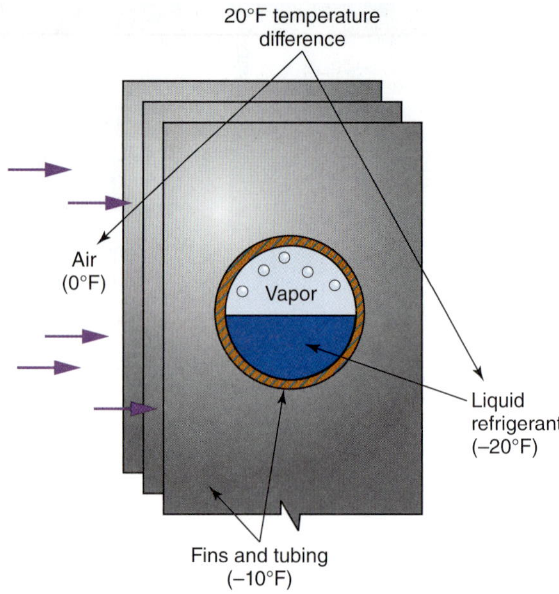

Goodheart-Willcox Publisher

Figure 13-4. A difference in temperature must exist between the refrigerant and the air passing through the evaporator for proper heat flow from hot to cold. Typically, the difference is about 20°F (11°C): the fins and evaporator tubing are 10°F (5.5°C) colder than the air, and the refrigerant, in turn, is 10°F (5.5°C) colder than the tubing and fins.

The evaporator ***temperature difference (td)*** must exist for heat to transfer from hot to cold, as specified by the Second Law of Thermodynamics. This temperature difference only exists when the system is running—otherwise known as the ***on cycle***. As the temperature of the air over the evaporator becomes colder, the evaporator and the refrigerant also become colder, thus maintaining the 20°F (11°C) temperature difference.

When the temperature of the incoming air reaches the desired lower temperature level, an air-sensitive thermostat turns the system off and stops the refrigeration process. However, in refrigerators, the evaporator fan runs constantly, circulating air through the evaporator even when the other system components are off. The continuous airflow quickly warms the evaporator and refrigerant to the same temperature as the air. Therefore, no temperature difference exists during the ***off cycle***. Refrigeration cannot be accomplished during the off cycle because no temperature difference exists.

The system is turned on when the rising air temperature reaches the ***cut-in*** (turn-on) point of the air-sensitive thermostat. Thus, the thermostat controls the temperature of the air inside the cabinet by controlling the on-and-off cycling of the system. See Chapter 18, *Working with Metering Devices*, for additional information on evaporator temperature differences.

Suction Line

The ***suction line***, usually made of copper tubing, connects the evaporator outlet to the compressor intake. These connections may be accomplished with brazed joints, flared fittings, or press fittings. The suction line is normally insulated to prevent moisture-laden air from contacting the cold tubing and condensing. Insulation eliminates moisture or frost problems, **Figure 13-5**.

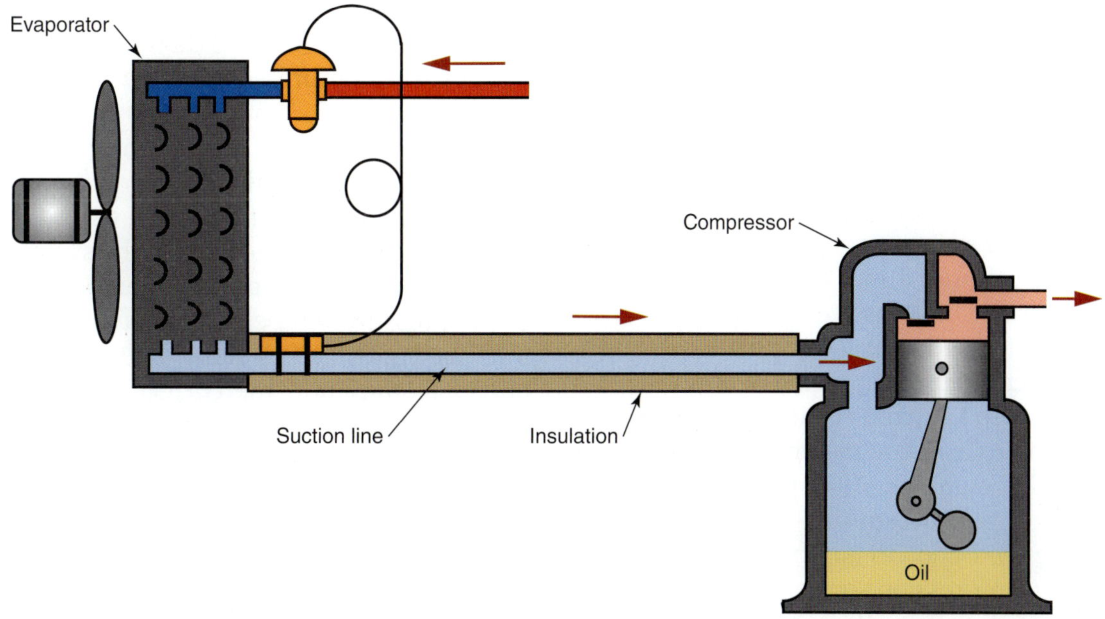

Goodheart-Willcox Publisher

Figure 13-5. The suction line connects the evaporator and the compressor. This line is insulated to prevent condensation from forming on the line and freezing. The suction line should contain only low-pressure, low-temperature vapor. Liquid refrigerant could damage the compressor.

The refrigerant inside the suction line is a superheated vapor, even at this low temperature. This is another reason the suction line is insulated, to minimize the extra heat picked up from the ambient air around the line. The compressor draws the refrigerant vapor through the copper suction line, similar to someone sipping soda through a straw. The drawing or suction action of the compressor creates the low pressure needed in the evaporator to provide a low boiling point for the liquid refrigerant.

The condition of the refrigerant inside the suction line is a *low-temperature, low-pressure, superheated vapor*. The superheat is obtained in the refrigerant's final pass through the evaporator and all the way through the suction line. Care must be exercised to prevent liquid refrigerant from entering the compressor where it can cause severe damage.

Compressor

The ***compressor*** is a vital part of the system and probably the least understood. Compressors often are unjustly condemned as being *inefficient* when the problem actually occurs elsewhere. The technician must have a thorough understanding of how the compressor operates.

The compressor is one of two division points that separate the ***low-pressure side*** (low side) of the system from the ***high-pressure side*** (high side), **Figure 13-6**.

The purpose of the compressor is twofold. First, it must remove vapor from the evaporator (by suction) to maintain a low boiling-point pressure inside the evaporator. Second, it must compress (squeeze) this low-temperature, low-pressure gas into a small volume, which results in a high-temperature, high-pressure gas.

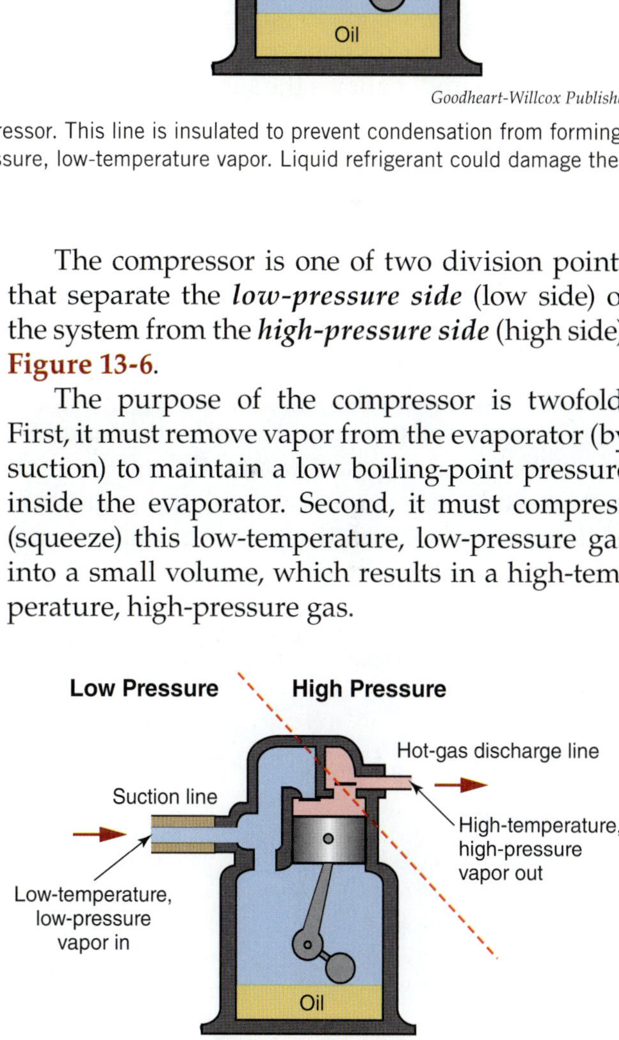

Goodheart-Willcox Publisher

Figure 13-6. The compressor is one of the division points between pressure regions in the refrigeration system. Low-temperature, low-pressure superheated vapor enters the compressor where it is compressed (squeezed) into a smaller volume. Refrigerant discharges from the compressor as a high-temperature, high-pressure superheated vapor.

Reciprocating Compressor

Reciprocating means moving first in one direction, then back in the opposite direction, such as back and forth or up and down. This is the operating principle of the reciprocating compressor widely used in refrigeration systems, **Figure 13-7**. The reciprocating compressor operates much like an automobile engine. It has two or more pistons driven by a crankshaft that causes the pistons to make alternating suction and compression strokes. At the top of each piston cylinder are two valve reeds: a suction valve reed and a discharge valve reed.

The downstroke of the piston creates a partial vacuum in the cylinder. This permits the low-pressure gas in the suction line to force open the suction valve reed and flow into the cylinder. On the upstroke (compression stroke), increasing pressure in the cylinder allows the suction valve reed to close. The vapor is compressed until the pressure within the cylinder overcomes the higher pressure in the hot-gas discharge line that holds the discharge valve reed closed. This occurs at the top of the stroke, after the gas has been compressed to a much smaller volume and a much higher pressure and temperature. The discharge valve reed opens, allowing compressed refrigerant to flow into the hot-gas discharge line.

The increased pressure created by the compressor pushes the refrigerant around the refrigeration system. Both the vacuum (suction) on the low side of the compressor and the pushing action on the high side are important qualities in a compressor. A malfunction on either side will greatly affect the system. The compressor *must* perform both duties simultaneously to have a properly operating system. For this reason, most compressors have at least two pistons. While one piston is on the downstroke, the other piston is on the upstroke.

The operating pressures of the system must be maintained to ensure proper operation. The boiling point (pressure-temperature) inside the evaporator must be controlled to achieve the desired cooling effect. To maintain the correct boiling point, the compressor must be able to remove vapor from the evaporator at the same rate the liquid is boiling off (changing state to a gas). If the temperature-pressure inside the evaporator is permitted to rise, the boiling point also increases (Charles' Law), reducing the cooling effect. Of course, the opposite is also true. Removing the vapor too fast (and lowering its boiling point) causes the evaporator to become too cold.

The compressor is the most expensive and critical component of the system. The compressor must be perfectly matched to other system components to balance the effect of the refrigerant circulating within the system. The refrigerant must be pushed through the high side of the system at the same rate the vapor is being removed from the evaporator by the vacuum action of the compressor. If the refrigerant fails to circulate properly, problems will develop and can be traced to one of two causes: no refrigerant in the system or poor circulation.

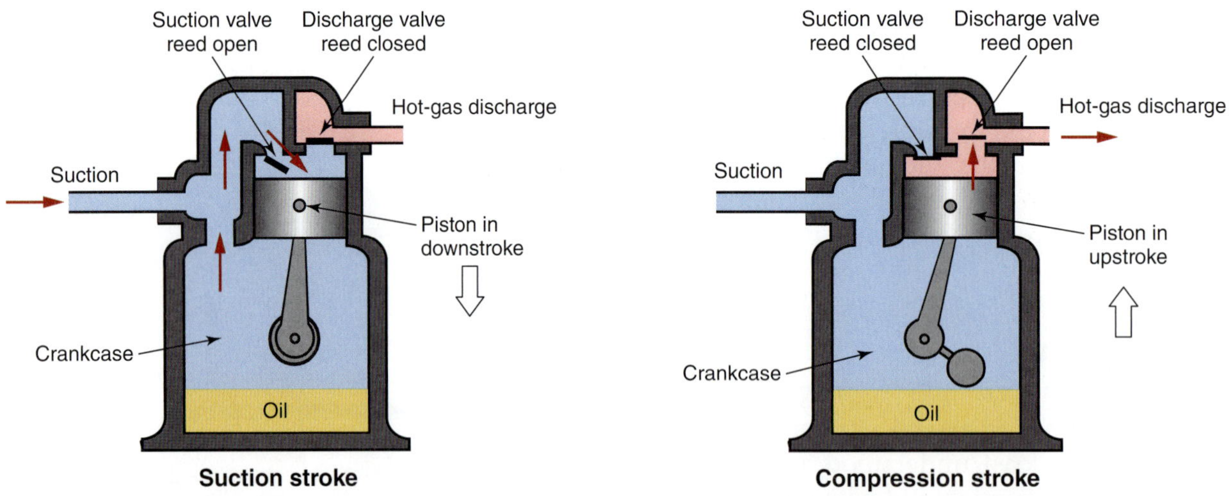

Goodheart-Willcox Publisher

Figure 13-7. On the piston downstroke (left), low-pressure, low-temperature refrigerant vapor is drawn into the piston cylinder. High pressure in the discharge line holds the discharge valve reed closed. As the piston rises during the piston upstroke (right), it compresses the vapor. This increases cylinder pressure to close the suction valve reed and eventually open the discharge valve reed and force the high-pressure, high-temperature vapor to flow from the cylinder into the hot-gas discharge line.

Hot-Gas Discharge Line

The refrigerant being discharged by the compressor is in the form of a high-pressure, high-temperature gas. The discharged hot gas travels through copper tubing to the condenser. See **Figure 13-8**. The copper tubing is called the ***hot-gas discharge line***.

The hot-gas discharge line is smaller than the suction line because the gas has been compressed to a smaller volume and higher pressure. While the suction line is cold or cool, the hot-gas discharge line is hot. Due to heat generated by compression, the discharge gas contains about 100°F (55°C) of superheat. Since the line must carry the superheated gas at high pressure (from 100 psig to 350 psig or 690 kPa to 2415 kPa), it *must* be leakproof.

Condenser

The ***condenser*** is a heat exchanger, somewhat like the evaporator. While the evaporator is designed to *absorb* heat, the condenser is designed to *release* heat. The sole purpose of the condenser is to remove heat from the superheated refrigerant vapor, causing the vapor to condense (change state) back to a liquid, as shown in **Figure 13-9**. Only liquid should leave the condenser. Most condensers are cooled by the flow of ambient (surrounding) air through the fins, but some are water-cooled. The differences between air-cooled and water-cooled condensers are explained later in this chapter.

The high pressure created by the compressor raises the saturation (condensing) point of the refrigerant to the point where ambient air can be used to:

- Remove superheat to the saturation point (sensible heat, lowering temperature).
- Remove enough additional heat to cause the refrigerant to condense back to a liquid (latent heat).
- Remove enough additional heat to partially subcool the liquid before it leaves the condenser (sensible heat, lowering temperature again).

When refrigerant passes through the first part of the condenser, its superheat is quickly removed (desuperheated), lowering the temperature of the high-pressure gas to the saturation point. Continued removal of heat causes a change of state from gas to liquid. In the final passes of the condenser tubing, the refrigerant is in liquid form, permitting the liquid to fall below saturation point, being

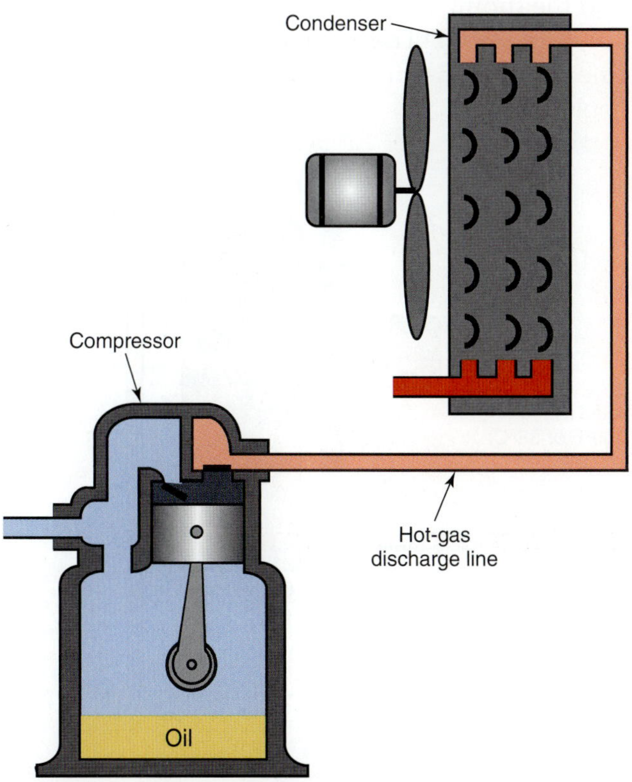

Goodheart-Willcox Publisher

Figure 13-8. The hot-gas discharge line connects the compressor to the condenser. The gas it carries is both high-temperature and high-pressure, so the connections must be leakproof.

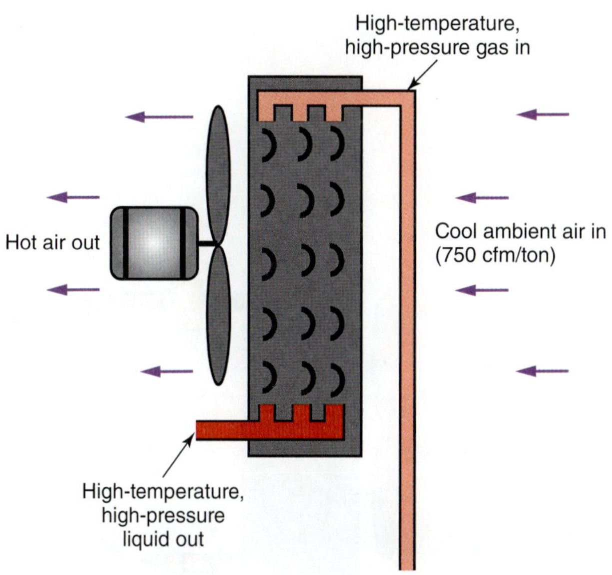

Goodheart-Willcox Publisher

Figure 13-9. The job of the condenser is to remove heat from the refrigerant vapor and change it back to a liquid state.

subcooled, before it leaves the condenser. This subcooling lowers the temperature of the liquid refrigerant. At the condenser outlet, the liquid is still at high pressure, but the temperature may be lukewarm (saturation temperature or subcooled), not hot.

The main objective of the condenser is to have all refrigerant in liquid form in the final pass of the condenser tubing. To illustrate, consider ambient air temperature of 70°F (21°C) entering the condenser. On a running system using R-134a, the compressor raises the saturation point about 30°F (17°C) above ambient air temperature, or 100°F (38°C). The saturation pressure for R-134a at 100°F (38°C) is 124 psig (856 kPa). (These values are pulled from the pressure-temperature chart found in Chapter 12, *Temperature and Pressure*.)

The temperature of the highly superheated gas at the compressor discharge is about 200°F (93°C). Some superheat is ***dissipated*** (lost) as the gas travels through the uninsulated hot-gas discharge line, so the gas arrives at the condenser at about 150°F (65°C). Ambient air at 70°F (21°C) passing through the condenser first removes all superheat (sensible heat) (about 50°F or 28°C), quickly lowering the temperature of the gas to the saturation point of 100°F (38°C).

Ambient air continues to pass through the condenser, removing additional heat (latent heat) and causing the gas to condense back to a liquid. The final lengths of condenser tubing should contain all liquid, which is easily subcooled by at least 10°F or 5.5°C (sensible heat). Including subcooling, the temperature of the liquid leaving the condenser is about 90°F (32°C) at 70°F (21°C) ambient air temperature. Subcooling in the final condenser passages makes the system more efficient.

During the refrigerant's change of state (and subcooling) in the condenser, the discharge pressure remains 124 psig (856 kPa). Although heat has been removed, pressure gauges continue to indicate saturation pressure. Pressure gauges cannot read superheat or subcooling. Remember, at saturation (boiling point) there can be all gas, a mixture of liquid and gas, or all liquid, depending upon how much heat has been removed.

Liquid Line

The copper-tubing ***liquid line*** connects the condenser outlet to the refrigerant metering device, **Figure 13-10**. The liquid line should contain only *liquid* refrigerant subcooled by about 10°F (5.5°C). The liquid refrigerant is still under high pressure, and the tubing is slightly warm to the touch (about 100°F or 38°C). The refrigerant inside the liquid line is a high-temperature, high-pressure, subcooled liquid.

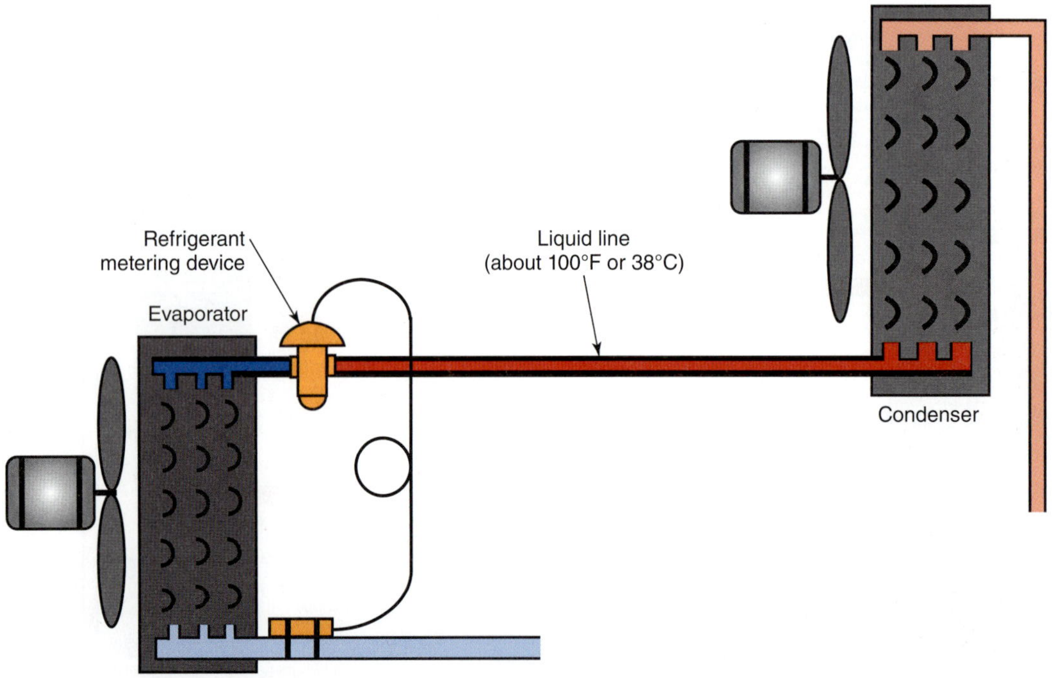

Goodheart-Willcox Publisher

Figure 13-10. The liquid line connects the condenser to the refrigerant metering device. Refrigerant leaving the condenser is a lukewarm liquid under high pressure.

Refrigerant Metering Device

The ***refrigerant metering device*** is a device that plays a key role in the efficient and automatic operation of the system, **Figure 13-11**. The metering device controls the flow of liquid refrigerant into the evaporator to ensure that all liquid is boiled off before it enters the suction line. Allowing liquid refrigerant to reach the compressor can result in compressor damage or even failure.

The metering device is the second division point between the high- and low-pressure sides of the system. Recall that the compressor is the first division point. The metering device works like a valve to automatically regulate the amount of liquid refrigerant entering the evaporator. At the *inlet* of the metering device, the liquid is in a high-temperature, high-pressure, subcooled condition. At the *outlet* of the metering device, as a result of a pressure drop across it, the liquid is in a low-temperature, low-pressure, saturated condition.

The amount of low-pressure liquid entering the evaporator is critical. The liquid must be correctly metered into the evaporator, so it boils off before it reaches the compressor. In other words, the amount of liquid entering the evaporator must be controlled to match the speed of evaporation. When the liquid boils rapidly, more liquid must pour into the evaporator. When the boiling process is slow, the valve must restrict the amount of liquid entering the evaporator.

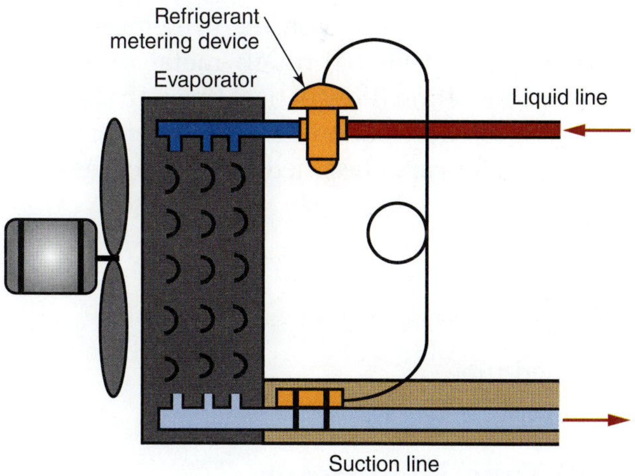

Goodheart-Willcox Publisher

Figure 13-11. The refrigerant metering device acts as a valve to meter the flow of liquid refrigerant into the evaporator. Metering is necessary to prevent liquid refrigerant from reaching and damaging the compressor.

There are different types of refrigerant metering devices such as capillary tubes, fixed bore, thermostatic expansion valves (TEVs), and electronic expansion valves (EEVs). While each type may operate differently, they all function to meter the flow of refrigerant into the evaporator. Exactly how each type operates is explained in Chapter 18, *Working with Metering Devices*.

Condition of Refrigerant in Each Component

A technician must understand the condition of refrigerant within each of a refrigeration system's components. The system illustrated in **Figure 13-12** is the basis for all compression refrigeration systems, regardless of how large or small.

Since refrigerant circulating within a system cannot be seen, thermometers and pressure gauges are used to check system operation. Knowing whether the refrigerant is superheated or subcooled can also aid in verifying correct system operation. These are the only ways that the technician can know for sure what is going on inside the refrigeration system. The technician must know where to obtain these readings and how to diagnose problems correctly. Refer to **Figure 13-12** to locate the following components:

1. **Evaporator:** Component where heat is absorbed.
   A. Inlet: Low-temperature, low-pressure, saturated liquid.
   B. Outlet (suction line): Low-temperature, low-pressure, slightly superheated gas.
2. **Compressor:** Division point between high and low pressure.
   A. Inlet: Low-temperature, low-pressure, slightly superheated gas.
   B. Outlet (hot-gas discharge line): High-temperature, high-pressure, highly superheated gas.
3. **Condenser:** Component where heat is discarded.
   A. Inlet: High-temperature, high-pressure, highly superheated gas.
   B. Outlet (liquid line): Moderate-temperature, high-pressure, subcooled liquid.

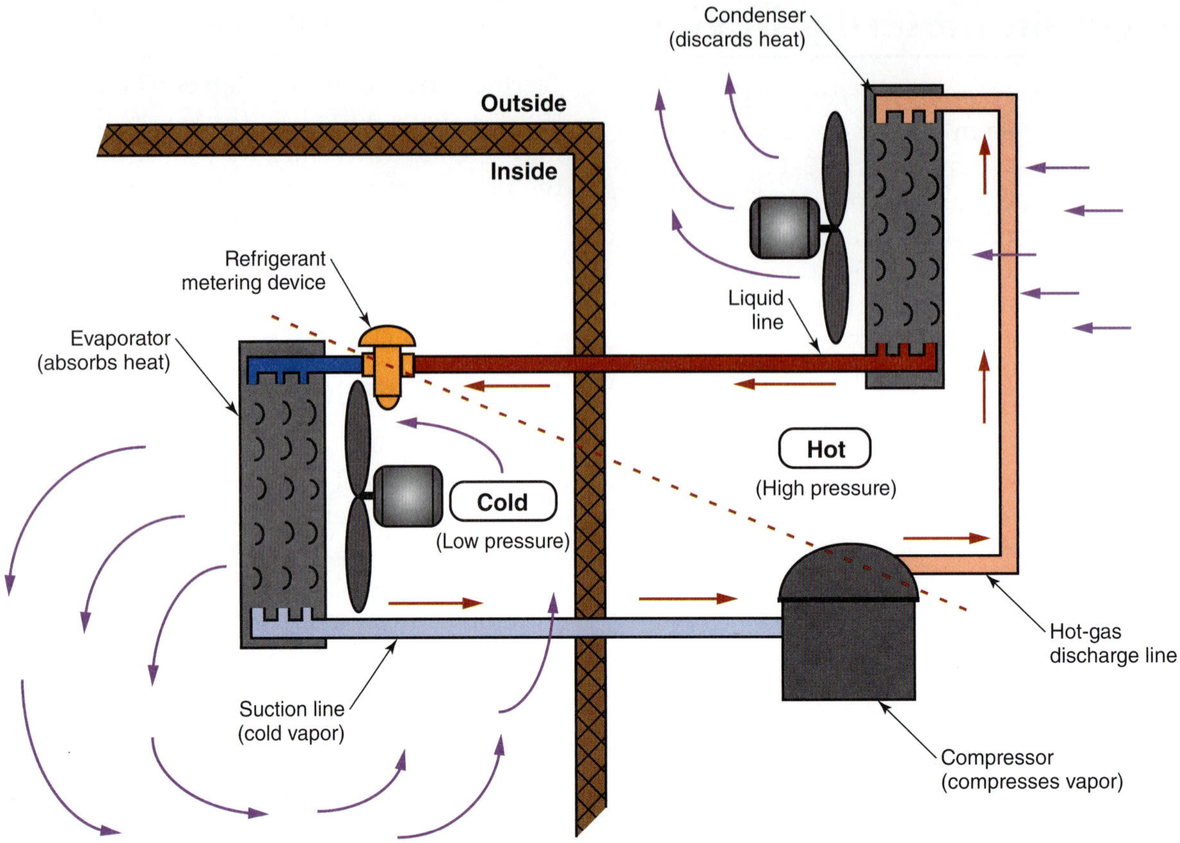

Goodheart-Willcox Publisher

Figure 13-12. A basic refrigeration system. The diagonal dashed line represents the boundary between the low-pressure (cold) side of the refrigeration system and the high-pressure (hot) side.

4. **Refrigerant metering device:** Division point between high-pressure and low-pressure sections of the system.
   A. Inlet: Moderate-temperature, high-pressure, subcooled liquid.
   B. Outlet: Low-temperature, low-pressure, saturated liquid (with some refrigerant vaporizing into what is called *flash gas*).

13.2 Variations in Basic Refrigeration System Components

All refrigeration system components serve their intended purposes but may vary in size, style, and shape. These differences are necessary to save cost or space or to improve overall system efficiency. Variations in major components are described in the following sections.

Evaporators

Evaporators are manufactured in many shapes and sizes to fill specific design or operational needs. They can operate on the principles of conduction (flow of heat through metal or another solid) or convection (flow of heat through a moving liquid or gas). Convection can be *natural* or *forced*.

Evaporators are classified into five types:

- Shell
- Shelf
- Wall
- Plate
- Finned-tube

Shell-Type Evaporator

The shell-type evaporator was commonly used as the freezing compartment in domestic refrigerators that did not have automatic defrost, **Figure 13-13**. The evaporator took the form of a

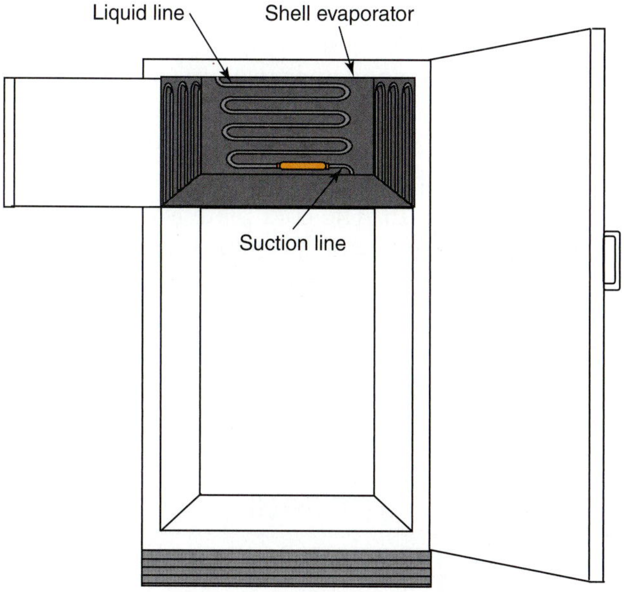

Goodheart-Willcox Publisher

Figure 13-13. The shell-type evaporator is commonly used for the freezing compartment of manual-defrost domestic refrigerators. The aluminum shell is easily damaged, so care must be exercised when defrosting.

metal box that served as the freezer compartment. Food is placed inside the shell (or box) for freezing.

The shell-type evaporator is made of thin aluminum sheet. The refrigerant travels through passages built into the shell during the manufacturing process. Since aluminum is a good conductor of heat, the shell-type evaporator operates on the principles of conduction and natural convection.

The main problem with this evaporator is its thin aluminum construction. The metal is easily punctured, resulting in loss of refrigerant. Repairs are difficult, time-consuming, and expensive.

Frost can accumulate on the evaporator and act as an insulator, preventing heat from getting to the evaporator. Proper defrosting is important, but the use of a sharp instrument is forbidden because of the danger of puncturing the evaporator. Defrosting is accomplished by turning the refrigerator off and placing a pan of hot water inside the evaporator. After the frost and water are removed, recovery to normal operating temperature is rapid.

The shell-type evaporator includes an ***accumulator***, a small reservoir used as a safety device to trap any liquid refrigerant that did not evaporate during passage around the evaporator. Liquid flowing into the accumulator evaporates before being drawn into the suction line.

Shelf-Type Evaporator

The shelf-type evaporator has been used extensively in domestic upright freezers, **Figure 13-14**. The evaporator is made of aluminum tubing that forms the shelves inside the freezer. Additional wires are soldered to the aluminum tubing to increase the strength and evaporator surface area of each shelf. Food containers are placed upon these shelves. Heat transfer is accomplished by conduction and natural convection.

Because warm air rises, one "shelf" of tubing, called the *ceiling coil*, is located in the uppermost part of the cabinet. Most upright freezers have three or four shelves, not counting the ceiling coil. The refrigerant goes to the ceiling coil first, then down through each of the shelves to the bottom. This is called a ***series connection*** because the refrigerant must travel through one shelf before going to the next. The refrigerant in this type of evaporator flows through each shelf's coil in a series.

The upright freezer is popular because it is easy to select frozen foods from its organized shelving. The drawback is that every time the door is opened, cold air spills out the bottom, and warm, moisture-laden air enters at the top. Frost readily accumulates on the ceiling coil and

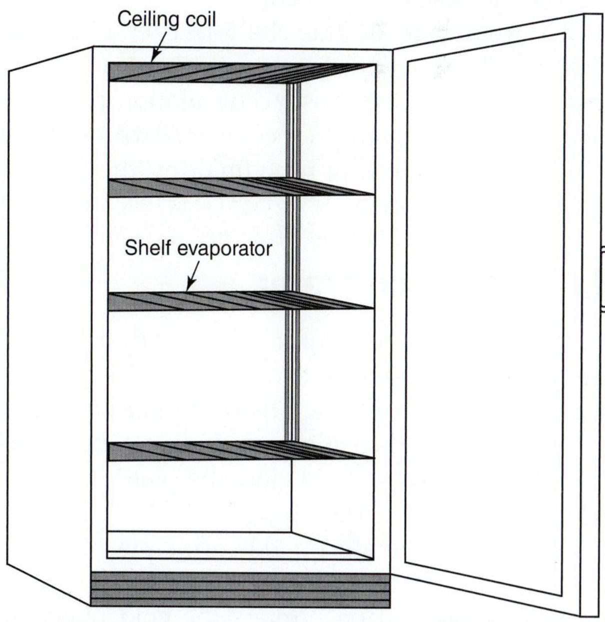

Goodheart-Willcox Publisher

Figure 13-14. The shelf-type evaporator is found in upright domestic freezers. The refrigerant tubing forms the shelves. A ceiling coil cools air that rises to the top of the freezer.

upper shelf. The excess frost acts as an insulator and prevents heat from getting to the evaporator. Manual defrosting must be done about once a year. Defrosting is accomplished by turning the freezer off, removing any packages of frozen food, placing one or two pans of hot water inside the freezer, and closing the door. Heat from the hot water causes the temperature inside the freezer to rise rapidly, melting or loosening the accumulated frost, which is removed by hand. Like the shell-type evaporator, the use of a sharp instrument on a shelf-type unit is forbidden. The sharp instrument might puncture the aluminum evaporator, permitting the refrigerant to escape. A puncture in the aluminum tubing is expensive to repair.

Wall-Type Evaporator

The wall-type evaporator is commonly used in chest-type domestic freezers. The evaporator tubing is made of steel or aluminum and is firmly attached to the outside surface of the inner cabinet liner.

This arrangement provides a smooth inside surface with uniform cooling throughout the cabinet. The rings of tubing go around the sides only, not across the bottom or lid. The refrigerant travels through the top ring first and continues to the lowest ring. Heat transfer is accomplished by conduction and natural convection.

An advantage of the chest freezer over the upright is that cold air does not spill out when the door is opened. Less frost-accumulation requires less frequent defrosting (once every two or three years) and less running time (making the freezer more energy efficient).

Plate-Type Evaporator

The plate-type evaporator is usually made of steel, and the refrigerant travels through passages formed in the cold plates. The cold plates are normally hung from the ceiling, with drip pans made from sheet metal located below the plates to catch and drain off the condensate.

The plates operate on the principle of natural convection (hot air rises and cold air falls). As warm air rises and encounters the cold plates, it gives up its heat to the plates and sinks toward the floor. This method creates constant, slow air movement that keeps temperature levels quite uniform.

The cold plate system is not common, but variations of it are used in domestic applications. The cold plate was the forerunner of the forced convection evaporator discussed next.

One type of cold plate is used in some domestic refrigerator-freezer combinations. Made of aluminum, it is located in the upper back section of the refrigerator compartment (not the freezer compartment), **Figure 13-15**. The cold plate is connected in *series* with the freezer evaporator. It cannot receive any liquid refrigerant until the freezer is satisfied. Then, the leftover liquid reaches the refrigerator cold plate. In this system, the cold plate functions as both an evaporator to cool the refrigerated section *and* as an accumulator to prevent liquid floodback.

Finned-Tube-Type with Forced Convection

The finned-tube-type with forced convection evaporator is the most widely used in domestic, commercial, and industrial applications. The introduction of the finned-tube, forced-convection evaporator made possible remarkable advances in the field of refrigeration and air conditioning. The ***forced convection*** evaporator has a mounted electric fan to increase airflow through the evaporator. The system has made it possible to quickly accomplish large amounts of refrigeration (Btu/hr) with a small evaporator, **Figure 13-16**.

The surface area of the evaporator tubing is significantly increased by the addition of fins. Fins are usually made of aluminum and are securely bonded to the tubing. Spacing the fins varies the capacity of the evaporator and compensates for depth. The fins must be kept straight and equally spaced, else airflow is reduced which would reduce the cooling capacity of the evaporator. Most evaporators are designed for a specific airflow in cubic feet per minute (cfm). The design airflow should not be changed. Changing the capacity of the evaporator affects the operation of other system components which are sized and selected according to the original capacity.

Suction Line

The correct size and proper installation of the suction line is important for efficient operation of a refrigeration system. The size of the suction line (as well as the liquid line) is determined by several factors involving the condition of the gas it contains. Factors include velocity, pressure, volume, density, and pressure drop. The size of the copper

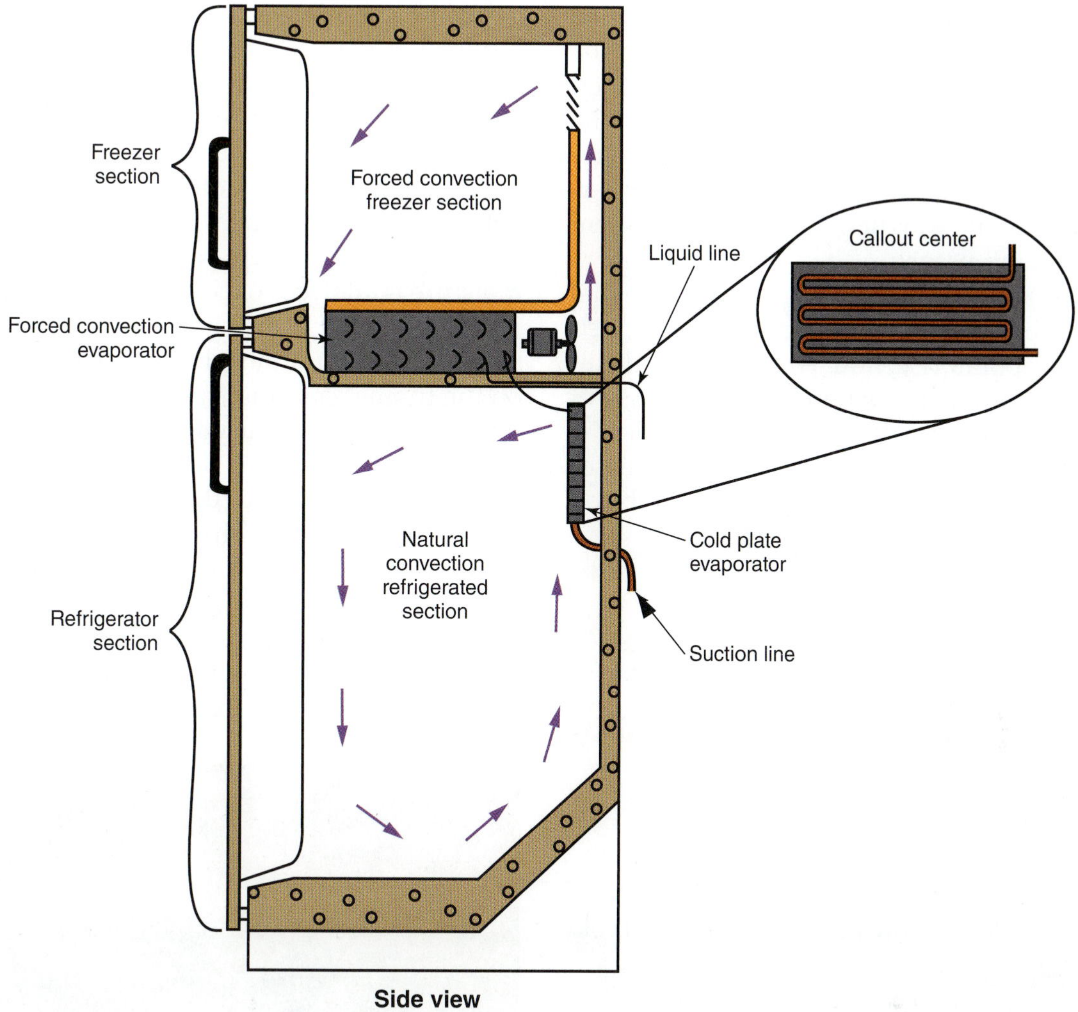

Goodheart-Willcox Publisher

Figure 13-15. A variation of the cold plate principle is used in some domestic refrigerator-freezer units. The plate is located at the back of the refrigerator section and connected in series with the freezer finned-tube-type evaporator.

You Touch Pix of EuToch/Shutterstock.com

Figure 13-16. The finned-tube-type with forced convection evaporator can quickly accomplish large amounts of refrigeration and air conditioning (Btu/hr) with a small evaporator.

tubing will vary according to the size of the system, ranging from 1/4″ to 6″ (6 mm to 152 mm) OD.

The gas-carrying capacity of a suction line is closely calculated to match the flow of gas from the evaporator to the compressor. A long suction line or one with many bends or fittings will lead to a ***pressure drop***. An excessive pressure drop (greater than 2 psi or 14 kPa) in the suction line is equivalent to operating the compressor at a lower pressure, which reduces compressor capacity.

Due to the compressor's lubrication design, some oil tends to circulate through the system along with the refrigerant. It is important that this oil returns to the compressor and does not become trapped somewhere in the system. Oil return can be improved by slanting the suction line downward from the evaporator, so oil drains naturally into the compressor. All horizontal suction lines should slope toward the compressor at the rate of 1/4″ for every 10′ of tubing. A low spot in the suction line functions as an oil trap, **Figure 13-17**. Oil accumulates in such a trap and decreases the efficiency of flow. As the low spot becomes filled with oil, the vapor builds up a pressure against the slug of oil, and the compressor lowers the pressure on the other side. As soon as the pressure difference overcomes the weight of the oil, the oil is sent back to the compressor.

Oil trapping is an advantage in commercial systems where the condensing unit is located above the evaporator section. An oil trap is deliberately installed in the suction line at the base of a ***vertical riser***, **Figure 13-18**. The vertical riser is usually reduced-diameter tubing, which increases the velocity of the refrigerant to help sweep the oil up the riser and back to the compressor.

Suction Line Insulation

Suction lines are insulated for two reasons:

- To limit superheating of gas that would reduce compressor efficiency
- To eliminate condensation that could cause frost and ice problems

The superheated gas inside the suction line is in a low-temperature, low-pressure state. This cold gas cools the compressor, so the amount of superheat gained through the copper tubing suction line must be limited. High superheat at the compressor inlet reduces compressor efficiency and causes damage due to excessive discharge superheat.

Insulating the suction line eliminates contact between moisture-laden atmospheric air and the cold copper tubing, preventing condensation. If moisture is allowed to condense on the copper tubing, it will drip. If the suction line temperature is below 32°F (0°C), atmospheric moisture will freeze on the tubing, creating frost and ice problems.

Suction lines are usually insulated with black, closed-cell, vaporproof flexible tubing insulation. It is available in 6″ lengths, with wall thicknesses of 3/8″, 1/2″, 3/4″, or 1″. Insulation is available with inside diameters (ID) ranging from 3/8″ to 4″ or larger.

In new installations, the insulation is slipped onto copper tubing and over elbows, **Figure 13-19**. For existing systems, insulation is slit lengthwise

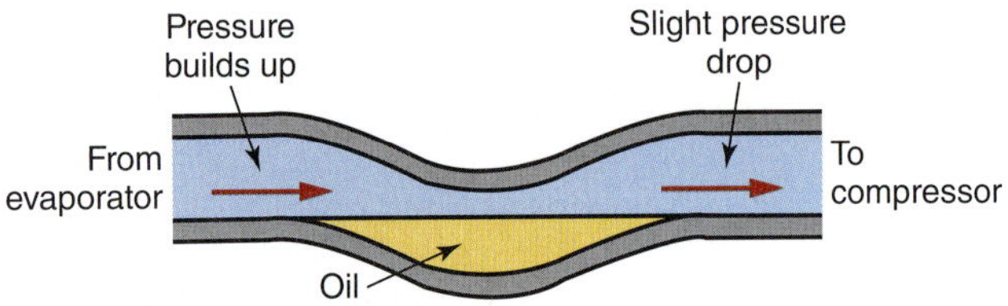

Figure 13-17. A low bend in the suction line functions as an oil trap. As the oil accumulates, vapor pressure builds up on the evaporator side of the trap. Eventually, it pushes the slug of oil to the compressor.

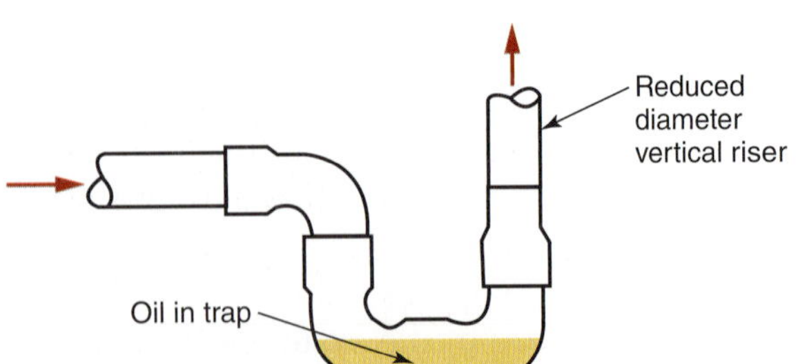

Figure 13-18. An oil trap in a commercial installation. The vertical riser uses reduced-diameter tubing to increase refrigerant velocity and help sweep oil back to the compressor.

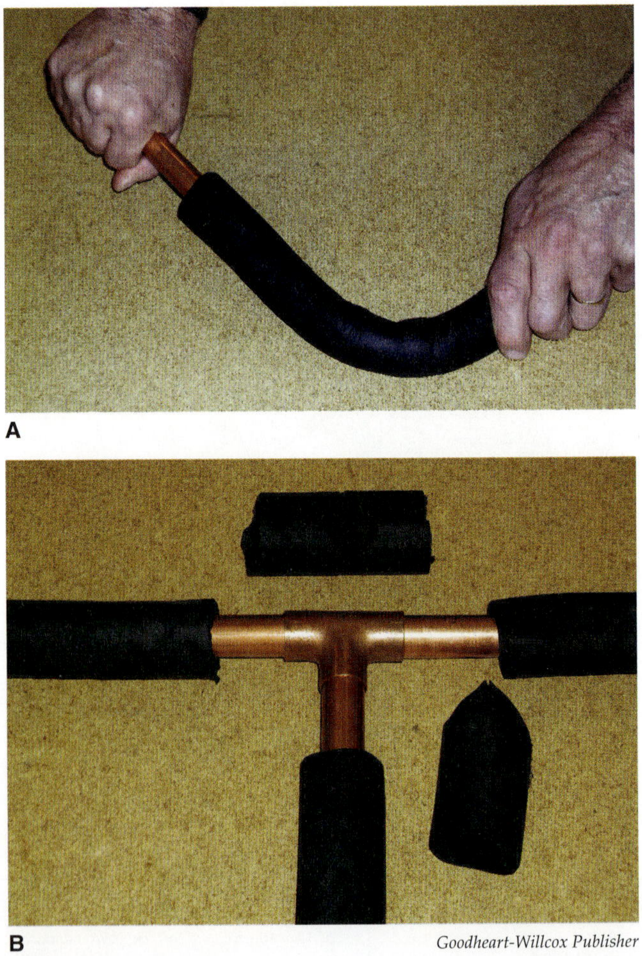

Figure 13-19. Installing insulation. A—Insulation is slipped onto tubing and over elbows in new installations. B—For tees and other fittings, the material is cut, then joined with adhesive.

with a sharp knife and slipped around tubing and fittings, **Figure 13-20**. A special adhesive is applied to seal all slits, seams, and butt joints.

Compressor Designs

The compressor is the "heart" of the refrigeration system and is driven by an electric motor. The electric motor can be mounted outside the compressor or inside the same housing as the compressor.

There are three basic "designs" of compressors, **Figure 13-21**:

- **Open.** This type of compressor is either belt-driven or directly driven (using a coupling) by an external electric motor.
- **Semihermetic.** The electric motor and compressor are enclosed in a forged iron body that is sealed and bolted. The motor shaft is

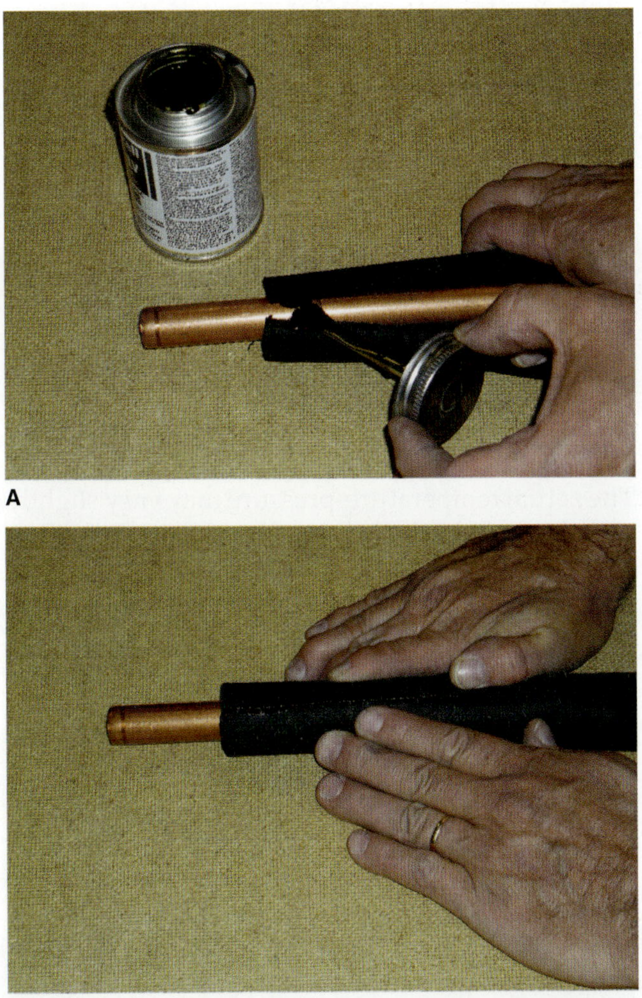

Goodheart-Willcox Publisher

Figure 13-20. Insulating existing installations. A—Insulation is slit to fit over tubing and fittings, then adhesive is applied. B—Adhesive bonds seams quickly when they are pressed together.

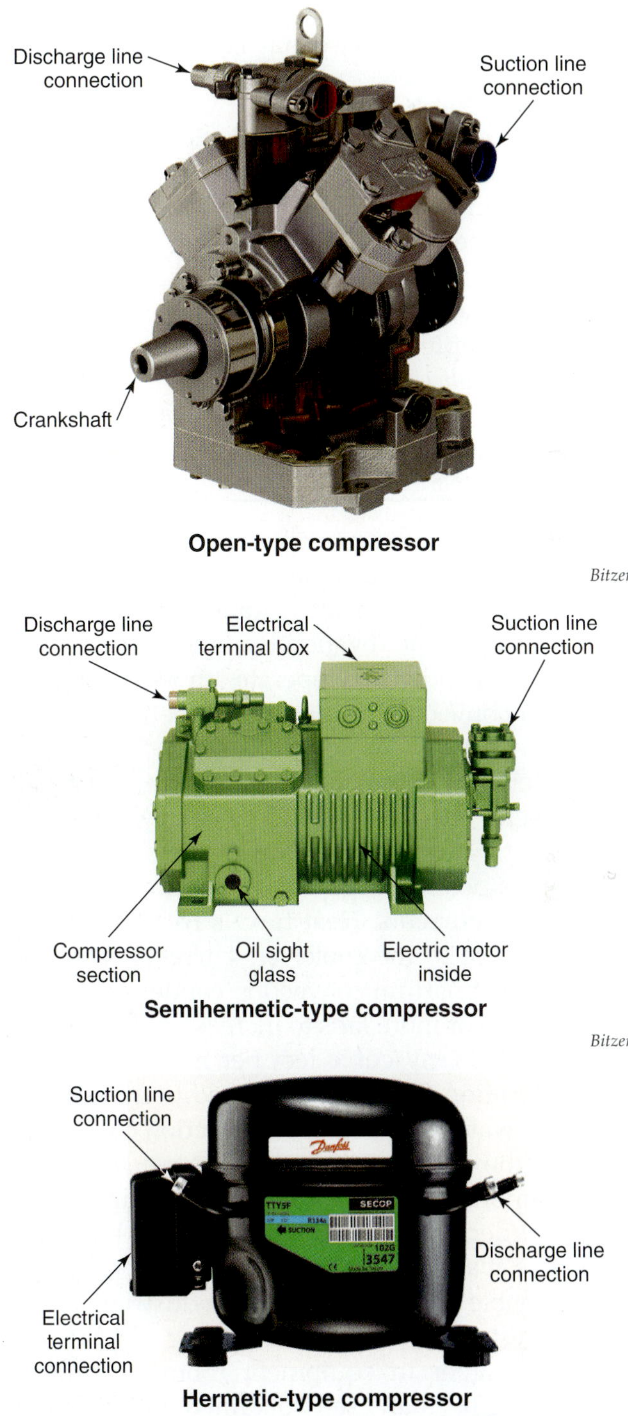

Bitzer

Bitzer

Danfoss

Figure 13-21. Basic styles of reciprocating compressors. The open-type compressor driven by an external electric motor. The semihermetic-type compressor with the motor and compressor inside bolted housing. The hermetic-type compressor with the motor sealed inside a welded metal shell.

directly connected to the compressor crankshaft. These may be disassembled and opened for service.
- **Hermetic.** The motor and compressor are sealed within a steel body that is welded together. Like the semihermetic type, the hermetic compressor is directly driven from the motor shaft. These may *not* be opened for service.

Both open and semihermetic compressors are repairable. Since hermetic compressors are sealed within a welded enclosure, they cannot be repaired. These three "designs" of compressors can be broken down into different types. Descriptions of these various types of compressors are found in Chapter 20, *Compressors* and Chapter 21, *Compressor Lubrication and Accessories.*

Condensers

Condensers are manufactured in a variety of shapes, sizes, and styles. Condensers remove heat absorbed by the refrigerant during its passage through the evaporator and compressor. Air-cooled condensers operate on the principle of forced convection or natural (sometimes called *static*) convection.

Forced convection permits the use of smaller condensers because increased airflow removes heat faster. Natural convection condensers remove heat by exposing large metal surfaces to slow-moving air currents. Heat travels from the warm metal surface to the cooler air currents. The disadvantage of natural convection condensers is the need for one or more fans to increase airflow to the required 750 cfm (cubic feet per minute) per ton of refrigeration. The temporary addition of fans or spraying water on a natural convection condenser changes the condenser's name. However, these are only temporary "fixes" for hot ambient air.

A condenser must be located where adequate airflow is available and the discharge of heat is not objectionable. Commercial and industrial refrigeration systems locate the condensing unit outside the building or in equipment rooms. Domestic systems are normally self-contained with the condensing unit located with the unit but still attached to the refrigerated space through tubing.

The condenser must be able to remove heat from the refrigerant at the same rate that heat is being absorbed during passage through other system components. The main source of heat to be removed by the condenser comes from the evaporator. Additional heat is absorbed by the cool refrigerant vapor when passing over the compressor motor windings and from the heat of compression in the compressor.

To achieve proper condenser operation, the compressor must increase the refrigerant pressure to a saturation temperature 30°F–35°F (17°C–19°C) higher than the ambient air entering the condenser. By raising the temperature-pressure relationship, the saturation point (condensing point) of the refrigerant is increased. This makes it possible to use ambient air or water as a cooling medium to condense the refrigerant to a liquid. The compressor is designed to maintain the 30°F (19°C) temperature difference during the running cycle, regardless of changing ambient temperature. If the ambient temperature goes up, the discharge pressure also goes up. Likewise, if the ambient temperature goes down, the discharge pressure also goes down.

For example, on a hot summer day, the ambient temperature may reach 95°F (35°C). The condensing temperature (saturation point) would be about 125°F (52°C), and the condensing pressure would be close to 185 psig (1277 kPa).

This method of determining discharge pressure applies to most air-cooled condensers. A high-efficiency condenser is able to reduce the temperature difference by about 20°F (11°C). These calculated differences are not precise, but they *are* guidelines for evaluating equipment performance. The actual temperature-pressure may vary slightly above (or below) the calculated pressure. The condensing temperature-pressure relationship is dependent upon several factors, but wide differences between actual pressure and calculated pressures should be investigated. Also, remember to use dewpoint pressures when calculating discharge pressures for zeotropic refrigerants (fluids with two or more components that have different vapor pressures and boiling points) in the 400-series. Zeotropic refrigerants are discussed in Chapter 15, *Refrigerants.*

Water-cooled condensers are more efficient than air-cooled condensers, and the discharge temperature-pressure corresponds to a temperature about 20°F (11°C) higher than the temperature of the water exiting the condenser.

Condenser Cooling Stages

As noted earlier, the refrigerant is cooled in three stages as it circulates through the condenser:

- First, the superheat of the gas is removed, cooling the gas down to the saturation point. (Sensible heat is removed.)
- Second, the gas is condensed to a liquid. (Latent heat is removed.)
- Third, the liquid is subcooled to a point below its saturation point. (Sensible heat is removed.)

As soon as the superheat of the gas is removed, the saturation point is achieved. Further heat removal will cause the gas to condense back to a liquid. The condenser's job is to remove enough heat so that all the gas condenses, leaving only liquid during the final passages where subcooling occurs.

Air-cooled condensers are easy to install, inexpensive to maintain and, if sized properly, satisfactory in all regions. Water-cooled condensers are efficient but are expensive to operate and keep clean. Water-cooled condensers are used occasionally where cool ambient air is not available.

Domestic Condensers

Domestic refrigeration commonly uses three types of air-cooled condensers:

- Finned-tube (forced convection)
- Wire static (natural convection)
- Wall static (natural convection)

Finned-Tube, Forced Convection Condenser

The finned-tube, forced convection condenser is the most commonly used condenser in domestic refrigeration. It is made of steel tubing with external aluminum fins that provide a large and efficient heat transfer surface. Heat transfer is accomplished by forcing large quantities of air through the compact tubing and fin assembly. The amount of air flowing through the condenser should be about 750 cfm per ton of refrigeration. Airflow is critical for proper operation of the condenser. **Figure 13-22** shows a typical finned-tube, forced convection condenser used in a domestic refrigerator.

Jesse33/Shutterstock.com

Figure 13-22. The external aluminum fins on finned-tube, forced convection condensers provide a large and efficient heat transfer surface.

The forced convection condenser is generally preferred for commercial applications. An adequate supply of fresh air must always be available. To achieve their compact size, these condensers are normally constructed with a small face (or surface area) and a depth of several rows of tubing. As the air is forced through the condenser, it absorbs heat from the refrigerant inside the tubes.

Most forced convection condensers use a fan that draws, rather than blows, air through the condenser. The draw-type fans result in more uniform airflow, which is preferred because even air distribution increases condenser efficiency.

Air-cooled condensers must be kept clean. The condenser fan draws into the condenser such foreign material as lint, dust, dirt, or leaves. Foreign material acts as an insulator and prevents proper airflow.

A dirty condenser causes the compressor to work harder. The condensing pressure-temperature must be raised to compensate for the reduced condenser capacity. This results in excessive discharge pressures that will seriously affect the operation of the system. Cleaning the condenser is an easy, but often neglected, task. Cleaning is done with a brush, vacuum cleaner, or by other appropriate means.

Wire Static Condenser

The wire static (natural convection) condenser is used on some domestic refrigerators, **Figure 13-23**. This condenser consists of rows of steel tubing with lengths of wire attached to the tubing. The wire increases the surface area of the tubing, permitting better heat transfer.

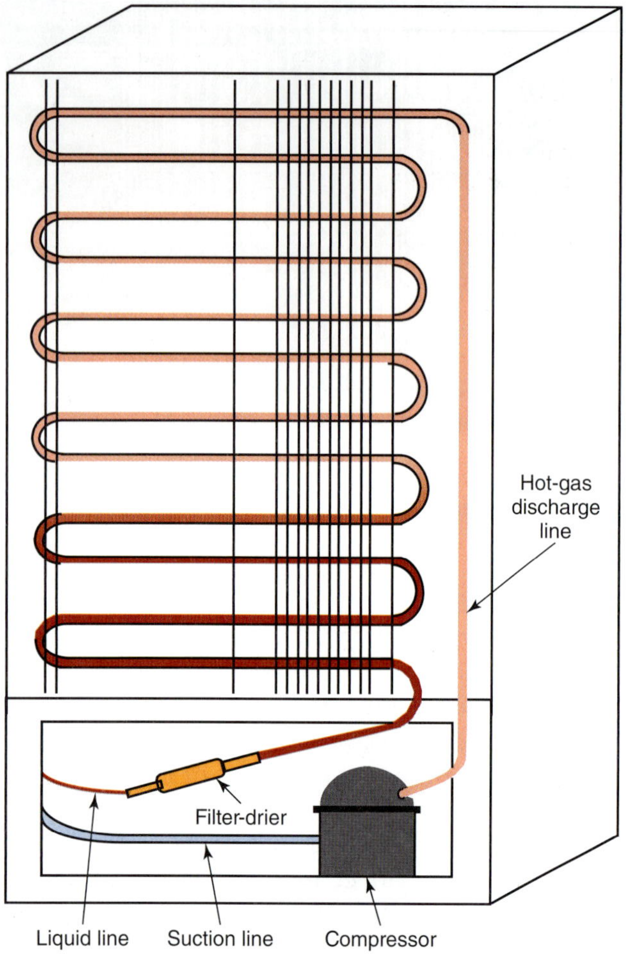

Figure 13-23. The wire static condenser is generally used for domestic refrigerators. Sufficient air circulation space must be provided around the condenser for efficient cooling.

The condenser is mounted on the back of the refrigerator, so the appliance should not be placed tightly against a wall. A space of at least 4″ to 6″ (10 cm to 15 cm) must separate the back of the refrigerator from the wall to allow for natural convection airflow over the condenser.

Wall Static Condenser

The wall static condenser is efficient, hidden from view, and space-saving. It is used on both chest and domestic upright freezers. The condenser tubing is securely fastened to the inside surface of the outer shell, **Figure 13-24**. A layer of insulation separates it from the inner shell (upon which the evaporator is mounted). The unit is actually a box inside a box, with insulation between the two boxes. Of course, the inner box is cold, while the outside of the freezer is warm to the touch. The insulation between the two boxes retards heat flow between the two temperature levels. When installing these freezers in the home, allowance must be made for proper airflow around the outside of the cabinet.

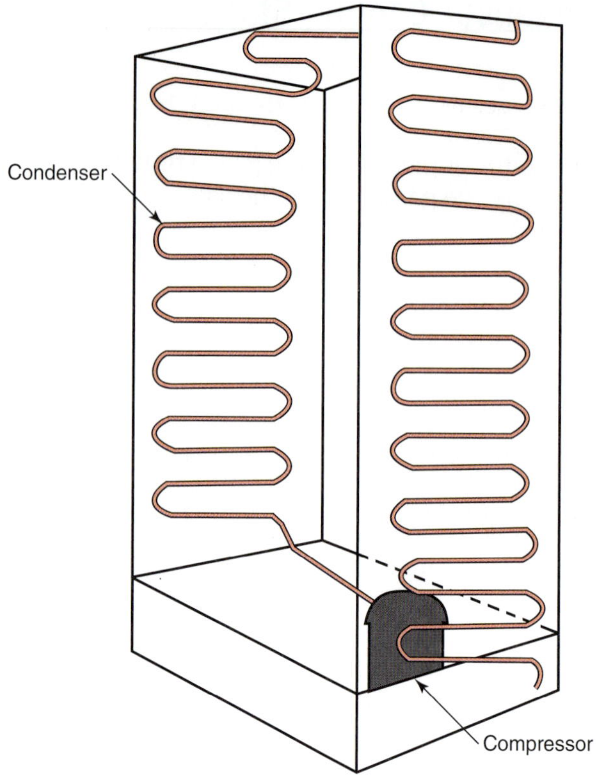

Figure 13-24. The wall static condenser is mounted on the inner surface of the outer wall of a chest or upright freezer. A layer of insulation separates it from the evaporator mounted along the inner wall of the refrigerated compartment. It is an efficient condenser design.

Commercial Condensers

Forced convection, air-cooled is the most commonly used type of condenser in commercial systems, although water-cooled systems are not unusual. Air-cooled units are cheaper to operate and have fewer service problems than do water-cooled units. Commercial air-cooled condensers vary greatly in size and style. Some outdoor units are mounted vertically. Others are mounted horizontally with one or more electric fans to draw air through the condenser, **Figure 13-25**.

Water-Cooled Condensers

At one time, water-cooled condensers were common. However, they have become costly to operate because the water is usually discharged into a sewer drain after passing through the condenser. Water and sewage service rates have

Rooftop condensing unit

Goodheart-Willcox Publisher

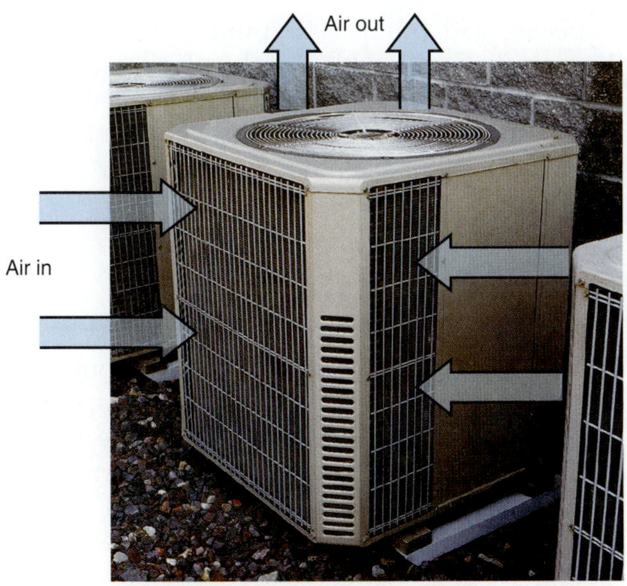

Air-cooled condensing units

Courtesy of Sporlan Division - Parker Hannifin Corporation

Figure 13-25. The rooftop unit is equipped with several electric fans for forced convection. The series of air-cooled condensers draw in ambient air from the sides and exhaust it out the top.

increased with the growing demand on facilities. As a result of increasing costs, many water-cooled condensers are being replaced with air-cooled condensers.

Water-cooled condensers are efficient at cooling refrigerant, but they also present certain problems related to water conditions. Lime and scale deposits occur inside the condenser tubing, which tends to insulate the tubing and reduce efficiency. Scale deposits can severely restrict or even completely stop waterflow. Water-cooled condensers are available in three types, suitably named for their designs:

- Tube-in-a-tube
- Tube-in-a-shell
- Coil-in-a-shell

Tube-in-a-Tube Condenser

The tube-in-a-tube condenser has one tube located inside another tube, **Figure 13-26**. Cooling water circulates through the inside tube, and hot gas discharges through the space between the inner and outer tubes. The *counterflow* principle is applied for maximum efficiency, meaning the water flows in one direction while the hot gas flows in the opposite direction. The warmest water is next to the warmest refrigerant, and the coolest refrigerant is next to the coolest water.

Tube-in-a-Shell-Condenser

This condenser is a cylinder made of steel with copper tubes running from end to end inside it, **Figure 13-27**. Water circulates through the copper tubes and condenses the hot vapors inside the cylinder to a liquid state. The bottom part of the cylinder serves as a storage tank for excess liquid refrigerant.

The tube-in-a-shell condenser has some distinct advantages. It is compact, usually located under the compressor frame, and needs no cooling fan. Manifolds at each end of the condenser control the flow of water through the copper tubing. When these manifold ends are removed, the water tubes are accessible for easy removal of scale deposits.

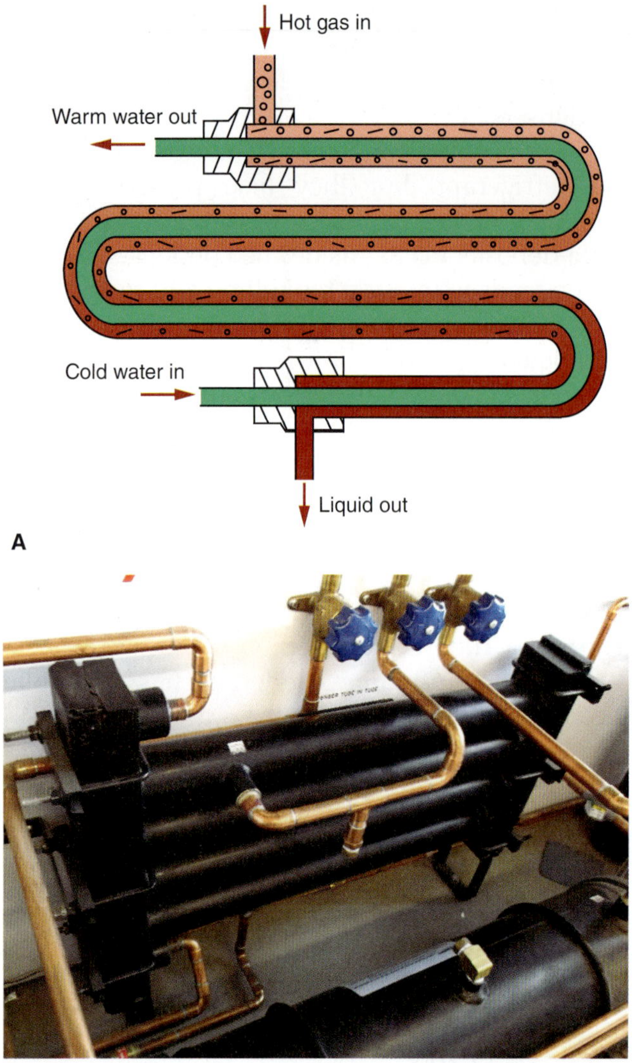

Goodheart-Willcox Publisher

Figure 13-26. Tube-in-a-tube condenser. A—Water flows through the inside tube. The water flows in the opposite direction as refrigerant flow (counterflow). B—A typical tube-in-a-tube condenser.

Cleaning scale deposits from water-cooled condensers is usually done with a wire brush attached to a rod turned by an electric drill. The brush is run back and forth through the water tubing to remove the scale. Removing the scale deposits may reveal small corrosion pits that have penetrated the copper tubing, allowing water to enter the refrigeration system. Cleaning water-cooled condenser tubing is always a gamble.

Coil-in-a-Shell Condensers

These condensers are composed of a shell that contains a coil of tubing around its inside. The coil circulates cooling water. The shell holds a high-pressure refrigerant that condenses and subcools as the coil's circulating water absorbs heat. Coil-in-a-shell condensers can only be cleaned with

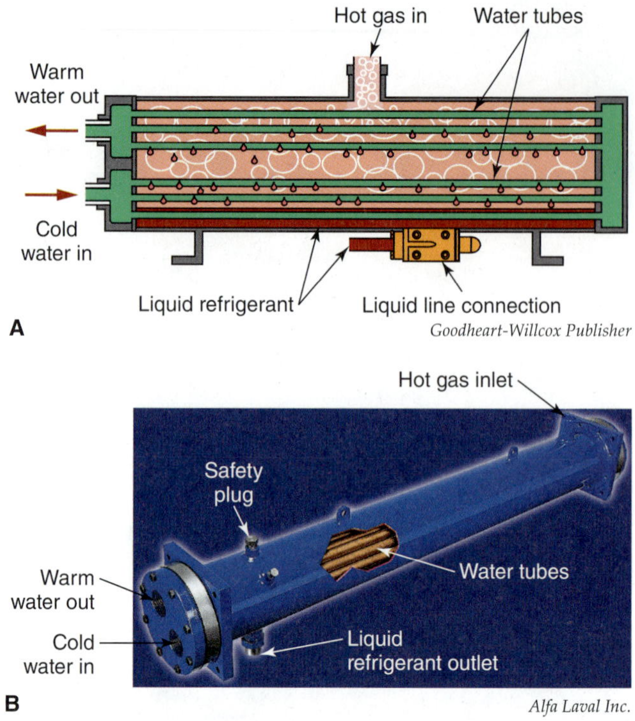

Alfa Laval Inc.

Figure 13-27. Tube-in-a-shell condenser. A—Water flows through the tubes inside the cylinder, absorbing heat from the refrigerant gas and condensing it to a liquid. B—A typical condenser of the tube-in-a-shell type.

a special acid pump. The pump circulates acid through the water tubing to dissolve scale deposits. The system is then flushed with fresh water and placed back in service. There is also a tube-in-a-coil condenser design not discussed here as this design is not common.

Water-Regulating Valve

The amount of water traveling through the compressor *water jacket* (casing) and the condenser is controlled by a water-regulating valve located at the inlet to the water jacket, **Figure 13-28.** The water-regulating valve is designed to control the head pressure by governing the flow of cooling water to the condenser. The bellows and diaphragm assembly at the bottom of the valve is connected by a capillary tube to the high-pressure side of the compressor. The capillary tube transfers the head pressure to the bottom of the water-regulating valve. As the head pressure increases, the valve opens more and permits more waterflow. Likewise, as the head pressure decreases, the waterflow decreases. In this way, pressure-responsive water-regulating valves are able to use the proper amount of water based on the cooling load of the system.

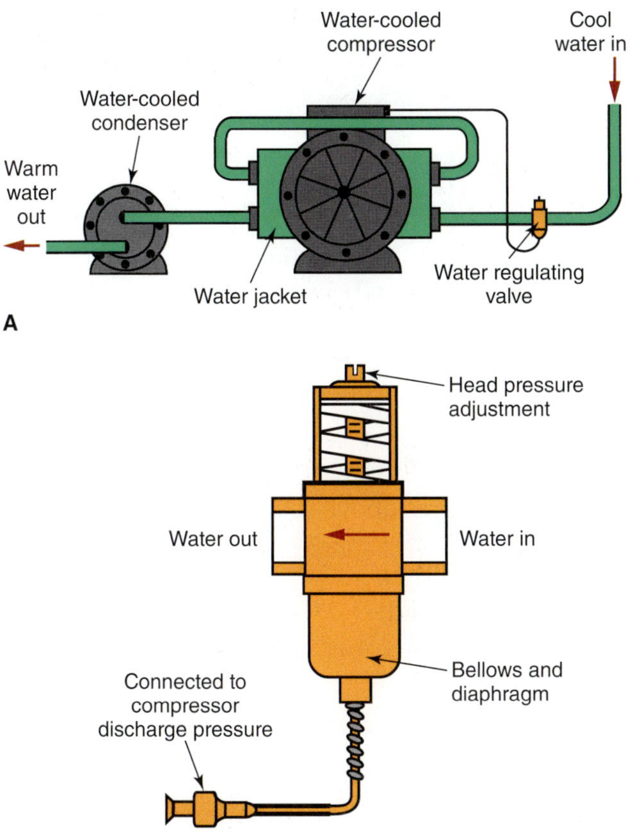

Goodheart-Willcox Publisher

Figure 13-28. Water-regulating valve. A—The valve is located at the inlet of the compressor water jacket where it controls water flow. B—A capillary tube transmits the compressor discharge pressure to the bellows and diaphragm of the valve to regulate the water flow. As head pressure increases, the valve opens more to increase the flow of water.

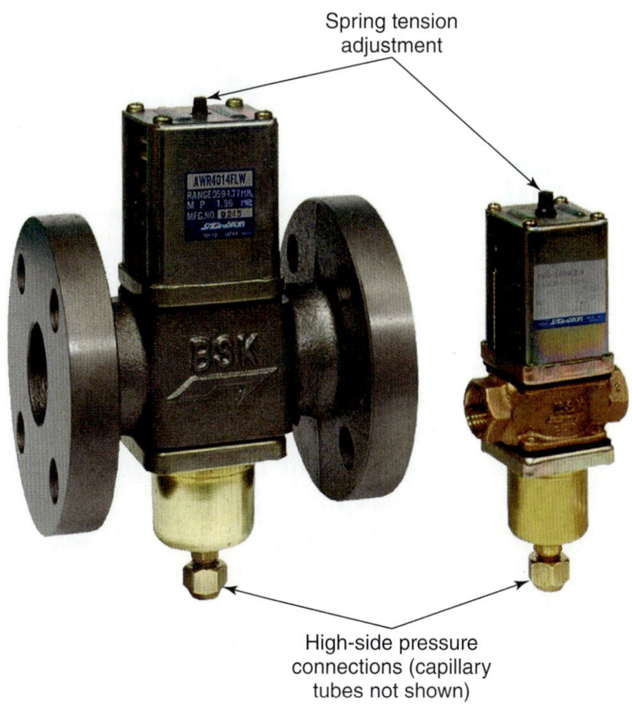

Danfoss

Figure 13-29. A spring-loaded adjustment at the top of the water-regulating valve permits the technician to precisely control head pressure.

The spring tension on top of the water-regulating valve is adjustable and permits the technician to precisely control the head pressure setting, **Figure 13-29**. The water valve automatically adjusts the amount of waterflow according to the head pressure. Thus, constant head pressure is maintained while the system is running. When the compressor stops, the head pressure falls below the valve setting, and the waterflow stops during the off cycle. The water-regulating valve is reliable; once it is set, it seldom requires further attention.

On water-cooled systems, the normal head pressure settings should correspond to a saturation temperature of 100°F (38°C). The setting is comparable to an air-cooled system operating at an ideal ambient temperature of 70°F (21°C). To save on water, a higher head pressure may be desirable.

Dirty tubes can affect the efficiency of a water-cooled condenser. Check the head pressure gauge reading. It should correspond to about 20°F (11°C) above the temperature of the water leaving the condenser. For example, if the water temperature leaving the condenser is 80°F (27°C), the head pressure measurement should correspond to 100°F (38°C) on a P/T chart for the type of refrigerant used in the system. If the saturated head pressure is higher than calculated, the condenser tubes probably need to be cleaned.

Chapter Review

Summary

13.1 The evaporator is a heat-exchanging device located inside the area where heat is to be removed. The suction line connects the evaporator outlet to the compressor intake. The compressor removes vapor from the evaporator to maintain a low boiling-point pressure inside the evaporator and compresses this vapor into a high-temperature, high-pressure gas. The hot-gas discharge line carries refrigerant from the compressor to the condenser. The condenser is a heat exchanger designed to remove heat from superheated refrigerant vapor, causing the vapor to condenser back to a liquid. The liquid line connects the condenser outlet to the refrigerant metering device. The metering device controls the flow of liquid refrigerant into the evaporator.

13.2 Evaporators are classified into five types: Shell, shelf, wall, plate, and finned-tube. Each type has advantages and disadvantages with their intended application.

13.2 The size of the suction line is determined by the condition of the gas it contains, velocity, pressure, volume, density, and pressure drop. The suction line is insulated to limit superheating of gas and eliminate condensation from occurring.

13.2 Compressors are available in open, semihermetic, and hermetic designs. Condensers are available as air-cooled and water-cooled types. Domestic refrigeration typically uses three types of air-cooled condensers: finned-tube, wire static, and wall static. Commercial refrigeration typically uses air-cooled condensers but water-cooled condensers are not unusual.

13.2 The water-regulating valve controls the amount of water traveling through the compressor water jacket (casing) and the condenser.

Know and Understand

1. The _____ is a heat-exchanging device located inside the area where heat is to be removed in a refrigeration system. (13.1)
   A. evaporator
   B. condenser
   C. compressor
   D. metering device
2. Temperatures above saturation are referred to as _____. (13.1)
   A. subcooled
   B. superheated
   C. condensed
   D. latent
3. *True or False?* In refrigerators, the evaporator fan runs constantly, even when other system components are off. (13.1)
4. The suction line connects the evaporator outlet to the _____. (13.1)
   A. condenser intake
   B. compressor intake
   C. metering device
   D. liquid line
5. The compressor must remove vapor from the evaporator to maintain a low boiling-point pressure inside the evaporator and it must compress _____. (13.1)
   A. low-temperature, low-pressure gas into a low-temperature, low-pressure liquid
   B. low-temperature, low-pressure gas into a high-temperature, high-pressure liquid
   C. low-temperature, low-pressure gas into a high-temperature, high-pressure gas
   D. low-temperature, low-pressure gas into a high-temperature, high-pressure liquid
6. *True or False?* The hot-gas discharge line is larger than the suction line because the gas is at a higher volume and lower pressure. (13.1)

7. In what condition is the refrigerant as it leaves the condenser? (13.1)
   A. Low-temperature, low-pressure superheated gas
   B. Moderate-temperature, high-pressure subcooled liquid
   C. High-temperature, high-pressure superheated liquid
   D. High-temperature, high-pressure highly superheated gas
8. The condition of refrigerant inside the liquid line is a _____. (13.1)
   A. low-temperature, low-pressure superheated gas
   B. moderate-temperature, high-pressure subcooled liquid
   C. high-temperature, high-pressure subcooled liquid
   D. high-temperature, high-pressure highly superheated gas
9. The two division points between the high- and low-pressure sides of a refrigeration system are at the compressor and at the _____. (13.1)
   A. condenser
   B. liquid line
   C. evaporator
   D. refrigerant metering device
10. The shell-type evaporator uses a(n) _____ to trap any liquid refrigerant that did not evaporate during passage around the evaporator. (13.2)
    A. shelf
    B. accumulator
    C. plate
    D. separate tube
11. What is the most widely used type of evaporator used in HVACR? (13.2)
    A. Finned-tube type with forced convection
    B. Shelf-type
    C. Shell-type
    D. Wall-type with natural convection
12. A long suction line with one or many bends or fittings are needed to create a(n) _____ in the suction line. (13.2)
    A. pressure increase
    B. pressure drop
    C. change of state
    D. accumulation of oil
13. High superheat at the compressor inlet reduces compressor efficiency and causes damage due to _____. (13.2)
    A. excessive condensation
    B. excessive discharge subcooling
    C. not enough superheat applied
    D. excessive discharge superheat
14. Most compressors are repairable, except for _____ compressors due to their enclosed design. (13.2)
    A. open
    B. semihermetic
    C. hermetic
    D. reciprocating
15. When liquid is subcooled in the condenser to a point below its saturation, then _____ at this stage. (13.2)
    A. sensible heat is added
    B. sensible heat is removed
    C. latent heat is added
    D. latent heat is removed
16. *True or False?* Finned-tube, forced convection condensers are generally preferred for commercial applications. (13.2)
17. The _____ condenser applies the counterflow principle with water and hot gas. (13.2)
    A. tube-in-a-shell
    B. coil-in-a-shell
    C. tube-in-a-tube
    D. tube-in-a-coil
18. The _____ is designed to control the head pressure by governing the flow of cooling water to the condenser. (13.2)
    A. metering device
    B. condenser
    C. water coil
    D. water-regulating valve

Apply and Analyze

1. What is the purpose of the evaporator? (13.1)
2. Explain the importance of having a temperature difference between the evaporator and the air passing through the evaporator. (13.1)
3. Describe two purposes of the compressor. (13.1)

4. Explain how the finned-tube, forced convection evaporator can provide large amounts of refrigeration. (13.2)
5. Why are suction lines insulated? (13.2)
6. Describe the three stages of condenser cooling. (13.2)

Critical Thinking

1. Describe the flow and condition of refrigerant in each system component throughout the entire refrigeration cycle. Begin at any point in the cycle. (13.1)
2. What is the effect of an excessive pressure drop in the suction line? (13.2)

Emerson Climate Technologies

14 Other Refrigeration System Components

LEARNING OUTCOMES

After studying this chapter, you will be able to:

14.1 Identify condensing and evaporator units.

14.2 Describe the use of liquid receivers in large commercial systems.

14.3 Recognize where service valves are located and how to operate them.

14.4 Explain how a filter-drier protects a system from moisture buildup.

14.5 Describe the purpose of a sight glass and moisture indicator.

14.6 Explain the function of a suction line-liquid line heat exchanger.

14.7 Recognize how an accumulator's design prevents liquid refrigerant from entering the compressor.

TECHNICAL TERMS

absorb
accumulator
adsorb
aspirator hole
backseated
condensing unit
cracked
desiccant
dip tube
discharge service valve (DSV)
evaporator unit
filter-drier
frontseated
heat exchanger
hydrostatic expansion
liquid receiver
liquid receiver service valve (LRSV)
moisture indicator
noncondensable
overcharge
process stub
pumpdown
Schrader valve
sight glass
suction service valve (SSV)
undercharged

Introduction

As described in Chapter 13, *Basic Refrigeration Cycle and System Components*, the basic refrigeration system consists of four basic components: evaporator, compressor, condenser, and refrigerant metering device.

Refrigeration systems are designed to remove heat faster than heat can enter a cabinet or room. The more heat leakage, or heat load, the bigger the system must be. Other than size, all systems are fundamentally the same. Those that appear more complicated simply have additional components to make them more efficient and serviceable.

To understand the operation of any system, knowledge of how individual components work and affect the system is needed. A malfunction by one component will impact the operation of other components. Different components, or combinations of components, make the system more efficient, versatile, and serviceable. For example, many accessory components facilitate checking system operation and performing repairs. Some components keep the system clean. Others provide a means for isolating sections for repairs. It is important to understand the purpose, theory, and operation of these components. Being able to identify them and understand their functions will help troubleshoot and service a system.

14.1 Condensing and Evaporator Units

Refrigeration system components can be grouped by function into condensing and evaporator units. The ***condensing unit*** includes the equipment necessary to reject the absorbed heat and reclaim the refrigerant vapor by condensing it back into a liquid. The condensing unit consists of the compressor, condenser, hot-gas discharge line, condenser fan, electrical panel box, and some accessory components, **Figure 14-1**.

The ***evaporator unit*** includes the equipment necessary to absorb heat and vaporize the liquid refrigerant. It consists of the evaporator, refrigerant metering device, evaporator fan, and some accessory components. The suction and liquid lines connect the evaporator unit with the condensing unit to complete the system.

The condensing unit must be located outside the refrigerated space, positioned so ambient air can cool the condenser. The evaporator absorbs heat from the refrigerated space, while the condenser discharges that absorbed heat into ambient air. Therefore, these two *heat exchangers* must be well separated from each other.

The suction and liquid lines connecting the two units may be either long or short. For example, on a domestic refrigerator-freezer, they would be quite short. The evaporator unit is in the freezer compartment, and the condensing unit is below the refrigerator section, **Figure 14-2**. This is a *self-contained* system because the condensing and evaporator units are located within the same cabinet.

In a remote or *split* system, the condensing and evaporator units are not in the same cabinet. While the evaporator is located inside the area to be cooled, the condensing unit is likely to be on the building's roof or outside at ground level. The condensing unit must be located where it is acceptable to discharge heat. The suction and liquid lines connecting the two units may be quite long. A residential central air-conditioning system is an example of a split system, **Figure 14-3**. Commercial split systems are essentially the same, but the components are larger.

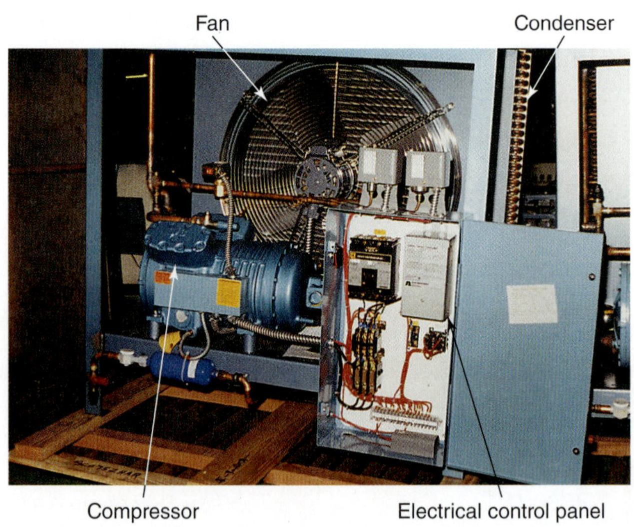

Dunham-Bush

Figure 14-1. A condensing unit for a commercial installation, consisting of a compressor, condenser, fan, and electrical control panel. The unit may be located far away from the evaporator unit.

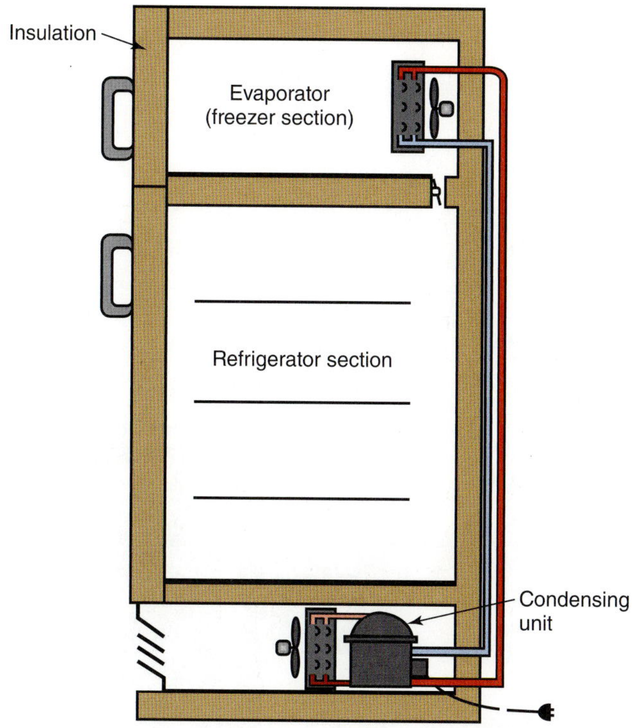

Goodheart-Willcox Publisher

Figure 14-2. A domestic refrigerator-freezer is a self-contained unit.

14.2 Liquid Receiver

The ***liquid receiver*** is an important accessory in large systems. It is installed in the liquid line and serves as a storage tank for excess liquid refrigerant, **Figure 14-4**. Ideally, the refrigerant should boil off inside the evaporator at the same rate it is being changed to a liquid state in the condenser. Such a balance is generally achieved in domestic systems, which operate at a steady rate. However, larger commercial systems require varying amounts of refrigerant at different times, depending on the heat load on the evaporator. Therefore, a reserve of refrigerant must be available.

Liquid receivers are not used in domestic refrigeration systems. The cost is prohibitive, and domestic systems use a capillary tube for the refrigerant metering device, which is not designed to work properly with liquid receivers. The amount of refrigerant in domestic systems is critical. Because excess refrigerant accumulates in the condenser and reduces its capacity, even a 1/2 oz ***overcharge*** (excess of refrigerant) will cause high head pressure in such systems.

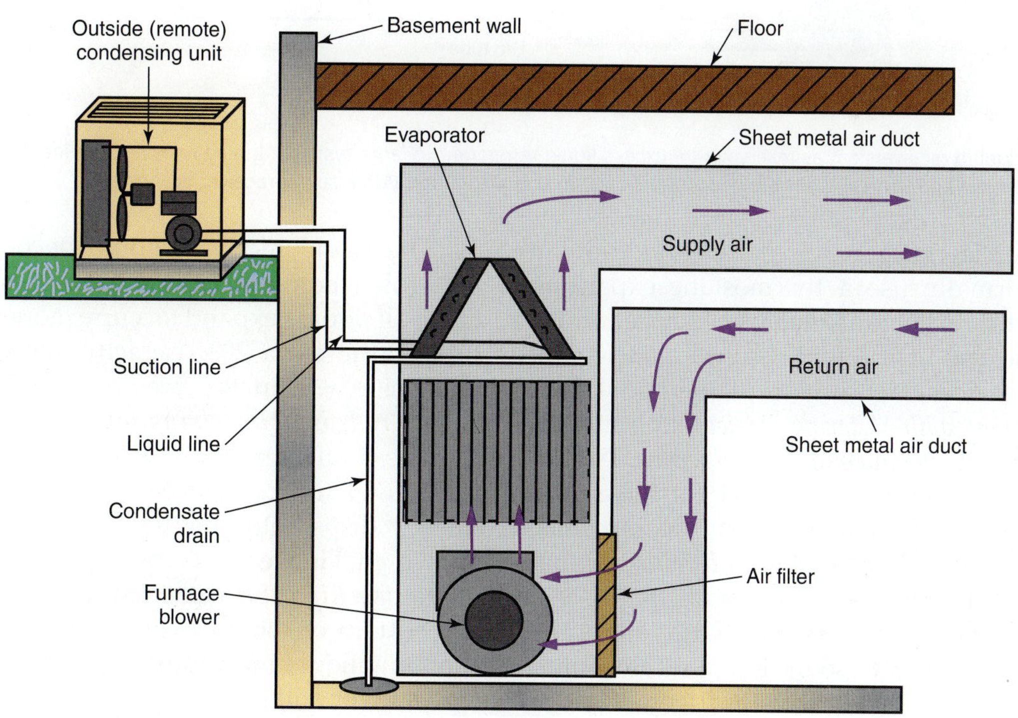

Goodheart-Willcox Publisher

Figure 14-3. A residential central air-conditioning system is a typical split system. The evaporator and condensing units are located separately but connected by copper tubing.

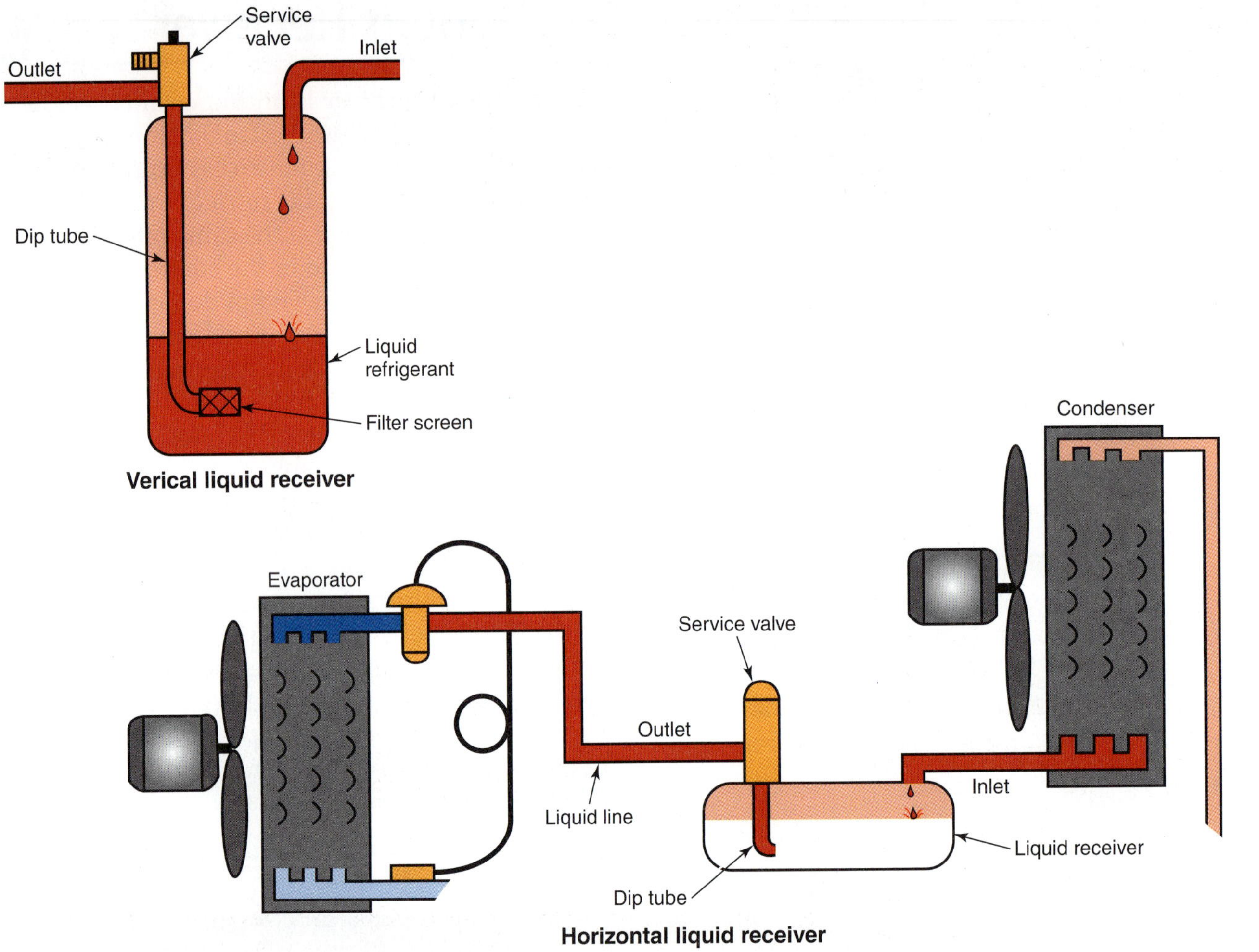

Goodheart-Willcox Publisher

Figure 14-4. Liquid receivers act as reservoirs for excess liquid refrigerant in large systems. Liquid receivers are located in the liquid line. Both horizontal and vertical types are used, depending on the application. A horizontal receiver is shown.

Commercial refrigeration and air-conditioning systems normally use a thermostatic expansion valve for the refrigerant metering device, which permits the use of a liquid receiver. Refrigerant metering devices are fully explained in Chapter 18, *Working with Metering Devices*. Commercial systems allow for some excess refrigerant. The liquid receiver acts as a storage place for the excess. The amount of refrigerant in such a commercial system is not critical unless it is ***undercharged*** (has insufficient refrigerant).

Commercial systems require large quantities of refrigerant to operate, so the liquid receiver must be large enough to hold all the refrigerant in the system plus any excess. All refrigerants are expensive, and the ability of the liquid receiver to hold the entire refrigerant charge is desirable when the system must be opened for repairs.

The liquid receiver should be no more than 80% full when holding the entire system charge. Since all liquids expand (occupy more space) when heated, the extra 20% capacity allows for liquid expansion within the receiver. This principle is called ***hydrostatic expansion***.

The liquid receiver is usually located close to (or below) the condenser. Liquid from the condenser drips into the liquid receiver inlet. As shown in **Figure 14-4**, the liquid receiver outlet contains a ***dip tube*** that extends to about 1/2″ from the bottom of the receiver. The dip tube ensures that only liquid (no vapor) enters the liquid line at the receiver outlet.

Some large receivers are equipped with two service valves, one at the inlet and one at the outlet. These receivers also contain a fusible plug or pressure relief valve as a safety device, **Figure 14-5**.

The plug or valve relieves any sudden increase in pressure down to a safe level. Such an increase might occur in a fire, for example. The spring-loaded pressure relief valve opens automatically when a specified pressure is reached in the receiver, **Figure 14-6**. The pressure at which the valve opens is adjustable within a given range. The fusible plug is a relief device designed to melt at a specified temperature, **Figure 14-7**. It releases the refrigerant if an unsafe temperature (and corresponding unsafe pressure) is reached.

Some large receivers are equipped with a dial-type gauge or a liquid level sight glass to indicate how much liquid refrigerant is in the receiver, **Figure 14-5**. As noted, liquid level in the receiver should never be more than 80% of its capacity when holding the entire charge of the system.

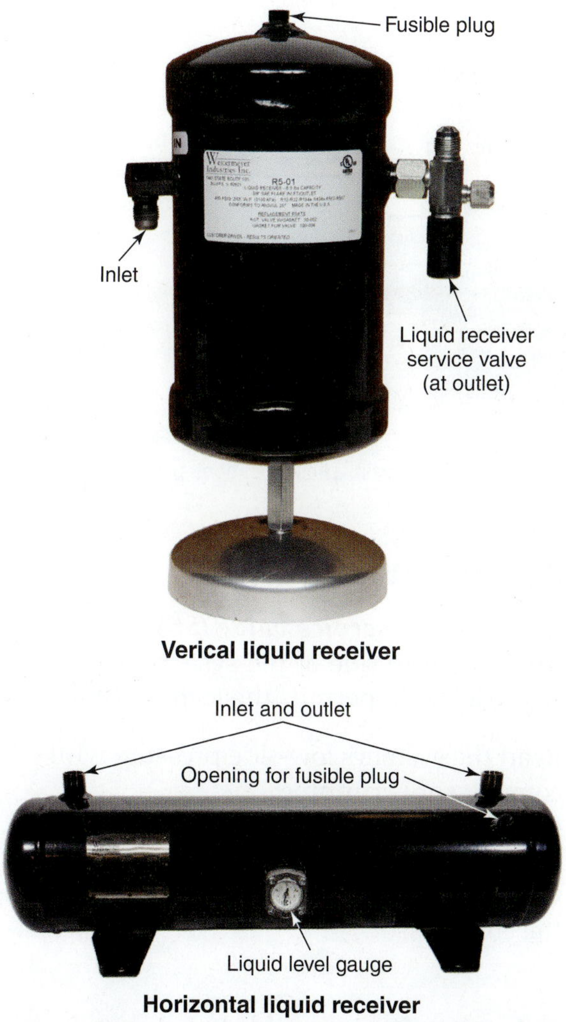

Westermeyer Industries, Inc.

Figure 14-5. The vertical receiver has a fusible plug installed. The horizontal receiver has a provision for either a fusible plug or relief valve.

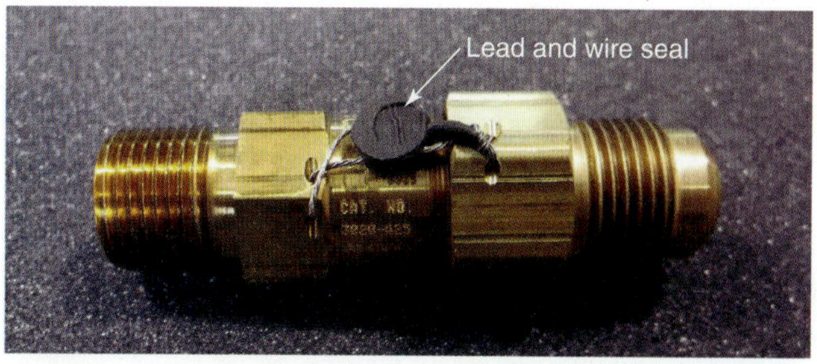

Image courtesy of United Refrigeration

Figure 14-6. The spring-loaded pressure relief valve opens at a specified pressure, then closes again when pressure drops below that level. The lead and wire seal prevents tampering with the valve pressure setting.

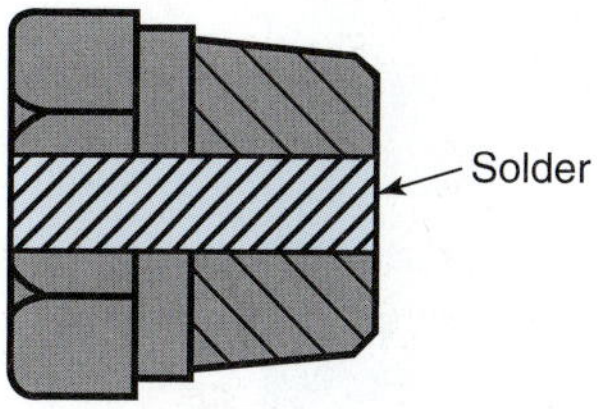

Goodheart-Willcox Publisher

Figure 14-7. A fusible plug has an opening sealed with solder that melts at a specified temperature, usually 165°F–210°F (74°C–99°C).

14.3 Service Valves

A service valve is an accessory that serves no operating function but is indispensable when work must be performed on the system. Service valves are provided on residential HVAC systems and commercial refrigeration systems for troubleshooting and making repairs. They are not factory-installed on domestic refrigeration systems but often are installed by technicians when servicing is needed. Domestic refrigeration compressors are equipped with service stubs to attach access fittings to for servicing. The valves make it possible to block off certain sections of the system for servicing or for reading operating pressures using a manifold gauge. The service technician must fully understand where service valves are located and how they operate.

Service Valve Locations

Service valves on commercial systems are normally placed in three strategic locations, **Figure 14-8**. The valves provide access for pressure readings and can be used to control system operation and isolate sections to diagnose problems.

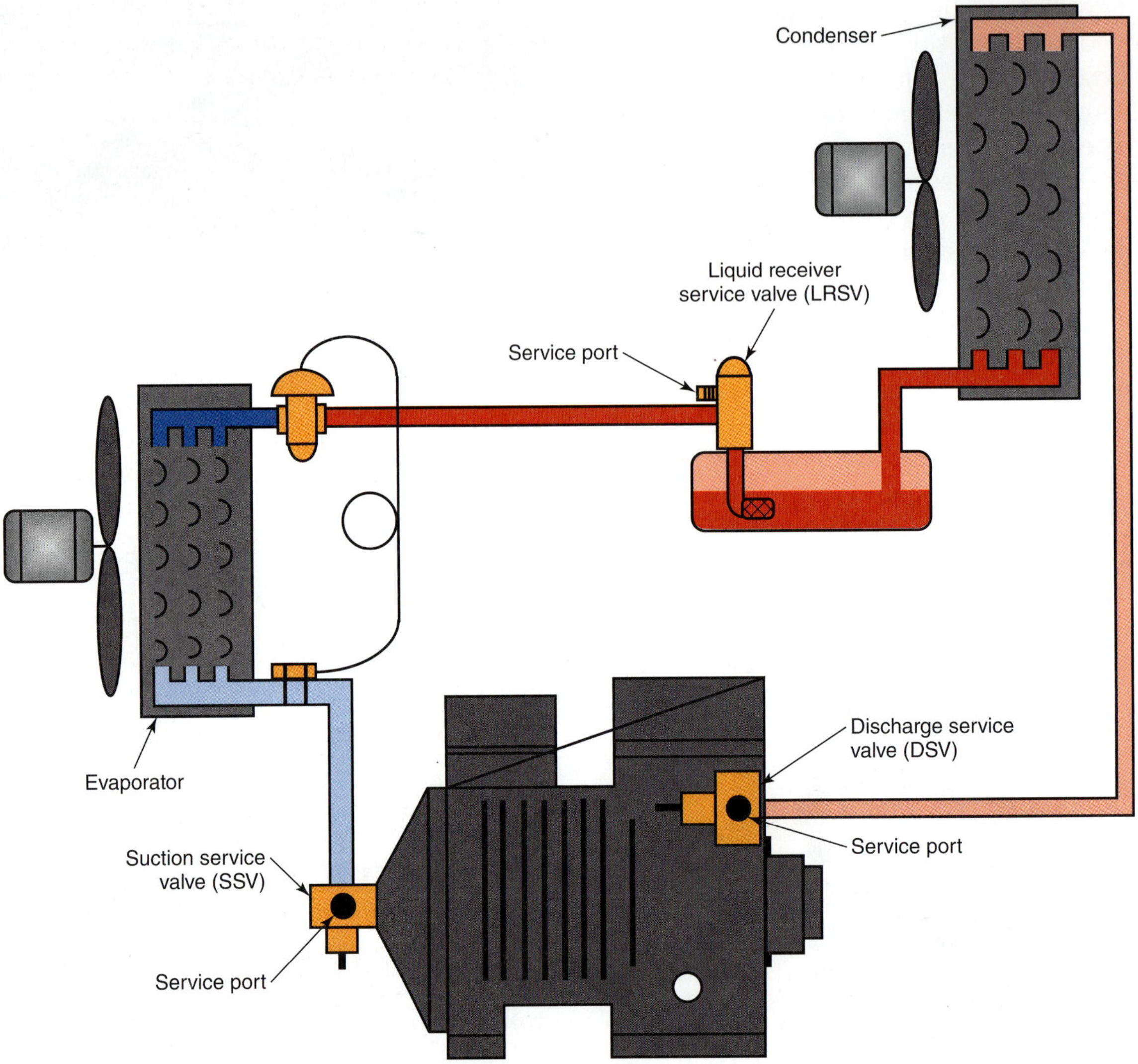

Figure 14-8. Service valves are normally located at three specific places (see the service ports). These locations permit easy access to the system for servicing.

Failure to use these valves properly may result in wrong diagnoses, excessive service time, and unnecessary loss of refrigerant.

SAFETY NOTE

Leaking Refrigerant

Leaking refrigerant can cause instant frostbite, causing bad burns to the hands or exposed flesh. Gloves and safety glasses should always be worn when working with refrigerants. Gloves can help protect your hands from small refrigerant leaks but with any leak, large or small, *never* use your hands to try and stop the leaking refrigerant. The gloves can freeze to your hands and would be painful to remove.

Suction Service Valve

The ***suction service valve (SSV)*** is located on the low-pressure side of the system in the suction line. The location permits the technician to:

- Read the system's low-side pressure while the compressor is running.
- Restrict the flow of refrigerant entering the compressor.
- Completely stop the refrigerant from entering the compressor.

Discharge Service Valve

The ***discharge service valve (DSV)*** is located on the high-pressure side of the system. The location permits the technician to:

- Read the system's high-side pressure while the compressor is running.
- Restrict the flow of gas leaving the compressor.
- Completely stop the gas from leaving the compressor.

CAUTION

Frontseating a Discharge Service Valve

Never frontseat (close) a discharge service valve while the compressor is running. Discharge pressure will rapidly reach extremely high, unsafe levels.

By having a service valve on each side of the system , it is possible to obtain high- and low-side pressure readings at a single location.

Liquid Receiver Service Valve

The ***liquid receiver service valve (LRSV)*** is located in the liquid line, at the liquid receiver outlet. The LRSV is connected to the liquid receiver outlet, and the liquid line is connected to the valve. The location permits the technician to:

- Read the system's high-side pressure while the compressor is running.
- Restrict the flow of refrigerant leaving the liquid receiver.
- Completely stop the liquid refrigerant from leaving the liquid receiver (while the compressor is running for system pumpdown).

Some large commercial systems include another service valve located at the liquid receiver *inlet*. This provides a method to isolate the liquid receiver from the rest of the system by closing (frontseating) both liquid receiver service valves. Closing both valves traps most of the refrigerant inside the liquid receiver.

An LRSV at a liquid receiver's *outlet* is the more common of the two service valves; it is commonly called a *king valve*. An LRSV on a liquid receiver's *inlet* is commonly called a *queen valve*.

Refrigeration Service Wrenches

Service valves vary in size along with the size of the system. Valve stem sizes vary as well, requiring refrigeration service wrenches of different sizes. The ratchet-type wrench with different sized openings is the most popular, **Figure 14-9**. It fits most valve stems and is easily reversed. This tool is a must in every technician's toolkit. Never use any other tool to open or close a service valve, as this could damage the valve stem and make it hard to open or close.

Suction Service Valve Operation

All service valves are designed to function in the same manner, regardless of location, size, or shape. All service valves have three openings. The valve stem can be turned all the way in one direction to block the service port. It could then be turned all the way in the *other* direction to block the inlet or outlet, depending on the valve and how it is connected. In this way, the valve stem controls the inlet (or outlet) and the service port, **Figure 14-10**. As shown, the suction line connects to the suction service valve's inlet and the compressor at the valve's outlet.

Suction Service Valve Backseated

With the valve stem in the ***backseated*** position—valve stem screwed all the way to the left, counterclockwise—the flow of refrigerant from the suction line to the compressor is unrestricted; however, the service gauge port is blocked off (closed), **Figure 14-11**. Backseated is the normal position of the suction service valve's stem when the system is operating.

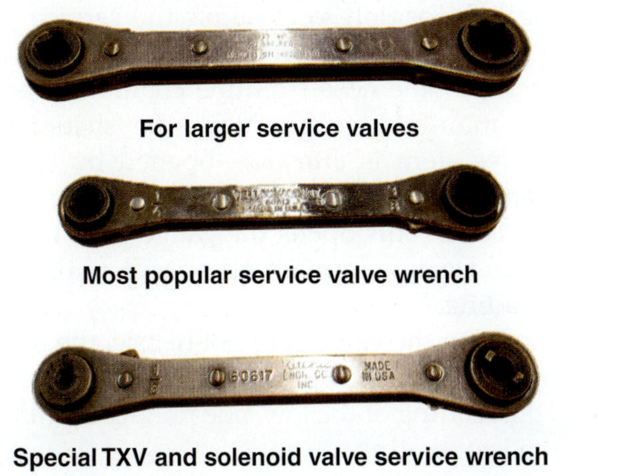

Figure 14-9. The most popular type of service valve wrench is a ratchet-type. These have different size openings, and their direction turn is easily reversed.

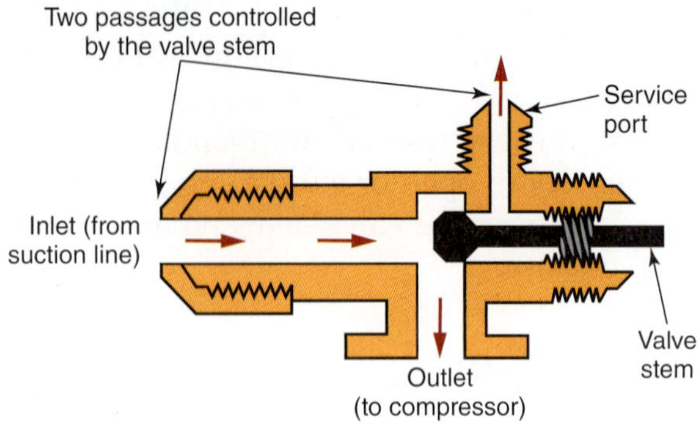

Goodheart-Willcox Publisher

Figure 14-10. Service valves have three openings: the inlet, the outlet, and the service port opening. The valve stem is able to close the service port and either the inlet or outlet, depending on how the valve is connected into the system.

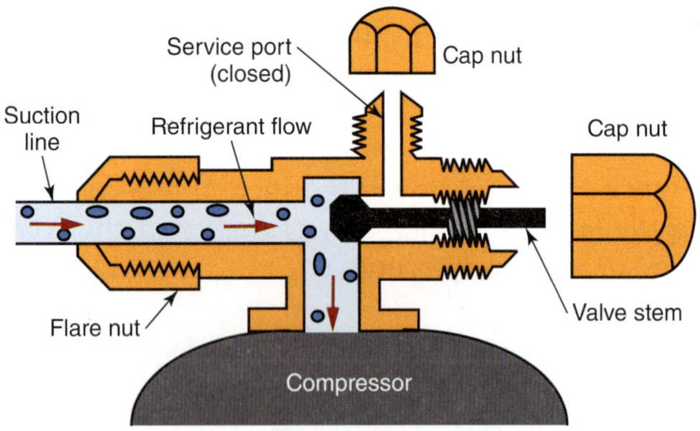

Goodheart-Willcox Publisher

Figure 14-11. Suction service valve with stem in the backseated position. This is the normal operating position of the valve stem.

Suction Service Valve Cracked

To obtain a pressure reading at the suction service valve, the valve stem must first be in the backseated position to close off the service port. Then, remove the cap from the service port and connect the blue hose from the compound gauge on the manifold gauge. Next, the suction service valve stem is ***cracked***—opened by turning the stem about one or two turns to the right. See **Figure 14-12**. This opens the gauge port, and the compound gauge immediately registers the pressure reading.

Cracking the valve does not restrict the flow of refrigerant to the compressor. The procedure permits the reading of the low-side pressure while the system is operating.

Suction Service Valve Frontseated

When the SSV is ***frontseated***—valve stem screwed all the way to the right, clockwise—the inlet opening is closed off. See **Figure 14-13**. This frontseated position prevents refrigerant vapor from entering the valve. However, the gauge port is open to the compressor. Remember that the compressor opening cannot be closed by the valve stem. An SSV's valve stem controls only the vapor inlet and the gauge port opening.

If the SSV is frontseated while the compressor is running, the compressor will immediately show a good vacuum on the compound gauge because the suction vapor cannot enter the compressor. This procedure is used to check the operation of the compressor. Failure of the compressor to pull a good vacuum when the SSV is frontseated indicates a problem with the compressor valve reeds.

CAUTION

Adjust Service Valve Stems

Never adjust a service valve stem unless the cap or a manifold gauge hose is attached to the service port. An open port can allow the refrigerant to escape or the compressor to draw moisture-laden atmospheric air into the system.

Service valves on small commercial units have a built-in 1/4″ (6.4 mm) male flare fitting at the gauge port for attaching a hose from the manifold gauge. Larger service valves may have one or two openings at the gauge port, plugged with a 1/8″ (3.1 mm) male pipe plug. To attach the gauge hose to the service valve, remove the plug and install a 1/8″ (3.1 mm) MPT × 1/4″ (6.4 mm) MFT half-union, as shown in **Figure 14-14**. All technicians should keep half-unions readily available.

Discharge Service Valve Operation

Just like the suction service valve, the discharge service valve has three openings and three possible valve stem positions, **Figure 14-15**.

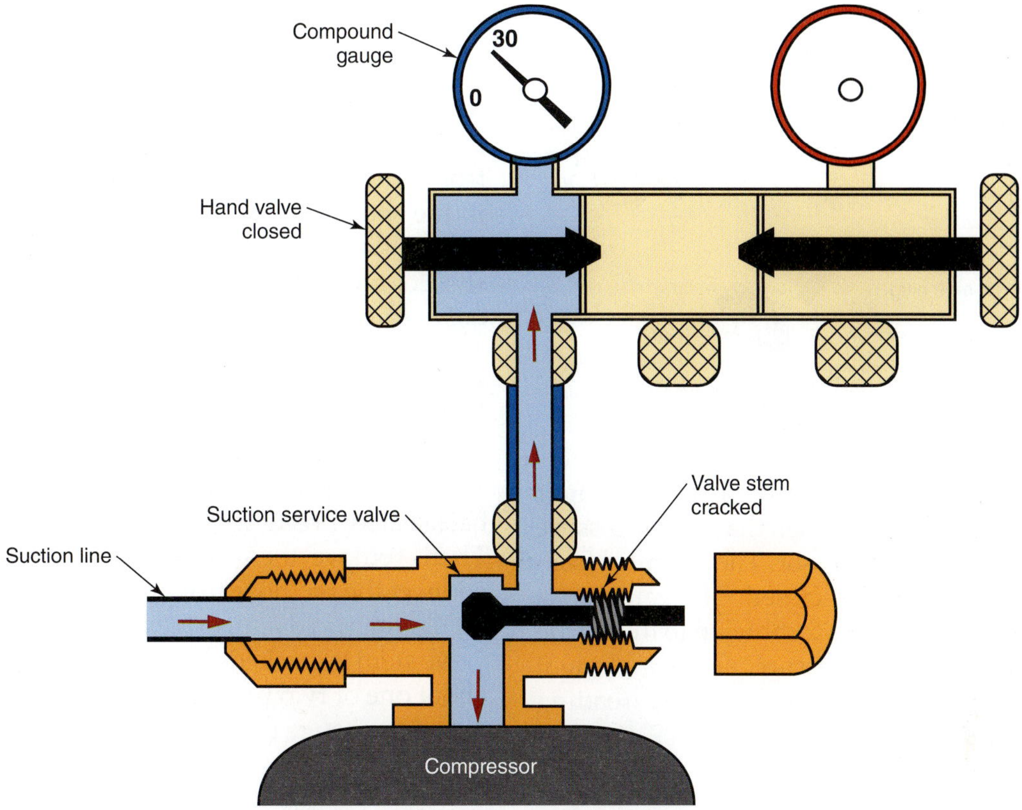

Goodheart-Willcox Publisher

Figure 14-12. The suction service valve with the stem in the *cracked* position permits the technician to read the low-side pressure while the system is in operation.

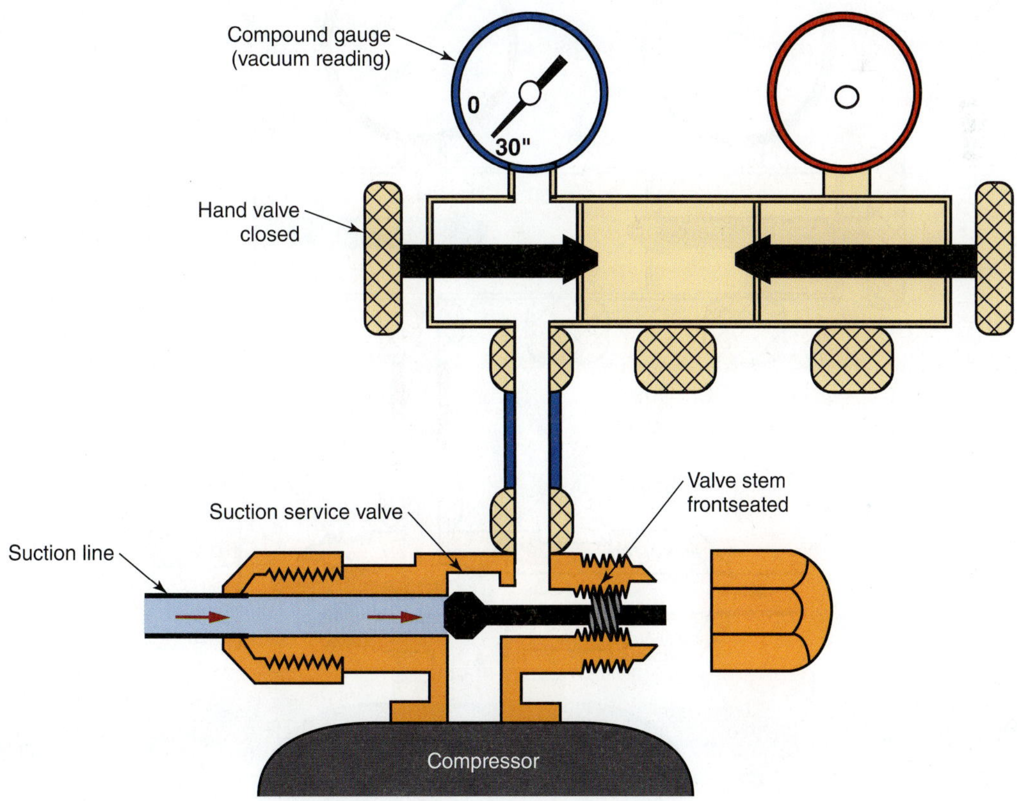

Goodheart-Willcox Publisher

Figure 14-13. The suction service valve with the stem in the frontseated position permits the technician to take a compressor vacuum reading. The inlet is closed, so no refrigerant vapor can reach the compressor.

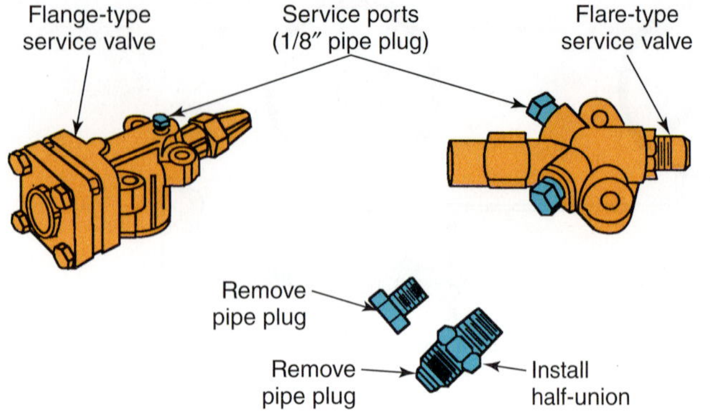

Goodheart-Willcox Publisher

Figure 14-14. Replacing pipe plugs on valves with half-unions.

However, the DSV controls high-pressure vapor leaving the compressor rather than low-pressure vapor entering the compressor. The opening from the compressor cannot be closed. The gauge port opening is available for the technician to install the red hose from the high-pressure gauge, making it possible to obtain a discharge pressure reading while the system is running.

Discharge Service Valve Backseated

The discharge service valve (DSV) is backseated for normal operation. This position stops the high pressure of the compressor discharge from putting any strain on the Schrader valve core in the service port. Though these are capped, it is best to keep the valve stem backseated. The DSV should be backseated when connecting the red hose from the manifold gauge.

Discharge Service Valve Cracked

To take a pressure reading at the discharge service valve, the valve stem must first be in the backseated position to close off the service port. Then, remove the cap from the service port and connect the red hose from the high-pressure gauge on the manifold gauge. Next, the discharge service valve stem is cracked—opened by turning the stem about one or two turns to the right, clockwise. This opens the gauge port, and the high-pressure gauge registers the pressure reading.

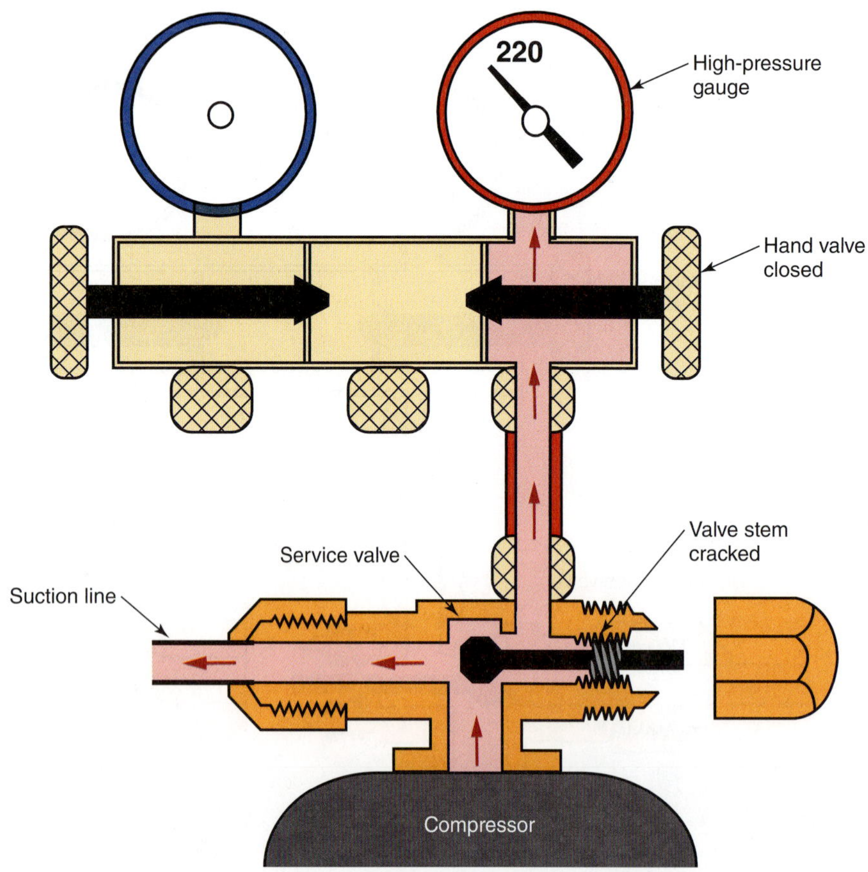

Goodheart-Willcox Publisher

Figure 14-15. The discharge service valve with the valve stem in the cracked position permits the technician to read the discharge (high-side) pressure while the system is in operation.

SAFETY NOTE

Discharge Service Valve Positions

The discharge service valve should *never* be frontseated (screwed all the way in) when the compressor is running. Frontseating a DSV will close the hot-gas discharge line. If the compressor is running, the high-pressure gas cannot escape, so high pressure will develop rapidly in the service valve. This extreme pressure may blow the hose off the valve or cause the compressor to shut off on a safety control. Only frontseat a DSV for a particular service procedure when the system is *off*. Remember to backseat the DSV before returning the system to service.

Discharge Service Valve Frontseated

If a compressor is shut off for replacement, both the discharge service valve and the suction service valve are frontseated to isolate the compressor from the system. Both service valves are unbolted from the compressor, then the old compressor is removed, and then the service valves are rebolted to the new compressor after it is installed. This procedure saves the refrigerant in the system, except for a small amount of gas trapped inside the old compressor which needs to be recovered. Refrigerant recovery will be discussed in Chapter 16, *Refrigerant Recovery and Recycling*.

Gaskets for each new compressor are used between the compressor body and the service valves. Old gaskets should be removed with a pocketknife, and the mating surfaces cleaned. Be careful not to scratch or cut the mating surfaces. Coat each new gasket with refrigeration oil before installing it. The oil expands the gaskets slightly so they can be squeezed between the mating surfaces for a leakproof fit.

Liquid Receiver Service Valve Operation

The liquid receiver service valve is welded or screwed onto the liquid receiver outlet, and the liquid line is connected to the valve with a flare nut. This service valve operates just like the suction and discharge valves, except it controls the liquid leaving the liquid receiver. As shown in **Figure 14-16**, the valve is backseated during normal operation. The valve can be cracked to read high-side pressures at the receiver valve during normal operation or frontseated to completely stop liquid refrigerant from leaving the receiver.

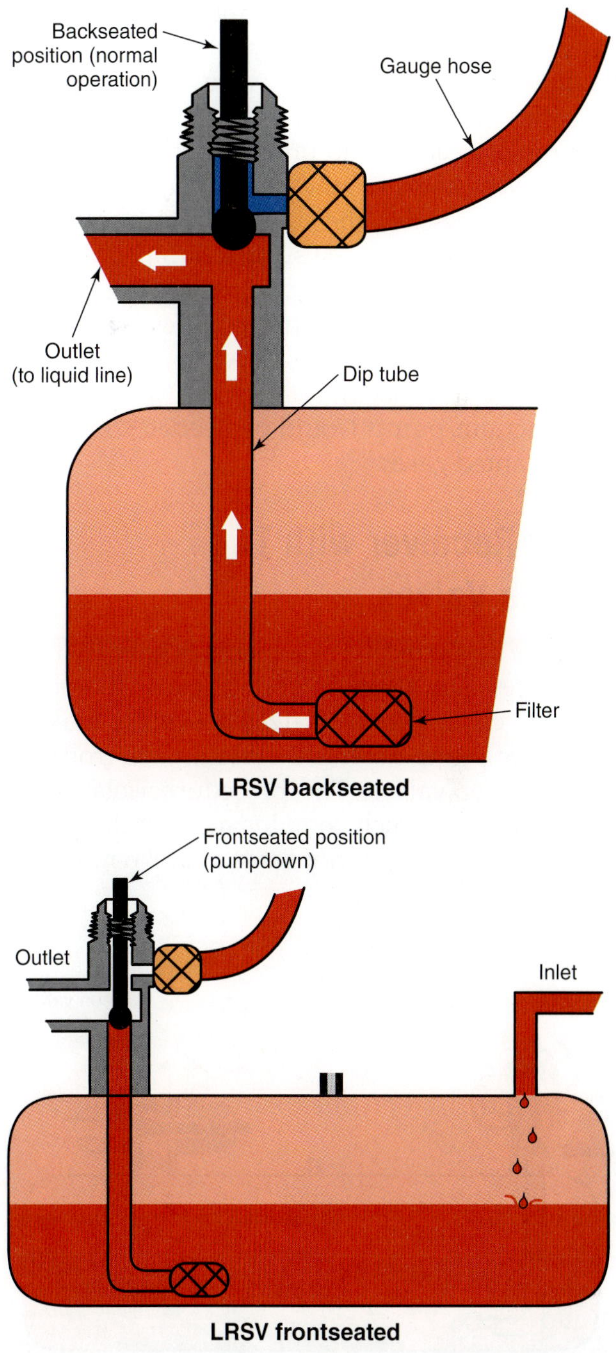

Figure 14-16. The liquid receiver valve is backseated for normal operation. The valve is frontseated to pump down the system.

Pumping Down the System

The technician often has to perform a ***pumpdown*** or manually *pump down the system.* This means removing all refrigerant from the low-pressure side and pumping it over to the high-pressure side of the system. Pumping down is accomplished by frontseating the liquid receiver service valve while the system is running. The compressor removes refrigerant from the low-pressure side of the system all the way back to the liquid receiver service valve. The compressor is then turned off. The valve reeds inside the compressor head should prevent refrigerant vapor from flowing back into the low-pressure side of the system. If necessary, the suction service valve can be frontseated after the system is pumped down.

Whenever a system is pumped down, its low-pressure side is usually in a good vacuum. The system should not be opened while in a vacuum because moisture-laden atmospheric air will rush in. A vacuum pump would be needed to remove the unwanted gases.

Liquid Receiver with Two Service Valves

Large liquid receivers sometimes have two service valves, one at the inlet and one at the outlet of the receiver, **Figure 14-17**. Two valves make it possible to pump down the system, then frontseat both receiver valves. The procedure isolates the liquid receiver, which contains most of the refrigerant in the system. Only some vapor remains in the condenser and the hot-gas discharge line. This procedure would be used to make repairs to the condenser, for example. Pumping down the system saves most of the refrigerant.

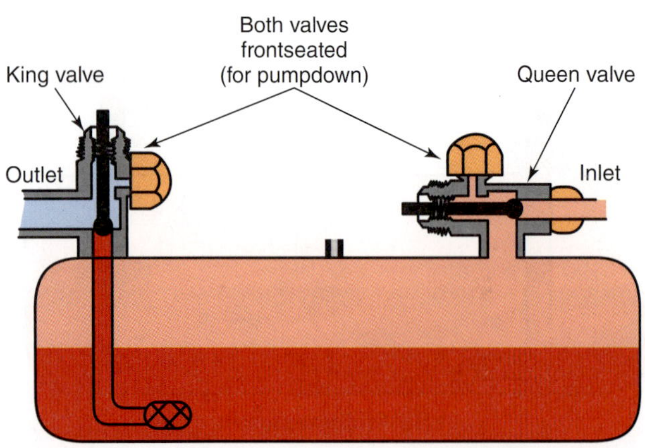

Goodheart-Willcox Publisher

Figure 14-17. Service valves at both the inlet and outlet allow isolation of the liquid receiver. Most of the system's refrigerant (in liquid form) can be trapped there. Noncondensables can be purged through the service port of the inlet valve.

The service valve on the inlet to the receiver also makes it easy to purge air from the liquid receiver. Whenever atmospheric air enters the system, it becomes trapped in the liquid receiver. Atmospheric air is considered a ***noncondensable***, meaning that it remains a vapor and cannot be condensed. Since atmospheric air remains a gas as it travels through the condenser, it eventually becomes trapped in the top of the liquid receiver.

A technician can remove noncondensables by turning the system off and frontseating a liquid receiver's inlet valve. Frontseating the inlet valve blocks off the valve's inlet from the hot-gas discharge line and opens the top of the liquid receiver to the service port. The noncondensables now can be removed through the service port. This procedure requires the use of a refrigerant recovery unit until all noncondensables are removed from the liquid receiver, and the head pressure returns to normal.

Service Valves for Domestic Systems

Manufacturers of domestic systems, such as refrigerators and freezers, seldom have factory-installed service valves. These systems are totally hermetic (sealed systems). To service these systems, a service technician must install service valves as needed. There are two basic types of valves installed by technicians: clamp-on valves and core-type valves.

Clamp-On Valves

These valves are known by various names—including *saddle valve* and *line-piercing valve*—but all are used in the same way. They straddle the tubing and clamp into place, then use a depressor pin to puncture a hole in the tubing, **Figure 14-18**. Gaskets and seals inside the valve prevent refrigerant from leaking around the punctured hole. The valve provides an access port for attaching a hose from the manifold gauge.

Once these clamp-on valves are installed on the suction or hot-gas discharge lines, they become a permanent addition to the system. They may also

A Sealed Unit Parts Co., Inc.

B Goodheart-Willcox Publisher

Figure 14-18. Saddle valves. A—The valve is clamped over the tubing to provide an access port for diagnosing system problems. Adapters permit use on different tubing sizes. B—The depressor pin punctures the tubing when the setscrew is tightened into the fitting. Then, the setscrew acts as a shutoff valve, and the cap seals the access port.

become a source of future problems due to possible leakage at the hole. Clamp-on valves should only be used for troubleshooting one of two problems:

- The compressor runs but does not compress properly.
- The system is entirely out of refrigerant due to a leak.

The pressures indicated on the manifold gauge will reveal which of the two problems exists. If the gauges reveal the *same pressure* on each side of the system, the compressor is defective (has bad valve reeds) and must be replaced. If the gauges reveal *no pressure* on either side of the system, the unit has a leak and has lost all refrigerant.

Both problems involve expensive repairs that require breaking into the hermetic system. The tube-piercing saddle valves are used for diagnosis only and should be replaced by ***Schrader valves*** (core-type) if repairs are authorized.

Schrader (Core-Type) Valves

One method for resealing a hermetic system after the refrigerant has been recovered using a clamp-on valve is to replace the clamp-on valve with a Schrader valve, **Figure 14-19**. Most hermetic compressors are factory-equipped with an extra copper tube, called a ***process stub***, that allows access to the suction pressure inside the hermetic shell. A Schrader valve is easily brazed to the end of the process stub.

SAFETY NOTE

Equipment Variances

Different refrigeration systems may require different methods of accessing and sealing the system. For instance, brazing Schrader valves on a system with flammable refrigerant could potentially lead to an explosion. Always refer to manufacturer literature and state and local building codes to make sure proper handling is observed.

Access to the high-pressure side is usually difficult because the hot-gas discharge line must be cut and a wrought copper tee installed to provide an opening for installing a valve. The valve is brazed to the branch opening on the tee.

Schrader valves have cores similar to those on automobile tires. Because of the special gaskets used, the valve core must be removed before brazing. It can be replaced after the joint has cooled. The hoses on the manifold gauge have a built-in device to depress the core when the hose is connected to the valve.

The seal cap for a core valve contains a rubber gasket to help prevent leaks. The cap should *always* be installed when the valve is not in use.

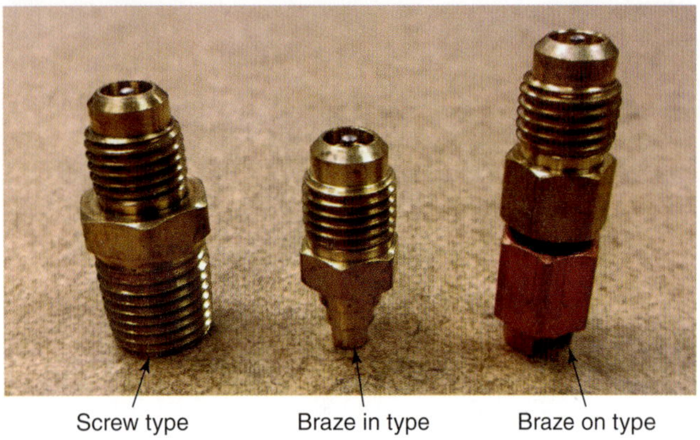

A

Image courtesy of Johnstone Supply

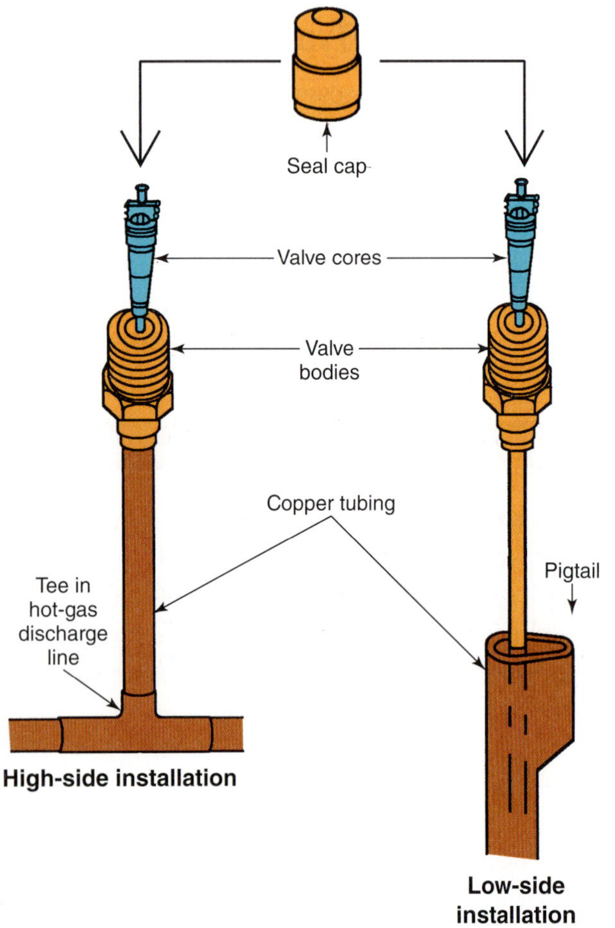

B

Goodheart-Willcox Publisher

Figure 14-19. Schrader valves are installed for permanent service access to domestic systems. A—Brazing-type and screw-type Schrader valves. B—Because heat will affect the gasket material, the cores must be removed while the valve body is being brazed to the tubing.

14.4 Filter-Driers

Regardless of the care used in evacuating and charging a system, it is best to assume that the system is *not* completely free of moisture. A ***filter-drier*** installed in the system will absorb the remaining moisture from the circulating refrigerant, **Figure 14-20**. It will also catch foreign particles circulating with the refrigerant.

On commercial systems, the filter-drier is normally installed in the liquid line, immediately after the liquid receiver. A second filter-drier may be temporarily installed in the suction line as well, **Figure 14-21**. On domestic systems, the filter-drier is typically located at the condenser outlet.

Moisture Problems

Water or moisture can sometimes enter a refrigeration system. This must be kept to an absolute minimum. Moisture is the primary factor in the formation of acids and sludge. This can lead to copper plating and corrosion. The service technician should always be alert to keep the moisture level of the refrigeration system as low as possible.

The main problems caused by moisture are:

- **Corrosion** that damages metal parts and adds contaminants to the system.
- **Formation of acid** that damages the motor windings. Since the motor windings are exposed to refrigerant, any acid that forms will cause a breakdown of the insulation and result in motor burnout.
- **Freezing of moisture** in the orifice of the refrigerant metering device or capillary tube. Frozen moisture can block the flow of refrigerant and stop the operation of the system.

Images courtesy of Johnstone Supply

Figure 14-20. Filter-driers remove residual moisture from a system and trap foreign particles circulating with the refrigerant.

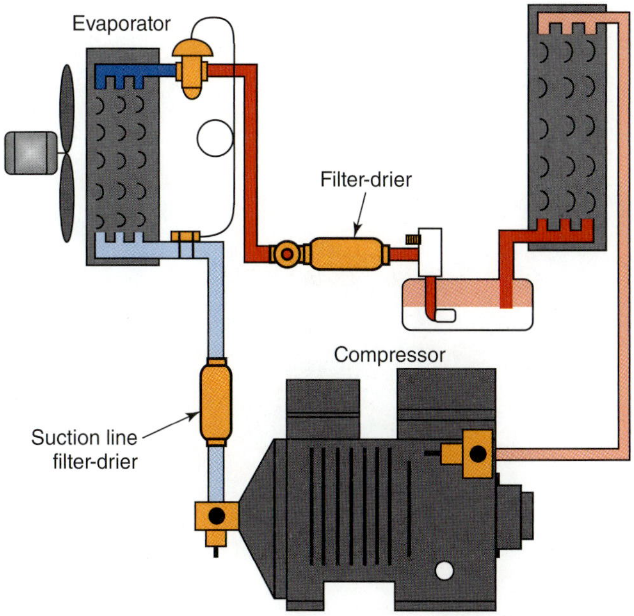

Goodheart-Willcox Publisher

Figure 14-21. On commercial systems, the filter-drier is usually installed in the liquid line, just after the liquid receiver. For additional protection, technicians sometimes install a second filter-drier, as shown, between the evaporator and the compressor in the suction line.

Other sources of problems within the system are dirt, sludge, rust, and foreign matter such as flux, copper or brass chips, and solder. These contaminants can damage piston cylinder walls or compressor bearings, or may plug capillary tubes and other refrigerant controls.

Filter-Drier Operation

Filter-driers in refrigeration systems work by bringing liquid refrigerant in contact with a substance that absorbs moisture in the refrigerant. The substance is called a drying agent, or ***desiccant***, and is usually capable of removing acid as well as moisture.

If properly sized, the filter-drier will not restrict the flow of refrigerant, even when the desiccant is full of moisture and no longer effective. Desiccants are extremely sensitive to moisture and must be protected against it until ready for use. The factory seal on a filter-drier should not be removed until just before installation. Most filter-driers are direction-sensitive, meaning they must be oriented to the direction of refrigerant flow. Direction-sensitive filter-driers have an arrow printed on the body to indicate the proper direction of flow, **Figure 14-22**.

Alco

Figure 14-22. Filter-drier may be designed for attachment to the liquid line with flare fittings or brazing. Note the arrow indicating the proper direction of refrigerant flow through the device.

Desiccants used in filter-driers include activated alumina, silica gel, and activated carbon, which ***absorb*** (soak up) moisture. They also include molecular sieves, which ***adsorb*** (collect substances on their surfaces in a condensed layer) contaminants from the refrigeration system. Desiccants are available in granular, bead, and block forms, **Figure 14-23**. Combinations of desiccants can be used in solid cores and have certain advantages over a single desiccant, such as absorption of a greater variety of contaminants.

Anytime a refrigeration or air-conditioning system is opened to the atmosphere, the filter-drier must be replaced. Trying to pull a vacuum on a system with an old filter-drier can release the absorbed moisture back into the system causing many problems.

On large commercial systems, a replaceable element filter-drier is used. The filter-drier is bolted

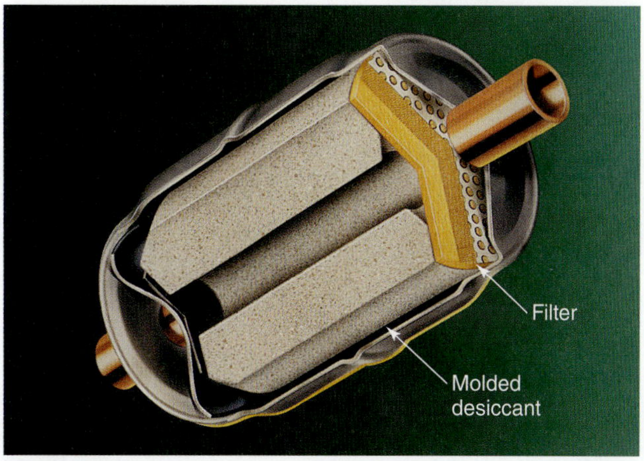

Sporlan Valve Co.

Figure 14-23. Desiccant granules can be molded into a core, as shown in this cutaway view.

together, and replacement cores are available in vacuum-packed metal containers, **Figure 14-24**.

Special acid-removing filter-driers are available for field installation in the suction line. Such filter-driers are used to clean the system and protect a new compressor following a burnout. A suction line filter-drier needs to be removed from the system after the system has been cleaned up; they are not designed to stay installed in a system and can cause problems later in the operation of the system.

Domestic Filter-Driers

Domestic systems (refrigerators and freezers) also have a filter-drier, but it is much smaller than those used on commercial systems. The body of the domestic filter-drier is made of copper and is brazed into the system at the outlet of the condenser, **Figure 14-25**. These filter-driers are normally direction-sensitive, but some small-capacity units are nondirectional.

Domestic systems need a filter-drier to ensure trouble-free operation. Due to the small amount of refrigerant circulating in the system, a single drop of excess moisture will cause a freeze-up. The components in the system are quite small. Any foreign material will cause severe problems. When a domestic (hermetic) system is opened for repairs, the filter-drier should always be replaced.

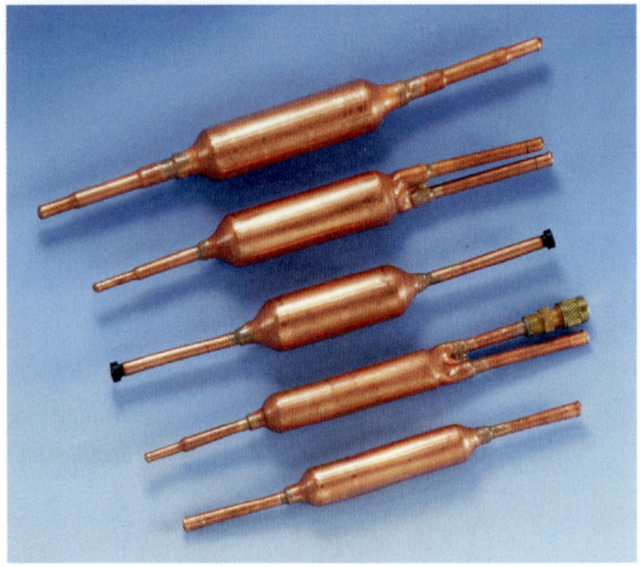

Sealed Unit Parts Co., Inc.

Figure 14-25. Small filter-driers intended for a domestic unit. The copper tubes at each end are brazed into the system at the condenser outlet.

Sporlan Valve Co.

Figure 14-24. The cutaway view (top) shows the installation of two cores (elements) in the drier shell. Access is by means of the bolted cover. Replacement elements are packed in sealed cans that exclude moisture. Typical cores are shown (bottom).

14.5 Sight Glass

The ***sight glass*** is a small window placed in the liquid line on commercial systems to provide a view of the flowing liquid, **Figure 14-26**. The sight glass serves as a valuable service aid. Visible bubbles in the liquid line can indicate problems within the system. Problems could be low refrigerant charge, low head pressure, insufficient subcooling, restrictions, or poor piping design. The sight glass is usually located close to the liquid receiver and immediately after the filter-drier, **Figure 14-27**. On long liquid line runs, the sight glass should be installed close to the refrigerant metering device.

The sight glass shows clear if the line is full of liquid and shows bubbles if the system is having problems. An occasional bubble is not unusual or harmful, but excessive bubbles indicate trouble in the system.

While the sight glass shows clear when full of liquid, it also shows clear when empty. To determine the correct situation, frontseat the liquid receiver service valve and pump down the system while looking through the sight glass. An empty system

Emerson Climate Technologies

Figure 14-26. Typical sight glass. The clear window allows the technician to "look inside" the system and diagnose problems by the appearance of the refrigerant. Sight glasses are available with various types of fittings to suit different system configurations.

shows no change. However, a full system appears full, then shows bubbles, and then appears empty.

Sight Glass with Moisture Indicator

Most sight glasses have a ***moisture indicator*** centered within the viewing window, **Figure 14-28**. The indicator is highly sensitive to moisture, gradually changing color to reflect the moisture content in the refrigerant. The indicator element is completely reversible and changes color as often as the moisture content of the refrigerant varies.

Some change in color takes place rapidly at the start-up of a new system or after the filter-drier is replaced. The system should operate for about 12 hours before determining (based on the moisture indicator) that another filter-drier change is needed. Drying of the refrigerant should continue until the indicator element changes to the proper color.

The amount of moisture in a refrigeration system must be kept to an absolute minimum to provide trouble-free operation. Every precaution must be taken to prevent moisture from entering the system during installation or service operations. Any moisture that does enter the system should be removed quickly.

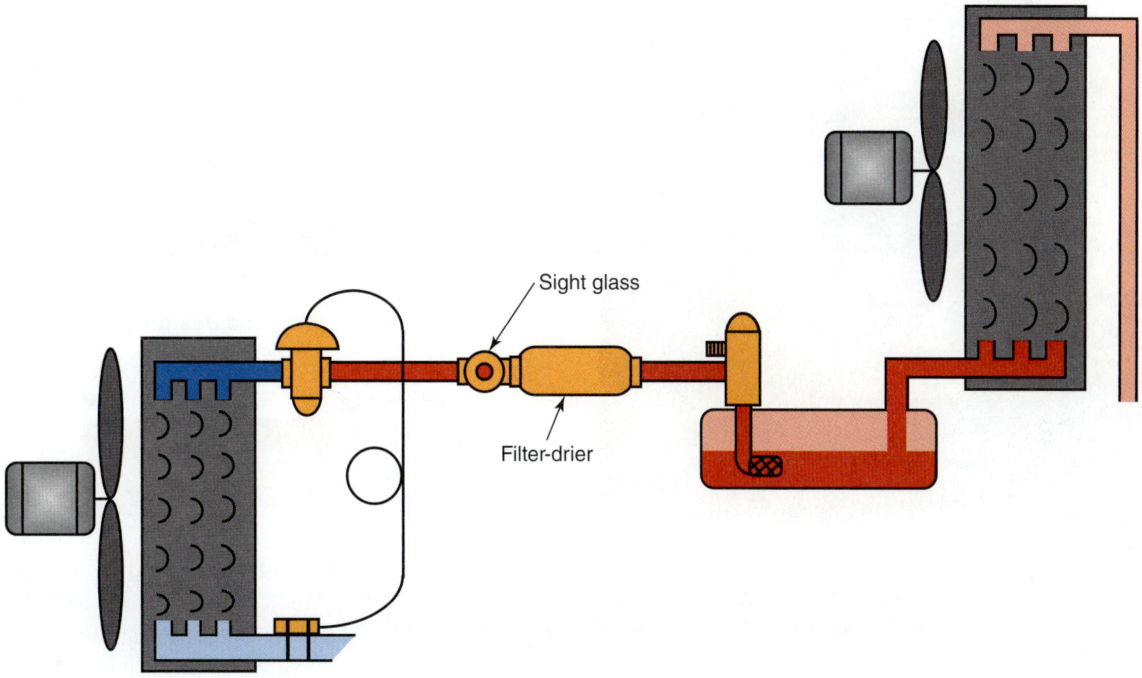

Goodheart-Willcox Publisher

Figure 14-27. The sight glass is normally installed immediately after the filter-drier in the liquid line.

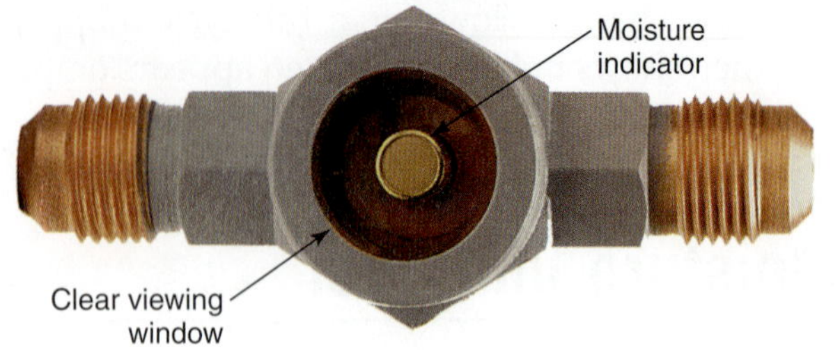

Sporlan Valve Co.

Figure 14-28. Typical sight glass with a moisture indicator in the center. The indicator gradually changes color to reflect the presence or absence of moisture.

A color reference code is often printed around the edge of the sight glass or on the front of the sight glass body, **Figure 14-29**. One manufacturer's color code varies from dark green (dry), to light green (caution), to bright yellow (very wet). Another manufacturer's indicator changes from dark blue (dry), to light blue (caution), to pink (very wet). A plastic or metal cap keeps the glass free of dust, dirt, and grease. The cap should always be replaced. Unlike commercial installations, domestic systems do not use a sight glass or moisture indicator.

The moisture indicator is chemically engineered for long life, accuracy, and reliability. The same indicator can be used for all common refrigerants. The indicator element will show a wet condition before installation, but that is normal and simply reveals ambient humidity. Most sight glasses are installed with flare connections, but brazed types are also available.

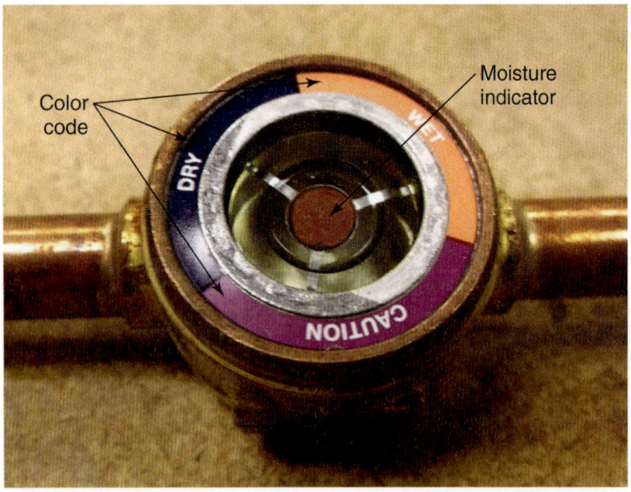

Images courtesy of Johnstone Supply

Figure 14-29. Colors the moisture indicator uses to show wet or dry conditions are displayed on the sight glass rim or body. In this example, the color code is on a label attached to the body. Colors for each condition vary among manufacturers.

14.6 Heat Exchanger

A ***heat exchanger*** is any device that transfers heat from one medium to another. However, in refrigeration, a heat exchanger generally describes a particular component that transfers heat from the warm liquid line to the cool suction line, **Figure 14-30**. A suction line-liquid line heat exchanger performs two mutually beneficial tasks:

- It subcools the refrigerant in the liquid line before it reaches the refrigerant metering device, improving the system's efficiency.
- It superheats vapor inside the suction line with heat removed from the warm liquid line. This prevents liquid refrigerant from reaching the compressor.

Many suction line-liquid line heat exchangers use the highly efficient tube-in-a-tube design. Other designs wrap the liquid line around the suction line for several turns before traveling to the refrigerant metering device. In domestic models, the small capillary tube used for the refrigerant metering device is soldered to the suction line for almost its entire length.

Suction line-liquid line heat exchangers are designed to perform a dual function in a system and should not be added into or removed from a system without proper engineering information.

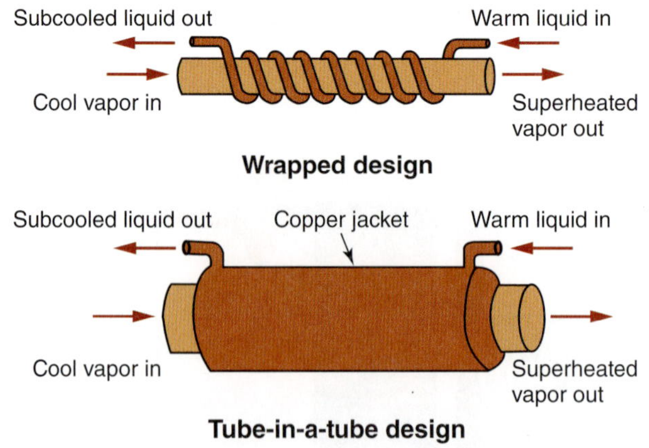

Goodheart-Willcox Publisher

Figure 14-30. A suction line-liquid line heat exchanger subcools liquid line refrigerant and superheats suction line vapor. Shown are two common designs.

14.7 Accumulator

An ***accumulator*** is a device installed in the suction line that keeps liquid refrigerant from entering the compressor, **Figure 14-31**. The accumulator is a cylinder that acts as a trap to collect liquid refrigerant and permits only vapor to exit. Liquid entering the accumulator must boil off inside the device before exiting as a vapor.

Domestic systems locate a small accumulator at the evaporator outlet, with the suction line coming out of the top of the accumulator. This prevents liquid from entering the suction line. Liquid entering the bottom of the accumulator remains inside until the liquid boils. Only vapor can exit and enter the suction line at the top of the accumulator.

An accumulator is generally an upright cylinder with two openings at the top: the *inlet* and the *outlet*, **Figure 14-32**. The suction line is connected to the inlet opening. Any liquid refrigerant entering from the suction line falls to the bottom of the cylinder. The outlet opening has a dip tube that goes to the bottom of the cylinder, makes a 180° bend, and extends upward toward the top of the cylinder where only vapor can enter the tubing leading to the compressor suction service valve.

Liquid refrigerant is trapped in the bottom of the accumulator until it evaporates. A small hole, called an ***aspirator hole***, along the side of the 180° bend permits small quantities of oil to enter the outlet tube and be drawn back to the compressor. Without the aspirator hole, the accumulator would trap oil and deprive the compressor of proper lubrication. Small amounts of liquid refrigerant may enter the aspirator hole. However, these droplets evaporate before reaching the compressor.

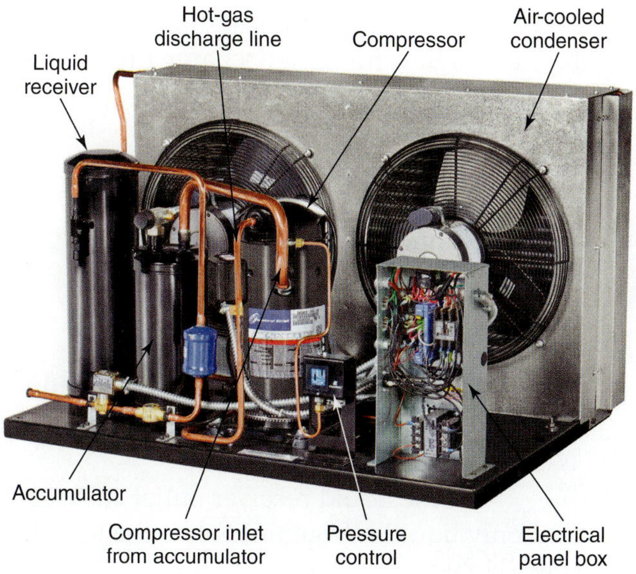

Emerson Climate Technologies

Figure 14-31. Commercial condensing unit showing the relationship of the accumulator and the compressor. The accumulator keeps liquid refrigerant from reaching the compressor.

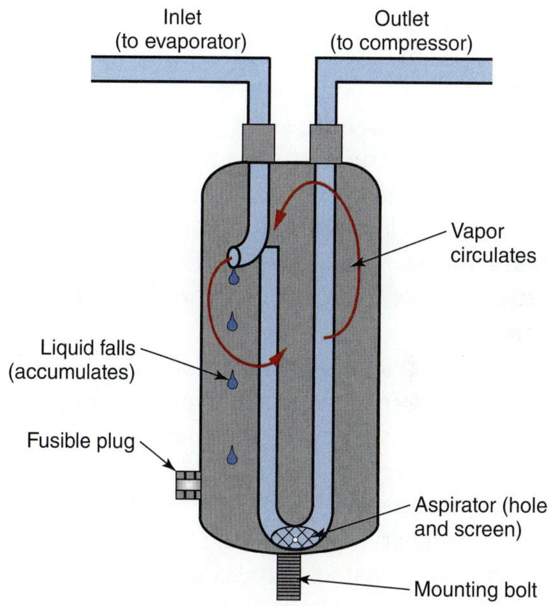

Goodheart-Willcox Publisher

Figure 14-32. Interior view of an accumulator. The 180° bend of the outlet tube permits only vaporized refrigerant to emerge from the accumulator.

Chapter Review

Summary

14.1 The condensing unit includes the equipment necessary to reject the absorbed heat and reclaim the refrigerant vapor by condensing it back into a liquid. The evaporator unit includes the equipment necessary to absorb heat and vaporize the liquid refrigerant.

14.2 Liquid receivers are installed in the liquid line and serve as a storage tank for excess liquid refrigerant. This allows a system to provide varying amounts of refrigerants depending on the heat load on the evaporator.

14.3 The suction service valve (SSV) is located on the low-pressure side of the system between the compressor and the suction line. The discharge service valve (DSV) is located on the high-pressure side of the system. The liquid receiver service valve (LRSV) is located in the liquid line, at the liquid receiver outlet. Clamp-on and Schrader valves are used on domestic systems to access them.

14.4 A filter-drier uses a desiccant to absorb moisture from refrigerant circulating in a system. On commercial systems, the filter-drier is normally installed in the liquid line, immediately after the liquid receiver. On domestic systems, the filter-drier is typically located at the condenser outlet.

14.5 A sight glass is a small window placed in the liquid line on commercial systems to provide a view of the flowing liquid. The behavior of the liquid refrigerant visible through the sight glass can help diagnose system problems. The moisture indicator is often included on a sight glass and changes color to reflect moisture content in system refrigerant.

14.6 The suction line-liquid line heat exchanger subcools refrigerant in the liquid line before it reaches the refrigerant metering device. It also superheats vapor inside the suction line with heat removed from the warm liquid line to prevent refrigerant from reaching the compressor.

14.7 The accumulator is located in the suction line and serves as a reservoir to trap liquid refrigerant, preventing it from entering the compressor.

Know and Understand

1. Which two copper lines connect the evaporator unit with the condensing unit? (14.1)
   A. Hot-gas discharge line, liquid line
   B. Suction line, hot-gas discharge line
   C. Accumulator, suction line
   D. Suction line, liquid line
2. In a split system, where might the condensing unit be located? (14.1)
   A. In the same place as the evaporator
   B. Outside the building
   C. Always below the evaporator
   D. Always above the evaporator
3. *True or False?* Liquid receivers are designed for use in commercial and domestic refrigeration systems. (14.2)
4. A _____ on the liquid receiver outlet ensures that only liquid refrigerant enters the liquid line. (14.2)
   A. pressure relief valve
   B. fusible plug
   C. dip tube
   D. dial-type gauge
5. Which of the following locations does *not* typically include a service valve? (14.3)
   A. Between the compressor and the suction line
   B. Connected to the hot-gas discharge line at the compressor
   C. At the liquid receiver outlet
   D. At the evaporator outlet

6. The discharge service valve allows the technician to do all of the following, *except* _____. (14.3)
   A. read the system's high-side pressure while the compressor is running
   B. completely stop refrigerant from entering the compressor
   C. completely stop refrigerant gas from leaving the compressor
   D. restrict the flow of refrigerant gas leaving the compressor
7. If the suction service valve is in the frontseated position while the compressor is running, then the compressor will show _____. (14.3)
   A. a low-side pressure reading
   B. a high-side pressure reading
   C. a vacuum reading
   D. both low- and high-side pressure readings
8. Pumping down a system involves frontseating the _____ service valve while the system is running. (14.3)
   A. liquid receiver
   B. discharge
   C. suction
   D. clamp-on
9. Where is a filter-drier located on a commercial system? (14.4)
   A. In the suction line only
   B. In the liquid line only
   C. In the liquid line, after the liquid receiver and in the suction line
   D. In the hot-gas discharge line
10. A filter-drier uses a(n) _____ to absorb moisture. (14.4)
   A. fan
   B. adhesive
   C. molecular sieve
   D. desiccant
11. Where is a sight glass normally installed? (14.5)
   A. In the suction line, before the compressor inlet
   B. In the hot-gas discharge line, at the compressor outlet
   C. In the liquid line between the metering device and the condenser
   D. With the liquid receiver valve
12. *True or False?* The same moisture indicator can be used for identifying moisture content with all common refrigerants. (14.5)
13. A suction line-liquid line heat exchanger subcools refrigerant in the liquid line before the refrigerant metering device, and _____. (14.6)
   A. subcools vapor inside the suction line
   B. superheats vapor inside the suction line with heat removed from the warm liquid line
   C. maintains temperature of the refrigerant in the liquid line
   D. subcools refrigerant before reaching the compressor
14. Domestic systems typically include an accumulator at the _____. (14.7)
   A. condenser outlet
   B. compressor outlet
   C. evaporator outlet
   D. refrigerant metering device

Apply and Analyze

1. List the refrigeration system components found in the condensing and evaporator units. (14.1)
2. Explain why a liquid receiver should be no more than 80% full when holding the entire system charge. (14.2)
3. Explain how to obtain a pressure reading at the suction service valve. (14.3)
4. Why do manufacturers of domestic refrigeration systems rarely install service valves on their systems? (14.3)
5. Explain the importance of keeping water or moisture out of a refrigeration system. (14.4)
6. How can a technician use a sight glass to know if a system's refrigerant charge is full? (14.5)
7. Explain the dual purpose of a suction line-liquid line heat exchanger. (14.6)
8. Explain how an accumulator prevents liquid refrigerant from entering the compressor. (14.7)

Critical Thinking

1. Name all 15 possible refrigeration system components discussed in this chapter, in order of refrigerant flow, beginning at the evaporator. (14.1)
2. List three problems that could cause bubbles to appear in a commercial refrigeration system's sight glass. (14.5)

Worthington Enterprises

15 Refrigerants

LEARNING OUTCOMES

After studying this chapter, you will be able to:

15.1 Recognize key developments in the history of refrigerants.

15.2 Explain the importance of Section 608 of the Clean Air Act.

15.3 Identify the different refrigerant classifications by their standardized name, safety and hazard classifications, and refrigerant cylinder.

15.4 Distinguish between pure component refrigerants, azeotropic mixtures, and zeotropic blends.

15.5 Explain the factors considered by HVACR engineers when selecting a refrigerant for use in a system.

15.6 Identify refrigerant applications within specific temperature ranges.

TECHNICAL TERMS

- Acceptable Exposure Limit (AEL)
- azeotrope
- azeotropic refrigerant
- chlorofluorocarbon (CFC)
- enthalpy
- fractionation
- halide refrigerants
- halogens
- high-temperature applications
- hydrocarbon (HC)
- hydrochlorofluorocarbon (HCFC)
- hydrofluorocarbon (HFC)
- hydrofluoro-olefin (HFO)
- inorganic refrigerants
- latent heat value
- low-temperature applications
- medium-temperature applications
- ozone depletion potential (ODP)
- P/T card
- refrigerant
- saturation
- standard conditions
- subcooled
- superheat
- temperature glide
- topping off
- ultralow-temperature applications
- zeotrope

Introduction

Refrigerants are the working fluids necessary in every refrigeration system. A ***refrigerant*** is any fluid or liquid that picks up heat by evaporating at a low pressure-temperature condition and gives up heat by condensing at a higher pressure-temperature condition. The purpose of a refrigerant is to move heat. Refrigerants absorb heat (by boiling) from an unwanted place and then carry the heat to a place where it can be discharged (by condensing into a liquid).

This chapter describes several common refrigerants and their basic properties and applications. To achieve low temperatures, it is necessary to select a refrigerant with a boiling point lower than the temperature being controlled. Remember that heat always flows from the *warmer* to the *colder* substance.

Factors other than boiling point must be considered when selecting a refrigerant for a specific application. These factors include toxicity, flammability, miscibility with oil, operating pressures, and Btus absorbed per pound. Some refrigerants can be used in a variety of applications, while others have more limited use.

The refrigeration and air-conditioning industry is in transition as new refrigerants are replacing older ones to protect the environment. Many refrigerants are currently available, and others are still being developed. It is beyond the scope of this textbook to describe all the available refrigerants. Such information is readily obtainable from manufacturers, supply houses, or the local library. Instead, this chapter concentrates on the composition, selection, identification, and application of the most common refrigerants. Sometimes a new or unusual refrigerant will be encountered, but the theories and principles described here apply to old and new refrigerants alike.

15.1 Brief History of Refrigerants

The following are key events in the development of refrigerants throughout the history of the refrigeration and air-conditioning industry:

- 1834: The first practical compression-cycle refrigeration system was built.
- 1850: The absorption machine was developed. It used water and sulfuric acid as the refrigerant.
- 1873: The ammonia compressor was introduced.
- 1876: The sulfur dioxide compressor was introduced.
- 1882: Thanks to Thomas Edison, the first electric power plant opened in New York.
- 1900: Use of the electric motor had become widespread.
- 1906: Willis Carrier patented his air-conditioning system, calling it an "apparatus for treating air." Stuart Cramer coined the term "air conditioning."
- 1915: Sulfur dioxide systems were common in domestic refrigerators. (Note: Sulfur dioxide is a toxic refrigerant and no longer used.)
- 1922: The first air-conditioned movie theater opened in Los Angeles, California (Grauman's Metropolitan Theater).
- 1928: Research to discover a "safe" refrigerant was begun by General Motors Corporation and DuPont Chemical Company. Researchers sought a revolutionary substance that would be nontoxic, nonflammable, odorless, noncorrosive, and nontoxic to humans. Experiments combined halogens, such as chlorine and fluorine, with hydrocarbon compounds (CFCs). The chamber of the US House of Representatives became air conditioned.
- 1929: The US Senate chamber became air conditioned.
- 1930: DuPont began manufacturing the first of many new refrigerants using halogen-hydrocarbon compounds. The first compound to be marketed (under the trade name *Freon-12*®) was *dichlorodifluoromethane*. This CFC refrigerant revolutionized the refrigeration and air-conditioning industry, leading to rapid growth and the discovery of many more refrigerants. Several chemical companies were involved in the research and development of refrigerants.
- 1930: The White House, the Executive Office Building, and the US Department of Commerce became air conditioned.
- 1945: The demand for room air conditioners increased, with 30,000 produced that year.
- 1957: The first rotary compressor was introduced, permitting refrigeration and air-conditioning units to be smaller, quieter, lighter, and more efficient than those with reciprocating compressors.

- 1977: New technology allowed heat pumps to operate at lower outdoor temperatures when heating on the reverse cycle (heating cycle).
- 1987: The Montreal Protocol was signed, establishing international cooperation for the phaseout of stratospheric ozone-depleting substances, including CFC refrigerants.
- 1990: The Clean Air Act (CAA) was passed by the Environmental Protection Agency (EPA), setting restrictions and penalties for venting refrigerants to the atmosphere. Section 608 of the Clean Air Act directed the EPA to enforce the regulations.
- 1995: At the end of this year, the US and many other industrialized countries ended production of CFC refrigerants.
- 2004: R-22 (HCFC) production was reduced to a level of 65% of production level in 1996.
- 2010: R-22 production was reduced to 25% of the 1996 level.
- 2010: Starting in January of this year, R-22 (HCFC) was phased out of new equipment.
- 2020: Remaining production and import of R-22 was banned. Future service of R-22 systems relies on recycled or stockpiled refrigerant.
- 2020: The American Innovation and Manufacturing (AIM) Act was signed. This legislation authorized the EPA to begin the phaseout of HFC refrigerants and transition to the use of low GWP and ODP refrigerant alternatives.
- 2022: The US ratified the Kigali Amendment, which is an international agreement to phase out the production of HFC refrigerants by 80%–85% by 2047.

15.2 The EPA and Refrigerants

Under Section 608 of the Clean Air Act, the Environmental Protection Agency (EPA) was given the authority to set guidelines on proper refrigerant safety and handling to protect the environment. First, they looked at a refrigerant's ***ozone depletion potential (ODP)***—potential to destroy the protective ozone layer in the stratosphere. The stratosphere is the part of the atmosphere extending 10 miles to 30 miles (16 km to 48 km) above the earth's surface. The ozone layer is an essential part of the stratosphere and serves as the earth's security blanket, surrounding the planet and blocking harmful solar radiation. Ozone depletion in the stratosphere is a serious matter. The ozone layer serves two important functions: it protects life on earth from harmful ultraviolet (UV) radiation and helps maintain stable earth temperatures. Chlorofluorocarbon (CFC) refrigerants were discovered to have high ODPs and as a result, were banned by the 1990 Clean Air Act.

As new refrigerants were developed with a lower ODP, they found that these new low ODP refrigerants contributed to global warming. ***Global warming***, also called the *greenhouse effect*, occurs when long-wave (infrared) radiation from the sun reaches the earth but cannot escape. Trapped radiation causes a gradual buildup of heat on the earth's surface. Several refrigerants are suspected of contributing to global warming. To express the risk of a refrigerant contributing to global warming, each one has been assigned a number called its ***global warming potential (GWP)***. Refrigerants that have a high GWP are said to contribute to global warming. The AIM Act and Kigali Amendment are enacting the phaseout of these high GWP refrigerants and transitioned the industry to using refrigerants with low ODPs and GWPs. **Figure 15-1** lists the ODPs and GWPs of various refrigerants.

These regulations resulted in many new non-ozone-depleting and low-GWP refrigerants and the introduction of synthetic lubricants for use with them. Many new refrigerants are continuing to be introduced, replacing those with high ODP and GWP.

HVACR service technicians must be knowledgeable in the use of these new refrigerants, lubricants, and associated products. New procedures are required for successful use of alternative refrigerants.

Widespread conversion from high ODP and GWP refrigerants in an existing system is not necessary unless there are strong reasons for doing so. A system that is leak-free and operating properly should be left alone. However, the availability of high ODP and GWP refrigerants for servicing existing equipment is becoming limited and expensive. Regulations are pushing customers and technicians to upgrade systems to equipment manufactured for use with low ODP and GWP refrigerants. These refrigerants will be discussed later in this chapter.

Ozone Depletion Potential (ODP) and Global Warming Potential (GWP)

Refrigerant	Type	GWP	ODP
R-12	CFC	10910	1
R-113	CFC	4800	0.8
R-22	HCFC	1810	0.05
R-401A	HCFC	2100	0.04
R-32	HFC	0	675
R-134a	HFC	0	1430
R-404A	HFC	0	3920
R-407C	HFC	0	1770
R-410A	HFC	0	2088
R-454B	HFO	0	466
R-1234yf	HFO	0	4
R-1234ze	HFO	0	0
R-290	HC	0	3
R-600a	HC	0	3
R-717	Inorganic compound	0	0
R-744	Inorganic compound	0	1

Goodheart-Willcox Publisher

Figure 15-1. Ozone depletion potential (ODP) is a measurement of a refrigerant's ability to destroy ozone molecules. Global warming potential (GWP) expresses the risk of a refrigerant contributing to global warming.

15.3 Refrigerant Classifications and Identification

There are many different refrigerants available for a variety of applications. As refrigerants continue to be developed with lower ODP and GWP, it is the HVACR service technician's responsibility to stay informed with these changes. This section discusses some of the most common refrigerant types used in the HVACR industry and how to identify them by their industry name.

Halide Refrigerants

For many years, the most widely used refrigerants contained one or more of the chemical substances called ***halogens***. Halogens include fluorine, chlorine, iodine, and bromine. When halogens are combined with a hydrocarbon compound—such as acetylene, methane, or ethane—they produce ***halide refrigerants***. Hydrocarbons transfer heat well but are extremely flammable, explosive, and somewhat toxic. However, when hydrocarbons are combined with halogens, the mixture is not flammable or explosive.

The manufacturing process for halide refrigerants is complicated. Each mixture is precise and must produce a new substance that has a specific boiling point and acts according to Charles' Law of gases. The molecule of the new compound also must be stable (not break down spontaneously). The combining of hydrogen with oxygen to form a compound such as water (H_2O) is a naturally occurring example. The halide refrigerants are classified into four groups according to their chemical makeup:

- Chlorofluorocarbons (CFCs)
- Hydrochlorofluorocarbons (HCFCs)
- Hydrofluorocarbons (HFCs)
- Hydrofluoro-olefins (HFOs)

These terms refer to manufactured (synthetic) fluids specifically designed for use as refrigerants.

Chlorofluorocarbons (CFCs)

Chlorofluorocarbon (CFCs) refrigerants are composed of chlorine, fluorine, and a hydrocarbon (such as methane, ethane, or propane). They are classified as CFCs due to their composition (chlorine-fluorine-carbon). Examples of CFCs are R-11 (trichloromonofluoromethane) and R-12 (dichlorodifluoromethane). The Greek prefixes *mono*, *di*, and *tri* indicate how many parts of a substance are used in the compound. "Mono" means one part, "di" means two parts, "tri" means three parts. The chemical name for R-12, dichlorodifluoromethane, describes the molecular structure of the compound. R-12 contains two parts chlorine (*di*chloro) and two parts fluorine (*di*fluoro), combined with methane. Therefore, the number 2 appears twice in R-12's chemical formula: CCl_2F_2. See **Figure 15-2**.

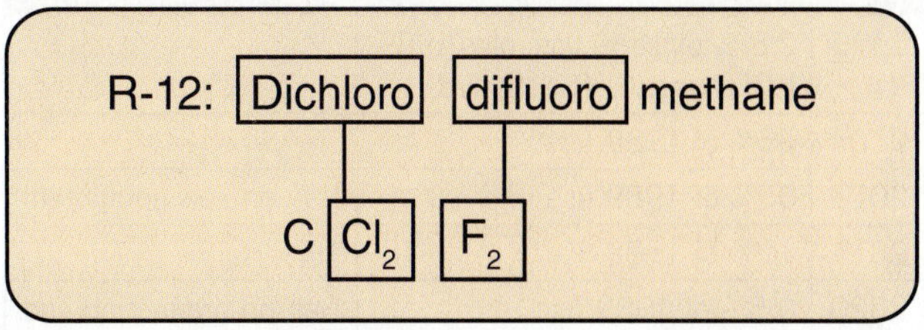

Goodheart-Willcox Publisher

Figure 15-2. The chemical name and chemical formula of a refrigerant describes its molecular structure. A molecule of R-12 (dichlorodifluoromethane) includes two chlorine atoms and two fluorine atoms combined with a methane molecule.

As mentioned earlier, all chlorofluorocarbon refrigerants (CFCs) were banned by the 1990 Clean Air Act due to their high ODP. Many alternative refrigerants with lower ODP levels are used in new equipment and offered for upgrading old equipment still using CFC refrigerants.

Hydrochlorofluoro-carbons (HCFCs)

Hydrochlorofluorocarbons (HCFCs) contain hydrogen atoms, causing the compound to be less stable in the atmosphere than CFCs. Therefore, HCFCs are considered less harmful to the ozone layer because the refrigerant molecules break up rather quickly. Only a small percentage of the chlorine in HCFCs can reach the ozone layer. The most widely used HCFC is monochlorodifluoromethane, designated R-22. R-22 is a widely used refrigerant for air-conditioning systems (other than automotive systems). Other examples of HCFCs are R-123, R-124, R-141b, and R-142b. R-123 is an acceptable alternative for R-11.

HCFC refrigerants have a low ODP level and are considered only slightly hazardous to the ozone layer. However, HCFCs are scheduled to be phased out of production by the Clean Air Act. Since January 2010, no new equipment may contain R-22. Total phaseout of HCFCs is scheduled for 2030, **Figure 15-3**. Current service of R-22 systems relies on recycled or stockpiled refrigerant.

Hydrofluorocarbons (HFCs)

Hydrofluorocarbons (HCs) contain no chlorine atoms and have an ozone depletion level of 0.0 (no chlorine). However, HFCs do have a high global warming potential (GWP). Refrigerants that have a high GWP are said to contribute to global warming. Research is underway to develop refrigerants that have low ODP and GWP. These refrigerants, or combinations of HFCs, are rapidly replacing CFCs. HFCs were developed to be interim retrofit gases, until new refrigerants could be developed with a lower GWP like HFO and HC refrigerants. HFCs are currently being phased out by the AIM Act and Kigali Amendment, so the use of HFCs is in decline.

HCFC Phaseout Schedule

Year	Production Levels	EPA Rule (HCFC-22)
1996	2.8 % of 1989 use, plus 1989 HCFC use, ODP-weighted	—
2004	65% of 1989 use	—
2010	35% of 1989 use	Ban on new equipment
2015	10% of 1989 use	—
2020	0.5% of 1989 use	Ban on production and consumption
2030	0% of 1989 use	—

Goodheart-Willcox Publisher

Figure 15-3. The schedule for total phaseout of HCFC production and use.

Hydrofluoro-olefins (HFOs)

Hydrofluoro-olefins (HFO) contain the same chemical elements as HFC refrigerants: hydrogen, fluorine, and carbon. The difference between HFCs and HFOs is the type of hydrocarbon compounds they contain. The hydrocarbon compounds in HFOs are classified as olefins. Olefins are hydrocarbon compounds that contain at least one double bond between carbon atoms. The hydrocarbon compounds in HFCs contain only single bonds between carbon atoms.

This difference in chemical structure impacts the GWP of the refrigerant. HFOs have a much lower GWP than HFCs. HFOs also have zero ODP. Due to their zero ODP and small GWP, many applications that currently use HCFC and HFC refrigerants are in the process of transitioning to HFO refrigerants.

Some HFO refrigerants are slightly flammable. This is a disadvantage of HFOs when compared with HCFCs and HFCs, which are nonflammable.

The use of HFOs is increasing while the use of HFCs is decreasing due to the difference in the GWPs of the refrigerant types. Examples of HFOs include R-1234yf and R-1234ze. R-1234yf is used in automotive air-conditioning systems. R-1234ze is being used in many applications, including residential and light commercial air-conditioning systems.

Hydrocarbon (HC) Refrigerants

Hydrocarbon (HC) refrigerants contain only two chemical elements: hydrogen and carbon. Some HC refrigerants are referred to as "natural refrigerants" because they contain naturally occurring compounds.

Like HFOs, HCs have zero ODP and low GWP. Due to the environmental concerns of using

HCFCs and HFCs, the use of HCs (and HFOs) is increasing.

A disadvantage of HCs is their high flammability, which creates concerns regarding leaks and equipment servicing. Safety precautions for working with flammable refrigerants are discussed later in this chapter.

Examples of HCs include R-600a (isobutene) and R-290 (propane). R-600a is being used in domestic refrigerator-freezers. R-290 is being used in many applications, including commercial air-conditioning, refrigeration, and freezer applications.

PRO TIP

Propane Refrigerant

Propane (R-290) refrigerant is not the standard propane used in gas grills and portable heaters. Standard propane is generally 65%–95% pure propane, with other contaminants and moisture present. Refrigerant-grade propane must be at least 98.5% pure propane and has limits on the amount of moisture that can be present. Never use "standard" propane as a refrigerant.

Inorganic Refrigerants

The 700-series of numbers is assigned to ***inorganic refrigerants***, such as ammonia or carbon dioxide, **Figure 15-4**. The last two numerals represent the atomic number of the substance. For example, the atomic number of ammonia is 17, so the refrigerant number is R-717. The atomic number of carbon dioxide is 44, so the number is R-744. Most inorganic refrigerants are considered "expendable," meaning the vapor is not recovered for reuse. The liquid is vaporized and released into the atmosphere.

Inorganic Refrigerants

Number	Name	Formula	Boiling Point (°F @ 0 psig)
R-702	Hydrogen	H_2	–423.0
R-704	Helium	He	–452.1
R-717	Ammonia	NH_3	–28.0
R-718	Water	H_2O	212.0
R-720	Neon	Ne	–410.9
R-728	Nitrogen	N_2	–320.4
R-732	Oxygen	O_2	–297.3
R-740	Argon	Ar	–302.6
R-744	Carbon dioxide	CO_2	–109.2

Goodheart-Willcox Publisher

Figure 15-4. Inorganic refrigerants are assigned numbers in the 700-series. These refrigerants are expendable.

Identifying Refrigerants

Trade terminology is often confusing because each refrigerant can be called by a different name. For example, R-22 can be called by the manufacturer's trade name (Freon-22®), by its chemical name (monochlorodifluoromethane), by its chemical formula ($CHClF_2$), by the ASHRAE "R" number (R-22), or by its composition (HCFC-22). The ASHRAE designation combined with the DuPont numbering system is the one most commonly used in the HVACR industry.

DuPont's numbering system for refrigerants came into general use in 1956. By that time, several other chemical companies were manufacturing halogenated refrigerants (those containing chlorine or fluorine). Each separate refrigerant chemical (such as dichlorodifluoromethane) was given a number, and each company produced these refrigerants under its own brand name. For example, DuPont's trademark for dichlorodifluoromethane was Freon-12®, while Allied Chemicals used Genetron-12®, and Virginia Chemicals used Isotron-12®.

The American Society of Heating, Refrigerating and Air Conditioning Engineers (ASHRAE) has standardized identification of all refrigerants. The system uses the DuPont numbering system, but it precedes each number by the letter "R" (for refrigerant), regardless of manufacturer. The ASHRAE "R" designation, combined with DuPont's numbering system, is most consistent in industry. The DuPont numbering system is explained in **Figure 15-5**.

Each refrigerant has its own special characteristics and applications. **Figure 15-6** lists the various names for several refrigerants. A few refrigerants are quite popular, while others are used only in special applications. The service technician must be able to recognize common and uncommon refrigerants and know where to find information about them.

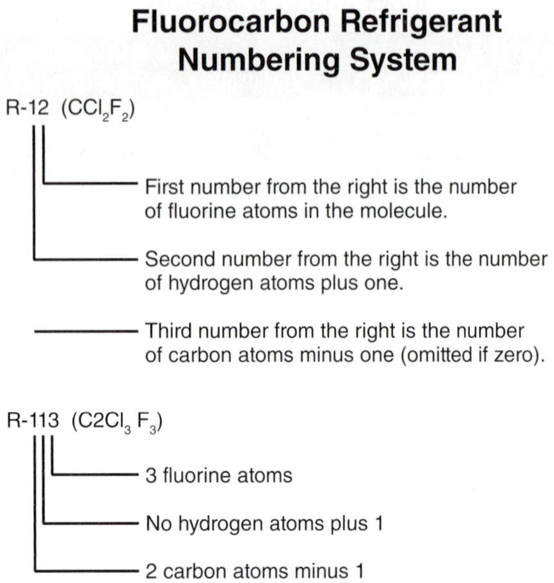

Goodheart-Willcox Publisher

Figure 15-5. The numbering system developed by DuPont has been adopted by all refrigerant manufacturers. The figure illustrates how the designations for R-12 and R-113 were determined.

Refrigerant Safety and Hazard Classifications

Along with assigning refrigerant numbers, ASHRAE also assigns the safety and hazard classifications to the refrigerants. **Figure 15-7** shows the two safety categories, flammability and toxicity, broken down into six classifications—A1, A2, A3, B1, B2 and B3—and two sub-classes—A2L and B2L.

According to industry standards, refrigerants are also classified into three groups, depending on their hazard potential. Group 1 refrigerants are considered the *least* hazardous. Group 3 refrigerants are the *most* hazardous.

The degree to which a refrigerant is hazardous can be measured by its ***Acceptable Exposure Limit (AEL)***. AEL is an airborne-exposure limit with a time-weighted average (TWA) to which workers may be exposed repeatedly during an 8- or 12-hour day, or 40-hour week, without adverse effects.

Refrigerant Names and Data

Number	Chemical Name	Formula	Type	Oil	Boiling Point (°F @ 0 psig)
R-11	Trichlorofluoromethane	CCl_3F	CFC	MO	74.9
R-12	Dichlorodifluoromethane	CCl_2F_2	CFC	MO	–21.6
R-22	Chlorodifluoromethane	$CHClF_2$	HCFC	MO	–41.4
R-123	Cichlorotrifluoroethane	$CHCl_2CF_3$	HCFC	MO/AB	82.2
R-23	Trifluoromethane	CHF_3	HFC	POE	–115.7
R-32	Difluoromethane	CH_2F_2	HFC	POE	–61.1
R-125	Pentafluoroethane	CHF_2CF_3	HFC	POE	–55.8
R-134a	Tetrafluoroethane	CF_3CH_2F	HFC	POE	–15.1
R-143a	Trifluoroethane	CH_3CF_3	HFC	POE	–53.7
R-152a	Difluoroethane	CH_3CHF_2	HFC	POE	–11.3
R-290	Propane	$CH_3CH_2CH_3$	HC	MO/AB/POE	–42.1
R-454B	Difluoromethane tetrafluoropropene	CH_2	HFO	POE	–59.6
R-600a	Isobutane	$CH(CH_3)_2CH_3$	HC	MO/AB/POE	10.9
R-1234yf	Tetrafluoropropene	$CF_3CF{=}CH_2$	HFO	POE/PAG	–20.9
R-1234ze	Transtetrafluoropropene	$CF_3CH{=}CHF$	HFO	POE	–2.2
R-717	Ammonia	NH_3	Inorganic compound	Depends on system	–28.0
R-744	Carbon dioxide	CO_2	Inorganic compound	POE	–109.2

Goodheart-Willcox Publisher

Figure 15-6. Each refrigerant has its own special characteristics and is identifiable by its ASHRAE "R" number, chemical name, and chemical formula.

Refrigerant Flammability and Toxicity Classifications

Flammability	Low Toxicity	High Toxicity
Higher flammability	A3 (Propane)	B3 (no known refrigerants)
Flammable	A2 (R-152a)	B2 (R-40)
Low flammability	A2L (R-32)	B2L (Ammonia)
Non-flammable	A1 (R-134a)	B1 (R-123)

Goodheart-Willcox Publisher

Figure 15-7. The refrigerant flammability and toxicity chart. Examples of a refrigerant in each classification are provided in parentheses below the classification name.

SAFETY NOTE

Refrigerant Release

Refrigerant vapor is almost odorless and is five times heavier than air. High accumulations of refrigerant will displace atmospheric air and prevent access to oxygen. Becoming light-headed (dizzy) is a warning that oxygen is not available.

If a large release of any refrigerant vapor occurs, the area should be evacuated immediately. Vapors may concentrate near the floor, displacing available oxygen. Once the area is evacuated, it must be ventilated using blowers or fans to circulate the air at floor level. Confined areas should be checked with a leak detector before entering back in. If vapors are inhaled at high concentrations, death by asphyxia, cardiac irregularities, and possible cardiac arrest could occur. Service technicians have died from entering tunnels and other areas where heavy concentrations of halide refrigerants had displaced the oxygen.

Refrigerant Cylinders

Another way to identify refrigerants is by checking the cylinders they are manufactured in. Since January 2020, all new refrigerant cylinders are a light gray-green color. The cylinders of flammable refrigerants are also marked with a red band around the cylinder. The type of refrigerant in a cylinder is clearly labeled.

Prior to January 2020, most refrigerant cylinders were color-coded for ease of identification and to help prevent accidental mixing of refrigerants. Some of these colored cylinders are still in use. Some refrigerant labeling and packaging incorporate the color coding still on the cylinder. Several common refrigerant cylinders are shown in **Figure 15-8**.

Sometimes, it is necessary to know the name or chemical formula of the refrigerant. All condensing units have a data plate displaying important information about the unit. The data plate may specify the refrigerant and the amount, such as "R-134a: 6.0 lb," "Tetrafluoroethane: 6.0 lb," or "CF_3CH_2F: 6.0 lb" See **Figure 15-9**.

15.4 Pure Refrigerants, Mixtures, and Blends

It is important to understand the differences between *pure* component refrigerants, azeotropic *mixtures* and zeotropic *blends*. When a refrigerant's

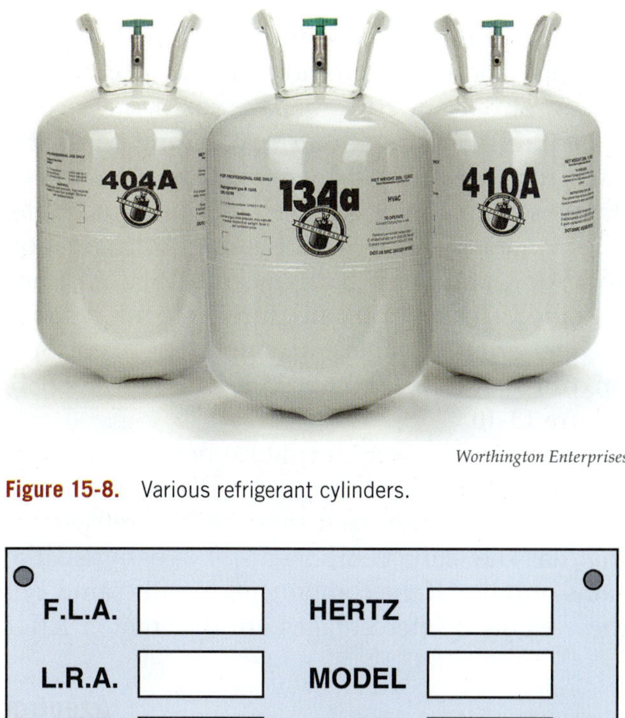

Worthington Enterprises

Figure 15-8. Various refrigerant cylinders.

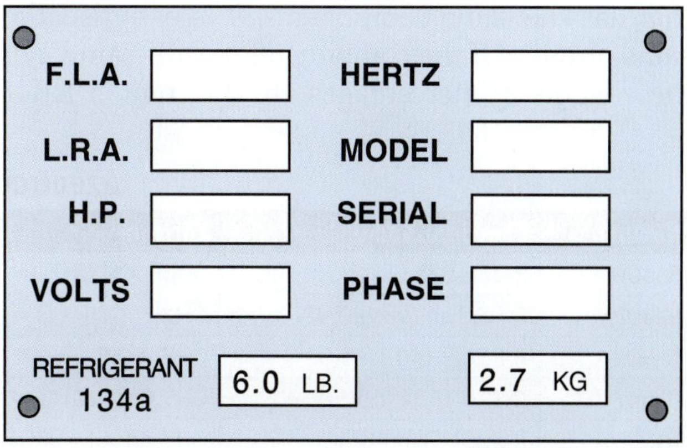

Goodheart-Willcox Publisher

Figure 15-9. Check the unit data plate for the type and amount of refrigerant.

entire composition consists of one molecule type, it is referred to as a single-component or "pure" refrigerant. A ***pure refrigerant*** is a single-component fluid that does not change composition when boiling or condensing. Examples of pure refrigerants are R-12, R-32, and R-134a. Pure refrigerants are generally the refrigerants of choice.

Azeotropic Mixtures

An ***azeotropic refrigerant*** is a mixture of two refrigerants that combine to form a third, unique refrigerant having its own individual characteristics. It is mixed in exact proportions, and its composition is always the same in both the liquid and vapor states. An *azeotrope* exhibits a single boiling temperature different from either of the refrigerants of which it is composed. Such a mixture must maintain a constant boiling point and act as a single refrigerant, strictly conforming to Charles' Law of gases.

Sometimes two pure refrigerants can be combined to form an azeotropic refrigerant mixture that cannot be separated into its individual components. These are commonly called ***azeotropes***. An azeotropic mixture behaves as a *single*-component refrigerant with a *single* boiling point at a particular pressure and temperature. Azeotropes do *not* separate in either the vapor or liquid state. The American Society of Heating, Refrigeration and Air Conditioning Engineers (ASHRAE) has classified all azeotropic mixtures into a 500-series of refrigerant numbers. Examples of azeotropic mixtures are R-500, R-502, R-507, and R-508B, **Figure 15-10**.

As manufacturers introduce new refrigerants within the series, it is assigned a consecutive number. If another manufacturer submits a refrigerant that has the same components, it is assigned the same number. If the components are the same, but the weights by percentages are different, a letter of the alphabet is added at the end of the number. Appropriate R-numbers de-emphasize individual manufacturer's brand names.

Two common azeotropic mixtures are R-500 and R-502, but these two refrigerants are classified as CFCs and have been eliminated because of their high ODP and GWP levels. They are included in the list of azeotropic mixtures in the table in **Figure 15-10** to show their history and unique qualities. Similar information is available for each refrigerant, but some have yet to develop a proven history.

SAFETY NOTE

Azeotropic Mixtures

Azeotropes are patented refrigerants that are mixed in a precise manufacturing process. The service technician should *never* attempt to make refrigerant mixtures.

Zeotropic Blends

When two or more single-component refrigerants are combined and do *not* form an azeotropic mixture, the result is a blended refrigerant. Blended refrigerants are commonly known as "blends" or ***zeotropes***. Most of the non-ozone-depleting refrigerants are classified as zeotropic blends, and ASHRAE has classified all zeotropes into a 400-series of refrigerant numbers. As noted, a zeotrope is a *blend* of refrigerants meaning it consists of two or three different refrigerants and may be described as a *binary* (two-part) or *ternary* (three-part) blend. These new refrigerant blends are designed to replace the ozone-depleting CFC and HCFC refrigerants. **Figure 15-11** lists many of the current blends, and the list continues to grow as more are being developed. Zeotropic blends are

Azeotropic Mixtures

Number	Chemical Name	Formula	Type	Oil	Boiling Point (°F @ 0 psig)
R-500	R-12/152a (73.8/26.2)	CCl_2F_2/CH_3CHF_2	CFC	Mineral Oil	–28.3
R-502	R-22/115 (48.8/51.2)	$CHClF_2/CClF_2CF_3$	CFC	Mineral Oil	–50.0
R-503	R-23/13 (40.1/59.9)	$CHF_3/CClF_3$	CFC	POE/AB/MO	–126.1
R-507	R-125/143a (50/50)	CHF_2CF_3/CH_3CF_3	HFC	Polyol ester	–52.1
R-508B	R-23/R-116 (46/54)	CHF_3/CF_3CF_3	HFC	Polyol ester	–126

MO = mineral oil, AB = alkylbenzene, POE = polyol ester

Goodheart-Willcox Publisher

Figure 15-10. Each new azeotrope is assigned a consecutive number in the 500-series.

Zeotropic Blends

Refrig. Number	Blend	Trade Name	Type	Lubricant Used*	Replaces	Boiling Point (°F @ 0 psig)
R-401A	R-22/152a/124	MP39	HCFC/HFC/HCFC	AB/POE	R-12/500	–16.5
R-401B	R-22/152a/124	MP66	HCFC/HFC/HCFC	AB/POE	R-12/500	–29.5
R-402A	R-125/290/22	HP80	HFC/HC/HCFC	MO/AB/POE	R-502	–53.6
R-402B	R-125/290/22	HP81	HFC/HC/HCFC	MO/AB/POE	R-502	–52.5
R-403B	R-290/22/218	69L	HFC/HC/HCFC	AB/POE	R-502/R-13	–57.1
R-404A	R-125/143a/134a	HP62/FX70	HFC/HC/HCFC	POE	R-502	–51.8
R-406A	R-22/142b/600a	GHG12	HCF/HCFC/HC	MO/AB	R-12	–22.0
R-407A, B	R-32/125/134a	KLEA60, 61	HFC/HFC/HFC	POE	R-502	–38.0
R-407C	R-32/125/134a	KLEA66	HFC/HFC/HFC	POE	R-22	–34.6
R-408A	R-125/143a/22	FX10	HFC/HFC/HCFC	MO/AB/POE	R-502	–46.3
R-409A	R-22/124/142b	FXS6	HCFC/HCFC/HCFC	MO/AB	R-12	–15.4
R-410A	R-32/125	AZ20	HFC/HFC	POE	R-22	–62.9

*Lubricants: MO = Mineral Oil, AB = Alkylbenzene, POE = Polyol Ester

Goodheart-Willcox Publisher

Figure 15-11. Examples of zeotropic refrigerant blends. New blends are continually being developed.

unusual because they do *not* have a specific pressure for each degree of temperature change. At a given pressure, the temperature may fluctuate as much as 16°F (9°C). The temperature difference is called *temperature glide* and is the result of *fractionation* (separation of parts).

Fractionation and Temperature Glide

All zeotropes exhibit fractionation and temperature glide. When both liquid and vapor states are present simultaneously (as in the evaporator or condenser), the composition of the liquid is different from that of the vapor. The components of the blend do not boil and condense as a single refrigerant. Instead, the most volatile component of the refrigerant blend boils first, the next most volatile boils second, and the next most volatile boils third. In other words, ***fractionation***, or the separation of parts, occurs when the most volatile component boils into a vapor more quickly than the less-volatile components.

Fractionation also occurs when the system develops a leak or when refrigerant is being charged into the system. Zeotropic refrigerants fractionate inside the charging cylinder when pressure is removed from the liquid. To prevent fractionation in the charging cylinder, zeotropic refrigerants are carefully charged into the system as a *liquid*. Charging the system as a vapor is possible when the entire contents of the charging cylinder are being emptied into the system.

The difference in composition of zeotropic refrigerants leads to a difference in boiling points for the liquid and vapor. There is a boiling *range* because the liquid begins boiling at one temperature and the vapor condenses at a different temperature. The temperature difference caused by fractionation between the evaporator inlet and outlet or between the condenser inlet and outlet is called ***temperature glide***. The amount of temperature glide varies from blend to blend and with system conditions. The old rules for setting superheat do *not* apply to zeotropic blends. Temperature glide for most zeotropic blends ranges from 1°F to 16°F (0°C to 9°C). Some zeotropic blends are referred to as *near-azeotropic* refrigerant mixtures, because the temperature glide is rather small (1 or 2 degrees).

Temperature glide can affect the system in several ways. First, the evaporator is no longer at a constant temperature, and the "set" temperature is the *average* temperature. The refrigerant metering device may sense the temperature at the evaporator outlet as too high. In some systems, temperature glide may cause frosting at the evaporator outlet. Evaporator frosting is a normal characteristic for refrigerants with glide. No problems should occur if the glide is recognized and proper adjustments are performed.

Topping Off

Topping off refers to the process of adding enough refrigerant to a system to bring it up to full capacity. A leak in a system using a blended refrigerant should not be topped off, because of fractionation. A leak on the liquid side of a system releases proportionate amounts of each refrigerant component. However, a leak from the low-pressure side tends to fractionate and leak disproportionate amounts of the lower-boiling components. Since there is no real way of knowing which refrigerant has leaked more, as a rule, a system with a blend should have the refrigerant charge fully recovered and new refrigerant charged back in. Refrigerant recovery and charging procedures are discussed in Chapter 16, *Refrigerant Recovery and Recycling* and Chapter 17, *Refrigerant Service Basics.*

15.5 Selecting a Refrigerant

The halide (halogenated) family of refrigerants was responsible for the tremendous growth of the refrigeration and air-conditioning industry. The chemical properties of these refrigerants have permitted their use under conditions that would be hazardous with more flammable or toxic chemicals. In many cases, one refrigerant may be used in several applications. The choice of refrigerant for a particular application depends upon several factors. A low boiling point is only one feature of a good refrigerant. Other important qualities include toxicity (safety), flammability, latent heat value, suction pressure, condensing pressure, vapor density, oil compatibility, corrosiveness, and cost.

The ***latent heat value*** is the amount of heat (in Btus) absorbed for each pound of liquid evaporated and is an important calculation. The latent heat value (Btu/lb) is different for each refrigerant. If the latent heat value for a refrigerant is high, less refrigerant is required, and system components can be smaller and less expensive. The latent heat of the refrigerant should be as high as possible, but other factors must also be considered. For example, ammonia has a high latent heat value (565 Btu/lb), but it is toxic to humans.

Selection of the best refrigerant for a particular system is often a compromise. For example, the pressure in the evaporator must be as high as possible for the boiling temperature desired. A high suction pressure results in a high-density vapor that sweeps oil back to the compressor. A refrigerant having low pressure (or a vacuum) in the evaporator results in a thin vapor, and oil return could be a problem.

Whereas the evaporator pressure should be as high as possible, the condensing pressure should be as *low* as possible. A low condensing pressure reduces the size of the condensing unit and decreases the load on the compressor. A high condensing pressure requires a high-capacity compressor and may require additional cooling for the compressor.

When selecting a refrigerant, the design engineer tries to achieve high capacity (Btu/lb) accompanied by low material and energy costs. System components are expensive to purchase and costly to operate. Therefore, the best choice is a refrigerant that has high capacity and low power requirements.

Standard Conditions

Each refrigerant is different, and it is important to select the best one for a given application. Many factors must be considered in the selection process, but the first step is to compare refrigerants under ***standard conditions***. Standard conditions have been established as 5°F (–15°C) evaporating temperature and 86°F (30°C) condensing temperature. See **Figure 15-12** for comparisons under standard conditions of several common refrigerants. Using standard conditions to compare refrigerants provides a quick method of eliminating unsuitable refrigerants. Final selection is made at actual

Thermodynamic Properties

Properties at 1 atmosphere (14.7 psia or 101 kPa)	R-22	R-32	R-410A	R-454B
Freezing point (°F)	–256	–212.8	–247	–241
Boiling point (°F)	–41.4	–61.06	–60.5	–58.9
Specific heat of liquid (Btu/lb/cu ft)	0.54	0.83	0.735	0.433
Specific volume (cu ft/lb) [suction gas]	0.27	0.36	0.35	1.12
Latent heat of vaporization (Btu/lb)	100.2	164.2	118.2	120.7

All the values in the table are at 77°F.

Goodheart-Willcox Publisher

Figure 15-12. Standard conditions are useful when comparing refrigerants. This table presents a side-by-side comparison of various refrigerants and their thermodynamic properties.

system operating temperatures. Refrigerant pressure-temperature charts provide this important information.

Saturation Point

At ***saturation***, a refrigerant can exist as a liquid, a mixture of liquid and vapor, or a vapor (all of which are at the same pressure and temperature). When a refrigerant is at the saturation point, the temperature-pressure relationship is predictable, **Figure 15-13**. The saturation point does not include superheat or subcooling. A vapor can be superheated to a temperature above the saturation point without affecting pressure. Likewise, a liquid can be subcooled below saturation without affecting pressure. A pressure reading reveals saturation temperature, but actual temperature may be different. For a given pressure, the actual refrigerant temperature may be higher or lower (if superheated or subcooled), **Figure 15-14**.

Pressure-Temperature Charts

Highly accurate pressure-temperature (P/T) charts help design engineers determine the best refrigerant for a particular application. P/T charts are useful for comparing refrigerants under various operating conditions. At a given temperature, the P/T charts automatically assume the refrigerant is boiling or condensing. At a given temperature, the charts reveal important information regarding the physical properties of the refrigerant: the correct saturation pressure (boiling point), the volume occupied by the saturated vapor (in cubic feet per pound), the density of the saturated liquid (in pounds per cubic feet), and the ***enthalpy*** (heat content) of both vapor and liquid (in Btu/lb). See **Figure 15-15** for a P/T chart of R-454B.

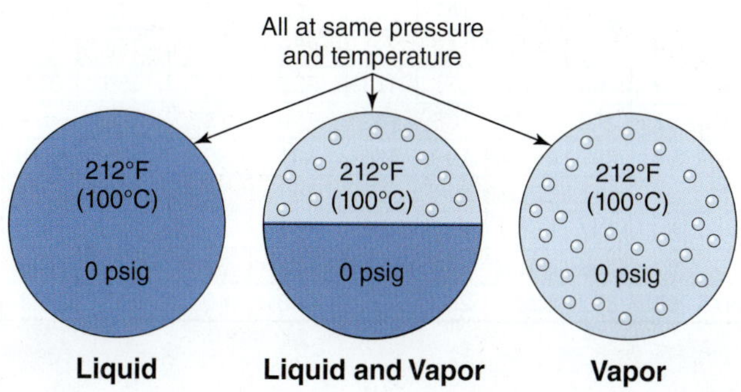

Goodheart-Willcox Publisher

Figure 15-13. At the saturation point (boiling point), a refrigerant can be a liquid, a mixture of liquid and vapor, or a vapor.

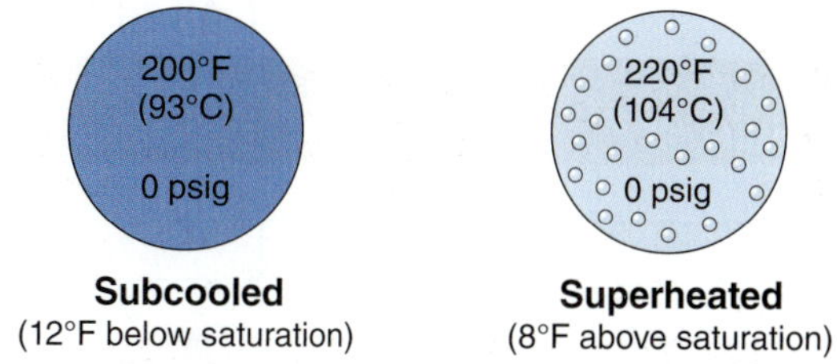

Goodheart-Willcox Publisher

Figure 15-14. A pressure reading can be converted to saturation temperature, but actual temperature may be different. Saturation charts reveal saturation temperature, not superheat or subcooling.

P/T charts are based on a single, pure gas that behaves according to Charles' Law. If another gas (or group of gases, such as atmospheric air) is permitted to enter the system, then Dalton's Law takes effect and the chart becomes useless. The condition is easily recognized because the system pressures greatly exceed the values shown on the P/T chart.

Service and installation technicians are primarily interested in the temperature-pressure relationship, but the charts are also used to select the best refrigerant for a system, to size suction and liquid lines, and to size system components, such as the evaporator, condenser, refrigerant metering device, and compressor. Proper sizing and selection of these components is important to efficient operation of the system.

Using Pressure-Temperature (P/T) Cards

Service technicians normally use the pressure-temperature relationship for checking system operation and troubleshooting system problems. Special pocket-size pressure-temperature charts, called ***P/T cards***, show the pressure and temperature relationship for several refrigerants. Most technicians keep a P/T card readily available. A typical P/T card is shown in **Figure 15-16**. Similar cards are available at no charge from refrigerant manufacturers or from local parts suppliers. Pocket P/T charts come in two different altitude levels: sea level (SL) and 5,000′ above SL.

P/T cards provide the information necessary to predict and control the behavior of each refrigerant. Service technicians are primarily interested in the pressure-temperature relationship, which differs for each refrigerant. Gauges are used to obtain refrigerant pressure. Then, the card is used

R-454B Pressure-Temperature Chart

Temperature (°F)	Pressure (psig)	Liquid Density (lb/cu ft)	Vapor Volume (cu ft/lb)	Liquid Enthalpy (Btu/lb)	Vapor Enthalpy (Btu/lb)
–40	9.5	75.46	2.9353	0	131.5
–30	12.7	74.44	2.3266	3.58	132.8
–20	24.3	73.40	1.8642	7.19	133.9
–10	34	72.33	1.5083	10.84	135
0	45.3	71.24	1.2312	14.53	136.1
10	58.6	70.11	1.0129	18.26	137
20	74	68.95	0.8391	22.05	137.9
30	91.7	67.76	0.6995	25.89	138.8
40	111.9	66.52	0.5863	29.79	139.5
50	135	65.23	0.4937	33.76	140.1
60	161	63.90	0.4174	37.81	140.6
70	190.4	62.49	0.3539	41.95	141
80	223.3	61.02	0.3008	46.19	141.2
90	260	59.46	0.2560	50.54	141.3
100	300.8	57.79	0.2180	55.04	141.1
110	346	55.99	0.1853	59.7	140.7
120	395.9	54.02	0.1572	64.57	140
130	450.9	51.83	0.1326	69.72	138.8
140	511.3	49.31	0.1108	75.26	137.1

Goodheart-Willcox Publisher

Figure 15-15. Properties of liquid and saturated vapor for R-454B.

Refrigerant Pressure-Temperature Chart

Temp in °F	Temp in °C	22	32	134a	410A	407C Bubble Liquid	407C Dew Vapor	1234yf	600a
–40	–40	0.5	11	14.8	11.6	2.7	4.6	11.5	21.6
–35	–37	2.6	14.4	12.5	14.9	5.1	0.9	8.9	20.4
–30	–34	4.9	18.2	9.9	18.5	7.7	1.6	6	18.9
–25	–32	7.4	22.3	6.9	22.5	10.6	3.9	2.8	17.4
–20	–29	10.1	26.8	3.7	26.9	13.7	6.5	0.4	15.6
–15	–26	13.2	31.7	0.6	31.7	17.2	9.3	2.3	13.7
–10	–23	16.5	37.1	1.9	36.8	20.9	12.3	4.4	11.5
–5	–21	20.1	42.9	4	42.5	25	15.7	6.7	9.2
0	–18	24	49.3	6.5	48.6	29.5	19.4	9.1	6.6
5	–15	28.2	56.1	9.1	55.2	34.3	23.5	12	3.8
10	–12	32.8	63.5	11.9	62.3	39.5	27.9	14.9	0.7
15	–9	37.7	71.5	15	70	45.2	32.7	18.1	1.3
20	–7	43	80	18.4	78.3	51.2	37.9	21.6	3.1
25	–4	48.8	89.2	22.1	87.3	57.7	43.5	25.4	5
30	–1	54.9	99.1	26.1	96.8	64.7	49.6	29.4	7.1
35	2	61.5	109.7	30.4	107	72.2	56.1	33.8	9.4
40	4	68.5	212	35	118	80.2	63.2	38.4	11.8
45	7	76	133	40.1	130	88.8	70.7	43.4	14.4
50	10	84	45.9	45.5	142	97.9	78.8	48.8	17.2
55	13	92.6	159.5	51.3	155	108	87.5	54.5	20.2

Temp in °F	Temp in °C	22	32	134a	410A	407C Bubble Liquid	407C Dew Vapor	1234yf	600a
60	16	102	174.1	57.5	170	118	96.8	60.6	23.5
65	18	111	189.5	64.1	185	129	107	67	26.9
70	21	121	205.8	71.2	201	141	117	73.9	30.6
75	24	132	223.2	78.8	217	153	129	81.3	34.5
80	27	144	241.5	86.8	235	166	141	89	38.6
85	30	156	260.9	95.4	254	180	153	97.2	43
90	32	168	281.3	104	274	194	167	106	47.7
95	35	182	302.9	114	295	209	181	115	52.7
100	38	196	325.7	124	317	226	196	125	57.9
105	41	211	349.6	135	340	242	212	135	63.4
110	43	226	374.9	147	365	260	229	146	69.3
115	46	243	401.4	159	391	279	247	157	75.4
120	49	260	129.3	171	418	299	266	169	81.9
125	52	278	458.6	185	446	319	286	182	88.7
130	54	297	489.4	199	476	341	307	195	95.8
135	57	317	521.8	214	507	363	329	209	103
140	60	337	555.7	229	539	387	352	223	111
145	63	359	591.4	246	573	412	377	239	120
150	66	382	628.8	263	608	438	403	255	128

All pressures in psi except underscored pressures in inches of mercury (vacuum).

Goodheart-Willcox Publisher

Figure 15-16. Technicians keep a P/T card readily available and refer to it often.

to convert the pressure reading to a corresponding temperature equivalent.

The service technician can raise or lower the pressure (according to the card) to achieve the desired refrigerant temperature. All refrigeration systems are designed to control the temperature of some substance, usually air. The air temperature is regulated by controlling the refrigerant temperature inside the evaporator. The saturation temperature of the refrigerant inside the evaporator is changed by raising or lowering the suction pressure. A colder evaporator (lower pressure) results in colder air temperature.

The service technician must control the pressure-temperature relationship inside the evaporator. The metal of the evaporator is about 10°F warmer than the refrigerant boiling inside. The temperature of the air in the evaporator is about 10°F warmer than the metal of the evaporator. Thus, during the run cycle the air temperature remains about 20°F warmer than the refrigerant, **Figure 15-17**.

Two pairs of reversible statements will help to remember the relationship between pressure and temperature in a refrigeration system:

- Refrigerant *temperature* can be controlled by controlling *pressure*.
- Refrigerant *pressure* can be controlled by controlling *temperature*.
- Refrigerant *pressure* is predictable by knowing *temperature*.
- Refrigerant *temperature* is predictable by knowing *pressure*.

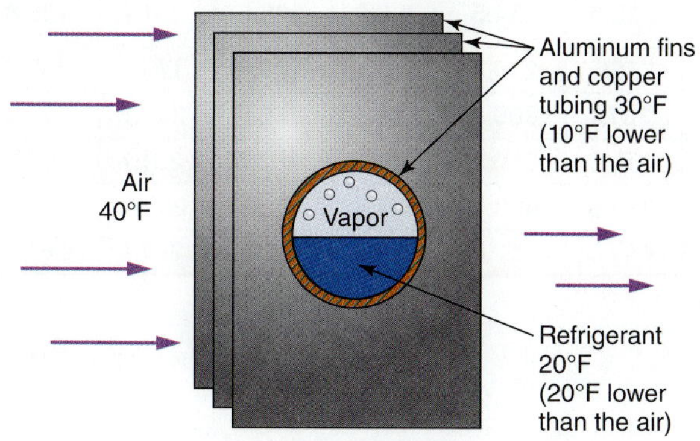

Goodheart-Willcox Publisher

Figure 15-17. To achieve proper heat flow, a temperature difference of 20°F is created between the airflow and the boiling refrigerant. The metal fins and tubing are 10°F colder than the air, and the refrigerant is 10°F colder than the metal fins and tubing.

Using P/T Cards for Zeotropic Blends

The P/T card for zeotropic blends is unusual. Instead of one pressure for each saturation temperature, it has two. **Figure 15-18** is a partial example of the P/T card used for blends.

The third column of the card shows the *saturated liquid pressure*, called the ***bubble point***. These numbers indicate the pressure for a given temperature in which the first bubble of vapor appears in a liquid refrigerant. The last column shows the *saturated vapor pressure*, called the ***dew point***. These numbers indicate the pressure at which the first drop of liquid condenses from a vapor. For example, R-409A has a bubble point temperature of –10°F (–23°C) when its pressure is 10 psig (69 kPa). It has a dew point temperature of about 5°F (–15°C) at approximately the same pressure (10.6 psig or 73 kPa).

P/T Cards and Temperature Glide

Temperature glide can be calculated by finding the difference between the dew point and bubble point temperatures at constant pressure. Actual measurements may differ slightly, depending on the state of the refrigerant and pressure loss through the evaporator or condenser.

Figure 15-19 shows a complete P/T card used by technicians when working with blends. These cards, or similar ones, are available free from the local refrigerant supplier or directly from the refrigerant manufacturer.

R-409A Pressure-Temperature

Temp. °F	Temp. °C	Liquid Pressure (Bubble)	Vapor Pressure (Dew)
–30	–34	0.9	7.9
–25	–32	2.9	4.8
–20	–29	5.1	1.4
–15	–26	7.4	1.2
–10	–23	10.0	3.2
–5	–20	12.9	5.5
0	–18	16.0	8.0
5	–15	19.3	10.6

Goodheart-Willcox Publisher

Figure 15-18. A sample of part of a P/T card used for a zeotropic blend.

Pressure-Temperature Card for Zeotropic Blends

°F	°C	401A (Liquid) Bubble	401A (Vapor) Dew	401B (Liquid) Bubble	401B (Vapor) Dew	402A (Liquid) Bubble	402A (Vapor) Dew	404A (Liquid) Bubble	404A (Vapor) Dew	407C (Liquid) Bubble	407C (Vapor) Dew	409A (Liquid) Bubble	409A (Vapor) Dew
−40	−40.0	*3.8*	*12.5*	*7.2*	*11.3*	8.5	7.1	5.1	4.7	3.3	*3.2*	*5.2*	*13.2*
−35	−37.2	*2.3*	*10.1*	*4.0*	*8.7*	11.3	9.9	7.6	7.2	5.7	0.3	*1.9*	*10.7*
−30	−34.4	*0.7*	*7.3*	*0.4*	*5.9*	14.5	13.0	10.3	9.9	8.3	2.3	0.9	*7.9*
−25	−31.7	1.1	*4.3*	1.8	*2.7*	17.9	16.2	13.3	12.9	11.1	4.6	2.9	*4.8*
−20	−28.9	3.1	*0.9*	3.9	0.4	21.6	19.9	16.6	16.2	14.3	7.1	5.1	1.4
−15	−26.1	5.3	1.4	6.3	2.3	25.7	23.8	20.2	19.8	17.7	9.8	7.4	1.2
−10	−23.3	8.9	3.4	8.9	4.4	30.1	28.1	24.1	23.7	21.4	12.9	10.0	3.2
−5	−20.6	10.3	5.6	11.7	6.7	34.8	32.8	26.3	27.9	25.5	16.2	12.9	5.5
0	−17.8	13.2	8.0	14.8	9.3	40.0	37.8	33.0	32.5	29.9	19.8	16.0	8.0
5	−15.0	16.3	10.6	18.2	12.0	45.6	43.3	37.9	37.5	34.7	23.8	19.3	10.6
10	−12.2	19.7	13.5	21.9	15.0	51.6	49.2	43.3	42.9	39.9	28.2	22.9	13.6
15	−9.4	23.3	16.7	23.3	18.3	58.0	55.5	49.1	48.6	45.5	32.9	26.8	16.8
20	−6.7	27.3	20.1	30.2	21.8	65.0	62.3	55.3	54.9	51.6	38.0	31.0	20.0
25	−3.9	31.6	23.8	34.9	25.7	72.4	69.6	62.0	61.6	58.1	43.6	35.5	24.0
30	−1.1	36.2	27.8	39.9	29.8	80.4	77.4	69.2	68.8	65.1	49.6	40.4	28.0
35	1.7	41.1	32.2	45.3	34.3	88.9	85.8	76.1	76.5	72.5	56.0	45.6	32.4
40	4.4	46.4	36.8	51.0	39.1	97.9	94.8	85.1	84.7	80.6	63.0	51.1	37.1
45	7.2	52.1	41.9	57.2	44.3	107.6	104.3	93.9	93.6	89.1	70.6	57.1	42.1
50	10.0	58.2	47.3	63.8	49.9	117.8	114.5	103.2	103.0	98.3	78.6	63.4	47.6
55	12.8	64.7	53.1	70.8	55.9	128.8	125.3	113.2	113.0	108.0	87.3	70.1	53.4
60	15.6	71.6	59.4	78.3	62.3	140.3	136.7	123.7	123.6	118.4	96.6	77.3	59.6
65	18.3	79.0	66.0	86.3	69.1	152.6	148.9	134.9	134.9	129.4	106.5	84.9	66.2
70	21.1	86.9	73.2	94.8	76.4	165.6	161.8	146.8	146.9	141.0	117.1	92.9	73.2
75	23.9	95.2	80.8	103.7	84.2	179.3	175.4	159.4	159.6	153.4	128.4	101.5	80.7
80	26.7	104.0	88.9	113.2	92.5	193.8	189.9	172.7	173.0	166.4	140.4	110.5	88.7
85	29.4	113.4	97.5	123.2	101.3	209.0	204.4	186.7	187.2	180.2	153.2	120.0	97.2
90	32.2	123.3	106.7	133.7	110.6	225.1	221.0	201.5	202.1	194.8	166.8	130.0	106.2
95	35.0	133.7	116.4	144.8	120.6	242.0	237.8	217.1	217.9	210.2	181.2	140.6	115.7
100	37.8	144.7	126.8	156.4	131.1	259.8	255.6	233.5	234.5	226.3	196.5	151.7	125.8
110	43.3	168.5	149.2	181.5	153.9	298.0	293.7	268.8	270.3	261.1	229.7	175.7	147.6
120	48.9	194.8	174.3	209.0	179.4	339.9	335.6	307.1	309.8	299.5	266.7	202.1	171.9
130	54.4	223.7	202.2	238.9	207.6	385.8	331.5	350.3	353.1	341.5	307.7	231.1	198.9
140	60.0	255.3	233.1	271.5	238.8	435.8	431.5	396.9	400.4	387.4	353.1	262.7	228.6
150	65.6	289.8	267.1	306.6	273.2	490.1	485.8	447.5	452.0	437.3	403.1	297.1	261.3

Italic figures indicate in. Hg vacuum.
Bold figures indicate psig.

Goodheart-Willcox Publisher

Figure 15-19. A complete P/T card for zeotropic blends.

Refrigerant Properties

As mentioned, the choice of refrigerant depends upon several factors. Properties of several refrigerants are discussed in this section, as well as hazards posed by their use.

R-134a

Chemical name: 1,1,1,2-Tetrafluoroethane
Chemical formula: F_3CCH_2F
Composition: HFC single-component refrigerant
Boiling point: –15°F (–26°C) at 0 psig
Class: Pure refrigerant
Safety class: A1

R-134a was the first commercially available fluorocarbon refrigerant that was not ozone-depleting. R-134a has characteristics similar to R-12 and was the preferred alternative refrigerant in most medium- and high-temperature applications where R-12 was used. It is scheduled for phaseout due to its high GWP. The overall stability of R-134a is makes it useful in virtually all applications where R-12 is suitable. R-134a is compatible with steel, copper, aluminum, and brass. The latent heat value for R-134a is 88.29 Btu/lb at 5°F (–15°C). See **Figure 15-20**.

R-134a

°F	Pressure psig	Volume Vapor cu ft/lb	Density Liquid lb/cu ft	Heat Content Btu/lb Liquid	Heat Content Btu/lb Vapor
–100	*28.09*	41.52	96.89	–17.93	87.24
–75	*25.06*	16.56	94.08	–10.47	90.76
–50	*18.73*	7.55	91.22	–2.99	94.24
–25	*6.92*	3.83	88.27	4.50	97.72
–20	*3.60*	3.41	86.47	5.71	98.81
–15	*0.05*	3.02	85.96	7.17	99.55
–10	1.93	2.69	85.44	8.64	100.28
–5	4.09	2.39	84.91	10.13	101.01
0	6.47	2.14	84.38	11.63	101.74
*5	9.07	1.92	83.85	13.14	102.47
10	11.92	1.72	83.31	14.66	103.19
15	15.11	1.55	82.76	16.19	103.90
20	18.40	1.40	82.21	17.74	104.61
25	22.07	1.26	81.65	19.30	105.31
50	45.62	0.78	78.75	27.27	108.73
75	78.42	0.50	75.63	35.57	111.95
*86	96.60	0.42	74.17	39.33	113.28
100	124.45	0.33	72.22	44.23	114.88
125	185.26	0.22	68.38	53.33	117.41
150	262.80	0.15	63.90	63.06	119.30
175	360.99	0.10	57.60	73.58	120.78

Italics indicate inches of mercury (vacuum).
* Indicated standard conditions:
5°F evaporating temperature
86°F condensing temperature

Goodheart-Willcox Publisher

Figure 15-20. Properties of liquid and saturated vapor for R-134a.

Because R-134a has a combustibility potential under pressure, leak-checking should never be done with a mixture of R-134a and air. Leak-checking can be performed safely with a mixture of R-134a and nitrogen. An R-134a leak may be detected by using soap bubbles, sniffers (halide torch and electronic), dyes, or ultrasonic detection.

R-134a is typically seen in several refrigeration applications including supermarket cases, walk-in coolers, vending machines, watercoolers, and home refrigerators. R-134a is also used in some centrifugal chillers.

Field tests show the capacity and energy efficiency of R-134a are like those of R-12. At evaporating temperatures below –10°F (–23°C), R-134a loses its attractiveness for several reasons:

- Capacity and efficiency are significantly lost compared to R-12.
- Compression ratios become high, reducing compressor reliability.
- Low-side pressures are in a vacuum, reducing system reliability.

R-134a is not a "drop-in" replacement for R-12, and it is not miscible with mineral oils (MO) or alkylbenzene (AB) lubricants. R-134a requires the use of polyol ester (POE) or polyalkylene glycol (PAG) lubricants. POE and PAG lubricants are manufactured in several variations. Most automotive original equipment manufacturers have chosen specific PAG lubricants for their systems. For non-automotive applications, most compressor manufacturers are recommending specific polyol ester lubricants. Check with equipment manufacturers for the recommended lubricants for their systems.

Chlorinated compounds (CFCs) should not be introduced into systems that use R-134a. When performing retrofits, the service technician should make every effort to eliminate residual contamination. The PAG lubricants used with R-134a are not compatible with mineral oil or CFC refrigerants. Tests have shown that up to about 0.5% of R-12 may be present in R-134a before decomposition of the PAG lubricant occurs.

A second concern regarding the mixing of R-12 and R-134a is that these two materials form a compound whose pressure is higher than the pressure of either component. The system pressure may be higher than expected and could result in performance problems. It is troublesome to separate such an azeotropic mixture, and recovery and reclamation efforts become difficult.

Compatibility of R-134a with desiccants can be a problem. Three common types of desiccant materials are used in making driers. They are molecular sieves, alumina, and silica gel. Under some conditions, the molecular sieve XH5, commonly used with R-12, is incompatible with R-134a. Molecular sieves XH7 and XH9 are recommended for use with R-134a. Drier manufacturers have developed driers and filters compatible with R-134a. Such driers can include all three types of drier materials. The solubility of water in R-134a is comparable to the solubility of water in R-22 (19.5 ppm).

R-22

Chemical name: Monochlorodifluoromethane
Chemical formula: $CHClF_2$
Composition: HCFC single-component refrigerant
Boiling point: –41°F (–41°C) at 0 psig
Class: Pure refrigerant
Safety class: A1

R-22 is a synthetic (manufactured) refrigerant that made its debut in the 1940s. R-22 was developed for low-temperature systems operating in the range of –20°F to –40°F (–29°C to –40°C). R-22 has high discharge pressure and high compression ratios when used for low-temperature applications. When used for high-temperature applications (air conditioning), R-22 has higher suction pressure, resulting in low compression ratios. R-22 is efficient when applied to high-temperature applications and used to dominate much of the air-conditioning market.

EPA regulations permitted the use of R-22 in new equipment until January 1, 2010. A total R-22 production ban became effective January 1, 2020. In the air-conditioning segment of the industry, substantial progress toward phasing out R-22 has been made; however, HCFCs can still be found in many systems. In high-temperature applications, the major refrigerant that has been used in place of R-22 in newer air-conditioning equipment has been R-410A. Starting in 2023, however, new units began replacing R-410A with R-32 or R-454B refrigerants.

R-22 is stable, nontoxic, noncorrosive, nonirritating, and nonflammable. It is more tolerant of moisture than most refrigerants and will absorb 19.5 ppm of water. Because of its strong attraction to moisture, R-22 requires the use of large amounts of desiccant. The latent heat value of R-22 is 93.2 Btu/lb at 5°F (–15°C). See **Figure 15-21**. An R-22 leak may be detected by using soap bubbles, sniffers (halide torch and electronic), dyes, or ultrasonic detection.

SAFETY NOTE

Phosgene Gas

Never expose refrigerants to open flames or glowing hot metal surfaces. At high temperatures, R-12 and R-22 decompose to form hydrochloric acid, hydrofluoric acid, and phosgene gas. Phosgene gas is toxic to the human system. It is wise to have a respirator that can filter out chemicals vapors as part of your toolkit.

R-22

°F	Pressure psig	Volume Vapor cu ft/lb	Density Liquid lb/cu ft	Heat Content Btu/lb Liquid	Heat Content Btu/lb Vapor
–100	*25.03*	18.43	93.77	–14.56	93.37
–75	*18.50*	8.36	91.43	–8.64	96.29
–50	*6.15*	4.22	89.02	–2.51	99.14
–25	7.39	2.33	86.48	3.83	101.88
–20	10.15	2.08	85.96	5.13	102.42
–15	13.17	1.87	85.43	6.44	102.94
–10	16.47	1.68	84.90	7.75	103.46
–5	20.06	1.52	84.37	9.08	103.96
0	23.96	1.37	83.83	10.41	104.47
*5	28.19	1.24	83.28	11.75	104.96
10	32.77	1.13	82.72	13.10	105.44
15	35.73	1.02	82.16	14.47	105.91
20	43.03	0.93	81.59	15.84	106.38
25	48.75	0.86	81.02	17.22	106.84
50	84.03	0.56	78.03	24.28	108.95
75	132.22	0.37	74.80	31.61	110.74
*86	158.17	0.32	73.28	34.93	111.40
100	195.91	0.26	71.24	39.27	112.11
125	277.92	0.18	67.20	47.37	112.88
150	381.50	0.12	62.40	56.14	112.73
175	510.72	0.08	56.14	66.19	110.83

Italics indicate inches of mercury (vacuum).
* Indicated standard conditions:
5°F evaporating temperature
86°F condensing temperature

Goodheart-Willcox Publisher

Figure 15-21. Properties of liquid and saturated vapor for R-22.

R-32

Chemical name: Difluoromethane
Chemical formula: CH_2F_2
Composition: HFC single-component refrigerant
Boiling point: –61°F (–52°C) at 0 psig
Class: Pure refrigerant
Safety class: A2L

R-32 currently mixed with other refrigerants to make R-410A. R-410A was the refrigerant of choice for all manufacturers of HVAC equipment until January 1, 2023, but will be phased out. R-32 was designed to replace R-410A, but it is not a drop-in replacement for R-410A systems. R-32 must be used in equipment designed for R-32. Additionally, R-32 has an ODP of 0 and has superior efficiency compared to R-410A. R-32 also offers high refrigeration capacity, thermal conductivity, low toxicity, and low pressure drop. Since R-32 is a single component refrigerant (not a blend or mixture), it is easy to recover and charge into a system.

Zeotropic Blend Properties

There are a variety of zeotropes available. Some are more widely used than others, but it is good to become familiar with them. Several zeotropic blends and their properties are described in the following paragraphs.

R-404A

Chemical name: Pentafluoroethane, Trifluoroethane, Tetrafluoroethane
Chemical formula: CHF_2CF_3, CH_3CF_3, CH_2FCF_3
Composition: Blend of R-125, R-143a, and R-134a
Boiling point: –52°F (–47°C) at 0 psig
Safety class: A1

R-404A was designed to serve as a long-term alternative to R-502 and R-22 in low- and medium-temperature commercial refrigeration applications. R-404A contains three HFCs and is scheduled for phasedown. It was a suitable retrofit refrigerant in such applications as supermarket freezer cases, reach-in coolers, display cases, transport refrigeration, and ice machines.

R-404A is not a "drop-in" replacement for R-502 since it is not miscible with existing lubricants used in R-502 systems. The mineral oil used with R-502 must be replaced with polyol ester lubricant to a residual mineral oil level of less than 5%. R-404A requires a polyol ester lubricant to ensure complete miscibility between oil and refrigerant. Miscibility is important for oil return to the compressor, especially in large systems with long runs of piping. Compressor manufacturers provide specific recommendations on the type or brand of lubricant to install in the field. Differences among polyol ester-based lubricants make it difficult to assume they are interchangeable. Always check with the compressor manufacturer prior to retrofit.

R-410A

Chemical name: Difluoromethane,
Pentafluoroethane
Chemical formula: CH_2F_2, CHF_2CF_3
Composition: Blend of R-32 and R-125
Boiling point: –61°F (–51°C) at 0 psig
Safety Class: A1

R-410A was the leading HFC refrigerant used to replace R-22, but R-410A is *not* a retrofit refrigerant. It must only be used in equipment designed for R-410A. This is because R-410A has a significantly higher operating pressure than R-22. For example, an R-22 air-conditioning system operating with an evaporator temperature of 40°F (4.4°C) would have a low side pressure of around 68 psig (469 kPa). A system using R-410A at that same evaporator temperature would have low side pressure of 118.8 psig (819 kPa). That is nearly double the pressure. It is for this reason that special manifold gauges are required to work on R-410A equipment.

R-410A has up to 60% greater capacity than current R-22 equipment. It is designed to be used in residential and commercial air-conditioning systems and heat pumps. R-410A is a near-azeotropic refrigerant mixture, because it has a small temperature glide. It is designed to use only polyol ester oils.

R-454B

Chemical name: Difluoromethane,
2,3,3,3-Tetrafluoropropene
Chemical formula: CH_2
Composition: Blend of R-32 and R-1234yf
Boiling point: –59°F (–51°C) at 0 psig
Safety class: A2L

R-454B offers higher-efficiency heating than R-32 and R-410A, requiring 5%–10% less refrigerant to operate. Its low GWP (466) makes it a lower impact alternative than other refrigerants available. The close operating pressures and temperatures of R-454B when compared with R-410A allow manufacturers to build equipment using parameters that are close to equipment that used R-410A. Despite being a blend of HFC and HFO refrigerants, R-454B is considered an HFO refrigerant due to its chemical makeup resulting in a lower GWP.

Superheat and Subcooling

A complete understanding of superheat and subcooling allows the technician to know the state of the refrigerant anywhere in the system.

Superheat

Superheat describes vapor with a temperature *above* its saturation point. Heat is added *after* the refrigerant changes its state from liquid to vapor. The temperature of superheated vapor is higher than the corresponding pressure reading on the P/T chart. Digital manifold gauges will read superheat and subcooling, but older analog gauges will not. When checking system pressures, the pressure reading shown on the manifold gauge will always be *saturation* pressure. If the refrigerant is superheated, the actual temperature of the vapor will be above saturation temperature. For troubleshooting purposes, superheat is normally checked at two specific places in the refrigeration system:

- At the evaporator outlet, which is referred to as *TEV (thermostatic expansion valve) superheat* or *evaporator superheat*. Always follow the manufacturer's instructions on the proper superheat needed.
- At the suction inlet to the compressor, which is referred to as *system superheat* or *suction superheat*.

Superheat for Zeotropes

To determine superheat for zeotrope blends, use *dew point* values. Superheat is the number of degrees *above* the dew point temperature for a specific pressure. For example, with a suction pressure of 8 psig (55 kPa), the dew point temperature is 0°F (–18°C). If an accurate temperature reading at the evaporator outlet reads 10°F (–12°C), then the superheat equals 10°F (6°C) *above* dew point.

Subcooling

Subcooled describes a liquid having a temperature *below* the saturation point. Heat is removed *after* the refrigerant changes its state from vapor to liquid. The temperature of a subcooled liquid does not correspond to its saturation pressure. Refrigerant in the condenser changes from vapor to liquid and then begins to subcool. Subcooling takes place in the bottom of the condenser and in the liquid line. Subcooling at the condenser outlet is normally 10°F (6°C) but can range from 5°F to 15°F (3°C to 8°C).

Superheat and subcooling are important because they are required for proper operation of the refrigeration system. In addition to controlling the saturation point in the evaporator and the condenser, the amount of superheat and subcooling within a system must also be controlled. **Figure 15-22** depicts an air-conditioning system with gauges and thermometers to illustrate saturation, superheat, and subcooling within an R-22 air-conditioning system.

Subcooling for Zeotropes

To determine subcooling, use *bubble point* values. Subcooling is the number of degrees *below* the bubble point temperature for a specific pressure. For example, the discharge pressure for R-409A is 152 psig at 100°F (1049 kPa at 38°C). If an accurate temperature reading at the condenser outlet reads 90°F (32°C), then the amount of subcooling is 10°F (6°C) *below* the bubble point.

PROCEDURE

Calculating Superheat or Subcooling

The amount of superheat or subcooling can easily be determined at any point in the system by using a manifold gauge and thermometer. Follow this four-step procedure:

1. Obtain a pressure reading with the manifold gauge.
2. Convert the pressure reading to refrigerant saturation temperature.
3. Using the thermometer, obtain an accurate temperature reading of the copper tubing at the test site.

(Continued)

PROCEDURE

Calculating Superheat or Subcooling *(continued)*

4. Find the difference in temperature readings (above or below saturation temperature). The difference is the amount of superheat or subcooling, in degrees Fahrenheit.

The example in **Figure 15-22** shows the pressure reading at the evaporator outlet is 60 psig. By referring to a P/T chart for R-22, the saturation temperature at 60 psig is 34°F. A thermometer reading taken at the evaporator outlet shows a temperature of 44°F. The difference in the readings (44°F – 34°F = 10°F) is the amount of evaporator superheat, 10°F.

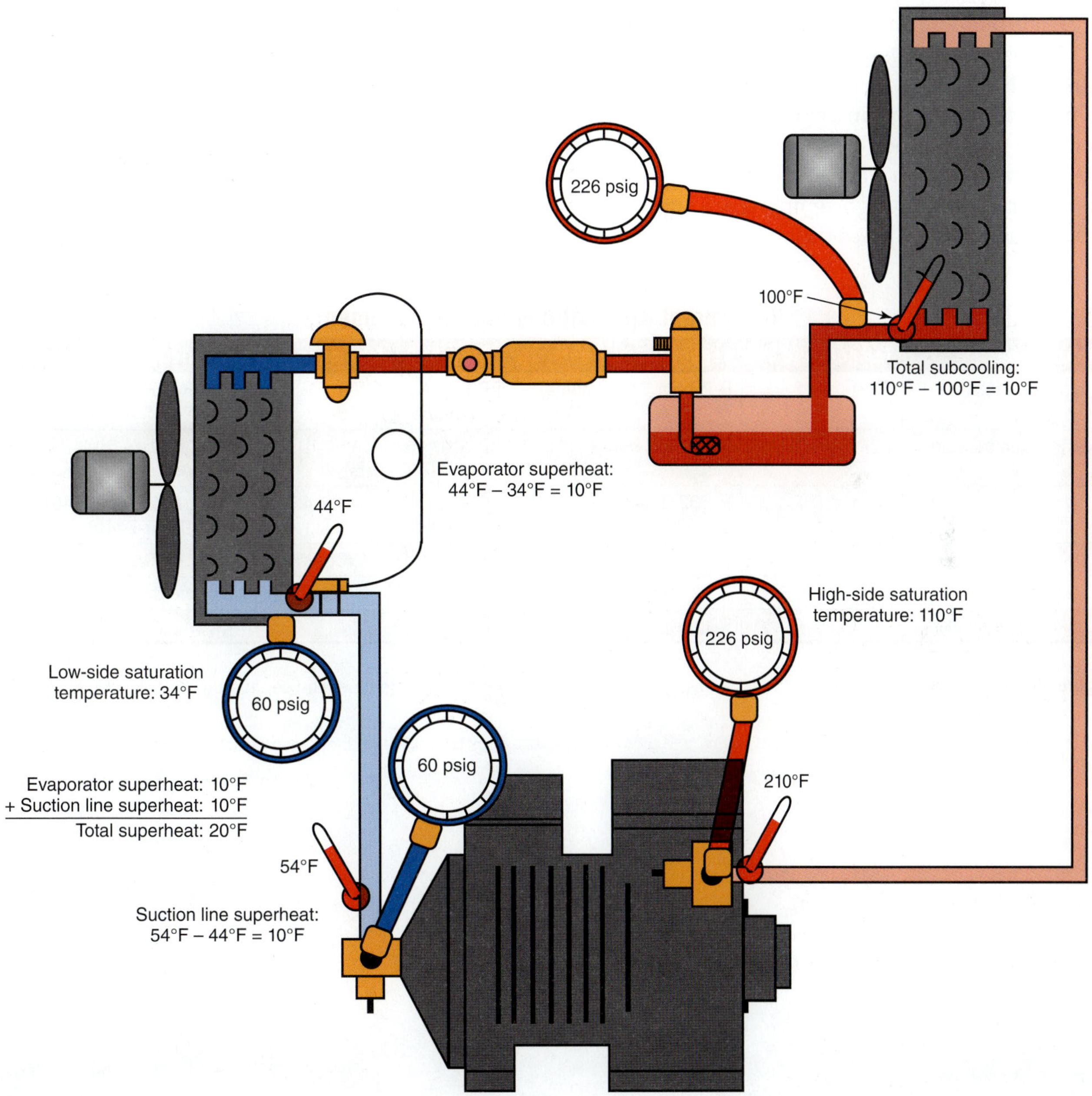

Goodheart-Willcox Publisher

Figure 15-22. This system uses R-22. Superheat or subcooling is checked by using a gauge and thermometer as shown. The difference between the thermometer and saturation temperature reveals the amount of superheat or subcooling.

15.6 Refrigerant Applications

For general purposes, refrigerants are categorized for operation within a specific temperature range or *application*. Some refrigerants operate best in high-temperature applications. Others are best in low-temperature applications. Still others are best at ultralow-temperature applications. **Figure 15-23** lists six applications (by temperature), the most popular refrigerants for each type of application, and their replacement refrigerants.

High-temperature applications primarily refer to air-conditioning systems (or comfort cooling). Air-conditioning systems are normally designed to maintain building interiors at Fahrenheit temperature levels in the 70s, which is considered high temperature for refrigeration systems. R-410A is a good refrigerant for high-temperature applications due to the low compression ratio but can be used for medium- or low-temperature applications with some equipment modifications.

Medium-temperature applications refer to domestic refrigerators, dehumidifiers, commercial walk-in coolers, watercoolers, dairy cases, meat cases, and produce cases. R-134a is an excellent medium-temperature refrigerant but is not practical for low-temperature uses. At low temperatures, R-134a must operate with the suction pressure in a vacuum and the Btu per pound (latent heat value) is quite low.

Low-temperature applications normally refer to frozen food cases, walk-in freezers, and ice cream storage. See **Figure 15-24**. R-507 is best for low-temperature applications because of such properties as low boiling point, azeotropic mixture, high suction pressure, low compression ratio, and low latent heat value.

Ultralow-temperature applications and cryogenic applications are special systems designed for use in research laboratories and industrial applications requiring extremely low temperatures.

Refrigerant Applications and Replacements

Application	Temperature	Legacy Refrigerant	Current Replacement	Possible Future Replacement
High temperature	55° to 90°F	R-22 (Air Conditioning)	R-22, R-410A (Air conditioning)	R-32, R-454B
Medium temperature	32° to 50°F	R-12 (Coolers)	R-401A, R-404A, R-407C, R-422D (Coolers)	R-290, R-407A, R-448A, R-449A, R-744
Low temperature	0° to –50°F	R-502 (Freezers)	R-401A, R-404A, R-422D (Freezers)	R-290, R-407A, R-448A, RR-449A, R-744
Ultralow temperature	–55° to –250°F	R-13 and R-503 (Cascade)	R-403B, R-744 (Cascade)	R-170, R-704
Cryogenics	–250° to –460°F	R-728 (Expendable)	R-728 (Expendable)	No ozone depletion
Low-pressure chillers	45° to 55°F	R-11 (Operates in vacuum)	R-123	R-513A, R-1234ze

Goodheart-Willcox Publisher

Figure 15-23. This table lists six application areas by temperature, popular refrigerants for each use, and their replacements.

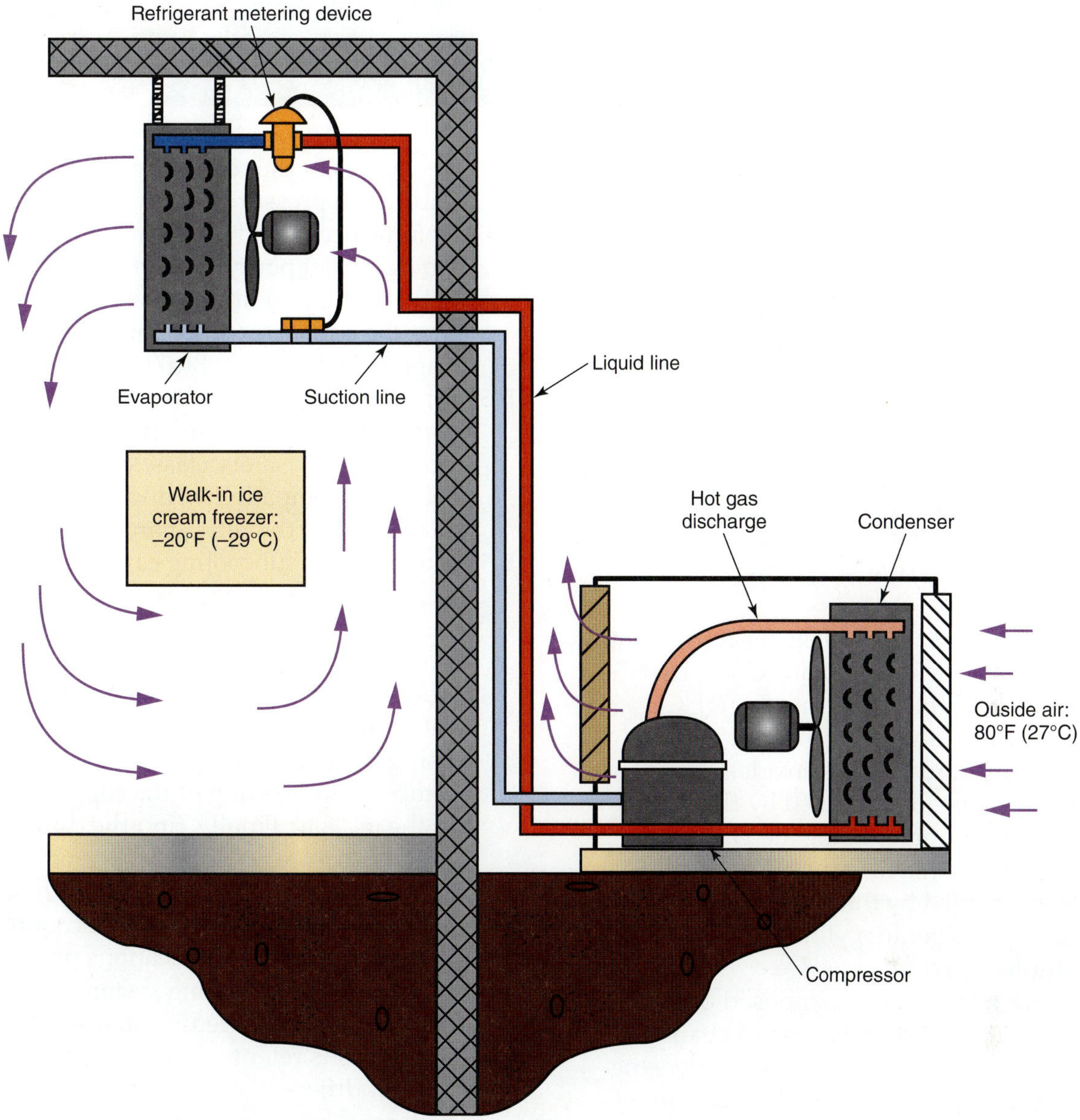

Goodheart-Willcox Publisher

Figure 15-24. Low-temperature application showing a walk-in ice cream freezer operating at –20°F (–29°C).

Chapter Review

Summary

15.1 The continual development of more efficient and environmentally safe refrigerants and advancements in technology have led to the refrigeration and air-conditioning systems seen today.

15.2 Section 608 of the Clean Air Act gives the EPA authority to set guidelines on proper refrigerant safety and handling. The EPA has targeted refrigerants with high ODPs and GWPs for phaseout to reduce the impact of HVACR systems on the environment.

15.3 The general classifications for refrigerants include halide refrigerants (CFCs, HCFCs, HFCs, and HFOs), hydrocarbon (HC) refrigerants, and inorganic refrigerants (ammonia and carbon dioxide).

15.3 Refrigerants are identified by the letter "R" and a number using the DuPont numbering system format. Refrigerants are also identified by their safety classification and the refrigerant cylinder they are manufactured in.

15.4 A pure refrigerant is composed of a single-component fluid that does not change composition when boiling or condensing. An azeotropic refrigerant is a mixture of two refrigerants that combine to form a third, unique refrigerant having its own individual characteristics. A zeotropic blend, or zeotrope, is a blend of two or more single-component refrigerants that behave differently than an azeotropic mixture.

15.5 The choice of refrigerant for a particular application depends upon several factors, including boiling point, toxicity (safety), flammability, latent heat value, suction pressure, condensing pressure, vapor density, oil compatibility, corrosiveness, and cost.

15.5 Pressure-temperature (P/T) charts help engineers compare refrigerants under various operating conditions. At a given temperature, the charts reveal important information regarding the physical properties of the refrigerant: the correct saturation pressure (boiling point), the volume occupied by the saturated vapor (in cubic feet per pound), the density of the saturated liquid (in pounds per cubic feet), and the enthalpy (heat content) of both vapor and liquid (in Btu/lb).

15.5 An HVACR service technician should be familiar with a refrigerant's composition, boiling point, safety class, system and lubricant compatibility, and other physical properties of the refrigerant.

15.5 Superheat or subcooling can be determined at any point in the system using a manifold gauge and thermometer. First, obtain a pressure reading with the manifold gauge. Then, convert the pressure reading to refrigerant saturation temperature and using the thermometer, obtain an accurate temperature reading of the copper tubing at the test site. Finally, find the difference in temperature readings (above or below saturation temperature). The difference is the amount of superheat or subcooling.

15.6 High-temperature applications primarily refer to air-conditioning systems (or comfort cooling). Medium-temperature applications refer to domestic refrigerators, dehumidifiers, commercial walk-in coolers, watercoolers, dairy cases, meat cases, and produce cases. Low-temperature applications normally refer to frozen food cases, walk-in freezers, and ice cream storage. Ultralow-temperature applications and cryogenic applications are special systems designed for use in research laboratories and industrial applications requiring extremely low temperatures.

Know and Understand

1. The first practical compression-cycle refrigeration system was built in _____. (15.1)
   A. 1834
   B. 1850
   C. 1873
   D. 1900

2. The _____ set restrictions and penalties for venting refrigerants to the atmosphere. (15.1)
   A. Montreal Protocol
   B. Clean Air Act
   C. AIM Act
   D. Kigali Amendment
3. The AIM Act targeted the phaseout of _____ refrigerants. (15.2)
   A. CFC
   B. HCFC
   C. HFC
   D. HFO
4. R-22 is an example of a _____ refrigerant. (15.3)
   A. CFC
   B. HCFC
   C. HFO
   D. HFC
5. One disadvantage of HFO refrigerants is that they _____. (15.3)
   A. have a high ODP
   B. have a high GWP
   C. are slightly flammable
   D. have limited applications
6. Following the ASHRAE and DuPont numbering system, how many hydrogen atoms are in R-32? (15.3)
   A. 0
   B. 2
   C. 3
   D. 4
7. *True or False?* HC-290 and R-290 refer to the same refrigerant. (15.3)
8. The cylinders of flammable refrigerants are marked with a(n) _____. (15.3)
   A. orange or red cylinder color
   B. red band around the cylinder
   C. tag near the data plate
   D. fire symbol in the middle of the cylinder
9. R-32 is an example of a(n) _____ refrigerant. (15.4)
   A. pure
   B. azeotropic mixture
   C. azeotrope
   D. zeotrope
10. Which is an example of an azeotropic refrigerant? (15.4)
    A. R-123
    B. R-1234ze
    C. R-454B
    D. R-502
11. Zeotropic blends do *not* have a specific pressure for each degree temperature change due to _____. (15.4)
    A. sublimation
    B. fractionation
    C. superheat
    D. subcooling
12. *True or False?* Zeotropic refrigerants should only be topped off if the system is below 20% charged with refrigerant. (15.4)
13. A refrigerant's latent heat value refers to the amount of _____ absorbed for each pound of liquid evaporated. (15.5)
    A. vacuum (in. Hg)
    B. contaminants
    C. heat (Btus)
    D. pressure (psig)
14. Standard conditions for refrigerants have been established as _____°F evaporating temperature and _____°F condensing temperature. (15.5)
    A. 5, 86
    B. 86, 5
    C. 32, 212
    D. 212, 32
15. Which of the following refrigerant information is *not* found on a refrigerant P/T chart? (15.5)
    A. Boiling point
    B. Pressure at a given temperature
    C. Temperature at a given pressure
    D. Chemical formula
16. The saturation temperature of refrigerant inside the evaporator is changed by raising and lowering the _____. (15.5)
    A. discharge pressure
    B. suction pressure
    C. liquid line pressure
    D. refrigerant metering device valve
17. Using **Figure 15-19** as a reference, calculate the approximate temperature glide of R-407C at 80 psig. (15.5)
    A. 5°F
    B. 10°F
    C. 15°F
    D. 25°F
18. *True or False?* R-454B has close operating pressure and temperatures when compared to R-410A. (15.5)

19. R-410A cannot be retrofit into R-22 systems mainly due to its _____. (15.5)
    A. high moisture absorption rate
    B. flammability
    C. higher operating pressure
    D. miscibility with R-22 lubricants
20. When calculating superheat for zeotropes, use _____ values. (15.5)
    A. bubble point
    B. dew point
    C. pressure above atmospheric
    D. vacuum pressure
21. Referring to the P/T chart in **Figure 15-16**, calculate the amount of subcooling for R-600a if the saturation temperature at 11.8 psig is 40°F and a temperature reading at the condenser outlet reads 32°F. (15.5)
    A. –8°F
    B. 5°F
    C. 8°F
    D. 12°F
22. Air-conditioning systems are categorized as a(n) _____-temperature application. (15.6)
    A. low
    B. medium
    C. high
    D. ultralow

Apply and Analyze

1. Explain why a refrigerant with a high ODP and high GWP is harmful to the environment. (15.2)
2. What is the basic composition of a halide refrigerant? (15.3)
3. Using the ASHRAE and DuPont numbering standard, list the chemical breakdown (number of fluorine, hydrogen, and carbon atoms) and safety classification of R-290. (15.3)
4. Explain the differences between azeotropic and zeotropic refrigerants. (15.4)
5. List some of the qualities that engineers consider when selecting a refrigerant. (15.5)
6. How can a service technician use a P/T chart be used to determine that a noncondensable is in the HVACR system? (15.5)
7. Give one example of each type of application: high-temperature, medium-temperature, and low-temperature. (15.6)

Critical Thinking

1. Review the brief history of refrigerants outlined in the text. Analyze the trends throughout history and speculate on what changes you think will occur in the future of refrigerants used in the HVACR industry. (15.1)
2. Refer to the P/T chart in **Figure 15-16** and determine the gauge pressures for R-22, R-134a, and R-410A at 20°F and for R-1234yf and R-600a at 10°F. (15.5)

JB Industries

16 Refrigerant Recovery and Recycling

LEARNING OUTCOMES

After studying this chapter, you will be able to:

16.1 Recognize venting and leak repair provisions enforced by the EPA.

16.2 Explain recovery, recycle, and reclaim procedures used when working with refrigerant.

16.3 Identify proper refrigerant disposal and record-keeping procedures.

16.4 State technician and equipment certification requirements.

16.5 Recognize recovery equipment and procedure requirements.

16.6 Perform recovery procedures using the passive recovery method and the active recovery method.

16.7 Describe how recycle units are designed to clean refrigerant.

16.8 Identify refrigerant capturing methods.

TECHNICAL TERMS

active recovery
capture
EPA certification
hazardous waste
low-loss fitting
motor vehicle air conditioning (MVAC)
passive recovery
pumpdown
R&R system
reclaim
recover
recovery cylinder
recycle
venting

Introduction

When working on HVACR systems, service technicians often must recover and recycle refrigerants for disposal or reuse. Part of an HVACR service technician's responsibility is to follow proper EPA guidelines to avoid releasing harmful amounts of refrigerant into the atmosphere.

16.1 Refrigerant Venting and Leaks

Venting refers to releasing *any* refrigerants to the atmosphere. The Clean Air Act of 1990 established regulations for refrigerant venting and is enforced by the Environmental Protection Agency (EPA).

Failure to comply with the EPA regulations could cost a technician and their company as much as $44,539 per day per violation. Service technicians who violate the Clean Air Act provisions may be fined, lose their certification, and be required to appear in federal court.

It is a violation of Section 608 of the Clean Air Act to:

- Falsify or fail to keep required records.
- Fail to reach the required recovery vacuum level prior to opening or disposing of appliances. An "appliance" refers to any refrigeration or air-conditioning system, regardless of size.
- Knowingly release (vent) regulated refrigerants while servicing, maintaining, disposing of, or repairing appliances.
- Service, maintain, or dispose of appliances designed to contain refrigerants without being appropriately certified.
- Fail to become certified (effective November 14, 1994).
- Fail to have an EPA-approved recovery device.
- Add nitrogen to a fully charged system for the purpose of leak detection and thereby cause a release of the mixture.
- Discard a disposable cylinder without recovering any remaining refrigerant. Doing this renders the cylinder useless.
- Fail to possess appropriate and approved recovery and recycling equipment with low-loss fittings.

Some state and local government regulations may contain additional or stricter regulations than those stipulated in Section 608 of the Clean Air Act. In case of conflict between the Clean Air Act and the Montreal Protocol, the more stringent law regarding the issue takes precedence.

Venting

The Clean Air Act prohibits individuals from knowingly venting ozone-depleting compounds used as refrigerants into the atmosphere.

Only four types of release are permitted:

- *De minimus* (minor) quantities of refrigerant released while making good faith attempts to recapture and recycle or safely dispose of refrigerant.
- Refrigerants emitted during normal operation of air-conditioning and refrigeration equipment, such as mechanical purging and leaks. However, the EPA requires the repair of substantial leaks.
- Mixtures of nitrogen and R-22 used as holding charges or leak test gases. However, a technician may not avoid recovering refrigerant by adding nitrogen to a charged system.
- Small releases of refrigerant resulting from purging hoses or from connecting or disconnecting hoses.

Refrigerant Leaks

Owners of equipment with charges over 50 lb are required to repair substantial leaks. Substantial leaks are defined as annual leak rates that exceed the following percentages:

- 30% of full charge for industrial process systems
- 20% of full charge for commercial refrigeration
- 10% of full charge for comfort cooling

When a substantial leak is discovered, the owner or operator must take corrective action. Either the leak must be repaired, or the owner or operator must develop a plan to retrofit or retire the equipment within one year. The repair must be made (or the plan must be developed) within 30 days of the discovery of the leak. If repair of an industrial system requires a process shutdown, the deadline for corrective action is extended to 120 days. Repair deadlines can also be extended if parts required for the repair cannot be readily obtained or if other regulations delay the repair.

After a substantial leak is repaired, initial testing is required to ensure that the repair was executed correctly. Additional inspections are required annually until the owner can demonstrate through leak rate calculations that the leak no longer exceeds the trigger amount. For systems with a charge over 500 lb, these follow-up inspections must be performed every three months.

16.2 The "Three Rs:" Recover, Recycle, Reclaim

The type of system, the condition of the refrigerant, and other factors should be considered when determining what to do with a system's refrigerant charge. There are three main procedures when dealing with a refrigerant charge:

- Recover
- Recycle
- Reclaim

Recover

Recover means to remove refrigerant in any condition from a system and to store it in an approved cylinder or external container. It is illegal to purge or vent refrigerants to the atmosphere. Recovery equipment is required to avoid releasing refrigerant to the atmosphere when making repairs or disposing of refrigeration equipment, **Figure 16-1**. Recovered refrigerant is normally returned to the same system after repairs are finished. Recovery does not address the quality or reuse potential of the refrigerant. Cleaning, processing, and testing are *not* part of the recovery process. Recovered refrigerant may contain acids, moisture, and oil.

Recovery must be performed:

- Before the system is opened for repairs
- Before disposing of any system or components containing refrigerant
- When removing excess refrigerant from a system
- Before pressurizing a system for leak detection with an inert gas

JB Industries

Figure 16-1. Refrigerant recovery equipment.

Recycle

Recycle means to clean refrigerant for reuse by removing moisture and contaminants from recovered refrigerant. Recycling is accomplished by repeatedly passing the recovered refrigerant through one or more filter-driers, **Figure 16-2**.

A recycle unit uses a liquid pump (not a compressor) to pull liquid from the bottom of a cylinder via the dip tube (liquid valve). The liquid refrigerant is circulated through a replaceable core filter-drier and returned to the top of the cylinder (vapor valve). Recirculation of the liquid refrigerant through the filter-drier removes moisture and contaminants, which may require several hours to complete.

The cleanliness of recycled refrigerant does *not* have to meet the specifications for new refrigerant, and chemical tests are *not* required. EPA regulations limit the use of recovered and/or recycled refrigerant to the *same system* or another system belonging to the *same owner*. Recovered or recycled refrigerant cannot change ownership by being sold or given away.

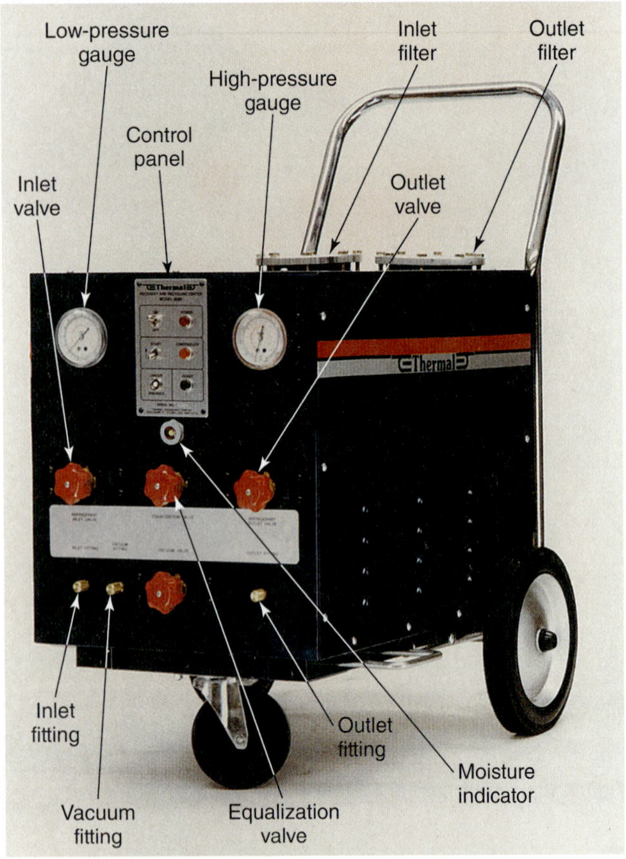

A

Thermal Engineering

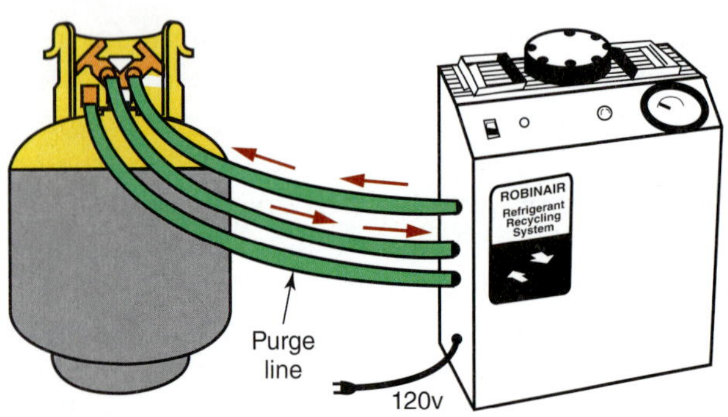

B

Robinair Mfg. Corp.

Figure 16-2. Recycle units and procedure. A—A recycle unit removes contaminants from a refrigerant to allow reuse. B—A recovered refrigerant charge is recycled outside of the refrigeration system. A liquid pump circulates liquid through filters and returns it to the cylinder.

Reclaim

Reclaim means to restore the recovered refrigerant to a level equal to new (virgin) product specifications as determined by chemical analysis. Refrigerant that is *recovered* or *recycled* can be returned to the *same system* or *other systems* owned by the *same person* without restriction. However, if a refrigerant is to change ownership, that refrigerant must be reclaimed. Reclaiming requires cleaning the refrigerant to the Air Conditioning, Heating, and Refrigeration Institute (AHRI) 700 standard of purity and chemically analyzing the restored refrigerant to verify that it meets the standard. Reclaiming is a process normally available only at a manufacturing or reprocessing facility, **Figure 16-3**.

16.3 Safe Disposal and Record-Keeping Requirements

Equipment that is typically dismantled on site before disposal (commercial and industrial equipment) must have the refrigerant recovered in accordance with EPA requirements for servicing. Equipment that typically enters the waste stream with the charge intact (auto air conditioners, domestic refrigerators, room air conditioners) is subject to certain safe disposal requirements.

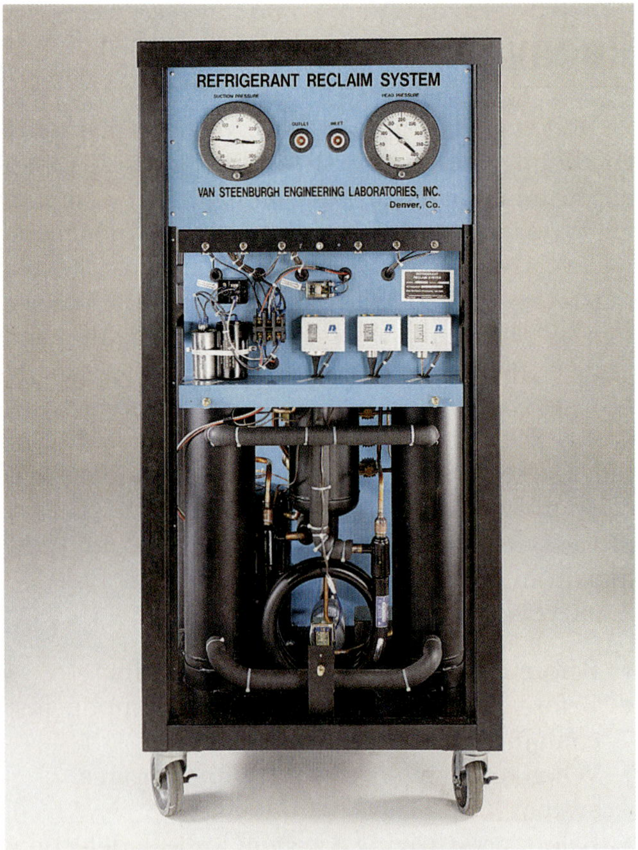

Van Steenburgh Engineering Laboratories, Inc.

Figure 16-3. A reclaim unit is used to restore refrigerant to its original purity.

The final person in the disposal chain (junkyard or landfill owner) is responsible for ensuring refrigerant is recovered from equipment before the equipment's final disposal. However, persons "upstream" may remove the refrigerant and provide documentation of its removal to the final person if it is more cost-effective.

Hazardous Waste Disposal

Recycled or reclaimed refrigerants are *not* considered ***hazardous waste*** under federal law. Likewise, used oils contaminated with CFCs are not classified as hazardous provided they are:

- Not mixed with other waste
- Subjected to CFC recycling or reclamation
- Not mixed with used oils from other sources

Individuals with questions regarding the proper handling of used oils should contact the EPA for guidance.

Record-Keeping Requirements

Technicians, appliance owners, refrigerant wholesalers, and reclaimers all must keep accurate records of services and transactions involving refrigerants.

Technicians servicing appliances that contain 50 or more pounds of refrigerant must provide the owner with an invoice indicating the amount of refrigerant added to the system. Technicians must also provide records of leak inspections and tests performed to verify repairs of leaks. Technicians must also keep proof of EPA certification(s) on them when repairing, servicing, or disposing of refrigerants.

When disposing of a system containing between 5 lb and 50 lb of refrigerant, technicians must keep records of the disposal, including the following information:

- Location of recovery
- Date of recovery
- Quantity of refrigerant recovered, by type
- Monthly totals of amounts recovered
- Amounts sent for reclamation or disposal, by type

Owners of appliances that contain 50 or more pounds of refrigerant must keep service records documenting the date and type of service (including leak inspections and repairs), as well as the quantity of refrigerant added. If a system leaks over 125% of its full charge in a year, the owner is required to submit a report to the EPA. The report must detail the corrective action taken to identify and repair the leak.

Wholesalers who sell HCFC refrigerants must retain invoices that indicate the name of the purchaser, the date of sale, and the quantity of refrigerant purchased. Since 1994, sale of these refrigerants has been restricted to technicians certified in refrigerant recovery.

Reclaimers must maintain records of the names and addresses of persons sending them material for reclamation and the quantity of material sent to them for reclamation. At the end of each calendar year, the reclaimer must submit an annual activity report to the EPA.

16.4 Certification Requirements

Not only do technicians need certification, so does the equipment the technicians are using. The following sections describe different technician certifications, testing information, and equipment certification.

Motor Vehicle Air Conditioning

Section 609 of the Clean Air Act requires all persons who service ***motor vehicle air conditioning (MVAC)*** units be certified in refrigerant recovery and recycling and proper use of approved equipment when performing service involving refrigerant. Servicing includes repairs, leak testing, and "topping off" air-conditioning systems low on refrigerant, as well as other vehicle repairs that require dismantling the air conditioner. MVAC certification can be obtained through a home-study program that can be taken online. The test is open-book.

Technician Certification

Section 608 of the Clean Air Act mandates that all persons who maintain, service, repair, or dispose of refrigeration and air-conditioning appliances (other than motor vehicle systems) must become certified by passing an EPA-approved test. The test is administered by an EPA-approved certifying organization and is monitored by an EPA-approved proctor. Each test contains a different set of questions. A passing score of 70% is required to obtain certification.

If EPA regulations change after a technician is certified, it is the technician's responsibility to comply with any future changes. Failure to comply with EPA regulations can cost a technician and their company as much as $44,539 per day, per violation.

There are four categories of EPA technician certification:

Type I

Technicians servicing small appliances must be certified in refrigerant recovery if they perform sealed-system service. The EPA definition of "small appliances" includes products manufactured, charged, and hermetically sealed in a factory with 5 lb or less of refrigerant. Persons recovering refrigerant during maintenance, service, or repair of small appliances (except motor vehicle air-conditioning systems) must be certified as either a Type I Technician or Universal Technician. Some EPA-approved organizations are offering Type I certification only, as an open-book, online test. The online test requires a passing score of 84%.

Type II

Technicians maintaining, servicing, repairing, or disposing of high-pressure or extremely high-pressure (R-134a, R-22, R-404A, R-410A, R-717) appliances (except small appliances and motor vehicle air-conditioning systems) must be certified as a Type II Technician or Universal Technician.

Type III

Technicians maintaining, servicing, repairing, or disposing of low-pressure (R-11 and R-123) appliances must be certified as a Type III Technician or Universal Technician.

Universal

Technicians maintaining, servicing, or repairing low- and high-pressure equipment, including small appliances (except motor vehicle air-conditioning systems) must be certified as a Universal Technician.

Test Format

Most certification tests contain all four sections (Core, I, II, III), and each section contains 25 multiple-choice questions. A minimum passing score of 70% (18 out of 25 correct) is required in *each section* (or group). A technician is certified in each group where a passing score is achieved. However, a technician *must* achieve a minimum passing score in the Core Section to receive any type of certification. A minimum passing score in all four sections earns a technician Universal certification.

- *Core Section* covers general knowledge about environmental impacts, rules and regulations of Section 608, the Clean Air Act, the Montreal Protocol, the "Three Rs," recovery devices, substitute refrigerants and oils, recovery techniques, dehydration, recovery cylinders, safety, shipping, and the refrigeration system.
- *Section I* contains questions pertaining to Type I certification such as recovery requirements and techniques and safety.
- *Section II* contains questions pertaining to Type II certification such as leak detection, leak repair requirements, recovery requirements and techniques, the refrigeration system, and safety.
- *Section III* contains questions pertaining to Type III certification such as leak detection, leak repair requirements, recovery requirements and techniques, leak repair requirements, recharging requirements, the refrigeration system, and safety.

The test is closed book, with no time limit for completion. Some personal information is required for the examination. The technician should be prepared to present:

- Picture identification
- Social Security number and date of birth
- Home address, including zip code

EPA-approved testing organizations are located throughout the United States. Check with a local air-conditioning and refrigeration parts supplier to obtain information about the testing organization(s) in the area. Most EPA-approved testing organizations provide the technician with a preparatory manual before taking the test.

A technician with an ***EPA certification*** means they have the required knowledge in refrigerant recovery techniques and agrees to abide by the EPA regulations prohibiting venting of refrigerants to the atmosphere. EPA certification does *not* mean you are qualified or knowledgeable in the proper repair, service, and installation of refrigeration and air-conditioning equipment. Job certification skills training is offered by industry organizations.

Equipment Certification

The EPA requires recovery units manufactured on or after November 15, 1993, be tested by an EPA-approved testing organization to ensure the equipment will achieve the required vacuum levels shown in the third column of **Figure 16-4**.

The EPA requires that persons servicing or disposing of air-conditioning and refrigeration equipment certify to the EPA that they have acquired (built, bought, or leased) recovery equipment and are complying with the applicable requirements. The certification must be signed by the owner of the equipment or another responsible officer and sent to the appropriate EPA regional office.

16.5 Recovery Requirements

Technicians are required by law to use recovery systems to evacuate air-conditioning and refrigeration equipment to established vacuum levels. If the recovery equipment was manufactured *before* November 15, 1993, the air-conditioning and refrigeration system must be evacuated to the levels indicated in **Figure 16-4**. If the recovery equipment is manufactured on or *after* November 15, 1993, the air-conditioning and refrigeration system must be evacuated to the levels indicated in **Figure 16-4** *and* the recovery equipment must be certified by an EPA-approved equipment testing organization. Because of the shift to lower GWP refrigerants (A2Ls), recovery equipment must be approved for service of A2L refrigerant and equipment.

Technicians repairing small appliances, such as household refrigerators, household freezers, and watercoolers, are required to recover 80%–90% of the refrigerant in the system, depending on the status of the system's compressor.

Exceptions to Recovery Requirements

If evacuation to the required levels is not attainable due to leaks, or if it would substantially

Recovery Requirements

Type of Appliance	Inches Mercury (in. Hg) Vacuum Using Equipment Manufactured Before 11/15/93	Inches Mercury (in. Hg) Vacuum Using Equipment Manufactured On or After 11/15/93
High-pressure appliance normally containing *less* than 200 pounds of refrigerant	0	0
High-pressure appliance normally containing *more* than 200 pounds of refrigerant	4	10
Medium-pressure appliance normally containing *less* than 200 pounds of refrigerant	4	10
Medium-pressure appliance normally containing *more* than 200 pounds of refrigerant	4	15
Very high-pressure appliance	0	0
Low-pressure appliance	25 mm Hg absolute	25 mm Hg absolute

Goodheart-Willcox Publisher

Figure 16-4. Required vacuum levels are based on the type of refrigeration appliance and the manufacture date of the recovery equipment.

contaminate the refrigerant being recovered, the person opening the system must:

- Isolate leaking components from nonleaking components, wherever possible.
- Evacuate nonleaking components to the required levels.
- Evacuate leaking components to the lowest attainable level without substantially contaminating the refrigerant. The level cannot exceed 0 psig (0 kPa).

Recovery Cylinders

Recovery cylinders are heavy-duty refillable cylinders that must meet Department of Transportation (DOT) specifications. The color code for recovery cylinders is a gray bottom with yellow shoulders and top. Recovery cylinders designated to contain A2L refrigerant must have a red band around the cylinder, **Figure 16-5**. Avoid mixing refrigerants in a recovery cylinder. Label each recovery cylinder permanently and clearly. Always carry at least one refillable recovery cylinder for each type of refrigerant commonly used. Otherwise, when using only a single recovery cylinder, it must be emptied and evacuated before each use.

> **CAUTION**
>
> **Damaged Cylinders**
>
> Do *not* use dented, rusted, gouged, or damaged cylinders. Examine the cylinder valve assembly for leakage, damage, or tampering.

Refillable cylinders must be hydrostatically retested and recertified every five years from the test date stamped on the cylinder shoulder, in accordance with DOT Title 49 CFR, Section 173.34(e) and 173.31(d). Do *not* fill a container that is out of date (five years or older). Return it empty and have it retested.

Recovery cylinders have two valves, **Figure 16-6**. One valve is marked "liquid" and the other "gas." The liquid valve has a dip tube that reaches to the bottom of the cylinder, making it possible to remove liquid without turning the cylinder upside down.

Some manufacturers require special recovery cylinders for use on their recovery units. The cylinders contain built-in safety devices to stop the recovery unit when the cylinder is full. The sensing devices are located inside the steel cylinder and consist of solid-state thermistors and float switches. The reliability of these automatic shutoff devices is questionable. All recovery cylinders should be inspected carefully and placed on an

Manchester Tank *Worthington Enterprises*

Figure 16-5. Recovery cylinders are easily identified by their color coding: yellow top and gray bottom. Recovery cylinders for A2L refrigerants have a red band around them.

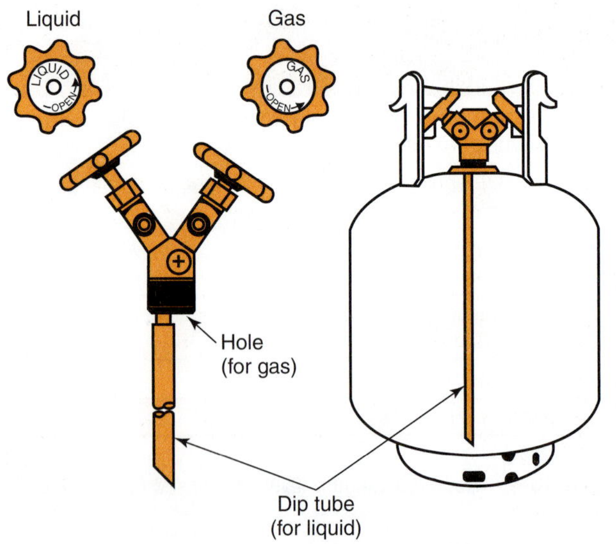

Goodheart-Willcox Publisher

Figure 16-6. Recovery cylinders have two valves: one marked "liquid" and the other marked "gas."

accurate scale before being filled. The technician should monitor the recovery process and stop the recovery when the required recovery vacuum has been achieved, when the scale shows the cylinder is full, or when leaks prevent recovery to the proper vacuum level.

Recovery cylinders are considered *full* at 80% of total cylinder capacity. The capacity rating of a recovery cylinder designates the maximum volume of liquid refrigerant (in pounds) the cylinder can safely hold. The capacity rating is stamped on the cylinder and sometimes reads "W.C." for *water column*. Cylinder manufacturers use water instead of refrigerant to determine cylinder capacity by weight. *Never* exceed a cylinder's weight capacity rating for any reason. Always use a refrigerant scale when recovering refrigerant into a cylinder, **Figure 16-7**.

Mastercool, Inc.

Figure 16-7. Electronic refrigerant scales allow accurate weighing of refrigerant. The service technician shown is recovering refrigerant from the condensing unit of an air-conditioning system.

SAFETY NOTE

Hydrostatic Pressure

All liquids expand when heated. The expanding liquid inside a cylinder creates hydrostatic pressure. Hydrostatic pressure in an overfilled refrigerant container can cause the cylinder to burst. At best, equipment could be damaged. At worst, someone could be killed by the explosion. Exercise great care when filling cylinders and avoid overfilling to prevent equipment damage and physical injury.

Recovery Equipment

Refrigerant recovery equipment is used to *avoid venting refrigerant into the atmosphere* when making repairs to a refrigeration system. Recovery equipment is essentially a refrigeration system designed to remove and store refrigerant in a *refillable* cylinder, **Figure 16-8**.

After recovery, refrigerant may be returned to the system from which it was removed or added to another system owned by the same person or company, without the refrigerant being *recycled* or *reclaimed*. While recovered refrigerant can simply be returned to the system from which it was originally withdrawn, it should be cleaned of any contamination before use in *another* system owned by the same person.

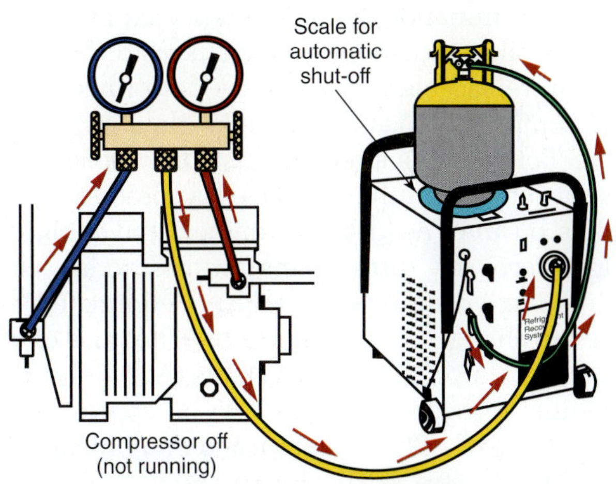

Goodheart-Willcox Publisher

Figure 16-8. Recovery equipment is a refrigeration system designed to remove and store liquid refrigerant in a refillable cylinder.

Some recovery machines are capable of removing liquid or vapor, while others remove vapor only. Removal of refrigerant in the liquid state is less time-consuming than removal of vapor. Regardless, a technician should always check system components to ensure all refrigerant has been removed. For example, an accumulator may contain extra refrigerant along a frost line that needs additional recovery time for the refrigerant to boil off or condense before being recovered.

Refrigerant recovery is accomplished by connecting a manifold gauge to the suction and discharge service valves, then connecting the center (yellow) hose from the manifold to the recovery unit. The system service valves are opened, the manifold gauge valves are opened, and the hoses are purged of atmospheric air. The center (yellow) hose serves as the suction line for withdrawing vapor from the system. A recovery unit includes an evaporator, compressor, condenser, various safety devices, and a separate refillable cylinder for liquid storage.

As mentioned earlier, most recovery units are able to remove refrigerant in the liquid or vapor state, though liquid removal is much faster. However, some recovery units are rated according to pounds of liquid recovered per minute, even when the unit only removes vapor. Older recovery systems contain compressors that use oil, so the recovery unit requires oil separators because some oil is removed from the system along with the refrigerant. Oil separators are not totally efficient. Consequently, some contaminated oil appears in the recovery cylinder. The contaminated oil, in turn, contaminates the recovery compressor oil. The contamination should be removed by replacing the oil in the compressor crankcase.

Oil replacement can only be performed when the compressor has an oil sight glass and drain plug. Before using a recovery unit to process a different refrigerant than the one last processed, the unit's oil may need to be drained and replaced. Most recovery units manufactured today contain oil-less compressors and have eliminated the need to change oil. See **Figure 16-9**. Because of the change to lower GWP refrigerants (A2Ls), the EPA requires recovery equipment to be approved for the service of A2L refrigerant and equipment.

The EPA also requires the use of ***low-loss fittings*** on recovery units and manifold hose ends. These fittings prevent loss of refrigerant to the atmosphere when connecting or disconnecting hoses from the recovery unit to a refrigeration system. The trapped refrigerant inside the recovery unit acts as a "holding charge" and prevents contamination by air. This is not a problem if the recovery unit is used for just one type of refrigerant. However, if it is used for different refrigerants, the recovery unit must be evacuated with a vacuum pump before use. Evacuation prevents contamination in the cylinder by two refrigerants.

16.6 Recovery Procedures

It is important to recover only one type of refrigerant into a recovery cylinder. Each recovery cylinder should be designated and properly marked for a specific refrigerant type. *Do not mix refrigerants in a recovery cylinder.* A mixture may be impossible to reclaim. If a reclaim facility receives a cylinder of mixed refrigerant, it may not be able to reclaim the mixture. The facility may agree to destroy the refrigerant, but it will typically charge a fee for the service.

The length and diameter of the hose connecting the recovery machine to the system affects the efficiency and time required for the recovery process. *Long hoses* and *small diameters* cause *excessive pressure drop* and *increased recovery time.* All refrigerants respond to a pressure-temperature relationship. Cold ambients result in low pressure, which results in a thin vapor and slow recovery rate. At low pressure, one pound of refrigerant vapor occupies several cubic feet of space. Likewise, high temperature results in high pressure, which results in a thicker vapor and faster recovery rate. At high pressure, one pound of refrigerant vapor occupies a much smaller volume.

With these interdependent pressure-temperature principles in mind, use short hoses with large diameters when recovering refrigerant. Also, to prevent high pressure from slowing recovery in the recovery cylinder, consider placing the recovery cylinder in a large bucket of ice and water. The low temperature of the cold water will cause lower

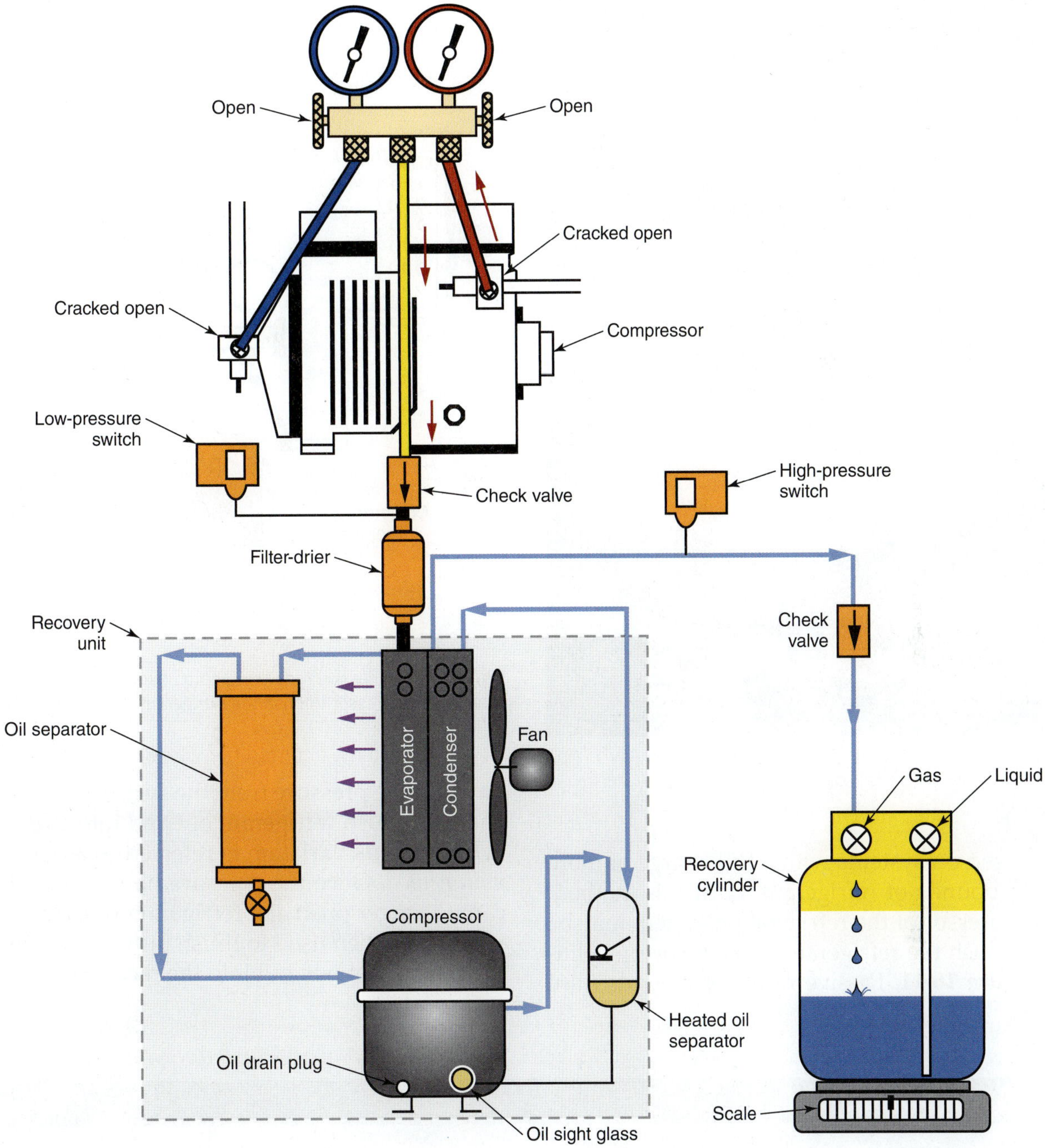

Goodheart-Willcox Publisher

Figure 16-9. Recovery units now come with oil-less compressors, eliminating the need to change the oil in a recovery unit.

pressure in the cylinder. This will allow liquid refrigerant to flow in more quickly and prevent excessive high pressure from building up inside the cylinder, **Figure 16-10**.

The two most common methods of recovering refrigerant from mechanical refrigeration systems are using passive recovery and active recovery.

Passive Recovery Method

The ***passive recovery*** method is used on a refrigeration system that has a higher static pressure than the recovery cylinder being used to store the charge in. This method is also referred to as the *system-dependent* method because it may

Ritchie Engineering – YELLOW JACKET Products Division

Figure 16-10. Recovery cylinder in an ice bucket for faster recovery time.

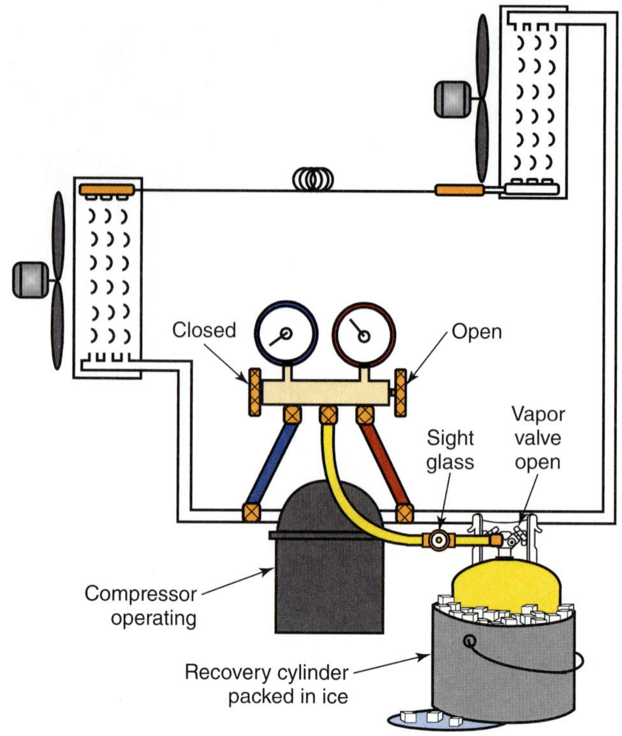

Goodheart-Willcox Publisher

Figure 16-11. Passive recovery method using a refrigeration system's compressor to push the refrigerant into the recovery cylinder.

not be used on systems with a full charge of more than 15 pounds of refrigerant. With this method, the compressor of the refrigeration system may be used to push the refrigerant into a recovery cylinder, **Figure 16-11**. Passive recovery may also use any positive pressure from the system's refrigerant charge to push refrigerant out and into a recovery cylinder. This variation is referred to as the *push-pull method*. Once the pressure has equalized and can no longer push any refrigerant out, a recovery machine is then used to finish pulling the remaining refrigerant out and into the recovery cylinder.

PROCEDURE

Recovering Refrigerant with the Passive Recovery Method

The following procedure demonstrates the passive refrigerant recovery method:

1. Connect the low-side manifold gauge hose to the suction service valve of the system. This will be for monitoring low-side pressure during recovery.
2. Connect the high-side manifold gauge hose to the liquid line service valve in the liquid line or on the receiver. This will be for recovering liquid refrigerant.
3. Place the recovery cylinder on a refrigerant scale and perform any scale operations necessary for monitoring the refrigerant being recovered, **Figure 16-12**.

SAFETY NOTE

Monitoring Refrigerant Recovery

Always use a refrigerant scale to measure the weight of the recovery cylinder, regardless of the type of recovery being used. Never overfill a recovery cylinder.

(Continued)

PROCEDURE

Recovering Refrigerant with the Passive Recovery Method *(continued)*

4. Install a sight glass between the hose attached to the center port of the manifold gauge. This will later connect to the recovery cylinder. The sight glass will show the flow of liquid refrigerant as it enters the cylinder during recovery.
5. Purge the air from the hoses and then connect the center hose with sight glass to the recovery cylinder.
6. Start up the refrigeration system.
7. Open the valve on the recovery cylinder and then open the high-side hand valve on the manifold gauge.
8. With the liquid line service valve opened, liquid refrigerant from the liquid line should flow through the manifold gauge and into the recovery cylinder.
9. When as much of the liquid refrigerant as possible has been removed, shut the hand valve on the manifold gauge and shut off the valve on the recovery cylinder. After the liquid has been removed, move on to the next method of recovery: the active recovery method.

JB Industries

Figure 16-12. Always use a refrigerant scale when recovering refrigerant.

Active Recovery Method

The ***active recovery*** method uses stand-alone recovery equipment and is used on refrigeration systems that do *not* have a working compressor or had refrigerant removed by the passive recovery method to an equalized pressure. See **Figure 16-13**. Active recovery should be used on systems that hold more than 15 pounds of refrigerant. Note that the active recovery method can be used on systems with less than 15 pounds of refrigerant, but the passive recovery method *cannot* be used on systems with more than 15 pounds of refrigerant.

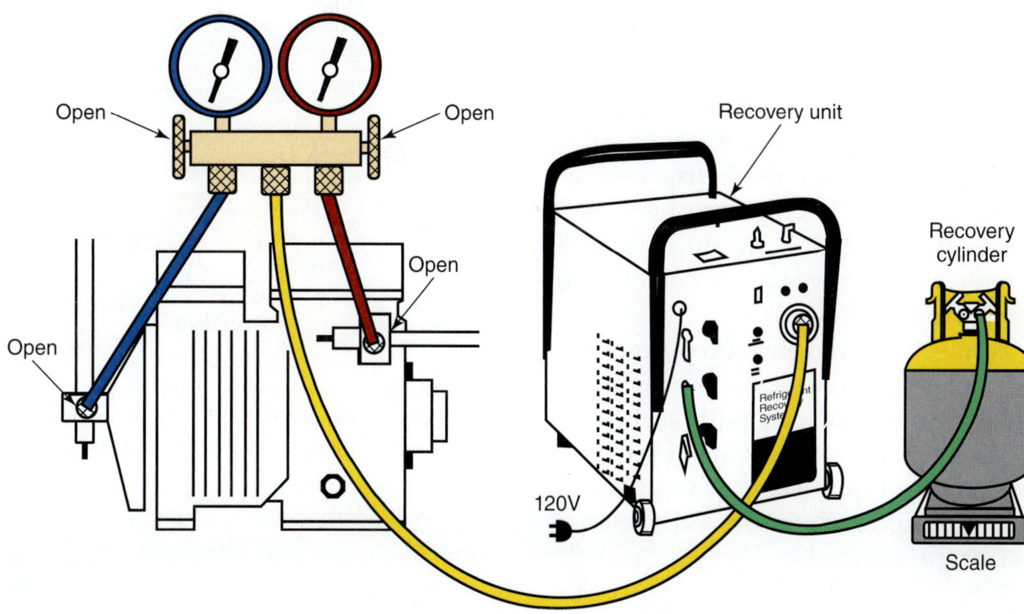

Goodheart-Willcox Publisher

Figure 16-13. Active recovery using a self-contained, independent recovery system. Refrigerant is withdrawn from the refrigeration system as a vapor. It is condensed into a liquid and then stored in a recovery cylinder.

PROCEDURE

Recovering Refrigerant with the Active Recovery Method

The following procedure demonstrates the active refrigerant recovery method:

1. Close both manifold gauge hand valves for now.
2. If recovering from refrigeration equipment: With the service valve initially closed (backseated), connect the manifold gauge hoses to the suction and discharge or liquid line service valves.
3. If recovering from refrigeration equipment: Crack open the service valves.
4. Connect the manifold gauge center hose to the recovery unit intake valve.
5. Connect a separate hose from the recovery unit output valve to the recovery cylinder vapor valve.
6. Carefully purge all hoses.
7. Place the recovery cylinder on a refrigerant scale and perform any settings necessary for monitoring the recovery process.
8. Open the recovery cylinder vapor valve, the recovery unit valves, and the manifold gauge hand valves.
9. Start the recovery unit and monitor the pressure gauges and refrigerant scale.
10. When an appropriate vacuum level has been reached, the recovery is complete. Check the system being recovered and the recovery equipment for appropriate vacuum level values.

In all methods of recovery, it is best to try to remove as much liquid refrigerant as possible before recovering the rest of the refrigerant as a vapor. If the refrigerant is only recovered as a vapor, the process will take a long time because the liquid refrigerant will have to first boil off into a vapor before it can be removed.

Recovery connections and start-up procedures may vary among manufacturers. Follow the manufacturer's instructions provided with each unit. Connections and procedures will vary if refrigerant is removed in the liquid state. Always use a recovery cylinder designated for use with the type of refrigerant being recovered.

16.7 Recycle Equipment

Recycle units remove contaminants from refrigerant prior to returning the refrigerant to the system from which it came or to other equipment belonging to the same owner. Recycle units contain a liquid pump that withdraws liquid refrigerant from the bottom of the recovery cylinder, circulates it through a large filter-drier, and returns the liquid to the top of the recovery cylinder. Most recycle units use one or more replaceable-core filter-driers for cleaning the refrigerant. It may take four to six hours of recirculation (including filter element changes) to remove sufficient moisture, acid, and particulate matter before the refrigerant is suitable for reuse. See **Figure 16-14** for a diagram of recycle unit operation.

The cleanliness of recycled refrigerant does *not* meet the specifications for *new* refrigerant, and chemical tests are *not* required. Some recycle units have built-in methods to determine the acid and moisture content of the recycled liquid refrigerant. Individual testing kits are available for performing acidity and moisture tests, if desired. Recycling is done at the jobsite or the local service shop. Recovery and recycle units are often housed in one system, called an ***R&R system***, that performs both operations.

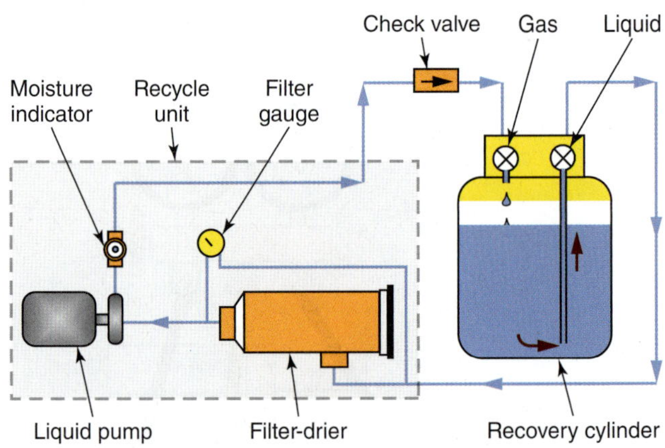

Figure 16-14. A pump draws liquid refrigerant from the bottom of the cylinder and circulates it through a replaceable-core filter-drier to remove moisture and contaminants from the refrigerant.

16.8 Capture Methods

In some repair situations, the refrigerant can be stored temporarily in another part of the system not in need of repair. For example, a system that has a liquid receiver with two service valves (inlet and outlet) can store the refrigerant while the area in need of repair is opened. See **Figure 16-15**. The procedure is not considered a recovery method because it does not remove the refrigerant to a container *outside* the system. It ***captures*** (traps) most of the refrigerant in the liquid receiver, with residual refrigerant residing in the condenser and hot-gas discharge line.

Capturing Refrigerant in the Liquid Receiver

Capturing refrigerant in the liquid receiver or the condenser is called a ***pumpdown***. On commercial HVACR systems, pumpdown is performed by closing the liquid receiver outlet valve (front-seated) while the compressor is running. When pumping down, sometimes the compressor shuts off on a low-pressure control before the pumpdown is completed. It may be necessary to use a "jumper" to electrically bypass the low-pressure safety control or otherwise force the compressor

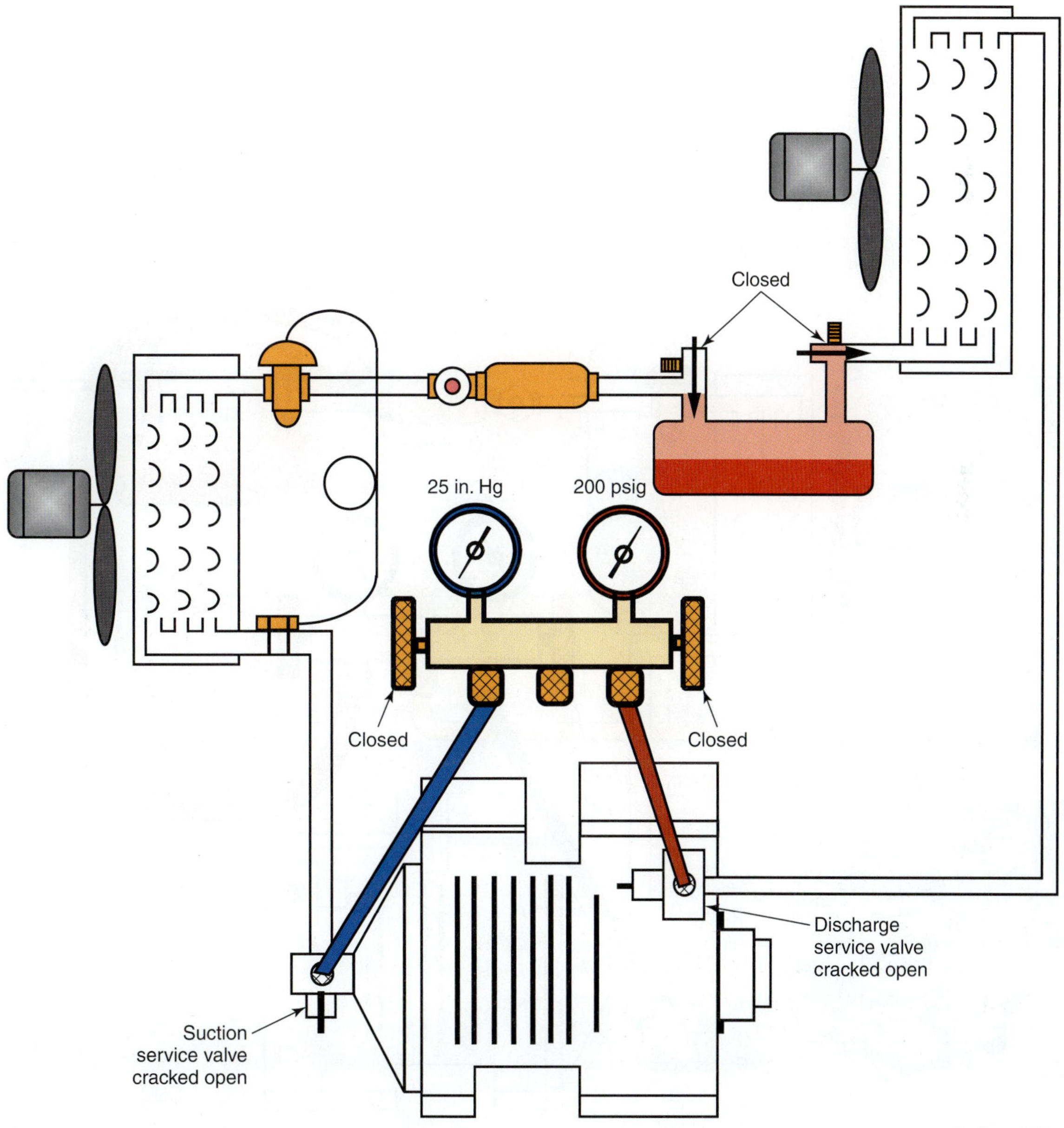

Goodheart-Willcox Publisher

Figure 16-15. A system that has a liquid receiver with two service valves (inlet and outlet) can be pumped down to capture most of the refrigerant charge while low-side repairs are made.

to continue running for a system pumpdown. On residential HVACR systems, pumping down refrigerant entails capturing the refrigerant in the outdoor unit by closing the discharge service valve, starting the unit, and allowing pressures on the manifold gauge to reach 0 psig, and then closing the suction service valve.

The compressor evacuates (removes) all refrigerant from the low-pressure side, including the liquid line from the discharge service valve, and stores it in the liquid receiver or condenser. Then, the compressor is turned off and the liquid receiver inlet valve is immediately closed (frontseated). The low-pressure side of the system is in a vacuum. The major portion of the system refrigerant charge is now stored in the liquid receiver or condenser, but residual high-pressure refrigerant remains in the condenser and discharge line. The suction service valve is also frontseated to prevent backflow of the refrigerant through the compressor valve reeds.

When a system is properly pumped down and the vacuum eliminated, the evacuated side of the system can be opened for minor, short-term repairs, such as changing a filter-drier. When repairs are completed, the part of the system that was opened needs to have a vacuum pulled before the service valves are opened and the system restored to service, **Figure 16-16**.

If the low-pressure side is left open to the atmosphere for too long, atmospheric air will enter the

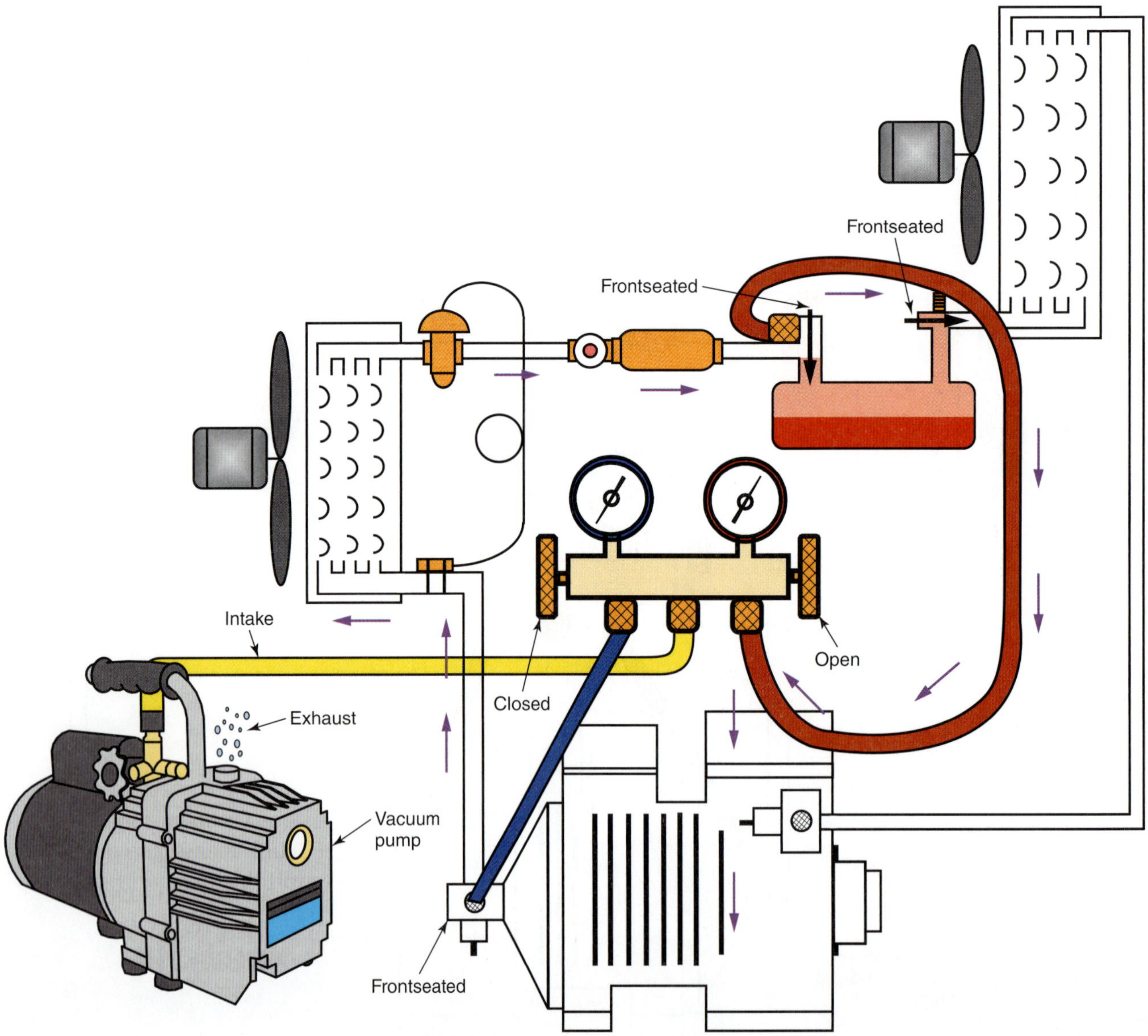

Goodheart-Willcox Publisher

Figure 16-16. If the system is left open to the atmosphere, air will eventually enter. The evacuation process for removing air from the low-pressure side is shown.

system, and vapor will leak back through the compressor valve reeds. The atmospheric air must be removed with a vacuum pump prior to releasing refrigerant from the receiver. Also, the lost refrigerant must be replaced. Evacuating air from the low-pressure side is done by connecting a manifold gauge to the receiver outlet service valve, with the valve in the frontseated position, so that the receiver is blocked but the gauge has access to the liquid line, **Figure 16-16**.

When in the pumpdown mode, repairs to the high-pressure side can be accomplished only after recovering refrigerant trapped in the hot-gas discharge line and condenser. The recovery process is performed at the discharge or liquid line service valve, **Figure 16-17**.

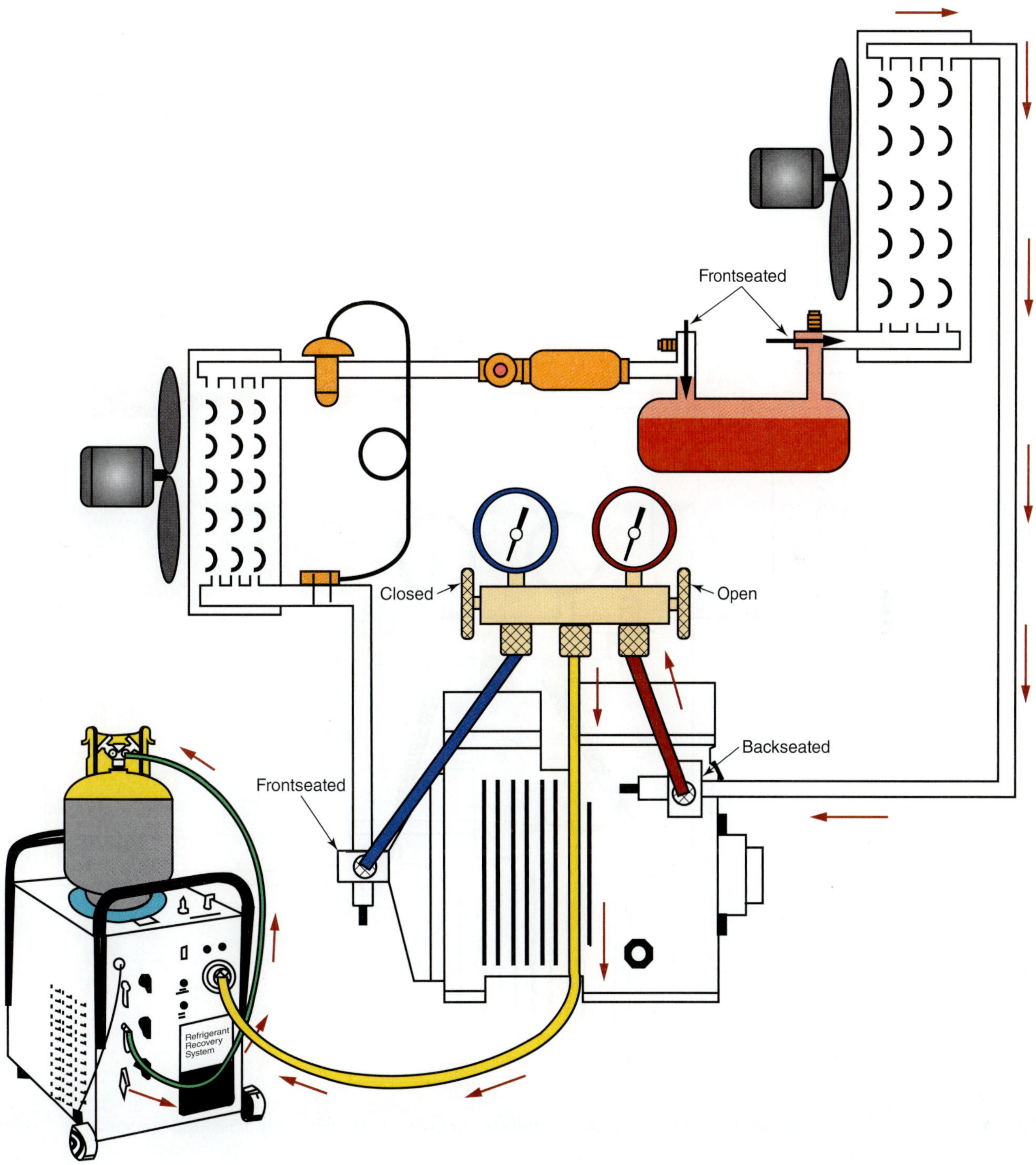

Goodheart-Willcox Publisher

Figure 16-17. After a pumpdown, vapor trapped in the hot-gas discharge line and the condenser must be removed by the recovery method.

Capturing Refrigerant in the Condenser

When a system does not have a liquid receiver, the condenser can be used as a storage area to capture refrigerant. A shut-off (isolation) valve must be available at the condenser outlet liquid line. The system is pumped down into the condenser after closing the shut-off valve at the condenser outlet liquid line. After pumpdown to the condenser, the compressor is shut off, and the discharge or liquid line service valve is frontseated. This captures all the liquid refrigerant in the condenser and the hot-gas discharge line. Using the condenser as a storage area does *not* cause high discharge pressure (as is commonly believed), unless the system is severely overcharged. The condenser will hold the normal system charge and not be completely full. Liquid occupies less space than vapor, and the vapor stops coming into the condenser *before* the condenser becomes full, **Figure 16-18**.

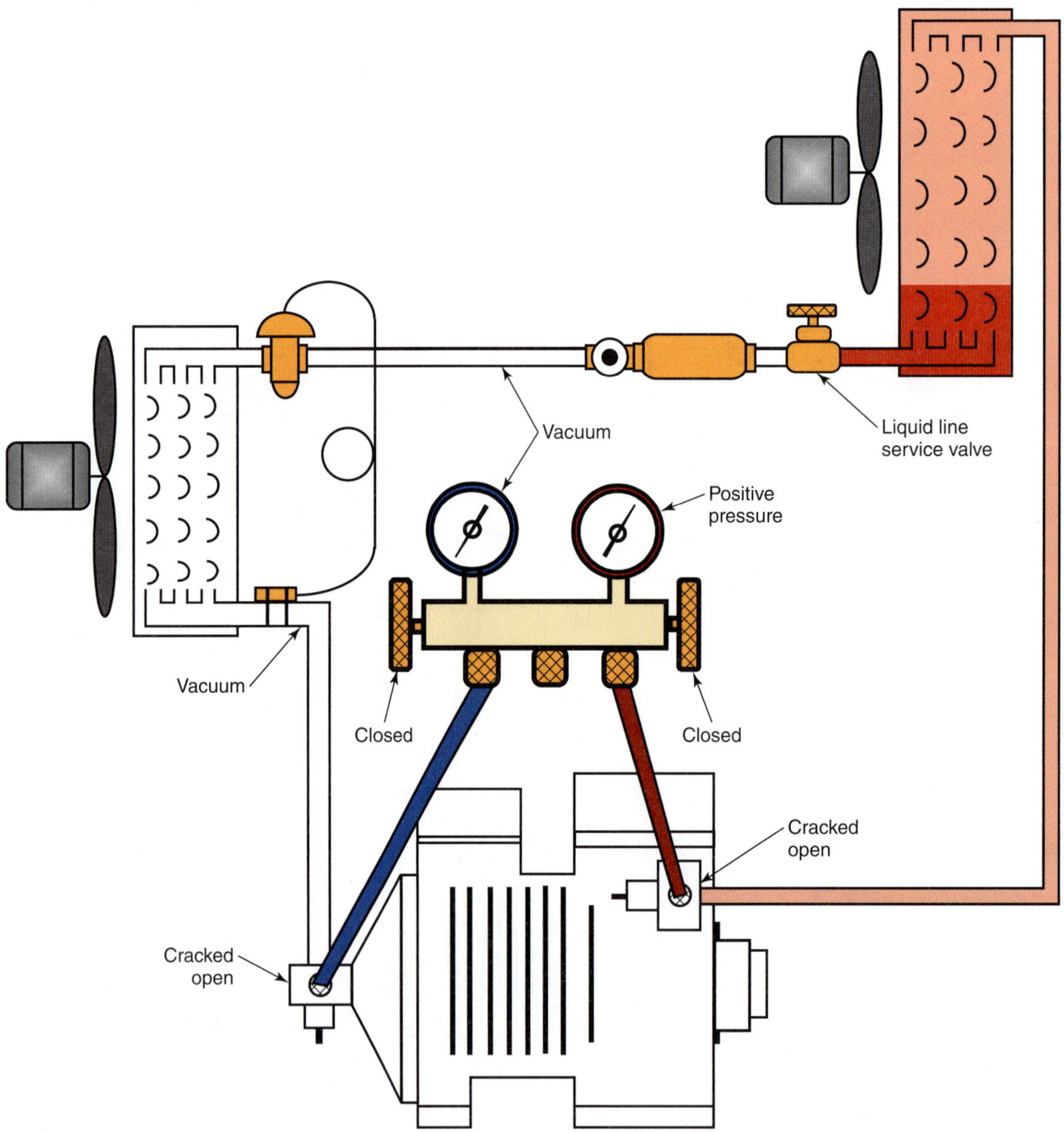

Goodheart-Willcox Publisher

Figure 16-18. The system is pumped down into the condenser after closing the shut-off valve at the condenser outlet.

Chapter Review

Summary

16.1 The EPA enforces strict venting and leak repair guidelines for technicians and equipment owners. Violation of these guidelines may result in fines, loss of certification, and legal consequences.

16.2 *Recover* means to remove refrigerant in any condition from a system and to store it in an approved cylinder or external container. *Recycle* means to clean refrigerant for reuse by removing moisture and contaminants from recovered refrigerant. *Reclaim* means to restore the recovered refrigerant to a level equal to new (virgin) product specifications as determined by chemical analysis.

16.3 Recycled or reclaimed refrigerants are not considered hazardous waste under federal law, but the EPA enforces guidelines for proper disposal and record-keeping of refrigerant service.

16.4 Section 609 of the Clean Air Act requires all persons who service motor vehicle air conditioning (MVAC) units be certified in refrigerant recovery and recycling and proper use of approved equipment. Section 608 of the Clean Air Act mandates that all persons who maintain, service, repair, or dispose of refrigeration and air-conditioning appliances (other than motor vehicle systems) must become certified by passing an EPA-approved test. Recovery equipment manufactured on or after November 15, 1993, must be tested by an EPA-approved testing organization to ensure the equipment will achieve required vacuum levels.

16.5 Technicians are required by law to use recovery systems to evacuate air-conditioning and refrigeration equipment to established vacuum levels. Recovery cylinders must meet Department of Transportation (DOT) specifications. Recovery equipment is used to avoid venting refrigerant into the atmosphere while making repairs to a refrigeration system.

16.6 The passive recovery method is used to recover refrigerant using a refrigeration system's compressor or any positive pressure from the system's refrigerant charge to push the refrigerant into a recovery cylinder. The active recovery method uses a recovery unit to move refrigerant into a recovery cylinder.

16.7 Recycle units contain a liquid pump that withdraws liquid refrigerant from the bottom of the recovery cylinder, circulates it through a large filter-drier, and returns the liquid to the top of the recovery cylinder.

16.8 Capturing refrigerant in the liquid receiver or condenser is called a *pumpdown*. It is performed by using the manifold gauge or by closing the liquid receiver outlet valve (frontseated) while the compressor is running. When a system does not have a liquid receiver, the condenser can be used as a storage area to capture refrigerant.

Know and Understand

1. It is a violation of Section 608 of the Clean Air Act to vent HFCs on or after _____. (16.1)
   A. January 1, 1990
   B. November 14, 1994
   C. November 15, 1995
   D. January 1, 1999
2. *True or False?* In case of conflict between the Clean Air Act and the Montreal Protocol, the more stringent law regarding the issue takes precedence. (16.1)
3. For a leak to be considered a substantial leak in a commercial refrigeration system containing 100 lb of refrigerant, at least _____% of the full refrigerant charge would need to have been lost. (16.1)
   A. 10
   B. 20
   C. 30
   D. 50

4. When refrigerant is removed from a system and stored in an approved cylinder or external container, this process is referred to as refrigerant _____. (16.2)
   A. recovery
   B. recycling
   C. reclaiming
   D. purging
5. *True or False?* Recovered refrigerant may contain acid, moisture, and oil. (16.2)
6. A recycle unit uses a(n) _____ to pull liquid refrigerant from the bottom of a cylinder via the dip tube. (16.2)
   A. compressor
   B. liquid pump
   C. solenoid valve
   D. electric motor
7. If refrigerant is removed from a system and is intended to change ownership, that refrigerant must be _____. (16.2)
   A. disposed
   B. recycled
   C. reclaimed
   D. retrofitted
8. CFC refrigerants may be considered hazardous waste if they are _____. (16.3)
   A. not mixed with other refrigerants
   B. mixed with oils from other sources
   C. subjected to CFC reclamation
   D. subjected to CFC recycling
9. If a system leaks over _____% of its full charge in a year, owners must submit a report to the EPA detailing the corrective action taken to identify and repair the leak. (16.3)
   A. 10
   B. 25
   C. 50
   D. 125
10. Section 609 of the Clean Air Act requires all persons who service _____ be certified in refrigerant recovery and recycling and proper use of approved equipment. (16.4)
    A. air-conditioning appliances
    B. motor vehicle air-conditioning units
    C. refrigeration systems
    D. commercial coolers
11. Type III EPA technician certification covers maintenance, service, repair, and disposal of _____. (16.4)
    A. high-pressure appliances
    B. motor vehicle air-conditioning units
    C. low-pressure appliances
    D. small appliances
12. A minimum passing score in all four sections of the EPA 608 certification exam earns a technician _____ certification. (16.4)
    A. Type III
    B. Universal
    C. Master Technician
    D. Job-Ready
13. *True or False?* Recovery cylinders are considered full at 100% of total cylinder capacity. (16.5)
14. The expanding liquid inside a recovery cylinder creates _____ pressure. (16.5)
    A. negative
    B. positive
    C. hydrostatic
    D. low
15. The EPA requires the use of _____ on recovery units and manifold hose ends to prevent loss of refrigerant to the atmosphere when connecting or disconnecting hoses from the recovery unit to a refrigeration system. (16.5)
    A. special hand valves
    B. a separate refrigerant accumulator
    C. low-loss fittings
    D. sealant
16. Passive recovery can be used on a refrigeration system that has a _____. (16.6)
    A. refrigerant charge of 15 pounds or more
    B. blended refrigerant only
    C. lower static pressure than the recovery cylinder being used to store the charge in
    D. higher static pressure than the recovery cylinder being used to store the charge in
17. Refrigerant removed by the active recovery method can remove refrigerant in the _____ state. (16.6)
    A. vapor
    B. liquid
    C. liquid or vapor
    D. solid

18. When recycling refrigerant, how long must the refrigerant normally recirculate through a filter-drier? (16.7)
    A. 30 minutes.
    B. 4–6 hours.
    C. At least 24 hours.
    D. No wait time is required.
19. R&R equipment refers to equipment that houses both _____ units in one system. (16.7)
    A. recycle and reclaim
    B. recovery and recycle
    C. reclaim and disposal
    D. recovery and reclaim
20. To perform a *pumpdown* means to capture refrigerant in the _____. (16.8)
    A. metering device
    B. condenser only
    C. liquid receiver only
    D. liquid receiver or condenser
21. After pumpdown to the condenser, the compressor is shut off, and the discharge or liquid line service valve is _____. (16.8)
    A. backseated
    B. frontseated
    C. cracked
    D. removed

Apply and Analyze

1. List the four types of refrigerant release that are allowed. (16.1)
2. Identify four situations when refrigerant recovery must be performed. (16.2)
3. Describe the EPA's regulations on the reuse of recovered and/or recycled refrigerant. (16.2)
4. A service technician is working on an appliance that contains 45 lb of refrigerant and finds that the equipment needs to be disposed of and replaced. Before disposing of the appliance, what information needs to be documented? (16.3)
5. List the requirements for a technician to obtain EPA 608 certification. (16.4)
6. Why is it typically quicker to recover refrigerant in the liquid state rather than as a vapor? (16.5)
7. What are the main differences between passive recovery and active recovery methods? (16.6)
8. When using the condenser to capture a system's refrigerant, does it cause high discharge pressure in the system? Why or why not? (16.8)

Critical Thinking

1. Speculate on the benefits of recovering and recycling refrigerant for reuse in systems. (16.2)
2. Why is it important to not overfill recovery cylinders? (16.5)

Mastercool, Inc.

17 Refrigerant Service Basics

LEARNING OUTCOMES

After studying this chapter, you will be able to:

17.1 Describe the reasons to perform a system evacuation, factors to consider before evacuating, and the two main methods of evacuation.

17.2 Explain the different types of refrigerant leak detection methods and how to verify when a system leak is corrected.

17.3 Recognize safe use and proper handling of refrigerant cylinders.

17.4 Predict and correct low-side and high-side pressures.

17.5 Describe procedures for vapor and liquid charging of refrigerant.

TECHNICAL TERMS

acid
air space
balance point
charging
contaminants
deep vacuum
disposable cylinder
electronic leak detector
electronic scale
halide torch
head pressure
heat load
hydrostatic expansion
liquid charging
low-side test pressure
permeation
pressure control
red iron oxide
setpoint
sludge
soluble
standing pressure test
temperature difference (td)
triple evacuation
ultrasonic leak detector
vacuum pump
vapor charging

Introduction

In addition to refrigerant recycling, recovery, and reclamation, technicians must also be able to perform system evacuations, check for leaks, recharge, and retrofit refrigerants to ensure proper operation and efficiency of the HVACR system. As always, practice safety first and follow manufacturer instructions when working with refrigerants and related equipment.

17.1 System Evacuation

For a refrigeration or air-conditioning system to operate at its peak efficiency, the refrigerant needs to be clean and dry. This is done by accomplishing the following important objectives before any refrigerant is charged into a system:

- Install all components of the system tightly.
- Locate and repair any system leaks.
- Perform a proper evacuation on the system to remove any moisture, air, or contaminants.
- Install a new filter-drier after opening the system.

Moisture, air, and contaminants are major causes of problems in compression refrigeration systems. Excess *moisture* in a refrigeration system can cause ice to form, restricting refrigerant metering devices and preventing proper refrigerant flow. Moisture may also cause rusting if other noncondensables are present, corrosion, acid formation, refrigerant decomposition, oil sludge, and system deterioration. *Air* in the refrigeration system causes high head pressure and high operating temperatures. Sludge, acid, corrosion, and other *contaminants* result in various system problems. Acid, for example, eats the insulation off motor windings, causing a burnout.

Moisture and Air Removal

The industry standard is to remove as much moisture as possible through proper evacuation methods. Water may be more ***soluble*** (easily dissolved) in some refrigerants than in others, but any excess water will exist as a separate liquid and cause considerable damage to a system. If the temperature is low enough, the water will freeze.

Moisture and air are the primary enemies of a refrigeration system. Proper evacuation and the use of driers are necessary for trouble-free operation. Evacuation is performed using a special pump called a ***vacuum pump***. As it operates, it creates a negative pressure and removes all moisture and noncondensables from a refrigeration system.

Thorough evacuation of a system should be performed whenever it has been contaminated due to improper procedures when hooking up manifold gauges or if exposed to atmospheric air when replacing components in the refrigerant circuit. Blowing the system out with an inert gas (nitrogen or carbon dioxide) might remove most of the air from the system, but then the inert gas must be removed as well. Blowing a system out does not remove moisture and air from trapped areas. While filter-driers remove small amounts of moisture and contaminants, they should *not* be relied on alone. Nothing is better than pulling a proper evacuation on the system.

Contaminants

In addition to moisture and air, ***contaminants***—such as brazing fluxes, metal chips, wax, acid, oil sludge, dirt, black oxides, and rust—can harm refrigeration systems. With hermetic and semi-hermetic compressors, the motor windings are exposed to contaminants in the system. Care is required to avoid contamination. Contaminants cause serious damage to the motor windings.

Sludge is caused by a chemical breakdown of oil that combines with other materials, such as carbon, metals, oxides, or salts. These compounds form a dark, gummy mass that blocks valves, oil ports, screens, and filters. Sludge problems can be traced to high compressor discharge temperatures and the presence of contaminants, including air.

Air in the system causes high discharge pressure (and high temperature). Atmospheric air is a source of oxygen and moisture. Oxygen and moisture contribute to the breakdown of the oil-refrigerant mixture. The breakdown forms ***acids***, which are corrosive chemical compounds. Acids are more corrosive in the presence of moisture than in a dry system, reacting quickly with metal parts and contributing (among other effects) to the formation of sludge. Therefore, it is important to keep systems as clean and dry as possible.

When contaminants are in a system, ordinary operating temperatures will produce corrosion. Oxygen from the moisture and air combine to form ***red iron oxide*** (rust). This leads, in turn, to the formation of iron salts and more water if acid is present.

Evacuation Factors

Before evacuation, a system's internal pressure must be reduced to atmospheric (0 psig). The vacuum pumps used to evacuate systems are *not* designed to handle pressures above atmospheric.

Proper evacuation of a system is performed with a two-stage, rotary-type vacuum pump. The vacuum pump draws all vapor (gas) out of the system, reducing its internal pressure to a low vacuum. The vacuum causes any moisture to boil, allowing it to be removed as a gas.

To make sure that as much air and moisture has been removed from the system that can be removed, the system needs to be pulled down to a vacuum level of 500 microns. Microns are a small measurement. See the following values for comparisons to other units of pressure:

- 500 microns = 0.5 mm Hg
- 25.4 mm Hg = 1 in Hg
- 1 in Hg = 25 400 microns
- 1 psia = 2 in Hg
- 1 psia = 51 715 microns
- 0 psig/14.7 psia (atmospheric pressure) = 760 209.4 microns

As shown in the relationship of these pressure units, a micron is extremely small compared even to psi. A system that is at atmospheric pressure (0 psig or 14.7 psia) contains 760 209.4 microns of pressure. To pull a vacuum on a system that is at atmospheric pressure, the vacuum pump must remove 759 709.4 microns!

While the proper evacuation of a refrigeration or air-conditioning system is one of the most important steps to having a clean and dry system, it is usually the one process that is not properly performed. Many technicians when asked why they do not perform a proper evacuation, their answer will almost always be "Time. It just takes too much time, and we need to get to the next job."

The time required for a vacuum pump to remove moisture and air from a system depends on several factors:

- Size of the system
- Amount of moisture present in the system
- Capacity of the vacuum pump being used
- Size and length of the connecting lines
- Condition of the vacuum pump oil

The larger the system, the longer it will take to achieve a proper evacuation. With the proper equipment and conditions, however, even a large system can be pulled down in a reasonable amount of time.

Note that time must be allowed for the pump to pull a vacuum in all parts of the system and for moisture to work its way out of the oil in the compressor crankcase.

A large quantity of moisture in a system cannot boil into a vapor immediately upon lowering the pressure. Similarly, an open pan of water does not immediately flash into vapor upon reaching the boiling temperature of 212°F (100°C). The boiling process requires time for the molecules to rearrange themselves in a new pattern. The boiling process must continue until all the molecules are rearranged. Stopping the vacuum pump too soon will leave moisture in the system.

The speed of moisture removal can be increased by applying a heat lamp to the compressor crankcase during the vacuum-drawing process. Severe moisture problems may require the use of heat lamps on other system components as well, **Figure 17-1**.

Evacuation of a refrigeration system should be performed using a vacuum pump of the proper size. Vacuum pumps are rated in cfm (cubic feet per minute). The most common sizes of vacuum pumps normally found in most supply houses and service vans are 5, 6, or 8 cfm units. Any one of these three sizes should cover most of the equipment that a technician will be working on. Additionally, when working with A2L flammable refrigerants, technicians need to make sure the vacuum pump is rated for use with A2L refrigerants.

When a technician does not pull a proper evacuation because it is taking too much time, the solution may be to use a bigger pump with a higher cfm rating to finish the job sooner. However, this is

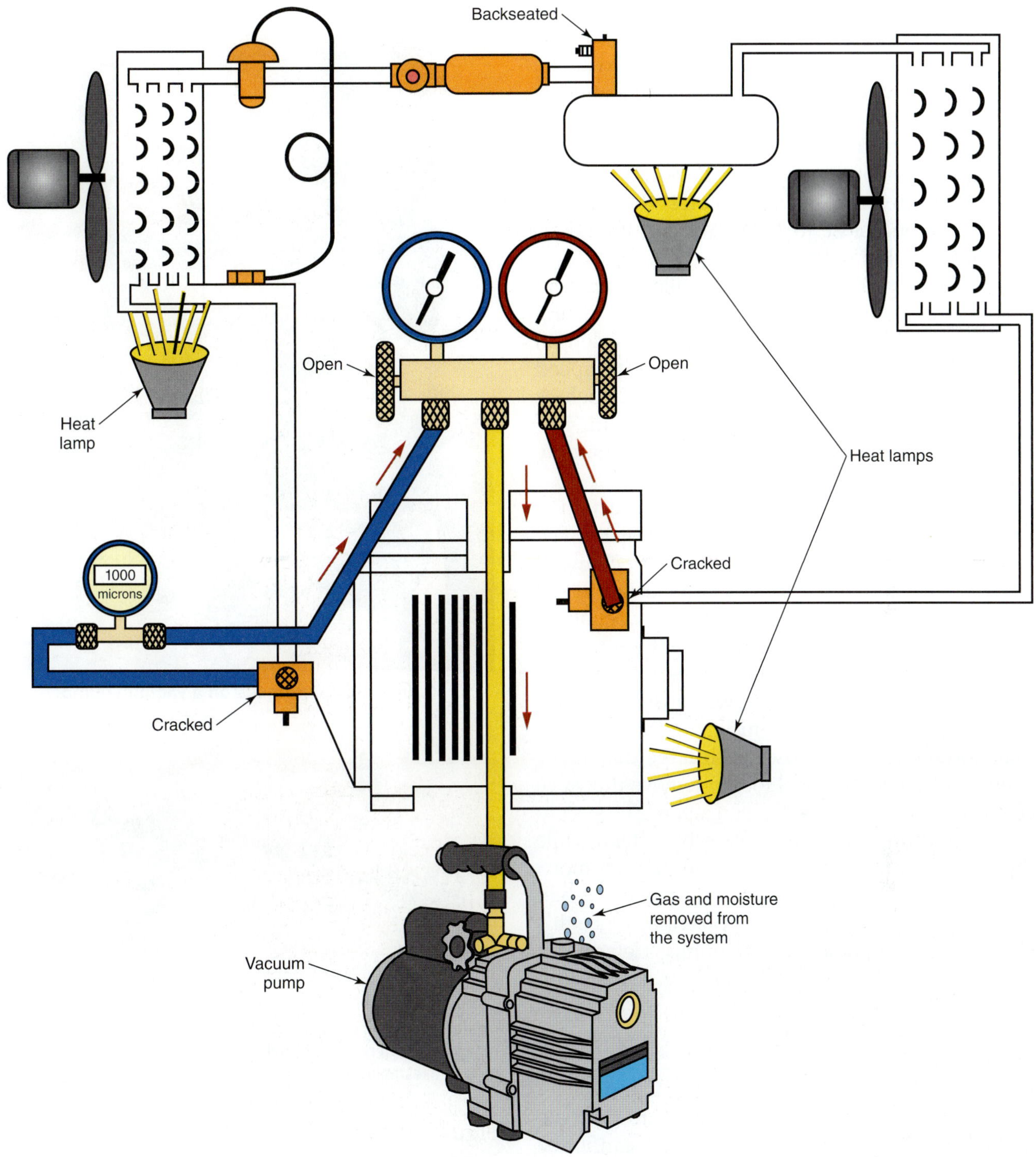

Goodheart-Willcox Publisher

Figure 17-1. Heat lamps may be used to drive moisture out of the system when evacuating. Moisture in the compressor oil is the last to be removed. Never use a torch or live steam for this purpose.

not always the right solution or the reason that the evacuation is taking too long. Another important factor to consider is the size and length of the connecting lines. Look at the following example.

A service technician is working on a 5-ton air-conditioning unit. The refrigerant has been recovered, and it is time to pull a vacuum. Using the regular hoses, manifold gauge, and a 5 cfm vacuum pump, the evacuation process takes too long to get the system to pull down. The chart shows that a 5 cfm pump will handle up to a 35-ton system, so a 5-ton system should take no time at all.

The problem is *not* the vacuum pump. The problem is the hoses. The service technician is using the regular 1/4″ hoses on the manifold gauge with the core depressors left in the hoses and the valve cores left in the service valves. With this setup, the 5 cfm pump will move less than 1 cfm through those hoses. If the technician removes the core depressors and valve cores, the vacuum pump will pull close to 1 cfm through the hoses. This is better, but not the best possible.

In addition to removing core depressors and valve cores, the solution should involve using the largest diameter hoses of the shortest length as possible. If a service technician wants to speed up the evacuation time, use large hoses, such as 3/8″ or 1/2″. Hoses that are 3/8″ or 1/2″ with 1/4″ and 3/8″ ends can be purchased at the local supply house. See **Figure 17-2**.

Another problem that slows down evacuation times is the condition of the vacuum pump oil. The big question is, "When should the oil be changed in the vacuum pump?" Most technicians will say to change the oil at the start of every evacuation. That is correct, but it should not stop there.

Continuing with the example earlier, the service technician realized what was causing the long evacuation time. So, the technician replaced the regular hoses with short hoses with large diameters, used a valve core removal tool to remove the valve cores, and started the evacuation again. The system started pulling down a lot quicker, but after a while the vacuum pump just stopped pulling anymore. The problem was that the vacuum pump oil had dropped to its vacuum sealing potential. Once this occurs, no matter how long the pump is left on, the vacuum level will never go lower. This is an indication that the vacuum pump oil needs to be changed. The real answer to the question of, "When should the vacuum pump oil be changed?" is as often as it takes to pull the system down to 500 microns.

The exact amount of time required to evacuate and dehydrate (dry out) a given system cannot be predicted because of these many factors. If the time to pull a proper evacuation cannot be predicted, then when does the technician know it has been achieved? A proper vacuum is achieved when a measurement of at least 500 microns is registered on a micron gauge. A compound gauge only shows vacuum in inches of mercury (in Hg). These gauges and units are *not* accurate enough to show a proper vacuum. A reading in microns is necessary. The only way to know when 500 microns has been reached is with a micron gauge. See **Figure 17-3**.

Large-diameter hoses on a vacuum pump

Valve core removal tools

Goodheart-Willcox Publisher

Figure 17-2. Using larger diameter hoses and using valve core removal tools to remove the valve cores can help to reduce the time it takes to pull a proper vacuum.

Figure 17-4 shows a micron gauge installed during an evacuation.

Two Methods of Evacuations

There are two methods of evacuation commonly used: deep vacuum and triple evacuation.

Deep Vacuum

The first method of evacuation is the ***deep vacuum***. When pulling a deep vacuum, the

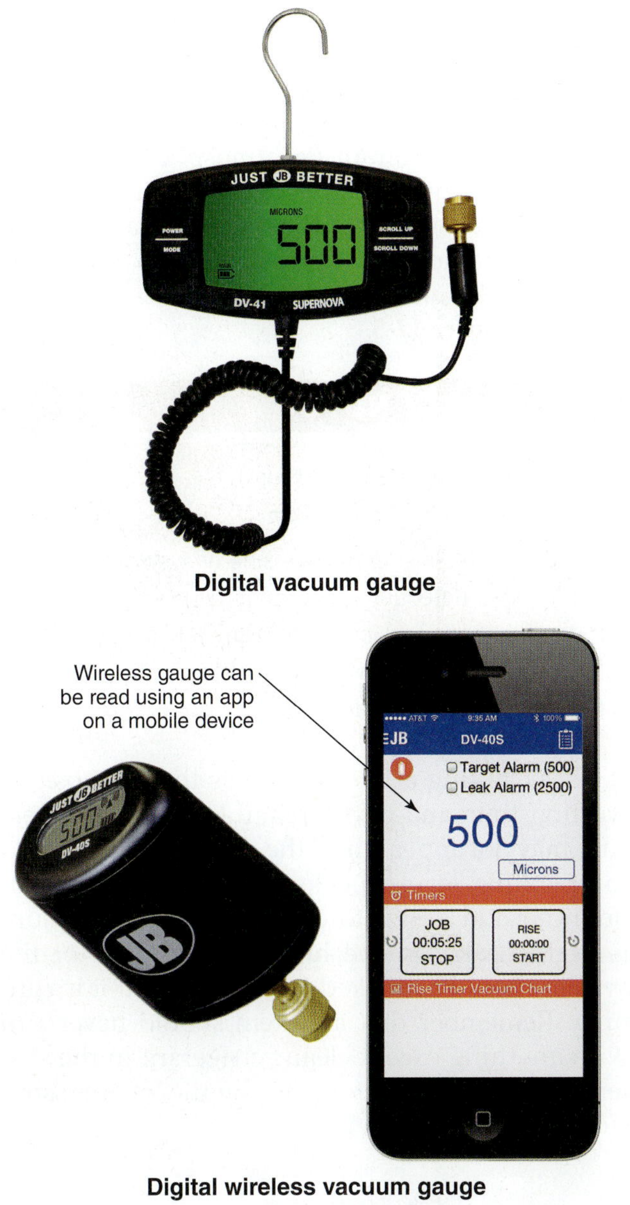

JB Industries

Figure 17-3. Different models of micron gauges.

goal is to pull the system down to a vacuum of 500 microns in one long pull. Using this method takes time to pull out air and moisture that might be trapped in different areas inside the system. This trapped moisture can only be removed after it has evaporated into a gas and then removed by the vacuum pump.

Note that a system can be pulled down to 500 microns and *still* have moisture present. When the system reaches 500 microns, isolate the system from the vacuum pump. This can be done with inline valves of the manifold gauge hand valves. Refer to the diagram in **Figure 17-1**. After isolating the system in a vacuum, watch the micron gauge for a while. If the reading stays at 500 microns for

Goodheart-Willcox Publisher

Figure 17-4. The micron gauge shown is installed in the hoses between the vacuum pump and the system.

15 minutes, the moisture has been removed from the system. But if the reading starts to rise and then stops, there is still moisture in the system. It will need to be pulled down to a minimum of 500 microns again and then valved off and tested again as previously.

A system may need to be tested several times in this way until all the moisture has been removed. If the reading keeps going up and does *not* seem to stop, there is a good chance that the system has a leak. In this case, find and repair the leak before pulling another deep vacuum.

Triple Evacuation

The ***triple evacuation*** method involves pulling three vacuums and injecting nitrogen into the system after the first and second vacuums. The reason for breaking the vacuum with nitrogen is that the inert gas will help to dry out the system allowing the vacuum pump to perform better.

Connect a vacuum pump, micron gauge, and nitrogen cylinder with pressure regulator to a refrigeration system that does *not* contain a refrigerant charge. Pull a vacuum down to a pressure reading of 1500 microns. Then isolate or valve off the vacuum pump. Use a dry inert gas like nitrogen to pressurize the refrigeration system to around 10 psig. Isolate the system and let it sit for about 10 minutes. Then, release the nitrogen from the system so the pressure reads 0 psig.

Pull a second vacuum of 1500 microns. Again, add nitrogen to the system. Let it sit for another 10 minutes. Then, once again release the nitrogen until 0 psig is measured. Pull a third vacuum down to a minimum of 500 microns. Isolate the refrigeration system with inline valves or the manifold gauge hand valves. Watch the micron gauge. If the pressure reading rises and then stops, turn the vacuum pump back on to remove the remaining moisture. Then retest it. Do this until the micron reading stops rising. If the pressure reading keeps going up and does *not* seem to stop, there is a good chance that there is a leak in the system. Fix the leak and perform the evacuation process again till it holds a steady vacuum.

17.2 Leak Detection

Refrigerants are expensive, and a system cannot operate properly without an adequate supply. Therefore, leakage in refrigeration equipment is a major problem for both manufacturers and service technicians. However, leak *detection* should be secondary to leak *prevention*. Vibration is probably the principal cause of leaks, but improperly made flared connections or poorly brazed joints are high on the list of contributing factors. Eliminating poor brazing techniques, the use of improper fluxes and brazing materials, and poor workmanship will result in fewer leaks.

Sometimes fluxes used for brazing can temporarily prevent a leak by plugging a tiny hole. However, when the system is pressurized, the flux or residue blows away and a leak develops. Using proper brazing techniques eliminates this problem.

Repairing a leak is not difficult, but *locating* the leak can prove time-consuming. Each piece of equipment presents specific problems and limitations. To say the leak cannot be found or to simply add refrigerant is *not* a solution. Good technicians are highly skilled in preventing leaks and in locating and repairing them. The following approved leak-detection procedures are discussed in this section:

- Bubble method
- Halide torch
- Electronic "sniffers"
- Ultrasonic
- Oil/dye

Bubble Method

The use of bubbles to pinpoint leaks is the oldest and least expensive leak-detection method available. It is also highly effective. Testing with the bubble method requires the system to be pressurized with nitrogen and possibly a trace amount of refrigerant. A recovery system is often used to remove the refrigerant (to atmospheric pressure) before pressurizing the system with nitrogen. If checking the low side of the system for a leak, then the system should be pressurized to equal the ***low-side test pressure*** listed on the nameplate of the unit, **Figure 17-5**.

SAFETY NOTE

Pressurizing Levels for Bubble Leak Detection

When pressurizing the low side of a system for bubble leak detection, *never* pressurize a system to the high-side test pressure. Only pressurize the low side of the system to the low-side test pressure found on the nameplate of the unit.

Leak testing at pressures less than 100 psig is wasting time and effort. Small leaks at low pressure may not appear until the pressure is increased to operating levels. Leaks are always more pronounced and easier to detect when the pressure is high. Leaks on the high-pressure side of the system should be looked for while the unit is running. Remember that a system should never run with anything except clean refrigerant in the system. It is best to look for leaks on the low-pressure side when the system is off.

Low-side test pressure

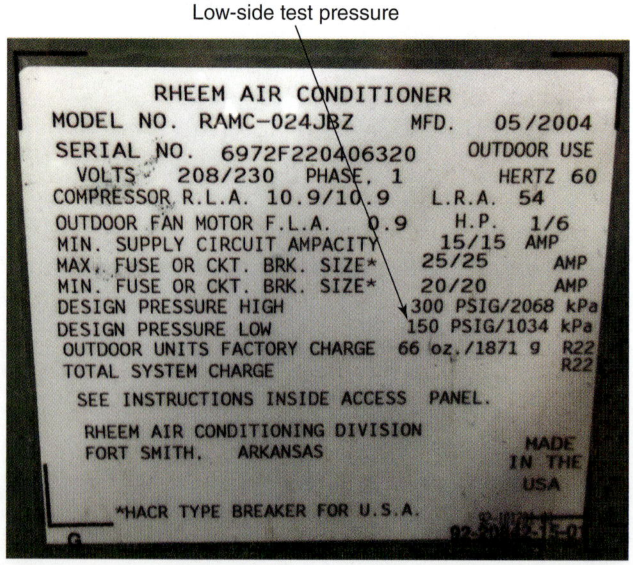

Goodheart-Willcox Publisher

Figure 17-5. Always check the nameplate for the low-side test pressure before pressurizing a system.

The bubble method is simple, but certain precautions must be observed for satisfactory results. The soap solution must be of the proper consistency, and the system must be properly pressurized. Apply the soap solution generously to the area being tested. Then wait for escaping gas to cause bubbles to appear. Do not forget to allow time for small leaks to appear. Locating large leaks is not difficult, but locating small leaks requires skill. The first thing most technicians do is apply bubble solution to the most likely spots, such as flare and brazed joints, fittings, and service valves.

Remember, vibration and poor craftsmanship are major causes of system leaks. It is not uncommon to find stress cracks in tubing bends or holes worn in tubing from rubbing together.

Do not forget to check the valve cores with bubbles when the hose connections are removed. It is common for the stem to get bent and stick open slightly, the core to loosen, or the gasket to be bad. Always replace the cap on a core valve. A valve core replacement tool makes it possible to replace a core under pressure without losing refrigerant.

Pressurizing the System

Nitrogen is often used to pressurize a system for leak detection. An inert gas like nitrogen is not harmful to the system, but the system must never be turned on while nitrogen is in it. A pressure regulator should always be used with such high-pressure gases to reduce them to a safe test pressure. See **Figure 17-6**. Again, never pressurize a system above the low-side test pressure found on the nameplate of the unit when locating a leak on the low side of the system. If a leak is suspected

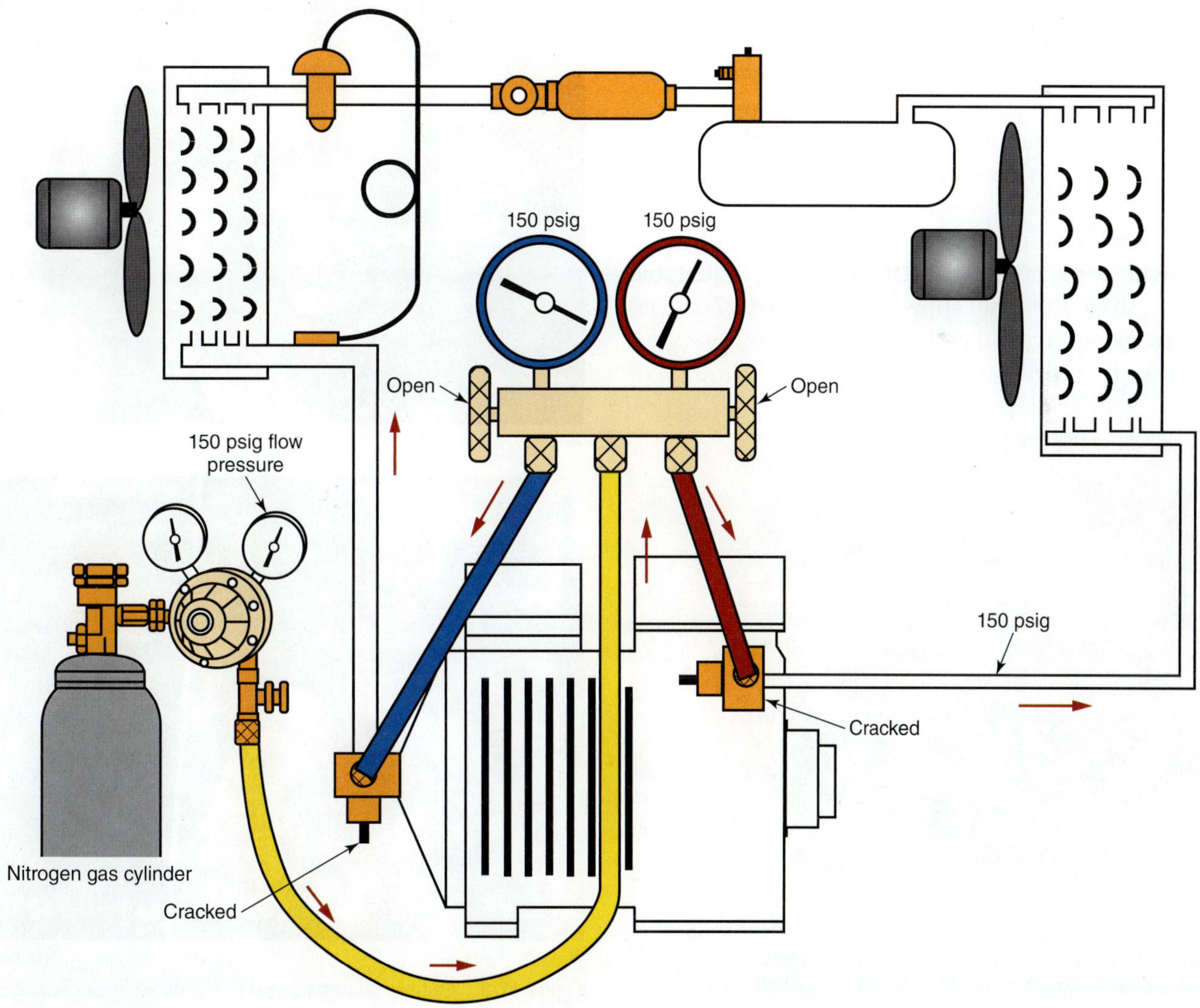

Goodheart-Willcox Publisher

Figure 17-6. Nitrogen is often used to pressurize a system for leak testing. This illustration is a typical example of leak testing a system with nitrogen and a detergent solution. The pressure causes bubbles to form, which pinpoint the location of the leak.

in the high side of the system, the high side can be isolated from the low side using service valves, then the system can be pressurized to the high-side test pressure found on the nameplate.

Detergent Solution

A special detergent solution for pressurized leak detection can be purchased from a local refrigeration supply house. This prepared leak indicator is available in plastic bottles with a dauber attached to the cap or a trigger spray, **Figure 17-7**. Detergent solution is highly effective for locating leaks.

An inexpensive detergent solution can be made from concentrated dishwashing liquid. As purchased, these detergents are too thick for proper use in leak detection. They must be thinned by mixing with water in a 1:1 ratio (*equal amounts* of water and detergent). Too much water will make the solution too *thin*, and it will not stick properly or blow bubbles. If the solution is too *thick*, more time will be needed for bubbles to appear (especially on small leaks). The solution must be of proper consistency to adhere (stick) to the surface but not so thick that spraying or bubble-blowing is difficult.

A handy applicator for detergent solution is a plastic trigger-spray bottle with an adjustable nozzle, like the one shown in **Figure 17-7**. The adjustable spray makes it easy to apply detergent solution in a fine stream to areas up to four feet away. A spray nozzle can eliminate a great deal of bending, climbing, crawling, and reaching to apply solution in hard-to-reach areas.

When leak testing, apply detergent solution *generously* to suspected leak sites to provide plenty of bubble-forming material. The solution can be easily wiped up with a rag after testing is completed. A large leak will form large bubbles quickly. Small leaks will form small bubbles slowly. See **Figure 17-8**.

Be sure to test the manifold gauge and hoses, as well as all system connections. Ordinary refrigeration hoses may allow gas to seep through their walls—a process called ***permeation***. Heavy-duty hoses do not allow refrigerant loss through permeation.

Goodheart-Willcox Publisher

Figure 17-7. Detergent solution for use in the bubble method of leak testing is available commercially, or it can be made from diluted dishwashing liquid. A trigger-spray bottle with an adjustable nozzle permits application in hard-to-reach areas.

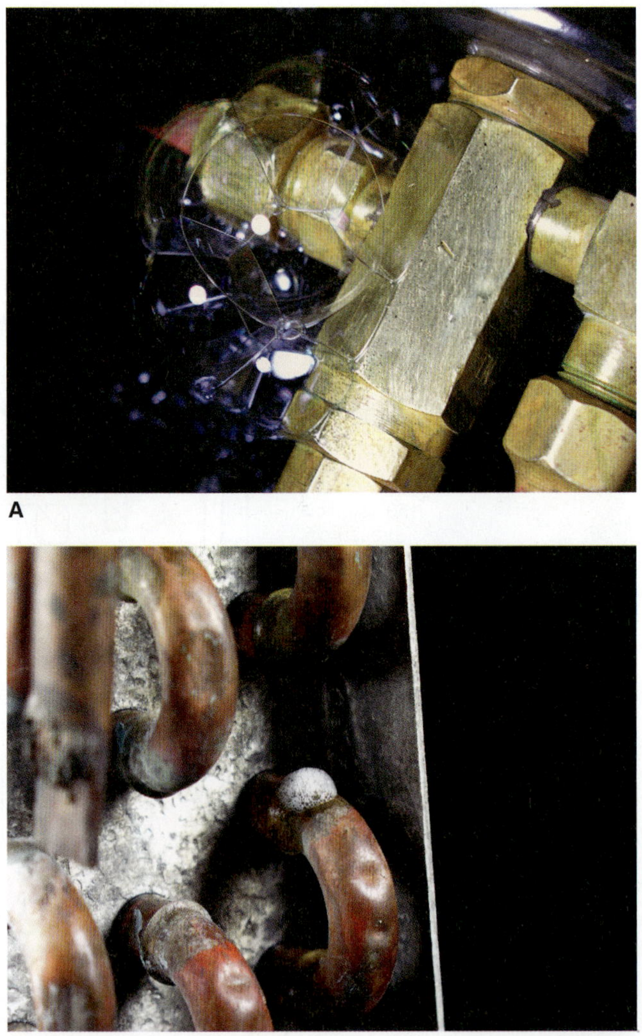

A

B

Refrigeration Technologies

Figure 17-8. Bubble testing for leaks. A—When a leak is large, bubbles will quickly appear. B—A froth of small bubbles will slowly form at the site of a small leak.

Halide Torch

The ***halide torch*** has been used for many years as a fast and reliable method for detecting halogenated refrigerant leaks, **Figure 17-9**. The torch detects *only* halogenated refrigerants, not other gases, such as nitrogen or carbon dioxide.

The halide torch operates on the principle that a flame will change color if it decomposes a halogen in the presence of copper. The essentially colorless flame of the halide torch will change to a color ranging from faint green to bluish green to bright blue, depending on the size of the leak (and the amount of halogen gas being decomposed in the flame). A halide torch is capable of detecting extremely small leaks, such as a leak of only one ounce of refrigerant per year.

Hydrocarbon fuels—such as propane, butane, acetylene, or alcohol—are burned in the torch to provide an almost colorless flame. The air is drawn in through a "sniffer hose" whose inlet nozzle is slowly passed over the area of the suspected leak. Leaking gas is drawn into the hose (along with atmospheric air) and carried to the flame. Color changes in the flame are visible through the viewing window of the torch. While halide torch leak detectors have been around for a long time, they are hardly ever used or seen anymore. Electronic leak detectors have largely replaced them.

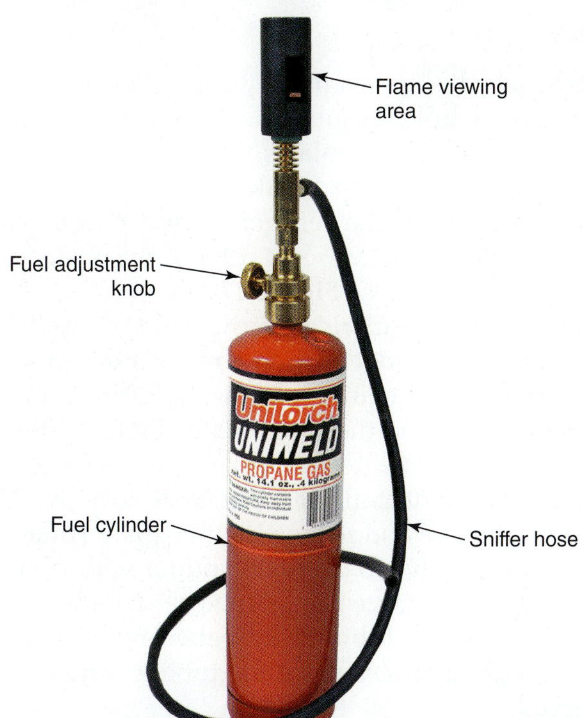

Uniweld

Figure 17-9. A typical halide torch. The sniffer hose draws in air from the suspected refrigerant leak site and directs it into the torch flame. Changes in flame color make it possible to detect even very tiny leaks.

SAFETY NOTE

Leak Detectors and A2L Refrigerants

Never use a halide torch leak detector when working on a system that contains an A2L refrigerant. The small flame could be enough to ignite the refrigerant and cause an explosion in the refrigerant lines. Always use an electronic leak detector rated for use with A2L refrigerants.

Electronic Leak Detectors

Small, hand-held electronic leak detectors ("sniffers") are extremely sensitive and easy to use. Most are powered by batteries, but some models must be plugged into an electric wall outlet for power. A typical ***electronic leak detector*** consists of a probe and a case containing the electronics, **Figure 17-10**. The unit may be equipped with visual indicators and/or audible signals used to notify the presence of halide refrigerant gas. When working with A2L refrigerants, make sure that the leak detector is rated for use with A2L refrigerants. When working on systems with hydrocarbon (HC) refrigerants—such as R-290 or R-600a—make sure to use an A3-rated leak detector.

To use an electronic leak detector, the system must have refrigerant in it. If the unit has refrigerant in it but is not up to the low-side test pressure, add some refrigerant to get the pressure up to leak check the low-pressure side. If the unit can run, turn it on to check the high-pressure side. If the

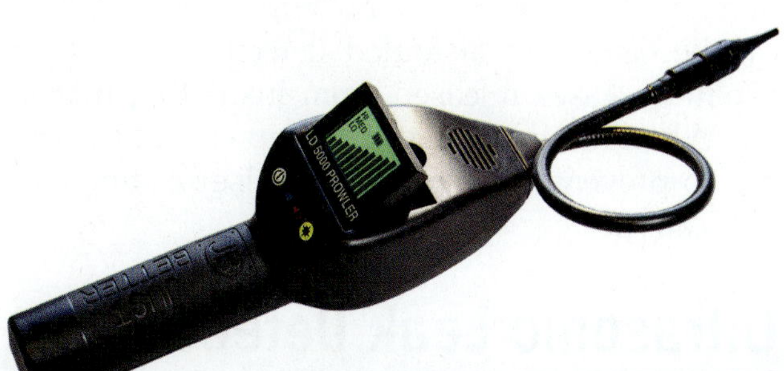

JB Industries

Figure 17-10. A highly sensitive electronic leak detector. This battery-operated model uses an audible tone to signal leaks.

system has lost its charge and has no refrigerant in it, add about 10 psig of any fluorinated refrigerant, such as R-134a or R-22 (regardless of the refrigerant the system holds). Then, pressurize the system up to the low-side test pressure with nitrogen. After the leak has been found, the refrigerant and nitrogen mixture can be vented out of the unit. After the leak is fixed, remember to pull a proper vacuum on the system before recharging.

When in operation, the instrument may emit a repeating, short "beep" signal. As the probe is passed over the site of the suspected leak, an air sample is pulled into the leak detector and across the sensor. The presence of even a minute trace of refrigerant causes the sound to change from a repeating tone to a long, steady tone.

PRO TIP

Electronic Leak Detector

When using an electronic leak detector, move slowly around the equipment when trying to locate a leak. It is recommended to move about an inch per second.

Like the halide torch, the electronic leak detector can give false readings because of refrigerant contamination of the air. Unlike the torch, however, the electronic detector can be recalibrated to compensate for a certain amount of atmospheric contamination. It will then look for levels above the atmospheric contamination, making leak detection easier. After narrowing down the location of the leak, pinpoint the leak location by applying bubble solution to the suspected leak location.

For accurate readings with an electronic leak detector, it is important to keep the sensing probe clean. Check the probe tip regularly for accumulations of lint, dirt, oil, or moisture. False readings can be caused by a fouled detector tip or from expander gases released from insulating materials. When not in use, keep the tip covered with a cap to prevent sensor fouling. False readings are also caused by soap solutions.

Ultrasonic Leak Detectors

Almost every leak produces a noise of some type, usually a high-pitched "hiss." An ***ultrasonic leak detector*** hears the "hiss" of a refrigerant leak and responds with audible and visual indications. It can pinpoint a leak regardless of contamination and works on pressurized systems or a vacuum. The ultrasonic detector is highly directional, which is an asset in confined spaces and a liability in open or windy areas.

Some leaks, such as those occurring in valves and other machined devices, do not produce an audible "hissing" noise. The sound level can be increased by spraying the area with plain water. Do *not* use bubble solution because it will muffle the sound. Be aware that false readings can be obtained from exterior noises (hissing sounds). These noises will adversely affect the ultrasonic leak detector and render it useless in that situation.

Oil as a Leak Indicator

Some refrigerant oil is always circulating through the system with the refrigerant. A certain amount escapes at the site of a refrigerant leak. A visual inspection of the system that reveals fresh oil indicates a system leak. Oil spills should always be cleaned up as a safety measure. Cleaning up old oil spills also makes it easier to spot new oil spills that indicate another leak.

Dyes

Special dyes are available for injecting into the compressor crankcase (oil). The dye mixes with the refrigerant oil and circulates throughout the system with the oil. When a leak occurs, oil and dye escape along with the refrigerant and produce a stain that glows a fluorescent color in the presence of ultraviolet light. See **Figure 17-11**.

Dyes are particularly effective in locating low-level leaks that cannot be found by other detectors. Furthermore, the equipment owner can see evidence of a leak, and the dye remains in the system for future detection.

It is important to allow adequate time between injecting the dye and inspecting for leaks. Time is needed for the dye to circulate and for very small leaks to produce a visible stain. The dye method cannot be used on tubing runs between walls and other enclosed spaces since the tubing cannot be observed.

Photo courtesy of Spectronics Corporation

Figure 17-11. Using a fluorescent tracer dye, system leaks can be detected using an ultraviolet light source.

Standing Pressure Test

Once leaks have been located and repaired, and if time permits, the system should be checked with a standing pressure test to determine whether all leaks were found. A ***standing pressure test*** involves raising system pressure to the low-side test pressure using nitrogen in a system that has had its refrigerant charge recovered.

After recovering the charge and pressurizing to the proper level, measure and record the system pressure and ambient temperature. When working with refrigeration equipment, backseat the service valves and remove the manifold gauge to prevent leakage there. The system is now pressurized and sealed. After several hours (overnight, if possible), reconnect the manifold gauge, open the service valves, and record new pressure readings and current ambient temperature. When working with air-conditioning equipment, give the pressure test at least 30 minutes for small residential units and upwards of 24 hours for large commercial equipment. If the pressures remain constant according to ambient temperature, the system has no leaks. Pressure loss indicates that an undetected leak remains. Note that even though nitrogen is an inert gas, extreme temperature changes will change the pressure.

17.3 Refrigerant Cylinder Handling and Safety

Cylinders used to store and transport refrigerants are made from either steel or aluminum and are available in several standard sizes. All but the smallest of cylinders must have a pressure relief device to protect against excessive pressure buildup. The US Department of Transportation (DOT) requires a relief valve or a fusible plug in any refrigerant cylinder more than 12″ (30 cm) tall and 41/2″ (11 cm) in diameter. On large steel cylinders, a fusible plug is normally located in the concave cylinder base. The DOT also requires all cylinders to be inspected every five years.

Steel Cylinders

Large cylinders containing 100–150 lb (45–68 kg) of liquid refrigerant are convenient when servicing systems that require large quantities of refrigerant. They also are less costly than smaller cylinders on a per-pound basis. Suppliers usually require a sizable deposit to ensure return of the empty cylinder. A steel protective cap screws down over the valve at the top of the cylinders when they are being moved or are not in use.

Steel cylinders are equipped with a dual valve that allows the technician to withdraw refrigerant as a vapor or as a liquid (by means of a dip tube). There is no need to invert these cylinders (turn them upside down) to withdraw liquid refrigerant, **Figure 17-12**.

A refrigerant cylinder should *never* be completely filled with liquid. Some space must be left for ***hydrostatic expansion*** (swelling of the liquid), which increases with a rise in temperature. Remember the principles of the Combined Gas Laws: In a confined volume, as temperature rises, pressure also rises.

A cylinder is considered "full" when liquid occupies 80% of its interior space. The remaining 20% allows for expansion within a normal range of temperatures without danger of rupturing the cylinder. In case of temperatures that might cause expansion too great for the available space (and present the danger of a cylinder bursting), the fuse plug in the base of the cylinder or a special safety valve will open and vent refrigerant to relieve the pressure. See **Figure 17-13**.

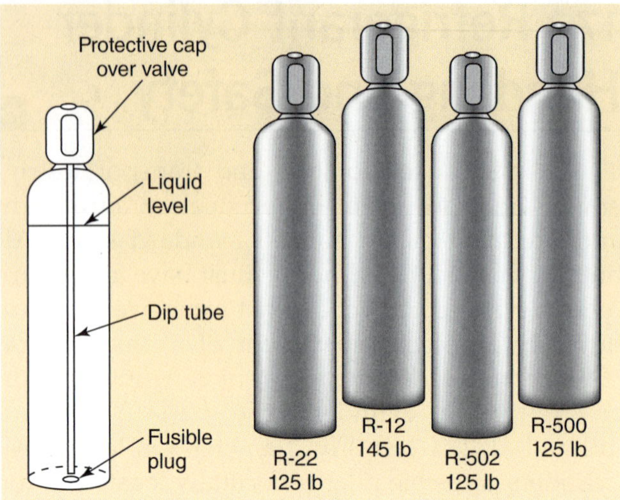

Goodheart-Willcox Publisher

Figure 17-12. Large steel refrigerant cylinders are returned to the supplier for refilling after use. A dual valve permits the technician to select refrigerant as either liquid (through a dip tube) or vapor while keeping the cylinder upright. To protect against excessive pressure, a fusible plug is installed in the cylinder bottom.

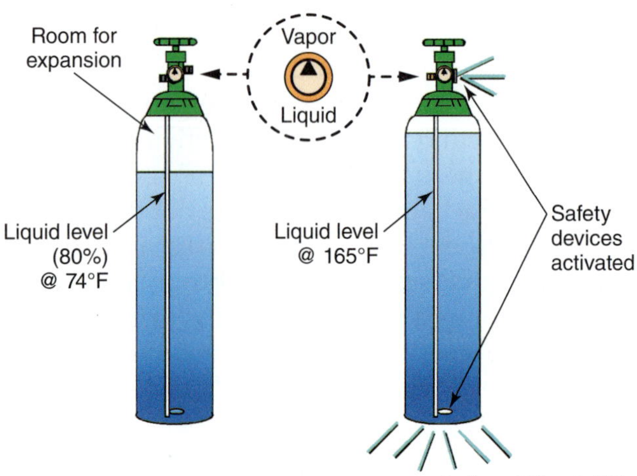

Goodheart-Willcox Publisher

Figure 17-13. When a cylinder is filled, at least 20% of cylinder volume must be left for hydrostatic expansion of the liquid refrigerant. If the temperature of the cylinder increases to 165°F (74°C), fusible plugs will melt or a safety valve will open to relieve pressure and prevent bursting.

Saturated conditions exist inside the cylinder, so the contents behave according to the pressure-temperature chart for the specific refrigerant involved. The space above the liquid is occupied by a dense refrigerant gas. A *balance point* is reached in which the pressure exerted on the liquid refrigerant by the gas molecules prevents any further *boiling off* or changing of liquid to the gaseous state.

When the vapor valve at the top of the cylinder is opened, gas molecules can escape. The leaving gas reduces pressure on the liquid, allowing it to boil and produce a steady flow of refrigerant gas. The flow of gas continues until the valve is closed or all the liquid has changed to a gas.

The heat required for the change of state ("boiling off" of gas) is drawn from the remaining liquid and the metal cylinder. As gas is released, both the temperature and the pressure of the remaining liquid drops. A continuous release of gas actually causes frost to form on the outer cylinder wall. Cylinder pressure may drop so low that it creates a problem when charging refrigerant into a system in vapor form. The cylinder pressure *must* be higher than the system pressure for vapor to flow from the cylinder into the system. See **Figure 17-14**.

To increase cylinder pressure in such a situation, the contents may be carefully warmed. The best and safest way to increase pressure in the cylinder is to place the cylinder in a container of warm water (80°F–110°F or 27°C–43°C) or use a specially designed cylinder heater, **Figure 17-15**.

SAFETY NOTE

Warming Refrigerant Cylinders for Charging

When warming a refrigerant cylinder for charging, only use gentle forms of heat transfer. *Never* heat a cylinder with a torch or use any other heating method that would produce a cylinder temperature greater than 125°F (52°C).

Disposable Cylinders

At one time, industry practice required technicians to transfer refrigerant from large steel cylinders to smaller, portable steel cylinders that held 25–30 lb (11–14 kg). Not only was this practice time-consuming and dangerous, but it also often resulted in contaminated refrigerant. The introduction of lightweight ***disposable cylinders*** has eliminated the need to transfer refrigerants. These throwaway cylinders are available in sizes ranging from 24 lb to 50 lb (11 kg to 23 kg), **Figure 17-16**.

Carrying handles on top of the cylinder also serve as protection against damage to the cylinder valve. The valve outlet is a 1/4″ male flare fitting for easy connection to the manifold gauge. To withdraw refrigerant *vapor*, the cylinder is used in the *upright* position. To withdraw *liquid*, the

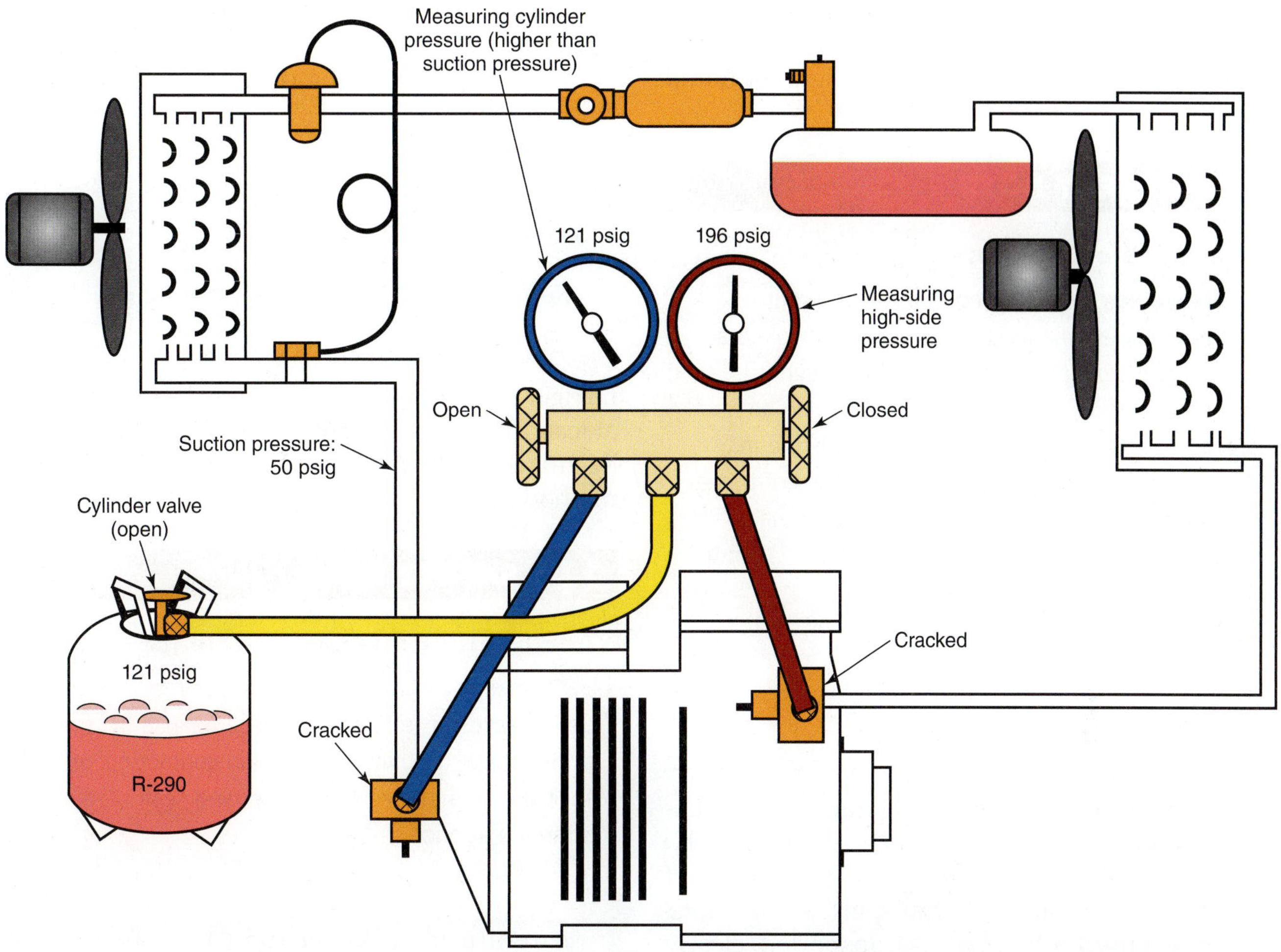

Goodheart-Willcox Publisher

Figure 17-14. If cylinder pressure is lower than or equal to system pressure, charging will stop. For refrigerant to flow into the system, cylinder pressure must be higher. If necessary, cylinder pressure can be increased by carefully heating the cylinder.

Ritchie Engineering – YELLOW JACKET Products Division

Figure 17-15. This refrigerant cylinder heater gently warms the cylinder for safe and effective heating.

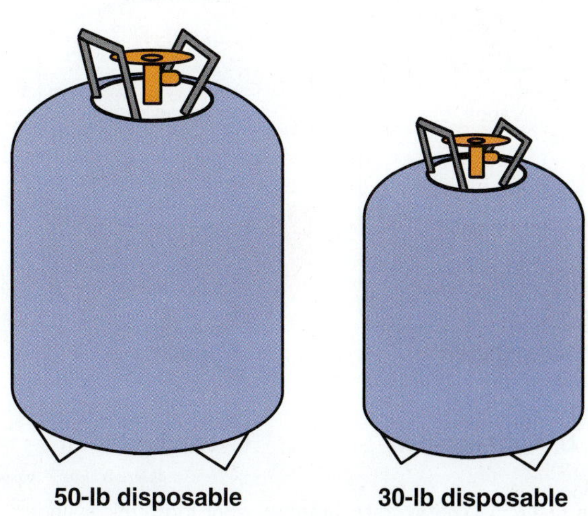

Goodheart-Willcox Publisher

Figure 17-16. Disposable cylinders are lightweight and intended to be discarded after use. An empty cylinder is considered hazardous waste until a hole is punched in it.

cylinder is *inverted* (upside down). See **Figure 17-17**. Some cylinders are equipped with a dip tube and dual-stem valve and can remain upright for liquid charging.

SAFETY NOTE

Disposable Cylinders

Federal law prohibits repurposing and refilling disposable cylinders with refrigerant. Violators are subject to fines and jail terms. Although it is not illegal, using empty disposable cylinders to transport compressed air is unwise. Moisture in compressed air could cause the thin metal of these cylinders to rust and eventually rupture. To help stop this, disposable cylinders are equipped with a check valve that will not allow refilling.

Cylinder Safety

When working with refrigerants, the technician must practice safety on the job every day. Safety-consciousness extends to the handling of cylinders that hold and transport refrigerants. Anytime a used refrigerant is transported, clearly label its container with a DOT classification tag to meet the requirements of the Code of Federal Regulation, Title 49. When moving a cylinder, be sure it is firmly strapped onto an appropriate wheeled device. *Never* roll a cylinder on its base or lay it down to roll it.

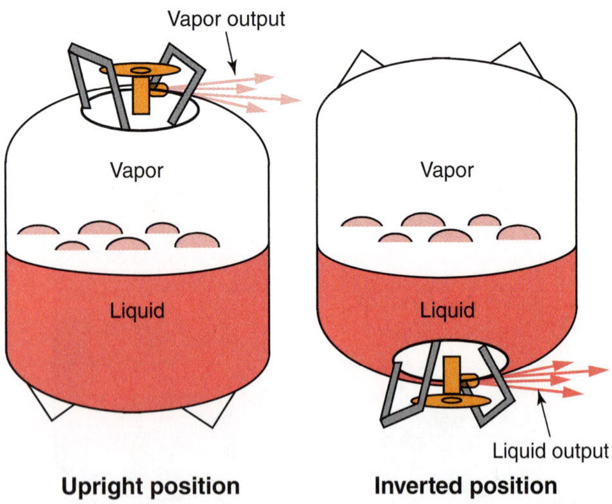

Goodheart-Willcox Publisher

Figure 17-17. To obtain refrigerant in vapor form, this cylinder is in the upright position. Inverting it will provide refrigerant in liquid form.

There are three major concerns when transferring refrigerants:

- Avoid contamination.
- Avoid venting to the atmosphere.
- Handle refrigerants safely.

Refrigerant cylinders must be the correct type. Always properly mark cylinders.

The cylinder used for holding transferred refrigerants must first be evacuated to a vacuum of at least 28 in. Hg. Different refrigerants should not be mixed in a cylinder because no economical method exists to separate refrigerants.

SAFETY NOTE

Accidental Refrigerant Release

Careless handling of cylinders can result in sudden releases of refrigerant that can cause frostbite, skin damage, or blindness. If skin contact occurs, soak the area in lukewarm water and seek medical attention at once.

A returnable cylinder must *never* be filled to more than 80% of the container volume. If the cylinder will be exposed to temperatures above 130°F (54°C), do *not* fill to more than 60% of the container volume. All liquids expand when heated. An expanding refrigerant liquid in an enclosed space determines the pressure in the cylinder. Overpressure safety devices provide some level of safety, but they do not eliminate all risk. Refer to Chapter 16, *Refrigerant Recovery and Recycling*.

Weigh cylinders on an accurate refrigerant scale. Inspect cylinders carefully before filling. Do not use dented, rusted, gouged, or damaged cylinders. Examine the valve assembly for leakage, damage, or tampering.

SAFETY NOTE

Refrigerant Cylinder Handling

Handle cylinders carefully. Do not drop them or bump them together. If the impact from a fall were to knock off a cylinder valve, the sudden release of refrigerant could turn the cylinder into a projectile.

Store refrigerant cylinders in a vertical position with their valves at the top. Become familiar with recovery equipment and apply all safety measures and instruction as prescribed by the manufacturer.

Refillable cylinders must be retested and recertified every five years. The test date must be stamped on the cylinder shoulder, in accordance with DOT Title 49 CFR Section 173.34(e) and 173.31(d). Retesting by visual inspection alone is *not* permitted. Do not fill a container that is out of date (five years or older). Outdated cylinders should be returned empty for retesting.

Disposable steel cylinders will oxidize and become so weakened by rust that the wall and seams can no longer tolerate pressure or contain gases. Discard disposable containers since they cannot be used for recovery or refilling. Refrigerant returned for reprocessing in improper cylinders will not be accepted. The penalty for transporting a refilled disposable cylinder is a fine up to $25,000 and five-years imprisonment. Disposable refrigerant cylinders should be stored in a dry location to prevent corrosion. *Never* allow used cylinders with residual refrigerant to sit at a jobsite; saturated vapor pressure will exist if even the smallest amount of liquid is present. An abandoned cylinder could explode under direct heat from any source. Before discarding a disposable container, recover any refrigerant remaining in the container. Then, for maximum safety, break off the cylinder valve and puncture the container before disposal as scrap metal.

Never use an empty disposable cylinder as a compressed air tank. The inside of the cylinder is untreated steel. When filled with compressed atmospheric air, oxygen and moisture contained in the air will form rust on the inside of the tank. It is not a question of "*Will* the cylinder burst?" but rather "*Where* and *when* it will burst?"

17.4 Controlling System Pressures

Since pure and azeotropic refrigerants are stable, pressure-temperature relationships are also stable and predictable. For ease of reference and use, these relationships have been compiled into pressure-temperature (P/T) or *saturation* charts. Zeotropic blends do *not* exhibit stable pressure-temperature relationships. Refer to Chapter 15, *Refrigerants*.

When working with systems that use one of the more common refrigerants (R-32, R-454B, R-1234yf, etc.), the technician does not even have to consult the pressure-temperature chart. The corresponding temperatures for these refrigerants are often printed on the face of pressure gauges. As shown in **Figure 17-18**, the faces of some gauges have several temperature scales. Often the outer scale shows pressure in psig. Inner scales show the corresponding temperatures (°F) for various refrigerants. The pressure-temperature relationship can be read directly from the gauge face. For other refrigerants, a P/T chart must be used to determine temperature for a given pressure.

Controlling Low-Side Pressures

The evaporator of an operating refrigeration unit absorbs heat from the space being cooled. The temperature of a product (such as ice cream in a home freezer) is controlled by the temperature of the air circulating around the product. The product gives up its heat to the air, which in turn gives up its heat to the evaporator. The evaporator temperature of a refrigerator or freezer is about 10°F (6°C) colder than the temperature of the air entering the evaporator. This air is often called the "return air."

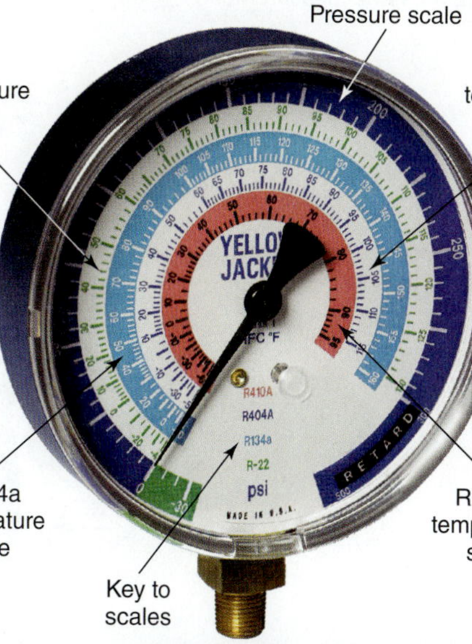

Ritchie Engineering – YELLOW JACKET Products Division

Figure 17-18. This compound gauge includes corresponding temperature scales for four widely used refrigerants. It allows relationships to be read directly from the gauge face, eliminating the need to consult a P/T chart. The temperature scales on this gauge are color coded.

The evaporator gives up its heat to the boiling refrigerant inside the tubing. The refrigerant is about 10°F (6°C) colder than the evaporator. This means that the refrigerant inside the evaporator is 20°F (12°C) colder than the return air. These temperature differences (air-to-evaporator and evaporator-to-refrigerant) must be maintained to accomplish heat removal. See **Figure 17-19**.

The amount of heat that must be removed per hour from an enclosed space determines the size of the evaporator that must be used. The correct amount of refrigerant must be boiled to maintain the necessary temperature inside the evaporator as well. When the air temperature is lowered to the desired level, the refrigeration system turns off. This eliminates the temperature differences, and heat removal stops, as shown in **Figure 17-20**.

Low-side pressure reveals *refrigerant temperature*, though this does not include superheat. When the system is not operating, the temperatures

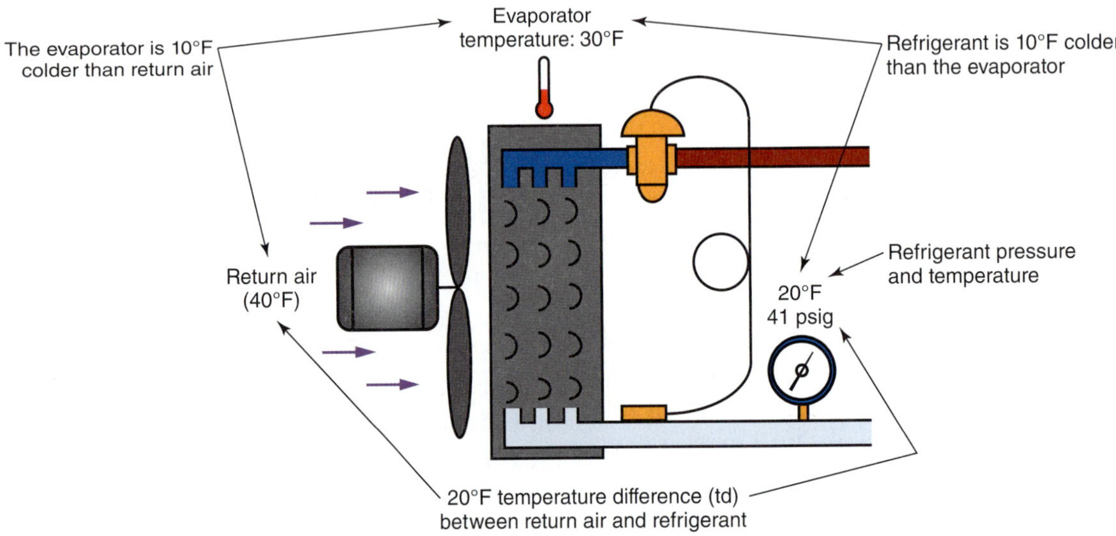

Goodheart-Willcox Publisher

Figure 17-19. Temperature differences are critical to accomplishing heat removal. The evaporator of an operating refrigeration unit will be 10°F colder than the air of the space being cooled, while the refrigerant inside the evaporator will be 10°F colder than the evaporator itself. As shown, if the return air temperature is 40°F, the temperature of the refrigerant inside the evaporator will be 20°F, based on its corresponding pressure measurement.

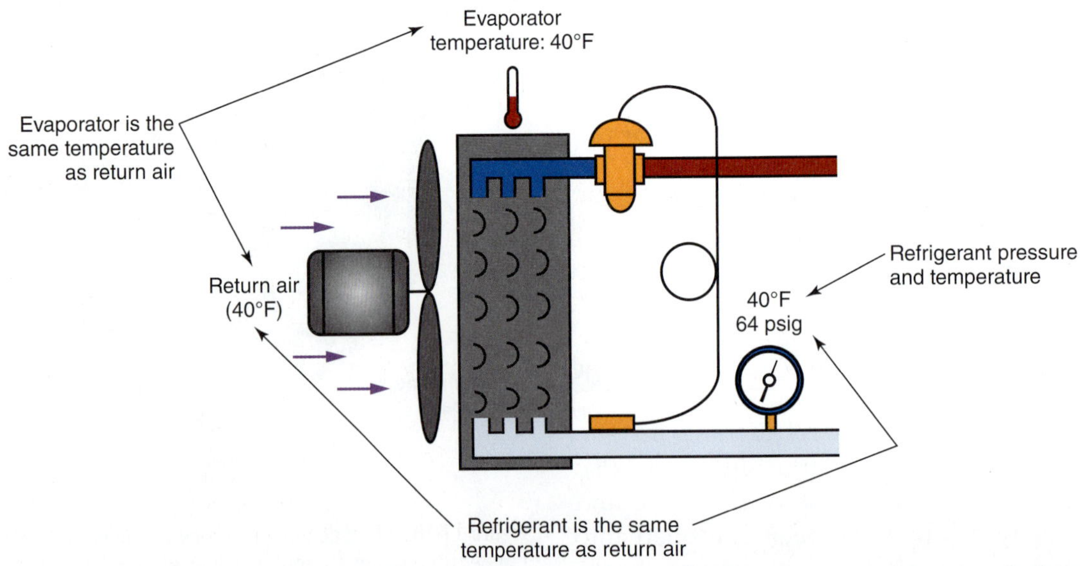

Goodheart-Willcox Publisher

Figure 17-20. When a refrigeration system is turned off, temperature differences quickly disappear, and heat removal stops. Note the air temperature and the refrigerant temperature. No heat can flow when a temperature difference does not exist.

of the refrigerant, the evaporator, and return air are all the same. This is because no ***temperature difference (td)*** exists among them.

For example, in a system charged with R-290 that has a low-side pressure reading of 63.9 psig (440.6 kPa), the refrigerant temperature at the evaporator outlet is 40°F (4°C). Since the system is not in operation, the temperature of the evaporator is also 40°F, as is the air temperature. Different refrigerants have different pressure readings at 40°F (68.5 psig for R-22 and 120 psig for R-32), but the principle remains the same.

A refrigeration system is designed to quickly *establish* and *maintain* a temperature difference once it begins operating. For example, in a system using R-290 with a low-side pressure of 36.3 psig (250.3 kPa), the refrigerant temperature is 15°F (–9°C). Thus, the temperature of the evaporator is 25°F (–4°C), and the return air temperature is 35°F (2°C). As the entering air becomes cooler, the evaporator and refrigerant temperatures drop accordingly. This relative temperature difference remains constant while the system is running. Pressure readings also decrease because of the lowered temperatures.

The temperature difference necessary for cooling is obtained only when the system is running. Once the system is turned off, the evaporator and refrigerant temperatures quickly rise and equal the airflow temperature. To maintain air temperature within a set range, a thermostat may be located within the area being cooled, known as the ***air space***.

During the "off" cycle, the air temperature typically is allowed to rise five or six degrees. When the air temperature reaches the highest ***setpoint*** (top of the range for which the thermostat is set), the system switches on. When the air temperature drops to the lowest setpoint, the thermostat switches the system off. Then the air begins to warm again, and the cycle repeats. An example is a walk-in dairy cooler with temperatures maintained in the range of 32°F to 38°F (0°C to 3°C). When air temperature rises to 38°F, the cooler's refrigeration system turns on. It runs until the temperature of the air falls to 32°F, then it shuts off.

Since the temperature differences allow low-side *pressures* to be predicted and controlled, many refrigeration systems are turned on and off by a ***pressure control***, rather than a thermostat. The pressure control functions as a switch, turning the system on and off as a result of changing low-side refrigerant pressure to maintain temperatures within the desired range.

Evaporator Airflow

To maintain proper temperature differences, the amount of airflow across the evaporator must be correct. Medium-temperature evaporators require a flow of about 1350 cfm per ton. Low-temperature evaporators require about 2000 cfm per ton. Airflow that is lower in volume than required results in a cold evaporator with low refrigerant pressure since the ***heat load*** (air to be cooled) cannot reach the evaporator in sufficient volume. Excessive airflow places too much heat load on the evaporator, resulting in high temperatures and pressures.

Air Conditioner Temperature Differences

The temperature differences designed for air conditioners are different from those designed for refrigeration systems. Because large volumes of very cold air are uncomfortable to people, air-conditioning evaporators are designed for an airflow of approximately 400 cfm per ton. The evaporator is designed to produce a temperature difference of 15°F to 20°F (9°C to 11°C) between the return air entering the evaporator and the supply air leaving the evaporator.

Although temperature differences for most refrigeration systems are calculated from *return air*, temperature differences for air-conditioning systems are calculated from *supply air*. In other words, the temperature of the evaporator is 10°F (6°C) colder than the air *leaving* it (supply air), rather than the air entering it (return air). As shown in **Figure 17-21**, the temperature difference between supply air (52°F or 11°C) and the evaporator (42°F or 6°C) is 10°F (or 5°C), while the temperature difference between return air (72°F or 22°C) and the evaporator temperature (42°F or 6°C) is 30°F (or 16°C).

A supply air temperature of 52°F (11°C) is considered the low limit for human comfort. As noted, this results in an evaporator temperature of 42°F (6°C) and a refrigerant temperature of 32°F or 0°C.

Allowing the *evaporator* temperature to fall to 32°F (0°C) in an air-conditioning situation can create cooling problems. At 32°F, moisture in the return air *freezes on the evaporator surface*. Ice buildup will restrict (or even totally block) airflow through the evaporator.

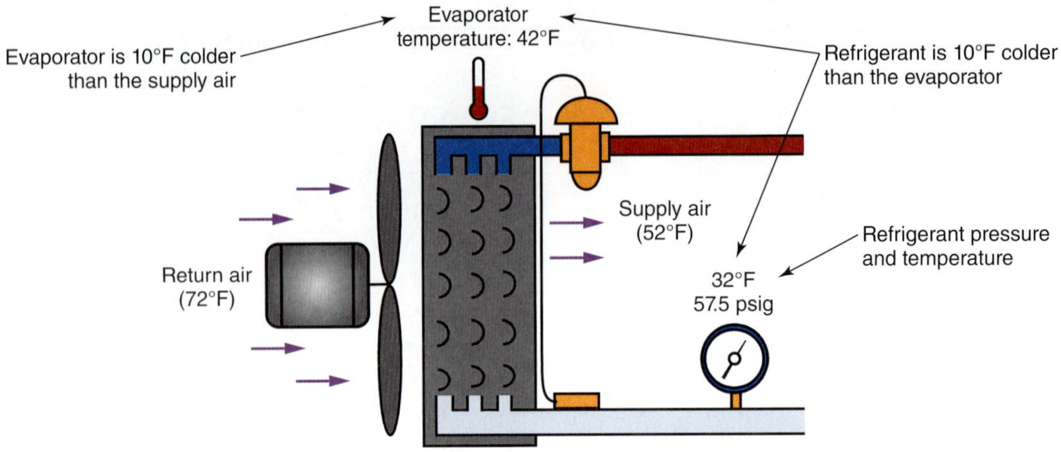

Goodheart-Willcox Publisher

Figure 17-21. In air-conditioning applications, a temperature difference of 10°F is maintained between the cooled air ("supply air") and the evaporator. For human comfort, air-conditioning systems are usually adjusted to provide a temperature no lower than 52°F.

Controlling High-Side Pressures

The condensing unit of a refrigeration system has a single task: to convert gaseous refrigerant back into a liquid form for reuse. The cooling medium is piped to the condensing unit, which is usually located outside in ambient air. Since heat moves from hotter areas to cooler ones, the gas must be *compressed* to a saturation point higher than the temperature of the ambient air. As the outside ambient air temperature goes up or down, the high-side pressure-temperature also goes up or down to maintain a temperature difference.

The system is designed so the high-side pressure rises to a saturation point where the refrigerant gas can be converted back into a liquid state. The rate at which the change of state takes place *must* equal the rate at which gas enters the condenser. This is called the ***balance point***. The condenser is sized to match system capacity. If the balance point changes, the high-side pressure-temperature will change accordingly.

Air-Cooled Condensers

The balance point of the condenser (high-side pressure) can be predicted. The compressor must compress the gas to a saturation temperature of anywhere between 15°F to 35°F (8°C to 19°C) above the temperature of the ambient air. With this information and the ambient temperature reading, a pressure-temperature (P/T) chart can be used to determine proper ***head pressure*** (pressure on the high side of the system).

For example, if the ambient temperature is 70°F (21°C), the temperature of the refrigerant (R-290, in this case) in the condenser should be between 85°F and 105°F (29°C and 41°C). See **Figure 17-22A**.

If the ambient temperature increases to 95°F (35°C), refrigerant temperature in the condenser should rise to around 110°F and 130°F (43°C and 54°C). Normal head pressure at an ambient temperature of 95°F would rise to around 200 psig and 259 psig (1379 kPa and 1786 kPa).

This method of predicting head pressure is applicable to most common refrigerants. However, some of today's high-efficiency units use closer to a 15°F to 25°F td design, instead of the 30°F to 35°F td seen in older units. With these 15°F to 25°F td systems, the service technician adds only 15°F to 25°F (8°C to 14°C) to the ambient temperature and uses that calculated value to determine the head pressure. See **Figure 17-22B**.

Water-Cooled Condensers

Some refrigeration systems use water-cooled condensers, which are more efficient than air-cooled condensers. Water-cooled condensers remove heat by conduction, providing better heat transfer. As shown in **Figure 17-23**, hot refrigerant vapor is drawn into the condenser while water is travelling in the opposite direction. Since the refrigerant is warmer than the water, heat will move from the incoming refrigerant to the water, until the refrigerant condenses into a liquid.

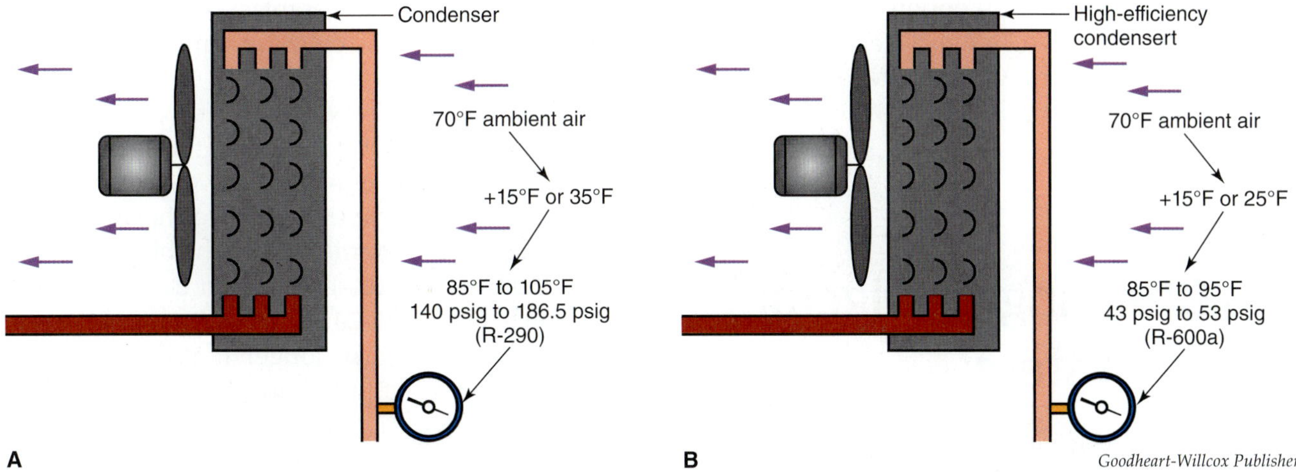

Goodheart-Willcox Publisher

Figure 17-22. A—In a balanced system, refrigerant temperature in the condenser will be anywhere from 15°F to 35°F higher than the temperature of the ambient air (depending on the efficiency of the system). This allows use of a pressure-temperature chart to predict head pressure. B—Adding 15°F to 25°F to the ambient air temperature allows the technician to determine head pressure in a high-efficiency system.

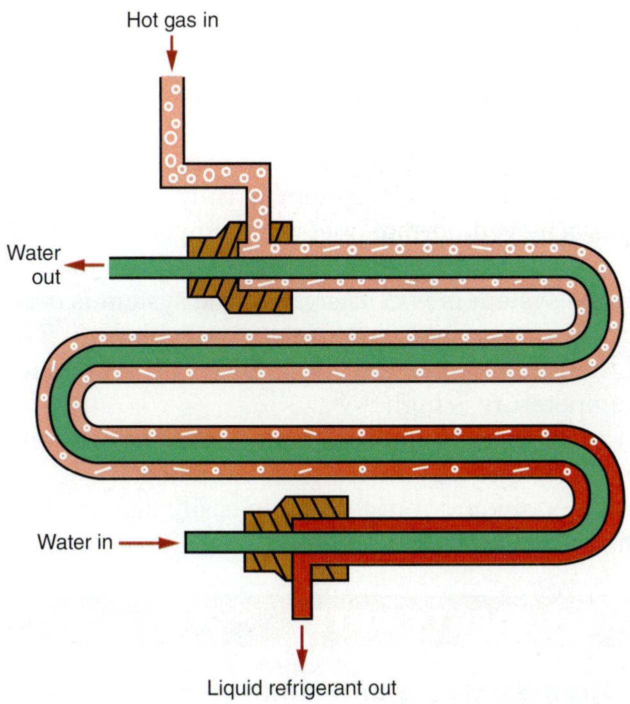

Goodheart-Willcox Publisher

Figure 17-23. Water-cooled systems transfer heat efficiently by means of conduction, so the refrigerant in the condenser is only 15°F to 20°F higher than the water leaving the condenser.

Excessive Head Pressure

As head pressure increases, the compressor works harder to achieve the balance point. Excessive head pressure should be avoided since it reduces system efficiency and causes compressor problems.

Higher-than-normal head pressures can usually be traced to one of five possible causes:

- **Air in the system.** Air is a mixture of noncondensable gases that behave according to Dalton's Law of Gases (pressures of each gas are added together to find total pressure). To achieve normal head pressure, the air must be removed.
- **Overcharge of refrigerant.** Any excess refrigerant collects in the condenser, reducing condenser capacity and raising the balance point.
- **Dirty condenser.** If dirt and debris is allowed to accumulate on an air-cooled or scale accumulate inside a water-cooled condenser, it will act as an insulator, reducing condenser capacity and raising the balance point.
- **Reduced air or water movement.** If there is no movement of a heat-transfer medium (air or water) past the condenser tubes, head pressure will rapidly rise to an excessive level. Head pressure safety controls prevent damage. Lack of movement in an air-cooled condenser can be caused by a burned-out fan or airflow blockage. In a water-cooled compressor, a closed water valve is a possible cause.
- **High suction pressure.** If the suction (low-side) pressure is above normal, the head pressure will also rise. High suction pressure and heat content require the compressor to achieve a higher balance point. System pressures correspond. If the high-side pressure goes up, so does the low-side pressure, and

vice versa. The only exception is when a compressor is defective, in which case an increase in suction pressure would *not* be accompanied by a corresponding increase in high-side pressure.

17.5 Charging Refrigerant into a System

In the context of a refrigeration or air-conditioning system, the term ***charging*** means to add refrigerant. Systems use different kinds and amounts of refrigerant, and the amounts are critical at times. For example, in some domestic refrigeration systems, a 1/2-ounce overcharge or undercharge can noticeably affect operation. In other systems, there is some room for error. For example, systems that use a liquid receiver can operate with an excess of refrigerant. As a rule, knowing *when to stop* adding refrigerant to a system is as important as knowing *how* to add refrigerant.

Single compound refrigerants can be charged into a system as a *vapor* or as a *liquid*; however, refrigerant blends must only be charged as a liquid. Always check if a refrigerant is a blend before charging a system. Before it can be charged, a system must be tested for leaks, dehydrated, and evacuated. Then, a manifold gauge can be connected to the suction service valve and the discharge service valve for charging. Both connections are important since they provide necessary pressure readings to monitor during charging.

Vapor Charging

Vapor charging involves charging refrigerant vapor into the low side of a system using the gauge port on the suction service valve. The service hose (normally yellow) from the manifold gauge is connected to a refrigerant cylinder, which must be in the *upright* position or use the "gas" valve to permit vapor charging. The cylinder valve is then opened, transferring control to the manifold gauge.

Cylinder pressure must be *higher* than suction pressure since refrigerant will not flow without a pressure difference. Sometimes, cylinder pressure must be carefully increased by warming the cylinder with hot water. As mentioned before, never use a torch to heat a refrigerant cylinder. Use warm water and never raise cylinder temperature above 125°F (52°C).

With both service valves in the cracked position, open the low-pressure manifold gauge hand valve. Refrigerant vapor will travel from the cylinder through the yellow (service) hose, through the left side of the manifold into the blue hose, and then into the compressor through the suction service valve, **Figure 17-24**.

Charging will take longer if it is done using Schrader valves rather than service valves. The passage for gas or liquid through a Schrader valve is much smaller than the passage through a service valve. Since the smaller passage restricts refrigerant flow, charging takes longer.

However, avoid charging too quickly. Fast charging may cause discharge pressure to rise rapidly. Do not exceed normal operating pressure. When charging with vapor, the process must be stopped periodically to check low-side operating pressure. The pressure should correspond to the conditions surrounding the evaporator.

Note that system operating pressures will not be normal until the system is fully charged and has achieved normal operating temperatures. *Both* high-side and low-side pressures will be too *low* if the system is *undercharged*. If the system is *overcharged*, the *high-side* pressure will be *high*. The low-side pressure will be high if the evaporator temperature is high.

Charging of small systems with hermetic compressors should always be done using the vapor method. Liquid refrigerant can easily damage hermetic compressors.

SAFETY NOTE

High-Pressure and Manifold Gauge Hand Valves

When the system is running during vapor charging, *never* open the right (high-pressure side) manifold gauge hand valve. Since the system's high-side pressure is greater than the cylinder pressure, the compressor will pump into the cylinder, **Figure 17-24**. This could contaminate the cylinder and cause the cylinder to rupture. The high-pressure gauge is connected to the discharge service valve *only* to allow monitoring of high-side pressure while charging.

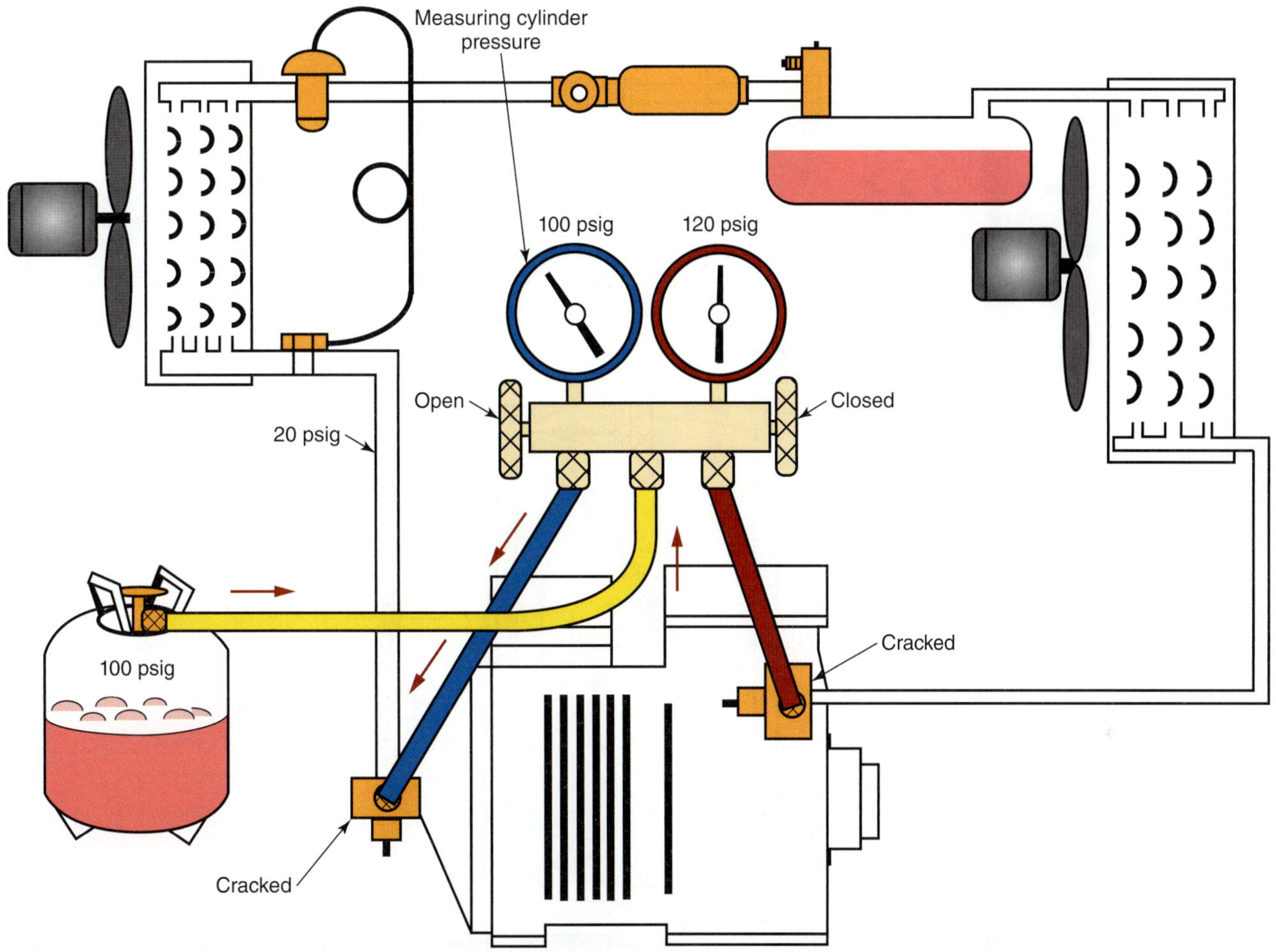

Goodheart-Willcox Publisher

Figure 17-24. During vapor charging, the refrigerant vapor flows from the cylinder through the hoses to the compressor. Cylinder pressure must be higher than suction pressure so the refrigerant vapor flows into the system.

Liquid Charging

Liquid charging involves charging refrigerant liquid into a refrigeration system. Charging a system with refrigerant in a liquid state is faster than vapor charging; however, liquid charging requires advanced skill. When liquid charging, the cylinder pressure remains constant since the liquid changes state to a vapor *inside the system*, rather than inside the cylinder. For liquid charging, the cylinder is usually inverted, as shown in **Figure 17-25**. Some cylinders, such as those used for refrigerant recovery, are equipped with a dip tube to allow liquid charging without inverting the cylinder.

Proper procedure for liquid charging entails *metering* (*throttling*) the refrigerant into the low-pressure side of the system by adjusting the manifold gauge hand valve. The low-pressure manifold gauge hand valve is cracked open so a pressure 10–15 psig (69–104 kPa) higher than suction pressure is maintained. Frost may form on the manifold body and the hose connected to the suction service valve.

Because low-side pressure increases as refrigerant is added to the system, the left manifold valve should be closed at regular intervals to check suction pressure. When the valve is closed, frost disappears. After pressure is checked, the valve should be cracked open once again, and the 10–15 psig (69–104 kPa) pressure difference again established.

Watch the high-side gauge to avoid excessive head pressure. If available, observe the sight glass for bubbles. When bubbles disappear from the sight glass and the two gauges register the correct pressure, the system is fully charged.

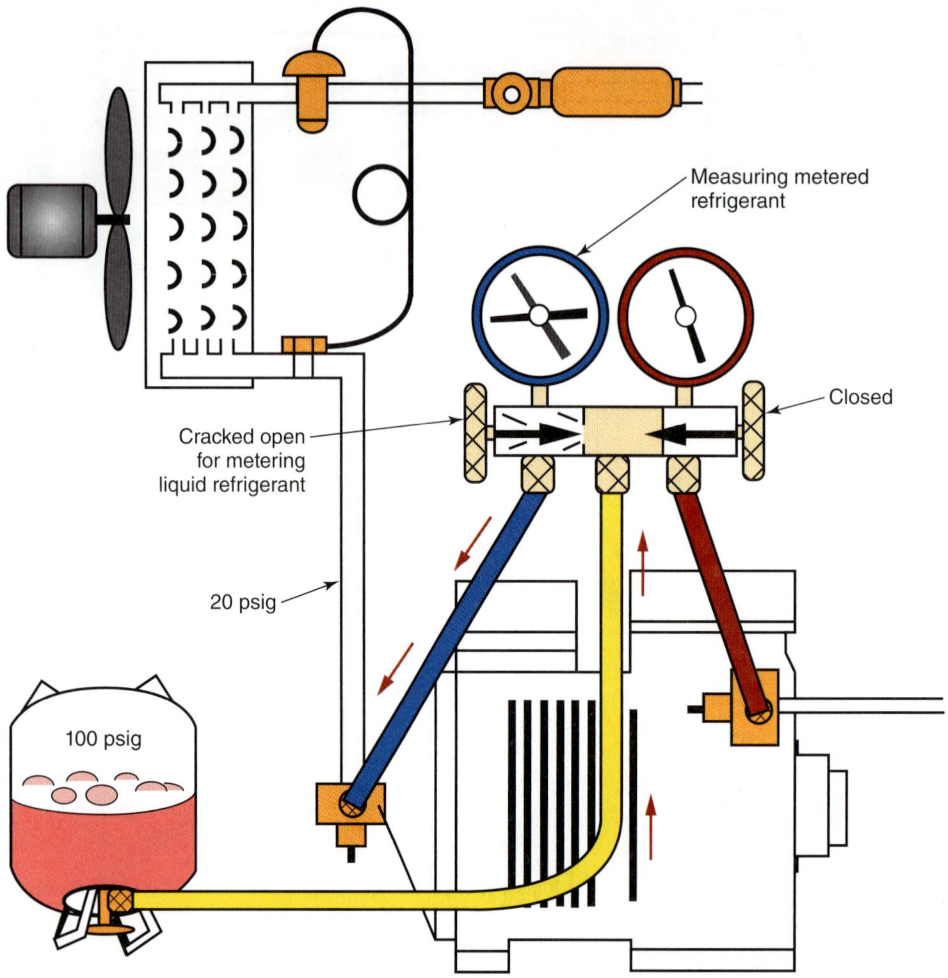

Goodheart-Willcox Publisher

Figure 17-25. When liquid charging, the technician must be careful to prevent liquid refrigerant from entering and possibly damaging the compressor. Gauge pressure should be 10–15 psig above actual suction pressure. Some types of cylinders can be left upright for liquid charging.

On larger systems, the liquid receiver service valve is usually used for liquid charging. The valve is frontseated, causing the system to pump down (evacuate the low side). When liquid is charged into the system through the LRSV valve, it travels to the expansion valve and the evaporator. The liquid becomes a vapor before it reaches the compressor, **Figure 17-26**.

Systems with liquid receivers are slightly overcharged on purpose. The technician must determine how much liquid is being stored in the liquid receiver. On large systems, it may be big enough to hold the contents of three or four 30-pound cylinders of refrigerant. Periodically, charging should be stopped and the liquid line sight glass checked for bubbles. When bubbles stop appearing, the liquid line is full, and the receiver dip tube (outlet) is covered with liquid. Some additional refrigerant is added to the system. The amount is determined by the technician's experience and judgment. The size of the liquid receiver and how much liquid it already contains must be considered.

Charging by Weight

Charging by weight is the most accurate and desired method of charging a system. The

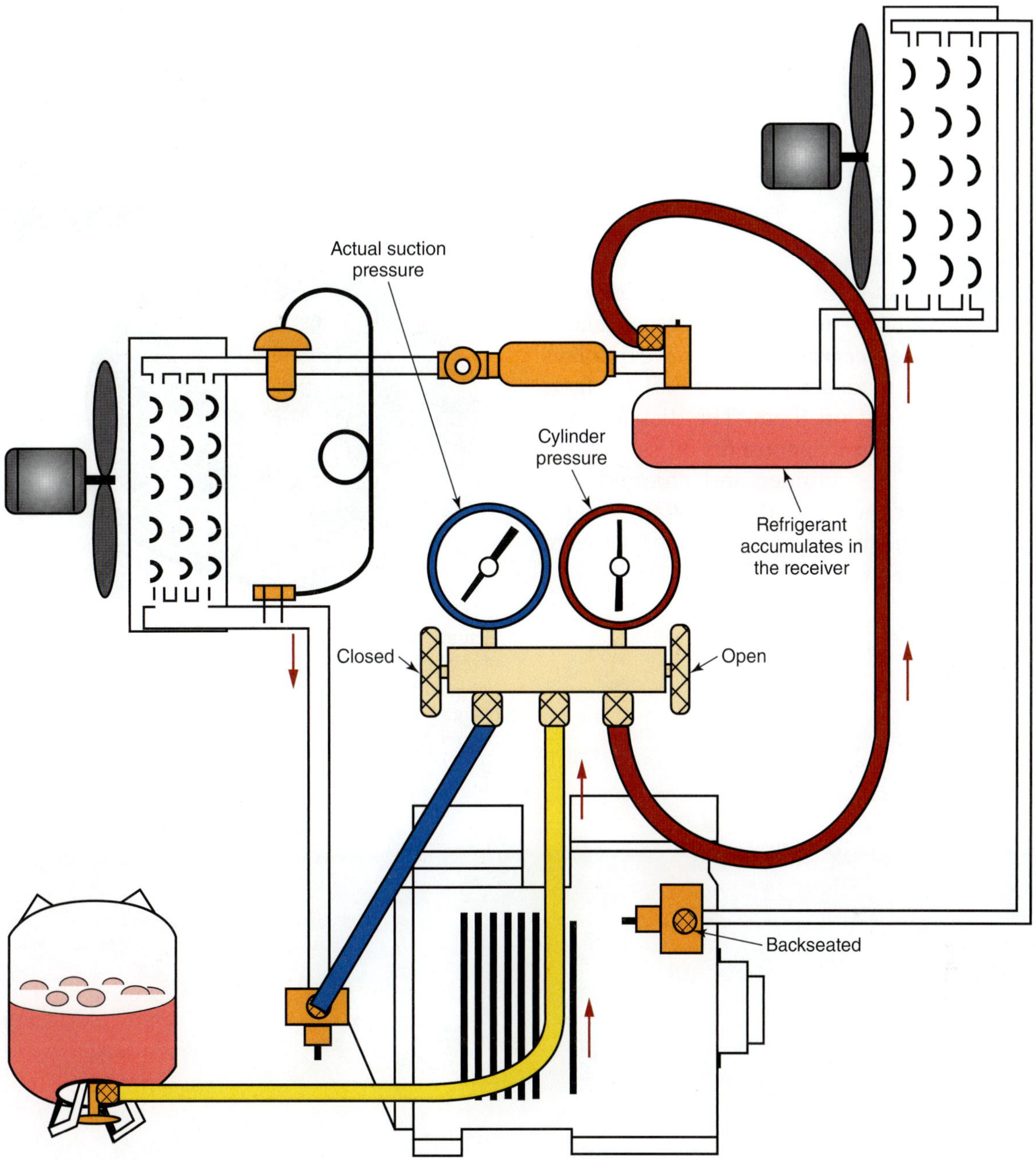

Goodheart-Willcox Publisher

Figure 17-26. For larger systems, liquid charging is often done through the liquid receiver service valve.

recommended charge is stamped on the data plate of the equipment, making it quick and easy to measure an exact amount of refrigerant and charge the system. Installation manuals will provide instructions on adding any additional charge due to line set lengths. The system must be thoroughly evacuated before charging.

Electronic Scales

Accurate, easy-to-use weighing devices are readily available for use by technicians, **Figure 17-27**. These ***electronic scales*** have displays that can be set to read zero when a full refrigerant cylinder is placed on them. The display then shows the amount of refrigerant (in ounces) being withdrawn from the cylinder and charged into the system. When the desired refrigerant weight is reached, close the manifold gauge hand valve to stop the process.

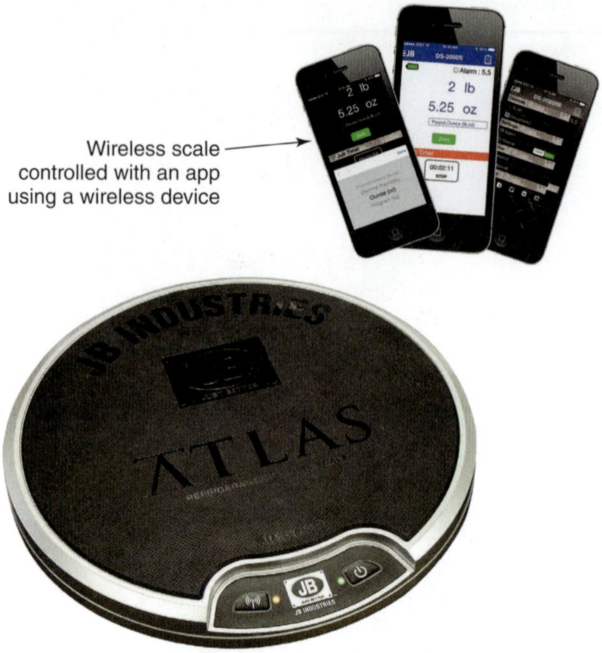

JB Industries

Figure 17-27. Electronic scales permit accurate measurement of refrigerant while charging. The cylinder is placed on the scale pad, and the weight readout is adjusted to zero. As refrigerant is charged into the system, the readout will keep a running total in pounds and ounces.

Chapter Review

Summary

17.1 Refrigeration and air-conditioning systems are evacuated to remove moisture, air, and contaminants from the system. Before evacuating, a system's internal pressure must be reduced to atmospheric using a vacuum pump. When a system is ready to be evacuated, technicians use either the deep vacuum method or the triple evacuation method.

17.2 Leaks can be detected in a refrigeration system using the bubble method, a halide torch, electronic and ultrasonic leak detectors, visual inspection of refrigerant oil, and special dyes. A standing pressure test is performed to verify that all leaks were corrected.

17.3 Refrigerant cylinders must never be completely filled with liquid; some space must be left for hydrostatic expansion. When transferring refrigerants, avoid contamination, avoid venting to the atmosphere, and safely handle the cylinders.

17.4 To control low-side pressures in a refrigeration system, the temperature difference must be maintained during operation. To control high-side pressures, the balance point of the condenser must be monitored by finding the head pressure. Higher-than-normal head pressures are caused by air in the system, an overcharge of refrigerant, a dirty condenser, reduced air or water movement, or high suction pressure.

17.5 Vapor charging of refrigerant involves charging refrigerant vapor into the low side of the system using the gauge port on the suction service valve. Liquid charging of refrigerant involves charging liquid refrigerant into the low side of the system by adjusting the manifold gauge hand valve. On larger systems, a sight glass and liquid receiver service valve can also be used for liquid charging.

Know and Understand

1. System evacuation is performed using a _____ to create a negative pressure in the system to remove moisture and noncondensables. (17.1)
   A. recovery cylinder
   B. vacuum pump
   C. special service valve
   D. water pump
2. In addition to air and moisture, refrigeration systems should be evacuated of any _____, which is caused by a chemical breakdown of oil that combines other materials. (17.1)
   A. flux
   B. acid
   C. sludge
   D. rust
3. *True or False?* The smaller the refrigeration system, the less time it will take to achieve a proper evacuation. (17.1)
4. Vacuum pump oil should be changed _____. (17.1)
   A. at the end of every evacuation
   B. every three evacuations
   C. as often as it takes to pull the system down to 500 microns
   D. at least twice during an evacuation
5. A technician is pulling a deep vacuum and notices the micron gauge reading start to rise and then stops. What should be the technician's next step? (17.1)
   A. Backseat the low-pressure valve on the manifold gauge.
   B. Disconnect the vacuum pump from the system.
   C. Find and repair any leaks in the system.
   D. Attempt to pull the system down to 500 microns again until the reading stays at 500 microns for 15 minutes.
6. The purpose of using an inert gas like nitrogen while performing a triple evacuation is to _____. (17.1)
   A. introduce moisture into the system
   B. dry out the system
   C. decrease system pressure
   D. locate any leaks

7. When using the ____ method for leak testing, refrigerant must be recovered from the system before pressurizing the system with refrigerant, inert gas, or a mixture of these gases. (17.2)
   A. vacuum
   B. bubble
   C. halide torch
   D. electronic leak detector
8. *True or False?* When checking the low-side of a system for a leak, a technician should *never* pressurize a system above its low-side test pressure. (17.2)
9. A(n) ____ detects the sound a refrigerant leak makes, can pinpoint a leak regardless of contamination, and works on pressurized systems or a vacuum. (17.2)
   A. electronic leak detector
   B. ultrasonic leak detector
   C. special dye
   D. halide torch
10. A standing pressure test involves ____. (17.2)
    A. charging a system with any refrigerant and nitrogen, then evacuating the unit
    B. charging a system with its appropriate refrigerant, then raising system pressure to the high-side pressure
    C. pulling a vacuum on a system with no refrigerant in it
    D. raising system pressure to the low-side test pressure using nitrogen in a system that has had its refrigerant charge recovered
11. Steel cylinders can charge refrigerant as a ____. (17.3)
    A. vapor only
    B. liquid only
    C. vapor or liquid
    D. superheated vapor
12. Which type of refrigerant cylinders should never be refilled? (17.3)
    A. Returnable
    B. Recovery
    C. Steel
    D. Disposable
13. A refrigerant cylinder used for holding transferred refrigerants must first be evacuated to a vacuum of at least ____ in. Hg. (17.3)
    A. 0
    B. 14.7
    C. 28
    D. 500
14. In a refrigeration system, evaporator temperature should be about ____°F colder than the return air. (17.4)
    A. 10
    B. 15
    C. 20
    D. 40
15. Boiling refrigerant inside the evaporator is about ____°F colder than the return air. (17.4)
    A. 10
    B. 15
    C. 20
    D. 40
16. *True or False?* When a refrigeration system is turned off, there is a temperature difference only between the refrigerant and the evaporator. (17.4)
17. Excessive airflow places a large heat load on the evaporator, resulting in ____. (17.4)
    A. low temperatures and pressures
    B. high temperatures and pressures
    C. increased system efficiency
    D. moisture and potential freezing in the evaporator surface
18. Temperatures for air-conditioning systems are calculated from the ____. (17.4)
    A. return air
    B. supply air
    C. air space
    D. ambient air
19. If ambient temperature is 78°F, the temperature of the refrigerant in the condenser should be within what range? The condensing unit is a high-efficiency, air-cooled unit. (17.4)
    A. 78°F–83°F
    B. 93°F–103°F
    C. 98°F–103°F
    D. 108°F–113°F
20. When vapor charging, ____ pressure must be higher than suction pressure since refrigerant will not flow without a pressure difference. (17.5)
    A. evaporator
    B. condenser
    C. cylinder
    D. discharge

21. What should the positions of the service valves and the manifold gauge hand valves be when vapor charging through the low-pressure side of the system? (17.5)
    A. Both service valves backseated, high-pressure valve open and low-pressure valve closed
    B. Both service valves frontseated, both manifold gauge valves open
    C. Both service valves cracked, low-pressure valve open, and high-pressure valve closed
    D. Suction service valve cracked, discharge service valve frontseated, low-pressure valve open, and high-pressure valve closed
22. *True or False?* When a refrigeration system is undercharged, both high-side and low-side pressures will be too high. (17.5)
23. When liquid charging, cylinder pressure _____. (17.5)
    A. increases
    B. decreases
    C. remains constant
    D. decreases then rapidly increases as charging nears completion

Apply and Analyze

1. What process can increase the speed of moisture removal from the compressor while pulling a vacuum? (17.1)
2. List the five main factors that determine the time required for a vacuum pump to remove moisture and air from a system. (17.1)
3. Explain the proper conditions for using an electronic leak detector. (17.2)
4. List three major concerns to be aware of when transferring refrigerants. (17.3)
5. On a high-efficiency, air-cooled system using R-600a as a refrigerant, what should the head pressure range be if ambient air temperature is 80°F? Consult the pressure-temperature chart in Chapter 15. (17.4)
6. What can happen if a system is vapor charged too quickly? (17.5)
7. Explain the flow of refrigerant during liquid charging through the liquid receiver service valve. (17.5)

Critical Thinking

1. Describe the type of vacuum pump needed to evacuate a 35-ton system that uses R-32 refrigerant. (17.1)
2. Explain how to perform a standing pressure test to verify that all leaks have been repaired in a refrigeration system. (17.2)
3. Speculate on why it is easy to overcharge a system when liquid charging. (17.5)

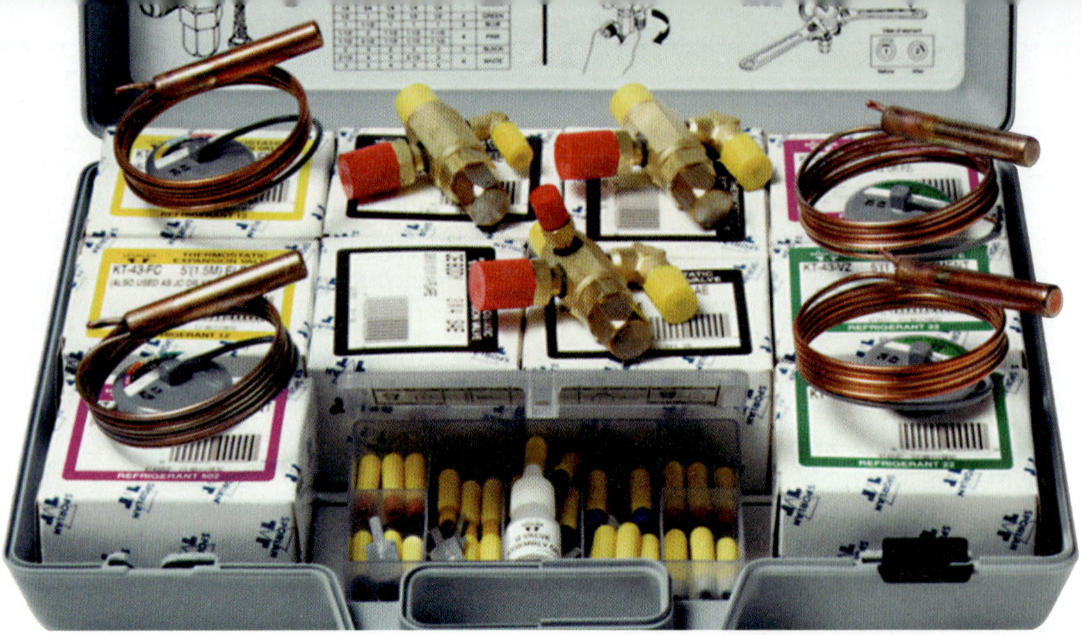

Courtesy of Sporlan Division - Parker Hannifin Corporation

18 Working with Metering Devices

LEARNING OUTCOMES

After studying this chapter, you will be able to:

18.1 Explain the operation of automatic expansion valves (AEVs), how to adjust and select AEVs, the disadvantages of AEVs, and how to troubleshoot them.

18.2 Select, install, and service capillary tubes.

18.3 Compare fixed bore (piston) metering devices with capillary tubes.

18.4 Install, adjust, and troubleshoot thermostatic expansion valves (TEVs).

18.5 Install and service electronic expansion valves (EEVs).

TECHNICAL TERMS

automatic expansion valve (AEV)
balanced port valve
bleedover
capillary tube
cross-charged
electronic expansion valve (EEV)
erratic
evaporator pressure drop
externally equalized
fixed bore (piston) metering device
flash gas
floodback
flooded start
hunt and surge
internally equalized
low evaporator load
migrate
modulates
overload
pressure drop
refrigerant distributor
refrigerant metering device
sensing bulb
starve
suction line pressure drop
thermostatic expansion valve (TEV)
undercharged

Introduction

A ***refrigerant metering device*** is the device that controls the amount of liquid refrigerant entering the evaporator. Since *all* liquid must boil off (vaporize or change state to a gas) inside the evaporator, the metering device has the task of metering into the evaporator the precise amount of liquid refrigerant needed to match the evaporator's boiling capacity. If not enough liquid is metered in, the evaporator will ***starve***. If too much liquid is metered in, ***floodback*** (liquid in the suction line) will result.

Operation of the refrigerant metering device must be automatic so the correct amount of refrigerant is supplied to the evaporator. If the evaporator starves because too little liquid is supplied, poor cooling will result from reduced system capacity. Also, pressures will be low because the compressor will be able to remove gas from the evaporator at a rate faster than the metering device feeding in liquid. Damage to the compressor can be caused by too much liquid being fed into the evaporator when there is not enough heat available to evaporate it all. When the excess liquid that did not evaporate flows through the suction line (floodback), it can be drawn into the compressor and severely damage it.

The refrigerant metering device acts as the division point between the high-pressure and low-pressure sides of the system. Subcooled liquid at high pressure enters the metering device, then exits as a low-pressure, saturated mixture of liquid and gas. This pressure drop across the metering device is necessary for proper system operation.

Every evaporator must have a refrigerant metering device. Systems with multiple evaporators must have a metering device for each evaporator. Once metering devices are installed and properly adjusted, they are reliable and virtually trouble-free. The five types of refrigerant metering devices discussed in this chapter are listed as follows:

- Automatic expansion valve (abbreviated AEV or AXV)
- Capillary tube
- Fixed bore (piston)
- Thermostatic expansion valve (abbreviated TEV or TXV)
- Electronic expansion valve (abbreviated EEV or EXV)

18.1 Automatic Expansion Valves (AEVs)

The ***automatic expansion valve (AEV)*** is a *pressure-type* control installed in the liquid line at the evaporator inlet, **Figure 18-1**. Most AEVs are installed with flare connections and are clearly marked to reflect refrigerant flow (*in* and *out*). The highly accurate valve will maintain a constant pressure in the evaporator anytime the system is running.

Since the AEV is designed to control pressure at its outlet, it responds to and controls the pressure inside the evaporator. When evaporator pressure drops below the selected level (by as little as 1 psig or 7 kPa), the valve opens wider, admitting more liquid refrigerant into the evaporator. If evaporator pressure increases above the desired level, the valve closes slightly to reduce the amount of liquid entering the evaporator. The AEV continually ***modulates*** (adjusts) refrigerant flow in this way to maintain constant evaporator pressure.

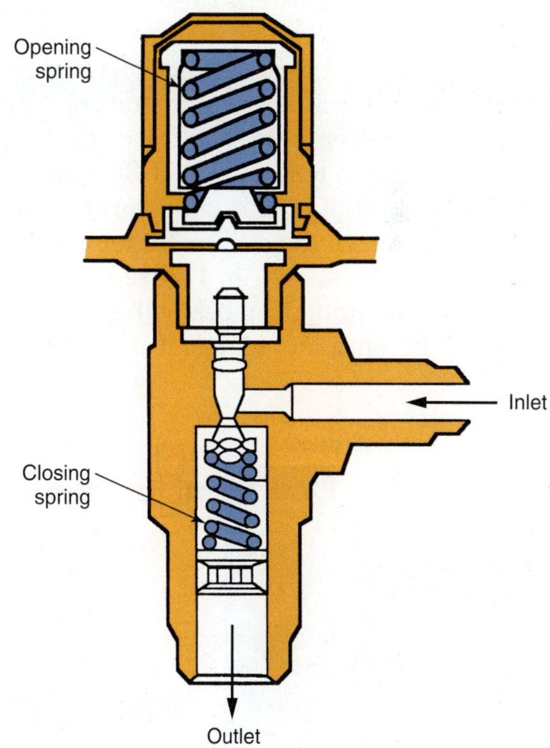

Goodheart-Willcox Publisher

Figure 18-1. A cutaway view of an automatic expansion valve operation. The adjustable opening spring in the top part of the valve forces the rod and ball assembly downward to open the valve. The closing spring moves the ball and rod assembly upward to close the valve.

When the compressor is off (not operating), the AEV is closed due to high evaporator pressure. It remains closed until the compressor cycles back on and reduces evaporator pressure to the valve's setting.

The desired pressure is set by turning an adjustment screw on top of the valve. The screw, protected by a metal cap, permits adjustment over a wide range of pressures. The technician must know the correct pressure setting for the boiling point of the refrigerant inside the evaporator. To set the pressure, the manifold gauge is installed to obtain a low-side pressure reading at the suction service valve. The automatic expansion valve is then adjusted until the proper pressure is registered on the compound gauge of the manifold gauge.

Determining Evaporator Pressure

The evaporator pressure is determined by the desired temperature of the product being cooled. Since the AEV is a constant pressure valve, the evaporator pressure and temperature remain constant. When an AEV is used as the refrigerant metering device, there is no temperature difference between the evaporator and the refrigerant.

For heat to flow, a temperature difference must exist. When the product (air) temperature is lowered enough to match the evaporator temperature, heat flow stops. As shown in **Figure 18-2**, product temperature cannot drop below the evaporator temperature because no temperature difference exists. In the same way, when the system is turned off, the temperature of the evaporator will match that of the ambient air, and no heat will flow.

The temperature-limiting feature of systems equipped with an AEV is desirable when cooling liquid products, such as milk or water. The evaporator is submerged in the liquid, and the AEV pressure is set to keep the evaporator temperature just above the freezing point of the liquid. Since heat flow ceases when there is no temperature difference, the liquid temperature can never drop below the evaporator temperature and cannot freeze.

This feature is illustrated in the milk cooling tank shown in **Figure 18-3**. The wall-type evaporator inside the tank is in direct contact with the milk, allowing excellent heat transfer by conduction. The system uses R-134a, so the AEV is set to maintain a pressure of 29 psig (197 kPa). At this pressure, R-134a produces an evaporator temperature of 33°F (0.5°C)—one degree above the freezing point of milk. The milk stays cold but cannot freeze.

How the AEV Works

The automatic expansion valve functions as a spray nozzle, breaking up the liquid refrigerant into fine droplets as it enters the evaporator. This makes it easier for the refrigerant to vaporize or change to a gas. As the gas is drawn into the compressor, pressure inside the evaporator is decreased. The AEV maintains a constant pressure level in the evaporator by metering in enough

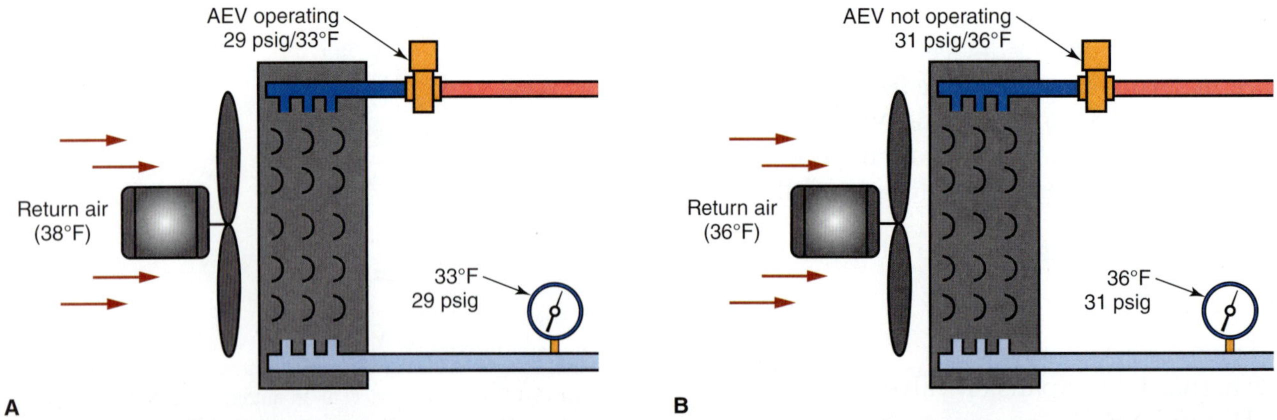

Goodheart-Willcox Publisher

Figure 18-2. Temperature when an AEV is running versus not running. A—When a system is running, the air temperature drops until it matches the evaporator temperature. Heat flow stops when the temperatures are the same. B—When a system is not running, all temperatures correspond to the air temperature (no heat flow).

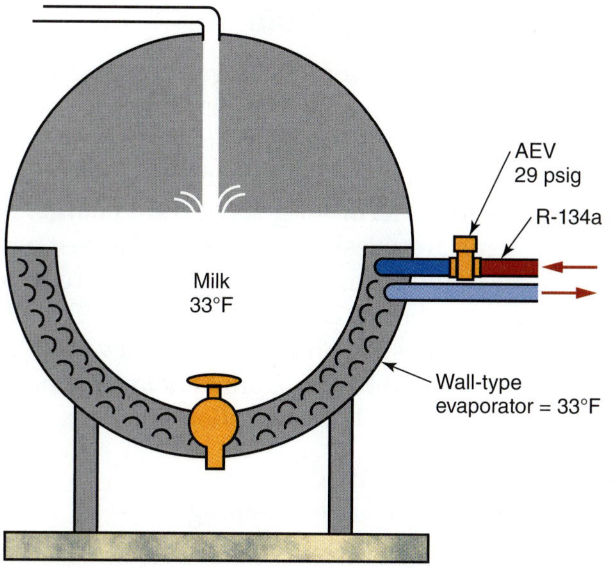

Goodheart-Willcox Publisher

Figure 18-3. The temperature-limiting feature of AEV systems allows liquids to be kept just above freezing. By holding the pressure of R-134a in the evaporator at 29 psig, the temperature of the milk stabilizes at 33°F.

liquid to balance the vapor drawn out by the compressor. When the compressor cycles off, pressure in the evaporator rises, and the AEV closes. It remains closed until the compressor cycles back on and lowers pressure in the evaporator.

Working forces in the AEV include:

- The range adjustment spring pressing downward on the diaphragm or bellows, which tends to open the valve
- The closing spring exerting an upward pressure on the push rod and ball assembly, which tends to close the valve
- Pressure on the diaphragm from the evaporator outlet, which tends to close the valve

Once set by the technician, spring pressure on the adjustment screw remains constant. As a result, the evaporator pressure determines how much the valve opens or closes. A rise in evaporator pressure will overcome spring pressure and force the valve toward a closed position. A drop in evaporator pressure will allow the spring pressure to open the valve wider. The AEV responds to slight pressure fluctuations, automatically modulating (partially opening or closing) to maintain a constant evaporator pressure.

Adjustment Procedure

A manifold gauge to monitor pressure readings is installed at the suction service valve to monitor evaporator pressure. With the system running, any adjustment of the AEV setting is immediately reflected on the compound gauge.

AEV setting adjustments are performed by removing the metal cap of the valve and turning the adjustment screw in or out. Turning the screw in (clockwise) increases the spring pressure on the diaphragm. More liquid enters the evaporator because higher evaporator pressure is needed to overcome the AEV spring pressure. Increasing the amount of refrigerant in the evaporator causes the evaporator pressure to rise. Conversely, turning the adjustment screw out (counterclockwise) has the opposite effect. It reduces spring pressure on the diaphragm, allowing it to open the valve more easily. This reduces the amount of liquid refrigerant entering the evaporator and lowers the evaporator pressure.

Controlling Product Temperature

A thermostat within the area of the product being cooled controls operation of an AEV system. When product temperature is lowered to the selected ("cut-out") point, the thermostat turns off the system's compressor. With the compressor turned off, pressure in the evaporator rises, closing the AEV.

During the "off" portion of the compressor cycle, the product temperature slowly rises. The evaporator temperature slowly rises, too. Once product temperature rises to the thermostat's cut-in point (typically 5°F or 3°C above the cut-out point), the compressor is turned on again.

When a compressor is cycled on, the AEV does *not* open immediately. It remains closed until the compressor reduces evaporator pressure to the valve setting. Adjustments to the AEV can be made only while the system is running since it is the only way a constant pressure can be maintained.

Selecting an AEV

The technician should select an automatic expansion valve with a capacity equal to the cooling capacity of the compressor at the design evaporator temperature. For example, a system with a 1-hp compressor running at an evaporator temperature of –10°F might have a cooling capacity of only 1/3 hp, so a 1/3 hp valve should be chosen. Using a valve that is too small will starve the evaporator. A valve that is too large will ***hunt and surge*** (fluctuate from fully open to fully closed) at start-up.

AEV Disadvantages

Increased heat load on the evaporator of an AEV-equipped system causes the evaporator pressure to rise, which causes the AEV to close or "choke down." The evaporator is starved under peak load conditions, and the time needed to bring the product temperature down to the desired level is increased. The AEV is best applied to equipment involving a fairly stable heat load, such as ice-making machines.

To control product temperature, the AEV system requires a thermostat. Since low-side pressure is constant on an AEV system, pressure-type motor controls cannot be used.

Systems using the AEV were once popular but are no longer common. They have largely been replaced by capillary tube or thermostatic expansion valve systems. The *operating principle* of the automatic expansion valve remains in wide use but not as the primary refrigerant metering device. Special valves using the AEV principle have different names and uses and are explained later in this book.

Troubleshooting the AEV

The automatic expansion valve (AEV) prevents evaporator pressure from dropping below a specified level. The compressor must be able to remove excess pressure from the evaporator. The AEV is a reliable system component; however, it *cannot* control evaporator pressure during the off cycle.

Undercharged System

A system that is ***undercharged*** suffers from a lack of refrigerant. Proper system pressures cannot be maintained without sufficient refrigerant in the system. Many AEV systems do not have a sight glass for determining refrigerant charge. In such systems, troubleshooting the AEV is best performed with the manifold gauge, taking pressure readings from both sides of the system. Trouble with the AEV is usually indicated when evaporator pressure becomes too low. Causes of low evaporator pressure may include the following:

- The AEV is defective.
- The AEV setting may be wrong.
- The system may be low on refrigerant (undercharged).

Before diagnosing a defective valve or incorrect setting, check to see if the system is undercharged. An AEV cannot feed an evaporator properly unless sufficient liquid is present. When the system is undercharged, a low-side pressure reading will be deceptive because the AEV will try to maintain a constant pressure. Its valve may be wide open, but sufficient liquid is not available to increase evaporator pressure, **Figure 18-4**.

An undercharged system must be checked for leaks. When found, the leaks must be repaired and tested. It will be necessary to change the filter-drier, then evacuate and recharge the system.

Erratic Valve Settings

A valve that is ***erratic*** does not maintain a particular setting for any length of time. If the refrigerant charge is correct, but the AEV requires periodic readjustment, then the valve should be replaced. An erratic valve is difficult to identify. A thorough examination of other possibilities is required before determining that a valve is erratic.

Contaminants

Erratic valve operation can also be caused by a foreign substance (ice, sludge, or wax) fouling the valve seat. Ice, sludge, or wax can be detected by turning off the system and warming the valve with a heat lamp. *Do not use a torch!* Heat will loosen or melt the foreign matter so it passes through the

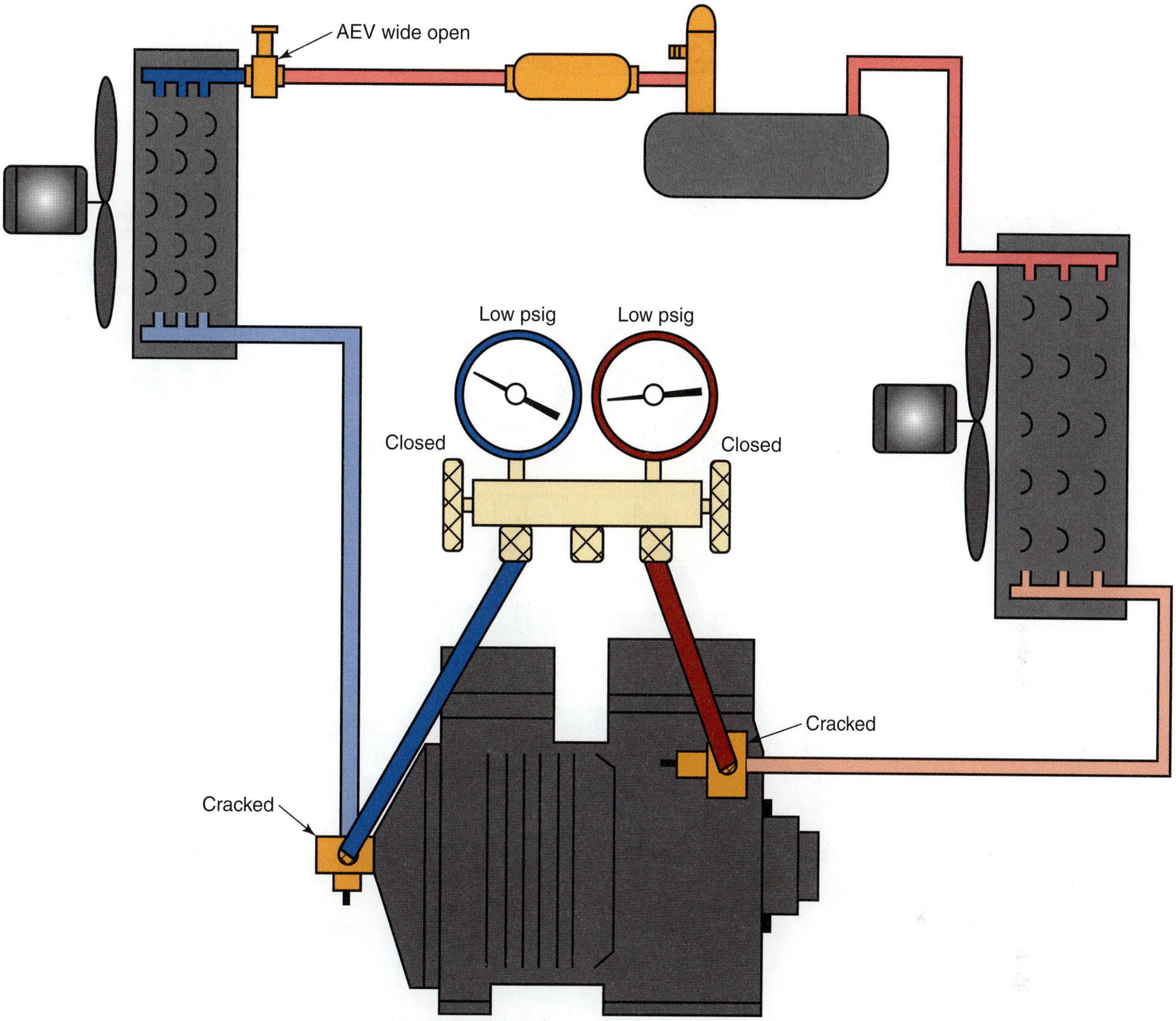

Goodheart-Willcox Publisher

Figure 18-4. When the system is undercharged, the AEV cannot feed the evaporator properly. The low-side pressure reading may be deceptive because the valve will try to maintain a constant pressure.

valve. Then, the AEV will work properly until the contaminants once again foul the valve seat. A contaminated system can be cleared up by changing the filter-drier as often as necessary to restore cleanliness.

Poor Cooling, Low Head, Normal Suction

The customer may complain, "The system is not cooling right. The product temperature is high and the system runs all the time." The gauges reveal low head pressure, but suction line pressure is acceptable. Is the system low on refrigerant? Maybe. Check for bubbles in the sight glass. The sight glass may show a full charge. Adding more refrigerant will raise the head pressure but will not correct the problem, **Figure 18-5**.

The compressor may have defective internal valve reeds or a broken piston rod. The compressor will run but is unable to draw gas and compress it properly, operating at half its normal capacity. The low-side pressure will be normal because the AEV is choked down to match the lower capacity of the compressor. The low rate of refrigerant removal from the evaporator will result in slow lowering of product temperature.

A defective compressor is usually indicated by *lower-than-normal* discharge pressure and *higher-than-normal* suction-line pressure. Troubleshooting the customer's problem is therefore difficult because the AEV maintains a normal suction pressure.

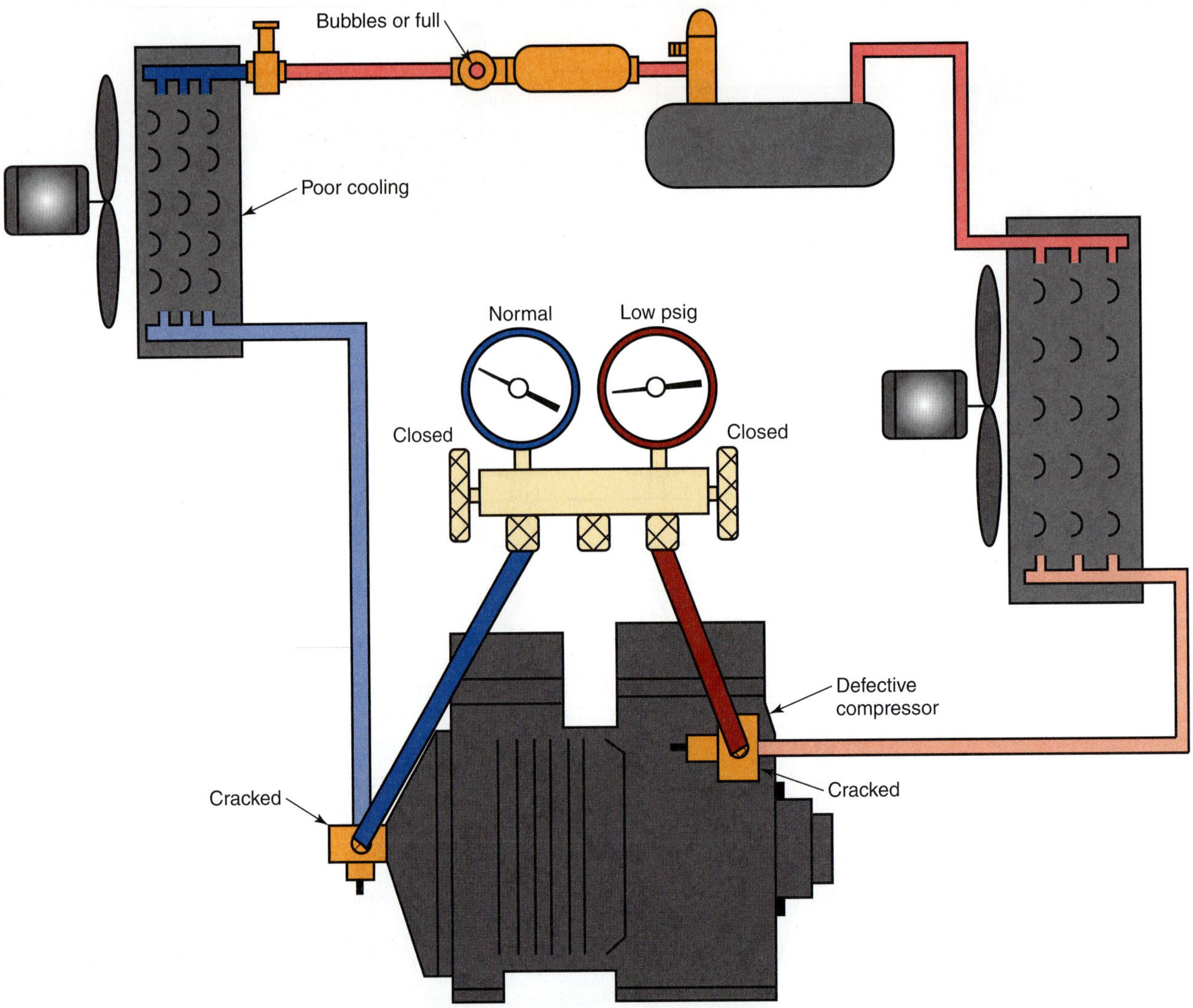

Goodheart-Willcox Publisher

Figure 18--5. Possible causes of low head pressure and poor cooling include undercharging and a defective compressor.

Poor Cooling, High Head, Normal Suction

Poor cooling with long running time is frequently caused by high head pressure. See **Figure 18-6**. Pressure drop across the AEV is excessive. It results in extra "flash gas," which increases evaporator pressure. High head pressure causes the AEV to throttle down, resulting in reduced cooling capacity and long running time. The AEV will maintain constant suction pressure, but head pressure will be higher than normal. Correcting head pressure will alleviate the problem, permitting the valve to feed subcooled liquid.

18.2 Capillary Tube Metering Devices

A ***capillary tube*** is a length of copper tubing with a tiny, accurately sized inner diameter (ID, or hole throughout its length). Capillary tubes are available in various sizes for several purposes, **Figure 18-7**.

The capillary tube is used in small refrigeration systems as a refrigerant metering device. The tube length and inner diameter are accurately calculated to restrict liquid refrigerant flow to control the amount of liquid entering the evaporator.

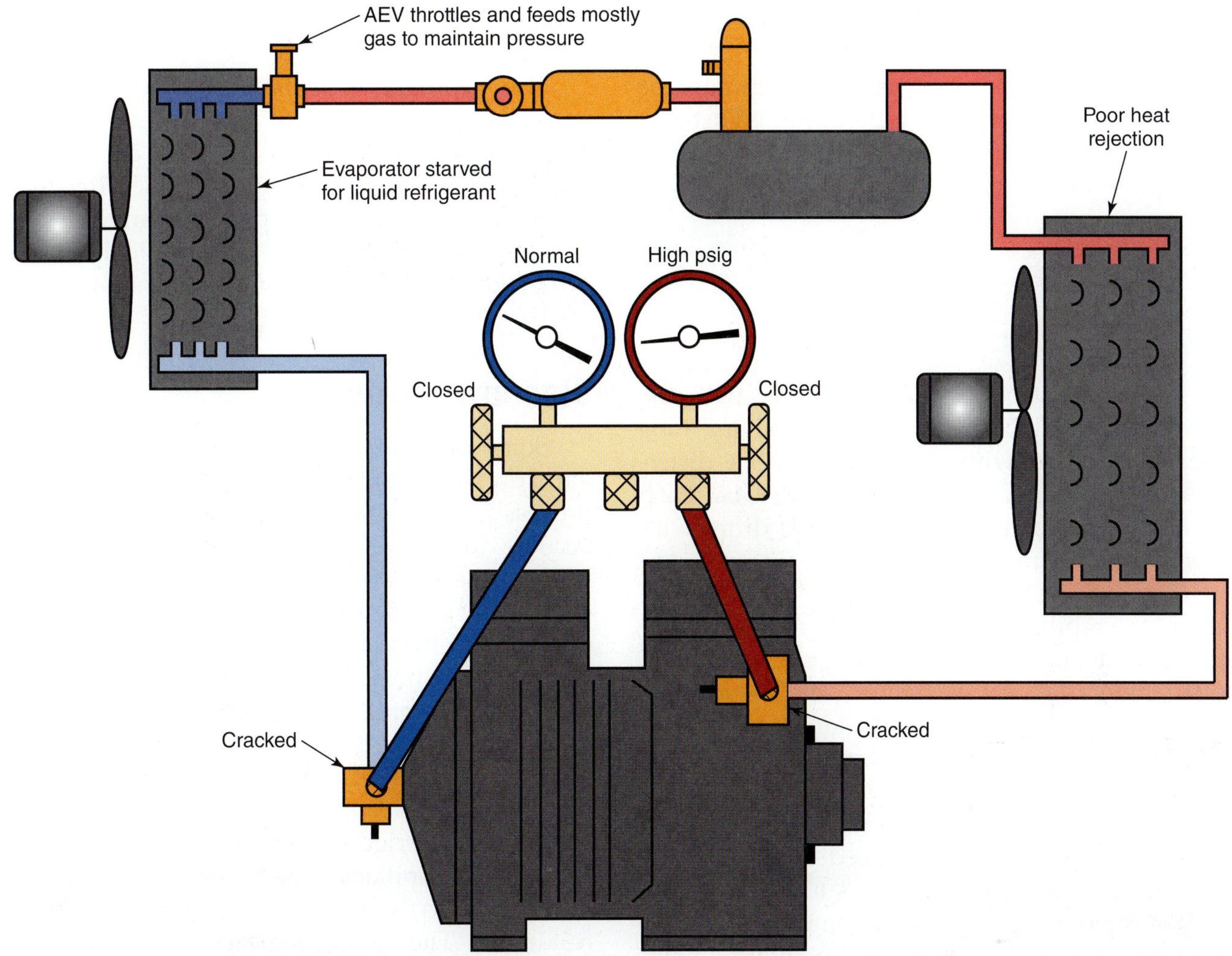

Goodheart-Willcox Publisher

Figure 18-6. High head pressure causes the AEV to throttle down, "starving" the evaporator by feeding more gas than liquid. The cooling capacity of the system is reduced significantly.

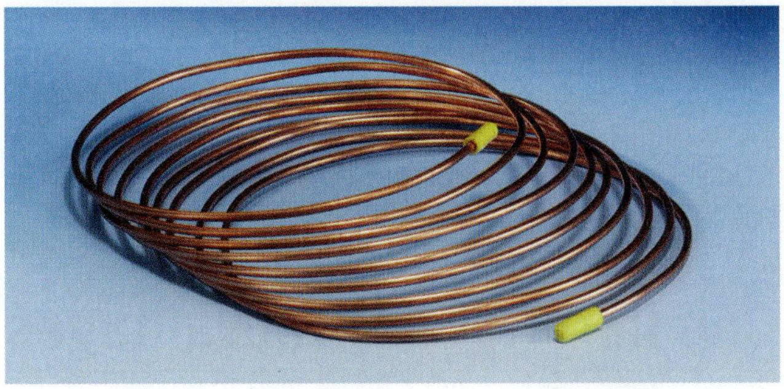

Sealed Unit Parts Co., Inc.

Figure 18-7. Capillary tubes come in different lengths and inside diameters.

In operation, one end of the capillary tube is connected to the condenser outlet and the other end to the evaporator inlet. In this way, the tube serves as both the liquid line and a metering device, **Figure 18-8**.

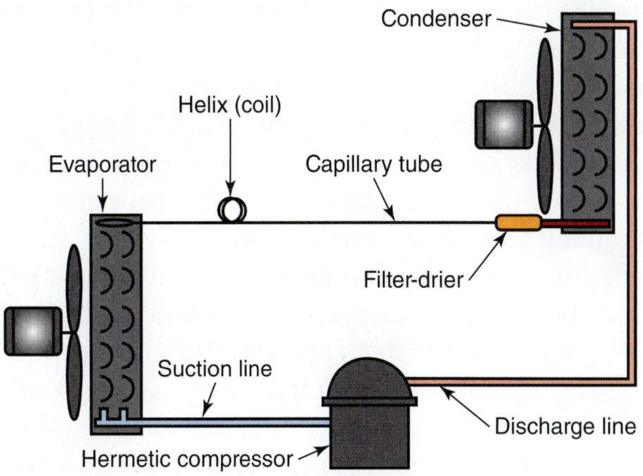

Goodheart-Willcox Publisher

Figure 18-8. In a capillary tube system, the capillary tube controls the flow of liquid refrigerant into the evaporator. It also serves as the liquid line to connect the condenser and evaporator. Capillary tube systems are often used on domestic and small commercial refrigeration systems.

Capillary Tube Selection

The capillary tube on a given system must be precisely sized, since the tube is *not* adjustable. Four variables should be considered when selecting capillary tubing:

- Length
- Inside diameter
- Refrigerant
- Tubing temperature

The capacity of the tube will vary with the refrigerant used. For example, a capillary tube selected for use with R-22 will have a different capacity if used with R-134a or R-404A. Each refrigerant has certain properties that differ with changing temperatures.

On many refrigeration coolers, the temperature of the capillary tube is controlled by soldering it to the suction line. This arrangement allows it to function as a heat exchanger. The suction line cools the liquid refrigerant inside the capillary tube, while heat from the capillary tube helps keep frost off the suction line.

The amount of liquid entering the evaporator must be balanced with the speed of boiling (rate at which the refrigerant changes to a vapor) and the gas-removing ability of the compressor. A perfect match must occur at the desired boiling point (pressure-temperature relationship). If too little liquid enters the evaporator to keep up with the compressor's gas-removing ability, the evaporator will starve. This results in a lower boiling point (suction pressure) and a lower cooling capacity. Excess liquid entering the evaporator raises the boiling point (pressure) and may permit liquid to enter the suction line. Liquid in the suction line (floodback) can cause severe compressor damage.

Selection and installation of the proper size of capillary tube is performed by the system manufacturer. Any replacement tube that is installed must be an *exact* duplicate. A replacement compressor installed in a capillary tube system also must be an exact replacement. The exact amount of refrigerant charge and every component used in a capillary tube system is critical. Therefore, always follow the manufacturer's recommendations.

Capillary Tube Advantages

The capillary tube is inexpensive and has no moving parts. In addition, it permits a small amount of high-side pressure to bleed over into the low-pressure side during the off cycle. ***Bleedover*** helps reduce the high-side pressure (and the starting load) on the compressor. As a result, the system can make use of a less costly compressor with low starting power.

Bleedover Process

Sometimes bleedover is referred to as a *balancing* of pressures during the off cycle. This balancing process does *not* mean the high and low pressures become identical. The pressures do not equalize because the system does not remain off long enough for condenser and evaporator pressures and temperatures to become equal. As the high-side pressure drops to ambient, the amount of flow through the capillary tube is reduced greatly due to bubbles of vapor. The evaporator is *much* colder than the condenser. The pressure in each will behave as described on the pressure-temperature chart for the refrigerant involved.

When a capillary tube system has shut off, a delay of two to five minutes must be allowed before restarting. The delay is necessary because the compressor cannot start under normal head pressure (that corresponds to temperatures 30°F–35°F or 17°C–20°C above ambient). An ***overload*** (electrical safety device) protects the compressor's motor until the head pressure is reduced to a value that corresponds directly with ambient temperature. *Cycling on the overload* should be avoided, as it is an indication that the motor is having problems. Motors and overloads are explained later in this book.

Capillary Tube Disadvantages

A capillary tube system has a fixed, nonadjustable capacity, so the amount of refrigerant is critical down to one-half ounce. Overcharging or undercharging the system will greatly affect operation. The amount of refrigerant in the system becomes the determining factor for proper control of system pressures.

Capillary systems are often self-contained, meaning the condensing unit is located within the cabinet. The cabinets are normally located inside a building where ambient temperatures are controlled (about 70°F or 21°C).

In a capillary tube system, an overcharge causes *high* head pressure, as excess liquid accumulates in the condenser. High head pressure, in turn, causes high suction pressure because a greater quantity of liquid is pushed through the capillary tube.

In a capillary tube system, an undercharge causes *low* head pressure, as the condenser is, in effect, "oversized." Since not enough liquid is being pushed through the capillary tube, low head pressure causes low suction pressure. Gas bubbles in the capillary tube compound the low-pressure problem because the bubbles tend to increase tube resistance.

Another disadvantage of the capillary tube system is the small passage through the tubing, which can easily be restricted or plugged by dirt, ice, or wax. To help guard against blockage, a filter-drier is always installed at the capillary tube inlet. When opening a capillary tube system for repairs, a thorough evacuation is critical for moisture removal. The filter-drier should always be replaced after service.

Working on Capillary Tube Systems

Capillary tube systems are normally hermetic (sealed) systems without service valves. In the manufacturing process, "pigtails" or process tubes are used to evacuate and charge the system. When charging is complete, the pigtail is pinched closed and cut off. The closed tubing end is then brazed to totally seal the system. For servicing, the technician must install saddle valves or Schrader (core-type) valves, **Figure 18-9**.

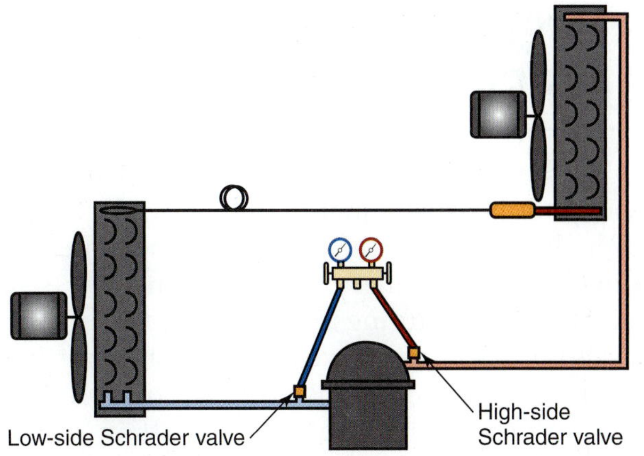

Goodheart-Willcox Publisher

Figure 18-9. Service valves are not provided on capillary tube systems. Schrader (core-type) valves or saddle valves must be installed at the points shown to obtain pressure readings.

A restricted (partly blocked) capillary tube or filter-drier causes low suction pressure and low head pressure. When the suction pressure is low, the load on the compressor is reduced and the condenser becomes "oversized." Subcooled liquid accumulates inside the condenser. Attempting to increase suction line pressure by adding refrigerant is a mistake, as it will cause excessive head pressure and compressor failure.

A restriction *always* causes a ***pressure drop***, resulting in flash gas and a cold surface at the location of the restriction. When a filter-drier's surface is cold (or shows frost), it has a restriction and must be replaced.

Another possible cause of low suction and head pressures is an *undercharge*. An undercharge indicates the system is losing refrigerant through a leak. Adding refrigerant will temporarily alleviate the problem but will not fix it. The only way to remedy a leak is to find and repair it, **Figure 18-10**.

As you can see, a restriction and an undercharge produce the same symptoms in the system. To determine which problem exists, add about 10% of the system charge into the system. If the pressures stay the same, the system has a restriction. If the pressures increase, the system is undercharged.

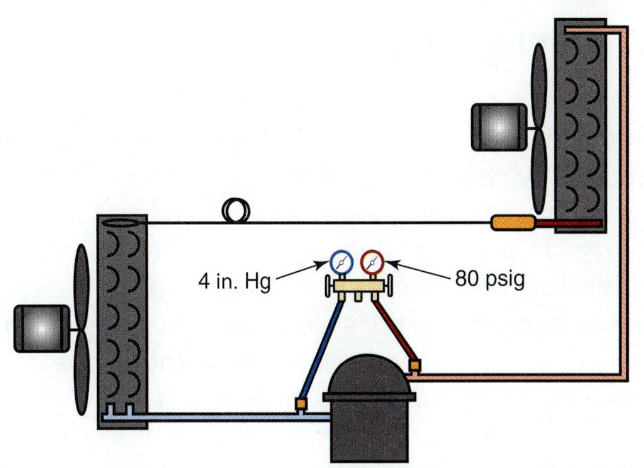

Goodheart-Willcox Publisher

Figure 18-10. Low pressure on both sides of the compressor indicates either a restriction or a leak. The technician must discover the problem causing the low pressure and fix it.

Charging a Capillary Tube System

A capillary tube system can be charged by slowly adding small quantities of refrigerant vapor to the low-pressure side. This must be done carefully because capillary tube systems are easily overcharged. Charging is done slowly to permit the system to stabilize after each addition of refrigerant. The evaporator temperature, suction pressure, and discharge pressure will not attain normal status until the system reaches normal operating temperatures.

An overcharge can be recognized by high discharge and suction pressures and by the presence of frost on the suction line. Correct the overcharge by slowly purging gas from the system using recovery equipment. The frost line will slowly retract until it disappears inside the cabinet.

Proper charging cannot be checked until the cabinet temperature is reduced to normal. During charging, the head pressure quickly rises to an above-normal point, while the suction pressure reading is in vacuum. Head pressure must not exceed 150 psig (1034 kPa) when using R-600a.

Vapor charging occurs faster than the condenser can liquefy incoming gas. Gaseous refrigerant in the capillary tube does not flow like liquid. System pressures should be allowed to stabilize after each addition of refrigerant. Cabinet temperature must be reduced to almost normal before the pressures become normal.

Weighing the Charge

An electronic scale can be used, as explained in Chapter 17, *Refrigerant Service Basics*, to quickly and easily charge a capillary tube system. The correct amount of refrigerant is measured and quickly charged into the system as a vapor. The procedure saves time, but the precise amount of refrigerant to be charged into the system must be known. The specified amount can usually be found on the data plate of the refrigeration unit. If necessary, check the service manual or contact the manufacturer to obtain the information.

18.3 Fixed Bore (Piston) Metering Devices

The ***fixed bore (piston) metering device***, like the capillary tube, is designed to operate on the principle of *restriction*. While the capillary tube is several feet long, the fixed bore is very small. See **Figure 18-11**. These often have flare connections and are installed in the liquid line at the evaporator inlet.

The brass body of the fixed bore contains a small hole (referred to as an *orifice* or *bore*) sized to match the equipment exactly. The manufacturer's recommendations must be followed. The fixed bore controls the amount of liquid entering the evaporator, and a mismatch could cause problems. The fixed bore is easily removed for cleaning or replacement.

The operating characteristics of a fixed bore and a capillary tube are identical, but the fixed bore has the advantages of being accessible and easily replaceable. These advantages account for its popularity among manufacturers of residential air-conditioning systems and heat pump units.

18.4 Thermostatic Expansion Valves (TEVs)

The ***thermostatic expansion valve (TEV)*** is the most common metering device, especially in commercial and industrial refrigeration and air-conditioning applications. The TEV can be mounted in any position at the evaporator inlet and is reliable, efficient, and virtually trouble-free, **Figure 18-12**. The operating principle and

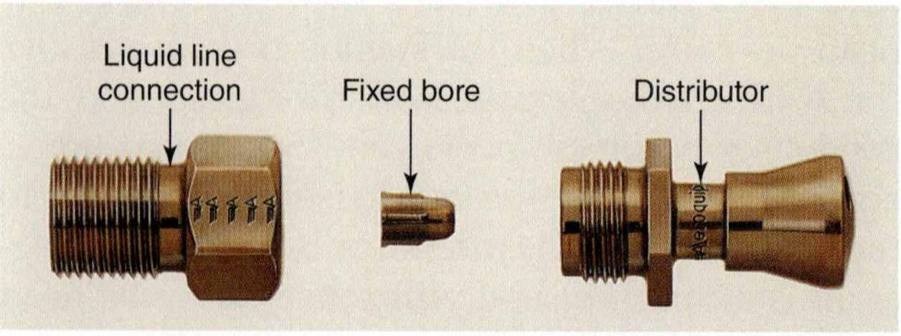

Aeroquip Corp.

Figure 18-11. The fixed bore is easily replaced. The assembly is located at the evaporator inlet.

Danfoss

Figure 18-12. Thermostatic expansion valves come in a variety of sizes and connections.

characteristics of the thermostatic expansion valve are often misunderstood. As a result, the valve is frequently and mistakenly blamed for other system troubles.

Controlling Superheat

Veteran technicians often refer to the TEV as a "superheat valve" because it maintains a constant level of superheat at the evaporator outlet. A constant level provides maximum efficiency under all types of load conditions. Too much superheat at the evaporator outlet starves the evaporator of liquid and reduces capacity. No superheat at the outlet permits liquid to enter the suction line.

The amount of superheat at the evaporator outlet varies according to the speed at which vaporization (boiling) takes place. The speed at which the liquid boils depends on the heat load placed on the evaporator. A high heat load causes the liquid to evaporate rapidly. A low heat load results in a slower change of liquid to gas.

Ideally, the evaporator should be well supplied with liquid refrigerant under all types of heat load conditions. All liquid refrigerant entering the evaporator *must* be converted to a superheated gas before leaving the evaporator.

How the TEV Works

The thermostatic expansion valve is designed to maintain a precise superheat at the evaporator outlet. This provides proper flooding of the evaporator with liquid refrigerant under all load conditions. The valve controls superheat by metering varying amounts of liquid into the evaporator. High superheat at the evaporator outlet causes the valve to open, while low superheat causes the valve to close.

As shown in **Figure 18-13**, the valve can be mounted in any position at the evaporator inlet, but the sensing bulb must be located on the suction line at the evaporator outlet.

Superheat at the location of the sensing bulb controls the amount of liquid flowing through the valve at the evaporator inlet. The desired amount of superheat (typically 10°F or 6°C) is set with an adjustable control on the valve body. The valve opens and closes slightly (modulates) in response to superheat changes. For example, if the sensing bulb detects 11°F (6.1°C) of superheat at the evaporator outlet, the valve will open enough to meter more liquid into the evaporator. If the sensing bulb registers 9°F (5°C) of superheat, the valve will close slightly to decrease liquid flow.

The advantage of the valve's modulating action is that there is a constant, properly flooded evaporator, despite changing heat load conditions.

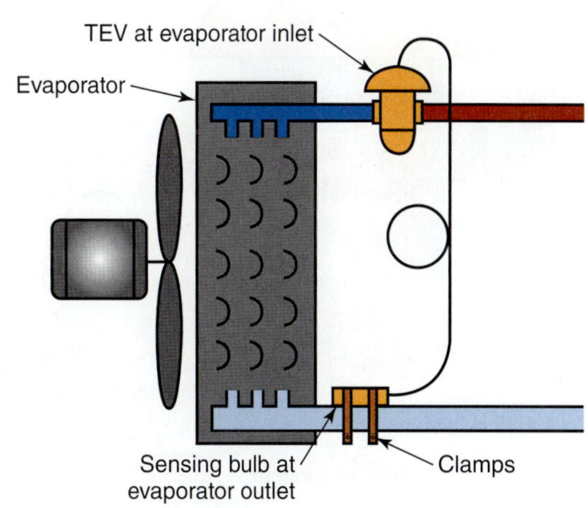

Goodheart-Willcox Publisher

Figure 18-13. The TEV is mounted at the evaporator inlet where it controls the flow of liquid refrigerant. The sensing bulb is mounted at the evaporator outlet to monitor changes in superheat.

The TEV opens and closes by moving the valve stem (pin) away from the valve seat. As shown in **Figure 18-14**, valve stem movement is controlled by four pressures exerted on working parts:

- Sensing bulb pressure (opens valve)
- Evaporator pressure (closes valve)
- Spring pressure (closes valve)
- High-side pressure (opens valve)

Sensing Bulb Pressure

The ***sensing bulb*** contains a special charge of refrigerant that is extremely sensitive to changes in temperature. The end of the sensing bulb is connected to a capillary tube that opens above the TEV's diaphragm. As the suction line temperature changes, the sensing bulb pressure increases or decreases accordingly. For example, a rise in suction line temperature would cause an increase in sensing bulb pressure that is transferred by means of a capillary tube to the diaphragm of the valve. The diaphragm is pressed downward, acting on push rods. The push rods exert a downward force on the pin carrier, moving the pin out of the seat to open the valve. See **Figure 18-15**.

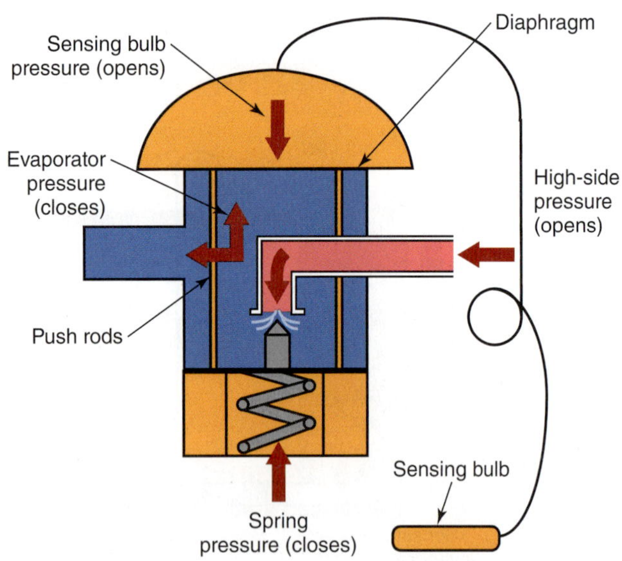

Goodheart-Willcox Publisher

Figure 18-14. While a system is operating, four pressures are acting on the TEV. Pressures tend to either open or close the valve by moving the stem (pin) into or out of its seat.

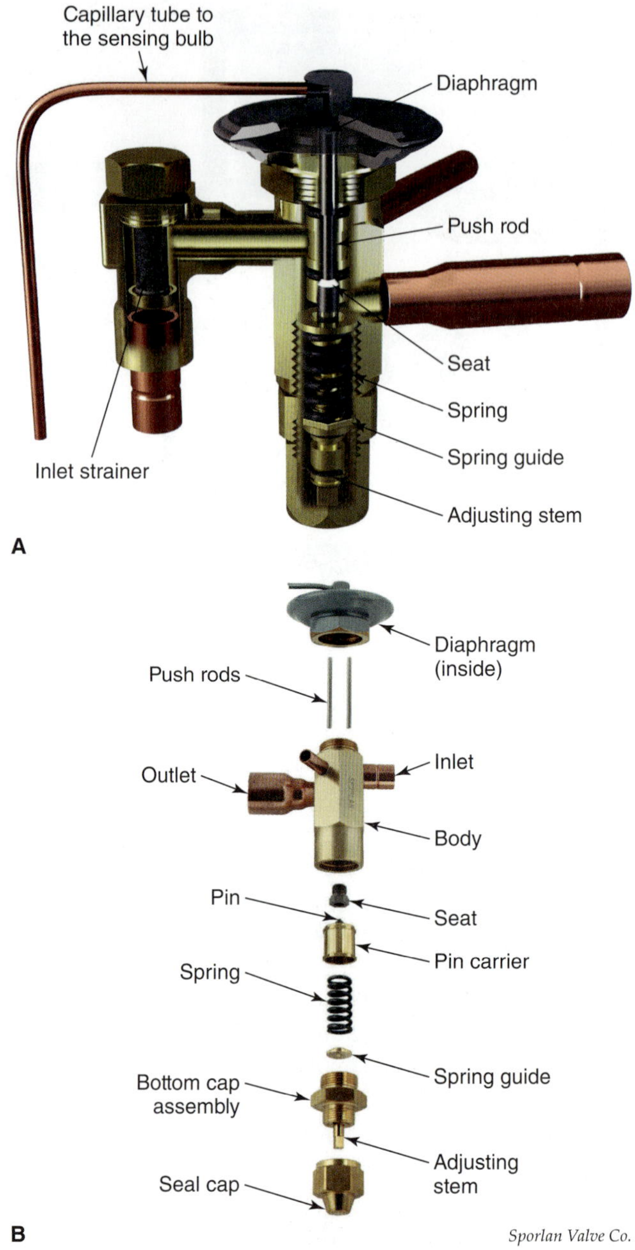

Sporlan Valve Co.

Figure 18-15. Thermostatic expansion valve. A—Cutaway view of a typical TEV. Varying pressures on the diaphragm and spring modulate operation of the valve. B—Exploded view of a TEV.

Sensing Bulb Charges

To provide accurate superheat control for different applications, different types of charges are used in the sensing bulb. In most cases, the bulb is charged with the same refrigerant used in the system. Different operating characteristics are obtained by varying the sensing bulb volume, which consequently determines the amount of its charge. Each valve is designed and charged for a particular application. A valve designed for a low-pressure application cannot be used for high-pressure applications. Sufficient refrigerant is charged into the bulb to ensure it contains some liquid under all temperature conditions.

Valves designed for use in low-temperature conditions are usually ***cross-charged***, meaning the bulb is charged with a refrigerant different from the one used in the system. The refrigerant used in the bulb is more sensitive to low temperatures.

Valves designed for high-temperature applications have sensing bulbs charged with little liquid. This provides a *pressure-limiting* feature, which means that when all the liquid has vaporized, a further increase in temperature produces almost no additional bulb pressure. In this way, the amount of pressure exerted upon the valve diaphragm is limited. Such valves are normally used in air-conditioning applications.

Sensing Bulb Mounting

Proper mounting of the sensing bulb is essential for good valve control. As shown in **Figure 18-16**, the bulb is clamped to the suction line at the evaporator outlet. It should be located on a horizontal section of the line, upstream from any trap. The sensing bulb should make good thermal contact along its entire length with the suction line. The bulb is normally secured to the suction line with two copper straps.

On suction lines 7/8″ (22 mm) or smaller in diameter, the bulb can be mounted on the top or on either side of the line. The bottom should be avoided. Any oil flow in the line will be along the bottom, which could prevent the bulb from sensing temperatures correctly. On suction lines larger than 7/8″ (22 mm), the sensing bulb should be mounted at the 4 o'clock or 8 o'clock position for the most accurate temperature readings.

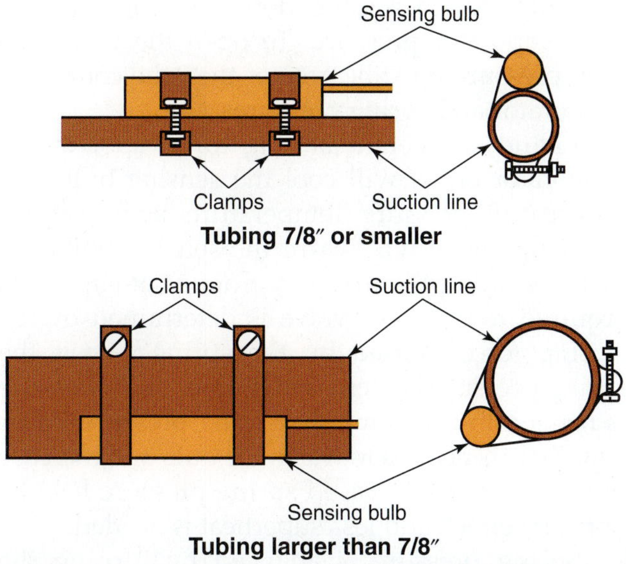

Goodheart-Willcox Publisher

Figure 18-16. On suction lines 7/8″ (22 mm) or smaller, the bulb should be clamped to the top or side of the line. Avoid bottom-mounting. On lines larger than 7/8″ (22 mm), mount the bulb at the 4 o'clock or 8 o'clock position.

Evaporator Pressure

Evaporator pressure acts on the bottom of the valve diaphragm, providing an upward movement that tends to close the valve (refer to **Figure 18-14**). An increase in evaporator pressure causes the valve to close until the compressor causes a reduction in evaporator pressure.

Slight changes in either evaporator pressure or bulb pressure make the diaphragm move up or down in small increments, causing corresponding changes in the size of the valve opening (modulation) and the flow of liquid refrigerant.

Internally Equalized TEV

In an ***internally equalized*** TEV, a special passage built into the valve body transfers evaporator pressure to the underside of the diaphragm. This type of valve is used when there is a pressure drop of *less than 2 psig (14 kPa)* through the evaporator, **Figure 18-17**.

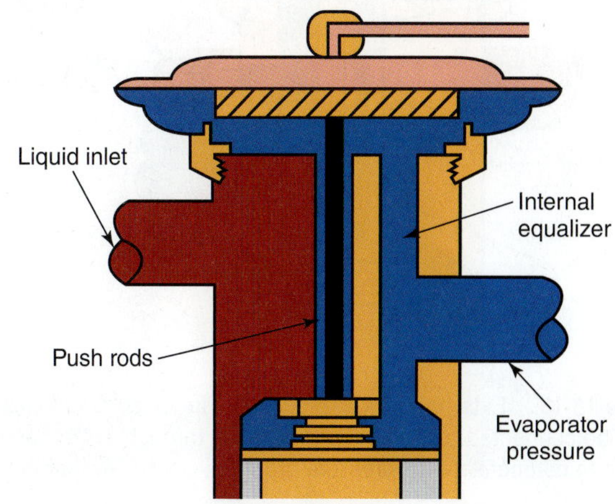

Goodheart-Willcox Publisher

Figure 18-17. Internally equalized valve. A special passage transfers evaporator pressure to the underside of the diaphragm.

Evaporator Pressure Drop

A pressure drop that develops in the evaporator due to the length of its tubing, the return bends in the tubing, and the vacuum (suction) action of the compressor is called ***evaporator pressure drop***. As a result, pressure at the evaporator inlet is higher than pressure at the evaporator outlet. A large evaporator may have a pressure drop of 6 psig (41 kPa) or more. For proper operation of the TEV, correct evaporator pressure on the underside of the diaphragm is essential. Evaporators that have a pressure drop *greater than 2 psig (14 kPa)* must use an externally equalized TEV.

Externally Equalized TEV

An ***externally equalized*** TEV is like the internally equalized type, but evaporator pressure on the underside of the diaphragm is supplied from the evaporator *outlet*, rather than the *inlet*. As shown in **Figure 18-18**, a separate passageway to the diaphragm is connected to a 1/4″ male flare fitting on the valve body. Tubing connected to this fitting is run to a point just downstream of the sensing bulb and inserted into the suction line, **Figure 18-19**. This transfers the correct evaporator pressure to the valve diaphragm.

An externally equalized TEV must *always* be connected to the suction line. It should *never be capped*, else the TEV may flood, starve, or operate erratically. An externally equalized TEV may be used with an evaporator that has very little pressure drop.

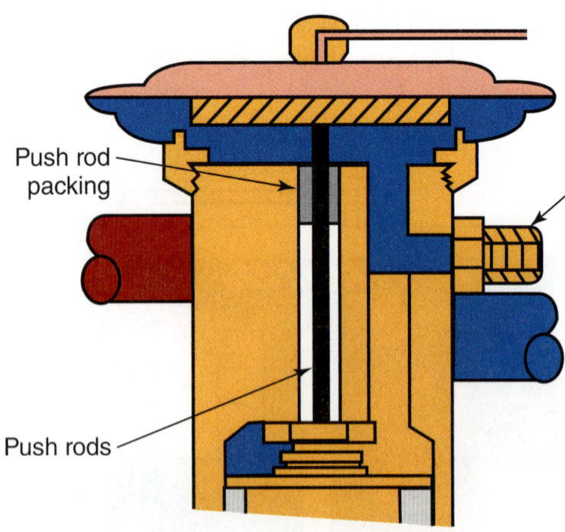

Goodheart-Willcox Publisher

Figure 18-18. Externally equalized valve. The external equalizer must be connected to the evaporator outlet through 1/4″ copper tubing to compensate for pressure drop through the evaporator.

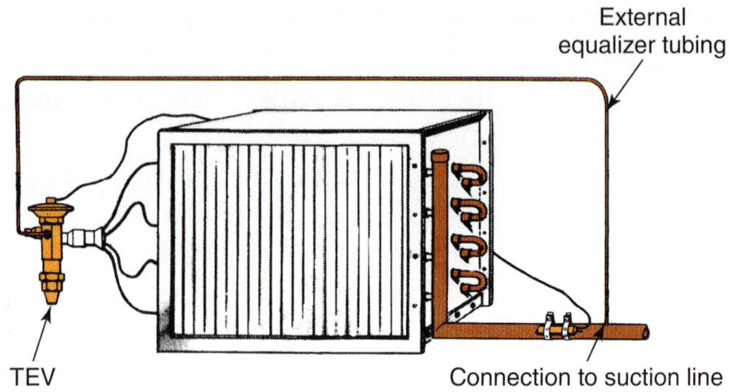

Sporlan Valve Co.

Figure 18-19. The tubing from an externally equalized TEV is connected to the suction line at a point immediately downstream of the sensing bulb.

Spring Pressure

The valve pin is mounted on a metal pin carrier that rides on top of a spring. Two push rods extend from the bottom side of the diaphragm to the top of the pin carrier. Upward pressure from the spring creates a closing force, pressing the pin into the valve seat. The force is also exerted on the push rods, which transmit it to the underside of the diaphragm.

Pressure *on top* of the diaphragm (exerted by expansion of refrigerant in the sensing bulb) opposes pressure on the bottom of the diaphragm resulting from a combination of spring pressure and evaporator pressure. To open the valve, the bulb pressure must be greater than the combined evaporator and spring pressures.

Liquid refrigerant boiling to a vapor at the evaporator outlet will cool the sensing bulb and reduce bulb pressure. Temperatures above saturation (superheat) will warm the sensing bulb and increase bulb pressure. The amount of superheat required to open the valve is determined by the closing force exerted by the spring. When the spring pressure is increased, higher bulb pressure is needed to overcome the spring pressure. More superheat is needed to warm the bulb and increase bulb pressure. Decreased spring pressure has the opposite effect, and less superheat is needed.

Spring pressure is adjusted by turning the valve stem. Turning the stem clockwise increases pressure. Counterclockwise rotation decreases pressure.

High-Side Pressure

High-pressure liquid refrigerant coming into the TEV tends to open the valve by pushing the pin away from the valve seat. This opening force can be offset by spring pressure. However, spring pressure is fixed, and wide changes can occur in head pressure due to changing ambient conditions at the condenser. Severe head pressure changes can result in an imbalance among the TEV's operating forces. Preventing wide changes in head pressure ensures more uniform operation of the TEV.

All thermostatic expansion valves are designed to operate at a certain *minimum* head pressure. For example, valves for use with R-290 require a minimum head pressure of 100 psig (690 kPa). Valves for use with R-22 require 200 psig (1379 kPa). TEVs cannot operate properly when head pressures drop below these minimums. Various methods used to control head pressures are discussed later in this book.

One way of dealing with varying head pressure is to install a ***balanced port valve***. Construction of the valve distributes head pressure between the valve pin and the push rods. By equalizing opening and closing forces, the valve design compensates for a wide range of head pressure changes.

TEV Superheat Adjustment

Basically, a TEV operates like any other valve: turning the stem in (clockwise) *closes* it. Turning it out (counterclockwise) opens it. Closing and opening the valve permits different amounts of liquid refrigerant to flow into the evaporator. While the valve stem physically changes the opening of the valve, it is adjusting the *superheat*.

Adjustments should be made gradually to avoid overshooting the desired setting. Make no more than one turn of the valve stem at a time. Then allow enough time for the system to stabilize (as much as 30 minutes may be needed) before making another adjustment.

Normal Superheat Settings

Normally, the TEV superheat setting is 10°F (6°C). High superheat settings reduce evaporator capacity since more evaporator surface is needed to produce the superheat. However, some superheat is needed for proper operation of the valve. In air-conditioning applications, the superheat setting can be as high as 15°F (8°C) before there is a noticeable loss of evaporator capacity. Many low-temperature systems require 5°F–6°F (2.8°C–3.3°C) of superheat. The lower superheat setting provides maximum capacity from the evaporator and makes use of the colder suction line vapor to cool the compressor.

For questions about the correct superheat setting for a particular system, consult the manufacturer. Recommendations from a manufacturer are normally the result of extensive laboratory testing.

When a valve is first installed, superheat adjustment is often necessary. Once properly installed and adjusted for superheat control, the TEV is virtually trouble-free.

Some thermostatic expansion valves are nonadjustable. These are factory-set for a specific superheat value determined by the manufacturer. If symptoms indicate a superheat adjustment is needed, carefully check for other system causes of incorrect superheat.

PROCEDURE

Measuring a TEV Superheat Setting

How well a TEV is performing is easily determined by measuring superheat at the sensing bulb location. For accurate temperature and pressure readings, the system must be running and in a stabilized condition. Measurement is made as follows:

1. Install the manifold gauge on the compressor service valves and take a pressure reading.
2. Obtain an accurate reading of the suction line temperature at the sensing bulb location.
3. Convert the pressure reading to a temperature value by consulting a P/T chart for the refrigerant in the system. This is the *saturation* temperature.
4. Subtract the saturation temperature from the actual temperature measured at the sensing bulb. The difference is *superheat*.

Allowing for Suction Line Pressure Drop

A pressure reading at the suction service valve will be lower than a pressure reading at the evaporator outlet. This is called ***suction line pressure drop***. It should be limited to about 2 psig (14 kPa) if proper line sizing and good piping practice were followed when installing the system. For an accurate superheat reading, the pressure drop must be considered. To do so, add 2 psig (14 kPa) to the pressure at the suction service valve *before* converting that pressure to a temperature reading.

Figure 18-20 illustrates how superheat is measured on a system using R-22. As shown, the temperature of the suction line at the sensing bulb is 52°F (11°C). Pressure at the suction service valve is 66 psig (455 kPa). To allow for pressure drop, 2 psi (14 kPa) is added for a total of 68 psig (469 kPa).

Converting 68 psig to a temperature reveals a saturation temperature of 40°F (4°C). Subtracting the saturation temperature from the actual temperature of 52°F (11°C) yields a superheat of 12°F (7°C).

Evaporator Temperature

The TEV does not control evaporator temperature or low-side pressure. It maintains a constant superheat, regardless of temperature or pressure. It is sized to match the capacity of the evaporator and compressor.

During operation, there is a difference of approximately 20°F (11°C) between the temperature of the product being cooled and the temperature of the refrigerant. The *evaporator* temperature falls midway between the two: 10°F (6°C) colder than the product but 10°F (6°C) warmer than the refrigerant.

As the system runs, the product temperature is reduced. The heat load on the evaporator is reduced as well. As the heat load is reduced, less liquid flows into the evaporator.

Since the compressor removes vapor from the evaporator at a fixed rate, less liquid flowing into the evaporator results in lower pressure inside the evaporator. Lower pressure means a lower boiling point for the refrigerant and a colder evaporator. Therefore, the difference of 20°F (11°C) between the product and the refrigerant is maintained. The superheat setting is maintained as well, regardless of evaporator temperature.

When product temperature reaches the desired level, the system is turned off by a thermostat or a low-pressure control. When the system is turned off, the compressor stops running and cannot remove vapor from the evaporator. Evaporator pressure rises quickly as the refrigerant remaining inside the evaporator warms up to match the product temperature. The increase in evaporator pressure is transferred to the bottom of the diaphragm in the TEV, closing the valve. The valve remains closed during the "off" portion of the cycle because the evaporator (closing) pressure is sufficient to overcome the bulb (opening) pressure.

Flash Gas

A certain small amount of liquid refrigerant that passes through the TEV orifice immediately evaporates into ***flash gas***. The sudden pressure drop (from high to low) causes some of the liquid to instantly boil (flash) into a gaseous state, or vapor. See **Figure 18-21**. The heat needed to cause this boiling action is drawn from the remaining liquid, which is cooled to the lower pressure. The

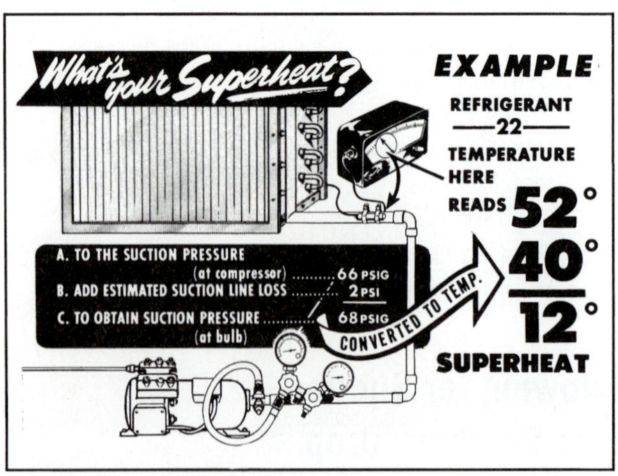

Sporlan Valve Co.

Figure 18-20. Including suction line pressure drop when calculating superheat.

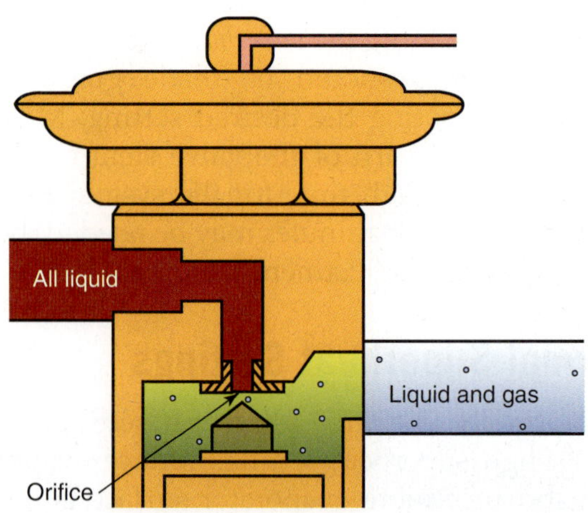

Goodheart-Willcox Publisher

Figure 18-21. A portion of the liquid refrigerant will "flash" into a gaseous state due to the pressure drop when it passes through the TEV orifice.

remaining liquid then boils at the proper evaporator saturation temperature.

The amount of flash gas is determined by the amount of pressure drop across the valve. This "loss" of liquid refrigerant is necessary as the refrigerant moves from a higher pressure, higher temperature part of the system to a lower pressure, lower temperature section (the evaporator).

The amount of liquid that flashes into a gas can be reduced by subcooling the liquid before it reaches the valve orifice. Subcooling the liquid makes the system more efficient. A heat exchanger can be installed to subcool the liquid, but many systems accomplish subcooling by simply having long, uninsulated liquid lines between the condenser and the TEV.

Refrigerant Distributor

On evaporators with multiple circuits, a ***refrigerant distributor*** directs liquid refrigerant through the different circuits of an evaporator. The distributor receives refrigerant directly from the thermostatic expansion valve and divides it equally among all evaporator circuits, **Figure 18-22**.

PRO TIP

Distributor Pressure Drop

Any system using a distributor *must* use an externally equalized TEV. This will help TEV operation to account for the pressure drop that occurs across the distributor.

In any refrigeration system, the refrigerant metering device is a division point where high pressure is changed to low pressure. On systems that do not include a distributor, all pressure drop is taken across the valve. When a distributor is used, some pressure drop occurs *across the distributor.* When brazing a distributor to a TEV, be careful to direct the torch flame away from the valve body, **Figure 18-23**. To prevent excessive heat

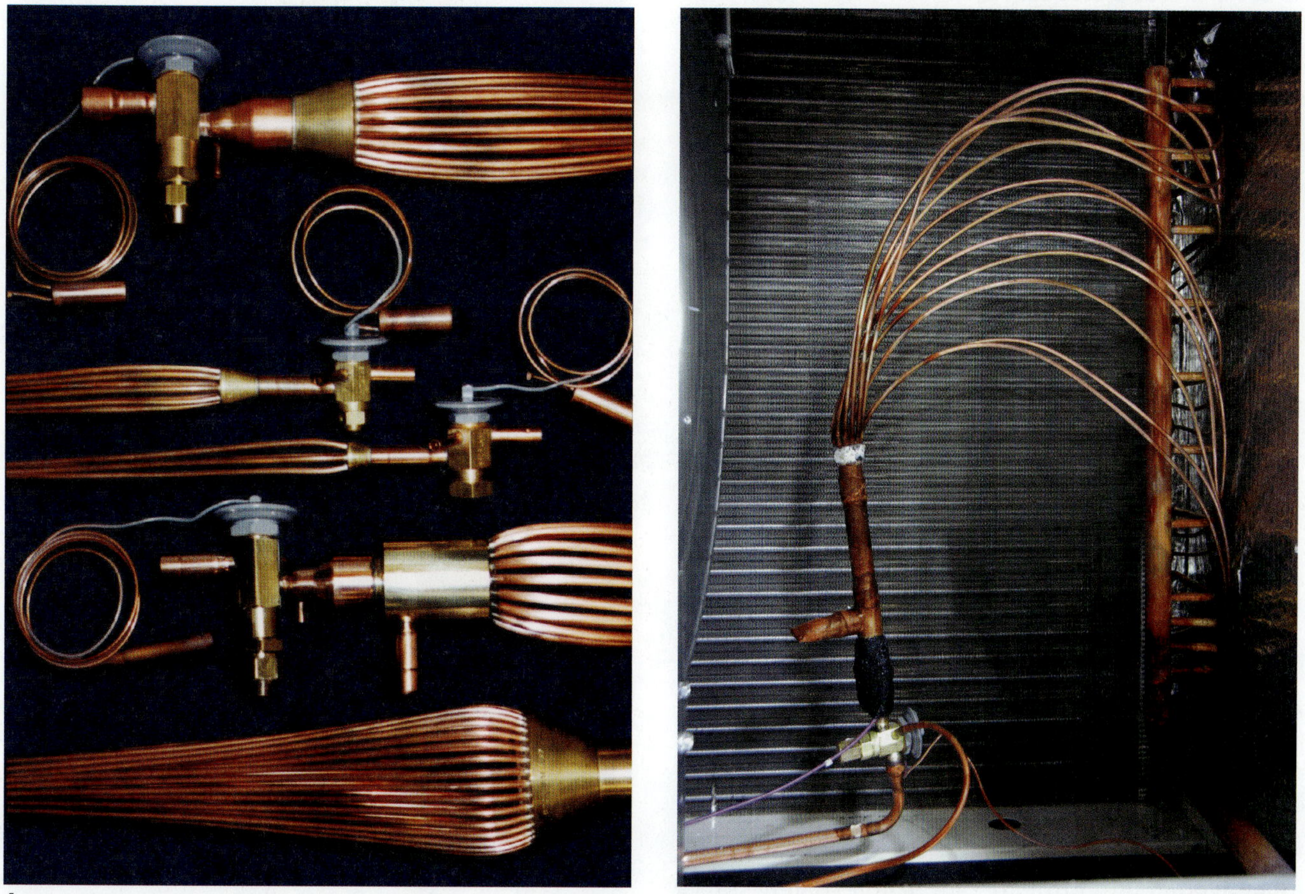

A B *Sporlan Valve Co.; ClimateMaster*

Figure 18-22. Distributors are used in large systems with multiple evaporator circuits. A—Distributors mounted directly to TEVs of different size. B—A distributor directing refrigerant into multiple evaporator circuits in a commercial heat pump.

Goodheart-Willcox Publisher

Figure 18-23. The flame of the brazing torch should be away from the TEV to prevent heat damage to its diaphragm. This distributor is piped further away from the TEV to protect it from the heat of brazing.

from damaging the valve diaphragm while brazing, wrap the valve body in a wet cloth.

Valve Mounting

A thermostatic expansion valve may be mounted in any position, but it should be installed as close to the evaporator inlet as possible. If a distributor is used, it is mounted directly to the outlet of the TEV. Many TEVs are installed with a flare or other mechanical connection, but some make use of brazed connections, **Figure 18-24**.

TEV Capacities

Thermostatic expansion valve capacities are rated in tons of refrigeration. Capacity varies based on these factors:

- Orifice size
- Pressure drop across the valve (liquid line pressure to evaporator pressure)
- Amount of flash gas (liquid temperature versus evaporator temperature)
- Refrigerant used

It is vital to use a TEV of the correct capacity. If an undersized TEV is used, the evaporator will starve, despite superheat adjustment. If the TEV is oversized, it will hunt and surge (alternately starve and flood the evaporator).

Flare connections

Braze connections

Chatleff connections

Sporlan Valve Co.

Figure 18-24. TEV mounting variations.

Number Codes

Printed on the body or diaphragm of each valve is a set of numbers (or letters and numbers) displaying the manufacturer's coded information about the valve. The codes are not secret, and information on how to read them is freely available from manufacturers. The code numbers are usually needed when ordering replacement parts.

Troubleshooting the TEV

Proper thermostatic expansion valve (TEV) performance is easily determined by measuring superheat at the evaporator outlet. So, checking superheat is the first step in diagnosing TEV performance. A starved evaporator results in high superheat. Excessive liquid flooding of the evaporator causes low superheat. Although both these symptoms point to improper TEV superheat control, the actual cause is usually elsewhere. Always check for proper installation of the sensing bulb and external equalizer.

Restricted Filter-Drier

A partially restricted filter-drier becomes cold and possibly frost-covered. A restricted filter-drier is caused by a screen plugged with contaminants, *not* by moisture. When a filter-drier has absorbed all the moisture it can hold, additional moisture passes through the filter-drier and freezes at the refrigerant metering device. Moisture *cannot* plug or restrict a filter-drier.

Improperly Seated Valve

Dirt, sludge, or foreign material lodged in a TEV's valve seat may prevent the TEV from seating properly. Liquid refrigerant may pass through the valve during the off cycle and fill the evaporator with refrigerant. If the seat leak is severe, the valve will feed too much refrigerant during the run cycle. A valve seat leak is detected by a gurgling or hissing sound during the off cycle. Also, the sight glass may indicate continued flow for a long period after shutdown. Disassemble the valve to determine if dirt or foreign material is responsible for the leaky seat. Replace the TEV if the seat is damaged.

Dead Sensing Bulb

If the sensing bulb loses its charge, the TEV cannot open. The capillary tube connecting the sensing bulb to the valve diaphragm is often a cause of trouble. For example, excess vibration causes the tube to become work-hardened and break off. Vibration is normally caused by airflow from the evaporator fan. Vibration of the coils against one another, or against the evaporator, can wear a hole in the soft capillary tube. The TEV must be replaced when the sensing bulb loses its charge. On some TEVs, only the sensing bulb and diaphragm assembly need to be replaced.

Low Evaporator Load

Low evaporator load describes a condition in which the heat load cannot reach the evaporator surface. Low evaporator load causes the evaporator to become too cold, **Figure 18-25**. Under a low-evaporator-load condition, a lesser amount of liquid is needed to maintain proper superheat, so the TEV restricts flow. Restricted flow causes low suction line pressure, a colder evaporator, and normal superheat. Low evaporator load can be caused by insufficient air over the evaporator, undersized blowers, dirty air filters, frost formation on the evaporator, or low temperature of the entering air. Correct the condition responsible for low heat load.

Liquid Migration

If a condensing unit is located outside and the ambient outdoor temperature is colder than the indoor evaporator, gaseous refrigerant will ***migrate*** (travel) to the compressor during the off cycle. This happens because the compressor is colder than the evaporator, and heat travels from hot to cold.

The cold ambient temperature also causes vapor inside the compressor to condense to a liquid. Upon restarting, the compressor must pump *liquid* rather than *vapor*. This is a condition called ***flooded start***. Flooded start can cause severe compressor damage. This situation is not caused by the TEV.

Any compressor that must operate while located in cold ambient temperatures should be equipped with a crankcase heater, **Figure 18-26**. The device is an electric heater, fastened to the

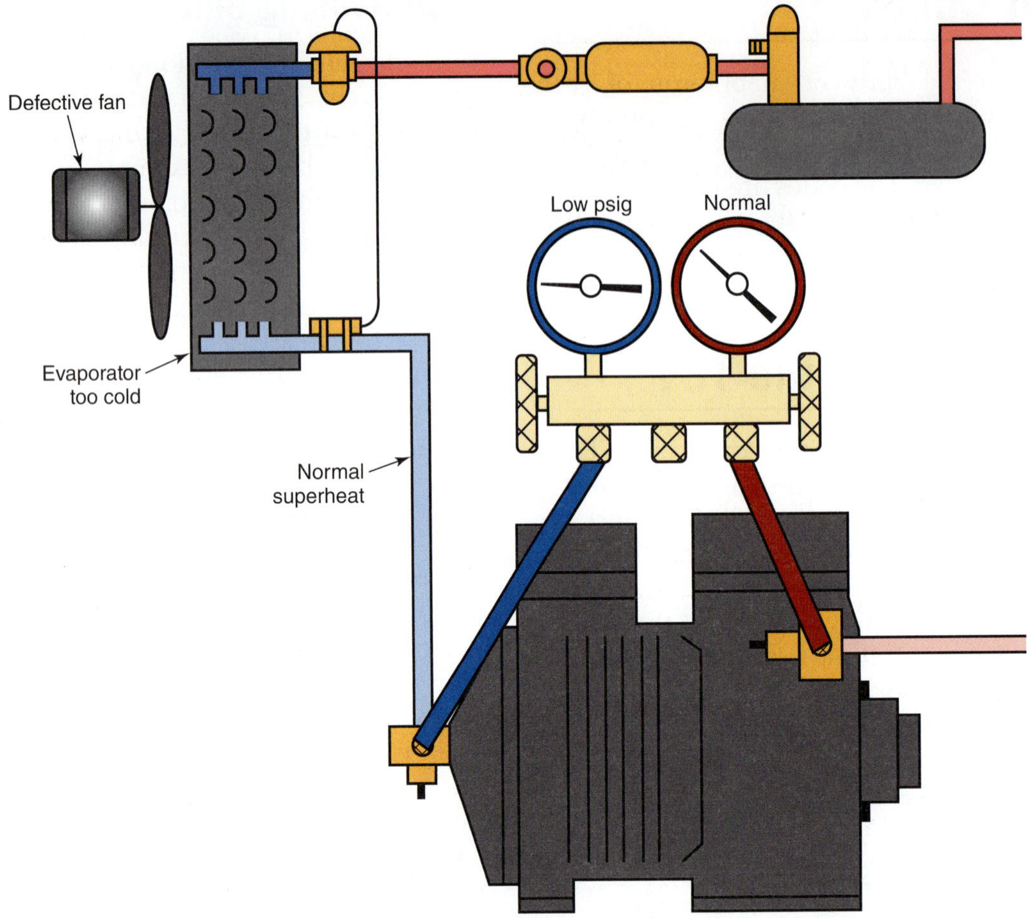

Goodheart-Willcox Publisher

Figure 18-25. A low evaporator load causes the evaporator to become too cold. A defective fan is one of the possible causes of insufficient airflow past the evaporator.

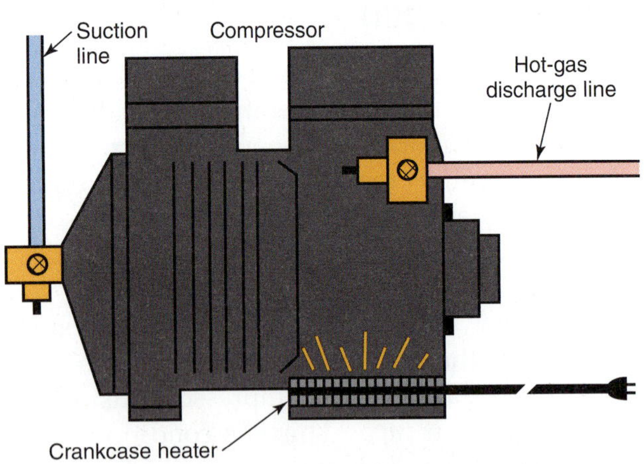

Goodheart-Willcox Publisher

Figure 18-26. An electrical heater at the bottom of the compressor warms oil in the crankcase under cold ambient conditions. This keeps condensation from forming and reduces the chances of having a flooded start by preventing liquid migration.

crankcase, that keeps the oil warm under cold conditions. Warm oil keeps condensation from forming in the crankcase and prevents liquid migration.

Most condensing units located in areas where cold ambient temperatures occur are equipped with a liquid line solenoid valve for automatic pumpdown. A pumpdown removes refrigerant from the low-pressure side before system shutdown. Pumpdown is another method of preventing a flooded start.

18.5 Electronic Expansion Valves (EEVs)

Electronic expansion valves (EEVs) are expansion valves operated by electronic controls,

Figure 18-27. The EEV's job is to control the flow of refrigerant into the evaporator with a valve similar to the TEV, except a small stepper motor is used to open and close the valve.

EEVs are used in many air-conditioning and refrigeration applications, including the following:

- Minisplit air-conditioning and heat pump systems
- Variable refrigerant flow (VRF) commercial air-conditioning and heat pump systems
- Standard split heat pump systems
- Commercial refrigeration
- Energy management systems
- Chillers
- Air handling unit air-conditioning coils

EEV Control System

An electronic expansion valve requires three components to operate, Figure 18-28:

- Sensor(s)
- Controller
- Electronic expansion valve

The sensors are attached to the suction line at the evaporator outlet. Many EEV systems include a temperature sensor (thermistor) and a pressure transducer. See Figure 18-29 and Figure 18-30. Some systems use only a single temperature sensor. Others use two temperature sensors: one at the evaporator outlet and one at the evaporator inlet. The pressure transducer converts the pressure reading into an electronic signal that can be used by the controller.

The sensors send signals to the controller, Figure 18-31. The controller is programmed with information about the system. For example, in a system with a temperature sensor and a pressure sensor, the controller uses the pressure signal to determine the pressure in the system. The controller is programmed with the P/T chart information for the refrigerant in the system, so the controller can translate the pressure reading to the saturation temperature. By comparing the saturation temperature to the actual system temperature being provided by the temperature sensor, the controller can determine the system superheat. The controller then compares the actual superheat to the design superheat. The controller then sends a signal to the EEV to provide more or less refrigerant to the evaporator.

The EEV receives a signal from the controller that either increases or decreases the size of the opening through which refrigerant passes. Thus, the EEV is dependent on the signal it receives from the controller to adjust and maintain correct system settings.

Electronic Expansion Valves (EEVs)

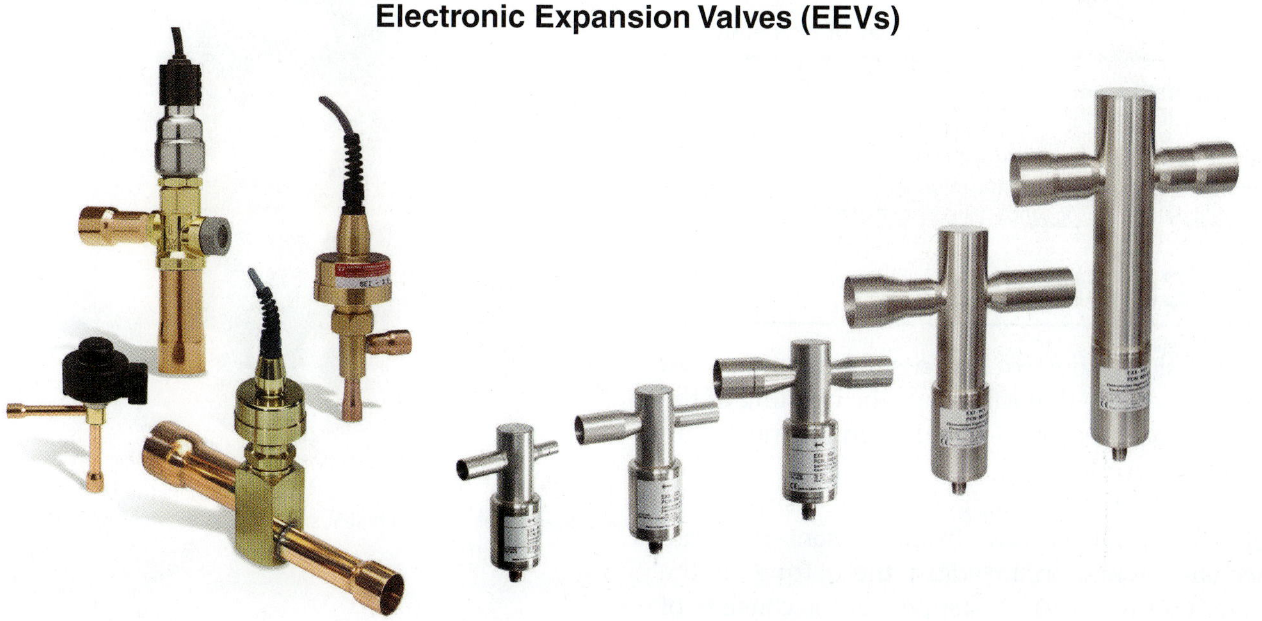

Courtesy of Sporlan Division - Parker Hannifin Corporation; Emerson Climate Technologies

Figure 18-27. Electronic expansion valves (EEVs) are available in a variety of types and sizes. The type of EEV used in a system is determined by the application and the size of the system (required capacity of the EEV).

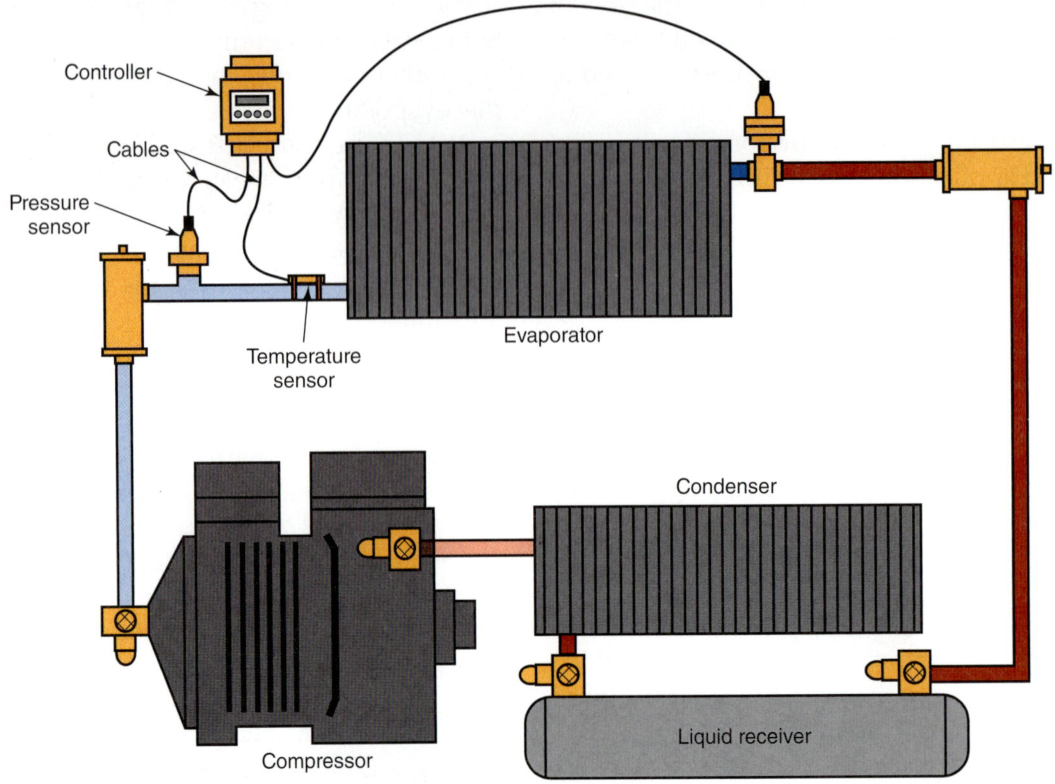

Goodheart-Willcox Publisher

Figure 18-28. An EEV system consists of the EEV, its controller, and the sensors that provide input to the controller.

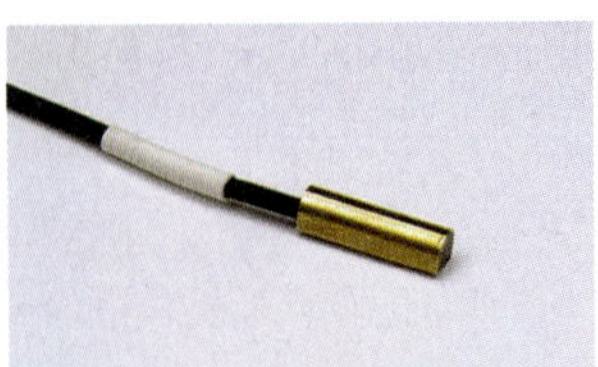

Temperature sensor

Cutaway of temperature sensor attached to suction line

Courtesy of Sporlan Division - Parker Hannifin Corporation

Figure 18-29. EEV systems use a temperature sensor (thermistor) to monitor the temperature at the evaporator outlet. The temperature sensor sends a signal to the controller.

EEV Construction

An electronic expansion valve contains an electrical device that can adjust the opening of the valve orifice based on a signal received from a controller, **Figure 18-32**.

In many EEVs, a stepper motor is used to convert the signal received from the controller into mechanical motion to adjust the setting of the valve, **Figure 18-33**. A stepper motor consists of

Courtesy of Sporlan Division - Parker Hannifin Corporation

Figure 18-30. A pressure transducer monitors the pressure at the evaporator outlet and sends a signal to the controller.

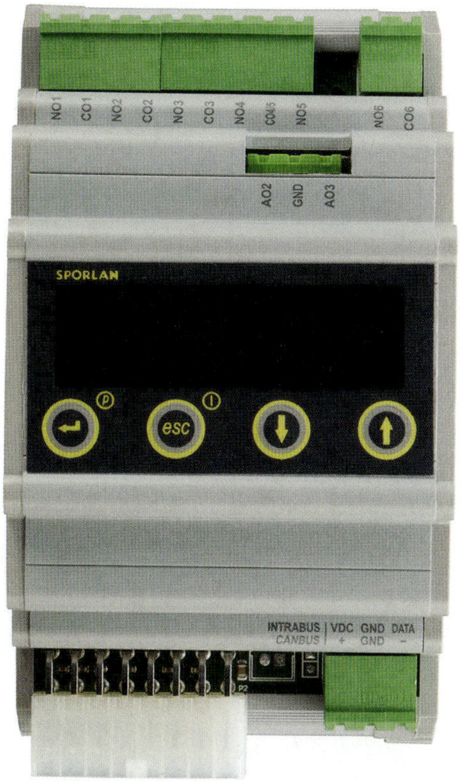

Courtesy of Sporlan Division - Parker Hannifin Corporation

Figure 18-31. In an EEV system, the controller receives signals from the sensors. Based on the signals it receives, the controller sends signals to the EEV to allow more or less refrigerant into the evaporator. Some controllers include a screen that can display information about the system conditions.

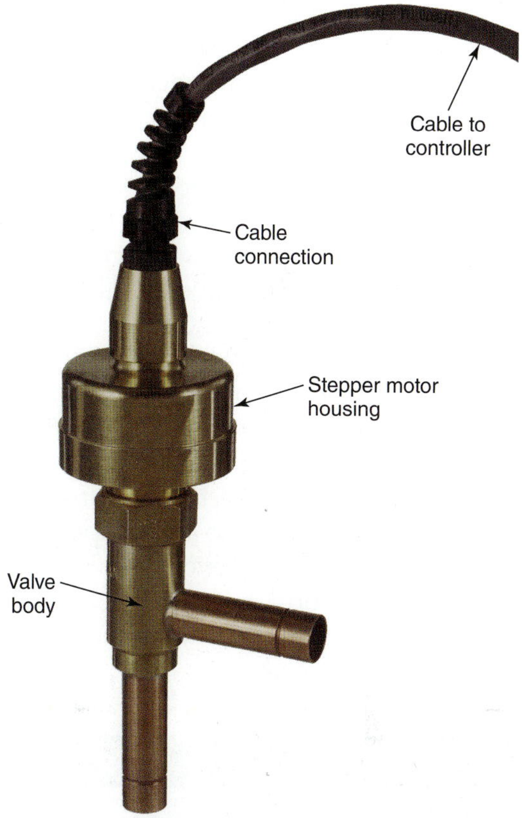

Courtesy of Sporlan Division - Parker Hannifin Corporation

Figure 18-32. External parts of an electronic expansion valve. A signal received from the controller causes the stepper motor to turn a discrete amount in either direction. This rotation of the motor causes a piston or plunger to move up or down, increasing or decreasing the size of the opening through which refrigerant flows.

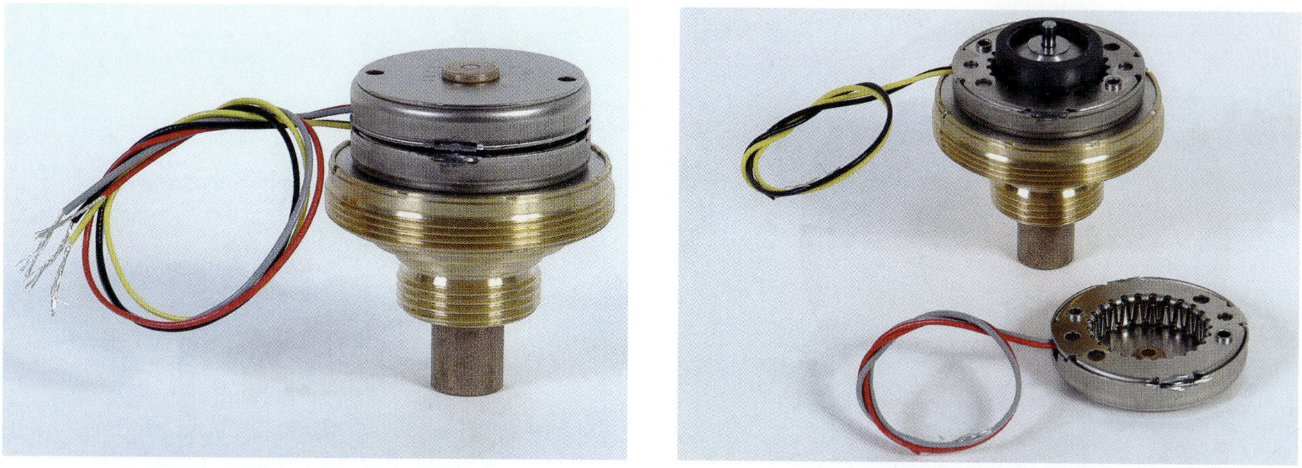

Stepper motor with housing removed

Partially disassembled stepper motor

Courtesy of Sporlan Division - Parker Hannifin Corporation

Figure 18-33. Stepper motors are common in EEVs.

a rotor (the part of the motor that turns) surrounded by many pairs of copper coils, **Figure 18-34**. The number of pairs of coils can range from 24 to 100. The number of coils determines the number of "steps" the rotor can take in one revolution. In some systems, the stepper motor can turn over 200 steps in one second.

The signal the EEV receives from the controller causes the rotor to turn one step. The rotor can turn in either direction, depending on whether the controller is increasing or decreasing the size of the opening. The rotation of the motor is converted to linear motion by a lead screw, **Figure 18-35**. The linear motion causes a plunger to move up or down, changing the size of the opening through which refrigerant flows into the evaporator. The plunger may be able to move a total of 0.25" to 0.50", depending on the size of the valve. **Figure 18-36** shows some of the internal parts of an EEV.

PRO TIP

EEV Types

The stepper motor EEV is the most common type of EEV, but older types of EEVs include heat motor, analog, and pulse types.

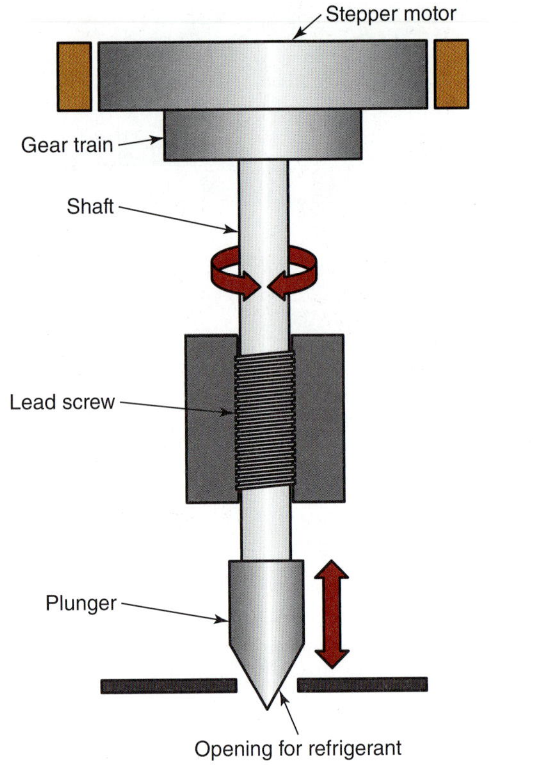

Goodheart-Willcox Publisher

Figure 18-35. The rotation of the stepper motor is transmitted through a gear train to the shaft. As the shaft rotates, so does the lead screw. The turning of the lead screw causes the plunger to move up or down, depending on the direction. The movement of the plunger increases or decreases the size of the opening.

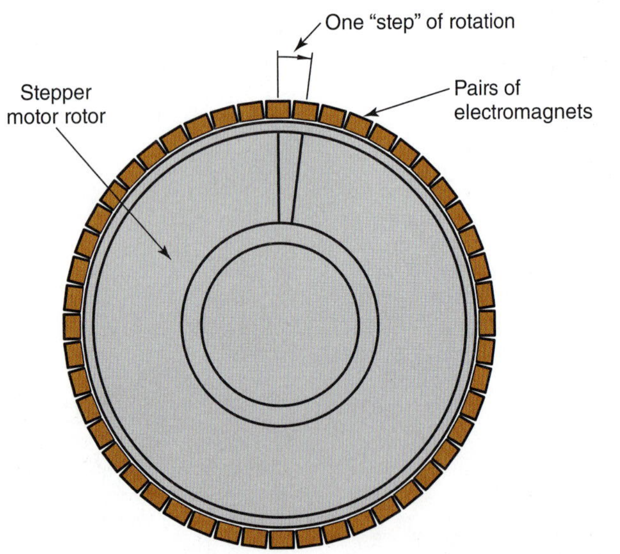

Goodheart-Willcox Publisher

Figure 18-34. A stepper motor turns in small increments, or steps. The motor can rotate in either direction. The size of each step is determined by the number of pairs of coils around the rotor.

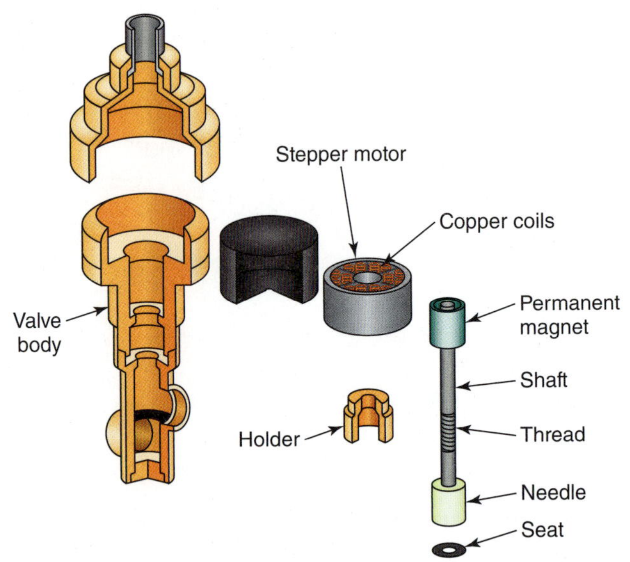

Goodheart-Willcox Publisher

Figure 18-36. Internal parts of an EEV.

Installing EEVs

The following are some general guidelines for installing an EEV. Always refer to and follow the manufacturer's installation instructions.

EEVs are often brazed to the refrigerant lines. Before brazing, remove the cable connection from the EEV if possible. Wrap the valve body in a wet rag to help keep it cool. When brazing, keep the flame away from the valve body and cables.

Attach the required sensors in the locations specified by the design. Run the cables from the sensors and from the EEV to controller. Use the controller installation instructions to determine which terminals each conductor connects to.

When the system is set up and you apply power to the controller, the controller will initialize the EEV. The controller will send signals to the EEV to fully close it. After initializing, the controller "knows" that the valve is in the fully closed position. The controller will then keep track of the plunger's position based on the signals it sends to the EEV.

Advantages and Disadvantages of EEVs

Using an EEV provides several benefits when compared with using other types of metering devices. Some of these benefits result from the more precise control of the refrigerant flow achieved with EEVs, along with the continuous monitoring of the system and adjustment of the EEV by the controller. Some of the benefits are as follows:

- EEVs provide more precise control over the flow of refrigerant into an evaporator than that of other metering devices. EEVs provide a more precise superheat in the evaporator.
- EEVs adjust quicker to changing load conditions. As a result, evaporator temperatures in EEV systems remain more consistent than in systems using other metering devices. This advantage is especially useful in more complex systems with varying load conditions.
- Smaller variation in evaporator temperature results in better system performance and improved energy efficiency. EEVs are estimated to be 30% more energy-efficient systems for certain system types.
- EEV use can help to reduce the stress on the compressor.
- Since components of the EEV system are connected with wiring, it is less problematic if the controller, EEV, and sensors are not located near one another.

Some disadvantages of the EEV include the following:

- The primary disadvantage of EEVs is their expense. However, on most systems, the added expense of an EEV will be offset over time by the energy savings achieved.
- HVACR technicians must have some knowledge of electronic controls to correctly service an EEV. EEVs require an advanced control system and thus a more experienced installer.

Given the similar functionalities between a TEV and an EEV, technicians and building owners may consider the following factors when deciding between which of these two metering devices to use in a system:

- Precision of the system
- Cost
- Energy efficiency
- Long-term reliability

Chapter Review

Summary

18.1 The automatic expansion valve (AEV) is a pressure-type control installed in the liquid line at the evaporator inlet. The AEV continually modulates (adjusts) refrigerant flow in this way to maintain constant evaporator pressure.

18.1 A manifold gauge and adjustment screw are used to adjust the AEV's spring pressure. An AEV with a capacity equal to the cooling capacity of the compressor at the design evaporator temperature should be selected.

18.1 Increased heat load on the evaporator of an AEV-equipped system causes the evaporator pressure to rise, which causes the AEV to close or "choke down." Some common problems with AEV systems include an undercharged system, erratic valve settings, contaminants, and poor cooling with low or high head pressures.

18.2 A capillary tube is a length of copper tubing with a tiny, accurately sized inner diameter. A capillary tube is selected based on its length, inside diameter, refrigerant used, and tubing temperature. To service capillary tube systems, pigtails are used to evacuate and charge these systems.

18.3 A fixed bore (piston) metering device is designed to operate on the principle of restriction, similar to the capillary tube. However, the fixed bore is smaller and more easily accessible and replaceable than capillary tubes.

18.4 The thermostatic expansion valve (TEV) is designed to maintain a precise superheat at the evaporator outlet by controlling the amount of liquid metered into the evaporator. The TEV is controlled by four pressures exerted on working parts: sensing bulb, evaporator, spring, and high-side pressures.

18.4 Turning the valve stem on a TEV clockwise closes it; turning it out counterclockwise opens it. Correct superheat settings, proper suction line pressure drop, evaporator temperature, TEV capacity, and other factors are important variables to consider when installing and servicing a TEV.

18.4 Some common problems with TEV systems include high superheat, restricted filter-drier, an improperly seated valve, dead sensing bulb, low evaporator load, and liquid migration.

18.5 An electronic expansion valve (EEV) is an expansion valve operated by a sensor, controller, and the valve itself. Many EEVs use a stepper motor to convert the signal received from the controller into mechanical motion to adjust the setting of the valve. EEVs provide more precise control over refrigerant flow into an evaporator compared to other metering devices.

Know and Understand

1. The AEV continually _____ refrigerant flow to maintain evaporator pressure. (18.1)
   A. maintains
   B. modulates
   C. stops
   D. increases
2. When an AEV is used as the refrigerant metering device, the temperature difference between the evaporator and the refrigerant should be _____. (18.1)
   A. 10°F
   B. 15°F
   C. 20°F
   D. nonexistent; the evaporator and refrigerant should be the same temperature
3. An AEV with a capacity that is too small for a system will _____. (18.1)
   A. flood the evaporator
   B. starve the evaporator
   C. flood the compressor
   D. decrease cooling time of the product
4. *True or False?* AEVs are the most commonly used metering device in refrigeration and air-conditioning applications. (18.1)

5. The _____ prevents evaporator pressure from dropping below a specified level. (18.1)
   A. compressor
   B. condenser
   C. automatic expansion valve
   D. capillary tube
6. Poor cooling with long running time is frequently caused by _____ pressure. (18.1)
   A. low head
   B. high head
   C. low suction
   D. high suction
7. When a capillary tube is installed along the suction line of a refrigeration cooler, then it can function as both a metering device and a(n) _____. (18.2)
   A. service valve
   B. compressor
   C. heat exchanger
   D. thermostat
8. During the bleedover process of a capillary tube system, the high- and low-side pressures will _____. (18.2)
   A. raise or lower until they both equalize
   B. both drop, then stabilize when the system is turned on
   C. behave as described on the P/T chart for the system refrigerant
   D. stay the same
9. The capillary tube works on the principle of _____ to cause a _____ between the high and low sides of the refrigeration system. (18.2)
   A. restriction, pressure drop
   B. superheat, pressure increase
   C. convection, temperature difference
   D. gravity, pressure drop
10. *True or False?* Since a capillary tube system is easily overcharged, refrigerant should be liquid charged into the low-pressure side of the system. (18.2)
11. A fixed bore metering device offers the advantage of being easily _____ when compared to the capillary tube. (18.3)
    A. adjusted
    B. replaceable
    C. charged
    D. paired with any refrigerant
12. The purpose of the TEV is to maintain a constant _____ when the system is running. (18.4)
    A. sensing bulb pressure
    B. evaporator temperature
    C. superheat
    D. head pressure
13. TEVs used in high-temperature applications generally have _____ refrigerant in the sensing bulb. (18.4)
    A. a little vapor
    B. a little liquid
    C. both liquid and vapor
    D. no
14. *True or False?* An externally equalized TEV must *always* be connected to the suction line. (18.4)
15. High-pressure liquid refrigerant coming into the TEV can be offset by _____ pressure. (18.4)
    A. evaporator
    B. spring
    C. sensing bulb
    D. low-side
16. The normal superheat setting of a TEV is _____°F. (18.4)
    A. 5
    B. 10
    C. 11
    D. 20
17. The suction line pressure drop in a TEV system should be limited to _____ psig. (18.4)
    A. 0
    B. 1
    C. 2
    D. 8
18. During operation of a TEV system, the evaporator should be _____°F warmer than the refrigerant. (18.4)
    A. 5
    B. 10
    C. 20
    D. 32
19. On large evaporators with multiple circuits, a(n) _____ is used to direct liquid refrigerant through the different circuits of an evaporator. (18.4)
    A. capillary tube
    B. automatic expansion valve (AEV)
    C. refrigerant distributor
    D. evaporator pressure regulator (EPR)

20. An undercharged system prevents the TEV from feeding properly and results in _____. (18.4)
    A. high suction pressure
    B. a cold evaporator
    C. low superheat
    D. high superheat
21. When a sensing bulb loses its charge, a technician should _____. (18.4)
    A. recharge it with R-22
    B. replace the capillary tube and recharge the bulb
    C. replace the TEV with an EEV
    D. replace the TEV or sensing bulb and diaphragm assembly
22. In an electronic expansion valve, _____ is typically used to open and close the valve. (18.5)
    A. evaporator pressure
    B. spring pressure
    C. a stepper motor
    D. a solenoid valve
23. *True or False?* EEVs provide more precise control over the flow of refrigerant into an evaporator than that of other metering devices. (18.5)

Apply and Analyze

1. Explain the working forces in an AEV. (18.1)
2. What symptom can indicate an undercharged AEV system? (18.1)
3. Name four variables that should be considered when selecting the capillary tube as a metering device. (18.2)
4. Explain what happens when there is a restriction in a capillary tube. (18.2)
5. What are some differences between a fixed bore and a capillary tube? (18.3)
6. List the four pressures that affect the operation of a TEV and how they control the valve. (18.4)
7. On suction lines 7/8″ (22 mm) or smaller, where should the sensing bulb of a TEV be mounted? (18.4)
8. What is the first step in diagnosing TEV performance? (18.4)
9. Explain a low evaporator load condition in a TEV system and how to correct the condition. (18.4)
10. Explain the function of each of the three major components in an electronic expansion valve (EEV). (18.5)

Critical Thinking

1. A technician is servicing a customer's chest freezer and finds the automatic expansion valve needs to be replaced. Should the technician replace the valve with a different metering device or stick with the automatic expansion valve for the replacement? What factors do the technician and customer need to consider when making this decision? (18.1)
2. When checking a TEV, explain the difference between system superheat and evaporator superheat and how the two calculations can be used as part of the troubleshooting process. (18.4)

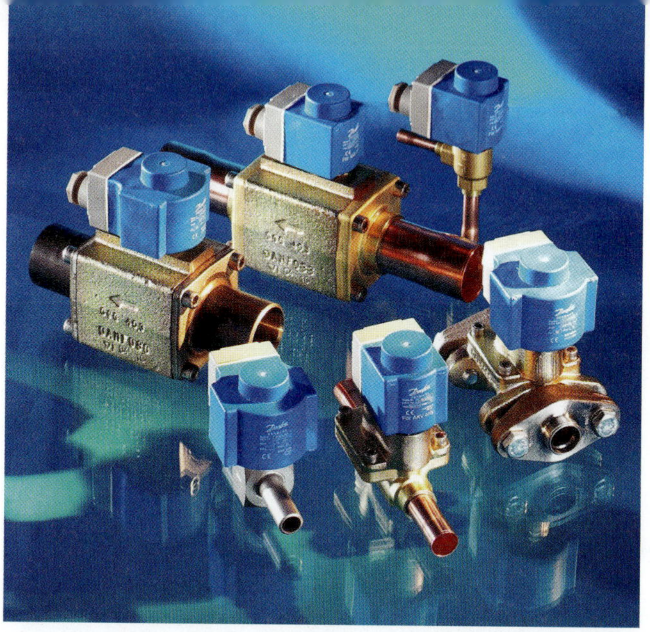

Danfoss

19 Special-Purpose Valves

LEARNING OUTCOMES

After studying this chapter, you will be able to:

19.1 Describe the uses of hand and check valves.

19.2 Identify applications of evaporator pressure regulators (EPRs) and how to troubleshoot these components.

19.3 Explain how solenoid valves operate, their installation process, and how to troubleshoot them.

19.4 Explain the operation, applications, and adjustment procedure for crankcase pressure regulators (CPRs).

19.5 Identify the application, location, operation, and troubleshooting of hot-gas bypass valves.

19.6 Explain the purpose of the desuperheating TEV.

19.7 Identify the two types of head pressure control valves, how they are used to solve low-head-pressure problems, and how to troubleshoot these valves.

TECHNICAL TERMS

adjustable head pressure control valve
capacity control
check valve
crankcase pressure regulator (CPR)
desuperheating TEV
evaporator pressure regulator (EPR)
false load
hand shut-off valve
head pressure control valve
hot gas bypass valve
multiplexed
multiplexed compressors
nonadjustable head pressure control valve
normally closed (NC)
normally open (NO)
pumpdown cycle
receiver pressure regulator
short cycle
solenoid valve
two-temperature system

Introduction

Many refrigeration systems use special accessory flow control valves to increase system efficiency, provide safeguards, facilitate repairs, or add system features. These special flow control valves are installed *in addition to* regular refrigerant metering devices. It is quite common to see one or more special purpose valves on a system. Each flow control valve serves a particular purpose, and a malfunction in a valve greatly affects operation of other system components. The operation and purpose of these valves are simple, but they can be misleading. This chapter explains these special-purpose valves and the functions they serve in the system. Knowledge of these valves will prevent time-consuming mistakes.

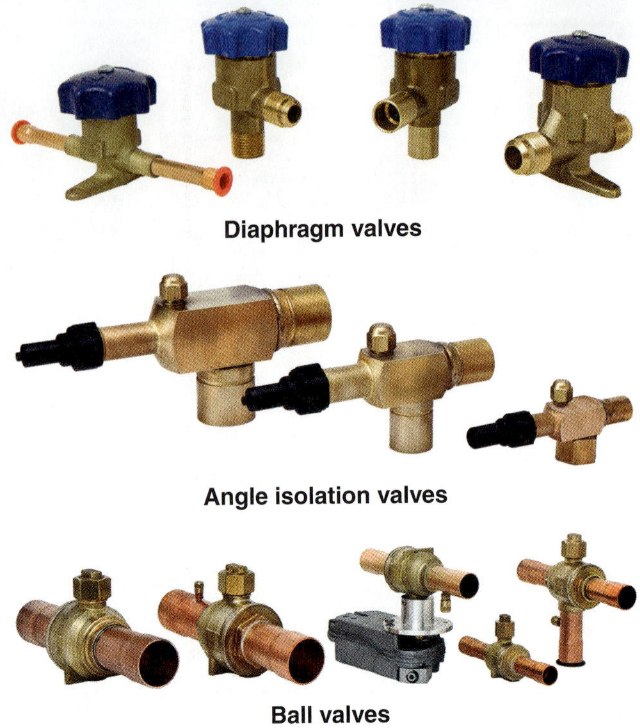

Mueller Industries, Inc.

Figure 19-1. Various manually-operated (hand) valves.

19.1 Hand Shut-Off and Check Valves

Hand shut-off valves are installed on commercial and industrial systems as a convenience for service technicians. Hand valves make it possible to quickly isolate sections of a system to perform a particular task, such as changing a filter-drier. For safety, local codes often specify the installation of hand valves in commercial and industrial refrigeration and air-conditioning systems, **Figure 19-1**.

Manually-operated valves must be sturdy and able to withstand frequent opening and closing without leaking. Pressure drop across the valve must be minimal. Hand valves are designed for different applications and can be brazed or flare-connected to copper tubing. The three main builds of hand valves include the following:

- Diaphragm-type
- Globe-type
- Ball-type

For proper operation, a hand valve must be installed according to the designed direction of flow through the valve body. An arrow on the valve body indicates correct refrigerant flow direction. Diaphragm-type hand valves are used for smaller lines, while ball and globe valves are used for larger lines.

The turning knob (handle) of a *diaphragm valve* is connected directly to the valve stem. *Ball valves* have a metal cap nut over the valve stem. The cap nut must be removed for access to the valve stem. *Globe valves* also have a cap nut over the valve stem, but the cap nut has extension handles for easy turning. The cap nut also has a recessed opening so it can be turned upside down and used as a wrench to turn the valve stem.

Hand valves can be installed in any line to evacuate and isolate specific sections of the system, **Figure 19-2**. Hand valves make it possible to perform repairs without loss of refrigerant.

Check Valves

Check valves are used in some refrigeration systems to prevent the flow of refrigerant in the wrong direction. Proper direction of flow is indicated on the body of these nonadjustable valves. If the liquid or vapor is flowing in the proper direction, the check valve will be open. A change of direction causes the valve to close tightly, preventing reverse flow, **Figure 19-3**.

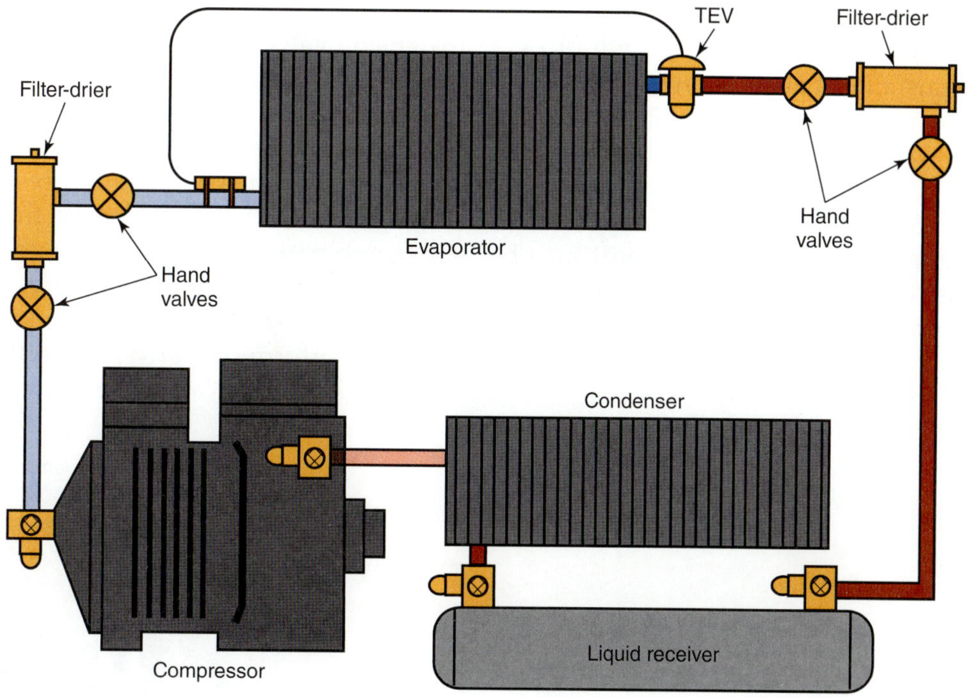

Goodheart-Willcox Publisher

Figure 19-2. Hand valves can be used to isolate and evacuate portions of the refrigeration system without loss of refrigerant. They are often used to permit easy changing of filter-driers.

Danfoss

Figure 19-3. The arrows on the bodies of these check valves show proper direction for flow.

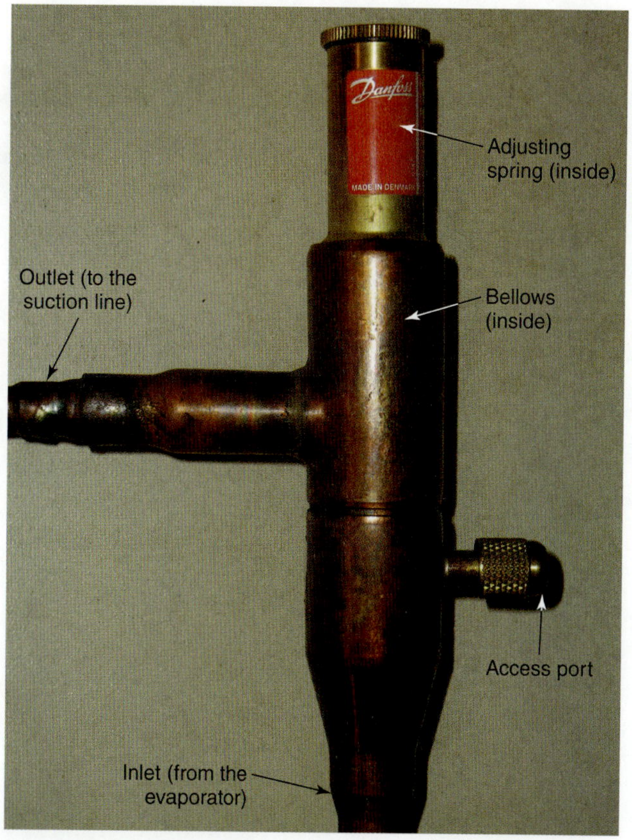

Goodheart-Willcox Publisher

Figure 19-4. The evaporator pressure regulator (EPR) valve keeps pressure at the evaporator outlet from dropping below a specified limit. An access port is available on the evaporator side to measure and monitor evaporator pressure.

19.2 Evaporator Pressure Regulators (EPRs)

Evaporator pressure regulators (EPRs) are special valves used to prevent evaporator pressure from falling below a set limit, **Figure 19-4.** Remember that an evaporator's pressure affects its temperature. EPRs make it possible to connect multiple evaporators in parallel but maintain different temperatures in each evaporator with EPRs.

An EPR is installed between an evaporator's outlet and the suction line, often by brazing. When installing EPRs with brazed connections, wrap the valve with a wet cloth to keep the valve body temperature below 250°F (121°C). The torch tip should be large enough to avoid prolonged heating of the connections. Overheating can be reduced by directing the torch flame away from the valve body. Be careful to avoid allowing molten alloy into the internal parts of the valve.

An EPR operates much like an automatic expansion valve. While the AEV controls pressure at its *outlet*, the EPR controls pressure at its *inlet*.

A TEV ensures proper evaporator flooding. An EPR provides its evaporator with a low-pressure limit. This prevents evaporator pressure (and temperature) from dropping below the EPR setting. The combination of valves ensures proper flooding of the evaporator *and* a low-pressure limit. The EPR can be installed anywhere in the suction line and still perform properly, **Figure 19-5**.

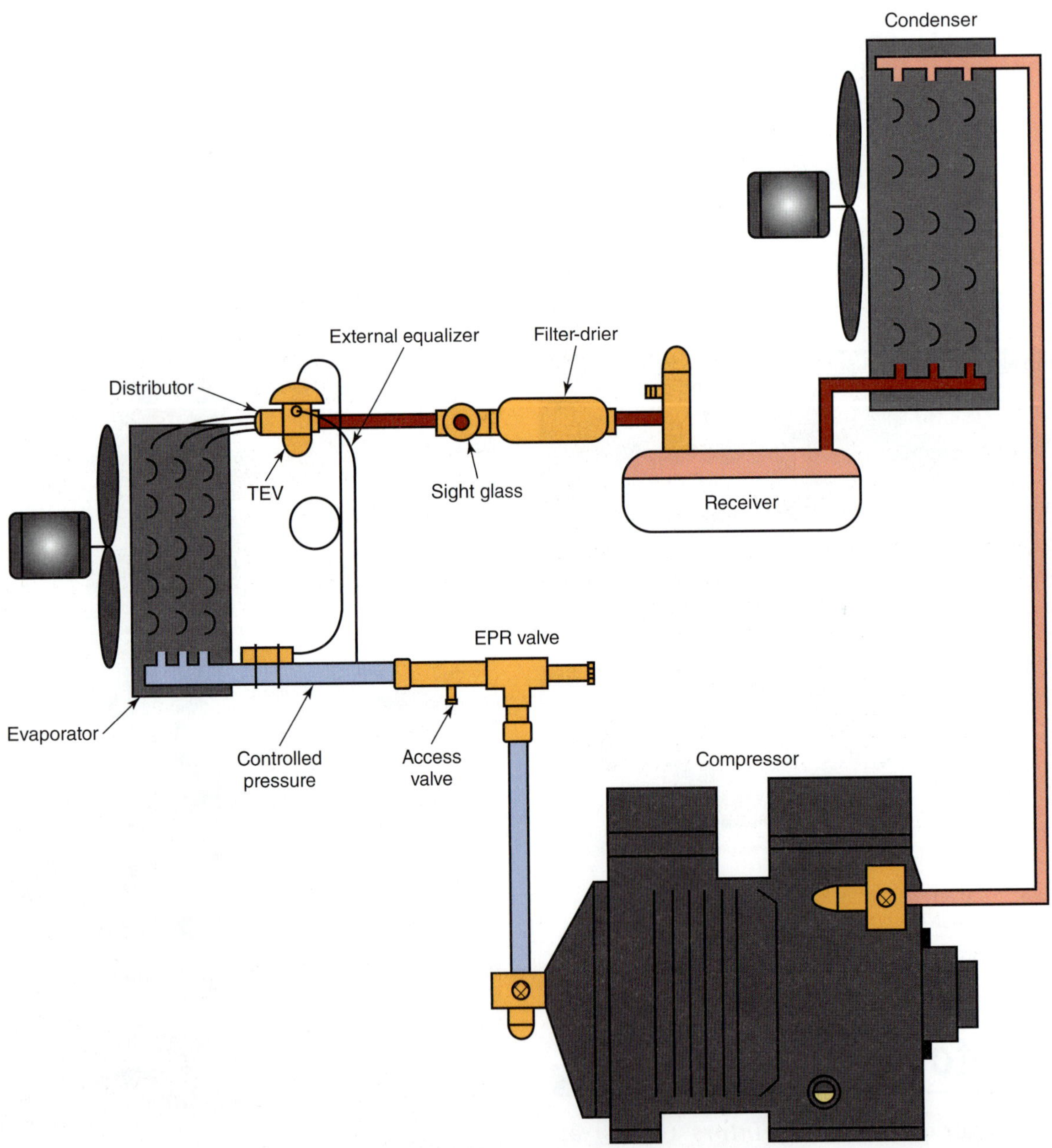

Goodheart-Willcox Publisher

Figure 19-5. An EPR valve is an inlet pressure regulator and can be installed anywhere in the suction line to prevent the evaporator pressure from going too low.

As noted, the primary function of an EPR is to prevent evaporator pressure from falling below a certain level. If the evaporator pressure rises above the valve setting, the EPR will open to release some pressure into the suction line. When the evaporator pressure drops to the valve setting, the EPR will close, preventing further pressure drop. The EPR is a "holdback" type of valve that automatically throttles vapor flow from the evaporator to maintain its setting. The valve modulates closely to the actual setting, constantly making slight adjustments to control evaporator pressure.

To adjust an EPR valve, remove the cap, and turn the adjustment screw with the proper size hex wrench (1/4" or 5/16"). A *clockwise (CW)* rotation increases the valve setting; a *counterclockwise (CCW)* rotation decreases the setting. To obtain the desired setting, a pressure gauge should be attached to the inlet side of the valve so the effects of any adjustments can be observed. Make small adjustments and allow adequate time between adjustments for the system to stabilize at the new setting.

EPRs are normally available in two pressure-adjustment ranges: 0–50 psig (0–345 kPa) and 30–100 psig (207–690 kPa). The 0–50 psig range is suitable for most applications. The 30–100 psig range is for applications using R-22 and R-404A with evaporating temperatures of 20°F (–7°C) and above. An access port or Schrader valve is provided on the EPR inlet (evaporator) side for connecting a pressure gauge. Inlet pressure readings are necessary when making pressure adjustments.

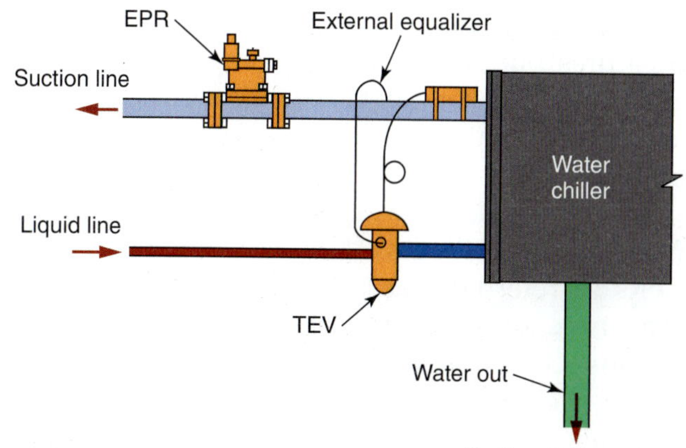

Goodheart-Willcox Publisher

Figure 19-6. An EPR valve is often used to prevent evaporator freeze-up. The EPR serves to prevent reduction of evaporator pressure below its low-pressure setting.

Adjustments to the EPR should be performed *before* checking TEV superheat and making any necessary adjustments.

Application of EPRs

A typical EPR application is a single evaporator system, such as a water chiller, where the EPR is used to prevent freezing of the water. See **Figure 19-6**. The valve's pressure adjustment is set to keep the evaporator pressure (which affects refrigerant temperature) above the freezing point of water. The EPR prevents evaporator pressure from dropping below this setting.

Another EPR application is the ***two-temperature system***. The system in **Figure 19-7**

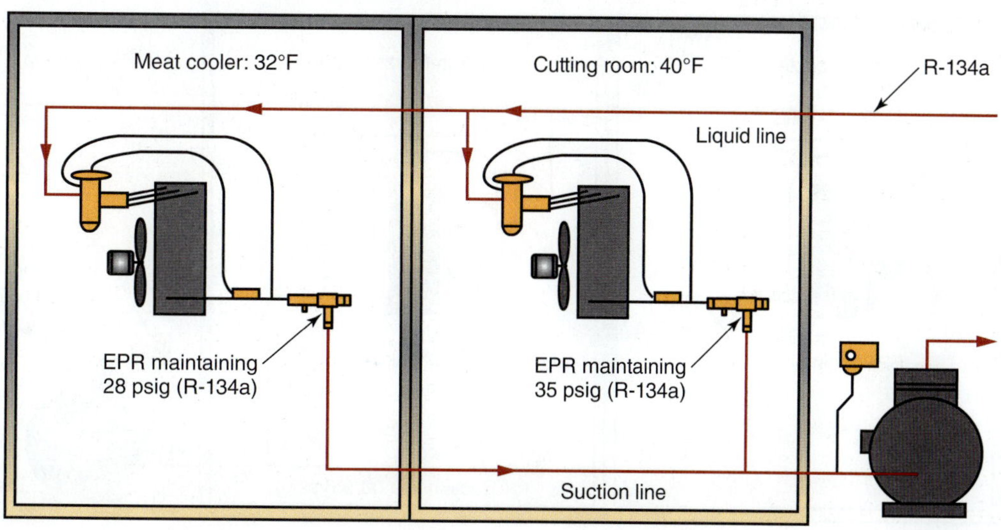

Goodheart-Willcox Publisher

Figure 19-7. A two-temperature system using two EPR valves. Using R-134a refrigerant, the meat cooler has a temperature of 32°F and a pressure at the access valve of 28 psig. The cutting room has a temperature of 40°F and a pressure of 35 psig.

has two evaporators operating at different temperatures but connected to a single compressor. The evaporator for the walk-in meat cooler is operating at 32°F (0°C), while the evaporator for the cutting room is operating at 40°F (4°C). An EPR is installed on each evaporator to prevent pressure from dropping too low. This system explains why EPRs are referred to as *two-temperature valves*. EPRs are a necessity when multiple evaporators share a single suction line controlled by one compressor or group of compressors.

An EPR is used when one evaporator operates at low temperature (a freezer) and another at medium temperature (a cooler), **Figure 19-8**. With two evaporators, it is unnecessary to use an EPR for the freezer (the lower temperature). The operating controls (thermostat or pressure control) are set for controlling the *freezer* (coldest), while the *cooler's* temperature (warmer) is controlled by the EPR.

Check valves are used on two-temperature applications in which two evaporators operating at different temperatures are connected to a single suction line. A check valve is installed in the suction line of the coldest evaporator, as shown in **Figure 19-8**. The check valve prevents warm suction gas from flowing into the freezer's evaporator during the off cycle. Warm suction gas entering the freezer's evaporator would create problems in the colder evaporator, possibly causing the system to ***short cycle*** (turn on and off rapidly).

Multiplexed Evaporators

EPRs make it possible to operate two or more evaporators at different temperature levels. These systems employ the use of a thermostatic expansion valve (TEV) to connect multiple evaporators in parallel to a single suction line leading to a single compressor or group of compressors. This setup is referred to as ***multiplexed***, **Figure 19-9**. Normally, such evaporators operate at nearly the same temperature since they have the same pressure, as regulated by the single suction line. However, EPRs make it possible to operate each evaporator at a different temperature by limiting their low pressure. There is no limit to the number of evaporators that may be connected to a single suction line if the compressor or group of compressors is sized for total refrigeration load.

An example of a multiplexed system is found in supermarkets. A line of four or five frozen food cases might be served by a single compressor or group of compressors located in an equipment room. Each case has its own TEV, evaporator, fans, and other components. Common liquid and suction lines connect the evaporators in the cases.

An advantage of multiplexed systems is the elimination of duplicated components, which saves money on materials. Multiplexed systems also save installation time, energy costs, and space.

A disadvantage of using multiplexed evaporators means a large compressor or group of compressors is needed. If using a single compressor that fails, replacement is time-consuming and

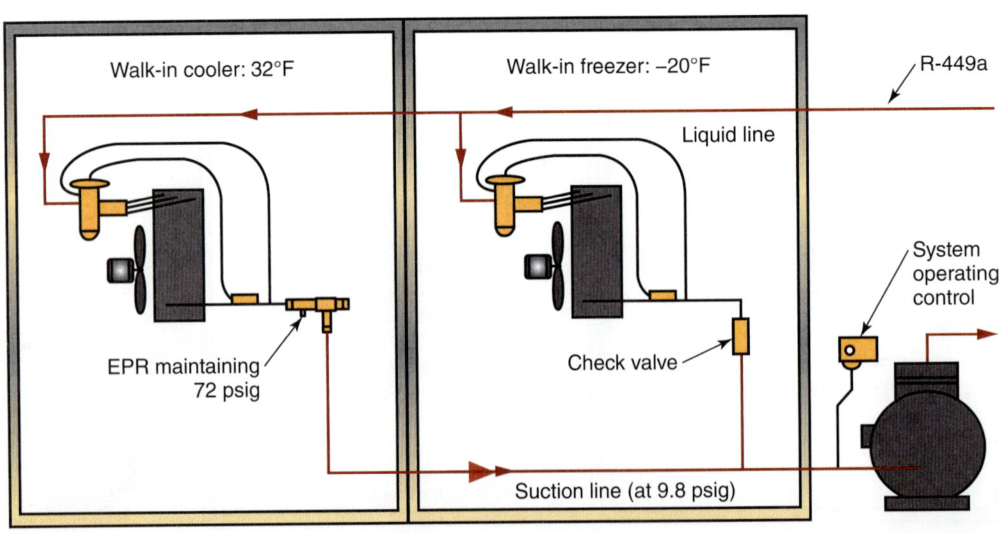

Goodheart-Willcox Publisher

Figure 19-8. A two-temperature system charged with R-449a using one EPR and the system operating control to regulate different temperatures in two areas. An EPR keeps the walk-in cooler at 32°F (at 72 psig), while the system operating control keeps the freezer at –20°F (at 9.8 psig).

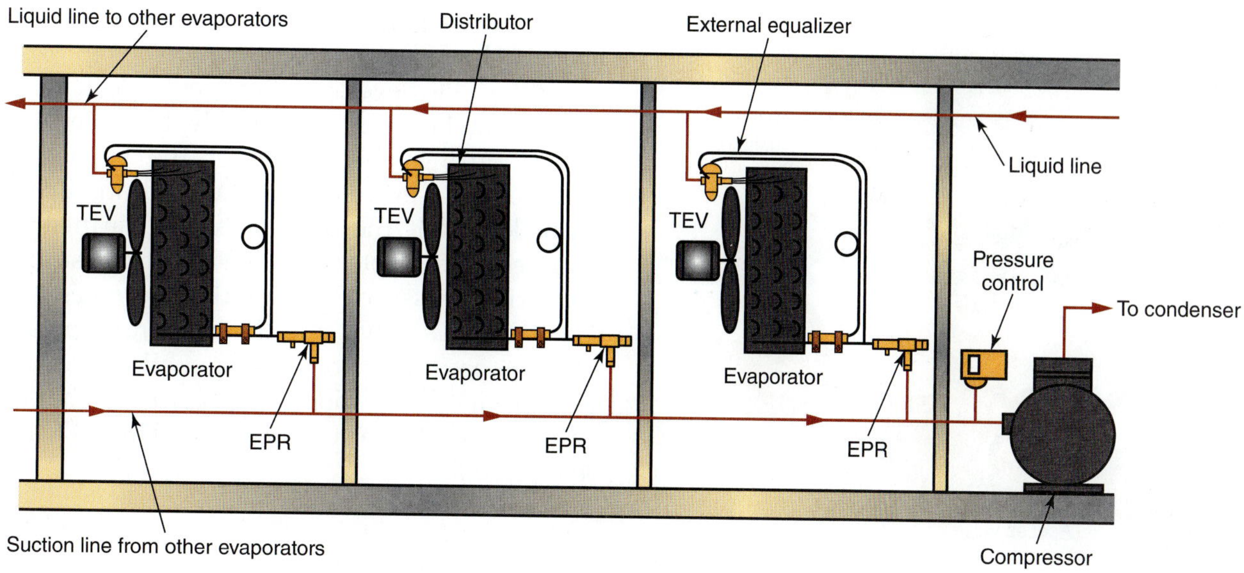

Goodheart-Willcox Publisher

Figure 19-9. Multiplexed evaporators (each with its own EPR valve) permit individual control of pressure (and temperature). Any number of evaporators can be connected to a single suction line and individually controlled by their EPR.

expensive and can lead to considerable loss of product during the repair time.

Compressor capacity under reduced load conditions often must be controlled. Under low-load conditions, a single large compressor reduces suction pressure too rapidly, causing short cycling to occur. Compressor cylinder unloaders may be required, or a hot-gas bypass valve may be installed to meter small amounts of hot gas into the suction line during low-load periods.

Multiplexed Compressors

The problems associated with *multiplexed evaporators* operated by a *single compressor* are easily overcome by using ***multiplexed compressors***. Rather than a single large compressor, multiple compressors are connected in parallel. The compressors are connected to a common suction manifold and a common discharge manifold. For example, instead of using a single 60 hp compressor, three 20 hp compressors are used. See **Figure 19-10**.

With multiplexed compressors, if one compressor fails, the others keep the system operational until repairs are completed. All compressors will operate under a full-load condition, which occurs when most of the EPRs are open. Large systems require large EPRs, such as the one in **Figure 19-11**.

The load is reduced when the EPRs throttle down, causing a reduction in suction line pressure. When the suction line pressure is reduced to a predetermined point, one of the compressors is turned off. Each compressor is cycled off in rotation, according to reducing suction line pressure. The individual pressure switches restart each compressor in response to increasing suction line pressure.

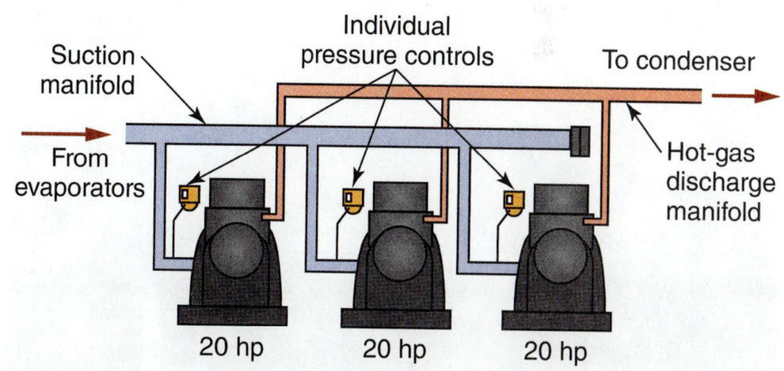

Goodheart-Willcox Publisher

Figure 19-10. Multiplexed compressors overcome the problems associated with multiplexed evaporators. Individual pressure controls cycle the compressors on as needed. If a compressor fails, the system will continue to operate with the remaining compressors.

Many supermarkets have multiplexed evaporators with EPRs and multiplexed compressors, such as the one in **Figure 19-12**. One rack with multiple compressors operates on R-449a and controls all the low-temperature evaporators, such as frozen food cases, walk-in freezers, and ice cream cases.

Some racks of four compressors operate with R-453a or R-22 and control all medium-temperature evaporators, such as walk-in coolers and dairy, produce, and fresh meat cases. On four-compressor systems like these, one compressor operates constantly because the load is never reduced to the

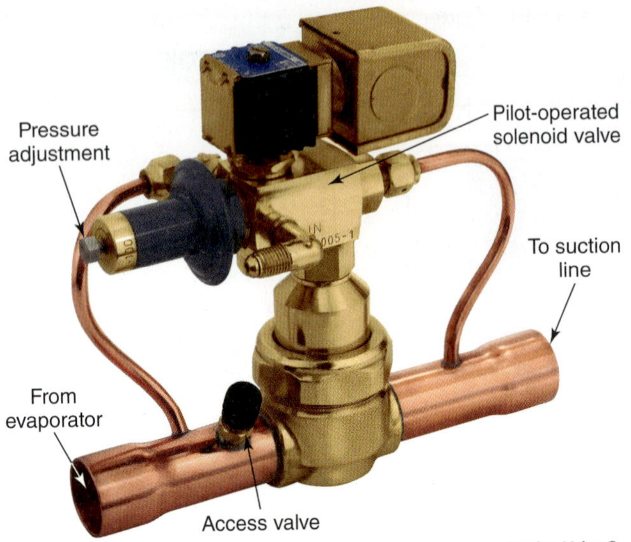

Sporlan Valve Co.

Figure 19-11. A large EPR used with large multiple-compressor installations.

Bitzer

Figure 19-12. A multiple-compressor system of the type used to meet the varied refrigeration needs of modern supermarkets.

point that the compressor can shut off. The other three compressors cycle on and off in response to changing load conditions signaled by pressure changes in the suction manifold.

Troubleshooting EPRs

Some EPRs are hermetic and cannot be disassembled for inspection and cleaning. They must be replaced if they become inoperative. If an EPR fails to open or close properly or if it will not adjust, alloy or other foreign material lodged in the port may be at fault. It is sometimes possible to dislodge these materials by turning the adjustment nut all the way out with the system running. Once cleared, the EPR can be reset.

If the EPR develops a refrigerant leak around the spring housing, it probably has been overheated during installation, or the bellows has failed due to severe compressor pulsation. When the bellows fails, the EPR closes until the inlet pressure becomes greater than the outlet pressure *plus* the spring pressure. The EPR then becomes a *pressure differential* valve, and the evaporator pressure changes according to varying suction pressures. This produces an evaporator pressure higher than desired and cannot be corrected. This EPR must be replaced.

19.3 Solenoid Valves

Solenoid valves are commonly used in refrigeration and air-conditioning systems. These valves serve a vital role in the automatic flow control of refrigerant, water, or other gases and liquids involved in the cooling process, **Figure 19-13**.

Solenoid Operating Principles

A ***solenoid valve*** is an electrically operated valve that is either fully open or fully closed, depending upon whether its electrical supply is on or off. Solenoid valves contain an electromagnet surrounding a plunger connected to a short valve

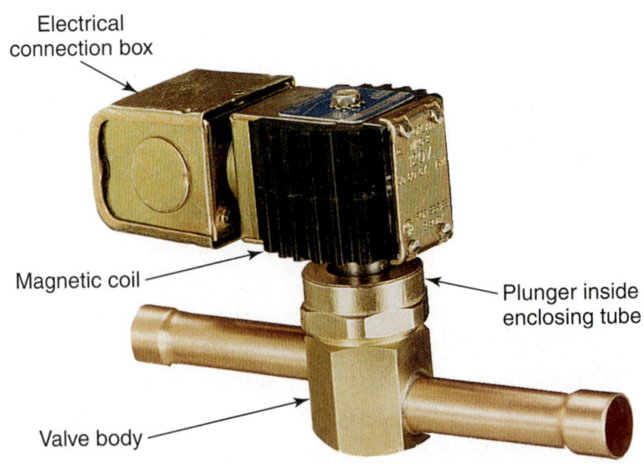

Sporlan Valve Co.

Figure 19-13. A typical solenoid valve. An electromagnetically operated plunger opens or closes the valve to regulate the flow of fluid. Valves usually have flare or braze-type connections.

stem. The electromagnet is a coil of insulated copper wire wrapped around a soft iron core with a round hole through the middle, **Figure 19-14**. Electricity flowing through the coil produces magnetism in the iron core. When the electrical flow is stopped, magnetism also stops. The magnetic coil is replaceable and available in different voltages.

The valve body section contains the valve seat, tubing connections, and a free-floating plunger connected to the valve stem, **Figure 19-15**. The plunger and valve stem are enclosed in a metal tube extending upward from the valve body. A special lock nut holds the enclosing tube to the valve body for a leaktight seal. The heavy plunger is free to move up and down in the enclosing tube, which fits into the hole in the electromagnet.

When the coil is magnetized by the flow of electricity, an opposing magnetic field is induced in the plunger. The plunger is pulled upward into the center of the magnetic field, **Figure 19-16**. Lifting of the plunger opens the valve. When the coil is de-energized, the plunger drops into the closed position. Sometimes, a small spring is located on top of the plunger to assist in closing.

Sporlan Valve Co.

Figure 19-15. An exploded view of the solenoid valve assembly. The magnetic coil, which surrounds the enclosing tube, has been removed for clarity.

Solenoid Installation

For proper plunger operation, a solenoid valve must be mounted in an upright position or at no more than a 45° angle. Special solenoid valves that use a closing spring (instead of relying on gravity) can be mounted horizontally (sideways). Most solenoid valves are ***normally closed (NC)***, meaning that they remain closed until energized. However, some solenoid valves are ***normally open (NO)***, so they remain open until energized, at which point they then close. Direction of flow is indicated by an arrow or the word "IN" on the valve body.

Before brazing, solenoid valves should be disassembled to protect internal components from heat damage. Merely removing the coil is not sufficient. Remove all the internal solenoid valve parts illustrated in **Figure 19-17**.

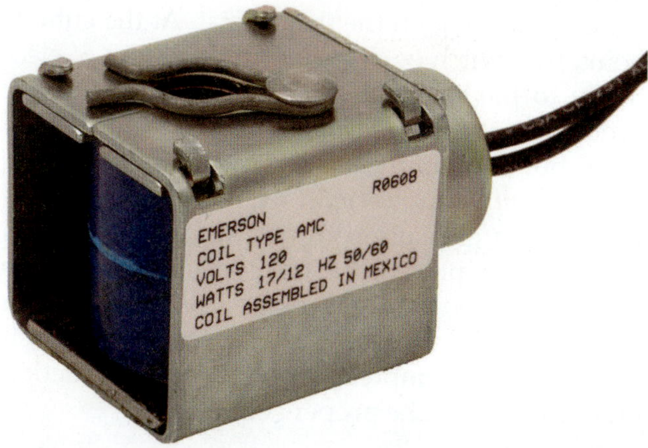

Emerson Climate Technologies

Figure 19-14. The magnetic coil of a solenoid has a hole in the middle for the plunger. Coils are replaceable.

PRO TIP

Brazing Valves to Tubing

When brazing solenoid valves (or any special-purpose valves) into a system, wrap the valve body with a *wet* cloth to keep the body temperature below 250°F (121°C) and prevent heat damage. The torch tip should be large enough to avoid prolonged heating of the copper connections. Always direct the flame away from the valve body and perform the brazing operation quickly and carefully. Be careful not to allow melted alloy to flow into the valve body.

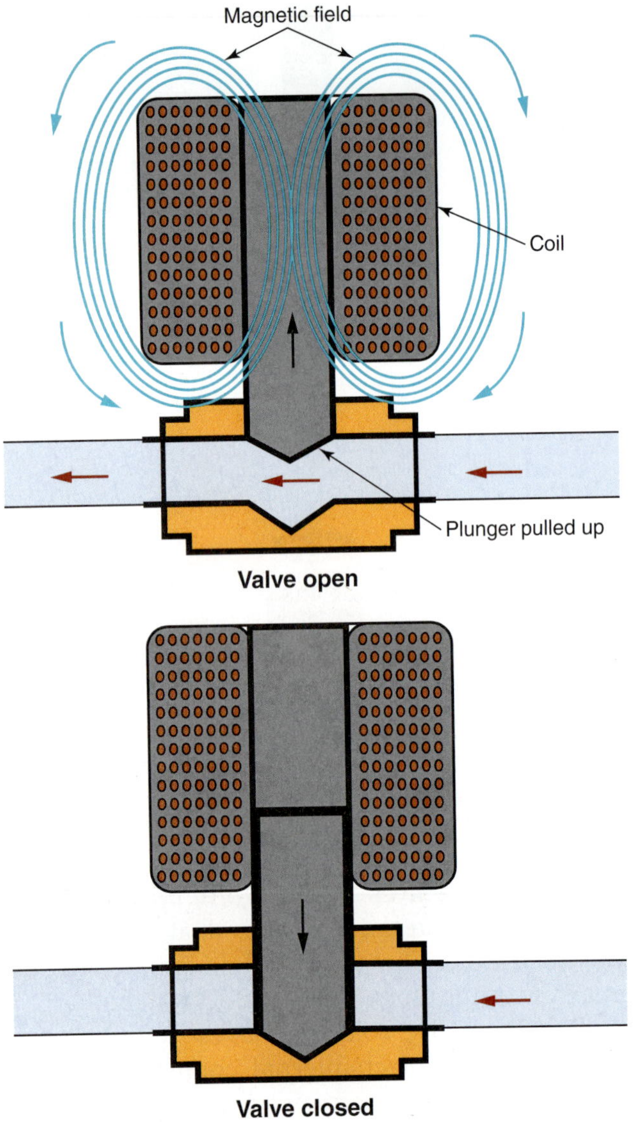

Goodheart-Willcox Publisher

Figure 19-16. Electricity energizes the coil, creating a magnetic field that pulls the plunger up into the enclosing tube, opening the valve. With no electricity flowing to the coil, the plunger is in the normally closed position, closing the valve and thus blocking flow.

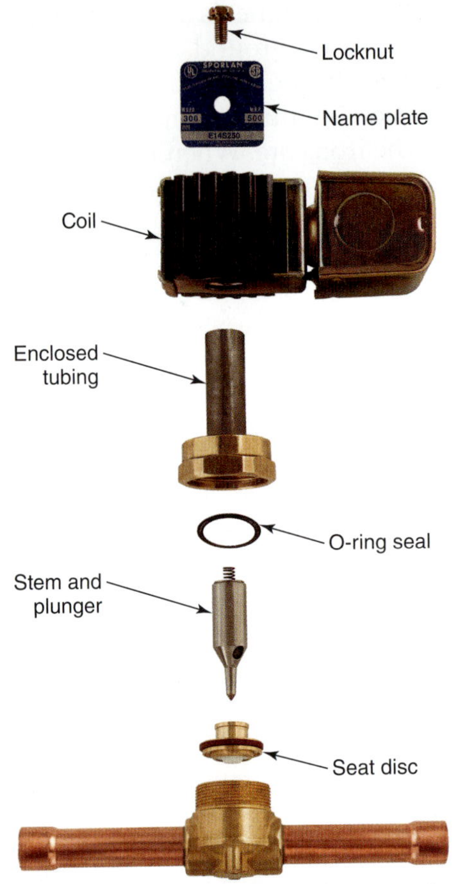

Sporlan Valve Co.

Figure 19-17. Internal parts of the solenoid, shown in this exploded view, should be removed before brazing. This will protect them from heat damage when the body is brazed to the tubing.

Solenoid Valve and Pumpdown Cycle

A common application of a solenoid valve is the automatic pumpdown cycle, **Figure 19-18**. A solenoid valve is installed in the liquid line between the sight glass and the TEV. Location may be just before the TEV (evaporator area) or just after the sight glass (condensing unit area).

A temperature control (thermostat) is mounted within the cooled space and electrically connected to the solenoid coil. When temperature within the space rises to the cut-in point on the thermostat, a switch closes and completes an electrical circuit to the solenoid coil. The solenoid coil is energized, opening the valve and allowing refrigerant flow.

The system continues to operate until the temperature in the space to be cooled is lowered to the cut-out point on the thermostat. At the cut-out point, the switch opens, stopping the flow of electricity to the solenoid coil. The solenoid plunger drops to the closed position. Refrigerant flow stops, but the compressor continues running.

On a ***pumpdown cycle***, the compressor is controlled by a pressure-sensitive switch that senses suction line pressure. The pressure control is adjusted to stop the compressor just before suction pressure reaches a state of vacuum. The switch also turns the compressor on again when suction pressure rises to the proper setpoint.

Pumpdown Run Cycle

When the solenoid coil is energized by the thermostat, the valve opens and permits liquid

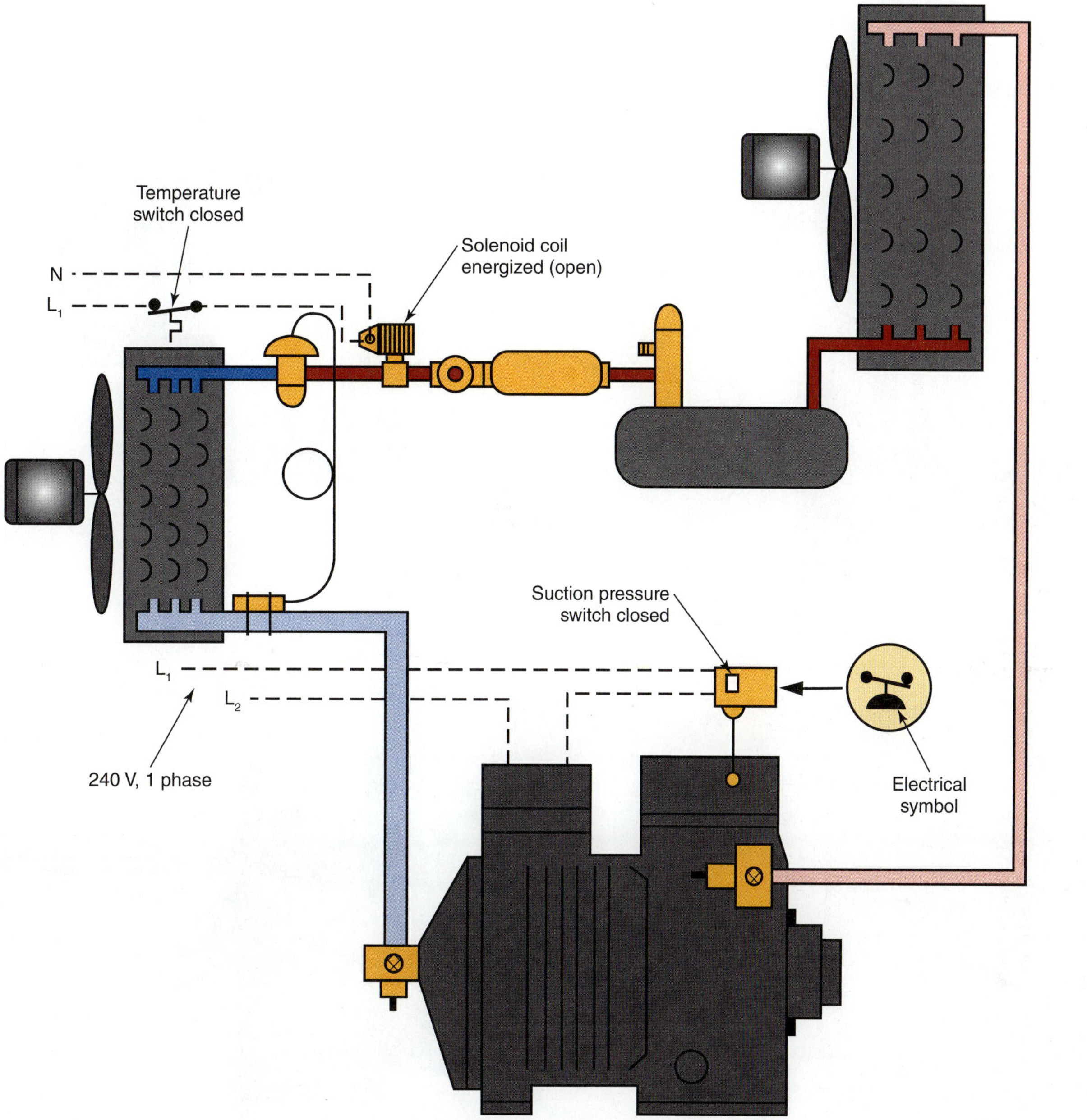

Goodheart-Willcox Publisher

Figure 19-18. In the pumpdown "run" cycle, the closed temperature switch energizes the solenoid coil, opening the valve and permitting refrigerant to flow. The suction pressure switch controls the compressor.

refrigerant flow to the TEV. Refrigerant flowing into the evaporator causes the suction pressure to rise. The pressure-sensitive motor control turns the compressor on. Normal refrigeration continues until product temperature is lowered to the cut-out point on the thermostat.

Pumpdown Off Cycle

When the thermostat switch opens, electrical flow to the solenoid coil stops, and the valve closes. Refrigerant flow to the TEV stops, **Figure 19-19**. The compressor continues to run until the low-side pressure is reduced to the cut-out point on the pressure-sensitive motor control (about 1–2 psig or 7–14 kPa). The compressor should stop before suction pressure reaches a vacuum, resulting in removal of all refrigerant from the low-pressure side following each run cycle. Some refrigeration and air-conditioning systems require such a pumpdown cycle to avoid high suction pressure at start-up. Pumpdown also ensures good oil return to the compressor and evacuates the evaporator prior to defrost. Otherwise, heat from electric defrost heaters would create high suction pressure at start-up.

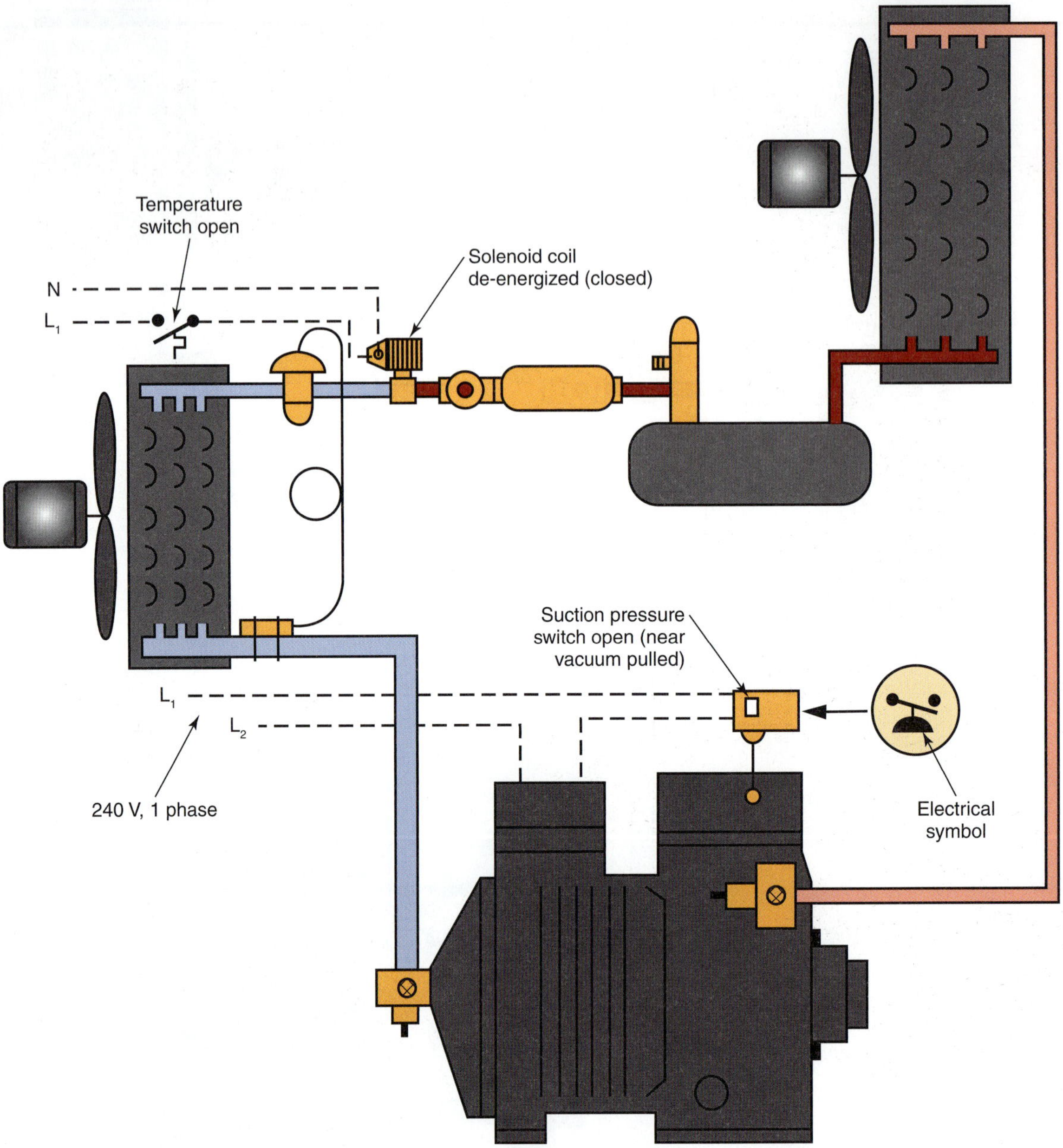

Goodheart-Willcox Publisher

Figure 19-19. In the pumpdown "off" cycle, the open temperature switch cuts current to the solenoid coil, closing the valve and stopping refrigerant flow. The compressor keeps running until the pressure switch opens (just before suction line pressure reaches a vacuum). This results in all refrigerant being removed from the low side of the system after each run cycle.

Troubleshooting Solenoid Valves

Troubleshooting a solenoid valve is primarily limited to determining whether the coil is energized. The coil cannot be energized unless proper voltage is applied to it. No voltage means no magnetism. Electricity and methods for checking voltage are explained in later chapters.

To quickly determine if the coil is energized, hold the tip of a small steel screwdriver close to the top center of the coil, **Figure 19-20**. If the coil is energized, magnetism will pull the screwdriver tip downward. If the coil is not energized, the screwdriver will not be pulled toward the coil. The magnetic field is not strong, so a *small* pocket screwdriver is required for proper detection.

Solenoid valves are noiseless during normal operation. A steady humming sound (an "alternating current" hum) indicates the coil is not correctly

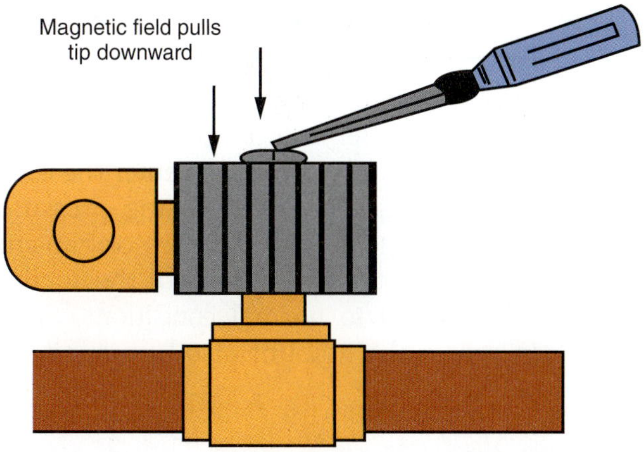

Goodheart-Willcox Publisher

Figure 19-20. A small steel screwdriver can be used to test for an energized solenoid coil. Magnetic force will pull the screwdriver tip downward if the coil is energized.

positioned or not properly anchored to the valve. Solenoid valves normally make a sharp "snap" sound when energized. This sound is caused by the plunger striking the top of the enclosing cylinder. To check solenoid operation, energize and de-energize the coil while listening for a snapping sound.

Coil burnouts are rare and normally caused by improper wiring or improper voltage. A burnout can occur if the plunger is not free to be pulled upward when magnetic force is exerted. The problem usually is caused by a damaged enclosing tube. A burnout also occurs if the coil is removed from the valve while energized.

Failure of the valve to open or close properly is usually caused by foreign material lodged in the valve seat. Failure can also be caused by a valve body that is deformed by overheating. A defective solenoid coil is easily replaced, but a defective valve body requires replacement of the entire valve.

19.4 Crankcase Pressure Regulators (CPRs)

Crankcase pressure regulators (CPRs) are used to protect the compressor from excessive suction pressure, which most often occurs at start-up. Like the EPR, the CPR is a pressure-limiting device. The CPR, however, is constructed to limit *outlet* pressure, **Figure 19-21**.

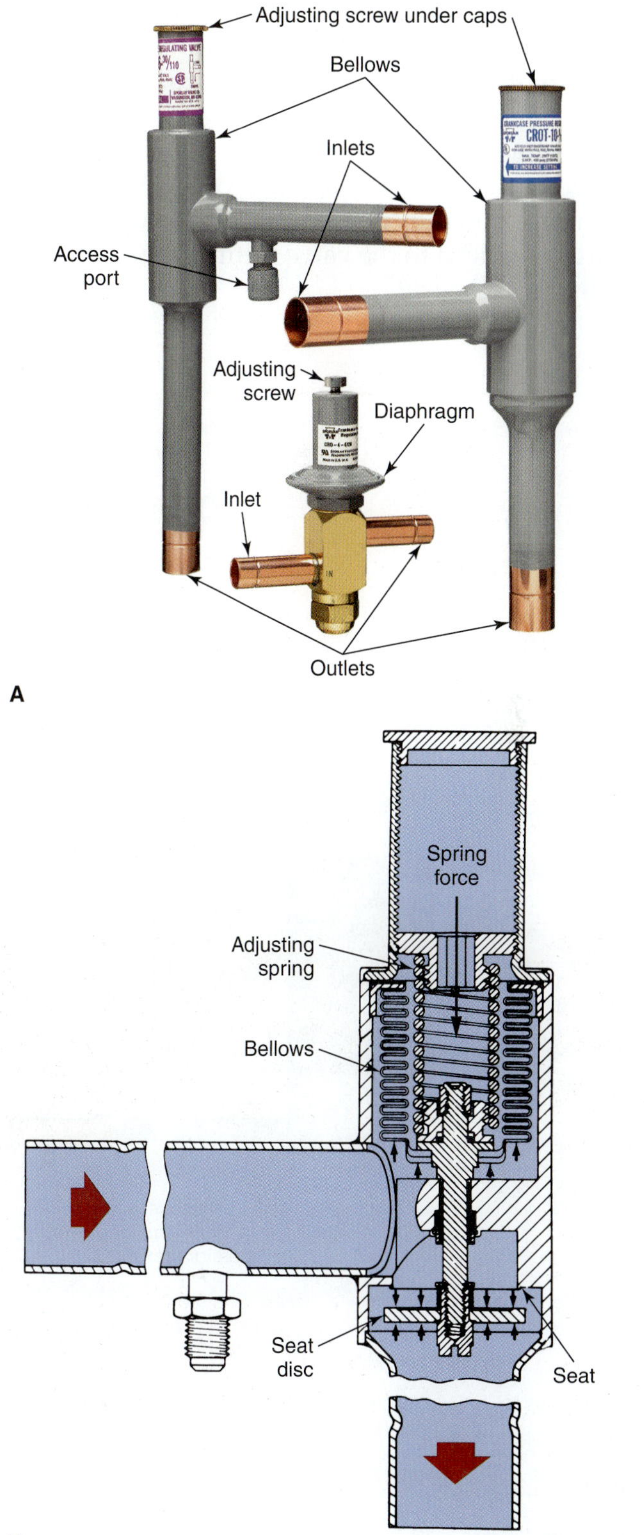

Sporlan Valve Co.

Figure 19-21. Crankcase pressure regulators. A—CRPs are available in different styles. The valve spring adjustment is usually under a cap. B—A cutaway view shows component locations.

The CPR is a holdback-type valve. As shown in **Figure 19-22**, it is mounted in the suction line close to the compressor. CPRs control or *limit* pressure leaving the valve, protecting the compressor from excessive pressure. When suction pressure exceeds the CPR setting, the valve throttles the pressure down to the setting. When suction pressure is reduced to the valve setting, the CPR opens fully.

CPR Applications

High suction pressure typically occurs after a defrost cycle when the evaporator is hot. If such high suction pressure is allowed to enter the compressor, it will cause high condensing pressure. This condition overloads the compressor, causing it to short cycle on the overload until the suction pressure is lowered to normal conditions. Short cycling increases the amount of oil pumped out

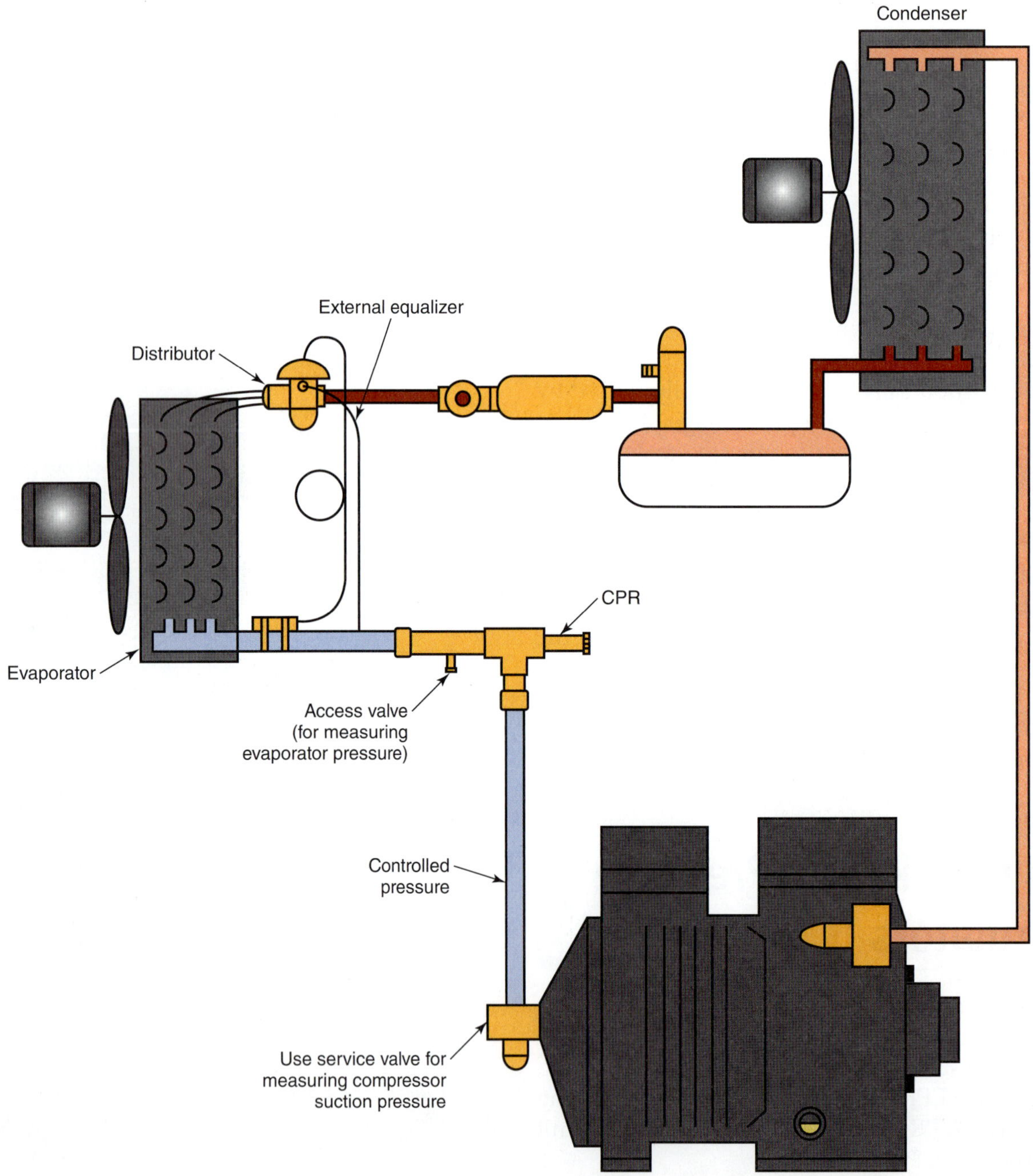

Goodheart-Willcox Publisher

Figure 19-22. The CPR is an outlet pressure limiter and may be installed anywhere in the suction line just before the compressor. It protects the compressor by limiting the pressure entering into it.

of the crankcase. Severe oil pumping causes broken piston rods, dry bearings, and damaged valve reeds.

CPR Adjustments

The CPR is adjusted by turning a screw located under the cover cap. When adjusting the valve, the manifold gauge must be installed on the suction service valve (SSV) to obtain accurate readings of pressure *entering the compressor.* Also a clamp-on ammeter should be installed to monitor the current draw of the compressor while setting the valve. The valve should be set to keep the compressor from pulling more current than the nameplate states. Most CPRs are set to limit entering suction pressure at 18–20 psig (124–138 kPa). Pressure adjustments can only be made when suction pressure (on the inlet side of the valve) *exceeds* the valve setting. Accurate adjustments cannot be made when the valve is fully open, a condition that occurs whenever low-side pressures are below the valve setting.

19.5 Hot-Gas Bypass Valves

Refrigeration and air-conditioning systems are usually designed to provide a specific capacity at maximum load conditions. Large systems must make provisions for operation at *reduced* load conditions to maintain temperature and humidity (air moisture) control. Hot-gas bypass valves are one method of controlling load variations. Some systems control capacity by compressor cylinder unloading or by using two or more compressors (multiplexing). On multiple compressor systems, individual compressors are turned off as the load and suction pressure is reduced.

It is often necessary to limit minimum evaporator pressure during periods of low load. This prevents frost or ice formation on the evaporator and avoids operating the compressor at low suction pressures. Without such ***capacity control***, the compressor would cycle on and off and rapidly reduce its lifespan. Such on-off control of air-conditioning systems permits wide temperature variations and does a poor job of controlling humidity. Human comfort requires control of both temperature and humidity.

Hot-Gas Bypass Valve Application

A practical and economical solution to the problem is to *bypass* a small portion of hot discharge gas directly into the low-pressure side. The amount of hot gas bypassed is controlled by installing a modulating control valve, called a ***hot-gas bypass valve***, **Figure 19-23**. This valve may also be called a *discharge bypass valve.* The valve opens on a decrease in suction pressure and can be set to automatically maintain a desired *minimum* evaporator pressure. Hot gas only bypasses when suction pressure is reduced to the valve setting. This method places a ***false load*** on the evaporator to maintain a minimum suction pressure.

Hot-Gas Bypass Valve Location

The hot-gas bypass valve is installed in a branch line off the hot-gas discharge line as close to the compressor as possible. Refer to **Figure 19-23B**. The hot gas is piped to the low-pressure side at the evaporator inlet or directly into the suction line. The evaporator inlet is the preferred location because the thermostatic expansion valve responds to the increased superheat leaving the evaporator and provides the liquid required for desuperheating. The evaporator also serves as an excellent mixing chamber for the bypassed hot gas and the liquid-vapor mixture from the expansion valve. Oil return from the evaporator is improved since the velocity in the evaporator is increased by the hot gas.

A solenoid valve should be installed ahead of the hot-gas bypass valve. The solenoid valve is energized only during the run cycle. This permits the system to operate with automatic pumpdown and guards against leakage during the "off" cycle.

Hot-Gas Bypass Valve Operation

Hot-gas bypass valves respond to changes in suction pressure. When suction pressure is above the valve setting, the hot-gas bypass valve remains closed. As suction pressure drops below the valve setting, the hot-gas bypass valve begins to open.

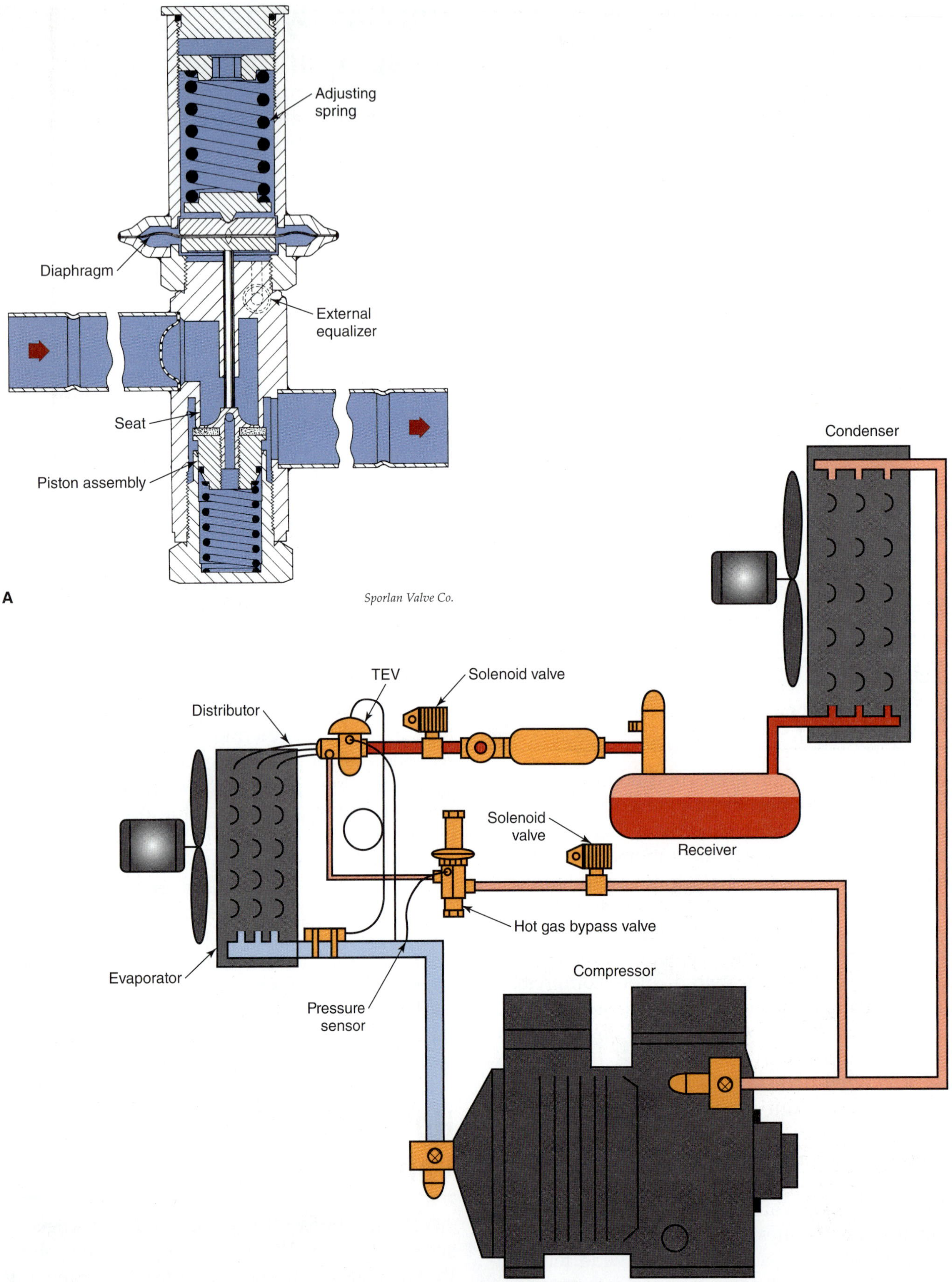

Figure 19-23. Hot-gas bypass valve. A—A cutaway view of an adjustable hot-gas bypass valve. B—Location of a hot-gas bypass valve.

As with all modulating valves, the amount of the opening depends on what is being controlled (in this case, the suction pressure). As the suction pressure continues to drop, the hot-gas bypass valve opens wider until the limit of the valve piston stroke is reached. However, in normal applications, the amount of pressure change is seldom sufficient to open the valve to the limit of its stroke.

Hot-gas bypass valves are usually rated at 6°F (3°C) temperature change from closed position to rated opening. A typical application would be a low-temperature compressor designed to operate at a minimum suction gas temperature on R-22 of –40°F (0.5 psig). The required suction temperature at normal load conditions is –30°F (5 psig). A hot-gas bypass valve would be selected to start opening at 3 psig (the pressure equivalent to –34°F) and bypass enough hot gas at 0.5 psig (–40°F) to prevent further decrease in suction pressure.

Many large air-conditioning systems use hot-gas bypass valves to prevent evaporator freeze-up and short cycling of the compressor. The hot-gas bypass valve begins opening at 60–61 psig (using R-22). See **Figure 19-24** for a valve used in this type of application.

Goodheart-Willcox Publisher

Figure 19-24. A hot-gas bypass valve is used in large systems to prevent evaporator freeze-up.

Troubleshooting Hot-Gas Bypass Valves

There are two possible problems with a hot-gas bypass valve: failure to open and failure to close. Failure to open is revealed by low suction pressure. Failure to close is revealed by high suction pressure.

Problems with the hot-gas bypass valve are often caused by foreign material lodged inside the valve. The valves usually can be disassembled for cleaning. Defective sensing elements are replaceable without changing the entire valve.

19.6 Desuperheating Thermostatic Expansion Valve

In applications where hot gas must be bypassed directly into the suction line, an extra thermostatic expansion valve is required. This valve is called a ***desuperheating TEV*** or *liquid injection valve,* **Figure 19-25**.

The purpose of this valve is to supply enough liquid refrigerant to cool the discharge hot gas to the recommended suction temperature. Cooling protects the compressor from becoming overheated by high suction gas temperatures. Hot gas entering the suction line must be desuperheated

Danfoss

Figure 19-25. A desuperheating thermostatic expansion valve is used in conjunction with hot-gas bypass valves when connecting into the suction line rather than into the evaporator.

to remain within the suction temperature limits specified by the compressor manufacturer. Again, a solenoid valve is installed in the liquid injection line upstream of the desuperheating TEV to permit an automatic pumpdown cycle and prevent leakage during the off cycle.

Proper mixing of the hot gas, liquid injection, and suction gas is important. The compressor must be always protected against high suction temperatures due to superheat and liquid floodback. Proper mixing of all of these in the suction line must be obtained. Several mixing methods are available, but the generally recommended method involves directing the discharge gas and liquid injection to enter the suction line at an angle of flow against the direction of suction gas flow, **Figure 19-26**.

Arranging a bypass directly into a suction accumulator is often a convenient way to obtain proper desuperheating of suction gas. In any event, introducing the hot gas and liquid into the suction line with separate connections is generally not recommended.

19.7 Controlling Head Pressure

Many commercial systems place the air-cooled condensing unit outdoors, often on the roof. Placing the unit outdoors conserves space inside the building and utilizes outdoor ambient air for condenser cooling. Outdoor units perform well, but cold winter temperatures present problems because of low head pressure. Many air-conditioning systems must operate during the winter months to provide controlled environments. In such applications, provisions must be made to maintain proper head pressure during periods when ambient temperatures are low.

The high-side pressure must be sufficient to maintain proper pressure drop across the thermostatic expansion valve. For proper operation of the TEV, certain minimum head pressures are required. Valves require different head pressures based on the refrigerant used:

- R-134a: 100 psig (690 kPa)
- R-22: 180 psig (1242 kPa)
- R-449A: 210 psig (1448 kPa)

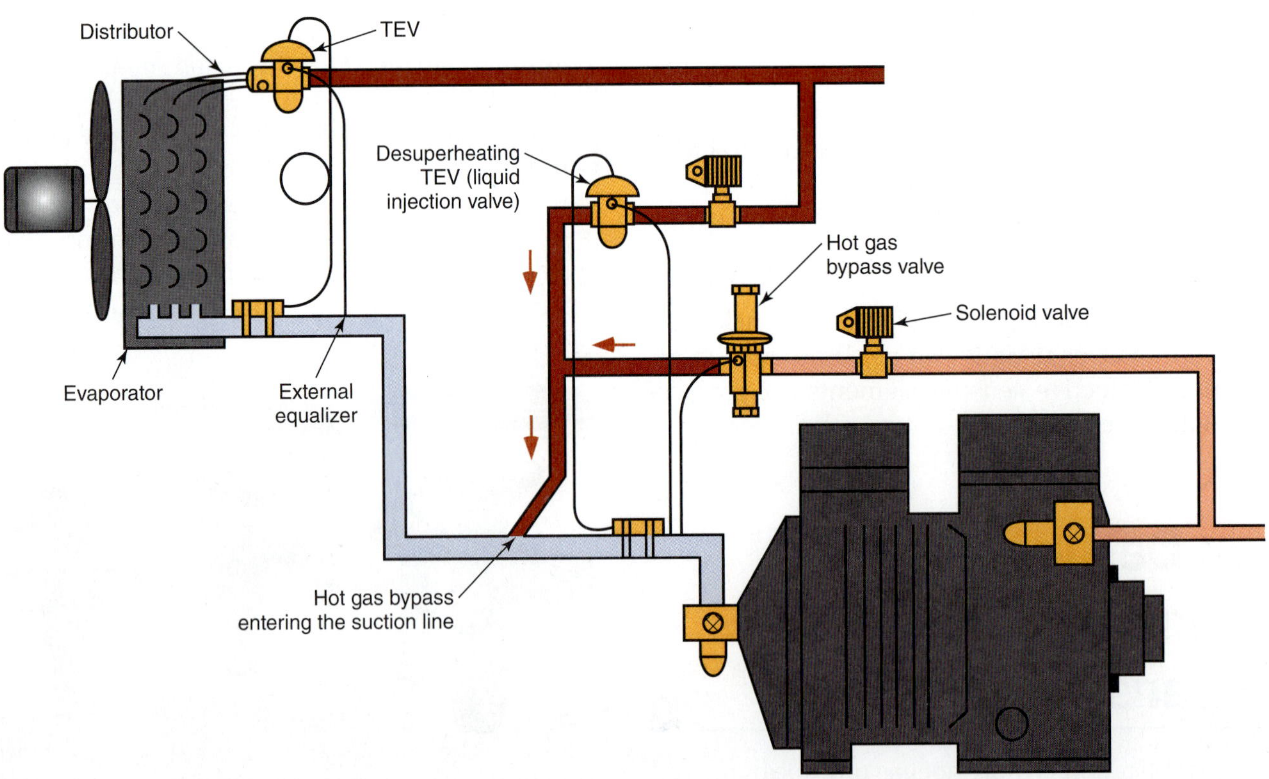

Goodheart-Willcox Publisher

Figure 19-26. An arrangement for proper mixing of liquid injection and discharge hot gas with suction vapor upon entering the suction line.

When head pressure drops too low, the expansion valve cannot feed the evaporator properly. When head pressure goes down, suction pressure also goes down, resulting in poor refrigeration or an ice-covered evaporator.

Low ambient temperature causes low head pressure on air-cooled condensing units, which can be observed with a manifold gauge measurement. The gauges reveal low pressures on each side, and the sight glass shows many bubbles (flash gas). The system contains the proper amount of refrigerant for summer operation but not for winter operation.

Air-cooled condenser capacity is selected to provide for efficient operation at 90°F (32°C) ambient airflow. Cold ambient airflow greatly increases condenser capacity, resulting in low head pressure. The low head pressure is not adequate to push liquid out of the liquid receiver and maintain correct pressure drop across the expansion valve. Liquid backs up in the receiver and remains there.

A *temporary* solution to the problem is adding enough refrigerant to partially fill the condenser with liquid. Filling the condenser with liquid reduces its capacity, causing head pressure to rise. However, when the outdoor ambient temperature rises, head pressure will rise drastically as a result of decreased condenser capacity. The system will cycle on the overload (or high-pressure safety control). The excess refrigerant must be removed from the system to return the condenser to normal capacity. Then head pressure will return to normal, until winter arrives again.

Two common types of ***head pressure control valves*** are used to solve low head pressure problems caused by low ambient temperatures. One method uses a single nonadjustable valve. The other uses two fully adjustable valves.

Nonadjustable Valves

The ***nonadjustable head pressure control valve*** is a three-way modulating valve that controls head pressure and is controlled by discharge pressure. It will maintain a head pressure of 100 psig (690 kPa) for R-134a, 180 psig (1242 kPa) for R-22, and 210 psig (1448 kPa) for R-449A. See **Figure 19-27**.

When head pressure drops to the valve setting, the valve will direct some (or all) hot gas to bypass the condenser and travel directly to the liquid receiver, **Figure 19-28**. The valve modulates in any position to control the amount of bypass that maintains proper head pressure. Under extreme conditions, the liquid receiver will act as the condenser as a result of cold ambient temperatures.

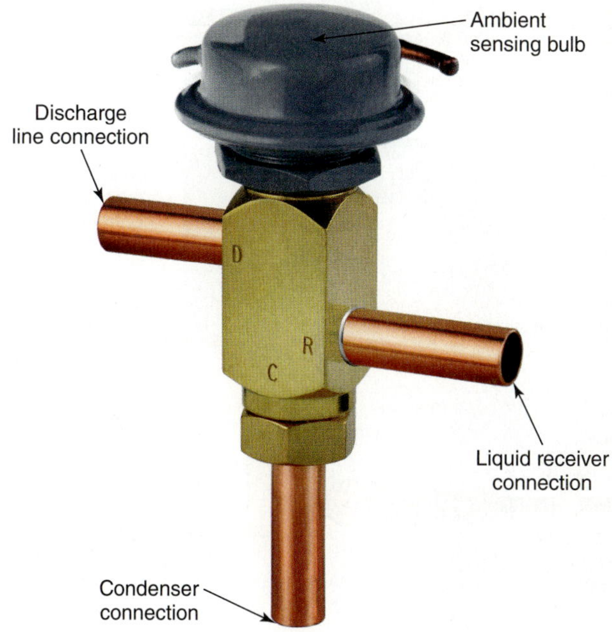

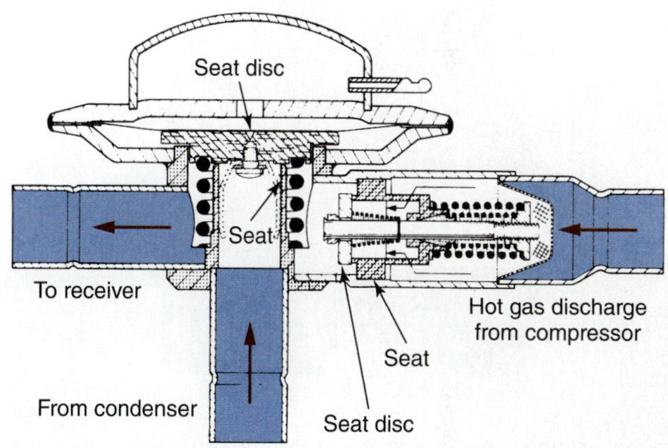

Sporlan Valve Co.

Figure 19-27. Nonadjustable head pressure control. A—A nonadjustable head pressure control valve. B—A cutaway view of the valve shows its components.

The refrigerant charge in the valve's dome exerts pressure on top of the diaphragm, which controls the seat disc. The downward diaphragm pressure tends to open the seat disc, permitting hot gas to bypass directly to the liquid receiver. The discharge (or condensing) pressure pushes upward on the diaphragm. The upward pressure tends to close the valve and prevent bypassing. Proper discharge pressure keeps the valve closed so the condenser is fully operational.

When discharge pressure falls below diaphragm pressure, the valve opens and allows discharge (bypass) gas to be metered into the liquid receiver. This creates a higher pressure at the

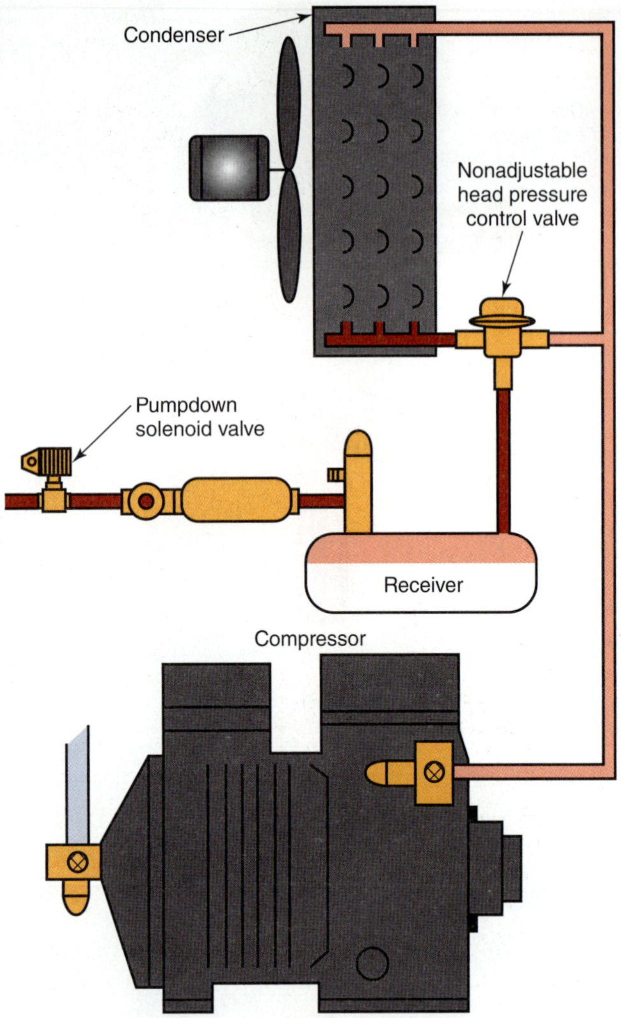

Goodheart-Willcox Publisher

Figure 19-28. When head pressure drops sufficiently, the valve allows hot gas to bypass the condenser and go directly to the liquid receiver.

condenser outlet. When the valve opens to permit hot gas to bypass, the condenser outlet line is restricted, causing liquid to back up inside the condenser. When the valve is fully open, the condenser outlet is fully closed. The advantage of this valve is that it can modulate and, thus, control the amount of liquid in the condenser, according to ambient temperature. Head pressure is maintained by controlling condenser capacity.

All head pressure control valves require the use of a liquid receiver large enough to hold the total charge for both summer and winter operation. Excess refrigerant is needed during cold weather to permit proper flooding of the condenser. These valves are only used on systems having a thermostatic expansion valve.

Manufacturers of outdoor condensing units usually include a head pressure regulating valve and an oversized liquid receiver. All other components are sized for normal operation. The liquid receiver must be oversized because good refrigeration practice requires the total system charge not to exceed 75% of receiver capacity. The excess charge needed for winter operation must be included in the receiver capacity.

Head pressure control valves should not be used on systems without a liquid receiver, or on systems where the receiver is too small. A lack of receiver storage space will cause liquid to back up in the condenser. When ambient temperatures begin to rise, discharge pressure will become excessively high due to liquid refrigerant reducing condenser capacity.

Adjustable Valves

The ***adjustable head pressure control valve*** works as a condenser pressure regulator, throttling liquid leaving the condenser. This causes liquid refrigerant to accumulate in the condenser and raise its pressure. This helps to maintain a set minimum pressure in the condenser.

An adjustable head pressure control valve is used in conjunction with a receiver pressure regulator to provide an improved system of head pressure control. A ***receiver pressure regulator*** is a differential valve that bypasses hot gas refrigerant from the compressor discharge line into the liquid receiver to maintain a set minimum pressure in the liquid receiver. See **Figure 19-29**. These two valves provide another method of maintaining constant receiver pressure during low ambient conditions. Adjustable head pressure valves can be used for common refrigerants such as R-22, R-134a, or R-449A, because they are adjustable over a range of 65 psig to 225 psig (4485 kPa to 15 525 kPa).

As shown in **Figure 19-30**, the adjustable head pressure control valve is located in the condensate line between the condenser and the liquid receiver. The receiver pressure regulator is located between the hot-gas discharge line and the liquid receiver, bypassing the condenser.

During periods of low ambient temperature, head pressure falls until it approaches the setting on the adjustable head pressure control valve. The valve then throttles down, restricting flow of liquid from the condenser. Liquid backs up in the condenser, reducing condenser capacity and raising condensing pressure.

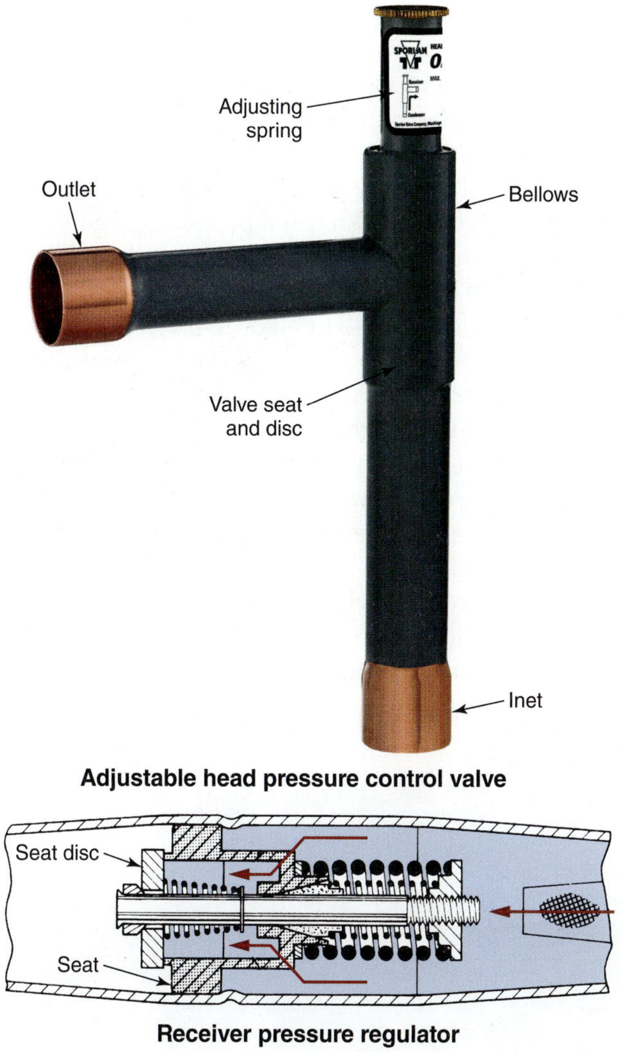

Sporlan Valve Co.

Figure 19-29. Two head pressure control valves.

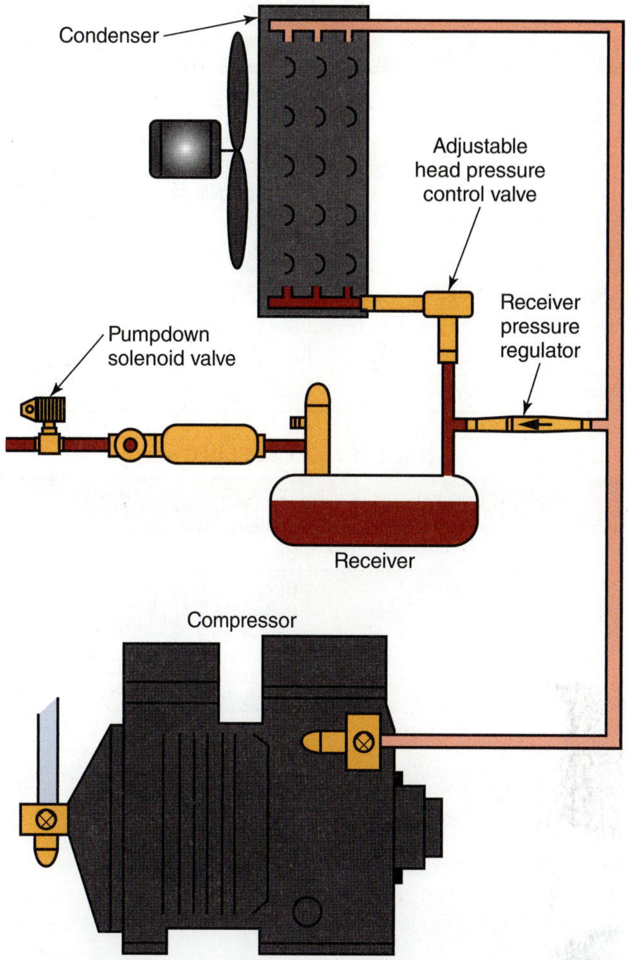

Goodheart-Willcox Publisher

Figure 19-30. The adjustable head pressure valve and the receiver pressure regulator work together to control receiver pressure.

It is actually liquid receiver pressure that must be maintained, *not* condenser pressure. Therefore, hot gas from the compressor bypasses the condenser through the receiver pressure regulator into the liquid receiver. The receiver pressure regulator is a differential valve that is preset for 20 psig and opens when the pressure difference between the liquid receiver and hot-gas discharge line exceeds that value. Hot gas flowing through the bypass line heats up the cold liquid being passed by the adjustable valve. As a result, the liquid is warm when it reaches the liquid receiver and has sufficient pressure to properly feed the expansion valve.

With a proper refrigerant charge, these two control valves modulate refrigerant flow automatically to maintain proper receiver pressure regardless of outside ambient temperatures.

Troubleshooting Head Pressure Controls

There are several possible causes for malfunctions in systems with head-pressure controls. These causes can be difficult to isolate from each other. Troubleshooting is not a "guessing game." Detailed information must be obtained to narrow the problem to a specific area or component. Operating temperatures and pressures must be known before system problems can be pinpointed.

Figure 19-31 lists the causes of the most common head pressure control malfunctions and recommended solutions to those problems.

Troubleshooting head pressure control valves to determine how the valve is feeding can sometimes be accomplished using your hands as tools.

Low Head Pressure

Possible Cause	Remedy
Low charge, cannot flood condenser	Add refrigerant
Low-pressure setting on valve	Increase setting
Valve fails to close due to foreign material stuck in valve	Open valve wide to pass material. If unsuccessful, replace valve
Valve will not adjust	Replace valve
Valve fails to close due to loss of element charge	Replace valve
Valve fails to open	Replace valve

High Head Pressure

Possible Cause	Remedy
Dirty condenser	Clean condenser
System overcharged	Purge until proper head pressure is obtained
Undersized receiver	Check receiver capacity
Air in system	Purge air from system
Valve fails to adjust or open	Replace valve
Bypassing hot gas when not required	Replace valve

Goodheart-Willcox Publisher

Figure 19-31. Possible causes and remedies for head pressure problems.

During very cold ambient temperatures, the head pressure control valve should completely bypass the condenser. In this case, the bypass line to the liquid receiver should feel warm. When ambient temperatures are high, the head pressure control valve should be fully closed, and the bypass line should feel cold. Also, with a proper charge in the system, the head pressure should not drop below 100 psig (690 kPa) for R-134a, 180 psig (1380 kPa) for R-22, and 210 psig (1448 kPa) for R-407A. In addition to checking the temperature of the copper tubing, always check system pressures and observe the liquid level in the sight glass.

A system low on refrigerant normally exhibits low head pressure. However, the head pressure control valve will bypass the condenser in its attempt to maintain proper head pressure. This condition is revealed by a very warm receiver and an empty or low liquid level in the sight glass. The valve is working properly, but the system suffers from an undercharge of refrigerant.

Chapter Review

Summary

19.1 Hand shut-off valves are installed on commercial and industrial systems to quickly isolate sections of a system to perform a particular task. Three main types of hand shut-off valves are the diaphragm-, globe-, and ball-types. Check valves are used in some refrigeration systems to prevent the flow of refrigerant in the wrong direction.

19.2 Evaporator pressure regulators (EPRs) are special valves used to prevent evaporator pressure from falling below a set limit. EPRs can be used in single or multiple (multiplexed) evaporator systems. Some EPRs are hermetic and cannot be repaired if they fail, and so must be replaced.

19.3 A solenoid valve is an electrically operated valve that is either fully open or fully closed, depending upon whether its electrical supply is on or off. When electricity is flowing, it moves through an electromagnet (copper coil) that surrounds a plunger connected to a short valve stem. When the coil is magnetized by the flow of electricity, an opposing magnetic field is induced in the plunger. The plunger is pulled upward into the center of the magnetic field, opening the valve. When the coil is de-energized, the plunger drops into the closed position.

19.3 For proper plunger operation, a solenoid valve must be mounted in an upright position or at no more than a 45° angle. A common application of a solenoid valve is the automatic pumpdown cycle, which stops the compressor just before suction pressure reaches a state of vacuum.

19.3 Troubleshooting a solenoid valve is primarily limited to determining whether the coil is energized. Failure of the valve to open or close properly is usually caused by foreign material lodged in the valve seat.

19.4 Crankcase pressure regulators (CPRs) are used to protect the compressor from excessive suction pressure, which most often occurs at start-up. The CPR is adjusted by turning a screw located under the cover cap.

19.5 Hot-gas bypass valves are used to control the amount of hot gas bypassed directly into the low-pressure side of a system. Hot gas only bypasses when suction pressure is reduced to the valve setting. The hot-gas bypass valve is installed in a branch line off the hot-gas discharge line as close to the compressor as possible.

19.5 There are two possible problems with a hot-gas bypass valve: failure to open and failure to close. Failure to open is revealed by low suction pressure. Failure to close is revealed by high suction pressure.

19.6 A desuperheating TEV is installed in the suction line to supply enough liquid refrigerant to cool the discharge hot gas to the recommended suction temperature.

19.7 Two common types of head pressure control valves are used to solve low-head-pressure problems caused by low ambient temperatures. One method uses a single nonadjustable valve; the other uses two fully adjustable valves. Depending on the malfunction, these valves exhibit conditions that can help a technician pinpoint the cause of the malfunction.

Know and Understand

1. A _____-type hand valve is connected directly to the valve stem by the handle.
   A. ball
   B. globe
   C. diaphragm
   D. check

2. Check valves are used to prevent the flow of _____ in the wrong direction. (19.1)
   A. air
   B. refrigerant
   C. moisture
   D. electricity
3. Evaporator pressure regulators (EPRs) are used to keep _____ pressure from falling below a setpoint. (19.2)
   A. evaporator
   B. condenser
   C. compressor
   D. metering device
4. *True or False?* The EPR can be installed anywhere in the suction line and still perform properly. (19.2)
5. A _____ valve is used on two-temperature applications in which two evaporators operating at different temperatures are connected to a single suction line. (19.2)
   A. globe-type hand
   B. ball-type hand
   C. check
   D. solenoid
6. Problems associated with multiplexed evaporators operated by a single compressor are easily overcome by using _____. (19.2)
   A. multiple TEVs
   B. EEVs
   C. hermetic compressors
   D. multiplexed compressors
7. Alloy or other foreign material lodged in an EPR port can be dislodged by _____. (19.2)
   A. turning the adjustment nut slightly with the system off
   B. turning the adjustment nut all the way out with the system running
   C. lightly heating the EPR with a torch
   D. increasing system pressure at the EPR until pressures normalize
8. A solenoid valve is commonly mounted in a _____. (19.3)
   A. downward position, at no more than a 45° angle
   B. upright position, at no more than a 45° angle
   C. downward position, at no more than a 90° angle
   D. upright position, at no more than a 90° angle
9. *True or False?* Most solenoid valves are normally open (NO). (19.3)
10. Solenoid valves used in automatic pumpdown cycles are typically installed in the _____. (19.3)
    A. suction line, at the evaporator outlet
    B. suction line, at the compressor inlet
    C. liquid line, between the sight glass and TEV
    D. evaporator
11. It can be quickly determined if a solenoid valve is energized or not by using a(n) _____. (19.3)
    A. brand new solenoid valve
    B. ohmmeter
    C. voltmeter
    D. small pocket screwdriver
12. Crankcase pressure regulators (CPRs) limit pressure entering the _____. (19.4)
    A. evaporator
    B. condenser
    C. compressor
    D. metering device
13. A common application of CPRs is to limit high suction pressure _____. (19.4)
    A. before a defrost cycle
    B. after a defrost cycle
    C. upon an automatic pumpdown start
    D. during normal operation
14. To accurately adjust a CPR, a manifold gauge should be used to obtain pressure readings _____. (19.4)
    A. at the metering device
    B. at the evaporator outlet
    C. at the compressor inlet
    D. at the compressor outlet
15. Which special purpose valve is commonly used for capacity control in reduced load conditions? (19.5)
    A. CPR valve
    B. Hot-gas bypass valve
    C. Solenoid valve
    D. Desuperheating TEV
16. A hot-gas bypass valve maintains a minimum evaporator pressure, which places a _____. (19.5)
    A. large load on the evaporator
    B. false load on the evaporator
    C. false load on the condenser
    D. large load on the compressor

17. The hot-gas bypass valve is installed in _____. (19.5)
    A. the hot-gas discharge line
    B. a branch off the hot-gas discharge line
    C. the compressor inlet
    D. the evaporator outlet
18. *True or False?* When suction pressure is above the hot-gas bypass valve setting, the valve remains closed. (19.5
19. *True or False?* A hot-gas bypass valve failing to close is caused by low suction pressure. (19.5)
20. The recommended method for mixing hot gas, liquid injection, and suction gas using a desuperheating TEV is to _____. (19.6)
    A. direct hot gas into the suction line in the same direction of suction gas flow, then introduce liquid injection against the direction of suction gas flow
    B. direct liquid injection into the suction line and restrict the hot gas
    C. direct hot gas and liquid injection into the suction line at an angle of flow against the direction of suction gas flow
    D. use separate connections to introduce hot gas and liquid injection into the suction line
21. Which of the following is a sign that head pressure is too low? (19.7)
    A. Improper feeding of the metering device to the evaporator
    B. A short cycling compressor
    C. A broken piston rod
    D. An ice-covered evaporator
22. All head pressure control valves require the use of a(n) _____ large enough to hold the total refrigerant charge for both summer and winter operation. (19.7)
    A. accumulator
    B. desuperheating TEV
    C. liquid receiver
    D. evaporator
23. A(n) _____ is a differential valve that bypasses hot gas refrigerant from the compressor discharge line into the liquid receiver to maintain a set minimum pressure in the liquid receiver. (19.7)
    A. receiver pressure regulator
    B. hot-gas bypass valve
    C. desuperheating TEV
    D. adjustable head pressure control valve
24. *True or False?* A system that exhibits low head pressure, a warm liquid receiver, and a low liquid level in the sight glass during operation likely has a malfunctioning head pressure control valve. (19.7)

Apply and Analyze

1. List two advantages of hand valves. (19.1)
2. How does an evaporator pressure regulator perform its primary function? (19.2)
3. Explain how EPRs are used to control the temperatures of multiple evaporators in multiplexed evaporator systems. (19.2)
4. What does it mean if a solenoid valve is normally closed (NC) or normally open (NO)? (19.3)
5. What sounds can you look out for when troubleshooting a solenoid valve? (19.3)
6. Explain how and when to adjust the pressure leaving a crankcase pressure regulator valve. (19.4)
7. Explain the operation of a hot-gas bypass valve. (19.5)
8. How does a desuperheating TEV protect the compressor from overheating? (19.6)
9. Explain how a nonadjustable head pressure control valve is used to solve low head pressure problems caused by low ambient temperatures. (19.7)

Critical Thinking

1. Do you think it would be better to have a multiplexed system with multiple evaporators connected to a single compressor or multiple compressors? Defend your answer. (19.2)
2. Why should adding refrigerant to partially fill the condenser with liquid only be considered a temporary solution to address low head pressure? (19.7)

Zero Zone, Inc.

20 Compressors

LEARNING OUTCOMES

After studying this chapter, you will be able to:

20.1 Identify the different types of compressors.

20.2 Compare open-type (external drive), hermetic, and semihermetic compressor designs, applications, and components.

TECHNICAL TERMS

access opening
clearance pocket
compliant
connecting rod
crankshaft
direct drive
discharge line stub
disc-type compressor valve
flywheel
full-floating
hermetic compressor
open-type compressor
piston
piston rings
reciprocating compressor
rotary compressor
scroll compressor
semihermetic compressor
stubs
suction line stub
valve plate
valve reed

Introduction

The compressor is a mechanical vapor pump that circulates refrigerant. It is the heart of any refrigeration system and is the most expensive mechanical component of the system. A compressor malfunction will greatly affect the operation of other components. Compressor problems are diagnosed without breaking into the compressor to examine individual parts. Problems involving the compressor are considered major because of the time and expense of replacement or repair. The technician must fully understand compressor types, designs, and performance to eliminate confusion and wrong diagnoses.

The purpose of the compressor is to remove heat-laden, low-temperature vapor from the evaporator and compress (squeeze) it into a much smaller volume. As the volume of the vapor is reduced, heat is concentrated in a smaller area, resulting in a much higher temperature.

20.1 Types of Compressors

The compressor is a mechanical device driven by an electric motor. It is categorized as one of three basic types based on how it compresses vapor:

- Reciprocating
- Rotary
- Scroll

PRO TIP

Screw and Centrifugal Compressors

There are other types of compressors (such as screw and centrifugal types) that are usually seen in specialized systems and industrial applications. Food packaging factories commonly use screw compressors in their refrigeration systems. Water chillers commonly use centrifugal compressors. These compressors are not discussed in this textbook.

Reciprocating Compressor

The reciprocating compressor is the workhorse of the industry. It is used for domestic and commercial applications ranging in size from fractional horsepower to the 100-ton to 150-ton range. *Reciprocating* means to move in one direction and then in the opposite direction (back and forth or up and down). The ***reciprocating compressor*** operates much like an automobile engine. It has one or more pistons driven from a crankshaft that is turned by an electric motor. These pistons move back and forth, making alternating suction and compression strokes in a cylinder equipped with suction and discharge valve reeds, **Figure 20-1**.

Rotary Compressor

Rotary compressors operate in a rotating or circular motion. The rotary compressor is a positive displacement unit and more efficient than a reciprocating compressor. See **Figure 20-2**. There are two types of rotary compressors. A *rotating vane* rotary compressor has vanes that rotate with the rotor. The *stationary blade* rotary compressor has stationary blades (*vanes*) that remain in place but slide in and out as the rotor turns around the inside of the drum.

In the domestic field, rotary compressors occasionally are used in fractional horsepower sizes. Recently, the rotary has gained some use in 1.5-ton to 5-ton residential air-conditioning systems. Large rotary compressors are sometimes used in low-temperature industrial systems.

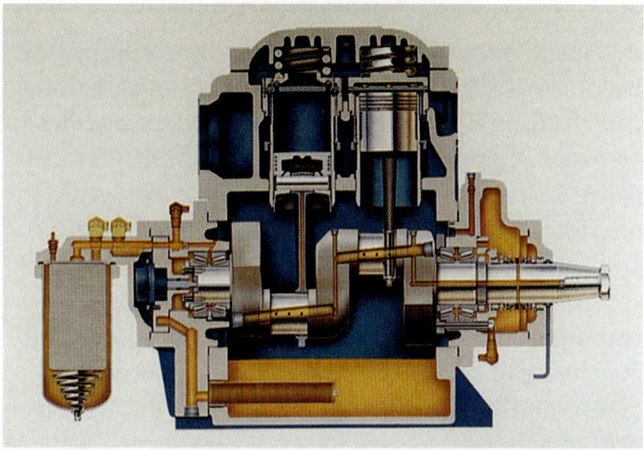

Vilter Mfg. Co.

Figure 20-1. This reciprocating compressor, shown in a cutaway view, is typical of those widely used in refrigeration applications.

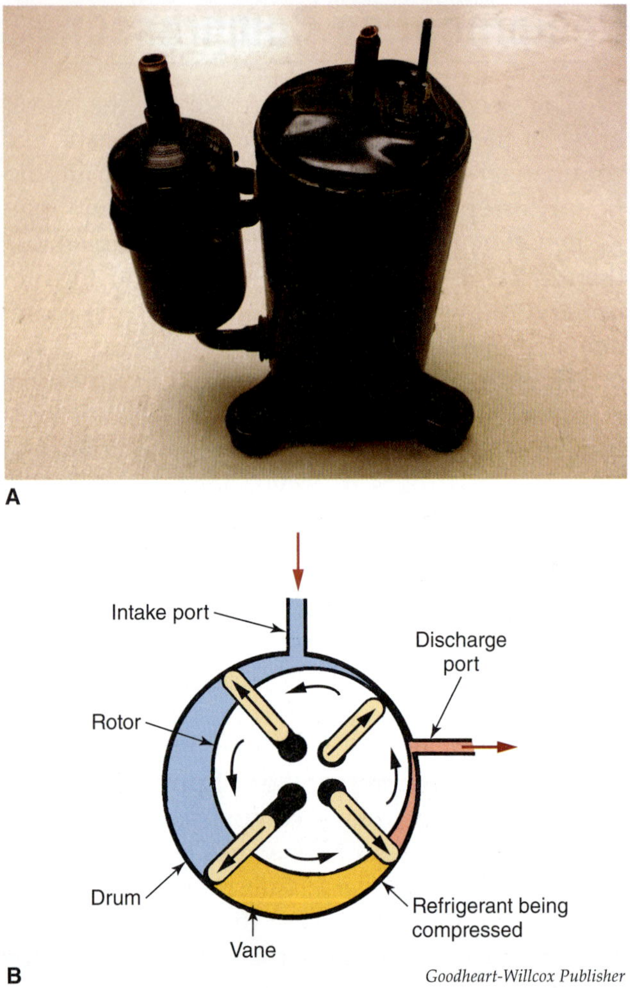

Goodheart-Willcox Publisher

Figure 20-2. Rotary compressors. A—Type of compressor used in many room air conditioners. B—A rotating vane rotary compressor has a rotor mounted on the shaft that is off center in the drum. As the rotor spins, refrigerant in the space between the vanes is moved and compressed into a progressively smaller volume before being discharged.

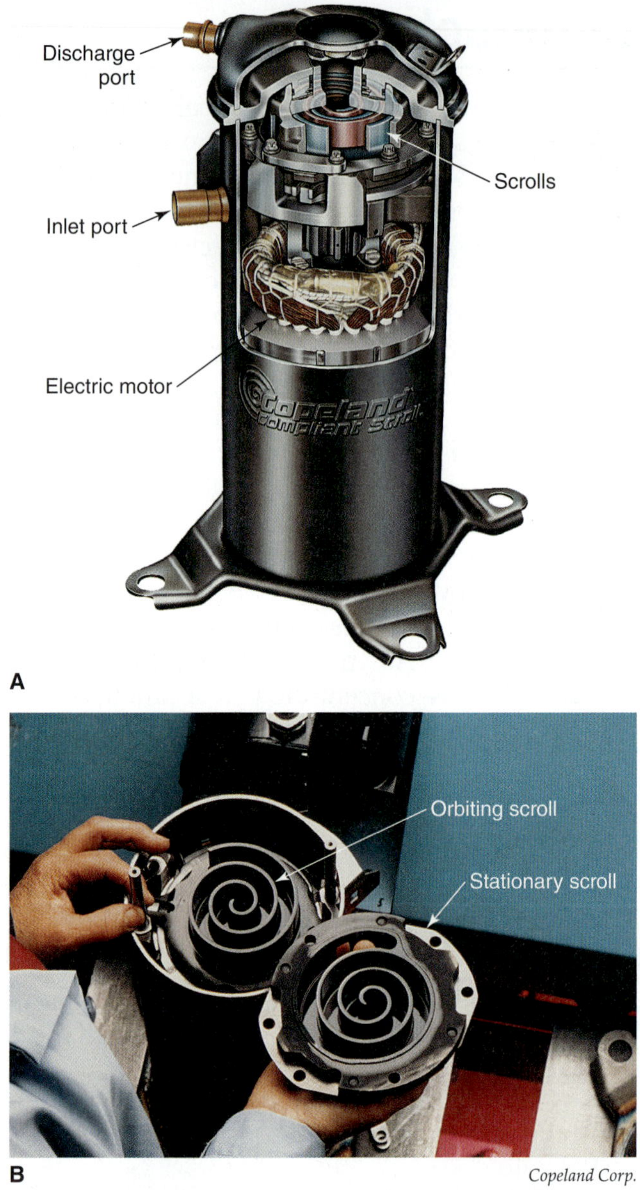

Copeland Corp.

Figure 20-3. Scroll compressors. A—Cutaway view of a typical scroll compressor. B—Compressor partly disassembled to show the two scrolls.

Scroll Compressor

The operation of the ***scroll compressor*** involves the orbiting of one scroll (spiral) that is intertwined with another scroll to form a series of crescent-shaped pockets. During compression, one scroll (the *fixed* scroll) remains stationary while the *orbiting* scroll turns (but does not rotate) within the space between it and the fixed scroll, **Figure 20-3**.

As shown in **Figure 20-4**, the orbiting motion traps gas in the pockets between the two scrolls. As the orbit continues, the pockets are reduced in size, compressing the gas and forcing it toward the center, where it is discharged. Several pockets of gas are compressed simultaneously during each orbit. Intake and discharge are continuous, providing smooth operation.

An advantage of the scroll compressor is that only two components, a fixed scroll and an orbiting scroll, are required to compress gas. These two components replace about 15 components in a typical piston-type reciprocating compressor. Scroll compressors require no valves, while piston-type compressors require a discharge and suction valve for each piston. The scroll compressor is 10% more efficient than a reciprocating compressor of the same size due to elimination of the valves. Scroll compressors also eliminate the need for accumulators and crankcase heaters and can be started under any load without special start kits.

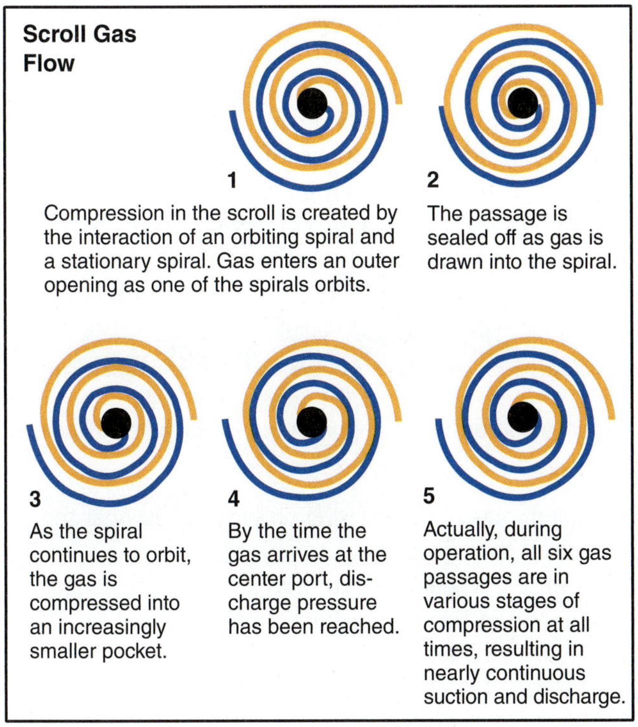

Copeland Corp.

Figure 20-4. Operation of the scroll compressor.

Unlike reciprocating compressors, which can be badly damaged by liquid slugging, scroll compressors have superior tolerance to liquid refrigerant. The scrolls are designed to be ***compliant*** (yielding). In other words, they will "give" or separate from each other in the presence of liquid. The pocket of liquid is simply pushed to the discharge opening, and the scrolls return to their original position. Compliant scroll compressors also can handle debris (foreign matter) without being damaged. Compliant scroll technology offers many advantages to meet the needs of air-conditioning and heat pump systems.

20.2 Compressor Designs

Compressor design deals with how a compressor is put together, not how it mechanically compresses vapor. There are three basic compressor designs:

- Open-type (externally driven)
- Hermetic (totally sealed)
- Semihermetic (accessible hermetic)

Open-Type Compressor

An ***open-type compressor*** refers to a compressor that is driven by an external source of power, usually an electric motor. The motor may be directly connected to the compressor, or it may use a V-belt to transmit the turning power. See **Figure 20-5**.

Open-type compressors are bolted together with the crankshaft extending through the crankcase. A source of power is connected to the external portion of the crankshaft. The most common method is to attach a ***flywheel*** (pulley) to the external crankshaft. An electric motor is connected to the compressor crankshaft pulley by one or more V-belts. Another, less-common method eliminates pulleys and belts by connecting an electric motor's rotor directly to the external crankshaft using a coupling. This open-type of compressor design is called ***direct drive***.

Although open-type compressors represent a small part of the total number of compressors sold today, a great number of them are still operating in commercial refrigeration applications.

Goodheart-Willcox Publisher

Figure 20-5. Many open-type compressors are driven by a V-belt connecting a pulley on the crankshaft with a pulley mounted on the motor shaft. A safety guard of metal mesh over the belt and pulleys is normally used.

Disadvantages of Open-Type Compressors

The primary disadvantage of open-type compressors is the need for a leakproof seal where the compressor crankshaft comes through the crankcase. Crankshaft seals are a common source of leaks. It is not unusual to observe traces of leaking oil directly under the crankshaft seal. A small leak is usually ignored, but a large leak requires field replacement of the crankshaft seal, **Figure 20-6**. The bearings of the electric motor must be oiled (or greased) every six months.

PROCEDURE

Checking Alignment of an Open-Type Compressor's Flywheel and Motor Pulley

To avoid unusual wear and tear of V-belts and to obtain good belt grip, the motor pulley and the compressor flywheel must be properly aligned. Alignment of the flywheel and pulley is checked by using a straightedge or a length of string. As shown in **Figure 20-7**, both edges of the flywheel and both edges of the pulley must touch the straightedge or string when held in a straight line. Proper alignment is obtained by moving the electric motor to conform with alignment of the flywheel. Check alignment using a string in this way:

1. Catch one end of a string under the belt near the top of the flywheel.
2. Turn the wheel by hand until the string crosses the rim at the 3 o'clock position, as shown in **Figure 20-7B**.
3. Stretch the string across the flywheel and the motor pulley in a straight line.
4. If the flywheel and the pulley are properly aligned, both edges of the flywheel and both edges of the pulley should barely touch the string.

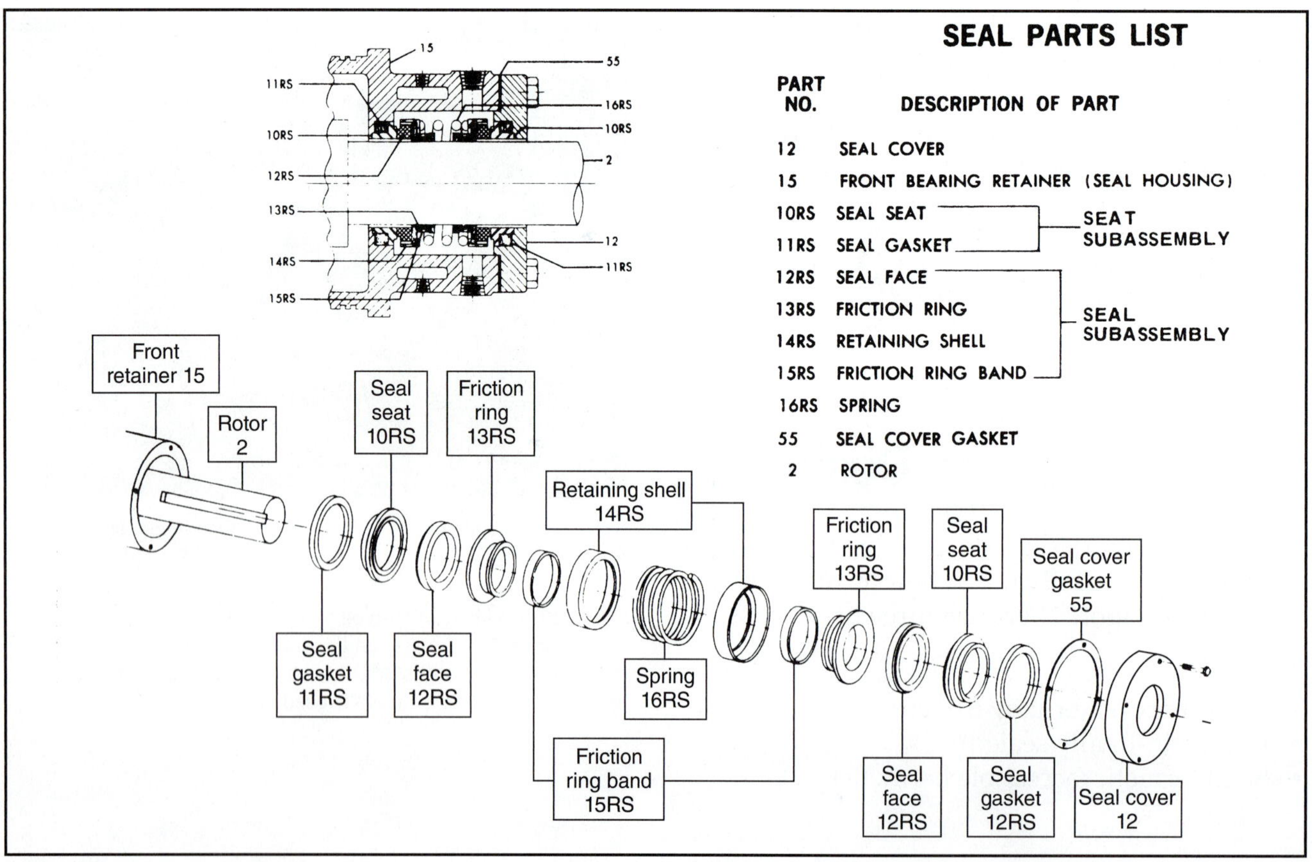

Fuller Co.

Figure 20-6. A double shaft seal like this one prevents leakage, when properly installed.

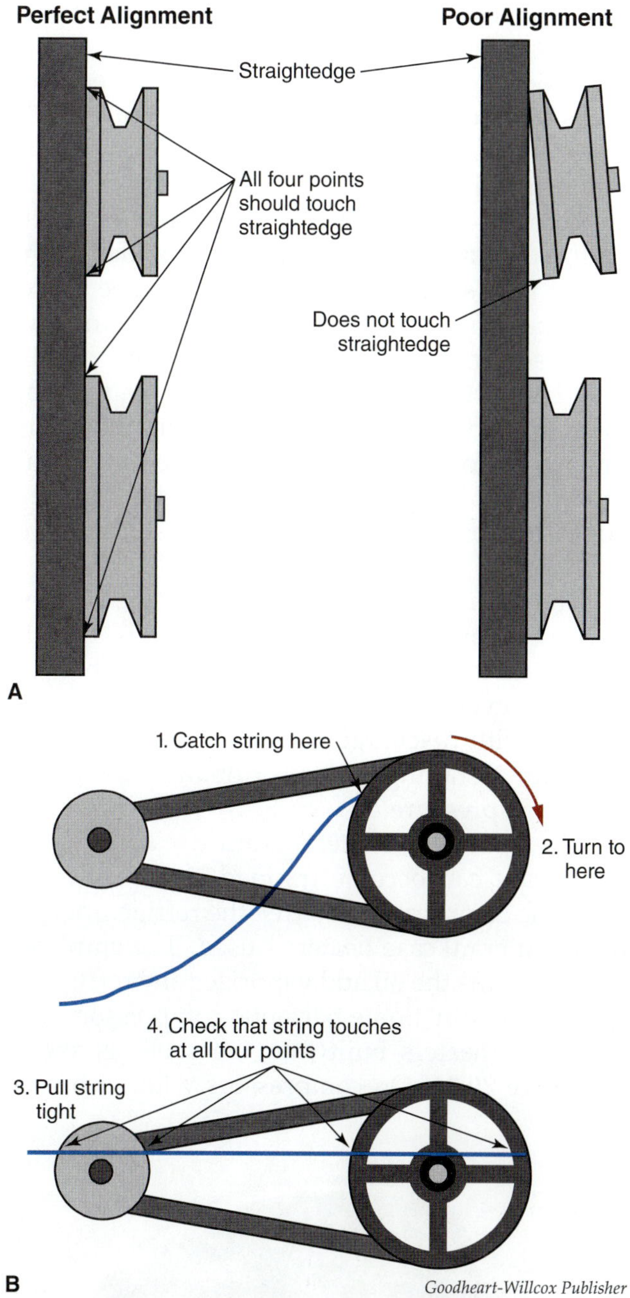

Goodheart-Willcox Publisher

Figure 20-7. A straightedge or a string can be used to check proper alignment of the flywheel and motor pulley. A—The straightedge method. B—The string method.

In the same way, a straightedge can be held against the flywheel and pulley to check alignment, if they are close enough together.

Advantages of Open-Type Compressors

Open-type compressors can be used with electric motors that have odd voltages or frequencies, an important factor in the world market. Open compressors are always field-serviceable. In case of a motor burnout, the motor is easily replaced.

The primary advantage of open-type compressors is flexibility. The speed can be changed, allowing a single compressor to operate with units of two or three horsepower sizes. By changing the size of the motor pulley or flywheel, the speed can be adjusted for use with different sizes of motors. By changing the speed, the same compressor can be used for high-, medium-, or low-temperature applications.

Hermetic Compressor

The word *hermetic* refers to being airtight, or totally sealed against all external forces. A ***hermetic compressor*** contains an electric motor that is connected directly to the compressor. The entire assembly is enclosed within a welded steel housing, **Figure 20-8**.

The hermetic compressor has several advantages. It is small, compact, almost vibration- and noise-free, has a refrigerant-cooled motor, and has positive lubrication.

Hermetic compressors include many types and styles of mechanical compression and differ

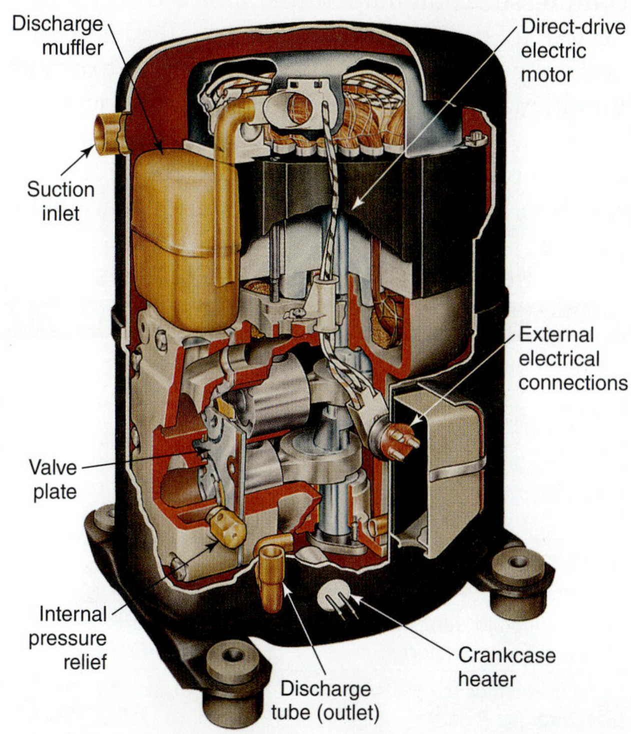

Tecumseh

Figure 20-8. A typical hermetic compressor. Since it is sealed, a hermetic compressor is replaced, rather than repaired, in the field.

widely from each other in appearance, application, and service.

Welded hermetics dominate the residential and light commercial market because of their performance characteristics and high reliability. In size, they range from fractional horsepower units for domestic refrigerators and freezers to 20-hp units for commercial air-conditioning systems.

Hermetics are engineered to perform a specific air-conditioning or refrigeration task. Using hermetic compressors within their design limits will yield favorable results. Asking them to perform outside their design limits will result in poor pumping efficiency and other problems. The key to selecting the right hermetic compressor for a particular job is the evaporator temperature of the system. Each hermetic compressor is designed for a specific evaporator temperature range, **Figure 20-9**. Each compressor is also designed for use with a specific refrigerant. Do not substitute refrigerants.

Hermetic compressors are *not* field-serviceable. In the event of a motor burnout, the entire hermetic compressor must be replaced. The refrigerant must be recovered from the system and a filter-drier installed to clean the system of contaminants.

Within a hermetic compressor, the motor and compressor combination (the motor directly connected to the compressor) is usually used in a vertical position and mounted on springs to absorb vibration. A small space exists between these two parts and the housing shell. This interior space is open to refrigerant gas drawn in through the suction line. The gas entering the interior space helps cool the motor and compressor. It also ensures oil returns to the crankcase.

The factory welds short copper-coated steel ***stubs*** to openings in the housing, **Figure 20-10**. One of these stubs—referred to as the ***suction line stub***—is for brazing of the suction line to the compressor. Another, smaller opening in the shell, called an ***access opening***, is provided with a 1/4″ or 3/8″ stub. A Schrader valve can be brazed onto the stub for evacuation and charging procedures during installation. The valve permits easy connection of the low-pressure gauge from the manifold gauge. After evacuation and charging are performed, the valve is closed off with a cap nut that has a rubber seal inside. If a valve is not installed on the access stub, the stub must be brazed shut.

The ***discharge line stub*** is welded where it passes through the housing and is directly connected to the compressor discharge. This procedure prevents the interior suction pressure from mixing with discharge pressure. No provision is made for installing gauges or obtaining access to discharge pressure.

On some low-temperature applications, refrigerant can become trapped in the oil inside hermetic compressors. To free the refrigerant from the oil, a crankcase heater is used. The crankcase heater warms the oil and vaporizes any refrigerant trapped within. Some hermetic compressors have crankcase heaters built into the shell, as shown in **Figure 20-11**. On compressors without heaters

Hermetic Compressor Applications

Application	Evaporator Temperature
Air conditioning	+32°F to +55°F (0°C to +13°C)
Heat pump (approved models)	–15°F to +55°F (–26°C to +13°C)
High evaporator temperature	+20°F to +55°F (–7°C to +13°C)
Medium evaporator temperature	–10°F to +30°F (–23°C to –1°C)
Low evaporator temperature (normal torque motor)	–30°F to +10°F (–34°C to –12°C)
Low evaporator temperature (high torque motor)	–40°F to +10°F (–40°C to –12°C)

Goodheart-Willcox Publisher

Figure 20-9. Evaporator temperature ranges for hermetic compressors used in different applications. A hermetic compressor is designed for use with a specific range of evaporator temperatures.

Maytag Corporation

Figure 20-10. Stubs extend through the welded shell of the hermetic compressor unit to permit necessary connections. The stubs are copper-coated for ease of brazing the connections.

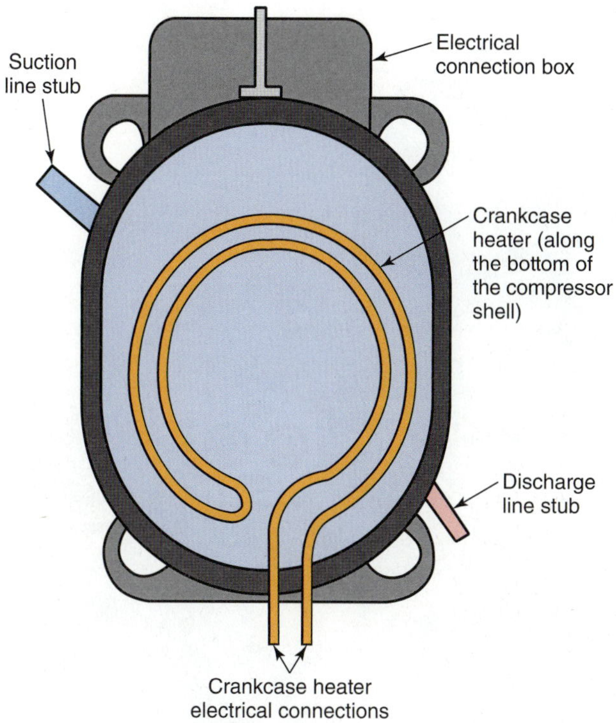

Goodheart-Willcox Publisher

Figure 20-11. This overhead view diagram of a hermetic compressor shows the crankcase heater coil along the bottom of the compressor shell.

built in, aftermarket crankcase heaters may be installed around the lower part of the compressor shell, **Figure 20-12**. These types of crankcase heaters are sometimes referred to as *belly band* crankcase heaters.

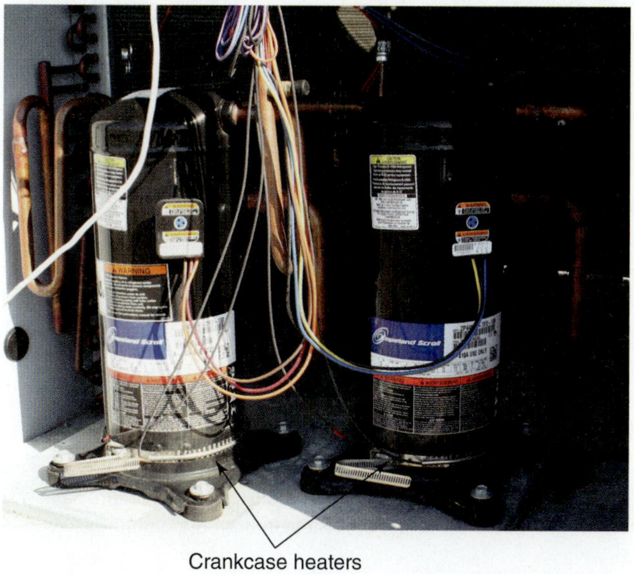

Goodheart-Willcox Publisher

Figure 20-12. These hermetic scroll compressors have aftermarket crankcase heaters wrapped around the lower part of their shells.

Semihermetic Compressors

The ***semihermetic compressor*** is a field serviceable version of a hermetic compressor. It contains an electric motor and compressor that can be accessed by opening the casing. The working parts of both the motor and the compressor are hermetically sealed within a common enclosure. The enclosure is bolted together, and gaskets are used to obtain leakproof seals between the mating parts, **Figure 20-13**.

Semihermetic compressors combine all the desirable features normally associated with hermetic compressors. The compressor is driven by an electric motor mounted directly on the compressor crankshaft. Such a design is compact, economical, efficient, and maintenance-free. The removable heads, end bells, and bottom plates allow access for field repairs in the event of compressor damage. The semihermetic compressor design was pioneered by Copeland Corporation and is widely used throughout the industry for light, medium, and heavy commercial systems.

Bolted semihermetics can be totally disassembled in the field and fitted with new parts as needed. This is a major advantage when dealing with large-tonnage machines where size and weight make total replacement undesirable. However, in many cases, a defective semihermetic is simply replaced and not repaired in the field. There are not enough technicians in the industry to handle the field repair of millions of compressor installations, nor can manufacturers and dealers stock all the parts needed for field repairs.

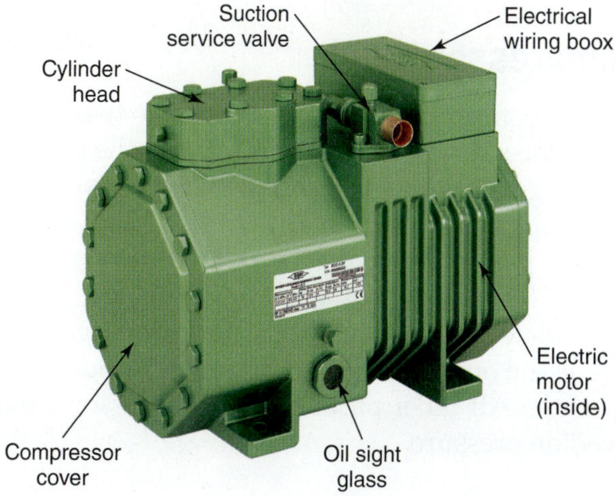

Bitzer

Figure 20-13. A semihermetic compressor. Unlike the welded hermetic compressor, a semihermetic compressor can be serviced in the field.

Fortunately, mass production, standardization, and improved quality have eliminated the need for field repairs in most cases. System reliability and compressor life expectancy have improved to the point where the failure rate for these devices is minimal. Even if the compressor is replaced instead of field repaired, the compressor should be opened up to determine what exactly the cause of the failure was. Occasional repairs are needed to replace a valve plate or an oil pump, but extensive overhaul is seldom required.

Compressor Construction

Although variations between manufacturers do exist, the general principles of construction and operation of semihermetic compressors are the same. The real difference is in quality, durability, and application.

Refrigerant vapor enters the compressor through the suction service valve, which is bolted onto an opening in the compressor body. The suction gas is permitted to scatter throughout the compressor body. The suction service valve is often located at the electric motor end bell to take advantage of the incoming cool suction gas for motor cooling.

Understanding how semihermetic compressor units are constructed is vital to effective troubleshooting and repair work. An accurate diagnosis must be performed *before* removing any bolts. The exploded view in **Figure 20-14** illustrates typical compressor construction details. Individual components may vary among models and manufacturers, but the basic method of assembly is similar.

Compressor Head

The compressor head is divided into two chambers. One side is suction pressure, and the other is discharge pressure. See **Figure 20-15**. Openings inside the compressor body create a pathway for suction gas to reach the suction chamber in the head. The compressor head rests on top of the valve plate. The head chamber for discharge pressure is the only part of the compressor under high pressure. All other parts of the compressor are at low-side pressure.

Goodheart-Willcox Publisher

Figure 20-14. Components of a typical semihermetic compressor, which can be disassembled for field service.

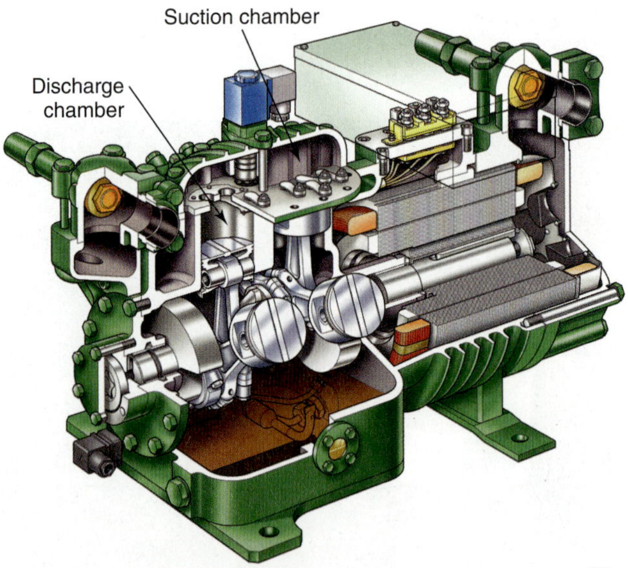

Bitzer

Figure 20-15. A cutaway view of a four-cylinder compressor shows the basic construction of a semihermetic compressor.

A special gasket is placed between the head and the valve plate, and another between the valve plate and the compressor body. Cap screws hold these parts together. The bolts extend through the head, gaskets, and valve plate and screw directly into the compressor body. The cap screws must be tightened slowly and evenly to achieve a leakproof connection.

Compressor Valve Plate

The ***valve plate*** is a specially milled steel plate between the head and the piston cylinders, **Figure 20-16**. The valve plate is a division point between the high-side pressure and the low-side pressure. The suction and discharge valve reeds are considered part of the valve plate.

Suction Valve Reed

The suction ***valve reed*** consists of a thin piece of steel located on top of the cylinder and under the valve plate. This thin reed is anchored at one end by two steel pins and the valve plate itself. The other end of the reed is permitted to move up and down slightly, due to a small depression cut into the cylinder wall. The piston is not permitted to rise above this small depression (or strike the valve plate), so a ***clearance pocket*** is formed at the top of the stroke, **Figure 20-17**.

The downstroke of the piston causes an area of lowered pressure (vacuum) below the reed. Since pressure of the refrigerant gas in the suction chamber is higher, the end of the reed is pushed down, opening the hole in the valve plate.

Refrigerant vapor flows through the hole in the valve plate, around the suction reed, and fills the cylinder cavity. On the upstroke of the piston, pressure inside the cylinder becomes higher than the pressure in the suction chamber of the head. This presses the suction valve reed firmly against the valve plate hole, trapping the refrigerant gas in the cylinder. Pressure inside the cylinder continues to rise as the piston moves upward, compressing the trapped gas into a smaller volume.

Discharge Valve Reed

The discharge valve reed is mounted on top of the valve plate, inside the discharge chamber of the head. It covers one or more holes through the valve plate, opening into the cylinder, **Figure 20-18**. High pressure inside the head chamber keeps the discharge valve reed closed. When the increasing cylinder pressure exceeds pressure in the head, the discharge reed is forced open, and the compressed gas flows into the head chamber. From the head

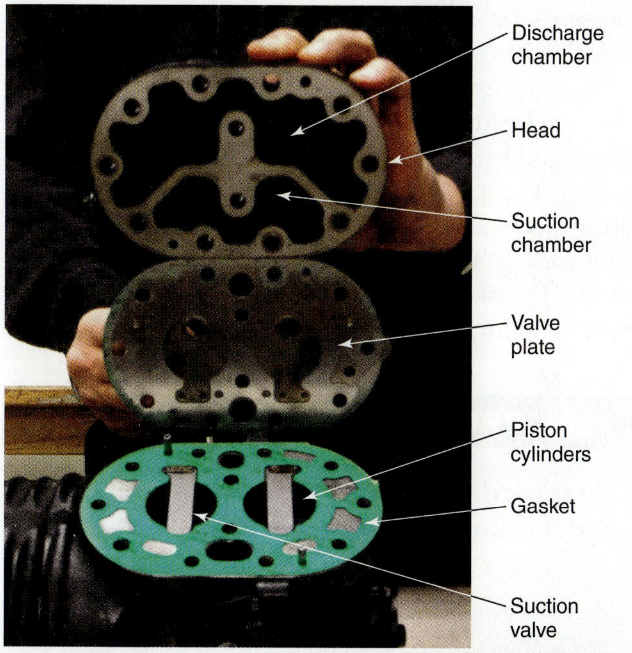

Goodheart-Willcox Publisher

Figure 20-16. The suction valve reed is mounted beneath the valve plate or on the cylinder wall and moves downward to admit refrigerant gas on the piston downstroke. Gaskets above and below the valve plate provide a tight seal.

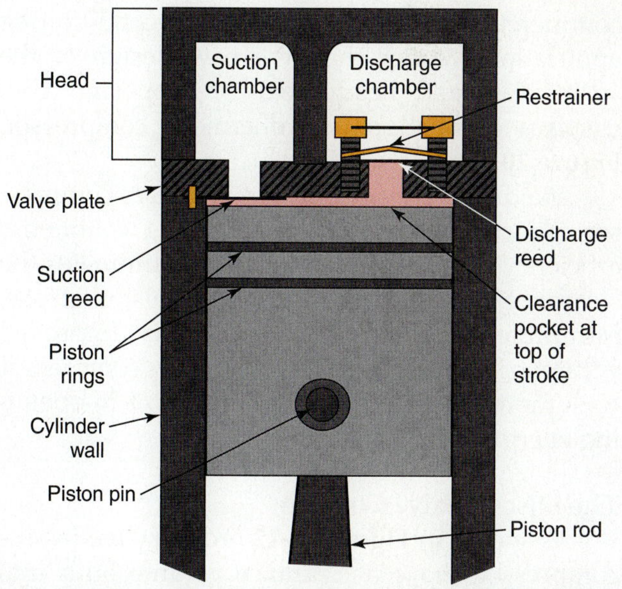

Goodheart-Willcox Publisher

Figure 20-17. The clearance pocket at the top of the cylinder avoids contact between the piston and the suction valve reed. Some compressed refrigerant gas remains in the clearance pocket after the discharge reed closes and is re-expanded as the piston moves downward.

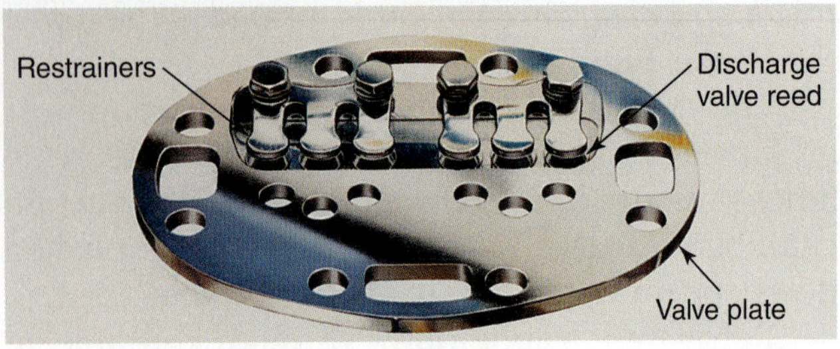

Carrier

Figure 20-18. Valve plate with discharge valve reeds and restrainers. The actual valve reed is a thin metal plate held in place by the bolts and restrainers (the thicker metal, three-fingered pieces shown). The restrainers prevent excessive flexing and damage to the reed from liquid slugging.

Goodheart-Willcox Publisher

Figure 20-19. The Discus® compressor is similar in appearance to compressors with conventional reed valves, but it has a radically different valve plate and discharge valve design.

chamber, the high-pressure and high-temperature gas travels through the discharge service valve and the hot-gas discharge line into the condenser.

The discharge reed usually has restrainers that help prevent the reed from bowing too far. These devices protect the discharge valve reed when the compressor is slugging oil or liquid refrigerant.

Each piston cylinder must be equipped with suction and discharge reeds. On larger cylinders, two or more of each may be used.

Disc-Type Compressor Valve

The ***disc-type compressor valve*** was designed by Copeland Corporation and marketed under the trade name Discus®. It has been used in new commercial refrigeration and air-conditioning applications. Other than the new design of the valve plate and head area, the Copeland Discus is a conventional semihermetic compressor, **Figure 20-19**.

The disc-type valve design increases compressor capacity up to 25% by improving volumetric efficiency. The valve plate virtually eliminates the clearance pocket at the top of the piston stroke, minimizing the re-expansion of gas. Improved volumetric efficiency provides the compressor with more capacity and saves up to 16% in operating energy costs.

The Discus Valve Concept

As shown in **Figure 20-17**, ordinary reed-valve compressors have a clearance volume built into the valve plate. High-pressure refrigerant gas is trapped in the discharge port area and re-expands during the piston downstroke. The volumetric efficiency of the compressor depends on the number and size of the discharge ports as well as the compression ratio.

Figure 20-20 shows the disc-type valve design. The valve plate is hollow and serves as a passage for the low-pressure suction gas. Conventional suction valve reeds control the flow of gas into the cylinder. The large discharge port is closed by the valve disc. A disc spring holds the valve disc against the bottom of the discharge port. This minimizes the clearance volume and improves the compressor's volumetric efficiency. In addition, disc-valve compressors have domed pistons, nearly eliminating the clearance volume above the piston.

Compressors with disc-type valves are extremely rugged and dependable, but they can be damaged by abuse in the same manner that a reed-valve compressor can be damaged. Proper application and installation procedures must be followed whenever a compressor replacement is made. Do not expect a compressor with disc-type valves to correct system problems.

PRO TIP

Displacement Considerations for Replacement Compressors

The disc-valve design results in increased capacity for a given displacement. When replacing a conventional reed-type compressor with a compressor that has disc-type valves, it may be necessary to select a unit with a smaller displacement.

A

Goodheart-Willcox Publisher

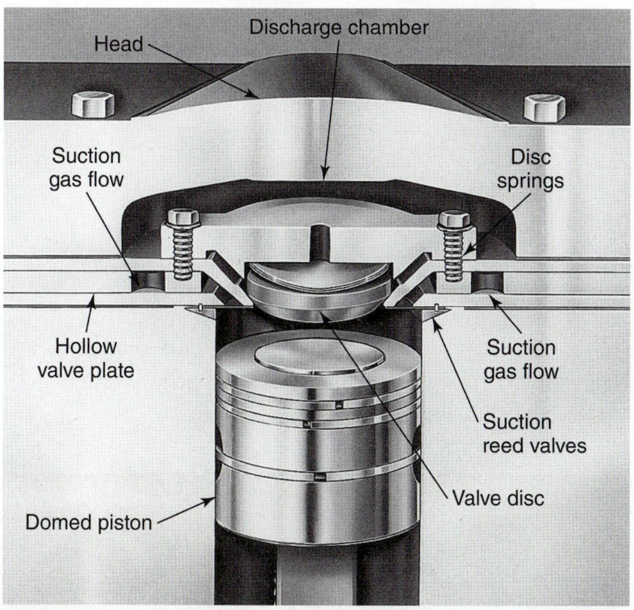

B

Copeland Corp.

Figure 20-20. A—Disc-type valve disassembled. B—This disc-type valve diagram shows how the hollow valve plate provides passage for suction gas and has conventional reed valves for gas flow into the cylinder. The large discharge valve disc is held in place by disc springs. The design improves volumetric efficiency by up to 25%.

Compressor Cylinder and Piston Arrangements

Compressors with one, two, or three pistons have just one valve plate and one head. Larger compressors with four or more pistons often use two valve plates and have two heads in a "V" arrangement, **Figure 20-21**. Six-cylinder compressors normally have three heads, and eight-cylinder compressors have four heads. It does not matter how many cylinders are involved. The compressors all work the same. The only difference is that the larger compressors have more capacity. A four-head, V-type compressor performs the same task as two separate single-head compressors that have the same-sized pistons as the four-head machine.

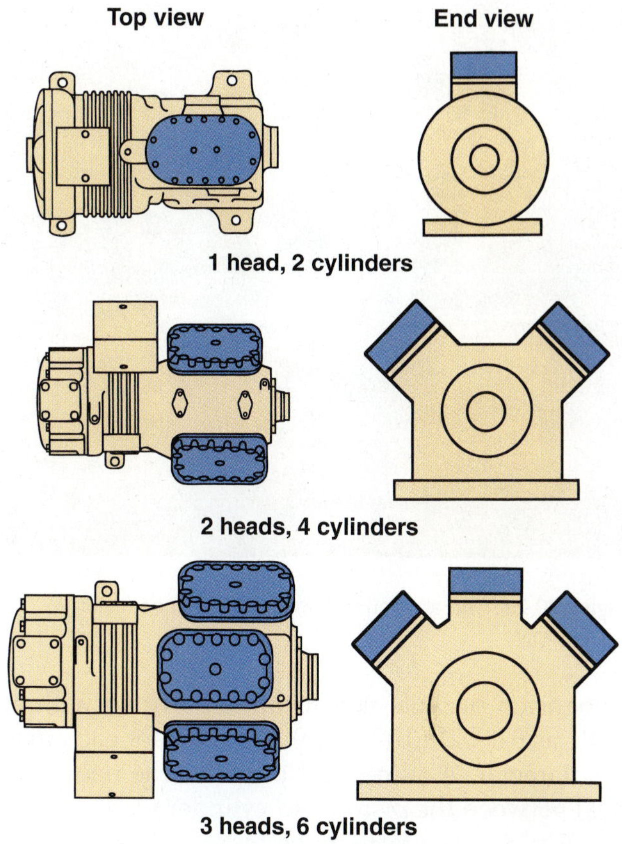

Copeland Corp.

Figure 20-21. Typical cylinder head arrangements for semihermetic compressors are shown in top and end views. Cylinders are usually grouped two per head.

Compressor Pistons

A ***piston*** is a device that moves up and down inside the cylinder to compress refrigerant gas. Most pistons are made from cast iron, but the small, high-speed, hermetic compressor pistons are made of die-cast aluminum. ***Piston rings*** wrap around the outside of pistons. They create a seal between the piston and the cylinder walls. This prevents any refrigerant from leaking past the piston. Piston rings are not used on compressors of less than 10 hp. See **Figure 20-22**.

In a compressor, the temperature of the piston rarely exceeds 250°F (121°C), so expansion of the piston or the cylinder wall is minimal. Minimal

Vilter Mfg. Co.

Figure 20-22. An aluminum piston for use in a hermetic compressor.

expansion permits the piston to be fitted with as little as 0.0002″ (0.005 mm) clearance for each inch in diameter. A film of oil provides the necessary seal between the piston and cylinder wall.

A piston pin fastens the piston to the connecting rod, **Figure 20-23**. The piston pin is accurately machined and ***full-floating*** (free to rotate within the connecting rod and the piston).

The ***connecting rod*** attaches the piston to the crankshaft. Connecting rods vary in design, depending upon the type of crankshaft. Some are clamped to the crankshaft, while others merely slide over the crankshaft eccentric.

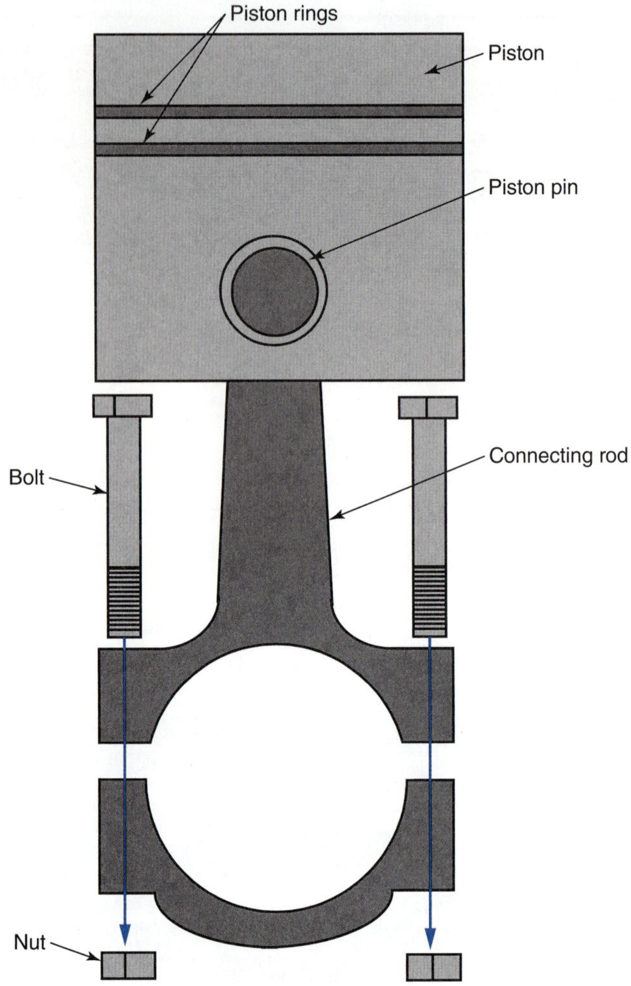

Goodheart-Willcox Publisher

Figure 20-23. Components of a typical compressor piston assembly.

Crankshaft

A ***crankshaft*** is used to change the rotary (circular) motion of the electric motor into a reciprocating (up-and-down) motion. There are various ways to accomplish the motion change using different crankshaft designs, such as the eccentric shaft, crank-throw, or Scotch yoke. Each of these crankshaft designs, however, has the same purpose: to cause the piston to move up and down, **Figure 20-24**.

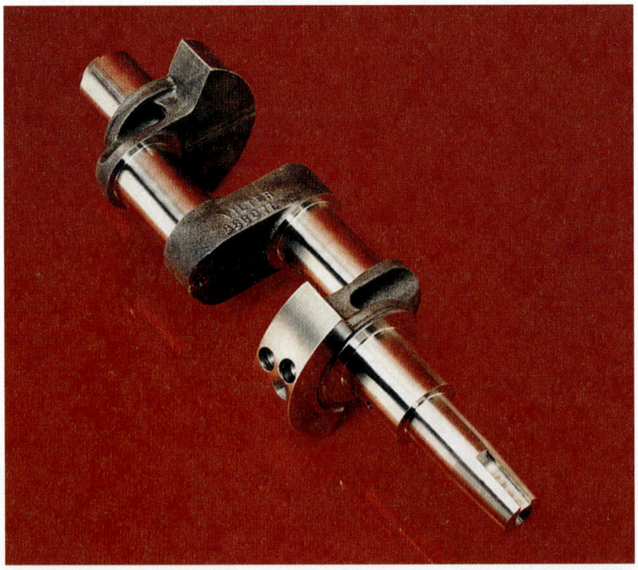

Vilter Mfg. Co.

Figure 20-24. A crankshaft for a two-cylinder compressor.

Chapter Review

Summary

20.1 The reciprocating compressor has one or more pistons driven from a crankshaft that is turned by an electric motor. These pistons move back and forth to provide compression.

20.1 A rotary compressor operates in a rotating or circular motion. There are two types of rotary compressors: rotating vane and stationary blade.

20.1 A scroll compressor provides compression by using a fixed scroll and an orbiting scroll that turns within the space between it and the fixed scroll.

20.2 An open-type compressor refers to a compressor that is driven by an external source of power, usually an electric motor. The primary disadvantage of open-type compressors is the need for a leakproof seal where the compressor crankshaft comes through the crankcase. The primary advantage of open-type compressors is flexibility.

20.2 A hermetic compressor is an enclosed steel assembly composed of an electric motor connected directly to the compressor. Hermetic compressors are used in residential and light commercial applications due to their performance characteristics and high reliability. However, hermetic compressors are *not* field serviceable.

20.2 A semihermetic compressor is a field serviceable version of the hermetic compressor in the sense that the electric motor and compressor can be accessed by opening the casing. Semihermetic compressors are composed of a compressor head with two chambers (suction and discharge), a valve plate with a suction valve reed and a discharge valve reed, cylinders, pistons, and a crankshaft.

Know and Understand

1. The _____ compressor operates in a rotating or circular motion and uses either rotating or stationary vanes to produce compression. (20.1)
   A. reciprocating
   B. rotary
   C. scroll
   D. centrifugal
2. The scroll compressor provides compression by _____. (20.1)
   A. back-and-forth motion of pistons
   B. using rotating augers
   C. trapping gas between pockets of fixed and orbiting scrolls
   D. using rotating vanes
3. *True or False?* The reciprocating compressor is 10% more efficient than a scroll compressor of the same size. (20.1)
4. The most common method of connecting an open-type compressor's external crankshaft to the motor is by attaching a(n) _____. (20.2)
   A. coupling to the crankshaft
   B. flywheel to the crankshaft
   C. Schrader valve
   D. special gasket
5. By changing the size of an open-type compressor's motor pulley, _____ can be adjusted. (20.2)
   A. refrigerant type
   B. refrigerant pressure
   C. speed
   D. alignment
6. The term *hermetic* refers to being _____. (20.2)
   A. field serviceable
   B. airtight
   C. efficient
   D. self-lubricating

7. What should a service technician do when a hermetic compressor's motor burns out? (20.2)
   A. Replace the entire compressor.
   B. Send the motor in to the manufacturer for service.
   C. Replace the motor.
   D. Lubricate the motor components and try to restart the motor.
8. Which of the following provides cooling to a hermetic compressor and motor? (20.2)
   A. Head fan
   B. The returning suction gas
   C. A water jacket
   D. They never need to be cooled.
9. What are the three welded studs on a hermetic compressor? (20.2)
   A. Shutoff valve, access opening, Schrader valve
   B. Suction line, access opening, discharge line
   C. Hot line, cold line, and medium-temperature line
   D. Suction line, crankcase heater access, discharge line
10. *True or False?* Even though semihermetic compressors are field serviceable, they are often replaced when defective due to lack of technicians and parts. (20.2)
11. The _____ is the only part of a semihermetic compressor under high pressure. (20.2)
    A. compressor body
    B. suction chamber
    C. discharge chamber
    D. motor end bell
12. The valve plate in a disc-type compressor valve virtually eliminates the _____. (20.2)
    A. high torque normally required for a semihermetic motor to start
    B. clearance pocket at the top of the piston stroke
    C. need for a suction chamber
    D. need for a discharge chamber
13. A _____ connects the piston to the connecting rod and is full floating. (20.2)
    A. piston ring
    B. crankshaft
    C. piston pin
    D. film of oil
14. *True or False?* On reciprocating compressors, the electric motor turns the crankshaft. (20.2)

Apply and Analyze

1. Explain the operation of each of the basic types of compressors: reciprocating, rotary, and scroll. (20.1)
2. List three advantages of scroll compressors. (20.1)
3. What might happen if a compressor is located outside in cold weather, and it does not have a crankcase heater? (20.2)
4. Explain how the disc-type compressor valve increases compressor capacity by up to 25%. (20.2)
5. What is the purpose of a compressor motor's crankshaft? (20.2)

Critical Thinking

1. Describe what might happen to a compressor's suction and discharge valves if the compressor experienced floodback? (20.2)
2. Imagine you are a service technician working on a small residential air-conditioning system and find that you will need to replace a defective compressor. How do you go about deciding which type of compressor to choose for the system? (20.2)

Sporlan Valve Co.

21 Compressor Lubrication and Accessories

LEARNING OUTCOMES

After studying this chapter, you will be able to:

21.1 Recognize relevant information of refrigerant lubricants such as proper handling, oil viscosity, miscibility, additives, and various types.

21.2 Identify lubrication methods and oil pressure safety controls and how these systems are used to achieve the proper oil level in a compressor.

21.3 Add oil to a compressor using three different methods.

21.4 Remove excess oil from a compressor using two different methods.

21.5 Perform an acid test on refrigeration oil.

21.6 Discuss the factors that can affect compressor capacity.

21.7 Recognize the factors that affect volumetric efficiency of a compressor and the methods to improve efficiency.

21.8 Describe how cooling is provided for compressors.

21.9 Explain how vibration absorbers and crankcase heaters can help prevent system malfunctions.

21.10 Recognize the information that can be found on compressor data plates.

TECHNICAL TERMS

- alkylbenzene (AB) lubricant
- clearance pocket
- compression ratio
- crankcase heater
- data plate
- density
- discharge port
- forced-feed system
- interstage pressure
- lubrication
- mineral oil (MO)
- miscibility
- naphthenic
- net oil pressure
- oil passages
- oil pressure safety control
- oil pump
- oil pump discharge pressure
- polyol ester oil (POE)
- radial
- slugging
- splash system
- stable
- tandem compressors
- two-stage compressor
- unloader
- vibration eliminator
- viscosity
- volumetric efficiency

Introduction

Lubrication is the process of providing oil or other friction-reducing material to critical points of mechanical devices to lessen wear and prevent overheating. Proper lubrication is vital to the efficient operation and long life of the compressor. Since some oil normally circulates along with the refrigerant, precautions must be taken to ensure that an adequate supply of oil is always in the compressor.

21.1 Refrigeration Lubricants

Many of today's refrigeration lubricants belong to a specially refined naphthenic group of mineral (petroleum) oils or are new synthetic oils. Both are different from the oil used to lubricate automobile engines and other motors.

PRO TIP

Compatibility and Lubrication Additives

In addition to using the specified refrigeration lubricant, some systems may require special oil additives to properly lubricate moving parts. Follow manufacturer recommendations in these special situations.

Lubricant Handling

Store refrigeration lubricant in sealed containers. Never leave lubricant exposed to atmospheric air. It will absorb moisture and other contaminants. Refrigeration lubricant is available in barrels or in one- or five-gallon containers. These containers may be plastic or metal, depending on the type of lubricant being held. Most service companies prefer the one-gallon containers for easier handling and contamination control, **Figure 21-1**.

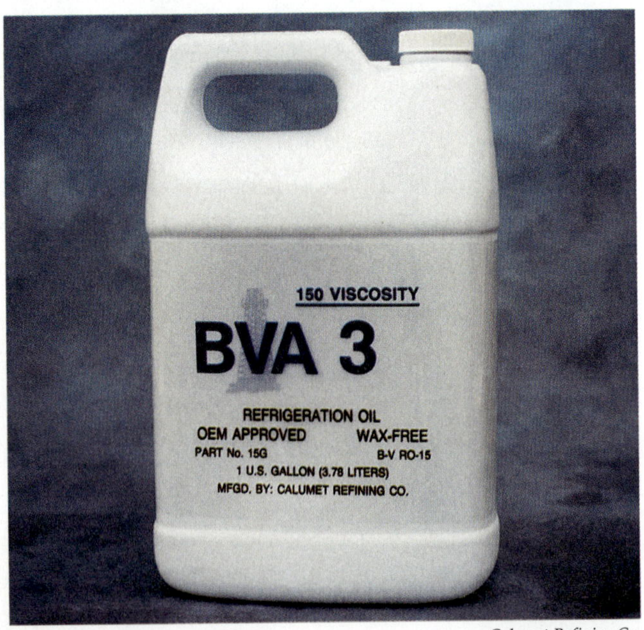

Calumet Refining Co.

Figure 21-1. One-gallon containers of refrigeration oil are preferred for ease of handling and storage. They also are less subject to contamination than larger containers with contents that might be only partly used for a given installation. The container shown holds 150 viscosity oil. Other viscosity oils are also available.

PRO TIP

Using Lubricants

To avoid potential problems, never use lubricant from an already opened container. Always use a new container of lubricant.

Oil Viscosity

Viscosity is the ability of oil to maintain good lubrication despite temperature changes. Choosing the proper viscosity depends on the temperature range to which the oil is exposed. *Higher* viscosity numbers indicate oil that will change little over an extreme range of temperatures. Oils with *lower* viscosity numbers may not flow properly or may even become solid at low temperatures.

Three viscosities are available for older refrigerants using naphthenic (mineral) oil: 150, 300, and 500. When choosing an oil viscosity for a specific application, *always follow the equipment manufacturer's recommendation*. If no recommendation is available, use these guidelines:

- For refrigerants with evaporating temperatures above –20°F (–29°C), use 150 or 300.
- For refrigerants with evaporating temperatures below –20°F (–29°C), use only 300.

Refrigeration Lubricant Requirements

Requirements for a good refrigeration oil are stringent. The oil must effectively lubricate and cool the compressor's moving parts. In a properly operating system, the oil will not decompose, wear out, or need replacement. Refrigeration oil must also be ***stable*** (not break down) and remain fluid in all parts of the system, even when in direct contact with hot motor windings, discharge valves, or cold evaporators. A good refrigeration oil must have these properties:

- **Low wax content.** Wax separating from the oil will clog valves and capillary tubes.
- **Good thermal stability.** The oil must withstand temperature extremes.
- **Low pour point.** The oil must remain fluid at low temperatures.
- **High dielectric strength.** The oil must resist electrical current flow. (Moisture reduces dielectric strength.)
- **Good miscibility.** The oil must readily mix with refrigerant.

Lubricant-Refrigerant Mixture

Refrigeration lubricants are highly soluble in liquid refrigerant. At normal room temperatures, they mix completely. However, lubricant and refrigerant vapor do *not* mix as readily. The ability of a lubricant to mix with refrigerant is called ***miscibility***.

Lubricant in the refrigeration system is never pure oil. It is really a lubricant-refrigerant mixture. The amount of refrigerant dissolved in lubricant depends upon pressure and temperature in each part of the system. An oil's lubricating ability and viscosity rating are reduced when it is mixed with refrigerant.

Oil Additives

Oil additives are sometimes used for better performance. Such additives can lower floc and pour points, improve stability, prevent foaming, improve viscosity, and prevent rust. Additives can be combined to provide special properties.

New refrigerants and oils have complicated the selection procedure. The old guidelines, using viscosity and temperature to select the proper oil, no longer apply. Mistakes in oil selection can be costly, so it is important to consult the manufacturer's recommendations before adding oil to a system.

Types of Refrigeration Lubricants

Refrigeration lubricants are either natural or synthetic. Mineral oil is natural, while the others are synthetic. Each group has different compatibilities and character traits.

Mineral Oil

Mineral oil (MO) is a refrigeration lubricant made by refining crude oil. It is used in CFC, HCFC, and HC systems. For years, it was the main refrigeration lubricant in use. However, it is *not* compatible with HFC refrigerants, such as R-134a. The mineral oil group of refrigerant lubricants is classified as ***naphthenic***. This means that these oils are *not compatible* with most of the new HFC refrigerants developed to lessen ozone depletion. Research has produced new synthetic oils that are compatible with new refrigerants. Two of the most popular families of synthetic oils are alkylbenzene (AB) and polyol ester (POE).

AB Lubricants

Alkylbenzene (AB) lubricants are manufactured from propylene and benzene. AB lubricants are used with CFCs, HCFCs, and blends including CFCs and HCFCs. Testing has shown that HCFC refrigerants perform better with alkylbenzene lubricants than with existing naphthenic mineral oils (MO). Mineral oils are *not* completely soluble in HCFC refrigerants. However, the HCFC refrigerants are soluble in a mixture of alkylbenzene and mineral oil if the mineral oil concentration is 20% or less.

POEs

Polyol ester oils (POEs) are a family of synthetic lubricants rapidly gaining popularity. They are compatible with all refrigerants (CFCs, HCFCs, HFCs, and HCs) and are miscible with mineral oil and alkylbenzene (AB) lubricants. Polyol ester oil is made from expensive base stock materials. Therefore, POEs cost more than other lubricants. However, some characteristics of POEs help offset its higher cost. For instance, polyol ester oil is backward-compatible with mineral oil, which means a compressor containing polyol ester oil can be installed in a refrigeration system that contains mineral oil.

Another advantage of polyol ester oil is that it can be designed to meet lubrication requirements equal to those of mineral oil. The viscosity of polyol ester oil varies less with temperature than mineral oil. Polyol ester oil is miscible with all refrigerants, and it has oil-return characteristics similar to mineral oil when used with conventional chlorine-containing refrigerants.

21.2 Compressor Lubrication

Two lubrication methods are utilized in reciprocating compressors: the *splash system* and the *forced-feed (oil pump) system*.

Splash System

In the ***splash system***, the crankcase is filled with enough oil to bring the level up to the bearings. As the crankshaft turns, the crank throw (eccentric) dips into the oil and splashes it around inside the compressor. Small dippers or scoops are sometimes added to help sling oil around to other parts, **Figure 21-2**.

The splash system is most common in small compressors of less than 3 hp. Hermetic compressors for use in domestic refrigerators, freezers, window air conditioners, and light commercial

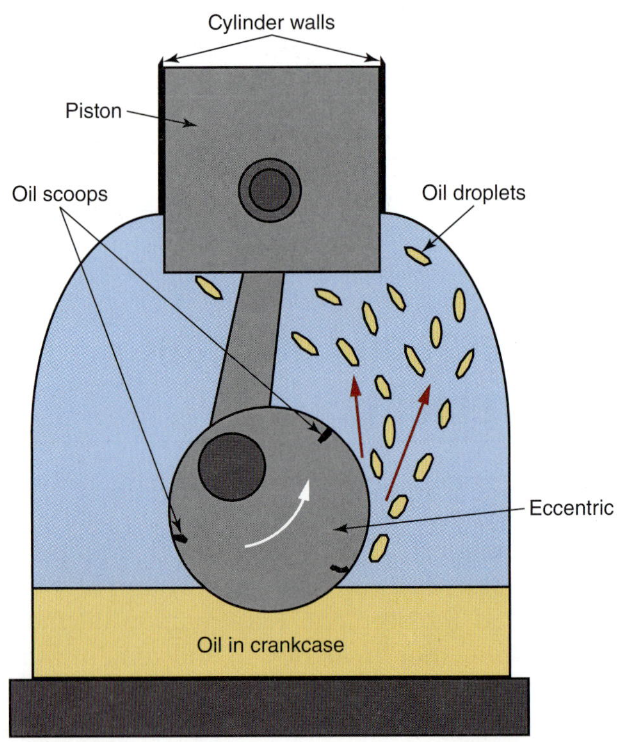

Goodheart-Willcox Publisher

Figure 21-2. Small oil scoops on the eccentric of the crankshaft splash oil onto compressor parts for lubrication and cooling.

systems are factory-assembled under controlled conditions. Although the amount of oil in these systems is critical, no method is provided to observe or check the oil level. When repairing hermetic systems, make every effort to prevent loss of oil. Any oil that is lost should be measured and replaced. If necessary, any remaining oil must be removed and a new factory charge installed.

Most welded hermetic compressors are installed in systems without service fittings. Adding oil to a welded hermetic unit usually involves cutting the suction line to allow the oil to be poured into the compressor. The suction line stub on a welded hermetic compressor opens directly into the shell of the compressor.

Forced-Feed (Oil Pump) System

The ***forced-feed system*** uses an oil pump to deliver oil under pressure to all bearing surfaces and other critical parts. The ***oil pump*** is a device operated by the compressor crankshaft that delivers oil through special ***oil passages*** drilled in the crankshaft, connecting rods, and other components. With this pressure system, the compressor gets better lubrication, permitting smaller bearing clearances. Oil pumps are usually provided on compressors of 3 hp or more.

The oil pump is normally mounted on the crankshaft bearing housing end of the compressor. As shown in **Figure 21-3**, the oil pump has a

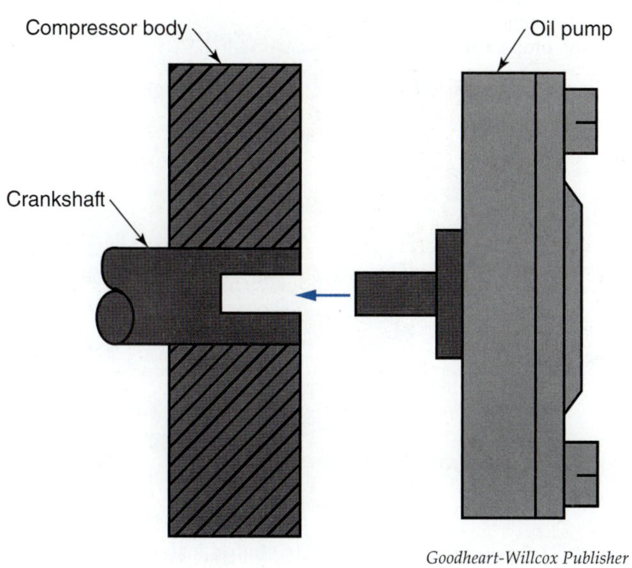

Goodheart-Willcox Publisher

Figure 21-3. The oil pump is bolted to the end bell of the compressor and driven by the crankshaft.

shaft with a flat end that fits into a slot cut into the crankshaft. The oil pump is driven by the turning of the compressor crankshaft.

The oil pump obtains oil directly from the crankcase and forces it through a hole in the crankshaft to the compressor bearings and connecting rods. The pump has a spring-loaded ball check valve that acts as a pressure-relief device. The valve allows oil to bypass directly to the crankcase if the oil pressure rises above its setting.

Oil pumps are normally equipped with a Schrader valve to attach the manifold gauge for oil pressure readings, **Figure 21-4**. Connecting to this valve, a technician can measure oil pump discharge pressure. However, there are three different pressures that must be considered when checking a compressor's oil pump.

An oil pump's intake is located in the crankcase, which is kept at low-side or suction pressure. Therefore, an oil pump's inlet is always subjected to crankcase or low-side pressure. This means that oil pump discharge pressure is based on crankcase pressure. ***Oil pump discharge pressure*** is the pressure at an oil pump's discharge. This value is measured on an oil pump's Schrader valve. It is composed of crankcase pressure and net oil pressure combined. ***Net oil pressure*** is the pressure added to oil by the pump (not including crankcase pressure). Therefore, to calculate net oil pressure, subtract crankcase pressure from oil pump discharge pressure:

$$\text{Net oil pressure} = \text{Oil pump discharge pressure} - \text{crankcase pressure}$$

For example, crankcase pressure is 60 psig (414 kPa), and the oil pump discharge pressure is 100 psig (690 kPa). To calculate net oil pressure, subtract crankcase pressure from oil pump discharge pressure.

$$\text{Net oil pressure} = \text{Oil pump discharge pressure} - \text{crankcase pressure}$$
$$\text{Net oil pressure} = 100 \text{ psig} - 60 \text{ psig}$$
$$\text{Net oil pressure} = 40 \text{ psig}$$

Be aware of the impact crankcase pressure has on net oil pressure. For instance, if a system has a low charge or the evaporator is otherwise starving for refrigerant, crankcase pressure can be lower than normal. This low suction pressure will result in lower-than-normal net oil pressure.

In normal operation, net oil pressure will vary depending on compressor size, temperature and viscosity of the oil, and the amount of clearance in

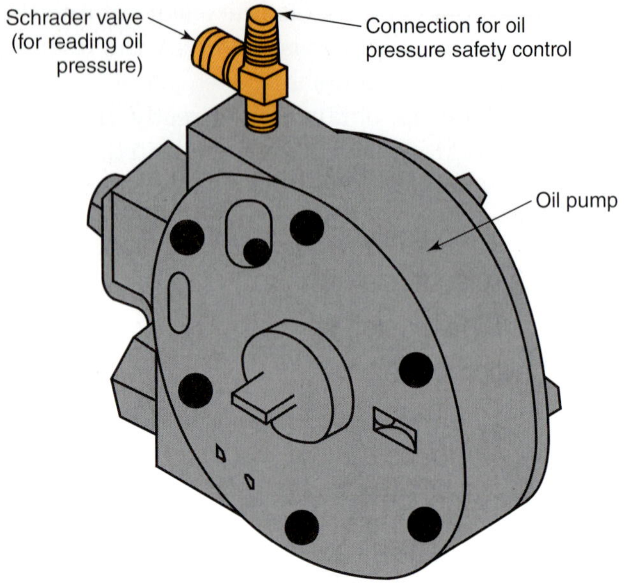

Goodheart-Willcox Publisher

Figure 21-4. A Schrader valve threaded into a tee fitting permits use of a manifold gauge for reading oil pressure at the pump.

the bearings. Acceptable oil pressures vary from one compressor manufacturer to another, but *minimum* oil pressures are standard.

Oil Level

For proper lubrication, an adequate supply of oil must always be maintained in the crankcase. Most semihermetic compressors have an oil sight glass (small window) on the side of the crankcase for determining the oil level in the crankcase.

Figure 21-5 illustrates proper oil levels in the widely used Copeland and Carlyle (Carrier) compressors. In Copeland units, the window should be half-full when the system is running and stabilized. For Carlyle "D" compressors, the oil level should also be half-full. For "E" compressors, the oil level should be from 1/8″ to 3/8″ up the sight glass. A higher level indicates an excessive amount of oil, which could result in slugging and possible damage to the compressor valves. On field-installed systems, it may be necessary to add or remove oil after the system stabilizes at its normal operating condition to maintain the proper level.

Oil Pressure Safety Control

A high percentage of all compressor failures are caused by lack of proper lubrication. Lack of lubrication can result from several factors:

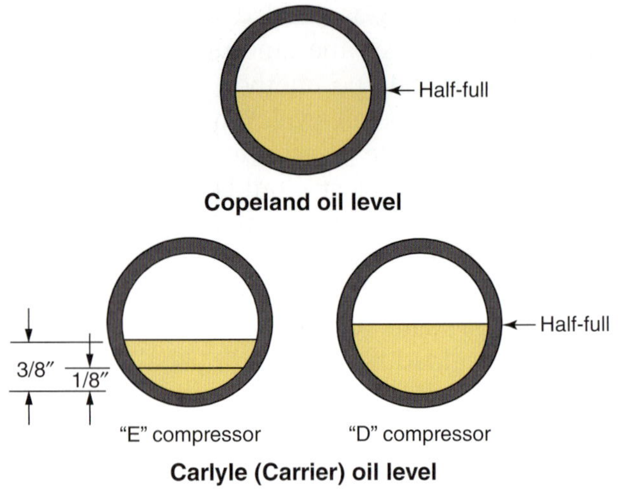

Goodheart-Willcox Publisher

Figure 21-5. Sight glass oil levels for two popular brands of compressors.

- Shortage of oil in the system
- Improper oil return from the evaporator ("logging")
- Shortage of refrigerant
- Refrigerant migration or floodback to the crankcase
- Failure of the oil pump
- Faulty operation of refrigerant flow controls

An ***oil pressure safety control*** is a device that protects the compressor against loss of oil pressure, **Figure 21-6**. Most compressor failures caused by loss of lubrication can be prevented by a safety control.

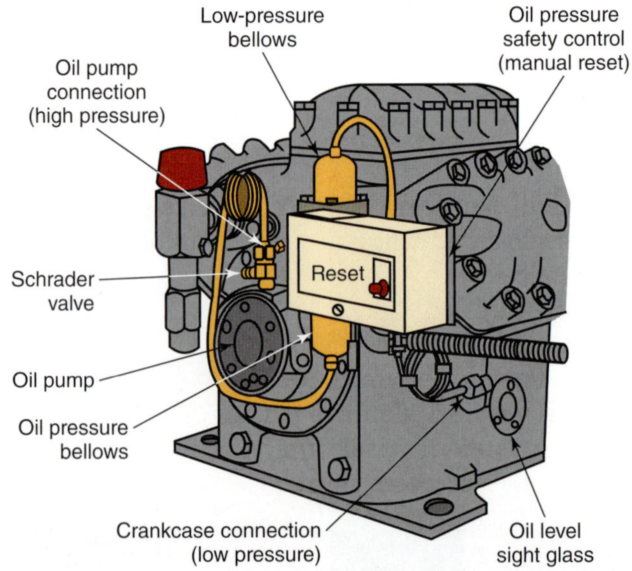

Goodheart-Willcox Publisher

Figure 21-6. An oil pressure safety control installed on a typical Copeland compressor.

Operation of an oil pressure safety control depends on the difference in pressure between the oil pump outlet (discharge) and the crankcase. The *top* of the control contains a low-pressure bellows. A capillary tube from the bellows is connected to the compressor crankcase, transferring crankcase pressure to the top of the control. The bellows on the *bottom* of the control has a capillary tube that must be attached to the oil pump discharge valve. This capillary tube transfers oil pump discharge pressure to the bottom of the control. Some oil pressure safety controls have the capillary tubes coming out of the bottom of the control and some controls are electronic and use a pressure transducer to check for pressures, **Figure 21-7**.

If oil pressure falls below safe limits, a switch inside the control opens, stopping the compressor. The compressor cannot restart until the manual reset button on the front of the control is pressed. Oil pressure safety controls have a time-delay feature. Time-delay options include 30-, 60-, or 120-second delays. A time-delay feature helps to avoid nuisance shutdowns caused by brief fluctuations in oil pressure during start-up.

The time-delay prevents a tripout for the specified number of seconds at start-up, regardless of oil pressure. When the time delay expires, net oil pressure *must* be at least 12–14 psig (83–96 kPa), else the control will stop the compressor. During a normal run cycle, the control prevents net oil pressure from dropping below 9 psig (62 kPa).

Tripping of the oil pressure safety switch is a warning that the system has been without proper lubrication for the set time-delay period. It is a clear indication that a lubrication problem exists and needs to be corrected. A well-designed system should not trip the oil pressure safety control. Once tripped, the safety control cannot be reset until it cools (about two minutes).

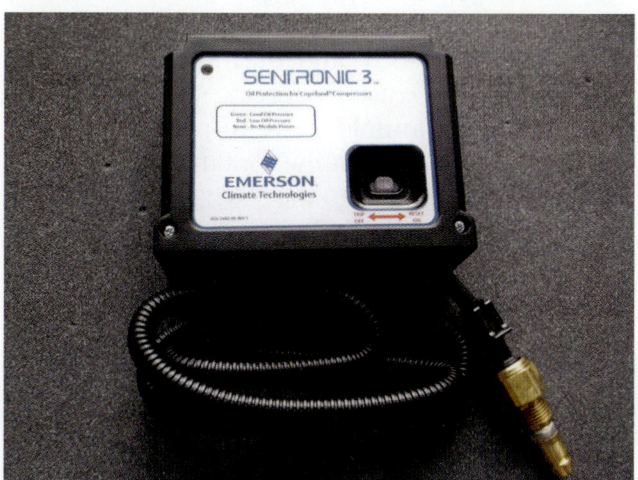

Image courtesy of United Refrigeration

Figure 21-7. This oil failure control uses an electronic pressure transducer to check oil pressure.

Compressor manufacturers often *require* the use of an oil pressure safety control. Eliminating or bypassing the safety control will void the compressor warranty. Operation of the safety control and proper wiring to the compressor electrical circuit are fully explained in Chapter 26, *Refrigeration Motor Controls.*

21.3 Adding Oil to a Compressor

There are three methods for adding oil to a compressor: *the open system method, the oil pump method*, and *the closed system method*. However, the third method is *not* recommended. It can introduce air into the system and cause damage to the compressor from slugging.

Open System Method

Semi-hermetic compressors are equipped with a removable crankcase plug for adding oil. If the system contains refrigerant, the crankcase can be isolated by stopping the compressor, frontseating the suction service valve, and reducing crankcase pressure to about 1 psig or 2 psig (7 kPa or 14 kPa). Remove the crankcase plug, and add the required amount of oil, **Figure 21-8**.

Refrigerant vapor in the crankcase is slightly positive at 1 psig or 2 psig, which is just above atmospheric pressure. This prevents entry of air and moisture during the oil-adding procedure. If desired, the crankcase can be purged by cracking the suction service valve off its seat for one or two seconds. After the oil has been added, replace the crankcase plug, backseat the service valve, and restore the system to operation.

Oil Pump Method

Most technicians purchase a small, hand-operated oil pump for adding oil to compressors. The hand pump is like a small bicycle tire pump, **Figure 21-9**. The pump permits the technician to add oil to an operating compressor through the

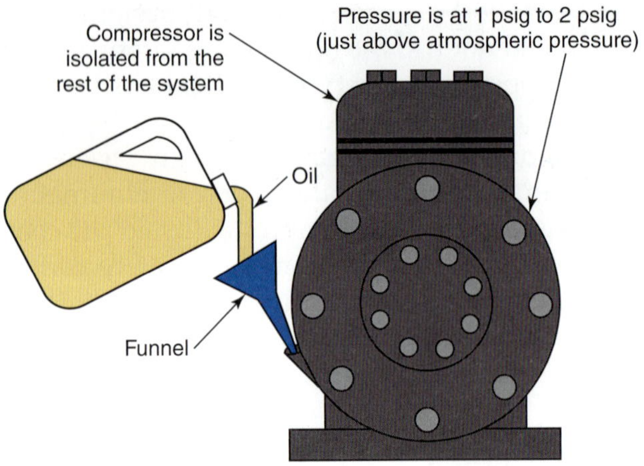

Goodheart-Willcox Publisher

Figure 21-8. The open system method of adding oil is used when the system is not running, the compressor is isolated, and at 1 psig or 2 psig (just above atmospheric pressure). Oil is added through the crankcase fill plug.

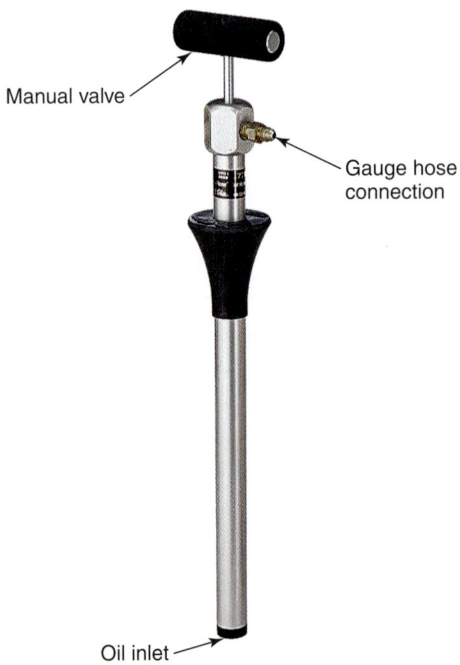

Ritchie Engineering – YELLOW JACKET Products Division

Figure 21-9. A small hand pump used to add oil to the compressor while the system is running. The pump is connected to the center hose of the manifold gauge.

manifold gauge and suction service valve (SSV), **Figure 21-10**. The hand pump contains a check valve to prevent backflow. The technician can easily develop sufficient pressure to overcome the operating suction pressure and add oil as needed. Adding oil while the system is *operating* is desirable

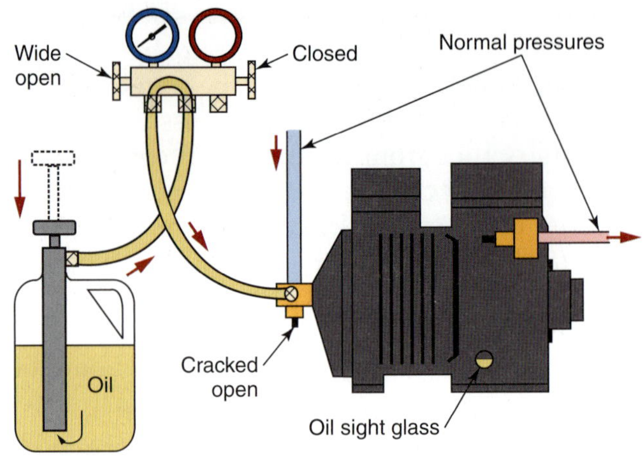

Goodheart-Willcox Publisher

Figure 21-10. While the system is running, oil is added using a hand pump connected to the suction service valve through the manifold gauge.

because the oil level can be observed through the oil sight glass. An accurate oil level cannot be determined when the system is off.

Closed System Method

Although the method is not recommended, oil may be drawn into the compressor through the suction service valve using the manifold gauge. *Only* try this on compressors where the suction service valve is located on the end of the compressor, *not* up by the heads. Connect the low-pressure gauge hose (blue) to the SSV and crack the valve open. Crack open the left manifold valve and vent a small amount of refrigerant through the hoses to purge them of air. Submerge the end of the center charging hose (yellow) in a container of refrigeration oil while venting it, **Figure 21-11**. Close the manifold valve after venting.

Frontseat the SSV and pull a vacuum in the compressor crankcase by forcing the unit to run. Open the left manifold valve to permit the compressor to draw oil from the container through the manifold gauge and into the SSV.

Be careful to keep the center hose end submerged in oil at all times. Otherwise, air will be drawn into the compressor. On smaller compressors, oil must be added slowly to prevent slugging. Continue adding oil until the proper amount has been drawn into the compressor.

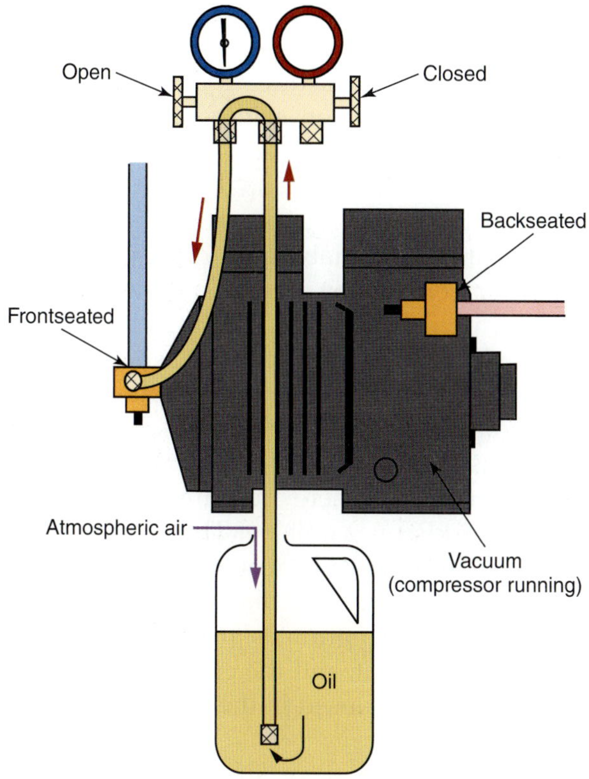

Goodheart-Willcox Publisher

Figure 21-11. Connections used to add oil to the compressor by the closed system method. This method is not recommended because of the danger of damaging the compressor or introducing air into the system.

PRO TIP

Adding Fresh Oil

Because of the hygroscopic properties of refrigerant lubricant (its tendency to absorb moisture from the atmosphere), it is a good practice to *not* add oil to a refrigeration or air-conditioning system from a container that has already been opened. Even when the container has been resealed with its cap, there is still a possibility that the oil has absorbed moisture, looks normal, and could then be introduced to the refrigerant system, causing many problems.

21.4 Removing Oil from a Compressor

Sometimes excess or severely contaminated oil must be removed from a compressor. Many compressors are equipped with a drain plug. Draining is performed with the compressor turned off.

Removing Excess Oil Using the Drain Plug

Install a manifold gauge and frontseat the SSV to reduce crankcase pressure to 1–2 psig (7–14 kPa). Stop the compressor, and frontseat the discharge service valve (DSV) to isolate the compressor. Carefully loosen the oil drain plug, but do *not* completely disengage the threads. Allow pressure to bleed off. Then drain oil to the desired level by seepage around the threads. Do *not* totally remove the plug. The oil seal at the drain hole and crankcase pressure prevent entry of air and moisture into the system.

After draining the excess oil, tighten the plug, open the service valves, and restore the compressor to operation. If the crankcase is totally drained because of oil contamination, tighten the plug, and refill with oil using one of the methods described earlier.

Removing Excess Oil through the Oil Fill Hole

Install a manifold gauge and reduce crankcase pressure to 1 psig or 2 psig by frontseating the SSV. Stop the compressor. Then frontseat the DSV to isolate the compressor.

Carefully loosen the oil fill plug and bleed off pressure before the threads are completely disengaged. Remove the oil fill plug and then insert a length of 1/4″ soft copper tubing so its end is near the bottom of the crankcase. Use tubing that is long enough to bend the other end down below the level of the crankcase, forming a siphon arrangement as shown in **Figure 21-12**. Wrap a clean rag tightly around the copper tubing to seal the oil fill opening. Crack open the SSV slightly to pressurize the crankcase to 4 psig or 5 psig (28 kPa or 34 kPa).

Crankcase pressure forces oil out through the copper tubing and into a container. The oil will continue to drain until the crankcase is empty, if desired. Refrigerant pressure in the crankcase also serves to prevent entry of air and moisture.

When oil removal is completed, remove the copper tubing and quickly reinstall the oil fill plug to prevent loss of refrigerant. Both service valves can be reopened, and the compressor restored to operation. If the crankcase is totally drained, refill before operating the compressor using one of the methods described earlier.

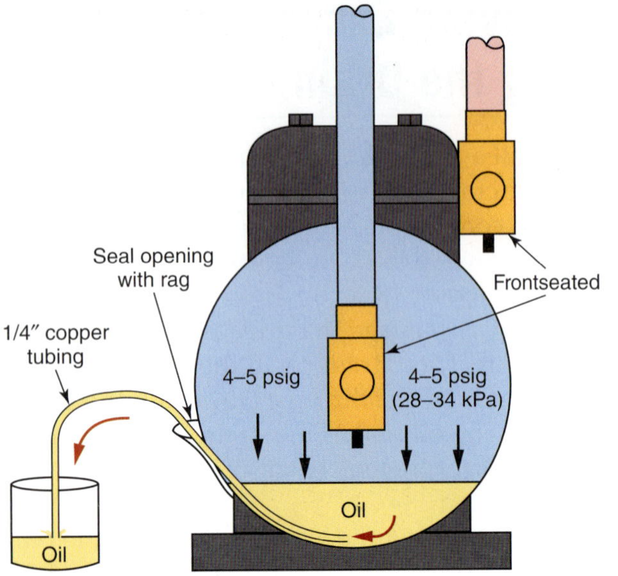

Goodheart-Willcox Publisher

Figure 21-12. Procedure for removing oil from the compressor through the crankcase opening. Sealing the opening around the tubing with a rag maintains the pressure difference necessary for removing the oil.

21.5 Testing Oil for Acid

Acid in a refrigeration system is picked up by the refrigeration oil, which acts as a scavenger. Acid test kits are available to check the oil for acid contamination, **Figure 21-13**. Each kit includes complete instructions for use.

The test is performed by obtaining a small (usually 1/2 oz) oil sample from the system being tested. The oil sample and a test solution are combined in specified amounts in a test tube or similar container. Then, the contents are thoroughly shaken to mix the ingredients. After the ingredients stand for a few minutes, the color is observed to determine whether acid is present and to what percentage of

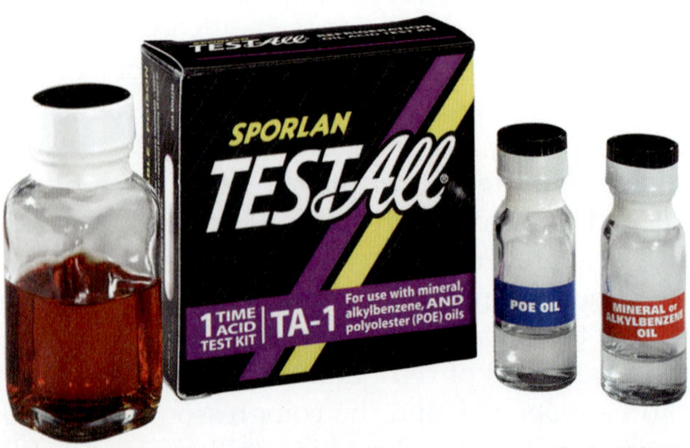

Sporlan Valve Co.

Figure 21-13. A kit used to test for acid contamination in refrigerant oil.

acid is present. The kit contains information on interpreting the color changes.

Periodic oil checks during a cleanup procedure will indicate when acid contamination has been reduced to a safe level. Cleanup is done by performing oil changes and by installing (and changing) acid-removing filter-driers until the test color reveals an acceptable acid level.

Carrier Corporation markets a kit called Totaltest™ that permits accurate readings of acid and moisture levels *without* shutting down or opening the system. The test instrument accepts disposable tubes containing indicator chemicals, **Figure 21-14**.

The instrument is attached to a service valve or Schrader fitting where a minimum pressure of 60 psig (414 kPa) is available. The chemical crystals in the disposable tube react to corrosive acid or any acid-causing moisture that might be present in the system. After 10 minutes, the tube is removed from the instrument, and the crystal color is compared to a chart enclosed in the kit.

21.6 Determining Compressor Capacity

A reciprocating compressor is designed as a positive displacement unit. Its capacity depends on the number and speed of the pistons and the volume of the cylinders. Since these compressors are normally driven by an electric motor, the speed is fixed at either 1725 rpm or 3450 rpm. Therefore, the amount of refrigerant pumped is calculated by the number of strokes per minute times the total cylinder volume. The capacity for any given compressor is limited.

Compressor capacity is determined by four factors:

- The *diameter* of the piston
- The *number* of pistons
- The *length* of stroke
- The *speed* (strokes per minute)

Capacity plays a large role in compressor selection for a particular refrigeration system. An oversized compressor (too much capacity) would rapidly remove refrigerant vapor from the evaporator and suction pressure would be too low. Likewise, an undersized compressor (low capacity) would not remove refrigerant from the evaporator fast enough; therefore, suction pressure would remain high for a prolonged length of time.

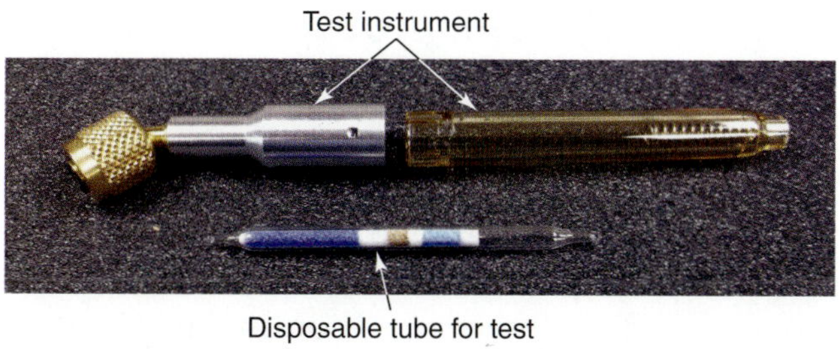

Goodheart-Willcox Pubilsher

Figure 21-14. The Totaltest™ system consists of an instrument that screws onto a service or Schrader valve, and a disposable tube filled with indicator chemicals that fits inside the instrument.

Refrigerant Density

Other factors influencing compressor capacity are the type of refrigerant being pumped and the temperature of the suction gas. The ***density*** of a gas (how closely the molecules are packed) varies with temperature. A low-temperature gas is "thin" and occupies more space (volume) per pound of refrigerant than does a denser, high-temperature gas. The compressor piston may have to make several strokes before it can move one pound of refrigerant. Since high-temperature gas is denser than low-temperature gas, one pound of high-temperature gas occupies less space than a low-temperature gas. The same compressor would require fewer piston strokes to move one pound of high-temperature refrigerant than it would for a low-temperature refrigerant.

The boiling point (*suction pressure*) in the evaporator should be kept as high as possible to require the least amount of compressor capacity. As suction pressure goes down, compressor capacity must go up. Low-temperature compressors require more capacity than medium- or high-temperature applications. For example, R-134a is *not* a good refrigerant for low-temperature applications because suction pressure is low (possibly in a vacuum), and the gas is too thin, requiring a high-capacity compressor. Using R-449A on a low-temperature application is preferable. R-449A has a higher suction pressure and much thicker gas, permitting use of a low-capacity compressor.

Clearance Pocket

The volumetric efficiency of a compressor varies with compressor design. If the valve reeds seat properly, the most important factor affecting a reciprocating compressor's efficiency is the ***clearance pocket***. A clearance space (or pocket) must exist at the top of the piston stroke to prevent the piston from striking the valve plate, **Figure 21-15**. Since the discharge valve reed is located on top of the valve plate, the ***discharge port*** (hole through the plate) adds some volume to the clearance space as well. At the top of the piston stroke, the clearance pocket remains filled with hot, compressed gas that does not exit.

As the piston starts downward on the suction stroke, the remaining high-pressure gas re-expands. As it expands, its pressure is reduced. Vapor from the suction line cannot enter the cylinder until the pressure of the re-expanding gas drops below suction pressure. From the standpoint of capacity, the first part of the suction stroke is actually lost. For this reason, the clearance pocket must be kept as small as possible to improve volumetric efficiency of the compressor.

On low-temperature applications, it is often necessary to reduce the clearance pocket to obtain the desired capacity. Low-temperature valve plates usually have smaller discharge ports.

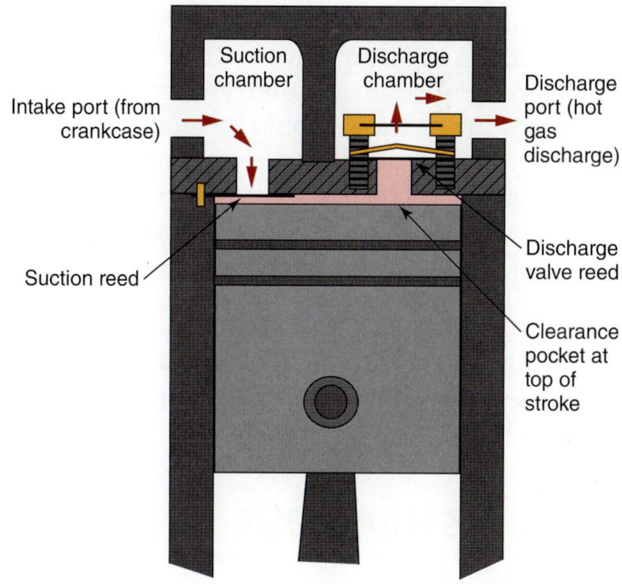

Goodheart-Willcox Publisher

Figure 21-15. The clearance pocket at the top of the piston stroke affects compressor efficiency. The hot gas remaining in the pocket after the discharge valve reed closes will re-expand as the piston moves down on the intake stroke, reducing the volume of new vapor that can be drawn in.

21.7 Volumetric Efficiency

The ***volumetric efficiency*** of a compressor is a comparison of the amount of gas *actually* pumped by the compressor to the amount of gas it *should* pump, according to piston displacement and length of stroke. The figure is expressed as a percentage. The actual volume pumped (expressed in cubic inches) is divided by the amount of gas the compressor should pump without any losses. For example, a compressor may be designed to pump 20 in^3 of vapor per stroke (piston displacement) but actually pumps 12 in^3. Therefore, volumetric efficiency is 60% ($12 \div 20 = 0.60$).

Compressor manufacturers try to keep volumetric efficiency as high as possible, but field problems can reduce efficiency.

High Head Pressure = Decreased Efficiency

As head pressure increases, efficiency decreases due to high-pressure gas left in the clearance pocket at the top of the stroke. Also, as suction pressure decreases, efficiency decreases, since the pressure of the gas left in the clearance pocket must be reduced further to drop below the lower suction pressure. This reduces the amount of incoming suction gas and lowers volumetric efficiency. Every effort should be made to operate the system within its designed capacity.

Compressor efficiency also depends on the size of ports or openings inside the compressor. Anything reducing the flow of gas through these ports will reduce compressor efficiency.

Each semihermetic compressor is designed to be most efficient for a particular evaporator temperature (high, medium, or low) and a particular refrigerant.

Compression Ratio

Compression ratio refers to the relationship between the low-side pressure and the high-side pressure. In other words, exactly how much compression is taking place. The compression ratio is figured by dividing the low-side absolute pressure (psia) into the high-side absolute pressure (psia).

$$\frac{\text{Head pressure (in psia)}}{\text{Suction pressure (in psia)}} = \text{Compression ratio}$$

Single-Stage Compressors

The majority of compressors in use today are the single-stage type. The compression ratio for single-stage compressors goes as high as 12:1. If the ratio needs to be higher than 12:1, a two-stage compressor is used. Lower compression ratios place less demand on the compressor, making it more energy efficient. See **Figure 21-16** for a comparison of single-stage and two-stage compressors.

The need for two-stage compressors can usually be avoided by selecting a refrigerant that operates below the 12:1 ratio. For example, a typical residential air conditioner using R-22 on a hot summer day (90°F or 32°C) has a head pressure (high side) of about 260 psig and a suction pressure (low side) of about 60 psig. Before calculating compression ratio, these gauge readings must be converted to absolute by adding 15 psi for atmospheric pressure:

$$260 + 15 = 275 \text{ psia (high-side absolute pressure)}$$

$$60 + 15 = 75 \text{ psia (low-side absolute pressure)}$$

$$275 \div 75 = 3.7 \text{ (approximately 4:1 compression ratio)}$$

Calculated to 3.7, this compression ratio is nearly 4 to 1 (4:1). Since it is below a 12:1 ratio, a single-stage compressor is sufficient, and a

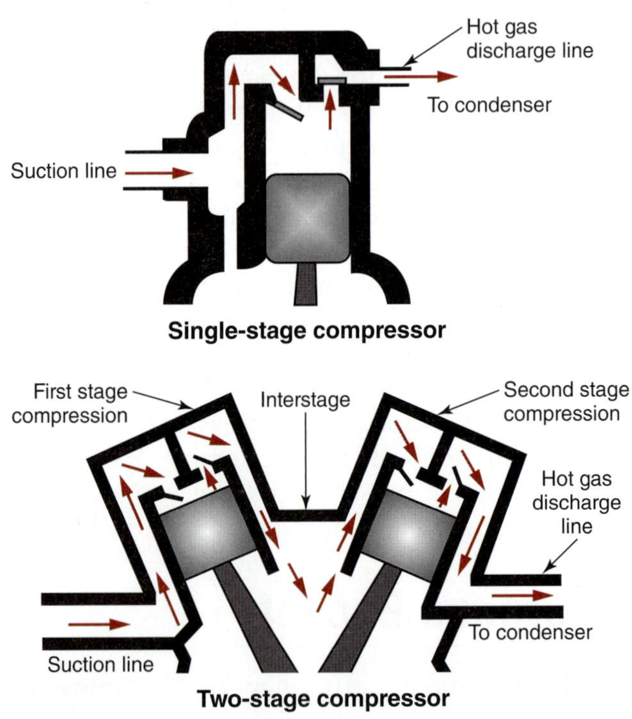

Figure 21-16. Cutaway views of single- and two-stage compressors.

two-stage compressor is unnecessary in this application.

Two-Stage Compressor Operation

Because of the high compression ratios needed in ultralow-temperature applications, ***two-stage compressors*** were developed. They provide increased efficiency when evaporating temperatures are in the range of –30°F to –80°F (–34°C to –62°C). **Figure 21-17** illustrates a typical two-stage system.

Two-stage compressors are divided internally into low (or first) and high (or second) stages. The three-cylinder models have two cylinders on low stage and one on high, while the six-cylinder models have four cylinders on low and two on high.

Refrigerant gas enters the low-stage cylinders *directly* from the suction line. The gas is compressed and discharged into the interstage manifold at ***interstage pressure***. The interstage pressure enters the motor chamber and crankcase, so the crankcase is at interstage pressure as well.

Interstage discharge gas is at high temperature (highly superheated). Liquid refrigerant is metered into the interstage manifold area by a desuperheating thermostatic expansion valve (TEV). The liquid refrigerant immediately boils to desuperheat the discharge gas. Desuperheating cools the motor and prevents excessive temperatures during second-stage compression.

Desuperheated refrigerant vapor at interstage pressure enters the suction ports of the high-stage cylinders, where it is compressed and discharged to the condenser.

Unloaders

Capacity can be controlled with unloaders. When large compressors must operate at a reduced load, unloaders are used to reduce compressor capacity, **Figure 21-18**. ***Unloaders*** are piston-operated valve assemblies that use a solenoid to control pressure that will hold open or closed suction valves on one or more cylinders in response to reduced suction pressure. With the suction intake blocked off, no refrigerant vapor can enter, so nothing is compressed and discharged into the high side. The pistons may still be pumping, but with nothing to pump, those unloaded cylinders perform no pumping action.

The unloader is controlled by an electric solenoid valve mounted directly onto the head and mechanically connected to the unloader. See

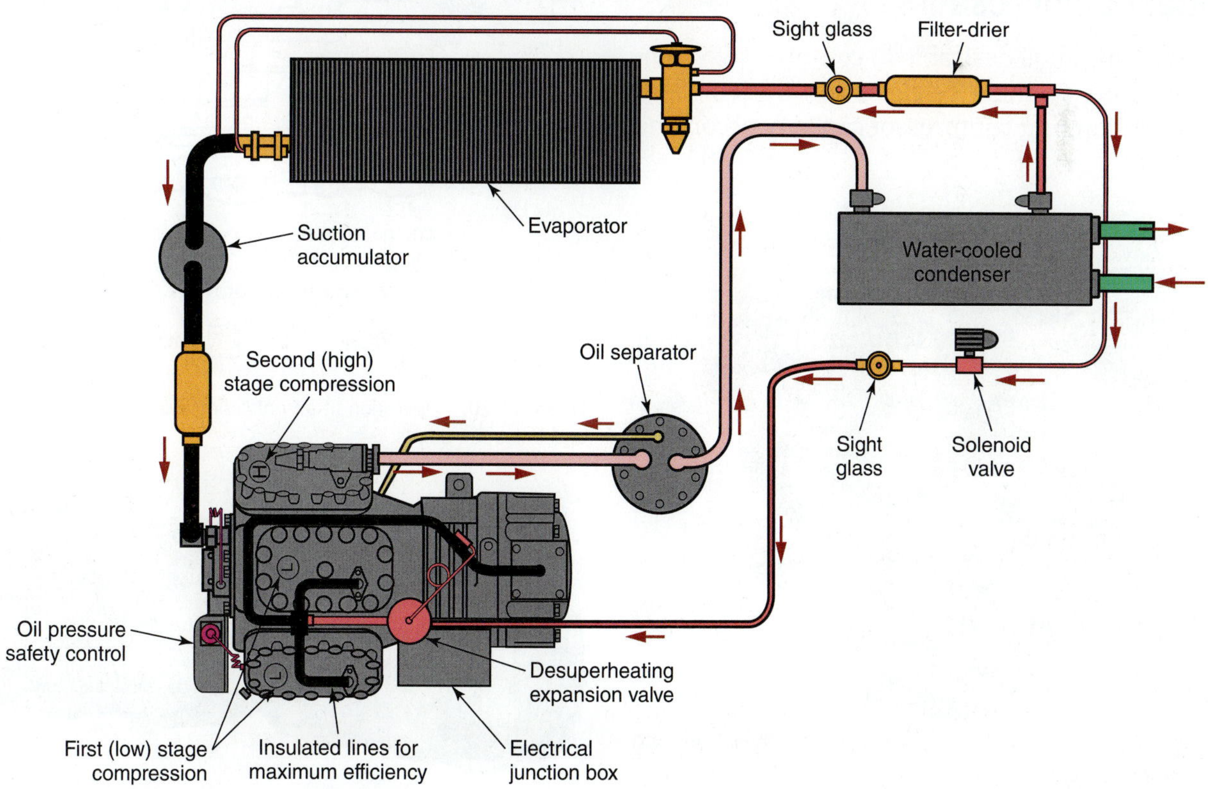

Goodheart-Willcox Publisher

Figure 21-17. A typical two-stage compressor used for low-temperature systems. Such compressors have high compression ratios and provide increased efficiency.

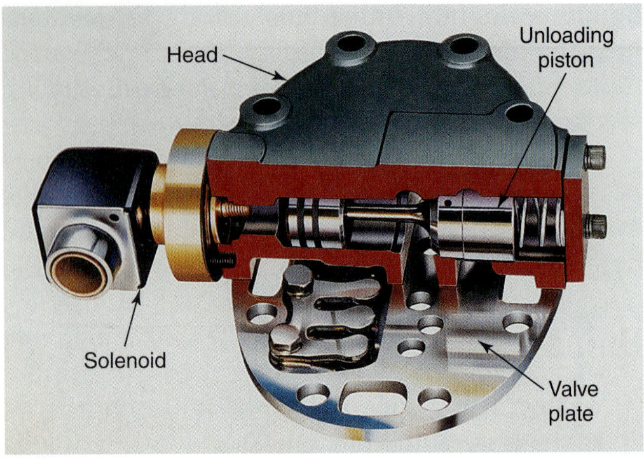

Carrier Corporation

Figure 21-18. Cutaway view of an unloader. The solenoid controls the unloading piston, which blocks or opens the suction manifold port. This controls capacity by taking a cylinder "on-line" or "off-line," as necessary.

Figure 21-19. Power supply to the solenoid is controlled by a pressure-sensitive switch operating on suction pressure. Figure 21-20 illustrates refrigerant flow through the unloader in both loaded and unloaded positions. Spring pressure or head pressure (as directed by the solenoid) control the position of the unloader's piston.

Tandem Compressors

Sometimes it is desirable to combine two compressors into a single refrigeration system as a means of changing compressor capacity according to reduced load, Figure 21-21. However, combining two compressors presents lubrication problems. Unless the pressures in the two crankcases are constantly equal, oil will leave the crankcase at a higher pressure level.

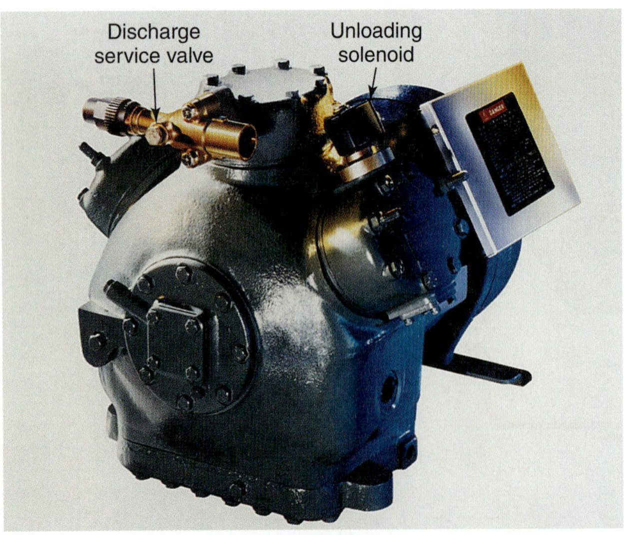

Carrier Corporation

Figure 21-19. The unloading solenoid is mounted directly onto the head of the cylinder it controls. This six-cylinder compressor has unloading solenoids on two cylinders for capacity control.

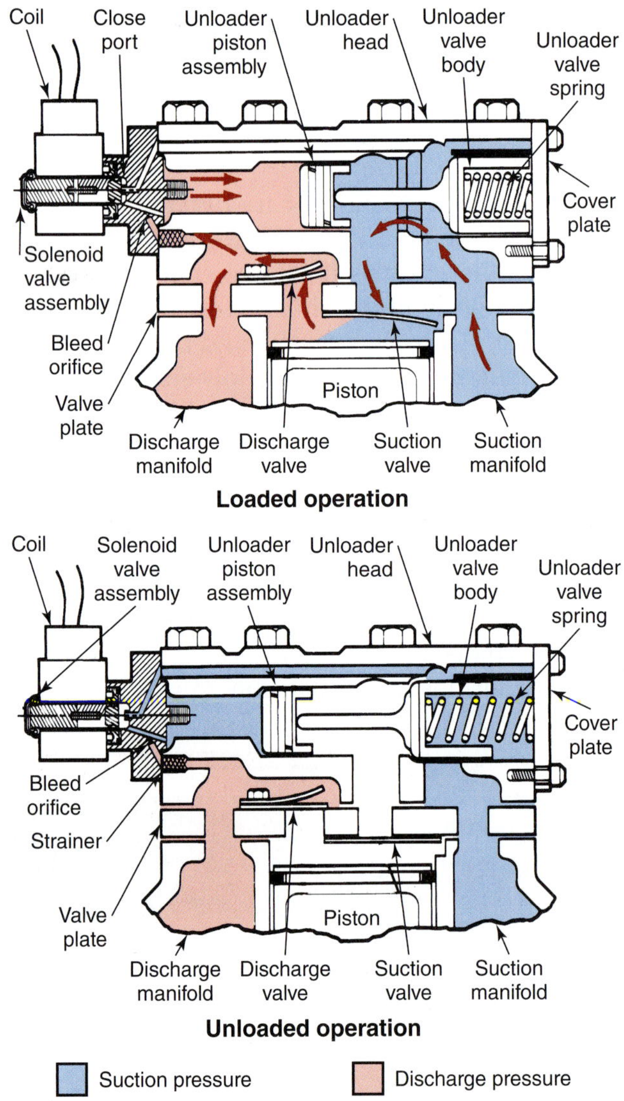

Carrier Corporation

Figure 21-20. Operation of an unloading system in the loaded and unloaded states.

Bitzer

Figure 21-21. A typical tandem compressor consists of two individual units joined with a connecting housing.

Tandem compressors were developed to overcome oil and vibration problems and still offer capacity control. The ***tandem compressors*** consist of two individual compressors with an interconnecting housing that replaces the individual end bells. The compressors can operate individually or together. Tandem compressors offer greater safety than a single compressor. If one compressor fails, the remaining one will operate until repairs are completed. Furthermore, staggered starting reduces electrical requirements. Tandem compressors provide simple, foolproof capacity control with maximum power savings. Tandem arrangement greatly simplifies system control.

21.8 Compressor Cooling

The hot discharge temperatures created at the top of the piston stroke by the heat of compression greatly affect efficient operation of the compressor. The highly superheated discharge gas temperature is directly related to the temperature of the incoming suction gas. Discharge temperatures above 325°F–350°F (163°C–177°C) will cause oil breakdown, resulting in sludge and acid. Discharge temperature must be kept below this range. Peak temperatures occur at the discharge valves. The temperature of the discharge line will be 50°F–100°F (28°C–55°C) *lower* than the temperature at the valve plate, which holds the discharge valves. The maximum allowable discharge line temperature is 250°F (121°C), measured six inches from the discharge service valve exit.

The temperature difference between gas at the discharge valve and the discharge line is absorbed by the metal parts of the compressor. This heat cannot be permitted to accumulate, so various mediums are used to remove it. Mediums used to cool compressors include the following:

- Air
- Refrigerant

Air for Compressor Cooling

Air-cooled compressors must have enough air blowing directly on the compressor body to cool the motor. Proper cooling can normally be accomplished by locating the compressor directly in the discharge airstream from the condenser fan. If the compressor cannot be in the condenser discharge airstream, cooling must be provided by another fan blowing air directly against the compressor. See **Figure 21-22**.

Refrigerant for Compressor Cooling

Refrigerant-cooled compressors are designed to provide a flow of cool suction gas around and through the motor. At temperatures below 0°F (–18°C), the incoming refrigerant gas is too thin (reduced in density) to provide sufficient cooling. In such cases, additional motor cooling by means of airflow is usually necessary.

21.9 Compressor Mounting

The compressor is usually mounted on four mounting bolts equipped with spring and washer assemblies. After installation, the top mounting nut is loosened to allow the compressor to float on the mounting springs, **Figure 21-23**. A space of about 1/16″ (1.6 mm) between the top nut and the rubber spacer is normal. This method of mounting reduces the transmission of noise and excessive movement during starting or stopping, when vibration of the motor-compressor is greatest.

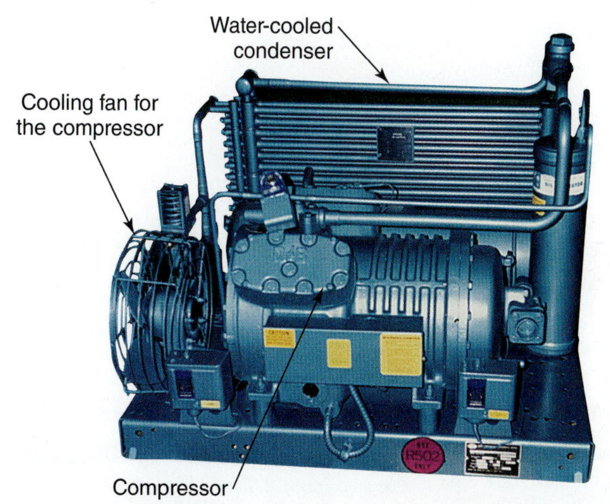

Goodheart-Willcox Publisher

Figure 21-22. On a water-cooled compressor, cool water circulates through a water jacket that surrounds the compressor body. After exiting the jacket, the water is piped to the condenser to provide further cooling action.

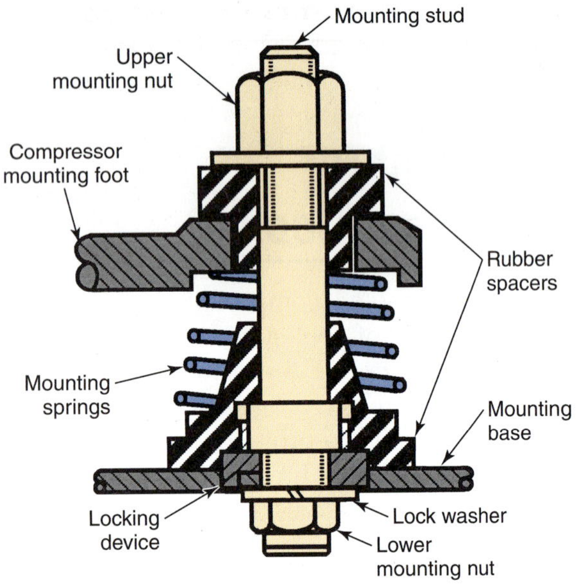

Goodheart-Willcox Publisher

Figure 21-23. The typical means of mounting a compressor, designed to minimize noise and vibration. Four such mounting assemblies are normally used.

Vibration Eliminators

Vibration eliminators prevent transmission of noise and vibration to refrigerant piping. Vibration of the piping causes leaks. Vibration eliminators are installed close to the compressor in both the suction and discharge lines. See **Figure 21-24**. On small units, the soft copper tubing is wrapped in a coil at the compressor to provide vibration protection.

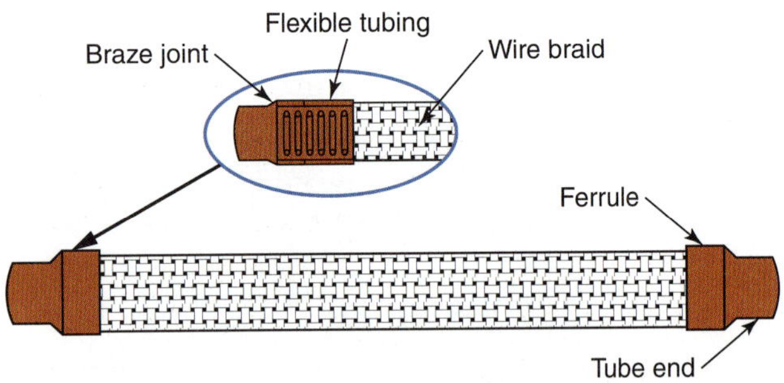

A

Goodheart-Willcox Publisher

B

Courtesy of Sporlan Division - Parker Hannifin Corporation

Figure 21-24. Vibration eliminators are made from flexible tubing and brazed into the suction and discharge lines at the compressor. Vibration of tubing can cause joints to leak. A—Illustration of a vibration eliminator. B—Photo of a vibration eliminator.

Installing Vibration Eliminators

Unless the vibration eliminator is properly installed, stress will cause it to fail or defeat its purpose. The metallic vibration eliminator is designed to adjust to movement in a ***radial*** (circular) direction. It must not be subjected to stress that causes *compression* (pushing together each end) or *extension* (pulling outward each end). Vibration eliminators should be installed as close to the compressor as possible, parallel to the crankshaft, **Figure 21-25**. The starting torque (turning power) of the motor tends to rock the compressor from side to side when starting. Parallel mounting to the crankshaft allows the vibration eliminators to easily adjust to the movement.

Internal connections of the metallic vibration absorbers are made with a brazing compound with a melting point of about 1300°F (704°C). To prevent damage to these internal joints, line connections should be made with a silver alloy with a melting temperature below 1200°F (649°C). Wrapping a wet rag around the vibration absorber before brazing is recommended. Brazing should be done quickly. Overheating may cause damage to the internal brazed connections.

Crankcase Heater

A ***crankcase heater*** is necessary when the compressor is surrounded by an ambient temperature that may become lower than the evaporating temperature. Due to low ambient temperature conditions, liquid refrigerant may migrate to the crankcase or vapor refrigerant may condense into liquid in the crankcase during the off cycle when the compressor is cold. Electric crankcase heaters keep the crankcase oil warm. Warming the oil helps prevent condensing of refrigerant vapor in the crankcase. Under mild conditions, the crankcase heater vaporizes liquid refrigerant and forces it out of the compressor.

Heat always travels toward colder-temperature areas. Therefore, if the compressor becomes colder than the evaporator, refrigerant will migrate (travel) to the compressor. Migration is assisted by the pressure difference between the evaporator and compressor crankcase.

Liquid refrigerant in the crankcase readily mixes with oil. Upon start-up, the crankcase pressure is rapidly lowered, and some liquid boils. This rapid boiling of the refrigerant in the oil can wash the oil off of bearing surfaces and reduce

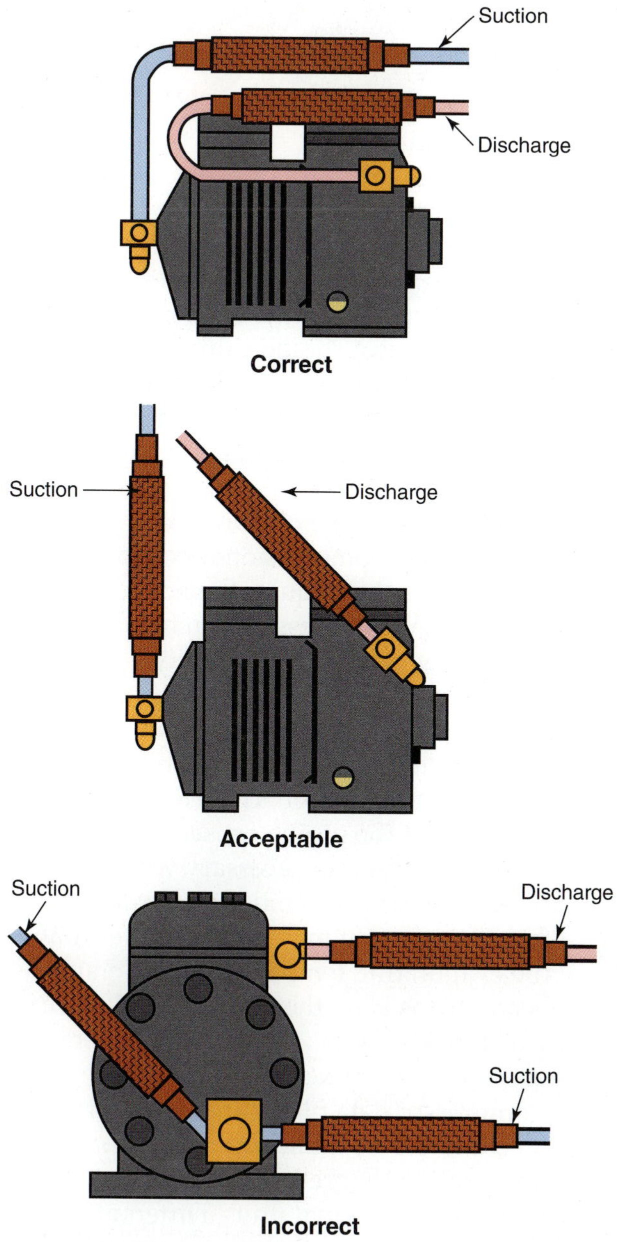

Goodheart-Willcox Publisher

Figure 21-25. In a correct installation, the vibration eliminators are installed parallel with the compressor crankshaft. Vibration eliminators should *not* be installed at a right angle to the crankshaft (as they are in the incorrect example).

the lubrication of the crankcase, resulting in flooded starts. An oil- and liquid-refrigerant mixture can be pulled into the cylinders, a condition called ***slugging***, which can cause severe damage to the discharge valve reeds in reciprocating compressors.

Slugging is prevented by eliminating liquid refrigerant in the crankcase. Elimination may require the use of a crankcase heater, suction accumulator, or pumpdown cycle.

The crankcase heater in some systems is energized continuously (never shut off). Other systems turn off the crankcase heater during the run cycle. An oversized heater may overheat the oil, while an undersized heater will be ineffective. Consult the compressor manufacturer or local dealer to determine the correct crankcase heater type and size.

PRO TIP

Crankcase Heaters for a Valid Warranty

Some heat pumps and air conditioners *require* a crankcase heater to maintain the compressor warranty. Always check manufacturer literature for such information.

21.10 Compressor Data Plates

The ***data plate*** mounted on each compressor contains coded information that reveals construction details of the unit. Knowing how to interpret the information (letters and numbers) can be useful to the technician. The data plate always contains the model number and serial number of the unit. These numbers *must* be used when ordering parts or obtaining an exact replacement. With these two numbers, the supplier or manufacturer has access to precise information concerning every detail of the unit. The data plate often includes useful electrical information pertaining to the motor compressor as well. Different manufacturers have unique ways of displaying unit information on the data plate.

PRO TIP

Data Plates and a Valid Warranty

The data plate is normally spot-welded or riveted to the unit. The manufacturer will usually *void* (not honor) the warranty if the data plate has been removed. Also, compressors that have the data plates removed cannot be returned for core credit.

Chapter Review

Summary

21.1 Refrigeration lubricant must not contain moisture or other contaminants. Always store lubricant in sealed containers.

21.1 Good refrigeration oil must have the correct viscosity for its application, stability, and good miscibility (ability to mix with refrigerant). Additives are added to refrigeration oil to improve performance.

21.1 Mineral oil (MO) is a naphthenic lubricant made from refined crude oil. MO lubricants are not compatible with newer refrigerants. Alkylbenzene (AB) and polyol ester (POE) oils are popular synthetic lubricants used with newer refrigerants. POE oils are compatible with all refrigerants and miscible with MO and AB lubricants.

21.2 In a splash system, the crankcase is filled with enough oil so that as the crankshaft turns, it can dip into the oil and splash it around inside the compressor. In a forced-feed system, an oil pump is used to deliver oil under pressure to all bearing surfaces and other critical parts. An oil pressure safety control is a device that can used on all systems to protect the compressor against loss of oil pressure.

21.3 In the open system method, semihermetic compressors are equipped with a removable crankcase plug for adding oil. In the oil pump method, a small oil pump is used to manually add oil to an operating compressor through the manifold gauge and suction service valve (SSV). In the closed system method, oil is drawn into the compressor through the SSV using the manifold gauge.

21.4 Excess oil is drained using either the drain plug or an oil fill hole.

21.5 Acid tests are performed to ensure refrigeration oil is not contaminated with acid.

21.6 Compressor capacity is determined by four factors: diameter of the piston, number of pistons, length of stroke, and the speed (strokes per minute). Refrigerant density and the size of the clearance pocket also affect compressor capacity.

21.7 The volumetric efficiency of a compressor is a comparison of the amount of gas actually pumped by the compressor to the amount of gas it should pump, according to piston displacement and length of stroke. High head pressure results in decreased volumetric efficiency. Higher compression ratios can cause compressor inefficiency, however, two-stage compressors can provide high compression ratios for applications that require them. Unloaders and tandem compressors can help with controlling compressor capacity.

21.8 Air and refrigerant are two common mediums used to cool compressors.

21.9 Vibration eliminators prevent transmission of noise and vibration to refrigerant piping. Vibration of the piping causes leaks. A crankcase heater is necessary when the compressor is surrounded by an ambient temperature that may become lower than the evaporating temperature. The crankcase heater helps keep the crankcase oil warm, which prevents condensing of refrigerant vapor in the crankcase.

21.10 Compressor data plates are mounted on compressors and contain coded information that reveal construction details of individual units. Different manufacturers have unique ways of displaying unit information on the data plate.

Know and Understand

1. *True or False?* Refrigeration lubricants are resistant to moisture and contaminants. (21.1)
2. Oil _____ is the ability of oil to maintain good lubrication despite temperature changes. (21.1)
   A. miscibility
   B. viscosity
   C. dielectric strength
   D. stability

3. Refrigeration lubricant that mixes well with refrigerant has good _____. (21.1)
   A. miscibility
   B. viscosity
   C. dielectric strength
   D. stability
4. Mineral oil (MO) is not compatible with _____ refrigerants. (21.1)
   A. CFC
   B. HCFC
   C. HFC
   D. HC
5. Polyol ester oils (POEs) are compatible with _____ refrigerants. (21.1)
   A. HFC
   B. HC
   C. HCFC
   D. all
6. The splash system is most common in _____. (21.2)
   A. low-temperature applications
   B. medium-temperature applications
   C. small compressors of less than 3 hp
   D. large compressors more than 12 hp
7. In a forced-feed system, the oil pump is normally mounted on the _____. (21.2)
   A. suction service port
   B. discharge service port
   C. crankshaft bearing housing end of the compressor
   D. oil pressure safety control
8. The oil pressure safety control operates on the difference in pressure between the _____ pressure and the _____ pressure. (21.2)
   A. crankcase, oil pump outlet
   B. suction line, discharge line
   C. compressor inlet, compressor outlet
   D. net oil, oil pump discharge
9. The closed system method of adding oil to a compressor should only be used on _____. (21.3)
   A. hermetic compressors
   B. reciprocating compressors
   C. compressors that have the suction service valve located on the heads
   D. compressors that have the suction service valve located on the end of the compressor
10. When removing excess oil through the oil fill hole, it involves use of a _____. (21.4)
   A. vacuum pump
   B. drain plug
   C. 1/4″ copper tubing to form a siphon arrangement
   D. special brazed-on service valve
11. *True or False?* When acid testing a system's refrigeration oil, the system must always be off. (21.5)
12. Low-temperature compressors require _____ than medium- or high-temperature applications. (21.6)
   A. lower suction pressure
   B. larger clearance pockets
   C. more capacity
   D. less capacity
13. *True or False?* The smaller the clearance pocket in a reciprocating compressor, the better the volumetric efficiency of the compressor. (21.6)
14. A compressor with a volumetric efficiency of 70% that actually pumps 24 in^3 of vapor per stroke is designed to pump _____ in^3. (21.6)
   A. 17
   B. 34
   C. 29
   D. 40
15. Calculate the compression ratio for a compressor with a discharge pressure of 247 psig and a suction pressure of 75 psig. Round up to the nearest whole number. (21.7)
   A. 2:1
   B. 3:1
   C. 5:1
   D. 8:1
16. *True or False?* If suction pressure remains constant but head pressure increases, the compression ratio would decrease. (21.7)
17. In two-stage compressors, liquid refrigerant is metered into the interstage manifold area by a(n) _____. (21.7)
   A. capillary tube
   B. TEV
   C. desuperheating TEV
   D. solenoid valve

18. A(n) _____ uses a solenoid valve and a piston mounted to a compressor's head to control compressor capacity. (21.7)
    A. EEV
    B. desuperheating TEV
    C. unloader
    D. additional suction service valve
19. Air-cooled compressors must have enough air blowing directly on the _____ to cool the motor. (21.8)
    A. discharge outlet
    B. suction inlet
    C. compressor body
    D. compressor head
20. On small compressors, _____ is used as vibration absorbers. (21.9)
    A. an additional mounting spring
    B. plastic tubing
    C. a small bearing
    D. the soft copper tubing wrapped in coil
21. A(n) _____ warms a compressor in cold ambient temperature to help prevent liquid refrigerant from condensing in or migrating to the compressor during the off cycle. (21.9)
    A. special lubricant
    B. crankcase heater
    C. evaporator
    D. superheating TEV
22. *True or False?* Liquid slugging can cause severe damage to the suction valve reed in reciprocating compressors. (21.9)
23. The data plate on a compressor always contains the model number and the _____ of the unit. (21.10)
    A. cost
    B. serial number
    C. anticipated life cycle
    D. compression ratio

Apply and Analyze

1. List five properties of a good refrigerant lubricant. (21.1)
2. Describe the differences in how splash systems and forced-feed systems provide compressor lubrication. (21.2)
3. What is the net compressor oil pressure if the suction pressure is 60 psig and the discharge oil pump pressure is 75 psig? (21.2)
4. Why is the closed system method of adding oil to a compressor not recommended? (21.3)
5. Explain how to remove excess oil using the drain plug. (21.4)
6. List the factors that determine compressor capacity. (21.6)
7. How does a "thin," low-temperature refrigerant gas affect compressor capacity differently from a denser, high-temperature gas? (21.6)
8. Describe how a two-stage compressor performs refrigerant compression. (21.7)
9. Where should vibration absorbers be installed to best provide vibration protection to a compressor and its components? (21.9)

Critical Thinking

1. A technician is troubleshooting an air conditioner with a hermetically sealed compressor. How would they check the oil pressure of the compressor? (21.2)
2. A technician checks the oil pressure on a semihermetic compressor by using the suction gas pressure entering the compressor and the crankcase pressure. Why will this method *not* work on a two-stage compressor? (21.7)

narai chal/Shutterstock.com

22 What Is Electricity?

LEARNING OUTCOMES

After studying this chapter, you will be able to:

22.1 Define various electrical terms and how they apply to the flow of electricity.

22.2 Describe the application of Ohm's Law in electrical circuits, including how to measure resistance and continuity using electrical meters.

22.3 Describe how to calculate and measure electrical power.

22.4 Identify characteristics of conductors and how to apply them to electrical applications, including wiring loads and switches.

22.5 Compare the flow of electrons in direct current and alternating current.

22.6 Explain the arrangement of conductors in three-phase power generation.

22.7 Explain the importance of the equipment grounding conductor in an electrical system and identify several examples of these conductors.

22.8 Identify various types of receptacles, switches, and disconnects.

TECHNICAL TERMS

- alternating current (ac)
- ammeter
- ampacity
- ampere
- atom
- busbar
- circuit
- conductance
- conductor
- contacts
- continuity
- coulomb
- counter-electromotive force (c-emf)
- current
- direct current (dc)
- disconnects
- double-pole, double-throw (DPDT) switch
- double-pole, single-throw (DPST) switch
- electrical load
- electrical power
- electricity
- electromotive force (emf)
- electron
- frequency
- fuse
- grounded
- ground fault
- insulator
- intensity
- line voltage
- lockout
- multimeter
- National Electrical Code (NEC)
- neutron
- normally closed (NC)
- normally open (NO)
- nucleus
- ohm
- ohmmeter
- Ohm's Law
- open circuit
- parallel
- phase
- pole
- potential difference
- proton
- reactance
- receptacle
- resistance
- resistive
- safety ground
- schematic
- series
- series-parallel
- short circuit
- single-phase load
- single-pole, double-throw (SPDT) switch
- single-pole, single-throw (SPST) switch
- stranded
- switch
- tagout
- three-phase power generation
- three-phase load
- throw
- valence electron
- volt
- voltage
- voltage drop
- voltmeter
- watt
- watt-hour meter
- wattmeter
- wire
- zero potential

Introduction

It is estimated that upwards 80% of HVACR service calls are related to an electrical problem. Understanding how electricity controls and operates heating and cooling equipment is essential for troubleshooting system problems and performing routine repairs. This chapter covers the fundamentals of electrical energy and how it relates to HVACR systems and equipment. An understanding of the relationship between Ohm's Law and Watts Law, and how voltage, current, resistance, and wattage are utilized are essential knowledge for troubleshooting systems.

22.1 Electricity Fundamentals

Electricity is a form of energy, just like light, heat, motors (mechanical), and batteries (chemical) are all forms or sources of energy. As was discussed in Chapter 11, *Basic Thermodynamic Principles*, the First Law of Thermodynamics states that energy cannot be destroyed, but it can be converted from one form to another. For example, heat can be used to produce electricity, and electricity can be used to produce light. The reverse is also possible. Electrical appliances and devices *convert electrical energy to another form of energy* to perform useful tasks.

What Is Electrical Energy?

Electricity is a form of energy in which electrons move from one atom to another atom. All matter is composed of ***atoms***, the smallest particle of any element. Although atoms are tiny, they are a vast source of potential energy. The center of an atom is called the ***nucleus***. The nucleus contains protons and neutrons. ***Protons*** are atomic particles that have a positive charge (+), and ***neutrons*** are atomic particles that have no charge (neutral). Orbiting around the nucleus are ***electrons*** carrying a negative charge (–). See **Figure 22-1**.

All electrons possess the same *negative charge*, while all protons possess an equal *positive charge*. Electrons orbit around the nucleus much like the planets orbit around the sun. Electrons travel in pathways called *shells*, and an atom can have as many as seven shells. Electrons traveling in shells located close to the nucleus are more tightly bound to the nucleus than those in the outer shells. The outer shell, called the *valence shell* or *ring*, can hold no more than eight electrons. Electrons located in the outer shell are called free, orbital, planetary, or ***valence electrons***. See **Figure 22-2**.

Electron Movement

The number of electrons in an atom's valence shell is meaningful in the study of electricity. Electrons orbiting close to the nucleus contain less energy than those orbiting farther away. The farther an electron orbits from the nucleus, the greater the electron's energy. If enough energy is added to an electron, it will move out of its orbit to the next outer shell. If enough energy is added to a valence electron, it will move out of its atom to another atom because no higher shell exists.

The energy released by the movement of free electrons from one atom to another is called *electricity*. Since valence electrons have the highest energy level, they are most easily set free. The amount of energy supplied to the valence shell is limited and is distributed equally among the electrons orbiting in that shell. The more valence electrons there are, the less energy each electron receives. A substance made up of atoms with 3 electrons or less in its valence shell is called a *conductor*. A substance made up of atoms with 5 to 8 electrons in its valence shell is called an *insulator*. Conductors and insulators will be discussed in more detail later in this chapter.

Atoms try to maintain an equal number of positive and negative charges (protons and electrons), **Figure 22-3**. The Law of Electric Charges states: *Like charges repel, and opposites attract.* When a substance with excess electrons contacts a substance lacking electrons, the excess electrons are attracted and flow to the atoms lacking electrons. To make electricity perform useful work, a constant and steady movement of electrons must be produced.

In 1798, Alessandro Volta, an Italian physicist, discovered and named the principle of *electromotive force (emf)*. Mathematicians further abbreviate emf with the letter *E*. However, most people refer to electromotive force as *volts*. Electromotive force

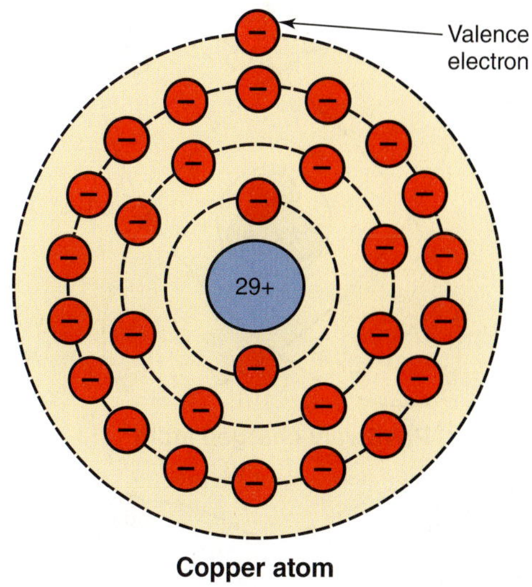

Goodheart-Willcox Publisher

Figure 22-2. Copper is a good conductor of electricity because it has one valence electron. All the energy in the valence ring is contained in this one electron.

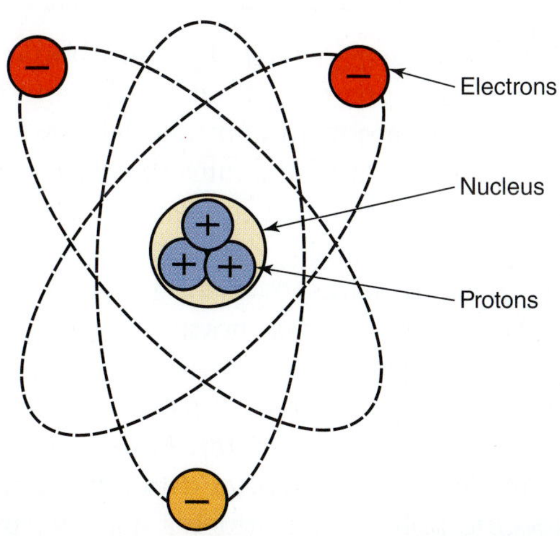

Goodheart-Willcox Publisher

Figure 22-1. Negatively charged electrons orbit around the nucleus of an atom. The nucleus contains neutrons (which have no electrical charge) and positively charged protons.

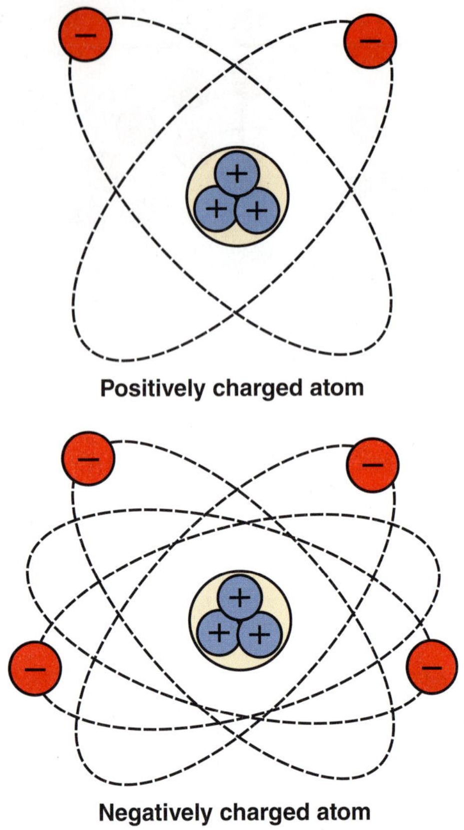

Goodheart-Willcox Publisher

Figure 22-3. The atom on top has three protons in its nucleus but only two electrons, giving it three positive charges and two negative charges. This makes the atom positively charged overall. With three positively charged protons and four negatively charged electrons, the atom on the bottom has an overall negative charge.

is *not* electricity. It is the driving force that causes electrons to move from one point to the other. To obtain electron flow (current), a potential difference must exist between two points (discussed in the next section). The flow of electrical current is much like the flow of water through pipes. The pipes serve as a "circuit" (pathway) for water traveling to a lower pressure. The force (voltage) is comparable to water pressure.

Potential Difference

When the electrical charge (positive versus negative) between two points is *not* in balance, the imbalance of electrons is called a ***potential difference***. With a potential difference, excess electrons have accumulated and are waiting for an opportunity to connect with atoms lacking electrons, **Figure 22-4**. The unit of measurement for potential difference is the ***volt***. Potential difference, also referred to as ***voltage***, is another way of describing the strength of the electromotive force

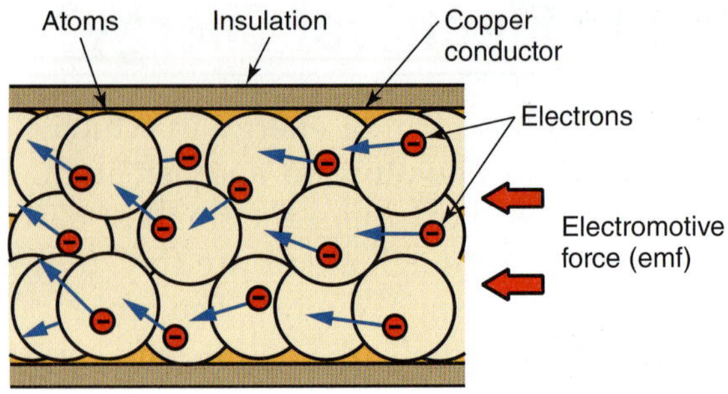

Goodheart-Willcox Publisher

Figure 22-4. Electrons already exist inside copper atoms. Electromotive force (voltage) causes the electrons to travel to the next atom, creating a domino effect throughout the length of a wire.

(emf) that causes electrons to move. To *perform work* (carry out a useful function), a potential difference must be maintained while electrons are moving.

A variety of energy sources can be harnessed to create a potential difference (electromotive force) between two points. Some of these sources are friction (static electricity), chemical energy (battery), thermoelectric energy (heat), photoelectric energy (light), and magnetic energy (generator).

PRO TIP

Electrical Terminology

The terms *potential difference*, *electromotive force (emf)*, *voltage*, and *electrical pressure* are often used interchangeably.

Electromotive Force

Electromotive force (emf) is the stimulus that causes free electrons to reel out of orbit and travel to an adjoining atom. In commercial power production, a power plant generates the electromotive force that causes electron movement. The power plant produces enough force to maintain high voltage through wires over long distances. The high voltage is reduced to normal levels before it reaches consumers.

The quantity of emf (volts) is measured with an instrument called a ***voltmeter***, **Figure 22-5**. Voltmeters measure the potential difference between two specific points. A voltmeter has two probes and detects a difference of potential between the probes. Some voltmeters provide accurate readings, while others make approximate readings. Voltmeters are used to check electrical circuits for

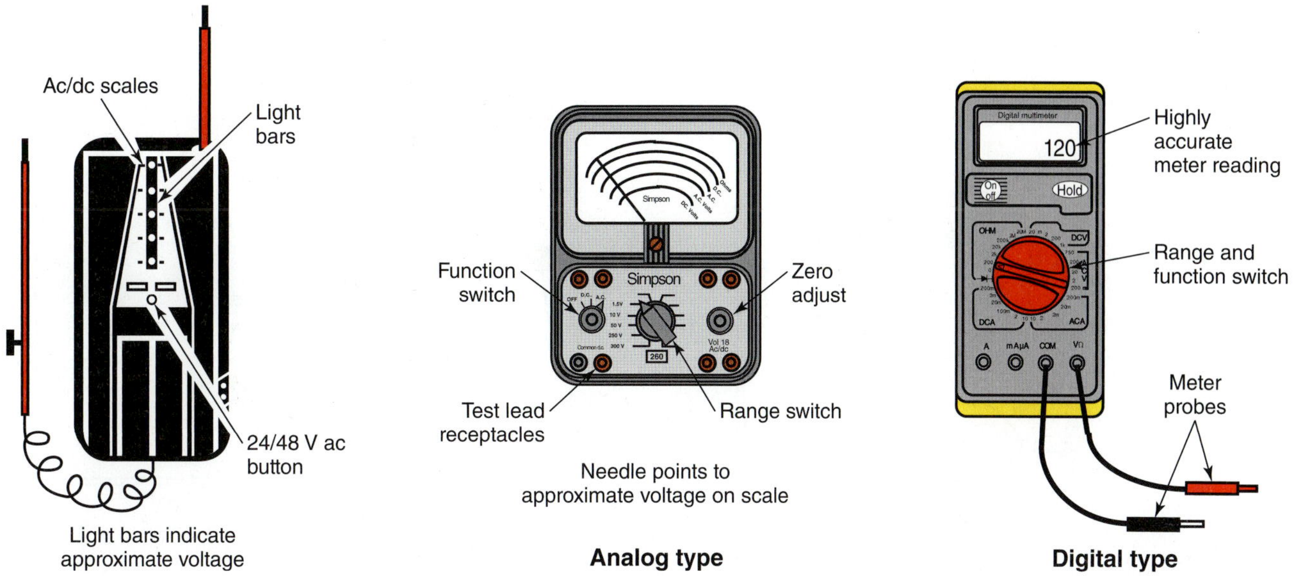

Goodheart-Willcox Publisher

Figure 22-5. Voltage testers are available in a variety of types, shapes, and capabilities. All voltage testers have two probes and a meter that indicates the potential difference (volts) between the two probes. A zero reading means there is no potential difference or voltage between the probes.

problems or to verify proper operation. To successfully troubleshoot electrical problems, the probes must be properly placed, and the meter readings correctly interpreted.

Electrical energy conversion devices (called *loads*) are designed for connection between a specific potential difference. Proper voltage must be supplied to force electron movement through a load. The load converts the electrical energy to another form of energy. The connecting wires supply the necessary electrons and provide a complete circuit, **Figure 22-6**.

Voltage is the force that causes electrons to move. Electrons cannot move unless they have a place to go. They cannot flow through an open switch or a broken wire. An opening in a circuit pathway acts like an open drawbridge. Traffic can flow through a closed drawbridge, but traffic stops when the drawbridge is open. Switches in electrical circuits act as "drawbridges" to start and stop the flow of electrons.

A potential difference between the two probes must exist for a voltmeter to register a reading. The voltmeter reads zero when voltage and polarity are the same at both probe locations. A voltmeter can also be used to check fuses and switches. **Figure 22-7** illustrates how a voltmeter can be used to discover an open switch or a blown fuse.

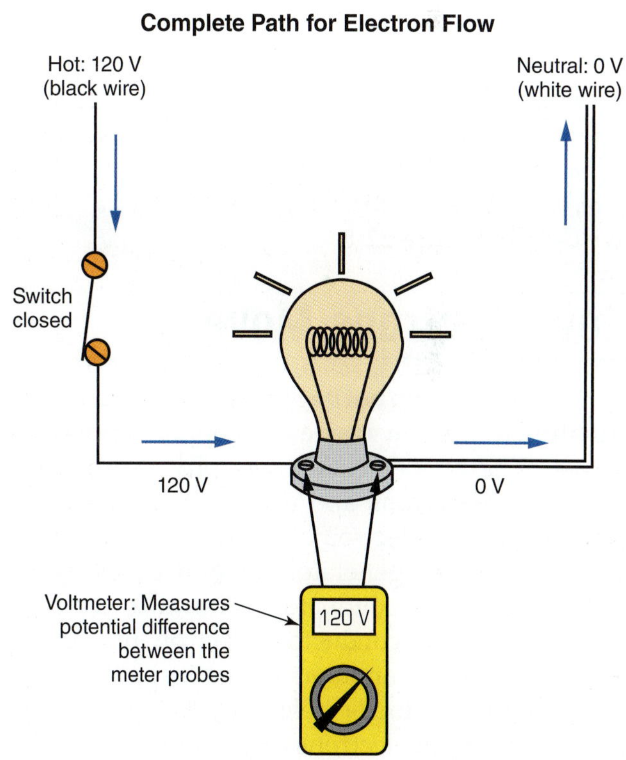

Goodheart-Willcox Publisher

Figure 22-6. An electrical load (such as a light bulb) must be part of a complete circuit (pathway) for electron flow. Voltage (electrical pressure) pushes the electrons through the filament (resistance) inside the light bulb. A potential difference must exist across the bulb for electrons to flow.

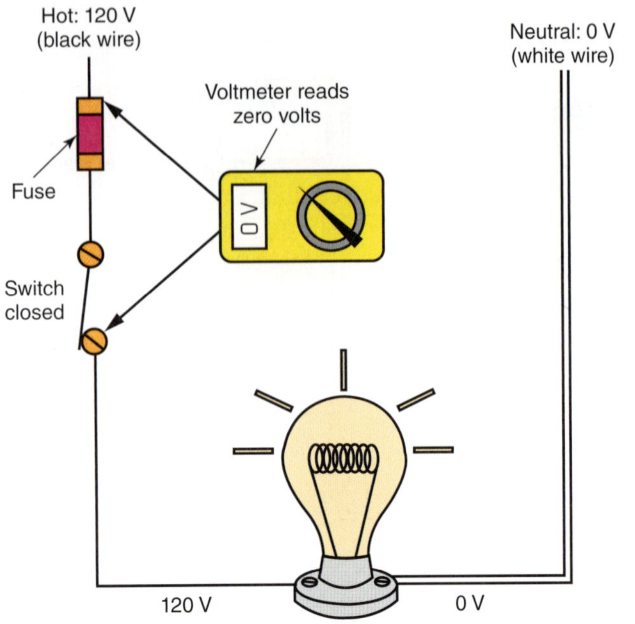

Goodheart-Willcox Publisher

Figure 22-7. The voltmeter reads zero when the potential is the same at both probes. No potential difference exists across a closed switch or an intact fuse, because these provide no resistance and are not electrical loads.

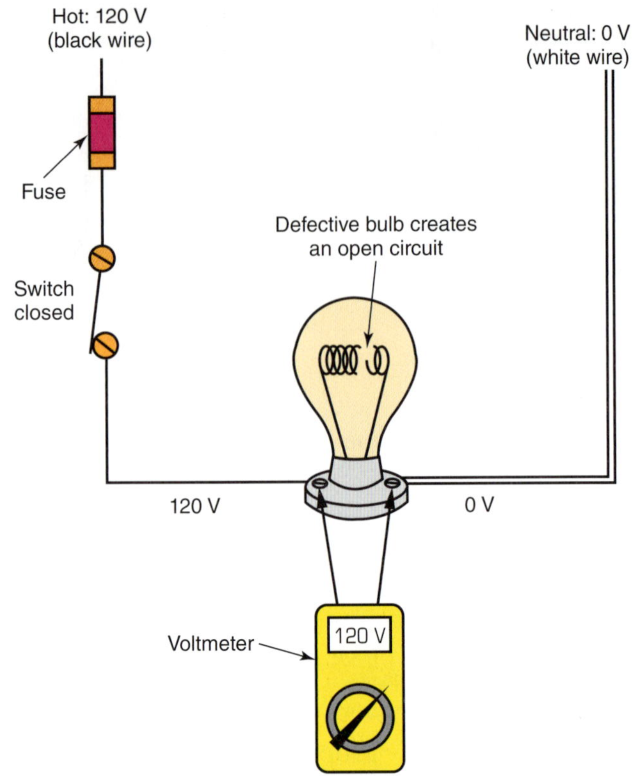

Goodheart-Willcox Publisher

Figure 22-8. A defective bulb causes an opening in the complete path between points of potential difference. Electrons cannot flow in an open circuit because a complete circuit, or pathway, does not exist.

When a load is connected to both sides of a potential difference, the load should operate. If the device (such as a light bulb) does not operate, it is defective. A voltmeter can quickly detect a defective device, **Figure 22-8**.

How Electrons Move

Power plants do not *produce* electrons. Electrons are already present inside the copper conductors (***wires***). The copper conductors provide the necessary free electrons and a pathway for electron movement. The electromotive force produced by the power plant generator forces free electrons inside the copper conductor to travel to the next atom within the conductor. Electron movement (from one atom to another) occurs throughout the length of the conductor (refer back to **Figure 22-4**). The movement of electrons in a conductor is rapid, creating a domino effect. An electrical impulse travels through a conductor at the speed of light (186,000 miles per second). It is this electron movement that produces electricity.

To illustrate the rapid movement from one atom to another, picture a long plastic tube filled with ping pong balls reaching more than 800 miles from New York City to Chicago. See **Figure 22-9**.

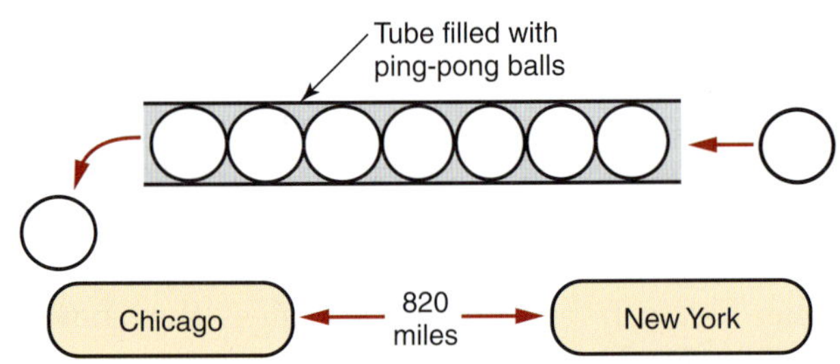

Goodheart-Willcox Publisher

Figure 22-9. Ping pong balls illustrate how electrical impulse travels at high speed over long distances. If a ball is inserted at one end of the tube, another ball immediately emerges from the other end.

If someone in New York City inserts another ball, a ball *immediately* pops out in Chicago. Each ball only moved a short distance, but the *impulse* of bumping each ball traveled at extremely high speed. If additional balls are rapidly inserted at New York City and permitted to fall out in Chicago, it would be an example of *direct current*. Electron flow would be in one direction. If additional balls were alternately inserted at each end, it would be an example of *alternating current*: movement would alternate first in one direction, then in the other direction. Direct current and alternating current are discussed in more detail later in this chapter.

Amperage

In 1836, a French physicist named Andre Ampere measured the number of electrons flowing past a given point in one second. The unit of measurement was called a *coulomb* of electricity. A ***coulomb*** equals 6.25×10^{18}, or 6,250,000,000,000,000,000 electrons flowing past a given point in one second. The term *coulomb* is now used only in textbooks as a measure of electron flow. The unit ***ampere***, named for the scientist, is the term most often used to describe the intensity of electron flow (current).

The following list gives fractional values for ampere:

Units of Current

1 ampere (A) = 1 coulomb/sec

1 milliampere (mA) = 1/1000 A

1 microampere (μA) = 1/1,000,000 A

1 ampere = 1000 mA = 1,000,000 μA

PRO TIP

Electrical Terminology

The words *ampere*, *amperage*, *amps*, *current*, and *flow* are all used to describe the movement of electrons. However, it should be noted that current and flow refer to the movement of electrons, while amps, amperage, and ampere refer to measurements of that flow. Current and flow are like the forward motion of a car, and amps are like the measure of the car's speed in miles per hour (mph).

Because of the vast amount of energy generated by electrons, their movement must be controlled to produce the desired amount of energy conversion. Toasters, light bulbs, and motors are examples of electrical loads. An ***ammeter*** measures the amount of electron flow. The clamp-on ammeter is common and is available with a needle pointer and scale (analog type) or a digital readout. See **Figure 22-10**.

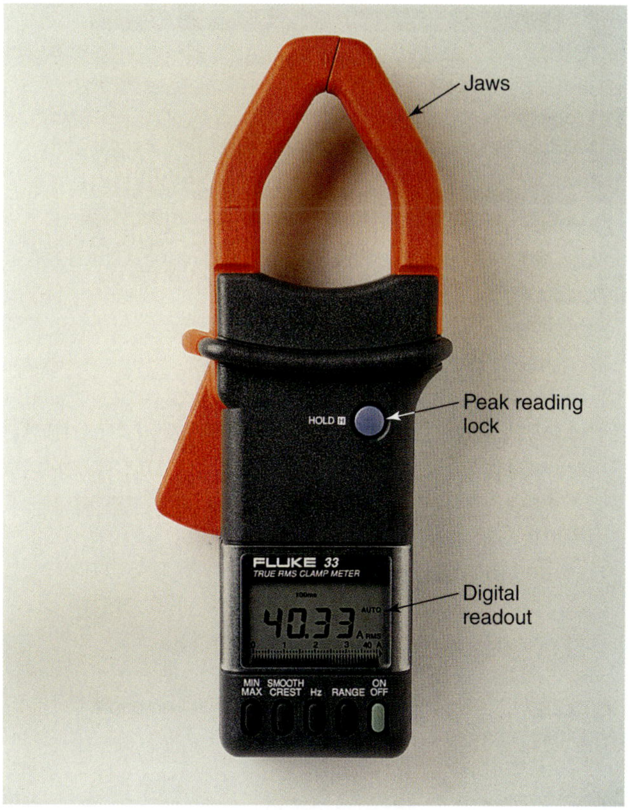

Fluke

Figure 22-10. A digital clamp-on ammeter.

When electrons flow through a conductor, a magnetic field is produced around the conductor. The strength of the magnetic field is determined by the intensity of electron flow. High current flow produces a strong magnetic field, while low current flow produces a weak magnetic field. When electrons are not flowing, no magnetic field exists around the conductor. See **Figure 22-11**.

When the jaws of an ammeter are clamped around a single conductor, the meter measures the ***intensity*** (strength) of the magnetic field. The ammeter converts the information to an amperage reading. See **Figure 22-12**. Because the ammeter measures intensity, the letter *I* is used in mathematical calculations to indicate intensity of electron flow.

The ammeter is designed to read the magnetic field around *one* wire. Do *not* clamp the jaws of the meter around two conductors at the same time. The two wires will be of opposite polarity because the currents are flowing in opposite directions. The two magnetic fields will cancel one another and give a false meter reading of zero.

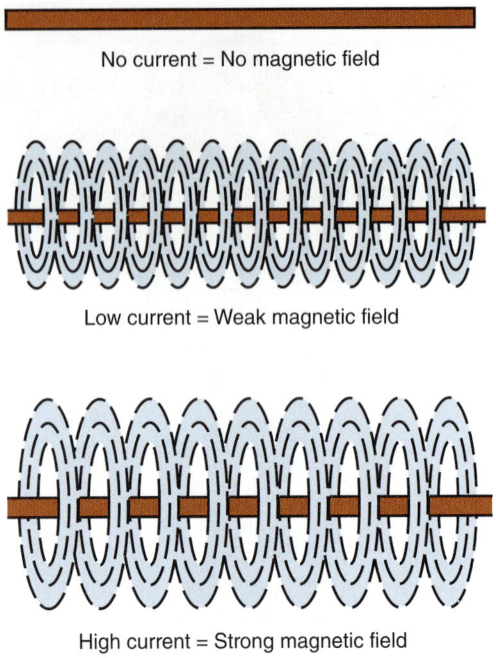

Goodheart-Willcox Publisher

Figure 22-11. The strength of a magnetic field around a conductor is determined by current flow.

Fluke

Figure 22-12. When the ammeter's jaws are clamped around a conductor, the meter reads the intensity of the magnetic field. The meter converts the information to an amperage reading.

PRO TIP

Taking Low Current Readings with an Ammeter

If a wire has too low of a current flow, then it will not be read by the ammeter. To overcome this problem, the jaws of the ammeter can be wrapped in wire to obtain a reading. When doing this, you must divide the reading by the number of wraps. For example, an ammeter that reads 20 A but has 10 wraps of wire around it would actually be 2 A (20 ÷ 10 = 2).

The clamp-on ammeter is a valuable tool for troubleshooting circuits. A multimeter, capable of reading volts, amperes, and resistance on several scales, is often used as well. Adapters and probes are included for easy conversion to another type of reading, and complete instructions are included with each ammeter.

Resistance

Resistance refers to anything opposing the flow of electrons. All materials offer some resistance to the flow of electrons, some more than others. Resistance converts electrical energy to another form of energy.

Resistance performs the energy conversion and controls the amount of electron flow. The amount of electron flow determines the rate of work performed by an electrical load. A load cannot function properly if the electron flow (amperage) is too high or too low.

Types of Resistance

There are two types of resistance: resistive and reactance. A resistance that is ***resistive*** remains constant (does not change). Examples of devices with fixed resistance are toasters and electric heaters. Each has a fixed (constant) resistance to controlled electron flow during operation.

With a fixed resistance, higher voltage increases current flow and lower voltage decreases current flow. Likewise, if resistance is increased, current flow decreases. If resistance is decreased, current flow increases.

Reactance is a form of resistance that resists a change in current or voltage. There are two forms of reactance depending on what is being resisted: inductive reactance and capacitive reactance.

With inductive reactance, a coil of wire produces a magnetic field that creates a voltage acting in direct opposition to the supply voltage. This *counter-voltage* causes an increased resistance that decreases current flow. Counter-voltage or back electromotive force (back-emf) is known as ***counter-electromotive force (c-emf)***. Motors, transformers (primary side), and solenoid coils are some devices that produce magnetic fields. The amount of counter-emf generated is determined by the strength of the magnetic fields. The c-emf acts as additional resistance, created whenever a device is operating. Counter-electromotive force is farther explained in Chapter 24, *Motors.*

With capacitive reactance, a capacitor is used to store voltage which can create a magnetic field that resists current flow. A capacitor is composed of two conductors with an insulating layer between them. Capacitors are also discussed in more detail in Chapter 24, *Motors.*

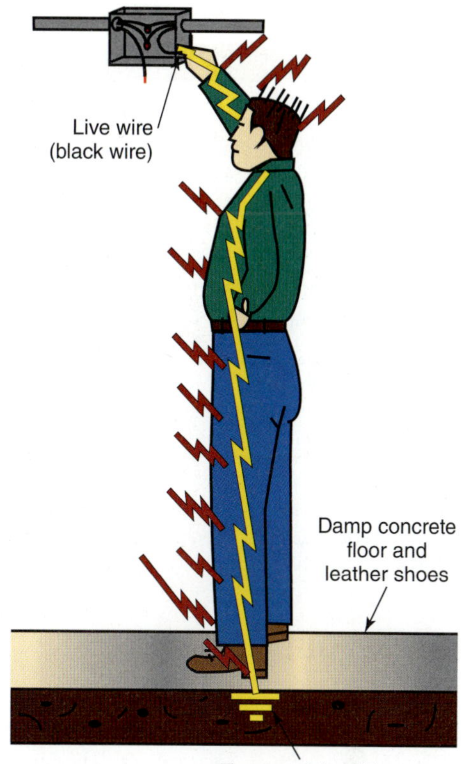

Figure 22-13. Always exercise care to avoid becoming part of an electrical circuit. Your body could provide a path for electron flow with painful (or even fatal) results.

Electrical Safety

Electrical ***circuits*** are composed of conductors (wires) and devices that control ***current***, which is the flow of electrons. Electrons should travel within the electrical circuit. Electrons on any other pathway are dangerous. Your body is a lower resistance for electrons to flow and could serve as a pathway for current to flow through your body, and to the ground. Never permit your body to become part of an electrical circuit! *Electricity is energy looking for someplace to go*, and if your body provides a pathway for a current to travel, it will indeed travel through your body.

You have heard it said that "electricity travels to the ground." The earth (ground) is an excellent conductor of electricity. The human body can easily complete an electrical pathway to the ground, becoming an electrical load if you make contact with an electrically live (hot) wire at the same time you make firm contact with the earth. Damp concrete is a good conductor to the ground, especially if the concrete is reinforced with steel rods. Wet shoes make a good connection to the ground, **Figure 22-13**.

Approximately 1000 people are killed by electricity every year. About one-half of these electrocutions occur on low voltage. Know that ordinary household current can kill in a fraction of a second. These deaths usually occur because people take chances.

An electrical shock may not be fatal, but it can cause you to fall off a ladder or scaffold and injure yourself. Never become overconfident or take unnecessary risks. Your life may depend on it.

Three factors determine whether or not an electrical shock is fatal:

- How much current (amperage) passes through your body.
- How long (duration) the current passes through your body.
- Whether the current passes through your heart.

A milliampere (mA) is 1/1000 of an ampere. About 100–200 mA of current will kill a person. The table in **Figure 22-14** shows the effects of various current values on the human body.

Effects of Current Values on the Human Body

Current Value	Effect
1 mA	Causes no sensation, nothing felt.
1–8 mA	Shock sensation, but not painful. No loss of muscle control.
8–15 mA	Painful shock. No loss of muscle control.
15–20 mA	Painful shock. Hand muscles contract. Cannot let go.
20–75 mA	Painful shock. Severe muscle contractions. Breathing is extremely difficult.
100–200 mA	Painful shock and *ventricular fibrillation* of the heart. This is like a pacemaker telling the heart to beat 3600 times per minute! Ventricular fibrillation is a fatal heart condition for which there is no known remedy or resuscitation. It usually means *death* because sophisticated equipment is required to quickly restore normal heartbeat.
Over 200 mA	Severe burns. Muscular contractions are so severe that chest muscles clamp the heart and stop it for the duration of the shock, preventing ventricular fibrillation. Artificial respiration should be administered immediately. In most cases, the victim can be revived if removed from the circuit quickly.

Goodheart-Willcox Publisher

Figure 22-14. Effects of current on the human body. Current values in the 1–8 mA range are considered safe, but keep in mind the three factors of a fatal electrical shock: how much current passes (amperage), how long the current passes (duration), and where the current passes (through the heart).

Lockout/Tagout

To prevent accidents during system maintenance or repair, lockout and tagout relevant circuits. According to OSHA, ***lockout*** is "the placement of a lockout device on an energy isolating device, in accordance with an established procedure, ensuring that the energy isolating device and the equipment being controlled cannot be operated until the lockout device is removed." This often involves placing a main switch in the "open" position and securing it with a key or combination lock.

According to OSHA, ***tagout*** is "the placement of a tagout device on an energy isolating device, in accordance with an established procedure, to indicate that the energy isolating device and the equipment being controlled may not be operated until the tagout device is removed." This often involves attaching a warning sign or tag on a main switch that informs others that repairs are in process, **Figure 22-15**. Together, a lockout/tagout (LOTO) procedure lets others know that this circuit may not be operated until the lockout/tagout devices are removed.

Goodheart-Willcox Publisher

Figure 22-15. Lockout/tagout (LOTO) procedures prevent the circuit or equipment from being energized while someone is working on it. Failure to follow tagout procedures could have serious, even fatal, results.

22.2 Ohm's Law

The relationship between voltage (E), amperage (I), and resistance (R) was discovered by the German scientist, George Ohm, at the beginning of the 19th century. The relationship, called ***Ohm's Law***, is used to predict and control the activity of electricity. The unit of resistance is called the ***ohm***.

Ohm's Law proves that $E = I \times R$ (voltage equals amperage times resistance). This mathematical formula can be depicted as a pie cut into three pieces, **Figure 22-16**. The horizontal line cut across the middle indicates *division*, and the vertical line cut from the middle indicates *multiplication*.

To use the pie, cover the item to be determined (the unknown). Then, perform the math operation indicated by the horizontal or vertical line. For

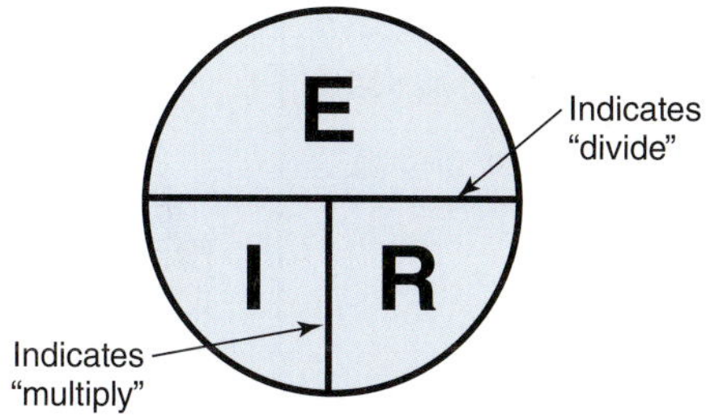

E = Voltage (I × R)
I = Amperage (E ÷ R)
R = Resistance (E ÷ I)

Goodheart-Willcox Publisher

Figure 22-16. The Ohm's Law "pie" makes it easy to determine the unknown value of an electrical circuit when the other two values are known. The horizontal line indicates division. The vertical line indicates multiplication.

example, to discover E, multiply I × R. To discover I, divide R into E. The English letter *R* or the Greek letter Ω (omega) indicate resistance (ohms). Ohm, the basic unit of resistance, is often too small for practical use.

Ohmmeter

An ***ohmmeter*** is an instrument for accurately checking resistance. Ohmmeters are available in analog or digital. However, the digital meters provide the most accurate readings. Ohmmeters measure resistance according to the voltage supplied by a battery inside the meter. Analog meters provide an adjustment knob for recalibrating the meter to compensate for a weak battery. Digital meters have automatic recalibration. See **Figure 22-17**.

SAFETY NOTE

Ohmmeters and Live Circuits

Never use an ohmmeter on a live circuit. It could result in equipment damage and an electrical shock.

Continuity

An ohmmeter can be used to check continuity. ***Continuity*** is a term that describes a complete circuit (path) for electron movement from one point to another. An ohmmeter measurement is taken and low to no resistance is checked. A continuity check can be used to prove that wires are not broken, fuses are not blown, or switches are not open in an electrical circuit (path of flow). See **Figure 22-18**.

Analog multimeter

Simpson Electric Co.

Digital multimeter

Fluke

Figure 22-17. An ohmmeter contains its own power supply (a small battery) and reads resistance to the battery voltage. Most technicians prefer a multimeter (like those shown) that can read resistance, voltage, or amperage. In analog multimeters, resistance values (in ohms) are read from the uppermost scale. In digital multimeters, resistance values are displayed on the screen.

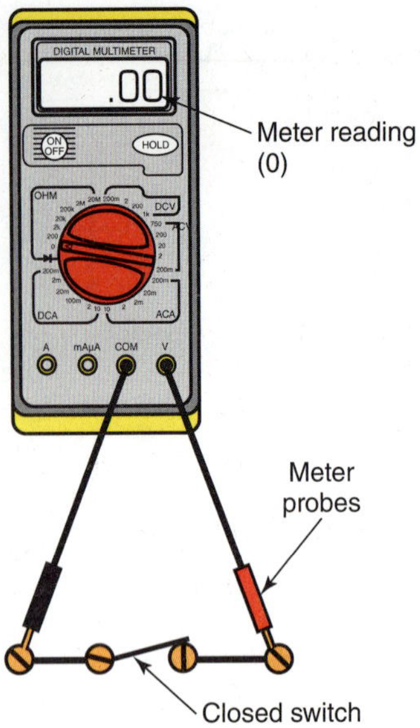

Goodheart-Willcox Publisher

Figure 22-18. A reading of zero (0 Ω) or low resistance proves continuity in a component or circuit. A complete path exists for electron movement. Continuity indicates no broken wires, open switches, or blown fuses.

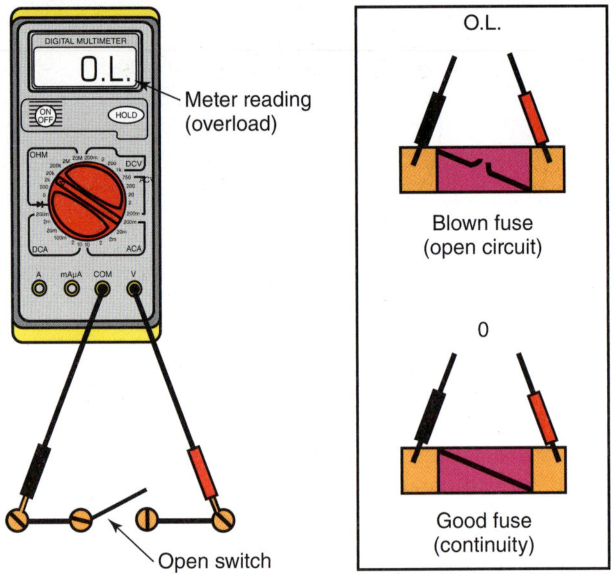

Goodheart-Willcox Publisher

Figure 22-19. Electrons cannot flow in an open circuit (a break in the electrical path). The ohmmeter indicates total resistance to electron flow by showing large numbers (999999), infinity (∞), or overload (OL).

Open Circuit

Electrons cannot flow in an ***open circuit*** (a circuit that is not complete or that contains a break in the path). The ohmmeter indicates an open circuit by registering unlimited resistance, because resistance is too high for the meter to measure it. Unlimited resistance is indicated by the symbol for infinity (∞). A digital meter shows unlimited resistance by displaying the letters OL, or "open line," because the meter cannot read such a large number (infinity).

An open circuit indicates an open switch, blown fuse, broken wire, or disconnection, among other possible failures. A failure *disconnects* the circuit, preventing electrons from flowing across the opening. A drawbridge is a good analogy. When the bridge opens (infinite resistance), traffic (electron flow) stops. The electrons must wait until the drawbridge recloses (zero resistance) before continuing their flow, **Figure 22-19**.

Short Circuit

The ohmmeter detects a ***short circuit*** by indicating zero resistance or a low resistance between two points that should have high resistance. In other words, a complete circuit (continuity) exists where none should, **Figure 22-20**.

A short circuit provides a "short cut" for unlimited electron movement (current flow). Some type of resistance is required to control current flow. However, in a short circuit there is none (zero resistance). Too much current flow causes wires to become hot, fuses to blow, fires to occur, and

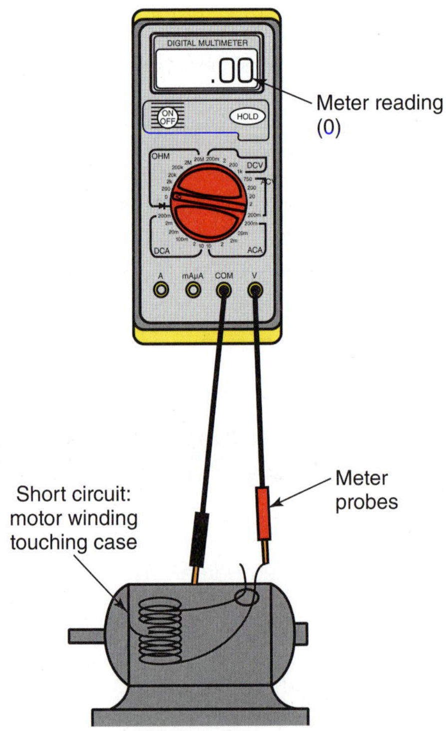

Goodheart-Willcox Publisher

Figure 22-20. If a short circuit exists, an ohmmeter shows no resistance between points where resistance should exist. A short circuit is a dangerous condition.

devices to explode. A short circuit is a dangerous situation because current flow is *unrestricted* and *unlimited.*

Multimeters

Electrical test meters can be combined into one instrument that performs multiple functions. The instrument is called a ***multimeter***. A multimeter is capable of measuring resistance, voltage, and current (amperage). To achieve proper results with multimeters, probes must be carefully placed and the settings correctly chosen. Always consult the manufacturer's instruction manual if you are unsure of proper probe hookup. Mistakes cause false readings and could result in meter damage.

22.3 Power and Watts

The English scientist James Watt (1763–1819) determined how to measure electrical power. ***Electrical power*** is the rate at which electrons are harnessed to perform a useful function. The function (or task) is measured in units called ***watts***. Watts (P for electrical power) are calculated by multiplying amperage (I) times voltage (E), expressed as:

$$P = I \times E$$

Similar to the formula for Ohm's Law, the formula for calculating power can be depicted as a pie cut into three pieces, **Figure 22-21**.

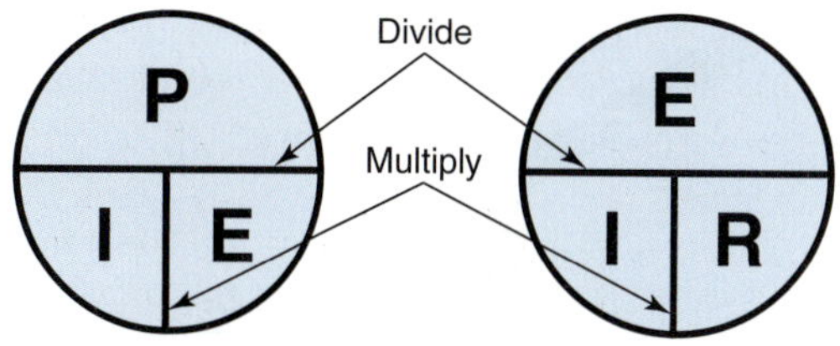

Goodheart-Willcox Publisher

Figure 22-21. Power can be determined by using a pie diagram like the one on the left. It is used in the same way as the Ohm's Law pie on the right. The horizontal line indicates division, and the vertical line indicates multiplication.

Units of Power

The basic unit of electrical power (the watt) is often too small for practical use. Instead, fractional or multiple values are used as follows:

Power = Watts (W)
1 watt (W) = 0.00134 horsepower (hp)
1 horsepower (hp) = 746 W
1 kilowatt (kW) = 1000 W
1 megawatt (MW) = 1,000,000 W

Wattmeter

A ***wattmeter*** measures the power (wattage) used by an appliance or house. The wattmeter is plugged into a power outlet, and the appliance is plugged into the wattmeter. Many electrical devices, such as light bulbs, window air conditioners, and resistance heaters, indicate a wattage number on the data plate or schematic. The wattage number tells the amount of power needed to operate the device. The wattage rating is often converted to terms that are more useful for troubleshooting.

The main purpose of any electric circuit is to carry electricity to electrical loads. The amount of energy (the rate of work) converted by the electrical loads depends on three factors: voltage, amperage, and elapsed time. A ***watt-hour meter*** measures the electrical energy used by a household or business. **Figure 22-22** shows the meter normally located where the electrical power supply wires enter the building.

The watt-hour meter measures how many thousands (kilo) of watts are used per hour (kWh). The watt-hour meter performs several functions at the same time. It measures voltage, amperage, and elapsed time. Then, the meter automatically multiplies the three quantities and records the number of kilowatt-hours used on its circular dials. The utility company sends a representative to read the meter dials every one or two months to record the number of kilowatt-hours used since the last reading. A lot of meters now are digital and send the power usage directly to the utility company.

22.4 Conductors

Electrical wires, also known as ***conductors***, provide the necessary free electrons and serve as the pathway for electron flow. Metals are conductors of electricity, but not all metals conduct

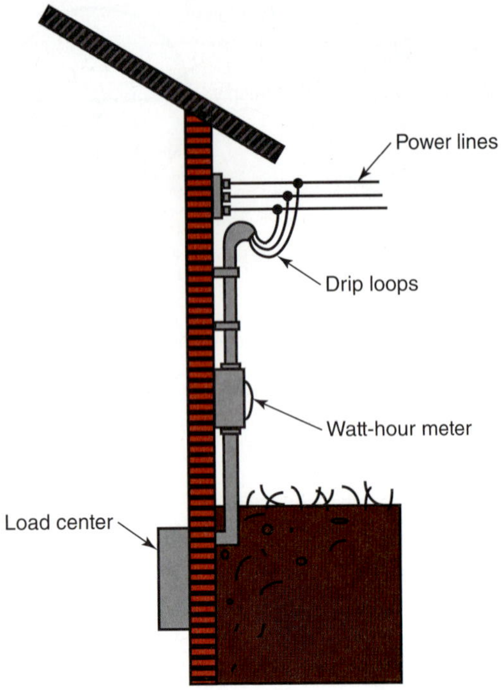

Goodheart-Willcox Publisher

Figure 22-22. The electric meter on a house or apartment building is a watt-hour meter that records a building's power consumption.

with equal ability. All metal conductors have an internal resistance that opposes the flow of electrons. While *resistance* is the opposition to flow, ***conductance*** is the allowance of current to flow.

A conductor that has high internal resistance automatically converts electrical energy to heat energy, so the conductor becomes hot. Internal resistance also causes a drop in voltage. The amount of internal resistance in a conductor depends on the type of metal from which it is made, the wire size and length, and the ambient temperature. Materials such as tungsten alloys, nichrome, and steel have high internal resistance. They heat up quickly, so they are often used for electric heating elements.

To be a good conductor, a material must have high conductance and low internal resistance. **Figure 22-23** lists the conductance and resistance values of different metals. Silver, copper, and aluminum are good conductors because they have high conductance and low resistance. Copper is the most common conductor. Copper wire bends easily, is strong, resists corrosion, and is easily joined. Aluminum is sometimes used because of its low cost. However, aluminum corrodes easily, and aluminum connections have a tendency

Conductance and Resistance of Metals

Metal	Conductance	Resistance
Silver	2207	1.00
Copper	2030	1.09
Gold	1471	1.50
Aluminum	1277	1.73
Tungsten	634	3.48
Brass	461	4.79
Iron	304	7.26
Nickel	263	8.39
Steel	235	9.39
Nichrome	34	64.91

Goodheart-Willcox Publisher

Figure 22-23. Conductance and resistance values of various metals. The higher the conductance value (in siemens) and the lower the resistance number (in ohms), the better the conductor. Note the very high resistance value for nichrome, which is used as a heating element for toasters and portable heaters.

to loosen. Loose connections become hot and can destroy insulation and cause fires. The high cost of silver keeps it from being widely used. Silver is limited to such applications as contacts in switching devices.

Voltage Drop

Voltage drop is a term that describes the loss of emf (voltage) between each of the two wires of the electrical source or across the electrical load. Voltage drop can be caused by an electrical load, a loose connection, or an undersized conductor. All conductors have some resistance to the flow of electrons. The type, size, and length of wire selected for a conductor is used to minimize the voltage drop through the length of the conductor. If the wires are too small, voltage drop within a conductor will cause the conductor to get hot and prevent the load from receiving its designed operating voltage. Any voltage lost between the power supply and the electrical load is voltage unavailable to the equipment.

All loads contain a specific resistance and are designed to operate within a specific voltage range. Most loads are designed to operate properly within ±10% of the rated voltage. Voltage drop should occur across the resistance of the load. Excess voltage drop is easily detected by an accurate multimeter, **Figure 22-24**.

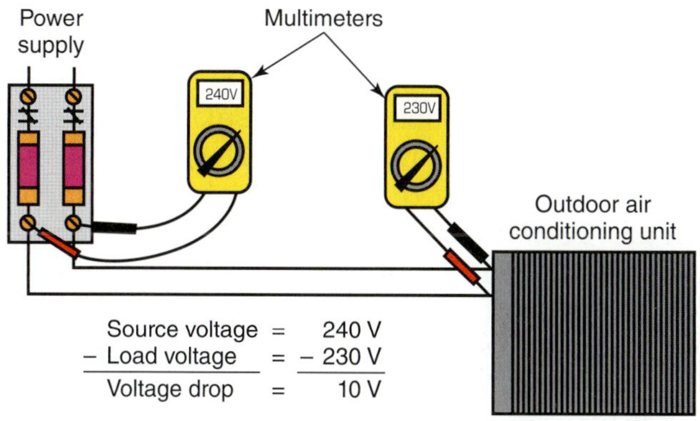

Goodheart-Willcox Publisher

Figure 22-24. Check voltage at the source and at the load. The difference between the two readings indicates the amount of voltage drop.

Wire Sizes

The amount of current that a conductor can safely carry without becoming overheated is limited. Limited current-carrying capacity is called ***ampacity***. The ampacity of a conductor depends on the wire size, length, location, type, and the quality and quantity of its insulation. An electrical conductor can be a single strand of solid wire or a group of wires twisted together. A group of wires twisted together this way are called ***stranded*** wires.

Wire sizes in the United States are established by the American Wire Gauge (AWG) system. The AWG lists wire sizes by number, **Figure 22-25**. The largest wire is #0000 (4/0), and #50 is the smallest. Larger and smaller sizes are available but are not commonly used. The most commonly used wire sizes range from #4 to #18. There is no problem with *oversizing* a wire. However, *undersizing* causes severe problems from voltage drop and overheating. **Figure 22-26** lists wire sizes, resistance, and ampacity for standard sizes of copper wire.

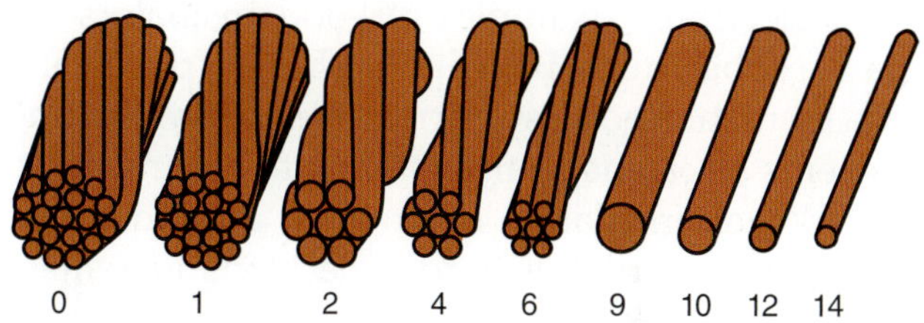

Goodheart-Willcox Publisher

Figure 22-25. Common sizes of solid and stranded wire. Conductors are stranded to improve their flexibility.

Dimensions, Typical Resistances, and Ampacity of Commercial Wire

Gauge No. (AWG)	Diameter Bare Wire (inches)	Ohms per 1000 ft. 70°F	Ohms per 1000 ft. 167°F	Current Capacity (in amperes) of Copper (TW and UF)	Current Capacity (in amperes) of Copper (RH, RHW, THHW, THW, and THWN)	Current Capacity (in amperes) of Aluminum (TW and UF)	Current Capacity (in amperes) of Aluminum (RH, RHW, THHW, THW, and THWN)
0000 (4/0)	0.460	0.050	0.060	195	230	150	180
000 (3/0)	0.410	0.062	0.075	165	200	130	155
00 (2/0)	0.365	0.800	0.095	145	175	115	135
0 (1/0)	0.325	0.100	0.119	125	150	100	120
1	0.289	0.127	0.150	110	130	85	100
2	0.258	0.159	0.190	95	115	75	90
3	0.229	0.202	0.240	85	100	65	75
4	0.204	0.254	0.302	70	85	55	65
6	0.162	0.400	0.480	55	65	40	50
8	0.128	0.645	0.764	40	50	30	40
10	0.102	1.020	1.216	30*	30*	20	25
12	0.081	1.620	1.931	20*	20*	16	18
14	0.064	2.570	3.071	15*	15*	—	—
16	0.051	4.100	4.884	10*	10*	—	—
18	0.040	6.510	7.765	5*	5*	—	—

*Load current rating and overcurrent protection must not exceed these figures.

Goodheart-Willcox Publisher

Figure 22-26. Wire size, resistance, and ampacity for the most common sizes of conductors. The copper and aluminum wire abbreviations in parentheses refer to the wire insulation. For example, "TW" wire is covered with thermoplastic and is water resistant.

When working with any electrical equipment, it is best to consult the ***National Electrical Code (NEC)***. The NEC is a national standard code that contains information on safe installation of electrical wiring and equipment. Always refer to the NEC for complete, accurate, and up-to-date information regarding ampacity and application of various types and sizes of electrical conductors approved for use in HVACR.

Insulation for Wire Coverings

Insulators are materials that make it difficult for electrons to set themselves free. Insulators have stable atoms, and their outer valence shells have eight electrons. Examples of good insulators are glass, rubber, mica, Bakelite™, asbestos, paper, and silk. Insulation acts as a barrier to electron flow and prevents electrons from traveling to an unwanted path. When insulation is wrapped around a conductor, the insulation forces electrons to remain inside the conductor.

All insulators can break down under moisture, heat, excess current flow, or chemical attack. Problems associated with breakdowns can be overcome by increasing the thickness and quality of insulation. Heat-resistant, moisture-resistant, and oil-resistant types of insulation are available. The type of insulation and covering determines where the conductor can be safely used. Always use care to avoid damaging the insulated covering on a wire.

PRO TIP

Wire Grease

When pulling conductors through metal conduit, use a special grease that makes pulling easier and helps prevent damage to insulation on sharp metal edges or turns.

Insulation Markings

The NEC requires all electrical conductors to be properly marked for easy identification. The markings appear on the outside surface of the wire's insulated covering. Markings contain:

- AWG size
- Type of insulation (coded)
- Voltage rating
- Testing agency (for example, UL or CSA)
- Number of conductors (for cable)

The most commonly used insulation types are TW, THHN, THW, and THWN. The more unusual types are reserved for special applications. **Figure 22-27** illustrates how insulated coverings are marked and lists codes and their applications.

Insulation Color

Insulation on conductors is color-coded to identify wiring. The NEC specifies the neutral wire be white or natural gray. The grounding conductor or safety ground wire must be bare copper, green, or green with yellow stripes. The current-carrying conductors (hot wires) are normally black or red, but hot wires can be any color except those designated for the neutral and ground wires.

Electrical Loads

An ***electrical load*** (or simply a *load*) is an energy-conversion device that converts electrical energy to another form of energy. Examples of loads are light bulbs, electric drills, and motors. The applied *voltage and resistance* determine the amount of electron flow (as described by Ohm's Law). *Wattage* measures the rate of energy conversion.

Manufacturers of electrical loads install the correct resistance to perform a particular job. Knowing how to connect these loads to the required voltage is critical. The voltage forces electrons through the resistance at a specific rate. If voltage is too high, the work is performed too fast. If voltage is too low, the work is performed too slowly. The amount of current flowing through the load is determined by the combination of voltage and resistance.

When connecting a load to a voltage source, a minimum of two conductors must be used and be connected from one side of the power source (L_1) to the other (L_2 or neutral), **Figure 22-28**. Electrons must be able to flow into and out of the load. The same amount of current flowing through the black wire also flows through the white wire. An amperage reading can be taken on either wire. The amount of electrons flowing within the circuit (amperage) is determined by Ohm's Law ($E = I \times R$). With a fixed amount of resistance, an increase in voltage results in an increase in amperage. Likewise, a decrease in voltage results in a decrease in amperage.

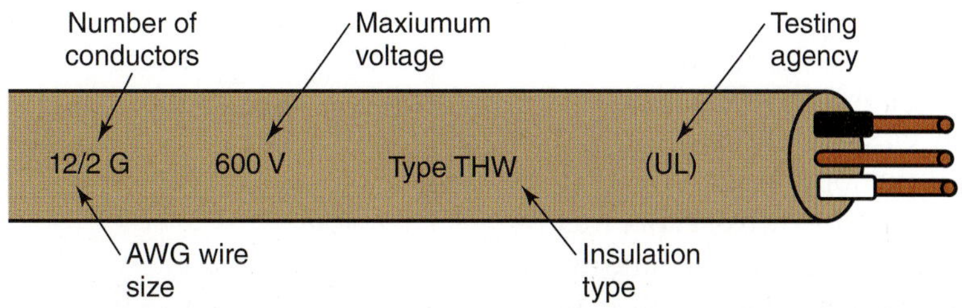

Type of Insulation	Letter Code	Maximum Temperature	Application
Asbestos	A	392°F	Dry locations only
Asbestos and varnished cambric	AVA	230°F	Dry locations only
Rubber	RH	167°F	Dry and wet locations
Heat- and moisture-resistant rubber	RHW	167°F	Dry and wet locations
Thermoplastic	T	140°F	Dry locations
Moisture-resistant thermoplastic	TW	140°F	Dry and wet locations
Heat-resistant thermoplastic	THHN	194°F	Dry locations only
Heat- and moisture-resistant thermoplastic	THW	167°F	Dry and wet locations
Underground feeder	UF	140°F	Moisture resistant
Varnished cambric	V	185°F	Dry locations only

Goodheart-Willcox Publisher

Figure 22-27. Manufacturers of electrical conductors print coded information on the outside of the insulation. The codes describe where the conductor can be safely used. Consult the National Electrical Code for a complete list of insulation types and approved uses.

Switches

Electrical switches should offer no resistance to the flow of electrons. Rather, switches stop and start the flow of electrons to an electrical load. When a switch is open, the pathway is broken, and electrons cannot flow to the load. Remember that when the drawbridge is open, all traffic stops until the drawbridge is closed. See **Figure 22-29**.

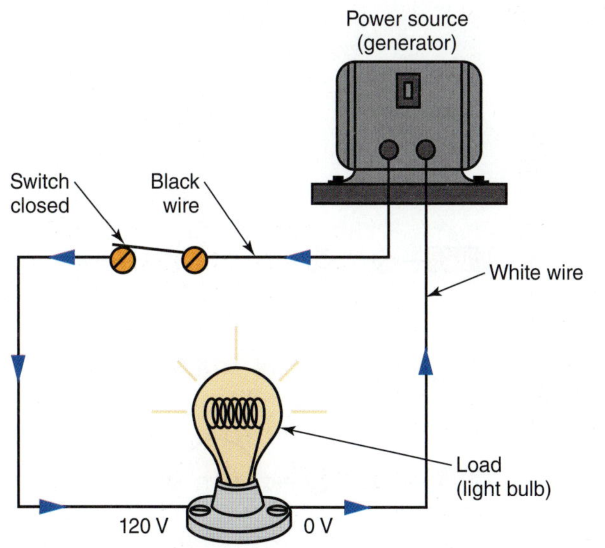

Goodheart-Willcox Publisher

Figure 22-28. A load, such as a light bulb, must be part of a complete circuit to function. The black wire has voltage and carries electrical power from the source to the load. The white wire has no voltage and carries electrical power from the load back to the source. The two wires provide the necessary pathway for electron flow.

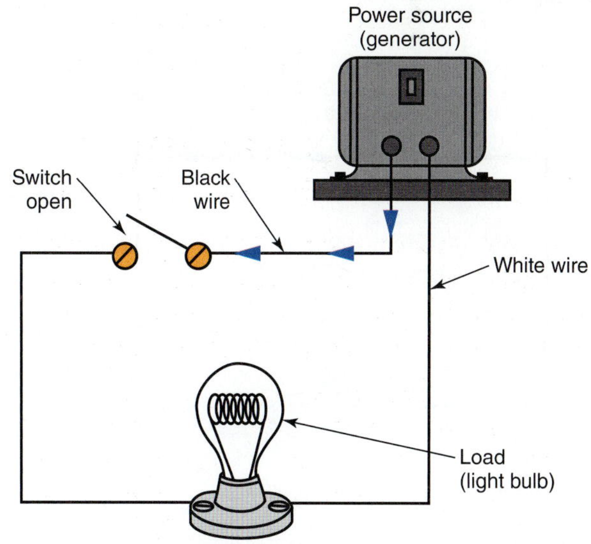

Goodheart-Willcox Publisher

Figure 22-29. A switch functions like a drawbridge to control the flow of electrons. When it is closed, the circuit is complete, and electrons can flow. When it is open, electron flow ceases because there is no longer a complete circuit.

Series Circuit

An electrical load cannot operate unless the circuit is completed for electrons to flow into and out of the load. For proper operation, a switch is connected in series with the load. In ***series*** means one after another, such as electrons having to travel through a switch before they reach a load. Switches are *never* connected from one power leg to the other without an electrical load between each leg. A connection with no load between power legs permits unrestricted flow of electrons and would be a short circuit, **Figure 22-30**. ***Fuses*** are devices installed to protect conductors and other circuit devices against unrestricted current flow. When a current above a fuse's rating flows, the fuse will blow, which breaks open the circuit to stop current.

Switches are always connected in series with a load, but often more than one switch controls a load. With multiple switches, the switches are usually connected in series with each other. Thus, electrons must travel through one switch before reaching the next switch and the next. All switches in series must be closed for the load to operate, but any switch can open and stop the electron flow, **Figure 22-31**.

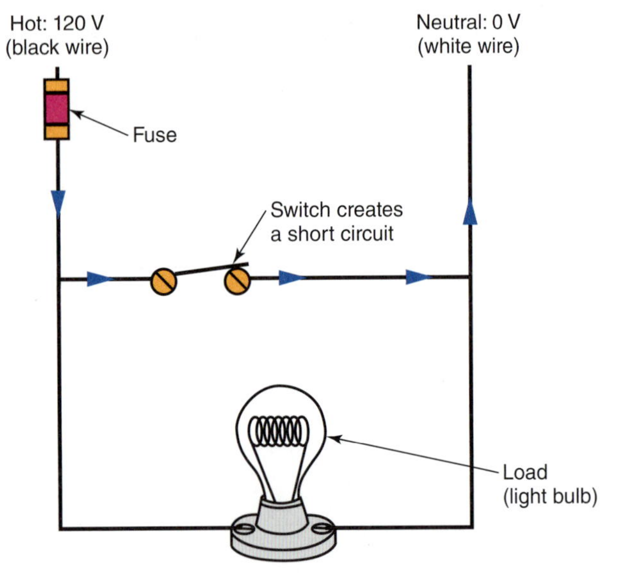

Goodheart-Willcox Publisher

Figure 22-30. A switch should never be connected in parallel with a load. Closing the switch creates a short circuit, permitting unrestricted current flow between power supply wires. Electrons will not travel through the light bulb (which has resistance) when they can take an easier path (with no resistance).

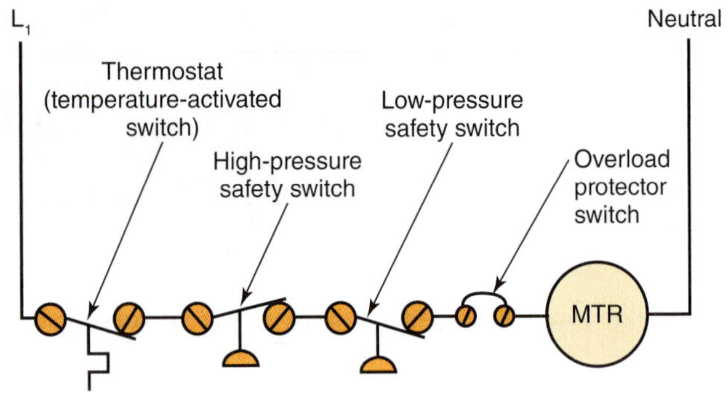

Goodheart-Willcox Publisher

Figure 22-31. In this series circuit, all four switches must be closed for the motor to run. If any switch opens, the motor will stop because electron flow has ceased.

Parallel Circuit

Electrical loads are connected in parallel, or independent from all others. ***Parallel*** means separate or multiple paths that are connected to the same input and output. Loads are designed for connection to a specific potential difference (voltage). When two loads are in series, they divide the total voltage drop, so that neither one gets enough voltage necessary for proper operation. The ultimate test of a parallel connection is when one load can be removed without affecting any others. It is common practice to connect more than one load to a supply circuit, but each load *must* be connected in parallel, **Figure 22-32**.

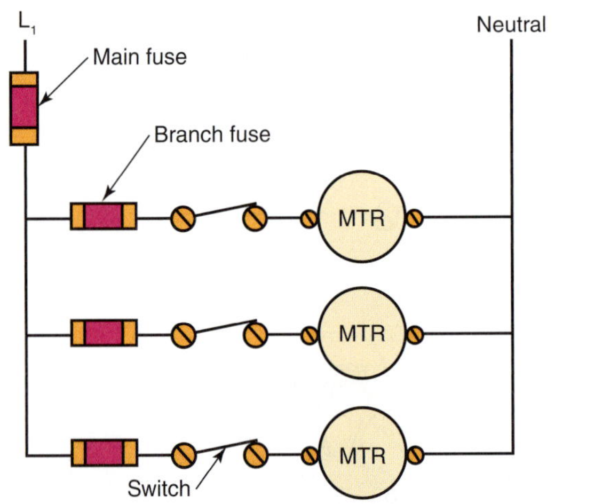

Goodheart-Willcox Publisher

Figure 22-32. These three loads are connected in parallel, so each branch line is independent of the others. If the switch is opened or a fuse blows on any branch line, it will affect only the motor on that branch. The others will continue to operate.

Series-Parallel Circuit

When more than one load is connected to a circuit, switches are required to control the individual loads. A separate switch can be installed to control each load, but a main switch is often installed in the main supply wire. A main switch serves to control power supply to all loads. The individual switches are connected in series with each load, and each load is connected in parallel with the other loads. The combination of series and parallel connections in a circuit is called ***series-parallel***.

Safety

Switches should be located in the black wire supplying power (voltage) to the load. When the switch opens, voltage stops at the switch, and the load is disconnected from the potential difference.

SAFETY NOTE

Properly Wiring an Electrical Load

If a switch is located in the neutral wire, electricity is available through the load. If the load becomes defective (ground fault), anyone touching the load could complete the circuit and be shocked, **Figure 22-33**.

22.5 Direct Current (DC) and Alternating Current (AC)

There are two basic types of electricity: *direct current* (which flows in only one direction) and *alternating current* (which reverses direction many times each second). Current continues to flow when a potential difference is maintained. If the polarity (+ vs. –) of the potential difference never changes, the current will flow in one direction. This is called ***direct current (dc)***.

Direct Current (DC)

A battery is the most common source of dc power. By converting chemical energy to electrical energy, it becomes a self-contained power source. To obtain continuous current flow, a constant voltage difference must exist between the negative and positive battery terminals. A "dead" battery exists when the chemical reaction stops (no voltage difference exists).

Other sources of direct current either exist in various forms or are manipulated from ac current via devices such as photovoltaic cells, diodes, and thermocouples. These are used in various HVACR applications and other industries.

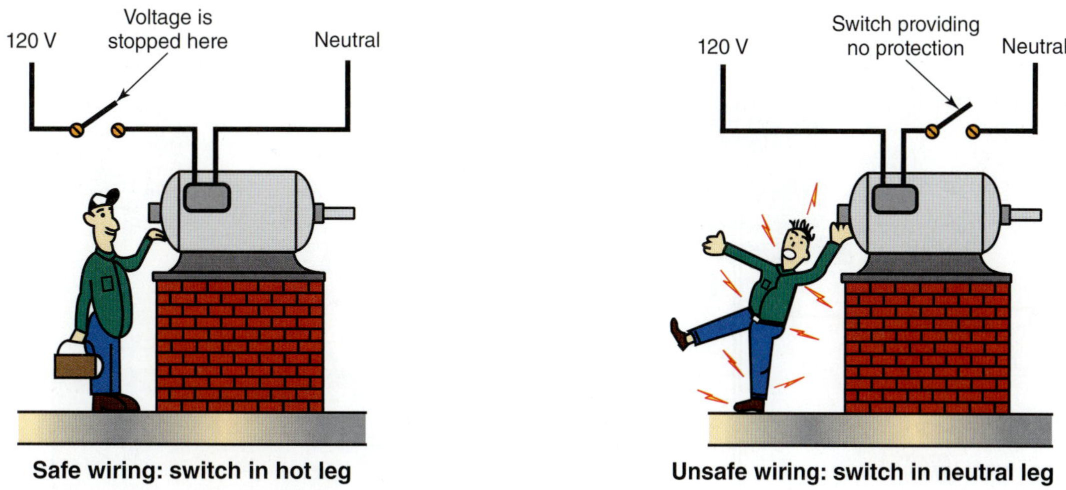

Goodheart-Willcox Publisher

Figure 22-33. For safety, a switch should always be located on the line (hot) side of the load rather than on the neutral side. When a switch is wired into the neutral side, a shock hazard exists.

Theories of How Electrons Flow

Two theories describe how electrons flow from one point to another. The most widely accepted theory is the *electron flow theory*, which states that electrons travel from negative to positive (magnetic attraction). The electron flow theory is used by most electricians and people working with ac current. The *current flow theory* states that current flows from positive to negative (difference of potential). The current flow theory is used by automobile technicians and people working with solid-state devices. Regardless of the theory involved, a *complete circuit* begins at the power source (positive or negative), typically travels through a switch and a load device, and returns to the other side of the power source, **Figure 22-34**.

Alternating Current (AC)

Alternating current (ac) flows first in one direction and then reverses to flow in the other direction. Since current normally flows from negative (–) to positive (+), the polarity of the power source must alternate from – to + and then + to –. The rapid alterations in magnetic polarity cause the electrons to move back and forth rapidly inside the conductor. The *movement* of electrons (not the direction of flow) produces electrical energy.

Alternating current (ac) is the name for reversing current. Power plants use alternators to produce a rapidly alternating polarity by rotating a conducting loop through a strong, stationary magnetic field. See **Figure 22-35**.

As the sides of the rotating conducting loop cut through the stationary magnetic lines of flux, a voltage (emf) is applied to the free electrons in the conductor. As the conducting loop rotates, each side cuts the magnetic field in opposite directions. Polarity changes as each side of the loop approaches the other magnetic pole (north and south). The result is alternating polarity and alternating current.

The AC Sine Wave

The ac sine wave illustrates the principle of alternating current, **Figure 22-36**. Voltage builds up at the magnetic north pole, falls back to zero, builds up again at the magnetic south pole, and again falls back to zero.

Electromotive force (voltage) drops to zero when the conducting loop is straight up and down and not cutting magnetic lines of flux. Voltage increases as more and more lines of flux are cut. Voltage is highest when the conductor is closest to the magnetic pole and is cutting lines of flux at right angles. Voltage drops off as the conductor travels away from the magnetic pole.

Cycles or Hertz (Hz)

To complete one cycle (or revolution), the loop passes one magnetic pole and then the other. Complete revolutions occur at the rate of

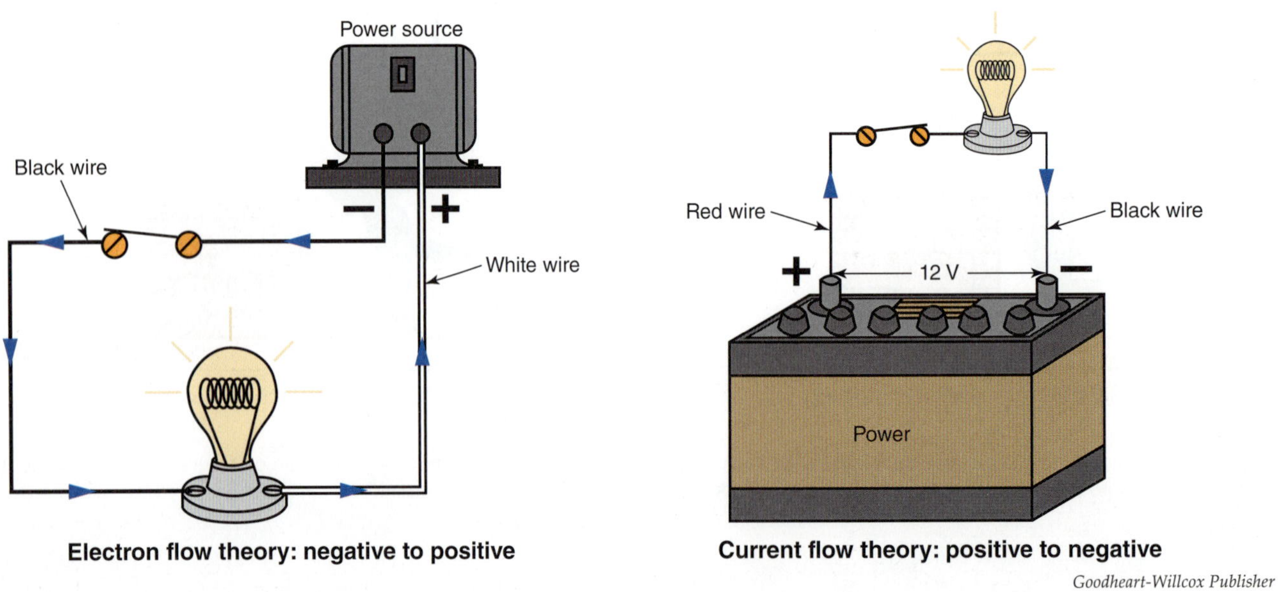

Figure 22-34. The two theories of how electrons flow.

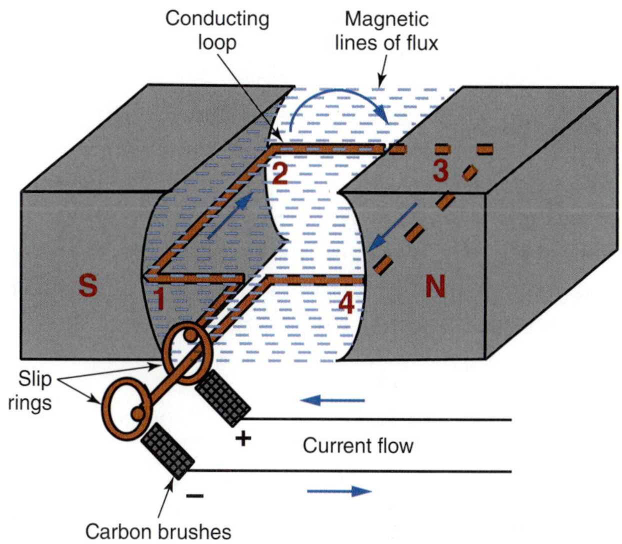

Goodheart-Willcox Publisher

Figure 22-35. Alternating current is produced by rapidly rotating a conducting loop (shown in simplified form) through a stationary magnetic field. Rotation of the loop causes the generated current to rapidly and regularly reverse polarity and direction. Polarity changes when the conducting loop rotates from one magnet to the other.

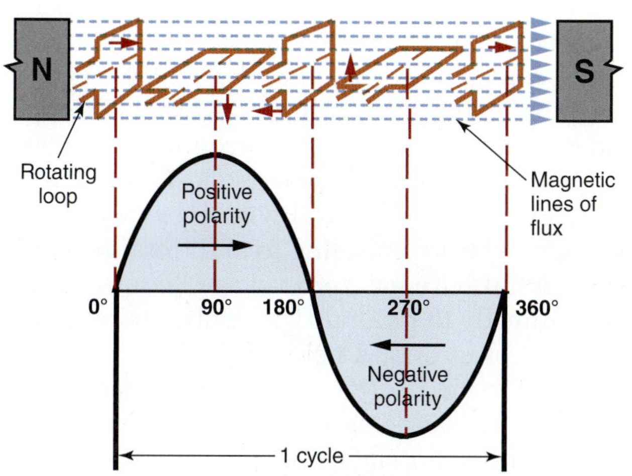

Goodheart-Willcox Publisher

Figure 22-36. As alternating current is generated by the rotating loop, the voltage builds to a peak, then declines to zero. Next, it reverses polarity and repeats the pattern in the opposite direction. The entire cycle is shown in a sine wave.

60 times (cycles) per second. Flow reversal occurs twice per cycle (1/2 cycle = north and 1/2 cycle = south). Flow reversal (120 times per second) is so rapid that a light bulb will not appear to flicker. The voltage and current flowing through the conductor are constant because of the speed of the cycles.

The ***frequency*** of alternating current is the number of complete cycles that occur in one second. In the United States and Canada, alternating current is produced at a frequency of 60 cycles per second. Most other countries use 50 cycles per second. The unit "cycles per second" is referred to as *hertz* (Hz) and is named for the German physicist Heinrich Hertz, one of the pioneers of alternating current.

22.6 Three-Phase Power Generation

In a power plant, a generator rotates three conducting loops at the same time. This means of producing electrical power is known as ***three-phase power generation***. The three conducting loops are spaced apart exactly 120° inside the generator. The loops are called ***phases*** or *legs*. The three phases are "out of step" along the sine wave. While one phase is positive, the second is negative, and the third is at zero. The three-way positioning is continuous as each loop takes its turn changing polarity from positive to negative to zero. Three-phase power (also referred to as *polyphase generation*) produces alternating current in three phases that are out of step with each other. See **Figure 22-37**.

The Greek letter phi (ϕ) is an abbreviation for phase. It may be used with a number to indicate the number of phases in a circuit. On a ***schematic*** (electrical circuit diagram) the letter *L* (standing for *line*) and a number identify each phase (L_1, L_2, L_3). See **Figure 22-38**. A potential difference exists between any two "hot" wires because of the

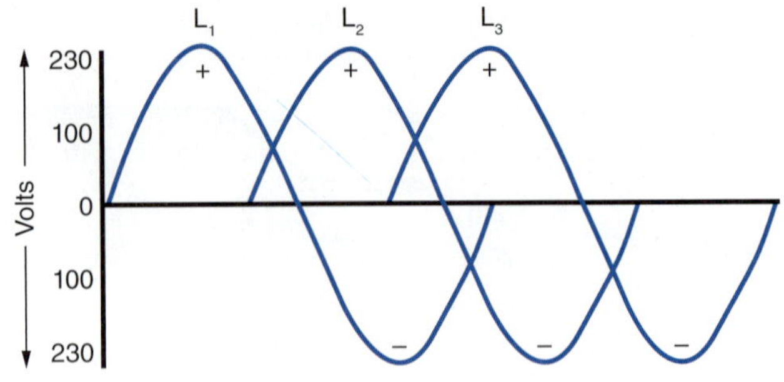

Goodheart-Willcox Publisher

Figure 22-37. In three-phase power generation, the three loops (and the three phases) are 120° apart, as shown by the sine wave.

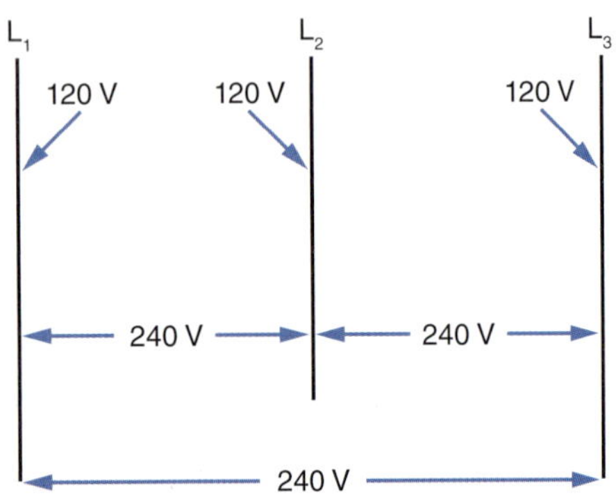

Goodheart-Willcox Publisher

Figure 22-38. In a three-phase system, voltage between any two wires is additive because the wires have different polarities. While one wire is negative, the other is positive. While one wire is pushing, the other is pulling, which doubles the voltage differential.

difference in polarity. As stated earlier, the normal color for hot wires is black or red. However, they can be any color except white or green.

Each hot wire has the same voltage but is different in polarity. When one wire is pulling (+), the other is pushing (–). Electrons flow according to polarity, but a voltage (120 V) exists on each wire. Therefore, the potential difference between any two hot wires (+ and –) is additive or the sum of their voltages. For example, 120 + 120 = 240.

Three-Phase vs. Single-Phase Loads

Many electrical devices are designed for connection to all *three* hot wires. Such devices are called ***three-phase loads***. Other devices are designed to operate with just *two* hot wires and are called ***single-phase loads***. Again, the voltage between the two hot wires is additive. The term "two-phase" is never used because it refers to an older method of power generation no longer used. Voltage higher than 120 V can be obtained only by using two of the three phases.

Many electrical devices operate with just one hot wire from a three-phase system and a second wire called the *neutral*. This method is also called *single-phase* because only one hot wire is used. A potential difference exists because the hot wire has voltage and polarity, while the neutral wire has zero voltage and no polarity.

Neutral Wire

The earth is a gigantic mass of elements and compounds that serves as an excellent conductor of electricity. Damp soil is a better conductor than dry soil. The earth is always at ***zero potential*** (no voltage) and can complete an electrical circuit.

The neutral conductor in a circuit has white or gray insulation and is connected to the earth. The neutral conductor is referred to as ***grounded***. The connection to the earth is typically made by the utility company. A bare copper wire travels from the transformer on the utility pole to a solid copper rod driven eight feet into the ground. The copper rod is called a grounding electrode, **Figure 22-39**.

Grounding the neutral wire provides a pathway for electrons traveling to and from the earth. The neutral wire has zero voltage because it connects directly to ground. The neutral wire should not be broken and is never fused. The electrical symbol for the neutral wire is the letter *N*. The neutral wire is used as a current-carrying conductor, but it has no voltage or polarity. A potential difference exists between the neutral wire (no voltage) and any black or red (hot) wire with 120 V. See **Figure 22-40**.

The NEC refers to the neutral conductor as the *grounded* conductor. Another wire (green) is reserved as the *grounding* conductor. The two wires serve entirely different purposes and should never be confused. The purpose and proper use of the grounding conductor (green wire) will be explained shortly. Single-phase loads designed to operate on 120 V require the use of one hot (L_1) wire and the grounded neutral (L_2) conductor. The hot (L_1) conductor, normally black in single-phase applications, supplies electrons, voltage, and magnetism, while the neutral (L_2) conductor, normally white in single-phase applications, supplies electrons at zero voltage and no magnetism. This provides a potential difference so electrons are

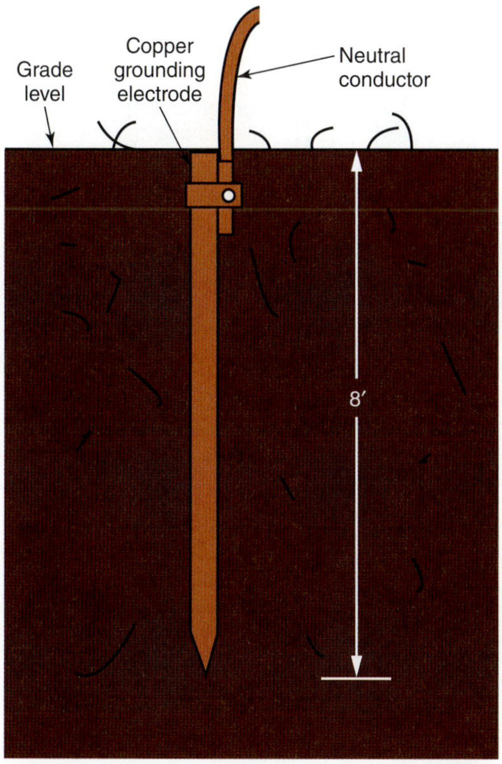

Goodheart-Willcox Publisher

Figure 22-39. A copper rod is driven eight feet into the earth by the power company and used to ground the neutral wire of the electrical supply system.

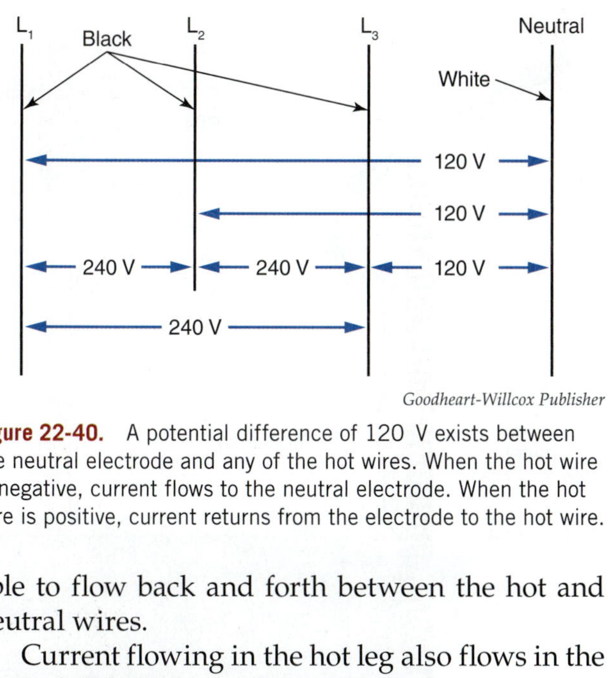

Goodheart-Willcox Publisher

Figure 22-40. A potential difference of 120 V exists between the neutral electrode and any of the hot wires. When the hot wire is negative, current flows to the neutral electrode. When the hot wire is positive, current returns from the electrode to the hot wire.

able to flow back and forth between the hot and neutral wires.

Current flowing in the hot leg also flows in the neutral. When the load is operating, an amperage reading can be obtained from either conductor. See **Figure 22-41**. The voltage and energy are used up by the electrical load, so the neutral wire remains at zero voltage. If either wire is disconnected from the load, the potential difference is removed and electrons cannot flow.

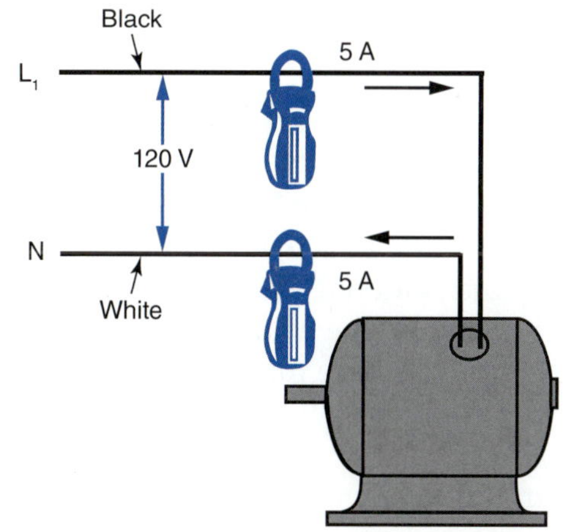

Goodheart-Willcox Publisher

Figure 22-41. Current flowing in the black wire also flows in the white wire. When the load is active, an amperage reading can be obtained from either wire.

Some single-phase loads are designed to operate on 240 V. Such loads require the use of *two* hot wires containing 120 V each (no neutral). The voltage between the two hot wires is additive (120 V + 120 V = 240 V). The two hot wires have the *same* voltage but opposite polarity. *Opposite* polarity (negative/positive) provides the necessary potential difference. Even though it is called single phase (1ϕ), the *higher voltage* can only be obtained by using two hot wires. See **Figure 22-42**.

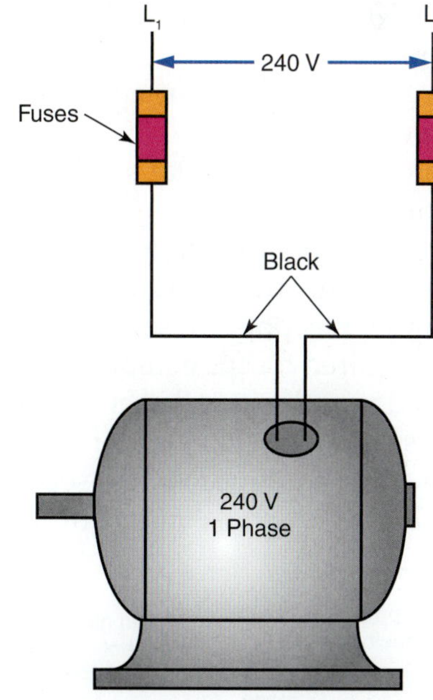

Goodheart-Willcox Publisher

Figure 22-42. A 240 V single-phase motor like this one uses two hot wires (but no neutral).

22.7 Equipment Grounding Conductor

The equipment grounding conductor, or ***safety ground***, is a wire added for safety. The wire is *required* by the National Electrical Code on all new electrical systems. The color code for this wire is green, or it can be bare (uninsulated) copper. The symbol for an equipment grounding electrode is shown in **Figure 22-43**.

Grounding conductors (green wires) from the local devices are connected to a busbar located inside the distribution panel. A ***busbar*** is a metal bar that serves as a common connecting device. A busbar has multiple screws for connecting wires, and it is securely bonded to the distribution box to obtain good metal-to-metal contact.

For residential wiring systems, the equipment grounding conductor (green) and the neutral conductor (white) share the same grounding busbar. The two busbars, located inside the distribution center (breaker box), are connected as shown in **Figure 22-44**. The breaker box is the only place where the neutral and safety grounds are joined. The NEC ruled that all new installations must keep the neutral and safety wires separated at all times.

The grounding busbar *must* be connected to a grounding electrode (also called a grounding rod). The grounding rod is a solid copper rod driven into the ground to make good contact with moist earth, **Figure 22-45**. Consult your local electrical inspector or the NEC regarding approved grounding techniques.

Commercial and industrial electrical systems require complete separation of the grounding electrode from the neutral electrode. As shown in **Figure 22-45**, the neutral is grounded by the utility company at the pole. A separate grounding electrode is required for the equipment grounding conductor and must be installed at the site.

The distribution panels of commercial installations are equipped with separate busbars, **Figure 22-46**. The equipment grounding

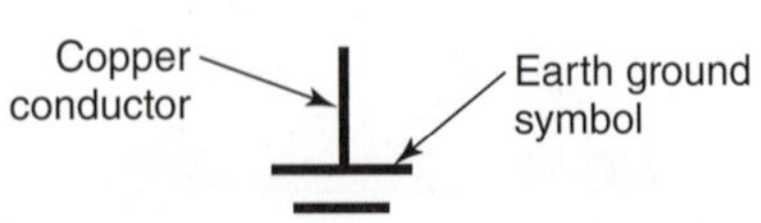

Goodheart-Willcox Publisher

Figure 22-43. A triangle symbol indicates the safety (earth) ground on wiring diagrams.

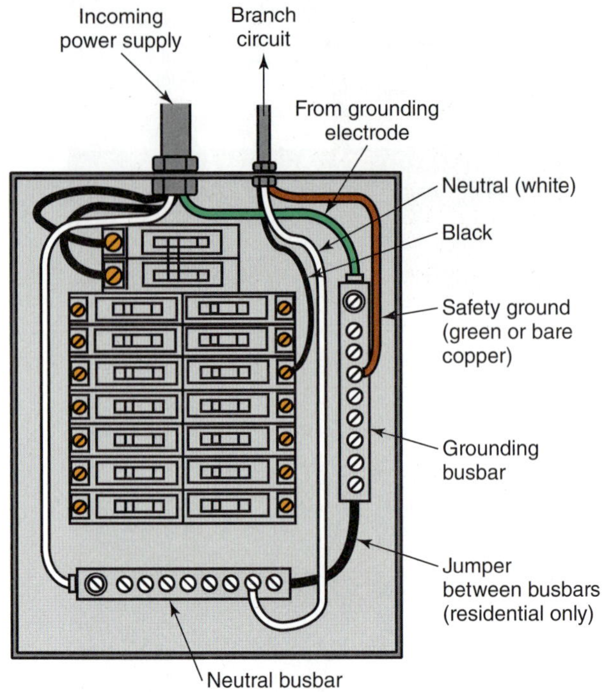

Goodheart-Willcox Publisher

Figure 22-44. For residential installations only, the neutral and equipment grounding busbars are connected to each other via a jumper, and both are connected to a grounding electrode.

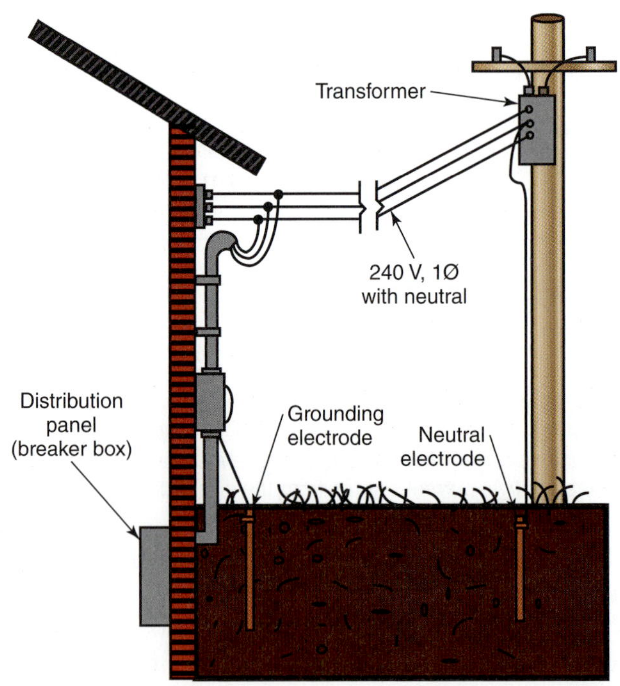

Goodheart-Willcox Publisher

Figure 22-45. The copper grounding rod, or electrode, is driven at least eight feet into the earth to make contact with moist soil.

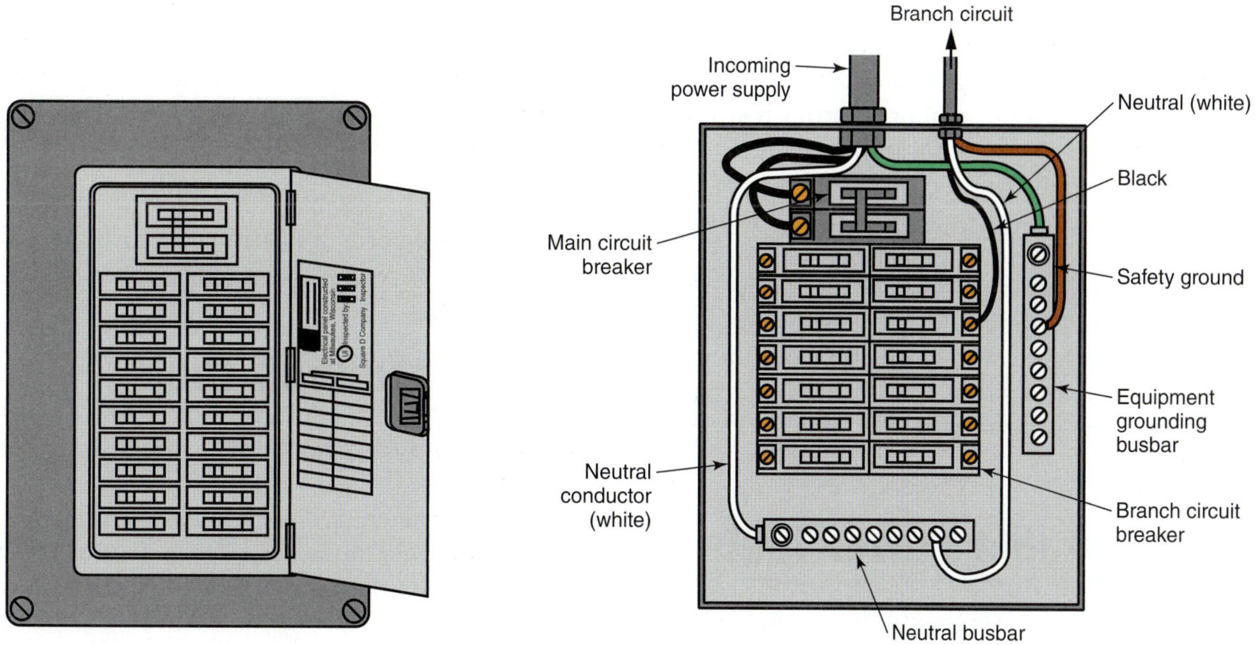

Goodheart-Willcox Publisher

Figure 22-46. In commercial and industrial applications, the neutral and equipment grounding busbars must remain separated at all times. The equipment ground is often connected to the steel framework of the building rather than to an electrode. The busbars are never connected to each other, as permitted in residential installations.

busbar is often connected to the steel beams used in constructing the building. The beams serve as a grounding electrode.

Purpose of the Safety Ground

The safety ground conductor (green or bare wire) is never connected to the normal electrical circuit. The safety ground conductor serves as a "safety valve" when a malfunction occurs in the electrical equipment. The safety ground wire is connected to the frame of a motor or appliance. Under normal conditions, the frame of a motor or appliance is safe to touch. However, an internal electrical problem may occur, permitting electron flow to the frame. Called a ***ground fault***, the flow of electrons makes the frame electrically live. Anyone touching the frame will receive a severe electrical shock. The body becomes a "load" by completing a circuit from the frame to ground, **Figure 22-47**.

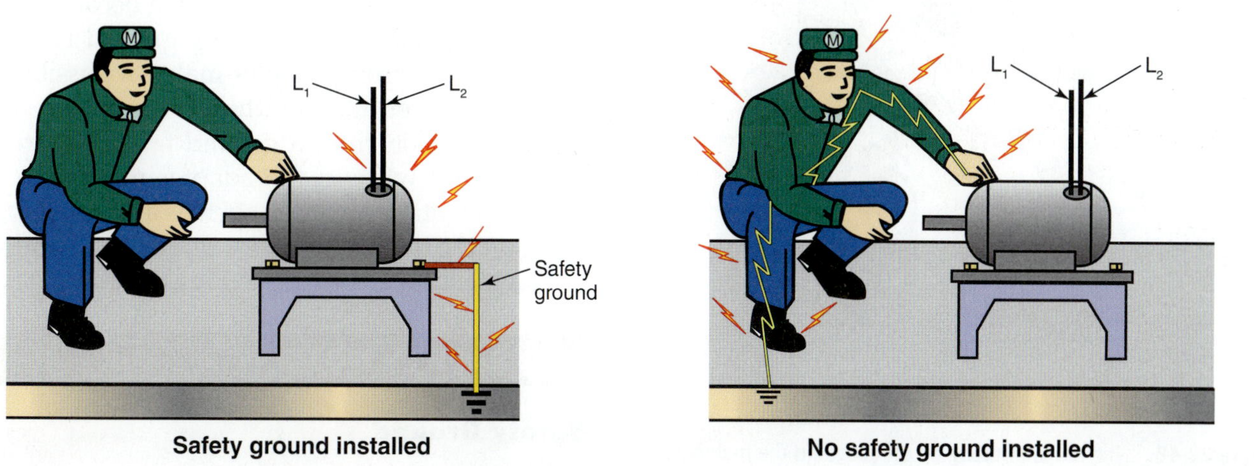

Goodheart-Willcox Publisher

Figure 22-47. An equipment ground provides a path for electron flow to ground in the event of a ground fault, preventing an electrical shock if the frame is touched. Without a safety ground, if the frame of a motor accidentally becomes electrically "live," touching the frame will complete the circuit to ground and make your body a load.

When a ground fault occurs, the safety ground wire provides an unbroken pathway for electrons to travel from the frame to the grounding electrode. The activity amounts to a short circuit, resulting in excessive current flow that will blow a fuse or trip a circuit breaker. So, the green wire provides an escape path for electrons when a ground fault occurs.

Because the safety ground wire has proven effective in preventing electrical injuries, it is required by the NEC. The third (rounded) prong on plugs and the matching hole in receptacles are reserved for the safety ground connection. See Figure 22-48.

Using Portable Power Tools

Caution must be exercised when using portable power tools. These handheld electrical devices can be dangerous if a ground fault occurs. Power tools are often constructed with a metal frame. Such tools should have a three-prong plug with the metal frame connected to the safety ground. If the frame becomes electrically "live" due to a ground fault, the third wire (safety ground) will safely carry the current to ground. The third wire protects the operator from becoming a pathway to ground. Excess current flow will blow a fuse or trip a circuit breaker.

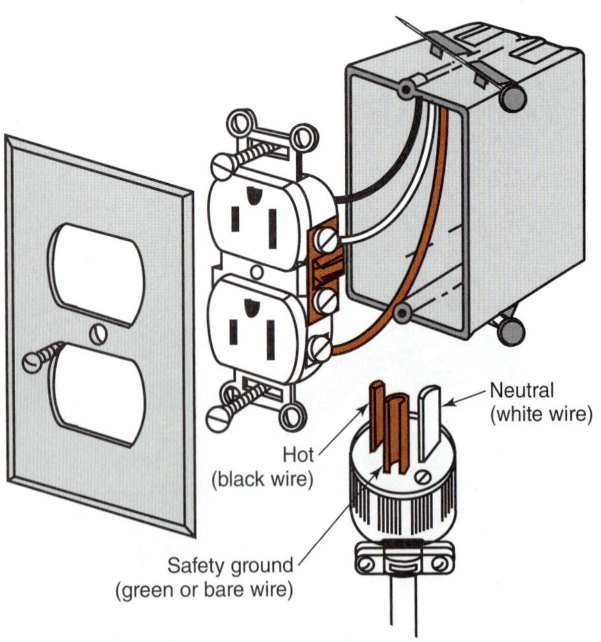

Goodheart-Willcox Publisher

Figure 22-48. The rounded prong on the plug and the matching socket in the receptacle are reserved for the safety ground. The receptacle should be installed with the ground uppermost to help prevent accidental shorting across the hot and neutral prongs.

SAFETY NOTE

Grounded Tools and Grounded Circuits

Always use properly grounded power tools and connect them to properly grounded circuits.

Double-Insulated Tools

Many power tools do not have the third (safety ground) prong on the male plug. Instead, the devices are supposed to be "double-insulated" to prevent a ground fault from the hot wire to the tool case. The insulating plastic is intended to protect the operator. However, a crack in the plastic case can permit moisture to enter the insulation and defeat its purpose. Always examine such tools for cracks or damage.

SAFETY NOTE

Power Tool Electrical Safety

Never use a double-insulated power tool that has a damaged case. Also, never use a power tool with a plug that has had the third (ground) prong removed.

Adapters

Many old-style wall receptacles are two-prong and do not accept a three-prong plug because they are not equipped with the safety ground feature. A three- to two-prong adapter makes it possible to connect a power tool to these receptacles. For proper protection from a ground fault, the green safety ground wire on the adapter must be connected to a suitable ground, such as a steel beam, a metal stake driven into the ground, or a similar good connection to earth. See Figure 22-49.

SAFETY NOTE

Safety Ground

Failure to connect the safety ground can result in death for the person whose body becomes a ground.

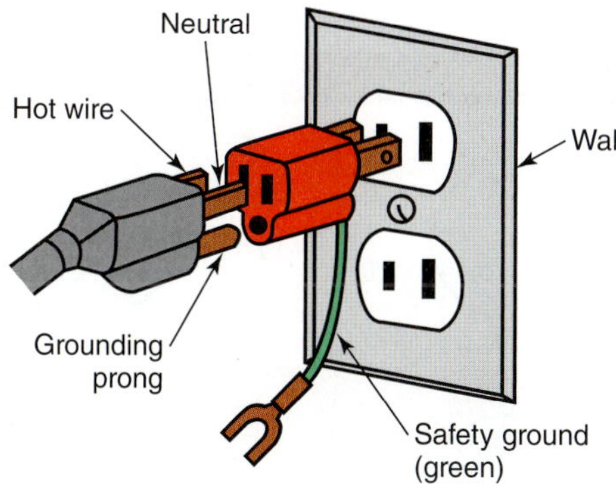

Goodheart-Willcox Publisher

Figure 22-49. When a three-to-two-hole adapter is used with a portable power tool, the green safety ground wire must be connected to a good ground. These adapters are sometimes called "suicide plugs" because of the fatal consequence of not connecting the safety ground wire to a good ground.

Most two-hole receptacles are *not* grounded, so ground fault protection is lost. Always use a voltage tester to determine if the receptacle is properly grounded. Zero voltage between the hot wire and a metal cover reveals no safety ground. A 120 V reading between the hot wire and the metal cover indicates a good safety ground.

If the receptacle box is not grounded, use another piece of wire to connect the adapter's green wire to a good ground. Taking the extra precaution of properly connecting the safety ground may save your life if a power tool becomes "shorted" to the metal case.

22.8 Power Circuit Devices

Circuits supplying electrical power to residential or commercial installations include *receptacles*, which allow easy connection and disconnection of equipment, and *switches* and *disconnects*, which control part or all of a circuit.

Receptacle Types

A ***receptacle*** is an electrical outlet including a fitting that allows easy connection and disconnection of electrical equipment. Various female outlet receptacles are available for equipment that has a flexible cord and a male plug. As a safety feature, each plug and receptacle are sized and styled for a particular voltage and amperage. For example, a 250 V plug cannot be inserted in a 125 V receptacle. Likewise, a plug rated for 20 A cannot be plugged into a 15 A receptacle.

The shapes and styles of plugs and receptacles are standardized by amperage and voltage. **Figure 22-50** shows some of the many types of receptacles.

Switch Types

As stated earlier in this chapter, a ***switch*** is a device used to control the flow of electrons to one or more electrical devices. All switches contain one or more sets of ***contacts*** that are opened or closed by the movement of a ***pole***. A switch may contain one or more poles. Movement of the pole is called the ***throw***.

Switches are rated to operate with a specific voltage and amperage. The specified ratings *must not* be exceeded. Switches for use in residential circuits are commonly rated 125 V at 10 A or 250 V at 5 A. Special-purpose switches may be rated for smaller or larger voltages and amperages.

The common household light switch is a ***single-pole, single-throw (SPST) switch***. It has one pole and makes (closes) or breaks (opens) one contact. The operation of an SPST switch, in its simplest form, is illustrated in **Figure 22-51**.

A ***single-pole, double-throw (SPDT) switch*** has one pole and operates two contacts, **Figure 22-52**. One contact is ***normally open (NO)***, and the other contact is ***normally closed (NC)***. There is no "off" position on this switch. Movement of the pole causes the NC contact to open and the NO contact to close.

A ***double-pole, single-throw (DPST) switch*** has two poles and a contact for each pole. Both contacts are either open or closed, making it possible to open or close two circuits at the same time. This switch is often used to control both hot wires on a 230 V, single-phase circuit. See **Figure 22-53**.

A ***double-pole, double-throw (DPDT) switch*** has two poles and two contacts for each pole. Power supply must be connected to the poles. The DPDT switch makes it possible to control a variety of loads from one location. Two contacts are NC, and two contacts are NO. See **Figure 22-54**.

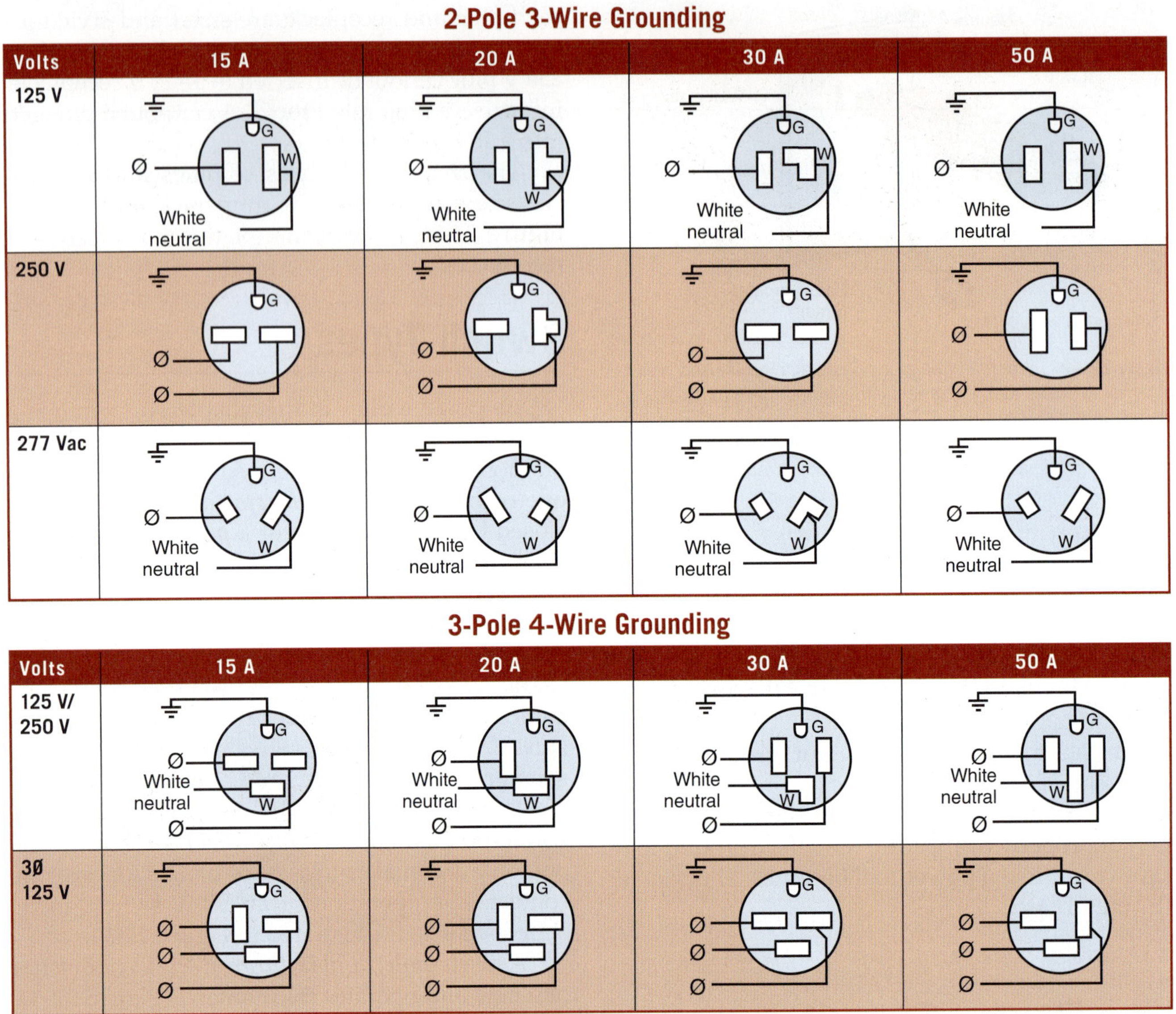

Goodheart-Willcox Publisher

Figure 22-50. Each type of plug and receptacle is sized and styled for a particular application, voltage, and amperage.

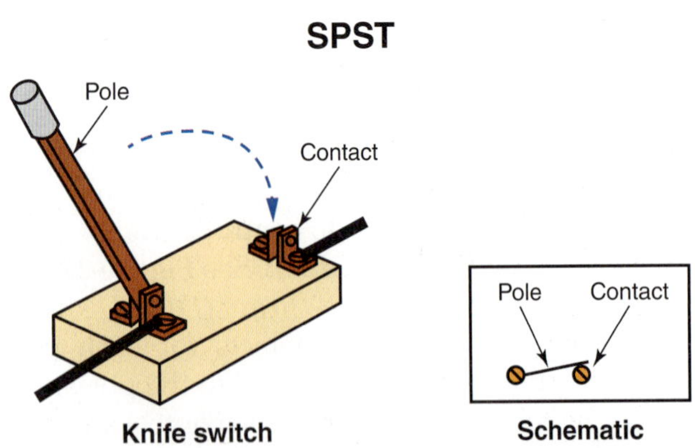

Goodheart-Willcox Publisher

Figure 22-51. The single-pole, single-throw switch (like the knife switch shown here) is the simplest type of switch. It is either on or off. The common household light switch uses a toggle mechanism but operates on the same principle: it either makes or breaks a contact.

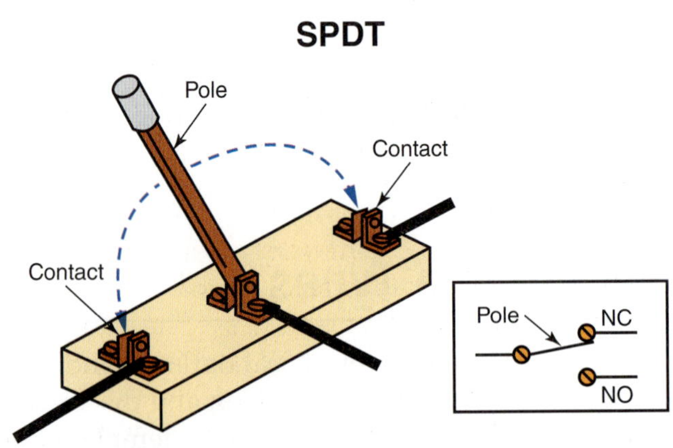

Goodheart-Willcox Publisher

Figure 22-52. On a single-pole, double-throw switch, the pole moves from one contact to another. One contact is normally open (NO), and the other is normally closed (NC).

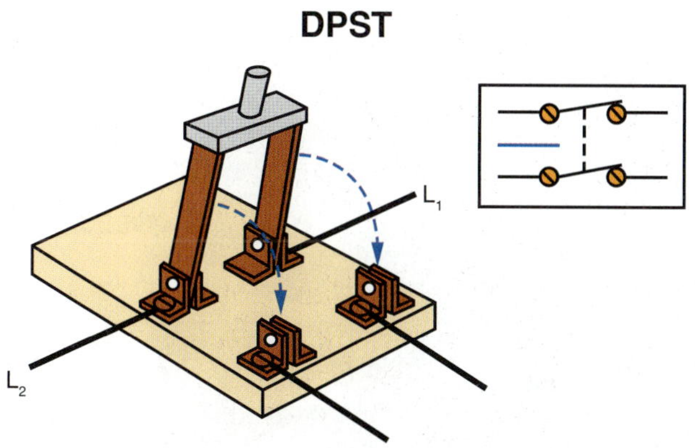

Goodheart-Willcox Publisher

Figure 22-53. A double-pole, single-throw switch can control two circuits or both hot wires in a 230 V, 1ϕ circuit.

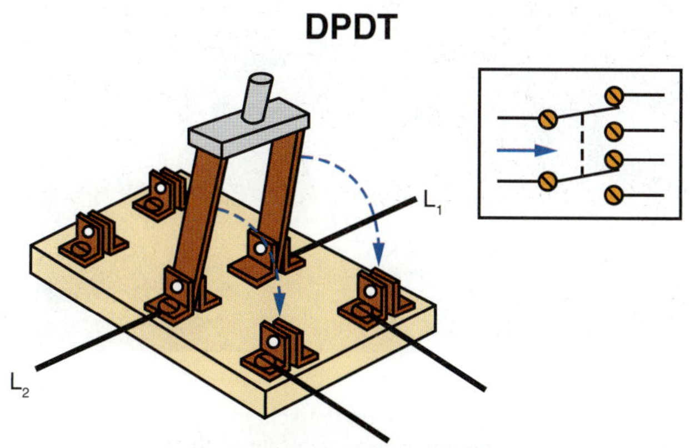

Goodheart-Willcox Publisher

Figure 22-54. A double-pole, double-throw switch can control a variety of loads. Two of the contacts are normally open (NO), and two are normally closed (NC).

Disconnects

Disconnects are heavy-duty, manually operated contacts for disconnecting large loads from a power supply. Disconnect switches are normally located close to the load for quick and easy access. A disconnect may have a two-pole or three-pole switch mounted inside a metal enclosure. Regardless of the number of poles, the switches are simultaneously controlled by a manually operated handle or by pulling the pull-tab out.

When a disconnect includes fuses, it is called a *fused disconnect*. The power supply, referred to as ***line voltage***, is connected to the terminals located inside the disconnect box, and are labeled L_1, L_2, in single-phase applications. In three-phase applications, disconnects include an additional terminal labeled L_3. These power supply terminal connections remain electrically live at all times, even when the disconnect is turned off. The switches are located between the power supply and the fuses. The *load* is connected to lugs provided at the other end of the fuses labeled T_1 and T_2 (and T_3 if it is a three phase). The arrangement permits safe replacement of fuses when the disconnect is off. See **Figure 22-55**.

SAFETY NOTE

Safely Pulling Fuses

Always use a voltage tester to be certain all switches open properly. Even though the switches are open, use an insulated fuse puller to remove or replace fuses.

Disconnects are rated and sized for a maximum amperage (such as 30, 60, or 100) and voltage. Each disconnect size requires a different fuse size. The fuse holders are designed to prevent using a too-large fuse. For instance, a 60 A fuse will not fit a disconnect designed for 30 A service.

SAFETY NOTE

Disconnect Switch Safety

Do *not* open the disconnect switches when the load is operating. Always stop the load before operating the disconnect switches. Under certain conditions (such as a short circuit), a sudden inrush of extremely high current could cause the disconnect box to explode. Always operate the disconnect handle with the left hand and stand to the side, facing away from the box in case of malfunction.

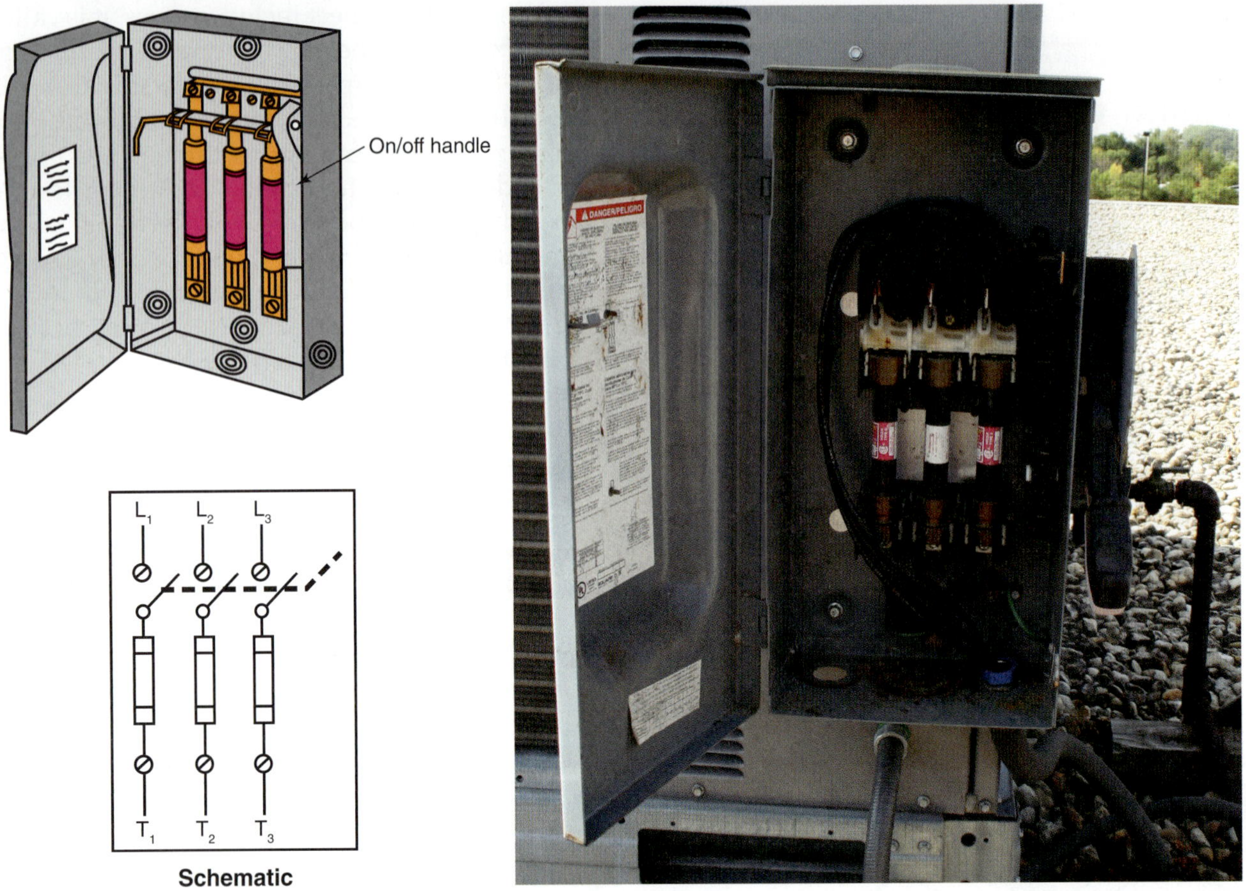

Goodheart-Willcox Publisher

Figure 22-55. In a fused disconnect, each line has a fuse located below the switch. The dotted line on the schematic indicates that all three switches open and close at the same time. An insulated fuse puller should be used to remove or replace fuses.

Chapter Review

Summary

22.1 Electricity is a form of energy in which electrons move from one atom to another atom. Atoms contain protons, neutrons, and electrons that determine an atom's charge. To make electricity perform useful work, a constant and steady movement of electrons must be produced.

22.1 When the electrical charge (positive versus negative) between two points is *not* in balance, the imbalance of electrons is called a *potential difference*. Electromotive force (emf) is the stimulus that causes free electrons to reel out of orbit and travel to an adjoining atom.

22.1 Electrons move through copper conductors, or wires, by electromotive force. This movement produces electricity.

22.1 The flow of electrical current is measured in amperes. Resistance is anything slowing the flow of electrons. There are two types of resistance: resistive and reactance.

22.1 How much current (amperage) passes through the body, how long (duration) the current passes the body, and whether the current passes through the heart determine whether or not an electrical shock is fatal. Never let your body become a pathway for electron flow. Using proper lockout/tagout procedures on circuits help prevent accidents during electrical repair work.

22.2 Ohm's Law refers to the relationship between voltage (E), amperage (I), and resistance (R) and how it can be used to predict and control the activity of electricity. An ohmmeter or multimeter can be used to measure resistance, check continuity, and detect open or short circuits.

22.3 Electrical power is the rate at which electrons are harnessed to perform a useful function. Power is measured in watts (P) and can be calculated using the following equation: $P = I \times E$.

22.4 Conductors provide the necessary free electrons and serve as the pathway for electron flow. Good conductors have high conductance, which is the number of amperes flowing in a conductor per volt of applied emf and is measured in siemens.

22.4 The type, size, and length of wire selected for a conductor is used to minimize the voltage drop through the length of the conductor. Wires are insulated with various materials to protect the wires from damage or malfunction.

22.4 An electrical load is an energy-conversion device that converts electrical energy to another form of energy. Electrical switches stop and start the flow of electrons to an electrical load.

22.5 Direct current is the flow of current in one direction due to a constant polarity of the potential difference. Alternating current first flows in one direction and then reverses to flow in the other direction. Both current movements produce electrical energy.

22.6 Three-phase power is produced using a generator that rotates three conducting loops at the same time. This produces higher voltages that many electrical devices are designed to operate on.

22.7 The equipment grounding conductor, or safety ground is a wire added to provide an unbroken pathway for electrons to travel to the grounding electrode in the case of a ground fault. Power tools should have a safety ground or be double-insulated to protect the user from electrical shock.

22.7 Adapters can be used to connect three-prong plugs to two-prong receptacles.

22.8 A receptacle is an electrical outlet including a fitting that allows easy connection and disconnection of electrical equipment. Various female outlet receptacles are available for equipment that has a flexible cord and a male plug.

22.8 A switch is a device used to control the flow of electrons to one or more electrical devices. There are single-pole, single-throw (SPST), single-pole, double-throw (SPDT), double-pole, single-throw (DPST), and double-pole, double-throw (DPDT) switches.

22.8 Disconnects are heavy-duty, manually operated contacts for disconnecting large loads from a power supply. When a disconnect includes fuses, it is called a *fused disconnect*.

Know and Understand

1. The movement of free electrons is called _____. (22.1)
   A. positive charge
   B. negative charge
   C. electricity
   D. valence flow
2. The imbalance of electrons between two points is referred to as _____. (22.1)
   A. neutral
   B. potential difference
   C. electromotive force
   D. negative charge
3. Electromotive force (emf) is measured with a(n) _____. (22.1)
   A. ohmmeter
   B. voltmeter
   C. wattmeter
   D. ammeter
4. The flow of electrical current is now measured in _____. (22.1)
   A. coulombs
   B. amperes
   C. volts
   D. ohms
5. An ammeter measures the _____ of a conductor's magnetic field and converts this information to an amperage reading. (22.1)
   A. revolutions
   B. intensity
   C. resistance
   D. potential difference between two points
6. *True or False?* The human body is at zero voltage. (22.1)
7. Resistance is measured with a(n) _____. (22.2)
   A. ohmmeter
   B. ammeter
   C. wattmeter
   D. voltmeter
8. Which of the following is *not* proved by Ohm's Law? (22.2)
   A. Voltage equals amperage times resistance.
   B. Amperage equals voltage divided by resistance.
   C. Resistance equals voltage times amperage.
   D. All of the following are proved by Ohm's Law.
9. When connected to an open circuit, a digital ohmmeter should read _____. (22.2)
   A. 0 Ω
   B. 1–100 Ω
   C. infinity or OL
   D. "E" for error
10. A multimeter is capable of measuring all of the following, *except* _____. (22.2)
    A. resistance
    B. voltage
    C. current
    D. frequency
11. Electrical power (the ability to do work) is measured in _____. (22.3)
    A. ohms
    B. watts
    C. volts
    D. siemens
12. The amount of energy (rate of work) converted by electrical loads depends on voltage, amperage, and _____. (22.3)
    A. resistance
    B. frequency
    C. length of the circuit
    D. elapsed time
13. The amount of internal resistance in a conductor depends on the type of metal from which it is made, the wire size and length, and the _____. (22.4)
    A. amount of power required
    B. ambient temperature
    C. elapsed time
    D. circuit load
14. *True or False?* Neutral wires are typically black or red. (22.4)
15. When connecting a load to a voltage source, a minimum of _____ conductor(s) must be used. (22.4)
    A. one
    B. two
    C. three
    D. four

16. The purpose of switches in an electrical circuit is to _____. (22.4)
    A. provide resistance
    B. protect against unrestricted current flow
    C. stop and start the flow of electrons to an electrical load
    D. disconnect an electrical load from the circuit
17. A complete circuit begins at _____. (22.5)
    A. the power source
    B. an electrical load device
    C. a switch
    D. the positively charged end of a device
18. The _____ of alternating current is the number of complete cycles that occur in one second. (22.5)
    A. amperage
    B. voltage
    C. wattage
    D. frequency
19. *True or False?* Devices designed to operate with just two hot wires are called *two-phase loads.* (22.6)
20. A neutral wire has _____ V because it connects directly to the ground. (22.6)
    A. 0
    B. 60
    C. 120
    D. 240
21. A _____ serves as a common connecting device for grounding conductors. (22.7)
    A. grounding electrode
    B. copper wire
    C. busbar
    D. neutral wire
22. When electrons flow to the frame of a motor or appliance, this condition is referred to as a(n) _____. (22.7)
    A. short circuit
    B. open circuit
    C. misfire
    D. ground fault
23. A _____ is an electrical outlet including a fitting that allows easy connection and disconnection of electrical equipment. (22.8)
    A. switch
    B. receptacle
    C. disconnect
    D. conduit
24. *True or False?* All four contacts can be normally closed at the same time in a double-pole, double-throw switch. (22.8)
25. Disconnect switches are normally located close to the _____. (22.8)
    A. load
    B. power supply
    C. fuses
    D. grounding electrode

Apply and Analyze

1. Explain how electron movement creates electricity. (22.1)
2. What is the difference between potential difference and electromotive force? (22.1)
3. If a series circuit has a voltage of 120 volts and an electrical load with a resistance of 25 ohms, what would the amp draw be on the circuit? (22.2)
4. Which five variables determine the ampacity of a conductor? (22.4)
5. Describe the concepts of direct current and alternating current. (22.5)
6. What is three-phase power generation? (22.6)
7. What is the purpose of the safety ground? (22.7)
8. Receptacles, switches, and disconnects are all rated to operate by which two electrical measurements? (22.8)

Critical Thinking

1. When using a digital multimeter, what functions may be used to help diagnose a faulty system? Explain each function and the pros and cons of using each function. (22.2)
2. While a system's condensing unit is operating, what effects might a set of pitted contacts have on electrical measurements? Also explain the cause of pitted contacts. (22.4)

Sealed Unit Parts Co., Inc. (SUPCO)

23 Power Transmission and Circuits

LEARNING OUTCOMES

After studying this chapter, you will be able to:

23.1 Explain how electrical power is transmitted for commercial, industrial, and residential applications.

23.2 Describe the operating principles of transformers and how to select transformers for an application.

23.3 Recognize the danger of overcurrent and the various circuit protection devices used to protect a circuit and people from overcurrent conditions.

23.4 Explain how to make good terminal connections.

23.5 Interpret electrical connections shown on pictorial, schematic, and ladder diagrams.

TECHNICAL TERMS

amperage interrupting capacity (AIC)
cartridge fuse
circuit protection device
crimp
delta
double-pole breaker
dual element
electrical symbols
full load amperage (FLA)
ground fault circuit interrupter (GFCI)
induced voltage
insulation
ladder diagram
load center
load rating
magnetic field
overcurrent
overload
oversizing
pictorial diagram
power factor
primary winding
schematic diagram
secondary winding
single-pole breaker
step-down transformer
step-up transformer
stinger leg
terminal
terminal connector
three-pole breaker
transformer
undersized
wire nut
wye

Introduction

Electrical power is normally generated as three-phase, 60-Hz, alternating current at about 26,000 V. A *step-up transformer* increases the voltage to 120,000 V (or more). In transmission lines, high voltage reduces current flow (I = V ÷ R). Reduced current flow permits the use of smaller-diameter wire over long distances. Power is often transmitted for hundreds of miles using steel towers to support the wires. The three-phase, four-wire system (with grounded neutral) is used. The neutral line grounds the steel towers and protects them from lightning. See **Figure 23-1**.

Substations are located at various points along the transmission circuit. At the substations, *step-down transformers* reduce voltage to 40,000 V for distribution through a particular region. Transformers at other substations *within* a region further step-down voltage to 13,200 V or 4800 V. Transformers for reducing voltage to individual homes and businesses are often located on utility poles but may be at ground level in protective enclosures. The final voltage varies from 120 V, single-phase to 480 V, three-phase, depending upon the needs of the customer.

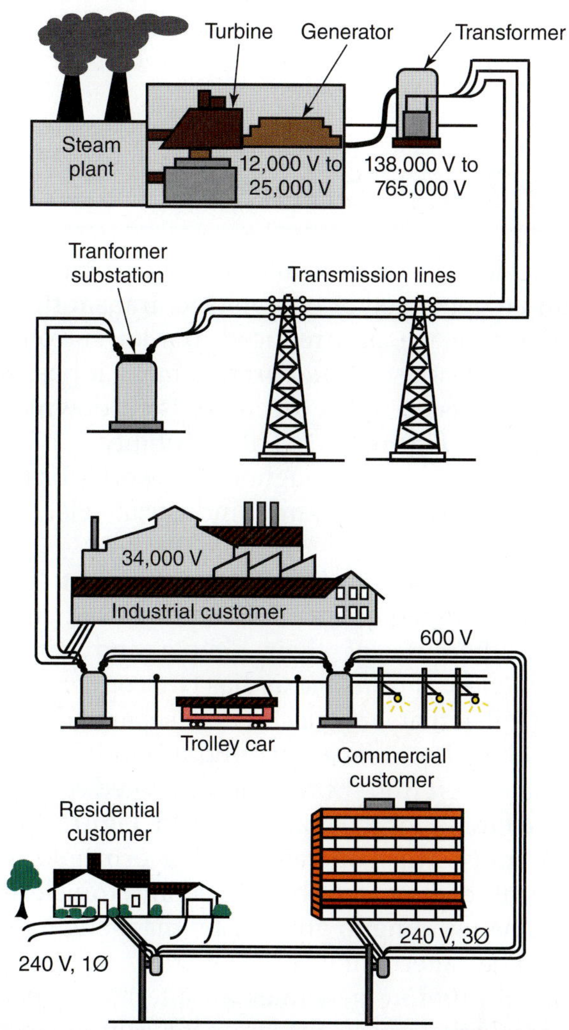

Goodheart-Willcox Publisher

Figure 23-1. The power transmission system steps up voltage to send electricity over long distances. Then, voltage is stepped down to meet the various needs of users connected to the distribution system.

23.1 Power Transmission

The power distribution system can supply electrical power in different configurations to meet the requirements of customers from large industrial plants to individual households.

Meeting Commercial and Industrial Electrical Needs

For commercial and industrial applications, three-phase alternating current is most common. Three-phase power is cheaper than single-phase and provides other advantages. For example, three-phase motors do not require starting components, and they offer better start and run characteristics. Three-phase is easier to understand and control. Easy access to 120 V, single-phase for lighting and receptacles is provided by including the neutral wire. **Figure 23-2** illustrates a typical 230 V, three-phase, four-wire system.

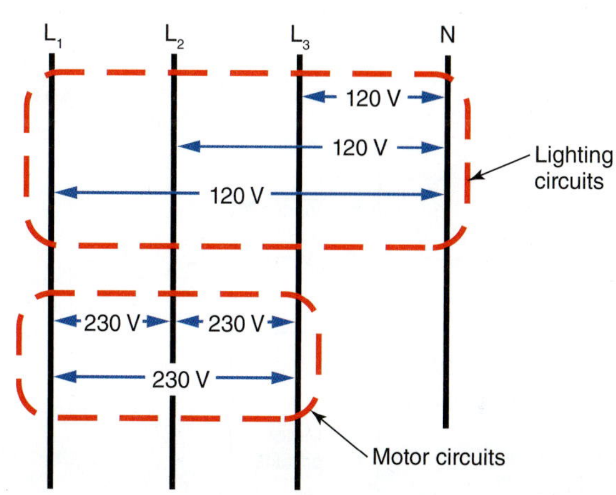

Goodheart-Willcox Publisher

Figure 23-2. Three-phase power is used for many commercial and industrial applications. Various hot and neutral wire combinations provide different voltages.

Three-phase motors use all three "hot" wires (no neutral). All three hot wires (L_1, L_2, and L_3) are required to obtain a three-phase supply. The voltage is measured between any two hot wires. Single-phase power is obtained from a three-phase supply by using just two hot wires, or one hot wire and the neutral. All voltmeters have two test leads that check the potential difference between any two hot wires or between one hot wire and the neutral.

An industrial customer, such as a factory, is supplied with a 480 V, three-phase, four-wire system (three hot and one neutral). The safety ground (grounding rod) is installed at the site. Much equipment inside the factory is operated by 480 V, three-phase motors. The lighting is operated on 277 V, single-phase (one hot leg and the neutral). See **Figure 23-3**. A step-down transformer can be used within the plant to reduce the 480 V, three-phase down to 240 V, three-phase. Voltage reduction provides additional options for use with smaller motors, incandescent lighting, or wall outlets. Large supermarkets also use this type of power system.

A commercial customer, such as a convenience store, is supplied with a 240 V, three-phase, four-wire system (three hot and one neutral). The safety ground (grounding rod) is added at the site. The three hot wires power refrigeration and air-conditioning equipment. Some equipment in the store operates on 208/240 V, single-phase, using two hot wires (and a safety ground). By using one hot wire and a neutral (with a safety ground), 120 V, single-phase is obtained for lighting circuits and receptacles. The single-phase loads must be equally divided among the three hot wires to balance the load on all three phases.

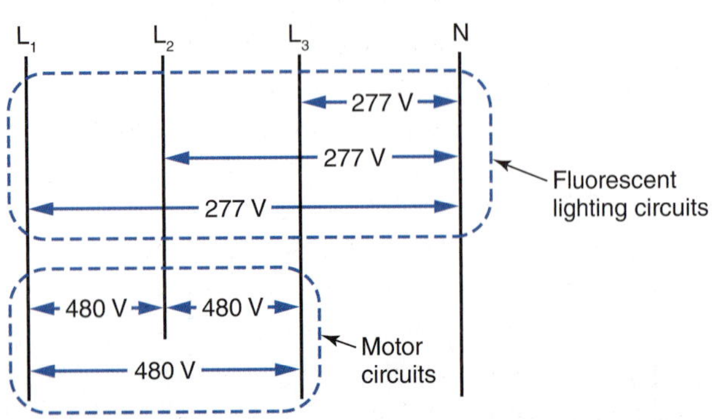

Goodheart-Willcox Publisher

Figure 23-3. A 480 V, three-phase, four-wire system is generally provided for customers with large power needs, such as industrial plants.

Meeting Residential Electrical Needs

Most residential units are provided with two hot wires and a neutral wire (240 V, single-phase). The safety ground is installed at the site, **Figure 23-4**.

Some home appliances, such as the water heater and central air-conditioning unit, require 240 V for operation. These appliances require two hot wires and the safety ground. Other appliances, such as the clothes dryer and electric range, require both 240 V and 120 V. *Dual voltage* appliances require two hot wires, the neutral, and the safety ground.

Residential lighting and receptacle circuits call for 120 V, single-phase connections. The connection is obtained by using one hot wire, the neutral, and the safety ground. The various 120 V circuits should be equally balanced between the two hot supply wires.

23.2 Transformers

A chief advantage of alternating current is that it can be generated at one voltage, transmitted at a higher voltage, and reduced to a lower voltage at the point of use. ***Transformers*** make it possible to increase (step up) or decrease (step down) the voltage. See **Figure 23-5**. This flexibility is highly desirable and used throughout the world for residential, commercial, and industrial electrical systems.

Transformers are 96% to 98% efficient. They have no moving parts and require little maintenance because of their simple and durable construction. A transformer has two copper windings (coils) that are electrically separated from each other. The coils are wound around a core of laminated soft iron sheets. The laminated core provides a circuit for the magnetic field, **Figure 23-6**.

A transformer operates on the principle that electrical energy can be efficiently transferred from one winding to another by magnetic induction. The alternating current flowing in the ***primary winding*** generates an alternating ***magnetic field*** that affects the laminated soft iron core. This magnetic field induces an alternating current in the ***secondary winding***. The magnetic field cuts across the turns of both windings; therefore, the same voltage is induced in each turn of the two

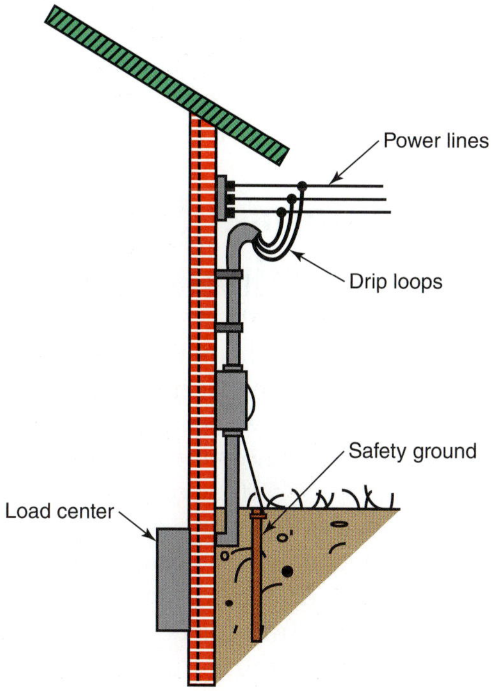

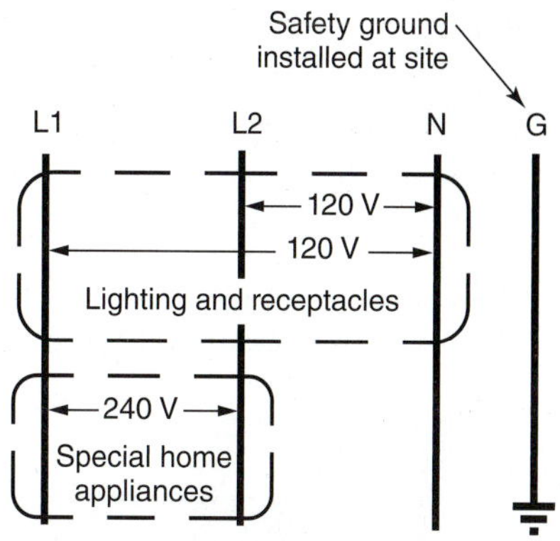

Goodheart-Willcox Publisher

Figure 23-4. A residential unit is typically supplied with a 240 V, one-phase, three-wire system. The safety ground is connected to a rod driven into the earth at least eight feet.

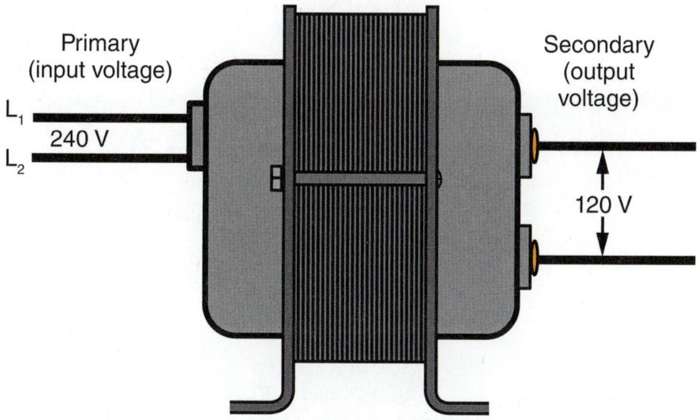

Goodheart-Willcox Publisher

Figure 23-5. A transformer is used to increase or decrease voltage. An electric power transmission and distribution system uses many transformers.

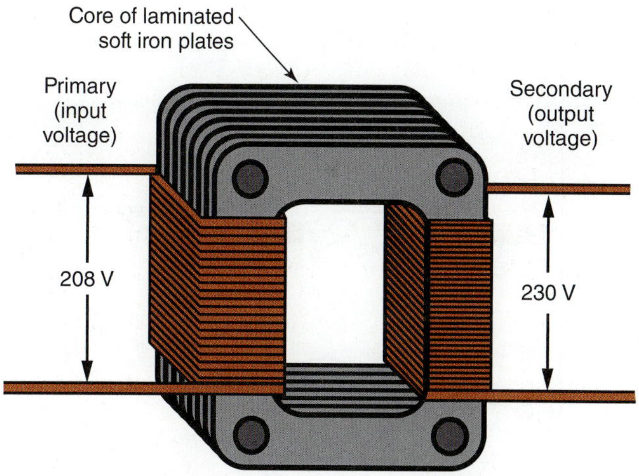

Goodheart-Willcox Publisher

Figure 23-6. Electrically separated windings transfer electrical energy through the iron core by magnetic induction. The differing number of turns of wire in the windings is responsible for the increase or decrease in voltage.

windings. The ***induced voltage*** is proportional to the number of turns in each winding.

Voltage at the secondary (output) winding is determined by the number of coils or wraps versus the number of coils in the primary (input) winding. Fewer coils in the secondary winding produce a lower output voltage (a ***step-down transformer***). More coils in the secondary winding increase output voltage (a ***step-up transformer***). See **Figure 23-7**.

Electrical energy conversion devices (electrical loads) are manufactured to tolerate a voltage of ±10%. However, many motors and other devices are limited to a 5% variation of voltage. A transformer is required to increase or decrease the supply voltage to match the load voltage specification. For example, a single-phase motor designed to specifically operate on 230 V cannot be connected to 208 V. A transformer is needed to change the power supply from 208 V to 230 V. As shown in **Figure 23-8**, the 208 V power supply is connected to

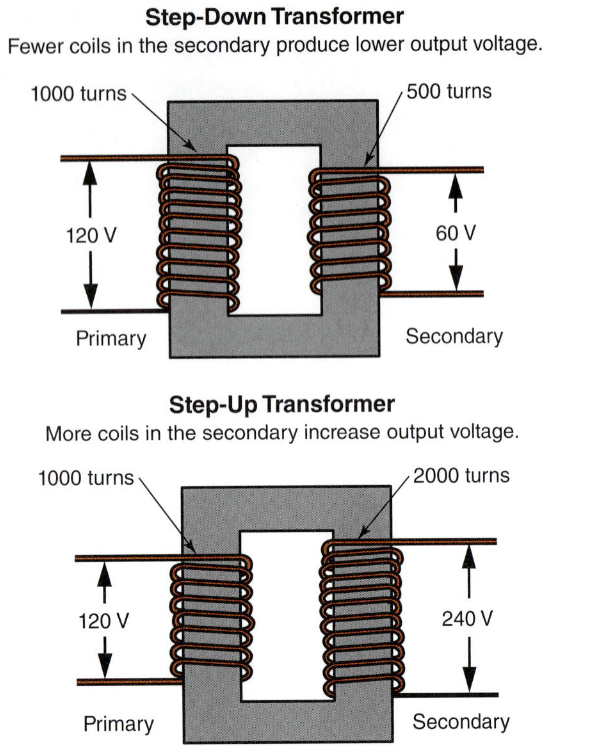

Goodheart-Willcox Publisher

Figure 23-7. Transformers can decrease (step down) or increase (step up) voltage.

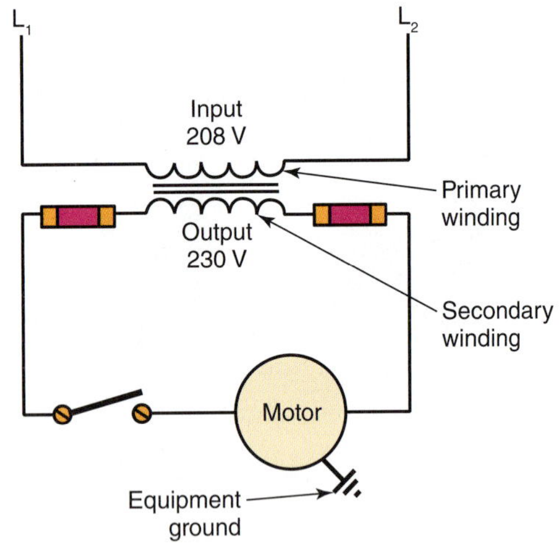

Goodheart-Willcox Publisher

Figure 23-8. A step-up transformer increases output voltage to the correct level for a motor.

the transformer's primary winding, and the motor is connected to the 230 V secondary winding.

Some single-phase transformers offer a choice of voltages and are called *multi-tap transformers*. The windings may have three or more connections (*taps*), which make it possible to use different input voltages or obtain different output voltages. Only *one connection* can be made on each winding at any time. The unused wires must be capped and sealed. Multi-tap transformers are convenient because one transformer can serve several applications. A wiring diagram or instructions are included on each transformer to ensure proper connections, **Figure 23-9**.

Transformer Symbols and Terminals

Electrical schematics commonly use similar symbols, shown in **Figure 23-10**. The left symbol, or similar variations, indicates a single-phase transformer. Shown in the center, three phase transformers consist of separate insulated windings for the three phases. The windings are on a three-legged core so three magnetic fields may be spaced 120° apart. On the right of the figure, three-phase transformers can be either delta- or wye-type. ***Delta*** is similar to a triangle (Δ), while ***wye*** (sometimes referred to as "star") is similar to the letter *Y*.

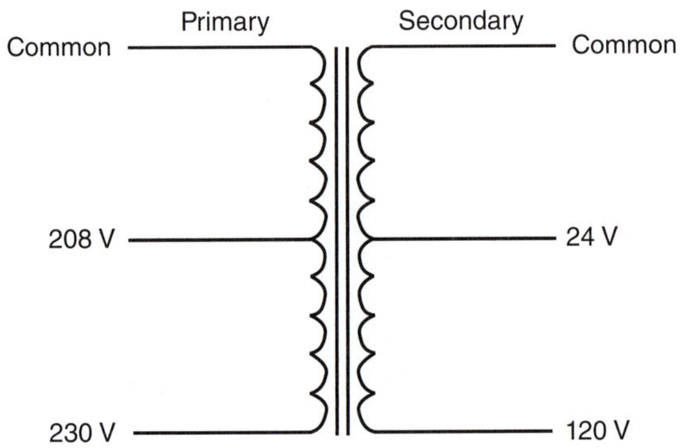

Goodheart-Willcox Publisher

Figure 23-9. Multi-tap, single-phase transformers provide several possible connections on both the input and output sides. Only one voltage connection per side can be made at a time. This input can be 208 V or 230 V. The output can be 24 V or 120 V.

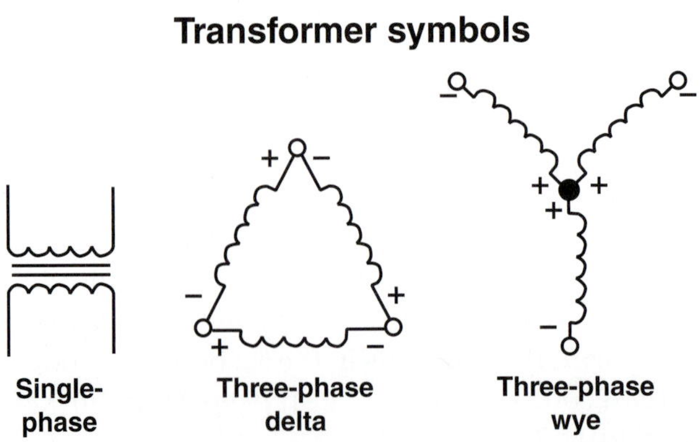

Goodheart-Willcox Publisher

Figure 23-10. Transformer symbols for use on electrical schematics.

Transformer ***terminals*** are labeled and tagged with letter and number combinations. Standard practice is to tag the transformer's primary terminals with H_1, H_2, H_3, and so on. The secondary terminals are tagged X_1, X_2, X_3, etc. (X_0 indicates the neutral terminal.) See **Figure 23-11**.

Transformer Ratings

In addition to the desired primary and secondary voltages, transformers are selected by the *amperage* that the secondary winding can safely carry. The secondary winding must be able to carry more than the minimum required amperage to accommodate the load(s) connected to it. ***Oversizing*** the transformer by 125% helps protect it from becoming overloaded.

Single-phase transformers are rated by VA (volt-amperes) at the secondary. VA = Volts × Amps (E × I). For example, a 48 VA transformer that has a 24 V secondary can safely carry two amperes of current (24 V × 2 A = 48 VA).

What size transformer should be selected if the secondary is to carry 20 A at 230 V? At minimum, 4600 VA (230 V × 20 A = 4600 VA). To protect the transformer from overload, a 125% oversizing is recommended. Therefore, a transformer rated at approximately 6000 VA should be chosen.

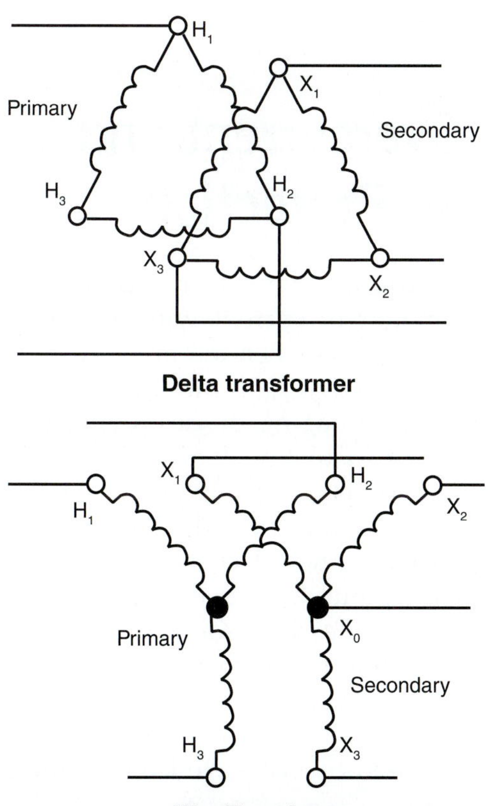

Goodheart-Willcox Publisher

Figure 23-11. Transformer terminals are identified by the letter *H* on the primary side and the letter *X* on the secondary side.

Transformer ratings of over 1000 VA are normally given in kVA (k = 1000). Therefore, the rating for a 6000 VA transformer is 6 kVA. Transformers with 1 kVA or higher ratings are always grounded by a separate cable or bus to a grounding electrode. The grounding electrode may be the steel beams supporting the building or a rod driven into the earth. A transformer rated under 1 kVA may be grounded to the cabinet or the conduit serving it. The cabinet or conduit is connected to the grounding electrode.

An overloaded or ***undersized*** transformer burns out because the secondary coil cannot carry the current. A minor overload causes a slow burnout, and a large overload causes a quick burnout. Transformers should be protected with fuses. *Oversizing* a transformer does no harm but is more expensive. When replacing a burned-out transformer, always check amperage in the secondary circuit to be sure there is adequate capacity.

Three-Phase Transformers

As noted earlier, three-phase transformers can be either wye- or delta-type, although the wye (star) is most popular. Three-phase transformers have a better ***power factor*** than single-phase transformers. A transformer's power factor is computed by multiplying VA × 1.73 and converting the answer to kVA.

Because electricity is produced in three-phases, a variety of transformer types (wye, delta, wye-delta, or delta-wye) must be used to produce the secondary voltages that match customer needs. Loads and other electrical equipment connected to the transformer secondary *must* match the voltage supplied by the transformer. **Figure 23-12** shows some commonly used voltages.

Some older installations use a "high-leg" system from a three-phase, delta transformer. Voltage readings from two of the hot legs to the neutral read 115 V. However, a reading from *one* of the hot legs to neutral registers 208 V. The higher-voltage wire is called the ***stinger leg*** (or wild leg) and cannot be used for 115 V circuits. The stinger leg is normally color-coded orange and should be tagged or marked for easy identification, **Figure 23-13**.

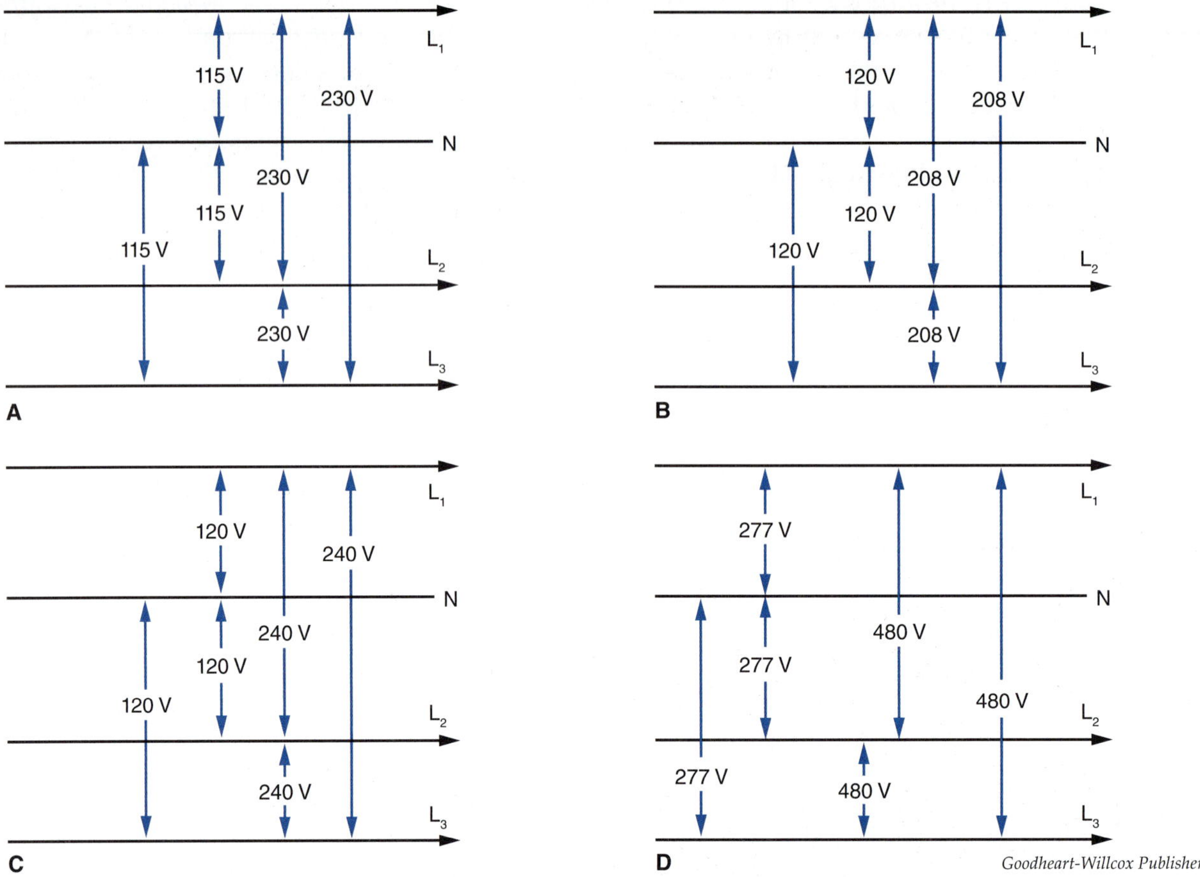

Figure 23-12. Transformers can provide a variety of voltages to meet different customer needs. A—230 V, three-phase, four-wire. B—208 V, three-phase, four-wire. C—240 V, three-phase, four-wire. D—480 V, three-phase, four-wire.

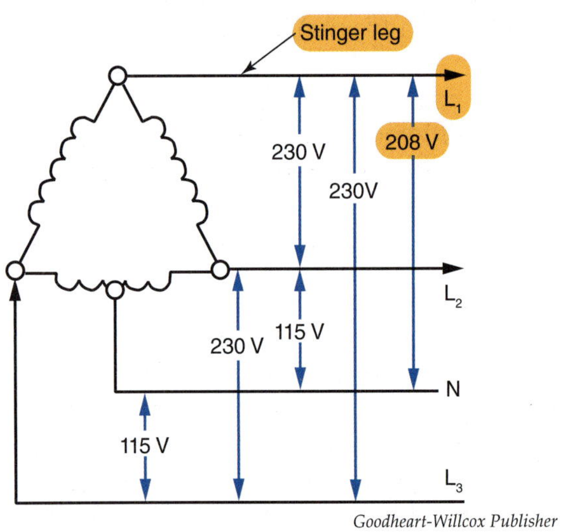

Figure 23-13. In a 230 V, three-phase, four-wire system with a "high leg," one leg provides higher voltage with the neutral wire than the other legs. In this case, the L_1 to neutral reading is 208 V. Readings from the other legs to neutral are 115 V. The high or "stinger" leg should be marked for easy identification.

23.3 Overcurrent and Circuit Protection

All conductors (wires) are sized to carry a limited amount of current without overheating. Overcurrent occurs when too much current flows through a wire. Overcurrent causes the wire to become hot, damaging or destroying the insulated covering. Insulation damaged by overcurrent presents a serious hazard. Exposed (bare) wire is dangerous to equipment and people. Overcurrent can be caused by such electrical problems as loose connections, ground faults (short circuits), defective resistance, or excessive loads.

Overcurrent in the form of an overload or short circuit is described as "current in excess of the normal flow for a given circuit." ***Overloads*** are typically caused by excessive loads in the circuit, loose connections, and dry motor bearings. An overload is usually confined to the circuit wiring. A ground fault is a type of short circuit that allows electrons to travel unrestricted through a path with low or no resistance to the frame of an appliance or device to make it electrically live. Other types of short circuits can occur from error or careless service practices. See **Figure 23-14**.

SAFETY NOTE

The Dangers of Overcurrent

Overcurrent resulting from a short circuit may exceed the normal current hundreds of times. High heat is generated rapidly and explosions may occur. A ground fault or short circuit is dangerous.

Circuit Protection

Every high-voltage electrical circuit must have some type of safety device, such as a fuse or a circuit breaker, to protect against overcurrent. Fuses and circuit breakers are manufactured in various shapes and sizes. They are designed to stop the flow of current when it exceeds safe limits. These ***circuit protection devices*** are rated in amperes and volts. Their amperage rating must not be greater than the ampacity of the wires being protected. Recall that ampacity is the amount of current a wire or load can safely carry.

Fuses and circuit breakers must be able to interrupt the extreme overcurrent created by a short circuit. However, each device has an *overcurrent interrupting limit* beyond which it can no longer interrupt current flow. If the overcurrent limit is exceeded, the device may violently arc or even explode and start a fire. The limit is called the ***amperage interrupting capacity (AIC)*** of the device. The AIC is much larger than the load current rating of a fuse or circuit breaker. The ***load rating*** is the current carrying-capacity during normal operating conditions.

All overcurrent protective devices are labeled with their normal load current rating and AIC for a given voltage. The amperage interrupting capacity of most circuit breakers is between 10,000 AIC and 20,000 AIC. Most fuses are rated at about 200,000 AIC.

Fuses

The purpose of a fuse is to detect excessive load current and open the circuit before danger arises. Fuses are typically located in the main power supply and in each branch circuit. A blown fuse in a branch circuit helps confine the problem to a specific area. Fuses and circuit breakers are used to protect circuits from overheating due to excessive current, **Figure 23-15**.

Zinc is a metal that has moderate internal resistance to the flow of electrons and a rather low melting point. For these reasons, it is often used as

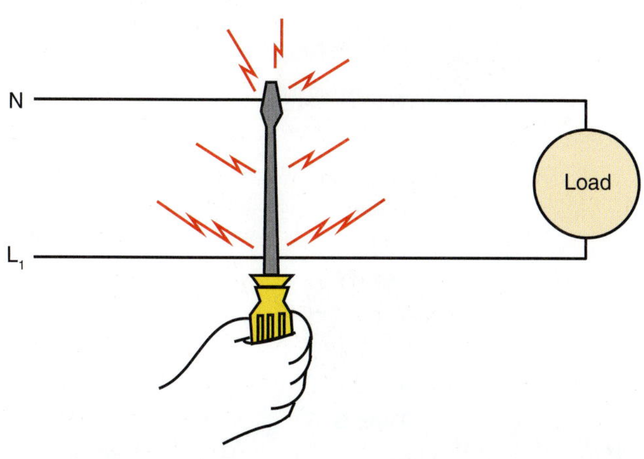

Goodheart-Willcox Publisher

Figure 23-14. Never work on a live circuit. A slip of a screwdriver or other tool can easily cause a short circuit. The resulting high overcurrent can have painful or even fatal results.

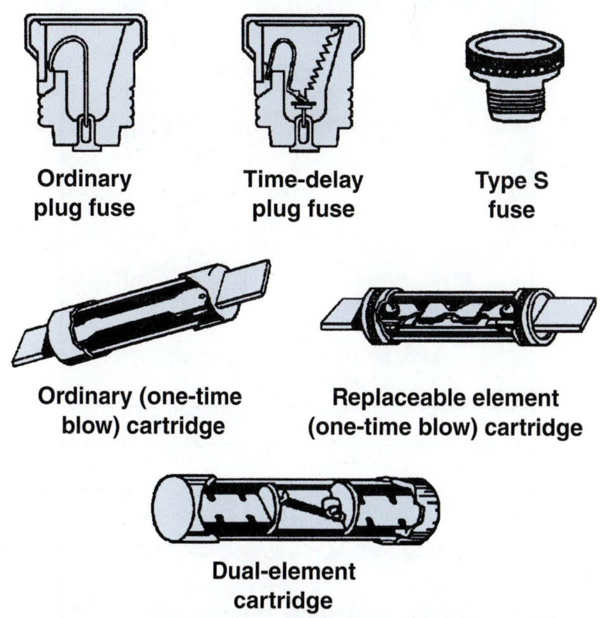

Goodheart-Willcox Publisher

Figure 23-15. Fuses have two basic shapes: plug and cartridge. These are available in many amperage ratings. Time-delay fuses allow short-duration overloads (such as a motor starting) without interrupting the circuit.

the connecting link (fusible element) in a fuse. See **Figure 23-16**. When excess current flows through the fuse, the link becomes overheated and melts (blows). The circuit opens and stops current flow.

The *amount* of excess current flow determines how fast the link melts. On a short circuit, the link is instantly vaporized. Overloads cause a slow buildup of heat, and the link melts slowly. A fusible link opens in 1 to 15 minutes on a 50% circuit overload. However, most fuses can withstand a 10% overload indefinitely.

A blown fuse indicates a serious circuit problem that must be corrected. The load amperage rating of a fuse must not exceed the ampacity (safe current level) of the feeder wire. Oversizing fuses is an invitation to disaster.

Plug-Type Fuses

Plug-type fuses (10 A to 30 A) are sometimes used for 120 V circuits. Three types of fuses are available: ordinary, time-delay, and Type S. See **Figure 23-17**.

Ordinary Plug Fuse

This type of fuse has no time delay. The link melts (blows) when the amperage rating is exceeded. The fuse prevents overheating of wiring in the circuit due to excess current flow, which could cause a fire. The fuse should open the circuit before the wire becomes hot.

Time-Delay Fuse

The time-delay, or ***dual element***, fuse is designed to permit an overload of short duration but to blow instantly if a short circuit occurs. Time-delay fuses are necessary when fusing a circuit containing an electric motor. They permit the momentary high starting current of the motor (about five times the normal running current) without opening the circuit. The high current is brief but occurs every time the motor starts. As the motor achieves full speed, counter-emf (increased resistance) fully develops, and the current drops to the normal running level.

An ordinary fuse used with a motor would have to be oversized to permit high starting

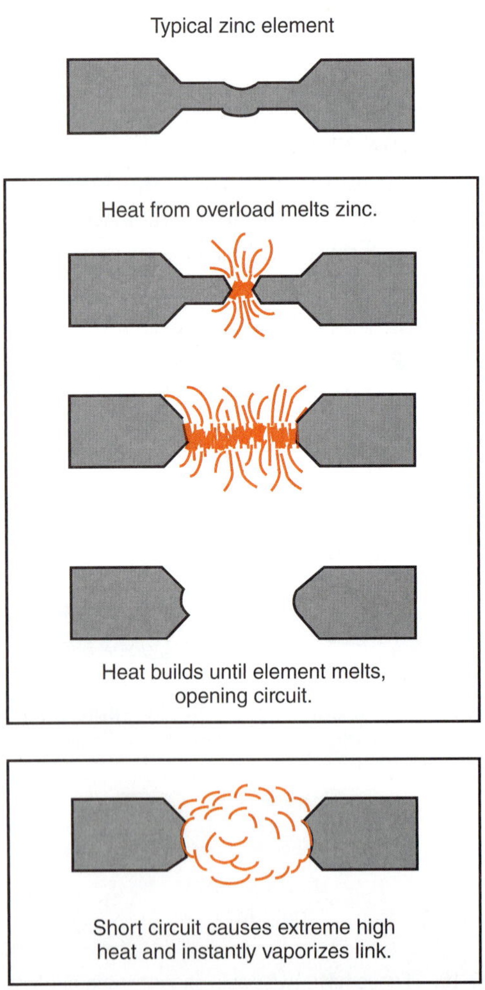

Goodheart-Willcox Publisher

Figure 23-16. The zinc element in a fuse reacts progressively to an overload and instantly to a short circuit.

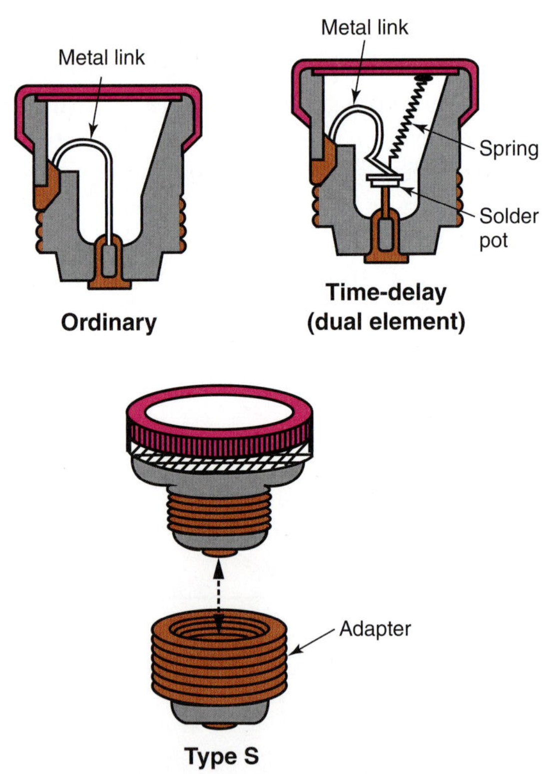

Goodheart-Willcox Publisher

Figure 23-17. Plug-type fuses are available as either ordinary or time-delay types. Type S is a special time-delay fuse with threads sized differently for each ampere rating. Adapters ensure a properly sized Type S fuse is used.

current. However, an oversized fuse is too highly rated to protect the wires during normal operation. If the fuse is sized for normal running current, it will blow every time the motor starts. The time-delay feature is the solution.

As shown in **Figure 23-18**, time-delay is accomplished by adding a spring and attaching one end of the link to a heat sink with a low-melting-point solder. When an overload occurs, heat is generated. If the condition continues long enough, the heat will melt the solder. The spring pulls the link away from the heat sink, opening the circuit. If a short circuit with high current occurs, the metal link will vaporize instantly and open the circuit.

No heat is generated if the motor starts quickly, so the solder does not melt. If the motor binds or does not start promptly, excess current flow generates enough heat to melt the solder. The delay feature permits the fuse to be sized according to the motor's *running* current, not the starting amps.

Time-delay fuses are sized according to the ***full load amperage (FLA)*** of the motor. FLA information is located on the motor data plate. A time-delay fuse is selected with a rating 25% greater than the motor FLA (*oversized* by 25%). For example, a motor with an FLA of 8 A should use a 10 A fuse.

Type S Fuse

This fuse has the time-delay feature but is made with threaded sections. The size of the threaded section varies according to the load amperage rating of the fuse. Type S fuses are used with an adapter that screws into the socket on the fuse panel and locks in place. The adapter is almost impossible to remove and prevents installing the wrong size fuse in the circuit. Refer to **Figure 23-17**. Type S fuses are commonly used in manufactured homes because of the greater risk of fire in such installations.

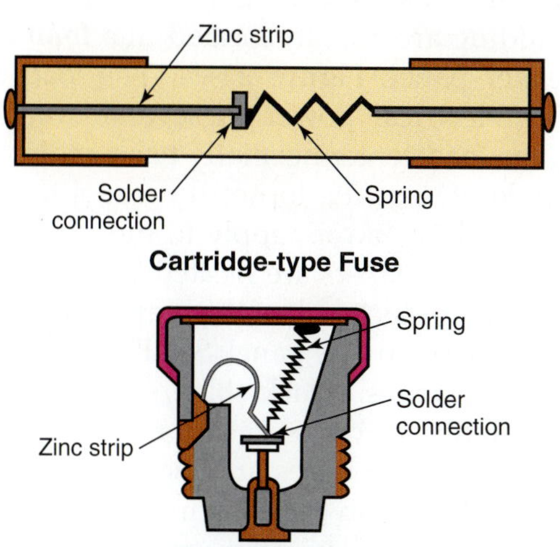

Goodheart-Willcox Publisher

Figure 23-18. Time-delay fuses permit brief overloads, such as the short period of high current flow when a motor starts. Prolonged overloads melt the solder connection, causing the spring to open the circuit.

Cartridge Fuses

Cartridge fuses are available as ordinary fuses or as time-delay, dual-element types, **Figure 23-19**. Round-end cartridge fuses are suitable up to 60 A. Knife-blade contacts are used for fuses over 60 A. It is important that the fuse ends make good contact in the fuse holder, **Figure 23-20**. Poor connections, which can produce high temperatures, or high air temperature around the fuse, reduce the

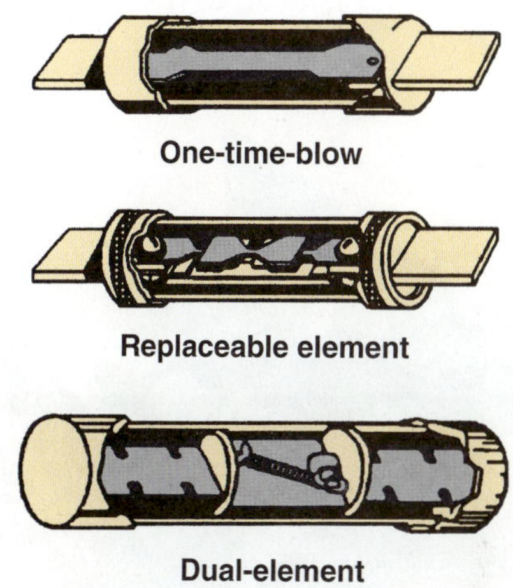

Goodheart-Willcox Publisher

Figure 23-19. Cartridge fuses are available in standard or time-delay (dual-element) types. In standard cartridges, either disposable (one-time-blow) fuses or those with replaceable elements can be used.

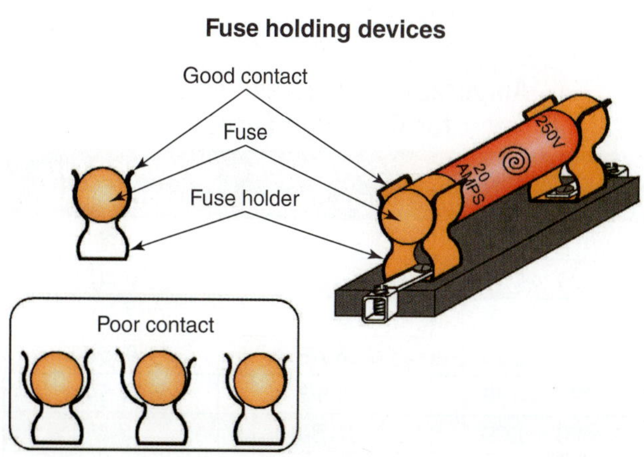

Goodheart-Willcox Publisher

Figure 23-20. A good mechanical fit between the fuse and fuse holder is important. Poor connections reduce the fuse's amperage rating.

fuse's amperage rating and cause needless blows and shutdown.

Amperage and voltage ratings determine the physical size of the cartridge fuse. Fuse-holding devices are also sized so fuses and holders match, **Figure 23-21**. The fuse holders limit the maximum fuse size. Such limitations help prevent oversizing fuses on circuits designed for a certain maximum amperage. For example, a cartridge fuse rated at 250 V with a rating from 1/10 A to 30 A only fits in a receptacle designed for a maximum of 30 A service at 250 V. See **Figure 23-22**.

Circuit Breakers

Circuit breakers are available as single-pole (one switch), double-pole (two switch), and three-pole (three switch) types. See **Figure 23-23**. The ***single-pole breaker*** is used to disconnect the black (hot) wire on 120 V, single-phase branch circuits. The ***double-pole breaker*** is used to disconnect both hot wires on 230 V, single-phase branch circuits. A ***three-pole breaker*** is used to disconnect all three hot wires on a three-phase circuit.

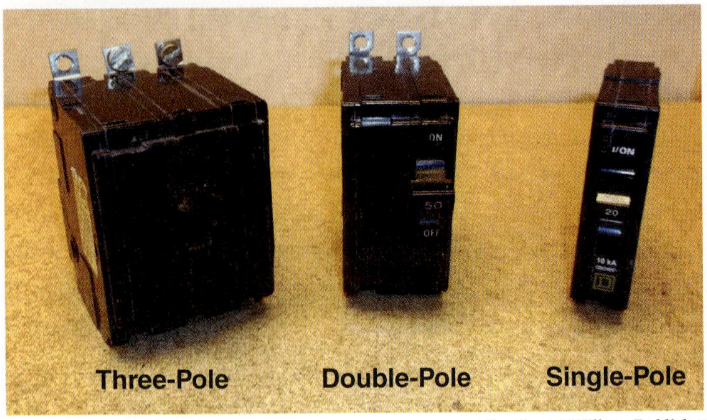

Goodheart-Willcox Publisher

Figure 23-23. Circuit breakers have one, two, or three switches, depending upon the application.

Goodheart-Willcox Publisher

Figure 23-21. Fuse-holding devices for cartridge fuses come in various forms. This fuse holder kit has metal clips that make a good mechanical connection.

Amperage Ranges and Lengths for Cartridge Fuses

Amperage Range	250-Volt Fuse Length (inches)	600-Volt Fuse Length (inches)
1/10 to 30	2	5
35 to 60	3	5 1/2
70 to 100	5 7/8	7 7/8
110 to 200	7 1/8	9 5/8
225 to 400	8 5/8	11 5/8
450 to 600	10 3/8	13 3/8

Goodheart-Willcox Publisher

Figure 23-22. Ampere ranges and lengths for 250 V and 600 V cartridge fuses. Fuse holders are sized to limit the maximum amperage fuses they can hold.

Circuit breakers perform the same job as fuses but can be reset after they open the circuit. When its load amperage rating is exceeded, the breaker *trips* (switches to the *mid-position*). All circuit breakers require a manual reset to the "on" position but first must be moved to the "off" position from the mid-position. Some older breakers trip to the "off" position, in which case the breaker can be reset by simply moving to the on position. A tripped circuit breaker indicates an overcurrent in the circuit. Failure to correct the problem will only result in another trip-out. Loose wire connections on the breaker can produce high temperatures just like the fuse and cause the breaker to trip.

Circuit breakers serving a building or portion of a building are usually located in a ***load center***, or breaker panel, **Figure 23-24**. The load center supplies electrical power to several branch circuits. The load center normally has a main circuit breaker located at the top of the panel to protect and disconnect power supply to the entire panel. Other circuit breakers are located below the main breaker and protect each of the branch circuits that obtain power from the panel. See **Figure 23-25**.

The hot wires (black, blue, or red) for the branch circuit are connected to a circuit breaker. All white (neutral) wires are connected to a neutral busbar which, in turn, is directly connected to a grounding electrode (earth ground). The safety ground busbar is directly connected to another grounding electrode. In some residential systems, however, the safety ground wires are connected to the neutral busbar in the main breaker panel for direct access to a single grounding electrode.

Goodheart-Willcox Publisher

Figure 23-24. A typical load center or breaker panel. The main circuit breaker disconnects the incoming electrical supply from the branch circuits. Branch circuit breakers protect individual circuits.

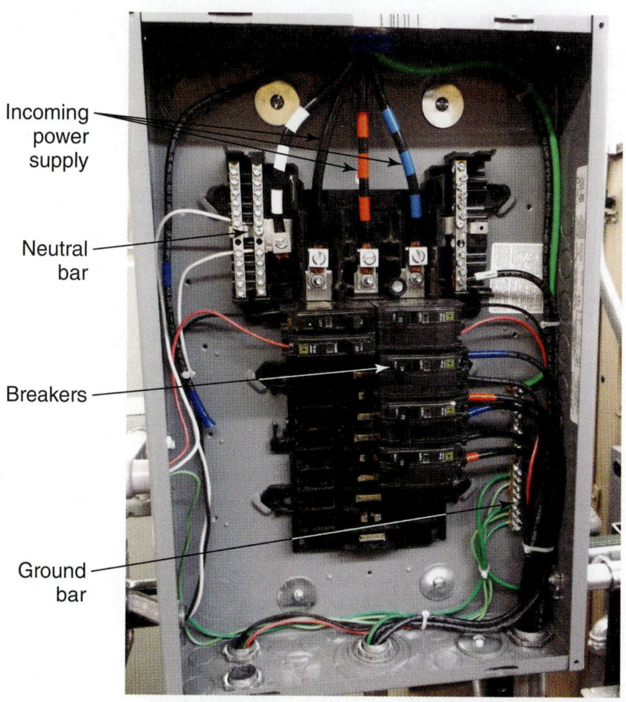

Goodheart-Willcox Publisher

Figure 23-25. The hot wire for each branch circuit is connected to the appropriate breaker, while the neutral wire is connected to a grounded neutral busbar. In some residential installations, the safety ground wire is also connected to the neutral busbar. For commercial installations, a separate safety ground busbar must be used.

This is the *only* place where the neutral and safety ground may be connected together.

Ground Fault Circuit Interrupters (GFCIs)

The importance of a good *safety ground system* cannot be overemphasized. Many people have lost their lives because of no safety ground or poor grounding methods. A ground fault can occur as a result of defective, worn, or misused equipment, especially when the equipment is operated in damp or wet areas. Examples include portable power tools, hair dryers, electric shavers, and kitchen appliances. Such equipment *does* fail. When it fails, it presents a dangerous situation to the operator.

Fuses and circuit breakers do *not* protect people using such equipment. The ***ground fault circuit interrupter (GFCI)*** is designed to perform a dual role: protect the circuit from overcurrent *and* protect people from potentially hazardous ground faults arising from the use of defective appliances or portable tools. See **Figure 23-26**. The National Electrical Code requires the use of GFCIs outdoors and in bathrooms, swimming pool areas, attached garages, motel guest rooms, and all countertop receptacles within six feet of the kitchen sink. GFCIs are available as circuit breakers and

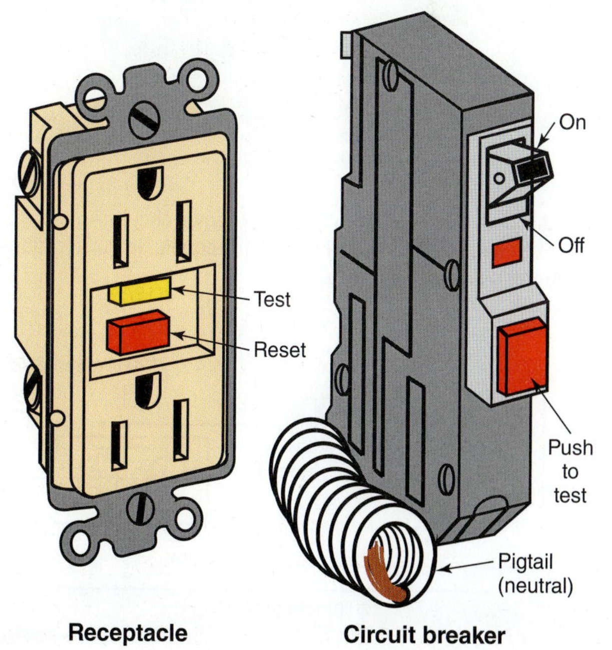

Goodheart-Willcox Publisher

Figure 23-26. Ground fault circuit interrupters protect people from potentially hazardous ground faults. GFCIs are available as receptacles or circuit breakers.

receptacles. Extension cords with built-in GFCI protection are available for use with portable tools.

How the GFCI Works

As shown in **Figure 23-27**, current normally travels to an appliance along the black (hot) wire and returns along the white (neutral) wire. The amperage in each conductor is the same. No current flows through the safety ground.

However, when a ground fault occurs, the normal current flow pattern (through hot and neutral wires) changes, because some current travels along the safety ground wire (or through a person's body) to ground. The flow creates an amperage *imbalance* between the hot and neutral wires.

The GFCI monitors current flow in both the hot wire and the neutral wire and is sensitive to any imbalance. Since an imbalance can only occur during a dangerous fault to ground, the GFCI opens the circuit whenever the imbalance reaches about 4 mA to 6 mA (milliamperes). See **Figure 23-28**.

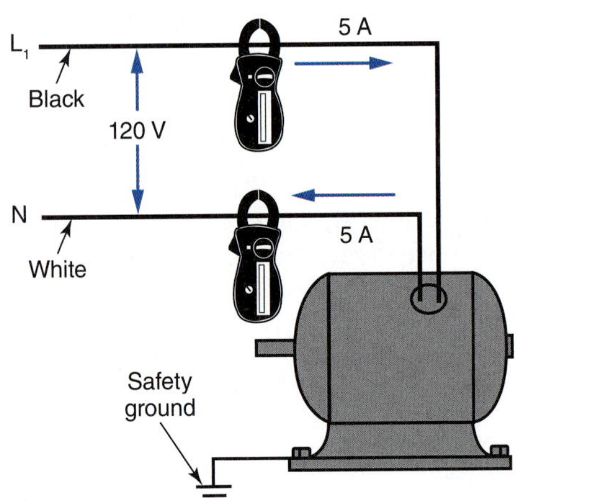

Goodheart-Willcox Publisher

Figure 23-27. Normal current flow in a motor circuit. Amperage in the hot wire (L_1) and neutral wire (N) is equal. No current flows through the safety ground wire.

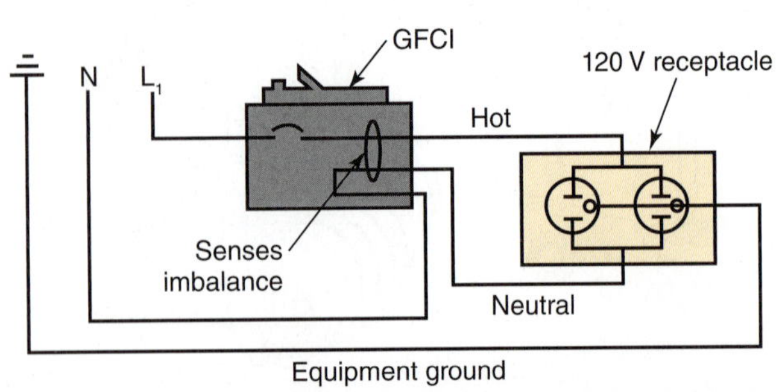

Goodheart-Willcox Publisher

Figure 23-28. A GFCI senses an amperage imbalance between hot and neutral wires as a result of a ground fault. It quickly opens the circuit to prevent injury to a person who might become part of the circuit.

As noted earlier, the GFCI serves a dual purpose: people *and* circuit protection. If the normal load amperage rating (15 A, 20 A, or 30 A) is exceeded, it will open the circuit. In this manner, it operates much like an ordinary circuit breaker.

23.4 Electrical Connections

Conductors are wires that carry electrical energy from place to place along a pathway. Wires connect to each other, terminating at various electrical devices, and are covered with ***insulation*** to prevent energy loss and unintentional flow of current. Where connections must be made, anywhere from 1/4″–1/2″ of insulation is removed from the wire ends, depending on application, allowing enough bare wire to make proper metal-to-metal connections.

Characteristics of a Good Connection

Properly made connections require a clean, tight contact between the conductor and the device terminals. The National Electrical Code does not permit splicing or connecting wires except inside a proper housing. Connections to an electrical device also must be made inside an approved housing or box. Connections between copper and aluminum can be made *only* with the use of an approved connector.

Poor connections are a constant source of electrical problems. A loose connection permits arcing and becomes a built-in resistance. Such poor connections cause voltage drop, energy loss, and overheating. The importance of good electrical connections cannot be overemphasized.

Making Terminal Connections

When connecting a solid conductor to a screw-type terminal, bend the wire end into a loop. Use long-nose pliers to form a circular loop to the right, **Figure 23-29**.

For good contact, the loop should encircle the screw shaft in a *clockwise* direction. The screw is inserted through the loop and turned clockwise to

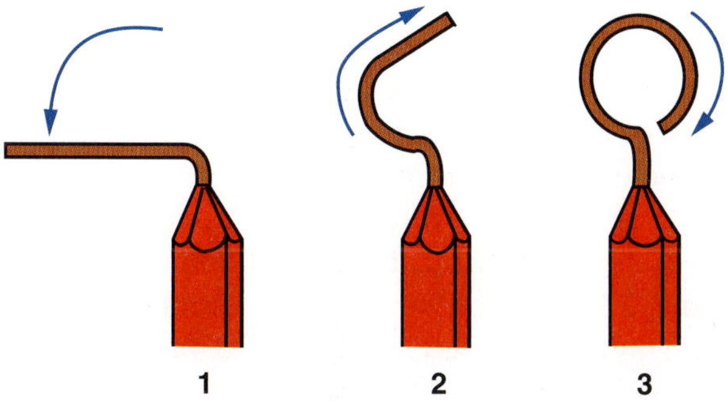

Goodheart-Willcox Publisher

Figure 23-29. A long-nose pliers is used to make a proper loop for wire connection to a screw terminal. The loop must be made in a clockwise (right-hand) direction.

tighten it. This causes the loop to be pulled inward for a tight connection. If the loop faces left (counterclockwise), tightening the screw will cause the loop to open outward, resulting in a poor and possibly unsafe connection. See **Figure 23-30**.

Terminal Connectors

Solderless ***terminal connectors*** that crimp onto wire ends are popular and easy to use. Types and sizes are available to fit most applications, **Figure 23-31**. When installed properly, terminal connectors avoid most problems arising from loose connections.

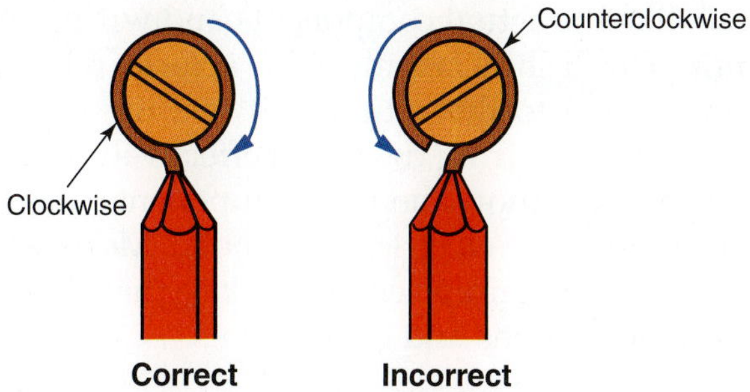

Goodheart-Willcox Publisher

Figure 23-30. The wire loop must encircle the screw in a clockwise direction so it is pulled inward as the screw is tightened. A counterclockwise loop tends to open as the screw is tightened.

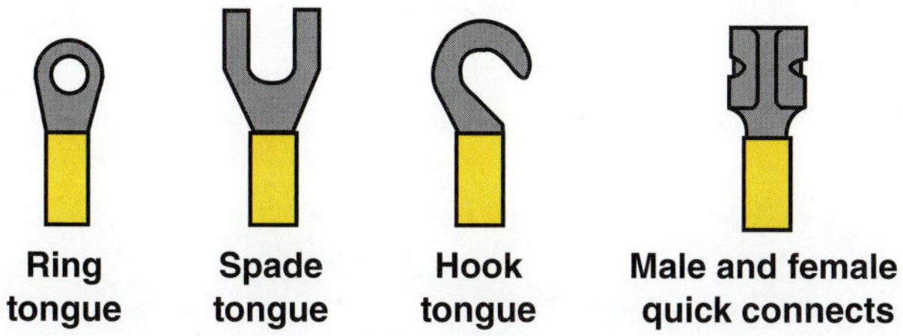

Goodheart-Willcox Publisher

Figure 23-31. Various types of terminal connectors are available for various applications. Terminal connectors almost eliminate loose connection problems.

To install a connector on a wire, remove about 1/4″ of insulation from the wire end. Then, insert the bare end into the connector. A special tool is used to ***crimp*** (squeeze) the connector onto the wire end, **Figure 23-32**. Two crimps usually achieve a secure bond between the wire and the connector. After crimping, make sure the connector does not pull off the wire. The connection is covered with a plastic insulator.

Solderless Screw-On Connectors

Solderless screw-on connectors are commonly called ***wire nuts***. They eliminate the need for soldering connections and are NEC-approved. Wire nuts are available in several sizes and made of plastic or Bakelite™ insulating material. To install a wire nut, remove about 1/2″ of insulation from the wire ends. Twist the bare ends together and screw the nut onto the twisted wires, **Figure 23-33**. Be certain the nut is tight and cannot fall off. If wire nuts become loose, the wires inside will separate.

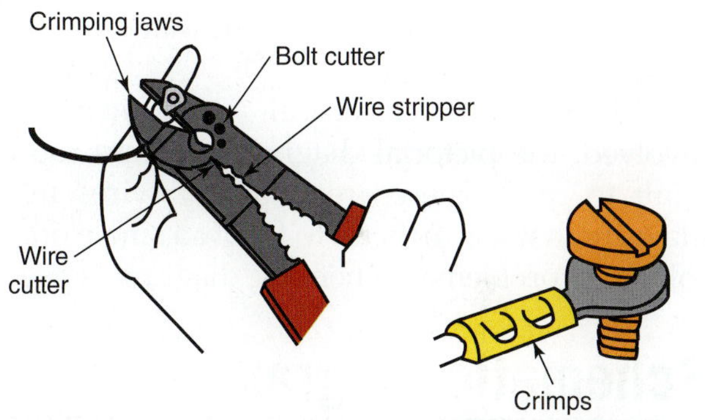

Goodheart-Willcox Publisher

Figure 23-32. A special multiple-use tool is used to crimp the connector to the wire end. As shown on the right, two crimps are usually made in the connector to ensure a good bond to the wire.

Goodheart-Willcox Publisher

Figure 23-33. A wire nut is installed by twisting the bare wire ends together, then screwing the nut onto the twisted wires. For a sound mechanical and electrical connection, the wire nut must be tight.

23.5 Wiring Diagrams

Manufacturers and organizations use different types of wiring diagrams for different purposes. Pictorial and schematic diagrams are two different types of illustrations commonly used to show electrical connections. Most manufacturers locate these important diagrams inside a cover panel on the appliance or other equipment. When making repairs and locating system problems, the technician must frequently refer to these diagrams to identify certain wires and component connections.

Pictorial Diagram

A ***pictorial diagram*** shows the physical appearance, component locations, wiring connections, and internal arrangement of a piece of equipment. See **Figure 23-34**. The components are arranged in the approximate positions they actually occupy inside the unit. Lines are then drawn from each component to indicate how the wires are connected in the circuit. These lines are sometimes labeled with the color or identification number of the wire used.

When many electrical components are involved, the pictorial diagram becomes too difficult to read. There are so many wires in the diagram that it is difficult to follow a single one. To solve this problem, a schematic diagram is used.

Schematic Diagram

A ***schematic diagram*** is an orderly method of presenting electrical circuits and components. The schematic diagram is usually referred to by technicians as simply the *schematic*. The schematic does not illustrate where the components are located or what they look like. It does illustrate the types of components involved and how they are connected in the circuit.

A standard set of ***electrical symbols*** is used for ease of communication in schematic diagrams, **Figure 23-35**. There are literally hundreds of different symbols used in electrical and electronic schematics, but technicians quickly learn to recognize those that are most commonly used.

Ladder Diagram

Another type of wiring diagram is a ***ladder diagram***. This is an electrical diagram that shows the logic of system operation. A ladder diagram is shaped somewhat like a ladder. The power supply lines are drawn parallel to form the side rails of a ladder, and loads are drawn between the power source lines, forming rungs. See **Figure 23-36**. Various switches and fuses are located on the rungs as needed to control and protect the loads. All components (motors, switches, fuses) in the diagram are indicated by electrical symbols.

Reading schematics and ladder diagrams requires more knowledge of electricity and electrical components than is needed to use a pictorial diagram. Schematics and ladder diagrams do not look like the actual components and wiring in the unit. The technician must be able to relate this information to the actual unit (via the pictorial). To use these wiring diagrams effectively, the technician must know the unit components and their corresponding electrical symbols. Most wiring diagrams contain a legend or key that describes the abbreviations and components.

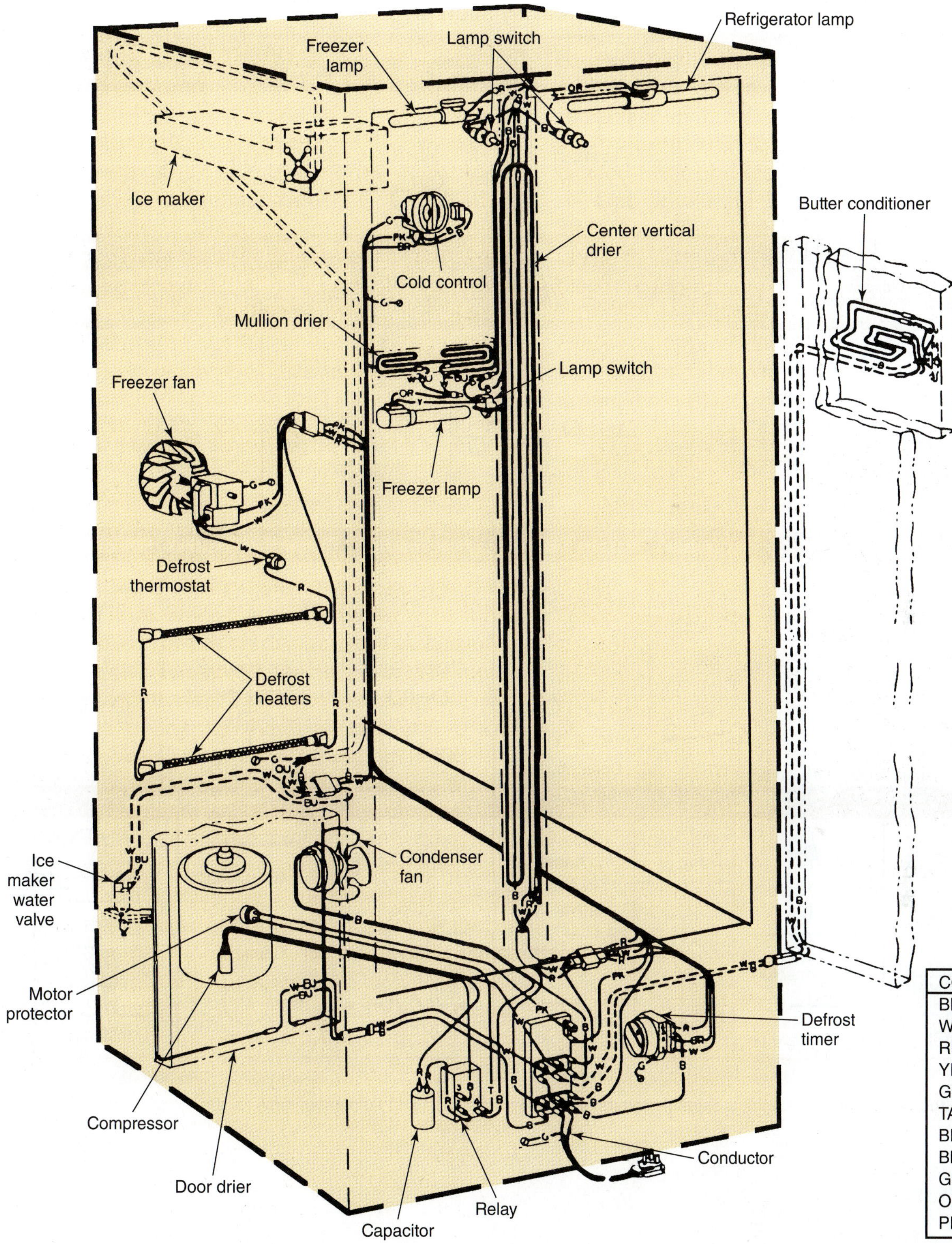

COLOR	SYM
BLACK	B
WHITE	W
RED	R
YELLOW	Y
GREEN	G
TAN	T
BROWN	BR
BLUE	BU
GRAY	GY
ORANGE	OR
PINK	PK

Frigidaire

Figure 23-34. On pictorial diagrams used for household appliances and similar types of equipment, the components are drawn realistically and in the approximate position they occupy inside the case. Note the key for wire colors at the lower right.

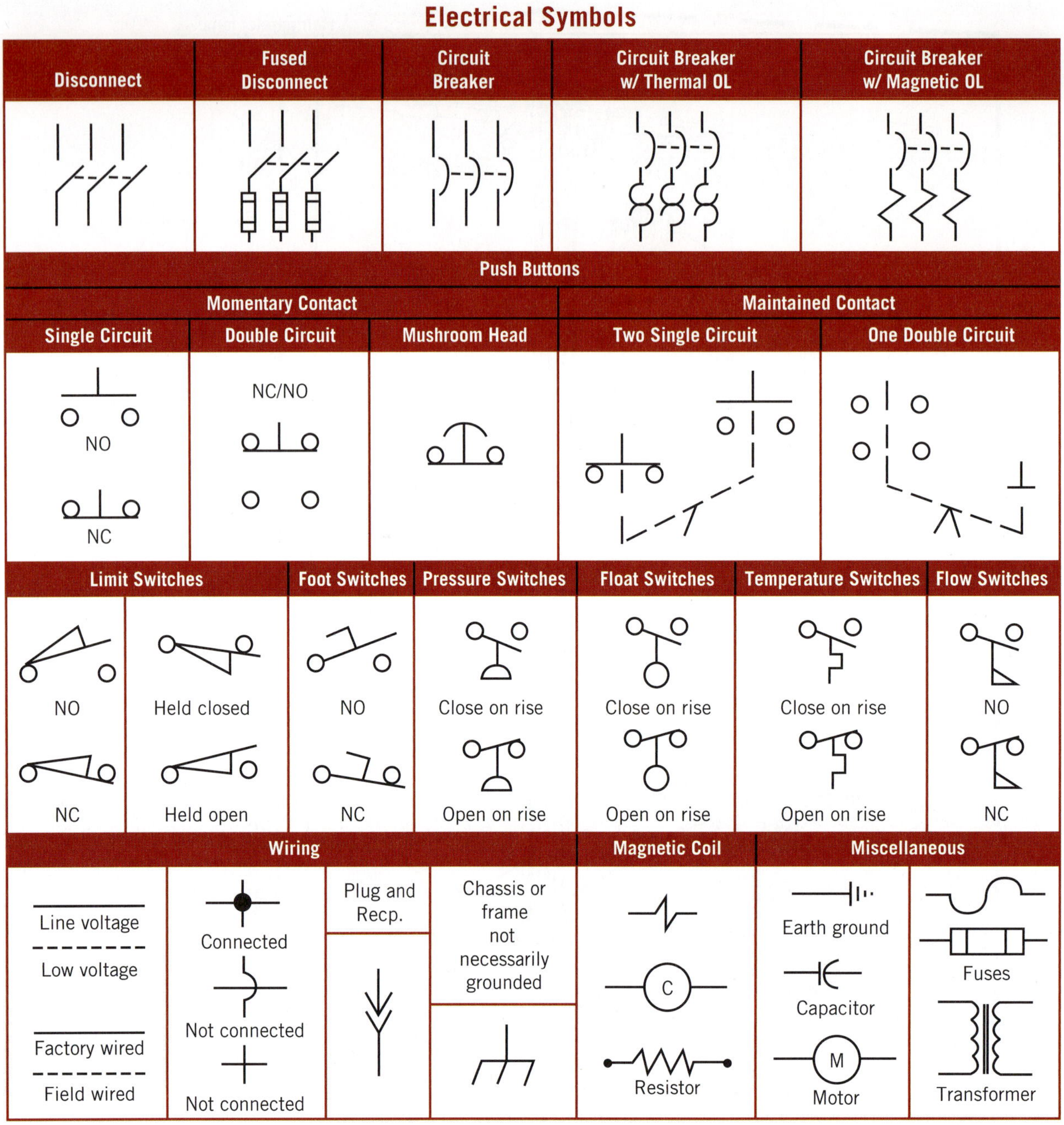

Goodheart-Willcox Publisher

Figure 23-35. A sampling of electrical symbols commonly used in schematic and ladder diagrams.

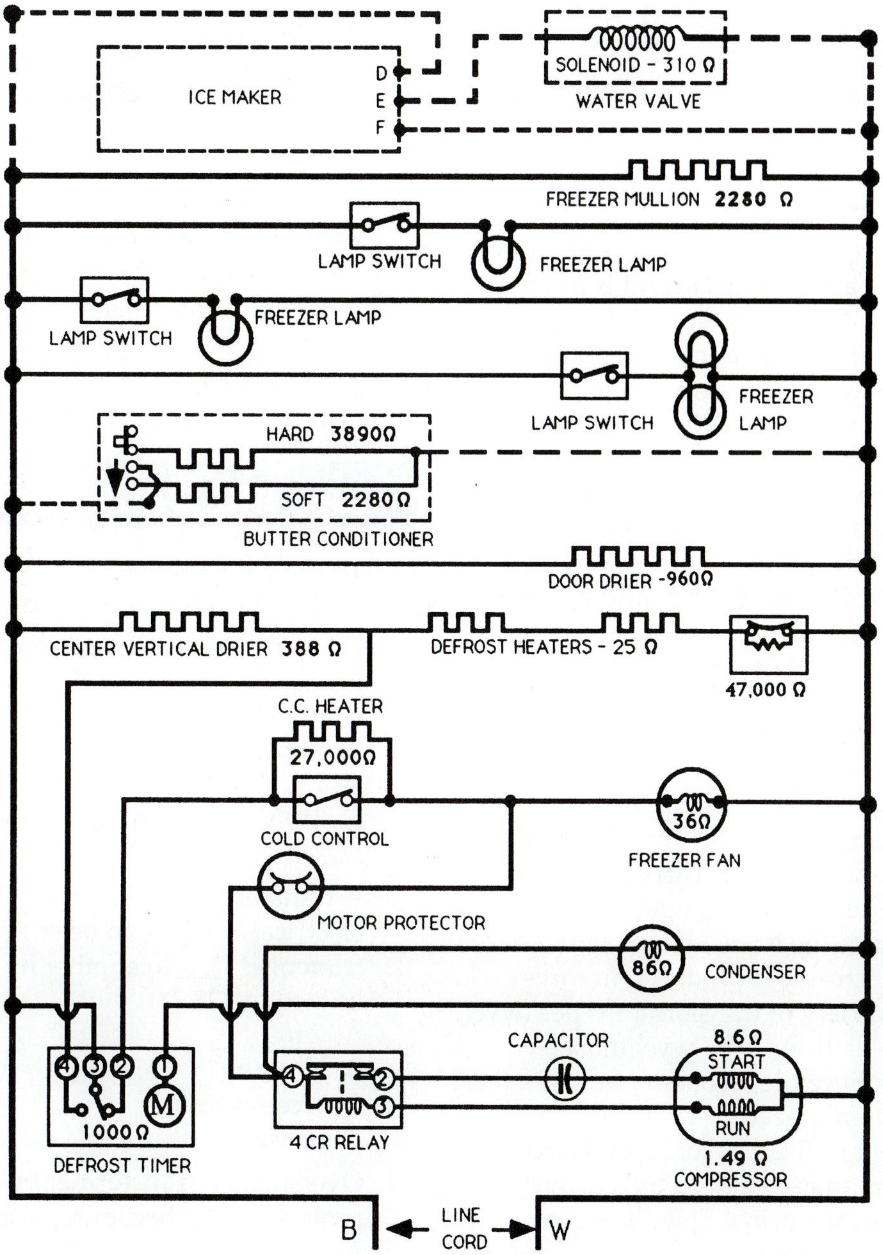

Frigidaire

Figure 23-36. A ladder diagram somewhat resembles a ladder with two vertical parallel rails and several horizontal rungs. This ladder diagram is for the refrigerator-freezer shown in **Figure 23-34**.

Chapter Review

Summary

23.1 For commercial and industrial applications, three-phase alternating current is most common for delivering electrical power. For residential applications, power is provided via two hot wires and a neutral wire with varying voltages depending on the application.

23.2 A transformer operates on the principle that electrical energy can be efficiently transferred from one winding to another by magnetic induction. Depending on the number of coils in the secondary winding, voltage can be either stepped up (more coils) or stepped down (fewer coils).

23.2 In addition to the desired primary and secondary voltages, transformers are selected by the *amperage* that the secondary winding can safely carry. The secondary winding must be able to carry more than the minimum required amperage to accommodate the load(s) connected to it. Because electricity is produced in three-phases, a variety of transformer types (wye, delta, wye-delta, or delta-wye) must be used to produce the secondary voltages that match customer needs.

23.3 Overcurrent in the form of an overload or short circuit is described as "current in excess of the normal flow for a given circuit." Overcurrent conditions can cause major damage to people and equipment near the circuit.

23.3 Circuit protection devices—such as fuses, circuit breakers, and GFCIs—are designed to stop the flow of current when it exceeds safe limits. Various types are available and are rated in amperes and volts.

23.4 Properly made electrical connections require a clean, tight contact between the conductor and the device terminals. Solderless connectors are available to provide tight connections.

23.5 A pictorial diagram shows the physical appearance, component locations, wiring connections, and internal arrangement of a piece of equipment. A schematic diagram is an orderly method of presenting electrical circuits and components. A ladder diagram is an electrical diagram that shows the logic of system operation with components indicated by electrical symbols.

Know and Understand

1. *True or False?* Single-phase power can be obtained from a three-phase system. (23.1)
2. What are the required supply voltage and phases for residential lighting and receptacle circuits? (23.1)
   A. 120 V, single-phase
   B. 120 V, three-phase
   C. 240 V, single-phase
   D. 240 V, three-phase
3. A transformer operates on the principle that electrical energy can be efficiently transferred from one _____ to another by magnetic induction. (23.2)
   A. wire
   B. winding
   C. receptacle
   D. phase
4. Oversizing a transformer by _____% helps protect it from becoming overloaded. (23.2)
   A. 10
   B. 25
   C. 50
   D. 125
5. Three-phase transformers have a better _____ than single-phase transformers. (23.2)
   A. efficiency
   B. power factor
   C. amperage rating
   D. protection from overcurrent
6. A ground fault is a type of _____. (23.3)
   A. short circuit
   B. overload
   C. circuit protection
   D. load center

7. The _____ of a circuit protection device is the current carrying-capacity during normal operating conditions. (23.3)
   A. amperage interrupting capacity (AIC)
   B. ampacity
   C. load rating
   D. voltage rating
8. *True or False?* A dual-element fuse should be used in circuits serving electric motors. (23.3)
9. The physical size of a cartridge fuse is determined by _____. (23.3)
   A. the space it is being placed in
   B. amperage and voltage ratings
   C. the number of loads in the circuit
   D. the number of phases in the circuit
10. Circuit breakers are normally located _____. (23.3)
   A. as close to the branch circuits as possible
   B. near the grounding electrode
   C. in a load center or breaker panel
   D. within an electrical application or load
11. *True or False?* When making a screw-type terminal connection, the wire should be bent into a loop that encircles the screw shaft in a counterclockwise direction. (23.4)
12. Wire nuts eliminate the need for _____. (23.4)
   A. wires
   B. soldering connections
   C. wire insulation
   D. a neutral wire
13. A standard set of _____ is used for ease of communication in schematic diagrams. (23.5)
   A. colors
   B. lines
   C. acronyms
   D. electrical symbols
14. Which of the following wiring diagrams shows the logic of system operation? (23.5)
   A. Pictorial
   B. Schematic
   C. Ladder
   D. Both pictorial and schematic diagrams

Apply and Analyze

1. Explain how electrical power is supplied to a convenience store. (23.1)
2. Explain how a transformer is used to increase or decrease voltage from alternating current. (23.2)
3. What size transformer should be selected if the secondary is to carry 20 A at 240 V? (23.2)
4. What is an operating difference between a fuse and a circuit breaker? (23.3)
5. Explain how to properly connect a solid conductor to a screw-type terminal. (23.4)
6. Explain the differences between pictorial, schematic, and ladder diagrams. (23.5)

Critical Thinking

1. Explain how a transformer and a single-phase motor can exhibit the same resistance in a circuit and how that similarity differs from a pure resistance. (23.1)
2. Explain why a three-phase motor does not need capacitors to start or run. (23.1)
3. Explain the operating differences in a motor when a winding-to-winding short occurs versus a short-to-ground. (23.3)
4. Explain the sequence of operation for the refrigerator-freezer shown in the ladder diagram in **Figure 23-36**. (23.5)

DiversiTech Corporation

24 Motors

LEARNING OUTCOMES

After studying this chapter, you will be able to:

24.1 Describe the operation and characteristics of induction motors.

24.2 Explain why start windings and overload protectors are used in single-phase induction motors.

24.3 Identify the five different types of single-phase induction motors and their unique design characteristics.

24.4 Recognize the two types of capacitors used in HVACR motors and their related ratings, safety practices, testing procedures, and application in changing motor types.

24.5 Explain how dual-voltage motors are designed to supply two voltages.

24.6 Explain how multi-speed motors are designed to run at different speeds.

24.7 Describe the operation and wiring of three-phase motors, including dual-voltage types.

24.8 Describe the operation and application of electronically commutated motors (ECMs).

24.9 Identify single-phase compressor terminals.

24.10 Troubleshoot induction motor problems.

24.11 Recognize various types of motor bearings and how to properly lubricate them.

24.12 Select and install fan blades used in refrigeration and air-conditioning systems.

TECHNICAL TERMS

- axial
- bearing
- bimetallic
- bleed resistor
- bushing
- capacitor
- capacitor-start, capacitor-run (CSCR) motor
- capacitor-start, induction-run (CSIR) motor
- centrifugal (squirrel cage) blade
- centrifugal switch
- common
- cycling on the overload
- dielectric
- dual-voltage motor
- electronically commutated motor (ECM)
- electrodes
- fractional horsepower
- full load amperage (FLA)
- hard start kit
- induced magnetism
- integral horsepower
- laminated
- lead end
- locked rotor amperage (LRA)
- microfarad
- motor overload protector
- multi-speed motor
- open capacitor
- open winding
- permanent split-capacitor (PSC) motor
- poles
- revolutions per minute (rpm)
- rotation
- rotor
- run capacitor
- run winding
- shaded-pole motor
- shading coil
- shaft end
- short cycle
- shorted capacitor
- shorted to ground
- shorted winding
- single-phasing
- slip
- split-phase motor
- squirrel cage rotor
- start capacitor
- start winding
- stator
- synchronous speed
- taps
- torque
- Vac
- zerk fitting

Introduction

Electric motors of the induction type are common in refrigeration, heating, and air-conditioning applications where they are used to drive compressors, fans, pumps, and other devices. Since a motor malfunction causes severe system problems, the technician must possess a good working knowledge of motors and their operation to properly troubleshoot a system and perform necessary repairs. Sometimes, the motor malfunction is a warning of other system problems that must be identified and solved.

There are six basic types of induction motors: five single-phase and one three-phase. Although methods of operation are similar, each type of induction motor serves a unique purpose. It is sometimes necessary to convert from one type of motor to another to achieve better operating characteristics.

24.1 Induction-Type Motor Operation and Characteristics

As mentioned, there are six different types of induction motors. However, the operation and characteristics of induction motors are generally the same. This section will discuss the general principles at work when an induction motor is in operation.

Induction Motor Operation

The induction motor uses ***induced magnetism***—magnetism generated by an electric current—to convert electrical energy into magnetic energy (a magnetic field). This magnetic field interacts with an existing magnetic field to cause mechanical energy (movement) in the form of a rotating rotor. As shown in **Figure 24-1**, a cylinder (***rotor***) is rotated inside a magnetic field produced by an assembly of stationary electromagnets (***stator***).

An electromagnet has two distinct advantages:

- Its core is magnetized only when current flows through the coil. When no current flows, there is no magnetic field.
- The polarity of an electromagnet can be changed by reversing the direction of the current flowing through the coil. Ordinary 60-cycle alternating current automatically reverses the north/south polarity of the stator poles 120 times per second.

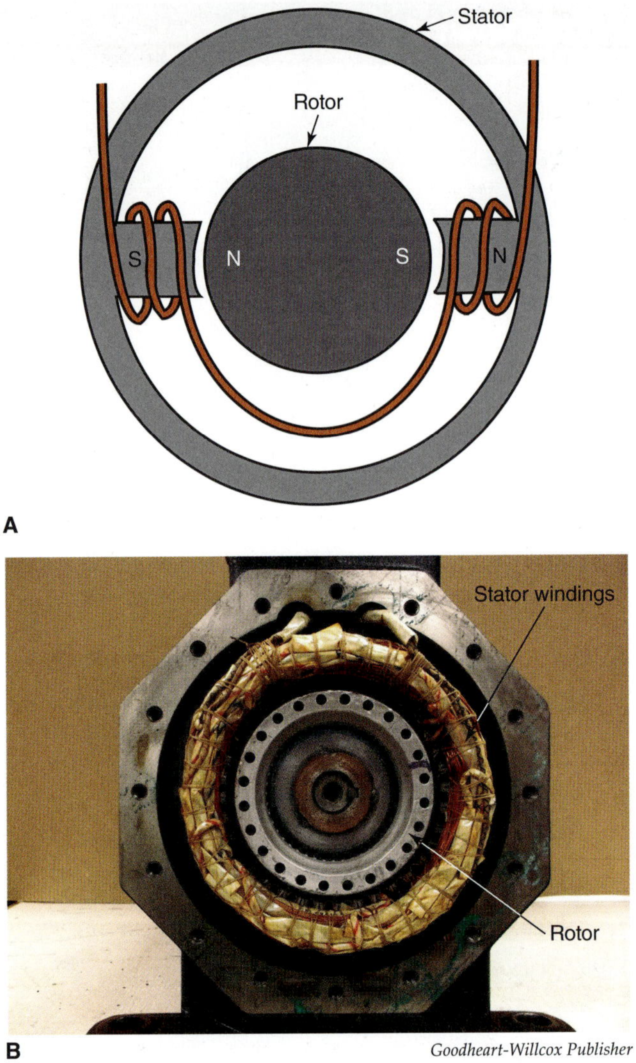

Goodheart-Willcox Publisher

Figure 24-1. Main parts of an induction motor. A—A simplified end view of an induction motor, showing the magnetic poles of the stator and rotor. B—A motor's rotor and the stator windings can be seen easily with the end bell of a semihermetic compressor removed.

As shown in **Figure 24-2**, two or more stationary electromagnets, referred to as ***poles***, are positioned at opposite sides of a circle inside the motor. These stationary poles have *opposite* magnetic polarity. When the alternating current flow automatically changes from positive to negative, the polarity of the electromagnets reverses as well (north to south or south to north). The polarity of the electromagnets changes constantly as the current alternates. By arranging the stationary electromagnets in a circle, a *rotating* magnetic field is produced.

A soft iron core (the rotor) is placed in the center of the stator's rotating magnetic field. The stator's magnetic field induces in the rotor a magnetic field opposite the polarity of the stator. Since opposite magnetic poles attract, the rotor is "locked" in an opposing position to the stator by their interacting magnetic fields. As the polarity of the stator's magnetic field moves, it also moves the rotor, because of the repulsion and attraction of magnetic poles. One magnetic field (the stator's) moves the other magnetic field (the rotor's). In other words, the stator rotates the rotor, **Figure 24-3**.

In other words, the rotor's south pole is attracted by the stator's north pole and repelled by the stator's south pole. The push-pull action is continuous, making the rotor spin as the stator poles change polarity. The rotor tries to catch up with the stator's rotating magnetic field.

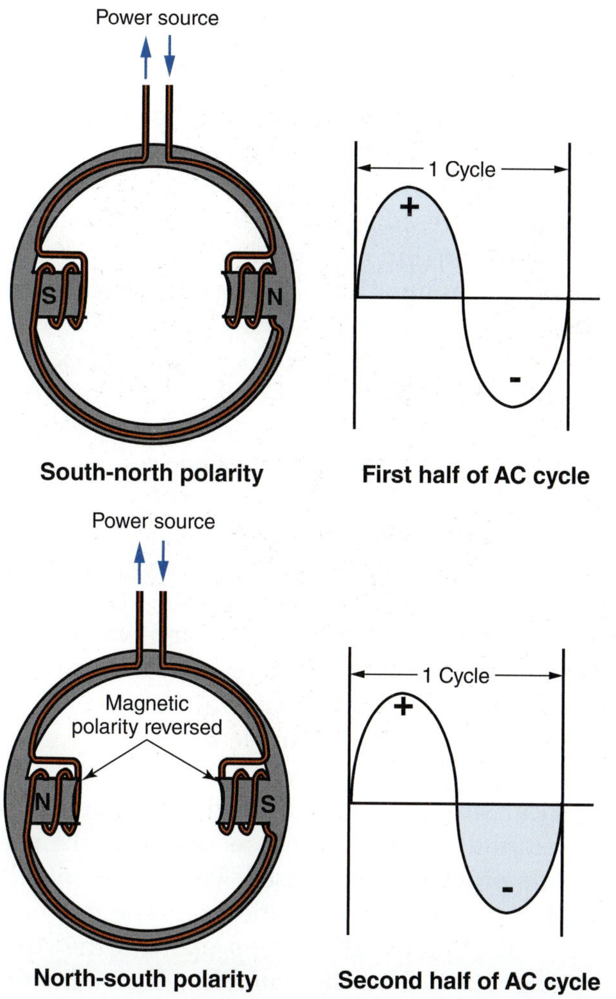

Goodheart-Willcox Publisher

Figure 24-2. Alternating current automatically reverses the polarity of the stator's magnetic poles during each cycle.

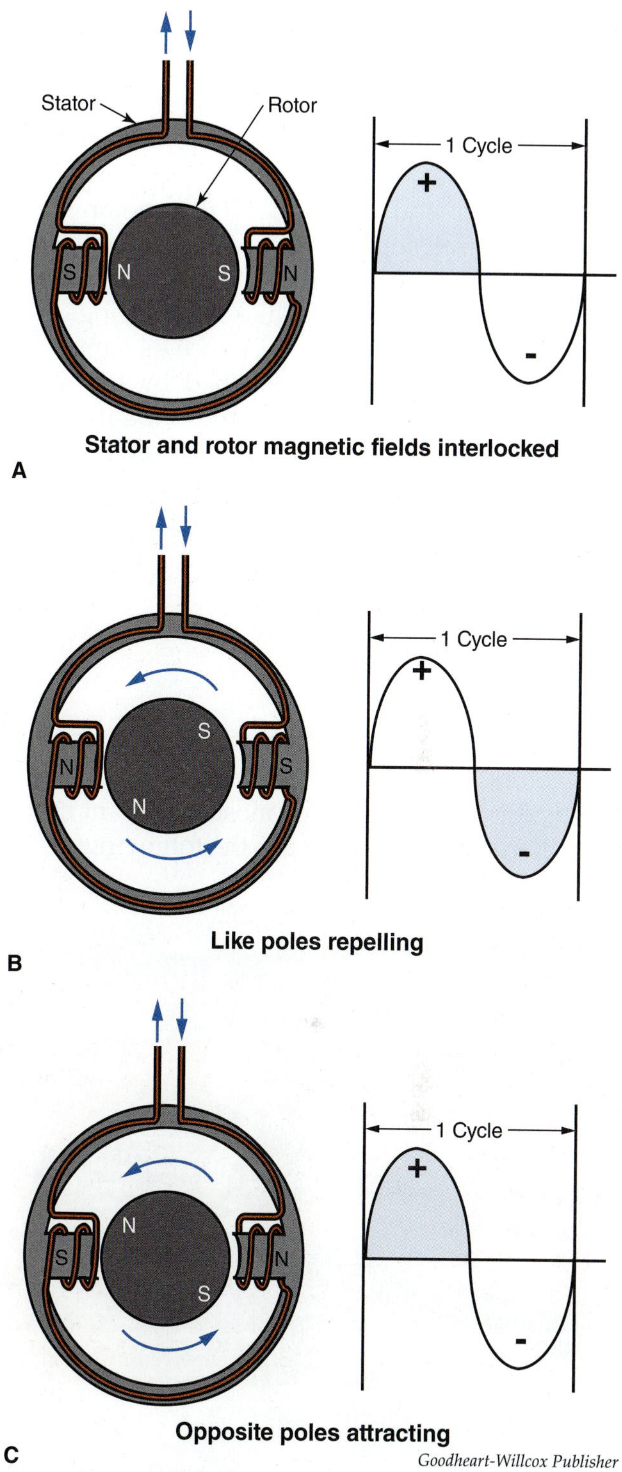

Goodheart-Willcox Publisher

Figure 24-3. Rotation of an induction motor's rotor is a result of the stator's changing magnetic field. A—Current flow through the stator induces opposite polarity in the rotor. B—Alternating current flow causes the stator polarity to reverse. The rotor rotates as its poles try to remain opposite those of the stator. C—The rotation continues as stator polarity once again changes.

Stator

Motors have stationary coils of copper wire (main windings) carefully wrapped around ***laminated*** (layered) cores of soft iron. As shown in **Figure 24-1B**, these coils and laminated cores are permanently mounted around the inside of the motor shell. The arrangement of coils and cores make up the stator.

The stator coils are wrapped to produce a north pole and a south pole. The poles are directly opposite each other, or 180° around the circle. Opposing polarity is produced by wrapping the first pole in one direction and the second pole in the opposite direction. The winding of the stator coil is called the ***run winding***.

The size (diameter) of copper wire used and the number of wraps in the coil determine the amount of resistance in the coil. The coil resistance limits the amount of current flowing through the wire. The amount of current, in turn, determines the strength of the magnetic field in the core.

Rotor

The ***squirrel cage rotor*** is the most common rotor type used in induction motors, **Figure 24-4**. Its body consists of a laminated soft iron cylinder enclosing and attached to the motor shaft. Instead of wires, copper bars are inserted into slots formed in the surface of the core. The copper bars are joined at their ends by thick copper rings. The effect resembles a round cage, hence its name.

The magnetic field established by the stator cuts across the closed loops in the rotor, inducing large currents. As a result of the induced currents, the rotor becomes a magnet and is turned by its interaction with the rotating magnetic field of the stator.

Induction Motor Characteristics

Motors are designated by their operational characteristics. Three of the most important motor characteristics to consider are the following:

- Motor speed
- Horsepower
- Efficiency

Motor Speed

Motor speed is determined by dividing the number of stator poles into the number of times the current changes direction (alternates) per minute. Because two alternations occur per cycle, 60-cycle current changes direction 120 times per second. Motors are rated in ***revolutions per minute (rpm)***, so alternations must be calculated on a per-minute-basis. Therefore, the current alternates 7200 times per minute (120 × 60 = 7200). See **Figure 24-5**.

For every alternation of current flow, the rotor travels from one stator pole to the next. With a two-pole motor, the rotor makes one complete revolution for every two alternations (one cycle). The speed of a two-pole motor is 3600 rpm (7200 ÷ 2). The speed of a four-pole motor is 1800 rpm (7200 ÷ 4), so it requires four frequency changes (two cycles) for the rotor to make one complete revolution. A six-pole motor rotates at 1200 rpm (7200 ÷ 6), so it requires six frequency changes (three cycles) for the rotor to make one complete revolution.

A motor's ***synchronous speed*** is based on the electrical supply's frequency and the number of poles in the motor winding, as shown in the rpm formula previously discussed. This represents the speed of the stator's magnetic field. The magnetic field of the rotor tries to keep up with the rotating magnetic field of the stator; however, a motor's

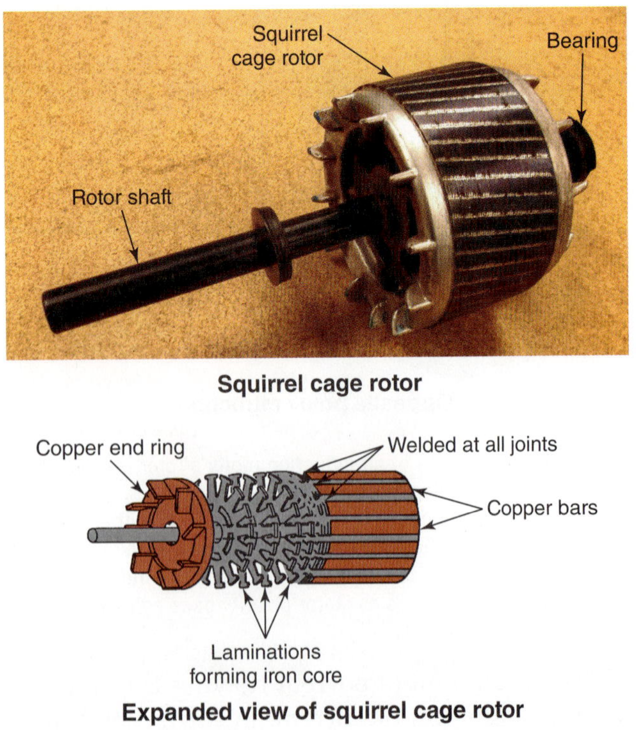

Goodheart-Willcox Publisher

Figure 24-4. A squirrel cage rotor consists of a laminated iron core with copper bars. The core surrounds the motor shaft.

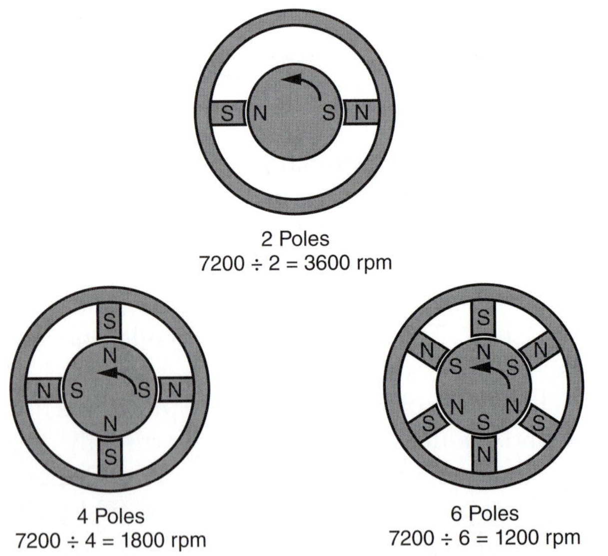

Goodheart-Willcox Publisher

Figure 24-5. Since 60-cycle current alternates 7200 times per minute, the number of stator poles determines motor speed. With two poles, the rotor turns 1/2 revolution for each current alternation (3600 rpm). With four poles, the rotor turns 1/4 revolution per alternation (1800 rpm). With six poles, motor speed drops to 1200 rpm as the rotor turns only 1/6 revolution per alternation.

actual speed is slower than its synchronous (stator field) speed. The rotor tends to lag behind the stator's rotating field as the motor is loaded. The greater the load on the motor, the more the rotor lags behind the stator's rotating field. Synchronous speed is nearly achieved when there is no load on the motor.

With a motor running at full load, the rotor turns at about 4% to 5% below synchronous speed. The percentage of relative motion compared to synchronous speed is called ***slip***, **Figure 24-6**.

Horsepower

Motors are rated by the amount of ***torque*** (turning power) they can produce. Turning power is rated in horsepower. The traditional English measure of horsepower is the amount of power required to lift 550 pounds one foot in one second.

In the United States, horsepower is calculated using the power needed by the motor to produce torque. Torque-producing power is rated in watts. One horsepower is equal to 746 watts. Thus, a five-horsepower motor uses 3730 watts ($5 \times 746 = 3730$). When the wattage and voltage are known, Ohm's Law makes it possible to determine the amount of amperage drawn under full load conditions.

Motors rated less than one horsepower are called ***fractional horsepower*** motors. Motors of one or more horsepower are called ***integral horsepower*** motors.

Efficiency

Electric motors are not 100% efficient. Like any machine, some power loss occurs in the motor when it operates. The efficiency of a fractional horsepower motor may be less than 50%, but an integral motor may be over 90% efficient.

Power loss in a motor is converted to heat energy. Motor cooling dissipates (gets rid of) the heat. Open-type motors are cooled by fans that draw air through openings in the motor frame and blow it over the windings. Sometimes, fan blades are attached to the rotor shaft inside the motor to blow air through the windings, **Figure 24-7**. For effective cooling, openings in the motor ends must be kept free of dirt and lint.

Motors located inside hermetic and semihermetic compressors are cooled by returning cool suction gas. Sometimes, additional cooling is provided by an externally mounted fan or liquid injection.

Motor Speeds

Number of Poles	Synchronous Speed	Actual Speed
Two-pole motor	7200 ÷ 2 = 3600 rpm	3450 rpm
Four-pole motor	7200 ÷ 4 = 1800 rpm	1725 rpm
Six-pole motor	7200 ÷ 6 = 1200 rpm	1150 rpm

Goodheart-Willcox Publisher

Figure 24-6. A comparison of the synchronous speed of various motors with the actual speed under load.

24.2 Single-Phase Motors

Single-phase induction motors are typically used in small business and residential applications. Single-phase motors do not start with the run (main) winding alone. To accomplish automatic starting, an additional field (stator) winding is required.

Start Winding

Called the ***start winding***, the additional winding establishes another magnetic field in the stator that is "out of step" with the main winding. As shown in **Figure 24-8**, the start winding is placed at a 90° angle to the main winding.

At start-up (when the main switch closes), current flows through both the start and run windings. The start winding has much smaller-diameter wire and more turns in the coil than does the run winding. The smaller diameter wire makes the start winding have high resistance.

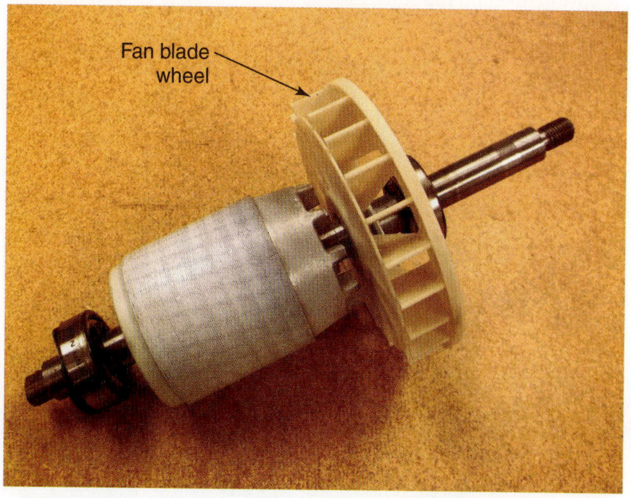

Goodheart-Willcox Publisher

Figure 24-7. To provide cooling in an open-type motor, fan blades may be attached to the same shaft as the rotor.

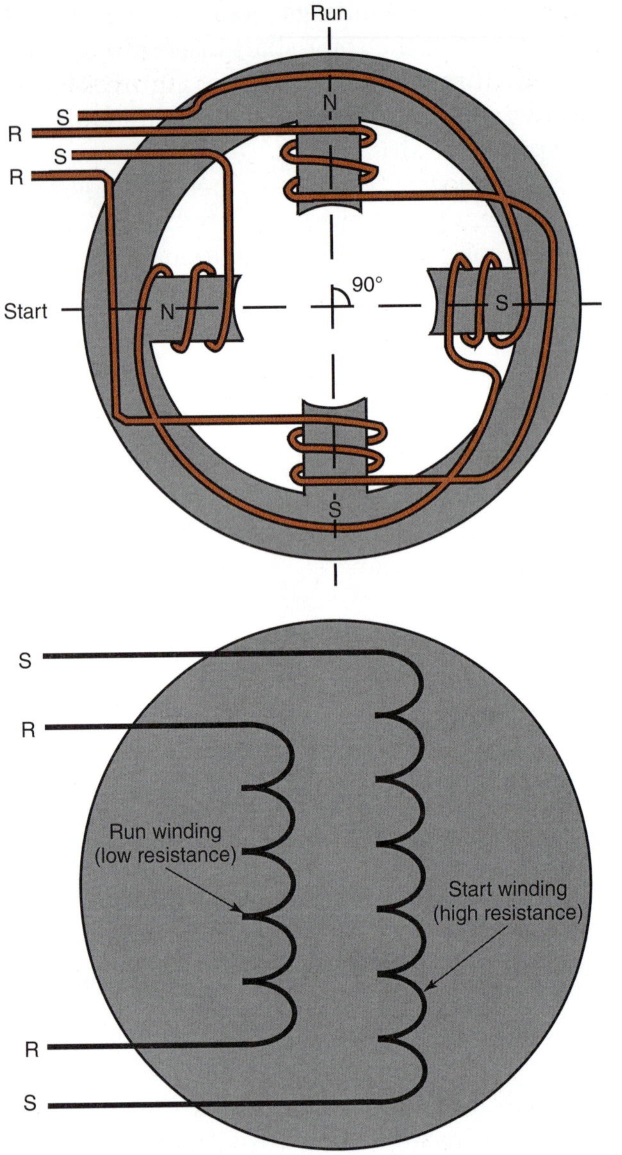

Goodheart-Willcox Publisher

Figure 24-8. The start winding in this two-pole motor is placed at a 90° angle to the run winding. The magnetic fields are out of step, which keeps the rotor turning.

Magnetism builds up faster in the run winding because it has less resistance. These two magnetic fields, one behind the other, create a turning force on the rotor.

Direction of Rotation

Rotation is the circulation motion of the motor shaft. The direction of rotation is determined by the start winding. Rotation to the right is clockwise (CW). Rotation to the left is counterclockwise (CCW). To reverse rotation, simply reverse the two connections to the start winding.

On open-type motors, the electrical connections are located at the ***lead end*** of the motor, which is where the electrical wires (leads) are located. The rotor shaft exits the opposite end of the motor (the ***shaft end***), **Figure 24-9**. For most types of motors, the direction of rotation (CW or CCW) is identifiable as viewed from the *shaft end* toward the lead end. Motors made by General Electric are an exception. Direction of rotation is determined by viewing from the lead end toward the shaft end in GE motors. Direction of rotation should always take into account which end of the motor is being viewed.

Disconnecting the Start Winding or Start Capacitor

The sole purpose of the start winding is to get the rotor moving (start the electric motor). After the motor starts, damage may occur to either the start winding or the start capacitor if not disconnected from the circuit. This is normally done once the motor reaches about 75% of normal speed.

Methods used to disconnect the start winding or capacitor from the circuit vary based on the type of motor. These methods are discussed later in this chapter.

The run winding and the start winding are two separate loads and must be connected in

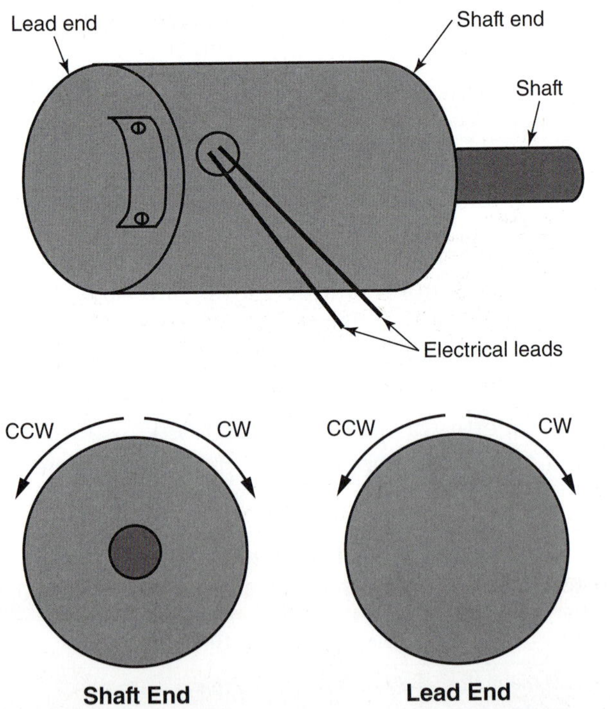

Goodheart-Willcox Publisher

Figure 24-9. When describing the rotation of a motor shaft, the end of the motor from which the rotation is viewed must be noted.

parallel, **Figure 24-10**. A main switch controls the power supply to the motor and is connected in series with both windings. The main switch can be installed at any remote location.

Locked Rotor Amperage (LRA)

At start-up, little or no back-emf is generated by the motor. Amperage flow is determined by the resistance of the motor windings. Starting current flow is typically five times higher than normal running amperage. The high current flow during start-up is called ***locked rotor amperage (LRA)***. As the motor picks up speed, back-emf is generated and rapidly increases the motor's resistance to current flow. As noted, the start winding is disconnected at about 75% of normal speed. Consequently, amperage flow is substantially reduced. The motor operates on the run winding alone. Amperage flow is now determined by the resistance of the run winding combined with the back-emf generated by the motor.

Full Load Amperage (FLA)

Immediately after the start winding is disconnected, the resistance of the run winding is increased by the back-emf generated by the motor. The operating ("running") amperage is called ***full load amperage (FLA)***. Induction motors usually operate at less than FLA because the motor rarely works at the fully loaded condition. An overload occurs when the amperage flow exceeds the FLA rating on the motor data plate. Except for the temporary high start-up amperage, any amperage that exceeds the FLA rating is converted to heat energy.

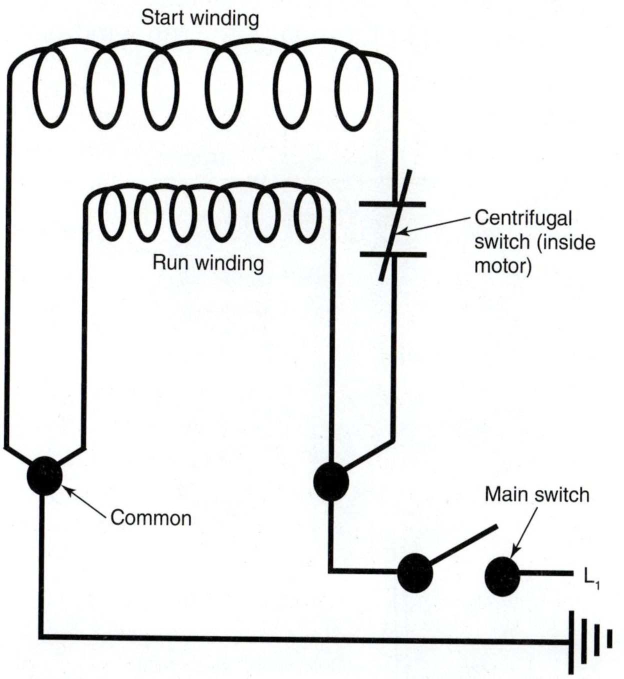

Goodheart-Willcox Publisher

Figure 24-10. The run and start windings are separate loads and connected in parallel. A centrifugal switch is connected in series with the start winding.

Overload Protectors

The induction motor is designed to convert electrical energy to mechanical energy. If the rotor does not turn normally, excess amperage flow will result. The motor will convert the excess amperage to heat energy, which will burn out the motor windings. The two most common causes of motor problems are:

- A locked rotor (rotor unable to turn)
- An overload (rotor turning too slowly)

The ***motor overload protector*** (usually referred to simply as the *overload*) is a device that protects the motor windings from damage caused by overheating due to overloaded conditions or poor ventilation.

The most common form of motor protection uses a "snap-acting" bimetallic disc to make and break a set of contacts, **Figure 24-11**. The snap action is obtained by fusing together two thin, circular discs of different metals (***bimetallic***)—usually steel on top and copper on the bottom. The bimetallic disc is anchored at the center. Since copper expands faster than steel, a rise in temperature causes the disc to expand. The outer edge curls (bows).

The movement of the bowed edge opens a set of stationary contacts. When the bimetallic disc cools, it snaps back to its original (closed) position. Excessive heat from any source will affect the overload and cause it to open the circuit supplying power to the motor.

Cycling on the Overload

If an overload condition is not corrected, the bimetallic disc will trip repeatedly, a malfunction known as ***cycling on the overload***. After each trip, more and more time is required for the bimetallic disc to cool. Eventually, the bimetallic disc could require up to an hour to cool and reset. Cycling on the overload is easy to hear. The motor makes

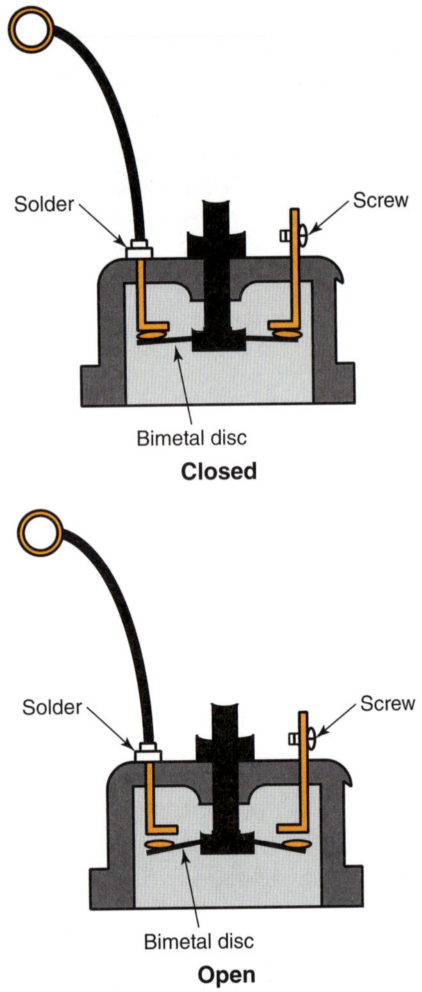

Goodheart-Willcox Publisher

Figure 24-11. A bimetallic disc of steel and copper is used in many overloads. In normal operation, the disc holds the contacts in the closed position. When sufficiently heated by an overload, the differing expansion rates of the metals cause the disc to curl in one direction. This opens the contacts and breaks the motor circuit. When the motor cools sufficiently, the disc returns to its normal shape, closing the contacts again.

a noise described as "hmmm...click." The humming noise is made by the motor trying to start, and the sharp "click" is made when the bimetallic disc snaps open.

Cycling on the overload indicates a condition that *must* be corrected. The overload protector is sensitive to three sources of heat:

- Excess amperage flow (usually an electrical problem related to the start winding and start components).
- Excess motor heat (usually resulting from dry or badly worn bearings).
- Excess compressor heat (low on oil or compressor cooling is not working).

Tripping the overload protector is a warning that the motor is in danger of a burnout. It is important to locate and repair the cause of an overload.

Two-Wire Overload Connections

The terminals on the overload protector are numbered 1, 2, and 3. The contacts are located between numbers 1 and 2. A small resistance heater wire is located between numbers 2 and 3. The resistance heater makes the overload react more quickly to overheating, **Figure 24-12**. Connections to the overload are limited to Terminals 1 and 3, which simplifies making electrical connections and prevents bypassing the heater. The heater is internally connected in series with the contacts.

One side of the supply voltage is connected to Terminal 1 so the current must flow through the contacts *and* the small heater to reach Terminal 3. A wire from Terminal 3 is connected to the *common* on the motor windings. ***Common*** describes the wire that connects one end of the run winding to one end of the start winding. Current flowing through the common supplies power to both windings, hence it is "common" to both windings, **Figure 24-13**.

Three-Wire Overload Connections

On open-type motors, a three-wire connection is often used for the overload protector. Power supply is connected to Terminal 1, the run winding to

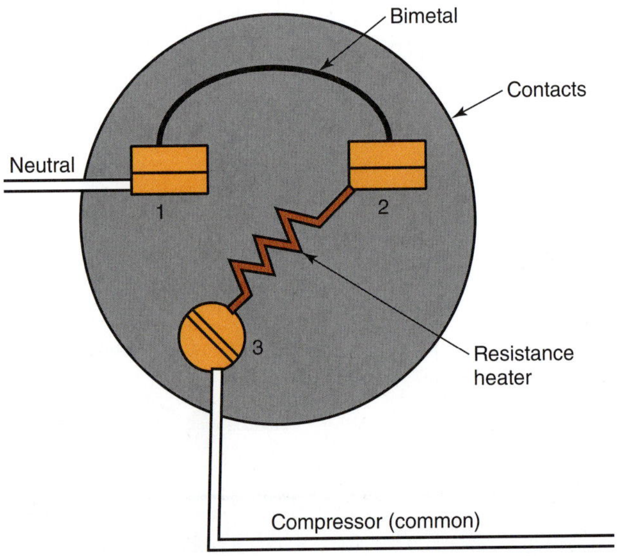

Goodheart-Willcox Publisher

Figure 24-12. A resistance heater wire, connected in series with the bimetallic element, makes an overload react more quickly.

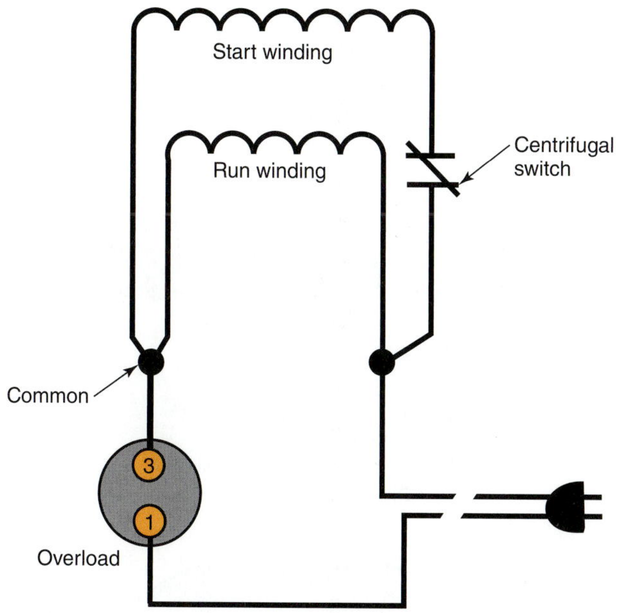

Goodheart-Willcox Publisher

Figure 24-13. The overload is in series with the common connection to the run and start windings, allowing the overload to protect against excess current flow through either winding.

Terminal 2, and the start winding to Terminal 3. See **Figure 24-14**.

The heater inside the overload is connected in series with the start winding. This connection makes the overload respond quickly to any fault in the start winding, the source of many motor problems. During normal operation, the *run* winding is protected by the bimetallic disc only. However, if the bimetallic disc opens the contacts between Terminals 1 and 2, the power supply to both windings will be interrupted.

Internal Overload Protection

Many large hermetic and semihermetic compressors have an overload protector buried inside the motor windings. This overload can sense the current draw of the motor and the temperature of the windings much better than external overloads. Internal overload protection is installed at the factory and is not accessible to the technician, **Figure 24-15**.

The internal overload is a bimetallic element that will open the circuit and stop the motor if the winding temperature reaches 200°F–250°F (93°C–121°C). It will close again when the winding temperature drops to 150°F–175°F (66°C–79°C). Depending on ambient temperature conditions, it may take one or two hours before the protector cools enough to close the contacts. Cooling the unit with forced air or ice speeds up the process. Unless an ohmmeter test confirms an open motor winding, do not condemn a motor until sufficient time has passed for the overload to reset.

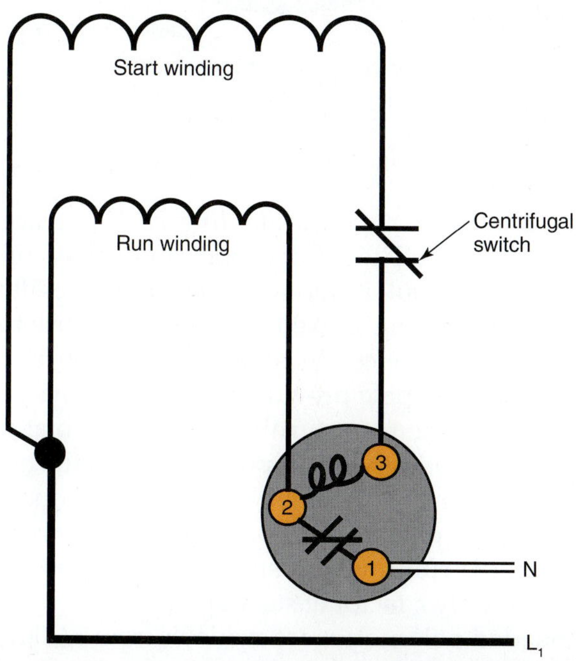

Goodheart-Willcox Publisher

Figure 24-14. Three-wire connections to the overload use all three terminals. The heater is connected in series with the start winding so the overload reacts quickly to any problem with the start winding.

24.3 Types of Single-Phase Induction Motors

In HVACR applications, five types of single-phase induction motors and three-phase induction motors are in general use. The type of motor for a particular application is determined by the location and type of load. Each motor is available in

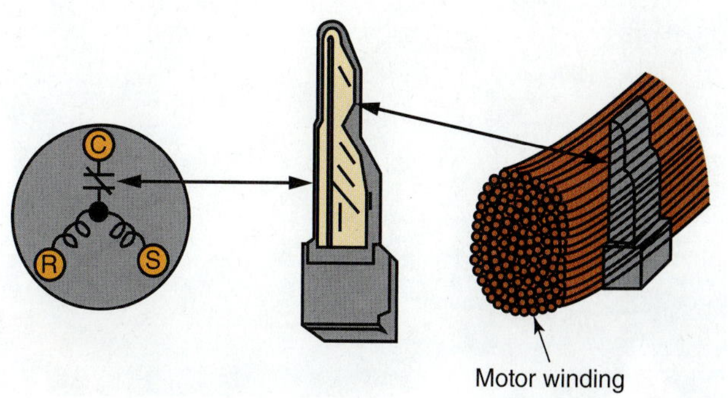

Goodheart-Willcox Publisher

Figure 24-15. An internal overload built into the motor windings is not accessible to the technician.

various horsepower ratings. Single-phase induction motors are described in the order of turning power (torque), from the least to the greatest. The three-phase motor is described later in this chapter.

The types of single-phase induction motors are:

- Shaded-pole
- Split-phase
- Permanent split capacitor (PSC)
- Capacitor-start, induction-run (CSIR)
- Capacitor-start, capacitor-run (CSCR)

Shaded-Pole Motor

The ***shaded-pole motor*** is a uniquely designed, low-cost motor used to operate small fans, such as the evaporator and condenser fans on domestic and light commercial refrigeration units, **Figure 24-16**. The shaded pole motor has a low starting torque, but it can carry a full load once it is up to speed. The capability of carrying a full load makes the shaded-pole motor ideal for operating small fans. At start-up, the only load on the motor is the weight of the fan blades. Full load is not reached until the blades are turning at operating speed and moving volumes of air.

Construction

Construction of the shaded-pole motor is different from construction of other induction motors. As shown in **Figure 24-17**, a slit is cut at the edge of each stator pole, producing a small prong.

Goodheart-Willcox Publisher

Figure 24-16. A shaded-pole motor used to operate a condenser fan on a light commercial refrigeration unit.

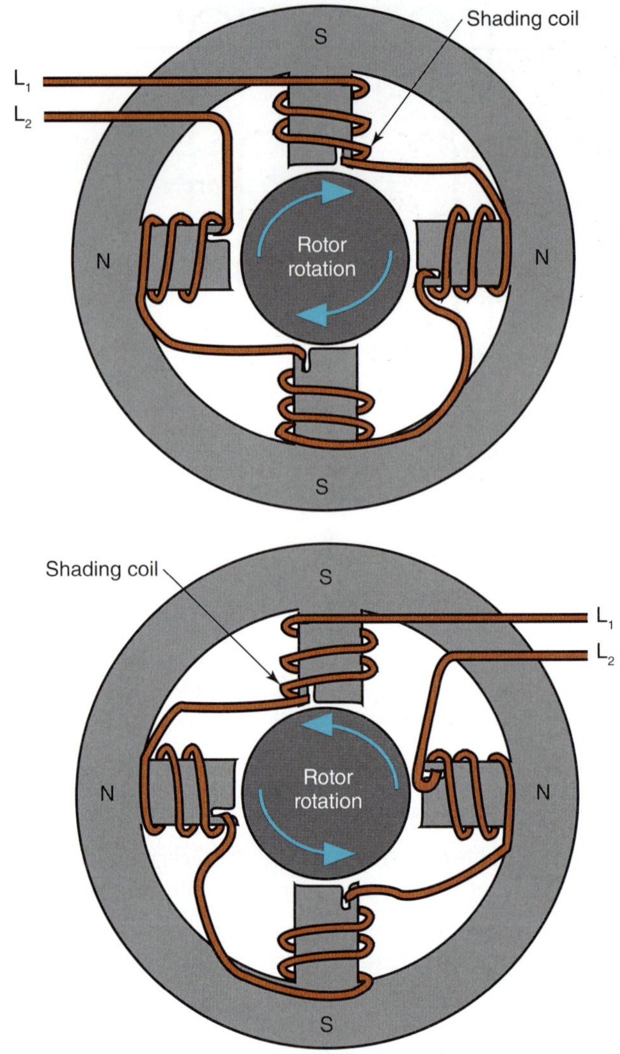

Goodheart-Willcox Publisher

Figure 24-17. The shaded-pole motor uses small "shading coils" on each pole to provide the necessary impetus for starting. Direction of rotation is determined by location of the shading coils on the poles.

A copper band, called a ***shading coil***, is placed over the prong. The shaded pole produces a tiny magnetic field out of step with the main poles. It is *not* a start winding in the usual sense because the shading coil is not connected to a power source.

The shaded pole provides just enough magnetism to start the rotor turning. Then, the rotating magnetic field in the main poles provides the necessary torque to bring the rotor up to full speed. The shaded pole is constantly energized when the motor is running, but such a small amount of energy is only a factor at start-up.

Shaded-pole motors are available from 1/100 hp to 1/6 hp. If a shaded-pole motor malfunctions, it is usually cheaper to replace it than repair it. Making electrical connections is simple because only two wires are involved. A power supply is connected to each end of the motor winding.

Reversing Rotation

The direction of rotation of the shaded-pole motor is determined at the factory by placing the shaded poles on one side or the other of the main poles. A left-side position produces a CCW rotation, and a right-side position produces a CW rotation. The placement of these poles cannot be changed in the field. Reverse rotation can be achieved only if the rotor can be physically removed and reversed (reinstalled so the shaft exits the opposite end of the motor). Rotor reversal is not always possible.

Split-Phase Motor

In the ***split-phase motor***, a start winding is installed in the stator to produce a magnetic field out of step (phase) with the main winding. The start winding is rather delicate because it is made of a small wire. Since it is such a small wire, it has higher resistance. The start winding also has more wraps per coil. The start winding assists the motor in getting started but begins to interfere with the magnetism of the run winding at about 75% of normal speed. The conflict of magnetic fields will cause the motor to burn out if the start winding is not promptly taken out of the circuit.

Application

The split-phase motor has low starting torque but good running torque. Open-type split-phase motors are popular for fans or blower motors where the load is too large for a shaded-pole motor. Split-phase motors are popular in hermetic compressors on systems using a capillary tube, such as domestic refrigerators and window air conditioners. During the off cycle, the capillary tube permits some high-side pressure to bleed over (unload) to the low-pressure side of the system. Bleedover helps reduce head pressure and makes it possible to use a less-costly motor with low starting torque for the compressor. To allow time for head pressure to drop, a capillary tube system should not be restarted for two to five minutes after it is stopped.

Open-Type Split-Phase Motor

On open-type motors, a terminal board for making electrical connections to the motor windings is located just inside the end bell (lead end). See **Figure 24-18**. The manufacturer connects the motor windings to the terminal board with quick-connect (push-on) connectors. The technician must connect the power supply voltage to the correct terminals on the board. Power supply terminals are typically the threaded-bolt-and-nut-type and labeled L_1 and L_2. See **Figure 24-19**.

One power supply wire (L_1) is connected to Terminal 1. Terminal 1 of the overload is connected at the factory to L_1. Terminal 3 of the overload is factory-connected to a pair of push-on-type terminals. The two push-ons connect one end of the run winding and one end of the start winding to the overload. The overload, in turn, is connected to

Goodheart-Willcox Publisher

Figure 24-18. On open-type motors, the terminal board is located at the lead end. Bolt-and-nut-type connections for the power supply wires are most common.

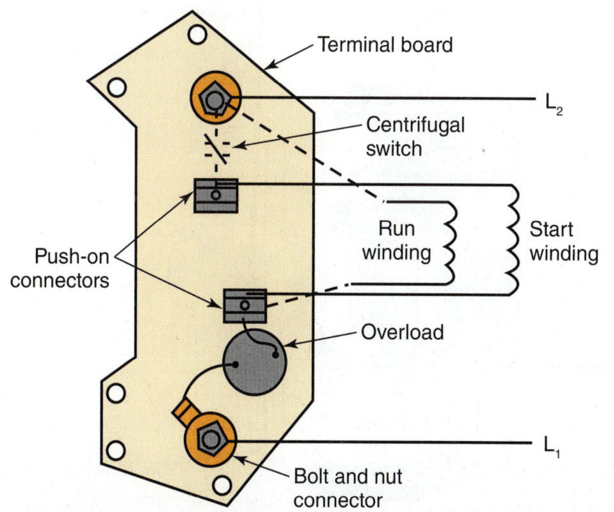

Goodheart-Willcox Publisher

Figure 24-19. Factory and field connections to the terminal board of an open-type motor. The factory connections are made with push-on connectors (the dashed lines represent connections made behind the terminal board).

one side of the power supply. The other side of the run winding is connected to the other side of the power supply at L_2. Connections to the run winding are now complete because it is connected from one side of the power supply to the other (parallel).

Connections for the start winding must include a centrifugal switch to disconnect the winding after start-up. The terminal board contains the electrical switch, which must be connected in series with one side of the start winding and power supply (at L_1). Connections are performed at the factory. Terminal board design and connections vary from one manufacturer to another, but the principles remain the same.

Contacts on a ***centrifugal switch*** are normally closed (NC) when the motor is off (not running). A pair of spring-loaded weights and a connecting linkage are attached to the rotor shaft on the inside of the motor. When the motor starts and achieves 75% of operating speed, centrifugal force causes the weights to swing outward. The movement of the weights, transmitted by the linkage, opens the switch, which disconnects the start winding from the circuit. When the motor stops rotating, springs pull the weights back to their normal position and the switch closes, ready for the next start-up. See **Figure 24-20**.

Reversing Rotation

Most manufacturers bring both wires from the start winding to the accessible part of the terminal board. The wires are different colors (identification is on the cover plate). Motor rotation is reversed by switching (reversing) the two terminal board connections. Switching the connections reverses current flow through the start winding, and the rotor follows the direction of current flow.

Hermetic Split-Phase Motor

Electrical connections to split-phase motors in hermetic compressors are made in the same way as open-type motors. The factory makes the necessary motor connections internally to three insulated terminals extending through the hermetic shell. The three terminals are: common (one end of each winding), start (other end of start winding), and run (other end of run winding). Because arcing occurs as the contacts open or close, the switch for disconnecting the start winding is located outside the shell. Connections outside the shell are made by the technician, **Figure 24-21**.

One leg of the power supply (usually the white wire) is connected to the overload at Terminal 1. Terminal 3 from the overload is connected to the common terminal on the compressor. On hermetic compressors, the overload is always connected in series with the common.

The other leg of the power supply (black wire) is connected to the relay and continues to the run terminal on the shell. The start winding is connected to a switch inside the relay. At this point, it is only necessary to understand the principles involved with these connections. The types of relays,

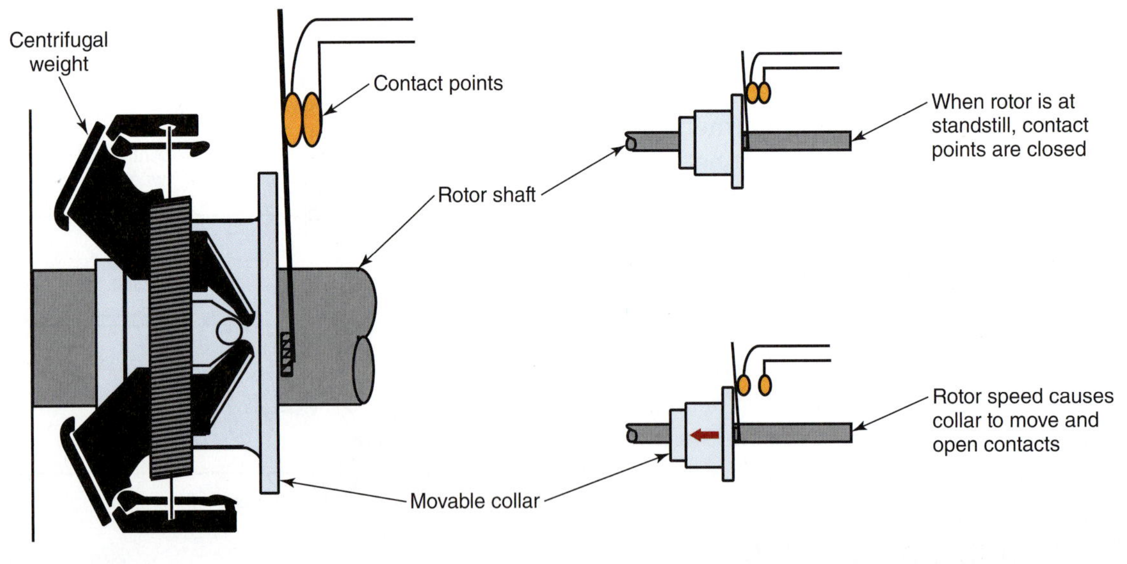

Goodheart-Willcox Publisher

Figure 24-20. A centrifugal switch opens when the motor reaches 75% of operating speed, disconnecting the start winding. As rotor shaft speed increases, the weights swing outward, sliding the movable collar along the shaft. The movement of the collar is transmitted through mechanical linkage, opening the contacts.

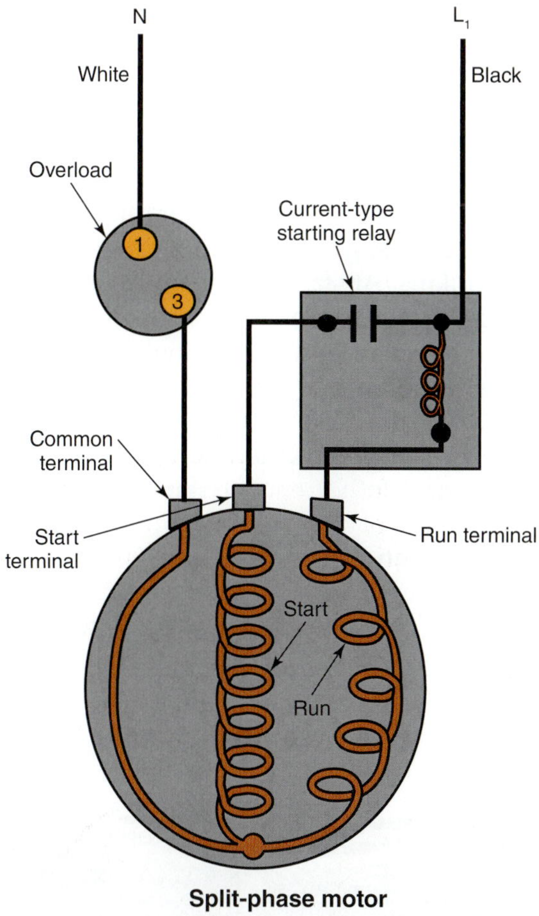

Goodheart-Willcox Publisher

Figure 24-21. Connections for the hermetic split-phase motor are made to the three insulated terminals (Common, Start, and Run) extending through the shell of the compressor.

their operation and connection are explained in Chapter 25, *Electromagnetic Control Devices.*

Permanent Split-Capacitor Motor

The ***permanent split-capacitor (PSC) motor*** is a split-phase motor with a capacitor in series with its start winding. It uses this design change to increase running torque, **Figure 24-22**. The PSC motor is used where *starting* load is low but *running* load is high. The high running load would cause too much slip in a split-phase motor. To overcome excessive slippage, a method was devised that uses the start winding to assist the run winding.

By connecting a capacitor in series with the start winding, the magnetic field of the start winding is reduced so it does not interfere with the run winding, **Figure 24-23**. The capacitor shifts the phase of the power in the start winding. By eliminating the conflict of magnetic fields, the start winding can assist the run winding. A controlled use of the start winding eliminates the need to disconnect the start winding after start-up.

As the name *permanent split capacitor* indicates, the capacitor is a permanent part of the circuit. Since the start winding is not disconnected, the PSC motor has no need for a centrifugal switch or relay.

Image courtesy of Johnstone Supply

Figure 24-22. An open-type PSC motor. The run capacitor is used to create better running torque.

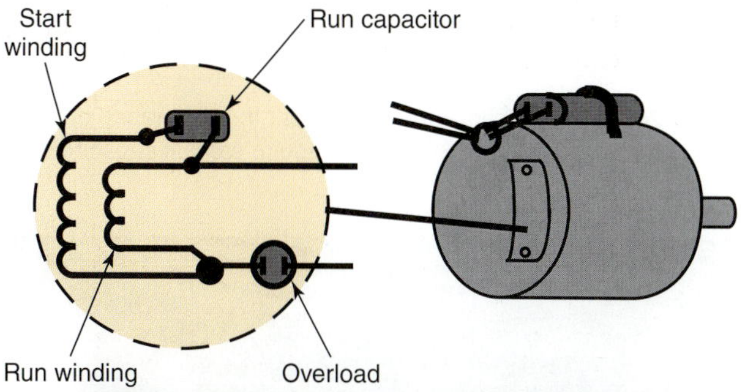

Goodheart-Willcox Publisher

Figure 24-23. In a PSC motor, the run capacitor is wired in series with the start winding. It shifts the phase of the electrical power, which shifts the start winding's magnetic field.

Run Capacitor

The capacitor on PSC motors is called a ***run capacitor*** because it is designed and used to create better running torque. However, as shown in **Figure 24-23**, the run capacitor is connected in series with the start winding.

The run capacitor is usually an oval-shaped, metal cylinder with push-on-type electrical connections on its top, **Figure 24-24**. Capacitor sizes vary and are prescribed by the motor manufacturer to match the operating characteristics of the PSC motor windings. When capacitor replacement is necessary, always use an *exact* replacement. Capacitors are described in greater detail later in this chapter.

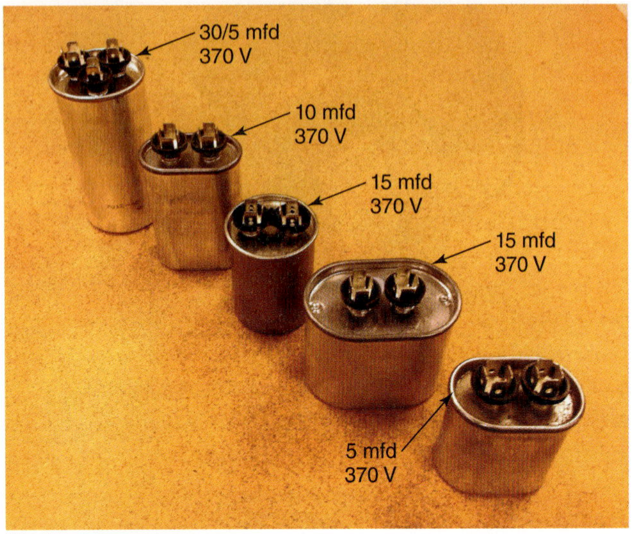

Goodheart-Willcox Publisher

Figure 24-24. Run capacitors are similar in appearance but vary in size because various PSC motors require various capacities.

Open-Type PSC Motor

Open-type PSC motors, like the one shown in **Figure 24-22**, are popular for operating large fans and blowers. They are also used for condenser and evaporator fans. At start-up, the motor load is low. At full speed, the fan blades are fully loaded by the weight and volume of the air being moved. Such a large load would cause a split-phase motor to slip too much, but the PSC motor solves the slippage problem by using both the run winding and the start winding.

Reversing Rotation

The rotation of a PSC motor is reversed by simply switching the two wires connected to the start winding if the motor supplies them outside the motor, **Figure 24-25**.

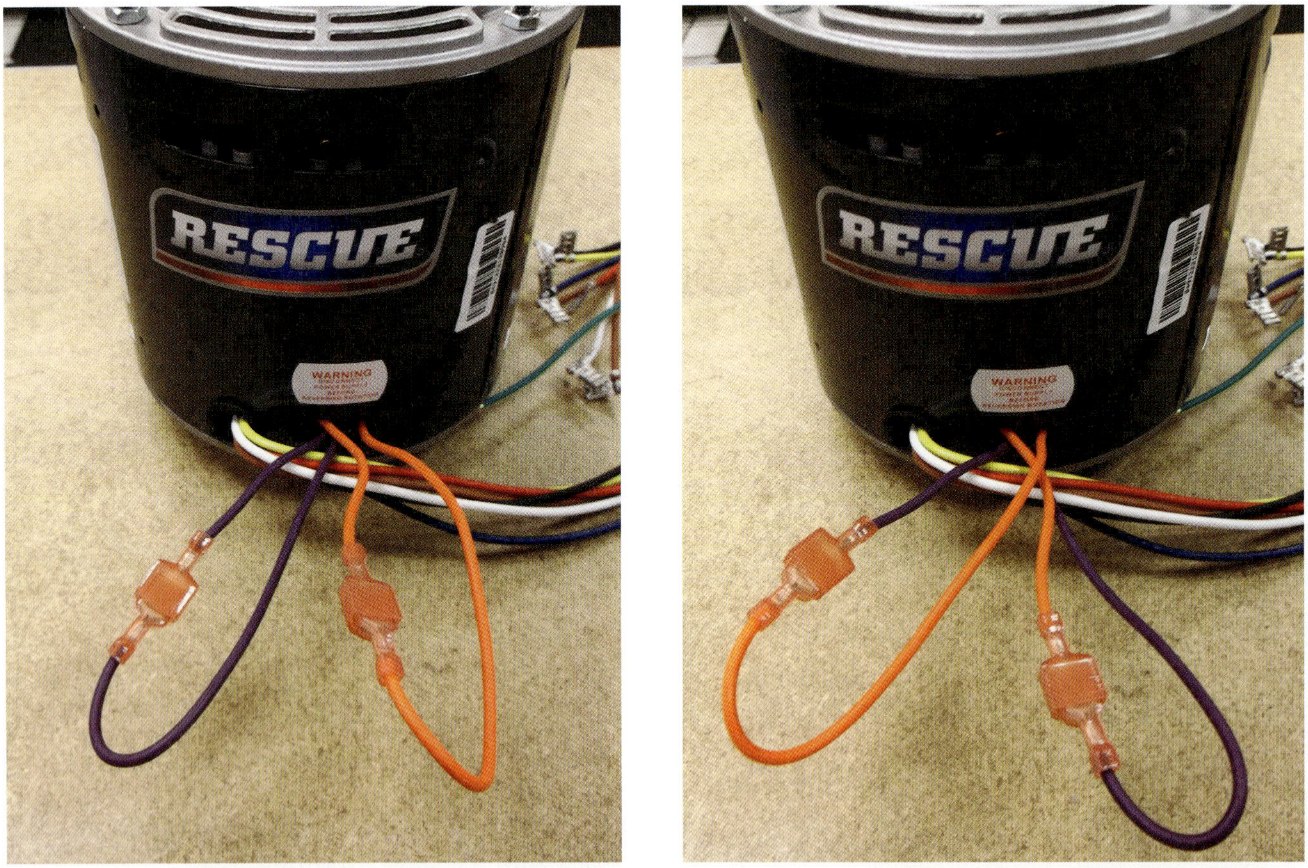

Regular rotation wiring

Reverse rotation wiring

Image courtesy of Johnstone Supply

Figure 24-25. Wires supplied on a PSC motor are able to change rotation based on how they are connected. For regular rotation wiring, purple is connected to purple and orange is connected to orange. For reverse rotation wiring, purple is connected to orange and orange is connected to purple.

Capacitor-Start, Induction-Run Motor

The ***capacitor-start, induction-run (CSIR) motor*** is used when a motor must start under a heavy load, but once started, the motor can operate with just the run winding. The open-type CSIR motor is found where V-belts connect the motor to a load, such as an open-type compressor or a centrifugal (squirrel cage) blower.

The capacitor-start, induction-run motor is actually a split-phase motor with a special capacitor that sets off the phase in the start winding to increase starting torque, **Figure 24-26**.

Start Capacitor

The capacitor on the CSIR motor is a ***start capacitor***. These are designed and used to create better starting torque than a run capacitor. A start capacitor usually appears as a black plastic cylinder with a hard cardboard top that houses the electrical push-on-type connections, **Figure 24-27**.

The start capacitor is not as durable in construction as a run capacitor because it does not stay in the circuit during motor operation. It gets the motor started, then is disconnected along with the start winding. Typically, a start capacitor can withstand about 20 starts per hour before suffering internal damage.

The start capacitor stores electrons during one current alternation, then discharges the electrons into the start winding when the current alternates again. This delays the phase of the current and the magnetic field generated by the start winding. The result is the necessary torque (turning power) to start the motor.

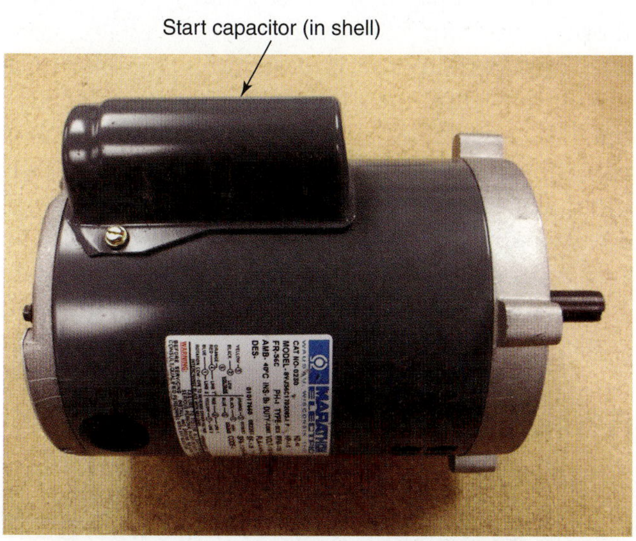

Goodheart-Willcox Publisher

Figure 24-26. A CSIR motor uses a capacitor to shift the phase of the power (and thus magnetic field) in the start winding. As a result, the motor is better able to start under heavy load.

Goodheart-Willcox Publisher

Figure 24-27. Start capacitors are available in various sizes, depending on motor requirements. They are less durably constructed than run capacitors because start capacitors are used only briefly, whenever the motor is started.

The size (capacity rating) of the start capacitor determines the amount of electrons discharged into the start winding. Capacitor size varies according to the horsepower rating of the motor and the load placed upon it. Be sure to use an *exact* replacement for a start capacitor.

Open-Type CSIR

Electrical connections for the CSIR motor are similar to electrical connections for the split-phase motor. The only difference is the addition of a start capacitor, **Figure 24-28**. On this type of motor, the start capacitor, centrifugal switch, and

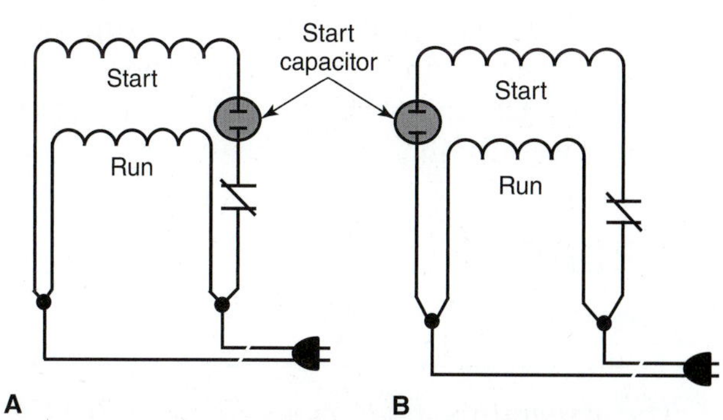

Goodheart-Willcox Publisher

Figure 24-28. The order of components in the start circuit does not matter as long as they are wired in series. A—The start capacitor is between the centrifugal switch and the start winding. B—The start winding is between the switch and the capacitor.

start winding are connected in series. The order in which the components are connected makes no difference; however, the components must be in series so the switch can disconnect them from the circuit.

Reversing Rotation

On the open-type CSIR motor, rotation is reversed by swapping the two wires that control current flow to the start winding. Information for identifying these wires is located on the motor's data plate or inside the cover plate.

Hermetic CSIR Motor

The hermetic CSIR motor is used when the compressor is required to start under high head pressure and high load conditions, such as in systems using automatic and thermostatic expansion valves.

Hermetic Electrical Connections

The CSIR hermetic compressor looks like other hermetic compressors, except the start capacitor is mounted immediately above (or beside) the electrical terminal box to simplify wiring connections. See **Figure 24-29**.

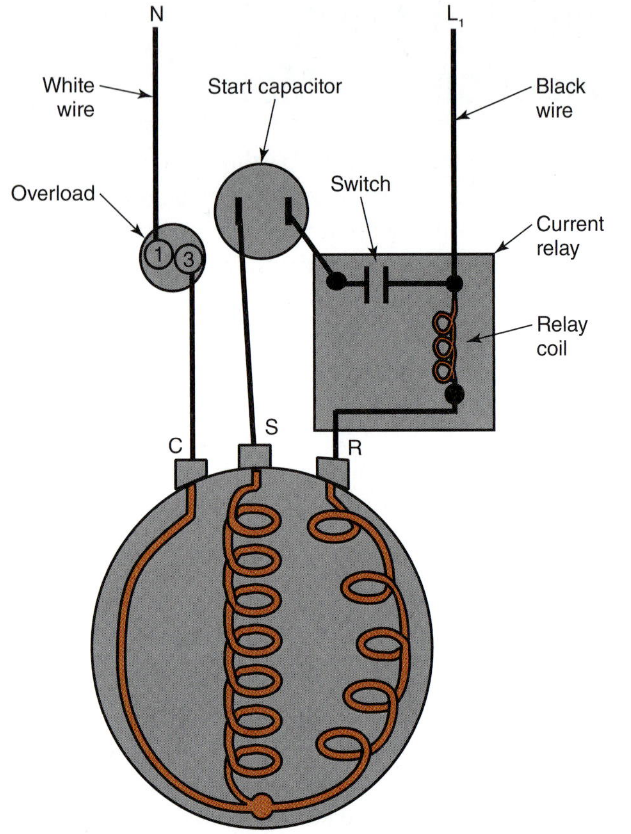

Figure 24-29. Electrical connections for a CSIR motor in a hermetic compressor. The start winding, start capacitor, and relay switch are in series.

One leg of the power supply (white wire) is connected to Terminal 1 of the overload. Terminal 3 of the overload is connected to the common terminal on the hermetic compressor. The other leg of the power supply (black wire) is connected to a relay terminal that splits to connect to the relay's switch on one side and the relay's coil on the other. The relay coil connection leads to the compressor's run terminal. The relay switch connection leads to the start capacitor and then to the start winding terminal on the compressor.

These connections place the overload in series with the common terminal (and both windings). The run and start windings are connected in parallel. The switch and start capacitor are in series with the start winding.

Capacitor-Start, Capacitor-Run Motor

The ***capacitor-start, capacitor-run (CSCR) motor*** runs as a CSIR during start-up and as a PSC for the rest of normal operation. It is another improvement on the split-phase motor. The most powerful of the single-phase induction motors, it provides both high starting torque and high running torque. The CSCR motor is used when the motor must start under a heavy load and carry a heavy load during the run cycle. The motor is typically used in small commercial systems with a thermostatic expansion valve.

The CSCR motor is essentially a split-phase motor that combines the operating principles of the PSC and CSIR motors, using both start and run capacitors. At start-up, it operates as a CSIR motor because of the start capacitor. After start-up, a relay switch disconnects the start capacitor. The run capacitor *remains in the circuit* and keeps the start winding energized. The motor continues to run as a PSC.

Hermetic CSCR Electrical Connections

One leg of the power supply (white wire) is connected to the overload at Terminal 1, **Figure 24-30**. A wire from Terminal 3 of the overload is connected to the common terminal. The other leg of the power

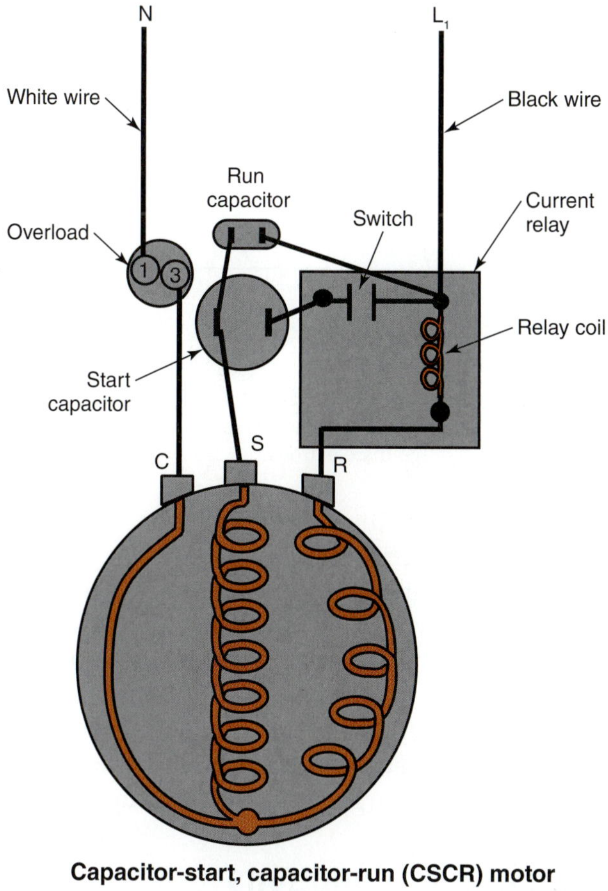

Goodheart-Willcox Publisher

Figure 24-30. Electrical connections for a hermetic compressor with a CSCR motor. The start winding, start capacitor, and relay switch are in series. The run capacitor is wired in parallel with the start capacitor and relay switch.

supply (black wire) is connected to the relay, which branches through the capacitors to the start terminal or to the run terminal through the relay's coil. Overall, the run and start windings are connected in parallel, and the overload is in series with common (so both windings).

PRO TIP

Wire Colors

Note that wire colors may vary from manufacturer to manufacturer. Always refer to manufacturer instructions for the wire colors used on their equipment.

Connections from the relay and capacitors to the start winding are critical but really quite simple. First, connect the start capacitor in series with the relay switch and start terminal by connecting a wire from the relay switch terminal to one side of the start capacitor. Another wire is connected from the other side of the start capacitor to the start terminal. These connections are exactly like those on the CSIR motor.

The run capacitor is connected to provide another circuit for the electrons to reach the start winding. This circuit must *bypass* the switch and start capacitor. The connections are called *series-parallel* because each capacitor is *in series* with the start winding but the capacitors are *in parallel* with each other. Such a connection permits each capacitor to act independently and perform its duty to the start winding. At start-up, both capacitors are energized. When the relay switch opens, only the start capacitor is disconnected. The circuit for the run capacitor is still available to control the phase of the current going through the start winding.

24.4 Capacitors

A ***capacitor*** is an electrical device that can store an electrical charge. It can be used to improve the operating characteristics of single-phase induction motors by changing the phase of the electrical signal, which affects the magnetic field produced by the start winding. **Figure 24-31** shows the two types of capacitors:

- **Start capacitor.** This is a dry-type capacitor intended for intermittent operation. It is fragile and only withstands 20 starts per hour.
- **Run capacitor.** This is an oil-filled capacitor intended for continuous operation. The oil serves to dissipate any heat buildup.

Start Capacitor

A start capacitor consists of two layers of aluminum foil, called ***electrodes*** or *plates*, separated by an insulating layer of specially treated paper. The foil and paper layers are about three inches wide and several feet long. The layers are tightly rolled into a cylinder and placed inside a black plastic case. A cardboard cover seals the top, and two electric terminals (usually push-on-type) are anchored to the cover. One aluminum layer is connected to each of the terminals. See **Figure 24-32**.

The large surface area of the aluminum layers serves as a "storage tank" for electrons. The insulating paper between the layers of foil prevents

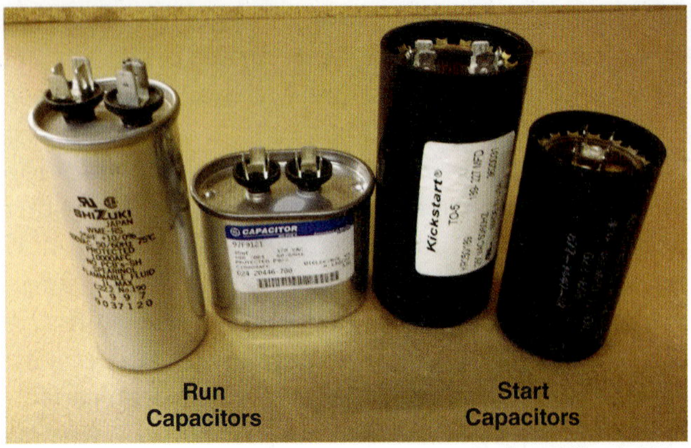

Goodheart-Willcox Publisher

Figure 24-31. In HVACR motor control, capacitors are of two types: run and start. Run capacitors remain in the circuit as long as the motor runs. Start capacitors are energized only during motor start-up.

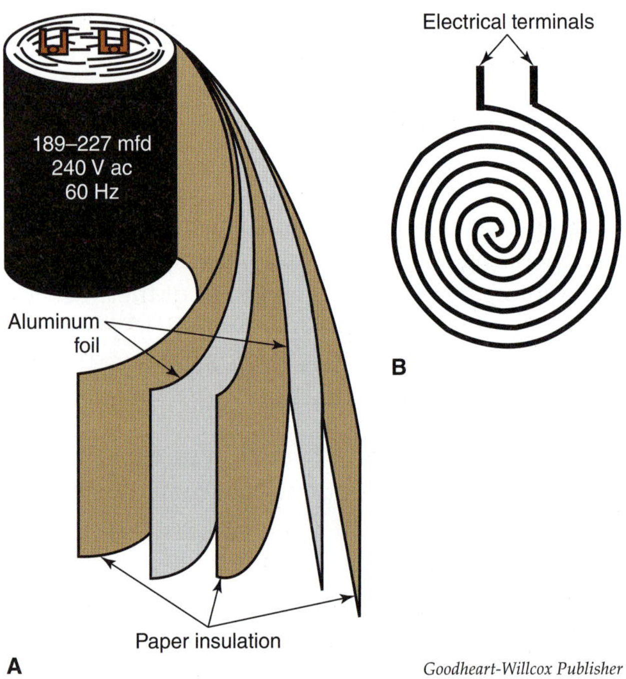

Goodheart-Willcox Publisher

Figure 24-32. Construction of a capacitor. A—Long strips of aluminum foil, separated by strips of paper insulation, are wound into a tight coil. B—An electrical terminal is attached to each of the aluminum strips (electrodes or plates).

electrons from traveling from one foil layer to the other. When connected to an alternating current circuit, the applied voltage forces one side of the capacitor to fill with electrons while the other side discharges electrons. When current flow alternates (reverses), the empty side fills up while the full side discharges, **Figure 24-33**.

Electrons do not travel *through* the capacitor. They simply accumulate on one side of the capacitor until the current alternates. Then they rush out of the capacitor and return to where they came from. Electrons are constantly rushing into and

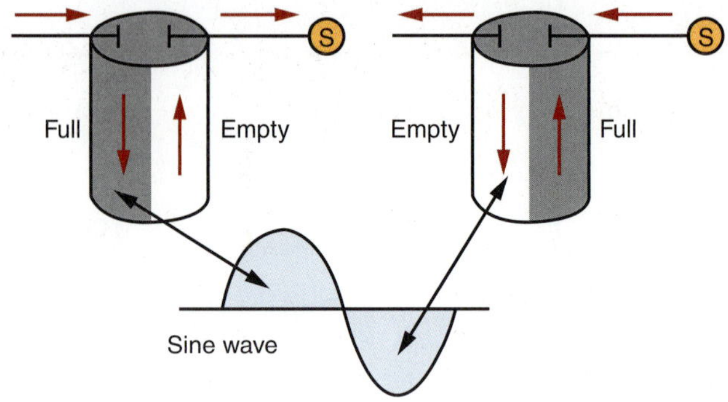

Goodheart-Willcox Publisher

Figure 24-33. Operation of a capacitor. When connected to an alternating current, the two foil electrodes alternately fill with electrons and discharge the electrons.

out of each side of the capacitor. When one side is full, the other side is empty.

When the start capacitor is connected in series with the start winding, one side of the capacitor is discharging its electrons into the winding. Discharging increases current flow in the start winding, strengthening the magnetic field, and producing better starting torque. The other side of the capacitor discharges back into the supply wire.

Capacitor Malfunctions

Both sides of the capacitor must be able to function, else the movement of electrons stops (no current flow). The movement of electrons into and out of the capacitor causes the capacitor to become hot. If a start capacitor is not removed from the circuit quickly, the capacitor will suffer a burnout or become shorted. A ***shorted capacitor*** is one in which the aluminum coils touch, such as a failure or burn-through of insulation. In this case, the capacitor is acting like any conductor, no longer as a capacitor. A burnout, or ***open capacitor***, occurs when one of the terminals separates from its coil. An open capacitor acts like an open circuit, stopping current through it.

The start capacitor is a frequent source of motor problems. Anything that causes the motor to ***short cycle*** (turn on and off frequently) will cause the start capacitor to burn out. If the motor tries to start and cannot, the capacitor will remain in the circuit and burn out.

Bleed Resistor

Sometimes a 20,000 Ω, 2 W, solid-state resistor is soldered in place to connect the capacitor terminals, **Figure 24-34**. Such a resistor does not

Goodheart-Willcox Publisher

Figure 24-34. A bleed resistor is a safety feature that prevents a possible electrical shock from touching the terminals of a charged capacitor. It also helps reduce arcing of relay contacts.

interfere with normal capacitor operation. During the off cycle, the resistor permits electrons trapped on one side of the capacitor to slowly migrate (bleed over) to the empty side until the two sides are equal. Called a ***bleed resistor***, it serves a twofold purpose:

- It protects the technician by preventing an electrical shock that could occur from touching the terminals of a charged capacitor.
- It prevents the excessive arcing that would otherwise occur when relay contacts close after a short run cycle. Excessive arcing can cause relay contacts to weld together or destroy the contacts.

Run Capacitor

The run capacitor is constructed like the start capacitor, but with some important differences. The run capacitor is better-built and is almost trouble-free because its capacitance rating is low, and it must remain in the circuit. The capacitor case is heavy gauge metal, which helps dissipate heat. The aluminum foil strips are shorter and thinner, and the coil is wound less tightly, limiting the amount of electron movement in the capacitor. Most distinctive is that the run capacitor is filled with a ***dielectric*** (insulating) oil that helps transfer heat from inside the capacitor to the metal case.

Manufacturers of outdoor condensing units for residential air-conditioning systems sometimes install a run capacitor with three terminals. Having three terminals indicates that it is actually two run capacitors in one case. These run capacitors are known as *dual run capacitors*. The rating for each capacitor is listed on the case. One terminal of the capacitor is identified as common to each capacitor and is connected to power supply. Another terminal is marked F (fan) and is connected to the start winding of the fan motor. The other terminal, marked H (hermetic), is connected to the start winding of the compressor.

Capacitor Ratings

Both start and run capacitors have two important ratings: ***Vac*** (volts, alternating cycle) and ***microfarads*** (mfd or µf).

The voltage rating of a capacitor indicates the nominal voltage at which it is designed to operate. Using a capacitor at voltages *below* its rating does no harm. However, excessive voltage causes severe damage. Run capacitors must not be subjected to voltages over 10% of their nominal VAC rating. Start capacitors must not be subjected to voltages over 30% of their nominal VAC rating. Start capacitors are normally rated at 110 VAC or 250 VAC, and run capacitors are normally rated at 370 VAC or 440 VAC. The 370 VAC or 440 VAC ratings allow for the back-emf generated by the start winding.

The amount of energy that a capacitor can store is expressed in *microfarads*. The rating tells how much capacitance the capacitor can store and then deliver to the start winding. Start capacitors have microfarad ratings much higher than those of run capacitors. Run capacitors are available from 1.5 mfd to 50 mfd, while start capacitors range from 75 mfd to 600 mfd.

Capacitors are selected according to the size of the motor and the load placed on the motor. A replacement capacitor should have the same microfarad and voltage ratings as the capacitor being replaced. If an exact replacement is not possible, the new capacitor's microfarad rating should be 5% to 10% *greater* than the one being replaced.

Capacitor Safety

The start winding of a motor can be damaged by a shorted and grounded run capacitor. Damage can be avoided by properly connecting the run capacitor. The electrical terminal attached to the aluminum foil closest to the metal case is identified with a special marking, **Figure 24-35**. The power supply should always be connected to the identified terminal, which is the terminal most likely to short to the metal case and be grounded in the event of a capacitor breakdown. If the capacitor should break down and become grounded, the fuse or breaker in the power supply line will blow before the motor is damaged.

SAFETY NOTE

Run Capacitors and PCB

Older run capacitors used polychlorinated biphenyl (PCB), a dangerous fluid, as the dielectric oil. If the metal shell is accidentally pierced or opened, do *not* touch the fluid or breathe the fumes. Disposal of such run capacitors should be done by properly trained technicians.

Testing Capacitors

One method of testing a capacitor is to use an ohmmeter. Before testing, the capacitor must be disconnected from the circuit and discharged by shorting across the terminals. This should be done with a 20kΩ, 2 W, solid-state resistor.

Touch the two leads from the ohmmeter to the capacitor terminals. If the capacitor is good, the meter battery (9 Vdc) will slightly charge one side of the capacitor. When the leads are reversed from one terminal to the other, one side of the capacitor discharges into the meter and the other absorbs electrons from the battery. A digital ohmmeter shows a slight resistance reading, then returns to zero. The resistance reading is slight because the meter battery is weak compared to normal line voltage. Increasing meter resistance can increase resistance for easier readings, **Figure 24-36**.

If the ohmmeter reads infinity and stays there, the capacitor is open. If the meter reads 0 and stays there, the capacitor is shorted. See **Figure 24-37**. An open or shorted capacitor is faulty and must be replaced.

An ohmmeter test will not reveal a weak capacitor. Low mfd capacity can only be discovered by using a tester specifically designed to test ac capacitors, **Figure 24-38**. These meters are accurate, inexpensive, and simple to operate. Most multimeters now have a capacitor checker built into them, which is used to get the actual mfd reading.

The most accurate way to test a capacitor is to leave it in the circuit and use a multimeter set to read voltage. To test using a multimeter, first place a meter lead on one terminal and the other meter lead to the other terminal and record the voltage. Then, using the amp probe on the multimeter, clamp the probe around one of the wires from the terminals and record the amperage reading. Next, take the amperage reading and multiply it by 2652 and divide that answer by the voltage reading to obtain the microfarad (mfd) reading, **Figure 24-39**. This step is reflected in the following formula:

$$\text{mfd} = \frac{\text{Amps} \times 2652}{\text{Volts}}$$

This mfd reading should be compared to the mfd rating set for that capacitor, and if it falls below that range or more than 5%–10% greater than that range, then the capacitor is faulty and should be replaced.

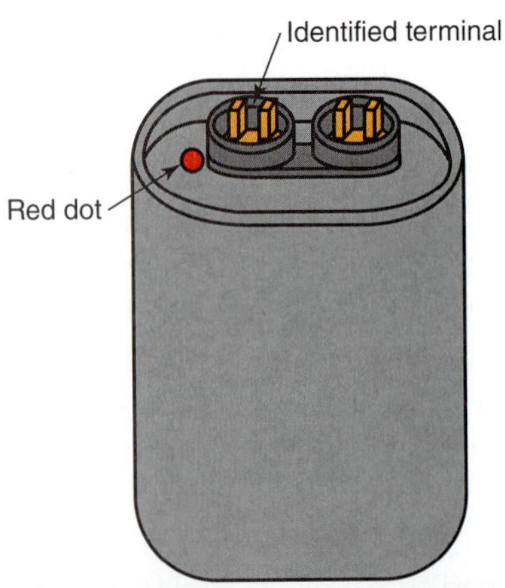

Goodheart-Willcox Publisher

Figure 24-35. The terminal connected to the outer aluminum strip is marked on a run capacitor. The power supply connects to this lead to help protect the motor. Some manufacturers use a dot, while others prefer an arrow or a dash.

Changing Motor Types

It is possible to add start components to a hermetic compressor's PSC motor to convert it to a CSCR motor. The procedure is sometimes necessary when the compressor's PSC motor has

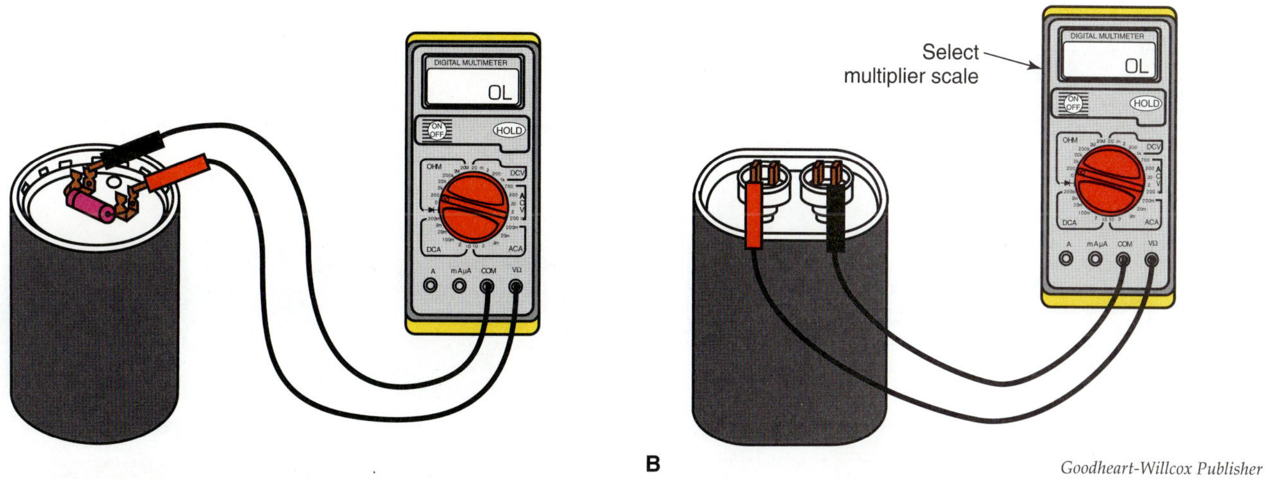

Goodheart-Willcox Publisher

Figure 24-36. Checking capacitors with an ohmmeter. A—Reversing leads should cause the needle of an analog meter to deflect slightly, then return to zero. A digital meter will briefly show a low resistance reading, then return to OL (overload). B—The meter reading can be increased for easier viewing by selecting the µ10,000 resistance scale.

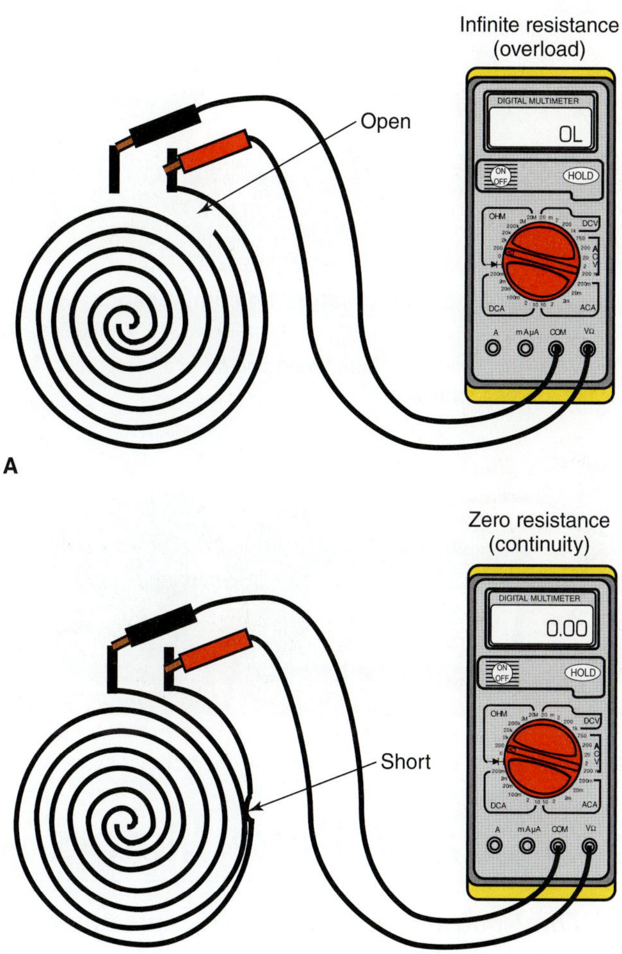

Goodheart-Willcox Publisher

Figure 24-37. Capacitor defects. A—An open (burned out) capacitor causes the meter to show an infinite resistance reading. B—A shorted capacitor shows zero resistance (continuity).

Goodheart-Willcox Publisher

Figure 24-38. A tester made specifically for use with capacitors helps the technician quickly diagnose problems.

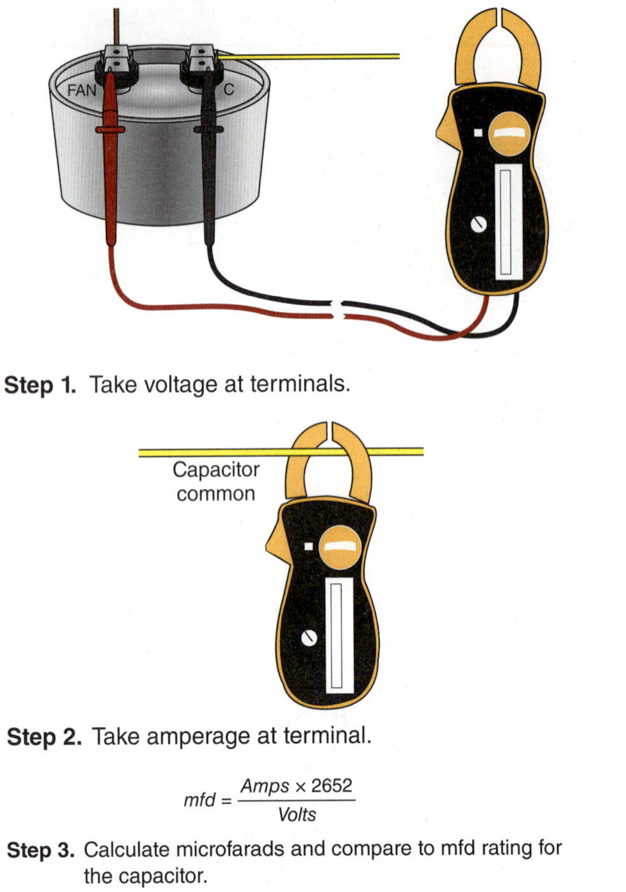

Goodheart-Willcox Publisher

Figure 24-39. Using the voltage and amperage settings on a multimeter, a technician can calculate the microfarad rating of a capacitor. This method is considered the most accurate way to test a capacitor.

difficulty starting, which is not unusual with residential central air-conditioning systems.

Selecting a properly sized capacitor and relay is critical for proper operation. Sizing information is available from the local dealer or compressor manufacturer. Most manufacturers of compressors using PSC motors furnish a ***hard start kit*** for their units' motors. The kit consists of a start capacitor and relay, along with complete instructions for installation. See **Figure 24-40**. The model and serial numbers of the motor/compressor are needed to obtain the correct kit. **Figure 24-41** shows a hard start kit designed to work on multi-sized compressors. This kit should be used as a temporary fix. If the compressor needs a hard start kit, obtain the one from the compressor manufacturer that is specifically designed for that compressor.

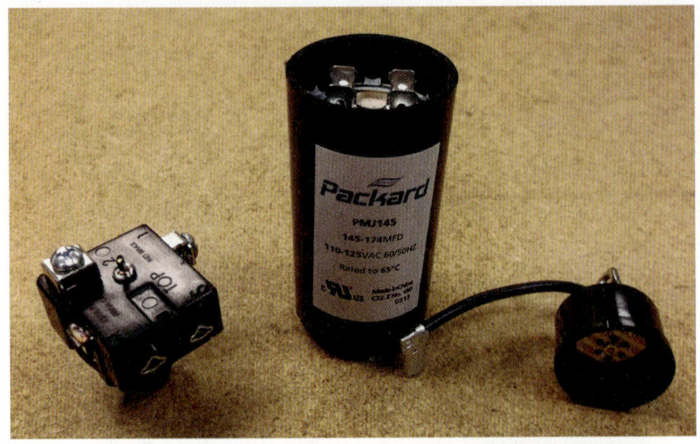

Hard start kit for fractional hp compressors

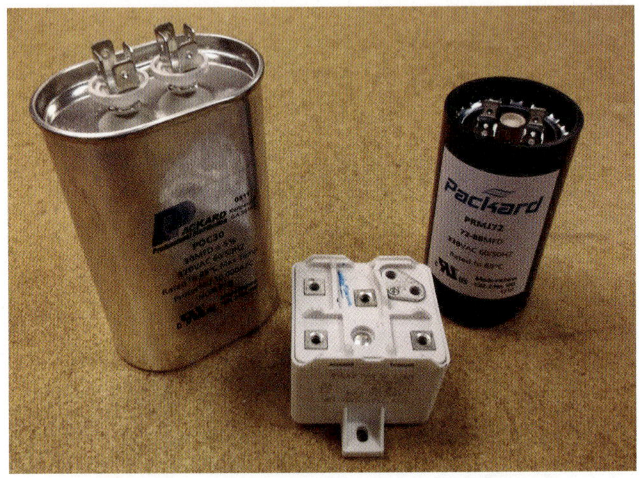

Hard start kit for larger hp compressors

Images courtesy of Johnstone Supply

Figure 24-40. Hard start kits for adding on to a compressor for greater starting torque.

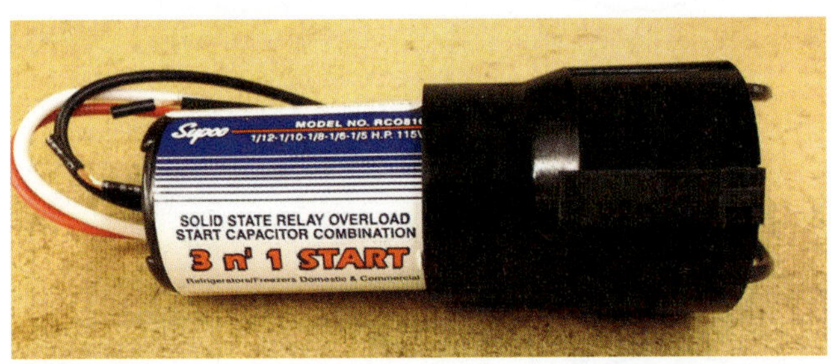

Goodheart-Willcox Publisher

Figure 24-41. A multi-compressor quick add-on hard start kit.

24.5 Dual-Voltage Motors

Dual-voltage motors are manufactured to operate on either of two supplied voltages. For example, the voltage rating on the motor data plate may read, *115/230 V*, indicating the motor can be connected to either a 115 V supply or a 230 V supply. Dual-voltage motors have two run windings that must be connected properly, as shown in **Figure 24-42**. For lower voltage, the run windings

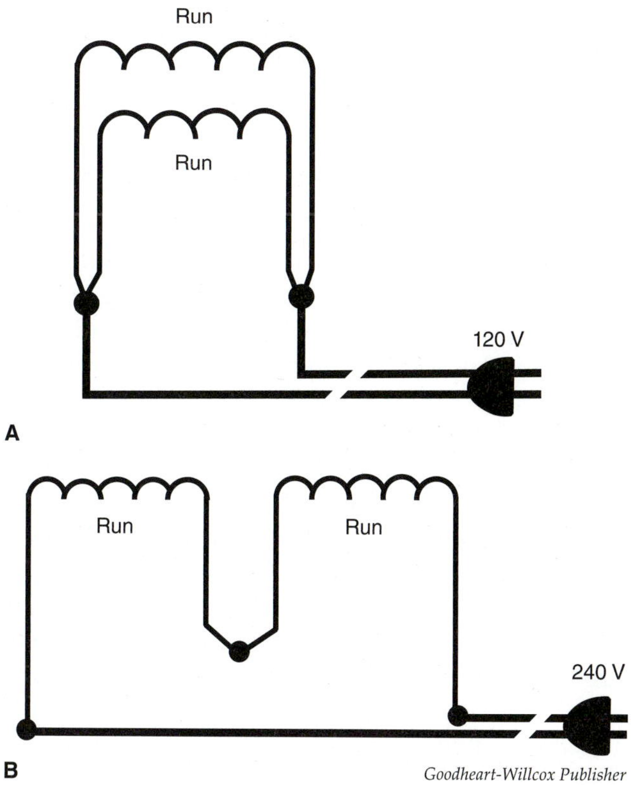

Goodheart-Willcox Publisher

Figure 24-42. Dual-voltage motor run winding connections. A—For the lower voltage, the windings are connected in parallel (for lower resistance). B—For high-voltage operation, the windings are connected in series (for high resistance).

are connected in parallel (providing lower resistance). For higher voltage, the windings are connected in series, doubling the resistance.

The magnetic fields of the two run windings *must* rotate in the same direction. If one field is "left" and the other is "right," opposing polarities will cause the rotor to lock. The factory connects the motor windings to terminals located in the terminal box.

Most dual-voltage motors are equipped with quick connect terminals for ease of change from one voltage to the other. Instructions for placement of these terminals are inside the terminal box cover. **Figure 24-43** depicts the proper placement of terminals on an open-type dual-voltage motor.

Because the 230 V power supply can vary from 208 V to 230 V depending on location, a lot of motors have been designed to run within that range. The voltage listed on their data plate could read *208–230 V*. This type of motor does not need to have its wiring changed. Some dual voltage motor's data plate could show *115/208–230 V*. This motor would need to be wired one way for 115 V and another way for 208–230 V.

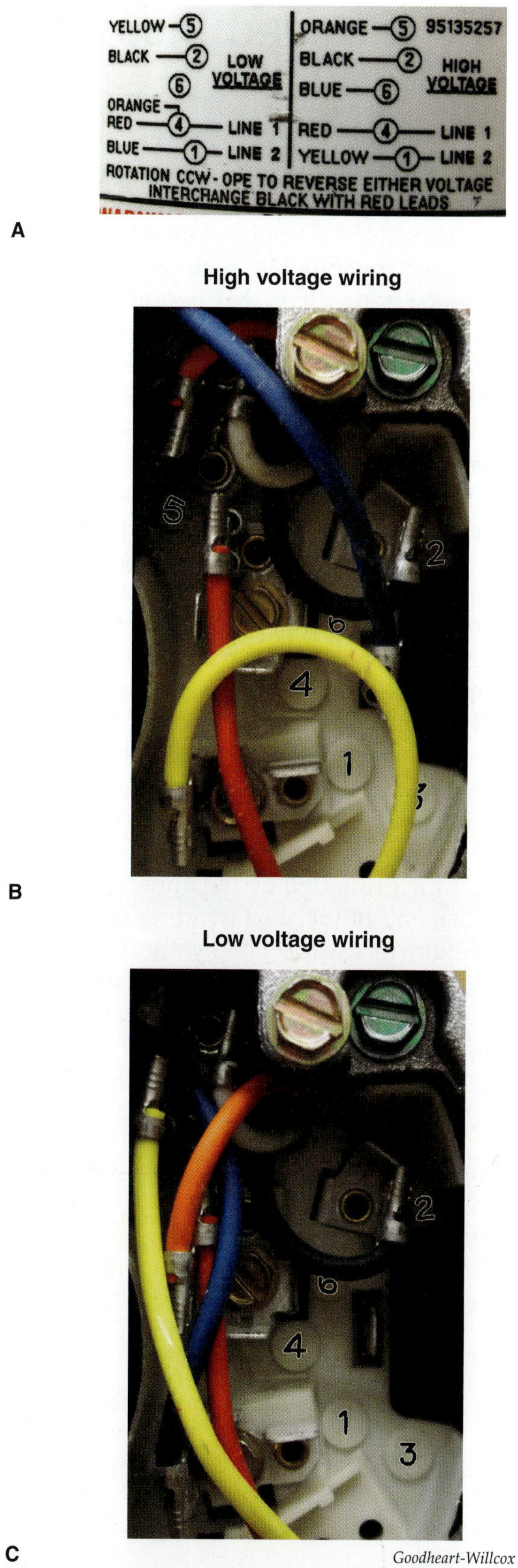

Goodheart-Willcox Publisher

Figure 24-43. Changing a dual-voltage motor from one voltage to another. A—Motor wiring diagram on data plate showing factory-installed internal connections at the terminal board. B—For high voltage, the wires connect the run windings in series. C—For low voltage, the wires connect the windings in parallel for lower resistance.

24.6 Multi-Speed Motors

Multi-speed motors are common in heating, cooling, and refrigeration systems. In contrast to the two wires used in single-speed motors, multi-speed motors have several wires. A ***multi-speed motor*** is necessary in applications that require different speed operations, such as a window air conditioner with a fan that operates at high, medium, or low speeds. Many furnace blower motors are multi-speed. Multi-speeds allow more air to be moved during the cooling season (high speed) than during the heating season (low speed).

Motor speed is determined by the number of poles in the run winding. The multi-speed motor has different ***taps*** (wires) that make it possible to select the number of poles being used, **Figure 24-44**. A high-speed tap uses two poles (3600 rpm), medium speed uses four poles (1800 rpm), and low speed uses all six poles (1200 rpm). One leg of the power supply is connected to the common terminal and the capacitor. The other side of the power supply is connected to the tap for the desired speed. Unless a selector-type switch is used, each unused tap is capped with a wire nut and sealed with electrical tape to prevent the motor windings from becoming grounded by the unused wires.

Only one end of the start winding is available for connection to a capacitor by the technician. The other end of the start winding is wired at the factory. A power supply that is parallel to the start winding is provided, regardless of the speed (tap) selected. Most manufacturers color-code the external wires for making connections to multi-speed motors. See **Figure 24-45** for the color codes.

24.7 Three-Phase Motors

Three-phase induction motors are common in commercial applications because they are smaller and more powerful than single-phase motors of equal horsepower. Furthermore, they are cheaper to operate, have fewer problems, and are easier to connect and control.

Three-phase alternating current induction motors offer high starting torque and high running torque without the need for a start winding, capacitors, or relays. Only the three "hot" wires (L_1, L_2, and L_3) are needed for proper operation, but the ground (green) wire is included for safety. The green wire is connected to the motor frame to provide an escape for electrons if the motor becomes *shorted to ground* (grounded).

Coding for Multi-Speed Motors

Tap	Color
Common	White
High	Black
Medium	Yellow
Low	Red
Capacitor	Brown (2 wires)

Goodheart-Willcox Publisher

Figure 24-45. Colors used by manufacturers to code wires used for making external connections to multi-speed motors.

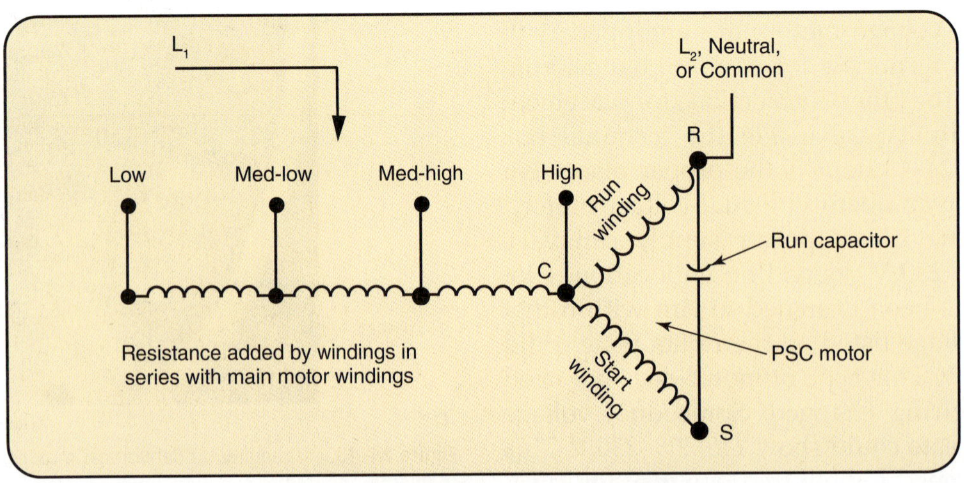

Multi-speed, single-phase PSC motor

Goodheart-Willcox Publisher

Figure 24-44. Different numbers of poles are used for each speed on a multi-speed motor. Connections are made to the common terminal and the proper tap for the desired speed. A selector switch may be installed for ease of changing from one speed to another.

Method of Operation

Three-phase motors have three pairs of stator poles (windings), one pair for each supply wire. Each winding should have identical resistance. As shown in **Figure 24-46**, each pair of windings has a north pole and a south pole directly opposite each other. The stationary poles are equally spaced around a circle, exactly 60° apart (6 × 60 = 360°). In this way, three-phase motors have "one pole per phase."

With three-phase alternating current, the three power–supply wires take turns changing polarity from north to south to zero. See **Figure 24-47**. Changing of polarity in the supply wires produces a strong rotating magnetic field. The alternating zero does not produce polarity in the stator poles, which permits the other two poles to produce the rotating push-pull effect on the rotor. Because the zero pole is constantly rotating around the stator poles, the rotor chases the rotating magnetic fields.

The alternation of north, south, and zero in a three-phase motor produces a strong starting torque and a strong running torque. If the motor is running and one wire is disconnected from the circuit or one phase is lost, the motor may continue to run as a single-phase motor (a condition called ***single-phasing***). However, this condition is bad for the motor. It will pull higher current on the other two phases and can quickly result in motor failure.

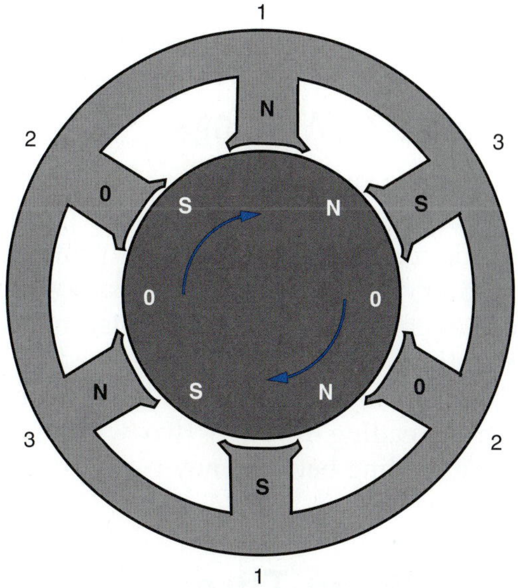

Goodheart-Willcox Publisher

Figure 24-47. The three poles change polarity from north to south to zero, producing a rotating magnetic field for the rotor to "chase."

Single-phasing of a three-phase motor is the most common failure of a three-phase motor.

Changing Rotation

Direction of rotation is determined by direction of the rotating zero, where no polarity is generated. A three-phase motor is easily reversed

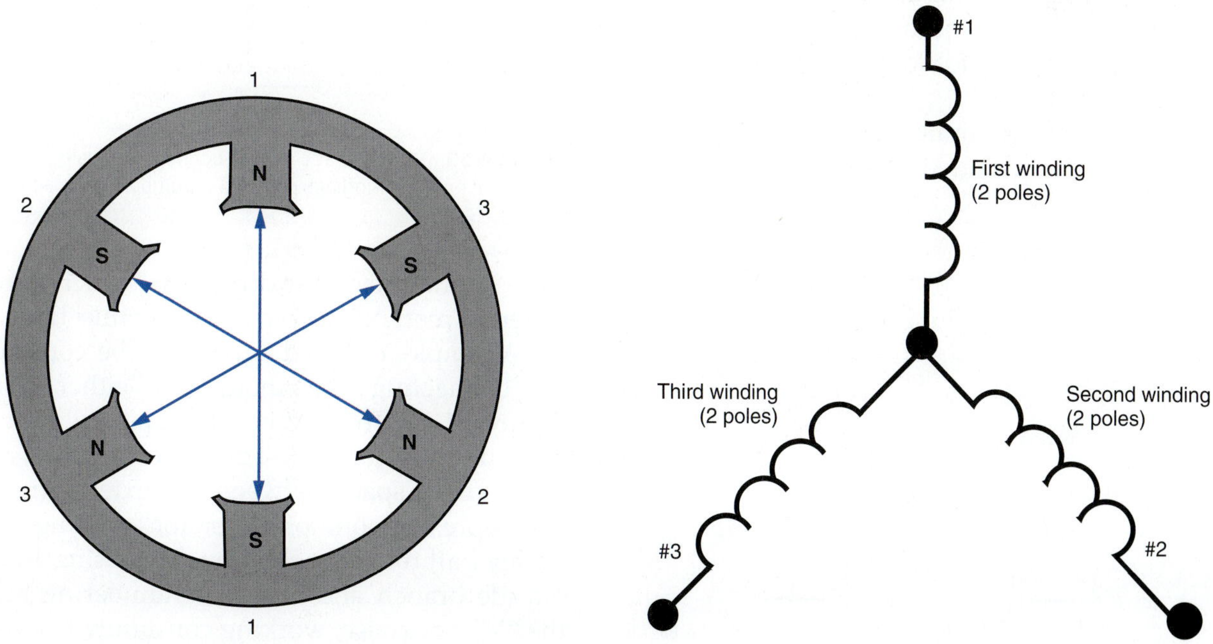

Goodheart-Willcox Publisher

Figure 24-46. The three windings, or stators, of a three-phase motor each have a north pole and a south pole. The stators are equally spaced around a circle.

by changing any two supply wires, causing the zero to rotate in the opposite direction.

Resistance of Windings

All the windings in a three-phase motor are identical in resistance. One end of each winding is factory-connected inside the motor, providing an ideal location for the internal overload. When the overload opens, all three wires are disconnected at the same time, **Figure 24-48**. When reading resistance between any two motor leads, the ohmmeter is actually reading through two windings. The resistance reading between any two wires should be the same.

Dual-Voltage Three-Phase Motors

Many three-phase motors are dual voltage. For example, a motor data plate may specify the voltages of 220/440 V can be applied. However, instead of three wires to connect, this dual-voltage motor has *nine* or more wires that are tagged and numbered for identification. *Never remove these numbers!* See **Figure 24-49**.

A dual-voltage three-phase motor contains an extra set of three poles. In addition to the three original wires, two wires for each of these extra poles are connected to wiring terminals in the motor's electrical box, **Figure 24-50**. All nine wires must be connected. The motor windings can be connected in parallel for low voltage and in series for high voltage.

Instructions for connecting the nine wires for each voltage are included on the motor data plate.

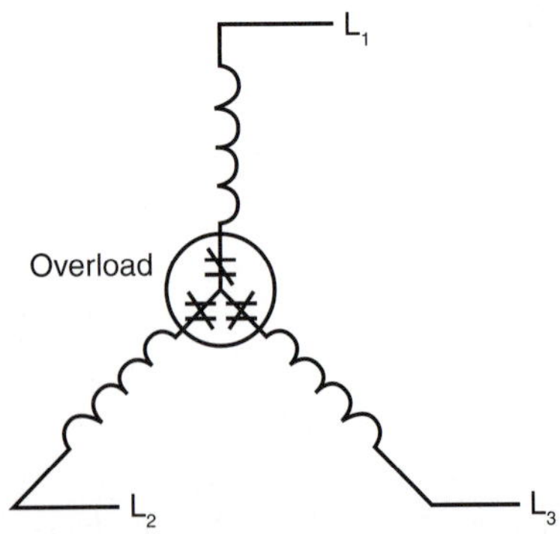

Goodheart-Willcox Publisher

Figure 24-48. All three windings are connected to the internal overload so when the overload opens, all the windings are disconnected.

Goodheart-Willcox Publisher

Figure 24-49. This dual voltage three-phase motor has 11 wires. Each wire has its number on its insulation as shown.

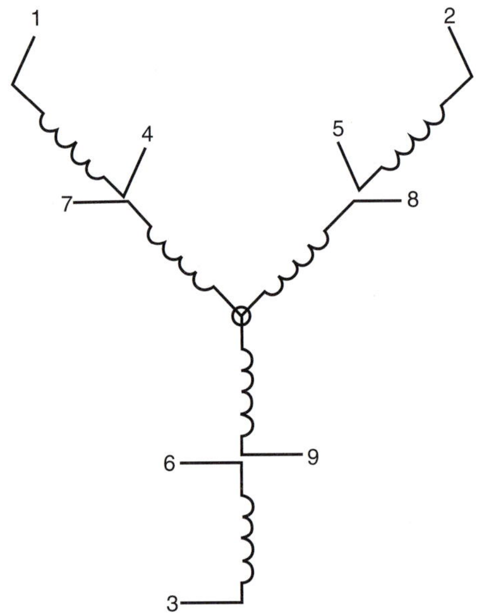

Goodheart-Willcox Publisher

Figure 24-50. A schematic of wye-connected, dual-voltage, three-phase motor windings shows the numbering for all nine wires.

Refer to manufacturer directions, which may use different methods to explain the connections.

A simple Y-type diagram can be constructed for connecting the motor for either voltage, **Figure 24-51**. To do this, start by drawing a "Y," and put extensions on each of the branches, allowing a small space between the extensions. Each line represents one of the stator windings. Next, number all the ends. Start by numbering 1 on any outside branch and continue numbering around the "Y" clockwise, working constantly toward the inside, until all ends are numbered.

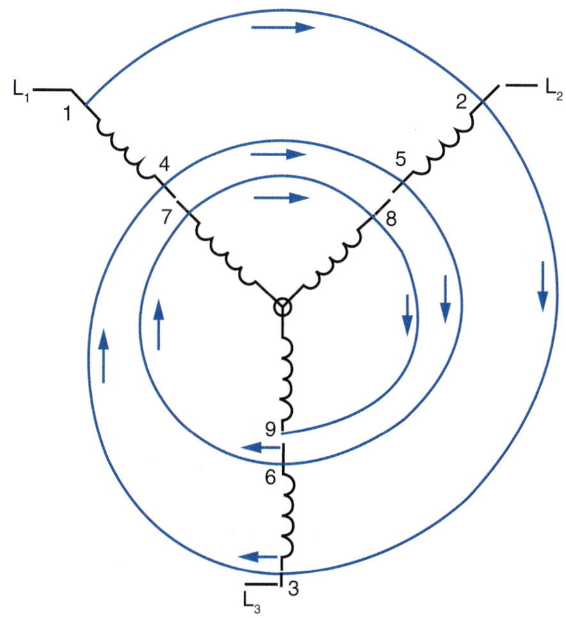

Goodheart-Willcox Publisher

Figure 24-51. Constructing a "Y" diagram and numbering the wires on a dual-voltage, three-phase motor. The spiral arrow shows the proper sequence for numbering the ends of the windings.

The three incoming power supply wires are always connected to motor wires l, 2, and 3. Except for direction of rotation, it does not make any difference which supply wire is connected to which terminal since the supply wires are the same.

High-Voltage Connection

Making the high-voltage connection is easy because the windings are wired in series, and the branches in the diagram are simply connected to form a larger "Y," **Figure 24-52**. Wires 4 and 7 are connected. Wires 5 and 8 are connected. Wires 6 and 9 are connected. The incoming power supply wires are always connected to wires 1, 2, and 3. For high voltage, there are two wires at each connection, and the current flows through four windings before exiting through another supply wire. Each wire connection should be made with a wire nut and secured with electrical tape.

Low-Voltage Connection

The parallel connections up for low voltage are more complicated than for high voltage, but the "Y" principle is still used. As shown in **Figure 24-53**, L_1 is connected to wires 1 and 7. L_2 is connected to wires 2 and 8. L_3 is connected to wires 3 and 9. Wires 4, 5, and 6 are connected to each other.

For low voltage, there are three wires at each connection, and the current flows through two windings before exiting through another supply wire.

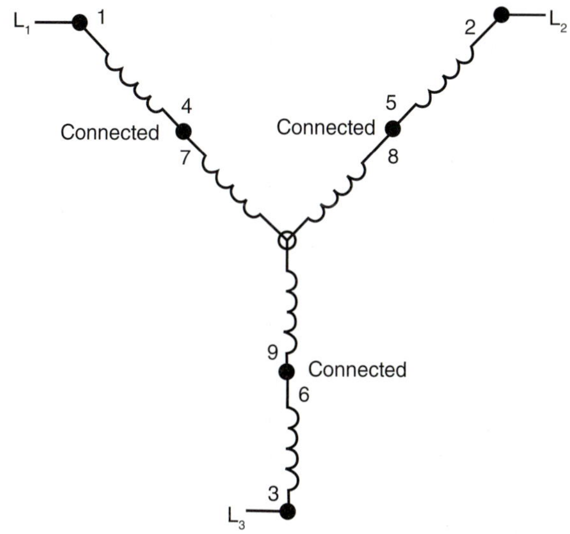

Goodheart-Willcox Publisher

Figure 24-52. The high-voltage connection places the windings in series. Wires 4 and 7 are connected, as are wires 5 and 8 and wires 6 and 9. Power supply wires always go to 1, 2, and 3.

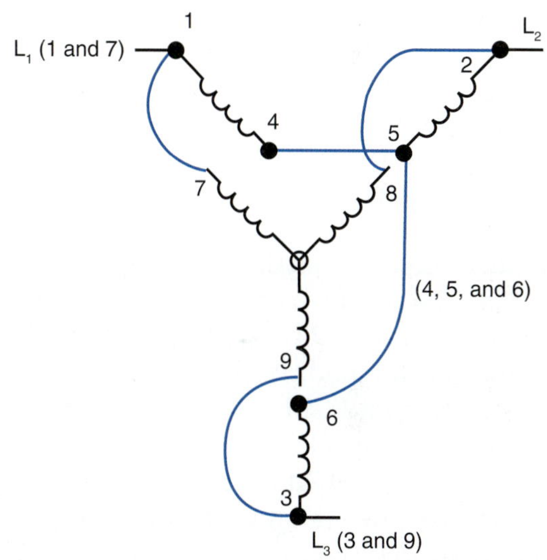

Goodheart-Willcox Publisher

Figure 24-53. The low-voltage connection places the windings in parallel. There are three wires at each connection, compared to two wires in the series connection for high voltage.

Resistance Check of Windings

The motor windings on a three-phase motor can be checked with an ohmmeter. If a resistance reading of zero occurs, the motor is shorted. If a reading is obtained from any motor lead to ground, the motor is grounded. A reading of infinite resistance (full scale deflection on an analog meter) across any two ends of a winding means the winding is open. In each of these cases, the motor must be rewound or replaced.

Depending on the size of the motor, resistance readings on three-phase motor windings range from 1 Ω to 50 Ω. The larger the motor, the smaller the resistance. On a single-voltage, three-phase motor, all windings have the same resistance. On a dual-voltage three-phase motor, the three "extra" windings have one-half the resistance of the main windings.

24.8 Electronically Commutated Motors (ECMs)

Electronically commutated motors (ECMs) are brushless dc three-phase motors with integrated electronic controls. ECMs provide more precise control of the motor speed and torque when compared with other ac motors. ECMs can be separated into two categories: *constant torque* where a multispeed motor maintains a constant torque and *constant volume* where a variable speed motor keeps constant airflow.

ECMs consists of two parts: a motor and a control module, **Figure 24-54**. On some ECMs, the motor and control module can be separated. When troubleshooting an ECM, a technician can disconnect the control module and then check the motor operation using an ECM testing tool.

ECMs are powered by standard ac voltage. Some ECMs can be configured to use either 120 V, 240 V, or 277 V. The electronics in the control module convert the ac voltage to dc voltage using an internal rectifier. The dc voltage is then supplied to the motor.

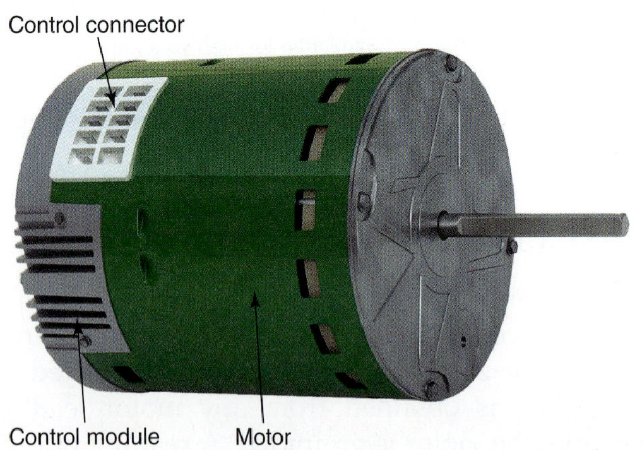

DiversiTech Corporation

Figure 24-54. An ECM has two components: a control module and a motor. This ECM could be used for an indoor blower on a residential or light commercial system.

The motor component of an ECM is a brushless dc motor. The rotor consists of permanent magnets. The stator consists of several windings, often six.

Most dc motors use brushes and a commutator to change the polarity of the current in the windings. In an ECM, the control module uses electronics to energize and de-energize each winding in sequence to control the speed and torque of the motor. This electronic control allows for a wide variation in motor speeds.

Some ECMs are programmed by their manufacturer for the specific type of equipment in the system. When replacing a manufacturer-specific ECM motor, the replacement motor must also be programmed for the specific equipment.

ECM motors can be programmed to provide constant air volume, constant torque, or constant speed. For example, an ECM powering an indoor blower is often programmed to maintain constant air volume. As it is operating, the ECM monitors motor speed and torque, and adjusts the motor speed based on changes in static pressure. For example, if a filter becomes dirty, static pressure increases and the volume of airflow decreases. This increases the torque on the motor. The ECM recognizes this change and increases the motor speed to maintain the design airflow.

ECMs provide several benefits when compared to similar ac motors:

- ECMs can adjust to changes in the system and maintain a constant volume of airflow, improving system performance.
- ECMs on blowers can be programmed to run at 50% for an initial few minutes to remove humidity more effectively, improving the performance of the system.
- ECMs use significantly less energy than comparable ac motors. ECMs have soft start capability, which eliminates a large inrush of starting current. In low-load conditions, an ECM can reduce motor speed and use less energy.
- ECMs make less noise than similar ac motors during operation.
- ECMs generally have ball bearings and require less maintenance than ac motors with sleeve bearings.

ECM motors also have some disadvantages relative to comparable ac motors. First, ECMs are much more expensive than ac motors. Second, servicing an ECM can be more difficult than servicing an ac motor and may require special equipment. Finally, obtaining a replacement ECM programmed for the specific equipment can be more difficult than obtaining a more generic ac motor replacement.

24.9 Identifying Single-Phase Compressor Terminals

Most manufacturers of single-phase, hermetic compressors mark the three electrical terminals: C, S, and R (Common, Start, and Run). If these terminals are not marked, they can be identified by checking the resistance of the windings.

Electrical power supply to the motor must be disconnected and the connections to the three motor terminals opened. This isolates the motor windings and prevents backfeed from other electrical circuits that would cause a false ohmmeter reading.

The first test with the ohmmeter is to determine across which two terminals have the highest resistance. Three separate readings are required to cover all the possible combinations. The two terminals that produce the highest reading—18 Ω in **Figure 24-55**—are the start and run terminals, because the meter is reading through both windings (in series). By process of elimination, the high reading means the other terminal is the common terminal.

With the common terminal identified, read from common terminal to each of the other two terminals. Common-to-start shows more resistance than common-to-run. The ohm readings will vary from compressor to compressor and from manufacturer to manufacturer. The proper reading is the math that adds up. When the measurement across the common-to-run terminals and the measurement across the common-to-start terminals is added together, it should equal the measurement across the run-to-start terminals, **Figure 24-56**.

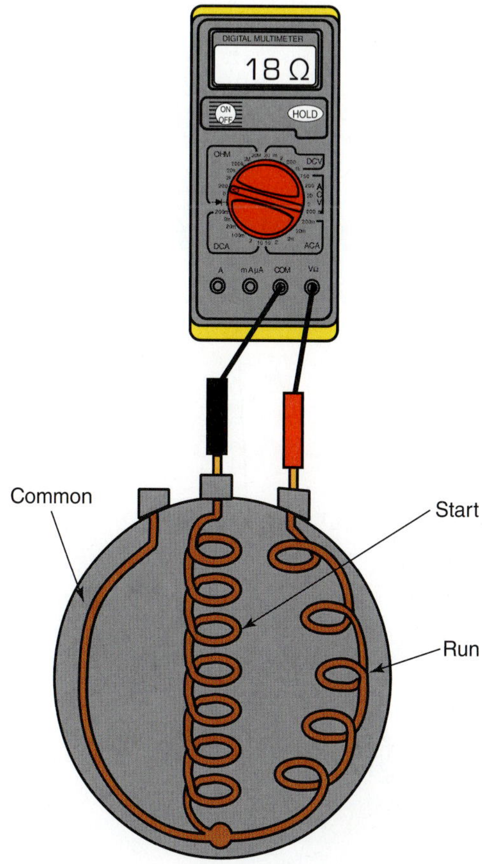

Goodheart-Willcox Publisher

Figure 24-55. The highest of the paired terminal resistance readings is across the start and run terminals, measuring the resistance of both windings. By process of elimination, the remaining terminal is the common.

24.10 Identifying Motor Problems

The ohmmeter is used to check motor windings, which is a process called "ringing out" a motor. All electrical connections to the motor must be removed to prevent false readings and possible damage to the meter.

Grounded Windings

Each motor terminal must be checked to determine whether it is ***shorted to ground***. A grounded winding is one that has developed an internal electrical short to the device's frame. Use one of the ohmmeter probes to scratch through paint or oxidation on the steel shell of the hermetic unit to provide a good electrical connection. Connect the other ohmmeter lead to one of the motor terminals (C, S, or R). The ohmmeter should read infinity or

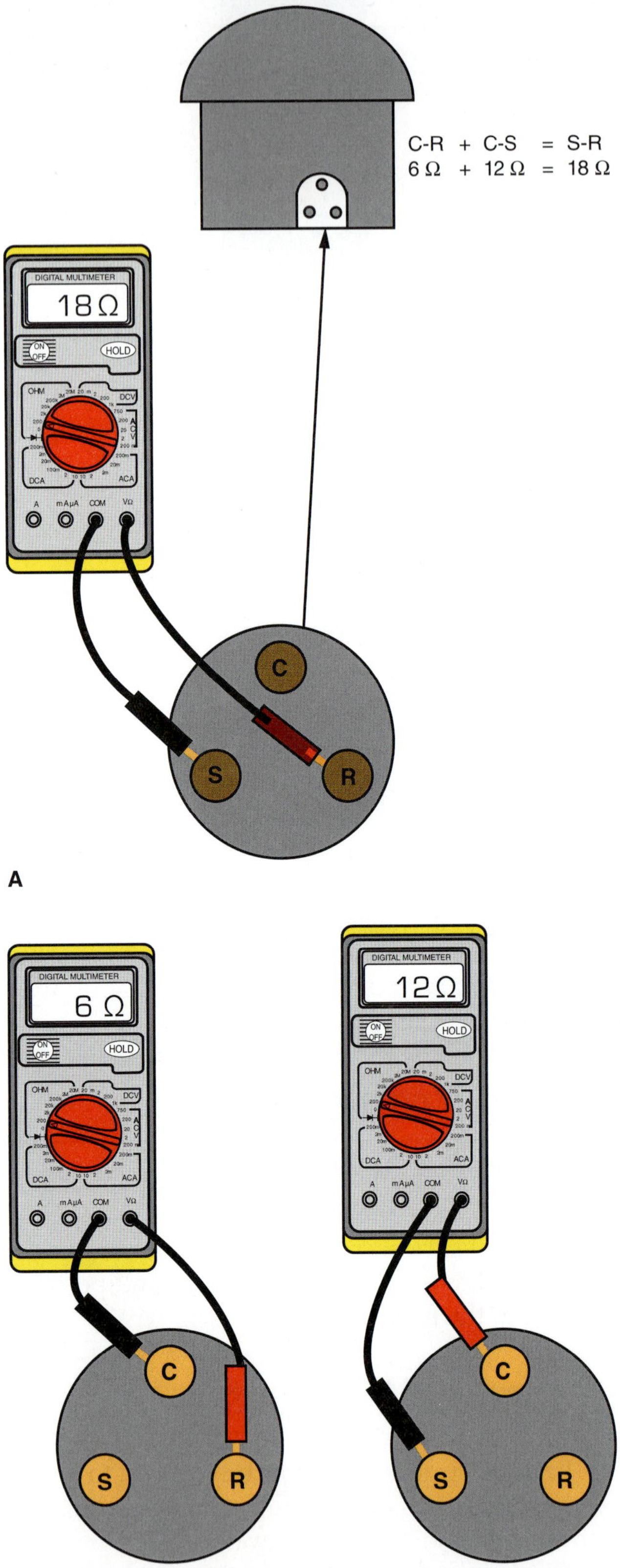

Goodheart-Willcox Publisher

Figure 24-56. Identifying terminals by measuring resistance. A—The common terminal is the one remaining when the highest terminal pair (run-to-start) reading is obtained. B—A reading from common-to-run yields the lowest reading. C—The common-to-start reading falls between the other two.

OL (unlimited resistance). A reading of zero resistance, or less than infinite resistance, indicates a winding has become shorted to ground. Repeat the test for each motor terminal. A "grounded" hermetic is not repairable and must be replaced. See **Figure 24-57**.

A motor with a winding shorted to ground will normally show itself by a blown fuse or a tripped breaker.

Open Windings

An ***open winding*** refers to a broken wire in the motor winding that is *not* touching the motor shell. To locate or determine an open winding, the motor must be checked with an ohmmeter. Check each winding by touching the meter probes to two terminals and recording the reading. Repeat for each pair of terminals until all combinations have been checked. Since there is no complete path (circuit) for current flow if the circuit is open, the resistance indicated by the meter would be infinity (OL), **Figure 24-58**. A broken wire can occur in either winding or in the common connection. The condition is not repairable in the field, so the hermetic must be replaced.

Shorted Windings

A ***shorted winding*** occurs when two wires on the same winding short out or touch each other, shortening the path of the current flow through that winding. This causes the winding to pull more amperage causing the fuse or breaker to trip. With power disconnected from the motor,

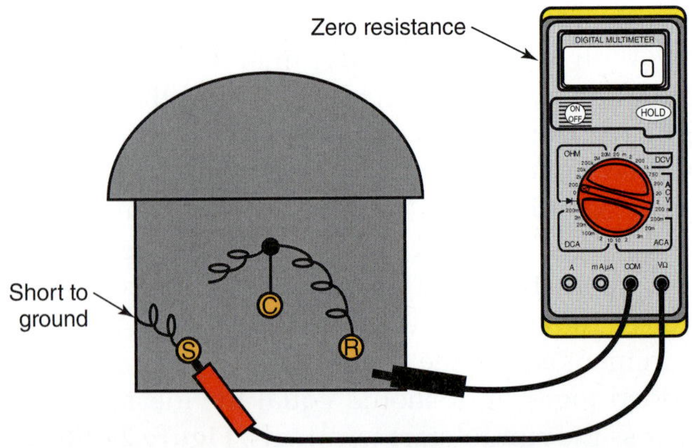

Goodheart-Willcox Publisher

Figure 24-57. A zero resistance reading when the ohmmeter probes are touching the hermetic case and a motor terminal indicates a short to ground. The hermetic must be replaced.

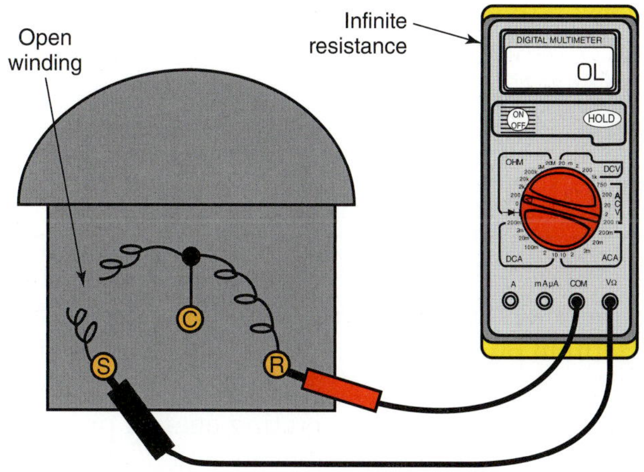

Goodheart-Willcox Publisher

Figure 24-58. An infinite resistance reading when the meter probes are touching two motor terminals indicates an open winding. This hermetic compressor must be replaced.

use an ohmmeter to check the winding resistance as discussed earlier. If the math of the resistance measurements does not add up, there is a problem. If the meter shows an infinite resistance reading, there is an open winding. If the meter shows a reading of zero resistance or a one that does not add up, the winding is shorted.

24.11 Motor Bearings and Lubrication

All motors have a ***bearing*** (lubricated support for a rotating shaft) inside the end bell (cover) at each end of the rotor. The rotor shaft is inserted through the bearing, which holds the shaft in perfect alignment. Alignment is *critical* because the air gap between the rotor and the stator is only a few thousandths of an inch, and the gap must be equal on all sides. The end bells are held in position by four bolts that extend through the stator shell from end to end.

A primary cause of motor failure is dry bearings, which cause the rotor to drag, lock up, or wear out. Dry bearings also cause excess amperage draw, while worn bearings cause a noisy motor.

Permanently Oiled Bushings

Many bearings for fractional horsepower motors are permanently lubricated. The bearing is actually a ***bushing*** (sleeve around a shaft) made of porous bronze that is saturated with oil at the factory. The bushing is considered permanently lubricated and is used in fan motors and other low-horsepower applications. No additional lubricating of the bearing is required for the life of the motor.

Oil-Type Bearings

Oil-type bearings should receive three to six drops of oil in each oil port every six months. Avoid over-oiling, since excess oil may accumulate in the stator and collect dirt. The oil ports are located at each end bell of the motor for direct access to the bearing, **Figure 24-59**. The oil ports are normally capped with rubber or plastic plugs to keep dirt out of the bearing. The plugs are easily removed for adding oil and should always be replaced. Regular 10W30, nondetergent automotive oil is used to oil motor bearings.

Grease-Type Bearings

Most large horsepower motors have ball bearings lubricated with grease rather than oil. A grease fitting—called a ***zerk fitting***—is installed on top of each bearing, and a removable plug is located at the bottom of each bearing, **Figure 24-60**. An ordinary grease gun is used to pump high-grade, medium grease to the bearings every three months. The bottom plug must be removed before pumping grease into the bearing. If it is not removed, pressure can rupture the diaphragm that holds grease in the bearing. A ruptured diaphragm will permit grease to enter the rotor area, resulting in a dry bearing.

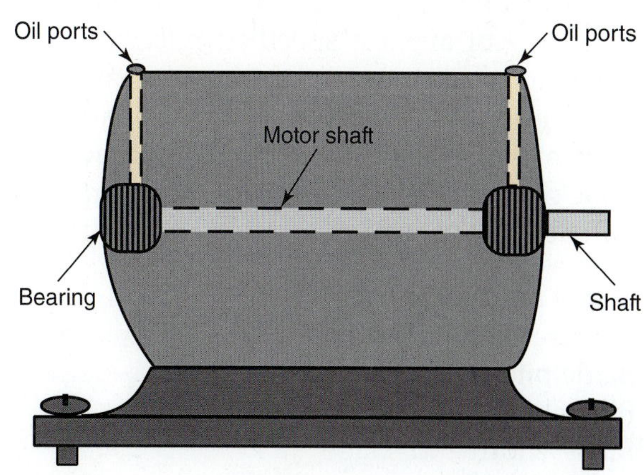

Goodheart-Willcox Publisher

Figure 24-59. Oil ports are located atop the bearings at each end of the motor. Periodic lubrication is needed to prevent motor damage.

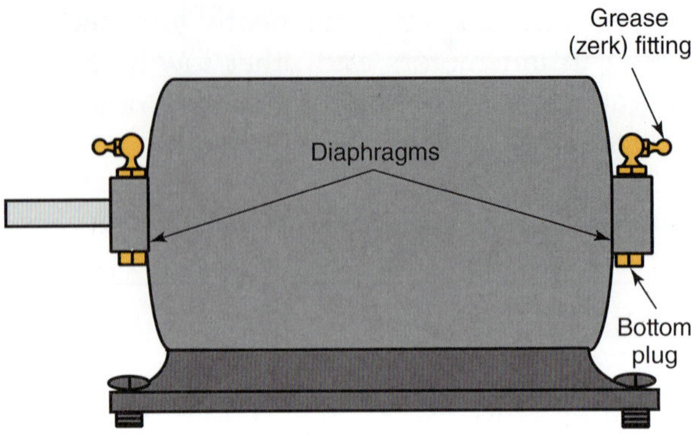

Goodheart-Willcox Publisher

Figure 24-60. Grease fittings are used to force lubricant under pressure into a bearing. The bottom plug must be removed when adding grease to prevent rupturing the diaphragm and allowing grease to leak out of the bearing.

Removing the bottom plug prevents any pressure on the diaphragm because new grease forces old grease (if any) out the bottom port. Continue pumping until new grease appears at the bottom port. Usually, only two or three pumps on the grease gun are required. If the bearing requires more pumps on the gun, the diaphragm has probably been ruptured, and grease is entering the motor. A ruptured diaphragm is revealed by a grease stain or puddle immediately under the motor. The bottom plug should always be reinstalled after greasing the bearing. A motor with a ruptured diaphragm must be greased frequently to delay or prevent bearing failure.

24.12 Fans and Blowers

Fans and blowers are essential parts of most refrigeration and air-conditioning systems. The purpose of any fan or blower is to move a specific volume of air at a specified *velocity* (speed). The amount of air moved (volume) is measured in *cubic feet per minute (cfm)*. The volume of air that a fan can move is determined by several factors. If any one of these factors changes, the volume of air moved also changes:

- Speed of rotation
- Number of fan blades
- Blade pitch (angle, in degrees)
- Blade length
- Blade width and shape

In HVACR applications, there are two general types of fan blades used to move air:

- ***Axial*** (propeller) blades are used for fans.
- ***Centrifugal (squirrel cage) blades*** are used for blowers.

Axial Blade Fans

There are two types of propeller (axial) blades: round and square, **Figure 24-61**. Each type serves a definite purpose. The round blade is best suited for free-air applications, such as pedestal fans. It does not perform well operating against pressure. To obtain proper airflow, the movement of inlet air and discharge air must be unrestricted. The round blade has excellent efficiency in free-air conditions and is almost noiseless because the air enters and exits at different angles along the blade length.

Square fan blades

Round fan blades

Goodheart-Willcox Publisher

Figure 24-61. Square blades and round blades are used for different applications. Square blade fans are used in most air-conditioning applications.

The square-type propeller blade is designed to operate against medium and high pressures in such applications as evaporators and condensers. The discharge blade tip is at full diameter, providing an air seal at the corners. The square blade fan is efficient when placed in a cowling or orifice that surrounds the blade tips. A gap of about 1/4″ (6 mm) is permitted between the blade tips and the orifice.

Fan Blade Rotation

Proper rotation of a fan blade is determined at the factory and cannot be changed in the field. Most propeller-type blades are mounted directly onto the motor shaft, so the rotation of the fan blades must match the rotation of the motor.

Rotation of fan blades is said to be clockwise (CW) or counterclockwise (CCW) as viewed from the *discharge* air side. Rotation for draw-type fan blades (those that pull air back over the motor) must be viewed from the motor side. Rotation for blow-type fan blades must be determined from the front of the blades, with the motor behind.

Most fan blades are not designed to be reversed (mounted backward). Reversing a fan blade causes severe reduction in the volume of air moved, increases motor and fan blade fatigue, and causes system problems. It is always best to replace a fan blade with an exact duplicate, but some allowance is permitted. The pitch can vary two degrees either way without trouble. Substituting a four-blade fan for a three-blade fan adds little to the volume of air moved.

A motor and fan must be properly matched to obtain desired results. Motor speed and direction of rotation are important. The motor shaft must precisely fit the hole in the fan hub to prevent slipping or wobbling. The fan is usually anchored to the motor shaft by one or more setscrews through the hub between the blades and the motor.

When purchasing fan blades, the supplier needs to know the direction of rotation, number of blades, degree of pitch, hub hole size, blade type (square or round), and diameter (from blade tip to blade tip).

Centrifugal Blowers

Centrifugal (squirrel cage) blade blowers are used in applications where airflow is controlled and directed by ducts, such as furnaces and central air-conditioning units. Excess airflow causes air pressure (called *static pressure*) to build up inside the duct. Static pressure ensures even distribution of air to all vents. As duct pressure increases, the volume of air delivered by the blower is reduced. The centrifugal blower is sized to deliver enough airflow to maintain proper static pressure levels.

Centrifugal blowers are measured by blade length and the diameter across the blade wheel. Some blowers are mounted directly on the motor shaft (direct drive), while others are belt-driven, **Figure 24-62**. The direction of rotation is determined by the curvature of the blades. Forward-curved blades are designed to cup incoming air and throw it forward by centrifugal force.

Direct drive centrifugal blower

Belt-driven centrifugal blower

Goodheart-Willcox Publisher

Figure 24-62. Blowers in heating and air-conditioning installations may be direct drive or belt-driven.

Chapter Review

Summary

24.1 An induction motor uses induced magnetism to convert electrical energy into magnetic energy, then magnetic energy into mechanical energy in the form of a rotating rotor. The magnetic field is created by an assembly of electromagnets collectively known as the stator.

24.1 Motor speed, horsepower, and efficiency are three important characteristics to consider when selecting an HVACR motor. Motors are rated in revolutions per minute (rpm) for speed and horsepower for torque (turning power).

24.2 A single-phase induction motor uses a start winding to establish a stator that is "out of step" with the main winding to accomplish automatic starting of the motor. The motor overload protector is used to protect motor windings from damage caused by overheating due to overloaded conditions or poor ventilation.

24.3 The five types of single-phase induction motors are shaded-pole; split-phase; permanent split capacitor (PSC); capacitor-start, induction-run (CSIR); and capacitor-start, capacitor-run (CSCR).

24.3 The shaded-pole motor uses a stator pole with a shading coil placed over it to produce a tiny magnetic field out of step with the main poles. This is done without a start winding, giving it the ability to operate with low starting torque and carry full load once up to speed.

24.3 The split-phase motor uses a start winding installed in the stator, giving it good running torque that is useful for larger loads. The permanent split-capacitor (PSC) motor is a split-phase motor with a capacitor in series with its start winding. This design increases running torque.

24.3 A capacitor-start, induction-run (CSIR) motor uses a start capacitor to start under a heavy load, but once started, operates using only the run winding. A capacitor-start, capacitor-run (CSCR) motor runs as a CSIR during start-up and as a PSC for the rest of normal operation, using both start and run capacitors.

24.4 A capacitor is an electrical device that can store an electrical charge. Start and run capacitors are used to improve the operating characteristics of induction motors. Both start and run capacitors are rated in Vac (volts, alternating cycle) and microfarads (mfd or µf).

24.4 Motor damage can be avoided by properly connecting the run capacitor to the correct terminal. Shorted or open capacitors can be detected by replacing a suspected faulty capacitor or testing with an ohmmeter.

24.4 Hard start kits are available for motors that have difficulty starting.

24.5 Dual-voltage motors have two run windings, which when connected properly, can supply two different voltages.

24.6 Multi-speed motors use different taps (wires) to select the number of poles in the run winding, which can be used to adjust motor speed.

24.7 Three-phase alternating current induction motors offer high starting torque and high running torque without the need for a start winding, capacitors, or relays. Three "hot" wires and a ground wire are used. Three pairs of stator poles (windings), one for each supply wire, are used to produce a strong magnetic field, which converts to strong starting and running torque.

24.7 Dual-voltage three-phase motors contains an extra set of three poles with two wires for each of these extra poles that supply low and high voltages.

24.8 Electronically commutated motors (ECMs) are brushless dc motors with integrated electronic controls. ECMs provide more precise control of the motor speed and torque when compared with ac motors. ECMs are also more responsive to changes in the system.

24.9 Most single-phase, hermetic compressors have their three terminals marked: C, S, and R (Common, Start, and Run). If these terminals are not marked, they can be identified by checking the resistance of the windings with an ohmmeter.

24.10 A grounded winding is one that has developed an internal electrical short to the device's frame. An open winding refers to a broken wire in the motor winding that is *not* touching the motor shell. A shorted winding occurs when two wires on the same winding short out or touch each other, shortening the path of the current flow through that winding. An ohmmeter can be used to check for all of these conditions.

24.11 All motors have a bearing inside the end bell at each end of the rotor. A primary cause of motor failure is dry bearings. Permanently oiled bushings do not need to be lubricated, but oil- and grease-type bearings require lubrication with oil and grease respectively.

24.12 Round axial blades are used in free-air applications (pedestal fans) and square axial blades are used in medium- and high-pressure applications (evaporators and condensers). Centrifugal blades are used in applications where airflow is controlled and directed by ducts (furnaces and central air-conditioning units).

Know and Understand

1. What are the energy conversions that occur in an induction motor? (24.1)
   A. Thermal energy to magnetic energy to mechanical energy
   B. Electrical energy to chemical energy to mechanical energy
   C. Electrical energy to magnetic energy to mechanical energy
   D. Electrical energy to thermal energy
2. A motor's main windings are located in its _____. (24.1)
   A. rotor
   B. stator
   C. end bell
   D. shaft
3. *True or False?* Motor speed is determined by dividing the number of stator poles into the number of times the current changes direction (alternates) per minute. (24.1)
4. Calculate the horsepower of a motor that uses 373 watts to operate. (24.1)
   A. 1/2 hp
   B. 1 hp
   C. 2 hp
   D. 5 hp
5. The direction of rotation for the motor shaft is determined by the _____. (24.2)
   A. rotor
   B. run winding
   C. start winding
   D. operating voltage
6. The start winding must be disconnected at about _____% of normal motor speed. (24.2)
   A. 25
   B. 50
   C. 75
   D. 80
7. *True or False?* The two most common causes of motor problems are failed windings and motor overload. (24.2)
8. Excess compressor heat can likely be attributed to _____. (24.2)
   A. an electrical problem in the start components
   B. a poorly lubricated compressor
   C. worn bearings
   D. a locked rotor
9. In a _____ motor, a start winding is installed in the stator to produce a magnetic field out of step (phase) with the main winding. (24.3)
   A. shaded-pole
   B. permanent split-capacitor
   C. split-phase
   D. capacitor-start, induction-run
10. The PSC motor includes a _____ in series with its start winding. (24.3)
    A. bimetallic disc
    B. relay
    C. start capacitor
    D. run capacitor
11. A CSIR motor is basically a split-phase motor that includes a _____ in series with its start winding. (24.3)
    A. bimetallic disc
    B. relay
    C. start capacitor
    D. run capacitor

12. A start capacitor consists of two layers of _____. (24.4)
   A. poles
   B. electrodes
   C. dielectric oil
   D. thermal insulation
13. When a start capacitor is wired in series with the start winding, one side of the capacitor is discharging its electrons into the winding, resulting in _____. (24.4)
   A. lower running torque
   B. higher running torque
   C. lower starting torque
   D. higher starting torque
14. *True or False?* Electrons travel through a capacitor. (24.4)
15. A capacitor _____ occurs when one of the terminals separates from the coil. (24.4)
   A. short
   B. burnout
   C. overload
   D. restart
16. Capacitors are selected according to _____. (24.4)
   A. motor size and load placed on the motor
   B. motor voltage and motor type
   C. single-phase or three-phase motor capability
   D. motor speed
17. A hard start kit generally consists of a start capacitor and _____. (24.4)
   A. run capacitor
   B. blower
   C. relay
   D. dielectric oil
18. *True or False?* Single-phase dual-voltage motors have two separate run windings. (24.5)
19. A multi-speed motor has different _____ that make it possible to select the number of poles being used, which determine motor speed. (24.6)
   A. capacitors
   B. relays
   C. windings
   D. taps
20. A three-phase motor uses _____ to provide high starting and running torque. (24.7)
   A. start and run windings
   B. start and run capacitors
   C. three supply wires
   D. relays
21. In a dual-voltage three-phase motor, the windings should be connected in _____ for low voltage and in _____ for high voltage. (24.7)
   A. series, parallel
   B. parallel, series
   C. parallel, series-parallel
   D. series, series-parallel
22. A technician uses an ohmmeter to measure resistance a three-phase motor winding. The ohmmeter reads 0 Ω. What does this reading tell the technician about the motor? (24.7)
   A. The motor is in proper operating condition.
   B. The motor is grounded.
   C. The motor is shorted.
   D. The motor has an open winding.
23. An electronically commutated motor (ECM) consists of a motor and _____. (24.8)
   A. relays
   B. a control module
   C. sensors
   D. a metering device
24. *True or False?* Electrical power supply to a single-phase compressor motor must be disconnected and the connections to the three motor terminals opened before taking a resistance measurement. (24.9)
25. A motor with a winding shorted to ground is indicated by a _____. (24.10)
   A. broken wire
   B. blown fuse or breaker
   C. an ohmmeter reading of infinite resistance or OL
   D. smoky smell
26. Larger horsepower motors often have _____. (24.11)
   A. permanently oiled bushings
   B. oil-type bearings
   C. grease-type ball bearings
   D. self-lubricating bearings
27. *True or False?* Fan blades can be mounted backward to increase the volume of air moved. (24.12)

28. Centrifugal blade blowers are commonly used in _____. (24.12)
    A. pedestal fans
    B. evaporators
    C. furnaces
    D. high-pressure applications

Apply and Analyze

1. Explain how an induction motor operates using induced magnetism. (24.1)
2. What are the two most common causes of motor problems and how can they be prevented? (24.2)
3. How is a shaded-pole motor different from other types of induction motors? (24.3)
4. How does a permanent split-capacitor motor provide higher running torque? (24.3)
5. Why is a bleed resistor used to connect capacitor terminals? (24.4)
6. List two methods of testing a capacitor for proper operation. (24.4)
7. What would happen if the two run windings in a dual-voltage motor were to rotate in opposite directions? (24.5)
8. How is motor speed determined in a multi-speed motor? (24.6)
9. How is the direction of rotation changed on a three-phase motor? (24.7)
10. Explain how to make the low-voltage connection in a three-phase motor. (24.7)
11. Is an ECM an ac motor or a dc motor? Explain your answer. (24.8)
12. List three advantages and three disadvantages of an ECM motor compared with an ac motor. (24.8)
13. How can a technician determine single-phase compressor terminals when they are not marked? (24.9)
14. What is an open winding? (24.10)
15. What is the purpose of a motor bearing? (24.11)
16. What characteristics of a fan blade affect the volume of air that can be moved? (24.12)

Critical Thinking

1. Explain why it is important to turn the power off and remove the capacitor from the circuit before testing it with an ohmmeter. (24.4)
2. Imagine you are a technician checking the resistance of motor windings on a compressor. Between Terminals 1 and 2, the ohmmeter reads 4.5 Ω. Between Terminals 2 and 3, the ohmmeter reads 20 Ω. Between Terminals 1 and 3, the ohmmeter reads 16 Ω. Based on these readings, determine which terminals are the Common, Start, and Run terminals. (24.9)
3. Imagine you are a technician checking the resistance of motor windings on a compressor. Between Terminals 1 and 2, the ohmmeter reads 2.5 Ω. Between Terminals 2 and 3, the ohmmeter reads 2.5 Ω. Between Terminals 1 and 3, the ohmmeter reads 2.5 Ω. Based on these readings, determine which terminals are the Common, Start, and Run terminals. (24.9)

Goodheart-Willcox Publisher

25 Electromagnetic Control Devices

LEARNING OUTCOMES

After studying this chapter, you will be able to:

25.1 Identify various types of control devices, how they operate, and how to troubleshoot them.

TECHNICAL TERMS

- across-the-line starter
- armature
- contactor
- control circuit
- current relay
- cycling on the overload
- electromagnet
- electromagnetism
- ferrous
- jumper wire
- magnetism
- pickup voltage
- positive temperature coefficient (PTC) thermistor
- potential relay
- PTC relay
- relay
- solenoid valve
- start relay
- thermistor

Introduction

Magnetism is the ability of a natural material or a field of force generated by an electric current to attract certain metals. An important magnetic principle is that a magnetized object can induce magnetism in certain other metals. ***Ferrous*** metals, which are metals containing iron, will readily accept magnetism. This means a ferrous metal will become a magnet if placed within a magnetic field (lines of flux). This is called *induced magnetism*. Soft iron is easily magnetized, but quickly loses its magnetism when the magnetic field is removed. This feature makes soft iron ideal for making ***electromagnets*** (magnets that can be turned on and off). Electromagnets are used in many electrical devices, such as motors, relays, and solenoids.

Electromagnetism is the generating of a magnetic field around a conductor (wire) as a result of current flowing through the wire. When the wire is wrapped into the form of a coil, the magnetic field becomes stronger. If the wire is coiled around a soft iron rod (that acts as the coil's core), the wire's magnetic field will induce magnetism in the iron, **Figure 25-1**. Magnetism in the core is controlled by starting or stopping current flow in the coil. Turning off the electrical supply to the coil kills the magnetic field in both coil and core. This electromagnetic principle is used in operating solenoid valves, relays, contactors, and line starters.

25.1 Control Devices

Electromagnetism is used to operate a number of control devices used in various applications. These applications include opening and closing of valves, motor starting, and circuit control.

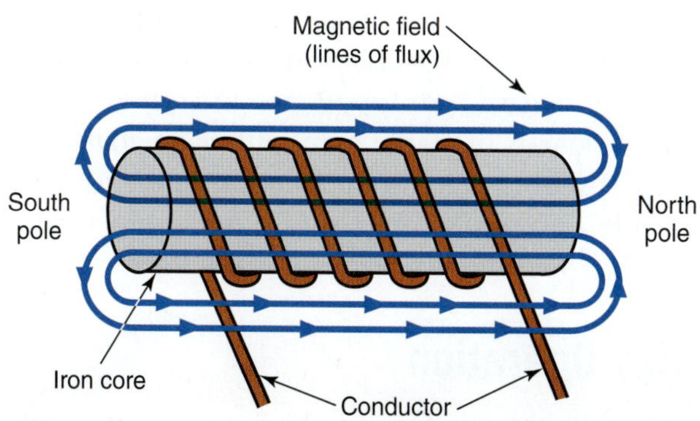

Goodheart-Willcox Publisher

Figure 25-1. When current flows through a wire coil wrapped around an iron core, the iron is magnetized. The magnetic field can be turned on or off by turning the electric current on or off.

Solenoid Valves

Solenoid valves use electromagnetism to electrically control the opening and closing of valves, **Figure 25-2**. A solenoid valve uses an electrical coil to control a movable iron core (plunger) that is used to open or close the valve. The magnetism generated by the coil induces opposite polarity in the iron plunger. Because opposite magnetic poles attract, the plunger is lifted into the coil's magnetic field and the valve is opened. The valve remains open as long as the coil is energized. When the coil is de-energized, magnetism disappears and the plunger returns to its closed position, **Figure 25-3**.

Solenoid valves are frequently used on applications where automatic control of fluids is desired.

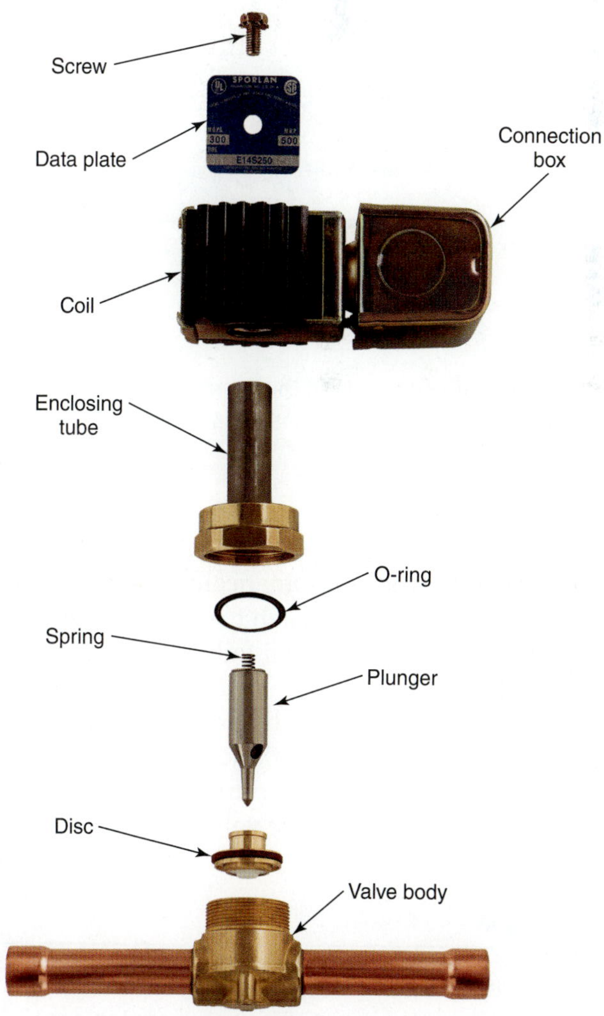

Courtesy of Sporlan Division – Parker Hannifin Corporation

Figure 25-2. A typical solenoid valve has the coil mounted on top of the valve body.

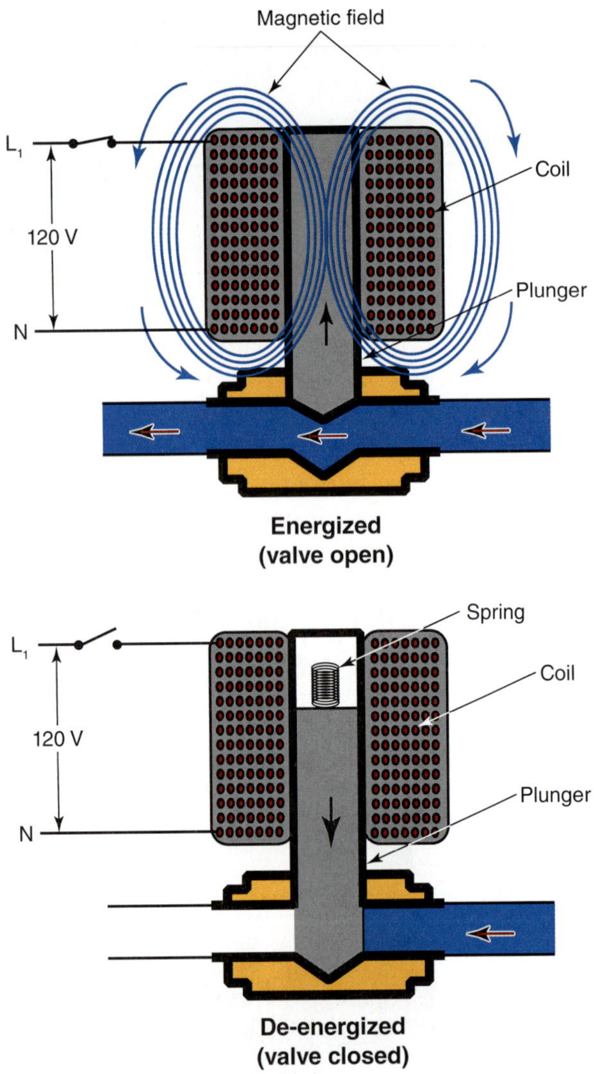

Goodheart-Willcox Publisher

Figure 25-3. When the solenoid coil is energized, the opposing polarities pull the plunger upward, opening the valve. The valve will remain open until the coil is de-energized.

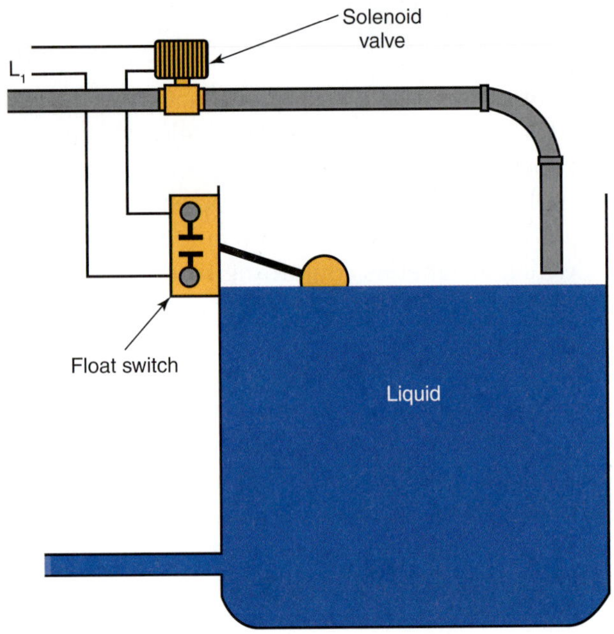

Goodheart-Willcox Publisher

Figure 25-4. A float switch controls current flow to the solenoid valve. The valve is opened and closed as needed to automatically maintain the liquid level in the tank.

Some type of switch is required to control power supply to the solenoid coil. **Figure 25-4** shows how proper liquid level is automatically maintained in a tank by using a float switch to control a solenoid valve. When the liquid level falls, the float switch closes. This energizes the solenoid coil, opening the valve and permitting liquid to flow into the tank. When the proper liquid level is reached, the float switch opens, cutting off power to the solenoid valve and allowing the valve to close.

Relays

An internal centrifugal switch cannot be used within a hermetic compressor as a motor control, due to arcing of the switch contacts. The arcing would cause breakdown of oil and refrigerant into acid and destroy the motor windings.

A ***relay*** is a device that opens or closes its electrical contacts when it is energized. It is basically an electrically operated switch. Relays can be designed to meet different system needs and applied to a variety of applications. For example, a ***start relay*** is a relay that is designed and used to start and stop a single-phase motor. It automatically disconnects the start winding and/or the start capacitor of a single-phase motor for normal starting and stopping of the motor. There are several types of start relays, but two of the most common are current and potential. Each type is easily recognized. The method of operation and the electrical connections of the two are different.

Most relays open or close switch contacts using the same electromagnetic principle as the solenoid. A coil of wire is wrapped around an iron core located near a movable arm called an ***armature***. An electrical contact is installed at one end of the movable arm, and a matching fixed contact below the arm.

Relay Operation

As shown in **Figure 25-5**, when the relay coil is energized, magnetism is induced in the core and the armature is pulled downward. This causes the contacts to close. When the relay coil is

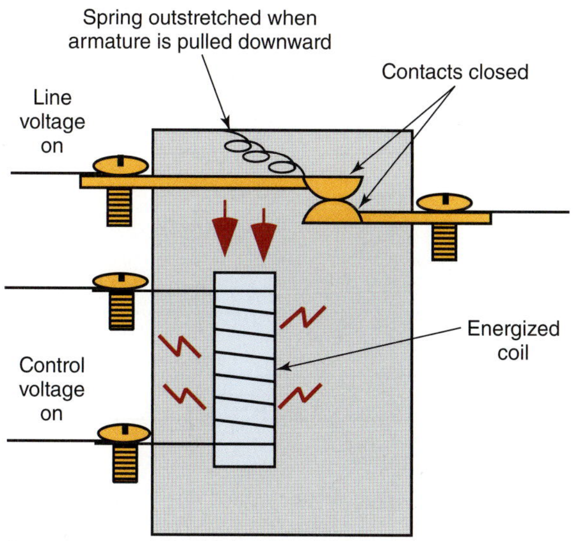

A

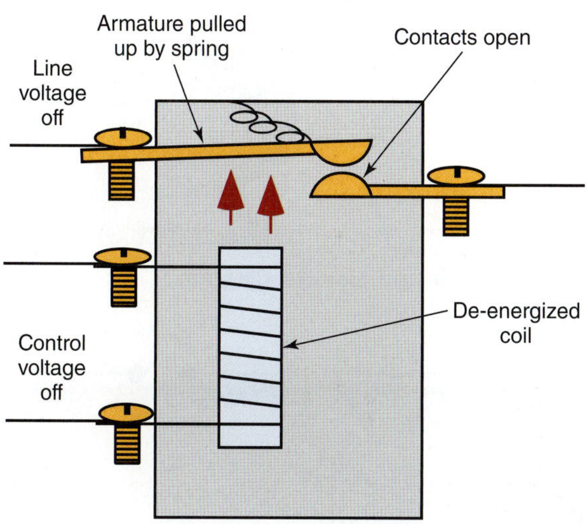

B

Goodheart-Willcox Publisher

Figure 25-5. Relay operation. A—When the coil is energized, the armature is pulled downward. This closes the contacts and completes the circuit. B—When the coil is de-energized, a spring pulls the armature upward again. This opens the contacts.

de-energized, a spring returns the arm to its original position, opening the contacts.

The electrical circuit to the coil is entirely separate from the circuit through the contacts, **Figure 25-6**. This separation permits use of a low-voltage (24 V) circuit to open or close a line-voltage (120 V) circuit. In such an application, the low-voltage circuit is referred to as a ***control circuit***.

Relay contacts have low current ratings, with a maximum of 10 A being considered normal. Current flow in the relay coil circuit (control circuit) is even lower (often less than 0.25 A). The low voltage (24 V) used in the control circuit is safer for the technician, permits the use of small gauge wiring, and is less likely to cause a fire. Also, switches in

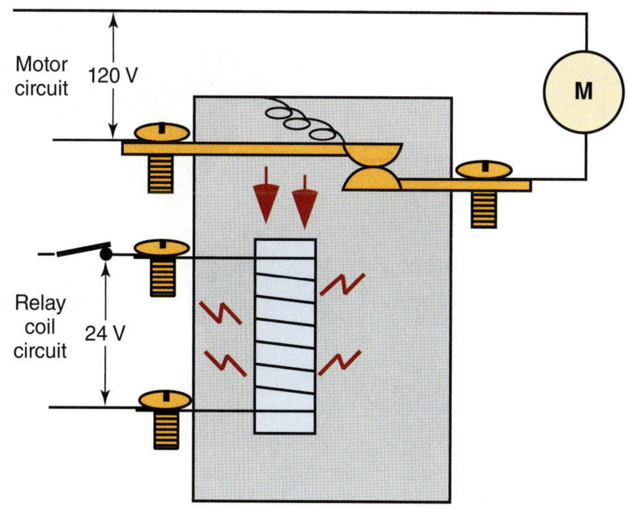

Goodheart-Willcox Publisher

Figure 25-6. The 24 V circuit for the relay coil is electrically separate from the 120 V circuit that controls the motor. The use of a low-voltage circuit to control a higher-voltage circuit is common in appliances and in heating and cooling equipment.

the control circuit last longer due to less arcing of the contacts.

The relay case often has a schematic (electrical drawing) or symbols for electrical components that reveal the internal connections and operation of contacts and coil terminals, **Figure 25-7**. The relay data plate reveals coil voltage and amperage rating of the contacts. Relays often contain more than one set of contacts. On a schematic, relay contacts are always shown in the state they assume when the coil is de-energized: normally open (NO) or normally closed (NC). All sets of contacts will change position at the same time when the coil is energized, **Figure 25-8**.

A relay is often used to control an electrical load from a remote location. For example, a 120 V motor located in a basement could be controlled from the building's second floor by using a

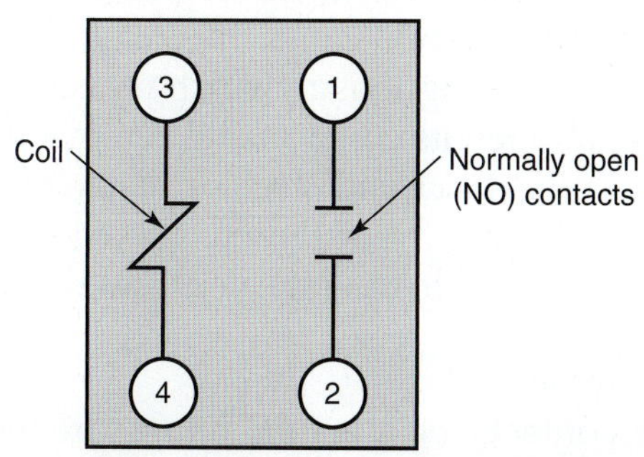

Goodheart-Willcox Publisher

Figure 25-7. Symbols like these are often used to depict relay components on schematics.

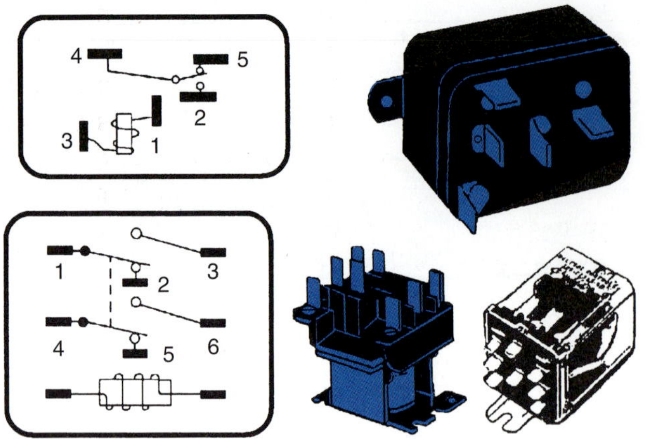

Goodheart-Willcox Publisher

Figure 25-8. A relay can have several contacts to control different devices simultaneously.

relay. The relay coil could be 24 V, permitting the use of small gauge wiring in the control circuit. **Figure 25-9** shows both pictorial and schematic views of such a system. Schematic drawings are often used to illustrate electrical circuits. Note that schematics do *not* show the *size* or *length* of wires.

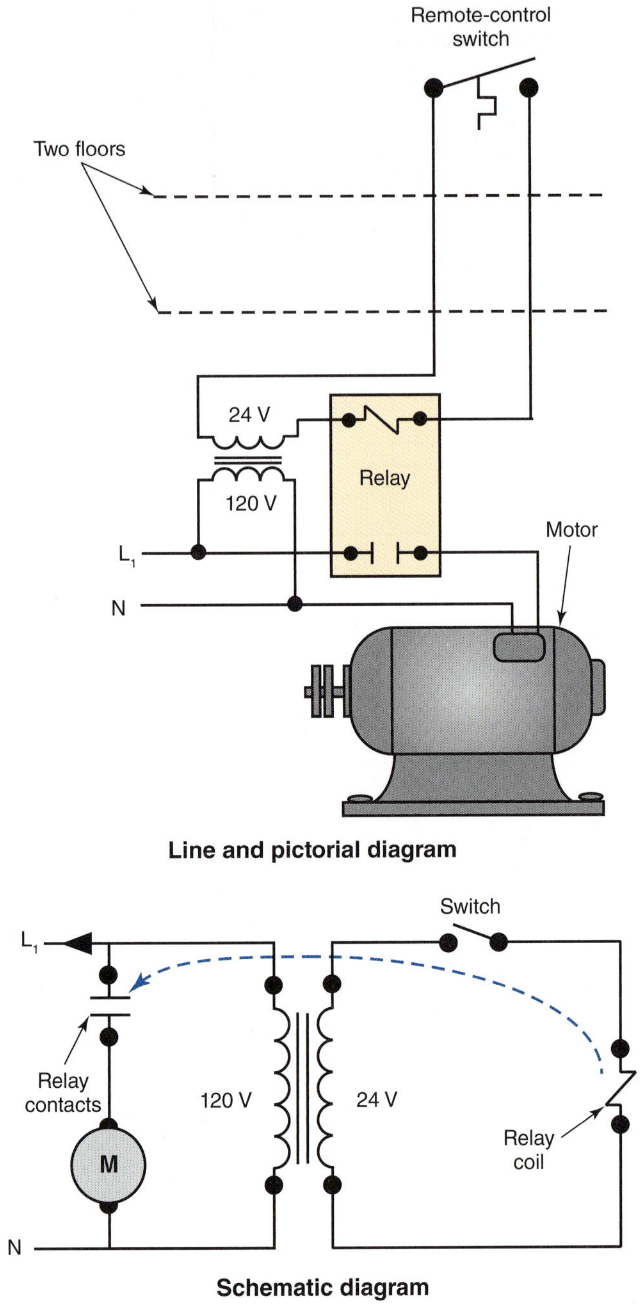

Goodheart-Willcox Publisher

Figure 25-9. A relay can use a low-voltage control circuit to operate a distant device. The line and pictorial diagram show a motor located two floors below a remote-control switch. When the switch is closed, it energizes the 24 V relay coil, which closes the contacts in the 120 V motor circuit.

Current Relays

A ***current relay*** is a start relay that operates on the current flowing through a single-phase motor's run winding. When a single-phase motor starts, current through the run winding is very high. The rotor is not yet moving, and the resistance of the run winding is very low. Since a current relay's coil is wired in series with the run winding, the same high current flows through the current relay's coil and the run winding. This high current actuates the current relay to close its contacts. This allows current to flow through the motor's start winding, which gives the motor the torque to start turning. As the rotor begins to turn, the current through the current relay and run winding will decrease as a result of increased resistance created by back-emf. The lower current will cause the current relay coil to produce less magnetism, which moves back the armature to open its contacts and remove the start winding from the circuit. This principle of temporary initial high-current flow to the run winding at start-up is used to operate a current relay.

Relay Contacts

The contacts on current relays are *normally open (NO)*, with the plunger located below them. The plunger must move upward to close the contacts and complete the circuit to the motor's start winding. Once the start winding is energized, the rotor quickly speeds up. As it does so, back-emf is generated and current flow through the run winding decreases. When the rotor speed reaches 75% of normal, amperage through the run winding drops and the magnetic field generated by the relay coil weakens. The plunger is pulled downward by gravity, opening the contacts to the start winding, **Figure 25-10**.

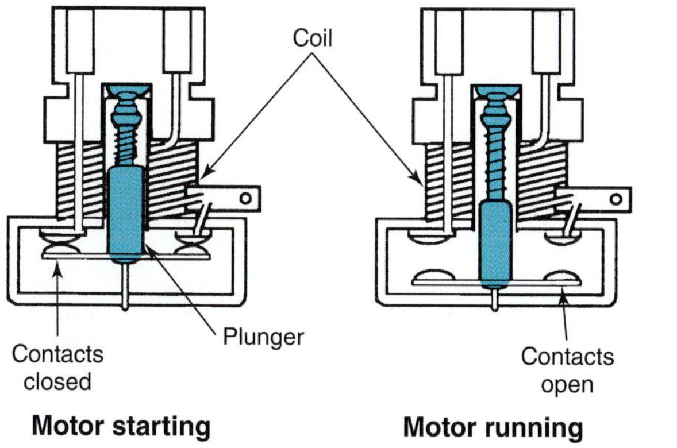

Goodheart-Willcox Publisher

Figure 25-10. When the coil of the current relay is energized, the plunger is pulled upward and closes the contacts to energize the motor's start winding. As motor speed increases, current flow to the coil decreases and the magnetic field weakens. This allows the plunger to drop down, which opens the contacts.

The contacts used on current relays must be durable, because of the arcing that takes place as they open and close under high-current conditions. Arcing can burn out contacts or weld them together.

Push-On Current Relays

Most current relays are designed for mounting directly onto the compressor terminals at S and R, **Figure 25-11**. These are push-on type relays, with the mounting procedure assuring an upright position. Electrical connections are simplified because the relay makes direct contact with the terminals.

Normally, this type of relay is used on split-phase hermetic compressors. Connection is easy, with the relay simply being pushed onto the compressor's S (Start) and R (run) terminals. The relay contacts are between S and R, while the coil is connected to the R terminal on the compressor and the L (line) terminal on the relay housing. The technician merely connects the black (hot) wire to the L terminal. The overload is mounted inside the terminal box and is already connected by a short wire to the compressor's Common terminal. The technician connects the neutral (white wire) side of the power supply to the overload.

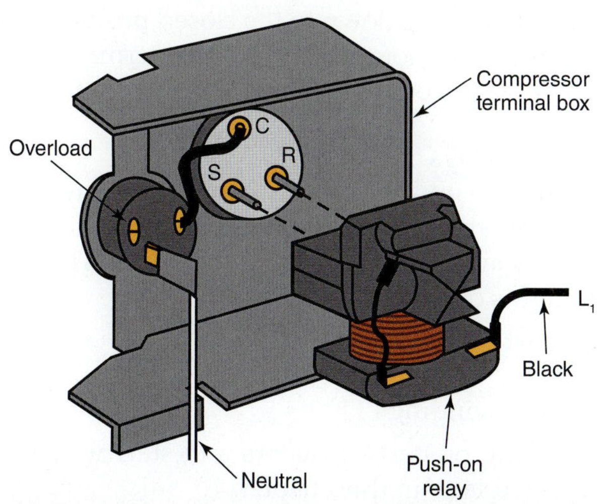

Goodheart-Willcox Publisher

Figure 25-11. Installation of a push-on current relay is done by simply pushing it onto the S and R terminals of a compressor. This mounting method helps assure correct positioning.

Current relays are position sensitive. A current relay must be mounted both upright *and* right-side-up. If the relay is not upright, the plunger will not be able to rise and fall. If mounted upside-down, the contacts will remain closed because the plunger cannot fall back to the open position.

Testing Current Relays

A current relay is checked for proper operation by using the resistance (ohms) setting of a multimeter. For safety and proper testing, the relay must be disconnected from the circuit and removed from its mounting. To check the contacts, first hold the relay in the upright position and place the test leads at Terminals M and S. When upright and de-energized, the relay contacts should be in the *open* position, giving a resistance reading of infinity (OL). If a zero resistance reading is obtained, the contacts are shorted by being welded together and the relay must be replaced.

If an infinity reading was obtained in the upright position, turn the relay upside down. The multimeter should now read zero resistance, indicating that the contacts have closed.

Current Relay Replacement

Fractional horsepower single-phase motors rated at less than 1/2 hp use current relays. Single-phase motors with ratings of 1/2 hp and above must use *potential-type relays*, which differ considerably from current relays.

Current relays are selected according to motor horsepower, and replacement must be exact for proper operation. Manufacturers place an identification number on each relay, so exact replacement is merely a matter of specifying the manufacturer's name and the relay identification number. If the number is missing or unreadable, provide the electrical supplier with the model number and serial number of the compressor where the relay was found. A cross-reference listing will allow the supplier to identify the correct replacement relay.

Potential Relays

A ***potential relay*** (also known as a *voltage relay*) is a start relay that operates on the back-emf (potential) that is generated by the start winding. Its normally closed contact allows the start winding to energize and provide the torque necessary to start the motor. Back-emf that is produced energizes a potential relay's coil and actuates to open its normally closed contact and remove electrical power from the start winding. The speed of a single-phase motor's rotor determines the amount of back-emf generated by the start winding. As speed increases, so does the back-emf. A potential relay does the same job as a current relay but in a different way.

Terminals on potential relays are identified by numbers, instead of the letters used on current relays. **Figure 25-12** shows the active terminals, which are conventionally labeled as 1, 2, and 5. The contacts are located between Terminals 1 and 2. The coil is between Terminals 2 and 5. This information is important to remember, since not all potential relays show their wiring diagram on the case. Most potential relays have dummy terminals labeled 4 and 6, which are included as a convenience for splicing wires. Use of these terminals to connect a thermostat in series with the run winding is shown in **Figure 25-13**.

Some potential relays are position-sensitive and others are not. A relay that *is* position-sensitive will have some indication (usually an upward-pointing arrow) printed on the case. The difference between the two types is in how the armature is moved to return the contacts to their normally closed (NC) position. Relays that are not position-sensitive use spring tension to move the armature and reclose the contacts. The position-sensitive types have a weight on the contact end of the arm. When the coil is de-energized, the weight pulls the contacts down to the closed position. The relay must be mounted upright to permit proper closing of the contacts.

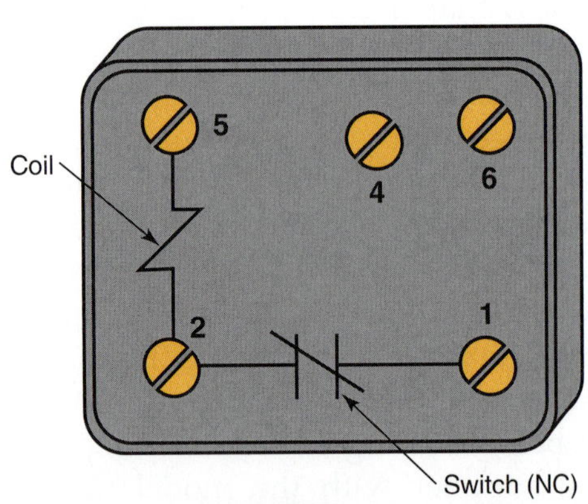

Goodheart-Willcox Publisher

Figure 25-12. Active terminals on the potential relay are labeled as 1, 2, and 5. Terminals 4 and 6 are dummies that can be used for wire splicing.

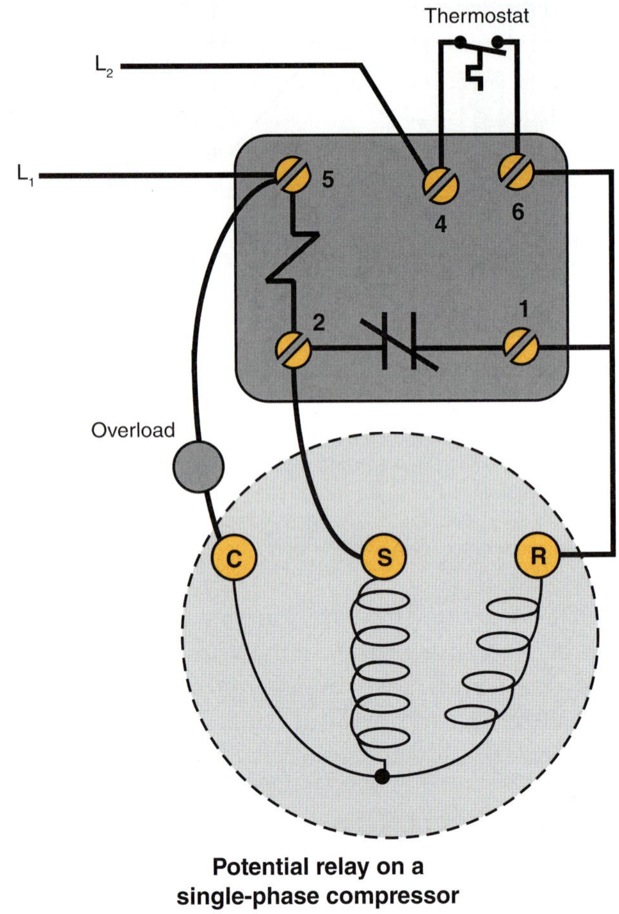

Goodheart-Willcox Publisher

Figure 25-13. A thermostat is wired in series with the run windings, using dummy Terminals 4 and 6 of the potential relay.

Testing Relays

When a motor fails to start, the relay, start capacitor, and start winding are checked to identify the source of the problem. Components usually fail in this order:

1. Start capacitor
2. Relay
3. Start winding

Testing of start capacitors and start windings is covered later in this chapter.

Another means of checking a relay is with the use of a shop-made test cord. This procedure is described later in this chapter.

To check a current relay's coil, place the multimeter test leads at Terminals M and L. A reading that shows a *small* resistance, such as 1 Ω, means the coil has continuity. If an infinity reading is registered, the coil is open and the relay should be replaced.

Potential Relay Operation

The potential relay uses a normally closed (NC) set of contacts that is *opened* by the energizing of the coil, **Figure 25-14**. The contacts control the power supply to the start winding of the motor. Since resistance of the coil is high, it will take a moderately high voltage to overcome resistance to energize it. Coil resistance is precisely measured so that, when the rotor reaches 75% of normal speed, back-emf will be sufficiently high to energize the coil. This point is the ***pickup voltage***. Coils are available with pickup voltages in different ranges, **Figure 25-15**.

When the coil is energized, it pulls downward on one end of the armature. The other end of the armature pushes the electrical contacts open, cutting off current flow to the start winding. The relay coil will hold the contacts open as long as motor speed is high enough to generate sufficient back-emf.

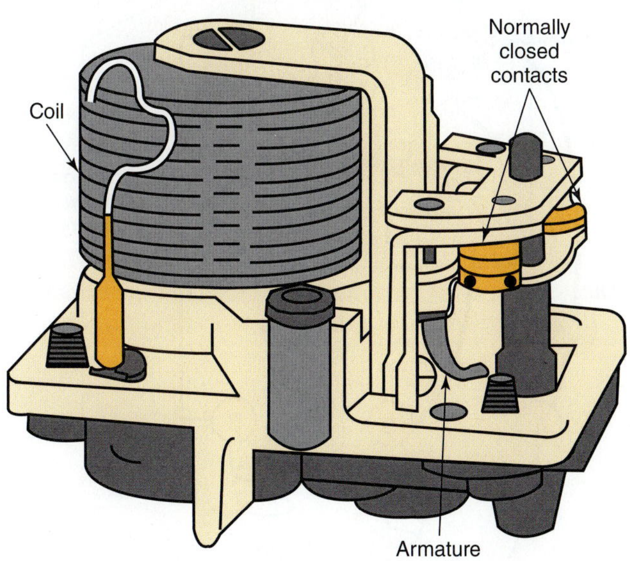

Goodheart-Willcox Publisher

Figure 25-14. The contacts of a potential relay are normally closed. The coil is energized when speed of the motor is fast enough to generate a predetermined amount of back-emf. The magnetic field generated by the coil moves the armature, which opens the contacts and takes the start windings out of the circuit.

Potential Relay Values

Coil No.	Pickup Voltage	Drop Out Voltage	Continuous Voltage	Coil Ohms (Approx.)
1	139–153	15–55	130	760
2	140–153	20–45	170	1400
3	159–172	35–77	256	3320
4	261–290	50–100	336	5180
5	280–310	50–100	395	7150
6	299–327	50–100	420	10000
7	323–352	60–135	495	11950

Goodheart-Willcox Publisher

Figure 25-15. This table shows the pickup voltage ranges and other information on different relay coils that are available.

Terminal Connections

Since the potential relay functions on back-emf generated by the start winding, the coil must be connected to both ends of the start winding. One end of the relay coil is connected to the start winding at Terminal 2, and the other end through the Common at Terminal 5, **Figure 25-16**.

The potential relay is the center of electrical activity, with many devices connected to or through it. These usually include such devices as the overload, system thermostat, condenser fan, and start capacitor.

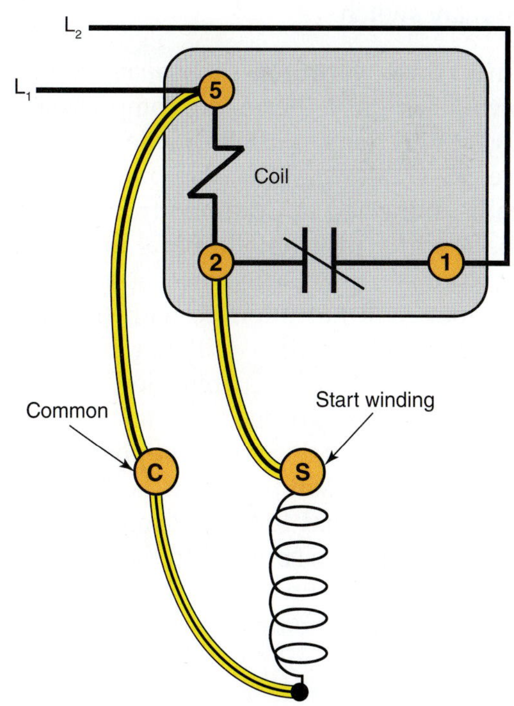

Goodheart-Willcox Publisher

Figure 25-16. A potential relay's coil must be connected to both ends of the start winding. It is connected through the Common at Terminal 5 and (on the other end) at Terminal 2.

SAFETY NOTE

Connecting Electrical Devices

When connecting electrical devices, the connections *must be correct*. Electricity does *not* permit mistakes. When making several connections, the best course is to consider and connect one wire at a time, while double-checking each circuit as it is worked on.

Connecting Hermetic Compressors

There are several methods used to make connections to a relay and compressor terminals. If extra terminals are available for multiple connections, a fair amount of wire can be saved. No matter what method is used, the basic principles remain the same:

- The power supply is always connected to the Common and Run terminals.
- The overload is wired in series with Common.
- The potential relay's coil is connected to both ends of the start winding.
- The start capacitor is in series with the relay switch.
- The run capacitor must be connected to bypass the relay switch.

Only three terminals (*Common, Start, Run*) are available on a hermetic compressor. Most wiring connections must be made on the relay. **Figure 25-17** shows how the power supply is typically connected to Common from Terminal 5. The back-emf wire from Terminal 5 to Common serves a dual purpose. The power supply traveling to Common will not interfere with the back-emf circuit. The voltage also will not influence the overload (OL), since the current flow through the back-emf circuit is extremely small.

The other side of the power supply must be connected to the run winding. This is done through the dummy Terminal 4 to allow connection of the system thermostat in series with the run winding. The thermostat is connected to Terminals 4 and 6, **Figure 25-18**. This temperature-activated switch will control power supply to the compressor at run.

If the compressor's motor is a split-phase type (one without capacitors), a jumper wire must be installed from Terminal 6 (power) to Terminal 1 (switch) on the relay, **Figure 25-19**. A ***jumper wire***

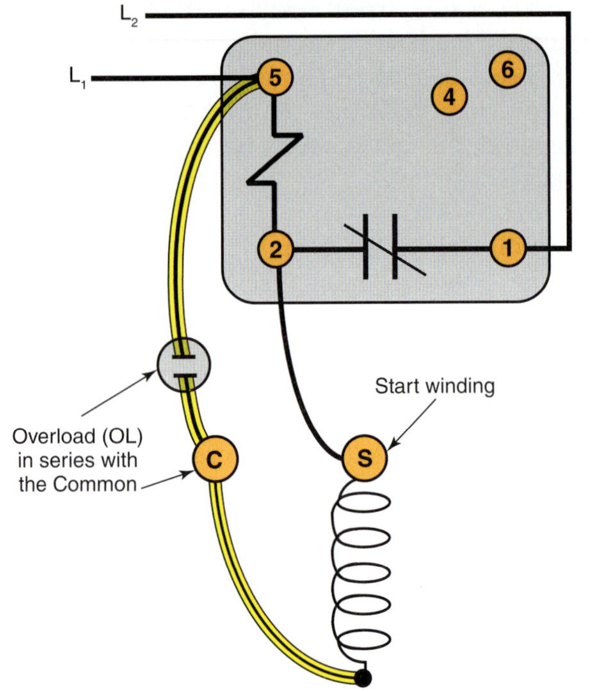

Goodheart-Willcox Publisher

Figure 25-17. The power supply is wired to Common at Terminal 5. Note the overload wired in series with Common.

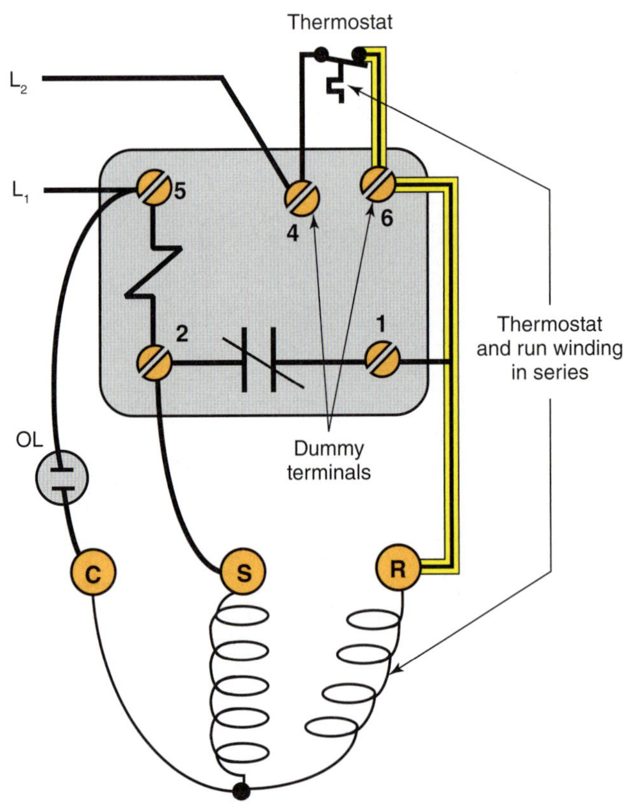

Goodheart-Willcox Publisher

Figure 25-18. The thermostat is wired in series with the run winding and one side of the power supply. It is wired using dummy Terminals 4 and 6.

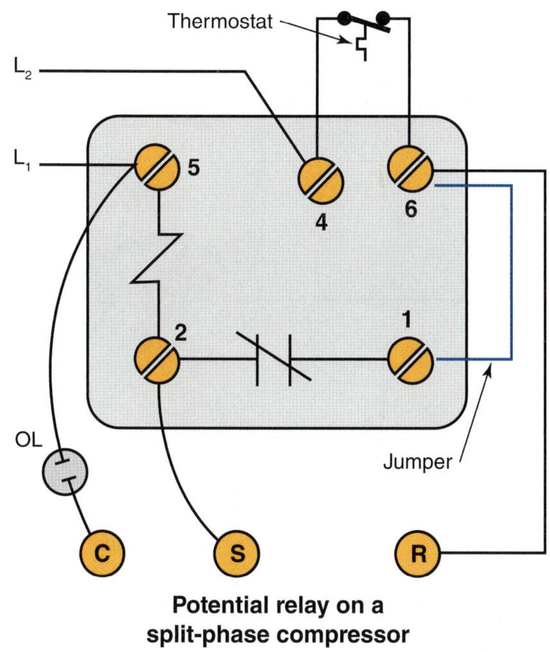

Goodheart-Willcox Publisher

Figure 25-19. On compressors with a split-phase motor, a jumper wire connects Terminals 6 and 1 on the relay. This provides a parallel connection between the power supply and the relay switch.

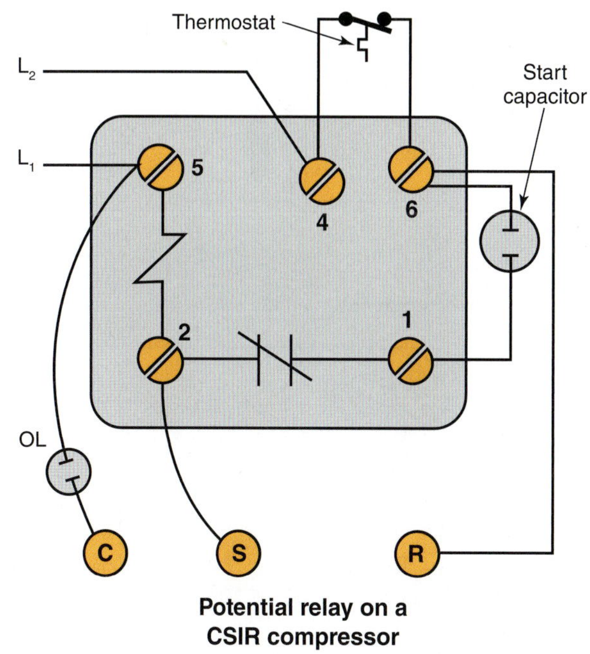

Goodheart-Willcox Publisher

Figure 25-20. On a compressor with a CSIR motor, a start capacitor is connected between Terminals 6 and 1 so that it is in series with the relay switch and start winding.

is a short wire installed to complete a connection from terminal to terminal. The jumper wire completes the parallel connection from one end of the start winding to the other side of the power supply. The relay switch is in series with the start winding. The relay switch is normally closed (NC) at start-up and will not open until the motor reaches 75% of normal speed. When the switch opens, the start winding is disconnected from the power supply. The back-emf circuit remains intact whether the relay switch is open or closed.

If a hermetic compressor contains a *capacitor-start, induction-run (CSIR)* motor, the jumper wire is used to install a *start capacitor* between Terminals 6 and 1, **Figure 25-20**. This places the capacitor in series with the switch, so both the capacitor and the start winding are disconnected from the power supply when the relay's switch opens.

A *capacitor-start, capacitor-run (CSCR)* motor has a run capacitor wired in parallel with the relay switch so it bypasses the start capacitor and switch. A typical method of wiring these capacitors into the circuit is shown in **Figure 25-21**. When both start and run capacitors are wired as shown, opening the relay switch will take just the start capacitor out of the circuit. The run capacitor will keep the start winding slightly energized to assist the run winding under loaded conditions.

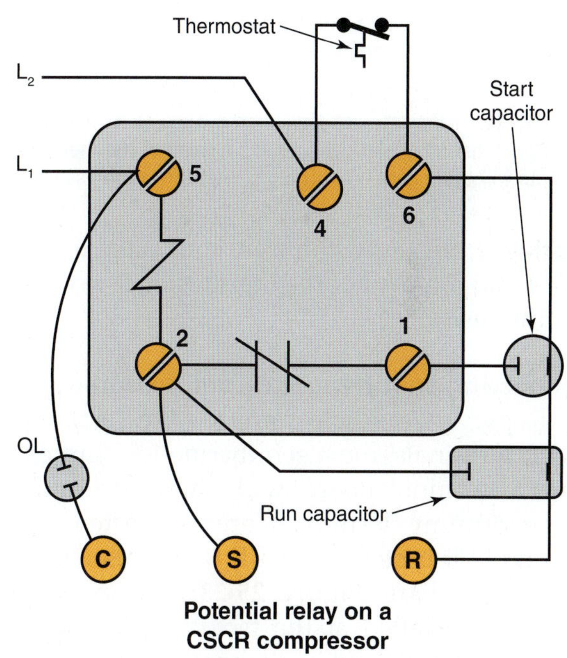

Goodheart-Willcox Publisher

Figure 25-21. This is the method used to wire capacitors to a relay controlling a compressor with a CSCR motor. When the relay switch opens, the start capacitor is taken out of the circuit.

Condenser Fan Connections

Since the condenser fan is usually cycled on and off with the compressor, it is wired to the relay, **Figure 25-22**. The condenser fan is connected to Terminals 5 and 6 (controlled by the thermostat)

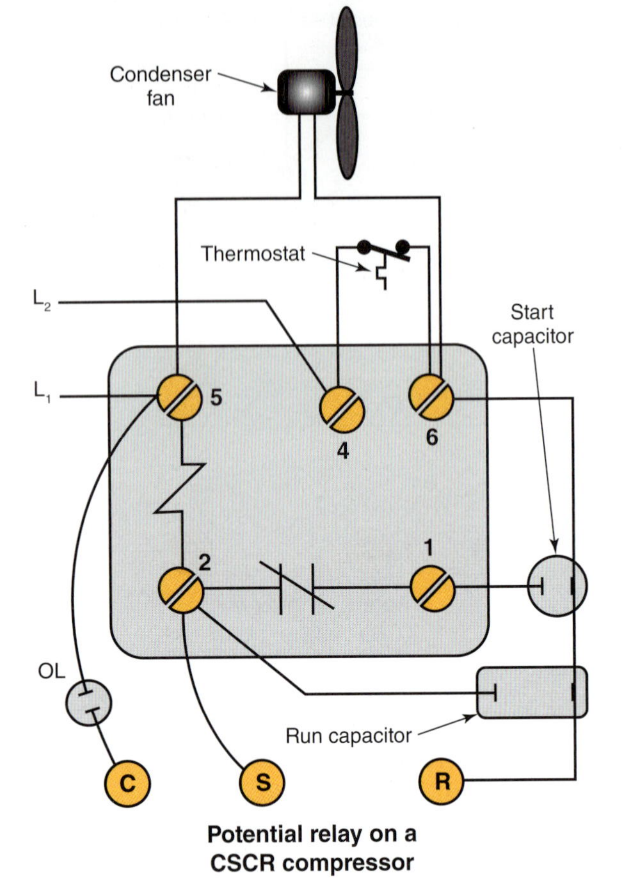

Goodheart-Willcox Publisher

Figure 25-22. When the condenser fan is wired to Terminals 5 and 6 of a potential relay, the opening and closing of the thermostat contacts will cycle it on and off along with the compressor.

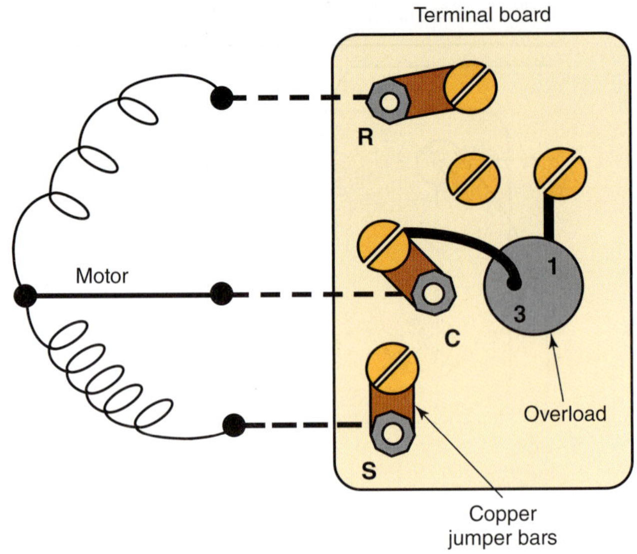

Goodheart-Willcox Publisher

Figure 25-23. The C, S, and R terminals are connected to dummy terminals with copper jumper bars. The dummy terminals should be used for all connections.

to achieve this. If the fan were connected to Terminals 4 and 5, ahead of the thermostat, it would run continuously.

Connecting Semihermetic Compressors

To prevent possible damage to the actual C, S, and R terminals, most semihermetic compressors have a terminal board with dummy terminals. These dummy terminals are connected to the nut-and-bolt type C, S, and R terminals with copper jumper bars, **Figure 25-23**. The overload is normally installed at the factory. The relay and capacitors are housed in a separate metal box, usually called a *power pack*. The wires connecting the terminal board and power pack are carried in a flexible metal conduit.

Figure 25-24 shows the common method for wiring a potential relay and capacitors to the terminal board. The wiring is summarized as follows:

- The overload is in series with Common.
- The relay coil is connected between Common and Start.
- The relay switch and start capacitor are in series with Start.
- The run capacitor is in series with Start but bypasses the relay switch.
- The system thermostat is in series with one power supply line.
- The condenser fan is connected in parallel with the compressor, after the thermostat.

PTC Relays

A ***PTC relay*** is a relay that uses a positive temperature coefficient thermistor as a sensor and switching device. They are used primarily in residential air conditioning and have proven to be convenient and reliable. PTC relays can be used to help resolve hard starting problems in a compressor with a permanent split capacitor (PSC) motor. It is connected in parallel with the run capacitor terminals., **Figure 25-25**.

A PTC relay contains a positive temperature coefficient (PTC) thermistor. A ***thermistor*** is a resistor that varies its resistances with temperature. A ***positive temperature coefficient (PTC) thermistor*** reacts rapidly to an increasing temperature change (having a "steep slope") by greatly increasing its resistance to current flow. The PTC relay will initially allow the large current needed for starting to flow to the motor's start winding. However, in a period of less than one second, heat from that large current flow will cause the PTC resistance to increase from about 50 Ω to approximately

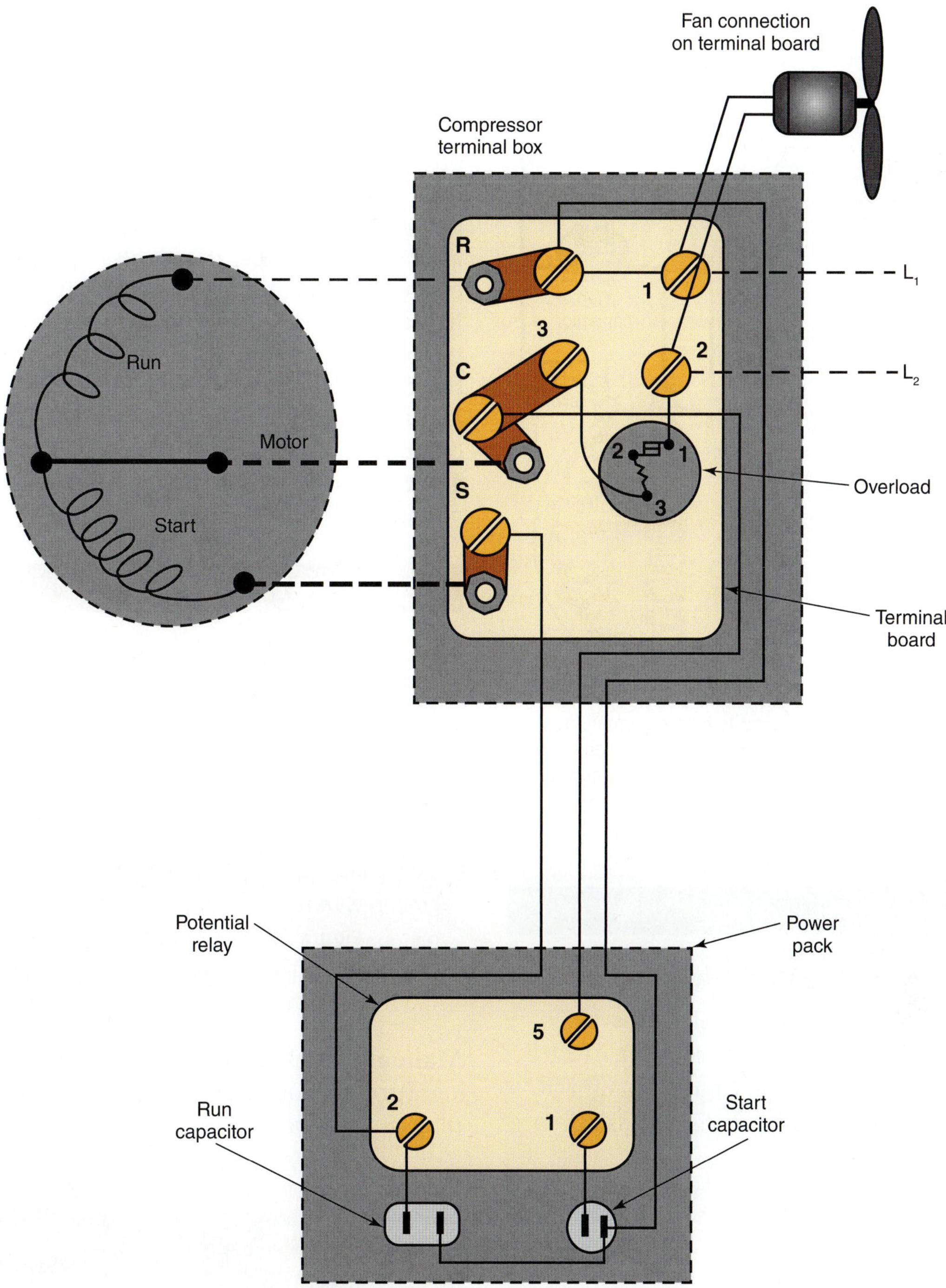

Goodheart-Willcox Publisher

Figure 25-24. Wiring of potential relay and capacitors to compressor terminal board.

80,000 Ω. This high resistance effectively stops current flow through the relay, allowing normal current flow through the run capacitor and motor winding. A hard start kit can be created by adding a start capacitor in series with a PTC relay. Advantages of a hard start kit include:

- Low cost
- Simple two-connection installation
- Wide applicability (usable on all single-phase PSC compressors up to 240 V and motors up to 5 hp)
- Reliability and long life
- Mounting flexibility (not position-sensitive)
- Increased torque (up to 300%)
- Capability of further increasing torque by wiring second kit in parallel to first

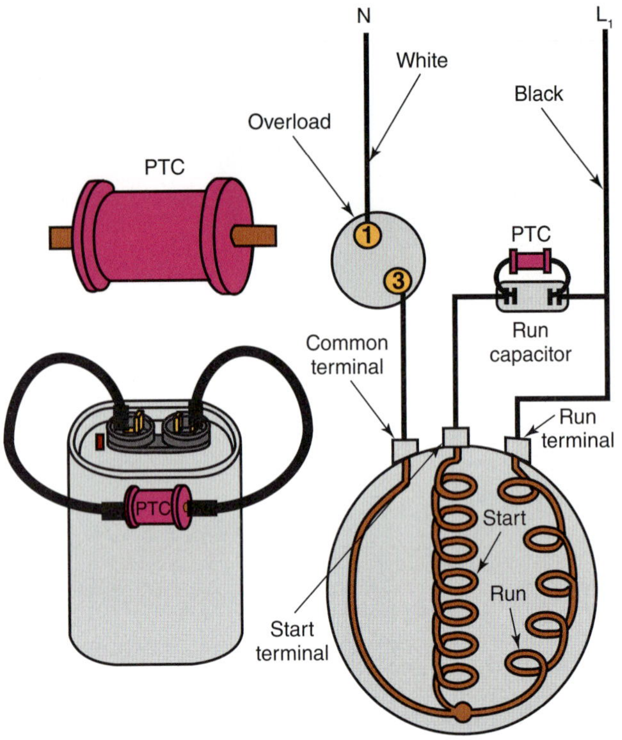

Goodheart-Willcox Publisher

Figure 25-25. PTC relays can be used to help a permanent split capacitor (PSC) motor to resolve its hard starting problems. A positive temperature coefficient thermistor in the relay controls current flow to the windings.

PRO TIP

Time Delays with PTC Relays

A system that uses a PTC relay should have some type of time delay control to allow the PTC to cool down after the compressor has shut off and before it starts again. This means the PTC is hot right after the compressor stops running so should not be worked on until it cools down.

Troubleshooting Relays

If a system is not running, always check the incoming power supply before making other electrical tests. Power supply to a potential relay can be checked by touching the leads of a multimeter to Terminals 4 and 5. If no voltage reading is obtained, check the fuse or breaker for the circuit serving the system. The fuse may be blown or the breaker tripped. If correct voltage is present at Terminals 4 and 5, move the test lead from Terminal 5 to Terminal 6. If there is no voltage present at Terminals 5 and 6, the system thermostat is open.

Shop-Made Test Cord

A simple test cord with two switches can be constructed by a technician, **Figure 25-26.** This test cord can be used along with a clamp-on ammeter to perform many troubleshooting procedures on small hermetic and semihermetic compressors. It is especially useful for bypassing start components and manually operating compressors. Among other uses, it can also be used to safely check the switch on potential relays. **Figure 25-27** lists the materials and shows the schematic needed to construct the test cord. The test cord can easily be converted to operate on 230 V by constructing another length of SJ cord with a 120 V female plug on one end and a 230 V male plug on the other end. The test cord will operate the same way with either voltage.

SAFETY NOTE

Include Safety Devices

When making test cords, exercise extreme care and obey all electrical safety precautions when using this test device, since it contains no built-in safety devices. Always connect a safety ground wire to the device being tested. Install a ground fault circuit interrupter (GFCI) and an additional safety ground wire from the switch box to the device being tested. **Figure 25-27** shows the principles of use and does not include recommended safety features.

Manual Compressor Test-Cord Start

Turn off both test cord switches and insert the plug into a power receptacle. Disconnect wires

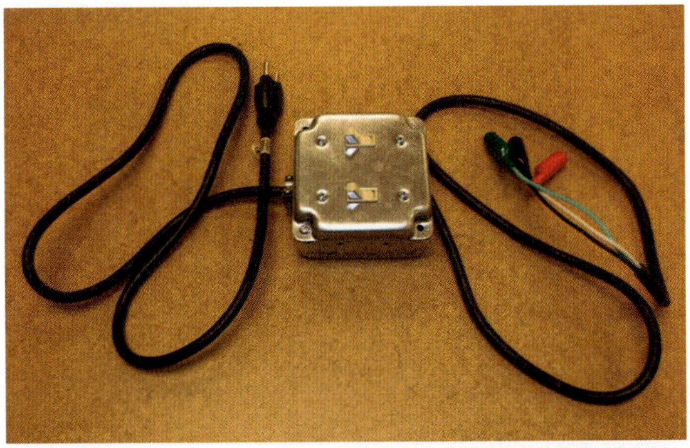

Goodheart-Willcox Publisher

Figure 25-26. When servicing hermetic and semihermetic units, a shop-made test cord is useful for many troubleshooting procedures.

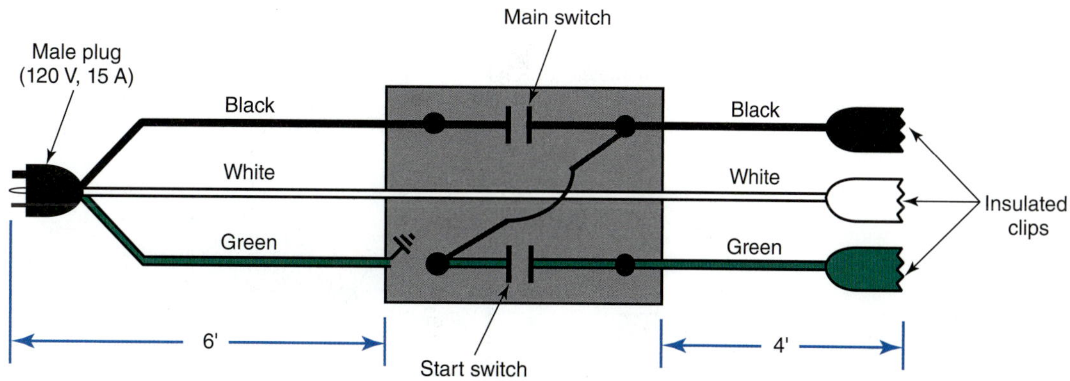

Test Cord Materials List
1–Male plug (120 V, 15 A)
1–Metal box (4" square) with 1/2" knockouts
1–Raised box cover with dual switch openings
1–10' three strand, 14 Ga. SJ cord
2–SPST 20 A toggle-type light switches
2–NMC box connectors (1/2")
3–Bulldog or alligator clips (with rubber insulated covers)

Goodheart-Willcox Publisher

Figure 25-27. Wiring connections must follow the schematic diagram shown. Label all switches and clips to prevent confusion. If desired, components of different colors (such as brown and ivory) can be used for the main and start toggle switches. Power supply to the box is controlled by the main switch. The start switch controls power supply to the green wire, which is attached to a bulldog clip. The green wire is used for the start winding connection. It becomes a power supply wire only when both switches are closed.

from the compressor's C, S, and R terminals and replace them with the test cord clips: white to Common, black to Run, and green to Start. Affix a clamp-on ammeter around the wire to Common. To attempt start-up, turn on both test cord switches at the same time to supply power to both the start and run windings. After about one-half second to one second, open the test-cord "start" switch to cut off the power being supplied to the start winding. The motor should continue to run and pull normal amperage.

Use a clamp-on ammeter to observe amperage flow while operating the switches. If necessary, a start capacitor can be easily connected in series with the green clip and the start winding.

Checking Potential Relays

When the test cord is used to check a potential relay, begin with both switches placed in the off position. Initially keep the test cord unplugged from a power source.

Turn off *system* power supply and check that it is off with a voltmeter. Remove the wire from relay Terminal 1. Place the green clip onto Terminal 1, **Figure 25-28**. Connect the test cord's black clip to relay Terminal 2. Close the test cord's "start" switch to provide a circuit from the "start" clip to the "run" clip with the start switch in series between them. Place a clamp-on ammeter around the wire to Common for observing amperage flow during the start-up procedure.

Observe the ammeter while turning on the *system* power supply (test cord's main switch) and (after about one second) turning the test cord's start switch to the off position. If the compressor starts and runs properly, the potential relay is defective and should be replaced.

Checking PTC Relays

A method for testing PTC relays uses just an ammeter for troubleshooting problems. The clamp-on ammeter is wrapped around one of the wires connecting the PTC relay to the run capacitor and used to read current flow. At start-up, the current reading should be high, then quickly drop to normal operating level as the PTC relay opens its contacts to disconnect the start winding. If there is no current through the PTC relay or if current does not drop (indicating that the relay contacts have *not* opened), the PTC relay is defective and should be replaced.

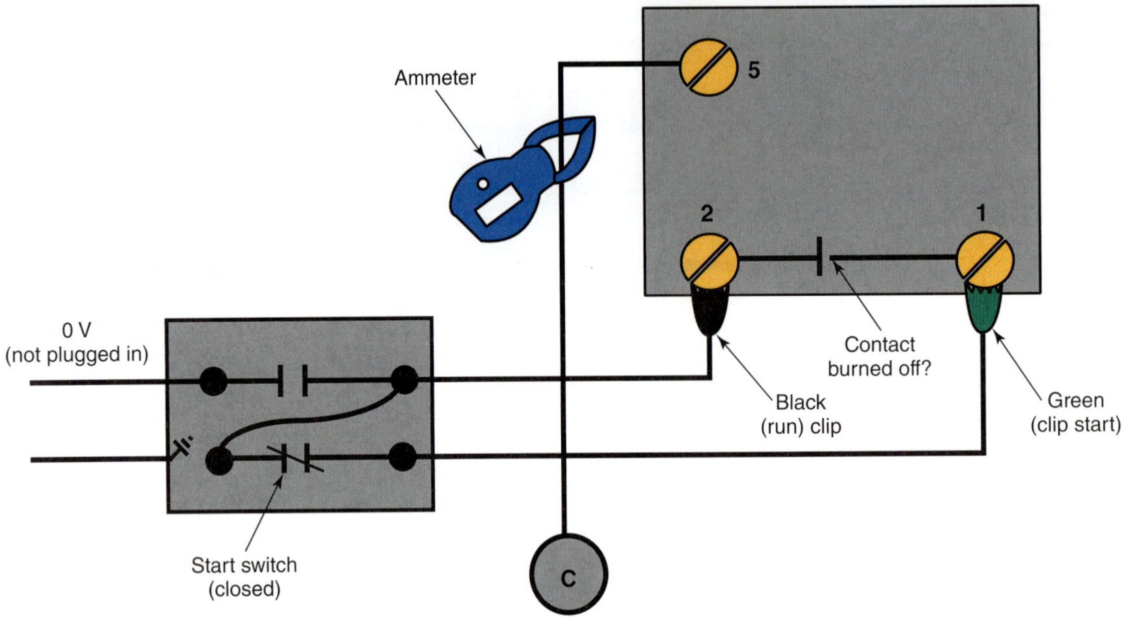

Goodheart-Willcox Publisher

Figure 25-28. Using the test cord to check the start relay contacts. Closing the "start" switch provides a circuit from Terminal 1 to Terminal 2 with the switch in series between them.

Faulty Run Capacitor

If a compressor's motor starts okay and runs for a short time before shutting off on the overload, the run capacitor may be suspected. However, the run capacitor seldom proves faulty, and high head pressure can cause the same symptoms. Install a manifold gauge and check for high head pressure before checking the run capacitor.

If the relay checks out properly, but the compressor fails to start, the problem is in the *compressor.* It may have a faulty start winding, be physically binding, or locked up.

Open Start Winding

A compressor's start winding may be open. This can be checked for continuity with a multimeter. A reading of infinite resistance (OL) will be obtained if the circuit is open.

SAFETY NOTE

Power Off for Ohmmeter Measurements

To protect a meter from damage when measuring resistance, all power to the circuit being tested *must* be off. Remove all wires connected to the compressor terminals. Then, use a voltmeter to check that no voltage is being applied to the circuit. It is then safe to test for continuity using the ohmmeter setting.

Cycling On the Overload

It is quite possible for the compressor's motor to be ***cycling on the overload*** while the condenser fan is operating. This condition is marked by operation of the overload each time the motor attempts to start, resulting in a "hmmmmm...click, hmmmmm...click" sound pattern. This provides the following items of information:

- The overload is doing its job of protecting the compressor from damage.
- The power supply is adequate and the thermostat is closed, else the fan would not be running.

This condition could be caused by high head pressure, which should be checked with a manifold gauge. If the head pressure is normal, the problem is in the start components or the compressor's motor windings.

To pinpoint the problem's location, use a clamp-on ammeter to check each circuit connected from the relay to the compressor. The *order* in which checks are performed is a matter of experience and the component locations.

Motor Shorted to Ground

Disconnect power and check that no voltage is present using a voltmeter. Then, use a multimeter set to read ohms to check the windings and confirm the shorted-to-ground diagnosis. If there is a short to ground (ground fault or grounded) situation, the compressor must be replaced.

Making Bench Tests

The test cord described earlier can be used for bench testing many electrical devices for proper operation. The test cord is used to supply power to the device for the test. For use in bench testing, both test cord switches are turned off.

The Common and Run clips are attached to the device being tested. A clamp-on ammeter is then clamped around either wire to permit the technician to observe the current during start-up, **Figure 25-29**. Then, the plug is connected to a power receptacle and the switches actuated. Always wait to plug in the test cord until ready to turn on the switches. This will prevent accidentally turning on the circuit too soon. Observe the clamp-on ammeter readings to determine if the device being tested is defective. Keep one hand on the main switch so it can be quickly turned off if necessary.

SAFETY NOTE

Adding a Safety Ground to Devices for Testing

When testing a compressor or other device using a test cord, remember to connect a ground wire to the device being tested for safety.

Contactors

A ***contactor*** is an electrical control device that uses an electromagnetic coil to control *heavy-duty* contacts that are opened or closed at the same time. The contacts are normally open (NO) and close when the contactor's coil is energized. A 2-pole contactor has two separate sets of contacts, and a 3-pole contactor has three sets of contacts. A three-phase motor uses a 3-pole contactor to simultaneously control all three hot legs, **Figure 25-30**.

The major difference between a relay and a contactor is the size of the contacts and the current ratings. Contactors are used to control power to large loads. Higher current to large loads requires the use of heavy-duty contacts. Some typical contactors are shown in **Figure 25-31**.

An insulated movable armature is located close to a contactor's iron core. This armature carries the movable contacts, which are held away from the stationary contacts by means of a spring, **Figure 25-32**. When the coil is energized, magnetism overcomes the spring pressure, snapping the contacts into the closed position. These snap-action contacts are used because the contacts must open and close quickly to reduce electrical arcing that always occurs under load conditions.

These devices are rated according to the maximum current through the contacts for a specific voltage. The current rating is usually expressed in full load amperage (FLA), which is an inductive load rating. The same contactor will tolerate a higher current if the load is resistive. Specific ratings are listed on the contactor data plate. An oversized contactor is perfectly acceptable, but an undersized contactor will result in burned-out contacts.

As a result of arcing when the contacts open and close, it is normal for contacts to become pitted and burned. Use of a file or sandpaper to clean the contacts is *not* recommended, because this destroys the surfaces of the contacts and increases arcing. If contacts become pitted and burned, it is best to replace them. Replacement contacts are usually available from a local supplier.

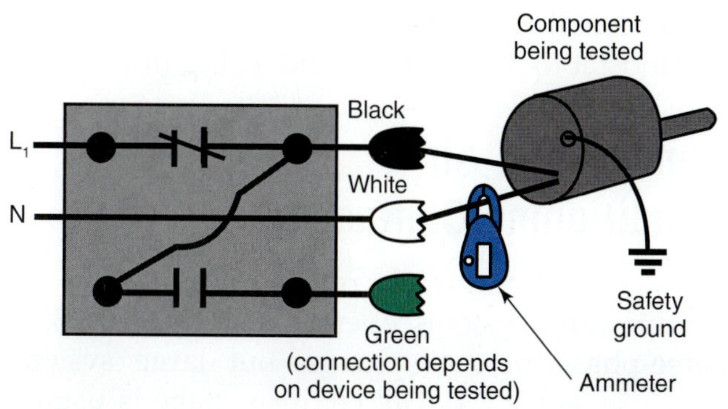

Goodheart-Willcox Publisher

Figure 25-29. Using the test cord for bench-testing an electrical component. The cord provides a 120 V (or 230 V with adapter cord) power supply.

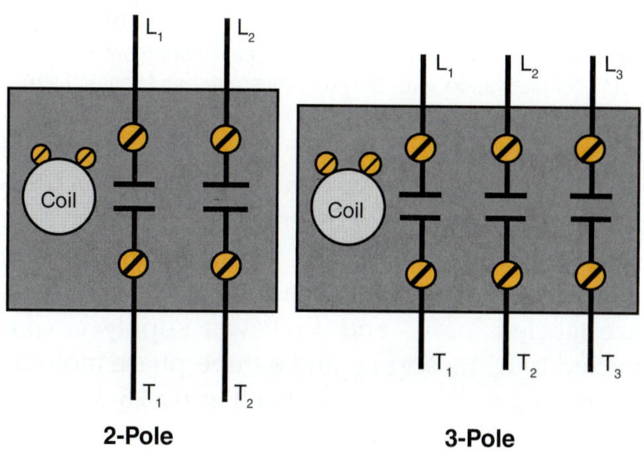

Goodheart-Willcox Publisher

Figure 25-30. Schematic views of 2-pole and 3-pole contactors.

2-pole contactor

3-pole contactor

Images courtesy of Johnstone Supply

Figure 25-31. Typical contactors.

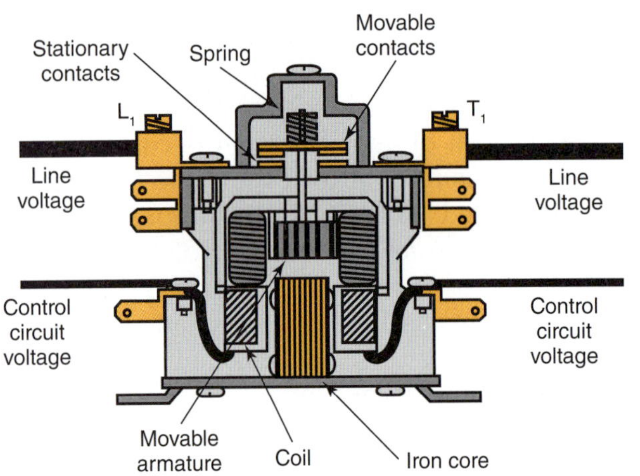

Goodheart-Willcox Publisher

Figure 25-32. Cutaway view of a contactor. Spring pressure holds contacts apart until the coil is energized and snaps them closed.

The numbering system for contactors determines direction of current flow through the contacts. The line power (inlet) terminals are labeled L_1, L_2, and L_3 and the load (outlet) terminals are labeled T_1, T_2, and T_3. Power supply is connected to L_1, L_2, and L_3, and a three-phase motor is connected to T_1, T_2, and T_3. See **Figure 25-33**.

Residential Air-Conditioning Contactors

Residential air conditioning uses a 230 V, single-phase circuit to operate the outdoor condensing unit. Power supply is typically obtained from the main breaker panel and is connected to the outdoor unit at the 2-pole (L_1 and L_2) contactor. The compressor and the condenser fan are typically both connected to T_1 and T_2 of the contactor, **Figure 25-34**.

The contactor coil and thermostat operate on 24 V, so a transformer must be used to provide the low control voltage. The low voltage permits the use of small (18 gage) wire to connect the indoor thermostat to the outdoor condensing unit.

Commercial Air-Conditioning Contactors

Most commercial refrigeration and air-conditioning systems operate on 208 V or 230 V three-phase power systems, but large systems may use 440 V. Control circuit voltage is usually

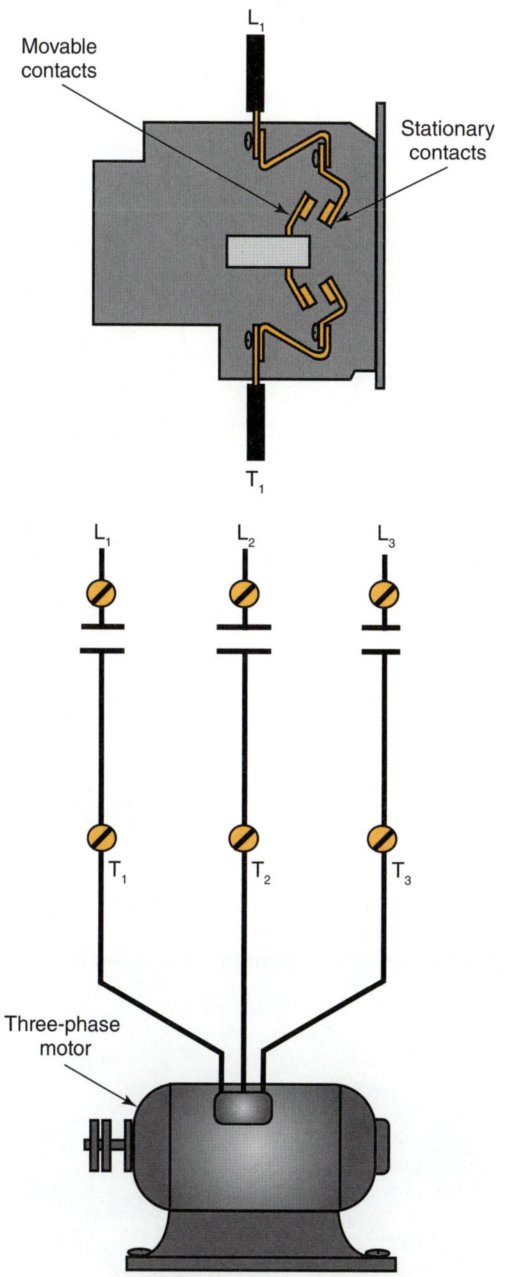

Goodheart-Willcox Publisher

Figure 25-33. Contactor terminals are labeled where the device is connected. The letter *L* is used for the line power side and *T* for the load side. All contacts open and close at the same time.

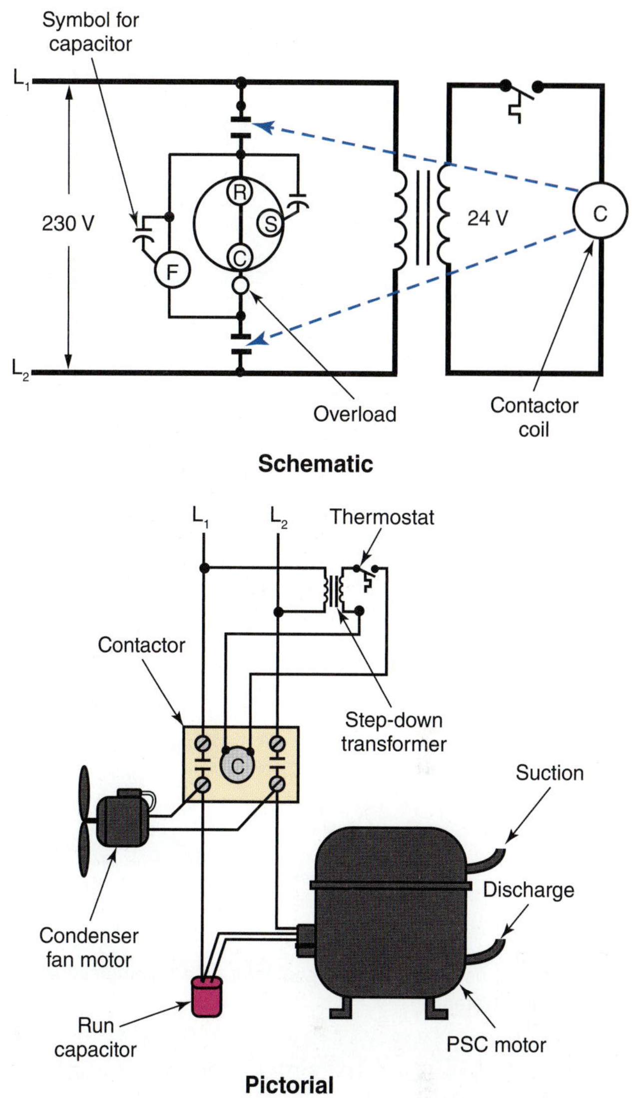

Goodheart-Willcox Publisher

Figure 25-34. Typical wiring for a residential system, using a two-pole contactor.

120 V, since this allows use of standard switching devices and wiring, **Figure 25-35**. Safety switches and thermostats are connected in series with the 120 V control circuit that supplies power to the contactor coil. If the thermostat or a safety switch opens, the contactor coil is de-energized, opening the main contacts and cutting off the power supply to the compressor's motor.

Commercial refrigeration systems may use a 230 V, single-phase control circuit. Power supply for the control circuit is obtained from any two supply wires connected to the contactor at L_1, L_2, and L_3. See **Figure 25-36**. The two control circuit wires are sometimes equipped with fuses to protect the control circuit wiring and the contactor coil. A blown fuse can narrow a problem down to the control circuit.

Contactor Coils

An electromagnetic coil is located inside the contactor, but it is *not* electrically connected to the main contacts. The contactor coil is electrically isolated from the contactor contacts. Because the coil is a separate device, the coil voltage is often different from the voltage at the main contacts. The coil has its own terminals for making electrical connections, **Figure 25-37**.

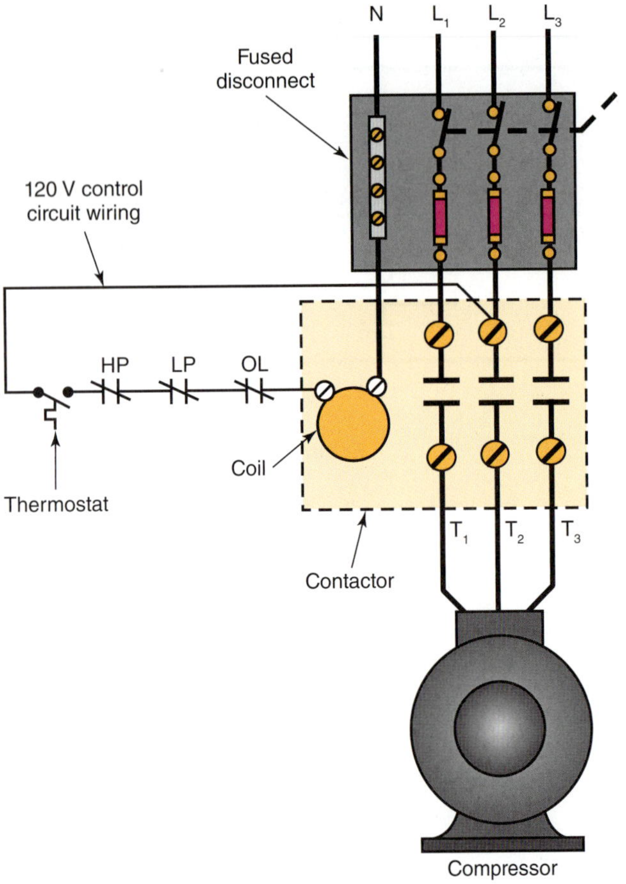

Goodheart-Willcox Publisher

Figure 25-35. A 230 V, three-phase system with a 120 V control circuit used for a 3-pole contactor on a commercial air-conditioning system.

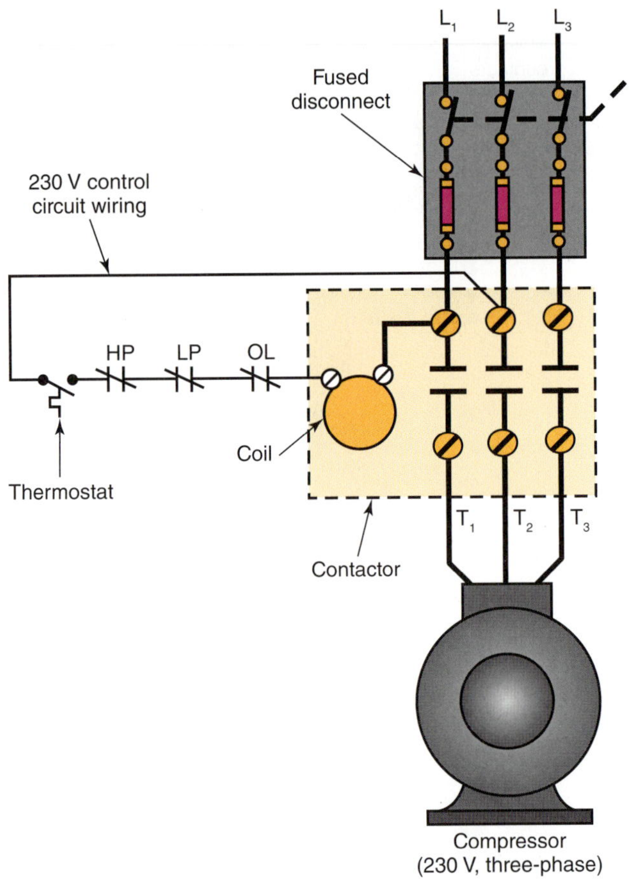

Goodheart-Willcox Publisher

Figure 25-36. Commercial refrigeration system using a 230 V, single-phase control circuit connected across two incoming power lines.

The coil is used to control the main contacts. Energizing the coil causes the main contacts to close simultaneously. De-energizing the coil causes the main contacts to open. Power supply to the coil can be controlled automatically or manually. As noted earlier in this chapter, the separate electrical circuit to the coil is called the *control circuit*, **Figure 25-38**.

Because the coil is electrically separate from the main contacts, it is common practice to use lower voltage to operate the coil. This permits easy control of a large motor operating on high voltage and amperage. Lower voltage and current in the control circuit is safer and permits the use of standard switching devices. As shown in **Figure 25-39**, a step-down transformer is used to obtain the lower control circuit voltage.

Any number of contacts and safety control switches can be located in the control circuit, **Figure 25-40**. The contacts for these controls are connected in series with the coil and may include overloads, thermostats, pressure controls, fuses, limit switches, flow controls, or other devices.

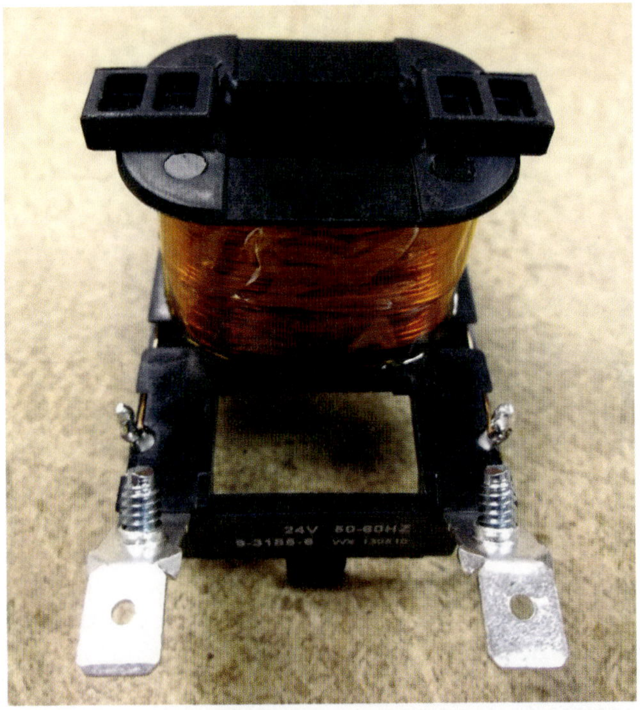

Images courtesy of Johnstone Supply

Figure 25-37. A replacement coil for a contactor. Note the terminals used with push-on connectors for ease of replacement.

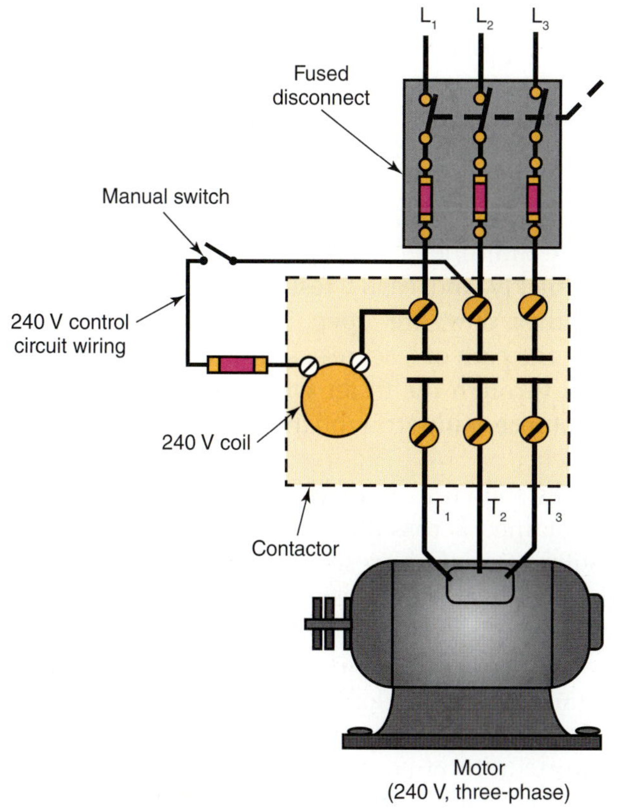

Goodheart-Willcox Publisher

Figure 25-38. Power supply to the contactor coil can be controlled manually or automatically.

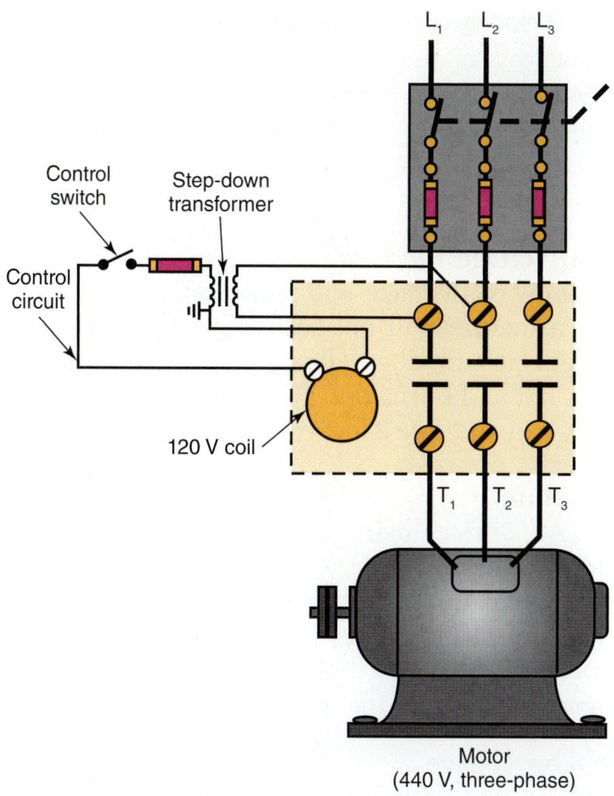

Goodheart-Willcox Publisher

Figure 25-39. A step-down transformer is used to obtain the lower voltage used for the control circuit.

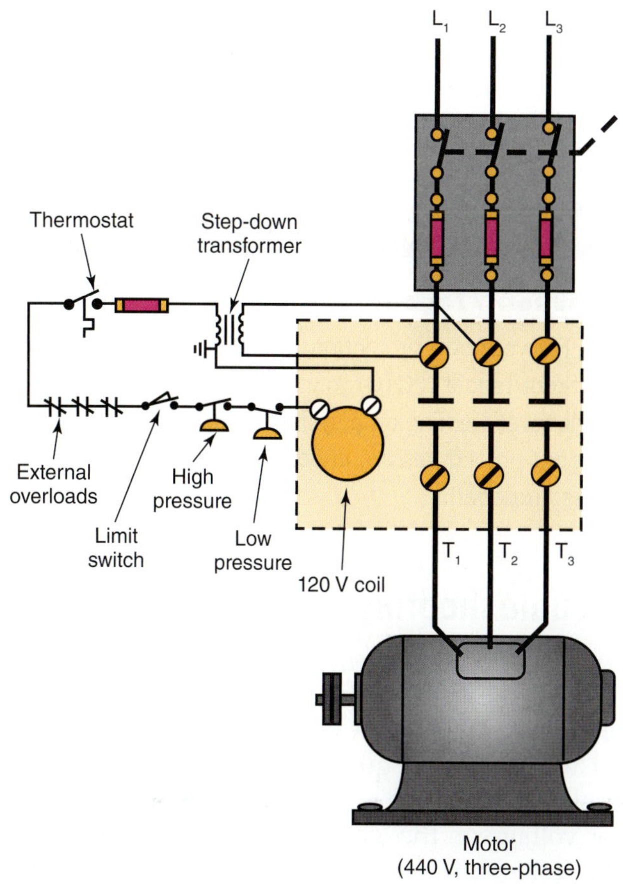

Goodheart-Willcox Publisher

Figure 25-40. All safety switches and overloads connected to a control circuit must be closed to energize the contactor coil.

Before the coil can be energized, *all* switches in the control circuit must be closed. Opening *any* switch will disconnect power supply to the coil and stop operation of the motor.

Although a step-down transformer is often used, control circuit voltage may be obtained from a *separate* power source. Thus, turning off the motor's fused disconnect may *not* de-energize the control circuit. This situation is not unusual and can be dangerous.

SAFETY NOTE

Voltage Testers

Never work on a circuit until you are certain the circuit is dead and cannot be accidentally re-energized while repairs are being made. Always check the circuit with your voltage tester before and after de-energizing it.

The main switch (disconnect) must be properly tagged and locked to inform others that work is being performed on the circuit. Lockout/tagout (LOTO) is a safety requirement in most industrial settings.

SAFETY NOTE

Lockout/Tagout

For your own protection, you should make lockout/tagout (LOTO) your standard procedure whenever working on a system. Proper lockout/tagout procedures are important and prevent injuries and deaths.

Troubleshooting Contactors

If the control circuit is not fused (and some are not), the entire system must be checked to locate electrical problems. A voltage tester or a multimeter will provide a quick method of locating these problems. With the power supply turned on, check for voltage on the load (L) side of the contactor. If the tester shows no voltage at these terminals, the main power source is off. If there is power missing from any legs while power remains on others, it could indicate blown fuses, bad disconnects or bad breakers.

If power supply to the contactor between L_1 and L_2, and between each terminal to ground are giving normal readings, then your issue may be inside the contactor. Next, move to the T-side of the contactor and check for proper voltage between T_1 And T_2 and again from each terminal to ground. If one of these terminals is dead and the others are giving proper readings, then the corresponding contact inside the contactor is burned out.

Control Circuit Problems

If power supply to the contactor is okay, but no voltage is available on the terminal (T) side of the contactor, the problem is in the control circuit. Check for voltage at the coil. The coil may be burned out, or a control circuit switch is open, either of which means the coil is not energized.

To check the contactor coil, test for voltage at the coil terminals. If the voltage is correct, the coil itself is defective and must be replaced. If voltage is not present at the coil, one of the control circuit contacts is open. All switches must be checked to determine which one is open.

Each switch serves a definite purpose for controlling power supply to the coil. Some of these controls are automatically reset and others are manually reset. When the open switch is located, that control pinpoints the problem to be corrected.

Voltage Test Method

A multimeter set to read volts can be used to quickly and easily test switches to determine whether they are open. Electricity will always seek the path of least resistance, so current will not flow through the tester when it can take a lower-resistance path straight through a fuse or closed switch.

As shown in **Figure 25-41A**, a 120 V reading is obtained across an open switch. Since an open switch has *infinite* resistance, electricity will flow through the lower resistance of the meter and provide a reading.

As shown in **Figure 25-41B**, a 0 V reading is obtained across a closed switch. In most cases, this means that the component is closed or there is continuity in the circuit.

SAFETY NOTE

Voltage Testing Contactors

A reading of 0 V across a switch or fuse or control does not *always* mean that the component must be closed. It just means that there is no difference of potential across the component. There could still be voltage at the control.

Coil Voltage Too Low

Low coil voltage will reduce magnetic pull, so the armature will not seat properly. This causes excess current, due to loss of back-emf, so the coil gets hot and burns out. "Chattering" of the armature may indicate a low coil voltage condition.

AC Hum

All devices using a magnetic field will produce a characteristic hum. This hum is caused by the changing magnetic fields, which produce mechanical vibrations. In addition to hum, noisy operation of contactors, relays, and line starters may be caused by the following:

- Low voltage
- Dirt, rust, or metal filings on magnet faces
- Inability of the armature to seat properly
- Binding of moving parts

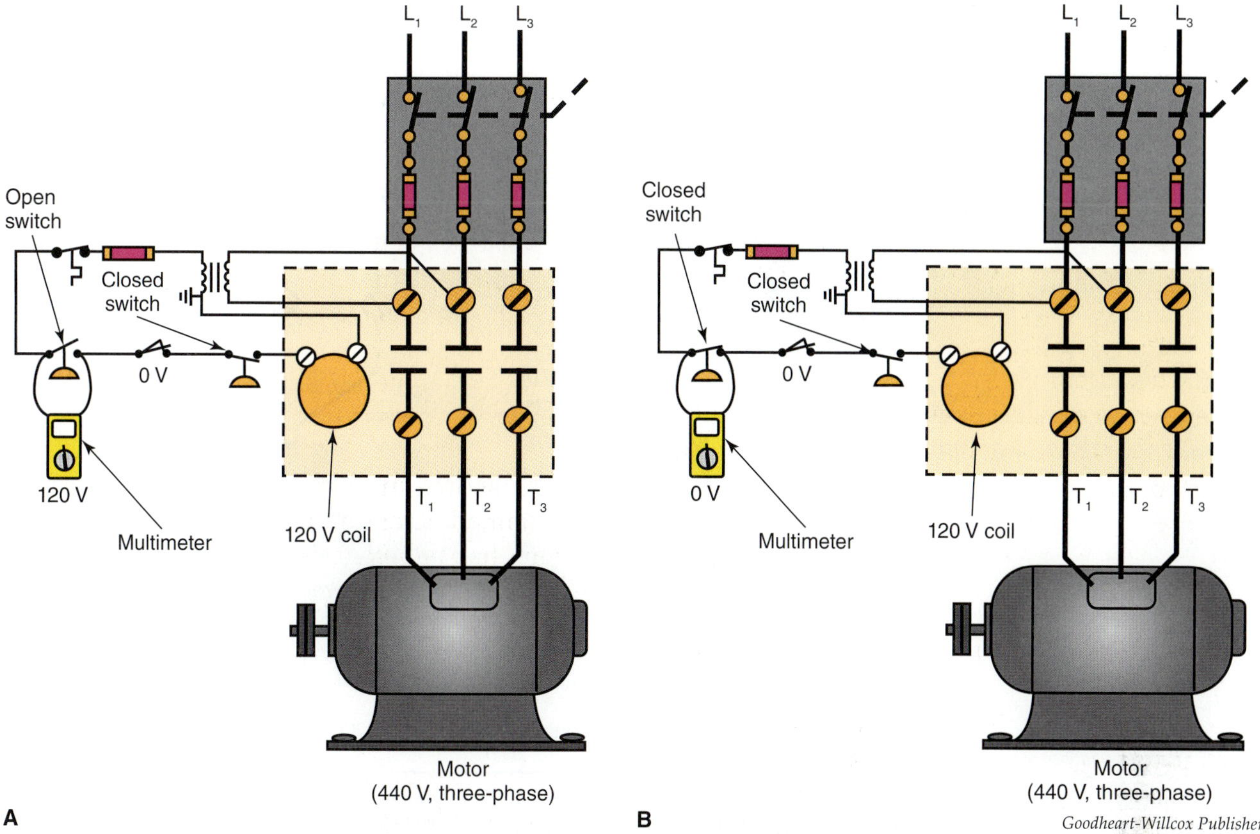

Goodheart-Willcox Publisher

Figure 25-41. A multimeter can locate an open switch or blown fuse by showing the voltage across the device. A—An open device will register 120 V. B—A closed switch or intact fuse may register 0 V.

Across-the-Line Starters

An ***across-the-line starter*** is a contactor that has built-in overload protectors. It is used to operate and protect three-phase motors. Across-the-line starters are also referred to as *line starters* or *magnetic starters*. Such motors are frequently chosen for belt-driven applications, such as operating compressors, large condenser fans, exhaust fans, cooling towers, or large air-conditioning blowers. The built-in overloads protect the motor against excess amperage (current), **Figure 25-42**.

The overload protectors are normally mounted just below the contactor and are often installed and prewired at the factory. Overloads are more accurate than fuses, providing excellent motor protection. One overload is needed to protect each phase supplying power to the motor.

The overloads operate on a *thermal* principle, with an element that is sized to permit a specific amperage flow. Excess amperage will cause the element to become hot and operate a trip mechanism that opens a set of contacts. This cuts off the power supply to the contactor coil and stops the motor, **Figure 25-43**. A manual reset is provided

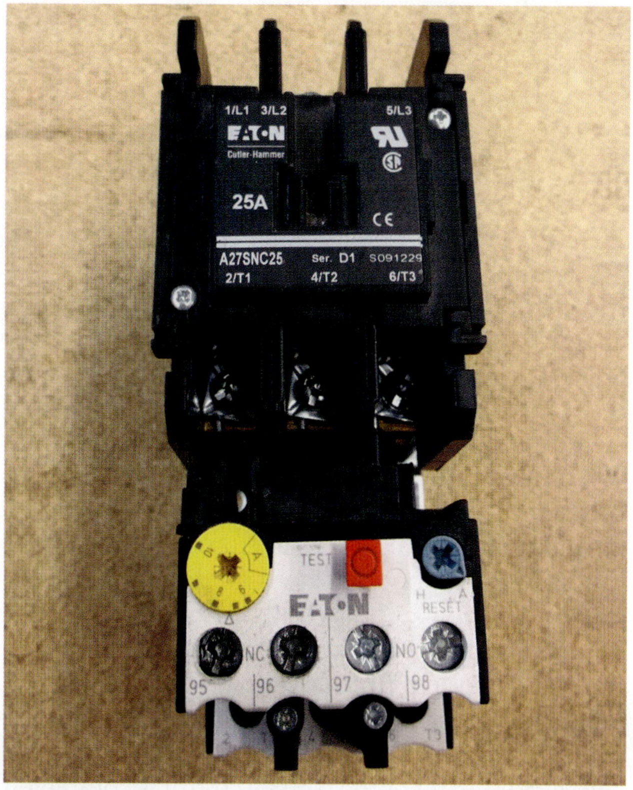

Goodheart-Willcox Publisher

Figure 25-42. A typical across-the-line starter. Most line starters have overload protectors mounted just below the contactor.

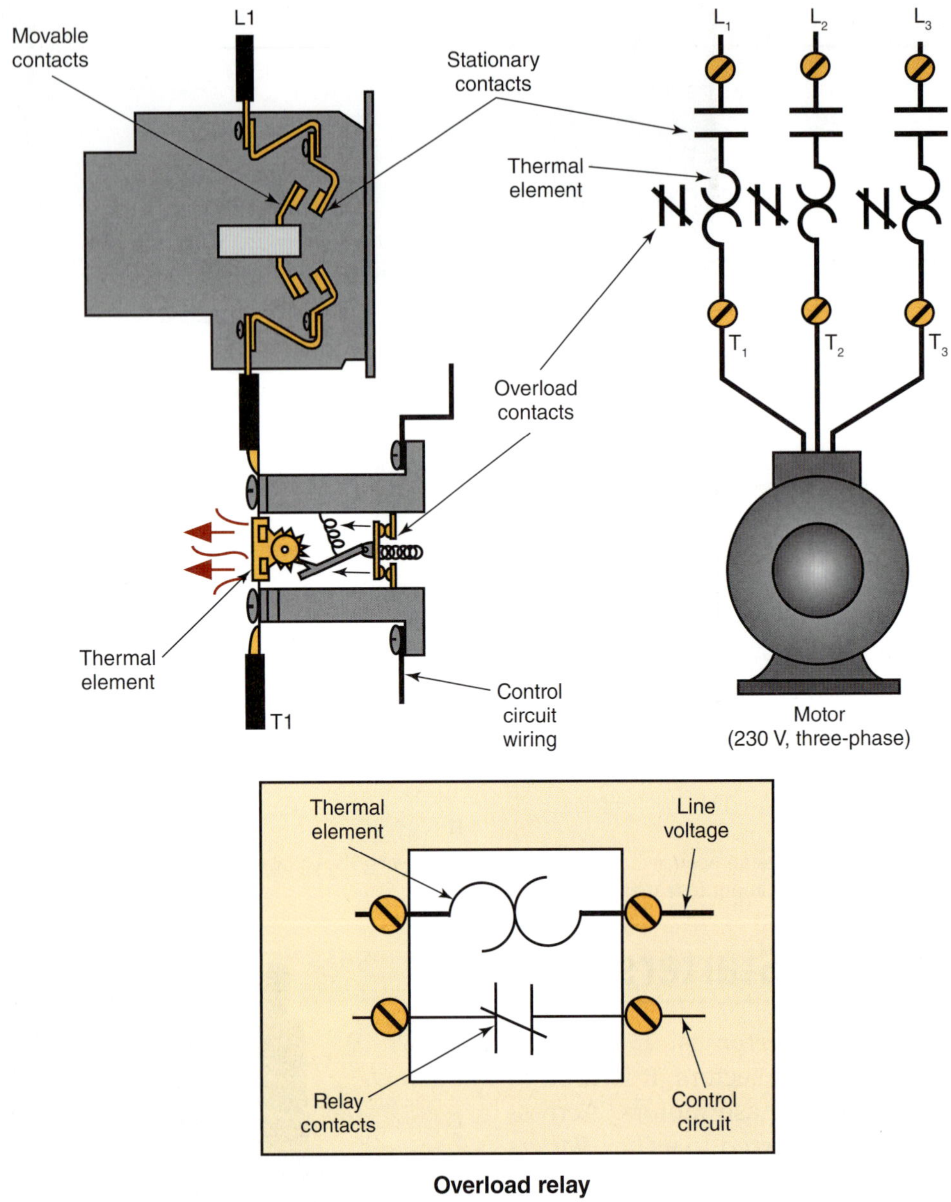

Goodheart-Willcox Publisher

Figure 25-43. The across-the-line starter provides thermal overload protection. Excess amperage will cause the thermal element to become hot and open the overload contacts. This cuts off power to the contactor coil, which opens the contacts and stops the motor.

to reclose the overload contacts after the thermal element cools off. Thermal elements, which are sometimes referred to as *heaters*, are available in different amperage capacities. They are sized to provide accurate motor protection at slightly above the full load amperage (FLA) rating of the contactor, **Figure 25-44**.

Each of the three overload contacts is connected in the control circuit. As shown in **Figure 25-45**, the factory connects a wire from L_1, through each overload switch to one terminal on the contactor coil. Field wiring of the control circuit is completed by connecting the other side of the coil through the various control circuit switches and back to either L_2 or L_3. This places the thermostat and all overload switches in series with the coil.

Some older across-the-line starters are equipped with just two overloads, instead of three. They are based on the principle that if any one supply wire becomes overloaded, another will also be overloaded. Thus, only two overloads are needed to protect the motor. However, current electrical codes require three overloads for a three-phase motor. Some across-the-line starters use three thermal elements, but only *one* set of contacts. Any thermal element can open the contacts. A manual reset button is used to reclose them, **Figure 25-46**.

A

B

Images courtesy of American Electric Co.

Figure 25-44. A—An across-the-line starter mounted in a control box. B—A typical replaceable thermal element for an across-the-line starter.

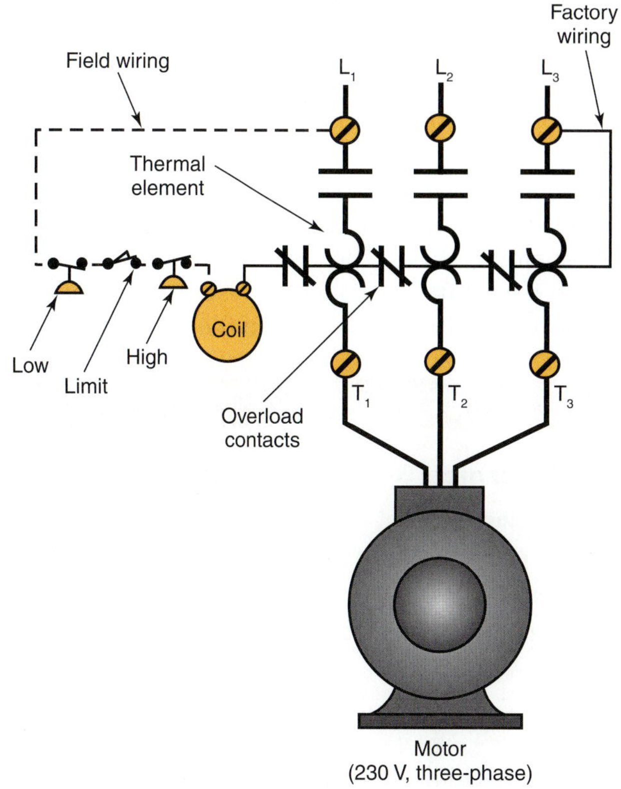

Goodheart-Willcox Publisher

Figure 25-45. Wiring of the overload contacts is usually completed in the field by incorporating the desired control devices in the circuit.

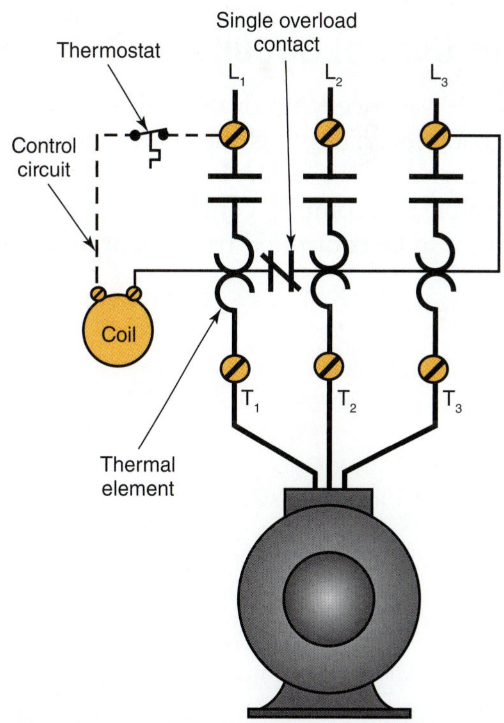

Goodheart-Willcox Publisher

Figure 25-46. Some across-the-line starters may have three thermal elements, but only a single set of overload contacts. Any thermal element can operate the contacts if an overcurrent problem occurs.

Troubleshooting Across-the-Line Starters

Troubleshooting across-the-line starters is performed as described under contactors. The only difference is the overloads. If an overload switch opens, power supply to the coil stops and the motor stops. When the overload cools, the reset button is manually pushed to reset the contacts to the closed position, so power to the coil is restored.

SAFETY NOTE

De-Energize Before Resetting

Pressing the reset button with the control circuit energized will start the motor immediately, while your hand is inside the starter box. Always de-energize the control circuit before you press a reset button.

The thermal elements and overload switches are reliable and rarely fail. If the overload will *not* reset or it quickly opens again, the problem is due to excess amperage. This indicates a motor problem.

Motor Control Circuits

Some advanced motor control circuits can be quite complex. They include relays, lights, pushbuttons, and safety controls. **Figure 25-47** shows an example of a complex control circuit that allows the motor to be controlled from separate locations.

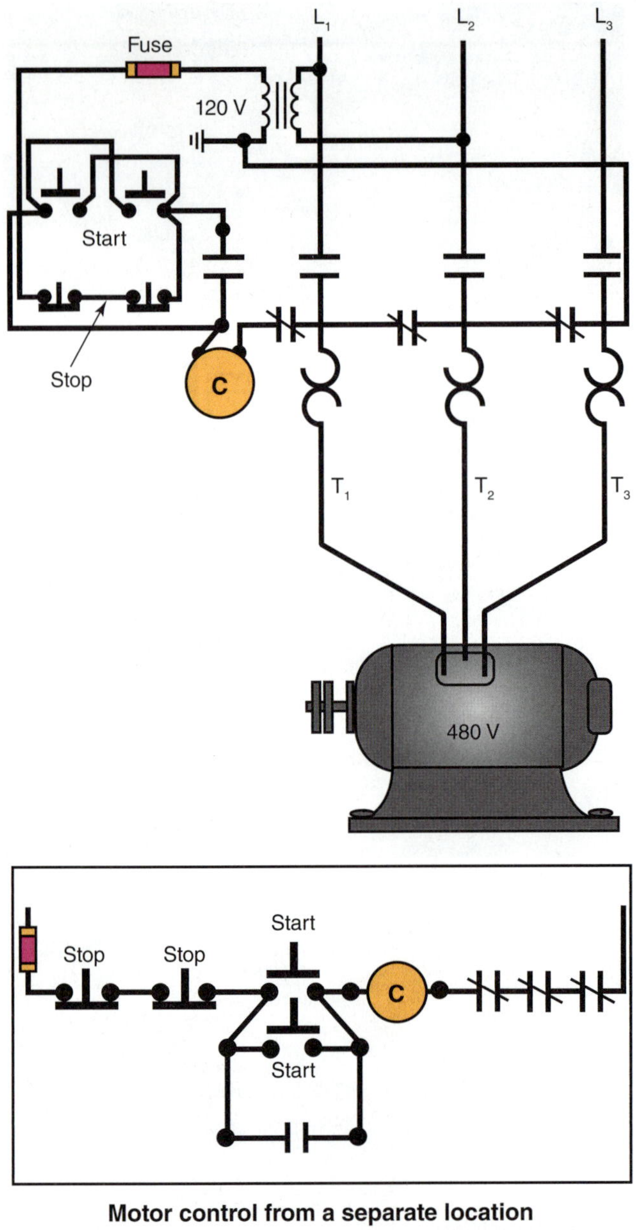

Goodheart-Willcox Publisher

Figure 25-47. With a more complex control circuit, a motor can be started and stopped from two or more locations. The circuit shown has two pushbutton control switches.

Chapter Review

Summary

25.1 A solenoid valve uses electromagnetism to electrically control the opening and closing of valves. Solenoid valves are frequently used on applications where automatic control of fluids is desired.

25.1 A relay is a device that opens or closes its electrical contacts when it is energized. Relays open or close switch contacts using an iron core and armature to open and close the contacts.

25.1 A start relay is designed to start and stop a single-phase motor. A current relay is a start relay that operates on the current flowing through a single-phase motor's run winding. A potential relay is a start relay that operates on the back-emf (potential) that is generated by the start winding.

25.1 A PTC relay is a relay that uses a positive temperature coefficient (PTC) thermistor as a sensor and switching device. A PTC thermistor reacts rapidly to an increasing temperature change by greatly increasing its resistance to current flow.

25.1 When troubleshooting relays, first make sure to power down the system and check incoming power supply before making other electrical tests. Test cords can be used to bypass start components and manually operate compressors. Be sure to include safety devices, such as GFCIs, when using test cords.

25.1 A contactor is an electrical control device that uses an electromagnetic coil to control heavy-duty contacts that are opened or closed at the same time. A voltage tester or multimeter provides a quick method of locating electrical problems in the control circuit.

25.1 An across-the-line starter is a contactor that has built-in overload protectors. It is used to operate and protect three-phase motors. Troubleshooting across-the-line-starters is similar to troubleshooting contactors, except overloads must be checked on across-the-line starters.

Know and Understand

1. A solenoid valve operates on the principle of _____. (25.1)
   A. heat transfer
   B. electromagnetism
   C. alternating current
   D. manual, mechanical movement
2. When a relay coil is energized, magnetism is induced in the core and the _____ is pulled downward. (25.1)
   A. armature
   B. float switch
   C. iron plunger
   D. relay switch
3. A current relay operates on _____. (25.1)
   A. low-current flow to the run winding at start-up
   B. high-current flow to the run winding at start-up
   C. back-emf generated by the start winding
   D. high-current flow at start-up through a start capacitor
4. Single-phase motors rated 1/2 hp and above must use _____ relays. (25.1)
   A. current
   B. push-on
   C. potential
   D. PTC
5. In a potential relay, the amount of back-emf generated is directly proportional to _____. (25.1)
   A. capacitor position
   B. armature position
   C. the number of dummy terminals used
   D. rotor speed
6. Motor components usually fail in this order: _____. (25.1)
   A. start winding, start capacitor, relay
   B. start capacitor, relay, start winding
   C. start winding, relay, start capacitor
   D. relay, start winding, start capacitor
7. *True or False?* A potential relay's contacts are normally closed (NC) and are opened by the energizing of the relay coil. (25.1)

8. Only _____ wiring terminals are available on a hermetic compressor. (25.1)
   A. two
   B. three
   C. four
   D. five
9. A PTC relay is connected _____. (25.1)
   A. In series with the run capacitor terminals
   B. In parallel with the start capacitor terminals
   C. In series with the start capacitor terminals
   D. In parallel with the run capacitor terminals
10. When checking a PTC relay with a clamp-on ammeter and there is no current running through the relay at start-up, then this indicates the _____. (25.1)
    A. PTC is operating properly
    B. PTC relay is defective and should be replaced
    C. start capacitor should be replaced
    D. run capacitor should be replaced
11. *True or False?* A resistance reading of 0 Ω at the start winding indicates the circuit is open. (25.1)
12. When testing a compressor using a test cord, be sure to connect a(n) _____ to the device being tested. (25.1)
    A. ground wire
    B. hot wire
    C. start capacitor
    D. contactor
13. The major difference between a relay and a contactor is the size of the contacts and the _____. (25.1)
    A. number of contacts
    B. current ratings
    C. type of armature used
    D. specific voltages
14. *True or False?* Contactors are rated according to their maximum current through the electromagnetic coil. (25.1)
15. Overload contacts are wired in _____ with a contactor's coil. (25.1)
    A. parallel
    B. series
    C. series-parallel
    D. parallel with main contacts, which are wired in series
16. When the power supply is okay, but no voltage is present at terminals T_1, T_2, or T_3 of a contactor, then the problem is in the _____. (25.1)
    A. main contacts
    B. rotor
    C. control circuit
    D. circuit devices
17. An across-the-line starter is essentially a contact with built-in _____. (25.1)
    A. start capacitors
    B. run capacitors
    C. overload protectors
    D. thermistors

Apply and Analyze

1. Explain how a solenoid valve operates. (25.1)
2. Why is a relay used as a motor control instead of an internal centrifugal switch within a hermetic compressor? (25.1)
3. Explain the difference(s) between a current relay and a potential relay. (25.1)
4. A service technician is on a service call where a single-phase compressor is not starting. The technician uses a test cord to bypass the potential relay and start the stuck compressor. The compressor starts and runs properly. Is the potential relay working properly or is it defective? Explain. (25.1)
5. Why is the contactor coil voltage different from the voltage at the main contacts? (25.1)
6. Explain how the overloads in an across-the-line starter protect a motor from excess amperage current. (25.1)

Critical Thinking

1. Imagine you are a service technician troubleshooting a locked-up hermetic compressor. You decide to first reverse motor rotation by switching the wires from the start and run windings, which does not work. Then, you reconnect the start and run windings in their original places and try starting the motor with the high-capacity start capacitor, which also does not work. Lastly, you use higher voltage to increase starting torque, which also fails to start the compressor. What should you do next? (25.1)

Johnson Controls, Inc.

26 Refrigeration Motor Controls

LEARNING OUTCOMES

After studying this chapter, you will be able to:

26.1 Explain thermostat operation, sensing elements, their use in domestic and commercial applications, and the process of troubleshooting thermostats.

26.2 Identify pressure-type motor control operation, settings, wiring configurations, and how to troubleshoot pressure-type motor controls.

26.3 Identify the function of low-pressure safety controls and common problems that lead to unwanted low pressure in a system.

26.4 Recognize different head-pressure safety controls and how they should be adjusted to maintain proper head pressure in a system.

26.5 Recognize and troubleshoot pumpdown cycles.

26.6 Install and troubleshoot oil pressure safety controls.

TECHNICAL TERMS

cut-in (CI)
cut-out (CO)
differential
dual-pressure motor control
erratic
fan cycle control
head-pressure safety control
logging
low-pressure motor control
net oil pressure
oil pressure safety control
oversized
pressure-operated switch
primary control
pumpdown cycle
riser
Rule of 10
secondary control
sensing element
setpoints
stage
temperature range
thermostat

Introduction

The two types of primary motor controls used to provide automatic operation of an HVACR system are thermostatic (temperature-sensitive) and pressure-type (pressure-sensitive). One or the other is found on every refrigeration system. These ***primary controls*** are used to start and stop the compressor when certain temperatures or pressures are reached. They include a switch that controls the power supply to the condensing unit. When the switch closes, the unit runs. When the switch opens, it stops.

A technician *must* understand the purpose and operation of these motor controls. Knowing the purpose of the control, how it operates, and how it is connected into the circuit makes troubleshooting easier and more accurate.

Thermostats are often used as the primary control for system operation. Other thermostats may function as secondary controls for individual operation of devices, such as the condenser fan, evaporator fan, solenoid valves, and defrost heaters. ***Secondary controls*** are safety devices used to shut off the system or turn on a safety mechanism if certain conditions exist.

Thermostatic Motor Controls

All refrigeration systems are ***oversized***. This means the system has more capacity than the minimum needed to permit it to lower the temperature to a certain point and then shut off. When temperature rises to a certain point, the system is turned on. Heat removal occurs only while a refrigeration system is running. A thermostat is used to cycle the system on and off. This maintains a temperature level in a desired range.

26.1 Thermostats

A ***thermostat*** is a switch that opens and closes its contacts according to temperature changes. Thus, a thermostat can be used to automatically turn a condensing unit on and off as set temperatures are reached. Thermostats are available in many different shapes and styles to meet different applications, **Figure 26-1**. While thermostats may look different, they all are essentially temperature-activated switches.

Thermostat contacts may *close* on rise in temperature or *open* on rise in temperature. The type that closes on rise is used in cooling applications. The type that opens on rise is used in heating applications. See **Figure 26-2**.

Image courtesy of Johnstone Supply

Figure 26-1. Thermostats are provided in a number of different forms, but all perform the same job of opening and closing a switch in response to temperature change. This thermostat shown can be found in domestic refrigerators and small commercial refrigerators.

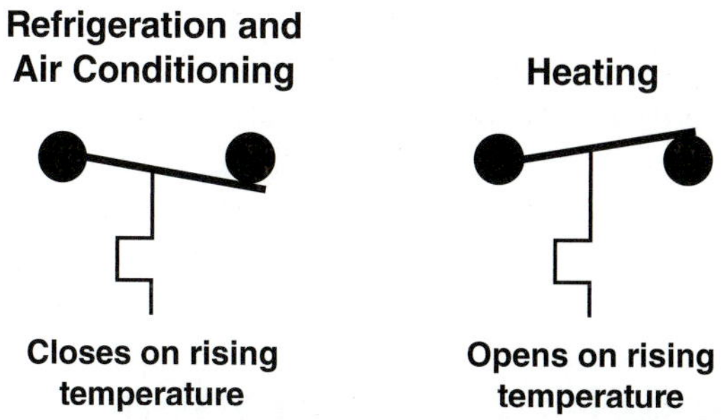

Goodheart-Willcox Publisher

Figure 26-2. Schematic symbols for thermostats. Those that close on a rise in temperature are used for cooling (refrigeration and air conditioning) applications, while those that open on a rise in temperature are used in heating applications.

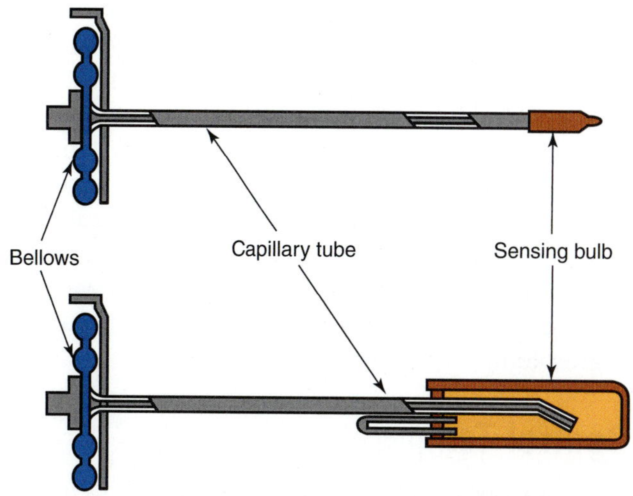

Goodheart-Willcox Publisher

Figure 26-3. As a refrigerant inside a capillary tube and sensing bulb expands or contracts in response to a temperature change, the thermostat bellows moves. This movement opens or closes a switch when the temperature change exceeds selected levels.

Thermostat Sensing Elements

The ***sensing element*** of a refrigeration thermostat consists of several parts, including a bellows (movable part), a capillary tube (connecting link), and a sensing bulb that holds a refrigerant, **Figure 26-3**. The refrigerant in the sensing element may be in a vapor or a liquid state. Different types of charges are used to achieve special operating characteristics.

Location of the sensing bulb determines what temperature is being controlled. Some sensing elements are designed for controlling air temperature, and others for evaporator temperature (evaporator-sensitive). Still others are submerged in liquid. Before replacing or adjusting a thermostat, the sensing bulb location must be known.

Sometimes the sensing bulb is omitted, and the capillary tube itself becomes the sensing element. The capillary tube will vary in length. The tube end may be straight or formed in the shape of a bulb, **Figure 26-4**. Capillary tubes without bulbs are generally used on such appliances as household refrigerators, window air conditioners, or domestic freezers. Twisted, coiled, and solid bulbs are often used in commercial refrigeration applications, such as food freezers or walk-in coolers.

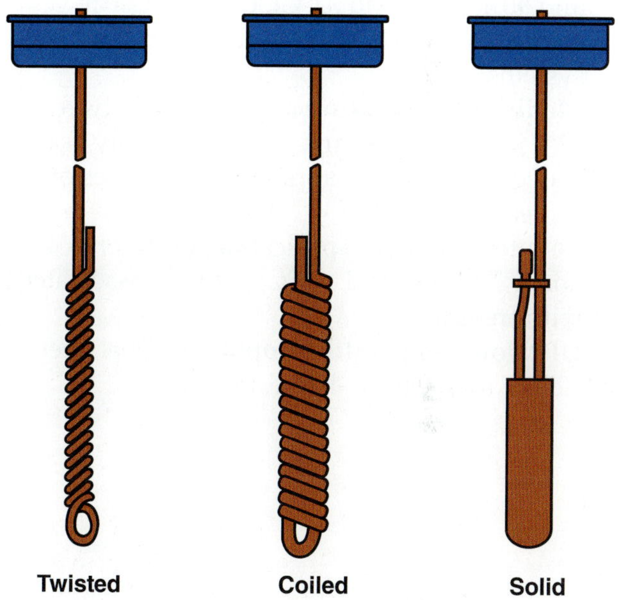

Goodheart-Willcox Publisher

Figure 26-4. On commercial refrigeration equipment, bulbs in twisted, coiled, or solid form are used.

Thermostatic Control Terminology

To understand motor controls, a technician must become familiar with the terms used to describe the operation of these controls. The setting value that causes the contacts to close is called the ***cut-in (CI)*** point. The setting value that causes the contacts to open is called the ***cut-out (CO)*** point. The difference between cut-in and cut-out is called ***differential***.

For example, a thermostat having a cut-in of 38°F (3°C) and a cut-out of 32°F (0°C) would have a differential of 6°F (38°F – 32°F = 6°F) or 3°C (3°C – 0°C = 3°C). Most domestic refrigeration thermostats have a fixed differential of 5°F or 6°F. Room thermostats for heating and air conditioning operate in a narrower range, with a fixed differential of 1.5°F–2°F.

Range Adjustment

The dial (or knob) on a domestic thermostat is a range adjustment, **Figure 26-5**. This range adjustment moves the cut-in and cut-out *equally* higher or lower. Turning the range adjustment to a colder setting lowers *both* ***setpoints*** (cut-in and cut-out), so the system will maintain the lower temperature level. Turning the range adjustment to a warmer setting will cause the system to maintain the higher temperature level.

Control Temperature Ranges

Controls are designed to operate within a specific ***temperature range***. These temperature ranges are rather general but fall into these five areas of application:

- **High-temperature application**, between 55°F and 90°F (13°C and 32°C). Usually associated with room heating and air-conditioning systems.
- **Medium-temperature application**, between 32°F and 55°F (0°C and 13°C). Normally associated with refrigerators and other types of coolers.
- **Low-temperature application**, between –50°F and 5°F (–45°C and –15°C). Usually associated with freezers.
- **Ultralow-temperature application**, between –250°F and –50°F (–157°C and –45°C). Associated with industrial and commercial fast-freezing systems.
- **Cryogenic application**, between –460°F and –250°F, or absolute zero (–273°C and –157°C). Usually associated with laboratory and research systems.

Domestic Refrigeration Thermostats

As noted, the range adjustment on domestic thermostats is usually limited to a few degrees either way (warmer or colder). A domestic thermostat is designed to operate with the range adjustment knob close to mid-position. It is poor practice to operate any system with the range adjustment set at either minimum or maximum. The thermostat cannot operate properly at either extreme. Typically, turning the knob all the way to the right is the warmest position and turns the thermostat off (contacts constantly open). Turning the knob all the way to the left moves it to the coldest position and the thermostat cannot open (contacts constantly closed).

Wiring Connections

While thermostats are used to control the power supply to the condensing unit, they are

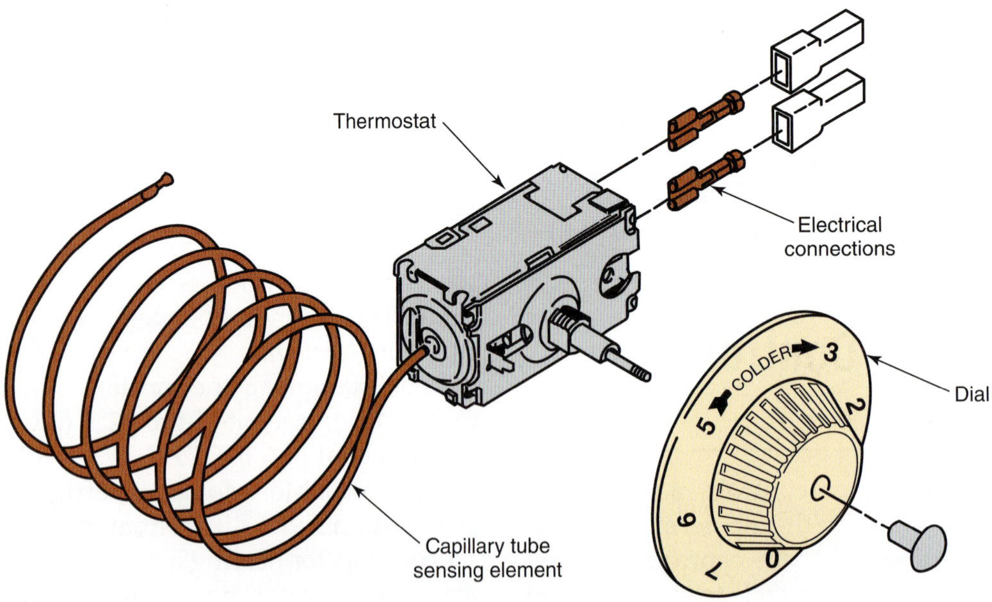

Ranco

Figure 26-5. Turning the dial on a domestic thermostat performs a range adjustment. The cut-in and cut-out temperatures become higher or lower, depending on the adjustment, but the differential does *not* change.

normally located at a point remote from the condensing unit. Connecting wires can be any length.

Thermostats normally have two terminals. Each is internally connected to one side of the switch. These terminals can be either push-on-type or screw-type. Wiring connections should be tight and well-insulated, since loose wiring connections are a primary cause of burned wires and component damage. Most thermostats include an extra (third) terminal, which is connected to the metal case and used for making the safety ground (green wire) connection.

For single-phase domestic systems, the thermostat switch is connected in series with one of the wires supplying power to the start relay. This permits the thermostat switch to disconnect the relay from the power supply. It is poor practice to connect the thermostat in the neutral wire. It makes troubleshooting more difficult and can cause safety problems.

Commercial Refrigeration Thermostats

Commercial thermostats are designed for a wide range of temperatures and have fully adjustable control settings, **Figure 26-6**. Two heavy-duty adjustment screws are located on top of the control for moving the pointers. A separate calibrated scale is provided for each pointer and is located directly below each screw. This permits the technician to control a wide temperature range, with the cut-in and cut-out easily adjusted to maintain desired temperatures.

Adjusting Controls

Commercial thermostat adjustments must be performed in proper order to achieve the desired results. When facing the control, the left screw changes the differential, while the right screw is the range adjustment. The range adjustment changes cut-in and cut-out equally. *Both* pointers will move when making range adjustments. The differential screw changes *only* the cut-out temperature. This screw moves only the pointer on the differential scale.

First, adjust the range screw until the desired cut-in temperature is indicated on the scale. Next, adjust the differential screw to obtain the desired differential setting. In a cooling application, the cut-out equals cut-in minus differential, **Figure 26-7**.

For example, consider a walk-in cooler that has an air-sensitive thermostat mounted on the inside wall to sense temperature of return airflow, **Figure 26-8**. Air temperature should be maintained between 32°F and 38°F (0°C and 3°C). The range adjustment screw is used to move the cut-in pointer to the 38°F mark on the range scale. The differential screw is then adjusted to move its pointer to the 6°F mark on the differential scale. With a differential of 6°F, the cut-out will be 32°F (0°C).

Some commercial motor controls have a cut-out screw and scale instead of a differential screw.

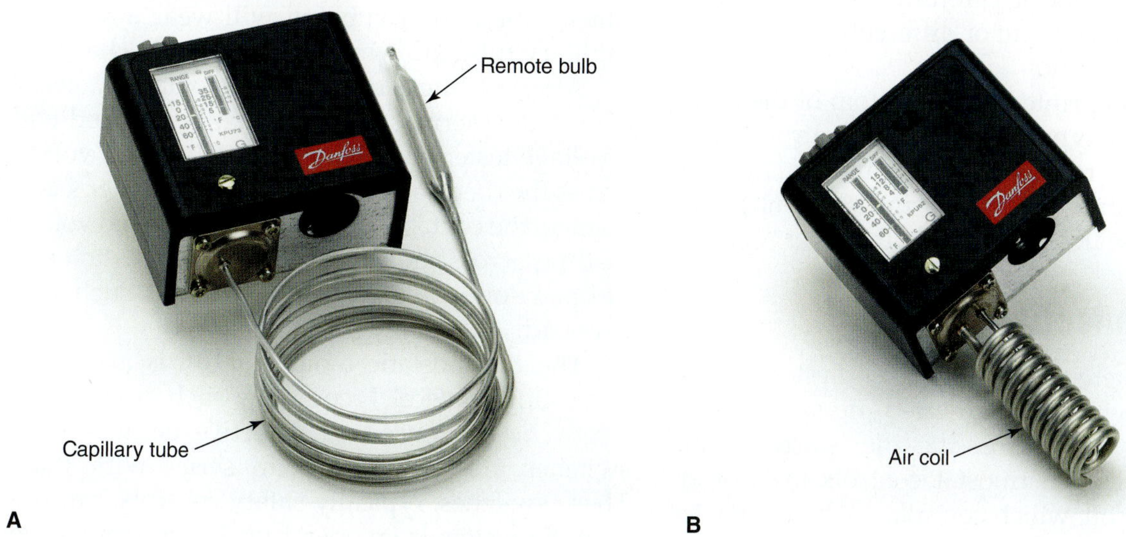

Danfoss

Figure 26-6. Typical commercial thermostats. Note the separate scales for setting cut-in temperature and differential. A—Thermostat with remote sensing bulb. B—Thermostat with an air coil for sensing.

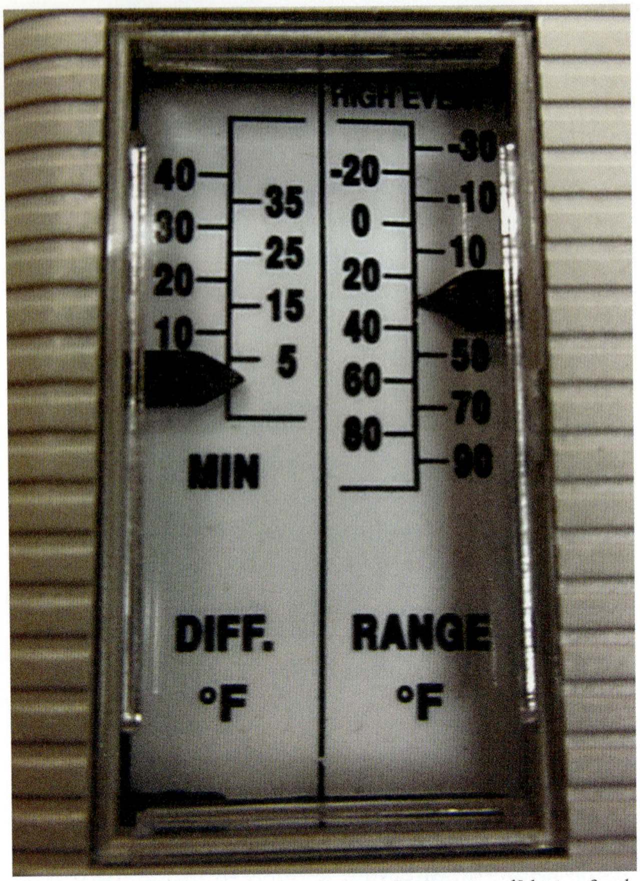

Image courtesy of Johnstone Supply

Figure 26-7. The range scale pointer indicates the cut-in temperature and is adjusted first. The differential pointer is then moved to set the desired cut-out temperature. On this thermostat shown, the cut-in temperature is 30°F, and the differential is 2°F, so the cut-out temperature is 28°F.

The cut-out screw directly establishes the cut-out point, according to the position of the pointer on the scale. Whenever dealing with a commercial thermostat, examine it carefully to determine if the left screw is a cut-out or differential screw. The settings are different.

Information printed at the bottom of the scale will identify the type of scale used. Controls using a differential screw will read "Cut-out is cut-in minus differential." Controls using a cut-out screw will label one scale "cut-out" and the other "cut in."

Wiring Connections

On commercial (three-phase) systems, the thermostat switch is connected in series with the contactor (or across-the-line starter) coil, **Figure 26-9**. The thermostat controls the power supply to the coil, which controls power supply to the compressor. The thermostat is located close to the area where the temperature is being controlled, due to limited length of the sensing element. Two wires connect the thermostat terminals to the contactor coil.

Disadvantage of Thermostatic Motor Controls

Since the thermostat operates according to the temperature of the sensing element, it may call for operation when the system is having problems. For example, a thermostat can require the system to operate when some or all refrigerant has been lost. If a leak occurs on the low-pressure side of the system, operation of the system will allow air and moisture to be drawn in and cause major problems. Relying strictly on thermostatic control may cause a system to self-destruct. Safety controls are often added to the system to keep it from operating if a problem develops.

Domestic systems, such as refrigerators, freezers, and window air conditioners, rely on the *compressor overload* to provide safety protection. However, prolonged cycling on the overload will result in compressor failure. Commercial installations normally include additional safety controls to protect the system. Such controls are explained later in this chapter.

Troubleshooting Thermostats

Thermostats are generally reliable. Most problems are related to improper settings, a sensing element that is not making good contact, or an element that has lost its refrigerant charge. Sometimes, vibration or rubbing will wear a hole in a capillary tube, allowing loss of the bulb charge.

After making certain that the contacts *should* be closed, the thermostat switch can be checked with a voltage tester or a multimeter. Check for voltage across the thermostat terminals. If the meter shows voltage, the contacts are open. If there is no voltage reading across the terminals, another switch may be open downstream of the thermostat, such as an overload.

The thermostatic motor control is a primary controlling device. Frequently, such controls are *incorrectly* diagnosed as defective because other switches are often wired in series with them. These switches, typically safety controls, can prevent the system from operating. For proper control by the thermostat, power supply should always be available at one (inlet) terminal of the thermostat.

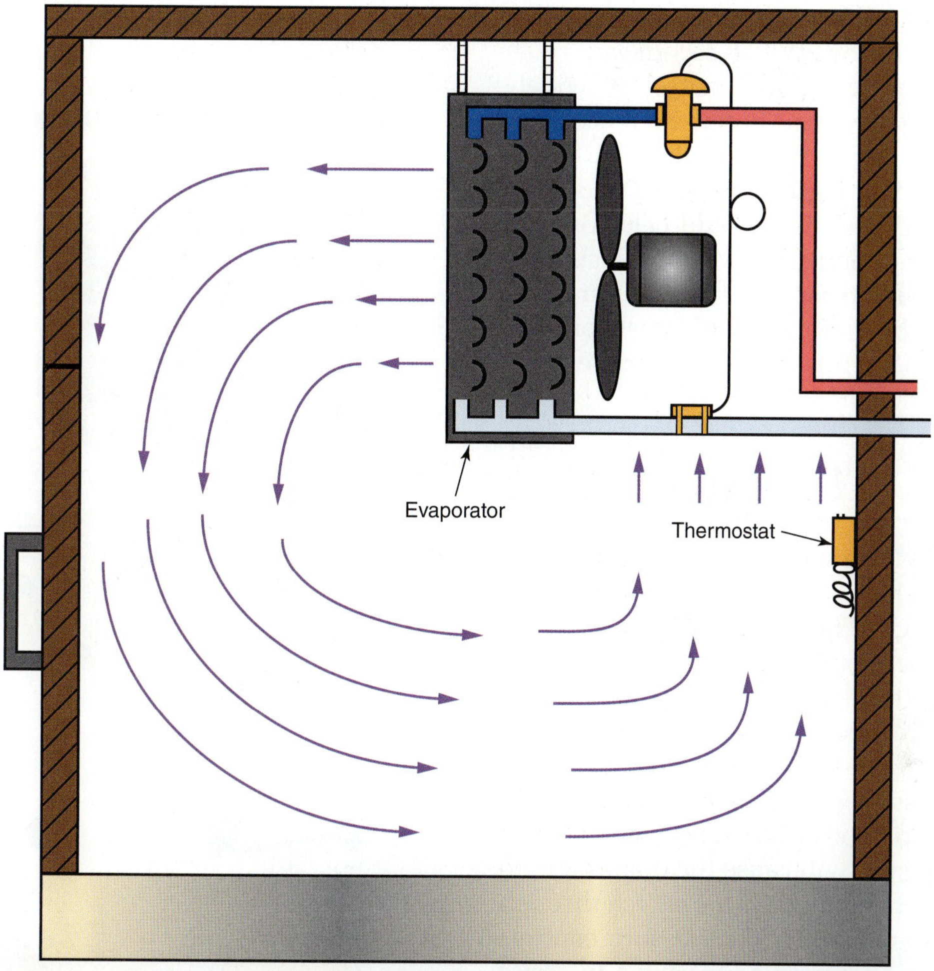

Goodheart-Willcox Publisher

Figure 26-8. The thermostat in this walk-in cooler is positioned to monitor the temperature of the return air. When properly adjusted, it will maintain the temperature inside the cooler in a range from 32°F to 38°F (0°C to 3°C).

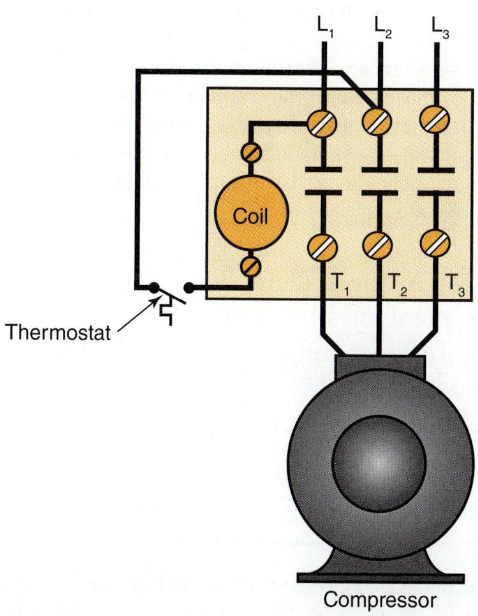

Goodheart-Willcox Publisher

Figure 26-9. The thermostat on a three-phase system is connected in series with the coil to control the power supply to the contactor.

However, sometimes an overriding control such as a defrost timeclock may prevent power from reaching the thermostat. Defrost systems and timeclocks are explained in Chapter 27, *Air and Defrost Cycles*.

26.2 Pressure-Type Motor Controls

Pressure-type motor controls operate using ***pressure-operated switches***. These are switches that respond to rising or falling system pressures. Like thermostats, pressure-operated switches are often used as the *primary control* of a system. They also may be used as a *secondary control* for safety to protect the system from high head pressure, loss of refrigerant, and low oil pressure. Troubleshooting these controls and making needed adjustments is an everyday task for the HVACR technician.

The pressure-type motor control is widely used on commercial systems because it is accurate, dependable, fully adjustable, and located at the condensing unit. See **Figure 26-10**. This location is convenient for the technician and helps prevent tampering with the control. However, adjustments are critical and vary from one system to another. Adjustments can be performed effectively only by a technician who is fully aware of the pressure-temperature relationship.

Principles of Operation

The pressure-type motor control includes an electrical switch that is opened or closed by pressure on a diaphragm, **Figure 26-11**. The diaphragm is a thin metal disc about 1 1/2″ in diameter, located in an assembly attached to the bottom of the control. A capillary tube connects the diaphragm assembly to the high- or low-pressure side of the system. This pressure connection is usually made on the compressor body. Either an increase or a reduction of pressure will cause the diaphragm to move. This movement operates the switch through a mechanical connection. Most pressure controls have two electrical terminals connected in series with the internal switch.

In terms of function, the pressure control is much like a thermostatic motor control. However, scales on thermostats are calibrated in degrees Fahrenheit (°F) or degrees Celsius (°C), while pressure controls are in pounds per square inch gauge (psig) or kilopascals (kPa). Product temperature can be maintained by controlling low-side pressures in the system. This is not difficult, but certain principles regarding low-side pressures must be fully understood. The technician must be able to use a *pressure-temperature (P/T) chart* to convert between temperature and pressure.

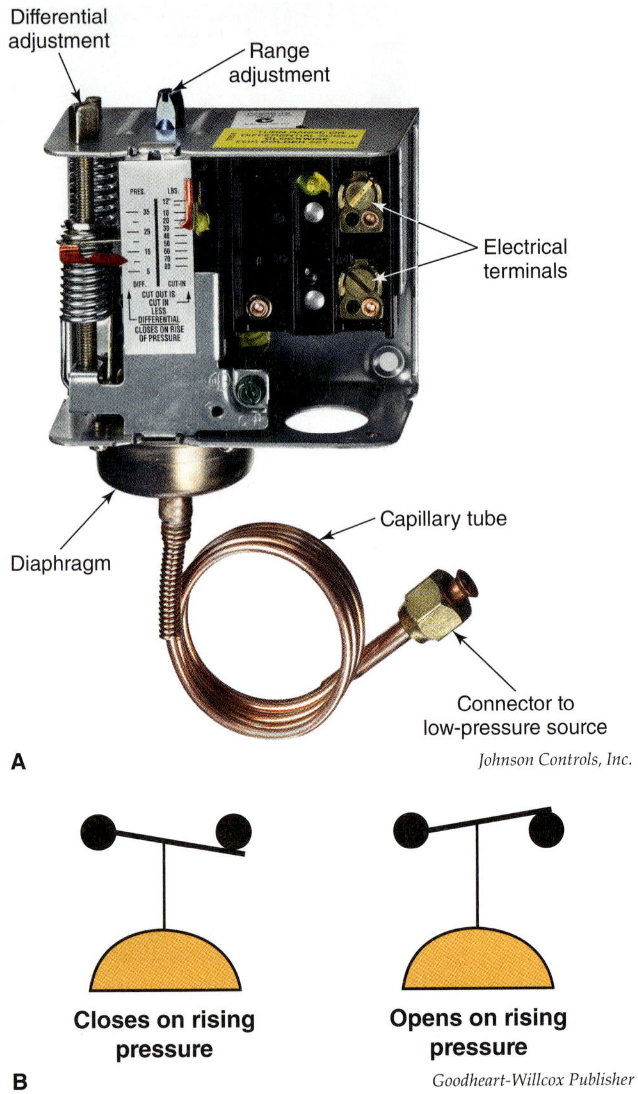

Johnson Controls, Inc.

Goodheart-Willcox Publisher

Figure 26-11. Pressure switch and symbols. A—Low-pressure switch with cover removed to show construction. The electrical terminals are in series with the internal switch that is opened or closed by movement of the diaphragm. This control permits setting of the cut-in pressure and a differential, as indicated on the scales. B—Schematic symbols for pressure switches. In refrigeration applications, thermostats that close a switch on a pressure rise are normally used.

Emerson Climate Technologies

Figure 26-10. A low-pressure control that opens or closes a switch when pressure causes movement of a diaphragm. This control has specific cut-in and cut-out settings, rather than using a differential to determine the cut-out pressure.

Low-Side Pressures

The sole purpose of any refrigeration system is to create a temperature difference so that heat can travel through the evaporator to be absorbed by the vaporizing refrigerant. This temperature difference can exist only when the system is running.

Once the system shuts off, the temperature differences quickly disappear. During the "off" cycle, refrigerant pressure (low-side pressure) equals the temperature of the air moving through the evaporator. As air temperature inside the cabinet increases, low-side pressure increases, as well.

During the "on" cycle, the operation of the system quickly establishes a temperature difference that permits heat to travel from the air to the evaporator, and then to the refrigerant inside the evaporator. The refrigerant is about 10°F (5.5°C) colder than the evaporator and the evaporator is about 10°F (5.5°C) colder than the air. This is called the ***Rule of 10***. Understanding the Rule of 10 can be helpful when troubleshooting a refrigeration system. As the temperature of the air becomes colder, the evaporator and refrigerant also become colder, thus maintaining the temperature difference during the run cycle.

Freezer Pressure Settings

As an example, consider a walk-in freezer using R-290 and a low-pressure motor control to maintain cabinet air temperatures that range from –10°F to –20°F (–23°C to –29°C).

Cut-In Point

The cut-in point is determined from the pressure-temperature (P/T) chart. No temperature difference exists during the off cycle. With the system off, cabinet air temperature slowly rises until air, evaporator tubing, and refrigerant are all –10°F (–23°C). At this temperature, the pressure for R-290 (as shown on a standard P/T chart) is 17 psig (117 kPa). To make this level the cut-in point, turn the range adjustment screw till it shows cut-in on the pressure switch scale at 17 psig (117 kPa).

At start-up, the air temperature is –10°F (–23°C), and the condensing unit quickly establishes the desired 20°F (11°C) air-to-refrigerant temperature difference: the evaporator being 10°F colder than the air and the refrigerant 10°F colder than the evaporator tubing. Therefore, the refrigerant temperature becomes –30°F (–34°C). A low-pressure gauge connected to the system would reflect this temperature difference by quickly dropping from the cut-in pressure of 17 psig (117 kPa) to 11 psig (76 kPa).

The system is designed to maintain the 20°F temperature difference during the run cycle. While the system is running, air temperature slowly drops from –10°F to –20°F (–23°C to –29°C). At the same time, low-side pressure slowly reduces from 17 psig to 11 psig (117 kPa to 76 kPa), and refrigerant temperature reduces from –30°F to –40°F (–34°C to –40°C).

Cut-Out Point

The cut-out point is determined by subtracting 20°F (11°C) from the desired cut-out *air* temperature. When the air temperature reaches –20°F (–29°C), refrigerant temperature is –40°F (–40°C), and low-side pressure is 1 psig (7 kPa). The cut-out pressure control is adjusted to read 1 psig. If the control has a *differential screw*, the differential is set at 16 psig (17 – 1 = 16). In metric terms, the differential would be 41 kPa (117 – 76 = 41).

Pressure Control Settings

A manifold gauge should be installed when adjusting pressure-type motor controls. Gauge readings are more accurate than the pointers and scales on the controls. To set the cut-in, do not wait for cabinet air temperature to reach ambient. Instead, use the manifold gauge to bleed pressure from the high side to the low-pressure side. This raises low-side pressure to the cut-in point rather quickly. This procedure is also used to double-check the cut-in setting.

The cut-out is quickly set by using the suction service valve (SSV) to restrict incoming suction gas. Frontseat the valve stem, then crack it open just enough to limit the incoming pressure at the desired cut-out point. Set the control (cut-out or differential) while the incoming low pressure is controlled. After the cut-out is firmly established and double-checked, backseat the suction service valve to remove the gauges and return the system to normal operation.

Cooler Pressure Settings

Consider a walk-in dairy cooler using R-290 and operating with cabinet air temperatures of 32°F to 39°F (0°C to 4°C). The system cut-in pressure is set for 60 psig (414 kPa), because no temperature difference exists when the system is off. The pressure-temperature for R-290 at 39°F is 60 psig (414 kPa).

Cut-out is determined by the temperature of the *refrigerant* when air temperature is 32°F (0°C). The refrigerant is about 20°F colder than air temperature, so the control would be set to cut out at 34 psig, which is the pressure equivalent of 12°F (–11°C). A differential screw would be set for 26 psig (60 – 34 = 26) or 179 kPa in metric terms, **Figure 26-12**.

Adjusting for Pressure Drop

When setting the cut-out on a pressure-type motor control, an allowance must be made for pressure drop through the suction line. The suction is strongest at the compressor when the piston makes its downstroke. Suction line pressure at the compressor is about 2 psig (14 kPa) *lower* than evaporator pressure.

Proper sizing of the suction line limits pressure drop to 2 psig. Pressure drop may be slightly higher if the suction line is long or contains many elbows, valves, and risers. A ***riser*** is a vertical section of tubing.

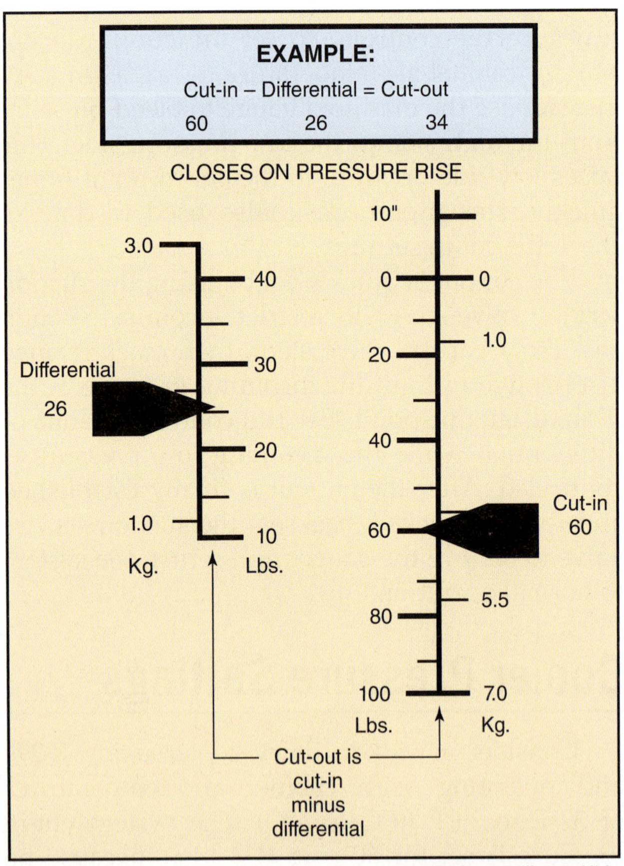

Goodheart-Willcox Publisher

Figure 26-12. Pressure switch control settings for a walk-in cooler using R-290. With a cut-in of 60 psig and a differential of 26 psig, the cut-out is 34 psig.

To allow for pressure drop, subtract 2 psig from the desired cut-out value just before setting the control. In the freezer example, cut-out was determined to be 1 psig (7 kPa). Thus, the cut-out setting on the motor control must be 1 psig (2 – 1 = 1) to compensate for pressure drop. In the dairy cooler example, the cut-out setting would be reduced from 34 psig (234 kPa) to 30 psig (207 kPa). Pressure drop affects only cut-out, because pressure drop disappears during the off cycle. Pressure drop cannot occur unless the compressor is running.

Allowing for Temperature Differences

The temperature difference of 20°F (11°C) between the air and refrigerant is a good guide. Some evaporators are more efficient than others and this temperature difference could be slightly more, or slightly less. After setting cut-in and cut-out on the pressure motor control, it is good practice to place an accurate thermometer inside the cabinet to check air temperature at both setpoints. Minor adjustments to the cut-out may be necessary to compensate for excess pressure drop or temperature difference through the evaporator.

Single-Phase Wiring Connections

The pressure-type motor control, as noted earlier, is a primary control. Its switch is used to control power supply to the condensing unit. These controls usually have two screw-type terminals, which are internally connected to the two sides of the switch. The wiring connections are made exactly like those for a thermostatic motor control, but the wires will be much shorter because a pressure control is always located close to the compressor. Wiring connections are usually performed inside the compressor terminal box, or inside the electric panel mounted on the condensing unit.

Three-Phase Wiring Connections

The ***low-pressure motor control*** is the preferred choice on commercial systems and is used to control power supply to the contactor coil, **Figure 26-13**. Both the contactor and the pressure motor control are located at the condensing unit, which makes wiring connections and troubleshooting easier. The pressure motor control switch is connected in series with one of the wires supplying power to the coil. The low-pressure motor control monitors suction (low-side) pressure at the compressor. When suction pressure at the compressor reaches the cut-out setting, the low-pressure motor control turns off the compressor. The cut-out pressure setting should correspond to the desired evaporator temperature minus suction line pressure drop.

Troubleshooting Pressure Motor Controls

Troubleshooting procedures for pressure controls are similar to those used with thermostats. The control is simply an electrical switch that operates according to low-side pressure at the compressor. The low-side pressures must be adequate for the control to operate properly.

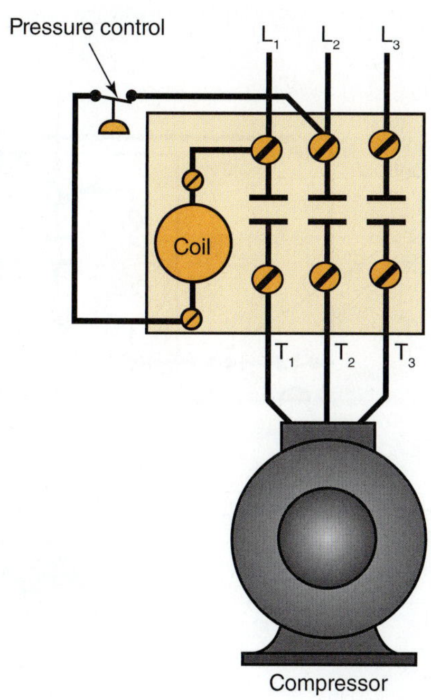

Goodheart-Willcox Publisher

Figure 26-13. The pressure control on a three-phase system is connected in series with the coil to control the power supply to the contactor.

System Low on Refrigerant

If the system is low on refrigerant, the low-pressure control will cause the system to short-cycle. The control switch opens and closes too quickly because the compressor is drawing gas out of the suction line faster than the expansion valve can feed liquid into the evaporator. Low-side pressure quickly drops to the cut-out point, the switch opens, then low-side pressure quickly rises to the cut-in point. Resetting the control does *not* solve the problem. Assuming that the system was initially commissioned with a proper charge of refrigerant, a likely cause of a low charge is a refrigerant leak. The leak must be located and repaired, and the system again charged to full capacity.

No Refrigerant in System

If a system loses all its refrigerant, the control contacts remain open. Low-side pressure cannot build up sufficiently to close the contacts and start the compressor's motor. When the control contacts are open, the voltage tester will reveal power supply (voltage) only at the inlet terminal.

Shutting off when the system is low on refrigerant is an advantage. If a leak occurs on the low-pressure side of the system, the control will shut the system off before it reaches a vacuum state, which would otherwise allow contaminants (such as air and moisture) to enter.

Defective Pressure Control

To properly diagnose a defective pressure motor control, the manifold gauge must be installed. While pressure controls are usually reliable, they sometimes become ***erratic*** (irregular) in operation. A control that is unable to maintain pressure settings must be replaced. This condition is usually caused by abuse to the control.

A common pressure control problem is a leak in the capillary tube that links the control diaphragm to the compressor. This leak is caused by vibration or by rubbing of the capillary tube against another metal object. Excessive vibration can cause the copper capillary tube to become work-hardened and crack or break off. Rubbing, especially against a steel object, will quickly wear a hole in the capillary tube. This occurs because steel is much harder than copper. The entire system charge can be lost by such a leak.

26.3 Low-Pressure Safety Control

The low-pressure safety control is used as a secondary control to shut off a system if the low-side pressure is reduced to unsafe levels. The low-pressure safety control is vital when the system uses a thermostat for temperature control. The thermostat is the primary control and the low-pressure safety control becomes a secondary (safety) control.

As noted, the low-pressure safety control will turn off the system when low-side pressure becomes too low. This is important because most compressors rely upon the suction gas to cool the motor windings. If this cool gas is not available, the compressor's motor will overheat and cycle on the overload. The low-pressure safety control is often used to protect the system against loss of refrigerant and protect air-conditioning systems from suffering an evaporator freeze-up.

Commercial low-pressure *safety* controls are similar to the *primary* low-pressure control. In fact, the same model control can often be used for either purpose, because they are selected according to the range and differential scales. Therefore, the same control can serve different purposes. **Figure 26-14** illustrates some typical scales.

The low-pressure safety control may have a 36″ or 48″ capillary tube with a flare nut for connection to the low-pressure side of the system. This connection is normally made at the compressor. The control contacts are designed to close on rising pressure and open once pressure falls to a certain point.

Wiring Connections

The contacts for the low-pressure safety control are connected in series with the thermostat contacts, **Figure 26-15**. The thermostat cycles the unit under normal operating conditions, with the low-pressure safety control acting as a back-up. The contacts on both switches must be closed to run the system, but either switch can open and stop operation.

Low-Pressure Controls

Range Scale	Differential Scale
12 in. Hg to 50 psig	5 to 35 psig
10 in. Hg to 100 psig	10 to 40 psig
50 in. Hg to 150 psig	10 to 40 psig

Goodheart-Willcox Publisher

Figure 26-14. This table shows that low-pressure controls are selected according to the range and differential scales available on the control. The same pressure control can often be used as either a primary or secondary control and with different refrigerants.

Pressure Settings

The low-pressure safety control must not be permitted to interfere with normal operation of the system, yet it must provide necessary protection. The control settings vary according to each system and refrigerant used. Pressure settings must protect the unit against loss of refrigerant, while the differential must be sufficient to prevent short-cycling.

When the system becomes low on refrigerant, the compressor quickly reduces low-side pressure. When the compressor shuts off, low-side pressure increases quickly, due to higher temperature at the evaporator. A close differential would permit the system to short-cycle during loss of refrigerant. The safety control is set to cut out just before the low-side pressure enters a vacuum. A wide differential places the cut-in at a point requiring more time for the low-side pressure to increase before

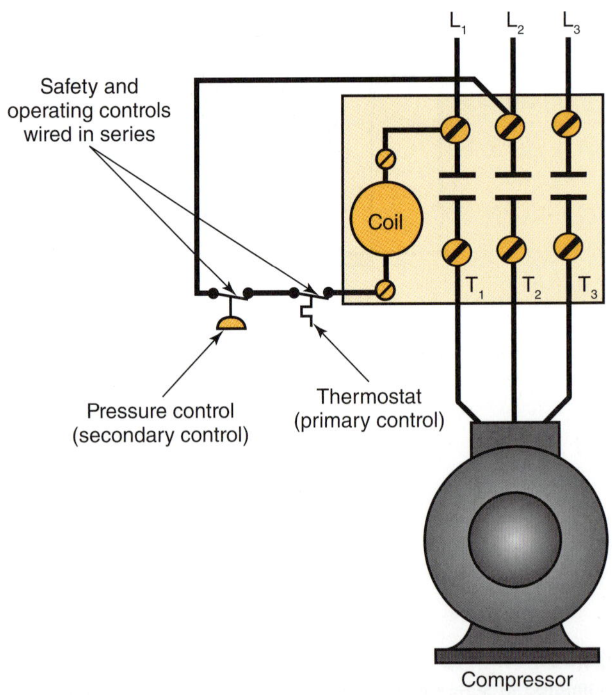

Goodheart-Willcox Publisher

Figure 26-15. In a thermostatically controlled system with a low-pressure safety control, the two controls are wired in series with the contactor coils. Both switches must be closed to operate the system. If either opens, the system will stop operating.

reaching the cut-in point. When trying to calculate the settings on a low-pressure safety control, it is best to think backward (from cut-out to cut-in).

For example, consider a walk-in cooler using a thermostat to control air temperatures between 32°F and 37°F (0°C and 3°C). The refrigerant used is R-290. If a leak occurs, the low-side pressure must be prevented from entering a vacuum. The low-pressure safety control cut-out should be established at a point just above a vacuum (between 2 psig and 5 psig, or 14 kPa and 34 kPa). The differential must be about 20–25 psig (138–172 kPa) to prevent short-cycling, so the cut-in should be set at about 30 psig (207 kPa). After calculating the proper setpoints, the control is adjusted by first setting the cut-in, then the differential.

Control settings of 25 psig (cut-in) and 3 psig (cut-out) do not interfere with normal operation of the system. The safety cut-in pressure is below the cut-in temperature of the thermostat, and the safety cut-out pressure is below the cut-out temperature. The corresponding low-side pressures for the temperatures at which the thermostat opens and closes are 34 psig (234 kPa) for cut-in and 14 psig (97 kPa) for cut-out.

Troubleshooting

Problems with the low-pressure safety control are rare. If the switch is bad, the control must be replaced. Most often, however, the problem is a capillary tube that is leaking due to vibration or rubbing. Corrective procedures for leaking capillary tubes are described earlier. The low-pressure safety control serves as a warning device that pressure is below safe limits. The actual problem is usually elsewhere, *not* with the control.

Some problems that will cause the system to have low pressure are:

- Low refrigerant level
- Evaporator freeze-up
- Evaporator fan not running
- Restricted airflow due to dirty return air filters
- Dirty evaporator due to missing air filters
- Moisture in the system, freezing at refrigerant control
- Faulty superheat setting on TEV
- Restriction at filter-drier
- Low head pressure due to low ambient temperature

26.4 Head-Pressure Safety Control

The ***head-pressure safety control*** is a device that will shut off a refrigeration system if head pressure exceeds proper limits. High head pressure results in poor system performance and damage to the compressor. High head pressure is common, and the safety control protects the system when it occurs.

The commercial high head-pressure safety control looks much like a low-pressure safety control, **Figure 26-16**. The head-pressure safety control is designed for the contacts to open on *rising pressure*. The capillary tube is connected to the high-pressure side at the compressor. Depending on the range scale, the same control can be used for R-134a, R-22, or R-507a. See **Figure 26-17**.

The head-pressure safety control is available with either automatic reset or manual reset. In automatic reset, the control contacts automatically close when head pressure drops to the cut-in setpoint. This means that the control turns the compressor on and off whenever head pressure

Image courtesy of Johnstone Supply

Figure 26-16. A head pressure safety control with a manual reset button. Note the scale shows only cut-out pressure.

Head-Pressure Safety Controls

Range Scale	Differential Scale
100 to 250 psig	20 to 100 psig
100 to 400 psig	35 to 150 psig
150 to 450 psig	35 to 150 psig

Goodheart-Willcox Publisher

Figure 26-17. This table displays the ranges and differentials available in head-pressure safety controls, which make it possible to use the same control with different refrigerants.

changes between unsafe and safe limits. If the condenser fan burns out, the automatic reset control permits the system to cycle on and off.

The automatic reset is common for commercial applications because it eliminates temporary nuisance shutdowns. These shutdowns usually result from high-heat-load conditions of brief duration. These result in temporary high suction pressure and high head pressure. Head pressure returns to normal when suction pressure is reduced to normal.

The manual reset is preferred by compressor manufacturers because the control must be manually reset to restore the system to operation. It is assumed that the person resetting the control will investigate and cure the problem that is causing the high head pressure. The manual reset control is often used for air-conditioning applications.

Head-Pressure Safety Control Settings

To calculate the safety pressure settings for a forced convection (fan-type) condenser, the technician must first know the normal head pressure during the hottest day of the year. For example, on a hot summer day, the ambient may reach 100°F (38°C). The normal head pressure is calculated by adding 30°F to 35°F (17°C to 19°C) to ambient. The lower temperature would be used for a clean condenser. The higher value is for a slightly dirty condenser. Use a pressure-temperature chart to convert the temperature to pressure. For example, on an R-134a system:

100°F + 30°F= 130°F (199 psig) (clean condenser)

100°F + 35°F= 135°F (214 psig) (slightly dirty condenser)

In this case, the head-pressure safety control is set to cut out at a pressure of 235–245 psig (1620–1689 kPa). This extra margin of safety protects the system and prevents nuisance trip-outs. The differential is set for 55–60 psig (379–414 kPa) to permit the system to cool down before automatic reset.

When the condenser becomes very dirty or if the condenser fan stops running, head pressure will rise rapidly to the cut-out. If the head-pressure safety control is *not* used, head pressure will continue to rise until the overload stops the compressor. The compressor will cycle on the overload until high head pressure is no longer a problem.

Wiring Connections

Wiring connections for the head-pressure safety control place the switch in series with the primary control, **Figure 26-18**. Safety control switches are usually connected into the same circuit as primary control, but act as a secondary control. Sometimes the head-pressure safety control is connected in series with the compressor overload. This method shuts the compressor off but permits the condenser fan to continue to operate. The two methods are equally effective for protecting the compressor.

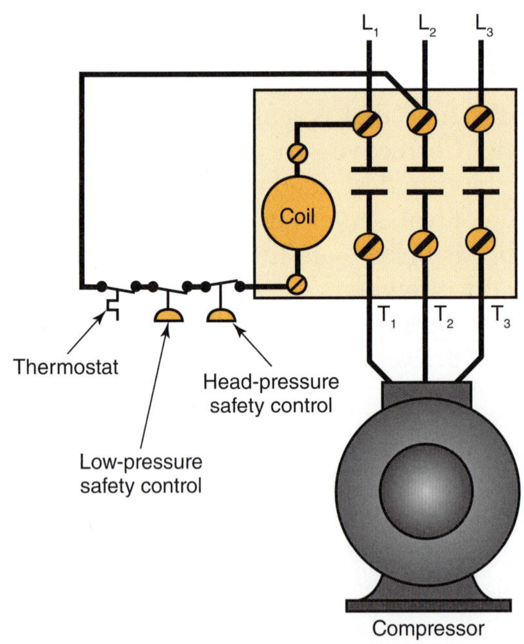

Goodheart-Willcox Publisher

Figure 26-18. The primary control (in this case, a thermostat) and secondary safety controls are usually wired in series with the contactor coil. If any switch opens, the power supply to the contactor coil will be cut off.

Dual-Pressure Motor Control

The ***dual-pressure motor control*** is a combination of a low-pressure control and a head-pressure (high-pressure) safety control, **Figure 26-19**. Typically, commercial systems used to have both a low-pressure control (either primary or safety) and a head-pressure safety control. Manufacturers recognized this and combined the two controls into a single unit.

The dual-pressure control saves time and money because double protection is provided with one control. Wiring connections are simplified because the control has only two terminals. The factory makes the internal connections that permit either switch to disconnect the power supply.

Each side of the control operates independently, with its own adjustment scale. If one side of the control becomes nonfunctional, the other side will remain operational. The low-pressure switch opens on falling pressure and the high-pressure switch opens on rising pressure.

The two capillary tubes must be properly connected to the compressor. The smaller bellows is connected to the high-pressure side, and the larger bellows is connected to the low-pressure side. Both sides of the control are fully adjustable by screws that are located directly above each scale.

Some head-pressure controls have only a single *range adjustment* screw, since the differential is factory fixed at 55 psig or 60 psig (379 kPa or 413 kPa). The proper cut-out pressure is selected by the range adjustment screw, while the cut-in is established by the fixed differential. This dual-pressure control is available with a manual reset for the high-pressure side.

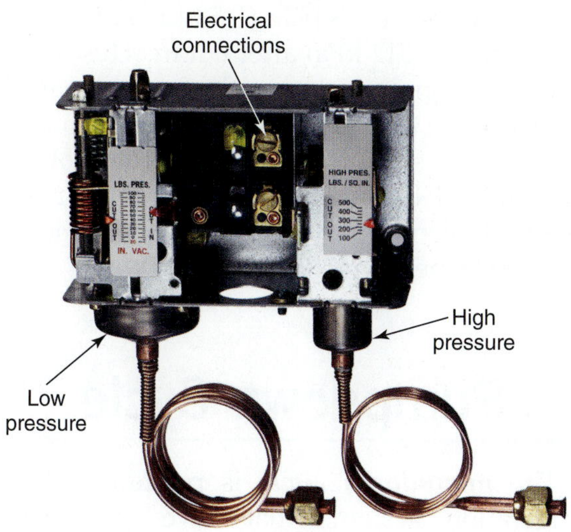

Johnson Controls, Inc.

Figure 26-19. A dual-pressure safety control for both high-pressure and low-pressure protection. On the high-pressure side, the differential is set by the factory, so only a cut-out pressure is set on the scale. The low-pressure control allows setting of both range and differential.

Condenser Fan Cycle Control

The ***fan cycle control*** is used to control the condenser fan or fans. When head pressure drops below acceptable limits, the fans are turned off. However, when pressure returns to normal, the control turns the fans back on. By turning off the condenser fan, heat from the high-side refrigerant will not dissipate as quickly. As the heat builds up, head pressure will correspondingly rise back to a more acceptable level. Remember that in a fixed volume (such as a condenser), temperature and pressure levels will correspondingly rise and fall.

Condenser fan cycling is a common method of maintaining head pressure during cold weather conditions. When the condensing unit is located outdoors, low ambient temperatures will cause low head pressure. Low head pressure results in poor system operation.

TEV Pressure Drop

Proper operation of the thermostatic expansion valve (TEV) depends on proper pressure drop across the expansion valve. This pressure drop is determined by subtracting normal suction pressure from normal head pressure. This pressure drop must be maintained else the TEV cannot feed refrigerant properly. When adequate pressure drop is not maintained, the compressor will remove refrigerant from the evaporator faster than the TEV can feed liquid into the evaporator. Thus, the evaporator pressure becomes too low, and the unit cycles off on the low-pressure control.

Head Pressure Limits

Head pressure for a system using R-134a should not fall below 100 psig (690 kPa). The minimum head pressure for R-22 and R-507a is 200 psig (1380 kPa). Head pressures should be maintained at or above these levels to assure proper system operation.

Control Contacts

The fan cycle control helps maintain proper head pressure by disconnecting the condenser fan when head pressures fall below an acceptable level. The fan remains off until the head pressure builds up to the cut-in point again. During low ambient, the condenser fan may not run or may cycle on and off. The contacts on this control must open on falling pressure and close on rising pressure, **Figure 26-20**.

Control Connections

The fan cycle control is located at the condensing unit and connected in series with one side of the fan's power supply. Sometimes two condenser fans are controlled by the same switch.

Large condensing units that use multiple fans will ***stage*** (operate in sequence) individual fans to cycle on or off at different pressures. These controls are set for a 10 psig (69 kPa) difference. One fan is cycled off at a fixed setpoint. If the pressure continues to drop, another fan turns off. This staging continues until all fans are off, or until head pressure stabilizes. Control settings also turn fans on as head pressure rises.

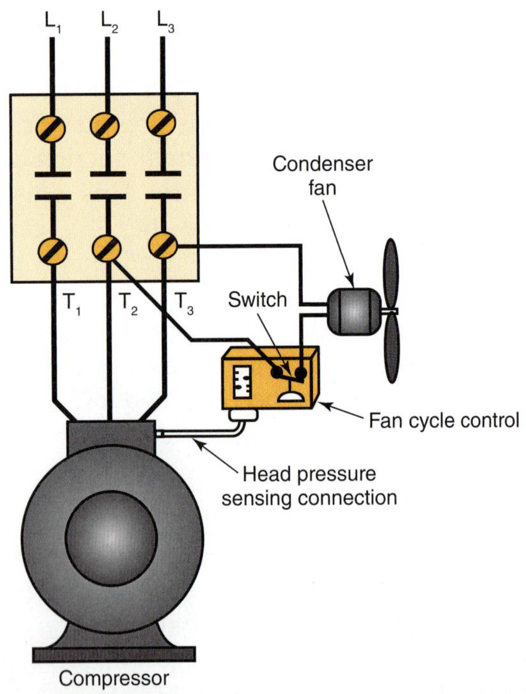

Goodheart-Willcox Publisher

Figure 26-20. The fan cycle control helps maintain proper head pressure by turning the condenser fan off and on as necessary.

Control Settings

Since systems using different refrigerants operate at different suction and head pressure levels, their fan cycle control settings also vary. The following sections discuss a few common refrigerants and their fan cycle control settings.

R-134a Systems

- Cut-in = 140 psig (966 kPa)
- Cut-out = 100 psig (690 kPa)

The 40 psig (276 kPa) differential keeps the fan from short cycling during periods of moderate ambient conditions. A delay exists at each setpoint while the pressure stabilizes. When the switch cuts out at 100 psig, head pressure drops to about 90 psig (621 kPa) because the fan requires time to stop. When head pressure rises to the 140 psig cut-in, time is needed for the fan to achieve full speed, so head pressure rises to about 150 psig (1035 kPa). Therefore, considering these delays, the actual differential is about 60 psig (414 kPa).

R-290 Systems

- Cut-in = 210 psig (1448 kPa)
- Cut-out = 178 psig (1227 kPa)

The 32 psig (220 kPa) differential permits head pressure to reach 168 psig (1158 kPa) after cut-out and 220 psig (1517 kPa) after cut-in. This differential prevents short-cycling of the fan and maintains head pressure within proper limits. The condenser fan must not short-cycle, because the motor will overheat and burn out. However, the fan needs to provide compressor cooling during moderate ambient temperatures.

26.5 Pumpdown Cycle

The ***pumpdown cycle*** is popular for many refrigeration systems that have the condensing unit located outdoors. The pumpdown cycle removes all refrigerant from the low-pressure side of the system prior to shutdown. This prevents liquid slugging and loss of oil from the compressor during start-up.

Principle of Operation

The pumpdown cycle uses a solenoid valve (electrically operated valve) in the liquid line and a thermostat controlling power supply to the solenoid coil, Figure 26-21. The solenoid valve can be located just before the TEV (evaporator location) or just after the sight glass (condenser location).

Power supply to the solenoid and thermostat is separate from the refrigeration system power supply. The air-sensitive thermostat, located inside the cooled space, controls power to the solenoid coil only. The solenoid valve controls refrigerant flow through the liquid line.

When cooling is required, the thermostat contacts close and energize the solenoid coil. The solenoid valve opens and permits refrigerant flow to the evaporator. The suction pressure rises, and a low-pressure control starts the compressor. When the thermostat is satisfied and opens, the solenoid coil is de-energized, allowing the valve to close. Refrigerant flow to the evaporator stops. The compressor continues to pump refrigerant from the evaporator and suction line until the suction pressure reaches the low-pressure control cut-out setting, Figure 26-22.

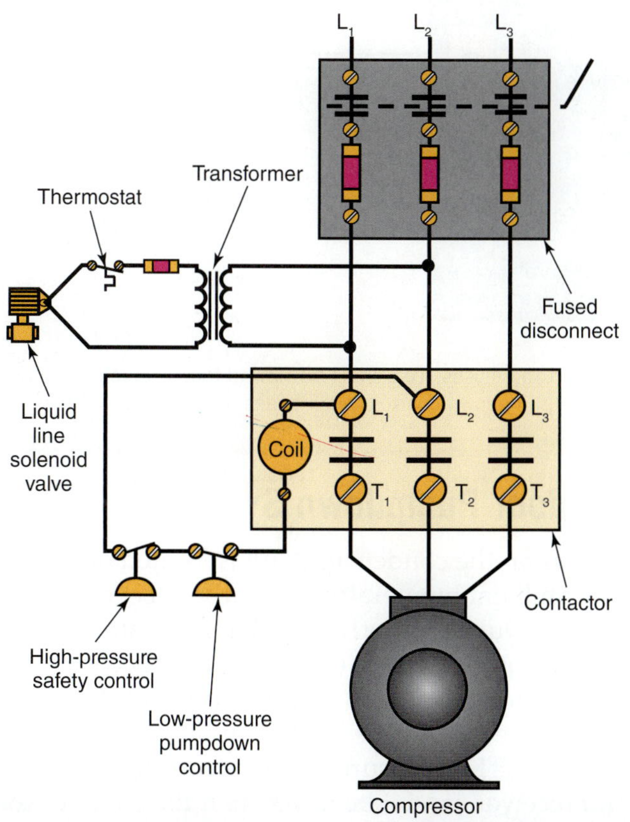

Goodheart-Willcox Publisher

Figure 26-21. The solenoid valve and the thermostat that controls it have a power supply isolated from the refrigeration system's power supply by a transformer.

In the pumpdown cycle, the compressor evacuates (removes) all refrigerant from the low-pressure side of the system. The refrigerant is stored in the liquid receiver (and condenser) after each run cycle. Since the liquid receiver and condenser are fully capable of storing this refrigerant, head pressure does not rise.

Advantages of the pumpdown cycle include:

- Preventing liquid migration to the compressor during the off cycle
- Eliminating high suction pressure at start-up
- Providing better oil return flow to the compressor crankcase
- Preventing refrigerant vapor from condensing in the crankcase during cold ambient

Pumpdown Pressure Control Settings

Pressure control settings for pumpdown are important and vary according to the system. At pumpdown, the cut-out should occur before suction pressure enters a vacuum (about 1–2 psig, or 7–14 kPa). The cut-in pressure setting is more difficult to determine. The cut-in depends upon whether the condensing unit is located indoors or outdoors.

Indoor Condensing Unit

For indoor condensing units, cut-in is determined from the lowest air temperature (thermostat cut-out). Subtract about 10°F (5°C) from the thermostat cut-in temperature and convert the result to pressure, using a pressure-temperature chart. This pressure becomes the cut-in on the pressure control.

For example, if the thermostat is set to control temperatures at 32°F to 38°F (0°C to 3°C), subtract 10°F from 32°F (thermostat cut-out). The result, 22°F (–6°C), is converted to pressure (for R-134a, 20 psig or 138 kPa). This becomes the cut-in for the pressure control.

Indoor Pumpdown System Operation

When air temperature reaches 38°F (3°C), the thermostat closes and energizes the solenoid coil.

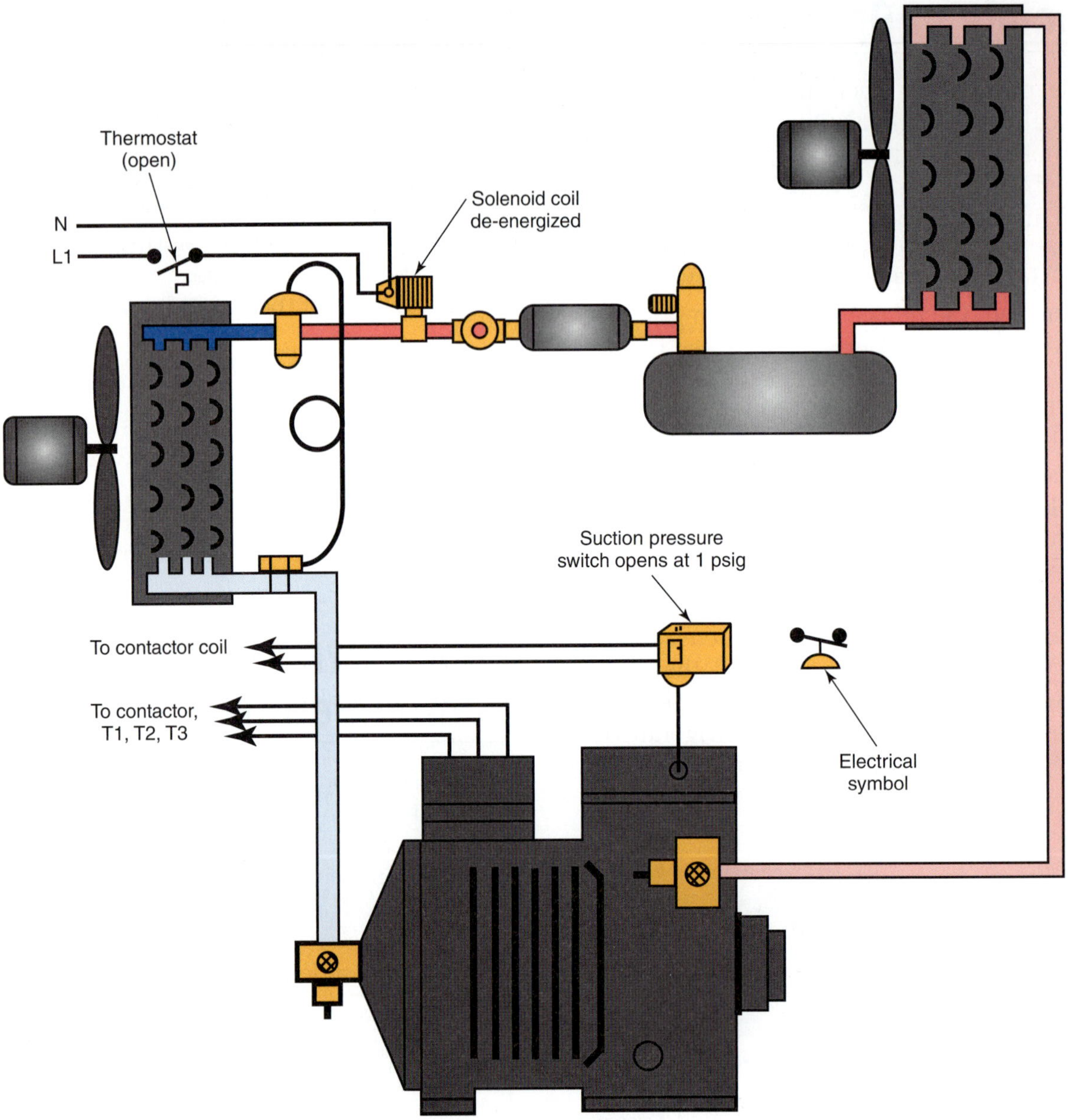

Goodheart-Willcox Publisher

Figure 26-22. When the thermostat opens, the solenoid valve is de-energized and closes. This shuts off flow to the evaporator through the liquid line. The compressor will continue to operate until the low-pressure pumpdown control switch opens. Refrigerant is stored in the condenser and liquid receiver.

Refrigerant flows to the evaporator. Low-side pressure quickly rises to 20 psig (138 kPa), and the low-pressure control closes. This starts the condensing unit and the system operates for a normal run cycle.

When air temperature is lowered to 32°F (0°C), the thermostat opens and de-energizes the liquid line solenoid. Refrigerant flow to the evaporator stops. The condensing unit continues to operate (pumpdown). Low-side pressure is quickly lowered to 1 psig or 2 psig, when the pressure control opens to stop the condensing unit.

Outdoor Pumpdown System

When the condensing unit is located outdoors, low ambient temperatures greatly affect the cut-in setting. During the off cycle, all refrigerant is stored in the liquid receiver (outdoors) and becomes the same temperature as outdoor ambient. Cut-in is determined from the coldest anticipated outdoor ambient. The pressure-temperature in a cold liquid receiver cannot be lower than the cut-in of the pressure control.

To overcome cold ambient problems, the cut-in setpoint *must* be equal to the coldest likely outdoor temperature. Failure to abide by this rule will result in repeat service calls when outdoor temperatures drop to low levels. The thermostat will energize the solenoid valve, but the compressor will not run because *refrigerant pressure* is lower than the control cut-in. **Figure 26-23** is a table of typical pressure control settings when outdoor ambient is below system operating temperatures.

Outdoor Pumpdown Cut-Out

The pressure control cut-out normally occurs before the system enters a vacuum. However, this is not always possible when outdoor ambient is low. The technician must adjust the control for cold ambient conditions. As shown in **Figure 26-23**, pumpdown systems using R-134a will frequently pull into a vacuum.

Pumpdown Differential

The differential for pumpdown cycles must prevent short-cycling after the initial cut-out. When the solenoid valve closes, the compressor rapidly removes refrigerant and reaches the cut-out point. Some gas remains in the low-pressure side. After shutdown, this gas residue causes low-side pressure to rise to the cut-in setpoint. The compressor cycles briefly until the low-side pressure is reduced to the cut-out. One brief recycle is not unusual and serves to remove residual refrigerant. Several short cycles before complete shutdown indicate a short differential.

Troubleshooting

A compressor recycling back on after shutdown may result from the following causes:

- A system that is low on refrigerant
- A differential set too close
- A leaking solenoid valve seat
- A leaking compressor discharge valve

These problems can be isolated, identified, and corrected using the following procedures.

Pumpdown System Low on Refrigerant

Install a manifold gauge on the compressor service valves and remove cover from the low-pressure control. Check control operation with a multimeter reading voltage or by observing the mechanical action of the control as it opens and closes.

Check operating pressures and the sight glass. The sight glass may show some bubbles immediately after start-up, but the bubbles should quickly disappear and the glass should indicate a full system. Excessive and continuous bubbles in the sight glass (even when the compressor is off), along with low pressures, indicate the system is low on refrigerant. A low charge often points to a system leak. The leak must be located and repaired, and the system recharged to full capacity.

Check the magnetism of the solenoid coil with a small screwdriver to determine if the solenoid is

Pressure Control Settings for Cold Ambient

Coldest Ambient	R-134a Cut-in	R-134a Cut-out	R-22 Cut-in	R-22 Cut-out	R-507 Cut-in	R-507 Cut-out
+30	22	1	40	1	45	6
+20	15	1	30	1	45	6
+10	10	0	25	1	40	6
0	4	5″	20	0	30	5
–10	1	8″	15	0	25	5
–20	5″	15″	8	0	18	5
–30	9″	15″	4	6″	14	0

Goodheart-Willcox Publisher

Figure 26-23. Table of pressure control settings for cold ambient. Ambient temperatures are in °F and cut-in/cut-out pressures in psig, except for vacuum readings, which are marked with ″ for in. Hg.

energized. If the solenoid is not energized, check power supply to the thermostat and determine if thermostat contacts are open or closed. Turning the thermostat to a lower temperature setting should energize the solenoid valve.

Differential Set Too Close

Remove the cover from the low-pressure control and check the scale settings against pressures observed on the manifold gauge. The control scale can be misleading or inaccurate, so adjustments should always be performed using manifold gauge measurements. Inaccurate or improper settings should be corrected to provide proper system operation.

Leaking Compressor Discharge Valve Reeds

When working on semihermetic compressors, the compressor discharge valve reeds that are leaking or broken are easily discovered by following this procedure:

1. Install a manifold gauge on the compressor service valves.
2. Frontseat the suction service valve to prevent any refrigerant from entering the compressor.
3. Force the compressor to run by using a jumper cord to bypass the low-pressure control.
4. Allow compressor to pull a good vacuum (20 in. Hg to 25 in. Hg).
5. Stop the compressor and observe the low-side gauge for rapid loss of vacuum. If it occurs, replace the compressor's valve plate.

Compressor discharge valve reeds generally show slight leakage, but a *rapid* loss of vacuum indicates faulty reeds. The compressor discharge valve reeds are a division point between the high-pressure and low-pressure sides of the system. If they are broken or are not seating properly, they will permit high pressure to leak back into the low-pressure side. This problem can be repaired in the field by a service technician using a valve plate replacement kit, which contains new valve reeds.

Leaking Pumpdown Solenoid Valve

A leaking pumpdown solenoid valve is rare. It is indicated by regular short-cycling of the compressor after pumpdown due to refrigerant leaking into the low-pressure side. The refrigerant leakage may result from leaking compressor discharge valve reeds or a leaking solenoid valve seat.

The problem is isolated by first checking the discharge valve reeds, as previously described. If discharge valve reeds are functioning properly (no loss of vacuum), the problem *must* be the solenoid valve.

Sometimes a Schrader valve is available in the suction line for checking suction pressures. If so, the solenoid valve seat can be checked by using this procedure:

1. Connect a compound gauge to the Schrader valve.
2. Force the compressor to pull a good vacuum, then frontseat the suction service valve. Stop the compressor. The suction service valve will prevent return pressure from the compressor.
3. Observe the compound gauge. It will reveal if the low-side pressure remains in a vacuum or rises to the cut-in point. If pressure rises, the solenoid valve seat is leaking. The defective valve must be replaced.

26.6 Oil Pressure Safety Control

The ***oil pressure safety control*** is used to protect the compressor when lubrication problems occur. Many compressor failures caused by lack of lubrication could have been prevented by using an oil pressure safety control. Warranties on compressors using an oil pump specify the unit *must* be equipped with an oil pressure safety control.

Loss of compressor lubrication can be caused by a variety of system problems:

- Shortage of oil in the system
- Oil remaining in the evaporator (called ***logging***) due to low refrigerant velocity
- Shortage of refrigerant
- Refrigerant migration or floodback to the crankcase
- Oil pump failure
- Faulty refrigerant control devices
- Short-cycling

Regardless of cause, the compressor must be protected against loss of oil pressure. Permitting the compressor to operate without proper lubrication will result in certain failure from broken piston rods, locked bearings, or similar causes. The oil pressure safety control is designed to stop the compressor before such damage occurs. It also warns the technician that an oil problem exists. An oil pressure safety control is standard equipment on any compressor equipped with an oil pump.

Control Connections

The oil pressure safety control is unlike any other type of control. It has two switches, two pressure-sensing bellows, and two capillary tubes for connection to the compressor. One capillary tube and bellows are tagged "low" and connected to the compressor crankcase (low-side pressure). The other capillary tube is tagged "oil" and connected to the oil pump (oil pump pressure). The tubes are tagged at the factory to prevent confusion when making these connections.

Two capillary tubes are needed because the control operates on the *difference* between low-side pressure and oil pump pressure. The oil pump produces a pressure that is higher than the low-side pressure. This pressure difference is called ***net oil pressure***. Net oil pump pressure is obtained by subtracting low-side pressure from oil pump pressure. Oil pump pressure *never* falls below low-side pressure, so it is necessary to subtract low-side pressure from oil pump pressure to obtain net oil pressure.

Most oil pumps are equipped with a tee fitting for making connections. The capillary tube (tagged "oil") from the safety control is connected to the #2 opening on the tee. The remaining tee branch (#3 opening) is fitted with a Schrader valve for obtaining oil pump pressure readings during operation. Two compound gauges are needed to check net oil pressure. One gauge is used to obtain low-side pressure at the suction service valve, and the other is connected to the oil pump Schrader valve to read oil pump pressure. The difference between the two pressure readings is net oil pressure.

Principle of Operation

The oil pressure safety control has two normally closed switches. One is used to control a small heater. The contacts controlling power to the heater are tagged T_1 and T_2. Operation of the heater causes a secondary (bimetallic) switch to open. The bimetallic switch contacts are tagged L and M. This switch is connected in series with the primary control and the contactor coil.

The two pressure-sensing bellows operate the switch at T_1 and T_2. This switch controls the power supply to the small heater. At start-up, power is supplied to both the compressor and the heater. When net oil pressure reaches 12 psig (83 kPa), the heater switch opens. This cuts off the power supply to the heater, but the compressor continues to run. During the run cycle, if net oil pressure drops to 9 psig (62 kPa), the switch will close and restore power to the heater.

If net oil pressure fails to achieve 12 psig after start-up (it is higher in normal operation), the switch at T_1 and T_2 remains closed and the heater gets hot. It requires about 120 seconds for the heater to produce enough heat to warp open the bimetallic switch connected between contacts L and M. When the bimetallic switch opens, power supply to the contactor coil is interrupted and the compressor stops. All oil pressure safety controls must be manually reset when they open.

Control Heater

The oil pressure safety control's heater provides a time delay required for the oil pump to establish oil pressure after start-up. It also prevents nuisance shutdowns due to brief shortages of oil pressure during the run cycle. The compressor is permitted to operate without proper lubrication for the amount of time required for the heater to get hot. The heater and bimetallic switch are carefully selected at the factory to provide a specific time delay. Oil pressure safety controls are available with time delays of 60 seconds, 90 seconds, or 120 seconds. This time delay is nonadjustable. Each compressor manufacturer specifies which delay is used on its product. The 120-second control is most common.

To provide for dual-voltage operation, there are two resistance heaters located inside each control. Only one is used when connecting the control for 120 V operation. The two resistance heaters are automatically connected in series for 240 V operation.

Wiring Connections

Power supply to the resistance heaters must be obtained from the bottom of the contactor

(T terminals, *not* L terminals). This prevents the heater from being energized during the off cycle. All other power supply connections are obtained from the top of the contactor (L terminals).

Heater Connections

Power supply to the heater is performed by connecting a wire from T_1 on the contactor to 240 V on the control. The other side of the heater obtains its power from the connections at the bimetallic switch. When the compressor contactor is closed, a circuit is made from T_1 on the contactor to the 240 V resistor connection, through the resistors, to the heater. From the heater, power travels through the switch at T_1 and T_2, through the bimetal contacts at L and M, through the low-pressure control, to L_2 at the top of the contactor.

Contactor Coil Connections

One side of the contactor coil is connected to L_1 on the contactor. From the coil, current travels to M on the bimetallic switch contacts, through the switch to L, and picks up the power supply traveling to the low-pressure control and up to L_2. The compressor cannot operate unless both the bimetallic switch at L-M and the low-pressure switch are closed. At start-up, if the net oil pressure does not build up sufficiently to take the resistance heater out of the circuit within 120 seconds, the heater will cause the bimetallic switch to open. This interrupts power to the contactor coil and stops the compressor.

Troubleshooting

To check the pressure settings on the oil pressure safety control, pump the system down by frontseating the liquid receiver service valve and forcing the compressor to run. When a vacuum of 10–15 in. Hg is obtained, shut off the compressor and briefly crack open the liquid receiver service valve to eliminate the vacuum. Disconnect both capillary tubes from the compressor. Leave the low-pressure capillary tube open to the atmosphere to sense "zero" pressure (0 psig, atmospheric pressure).

Connect the middle hose from the manifold gauge to a refrigerant cylinder. Connect the low-pressure gauge hose to the "oil" capillary tube. Gently open the refrigerant cylinder valve so a pressure between 20 psig and 60 psig (138 kPa and 414 kPa) is available in the manifold gauge's central chamber. Open the left manifold gauge hand valve until the low-pressure gauge indicates a pressure between 20 psig and 60 psig (138 kPa and 414 kPa). Use a multimeter to check for continuity between T_1 and T_2. The switch should be open.

Bleed the pressure off slowly and observe the compound gauge. The oil pressure safety control's switch should close at about 9 psig (62 kPa). Gradually raise the pressure again until the switch opens. This should occur at a pressure of between 12 psig and 14 psig (83 kPa and 97 kPa). The switch will make a "click" sound each time it opens or closes. If the oil pressure safety control's switch does not open and close within the specified pressure range, it should be replaced.

Chapter Review

Summary

26.1 A thermostat is a switch that opens and closes its contacts according to temperature changes. It uses a sensing element to detect temperature changes and responds based on the thermostat's established setpoints (cut-in and cut-out). These setpoints are set using a range adjustment dial or knob, which is common on domestic thermostats. Commercial thermostats have adjustment screws and are designed to adjust to a wide range of temperatures.

26.1 Most thermostat problems are related to improper settings, a sensing element that is not making good contact, or an element that has lost its refrigerant charge. After making certain that the contacts should be closed, the thermostat switch can be checked with a voltage tester or a multimeter.

26.2 Pressure-operated switches respond to rising or falling system pressures. They may be used as the primary control or as a secondary control of a system. These switches include an electrical switch that is opened or closed by pressure on a diaphragm.

26.2 Pressure control settings can be adjusted by using a manifold gauge to check pressure as the setpoints are changed. Pressure-type motor controls can be wired single-phase or three-phase, depending on the application.

26.2 Troubleshooting procedures for pressure controls are similar to those used with thermostats. The control is simply an electrical switch that operates according to low-side pressure at the compressor. A pressure control that is unable to maintain pressure settings must be replaced.

26.3 The low-pressure safety control is used as a secondary control to shut off a system if the low-side pressure is reduced to unsafe levels. Some problems that cause a system to have unsafe low pressure are: low refrigerant level, evaporator freeze-up or malfunction, restricted airflow due to dirty evaporator or air filters, moisture in the system, faulty superheat setting, filter-drier restriction, or low head pressure.

26.4 The head-pressure safety control is a device that will shut off a refrigeration system if head pressure exceeds proper limits. A dual-pressure motor control is available that combines the functions of low-pressure and high-pressure safety controls. A fan cycle control is used to control the condenser fan(s), maintaining head pressure to acceptable limits. These head-pressure safety controls should be adjusted according to the pressures and temperatures that correspond to the refrigerant(s) used in the system.

26.5 The pumpdown cycle uses a solenoid valve and thermostat to remove all refrigerant from the low-pressure side of a system prior to shutdown. This prevents liquid slugging and loss of oil from the compressor during start-up. A compressor recycling back on after shutdown may result from a system low on refrigerant, an improper differential setting, or a leaking solenoid valve seat or compressor discharge valve.

26.6 The oil pressure safety control is used to protect the compressor when lubrication problems occur. This control device has two switches, two pressure-sensing bellows, and two capillary tubes that connect to the compressor. Switches open and close based on net oil pressure maintained in the system.

26.6 Oil pressure safety controls can be checked by pumping the system down using the liquid receiver service valve and running the compressor. A manifold gauge and refrigerant cylinder are used to observe system pressures.

Know and Understand

1. Thermostats that open on rise in temperature are used in _____ applications. (26.1)
   A. cooling
   B. heating
   C. residential cooling
   D. industrial
2. *True or False?* All refrigeration systems are oversized in terms of capacity. (26.1)
3. Turning the range adjustment on a domestic thermostat to a colder setting lowers _____. (26.1)
   A. the cut-in setpoint
   B. the cut-out setpoint
   C. both setpoints
   D. the cut-in setpoint, and raises the cut-out setpoint
4. A commercial freezer operates at a range of 0°F (–18°C) to –5°F (–20.5°C). The motor controls used for this freezer are categorized as a _____-temperature application. (26.1)
   A. high
   B. medium
   C. low
   D. ultralow
5. It is poor practice to operate any system with the thermostat range adjustment set at _____. (26.1)
   A. minimum
   B. maximum
   C. minimum or maximum
   D. mid-position
6. For single-phase domestic systems, the thermostat switch is connected in _____. (26.1)
   A. the neutral wire
   B. series with the wires supplying power to the start relay
   C. parallel with the wires supplying power to the start relay
   D. parallel with the contactor coil
7. For three-phase commercial systems, the thermostat switch is connected in _____. (26.1)
   A. the neutral wire
   B. series with the contactor coil
   C. series with the wires supplying power to the start relay
   D. parallel with the contactor coil
8. After making certain that the contacts are closed, the thermostat switched can be checked for voltage _____. (26.1)
   A. at the overload
   B. at the power supply
   C. across the thermostat terminals
   D. at the safety controls
9. On commercial systems, pressure-type motor controls are conveniently located at the _____. (26.2)
   A. thermostat
   B. furnace
   C. condensing unit
   D. evaporator
10. During the "on" cycle, the refrigerant inside the evaporator is about _____°F colder than the evaporator and the evaporator is about _____°F colder than the air. (26.2)
    A. 5, 5
    B. 10, 10
    C. 10, 20
    D. 10, 30
11. The cut-out point is determined by subtracting _____°F from the desired cut-out *air* temperature. (26.2)
    A. 5
    B. 10
    C. 20
    D. 30
12. When setting system cut-out, how much pressure must be subtracted from the desired cut-out value to allow for pressure drop? (26.2)
    A. 1 psig
    B. 2 psig
    C. 5 psig
    D. 10 psig
13. The low-pressure motor control will cause a system to short-cycle if the system _____. (26.2)
    A. has a compressor burnout
    B. is incorrectly wired at the controls
    C. is low on refrigerant
    D. has no refrigerant
14. *True or False?* The low-pressure safety control becomes necessary when a thermostat is used for temperature control. (26.3)
15. The contacts for the low-pressure safety control are connected in _____. (26.3)
    A. series with other safety controls
    B. series with the thermostat contacts
    C. parallel with the thermostat contacts
    D. parallel with the power supply

16. *True or False?* The head pressure safety control switch is connected in parallel with the primary control. (26.4)
17. A dual-pressure motor control has _____ connected to the compressor. (26.4)
    A. an adjustment scale
    B. three terminals
    C. two capillary tubes
    D. an automatic reset
18. A _____ is used to control condenser fan(s), which is a common method of maintaining head pressure during cold weather conditions. (26.4)
    A. dual-pressure motor control
    B. low-pressure safety motor control
    C. pumpdown cycle
    D. fan cycle control
19. What is the purpose of the solenoid valve on a pumpdown cycle? (26.5)
    A. To control refrigerant flow through the liquid line
    B. To turn the compressor on and off
    C. To signal to the thermostat when a setpoint temperature is reached
    D. To prevent a system from pulling into a vacuum
20. If a system continues recycling after a pumpdown and the system refrigerant, low-pressure control, and compressor discharge valve reeds have already been checked and no issues were found, where is the issue likely to be found? (26.5)
    A. In the metering device
    B. In the pumpdown solenoid valve
    C. In the head-pressure safety control wiring
    D. In the suction line
21. Which of the following statements is *true* regarding oil pressure safety controls? (26.6)
    A. Both capillary tubes are connected to the compressor crankcase and oil pump.
    B. Net oil pressure is found by subtracting high-side pressure from oil pump pressure.
    C. Oil pressure safety controls have one normally closed switch and one normally open switch.
    D. All oil pressure safety controls must be manually reset when they open.

Apply and Analyze

1. Explain what is meant by the following phrase: "All refrigeration systems are *oversized*." (26.1)
2. A walk-in freezer has an air-sensitive thermostat mounted on the inside wall to sense the temperature of return airflow. The air temperature should be maintained between –10°F and 0°F. Explain how to properly make the range adjustment at the thermostat for this system. (26.1)
3. Explain the operation of pressure-type motor controls. (26.2)
4. What is the "Rule of 10" and how does it apply to refrigeration system troubleshooting? (26.2)
5. Why is a low-pressure safety control important in a refrigeration system with a compressor that relies on suction gas to cool the motor windings? (26.3)
6. List two methods of wiring the head-pressure safety control for effective compressor protection. (26.4)
7. On a pumpdown cycle, what does the cooled-space thermostat control? (26.5)
8. A technician installs a manifold gauge on the compressor service valves, frontseats the suction service valve, and forces the compressor to run using a jumper cord. Once the compressors pulls into a vacuum (25 in. Hg), the technician observes the low-side gauge showing a rapid loss of vacuum. What should the technician do next? (26.5)
9. How can pressure settings on an oil pressure safety control be checked? (26.6)

Critical Thinking

1. Describe the purpose of a primary control in a refrigeration system. (26.1)
2. Imagine you are a technician servicing a floral reach-in cooler and you need the reach-in temperature to operate in the 35°F to 38°F temperature range. The reach-in cooler uses R-290 refrigerant. Using the Rule of 10, how would you set the low-pressure switch for box temperature control? (26.3)
3. If the heater for an oil pressure switch was wired to terminals L_1, and L_2, what would happen? (26.6)

Sammyvision via Getty Images

27 Air and Defrost Cycles

LEARNING OUTCOMES

After studying this chapter, you will be able to:

27.1 Explain the physical properties of air and how refrigeration and air-conditioning systems are used to account for these properties.

27.2 Describe the types of defrost cycles used in both domestic and commercial refrigeration systems.

TECHNICAL TERMS

atmospheric air
capacity control
condensate pan
condense
defrost cycle
defrost termination thermostat
defrost timeclock
dehumidifier
dew point
electric defrost
evaporator fan delay thermostat
fail-safe setting
freeze-up
hot-gas defrost
humidistat
humidity
initiate
mullion heater
off-cycle defrost
relative humidity
reverse-cycle defrost
saturated
terminate
time-initiated/pressure-terminated timeclock
time-initiated/temperature-terminated timeclock
time-initiated/time-terminated timeclock
water vapor

Introduction

Most refrigeration systems are designed to remove heat from air which, in turn, removes heat from a product or space. The physical properties of ordinary air can become complicated and must be controlled. Otherwise, serious problems will develop in the refrigeration system.

The temperature and moisture content (humidity) of atmospheric air are crucial factors to consider when controlling air, but filtering (cleaning) and circulation (movement) are crucial as well. Refrigeration and air-conditioning systems are designed to provide control of temperature, humidity, filtration, and circulation. This chapter explains the various properties of air and how systems are designed to control air in refrigeration systems.

27.1 Physical Properties of Air

Atmospheric air is a mixture of gases and ***water vapor*** (moisture). Air has weight, density, temperature, and specific heat. More energy is required to move air with less heat content because "cold" air is heavier than "warm" air. Density of air varies with atmospheric pressure and the amount of water vapor the air contains. At sea level, 1 lb of air occupies about 14 cu ft of space and has a density of 0.0725 lb/cu ft. The specific heat of *dry* air at sea level is 0.24 Btu/lb. The specific heat of *moist* air increases with the moisture content.

Humidity

The moisture content of air is described as ***humidity***. The amount of moisture, or water vapor, that air will hold depends on the temperature. Humidity directly influences human comfort. For example, dry air causes rapid evaporation of moisture through the skin and can make a person feel cold even in a moderate temperature. Moist (very humid) air does not absorb much moisture, so a person may feel uncomfortable due to lack of evaporation.

When *warm*, moisture-laden air touches a *cold* surface, it releases much of its moisture. The air temperature near the cold surface is lower than the surrounding air, causing the moisture to ***condense*** (change to a liquid state) on the surface. Water dripping from cold pipes is caused by condensation of moisture from the air onto the pipes. Insulation on the pipes prevents condensation by keeping the moist air from contacting the pipes. For this reason, cold suction lines on refrigeration equipment are insulated. If the suction line temperature falls below the freezing point of water, condensation will freeze on the lines. Frozen condensation on suction lines is easily confused with frost formed in liquid floodback situations.

Relative Humidity

Relative humidity is the amount of moisture in the air compared to the amount of moisture the air would hold if the air were ***saturated*** (filled to capacity, of moisture in this case). The ratio of actual water vapor content to the maximum possible water vapor content is expressed as a percentage. The relative humidity of saturated air is 100%, so air holding half the moisture it can hold has a relative humidity of 50%. Low relative humidity indicates dry air. High relative humidity indicates moist air. Warm air holds more moisture than cold air. As air temperature decreases, relative humidity goes up. When the relative humidity reaches 100%, moisture will be released (condensed out) from the air.

Dew Point

Dew point describes the point at which the air becomes saturated (holds all the moisture it is capable of holding) for a given temperature. Dew point is simply another way of describing 100% relative humidity. This is also the temperature where the air will start releasing its moisture as condensation.

Dehumidifiers

A ***dehumidifier*** is a refrigeration system designed specifically to remove moisture from air without affecting air temperature, **Figure 27-1**. A dehumidifier is useful for "drying out" a damp basement. The evaporator is usually a coil of tubing and is placed immediately in front of the condenser. The fan draws air across the evaporator tubing and the condenser.

When humid air contacts the cold evaporator coil, moisture condenses on the tubing and runs

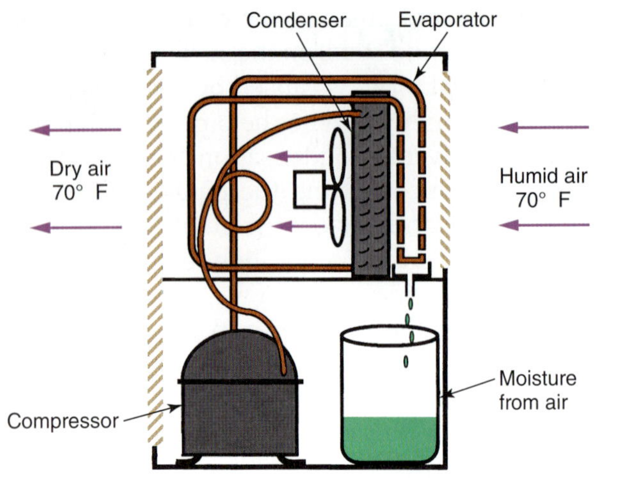

Goodheart-Willcox Publisher

Figure 27-1. A dehumidifier removes moisture from air without affecting the air temperature.

down into a catch pan. The catch pan drains into a basin, or it may be routed to a hose leading to a floor drain. Air flowing across the evaporator is cooled and gives up its moisture. Then it flows across the condenser and picks up heat.

A ***humidistat*** (moisture-sensitive switch) controls system operation. In mechanical humidistats, the sensing element is a nylon ribbon that stretches and contracts according to humidity. The refrigerant pressure-temperature relationship must be maintained to keep the evaporator temperature above the freezing point of water (32°F or 0°C). Moisture must *not* be allowed to freeze on the evaporator.

Humidity and Air Conditioning

All air-conditioning systems act as dehumidifiers because the evaporator lowers air temperature to below the dew point. Moisture collects on the cold evaporator and drips into a catch pan connected to a floor drain. The process of removing moisture from the air adds a significant heat load on the system. Humidity removal is a latent heat load. Latent heat (moisture) must be removed before sensible heat (temperature) can be lowered. High humidity can prevent an air conditioner from lowering temperature.

The temperature of an air-conditioning evaporator must be maintained above the freezing point of water to prevent moisture from freezing on the evaporator. Evaporator temperature is maintained by controlling suction pressure. Airflow across the evaporator (heat load) is limited by blower and duct sizes. During periods of high sensible and latent heat loads (high air temperature with high humidity), these limits may prevent the system from performing properly.

Many residential air-conditioning systems are sized for operation under normal conditions, but they fail to perform properly during periods of peak loads involving high temperature with high humidity. Such systems are undersized for operation during hot and moist summer days. However, oversizing to compensate for extreme conditions can cause the system to short-cycle during normal conditions which can lead to high humidity levels and shortened unit lifespan.

Most commercial air conditioning systems are sized to operate properly under peak loads but provide capacity control for use under low-load conditions. ***Capacity control*** is a controlling method for reducing the heat displacement ability of the system. It may be attained by using unloaders, bypass valves, or by cycling multiple compressors.

27.2 Defrost Systems

In refrigeration systems, moisture will freeze on the evaporator surface if the evaporator temperature is lower than the freezing point of water, 32°F (0°C). The accumulation of frost and ice restricts proper airflow and acts as an insulator, preventing heat transfer. An evaporator full of frost and ice cannot remove heat from the air because the air molecules cannot touch the evaporator surface. To remedy this problem, the evaporator must be defrosted. A ***defrost cycle*** is a sequence of operations used to remove frozen moisture from an evaporator and restore it to full efficiency.

Air-conditioning (comfort cooling) systems do not require a defrost cycle because their evaporator temperature is maintained above 32°F (0°C). However, refrigeration systems usually require a defrost cycle, because the temperature of the evaporator is below freezing.

Automatic defrost is typically controlled by a ***defrost timeclock***. A defrost timeclock uses a timer and two or more switches to automatically turn off certain system components (the compressor and evaporator fan) and turn on other components (the electric heater on the evaporator or hot-gas solenoid valve). After a certain length of time has elapsed, the timer resets the switches and restores the system to normal operation.

Defrost timeclocks can be divided into domestic and commercial types, **Figure 27-2**. Many domestic and commercial timeclocks are accessible and fully adjustable. Timeclocks either have electric motors for turning gears that open and close electrical switches or electronic circuitry that performs the same function. Various types of defrost timeclocks are available. Selection is determined by the type of defrost cycle used in the system.

Domestic timeclock

Goodheart-Willcox Publisher

Commercial timeclock

Sealed Unit Parts Co., Inc.

Figure 27-2. Defrost timeclocks. The starting time and duration of each defrost cycle can be set on commercial timeclocks.

Types of Defrost Cycles

Different methods are used to clear the evaporator of frost accumulation. Even though domestic and commercial defrost systems differ, they make use of similar defrost methods. There are four basic types of automatic defrost systems:

- ***Off-cycle defrost***. This commercial system automatically defrosts during each off cycle. After the compressor cycles off, the evaporator fan continues to run (all the time). The constant airflow helps to melt any frost or ice that accumulated during the daily run cycles. The thermostat settings (or the low-pressure control) prevent the compressor from operating until the evaporator temperature reaches 38°F (3°C). Some commercial systems use a timeclock to turn off the condensing unit. The timeclock can perform several defrost cycles each day.
- ***Electric defrost***. This defrost method is used on both domestic and commercial systems and is controlled by a timeclock that automatically turns off the condensing unit, stops the evaporator fan, and turns on electric resistance heaters fastened to the evaporator. The timeclock performs several defrost cycles each day.
- ***Hot-gas defrost***. Hot-gas defrost is primarily a commercial system. High-temperature gas from the hot-gas discharge line is piped directly to the evaporator inlet. A solenoid valve is located in the bypass line and controlled by a timeclock. At defrost time, the timeclock stops the evaporator fan and energizes the solenoid valve. The compressor operates, but hot gas bypasses the condenser and travels directly to the evaporator inlet. This supplies heat to melt frost on the evaporator.
- ***Reverse-cycle defrost***. This commercial defrost system was used for defrosting multiple-deck frozen food display cases in supermarkets. A timeclock turns the condensing unit off and activates a special relay that reverses rotation of the evaporator fans. The

reverse rotation draws warm ambient air into the canopy air ducts of the display cases. The warm air is blown through the evaporator to melt frost. Then the air is discharged back into the store through the cases' lower air ducts. Note that many of these multiple-deck frozen food cases have commonly been replaced with glass door reach-in cases.

Manual Domestic Defrost Systems

Some domestic freezers do not have automatic defrost. These must be *manually* defrosted on a regular basis.

PROCEDURE

Manually Defrosting a Domestic Refrigerator-Freezer

Manual defrost is best accomplished by these steps:

1. Turn off the thermostat or unplug the appliance.
2. Empty the freezer section of food items.
3. Place containers of hot water in the freezer section. Heat from the water quickly softens the frost and ice. Usually, large chunks of frozen material can be removed by hand.
4. Restart the system after frost is removed from the freezer section.
5. Return food items to the freezer section.

CAUTION

Manual Defrosting

Never use a sharp instrument for prying or chipping frost from a freezer compartment. It is easy to poke a hole in the thin aluminum evaporator. Repairing holes in the evaporator is difficult and expensive.

Automatic Domestic Defrost Systems

Domestic refrigerator-freezer systems with automatic defrost are sometimes called *frost-free.* An electric defrost system automatically performs the defrost and disposes of resulting water. The system is quite simple but differs from commercial systems.

Domestic Timeclocks

All automatic defrost (frost-free) refrigerator-freezers use a defrost timeclock to regulate frequency and duration of each defrost cycle. The clock is usually located behind the toeplate grille at the bottom front of the unit. Some are located in the rear compressor section or inside the refrigerator with the thermostat and lightbulb, **Figure 27-3.**

A mechanical defrost timeclock consists of a motor, gear assembly, switches, and a rotating cam. The camshaft is geared to the motor so it completes a certain number of revolutions per day. Revolutions per day vary according to manufacturers' specifications. There may be four (one every 6 hours), three (one every 8 hours), or two revolutions (one every 12 hours). At the end of each revolution, the switches are activated, and the system enters a defrost cycle. Duration (length) of the defrost cycle varies from 10 to 35 minutes. Average defrost length is 28–30 minutes.

Domestic defrost timeclocks are a critical element in proper system operation, but failure is rather common. Be sure to obtain an exact replacement. Defrost cycle frequency and duration recommendations are available from the equipment manufacturer.

The front of a mechanical defrost timeclock may contain a plastic screwhead or knob connected directly to the timer camshaft. The screwhead is designed for clockwise rotation and prevents counterclockwise movement, **Figure 27-4.** When turning the cam with a screwdriver, a sharp "click" sound is made at the beginning and end of the defrost cycle.

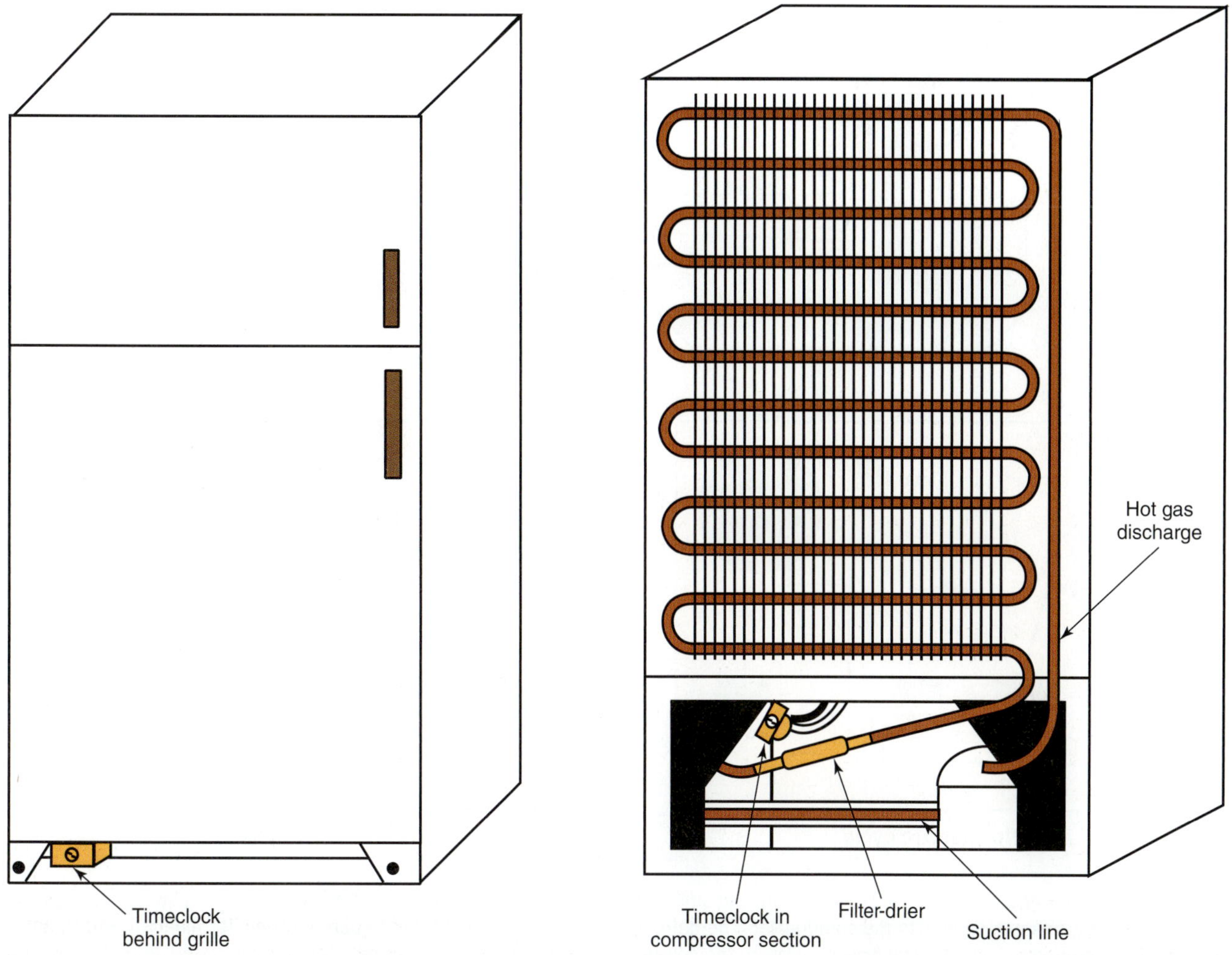

Goodheart-Willcox Publisher

Figure 27-3. Typical locations of defrost timeclocks on domestic refrigerators. A—In front, behind the toeplate. B—In the rear, in the compressor section.

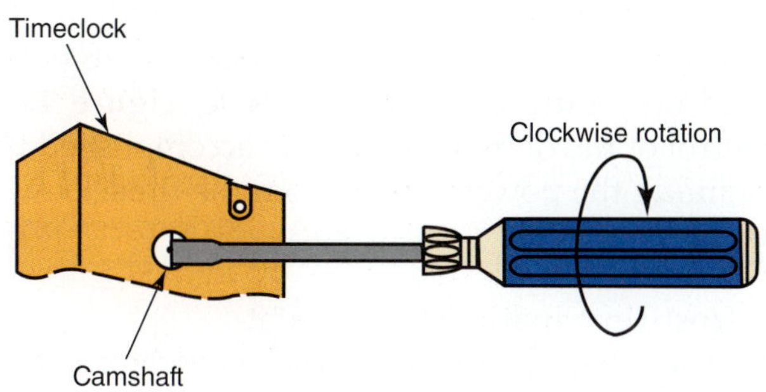

Goodheart-Willcox Publisher

Figure 27-4. When the camshaft screw is turned clockwise, a distinct "click" sound is made each time the defrost cycle switch opens or closes.

Most domestic defrost timeclocks have four terminals. The power supply is normally connected to Terminals 1 and 3. These supply power to the clock motor. Terminal 1 also supplies power to the switches. The normally closed (NC) contacts are between Terminals 1 and 2. The normally open (NO) contacts are between Terminals 1 and 4. During a defrost cycle, the normally closed (NC) switch opens, and the normally open (NO) switch closes. See **Figure 27-5**.

The switches and terminal numbers just described are the most common, but some exceptions can be found. For example, General Electric

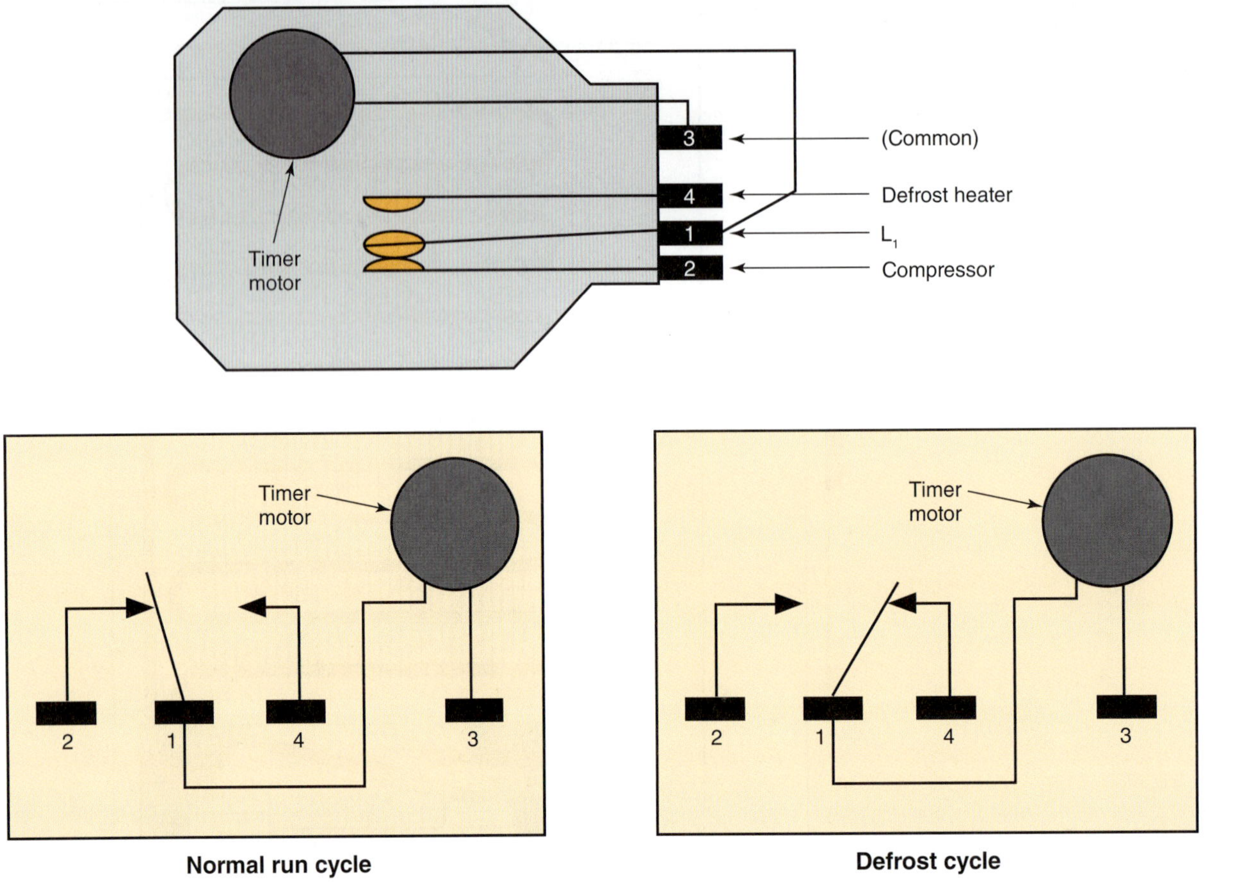

Goodheart-Willcox Publisher

Figure 27-5. The timer causes the switches to change position for the defrost cycle. As shown, the switch is normally closed between Terminals 1 and 2 to supply power to the compressor. The defrost cycle causes the contact to open between Terminals 1 and 2, and close between Terminals 1 and 4. The power supply to the compressor is cut off, and the defrost heater is energized.

and Hotpoint use a timeclock with the normally closed switch located between Terminals 1 and 4, and the normally open switch located between Terminals 1 and 2. When making a replacement, always check the appliance schematic against the new timeclock for proper terminal numbers and arrangement.

Domestic Electric Defrost Cycle

On an electric defrost system, power supply (hot leg) to the condensing unit is obtained from the timeclock at Terminal 2 because it is the normally closed switch. The other side of the power supply (neutral) can be obtained at Terminal 3 on the timeclock, but the neutral can also be obtained from another location. It is standard practice to place all switches in the hot wire and never disconnect the neutral (white) wire. It is then possible to tap into the neutral wire at any location to obtain that side of the power supply.

The evaporator fan also obtains power from Terminal 2. During a defrost cycle, power supply to Terminal 2 is open and stops operation of the condensing unit and evaporator fan, **Figure 27-6**.

One or more electric resistance heaters are attached to the evaporator. The heaters are energized only during the defrost cycle, **Figure 27-7**. Control of the defrost heater(s) is accomplished by obtaining the power supply from the timeclock at Terminal 4.

Defrost Termination Thermostat

If frost accumulation on the evaporator is less than normal, defrosting may be completed before the set amount of time on the timeclock expires. Such a situation could prove dangerous since the defrost heaters could become too hot and cause severe damage. A ***defrost termination thermostat*** is used to disconnect the defrost heater when the evaporator temperature reaches about 50°F (10°C). A defrost termination thermostat is a bimetallic disc inside a sealed container that looks somewhat like an overload. Two wires are provided for making electrical connections. The defrost termination thermostat is connected in series with the power

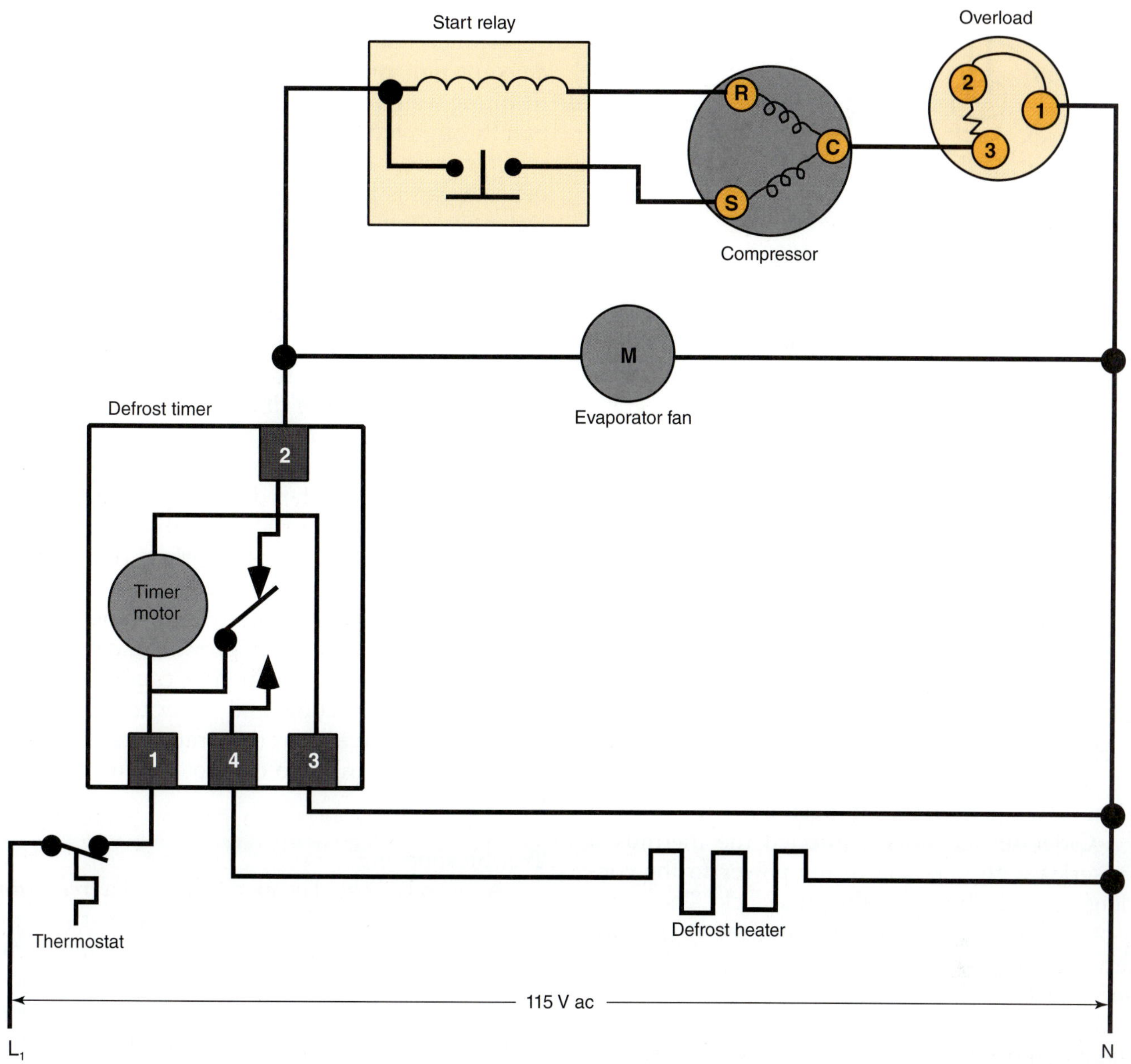

Goodheart-Willcox Publisher

Figure 27-6. Typical wiring diagram showing how the timeclock controls operation of a refrigeration system. The circuit to the compressor's motor and evaporator fan is through a normally closed switch that is first regulated by the thermostat. The defrost heater is connected to the normally open switch on Terminal 4.

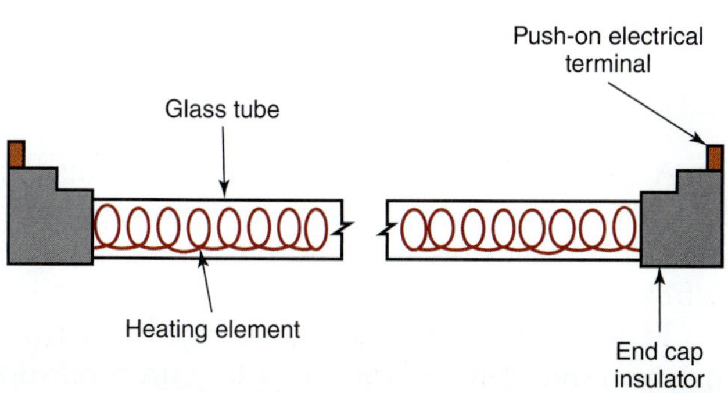

Goodheart-Willcox Publisher

Figure 27-7. Cutaway view of a defrost heater used in domestic applications.

supply from the timeclock to the defrost heater, **Figure 27-8**.

A defrost termination thermostat is clamped onto the evaporator tubing or mounted close to the evaporator. Once all frost has been removed, evaporator temperature rises quickly as the defrost heaters continue to produce heat. When evaporator temperature reaches about 50°F (10°C), the bimetallic disc inside the defrost termination thermostat warps and *opens* its contacts, cutting off the power supply to the defrost heater. The system must wait for the defrost timeclock to finish

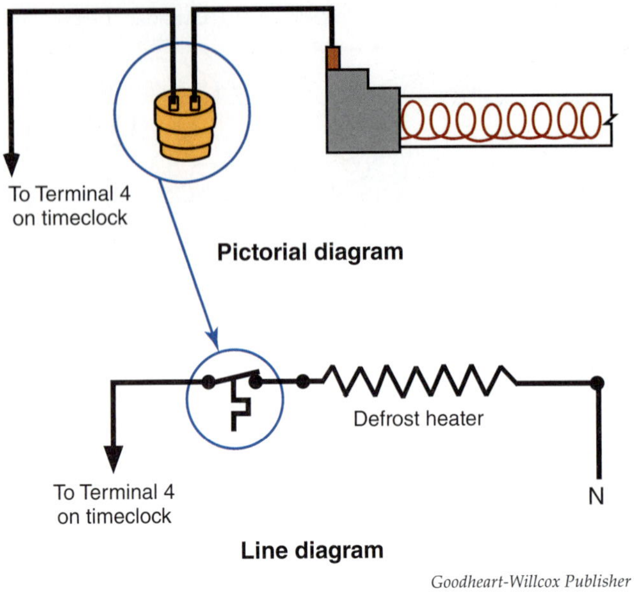

Figure 27-8. A defrost terminator thermostat is connected in series with the defrost timeclock and the defrost heater.

its timed cycle before normal operation resumes. When evaporator temperature drops to about 30°F (–1°C), the defrost termination thermostat's bimetallic disc contacts close, ready for the next defrost cycle.

Domestic Thermostat Connections

Older refrigerators connected the thermostat in series with one side of the power to the condensing unit, but *after* the timeclock. The timer motor operated continuously, and the evaporator fan shut off only during a defrost cycle. To conserve energy and increase the refrigerator's energy efficiency ratio (EER), newer refrigerators place the thermostat in series with the power supply to the timeclock. The timeclock, condensing unit, and evaporator fan can operate only during a run cycle when the thermostat is closed. In this case, the defrost cycle is based on accumulated running time, not on continuous time elapsed.

The newer wiring connections result in some energy savings because the clock and the evaporator fan are also stopped during the off cycle. The thermostat determines running time (on cycle) and is located inside the refrigerator section. The refrigerator door is opened more frequently than the freezer door, allowing heat to enter and causing the thermostat switch to close. The refrigerator section normally receives its cold air from an evaporator inside the freezer section. A small air duct controls air flowing from the freezer to the refrigerator section.

Defrost Water Disposal

All automatic defrost systems require a ***condensate pan*** under the evaporator to catch the resulting water. Water travels from the condensate pan through a drain line to a catch pan located in the compressor compartment. Sometimes the condensate pan is heated during defrost to keep the water traveling down the drain instead of freezing in the pan.

On domestic systems, the condensate water is drained to a removable catch pan within the compressor compartment. The catch pan rests on a metal plate heated by the hot-gas discharge line. The hot-gas discharge line is routed beneath the metal plate before traveling to the condenser. The heated plate causes defrost water to completely evaporate before another defrost cycle begins. The process serves two purposes: it disposes of condensate water and removes superheat from the hot gas as it travels beneath the plate.

The condensate drain line can become plugged with debris or algae that grows in damp, cold climates. A plugged drain line causes the condensate pan to overflow and water to accumulate and freeze. Once a year, the drain line should be cleaned by pouring two cups of hot (not boiling) water through it.

Troubleshooting

A severely frosted evaporator, called a ***freeze-up***, may signal trouble with an electric defrost system. A freeze-up causes a sharp rise in cabinet temperature. A customer might describe the condition in these terms: "The food in the freezer section is thawing, the milk is warm, and the unit runs constantly."

One of four malfunctions will usually cause a freeze-up:

- Defective evaporator fan
- Defective defrost timeclock
- Burned-out electric defrost heater
- Defective defrost terminator thermostat

A defective evaporator fan is not unusual, and a defective timeclock is rather common, too. A defrost heater occasionally burns out, but the defrost terminator rarely fails.

To begin troubleshooting an electric defrost system, remove the cover panel to gain access to the evaporator. The evaporator fan should operate whenever the compressor is running. If the fan is operating properly, use a screwdriver to switch the timeclock into defrost mode. With an electric defrost system, the compressor and the evaporator fan should both stop, and the defrost heater should be energized. Place a clamp-on ammeter around the wire supplying power to the defrost heater to check for a reading of 2 A to 10 A.

If the ammeter reading is in the 2 A to 10 A range, allow the system to defrost. The problem is a defective timeclock that cannot switch from run to defrost.

The timeclock can be checked with an ohmmeter. First, disconnect the power and then remove the timeclock from the unit. Resistance through the timeclock motor should be about 2400 Ω. Check continuity through the switches *while* rotating the cam with a screwdriver, **Figure 27-9**.

There should *only* be continuity between Terminals 2 and 4 during defrost mode. If continuity occurs during normal operating mode, the defrost heater is being energized during the run cycle. This would cause more heat to be added to the evaporator than the system can remove. The defrost heater will cycle on the defrost terminator thermostat, allowing temperatures to return to normal until the defrost terminator thermostat closes again. These symptoms may be incorrectly diagnosed as moisture restriction in the capillary tube or a defective compressor.

Visual inspection of the defrost heater usually uncovers a burnout or a wiring problem. An ohmmeter can be used to check proper resistance through each heater. The appliance schematic

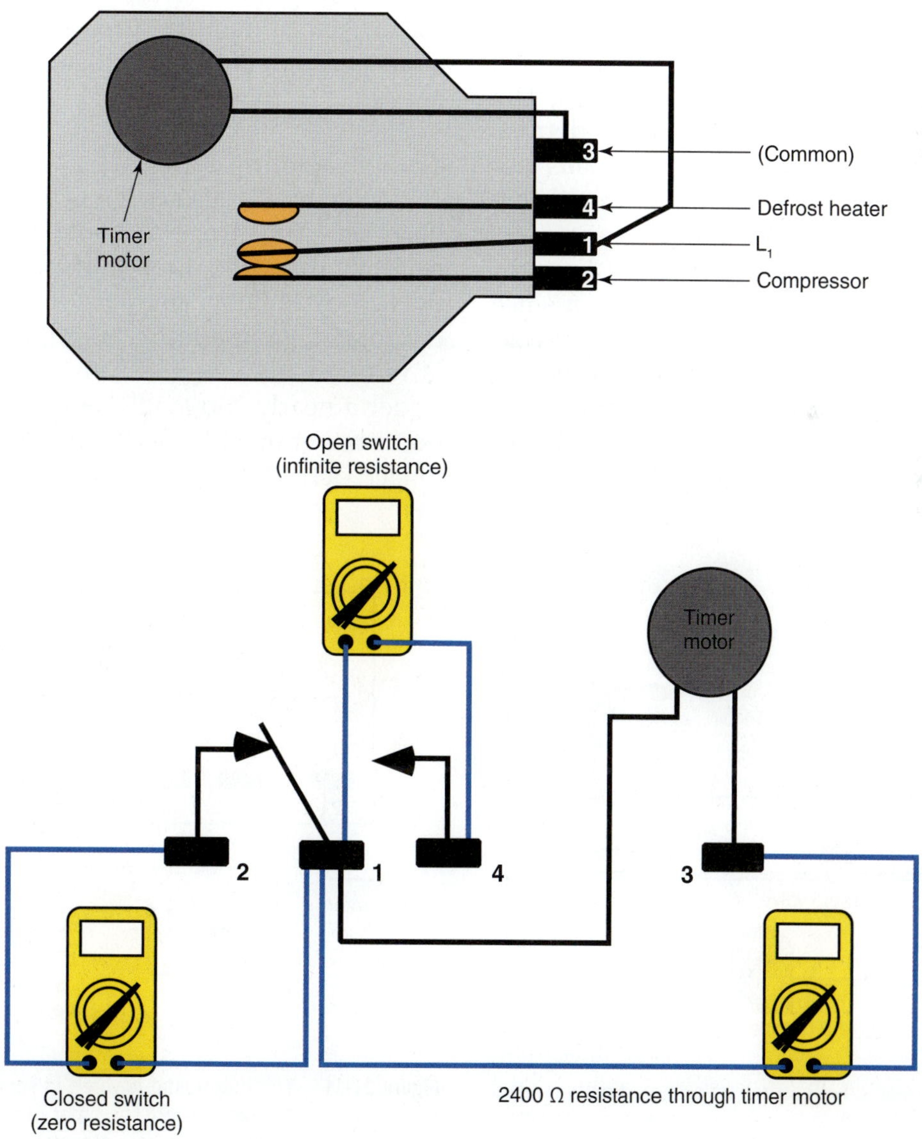

Goodheart-Willcox Publisher

Figure 27-9. An ohmmeter is used to check the defrost timeclock to determine whether it is defective. Before testing, turn off system power and disconnect all wiring from the timeclock.

normally shows the resistance value for each heater.

Hot-Gas Defrost Cycle

The hot-gas defrost cycle is different from the electric defrost cycle. For hot-gas defrost, hot gas is pumped directly from the hot-gas discharge line (compressor outlet) and delivered to the evaporator inlet through a hot-gas bypass line, **Figure 27-10**. The 1/4″ copper tube is factory-installed to bypass the condenser and refrigerant metering device. To control flow, a solenoid valve is located in the hot-gas bypass line within the condensing unit area. When the solenoid valve is energized and opens, hot gas takes the path of least resistance and travels directly to the evaporator inlet. When the solenoid valve closes, the hot gas is forced to travel through the condenser.

During defrost, the timeclock energizes the solenoid valve and stops the evaporator fan. The compressor continues to operate during defrost, producing hot gas. The hot gas quickly melts any frost accumulation as the gas travels through the evaporator and into the suction line.

Any system using hot-gas defrost *must* protect against liquid floodback to the compressor, since it is possible for hot gas to condense in the cold evaporator and enter the suction line as liquid. Protection against liquid floodback is usually provided by an accumulator. On domestic systems, the accumulator is a chamber located on the evaporator outlet.

Wiring Connections

For domestic hot-gas defrost systems, the timer must control the solenoid valve *and* the evaporator fan. The timeclock must have a set of normally open contacts (for the solenoid valve) and a set of normally closed contacts (for the evaporator fan). The thermostat is connected in series with the power supply to the timer and the power supply to the compressor. The timer must *not* control the compressor, **Figure 27-11**.

The timer operates on accumulated running time and cannot place the system in defrost unless the thermostat contacts are closed. This ensures the compressor is operating and producing hot gas when the timeclock calls for defrost. Frequency and duration of defrost are controlled by the timer (*time-initiated/time-terminated*). The required number and length of defrost cycles is determined by the appliance manufacturer, so the clock is non-adjustable. A defrost terminator thermostat is not needed.

Troubleshooting

When an R-600a system is in the defrost cycle, head pressure drops to about 52 psig (358 kPa) because the condenser and capillary tube are bypassed. Suction pressure rises according to evaporator temperature. Compressor amperage is reduced because the compressor is operating under a nearly "no load" condition. Defrost must occur rather quickly (5–10 minutes) because a prolonged defrost produces little hot gas.

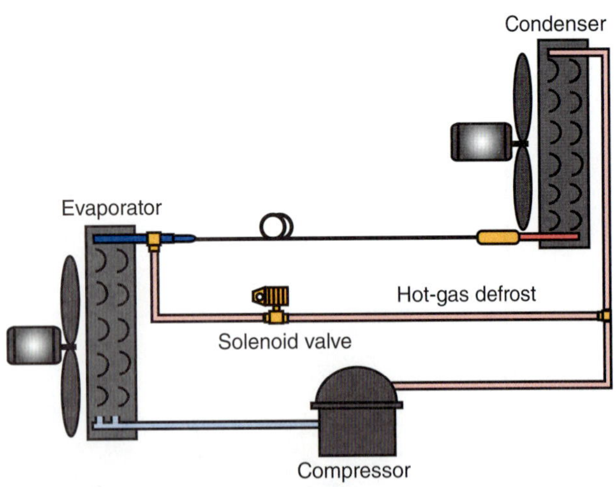

Goodheart-Willcox Publisher

Figure 27-10. A domestic hot-gas defrost system directs hot gas from the compressor directly to the evaporator, where it quickly melts accumulated ice. The hot-gas defrost (bypass) line should be connected as close to the compressor as possible.

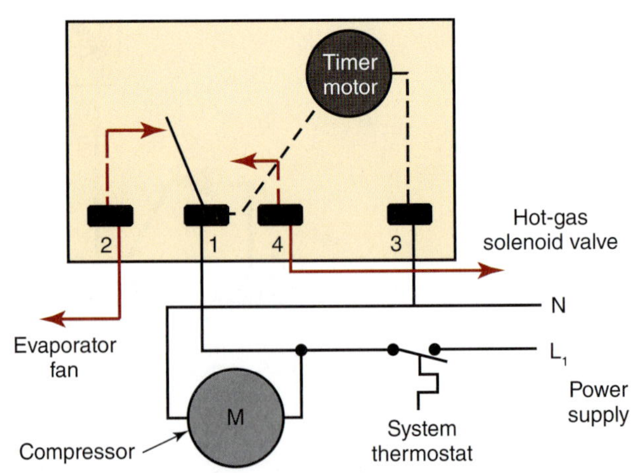

Goodheart-Willcox Publisher

Figure 27-11. The thermostat is wired in series with both the timer and the compressor. The timer must not control the compressor since the compressor provides the hot gas for defrosting the evaporator.

Problems with hot-gas defrost systems usually can be traced to a defective timer or to a low level of refrigerant. Low head pressure (because of low ambient temperature) cannot produce sufficient hot gas for defrost, so the evaporator eventually freezes up. However, a freeze-up seldom occurs with *domestic* hot-gas defrost systems since the appliance is inside, where the ambient temperature is typically about 70°F (21°C).

Domestic Accessories

Domestic refrigerator-freezers have become quite sophisticated as manufacturers have added accessories, such as ice makers and LCD displays with WiFi connectivity. These devices are not controlled by the timer. Consult the appliance schematic when servicing units with such devices.

Switches that turn interior lights on and off are standard accessories for most refrigerator-freezers. Sometimes, the door switch for the freezer section controls the light and the evaporator fan. When the freezer door is opened, the light is turned on and the evaporator fan is turned off. Stopping the fan keeps warm, moist air from being drawn into the freezer. The lighting circuits are connected in parallel and do not affect other system components.

Refrigerator-freezers also contain resistance heaters inside the door frames. Resistance heaters prevent condensation of moisture around the doors during periods of high humidity. Door frame heaters are often called ***mullion heaters***, or "anti-sweat" heaters. Sometimes a manual energy-saving switch is provided to turn off door heaters during the winter (low humidity) months.

A small heater is often placed inside the butter storage compartment. Called a "butter conditioner," it maintains the proper temperature to keep butter from getting too hard. These and other small resistance heaters are designed to operate continuously. **Figure 27-12** shows a complete pictorial diagram of a modern refrigerator-freezer. The same appliance is shown in ladder diagram form in **Figure 27-13**.

Commercial Off-Cycle Defrost

Off-cycle defrost is used on such light commercial systems as small walk-in coolers and self-contained cases and medium temperature cases. These systems may operate with R-134a and maintain an air temperature between 32°F and 38°F (0°C and 3°C). A low-pressure control or an air-sensitive thermostat controls the space temperature. The low-pressure control is set to cut-in at 38°F (33 psig) or 3°C (228 kPa). Cut-out is set at 12°F (13 psig, minus 2 psig for pressure drop) or –11°C (91 kPa).

During the run cycle, frost accumulates because the evaporator temperature is below 32°F (0°C). When air temperature is reduced to 32°F (0°C), the evaporator tubing is 22°F (–6°C) and the refrigerant is 12°F (–11°C).

After cut-out, air temperature, evaporator temperature, and refrigerant temperature quickly equalize at 32°F (0°C). The evaporator fan runs continuously. As air temperature increases, heat is absorbed by frost melting off the evaporator. The latent heat of melting ice (144 Btu/lb) helps maintain the cabinet temperature during the off cycle. The space temperature must be warmed to 38°F at 33 psig (3°C at 228 kPa) before the low-pressure control (or thermostat) will reach the cut-in point. All frost will melt before the evaporator warms up to 38°F (3°C).

Occasionally, an off-cycle defrost that uses an air-sensitive thermostat has an excessively long run cycle. Frost accumulation becomes extremely heavy and stops airflow over the evaporator. The evaporator becomes colder and accumulates more frost. Air temperature rises sharply due to lack of heat transfer.

Such heavy frosting of the evaporator usually occurs during normal working hours when the door of the cooler or case is opened frequently. After several hours, frost accumulation is so thick that little heat transfer is taking place. The system is unable to satisfy the air-sensitive thermostat. In such circumstances, defrost is accomplished by manually turning off the condensing unit for about two hours. The evaporator fan continues to operate and melt the frost accumulation.

Automatic off-cycle defrost is accomplished by installing an inexpensive timer that turns the condensing unit off for two hours each evening. Typically, the timer is set to accomplish defrost between midnight and 2:00 a.m. After defrost, the system reduces air temperature in the cooler to the desired level before normal working hours.

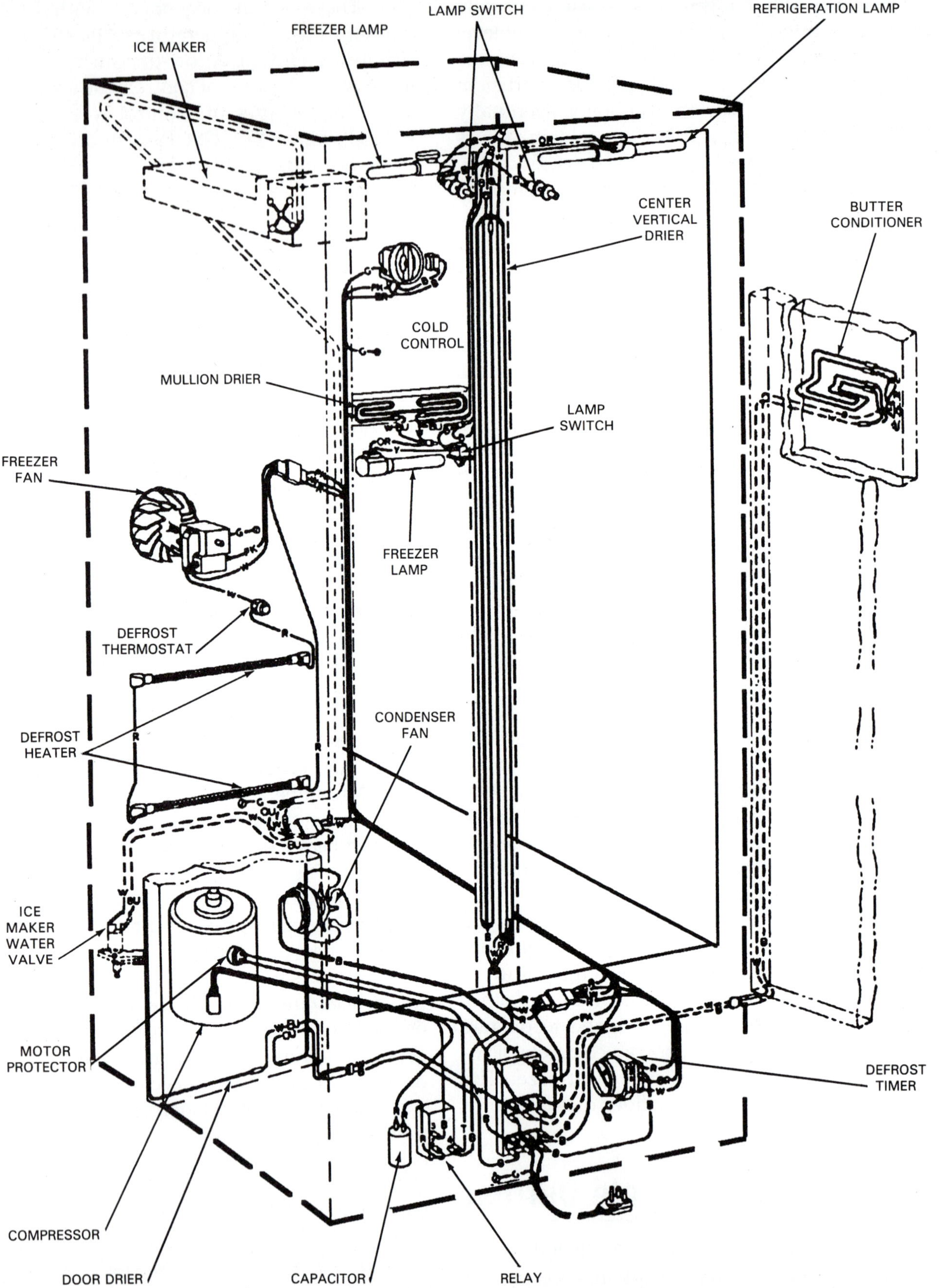

Frigidaire

Figure 27-12. A pictorial diagram of the electrical components of a typical side-by-side refrigerator-freezer unit.

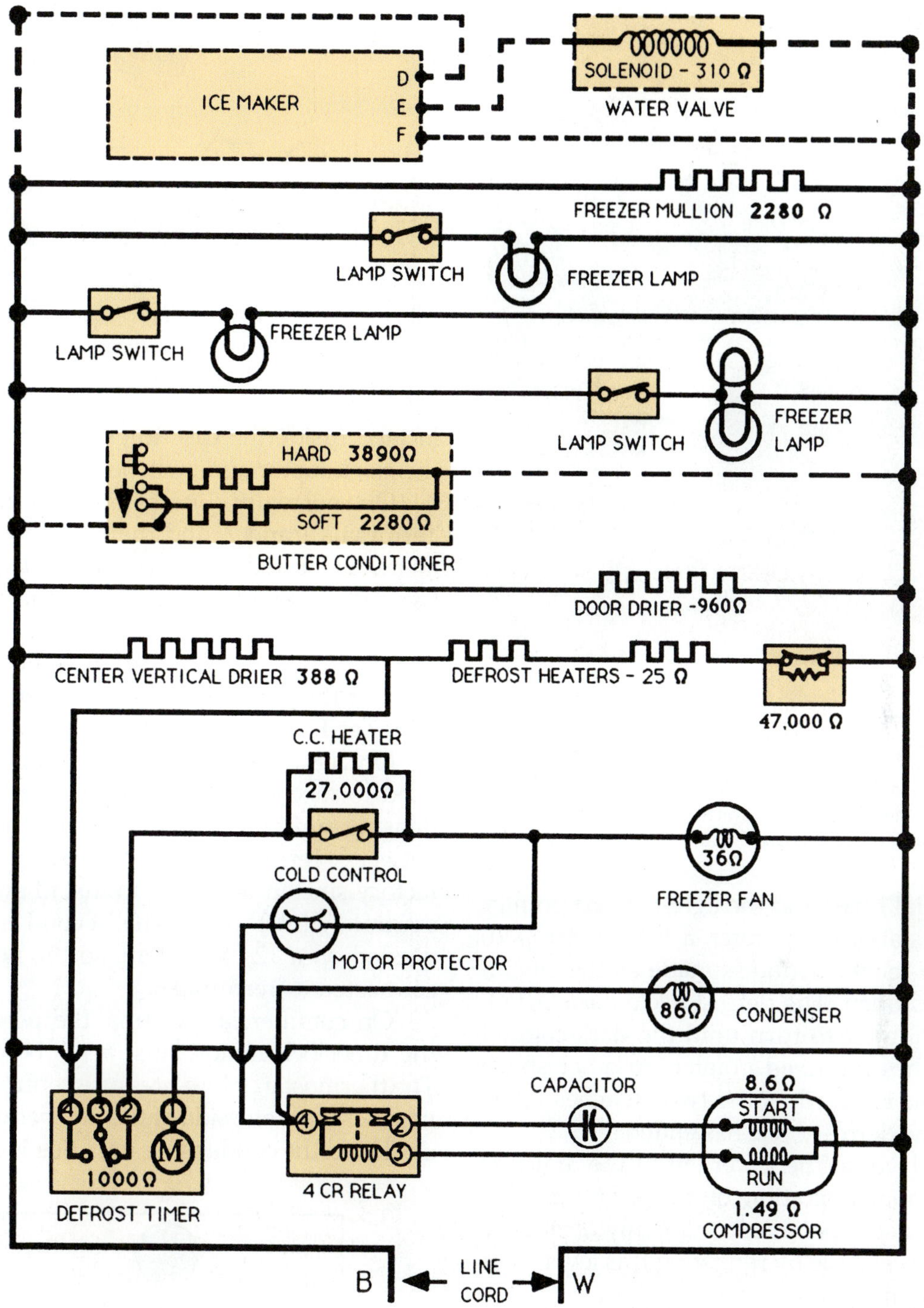

Frigidaire

Figure 27-13. A schematic ladder diagram of the electrical components of a typical side-by-side refrigerator-freezer unit.

Off-Cycle Defrost Timer

A typical timeclock used for off-cycle defrost is shown in **Figure 27-14**. Commercial refrigeration defrost timeclocks operate on a 24-hour basis. The large dial face is divided into two 12-hour sections. Half the dial face is black for easy identification of the 12 nighttime hours (6:00 p.m. to 6:00 a.m.). The dial face slowly rotates clockwise past the stationary pointer, which should indicate the correct time of day. The dial face can be rotated manually to reset the time.

Pairs of metal pointers (one light and one dark for each defrost cycle) attach with screws to the outer edge of the dial face. The light-colored pointer is anchored at the time the defrost should begin. The dark pointer is anchored at the time defrost should end. The minimum defrost period is two hours.

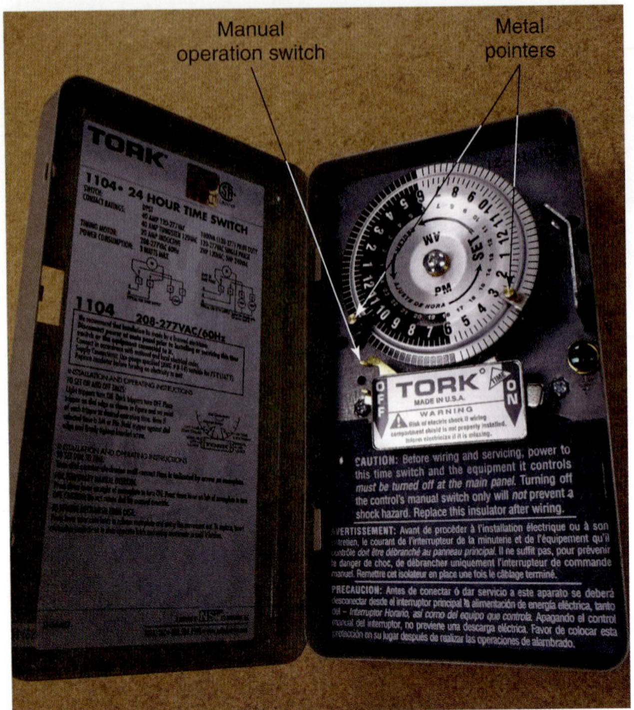

Image courtesy of Johnstone Supply

Figure 27-14. A commercial automatic off-cycle defrost timer. Metal pointers are attached to the dial face to start and stop the defrost cycle. This timer is set up for one defrost cycle in a 24-hour period that begins at 3:00 p.m. and ends at 10:00 p.m. The manual operation switch can be used to start and stop a defrost cycle.

As the dial turns and the light-colored pointer reaches the stationary pointer, a lever is tripped, and a sharp "click" sound is made as the clock's two switches open. The defrost cycle has started. The dial continues to turn until the dark-colored pointer reaches the fixed pointer. Another sharp "click" sound is heard as the two switches close, indicating the defrost cycle has ended.

A small lever at the lower left of the dial face permits manual operation of the switches. The technician must remember to manually ***terminate*** (end) the defrost cycle by returning this lever to its proper position when finished.

Some refrigeration cases require less time to defrost but need to defrost more times a day. In this case, use a timer that can be set up for more than two defrost cycles a day.

Electrical Connections

Most commercial timer switches are rated to carry 40 A per switch. The timeclock is available in either 120 V or 208 V to 240 V configurations for ease of matching voltages. The steel case is equipped with standard knockouts for attaching electrical box connectors. An electrical schematic drawing, like the one shown in **Figure 27-15**, is

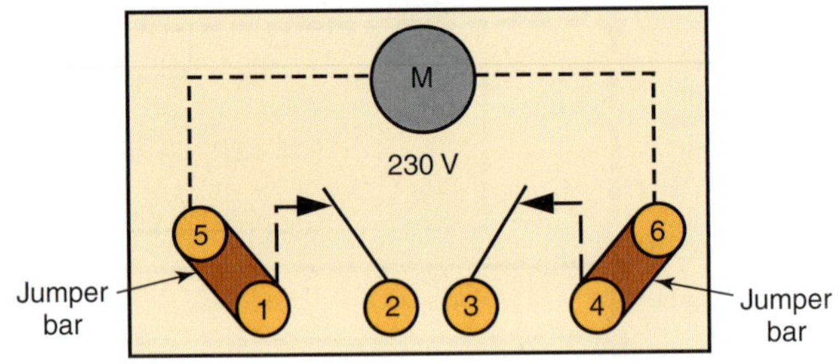

Goodheart-Willcox Publisher

Figure 27-15. Electrical schematic for the commercial automatic timeclock shown in **Figure 27-14**. The jumper bars are factory-installed to provide power to the timer motor from the power supply connected to Terminals 1 and 4.

located inside the case to explain connections for the clock motor and switches.

One normally closed switch is located between Terminals 1 and 2, and the other normally closed switch is located between Terminals 3 and 4. "The two power supply wires are connected to Terminals 1 and 4. This connection provides power to each switch, and the factory-installed jumper bars transfer the power supply to each side of the clock motor at Terminals 5 and 6.

On 208 V to 240 V systems, both switches are used to disconnect the power supply to the condensing unit. On 120 V systems, the jumper bar between Terminals 4 and 6 are omitted by the factory since it is unsafe to install switches in the neutral (white) wire. The second switch is not needed on a 120 V system, so the neutral wire is connected to Terminal 6.

On commercial systems, the power supply to the timeclock motor must *never* be disconnected. The thermostat or low-pressure control is connected in series with one side of the line between the timeclock and the condensing unit. See **Figure 27-16**.

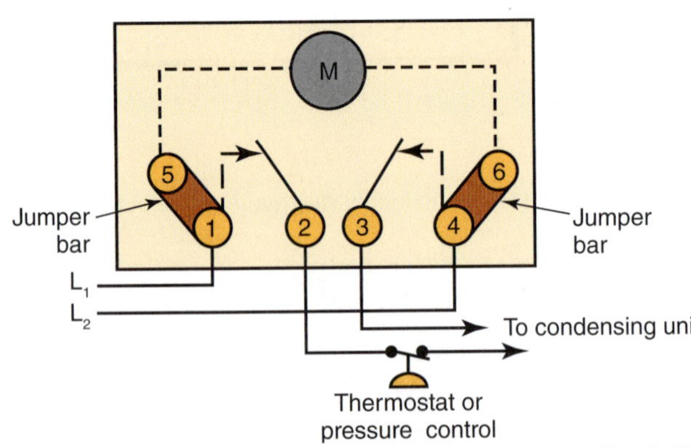

Goodheart-Willcox Publisher

Figure 27-16. The thermostat or low-pressure control is connected in series with one side of the line supplying power to the condensing unit. However, the thermostat does *not* control the timer. The timer controls both switches, opening or closing them at the same time.

Troubleshooting

The timer motor can be checked for proper rotation by noting the position of the pointer and waiting one-half hour to see if the dial has turned the proper distance on the scale. An ohmmeter can be used to check resistance of the timer motor (it should be between 1800 Ω and 2800 Ω). After disconnecting the power supply, remove the clock body from the metal case to visually inspect the clock parts and switches. The dial face can be rotated by hand to observe the switches opening and closing.

Commercial Electric Defrost

Electric defrost is popular for commercial refrigeration systems because the defrost period is short and easily controlled. A defrost sensor or timeclock is used to ***initiate*** (enter) the defrost cycle according to time of day, **Figure 27-17**. Termination can be accomplished by time, evaporator temperature, or suction pressure. Each means of termination has advantages and limitations. Commercial defrost timeclocks can be separated into three types, depending upon the method of termination:

- Time-initiated/time-terminated
- Time-initiated/temperature-terminated
- Time-initiated/pressure-terminated

Time-Initiated/Time-Terminated

The ***time-initiated/time-terminated timeclock*** is used to place the system in defrost according to time of day and terminate the defrost cycle after a specified length of time. The clock contains a normally closed (NC) switch and a normally open (NO) switch for controlling the compressor, evaporator fan, and defrost heaters. As shown in **Figure 27-18**, the large dial face is divided into two 12-hour sections, with nighttime hours (6:00 p.m. to 6:00 a.m.) in black. The dial face turns counterclockwise (CCW) and has a stationary pointer to indicate the actual time of day. Holes are drilled and threaded around the outer edge of the dial face at all even-numbered hours. Metal pins are provided to screw into the threaded holes to indicate the time of day the system will enter a defrost cycle. The number of pins determines how many defrost cycles occur each day.

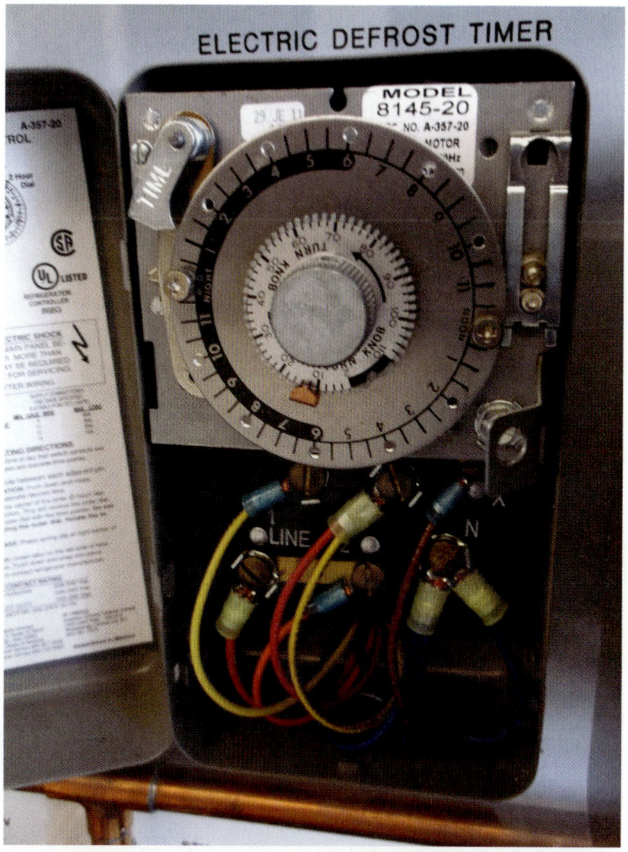

A Goodheart-Willcox Publisher

B Grasslin Controls Corporation

Figure 27-17. A—Commercial electric defrost systems use timeclocks like this one to initiate the defrost cycle. Termination can be by time or temperature. B—Electronic timers are also used in commercial defrost systems.

The large dial completes one rotation every 24 hours. When a pin nears the stationary pointer, it trips a lever and places the system in a defrost cycle. The tripped lever opens the NC contact and closes the NO contact.

The knob and small dial in the center of the large dial determine the duration (length of time) of the defrost cycle. The small dial is calibrated in

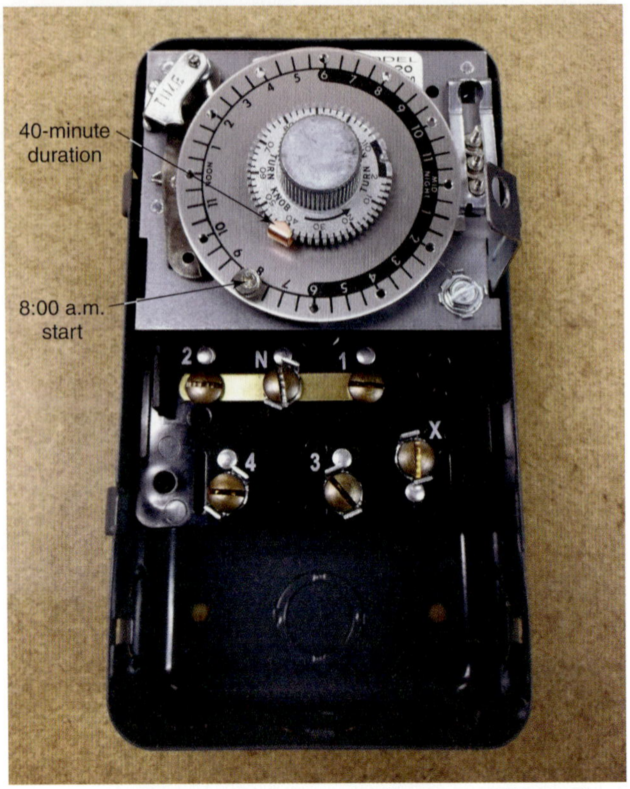

Image courtesy of Johnstone Supply

Figure 27-18. A timeclock used for time-initiated/time-terminated defrost. The timeclock setting provides one defrost cycle beginning at 8:00 a.m. (note pin on outer dial) and lasting 40 minutes (as indicated by pointer at 40 on notched inner dial). Several different cycles can be programmed, but all cycles must be the same length.

minutes and is sometimes referred to as the "fail-safe" dial. The outer edge of the small dial contains notches calibrated at two-minute intervals (up to 110 minutes). To set the length of the defrost cycle, the pointer is depressed and moved to the desired number of minutes (the termination point). When the specified number of minutes has passed, the pointer trips the switches back to their original positions.

The knob attached to the smaller dial is used to manually advance both dials simultaneously. The knob is turned counterclockwise (CCW) to set the correct time of day, place the system in defrost, or take the system out of defrost.

Defrost Requirements

The time-initiated/time-terminated defrost cycle cannot adjust itself to seasonal changes. Adjustments are required to compensate for summer (high humidity) versus winter (low humidity). Longer defrost periods are required during summer months when frost accumulation is heavier. Less frost occurs during fall and winter when humidity is lower.

The length of defrost must be sufficient to completely remove frost from the evaporator during any season. Partial defrosts cause a slow, steady accumulation of ice and result in evaporator freeze-up. Longer, but less frequent, defrosts are preferred over frequent short defrosts.

Electrical Connections

The timer just described is available in 120 V or 208 V to 240 V configuration, with contacts rated at 40 A. The power supply must be connected to Terminals N and X for continuous power supply to the clock motor. See **Figure 27-19**. A factory-installed jumper bar transfers the power supply from Terminal N to one side of the switches at Terminals 1 and 2. The switch between Terminals 1 and 3 is normally open, and the switch between Terminals 2 and 4 is normally closed.

Power supply to the defrost heaters or hot-gas solenoid is obtained at Terminal 3 (NO). The other side of the heaters or hot gas solenoid is connected at Terminal X (Common). The compressor and evaporator fan are connected to Terminals 4 (NC) and X (Common). Each switch is controlled by one side of the power supply. Terminal X is called *common* because all components use that terminal to obtain the other side of the power supply. See **Figure 27-20**.

In commercial systems, power supply to the timeclock is *never* interrupted. The primary motor

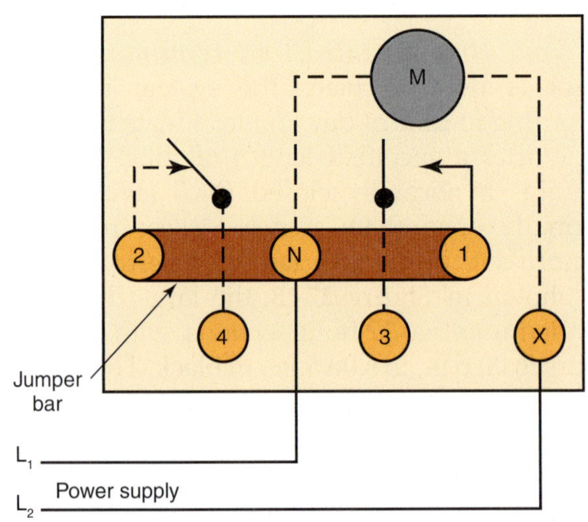

Goodheart-Willcox Publisher

Figure 27-19. Power supply to the timer is connected to Terminals N and X. A factory-installed jumper bar connects Terminals N, 1, and 2.

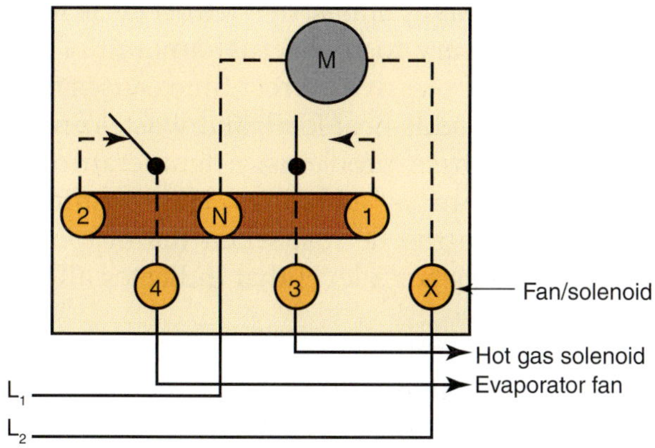

Goodheart-Willcox Publisher

Figure 27-20. Terminal X is the common terminal for the timer since it serves as one side of the power supply for all the timer-controlled components of the system.

control (thermostat or pressure control) must be connected in series between the normally closed (NC) contacts and the compressor. On three-phase electrical systems, Terminal 4 (NC contact) is used to control power supply to the contactor coil.

Troubleshooting

Trouble with the timeclock usually results from improper defrost time settings. Complete failure may be due to a burned-out timer motor or burned-out contacts on the switches. Resistance of the timer motor is about 2800 Ω. Proper operation of the motor and gears can be checked by using a pencil to mark the position of the dials, waiting about one-half hour, then rechecking the dials for proper rotation.

The entire timer body can be removed from the case for visual inspection. After disconnecting the power supply, remove the clock from its case. Manually rotate the dials by turning the knob counterclockwise and observing the switches for proper operation. Visual inspection will usually reveal any malfunction.

Field repairs to the timeclock and switches are *not* recommended. The devices should be replaced. A new clock is far less costly than the loss of product and sales due to continued defrost problems.

Evaporator Fan Delay Thermostat

On low-temperature systems (freezers), the evaporator fan must *not* be permitted to restart immediately after defrost. During defrost, the evaporator becomes very warm, about 55°F to 60°F (13°C to 16°C). If the fan is restarted as soon as the defrost ends, the fan will blow evaporator heat into the storage area. The heat will cause moisture to condense on the ceiling panels, drip down onto the product, and then freeze. An ***evaporator fan delay thermostat*** is used to prevent restarting of the fan until evaporator temperature drops to a set value (around 5°F or –15°C).

The evaporator fan delay thermostat has a bimetallic disc with normally closed contacts that open when the evaporator temperature rises to about 20°F (–7°C) and close when the temperature falls to about 5°F (–15°C). See **Figure 27-21**. The fan control setting (such as F-20 or F-38) is usually stamped on the control body. The letter *F* indicates fan control, and the number indicates the temperature at which the contacts open. This thermostat is connected in series with one side of the power supply to the evaporator fan. The evaporator fan delay thermostat is mounted or clamped directly to the evaporator on the return bends.

Troubleshooting

Failure of the fan delay thermostat is rather rare. However, when the control *does* fail, it is usually in the open position due to burned-out contacts. In this case, the evaporator fan can no longer operate, and an evaporator freeze-up occurs. To determine if an evaporator fan delay thermostat has failed, be certain the evaporator is cold enough for the thermostat contacts to close. Then jumper the evaporator fan delay thermostat to "make" (close) the switch. If the evaporator fan operates, the control is defective and should be replaced. If the evaporator fan does *not* operate after jumping the control, the switch is already closed, and the problem is something else (perhaps a burned out evaporator fan or wiring problem).

If an evaporator fan delay thermostat fails in the closed position, the timeclock will control the fan, and the fan will operate immediately after defrost (having no delay). Excessive heat and moisture following a defrost cycle indicate that an evaporator fan delay thermostat has failed in the closed position. The assumption can be tested by heating the fan control sensor with warm water to its cut-out point.

Time-Initiated/ Temperature-Terminated

Time-terminated defrosts are difficult to program because of such factors as seasonal changes, number of door openings, restocking of product, humidity, and temperature. The programmed length of defrost must allow for maximum frost

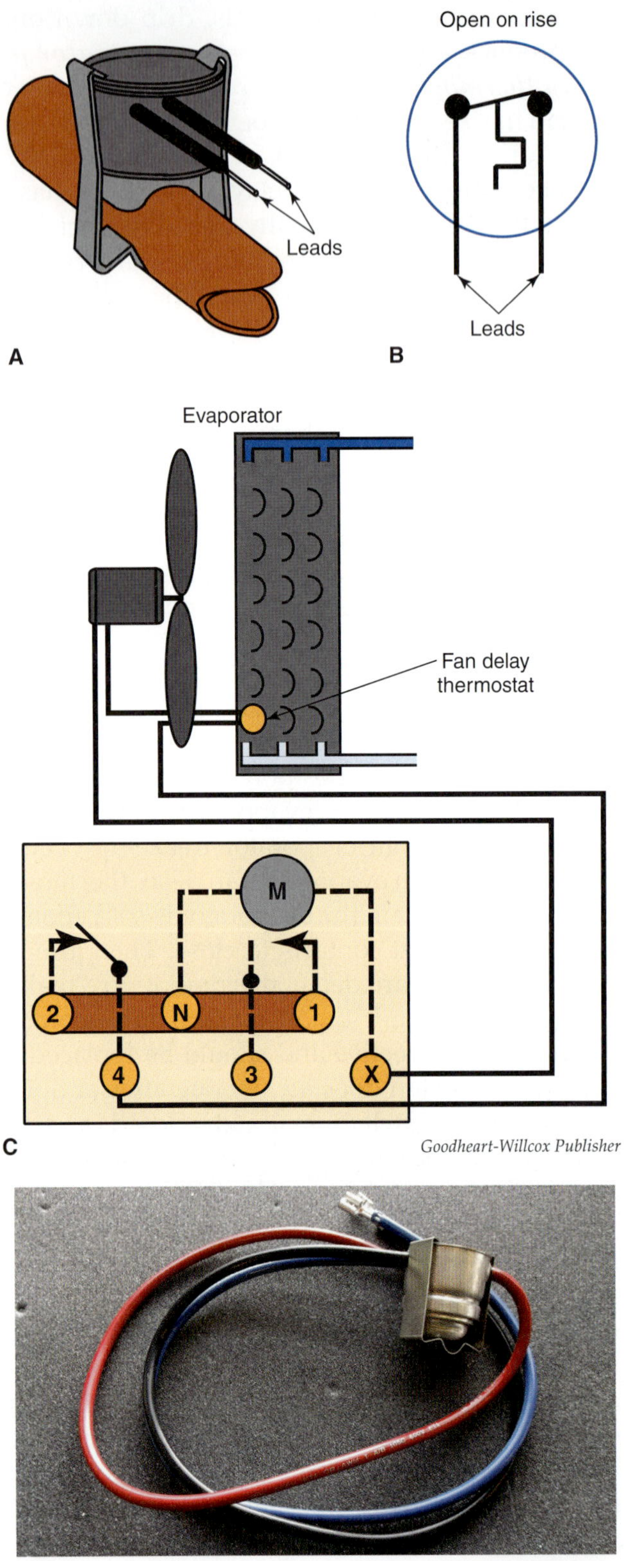

Goodheart-Willcox Publisher

Image courtesy of United Refrigeration

Figure 27-21. The evaporator fan delay thermostat restricts fan operation until temperatures have dropped sufficiently to avoid condensation problems. A—Mounting of the thermostat on evaporator tubing. B—Schematic symbol. C—Wiring connections to the timeclock. Note that the evaporator fan delay thermostat is wired in series with the evaporator fan. D—A combination fan delay and defrost termination thermostat clamps to the evaporator tubing. Note that it has three color-coded wires.

accumulation. Many times, the defrost cycle is longer than necessary for the actual amount of frost accumulation. Excessive defrost time causes moisture problems, adds heat load, and wastes energy. To overcome these problems, a temperature-termination feature is added to time-terminated systems to stop the defrost when the evaporator temperature reaches a level that indicates all frost has been removed.

Time-Initiated/Temperature-Terminated Timeclock

A ***time-initiated/temperature-terminated timeclock*** initiates defrost according to time of day, but termination is controlled by a thermostat mounted on the evaporator. The termination thermostat senses evaporator temperature. This defrost system has the advantage of automatically adjusting to seasonal changes and maintaining a minimum defrost time, eliminating guesswork when setting the defrost length.

The timeclock has a small solenoid with a plunger connected to the clock's trip lever. See **Figure 27-22**. Internal wiring connections prevent the solenoid from being energized except during a defrost cycle. When the solenoid is energized by the termination thermostat, the plunger trips the lever and ends the defrost cycle, regardless of the time remaining on the small dial. The timeclock initiates the defrost cycle, but the thermostat and solenoid end the cycle when evaporator temperature indicates the frost has melted.

Uninterrupted power supply to the timeclock is connected to Terminals 1 and N because the timer motor is internally connected to these terminals. The factory-installed jumper bar transfers power supply from Terminal 1 to Terminal 2. The normally open contact is located between Terminals 1 and 3 (controlling the defrost cycle). The normally closed contact is located between Terminals 2 and 4 (controlling the refrigeration cycle).

One side of the power supply to the small solenoid is factory-connected to Terminal 3 (NO switch). This prevents the solenoid from being energized, except during a defrost cycle when this switch is closed. The other side of the solenoid is factory-connected to Terminal X. The termination thermostat is connected between Terminals X and N, completing the solenoid circuit to the Common.

Fail-Safe Feature

When temperature (or pressure) is used to terminate defrost, the small dial on the timeclock

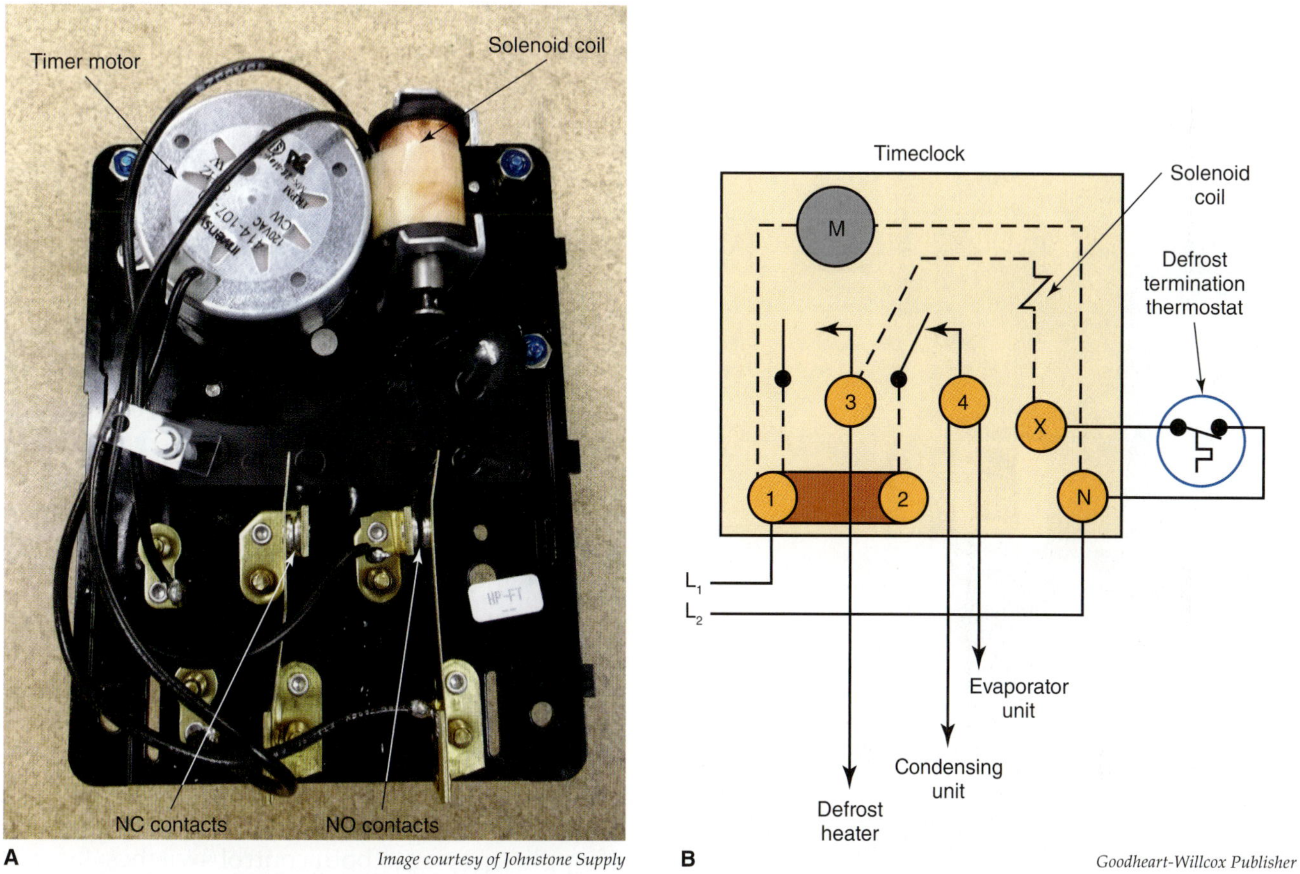

Figure 27-22. Time-initiated/temperature-terminated timeclock. A—Rear view of the timeclock, showing the solenoid that operates the trip lever. B—Schematic showing the electrical connections. The thermostat closes on rising temperature.

serves as a safety feature (the reason it is often referred to as a "fail-safe"). The ***fail-safe setting*** must allow sufficient time for the system to perform a complete defrost of maximum frost accumulation. If temperature or pressure termination components fail to operate properly, the fail-safe acts as a back-up feature by terminating the defrost cycle.

Defrost Termination Thermostat

The defrost termination thermostat contacts *close on rising temperature* (when evaporator temperature increases to the thermostat's setpoint). The setpoints (cut-in and cut-out) vary according to the application and the manufacturer's specification. On a "high limit" thermostat, the cut-in (L-50, L-55) is normally stamped on the control body. In this case, the letter *L* stands for limit control, and the number indicates the temperature at which the contacts close. This type of thermostat is nonadjustable and frequently found on evaporators inside walk-in freezers, coolers, and display cases.

The *adjustable-type* defrost termination thermostat is commonly used for commercial display cases, **Figure 27-23.** The thermostat is normally located behind the kickrail (toeplate) of the case to facilitate making wiring connections and adjustments. The sensing element is attached to the evaporator, usually located in the bottom of the case. A capillary tube linkage transfers bulb temperature to the control body.

Troubleshooting

Termination problems can occur if the clock solenoid burns out, or if the termination thermostat fails or is not properly adjusted. The customer may complain, "The system gets too hot during defrost, and the product tends to thaw out but refreezes during the run cycle." Such a complaint is often heard during the first heat wave of summer.

The technician must first determine the type of defrost clock used. A time-initiated/time-terminated timeclock may need manual readjustment. A time-initiated/temperature-terminated timeclock may be having solenoid or thermostat problems. The solenoid may be burned out and physically unable to trip the lever to terminate defrost. Visual inspection of the solenoid usually reveals a burnout, but an ohmmeter reading of *infinite resistance*

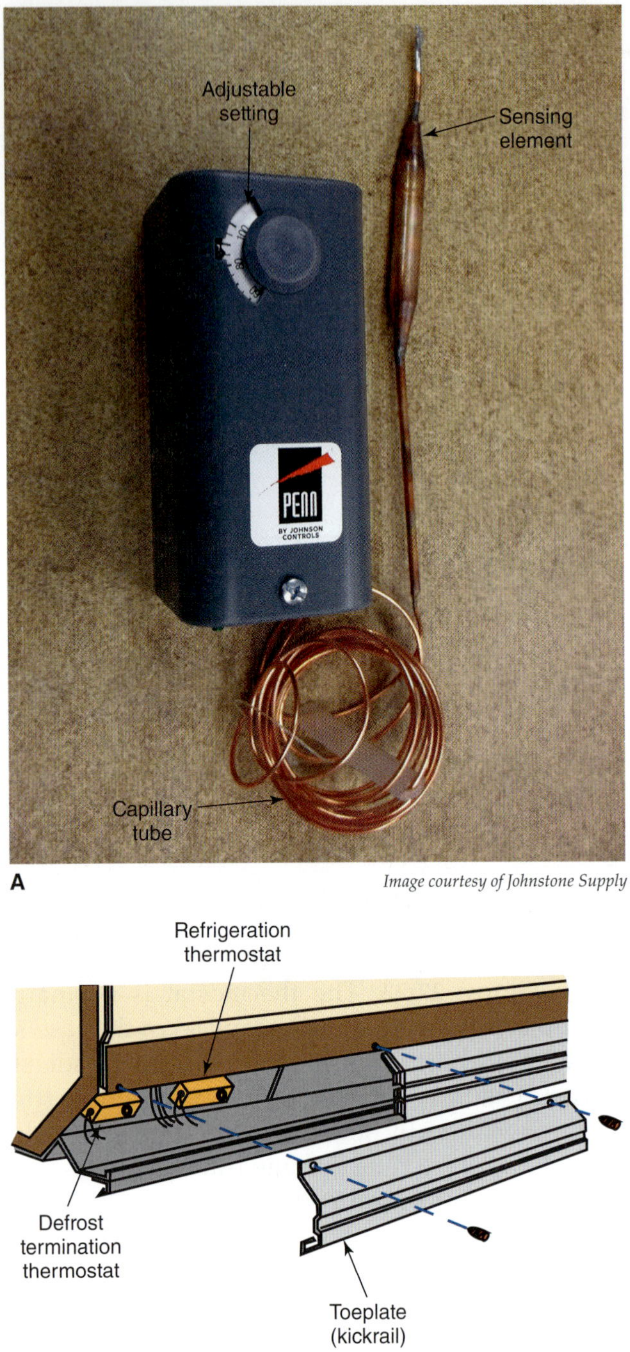

Figure 27-23. Adjustable defrost termination thermostat. A—Thermostat and sensing element. B—The thermostat is usually mounted behind the toeplate of the refrigerated display case.

will confirm it. If the solenoid is burned out and a replacement clock is not readily available, the fail-safe can be adjusted to operate the defrost in a time-terminated mode until a replacement is obtained.

If a timeclock's solenoid is functioning properly, the defrost termination thermostat can be checked by manually placing the clock in the defrost cycle. Then jumper the thermostat connections across timeclock Terminals X and N. If the clock switches out of the defrost cycle, the defrost termination thermostat is defective or improperly adjusted. If the clock *remains* in defrost, the clock itself is defective while the thermostat is functioning properly.

Combination Fan Delay and Defrost Termination

Sometimes the evaporator fan delay thermostat and defrost termination thermostat are combined into one component. The device is used for walk-in coolers and freezers where the evaporator and fan are readily accessible. The combination thermostat is easily recognized because it has three wires rather than two, **Figure 27-21D**.

The three wires are marked C (for common), F (for fan), and T (for terminator). Power supply to the control is obtained from the timeclock at Terminal N and is connected to Terminal C on the control. The connection supplies power from the timeclock to the movable pole inside the control, **Figure 27-24**. This connection is *common* because it supplies power to both control switches: fan and terminator.

Figure 27-25 illustrates a typical commercial defrost system using fan delay and defrost termination (combination) thermostat. One side of the power supply to the evaporator fan is connected to Terminal F on the combination thermostat and the other side to the normally closed switch on the timeclock (Terminal 4). The connection places a switch in each wire supplying power to the evaporator fan. A wire from Terminal T of the thermostat is connected to the timeclock at Terminal X

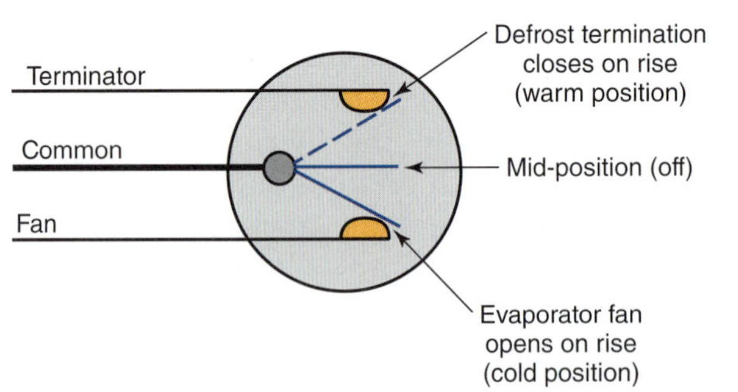

Figure 27-24. A movable pole inside the combination fan delay and defrost termination thermostat permits the thermostat to perform both functions. During normal system operation, the contact is in the fan position. In the defrost cycle, the pole moves to the mid-position as the temperature of the evaporator begins to rise. When evaporator temperature indicates all frost has melted, the pole moves to the defrost termination contact.

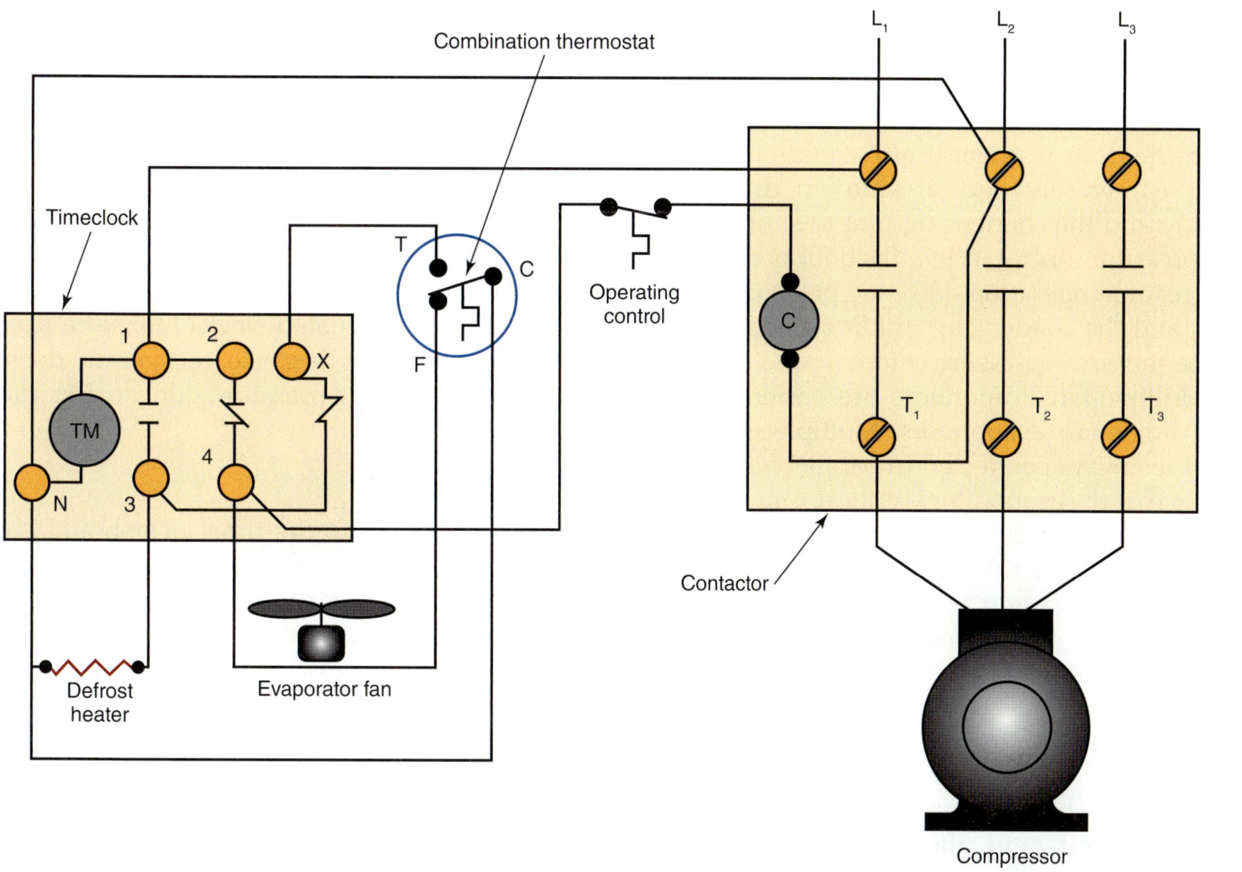

Goodheart-Willcox Publisher

Figure 27-25. Schematic for a system using a combination fan delay and defrost termination thermostat.

and serves as the defrost termination connection. The connection also places a switch in each wire supplying power to the small solenoid inside the timeclock.

During a normal run cycle, the evaporator fan operates because the timeclock switch is closed (between Terminals 2 and 4), and the combination thermostat's fan switch (between Terminals C and F) is closed (since the evaporator is cold). Both switches controlling power supply to the solenoid inside the timeclock are open (between Terminals 1 and 3 of the timeclock and also between Terminals C and T of the combination thermostat).

At the time of defrost, the timeclock stops the evaporator fan and condensing unit by opening the NC switch between Terminals 2 and 4 in the timeclock. The timeclock's NO switch between Terminals 1 and 3 closes to energize the defrost heater. Closing this switch also connects one side of the power supply to the timeclock's solenoid.

As evaporator temperature rises quickly, the movable pole in the combination thermostat warps to the mid-position where both control switches (F & T) are open. The combination thermostat remains in the mid-position until the evaporator temperature reaches 50°F to 55°F (10°C to 13°C). The temperature further warps the movable pole, closing the contact connected to the Terminal T. This completes the circuit to the timeclock solenoid, which terminates the defrost cycle.

The evaporator fan cannot restart until the evaporator temperature becomes cold enough to cause the combination thermostat's movable pole to warp back to the contact connected to Terminal F. This movement also opens the contact connected to Terminal T, disconnecting the solenoid.

Troubleshooting

Troubleshooting the combination thermostat is accomplished by checking each switch for proper operation. When checking evaporator fans, remember that there may be multiple causes for non-operation:

- The combination thermostat may be malfunctioning
- The fan motor may be defective (burned out)
- The cooling case may have a defective door switch

Many walk-in freezers have a door switch that turns off the evaporator fans when the door is opened. Stopping the fans prevents warm moisture-laden air from being drawn into the freezer when the door is open. It also causes the evaporator temperature and pressure to drop rather quickly and the condensing unit to shut off by its low-pressure control. When the door is closed and fans resume operating, low-side pressure quickly rises, and the condensing unit is turned on again by the same low-pressure control.

Multiple-fan evaporators are easier to check than single-fan evaporators. Multiple evaporator fans are always connected in parallel because airflow across the evaporator is critical. One fan motor may burn out, but the others continue to operate. If one fan stops functioning, a partial freeze-up will occur at the location of the malfunction. Power and control to evaporator fans is proven when at least one fan is operating.

Time-Initiated/Pressure-Terminated

The ***time-initiated/pressure-terminated timeclock*** is similar to the timeclocks discussed previously, except that termination is accomplished by rising pressure in the suction line, rather than by rising temperature in the evaporator. The clock is located at the compressor, with a 1/4″ copper tube linking the clock diaphragm to suction pressure at the compressor. During defrost, suction pressure rises. The rise in pressure moves a diaphragm mechanically linked to the timer lever. The appropriate rise in suction pressure causes the defrost cycle to terminate.

To initiate defrosting, the timeclock turns the compressor off, energizes electric defrost heaters, and stops any evaporator fans. During defrost, suction pressure reflects evaporator temperature according to the pressure-temperature (P/T) chart for the system's refrigerant. When suction pressure rises to the timeclock's setpoint, it pushes against the diaphragm to trip the clock lever, which terminates the defrost cycle.

The pressure required to trip the clock lever is adjustable. An adjustment screw (and scale) are provided on the clock. The scale is calibrated from 36 psig to 110 psig (248 kPa to 759 kPa), so the same clock can be used for R-134a, R-22, or R-507. Termination pressure settings for R-134a can be from 30 psig to 35 psig (207 kPa to 241 kPa), resulting in an evaporator temperature of 35°F to 40°F (2°C to 4°C). The pressure setting for R-22 is 85 psig (586 kPa), resulting in an evaporator temperature of about 50°F. The termination pressure setting for R-507 is 106 psig (730 kPa), resulting in an evaporator temperature of 49°F (9°C).

Frequency of defrost varies according to the system, but the fail-safe should be set for 45 to 60 minutes. This recommendation is approximate, but the setting should result in defrost being terminated by *pressure* rather than *time*.

Pressure-terminated clocks *cannot* be used on pumpdown cycles because refrigerant does not remain in the low-pressure side during the off (defrost) cycle.

Troubleshooting

Power supply to the timeclock should always be checked first. Checking for proper rotation of timer dials is standard procedure. Mark the location of the dials, and then check rotation against elapsed time. The clock can be removed from the case for visual inspection, if desired.

To check the accuracy of the calibrated dial (termination pressure), install a manifold gauge on the compressor service valves. Turn the clock dials to place the system in the defrost cycle. Slowly bleed some high-pressure gas through the manifold gauge and into the low-pressure side to raise the suction pressure to the termination point. Close manifold valves, and observe the suction pressure while listening for a clear "click" as the timeclock terminates. Always double-check the setting to be certain the clock is operating consistently, not erratically.

Commercial Hot-Gas Defrost

Commercial hot-gas defrost systems are rather common, and their operation is similar to the hot-gas defrost in domestic systems described earlier in this chapter. Hot gas is taken directly from the hot-gas discharge line (close to the compressor) and delivered directly to the evaporator inlet. See **Figure 27-26**. A solenoid valve is located in the hot-gas bypass line to control flow. When the solenoid valve is open, hot gas travels directly to the evaporator inlet, bypassing the condenser, liquid receiver, and refrigerant metering device. However, when the solenoid valve is closed, hot gas *must* follow its normal route through the condenser.

The defrost timer is used to control electrical circuits to the hot-gas solenoid valve and evaporator fan(s). The clock does *not* control operation

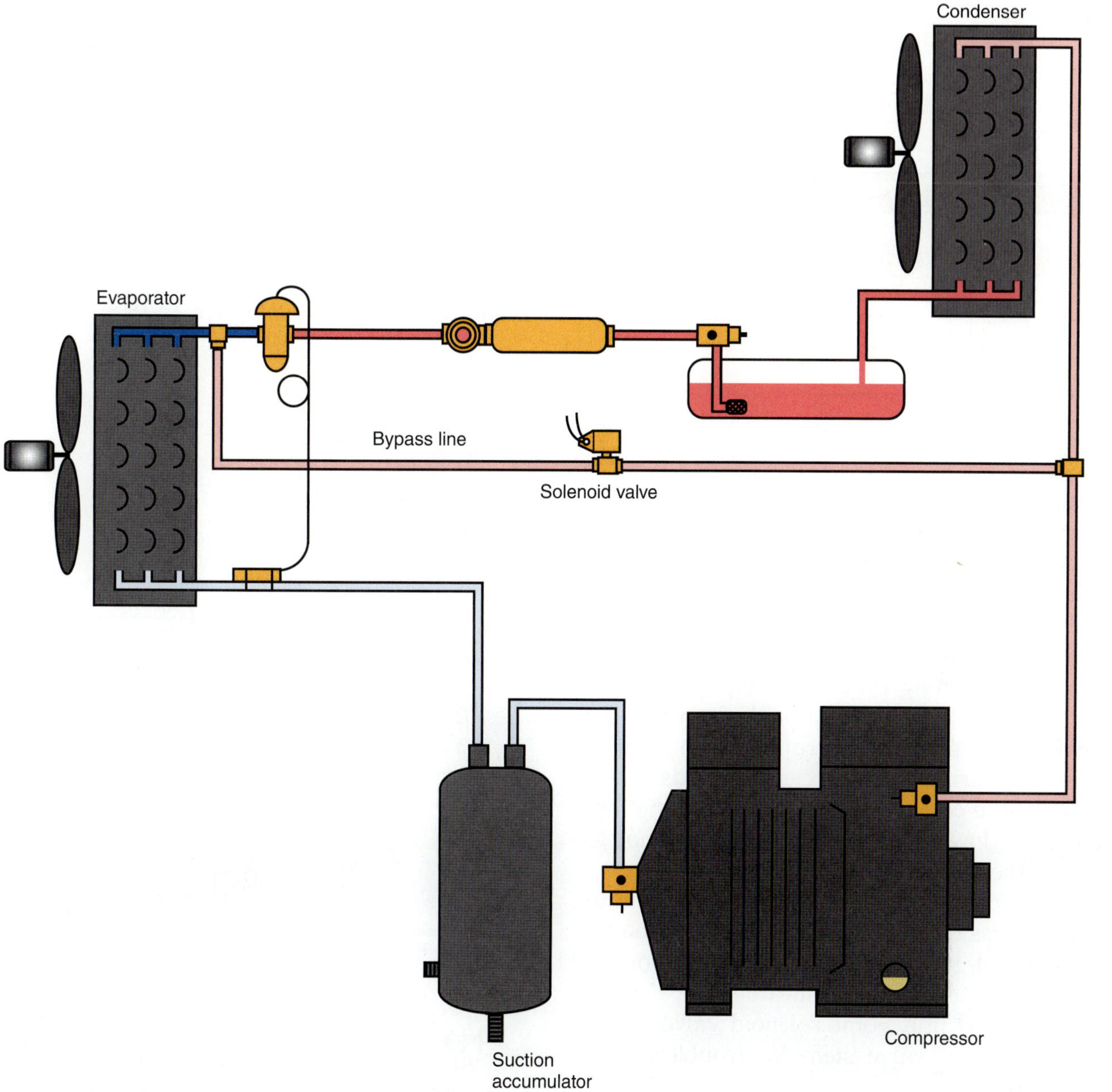

Goodheart-Willcox Publisher

Figure 27-26. A typical commercial hot-gas defrost system. The solenoid valve controls hot gas flow through the bypass line. An accumulator in the suction line is required to prevent liquid floodback, which could damage the compressor.

of the condensing unit since the compressor must operate to produce hot gas for the defrost cycle. See **Figure 27-27**. Complete defrost usually occurs within five minutes, with the fail-safe set for 10–20 minutes. Hot-gas defrost systems may require more frequent cycles because the length of defrost is determined by the amount of hot gas available.

Troubleshooting

The major disadvantage of hot-gas defrost is that the hot gas may condense back to a liquid as it enters the cold evaporator. Liquid floodback could occur and damage the compressor. All commercial hot-gas defrost systems are provided with an accumulator in the suction line to prevent liquid floodback during defrost.

Proper hot-gas defrost can be obtained only with sufficient head pressure. A condenser fan cycle control maintains head pressure and stops the condenser fan during defrost. Cold ambient temperatures drastically affect hot-gas defrost systems, so provision must be made to control the head pressure.

High suction pressure during the refrigeration cycle may indicate that the hot-gas solenoid valve

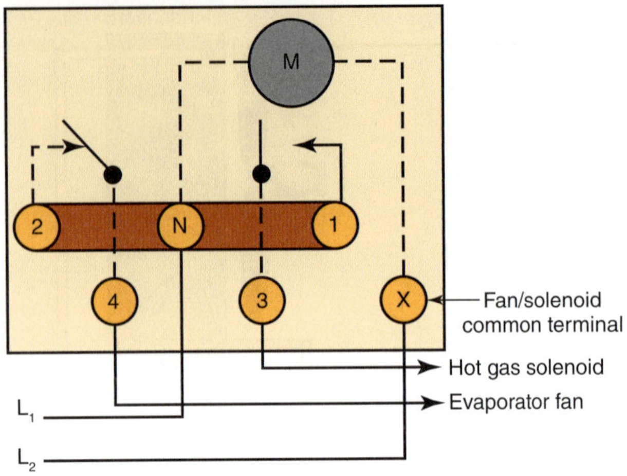

Goodheart-Willcox Publisher

Figure 27-27. In a hot-gas defrost system, the timeclock does *not* control the condensing unit. It only controls the hot gas solenoid and evaporator fans.

is leaking. The thermostatic expansion valve desuperheats the hot gas by permitting more liquid to enter the evaporator, resulting in high suction pressures and a prolonged run cycle. A leaking solenoid valve may produce a whistling sound, indicating gas is flowing through the valve seat. A leaking hot-gas valve permits gas to enter the evaporator during the off cycle. This leads to short cycling of the compressor, as it reacts to the rapid increase of low-side pressure caused by hot gas entering the low side through the leak.

To detect a leaking valve, feel the temperature of the hot gas defrost tubing some distance away from the hot-gas discharge line. Allow for heat from the hot-gas discharge line traveling along the hot-gas defrost line. Some solenoid valves have a manual shut-off valve stem for troubleshooting and isolation.

Defrost Contactors

The contacts on commercial timeclocks are normally rated at 40 A, but some electric defrost heaters require higher amperage. Timeclocks with heavier-duty contacts rated at 55 A per switch are available. The lug-type terminals accept wire sizes #4 to #14. Such heavy-duty timeclocks are available as time-terminated, temperature-terminated, or pressure-terminated models. Operation is identical to the timers previously described.

Three-phase systems often use a contactor to control the power supply to the defrost heaters. The contactor is located inside the control panel at the condensing unit and is mounted beside the compressor contactor. The timeclock controls both the coil on the compressor contactor and the coil on the defrost contactor. The defrost heater circuit is fused separately. See **Figure 27-28**.

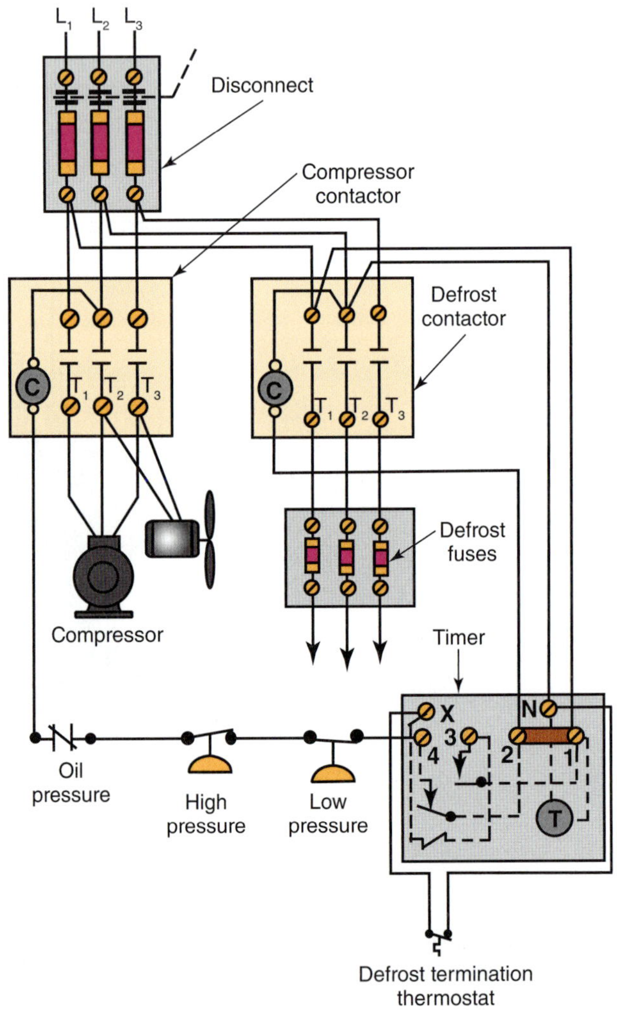

Goodheart-Willcox Publisher

Figure 27-28. Timeclocks are used to control several contactors. In this circuit, both the compressor contactor and the defrost contactor are controlled by the timer.

Chapter Review

Summary

27.1 Atmospheric air is a mixture of gases and water vapor (moisture). The moisture content of air is described as *humidity*. The amount of moisture, or water vapor, that air will hold depends on the temperature. Humidity directly influences human comfort.

27.1 Relative humidity is the amount of moisture in the air compared to the amount of moisture the air would hold if the air were saturated (filled to capacity, of moisture in this case). Dew point describes the point at which the air becomes saturated (holds all the moisture it is capable of holding) for a given temperature. All air-conditioning systems act as dehumidifiers because the evaporator lowers air temperature to below the dew point.

27.2 A defrost cycle is a sequence of operations used to remove frozen moisture from an evaporator and restore it to full efficiency. There are four types of automatic defrost systems: off-cycle, electric, hot-gas, and reverse-air. Some systems have manual defrost, which requires the use of hot water to soften frost and ice buildup.

27.2 All automatic defrost systems use a defrost timeclock to regulate frequency and duration of each defrost cycle.

27.2 Commercial systems may use off-cycle defrost with a timeclock, electric defrost with timeclocks and thermostats, or a hot-gas defrost system to control and execute the defrost cycle.

Know and Understand

1. The moisture content of air is described as _____. (27.1)
   A. condensation
   B. humidity
   C. wet bulb
   D. saturation
2. *True or False?* When relative humidity reaches 100%, any reduction of the temperature causes relative humidity to decrease. (27.1)
3. Which of the following statements about dehumidifiers is *not* true? (27.1)
   A. All air-conditioning systems act as dehumidifiers.
   B. Dehumidifiers serve as a latent heat load.
   C. Dehumidifiers only slightly decrease air temperature.
   D. Dehumidifiers use humidistats to control system operation.
4. Defrost cycles are necessary when evaporator temperature is below _____°F. (27.2)
   A. 0
   B. 32
   C. 45
   D. 78
5. A(n) _____ defrost system is controlled by a timeclock that automatically turns off the condensing unit, stops the evaporator fan, and turns on electric resistance heaters fastened to the evaporator. (27.2)
   A. off-cycle
   B. electric
   C. hot-gas
   D. reverse-cycle
6. The _____ defrost system uses warm ambient air to melt frost buildup on the evaporator. (27.2)
   A. off-cycle
   B. electric
   C. hot-gas
   D. reverse-cycle
7. Manual defrosting of a domestic refrigerator-freezer is done by using _____. (27.2)
   A. a chisel
   B. a heat gun
   C. a screwdriver
   D. containers of hot water
8. *True or False?* All automatic defrost (frost-free) refrigerator-freezers use a defrost timeclock to regulate frequency and duration of each defrost cycle. (27.2)

9. To prevent a defrost cycle from getting too hot, the defrost terminator thermostats are wired in _____. (27.2)
   A. parallel with the power supply from the timeclock
   B. series with the power supply from the timeclock
   C. series with the power supply from the neutral
   D. parallel with the system thermostat
10. A technician is troubleshooting a refrigerator-freezer experiencing a freeze-up. The refrigerator is equipped with an electric defrost system. The technician checks continuity between Terminals 2 and 4 during the refrigerator's normal operating mode and discovers continuity, what does this indicate? (27.2)
    A. The compressor is defective.
    B. There is a restriction in the capillary tube.
    C. The defrost heater is being energized during the run cycle.
    D. The issue is not with the defrost system.
11. The evaporator fan delay thermostat is mounted or clamped directly to the _____. (27.2)
    A. condenser tubing
    B. evaporator return bends
    C. metering device
    D. liquid receiver
12. What does the timeclock control on a domestic hot-gas defrost system? (27.2)
    A. the solenoid valve and evaporator fan
    B. mullion heaters
    C. the compressor
    D. the thermostat and metering device
13. *True or False?* Mullion heaters are often used in place of an automatic defrost system. (27.2)
14. In a commercial refrigeration system with off-cycle defrost, the evaporator fan runs _____. (27.2)
    A. continuously
    B. until the cabinet temperature reaches 32°F (0°C)
    C. until the cabinet temperature reaches 38°F (3°C)
    D. intermittently as needed
15. Commercial refrigeration defrost timeclocks operate on a _____-hour basis. (27.2)
    A. 1
    B. 8
    C. 12
    D. 24
16. *True or False?* The time-initiated/time-terminated defrost cycle can recognize seasonal weather changes and will automatically adjust the defrost cycle to compensate. (27.2)
17. A(n) _____ is used to prevent restarting of the fan until evaporator temperature drops to a set value. (27.2)
    A. anticipator
    B. relay
    C. evaporator fan delay thermostat
    D. fail-safe setting
18. The combination fan delay and defrost termination thermostat is easily recognized because it has _____ wires. (27.2)
    A. two
    B. three
    C. four
    D. six
19. A time-initiated/pressure-terminated timeclock links the clock diaphragm to _____. (27.2)
    A. discharge pressure at the compressor outlet
    B. suction pressure at the compressor
    C. liquid line pressure at the condenser
    D. the evaporator fan
20. Which of the following statements about commercial hot-gas defrost systems is *false*? (27.2)
    A. A solenoid valve is used to direct hot gas to the evaporator inlet.
    B. Hot-gas defrost systems may require more frequent cycles based on the amount of hot gas available.
    C. The defrost timer controls the operation of the condensing unit.
    D. A suction line accumulator helps prevent liquid floodback in hot-gas defrost systems.
21. *True or False?* Three-phase systems often use a contactor to control the power supply to the defrost heaters (27.2)

Apply and Analyze

1. A room is 10′ × 10′ with a 9′ ceiling. How much does the air weigh in the room? (27.1)
2. Explain how a dehumidifier can remove moisture from air without affecting air temperature. (27.2)
3. Why do air-conditioning (comfort cooling) systems *not* require a defrost cycle? (27.2)
4. What is unique about the evaporator fan motors used in a reverse-cycle defrost system? (27.2)
5. Why might it be dangerous if electric heat defrosting is completed before the set amount of time on the timeclock expires? (27.2)
6. What would happen if the evaporator fan restarted immediately after a defrost cycle in a freezer? (27.2)
7. Explain the operation of a domestic hot-gas defrost system. (27.2)
8. Describe how to properly troubleshoot an off-cycle defrost timer. (27.2)
9. What's the purpose of a timeclock in a commercial defrost system? (27.2)
10. What happens if an evaporator fan delay thermostat fails in the closed position? (27.2)
11. What are three common causes of malfunctioning evaporator fans? (27.2)

Critical Thinking

1. Imagine that the wiring diagram in **Figure 27-5** is your reference for a refrigerator-freezer with an automatic defrost system that you are servicing. While troubleshooting, you discovered that the defrost timeclock needed to be replaced. After replacing the timeclock, you plugged the refrigerator back in and the compressor and evaporator fan motor did not start and the freezer is getting extremely hot. Where did the error likely occur? (27.2)
2. A customer calls and says that their refrigerator is not properly cooling. They explain that they cannot hear the fridge's motor running or feel any airflow when they open the fridge door. What might be keeping the fan motor from working? (27.2)
3. Using the example in the previous question, image you are troubleshooting this refrigerator. First, with the refrigerator plugged in, you observe that the compressor is running and the evaporator is cold. Then, you use a multimeter to see if there is proper voltage going to the fan motor. The multimeter reads 0 V, which indicates there is no power going to the fan motor. What might be the problem? (27.2)

Goodheart-Willcox Publisher

28 Gas Heat with Air Conditioning

LEARNING OUTCOMES

After studying this chapter, you will be able to:

28.1 Identify the main components of air distribution in a forced-air system.

28.2 Describe the operation of the heating system and the various types of ignition systems used in gas-fired furnaces.

28.3 Explain the operation, components, and various types of thermostats used in HVAC systems.

28.4 Distinguish the operating principles of combination fan/limit controls and integrated control modules used in blower motors.

28.5 Recognize key steps in troubleshooting gas-fired furnaces, including being familiar with system sequence of operation and common problems and issues.

TECHNICAL TERMS

air filter
automatic gas valve
blower relay
blower section
cold anticipator
combination fan/limit control
combustion
crossover igniter
draft
fan switch
flame rectification
gas furnace
gas manifold
heat anticipator
hot surface igniter
hot surface ignition system
incomplete combustion
integrated control module
intermittent pilot ignition system
limit switch
pilot assembly
pilot flame
primary air
secondary air
spark igniter
standing pilot
standing pilot ignition system
thermocouple
thermostat
thermostat overshoot

Introduction

Most residential heating and cooling systems are the forced-air type, also known as *forced-convection* systems. Total comfort is achieved by automatically controlling air circulation (movement), temperature (heating and cooling), filtration (cleaning), and moisture content (humidity). During its operating cycle, a forced-air system heats or cools air and directs it to all areas of the conditioned space.

A ***gas furnace*** is a furnace that uses forced convection to distribute heated air that has been warmed by the combustion of gas. These are rated as to their efficiency of operation: 70%, 80%, or 90% efficiency. See **Figure 28-1**. Different efficiencies are possible through different designs and systems controls. A furnace can be in a basement, attic, crawl space, garage, or utility closet. Basic components common to residential gas-fired forced-air heating and cooling systems include air ducts, burners, blower, filter, evaporator, condensing unit, and thermostat.

A furnace contains a blower section (motor and fan) to circulate air in the home. Separate air ducts are used to control air movement to and from the furnace (supply and return).

Air conditioning is easily added to a gas forced-air heating system to remove heat and humidity from the air. In furnaces with air conditioning, the evaporator is installed inside the furnace plenum, and the condensing unit is located outdoors. The suction and liquid line connect the evaporator to the condensing unit, **Figure 28-2**. The furnace blower is used to circulate air during either heating or cooling cycles.

An air-sensitive thermostat is centrally located in the conditioned space to control system operation. The thermostat is designed to prevent the heating and cooling systems from operating at the same time.

28.1 Air Distribution System

A forced-air HVAC system uses an air distribution system that is composed of a network of ducts to control airflow to and from the conditioned space. Ducts carrying air *to* the conditioned space are called *supply air ducts*. Ducts bringing air *back* to the furnace for reconditioning are called *return air ducts*. A blower section (fan and motor) inside the furnace provides the forced airflow through the furnace and air ducts.

An air-conditioning evaporator (often called an A-coil because of its shape) is located inside the supply plenum. Supply air ducts connected to the supply plenum provide passage for conditioned air to reach the conditioned spaces in a building. The return plenum is a separate chamber located at the furnace air intake. Return air ducts provide passage for conditioned space air to return to the furnace for heating or cooling.

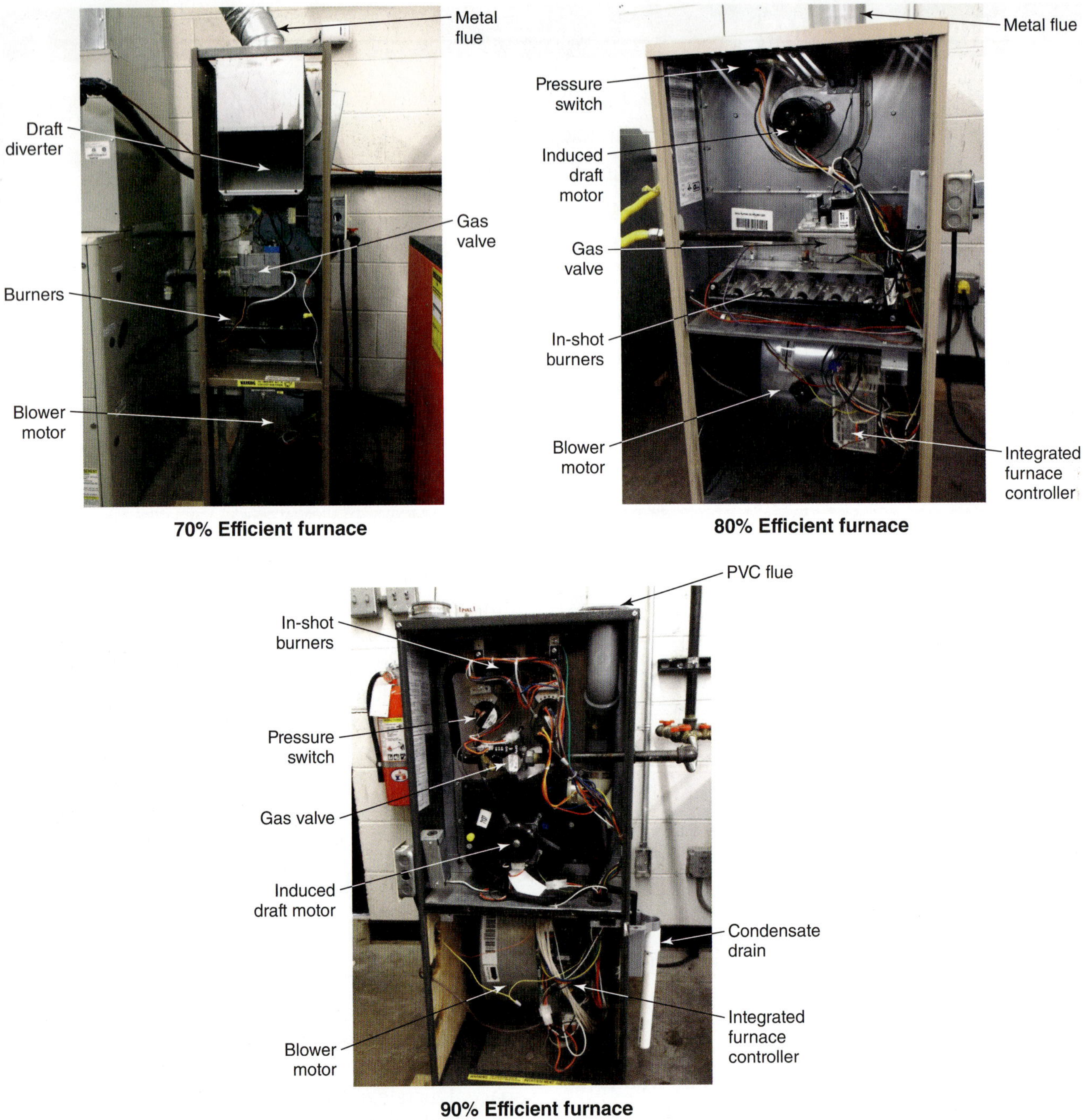

Goodheart-Willcox Publisher

Figure 28-1. Gas furnaces with varying efficiencies.

Supply air is released through outlets that are called *registers* or *diffusers*. Registers deliver air in a concentrated flow pattern and have adjustable vanes or dampers. Diffusers have stationary vanes and provide wider, fan-shaped patterns.

Inlets for return air are called *grilles*. Grilles are nonadjustable because return airflow to the furnace must be unrestricted. Ductwork and related components are discussed in more detail in Chapter 31, *Ductwork*.

Blower Section

The ***blower section*** is located inside the furnace cabinet. It consists of a motor and squirrel cage-type centrifugal fan. The fan draws air from the return air ducts and forces it through the furnace for conditioning (heating or cooling). Then, the conditioned air is pushed through the supply air ducts to individual rooms. There are two types of blowers: belt-drive (which were used in older

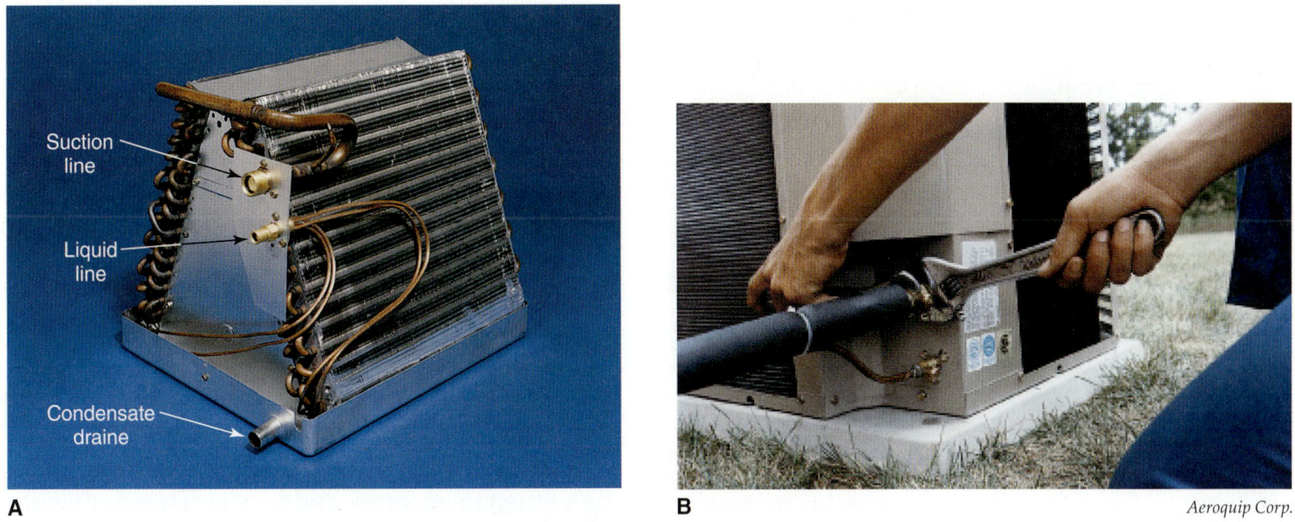

Aeroquip Corp.

Figure 28-2. Add-on air conditioning. A—The evaporator, often referred to as an A-coil because of its shape, is installed in the furnace plenum. B—The condensing unit is installed outside the building and connected to the evaporator by means of suction and liquid lines.

residential furnaces and are still used in commercial systems today) and direct-drive, **Figure 28-3**.

Air Filter

An ***air filter*** is a cleaning device that captures dust, lint, dirt, and other impurities from the circulating airstream of a furnace. The air filter is in the return-air side of the blower section. A fiberglass filter medium is commonly used as a screen for cleaning the air before it enters the blower. Filters should be replaced regularly. Certain special filters may not need replacement but require regular cleaning. A dirty air filter reduces system airflow, which significantly reduces system capacity. Dirty air filters can cause the furnace to overheat during the heating season or to freeze up the evaporator coil during the cooling season, **Figure 28-4**.

Disposable air filters consist of a rectangular fiberglass screen in a cardboard frame, **Figure 28-5**. Disposable air filters are available in many sizes and in two types: standard and pleated. A standard air filter traps large particles and should be changed monthly. A pleated air filter has finer openings than the fiberglass screen and is pleated to expose more surface area to circulating air. These can trap much smaller particles. An air filter should be sized to fit completely across the

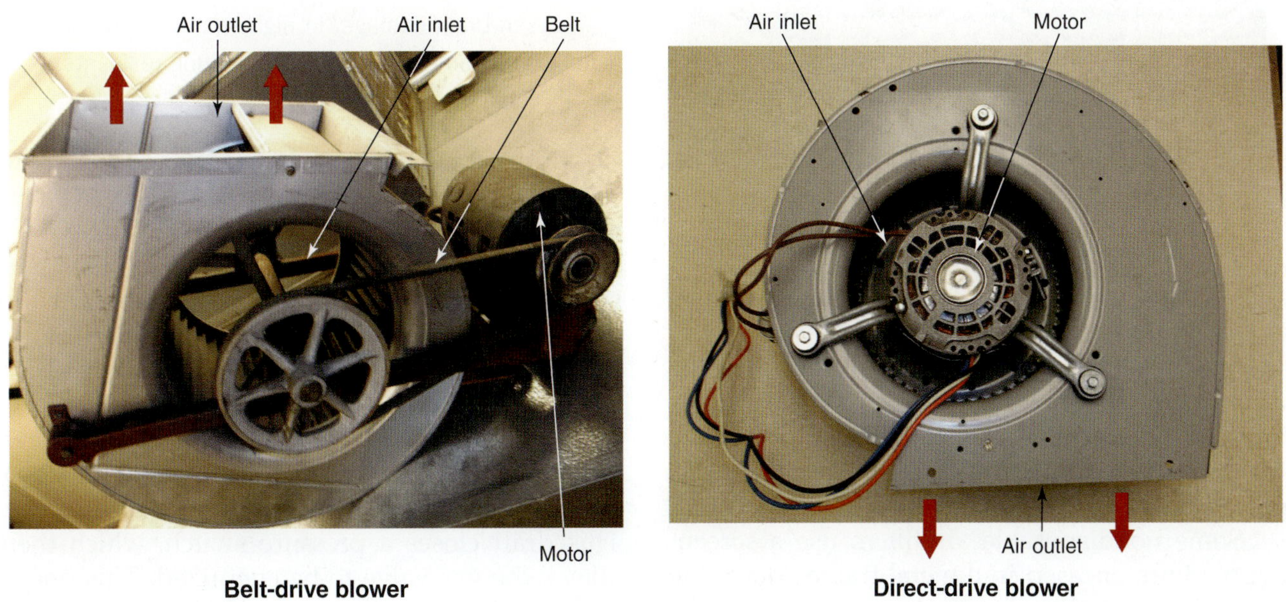

Goodheart-Willcox Publisher

Figure 28-3. Furnace blowers.

Robert Reasor/Shutterstock.com

Figure 28-4. A forced-convection furnace with frozen evaporator coils.

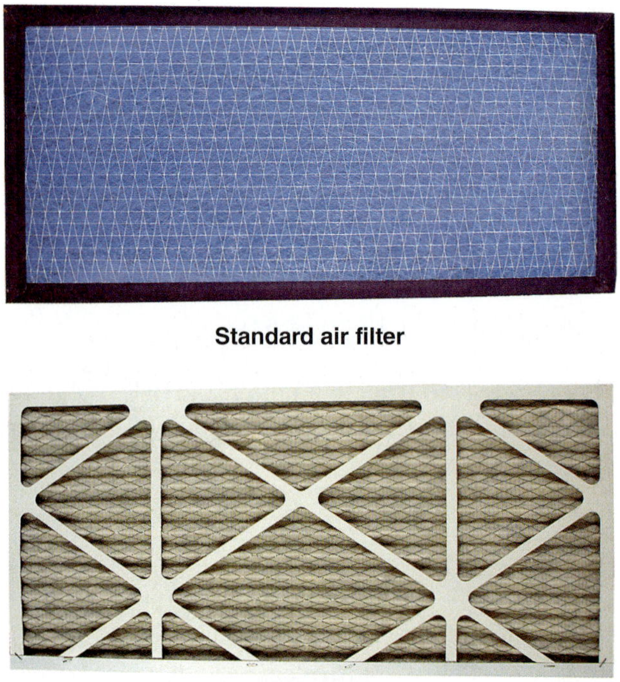

Standard air filter

Pleated air filter

Goodheart-Willcox Publisher

Figure 28-5. Standard air filters trap large particles. Pleated air filters have more surface area and can trap small particles, cleaning the air more effectively than standard air filters.

incoming airstream. Arrows on an air filter frame indicate the proper direction of airflow. If airflow is permitted to bypass the filter, then the filter is not doing its job.

Some nondisposable air filters use a screening medium encased in a metal frame. When the screen becomes dirty, it may be cleaned by washing or vacuuming before putting it back into use.

Never operate an HVAC system without an air filter. Dirt will gather on the evaporator, greatly reducing airflow. Cleaning the evaporator is a difficult and time-consuming task that can be avoided by proper filter replacement or cleaning. Dirt and lint can also plug fan motor ventilation and cause the motor to overheat, resulting in expensive repairs.

28.2 Heating System

Gas-fueled furnaces rely on ***combustion*** (burning of fuel) to generate usable heat. The combustion process produces waste products, such as carbon dioxide, water vapor, and small amounts of other chemical compounds. ***Incomplete combustion*** may produce such substances as carbon monoxide, aldehydes, and soot, which are dangerous to human health. Gases and other waste products resulting from combustion must be properly vented to the outdoors.

SAFETY NOTE

Proper Venting

Proper venting of gas-fired appliances (furnaces, water heaters, etc.) is of extreme importance. Carbon monoxide (CO) is a colorless, odorless, and poisonous gas. Breathing in enough carbon monoxide can cause headaches, dizziness, nausea, vomiting, body weakness, other flu-like symptoms, and even brain damage. In high amounts, carbon monoxide can asphyxiate unsuspecting residents.

The heating section of a furnace fueled by natural gas or liquefied petroleum (LP) gas consists of burners, a heat exchanger, and a venting system. The heat exchanger is usually divided into sections. Burners fit into special chambers at the inlet of a heat exchanger. One factor determining the efficiency of a furnace is the design of its heat exchanger, **Figure 28-6**.

On newer furnaces (80% to 90% efficiency), when a thermostat calls for heat, the combustion fan turns on, creating a draft through the furnace. This draft closes a pressure switch, which then allows the gas valve to be energized. This opens the gas valve and allows gas to flow to the burners, **Figure 28-7**.

70% furnace heat exchanger

80% furnace heat exchanger

90% furnace heat exchanger

Goodheart-Willcox Publisher

Figure 28-6. Heat exchangers transfer heat from combustion gases to recirculated room air, while keeping the two separated. Heat exchangers are designed differently for the different efficiency levels of furnaces.

Goodheart-Willcox Publisher

Figure 28-7. The air pressure switch determines whether a combustion blower is operating. After confirmation, the gas valve opens to allow gas to flow to the burners for ignition and combustion.

The gas in the burners is ignited, and the flames warm the heat exchanger. Hot gas products from combustion travel through the heat exchanger passageways. Air circulated by the blower travels around the outside of the heat exchanger passageways where it absorbs heat used to warm the building. Any product of combustion gases not used is directed to a vent. A vent pipe (flue) runs from the vent connection to the outside of a building or to a chimney, **Figure 28-8**. On 70% furnaces, the vent is connected to a box called a *draft hood* or *diverter*. The vent flue pipe runs from the diverter box to the chimney. On 80% and 90% furnaces, the flue pipe connects to a combustion blower that aids in the travel of flue gases through the heat exchanger.

A heat exchanger is designed to keep the recirculated room air separate from the combustion gases. The blower forces recirculated room air around and between the heat exchanger sections. The air is heated by conduction as it touches the hot surface of the sections. The heated air travels from the heat exchanger to the supply plenum and is delivered to living spaces. The name plate on the furnace will state what the desired temperature rise range is for that furnace.

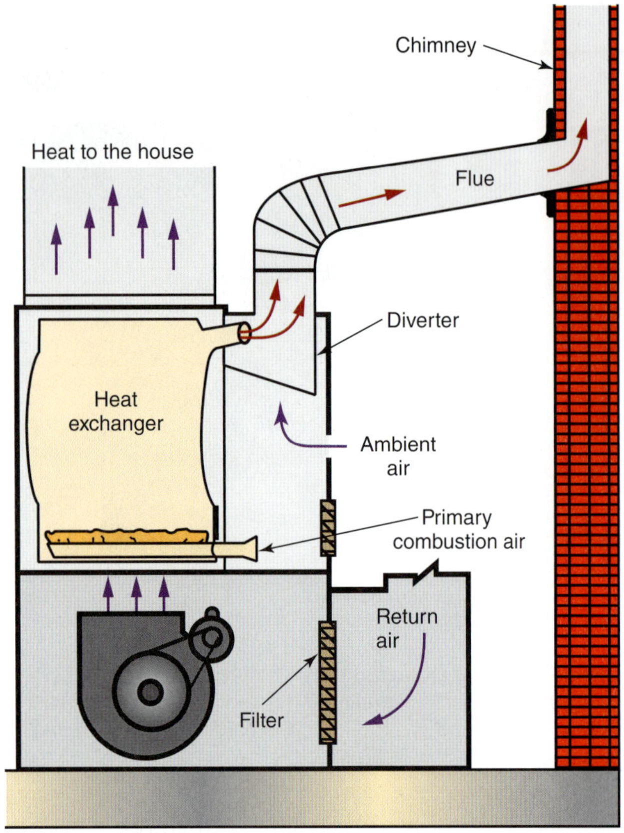

Goodheart-Willcox Publisher

Figure 28-8. A heat exchanger transfers heat from combustion gases to recirculated room air, while keeping the two separated. Hot waste gases are drawn up the flue and out of the chimney.

Gas Burners

The number of gas burners in a furnace depends on the size of the furnace. Gas burners vary in style and shape but essentially work the same: when the thermostat calls for heat, a mixture of atmospheric air and gas enters the burner and is ignited on the surface of the ports. Both primary and secondary air must be provided to the burner for complete and proper combustion.

For natural gas furnaces to have complete combustion, the ideal ratio of air to fuel is 2 parts oxygen to 1 part fuel. Because the air around us only contains about 2% oxygen, the ratio of air to fuel needs to be 10 parts air to 1 part fuel. But to help protect from the possibility of having incomplete combustion, excess air is added. Therefore, to have complete combustion, 15 cubic feet of air is needed for every 1 cubic foot of fuel.

Primary air is atmospheric air that is drawn into the burner by the jet of gas. The air is mixed thoroughly with gas inside the burner before delivery to the ports. ***Secondary air*** is air that is drawn into the burner chamber after ignition of the flame to help support complete combustion, **Figure 28-9**. On 70% furnaces, the natural convection (the rising of less-dense hot air) creates a ***draft***, or air movement, into the burner chamber, up through the heat exchanger, and out the flue. On 80% and 90% furnaces, a combustion blower forces (pushes) or induces (pulls) the combustion gases through

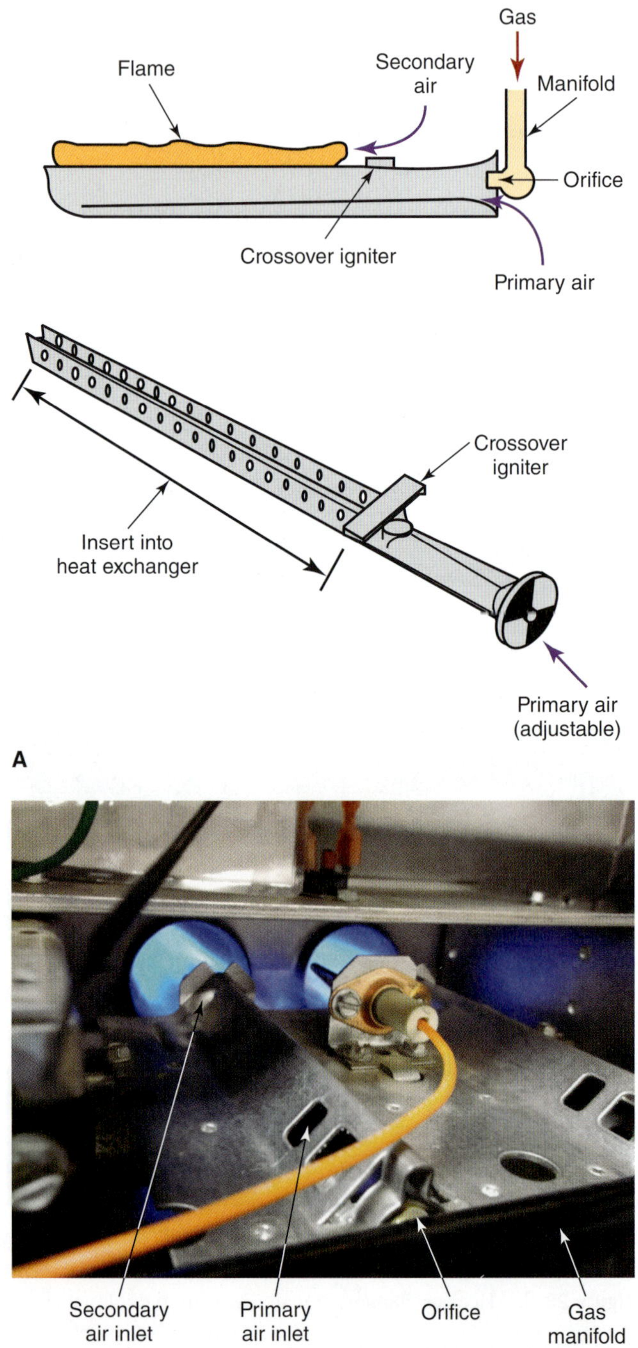

Goodheart-Willcox Publisher

Figure 28-9. A—Burners combine air and gas into a mixture that can be burned in a furnace. Primary air may be adjusted by rotating a shutter on the burner body. B—An in-shot burner used in a high-efficiency gas furnace.

the heat exchanger. Primary air is rarely sufficient for complete combustion. Secondary air provides the additional air needed for complete combustion of the fuel gas.

The amount of primary air on a 70% furnace is controlled by small, adjustable shutters located at each burner inlet. Secondary air cannot be controlled. A *yellow* gas flame indicates poor combustion. A *blue* flame indicates good combustion. Addition of secondary air results in a blue flame that is smaller, faster burning, and hotter than a yellow flame.

Gas Valve and Gas Manifold

The supply of fuel gas to a furnace is governed by a manual gas shut-off valve located just outside a furnace. After the shut-off valve, there is a drip leg in the gas line before the piping enters the furnace. The gas supply inside the furnace connects to a gas valve. An ***automatic gas valve*** is used to start and stop the flow of gas to the burners, **Figure 28-10**. This gas valve includes a solenoid valve that is operated by the heating system thermostat. The automatic gas valve connects to a pipe, called a ***gas manifold***, that distributes gas to the burners through an orifice. The orifice is a precision-sized hole drilled into a replaceable spud.

PRO TIP

Gas Burner Orifice

The orifice is dependent on the fuel source and needs to be changed if changing from natural gas to liquid petroleum (LP) gas or LP to natural gas. Kits to accomplish this task can be purchased through a local supply house.

On 70% furnaces, the gas valve contains a 24 V solenoid valve. When the solenoid coil is energized by the thermostat, the valve opens, and gas flows to the burners. See **Figure 28-11**. When the coil is de-energized, the gas valve closes, and the burner flame goes out, having no fuel to burn. The system thermostat controls electrical power supply (24 V) to the gas valve's solenoid. When the thermostat calls for heat, electrical power is supplied to the gas valve to open its solenoid valve.

On 80% and 90% furnaces, the gas valves can have multiple 24 V solenoids to control the gas flow to the manifold depending on the number of gas valve stages and the way the burners are lit. The gas valves are energized after the combustion blower air pressure switch proves that the combustion blower is on.

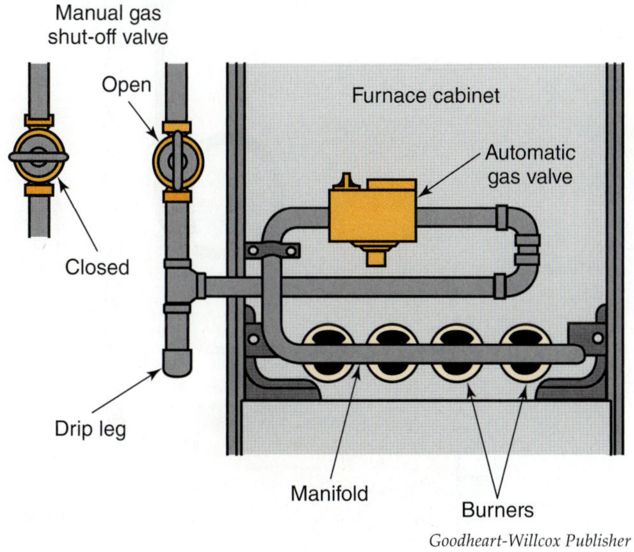

Goodheart-Willcox Publisher

Figure 28-10. An automatic gas valve controls the flow of gas from the supply to the manifold and the burners.

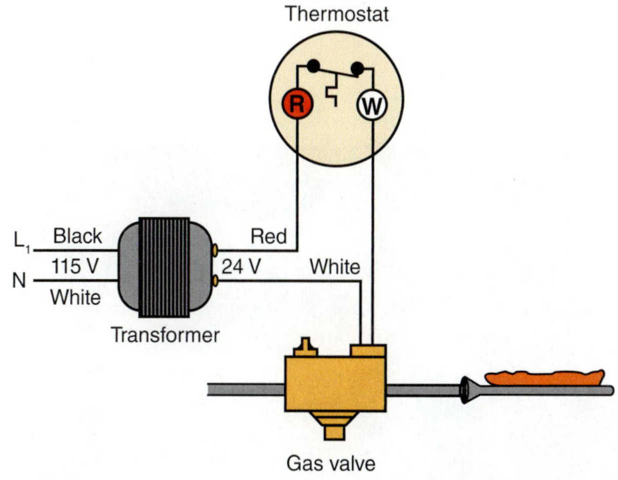

Goodheart-Willcox Publisher

Figure 28-11. The system thermostat controls power to the gas valve's solenoid. The control circuit operates on 24 V, opening and closing the gas valve's solenoid valve to regulate gas flow to the burners.

A gas furnace's burners are ignited by one of three types of ignition systems. One of these three ways will be found in any type of furnace:

- Standing pilot
- Intermittent pilot ignition
- Direct ignition

Standing Pilot Ignition System

A ***standing pilot ignition system*** uses a pilot flame to light its gas burners. This ignition system is controlled by a standing pilot combination gas valve, which is a combination gas valve having such features as manual shutoff, pilot light safety control, and an internal gas pressure regulator. See **Figure 28-12**. A manually operated knob on top of the gas valve has three positions:

- **On:** Normal operating position. The opening and closing of the gas valve solenoid are electrically controlled by the thermostat at Terminal W.
- **Off:** Used to manually shut off the gas supply to the furnace.
- **Pilot:** Used to light the pilot without opening the main valve. Depressing the knob in this position permits gas flow only to the pilot valve.

Goodheart-Willcox Publisher

Figure 28-12. A combination gas valve with manual shut-off, an internal pressure regulator, and pilot push button.

A ***pilot flame*** is a small gas flame used to ignite the flame on the main burners. A constantly burning pilot is called a ***standing pilot***. Gas is constantly supplied to a standing pilot through a pilot connection on a gas valve. A standing pilot is always burning and ready to ignite the main burners during a call for heat. If a pilot goes out, the pilot push button is used to re-ignite the pilot flame.

Gas flow to the main burners cannot be allowed when the pilot flame is out (not lit). Without a pilot flame, gas would collect inside the furnace and eventually explode when exposed to a spark or flame.

A ***thermocouple*** is a heat-sensitive safety device used to confirm that a pilot light is lit. A thermocouple prevents the main gas valve from opening if there is no pilot flame, **Figure 28-13**. It converts heat energy to electrical energy. The sensing element of the thermocouple is positioned within the pilot flame. Heat from the flame

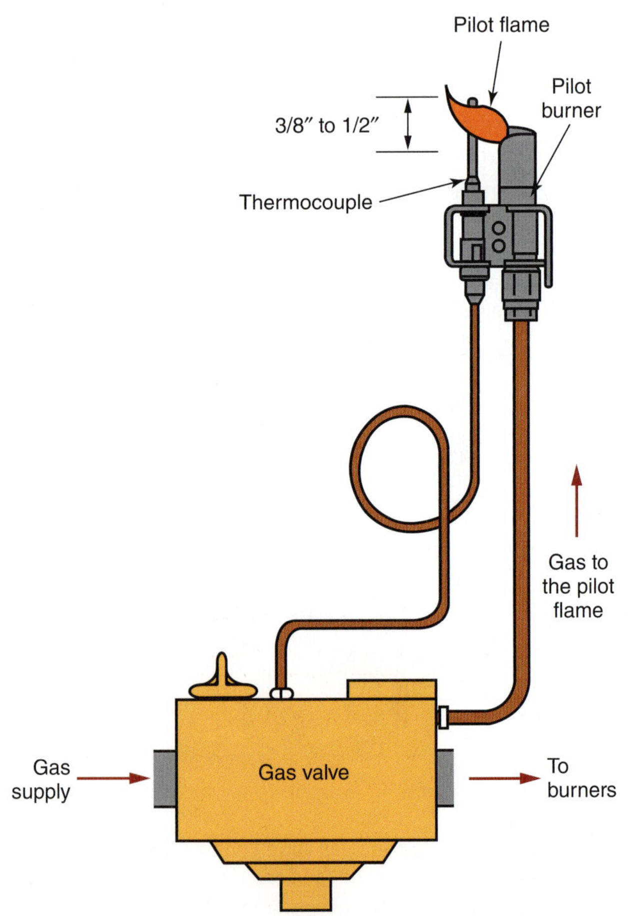

Goodheart-Willcox Publisher

Figure 28-13. In a gas furnace, a thermocouple acts as a safety device that confirms the existence of a pilot flame. Without the electrical signal generated by the thermocouple, the gas valve will not open to supply fuel to the burners.

generates a tiny amount of electricity (measured in millivolts) in the thermocouple. The electrical energy is conducted to the main combination gas valve as a signal that a pilot flame is available. If the pilot flame goes out, the electrical signal is *not* generated, and the main gas valve cannot open. This safety feature prevents gas flow to the burners when the pilot flame goes out.

A pilot light should have a soft blue flame and envelop (cover) between 3/8″ and 1/2″ (10 cm and 13 cm) of the thermocouple's tip (hot junction). Proper positioning of the thermocouple within the pilot flame is critical. A pilot flame is adjustable by a screw on the combination gas valve. Remember, a gas valve cannot open unless the thermocouple generates the proper signal. Minimum acceptable dc voltage is 18 mV.

A ***crossover igniter*** is a component used to ignite several main burners from a single pilot flame. It is a slotted length of metal that acts as a bridge between burners, **Figure 28-14**. Gas flowing to the main burners passes through the igniter slots until it comes in contact with the pilot flame. Properly positioning the burners will correctly align the crossover igniter. A poorly positioned burner will cause noisy and delayed ignition.

Intermittent Pilot Ignition System

An ***intermittent pilot ignition system*** lights a pilot flame just before lighting the main burners. It saves fuel by eliminating the continuously burning (standing) pilot flame. The pilot flame is ignited by an electric spark and burns only on a call for heat. If the pilot flame blows out during a call for heat, the system will automatically relight the pilot. An intermittent pilot ignition system uses a ***pilot assembly***, which consists of a pilot burner, a spark electrode, and a sensing probe, **Figure 28-15**.

When a thermostat calls for heat, the ***spark igniter*** produces an electric arc across its electrodes to automatically light the pilot flame. Then, the sensing probe confirms the presence of the pilot flame. Once the pilot flame is confirmed, through flame rectification or thermal detection, the spark igniter is de-energized, and the main gas valve is opened. The main burners are ignited by the pilot flame. When the thermostat is satisfied and opens its switch, the main gas valve is turned off. The flow of gas to the burners and the pilot flame stops.

Hot Surface Ignition System

In a ***hot surface ignition system***, the main gas valve is opened and the burners are immediately lit by an igniter. Hot surface ignition does *not* use a pilot flame at all. When there is a call for heat, the main burners are lit directly by either a spark igniter or a hot surface igniter. A ***hot surface igniter*** uses a special high-resistance element that becomes extremely hot when voltage is applied, **Figure 28-16**. It can be used to light the main burners directly.

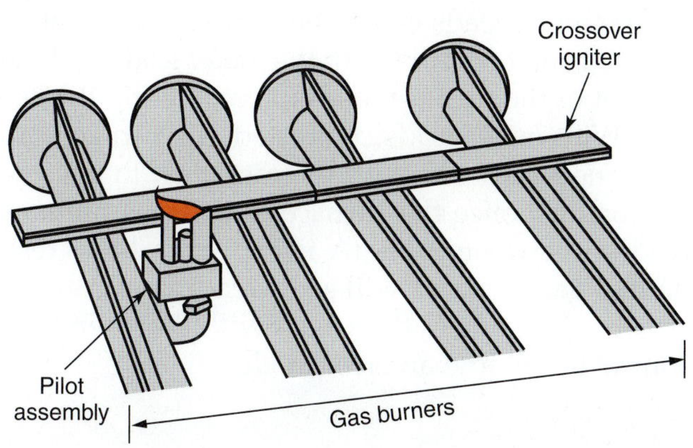

Goodheart-Willcox Publisher

Figure 28-14. A crossover igniter serves as a bridge between burners, allowing several burners to be ignited by a single pilot flame.

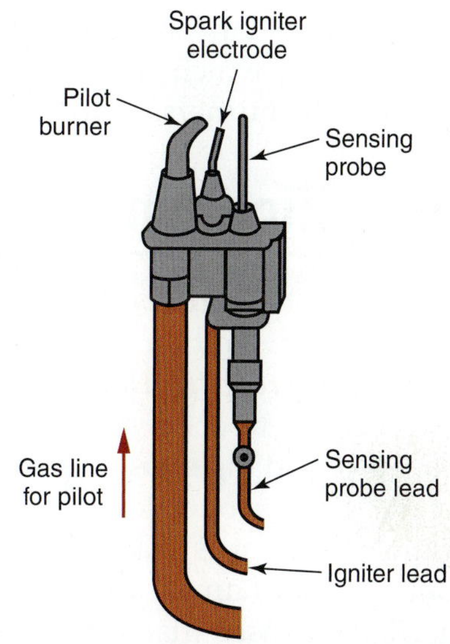

Goodheart-Willcox Publisher

Figure 28-15. When the thermostat calls for heat, an intermittent pilot ignition system lights the pilot flame, which then lights the main burners. By only lighting a pilot flame when there is a call for heat, an intermittent system eliminates the cost of fueling a constantly burning standing pilot.

Goodheart-Willcox Publisher

Figure 28-16. A direct ignition burner and hot surface igniter.

Once the main burners are lit, flame rectification proves that the burners are lit. If the burners are not proven to be lit by the control board, the gas valve will shut off gas flow as a safety feature.

Direct Spark Ignition System

A ***direct spark ignition system*** uses an electric spark to ignite its burners. Similar to a hot surface ignition system, no pilot light is used. The ignition control module sends voltage through two electrodes. The arcing that occurs between the electrodes is a result of current jumping from one electrode to the other, which creates the spark at the igniter to light the burners.

Flame Rectification

An essential safety feature equipped with all types of gas furnace ignition systems is flame rectification. ***Flame rectification*** is a process used on gas furnaces to confirm the presence of a flame and prevent dangerous operating conditions. Flame rectification operates by using a flame sensor or flame rod that is placed in the burner compartment. When a flame is established by the burners, the flame sensor (flame rod) senses the flame and signals to the control board to continue the sequence of operation.

When the gas ignites at the burners, an ionization process occurs which allows the flame to conduct electricity. An integrated control module or other flame sensing unit will send an ac voltage signal to a flame sensor located at the burners. This flame sensor is made of steel and is used as an electrode to transfer this voltage signal. The flame sensor is insulated so that it can be connected to the burner assembly without any electrical connection between the flame sensor and the ground of the furnace. These flame sensors will typically have 120 V or less applied to them, depending on the manufacturer. When the furnace starts the combustion process, whether it is lighting a pilot assembly or lighting a main burner (hot surface or direct spark ignition), the input voltage is found at the flame sensor. Once the flame is present, the ionized gas within the flame acts as a diode, allowing the ac current to only flow in one direction through the flame and into the furnace burner ground. This ground has a path back to the control module to complete the circuit path. Once the circuit is confirmed by the furnace's control module, then it will allow the flame to continue burning. Any disruption of the current path through the flame sensor will stop gas flow to the burners. In the process of rectifying the flame, ac current is converted to dc current. This current is read in dc microamps (μA) and can be a useful measurement when troubleshooting gas furnaces.

Technicians can measure the flame current by connecting a multimeter set to read dc microamps in series with the flame sensor path. With the power turned off to the furnace, the wire connected to the flame sensor can be removed. One meter lead is connected to the removed flame sensor wire and the other lead is connected to the flame sensor itself. The use of alligator clips for the meter leads can be helpful. It is imperative that no contact occurs with the meter leads and the cabinet of the furnace to avoid completing the circuit. With meter leads connected, the furnace can be turned back on. With the meter set to read dc microamps, once the furnace establishes flame, a reading between 2–25 μA is expected. Most residential gas furnaces will see a reading of 2–6 μA. Any reading of less than 2 μA will likely send the control board into an error mode.

28.3 Heating Thermostats

A ***thermostat*** is a switch that opens and closes according to changes in temperature. An HVAC system thermostat is located within the living space where temperature is being controlled. It turns heating and air-conditioning equipment on and off automatically to maintain specific temperatures within the space. Typically, only one thermostat is used, and it must be located where it can sense temperature uniformly. Cold walls and doorways must be avoided.

Low-Voltage Heating Thermostats

Thermostats for comfort heating and cooling are different from those used for refrigeration. Many heating and cooling thermostats are now designed to operate on 24 V because they are often located at a considerable distance from the heating and cooling equipment. Low voltage permits the use of small wires that are easily hidden from view. The 24 V system has other advantages. It produces little arcing of the switch contacts, resulting in long switch life. Also, both shock and fire hazards are virtually eliminated.

Power is supplied to the thermostat by the secondary winding of a transformer. A red wire is used to connect the power supply from the transformer secondary (24 V) to Terminal R on the thermostat. The thermostat becomes an automatic switching device that controls the 24 V power supply to various electrical components. Thermostats are divided into mechanical or digital types. Mechanical thermostats operate using either mercury or snap action. Digital thermostats include programmable or smart types.

Mechanical Thermostats

Mechanical thermostats are constructed with a small, coil-shaped, bimetallic (brass and steel) element. The coil is extremely sensitive to small changes in temperature. The end of the coil at the center of the spiral is anchored. As the coil winds and unwinds in response to small temperature changes, the free end moves in an arc, **Figure 28-17**.

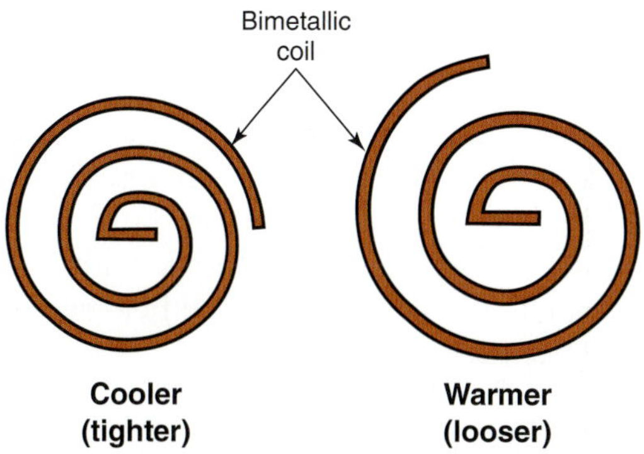

Figure 28-17. The coil is anchored at the center of the spiral so the outside end moves in an arc as the temperature changes. As the air cools, the coil becomes tighter. As the air warms, the coil becomes looser.

At the free end of the bimetallic element is a set of copper-coated contacts that close when there is a call for heat.

Thermostat Overshoot

When the room temperature has reached the desired level in heating mode, the thermostat opens and shuts off the gas valve. However, the blower motor continues to run and deliver heated air until the furnace cools down to 100°F (38°C), depending on whether the system uses a combination fan/limit switch or an integrated control module (discussed later in this chapter). As a result, room temperature may become too high. Delivering too much heat after the thermostat is satisfied is called *system overshoot* or ***thermostat overshoot***.

Heat Anticipator

To overcome thermostat overshoot, a small resistance heater—called the ***heat anticipator***—is mounted inside the thermostat near the bimetallic coil. Current flow through the heat anticipator during the heating cycle warms the bimetallic coil. Heat from the anticipator causes the thermostat switch to open just before room temperature reaches the desired level. The thermostat ends its call for heat, but the furnace blower continues to operate to cool the furnace to 100°F (38°C). The additional heat conveyed by the blower raises the temperature of the living space to plus or minus 1°F (0.5°C) of the desired temperature.

Heat Anticipator Operation

The heat anticipator is energized when the furnace is in heating mode. The small heater is factory-connected between the copper contact and Terminal W (the heat terminal) on the thermostat. This places it electrically in series with the gas valve in a gas furnace or with the heat sequencer in an electric furnace.

Heat Anticipator Adjustment

The heat anticipator has a slide with a pointer and a scale calibrated in tenths of an ampere, **Figure 28-18**. The pointer is moved to the scale reading that agrees with the amperage flow of the heating circuit. Such a low amperage draw can be measured with a digital clamp-on ammeter.

The digital ammeter is clamped around the thermostat wire coming from Terminal W of the thermostat to the furnace. The heat anticipator should be set to the same amperage to control override. For example, if the draw of the circuit is 0.45 A, the anticipator pointer must point to 0.45 on the scale. Matching amperages provide the additional heat necessary for precise control of room temperature.

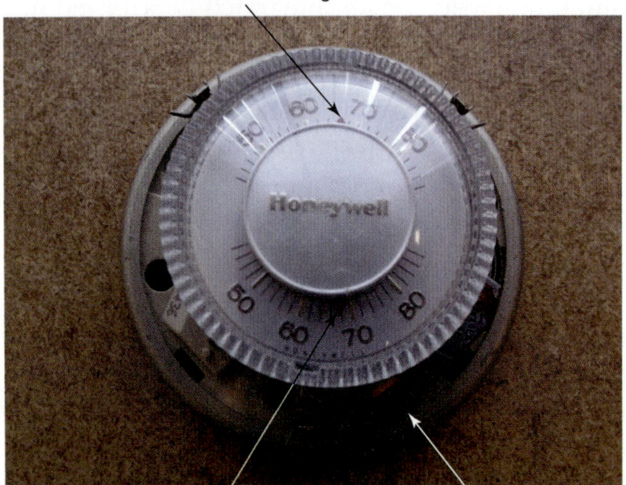

Goodheart-Willcox Publisher

Figure 28-18. The scale and pointer for a heat anticipator is visible by removing the thermostat cover. The pointer should be set to the scale marking equal to the amperage draw of the heating control circuit.

Figure 28-19 illustrates, in both pictorial and schematic forms, a complete gas heating system of a 70% furnace. **Figure 28-20** illustrates the wiring diagrams that might be found on an 80% or 90% furnace.

Most thermostats in the field today are digital thermostats and do not have heat anticipators. Instead, digital thermostats have the functions of the heat anticipator designed into the logic of the thermostat and no adjustments are needed. However, there are still older mechanical thermostats in use, so it is good to be aware of how to adjust the heat anticipator on these types of thermostats.

Heating and Cooling Thermostats

When heating and cooling systems are combined for year-round comfort, it is impractical to use a separate thermostat for each system. The different switching requirements are easily combined into a single thermostat and subbase that includes an extra set of contacts for the cooling switch.

Unwanted switching back and forth from heating to cooling is prevented by a three-position *manual* switch on the thermostat, labeled COOL-OFF-HEAT. The switch prevents the thermostat from energizing both systems at the same time or automatically switching back and forth between them, **Figure 28-21**.

When the manual switch is in the OFF position, both heating and cooling systems are turned off (unable to operate). When the switch is in the HEAT position, only the heating system is able to operate, and the cooling mode is disconnected. When the switch is in the COOL position, only the cooling system is operational, and the heating mode is disconnected.

Cooling Cycle

When the manual switch on the thermostat is set to COOL, and the cooling contacts close, 24 volts are supplied to Terminal Y on the thermostat. A yellow wire connects Terminal Y with the coil on the contactor in the outdoor condensing unit. A white wire connects the other side of the contactor coil to the neutral side of the 24 V transformer, **Figure 28-22**. When the contactor coil is energized, the outdoor condensing unit is turned on.

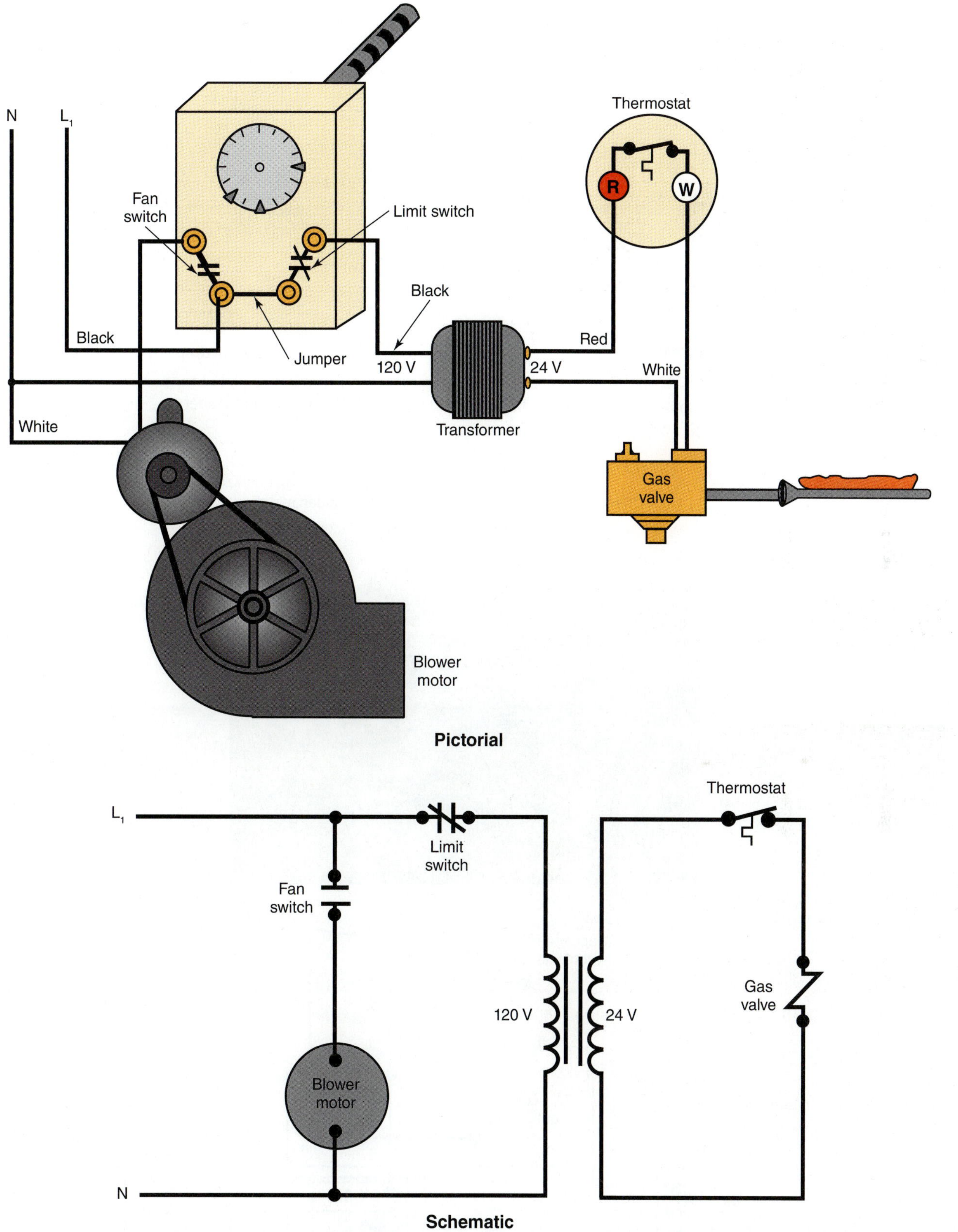

Goodheart-Willcox Publisher

Figure 28-19. A typical standing pilot gas heating system shown in both pictorial and schematic forms.

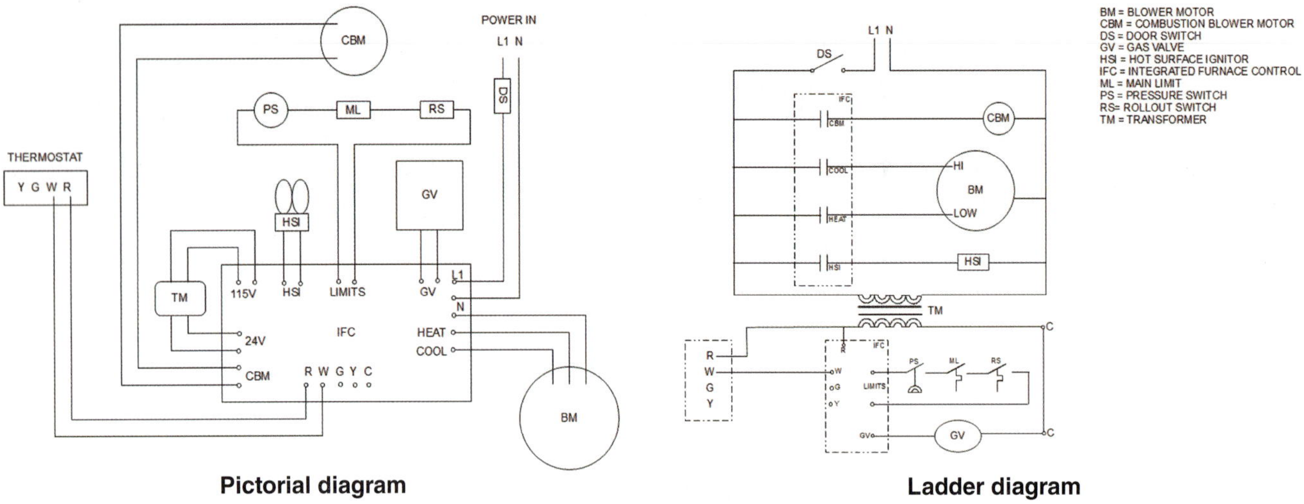

Goodheart-Willcox Publisher

Figure 28-20. A typical gas heating system on an 80% or 90% furnace shown in both pictorial and ladder diagrams.

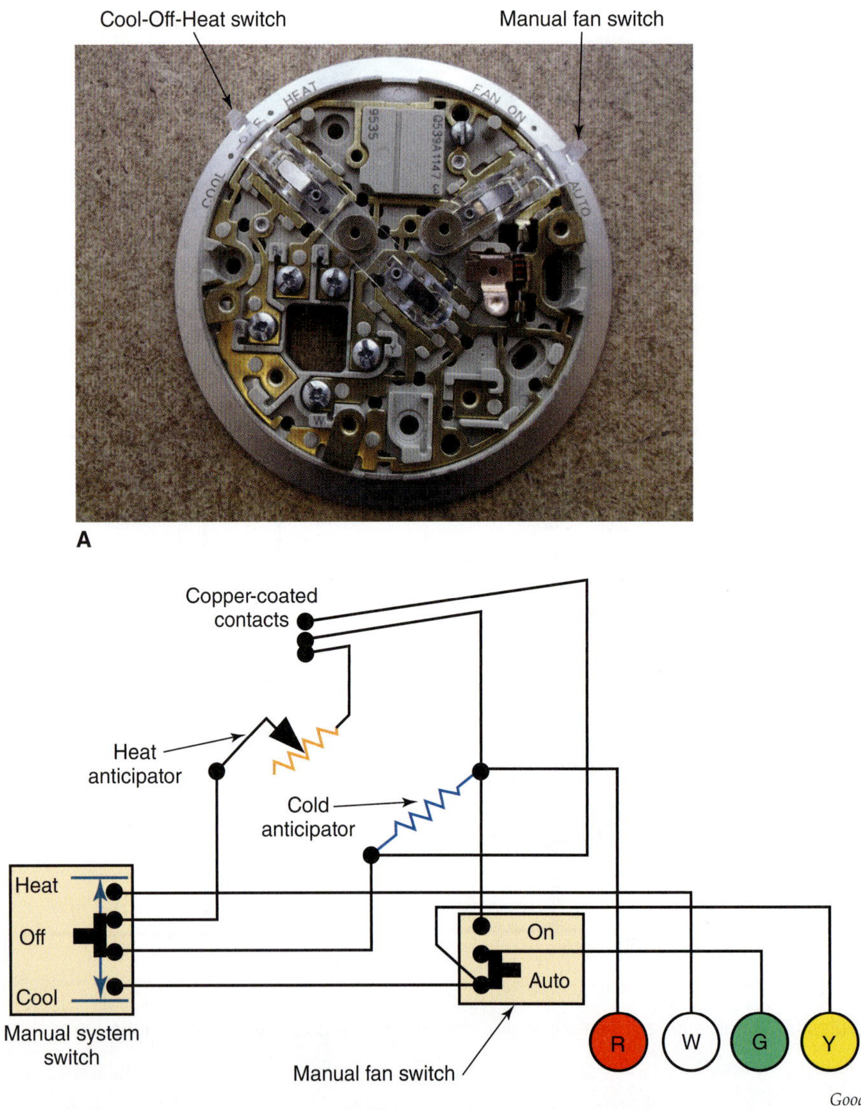

Goodheart-Willcox Publisher

Figure 28-21. Manual switching of modes. A—The COOL-OFF-HEAT manual switch is on the subbase of the thermostat. The fan switch is also in the subbase. B—Schematic showing connections for the manual-mode switches.

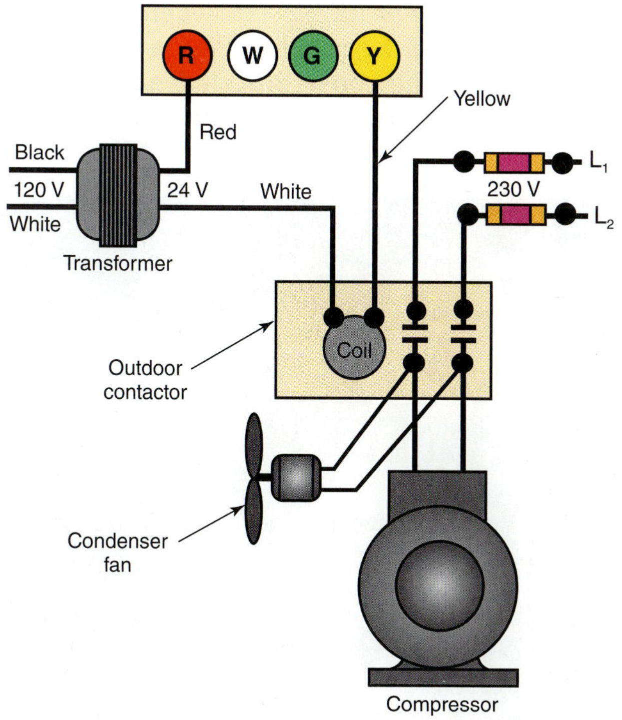

Goodheart-Willcox Publisher

Figure 28-22. When the cooling cycle is active, the circuit is completed from Terminal Y of the thermostat to the contactor on the outdoor condensing unit.

Cold Anticipator

A thermostat's cold anticipator does the same job as the heat anticipator, only in reverse. A ***cold anticipator*** must compensate for the delay between the call for cooling and the time when the system begins delivering cool air. The cold anticipator is a small, fixed resistor wired in parallel with the cooling switch contacts. When the contact switch is closed during a call for cooling, the cold anticipator is in parallel. Since the cold anticipator is in parallel and the circuit through the cooling contacts has much less resistance, the cold anticipator is bypassed. Remember that electricity always takes the path of least resistance.

When the cooling demand is satisfied, the contact switch opens and places the cold anticipator *in series* with the compressor contactor coil. Due to high resistance, a small current flows through both the anticipator and the compressor contactor coil. The amperage is *not* sufficient to energize the contactor coil, but it does cause the cold anticipator to heat up. The resistor's heat makes the bimetallic coil tilt just slightly in advance of the room temperature. The cooling equipment has the opportunity to get started and begin delivering cool air just before the room temperature reaches the thermostat setting.

Digital Thermostats

Most newly installed and replacement thermostats are digital thermostats, **Figure 28-23**. Digital thermostats are more accurate than mechanical thermostats because of their solid-state operation using temperature sensors, instead of using a bimetal strip to move a mercury bulb. Additionally, states have begun restricting the sale of mechanical thermostats that use mercury as a sensing element, which in part explains the shift from mechanical to digital thermostats.

Digital thermostats come in both non-programmable and programmable types. Programmable thermostats can be programmed to turn the temperature setting up or down at different times of the day. This allows for better energy saving from the furnace. Digital thermostat wiring connections are made just like those in mechanical thermostats. Digital thermostats do *not* have a heating or cooling anticipator. These adjustments to help with temperature overshoot are accomplished in the solid-state circuits of the thermostats. Most digital thermostats are powered by batteries installed in the thermostat. Some get their power from the furnace.

Smart Thermostats

Smart thermostats are programmable and use solid-state devices, like digital thermostats, but they are also equipped with "smart" technology that learns the heating and cooling patterns of the homeowner. Motion sensors are installed so the thermostat knows when there are occupants in a home or not. Another feature of smart thermostats is that they are connected to the internet, so they can

Provided courtesy of Goodman Manufacturing Company, L.P. All rights reserved.

Figure 28-23. Digital thermostats provide the capability to program different thermostat settings based on the day of week and time of day. Some offer wireless connectivity and can be monitored and adjusted using a mobile device.

provide other information such as the date, time, outdoor temperature, and humidity. Smart thermostats are also able to interact with devices, such as phones. This allows the homeowner to adjust the thermostat temperature from their phones, in or outside of the home. See **Figure 28-24**.

28.4 Blower Motor Controls

The blower motor is operated by either a combination/fan limit control or integrated control modules. Older furnaces use a combination fan/limit control, while newer furnaces use integrated control modules. Both controls use a special relay to energize the blower motor during the cooling cycle for furnaces with air conditioning. Technicians should be familiar with the operation of these controls.

Combination Fan/ Limit Control

Older furnaces use a ***combination fan/limit control***, which serves a dual purpose. It controls operation of the blower motor and provides protection against overheating of the furnace. The fan/limit control contains two electrical switches: one normally open (NO) and one normally closed (NC). Movement of a coiled bimetallic strip controls operation of the switches, **Figure 28-25**.

The bimetallic coil is inserted into the heat exchanger area to sense furnace temperature,

Google Nest

Figure 28-24. The Google Nest is an example of a smart thermostat. This thermostat is compatible with an app, where the homeowner can track and program heating and cooling usage.

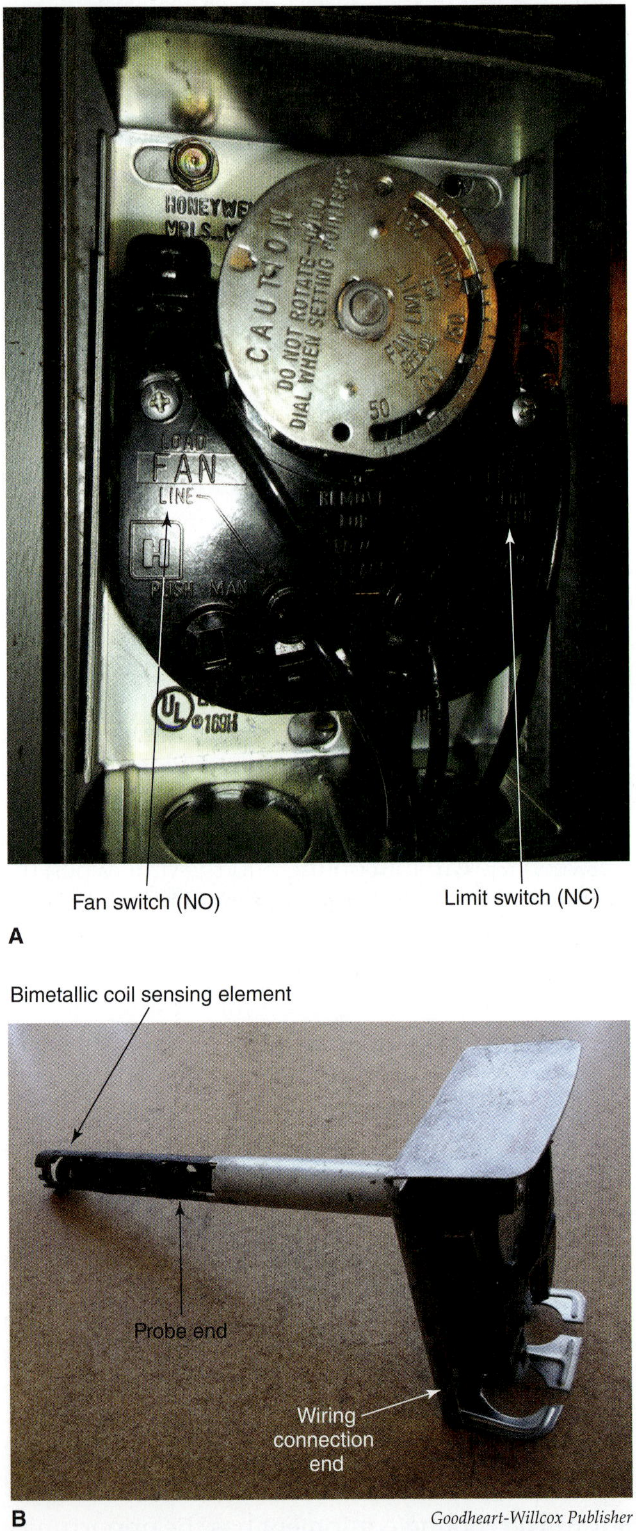

Goodheart-Willcox Publisher

Figure 28-25. A combination fan/limit control. A—The fan switch is normally open, while the limit switch is normally closed. B—The probe end is inserted into the furnace. The wiring connection end is mounted on the furnace firewall.

Figure 28-26. Heat causes the coil to twist in a circular movement. The movement controls the NO and NC switches. The amount of heat necessary to activate the switches can be adjusted by a calibrated dial located on the control face, Figure 28-27.

Fan Switch (Normally Open)

The normally open switch that operates the blower motor is called the ***fan switch***. When the furnace temperature reaches a selected setting on the dial face (usually 140°F or 60°C), the fan switch closes and turns on the furnace's blower motor. Air is pulled from the living space, forced through the heat exchanger where it is heated, and returned to the living space. This movement of air through the heat exchanger causes the furnace temperature to drop to about 125°F (52°C), where it remains during the heating cycle.

When the heated space reaches the desired temperature setting (typically 72°F or 22°C), the thermostat contacts open and stop current flow to the gas valve. Once current flow to the gas valve stops, the burner flame goes out. However, the blower motor continues to run because the furnace is still hot. Cooling of the furnace causes the bimetallic coil to twist in the opposite direction as when heated. The blower fan keeps running until the furnace temperature drops to about 100°F (38°C). At that time, the fan contacts reopen and stop the blower motor.

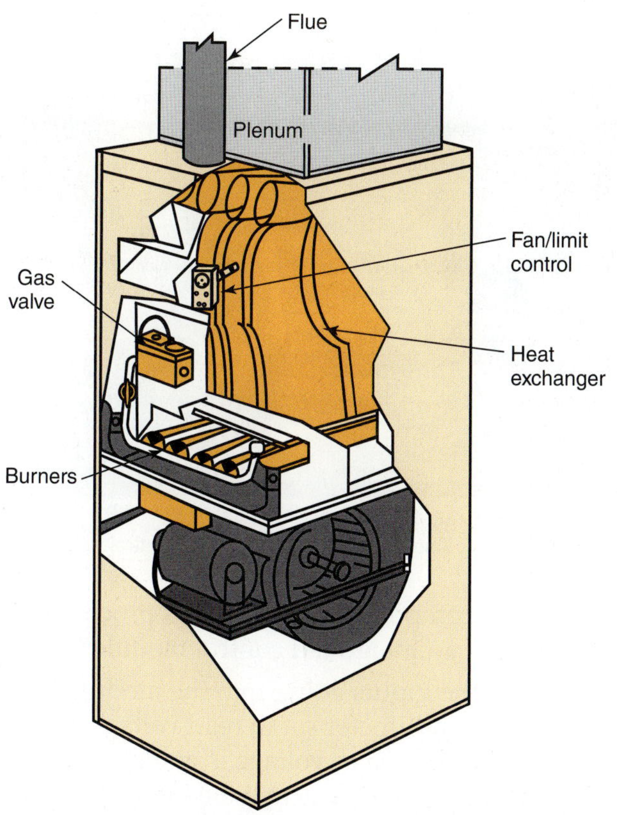

Goodheart-Willcox Publisher

Figure 28-26. The fan/limit control is positioned with the sensing element inserted into the heat exchanger to sense furnace temperature.

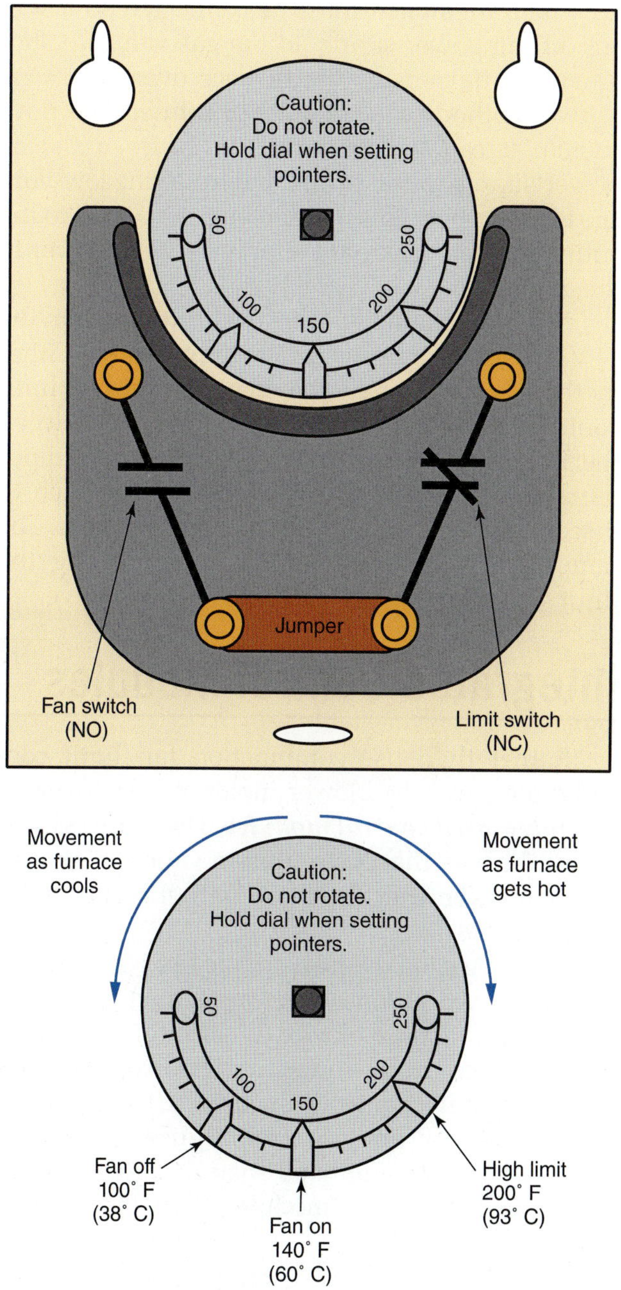

Goodheart-Willcox Publisher

Figure 28-27. A calibrated dial on the face of the fan/limit control is used to set the fan-on, fan-off, and high-limit temperatures. As the bimetallic coil sensing element is heated or cools, it twists one way or the other, rotating the dial.

Limit Switch (Normally Closed)

A combination fan/limit control's ***limit switch*** is a safety device that cuts off the power supply if the furnace becomes overheated. The furnace may overheat due to lack of airflow caused by blower fan motor failure, a broken V-belt, a dirty filter, or failure of the fan switch contacts. The limit switch, which is normally closed, will open if the furnace temperature reaches about 200°F (93°C).

The limit switch must be connected so it will cut off the power supply to the gas valve (to shut down the burners) if the furnace overheats. One of two methods can be used to connect the limit switch. It can be connected in series with the power supply to the primary wire of the low-voltage transformer. This method cuts off power to the entire low-voltage circuit supplying 24 V power to the gas valve.

The other method places the limit switch in the 24 V circuit supplying power to the gas valve. Either method is acceptable. The combination fan/limit control has a factory-installed jumper between the fan switch and the limit switch. The jumper remains in place when using the limit switch to disconnect line voltage to the transformer. The jumper must be removed when connecting the limit switch in the 24 V circuit to the gas valve.

Integrated Control Modules

Rather than use combination fan/limit controls to control the blower motor, newer furnaces use ***integrated control modules***. These are printed circuit boards that have built-in timers that turn on the blower after the main burner has been lit, **Figure 28-28**.

Goodheart-Willcox Publisher

Figure 28-28. This printed circuit control board is part of an integrated control module used in a furnace to control the fan cycle on and off times.

An integrated control module uses operational controls based on *time* (not *temperature*) to turn on and off the furnace blower. In addition to normal operating controls, integrated control modules also use safety controls, such as a temperature-operated safety limit control, **Figure 28-29**. This limit switch uses a bimetal snap disc to break the power to the control module when the furnace gets too hot.

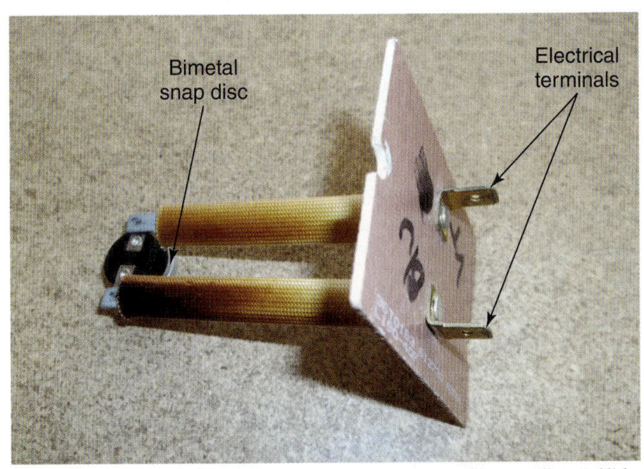

Goodheart-Willcox Publisher

Figure 28-29. A limit switch used with newer control boards.

Integrated Control Module Troubleshooting

To troubleshoot an integrated control module, you will use input-output testing and check all relevant components. Follow this procedure to troubleshoot an integrated control module:

1. Locate the inputs going into the module. Using a multimeter set to read voltage, test to see if the proper voltage is going to the module. These inputs may be low voltage to the board from safety switches.
2. Locate the outputs coming out of the module. Using the multimeter again, test to see if the proper voltage is coming out of the module or going out to safety switches. If you have

the proper input voltage but do not have the proper output voltage at the module, then the printed circuit board may be bad and need to be replaced.
3. If both the input and output voltage are correct at the module, check the module fuse to see if it has blown. Most modules have the fuse mounted on the module. Refer to **Figure 28-28**.
4. If the fuse has not blown, check the power supply wiring from the module to the furnace. The white wires should be tied together, and the black wires should be tied together.

Blower Relay

On 70% furnaces, the combination fan/limit switch turns on the blower motor when the furnace gets hot enough. Provisions must be made to energize the blower motor during the cooling cycle, as well. A ***blower relay*** accomplishes this task. The blower relay has a 24 V coil and one set of normally open contacts. When a blower relay's coil is energized by the thermostat, the relay contacts close to complete a 120 V circuit to the blower motor, **Figure 28-30**.

Another switch on the thermostat (the *fan switch*) controls power to the blower relay coil. The fan switch is labeled ON and AUTO (for automatic). When the thermostat calls for cooling and the fan switch is in the AUTO position, another 24 V circuit is completed to Terminal G inside the thermostat, **Figure 28-31**.

Terminal G is used to control the 24 V power supply to the blower relay coil. A green wire connects Terminal G on the thermostat to the blower relay coil. The other side of the relay coil is connected to the common side of the 24 V transformer.

When the manual fan switch is in the ON position, Terminal G receives a continuous 24 V power supply. The blower relay coil remains energized, and the blower motor runs continuously. The fan switch permits the homeowner to control blower operation independently of the COOL-OFF-HEAT switch. Sometimes air circulation only, without cooling, is adequate to satisfy the homeowner's needs.

A two-speed blower motor may be used to provide better air circulation. Low speed is used for the heating cycle because the temperature rise through the furnace is 55°F (125 – 70 = 55) or 30°C.

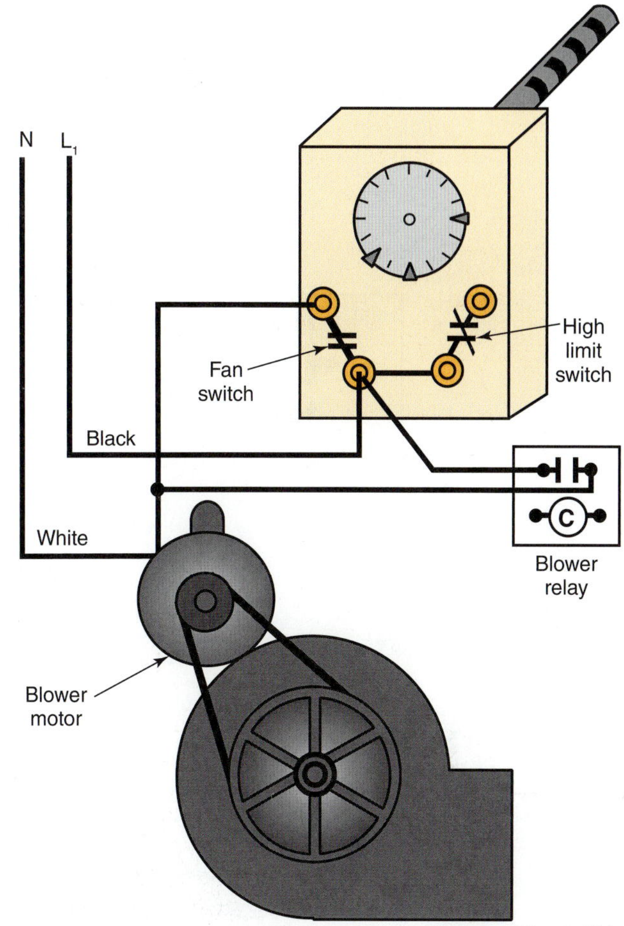

Figure 28-30. The blower relay controls the blower during the cooling cycle.

High speed is used for the cooling cycle because the temperature difference across the evaporator is 16°F to 20°F (9°C to 11°C). Electrical connections for the two-speed blower motor are accomplished by connecting its low speed (heating) terminal to the fan switch and its high speed (cooling) terminal to the blower relay switch, **Figure 28-32**.

Note that newer furnaces using integrated control modules do *not* use a blower relay. The blower in these systems is controlled by a relay on the integrated control module. **Figure 28-33** illustrates a typical gas-fired, forced-air furnace with air conditioning and a two-speed blower motor in both pictorial and ladder diagrams.

Electronically Commutated Motors (ECMs)

Electronically commutated motors (ECMs) are commonly used in gas furnaces. As discussed in Chapter 24, *Motors*, an ECM is a brushless dc

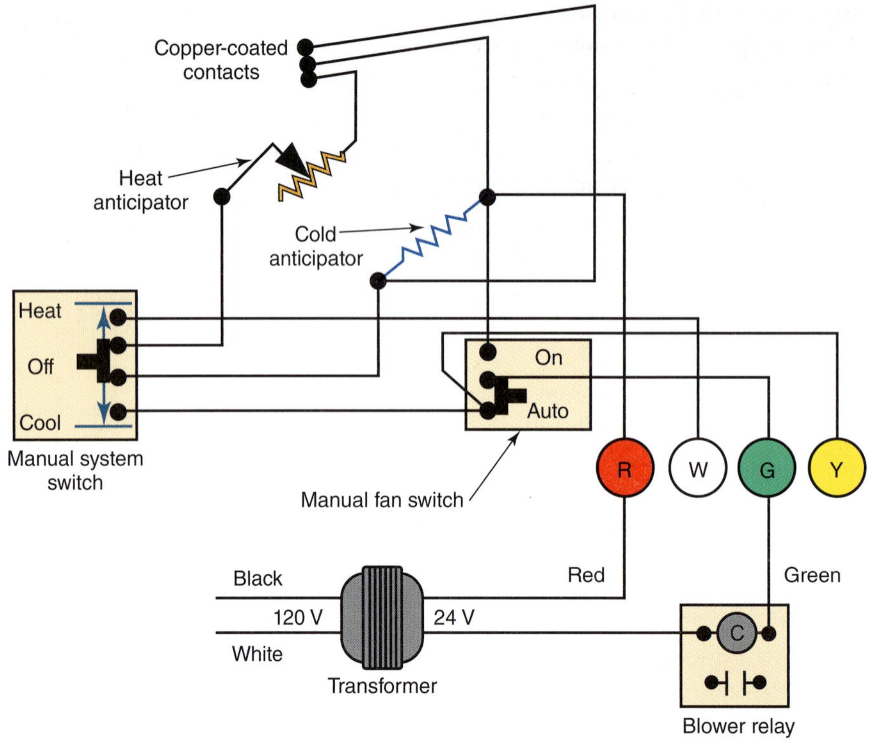

Goodheart-Willcox Publisher

Figure 28-31. The fan switch on the thermostat allows the selection of continuous fan operation or operation only during the cooling cycle.

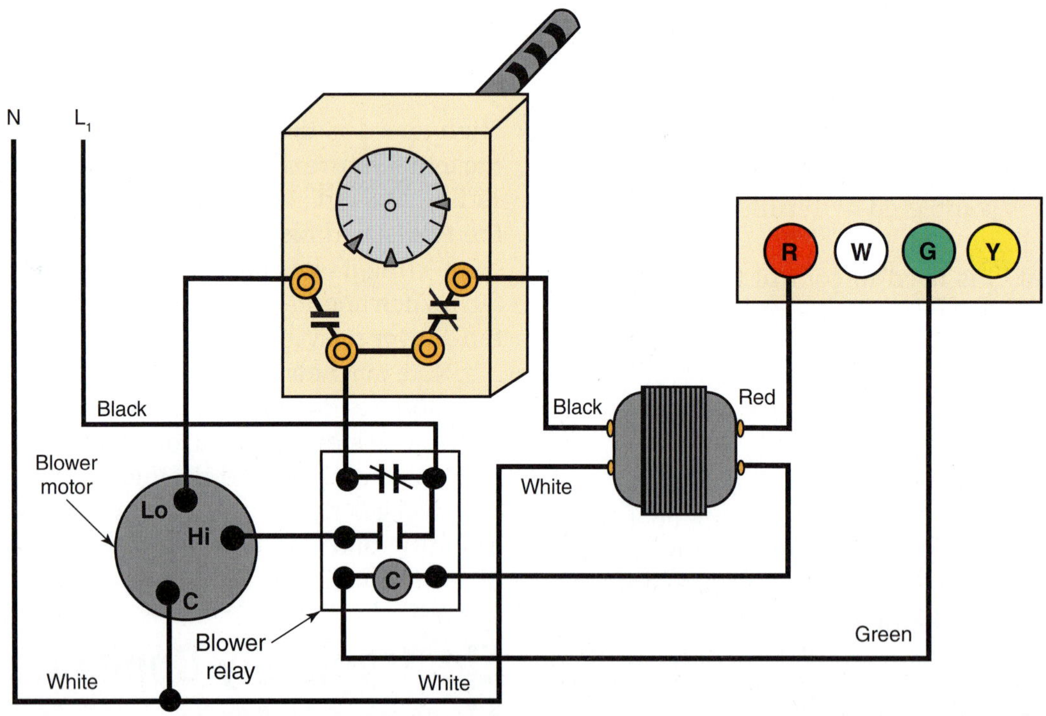

Goodheart-Willcox Publisher

Figure 28-32. On a 70% furnace, the low speed of a two-speed blower fan is energized during the heating cycle. High speed is energized during the cooling cycle.

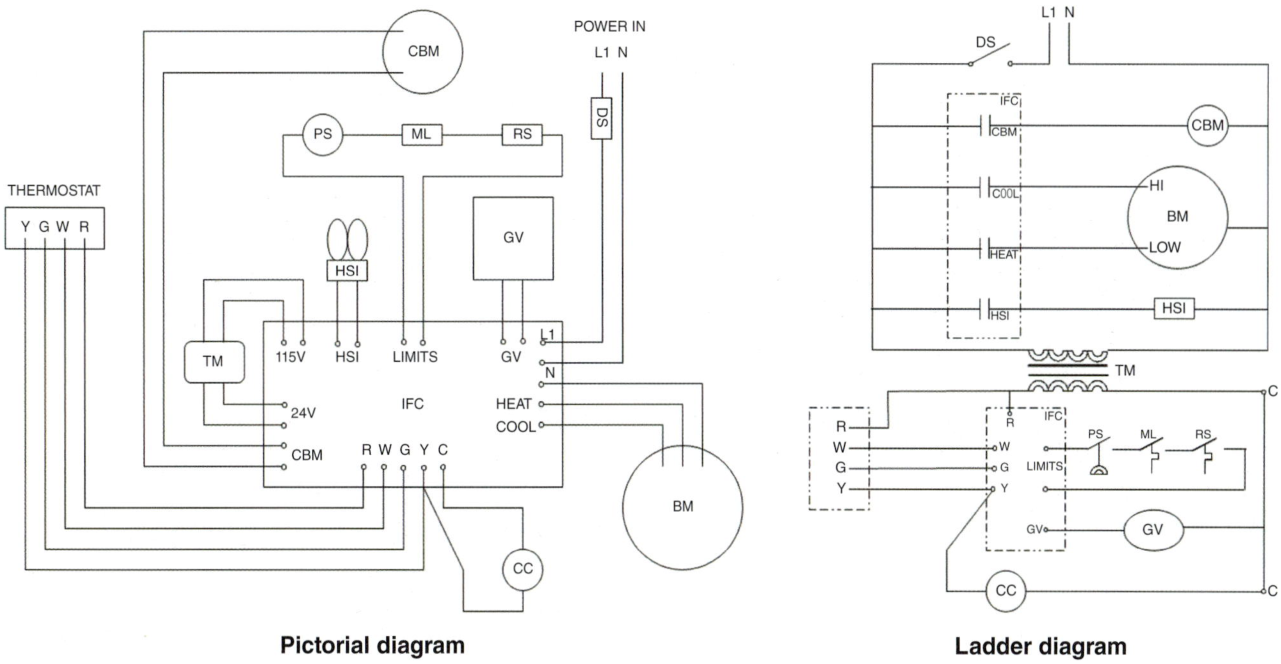

BM = BLOWER MOTOR
CBM = COMBUSTION BLOWER MOTOR
CC = COOLING CONTACTOR
DS = DOOR SWITCH
GV = GAS VALVE
HSI = HOT SURFACE IGNITOR
IFC = INTEGRATED FURNACE CONTROL
ML = MAIN LIMIT
PS = PRESSURE SWITCH
RS = ROLLOUT SWITCH
TM = TRANSFORMER

Goodheart-Willcox Publisher

Figure 28-33. A typical gas-fired heating system for an 80% or 90% furnace with air conditioning added is shown in both pictorial and ladder forms.

motor that uses electronic controls to regulate its speed, torque, or power output without external devices or sensors. The ECM in a gas furnace may be constant torque or constant volume and will adjust motor speed to maintain system design airflow. At installation, the ECM is programmed with these settings and must be set correctly to optimize efficiency.

ECMs have largely replaced permanent split-capacitor (PSC) motors used in furnace. ECMs do not work with blower relays, instead they are wired from a control board to the indoor blower. Another distinction in ECMs compared to PSC motors is that they are always supplied with power. Some ECMs require a low-voltage 24 Vac signal which tells the blower to run at a set speed. Others are signaled by the control board using dual in-line package (DIP) switches to control the blower speed. ECM blower motors are commonly programmed for air conditioning at 400 cfm per ton. In high humidity areas, 350 cfm per ton may be more common. Areas of low humidity may be programmed for 450 cfm per ton. Furnaces that utilize an ECM will need to be programmed for the specific airflow needs of the system.

28.5 Troubleshooting Gas Furnaces

Systematic troubleshooting skills must be developed. A service call tests a technician's knowledge, mechanical competence, and observational and interpretational skills. Good troubleshooting requires analyzing information accurately and drawing logical conclusions. Hasty decisions and inaccurate information lead to mislead conclusions. Following step-by-step procedures and a *process of elimination* usually pinpoints the problem. There are five important steps in systematic troubleshooting:

1. Know what the equipment *should* do (sequence of operation).
2. Know what the equipment *is* doing.
3. Know what the equipment is *not* doing.
4. Eliminate areas that *are* functioning properly.
5. Concentrate on areas that are *not* functioning properly.

For example, if a technician is working on a hot surface ignition furnace, they will know that the equipment should turn on when the thermostat is set to heat and the temperature should begin to rise. Then, the technician should observe the furnace draft inducer motor turn on and clear the flue chamber. This step will create a vacuum so the pressure switch can close and allow the board to verify that all the safety switches are closed. Once verified, the hot surface ignitor should turn on which will then open the gas valve and allow the gas to ignite. If the gas never ignites, the technician can eliminate every step up to the hot surface ignitor since those sequences flowed as they should have up to that point. Next, the technician can begin to pinpoint the issue more efficiently by focusing on the components involved with igniting the gas.

Sequence of Operation

To be able to troubleshoot furnaces, technicians must be familiar with the sequence of operation of different types of ignition systems. Recognizing how a system should be working and the order it should be working in can help a technician narrow down the problem of a malfunctioning system.

Standing Pilot Systems

1. In standing pilot systems, the pilot light should always be lit. When the thermostat calls for heat, the gas valve is turned to the ON position.
2. The gas valve will open, and the burners will light.
3. Next, you will hear a click, and the blower motor will turn on. The blower will then circulate heated air through the air distribution system.

Intermittent Pilot Ignition Systems

1. In intermittent pilot ignition systems, the pilot light should *not* always be lit.
2. When the thermostat senses that the temperature is below the desired setpoint, it sends a 24 V control signal to the furnace.
3. The draft inducer motor starts, creating a vacuum within the combustion chamber and then begins exhausting any combustible gases that are in the chamber.
4. The pressure switch then verifies that the inducer motor is running and that the exhaust system is free of any obstructions.
5. The igniter then receives a signal to spark and ignite the pilot light.
6. Once the pilot light is verified to be lit through flame rectification, the gas valve will open to allow the fuel gas to flow to the main burners.
7. The main burners will then light and after a delay to allow the heat exchanger to heat up, the blower motor will then start. A flame sensor will verify that the burners keep a flame. If at any time the flame goes out, the flame sensor will stop communicating with the control board thus stopping the signal to the gas valve which will then cut off gas flow.

Hot Surface and Direct Spark Ignition Systems

1. The thermostat sends a signal to the furnace to call for heat.
2. The draft inducer motor will start up creating a vacuum in the combustion chamber and begin exhausting any combustion gases that are in the chamber.
3. The pressure switch then verifies that a vacuum has been met and allows the board to send a signal through all safety switches to verify it is safe to move forward.
4. The hot surface igniter or spark igniter will turn on and glow red for a short period of time, usually less a minute.
5. The board will then send a signal to the gas valve to open. At this time, a burner flame should start inside the fire box or burner tubes to the heat exchanger. If the system has a spark ignitor, it will begin to spark at the same time the gas valve is opened.
6. Once the main burners are lit, a flame rod (sensor) rectifies the voltage and proves that the burners are lit.
7. In a short amount of time, the blower will turn on. The blower will then circulate heated air through the air distribution system.

Common Furnace Issues and Causes

Speaking with a customer usually helps narrow the problem to a specific area. Furnace issues typically fall into six categories:

- No heat
- Not enough heat
- Too much heat

- Noise
- Odor
- Excessive operating cost

A question or two to the homeowner, plus observation of the equipment, should aid in the process of elimination. By listening, observing, and testing, a technician will discover clues that point to the exact problem. The following are some of the symptoms and probable causes of furnace issues.

No Heat

Symptom: Burner does not start.

Possible causes:

- Wrong thermostat setting
- Defective thermostat
- No power supply to the furnace (Check fuses and circuit breakers)
- Gas valve turned off
- Defective high-limit control
- Roll out switch tripped
- Defective transformer (Check fuse, if any)
- Defective gas valve
- Hot surface igniter burned out

Symptom: Pilot flame is out.

Possible causes:

- Main gas valve turned off
- Plugged or restricted pilot orifice
- Pilot blows out due to draft through furnace
- Pilot blows out due to high gas pressure
- Pilot blows out due to delayed ignition

Symptom: Pilot flame will not stay lit.

Possible causes:

- Thermocouple loose at valve
- Thermocouple out of alignment at pilot
- Defective thermocouple

Not Enough Heat

Symptom: Burner short cycles.

Possible causes:

- Unit cycling on high-limit control
- Defective blower motor
- Dirty filter
- Loose or broken V-belt
- Wrong rotation of blower motor
- Restriction in return air system
- Low voltage at motor (causes high amps)

Too Much Heat

Symptom: Burner runs too long.

Possible causes:

- Wrong thermostat setting
- Heat anticipator set too high
- Thermostat not level
- Thermostat on cold wall or in draft
- Defective gas valve (stuck open)

Noise

Possible causes:

- Delayed ignition causing flashback
- Flashback after shutdown (excess primary air)
- Loose blower bearings
- Blower bearing dry and squeaking
- Blower wheel off-center
- Loose V-belt
- Loose running gear and mounts
- Filter caught in running gear
- Loose cabinet panels

Odor

Possible causes:

- Leak in gas line or pilot line
- Loose or cracked heat exchanger
- Leak in flue pipe
- Downdraft causing diverter spillage
- Negative pressure in furnace area
- Dirty air filters
- Stagnant water or sludge in humidifier

Excessive Operating Cost

Possible causes:

- Dirty air filter
- Poor combustion
- Insufficient home insulation
- Excessive air leakage (fireplace, vents)
- Excessive gas manifold pressure
- Oversize burner orifice
- Excessive flue draft
- Wrong fan control settings

Chapter Review

Summary

28.1 A forced-air HVAC system uses an air distribution system that is composed of a network of ducts to control airflow to and from the conditioned space. Inside the furnace cabinet is a blower section that consists of a motor and fan, along with an air filter to prevent impurities from entering the airstream.

28.2 Gas-fueled furnaces rely on combustion (burning of fuel) to generate usable heat. Gas burners, a gas valve, and gas manifold are responsible for generating that usable heat.

28.2 A standing pilot ignition system uses a pilot flame to light its gas burners. An intermittent pilot ignition system lights a pilot flame just before lighting the main burners. In a hot surface ignition system, the main gas valve is opened and immediately lit by an igniter. In a direct spark ignition system, an electric spark created by a spark igniter lights the burners on a call a heat.

28.3 A thermostat is a switch that opens and closes according to changes in temperature. Heating and cooling system thermostats are designed to operate on 24 V.

28.3 Thermostats are divided into mechanical or digital types. Mechanical thermostats detect changes in temperature using a bimetallic coil and mercury switch or copper-coated contacts. Mechanical thermostats are equipped with a heat anticipator to overcome thermostat overshoot. A cool anticipator is included for systems with air conditioning. Digital thermostats use solid-state operation and come in non-programmable and programmable types. Digital thermostats do *not* use heating or cooling anticipators.

28.4 The blower motor is operated by either a combination/fan limit control or integrated control modules. Older furnaces use a combination fan/limit control, while newer furnaces use integrated control modules and electronically commutated motors (ECMs).

28.5 When troubleshooting a gas furnace, a technician should know what the equipment *should* do (sequence of operation), what the equipment *is* and is *not* doing, eliminate areas that *are* functioning properly, and concentrate on areas that are *not* functioning properly. Furnace issues typically fall into six categories: no heat, not enough heat, too much heat, noise, odor, and excessive operating cost.

Know and Understand

1. Supply air is released into a conditioned space through outlets called _____. (28.1)
   A. grilles
   B. registers
   C. blowers
   D. filters
2. The blower section is located inside the _____. (28.1)
   A. supply air ducts
   B. return air ducts
   C. furnace cabinet
   D. draft hood
3. *True or False?* A furnace's heat exchanger freely mixes recirculated room air and combustion gases. (28.2)
4. Which of the following statements describes the correct ratio to achieve complete combustion in a gas-fired furnace? (28.2)
   A. 1 part oxygen to 2 parts fuel
   B. 10 oxygen to 2 parts fuel
   C. 2 cubic feet of air for every 1 cubic foot of fuel
   D. 15 cubic feet of air for every 1 cubic foot of fuel
5. *True or False?* A blue furnace flame indicates good combustion, and yellow indicates poor combustion. (28.2)

6. The automatic gas valve controls gas flow to the manifold, which distributes the gas into the _____. (28.2)
   A. burners
   B. flue
   C. fan
   D. igniter
7. A _____ is a heat-sensitive safety device used to confirm that a pilot light is lit in a standing pilot ignition system. (28.2)
   A. fuse
   B. thermocouple
   C. sensing probe
   D. bimetallic coil
8. A pilot assembly in an intermittent pilot ignition system is composed of the following, *except* a _____. (28.2)
   A. pilot burner
   B. spark electrode
   C. hot surface igniter
   D. sensing probe
9. The main purpose of flame rectification is to _____. (28.2)
   A. ignite the burner
   B. generate current
   C. confirm the presence of a flame
   D. convert dc current to ac current
10. Mechanical thermostats use a _____ to sense changes in temperature. (28.3)
    A. bimetallic coil
    B. heat anticipator
    C. solid-state temperature sensors
    D. solenoid valve
11. A heating thermostat can easily be used to accommodate a cooling system by adding a _____. (28.3)
    A. cooling anticipator
    B. subbase with contacts for the cooling switch
    C. blower relay
    D. digital thermostat
12. *True or False?* Digital thermostats are more accurate than mechanical thermostats. (28.3)
13. A combination fan/limit control's limit switch opens when furnace temperature reaches about _____°F. (28.4)
    A. 100
    B. 125
    C. 140
    D. 200
14. An integrated control module uses operational controls based on _____. (28.4)
    A. pressure
    B. temperature
    C. time
    D. homeowner preferences
15. The purpose of the blower relay is to _____. (28.4)
    A. improve blower motor heating efficiency
    B. energize the blower motor during the cooling cycle
    C. compensate for the delay between the call for cooling and the time when the system begins delivering cool air
    D. cut off power supply if the furnace becomes overheated
16. In high humidity areas, an ECM should be programmed to operate at _____ cfm per ton. (28.4)
    A. 350
    B. 400
    C. 450
    D. 500
17. *True or False?* In a direct spark ignition system, the sensing probe will verify if the pilot light is lit. (28.5)
18. A strong gas odor coming from a furnace may indicate all of the following, *except* _____. (28.5)
    A. a leak in the gas line
    B. dirty air filters
    C. a heat anticipator set too high
    D. stagnant water in the humidifier

Apply and Analyze

1. What are the main components of an air distribution system in a gas-fired system? (28.1)
2. Why is secondary air needed in the burner chamber of a gas-fired furnace? (28.2)
3. What is the purpose of a thermocouple in a standing pilot ignition system? (28.2)
4. List two differences between mechanical and digital thermostats. (28.3)
5. Explain the operation of the fan switch and the limit switch in a combination fan/limit control. (28.4)

6. In furnaces with integrated control modules, what is the main variable used to control blower operation? (28.4)
7. List the five important steps in systematic troubleshooting. (28.5)

Critical Thinking

1. Imagine you are a service technician working on a gas-fired furnace with standing pilot ignition. The homeowner reports that the furnace is noisy upon startup and takes a long time to heat the space. You set the thermostat to call for heat and observe the burners. You notice that the furnace has six burners and you only see five lights. What might be the problem? (28.2)
2. Imagine you are a service technician working on a gas-fired furnace with intermittent ignition. After a call for heat, you observe the pilot light, and the ignitor keeps trying to light the pilot, but the burners do not light. What might be the problem? (28.2)
3. Imagine you are a service technician working on a gas-fired furnace with hot surface ignition. After a call for heat you see and hear the combustion blower turn on, but the hot surface ignitor does not turn on. What might be the problem? (28.2)
4. Imagine you are a service technician working on a gas-fired furnace with air conditioning. You turn the thermostat to call for heat and the outdoor condenser starts running. What might be the problem? (28.3)
5. Imagine you are a service technician working on a gas-fired furnace with standing pilot ignition. The homeowner is reporting no heat. After a call for heat, you observe that the burner does not start, but the pilot flame is still light. What component is likely the cause of the problem? (28.5)

ClimateMaster

29 Electric Heat with Air Conditioning

LEARNING OUTCOMES

After studying this chapter, you will be able to:

29.1 Identify how the power supply and circuit protection devices are connected to an electric furnace.

29.2 Explain the purpose and operation of electric heating elements and related components in an electric heating system.

29.3 Describe the operation of two-stage electric heating and the thermostats used in these systems.

29.4 Explain the components used when adding air conditioning to electric furnace systems.

29.5 Recognize how to troubleshoot electric furnace systems.

TECHNICAL TERMS

electric furnace
electric heating element
feeder service
fusible link
heating-cooling thermostat
high-pressure safety control
limit switch
low-pressure safety control
outdoor thermostat
sequencer
two-stage heating system
two-stage heating thermostat

Introduction

An ***electric furnace*** is a furnace that uses forced convection to distribute heated air that has been warmed by electric heating elements. Electric heating involves 100% conversion of electrical energy to heat energy. Because no fuel is burned, there is *no need* for a chimney to vent combustion gases. This provides great flexibility in locating an electric furnace. Supply and return air ducts are installed from the furnace to the conditioned space.

This chapter explains the function of electric heating system components and their connections in electrical circuits. For proper servicing, it is important to understand the relationship of each component to the others in the system. The system must operate properly and automatically to safely maintain conditioned space temperature, **Figure 29-1**.

Goodheart-Willcox Publisher

Figure 29-1. Electric forced-convection furnace with air conditioning.

A schematic wiring diagram is presented step-by-step in this chapter to illustrate the components' relationships and connections. Since there are variations in wiring diagrams on different units in the field, a thorough understanding of both pictorial and schematic wiring diagrams is important.

29.1 Electric Furnace Power Supply and Wiring

Electric furnaces require 240 V, three-wire feeder service from the main disconnect or load center. ***Feeder service*** refers to wiring directly from the main disconnect or electrical load center to the electric furnace. This includes two hot wires and one ground wire. The feeder circuit to the furnace must be sized (and fused) in accordance with the National Electric Code (NEC). Electrical requirements are printed on the furnace wiring diagram. The 240 V furnace feeder wires are connected to a fused disconnect or circuit breaker. The disconnect or breaker must be within sight of the furnace and capable of locking in the open position. The fused disconnect or breaker is a convenient means of making sure power to the furnace is safely disconnected for service procedures.

Fuses and Circuit Breakers

The two hot wires and the ground wire are run from the load side (bottom) of the fused disconnect to a set of terminals on the fuse block within the electric furnace's enclosure. Some electric furnaces are equipped with a double-pole circuit breaker rather than a fuse block, **Figure 29-2**.

The feeder circuit wires must be enclosed in metal rigid or flexible conduit. The NEC requires these circuits to be grounded in the furnace (not at the disconnect). Most furnaces provide a ground connection in the control box next to the high-voltage terminals.

An electric furnace is supplied with 240 V single-phase electrical power with safety ground. When the electrical supply connections are made, the field-installed line voltage wiring is complete. The remaining 240 V circuits (inside the furnace) are factory-installed.

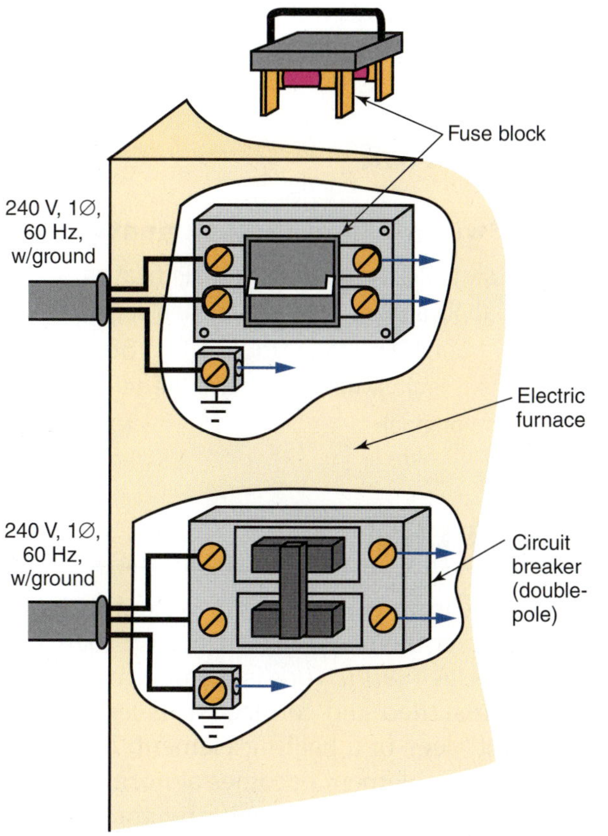

Goodheart-Willcox Publisher

Figure 29-2. Power supply connections for an electric furnace. Some furnaces use a fuse block, while others use a double-pole circuit breaker. Either method is acceptable.

29.2 Electric Heating Elements

The function of an ***electric heating element*** in an electric furnace is to generate heat for the conditioned space. The heating element wire is made of nickel and chromium and is often referred to by its trade name, *nichrome*. The wire has a built-in resistance to electron flow with the amount of resistance determined by its length and diameter. Resistance to electric current flowing through the wire causes it to get hot.

The nichrome wire is coiled and threaded through a metal frame with ceramic insulators to prevent metal-to-metal contact, which would cause a short circuit, **Figure 29-3**. The heating element used in furnaces is similar to the heating element used in toasters and electric clothes dryers. When 240 V is applied to the nichrome element, the current flow causes it to heat up. The heat is transferred to the air being circulated through the furnace by the blower.

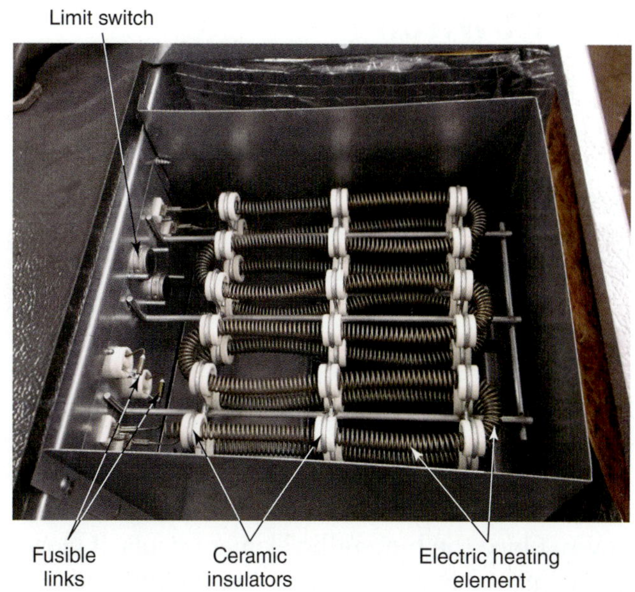

Goodheart-Willcox Publisher

Figure 29-3. A section of the coiled-wire resistance heating element typically used in electric furnaces. Ceramic insulators isolate the heating element from the frame to prevent metal-to-metal contact, which would result in a short circuit.

Furnace heating elements are normally sized at 5 kW or 5000 W. At 3.415 Btus per watt, the heat output of such an element is 17,075 Btu/hr.

$$5000 \text{ W} \times 3.415 \text{ Btus} = 17{,}075 \text{ Btu/hr}$$

To provide the necessary amount of heat for most applications, several electric heating elements may be required. A fuse or circuit breaker must be installed in each power supply leg to each element. Since a 5 kW element draws about 20 A, a 30 A fuse or breaker is required in each leg. The fuse or breaker monitors the amount of current flow in the heating element branch circuit. Excessive current flow will cause the fuse to burn out (open) or the breaker to trip (open), thus protecting the circuit wiring from damage or burning out.

CAUTION

Exact Fuse Replacement

When a fuse burns out, it must be replaced with the fuse size shown on the unit wiring diagram. Before replacing a fuse, always find and correct the fault that caused it to burn out.

Limit Switch

The purpose of a ***limit switch*** is to de-energize the heating element if the temperature exceeds the switch's setting. A limit switch uses a normally closed, bimetallic disc enclosed in a metal housing. The limit switch is mounted on the frame near the heating elements, **Figure 29-4**.

Normally, air being circulated through a furnace removes heat from the furnace to maintain the desired temperature. However, if airflow is not adequate, furnace temperature will rise rapidly. The bimetallic element will sense the excessive heat and cause the NC contacts to open. This de-energizes the heating element until the furnace cools down. Once the bimetallic element cools, it closes its NC contacts and permits the heating element to function again. The process is called "cycling on the limit switch." Dirty air filters and blower motor failures are the primary causes of a furnace overheating and cycling on the limit switch.

A limit switch has a temperature differential built-in between its "open" and "close" setpoints. The close setpoint is normally 35°F (19°C) cooler than the open setpoint. For instance, if a limit switch opens at 160°F (71°C), it will only close after the temperature has reduced to 125°F (52°C). The differential setting prevents the heating element from short-cycling the system.

Each heating element contains its own safety controls (limit switches and/or fusible links), which are unaffected by the operation of other elements. Therefore, safety limit controls are *not* needed in the control circuit. Heating elements are *not* position-sensitive and can be mounted in *any* position.

Goodheart-Willcox Publisher

Figure 29-4. The limit switch is located on the frame near an electric heating element. It has a bimetallic element that opens the electric circuit when the temperature exceeds a high limit.

CAUTION

Limit Switch Exact Replacements

Limit switches are sized for certain heating element applications. The various open setpoints may vary from 130°F to 190°F (54°C to 88°C). A replacement limit switch must have the same ratings as the original.

Fusible Link

Some heating elements also contain a fusible link to protect them against burnout if their limit switch fails. A ***fusible link*** is a safety device that is factory-installed and connected in series in one of the "hot" legs of a heating element. If the temperature of the element becomes abnormally high, the fusible link will melt, opening the line voltage circuit. Once the fusible link melts, it must be replaced with a new one.

Figure 29-5 is a schematic of the line voltage (240 V) circuit supplying power to three heating elements. The wiring is factory-installed inside the furnace cabinet.

Sequencers

The heating elements in electric furnaces are *step-started*. In other words, the heating elements are energized in *sequence* one or two at a time. Energizing all heaters at one time would place an overwhelming demand on the power supply. Devices for staging the elements on and off are called ***sequencers***, **Figure 29-6**. A sequencer uses a temperature-responsive bimetallic element for the normally open switch and includes a small heater to activate (close) the switch after a set amount of time.

Sequencers contain one or more switches and are used to stage heater elements on or off. The sequencer that controls the first heating element to be energized must have two switches: one to turn on the heating element and one to turn on the blower motor. Sequencers can also contain several switches to stage multiple elements as well as the blower motor, **Figure 29-7**.

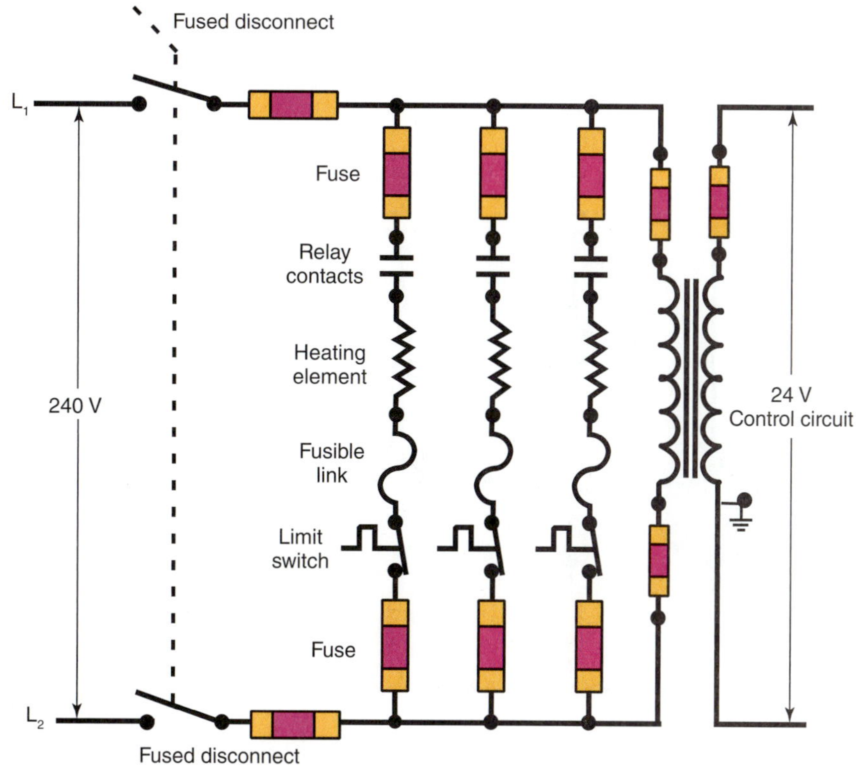

Goodheart-Willcox Publisher

Figure 29-5. Schematic of the line voltage connections for three heating elements.

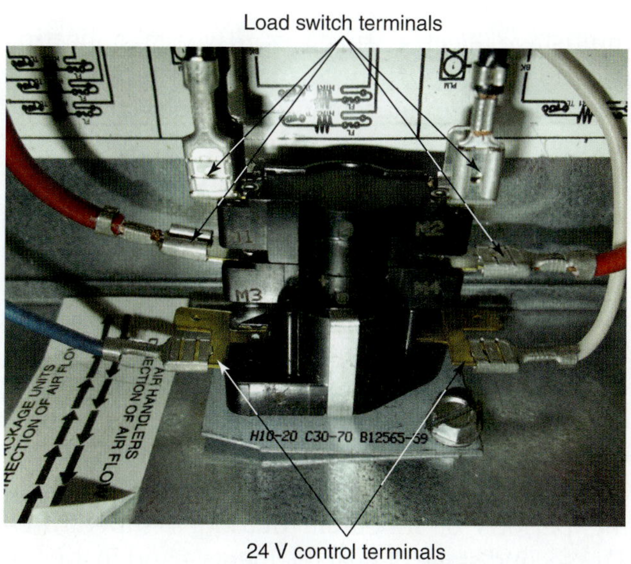

Goodheart-Willcox Publisher

Figure 29-6. A sequencer used to energize electric heating elements in sequence in an electric furnace.

Sequencers Controlling Multiple Heating Elements

When multiple heating elements are used, the sequence of operation for an electric furnace is as follows:

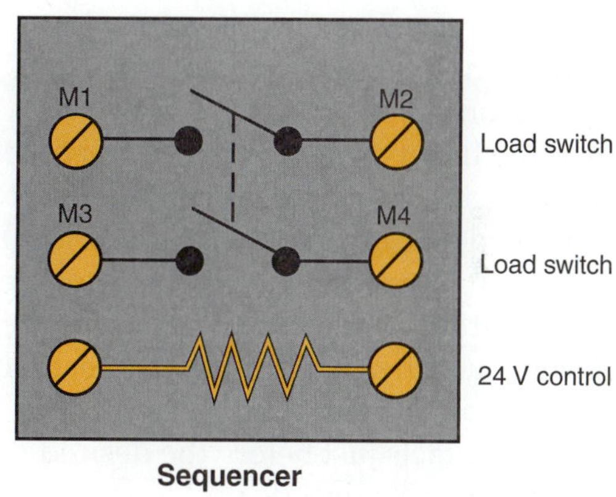

Goodheart-Willcox Publisher

Figure 29-7. A sequencer uses a small heater to close load switches.

1. Upon a call for heat from the thermostat, a 24 V circuit is completed to the small heater in the sequencer. After about 20 seconds, heat from the small heater causes the first set of contacts in the sequencer to close. One switch energizes the first heating element and the blower motor, **Figure 29-8**.
2. After a time delay, the second switch closes turning on the second heating element. After another time delay, if equipped, the third switch closes which turns on the third heating element. Thus, each heating element

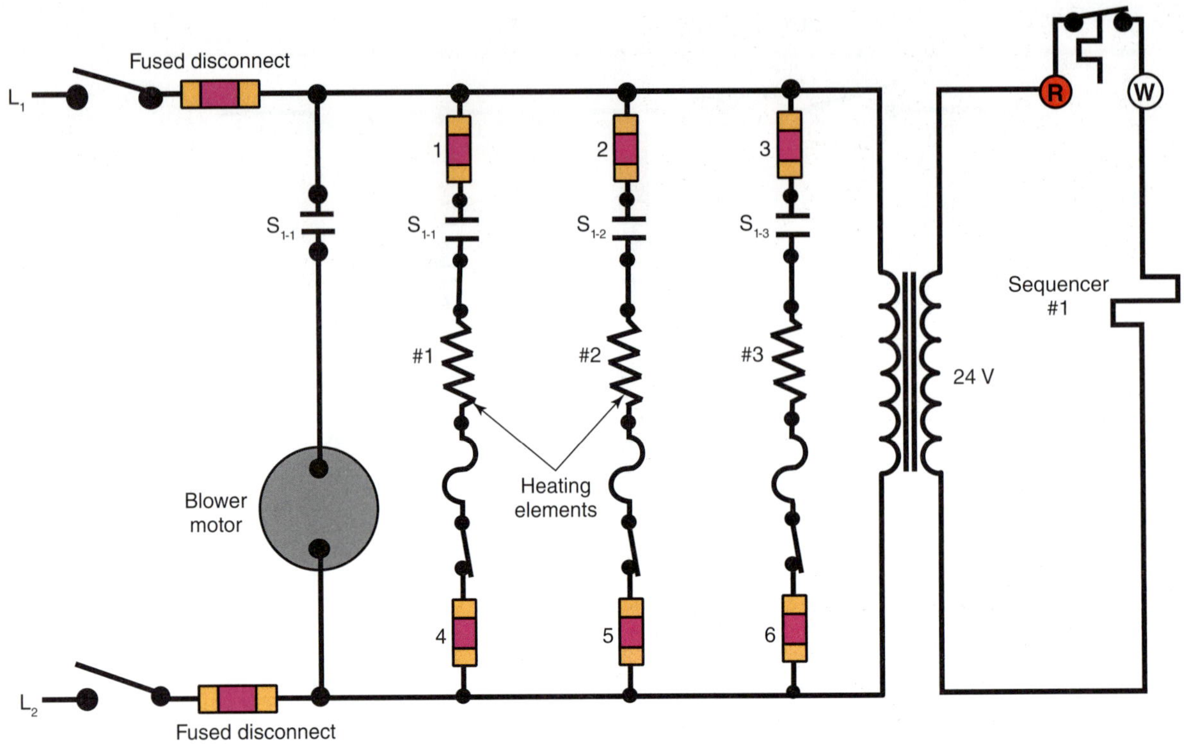

Goodheart-Willcox Publisher

Figure 29-8. When the thermostat calls for heat, a small resistance heater in the sequencer causes the normally open relay contacts (labeled S_{1-1} on this schematic) to close, energizing the first heating element and the blower motor. In turn, sequencer switches S_{1-2} and S_{1-3} energize the remaining two heating elements.

is staged to come on one after the other. The time interval between stages can be 15–45 seconds.

3. When the thermostat is satisfied and opens, the heating elements stage themselves off in reverse order. Time must be allowed for the small relay heater to cool. The heat anticipator inside the thermostat compensates for the delayed shutdown and additional heat. The heat anticipator causes the thermostat to open its contacts just before the desired room temperature is achieved.

29.3 Two-Stage Heating Systems

A ***two-stage heating system*** can produce two different heat outputs, depending on how much heat is desired or needed. Systems with more than one stage of heating include heat pump systems with gas heat as a secondary source (commonly referred to as *dual fuel systems*) or electric heat as a secondary source, two-stage gas furnaces, or multistage electric furnaces. Two-stage heating tries to match a furnace's heating ability to the actual heating requirement. This popular energy-saving feature is accomplished by controlling furnace capacity. In an electric furnace, the first stage switch of a two-stage heating thermostat energizes the first bank of electric heating elements. If more heat is needed than the first stage can supply, the thermostat will energize a second bank of heating elements.

Figure 29-9 illustrates the staging of heating elements in an electric furnace with five heating elements. The first stage of heat (elements 1, 2, and 3) operates when the thermostat calls for heat during periods of moderate outdoor temperatures. The second stage of heat (elements 4 and 5) is energized only during periods of extreme cold, when the first stage cannot satisfy the demand. Energy savings result because not *all* the elements are energized every time there is a call for heat.

The first stage of heat is always energized before the second stage of heat. The second stage is always de-energized before the first stage turns off. A control board or sequencer will keep the blower energized until all heat is dissipated from the furnace after shutdown.

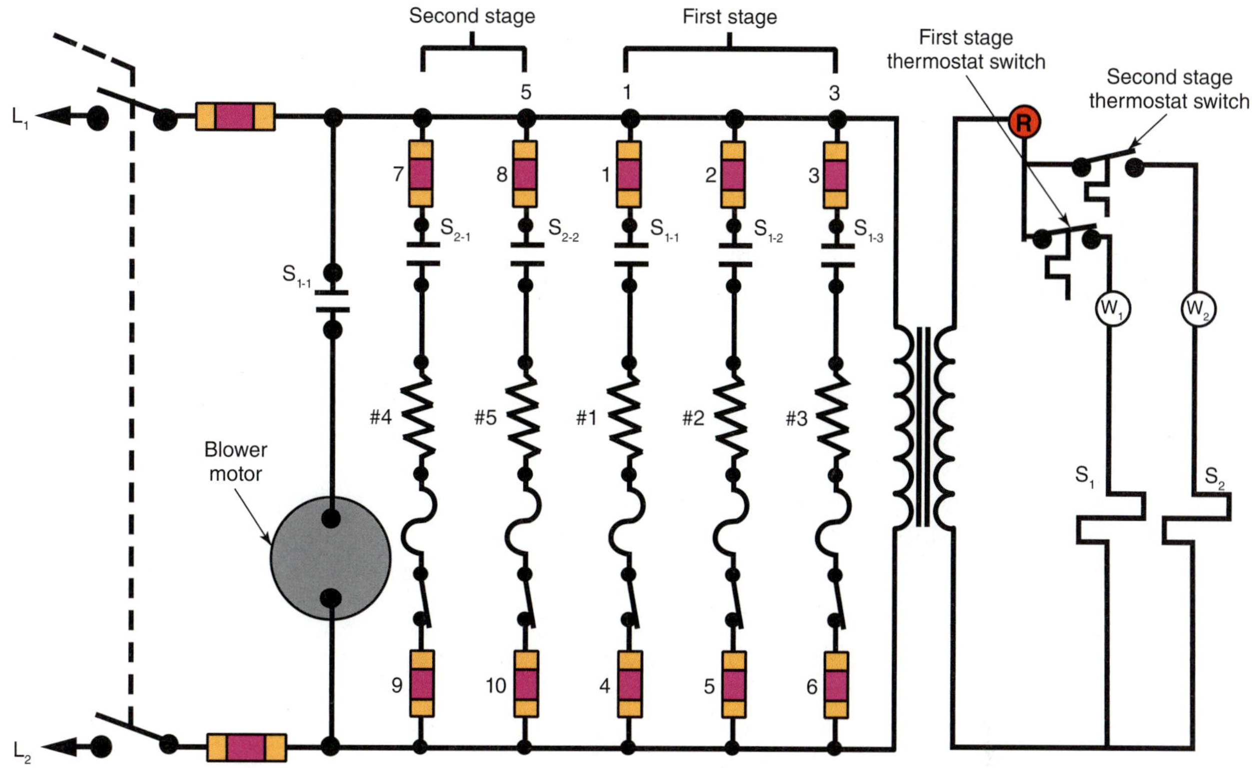

Goodheart-Willcox Publisher

Figure 29-9. An electric furnace with a thermostat controlling two heat stages. The first stage (heating elements 1, 2, and 3) is supplemented by the second stage (elements 4 and 5) when additional heating is needed.

Two-Stage Heating Thermostats

A ***two-stage heating thermostat*** has two copper-contact heating switches, which eliminate the need for separate thermostats. One contact switch controls the first stage of heat, and the second switch controls the second stage of heat. The contact switches open and close at different temperature setpoints. Separate heat anticipators are included for each switch, **Figure 29-10**.

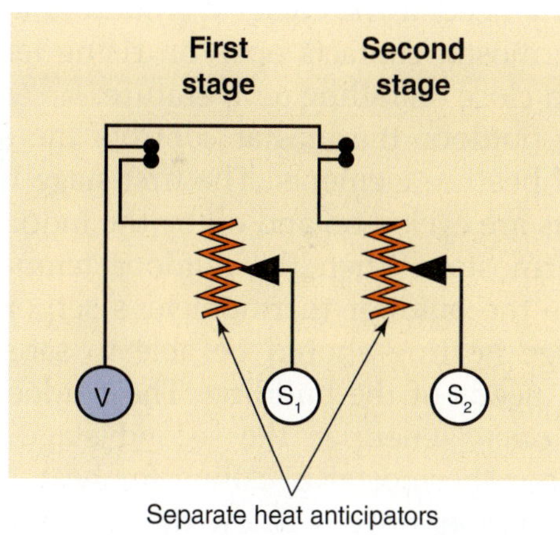

Goodheart-Willcox Publisher

Figure 29-10. A two-stage heating thermostat contains two copper-contact switches. One controls first-stage heating elements, the other controls the second-stage heating elements. Each switch is equipped with a heat anticipator.

The power supply from the transformer is connected to Terminal R on the thermostat, supplying 24 V power to each contact switch. One switch controls power to the W_1 terminal (first-stage heating), and the other switch controls power to the W_2 terminal (second-stage heating).

When room temperature drops to the thermostat setpoint, the thermostat contacts controlling power to W_1 close. First-stage heating is energized. If heat from the first stage is sufficient to prevent further temperature drop, the second stage remains off. However, if room temperature drops another 0.5°F–2°F, the contacts in the second switch will close. The second-stage heating elements are subsequently energized. When room temperature rises, the system de-energizes the stages in reverse order. The second stage shuts off first. Then the first stage shuts off.

Two-Stage Digital Thermostats

Two-stage digital thermostats are also available. These thermostats use sensors and sequential logic to provide the same function as electromechanical thermostats but provide more programmability and smart technology.

A two-stage digital thermostat controls a heating system with two signals of heat output to provide more comfortable and efficient heating. When the room temperature drops below the setpoint, the first stage of heat is activated. The thermostat uses the first stage for heating most of the time to maximize efficiency. If the room temperature continues to drop below a setpoint, the thermostat then uses the second signal to activate the second stage. The setpoint at which the second stage activates may be adjusted in the thermostat under economy or comfort settings. Second stage heat may be referred to as *auxiliary heat* or *emergency heat* on the thermostat depending on the thermostat model.

Outdoor Thermostat

Another method of staging two banks of heating elements is to use an outdoor thermostat in conjunction with a regular (single-stage) indoor thermostat. An ***outdoor thermostat*** monitors external air temperature, since more heat indoors is usually needed the colder the outdoor air gets. The thermostat contacts open on rising temperature and close on falling temperature.

The outdoor thermostat controls the second bank of heating elements. The first-stage heating elements are cycled on and off by the indoor heating thermostat. When the outdoor temperature is above the outdoor thermometer's setpoint, the first-stage heating should be able to satisfy the heating needs of the building. The outdoor thermostat *cannot* energize the second stage unless the indoor thermostat is calling for heat, *and* the outdoor temperature is below the setpoint of the outdoor thermostat. **Figure 29-11** shows how the indoor thermostat is wired between Terminals R and W. If the indoor thermostat switch is *not* closed, the outdoor thermostat does *not* receive power.

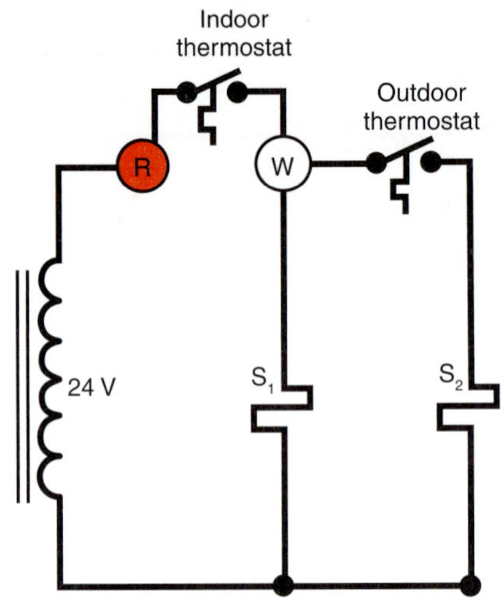

Goodheart-Willcox Publisher

Figure 29-11. An outdoor thermostat can be used in conjunction with an indoor thermostat to control a two-stage heating system, energizing the second-stage heating elements when outdoor temperature drops below the outdoor thermostat's setpoint.

29.4 Add-On Air Conditioning

Air conditioning is easily added to an electric forced-convection heating system. The evaporator is installed inside the furnace plenum or air handler, and the condensing unit is located outdoors. A suction line and a liquid line connect the evaporator to the condensing unit. The furnace blower is used to circulate air during heating and cooling cycles.

Cooling Blower Relay

A blower relay is required when adding air conditioning to a furnace. Blower controls for the heating cycle energize the blower when heat is produced. A separate relay—called a *blower relay*—must be used to energize the blower during the cooling cycle. Blower relays are normally provided with wiring diagrams that indicate which terminals are normally open (NO) and normally closed (NC). The coil on the blower relay is 24 V and controlled from Terminal G on the indoor thermostat. When a thermostat's manual fan switch is set for AUTO, Terminal G automatically receives 24 V when the thermostat calls for cooling. Then, the blower relay coil is energized and closes its switch. The blower relay switch is connected in series with the blower motor, **Figure 29-12**.

Setting a thermostat's manual fan switch to ON causes the blower relay coil to be energized continuously. On the continuous setting, the blower motor can operate any time, regardless of the thermostat setting.

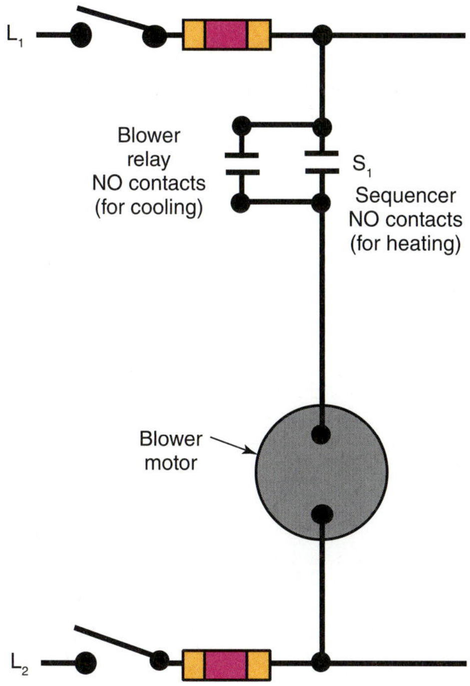

Goodheart-Willcox Publisher

Figure 29-12. The blower relay controls the blower motor during the cooling cycle.

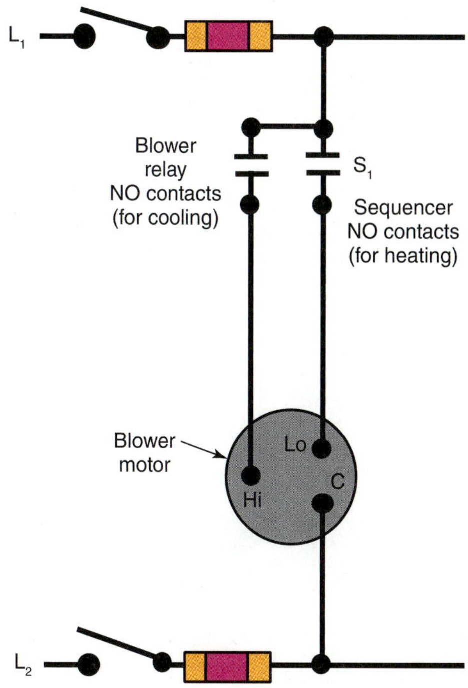

Goodheart-Willcox Publisher

Figure 29-13. A blower relay connected to a two-speed blower motor.

Two-Speed Blower Motor

A two-speed blower motor is often used to provide proper airflow through a furnace. Low speed (1025 rpm) is normally used for the heating cycle and high speed (1750 rpm) for the cooling cycle. **Figure 29-13** illustrates the wiring for a two-speed blower motor.

The blower relay often uses a single-pole, double-throw switch. One set of contacts is NO, and the other is NC, **Figure 29-14**. The NC switch is connected in series with the sequencer's NO contacts and the power supply to the blower motor for the heating cycle (low-speed). The NO switch is connected in series with the blower motor's high-speed windings, **Figure 29-15**.

The normally closed switch permits low-speed motor operation during the heating cycle but acts as a "lockout" when the blower operates at high speed (cooling). The normally open switch functions in just the opposite manner, permitting high-speed operation during the cooling cycle but acting as a "lockout" when the motor operates at low speed (heating). This arrangement prevents both speeds from being energized at the same time.

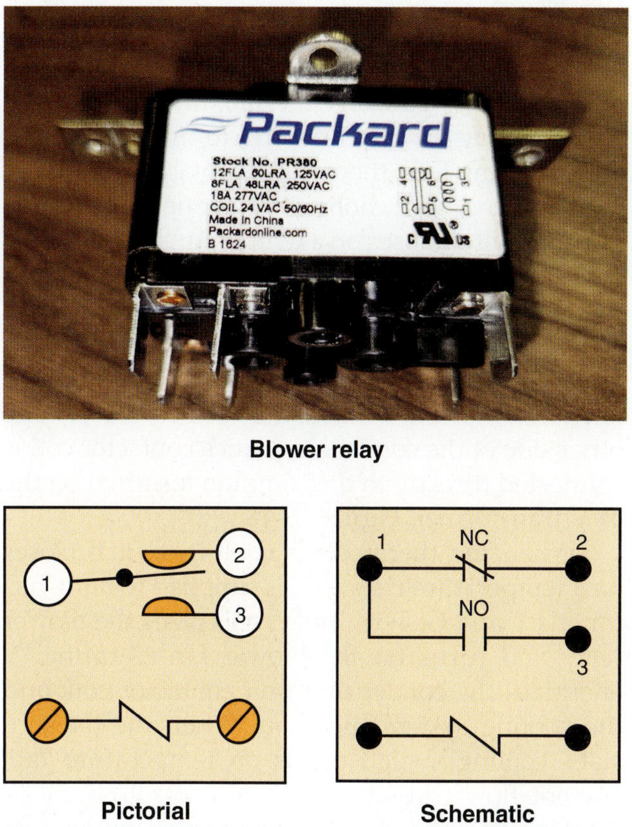

Goodheart-Willcox Publisher

Figure 29-14. A blower relay used with a two-speed blower motor has a single-pole, double-throw switch. As shown in both pictorial and schematic forms, one contact is normally closed, and the other is normally open. The NC contacts are in series with the blower's low speed (heating) winding.

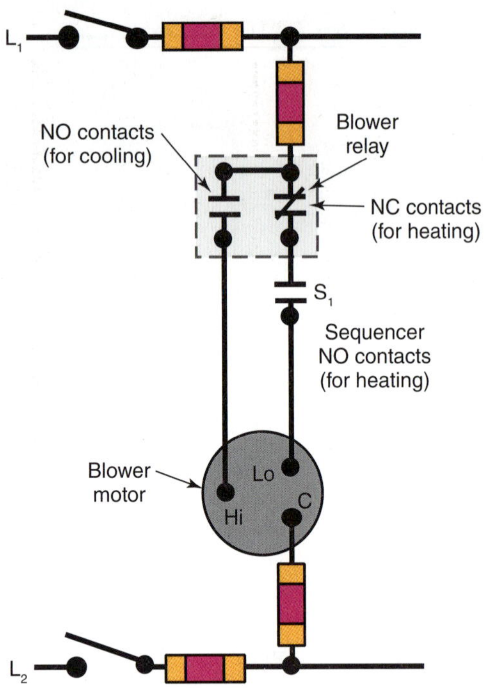

Goodheart-Willcox Publisher

Figure 29-15. Schematic of a blower relay with a single-pole, double-throw switch used with a two-speed blower motor.

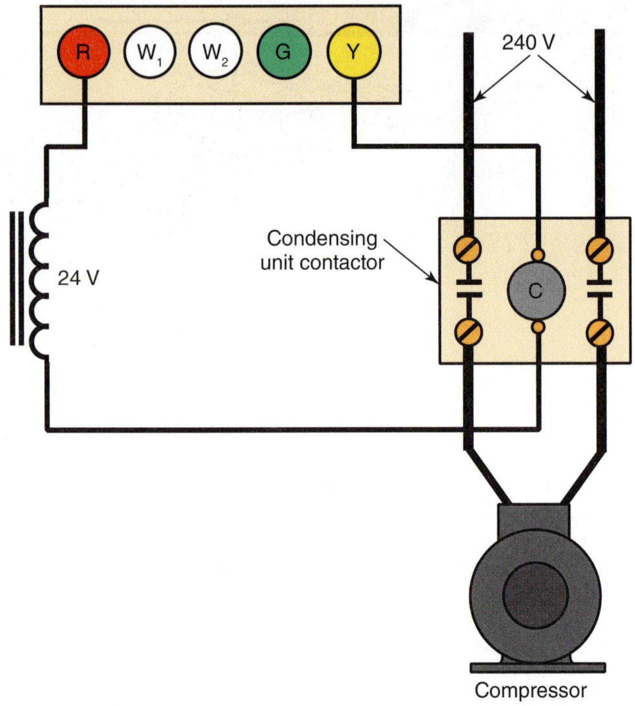

Goodheart-Willcox Publisher

Figure 29-16. A heating-cooling thermostat controls operation of the condensing unit contactor coil.

Heating-Cooling Thermostat

A ***heating-cooling thermostat*** is required when adding air conditioning to an electric furnace system. This thermostat provides control of both heating and cooling from one device. The heating switch closes on a temperature fall, and the cooling switch closes on a temperature rise. When the cooling switch closes on a temperature rise, 24 V is supplied to Terminal Y (yellow wire) on the thermostat. A wire is connected from Terminal Y to the coil on the condensing unit contactor. The other side of the condensing unit's contactor coil is connected directly to the common terminal on the 24 V transformer, **Figure 29-16**.

When the thermostat's cooling switch closes on a temperature rise, 24 V is supplied to both Terminals Y and G. Terminal G energizes the blower relay and turns on the blower fan. Terminal Y energizes the condensing unit contactor coil, and the cooling system turns on. When the thermostat's cooling switch opens on temperature fall, current flow through the 24 V control circuit to Terminals G and Y stops. The blower motor and compressor immediately stop running because the blower relay coil and the condensing unit coil are de-energized. Control boards may allow the blower motor to continue running for up to a minute, depending on the settings, after the call has been satisfied. The contacts in the high-voltage circuit are then opened.

High-Pressure Safety Control

Many residential air-conditioning units are equipped with a ***high-pressure safety control***, **Figure 29-17**. The high-pressure control is attached directly to the hot-gas discharge line. If discharge pressure exceeds the factory setting, an internal switch will open. The two factory-installed wires are connected to place the switch in series with the 24 V power supply to the condensing unit contactor coil.

The discharge pressure acts against a diaphragm balanced by a calibrated spring inside the control. If discharge pressure exceeds the setpoint, it overcomes the spring tension and opens the contacts. This stops the compressor. The safety control normally has a manual reset. When the control is tripped open, the service technician must first correct the cause of high head pressure, such as lack of condenser air (blocked or dirty coils), a refrigerant overcharge, or the presence of noncondensables (such as air) in the refrigerant circuit.

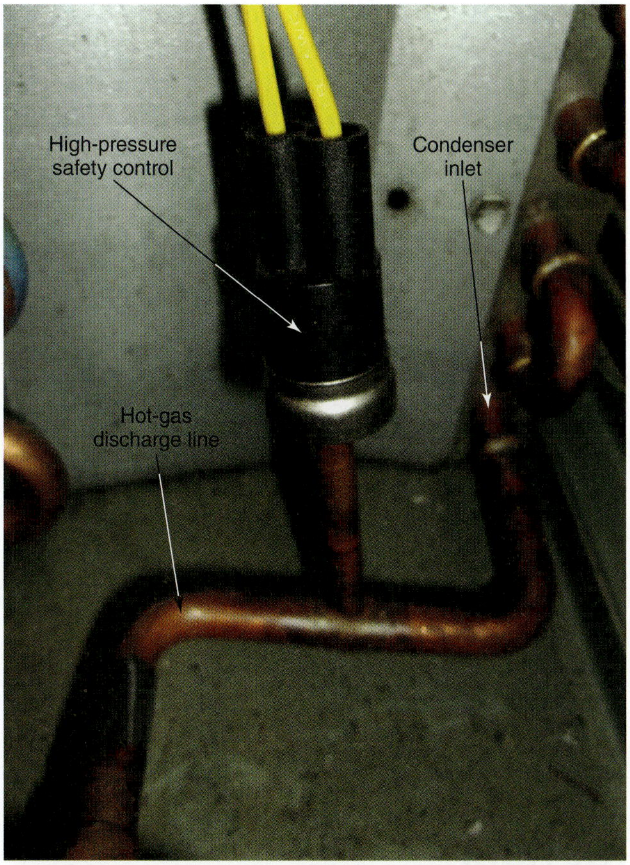

Goodheart-Willcox Publisher

Figure 29-17. A high-pressure safety control for residential air conditioning. Some switches have a manual reset button on top.

Low-Pressure Safety Control

Many residential air-conditioning units are also equipped with a ***low-pressure safety control***, **Figure 29-18**. The low-pressure safety control is attached directly to the suction line. Suction pressure works against a spring-loaded diaphragm that is mechanically linked to the normally closed contacts. If suction pressure falls below the control setpoint, the spring opens the contacts. Low suction pressure could be caused by dirty air filters, blower motor problems, a low refrigerant charge, or poor flow of refrigerant (a restriction).

The low-pressure safety control is intended to protect the compressor against low suction pressure. The control setpoints are established by the factory, and reset is automatic. Installation entails connecting two factory-installed wires to place the switch in series with the condensing unit contactor coil, **Figure 29-19**.

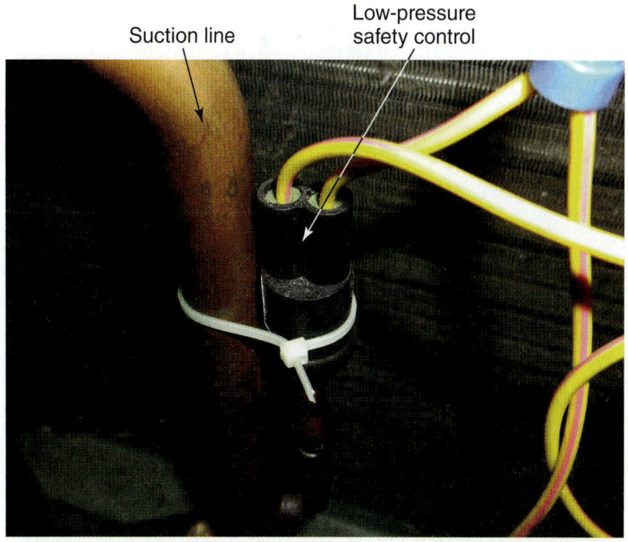

Goodheart-Willcox Publisher

Figure 29-18. A low-pressure safety control for residential air-conditioning installations.

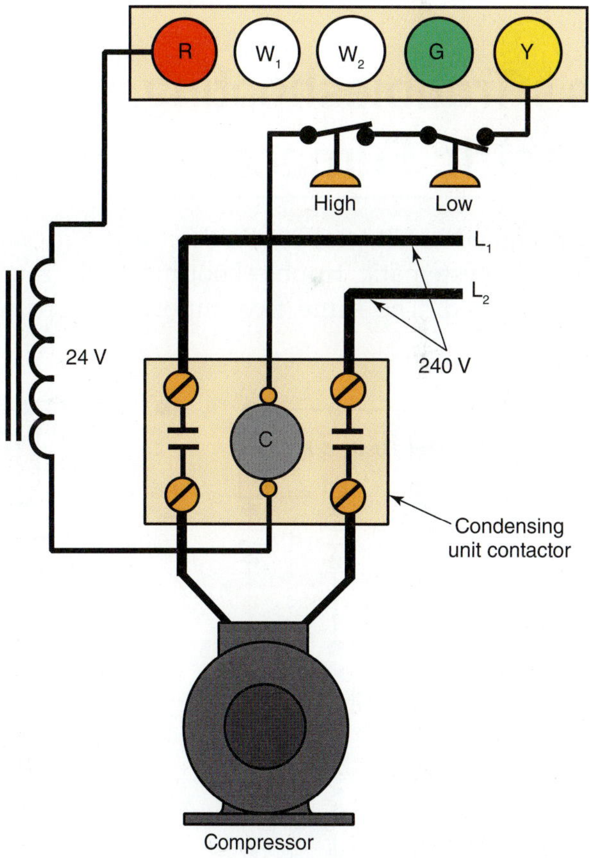

Goodheart-Willcox Publisher

Figure 29-19. High-pressure and low-pressure safety controls are connected in series with the condensing unit contactor coil.

Schematic and pictorial diagrams for a two-stage electric furnace and single-stage air-conditioning system are shown in **Figure 29-20** and **Figure 29-21**. Notice in these diagrams that second-stage heating cannot be energized unless first-stage heating is also energized. This is accomplished by the stage interlock contacts $S_{1\text{-}3}$ that are in series with thermostat Terminal W2 and sequencer coil S_2 shown in **Figure 29-20**.

Note that not all electric furnace manufacturers use stage interlock wiring practices. When interlocked, if the first stage sequencer goes bad, then the electric furnace will not be able to produce any heat at all. When interlocked, if the first stage cannot turn on, then the second stage also cannot turn on. However, when the stages are not interlocked, a bad first stage sequencer will only prevent the first stage from starting. The second stage heat may still be able to turn on and produce some heat if it is only controlled by the second stage thermostat.

29.5 Troubleshooting Electric Furnaces

As noted in Chapter 28, *Gas Heat with Air Conditioning*, systematic troubleshooting skills must be developed. The same five important steps in systematic troubleshooting discussed with gas furnaces apply to electric furnace systems as well:

- Know what the equipment *should* do (sequence of operation).
- Know what the equipment *is* doing.
- Know what the equipment is *not* doing.
- Eliminate areas that *are* functioning properly.
- Concentrate on areas that are *not* functioning properly.

The customer interview usually helps narrow the problem to a specific area. Electric furnace issues typically fall into six categories:

- No heat
- Not enough heat
- Too much heat
- Noise
- Odor
- Excessive operating cost

The following are some of the symptoms and possible causes of electric furnace issues.

No Heat

Possible Causes:

- Wrong thermostat setting
- Broken or loose thermostat wires
- Defective thermostat

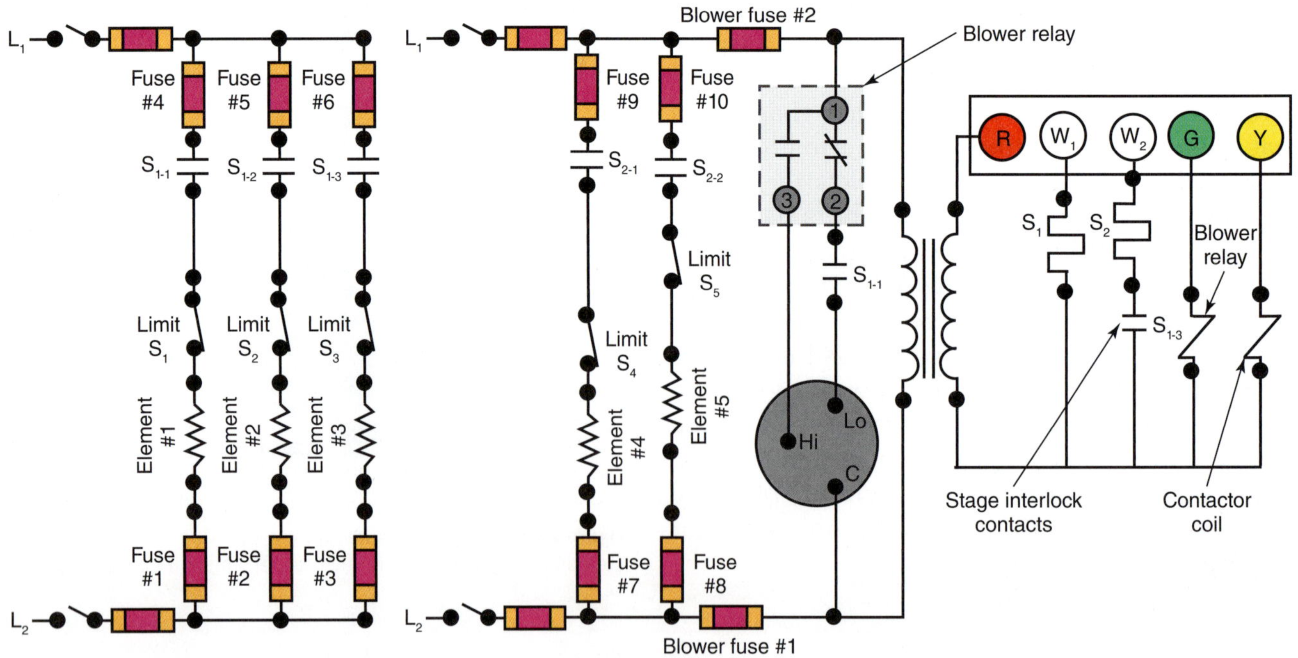

Goodheart-Willcox Publisher

Figure 29-20. A schematic diagram of a typical two-stage electric furnace with a single-stage air-conditioning system.

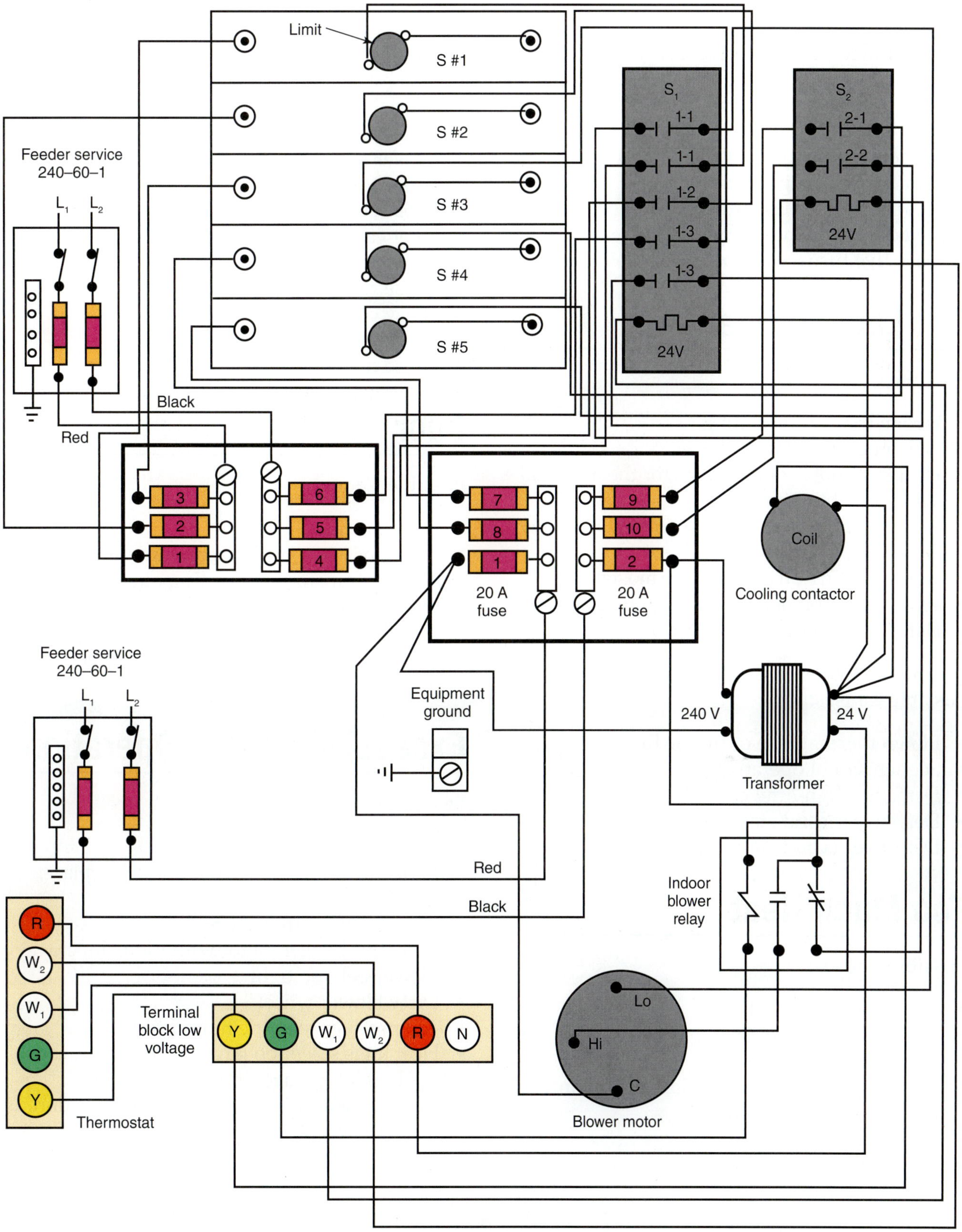

Goodheart-Willcox Publisher

Figure 29-21. A pictorial diagram of a typical two-stage electric furnace with a single-stage air-conditioning system.

- No power supply to furnace (Check fuses, circuit breakers)
- Low voltage at transformer secondary (less than 22 V)
- Low voltage at transformer primary (more than 10% drop)
- Defective transformer (Check fuse)
- Defective heating elements
- Blown thermal fuse or fusible link
- Defective sequencer
- Defective furnace contactor

Not Enough Heat

Possible Causes:

- Wrong thermostat setting
- Not enough airflow
- Heat strip not properly sized for the heat load
- Vibration at thermostat
- Thermostat influenced by warm air
- Defective thermostat
- Wrong setting on outdoor thermostat
- Wrong location of outdoor thermostat
- Defective outdoor thermostat

Symptom: Some heating elements not operating
Possible Causes:

- Blown thermal fuse or fusible link
- Sequencer not operating
- Defective high-limit control
- Defective second-stage sequencer
- Defective second-stage contactor

Too Much Heat

Possible Causes:

- Thermostat not level
- Thermostat in cold draft
- Defective thermostat
- Heat sequencer stuck in a closed position, leaving heat strips on at all times
- Control circuit shorted
- Too much airflow

Noise

Possible Causes:

- Loose blower bearings
- Blower bearing dry and squeaking
- Blower wheel off-center
- Loose V-belt
- Loose running gear and mounts
- Filter caught in running gear
- Loose cabinet panels
- Loose relay mountings
- Noisy contactor (Replace it)
- Humming transformer (Replace it)
- Loose ductwork

Odor

Possible Causes:

- Dirty air filter
- Stagnant water or sludge in humidifier
- Control transformer burning out
- Overheated wiring or overloaded circuit
- Dirty/dusty heating elements (proper preventative maintenance can stop this before it occurs.)

Excessive Cost of Operation

Possible Causes:

- Insufficient home insulation
- Excessive air leakage from home (fireplace, vents)
- Wrong fan control settings
- Defective blower motor with high amperage
- Blower belt too tight
- Dirty blower wheel
- Too much or too little humidity
- Heat kit is sized too large for the home, causing high amperage draw
- Heat sequencer is stuck in a closed position, keeping heat strips on at all times

Chapter Review

Summary

29.1 Electric furnaces require 240 V, three-wire feeder service from the main disconnect or load center. Some electric furnaces are equipped with either a fuse block or a double-pole circuit breaker.

29.2 Electric furnaces use electric heating elements to generate heat for the conditioned space. The limit switch is used to de-energize the heating element if the temperature exceeds the switch's setting. A fusible link is a safety device used to open the line voltage circuit if the temperature of the heating element becomes abnormally high.

29.2 Heating elements in electric furnaces are energized in sequence one or two at a time. Sequencers are devices used to stage the heating elements on and off in sequential order.

29.3 A two-stage heating system can produce two different heat outputs, depending on how much heat is desired or needed. Two-stage heating thermostats or outdoor thermostats can be used to stage two banks of heating elements in two-stage heating systems.

29.4 Air conditioning is added to an electric forced-convection heating system by installing an evaporator inside the furnace, and a condensing unit outdoors. A blower relay is used to energize the blower during the cooling cycle. A heating-cooling thermostat is used control both heating and cooling operation. High-pressure and low-pressure safety controls are used to protect the system from exceedingly high discharge pressures and dangerously low suction pressures.

29.5 When troubleshooting an electric furnace, a technician should know what the equipment *should* do (sequence of operation), what the equipment *is* and is *not* doing, eliminate areas that *are* functioning properly, and concentrate on areas that are *not* functioning properly. Electric furnace issues typically fall into six categories: no heat, not enough heat, too much heat, noise, odor, and excessive operating cost.

Know and Understand

1. Electric furnaces require _____ feeder service from the main disconnect or load center. (29.1)
   A. 120 V, two-wire
   B. 120 V, three-wire
   C. 240 V, two-wire
   D. 240 V, three-wire
2. Electric heating element wires used in electric furnaces are made of _____. (29.2)
   A. nickel
   B. copper
   C. nichrome
   D. ceramic
3. *True or False?* A fuse is required in each power supply leg to each electric heating element. (29.2)
4. The limit switch will de-energize an electric heating element if it senses excessively high _____. (29.2)
   A. current
   B. pressure
   C. temperature
   D. resistance
5. What device is commonly used to step-start electric heating elements? (29.2)
   A. Fusible link
   B. Limit switch
   C. Relay
   D. Sequencer
6. Two-stage electric heating results in energy savings because _____. (29.3)
   A. electricity is cheaper than gas
   B. electric heat provides more heat output than gas
   C. not all heating elements are energized when there is a call for heat
   D. heating is provided in half the time

7. Two-stage heating systems can operate with a single thermostat due to the _____ in the thermostat. (29.3)
   A. contact heating switches
   B. heat anticipators
   C. blower relay
   D. multiple power supplies
8. *True or False?* The outdoor thermostat can energize the second stage of heating elements without a call for heat if outdoor temperature is below the thermostat setpoint. (29.3)
9. A two-speed blower motor's blower relay often uses a _____ switch. (29.4)
   A. single-pole, single-throw
   B. single-pole, double-throw
   C. double-pole, single-throw
   D. double-pole, double-throw
10. The high-pressure safety control is attached directly to the _____. (29.4)
    A. suction line
    B. metering device
    C. hot-gas discharge line
    D. condensing unit
11. The low-pressure safety control is attached directly to the _____. (29.4)
    A. suction line
    B. metering device
    C. hot-gas discharge line
    D. condensing unit
12. *True or False?* All electric furnaces are wired with sequencers interlocked. (29.4)
13. A service technician is troubleshooting an electric furnace and discovers that two of the heating elements are not operating. Which of the following is *not* a likely cause of the issue? (29.5)
    A. Blown fusible link
    B. Defective high-limit control
    C. Dirty air filter
    D. Sequencer not operating
14. A humming transformer in the thermostat circuit should be _____. (29.5)
    A. sent back to the manufacturer for repair
    B. cleaned thoroughly
    C. replaced
    D. disconnected and rewired to a different power source

Apply and Analyze

1. What are two installation requirements for a fused disconnect or breaker used in an electric forced-convection heating system? (29.1)
2. List the sequence of operation for an electric furnace with multiple heating elements. (29.2)
3. Explain how an electric two-stage heating system can produce two different heat outputs. (29.3)
4. Why are two-speed blower motors often used in electric furnace systems with air conditioning? (29.4)
5. A service technician is troubleshooting an electric furnace that is not supplying enough heat. The technician discovers that the second stage of heating elements is not operating. Where is the cause of the issue *likely* to be found? (29.5)

Critical Thinking

1. List three causes of high head pressure and how a technician can correct those causes. (29.4)
2. Imagine you are a service technician troubleshooting an electric furnace. You set the thermostat to call for heat. You hear the sequencers click on and the blower comes on but it is not blowing heat. You discover that the sequencer is energizing the blower motor but not the heating elements. What could be the cause of the problem and how can it be fixed? (29.5)

Rheem Manufacturing Co.

30 Heat Pumps

LEARNING OUTCOMES

After studying this chapter, you will be able to:

30.1 Describe the operating principles of a heat pump, including how coefficient of performance, energy efficiency ratio (EER), balance point, and airflow calculations factor into heat pump design.

30.2 Recognize and explain the purpose of major heat pump components.

30.3 Describe the operation of thermostats and defrost controls used on heat pumps.

30.4 Read and interpret heat pump wiring schematics.

30.5 Describe the primary components of a ductless split system.

30.6 Recognize how to troubleshoot heat pump systems.

TECHNICAL TERMS

- air-source heat pump
- auxiliary heat
- balance point
- bypass line
- coefficient of performance (COP)
- ductless split system
- emergency heat
- energy efficiency ratio (EER)
- ground-source heat pump
- heat pump
- indoor coil
- low-ambient thermostat
- metering orifice
- outdoor coil
- outdoor defrost thermostat
- outdoor fan and defrost relay
- pilot valve
- reversing valve
- variable refrigerant flow (VRF) system

Introduction

A ***heat pump*** is a refrigeration system capable of reversing the flow of its refrigerant so that it can absorb and release heat. It does this by using unique heat pump components: an indoor coil, an outdoor coil, and a reversing valve. Heat pumps can be generally classified as air-source or ground-source (geothermal). An ***air-source heat pump*** relies on ambient air as a heat sink and as a heat source. A ***ground-source heat pump*** relies on the earth or a body of water as a heat sink and as a heat source. This chapter will focus on how an air-source heat pump works, explain the operation of various heat pump components, and discuss the application of heat pumps in residential units.

30.1 Residential Heat Pumps

Like a standard refrigeration system, an air-source heat pump has an ***indoor coil*** and an ***outdoor coil***. Both function as refrigerant-circulating heat exchangers. For the cooling cycle, the indoor coil functions as an evaporator, absorbing heat from the space, and the outdoor coil functions as a condenser, releasing heat out of the space. However, when the refrigerant flow is reversed for the heating cycle, the outdoor coil functions as an evaporator, absorbing heat from the space, and the indoor coil functions as a condenser, releasing the heat into the space.

The familiar window air-conditioning unit illustrates the operation of a heat pump. In the summer, the air conditioner absorbs heat from inside the room and discharges the heat to the outside. When winter arrives, the unit could be removed from the window, turned end-for-end, and reinstalled. In this position, the unit would then absorb heat from the outside air and discharge the heat into the room.

A heat pump is designed to accomplish refrigerant flow reversal using a four-way reversing valve to control the direction of refrigerant flow. During the cooling cycle, the high-temperature, high-pressure, superheated vapor that was absorbed from the indoor coil leaves the compressor and enters the reversing valve and is directed to the outdoor coil to release the heat. While in the heating cycle, once the refrigerant flows into the reversing valve, it is then directed to the indoor coil to release its heat, providing heat inside. This process is discussed further in the chapter.

A heat pump is *not* simply an air conditioner with reverse flow capability. It is a precisely engineered system designed to heat and cool with similar components; however, it looks like a conventional air-conditioning system, **Figure 30-1**.

There are many design factors carefully considered for heat pump systems. The following sections describe some of these factors.

Coefficient of Performance

Outside air always contains a certain amount of heat (measured in British thermal units, or Btus). Air does not lose *all* its *heat content* until it reaches absolute zero (–460°F or –237°C). A heat pump is designed to extract heat from outdoor air and to do this more efficiently than electric resistance heating elements produce heat. Electric resistance heat is considered 100% energy-efficient in converting electrical energy to heat energy. For every kilowatt of electricity used, an electric resistance heating system produces 3412 Btu of heat energy. A heat pump *improves upon* this energy efficiency.

The measurement of heat output (Btu) divided by its power input (Btu/watt) is called the ***coefficient of performance (COP)***. A heat pump with a COP of 3.00 produces three times as many Btu of heat for the same electrical energy input as electric resistance heating elements. For example, a heat pump that produces 50,000 Btu at 60°F (16°C)

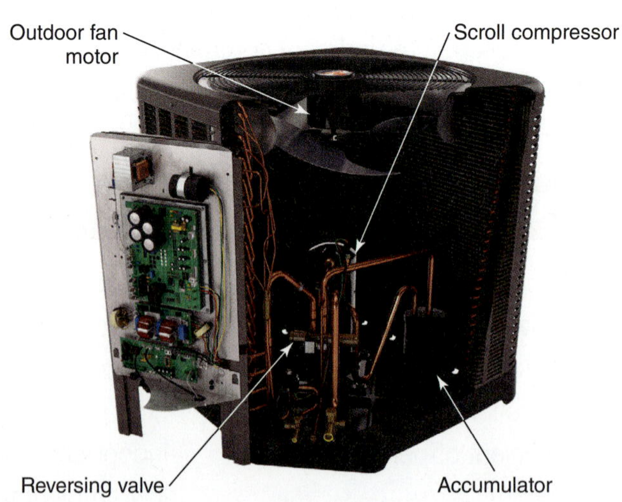

Rheem Manufacturing Co.

Figure 30-1. Cutaway view of a residential air-source heat pump.

outside air temperature uses about 4400 watts (4.4 kW) of electrical energy. Electric resistance heat using the same amount of electrical energy (4.4 kW) produces only 15,012 Btu. Therefore, at 60°F (16°C) outdoor air temperature, the heat pump has a COP of 3.32, meaning it has three times the energy-efficiency of electric resistance heat. However, as the outdoor temperature falls, less outdoor heat is available. Therefore, as outdoor temperature decreases, a heat pump's COP decreases (becomes less energy-efficient).

Energy Efficiency Ratio

Energy efficiency ratio (EER) is a term used to rate the energy efficiency of appliances. The EER is stated as the ratio of energy produced (output) to energy used (input). The EER provides the consumer with the relative operating cost of an appliance. The higher the EER, the more energy efficient the product is. In the preceding COP example, the heat pump produced 50,000 Btu with an input of 4400 W at an outdoor air temperature of 60°F (16°C). Dividing 50,000 by 4400 gives an EER of 11.36.

At an outdoor air temperature of 0°F (–18°C), heat pump efficiency is greatly reduced. Since energy input is the same (4.4 kW) and there is less heat available with an outdoor air temperature of 0°F (–18°C), a heat pump's heat output is reduced. With an output of only 16,000 Btu and an input of 4400 watts, the EER at 0°F (–18°C) becomes 3.64 (16,000 ÷ 4400 = 3.64).

Balance Point

Heat pumps require accurate calculations for building heat loss and gain. After heat loss and gain loads have been figured, the balance point of the heat pump is determined.

The ***balance point*** of a heat pump is the point at which the *heating ability* of the device equals the *heat loss* of the building, **Figure 30-2**. Balance point varies according to outdoor temperature, building design, building construction, and other factors, but it usually falls between 30°F and 40°F (–1°C and 4°C). When the balance point is reached, the heat pump runs continuously.

When outdoor air temperature is *above* the balance point of the system, the heat pump can readily make up for heat loss. When outdoor air temperature is *at or below* the balance point, auxiliary (supplementary) heat must be supplied. ***Auxiliary heat*** is another form of heat that augments a heat pump's heat production and may also be used in the defrost cycle. Auxiliary heat is also called *supplemental heat* and may come in the form of staged electric heating elements or a gas-fired furnace. When outdoor air temperature drops to a certain temperature (check each manufacturer's specifications for the set temperature), a heat pump is normally turned off, and the dwelling is heated by a conventional electric or gas furnace. At these low temperatures, when the heat pump shuts off and auxiliary heat alone warms a building, it is referred to as ***emergency heat***. Emergency heat will also provide heating if any issues arise with the heat pump that prevents it from running.

Airflow Standards

Ensuring proper airflow is a critical part of heat pump installation, as it is for other types of HVACR systems. Correct heat load calculations, duct designs, and airflow performance charts need to be adhered to so the system can operate as it is designed.

During the heating cycle, the temperature rise over the indoor coil is considerably less than the rise in a conventional furnace. Supply air temperature to a room may be about 105°F (41°C) when outdoor temperature is 30°F (–1°C). This provides a temperature rise of about 35°F (19°C) over the coil, compared to about 60°F (33°C) temperature rise for a gas furnace. The air seems cooler due to increased volume. The system's ability to maintain the desired room temperature is *not* affected. The system simply runs longer to satisfy the thermostat.

The outdoor coil receives much less air than is normal for a regular air-conditioning system. Heat rejection during the cooling cycle is less efficient. During the heating cycle, slower airflow permits better heat pickup. Some manufacturers may use either a variable speed or two-speed motor on the outside coil to deliver more air during the cooling cycle. Head pressure is lowered, and operation is more efficient.

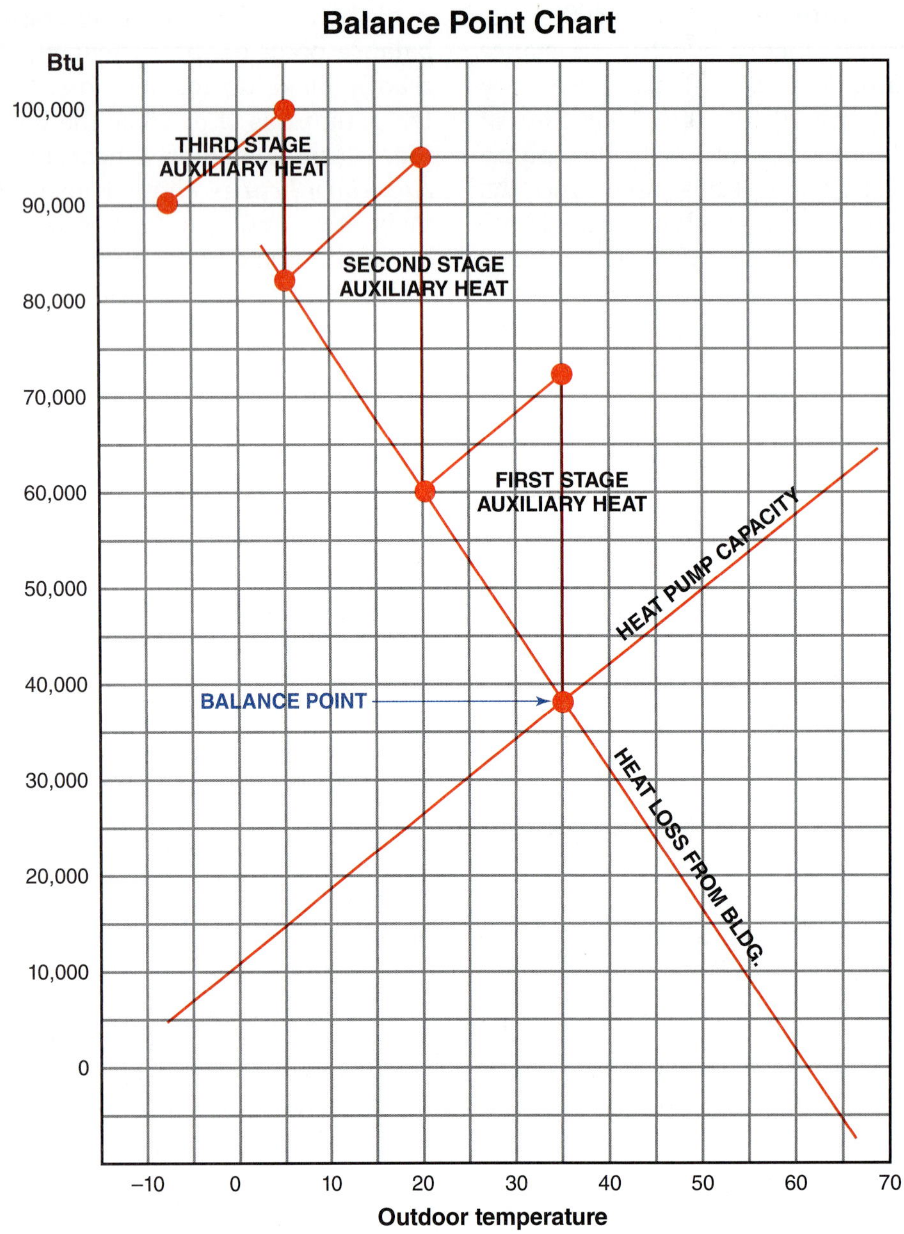

Goodheart-Willcox Publisher

Figure 30-2. The balance point of a heat pump is where heat pump capacity and building heat loss are equal.

30.2 Heat Pump Components

Although a heat pump looks much like a conventional air-conditioning unit, its components must operate equally well in both heating and cooling modes. Specially engineered heat pump components that can withstand adverse conditions are often required. A reversing valve, electrical controls, additional safety devices, and other components permit operation in both modes.

Reversing Valve

A ***reversing valve*** is a solenoid-operated, four-way valve that changes the direction of refrigerant flow through a heat pump, **Figure 30-3**. The reversing valve is located within the outdoor unit, immediately after the compressor. The valve body contains a piston-operated slide valve that is actuated (moved) by refrigerant pressures as directed by a solenoid-operated pilot valve. Energizing or de-energizing the 24 V solenoid pilot valve determines on which side of the piston pressure is

Goodheart-Willcox Publisher

Figure 30-3. A reversing valve makes it possible to alternate a heat pump's heating and cooling modes by reversing the flow of refrigerant. The valves are available in sizes to match the different heat pump capacities.

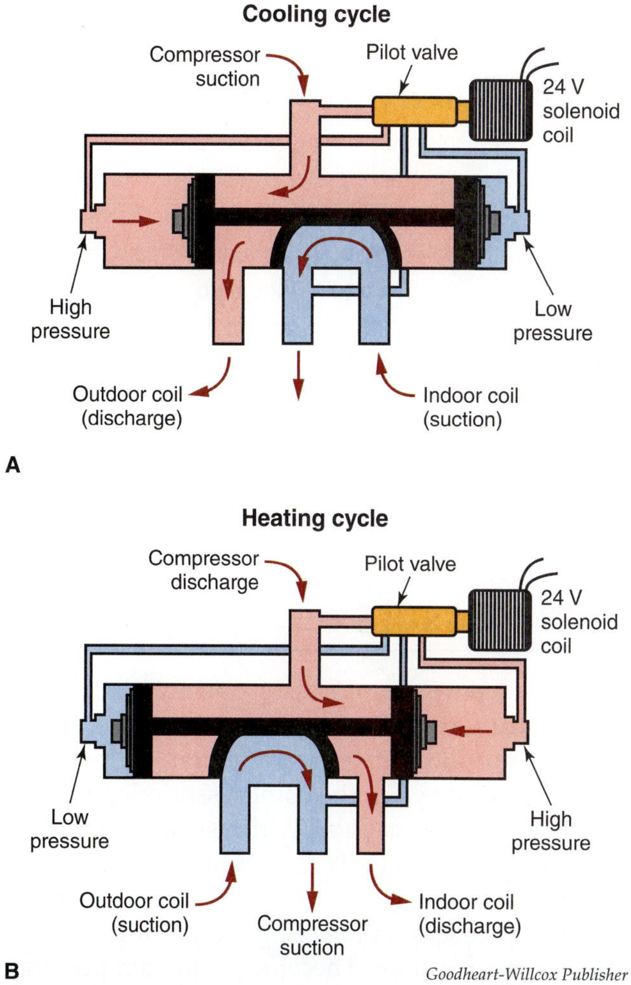

Goodheart-Willcox Publisher

Figure 30-4. Reversing valve operation. A—In the cooling cycle, high-pressure refrigerant is directed to the outdoor coil. B—In the heating cycle, high-pressure refrigerant is directed to the indoor coil.

exerted, which determines piston position and the refrigerant pathways through the reversing valve. Generally, three refrigerant-circulating tube connections are on one side of the valve body, and a single tube connection is on the opposite side.

As shown in **Figure 30-4**, the compressor discharges into the inlet tube. The middle tube on the *opposite* side of the reversing valve is connected to the compressor suction. Note that refrigerant flow through the compressor is always the *same*; it does *not* reverse. The remaining two tubes on the reversing valve are connected to the indoor coil and the outdoor coil.

The position of the piston-operated slide valve inside the reversing valve determines the direction of refrigerant flow to and from each coil. Flow reversal is accomplished by the slide valve connecting one of the outside tubes with the compressor suction tube of the valve. The valve position is controlled by energizing or de-energizing the solenoid coil of the pilot valve.

The reversing valve is pilot-operated. The 24 V solenoid coil and ***pilot valve*** control the pressure difference at each end of the reversing valve's piston. Four small tubes are connected to the pilot valve body. Two of these tubes are connected to the ends of the reversing valve body. Another tube is connected to the suction line where it leaves the reversing valve body. The final tube connects to the compressor discharge line. Note that there is a great pressure difference between a compressor's suction and discharge lines.

The pilot valve automatically controls pressure (high or low) at each end of the piston. Energizing or de-energizing the solenoid coil causes pressure on the piston ends to reverse. The side with high pressures pushes the piston to move the slide valve. Thus, system pressures are used to reverse piston position. Some heat pump systems energize the reversing valve during the cooling cycle, others during the heating cycle.

CAUTION

Fail-Safe Reversing Valve Position

Most heat pump reversing valves are installed to fail-safe into the heating mode position. This means that if power is not supplied to the reversing valve's solenoid, the valve position will automatically be in heating mode position (directing high pressure to the indoor coil and low pressure to the outdoor coil). Even if the solenoid coil burns out, the heat pump can still warm a building during cold weather. However, there are some manufacturers that set their systems to fail-safe in cooling mode. Always refer to manufacturer specifications for a given unit.

Compressors

A heat pump compressor operates as a high-temperature machine during the cooling cycle, which is normal. During low ambient temperature (in the heating cycle), the compressor operates as a low-temperature machine with a steadily reducing capacity. The possibility of liquid *floodback* to the compressor during the heating cycle is ever present. Compressors for heat pumps contain specially engineered features. Therefore, only compressors designated as "heat pump models" should be used as replacement compressors.

Because liquid floodback is probable, heat pumps are equipped with an accumulator in the suction line to protect the compressor. A crankcase heater is used to warm the compressor, which protects it against liquid migration (or condensation) during the off-cycle.

The compliant scroll compressor has greatly improved compressor efficiency and reliability, **Figure 30-5**. For heat pumps, it is the compressor of choice. The compliant scroll feature enables the compressor to pump liquid without damage, eliminating the need for a suction accumulator and a crankcase heater. The scroll compressor also has fewer moving parts than the reciprocating compressor, significantly reducing lubrication problems.

Copeland

Figure 30-5. The compliant scroll compressor design eliminates the need for an accumulator and a crankcase heater since the scroll can pump liquid without damage.

Refrigerant Metering Devices

Proper control of refrigerant entering the coils can be accomplished in several ways. Controls include thermostatic (and electronic) expansion valves with bypass lines and metering orifice flow controls.

Thermostatic Expansion Valves

Thermostatic expansion valves (TEVs) with bypass lines and check valves are popular on large heat pump systems, **Figure 30-6**. Because each coil operates as an evaporator, each coil requires a TEV to control the amount of liquid entering.

A ***bypass line*** is also needed for each coil. This is an alternate tubing path in parallel with the TEV that permits refrigerant to bypass the expansion valve when the TEV is not in use. A check valve is installed in the bypass line to control the direction of refrigerant flow. Refrigerant can only flow through the check valve and bypass line when its TEV is not in use. For example, when an outdoor coil is functioning as a condenser during the cooling cycle, the outdoor TEV is bypassed and thus not in operation. Only the indoor TEV is functioning during the cooling cycle.

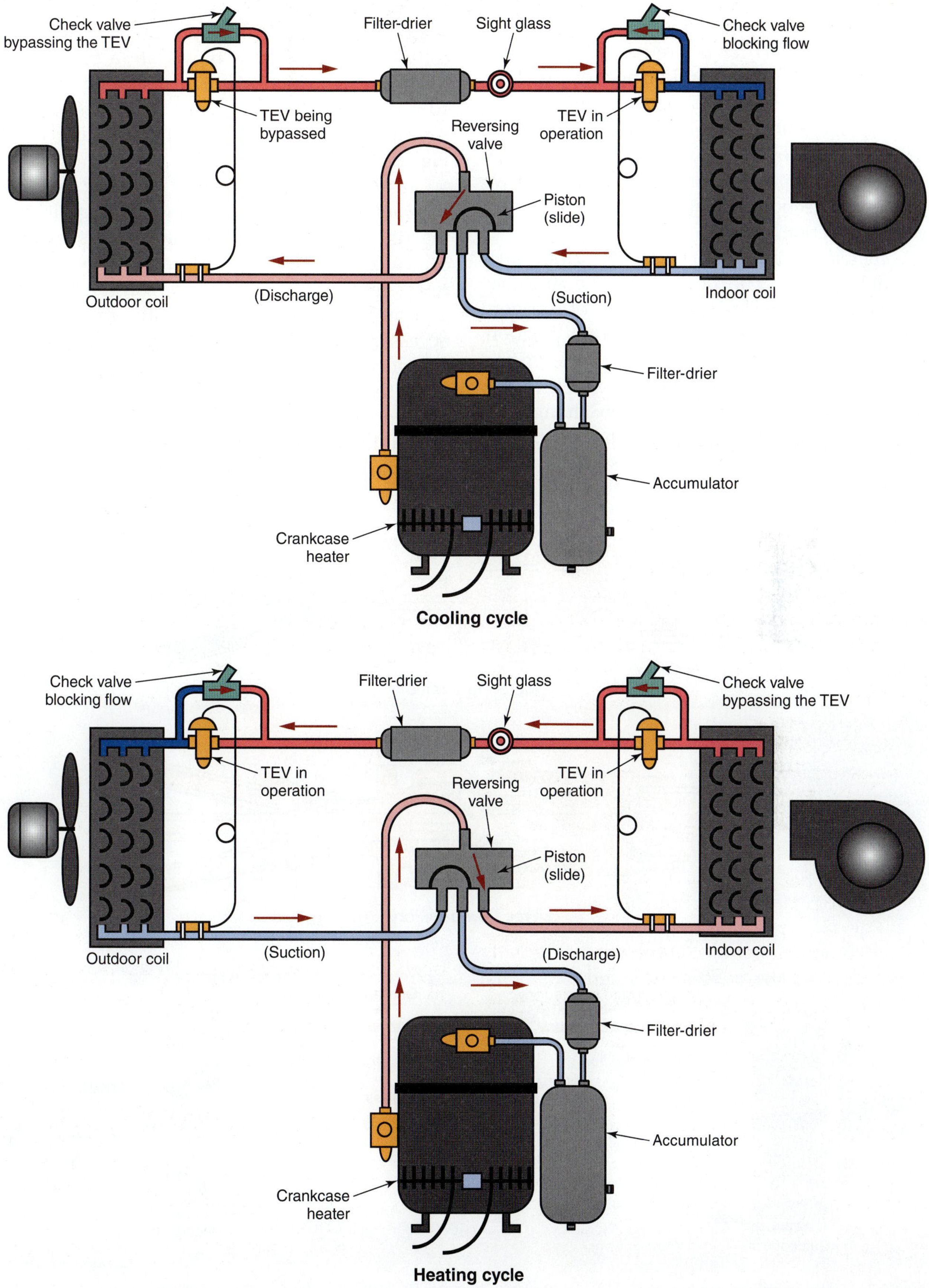

Goodheart-Willcox Publisher

Figure 30-6. A heat pump system using TEVs and bypass lines with check valves. Note how one check valve allows refrigerant to bypass the TEV in each mode.

Metering Orifice

The ***metering orifice*** is a refrigerant metering device that uses a piston and orifice for pressure drop and flow control. It operates much like a capillary tube, allowing refrigerant flow in both directions and providing a fixed pressure drop. A metering orifice meters (restricts) refrigerant flowing in one direction and allows free flow in the opposite direction.

A metering orifice contains an orifice and a movable piston. The position of the movable piston is determined by the direction of refrigerant flow. In one direction, the piston's position forces the refrigerant to flow through the orifice (metered position). When refrigerant flows in the other direction, the piston's position changes to allow refrigerant to flow freely without restriction or pressure drop. This option to restrict or allow free flow means that a heat pump will need to use two metering orifices (one on each refrigerant coil). When refrigerant flows *into* a coil through the metering orifice, it will be restricted for a pressure drop. When refrigerant flows *out* of a coil, it will flow freely through the metering orifice, **Figure 30-7**.

In the cooling mode, the metering orifice on the *outdoor coil* allows free flow through it, creating no pressure drop. Meanwhile, the metering orifice on the *indoor coil* meters the flow of liquid refrigerant, making the indoor coil function as an evaporator. The orifice creates the pressure drop necessary to maintain the proper boiling point in the indoor coil.

When in heating mode, the metering orifice on the indoor coil permits unrestricted reverse flow. As liquid refrigerant flows in the reverse direction, the metering orifice automatically moves to

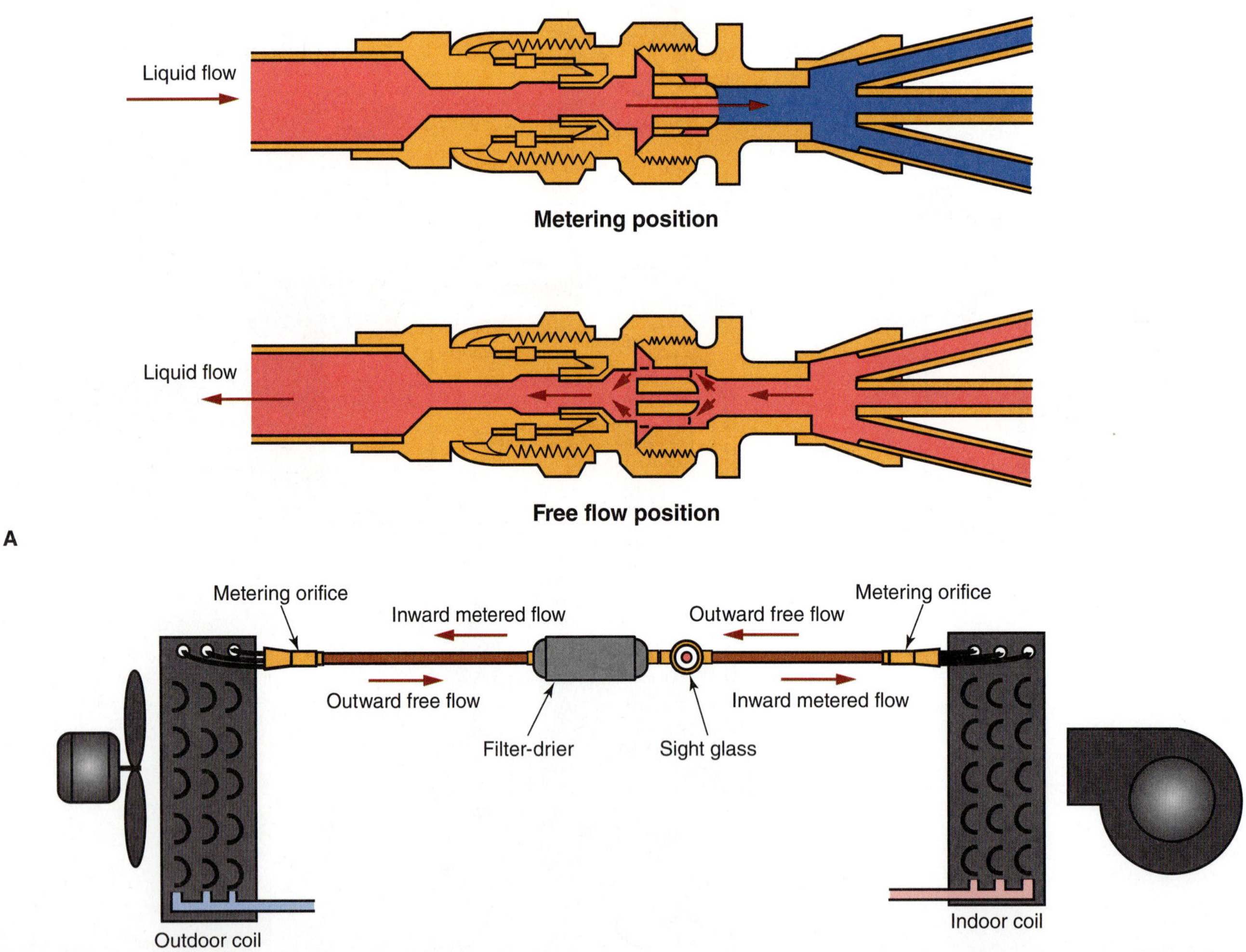

Goodheart-Willcox Publisher

Figure 30-7. Metering orifice-type flow control. A—As shown in the cutaway views, the metering orifice precisely meters refrigerant in one direction and allows unrestricted free flow in the reverse direction. B—One metering orifice is needed at each coil. Flow through the orifice into the coils is metered, and flow out of the coils is free flow. Systems with two metering orifices do not need any expansion valve, check valves, bypasses, or capillary tubes.

the free-flow position. The reverse-flow capability eliminates the need for a check valve and bypass tubing. The metering orifice on the outdoor coil restricts refrigerant to create the pressure drop for the outdoor coil to act as an evaporator.

On a heat pump, the indoor and outdoor coils differ in size and capacity. To compensate for the pressure difference this creates, a different-sized metering orifice is used for each coil. During the heating cycle, the outdoor coil becomes the evaporator and requires more restriction. A metering orifice may be easily removed in the field and replaced with a different-sized orifice to provide a different flow rate and corresponding pressure drop.

Heat Pump Filter-Driers

Most filter-driers are direction-sensitive and must be installed in the direction of refrigerant flow. Such filter-driers cannot be used on heat pumps because of reversing refrigerant flow. Special filter-driers that permit reverse flow are available for heat pumps.

PRO TIP

Charging a Heat Pump System

The amount of refrigerant in a heat pump is critical for efficient operation of the system. The most accurate and efficient method of charging the system is by weight. The proper type and amount (pounds and ounces) of refrigerant is indicated on the unit's data plate. Charging by exact weight can be performed year-round in either the heating or cooling mode.

30.3 Heat Pump Thermostats and Defrost Cycles

Thermostats used in heat pumps must be able to provide controls for both heating and cooling cycles. The thermostats may also control defrost cycles needed to melt frost and ice that may accumulate on the outdoor coil. Other methods of providing defrost are also used.

Thermostats

The thermostats used with heat pumps are similar to digital thermostats used with other systems, with a few wiring differences. Similar to other digital thermostats, heat pump thermostats use a thermistor to sense air temperature and signal the heat pump unit to turn on or off based on the temperature sensed. One important distinction with heat pump thermostats is they may list a minimum of two options for heat and at least one option for cooling. Thermostats will usually specify on the packaging, "Heat pump with auxiliary heat or heat pump without auxiliary heat." The two options for heat are for running the unit in heating mode and for any auxiliary or emergency heat. Another important distinction with heat pump thermostats is that the heat pump should always be the first stage of heat and will always wire to an O/B terminal (depending on manufacturer) in the thermostat subbase. Always refer to the thermostat's installation manual to confirm wiring and settings are correct. Heat pump thermostats may have additional settings in the "Mode" feature. Other than the standard "Heat-Off-Cool," these thermostats may be set up to offer an "Emergency Heat" mode to allow the option of turning on auxiliary heat if the heat pump loses function.

Auxiliary heat can also be used to provide additional comfort or be set up as an economy (Eco-mode) option. With these options, if the thermostat in heating mode senses that it is not raising the space temperature, then it may bring on the auxiliary heat to make up difference. The comfort and economy options allow the thermostat to decide at which set point the auxiliary heat will turn on.

Various terminals that may be found on the heat pump thermostat subbase include:

R = voltage from the transformer (hot, 24 V)
G = indoor fan
Y = compressor
O/B = reversing valve
W2 = auxiliary heat
E = emergency heat
C = common from transformer

Outdoor Thermostat

On systems that use a gas furnace as an auxiliary heat source, an outdoor thermostat may be used to prevent energizing of auxiliary heat stages until the outdoor temperature is low enough to require auxiliary heat. Once the thermostat calls for auxiliary heat, it signals the outdoor unit to turn off so that it does not cause unnecessary pressure buildup at the indoor coil while the gas furnace is running. The outdoor thermostat is connected in series with the indoor thermostat's second stage setting for heat. This connection ensures that the auxiliary heat can *only* be energized when *both* indoor and outdoor thermostat contacts are closed.

Low-Ambient Thermostat

Heat pump efficiency is severely reduced when outdoor temperature drops between 10°F and 15°F (–12°C and –9°C). At a certain point, heat pump efficiency drops to a level at which it would be more efficient to rely on auxiliary heat for comfort heating. A ***low-ambient thermostat*** is a thermostat located outdoors that stops compressor operation when the system should run using auxiliary heat rather than as a heat pump. The low-ambient thermostat has a set of NC contacts connected in series with the compressor contactor coil. It opens its NC contacts to prevent the compressor from operating during periods of low outdoor temperatures.

Defrost Cycles

Heat pumps require a defrost cycle to remove frost from the outdoor coil. The outdoor coil becomes an evaporator during the heating cycle and is about 10°F (5.5°C) colder than the ambient air. Whenever the coil temperature is below 32°F (0°C), frost and ice accumulate on the outdoor coil, Figure 30-8. The frost buildup must be removed periodically to maintain proper airflow.

The most practical way to remove ice buildup is to heat the coil. For a defrost cycle, the system is automatically switched to the cooling mode, even though the thermostat calls for heating. The outdoor coil is converted to a condenser. Frost quickly melts because the outdoor fan motor is turned off during defrost, allowing heat to accumulate and temperature to rise quickly in the outdoor coil.

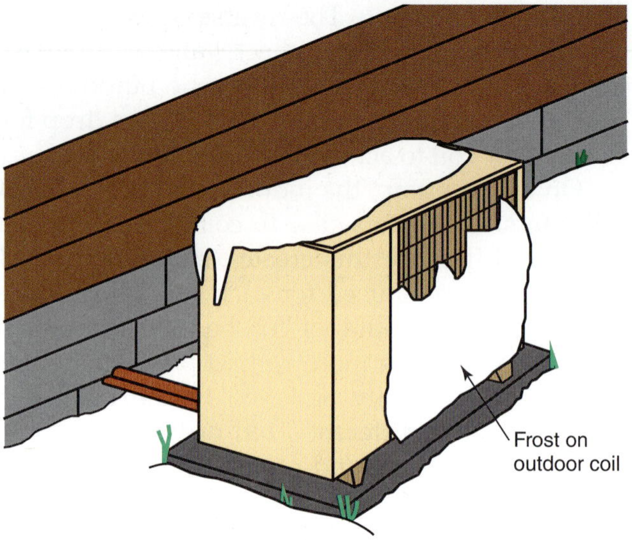

Goodheart-Willcox Publisher

Figure 30-8. During heating in low ambient temperatures, a defrost cycle is needed to remove frost accumulated on the outdoor coil of a heat pump.

Unfortunately, operating in cooling mode removes heat from the interior of the building during a defrost cycle. To solve this problem, the auxiliary heat is turned on during defrost so air delivered to the building interior is warm.

There are several ways to control the defrost cycle. A pressure differential switch can be used to sense increased pressure drop across the coil as frost forms. A temperature switch can be used to sense the decrease in coil temperature as frost forms. The most common method is the time-initiated/temperature-terminated defrost. In this system, a defrost timer runs whenever the compressor is operating. About every 90 minutes, the timer mechanically closes a switch to energize the defrost relay and initiate a defrost cycle.

A special thermostat called an ***outdoor defrost thermostat*** senses outdoor coil temperature. This thermostat's contacts open on rise (65°F or 18°C) and close on fall (32°F or 0°C). The outdoor defrost thermostat's contacts are connected in series with the defrost timer switch and defrost relay. The thermostat's contacts close when coil temperature drops to about 32°F (0°C). To initiate a defrost cycle, two conditions must be met: the timer must have completed its 90-minute cycle *and* the coil temperature must be 32°F (0°C). If one of these elements is lacking, defrost will not occur. The system will wait another 90 minutes to try again.

The defrost cycle is terminated by the outdoor defrost thermostat when coil temperature rises to 50°F–70°F (10°C–21°C). This temperature is set by the manufacturer. The defrost cycle may require 5–6 minutes when frost accumulation is heavy, typically when outdoor ambient temperature is about 40°F (4°C). At lower ambient temperature, outdoor air contains less humidity, and the defrost cycle may terminate in 3 minutes. If the outdoor defrost thermostat fails to do so, the clock will terminate the defrost cycle after 10 minutes.

PRO TIP

Defrost Vapor Cloud

While a heat pump is defrosting, a large vapor cloud may appear at the outdoor unit. It is normal and indicates that the defrost cycle is operating properly. When installing a system for first-time heat pump buyers, always inform them of this defrost cloud to prevent them from worrying or incurring the costs of an unnecessary service call.

30.4 Reading Heat Pump Schematics

Electrical circuits for controlling heat pumps vary greatly from model to model and from one manufacturer to another. Electrical control circuits range from simple to complex. Most heat pump schematics are complex since they may include a defrost cycle, safety controls, and accessory features.

The heat pump service technician must be knowledgeable of heat pump theory and able to read schematics. Manufacturer schematics explain how a unit's electrical components are connected. Understanding the purpose of these controls and circuits is the basis for effective troubleshooting.

Heat Pump Circuits

The following electrical schematics illustrate the general operating principles of heat pumps. The electrical circuits shown are common to many systems. Keep in mind that variations in types of components and control circuits are possible. Always check each unit's wiring schematic and pictorial representation for methods and sequences of operation.

The schematic in **Figure 30-9** begins with basic connections for a heat pump system. Components include the compressor, outdoor fan, transformer, thermostat, auxiliary heat relay, reversing valve, compressor contactor, and indoor blower relay. Additional components will be added to subsequent schematics.

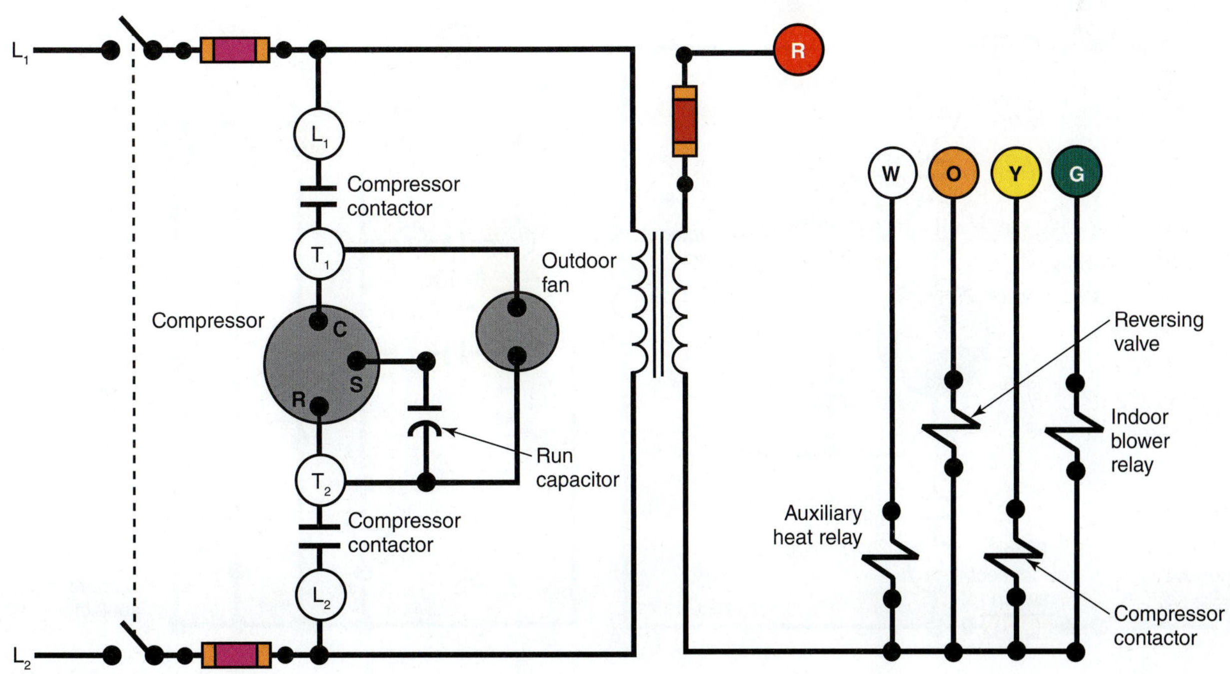

Goodheart-Willcox Publisher

Figure 30-9. Basic schematic for a heat pump.

Outdoor Fan and Defrost Relay

The ***outdoor fan and defrost relay*** combines two relays into one component, **Figure 30-10**. Several switches are controlled by the relay coil. The relay coil can be either 24 V or 240 V, depending on the type of defrost control circuit. When a defrost cycle is initiated, the relay coil is energized and initiates three actions:

- A NC contact opens and turns off the outdoor fan. Defrost is easier and quicker if cold air is not blowing over the outdoor coil.
- A NO contact closes, energizing the reversing valve coil. The system is switched to the cooling cycle. Hot gas is pumped to the outdoor coil.
- Another NO contact closes, completing a 24 V circuit to the auxiliary heat relay. The auxiliary heat is needed to offset the cooling effect of the indoor coil, **Figure 30-11**.

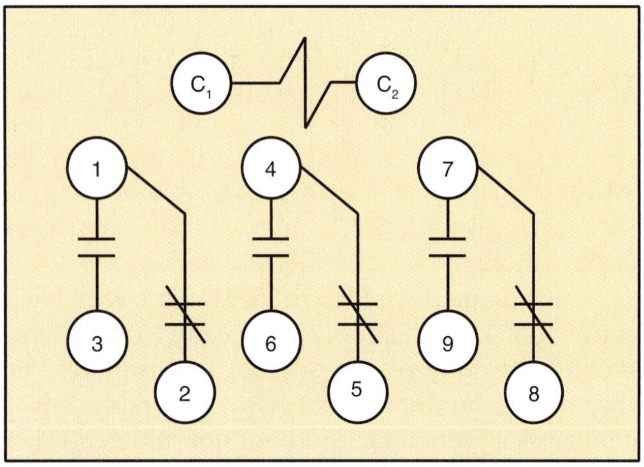

Goodheart-Willcox Publisher

Figure 30-10. Schematic of the combination outdoor fan and defrost relay.

Types of Defrost Controls

Several types of defrost controls in common use are the timeclock, solid-state defrost control, mechanical switch, and air differential switch. Each control accomplishes the same purpose but in a different manner. A defrost timeclock in combination with a defrost thermostat is illustrated next.

Defrost Timeclock

The defrost timeclock motor operates on 208 V to 240 V. A rotating cam controls the mechanical opening and closing of two switches, **Figure 30-12**. The timer motor is connected in parallel with the outdoor fan motor and runs whenever the compressor contactor is energized. Every 90 minutes, the timer closes the normally open set of contacts between Terminals 3 and 4, **Figure 30-13**. The outdoor defrost thermostat contacts are connected in series with the power supply to clock Terminal 4.

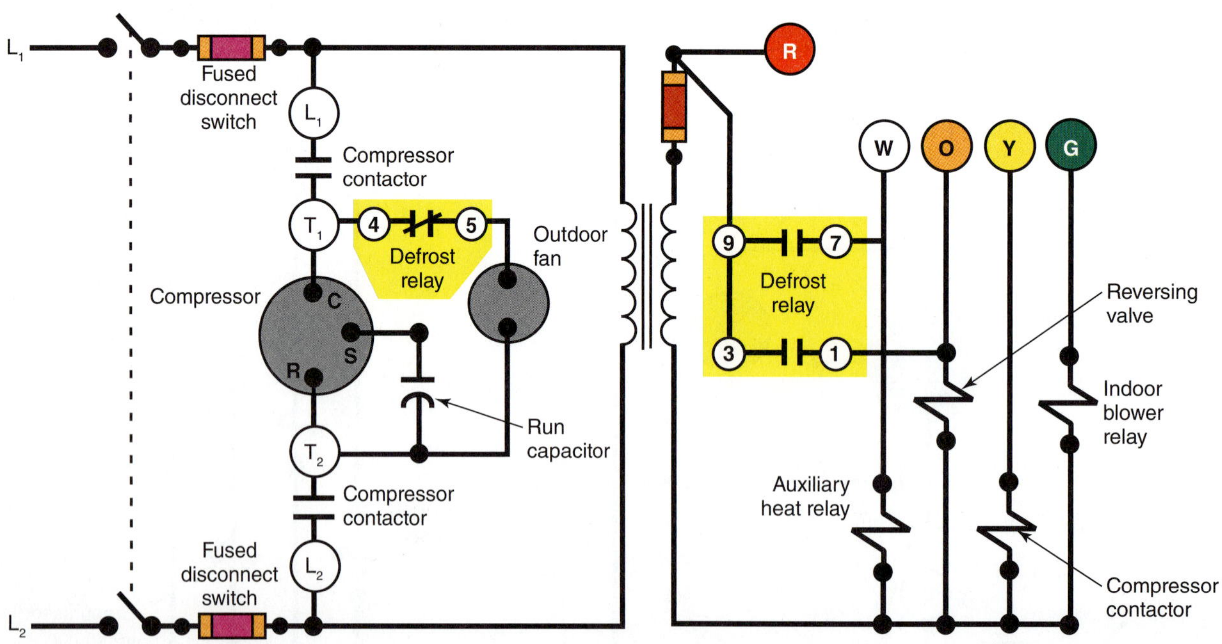

Goodheart-Willcox Publisher

Figure 30-11. Basic heat pump schematic with outdoor fan and defrost relay added. Notice how the NC contacts control the outdoor fan and how the NO contacts control the reversing valve and auxiliary heat.

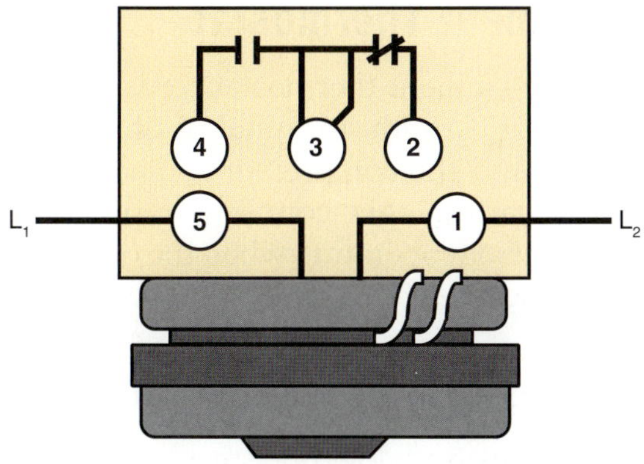

Goodheart-Willcox Publisher

Figure 30-12. To begin the defrost cycle, the timer closes the NO switch between Terminals 3 and 4 every 90 minutes. If the outdoor coil temperature is at or below 32°F (0°C), the defrost cycle will begin.

The outdoor defrost thermostat contacts do *not* close unless the outdoor coil temperature is at or below 32°F (0°C). A circuit to the defrost relay cannot occur unless *both* the timeclock (defrost timer) and the outdoor defrost thermostat contacts are closed.

The defrost cycle is normally terminated by the outdoor thermostat when the outdoor coil temperature reaches 50°F–70°F (10°C–21°C) as set by the manufacturer. If the defrost timer switch closes but the thermostat contacts are open, a defrost is *not* needed and *cannot* occur. The timer then proceeds through another 90-minute cycle. During a defrost cycle, the timer motor is de-energized along with the outdoor fan motor. Both the timer motor and outdoor fan are wired in parallel and controlled by the defrost relay.

Solid-State Defrost Control Boards

Most heat pumps now use solid-state defrost control boards, instead of the motor/cam type of control. **Figure 30-14** shows a solid-state defrost timer board used in some heat pumps. The operation is the same as was previously discussed, but this control combines both the defrost timer and relay into one unit.

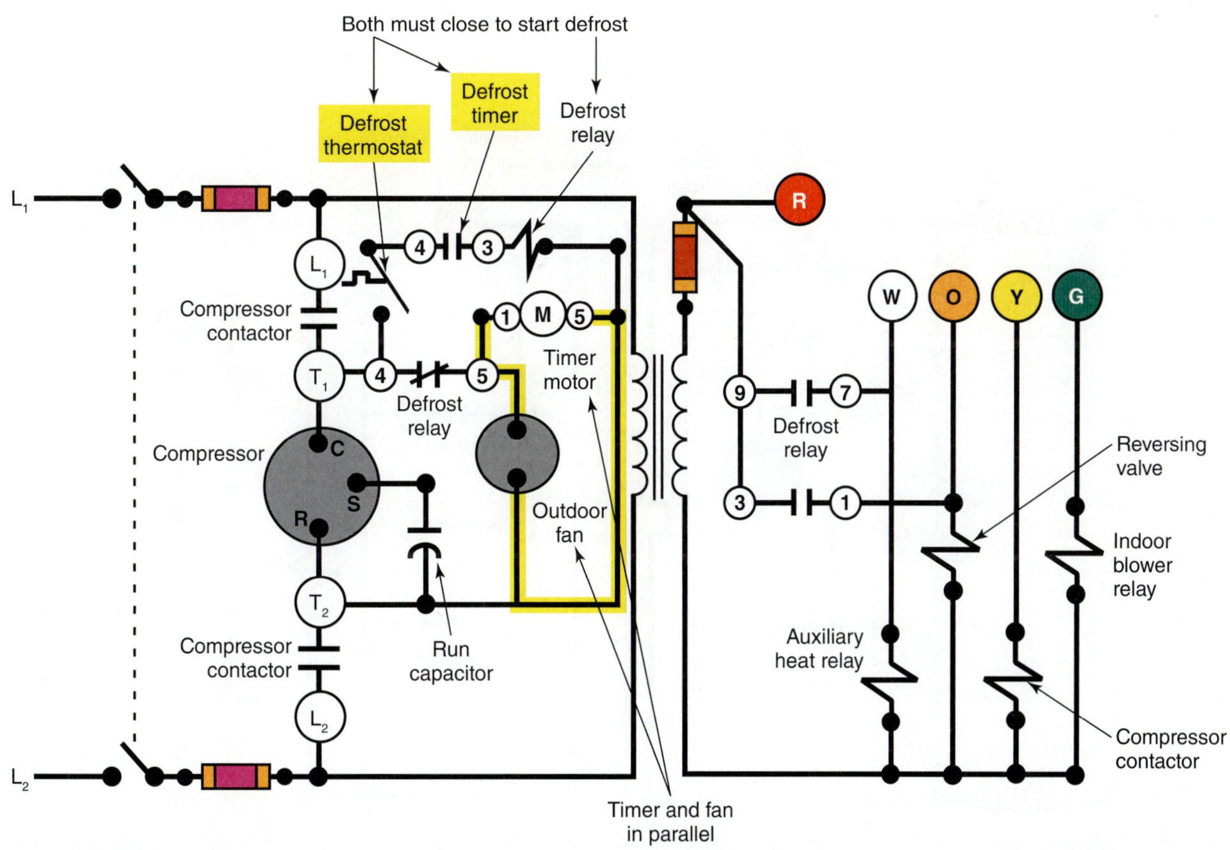

Goodheart-Willcox Publisher

Figure 30-13. Schematic with the defrost timer added. Both the timer contacts and the outdoor (defrost) thermostat contacts must be closed to begin the defrost cycle.

Goodheart-Willcox Publisher

Figure 30-14. Heat pumps now use solid-state defrost control boards that combine the defrost timer and relay.

Crankcase Heater

The crankcase heater is connected to the line side of the compressor contactor and is energized at all times, **Figure 30-15**. The crankcase heater helps prevent possible compressor damage caused by liquid slugging. Slugging typically occurs at startup.

Low-Ambient Thermostat

The low-ambient thermostat contacts are connected in series with Terminal Y on the indoor thermostat and the compressor contactor coil. As explained earlier, this connection prevents the compressor from operating when the outdoor temperature reaches 10°F to 15°F (–12°C to –9°C).

High- and Low-Pressure Controls

Both high- and low-pressure safety controls are connected in the 24 V circuit. Their NC contacts are connected in series with Terminal Y on the indoor thermostat, which controls the power supply to the compressor contactor coil. If either pressure control is open, the contactor coil cannot be energized to turn on the compressor.

Typical Heat Pump Schematic

Figure 30-16 is a schematic of a typical heat pump system. It shows the outdoor unit, indoor blower unit, auxiliary electric heat, and control

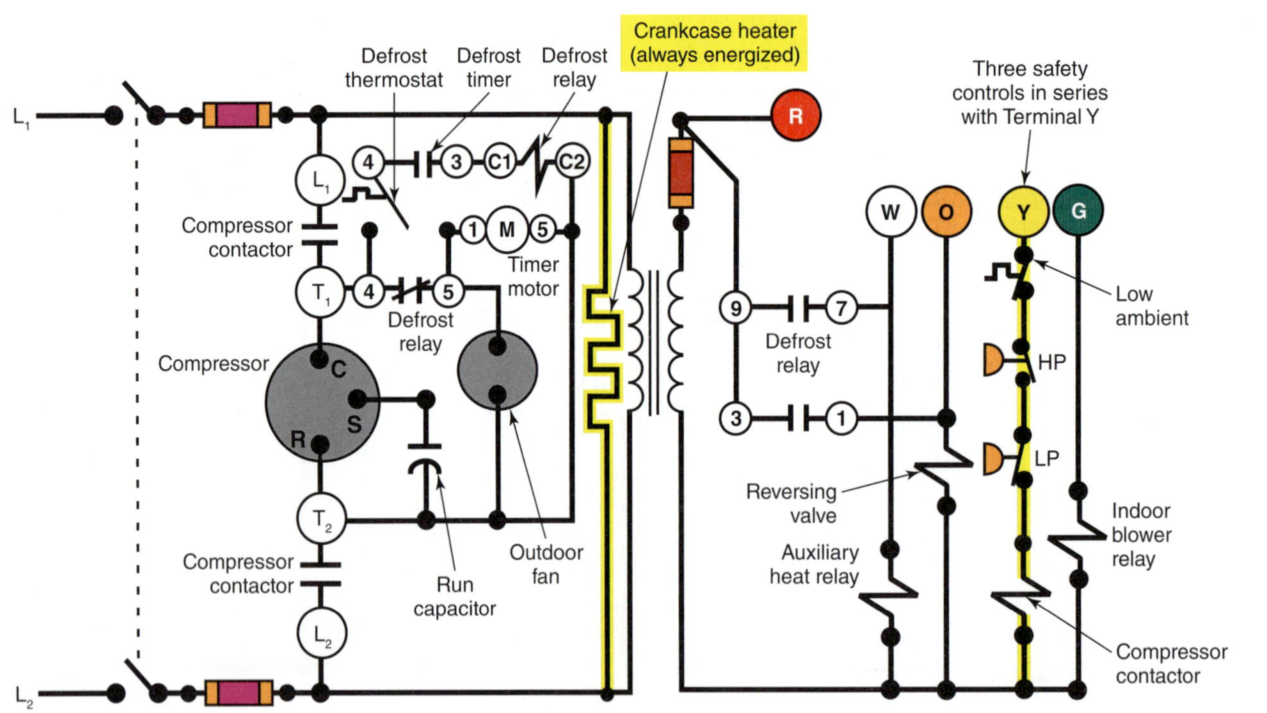

Goodheart-Willcox Publisher

Figure 30-15. The crankcase heater, low ambient thermostat, and high- (HP) and low-pressure (LP) safety controls have been added to the heat pump schematic.

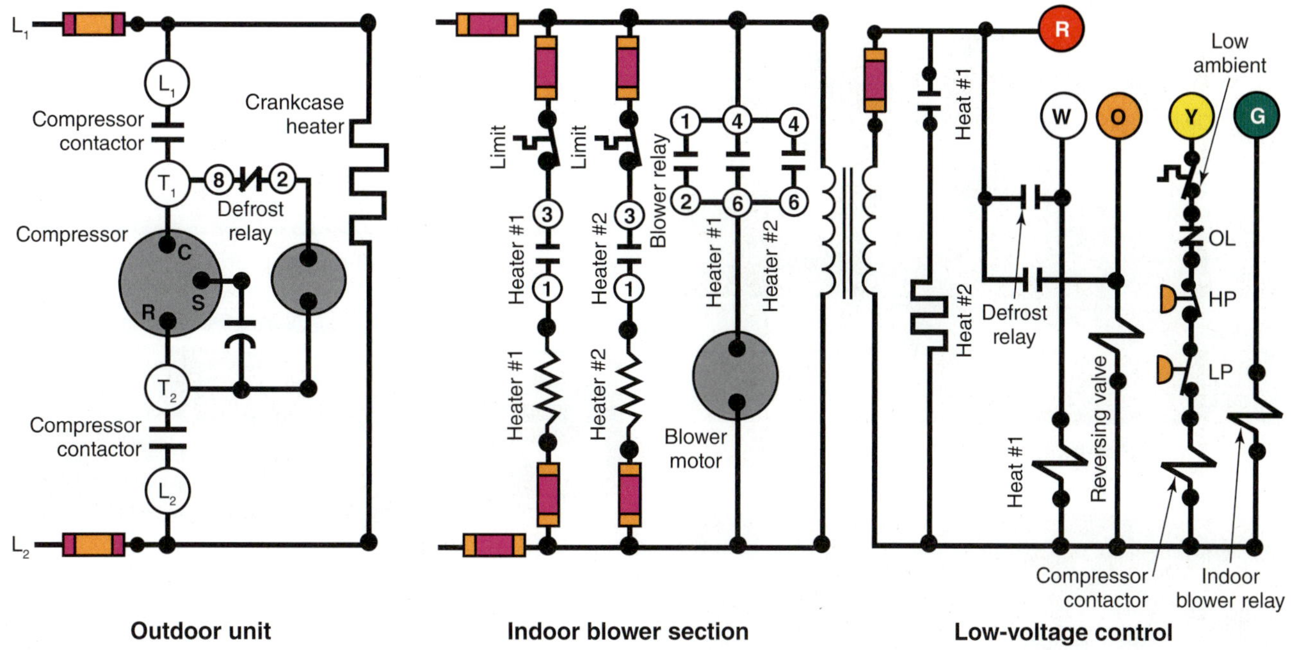

Goodheart-Willcox Publisher

Figure 30-16. A typical schematic for a complete heat pump system.

circuits. Separate power supplies (each with disconnect and fuses) are required for the outdoor and indoor units.

30.5 Ductless Split Systems

A ***ductless split system*** is a heat pump or air-conditioning system in which the evaporator (or indoor coil) and indoor blower can be located in the space to be conditioned. These systems eliminate the need for ductwork, **Figure 30-17**.

Like a conventional heat pump or air-conditioning system, a ductless system consists of an outdoor unit and an indoor unit. However, ductless systems can have multiple outdoor units and multiple indoor units. The units are connected by line sets.

The outdoor unit contains the compressor, condenser (or outdoor coil), and condenser fan, **Figure 30-18**, The electronic expansion valve (EEV) metering device is also found in the outdoor unit. Ductless systems use a variable-speed compressor, which allows the system to continually adjust the amount of refrigerant supplied to the evaporator. The ability to reduce the load on the compressor makes these systems more energy efficient than traditional split systems.

Mitsubishi Electric, HVAC Advanced Products Division

Figure 30-17. This ductless system consists of three single-zone units and one multizone unit. One of the outdoor units is connected to three indoor units.

Indoor units are installed in living spaces. The types of indoor units selected for an installation depend on both the interior design and the heating and cooling loads, **Figure 30-19**.

Indoor units contain an evaporator (or indoor coil), blower, filter, condensate pan, thermistor, and electrical connections. In a conventional ducted split air-conditioning system, air passes through ducts after it passes over the evaporator or indoor coil. As the air travels through the ductwork, it

Pad-mounted outdoor unit

Fujitsu General America, Inc.

Wall-mounted outdoor unit

C5 Media/Shutterstock.com

Figure 30-18. Outdoor units in ductless systems are normally pad-mounted or wall-mounted. For heat pump systems in cold regions, the outdoor unit is often raised off the ground to prevent snow from accumulating around the unit and blocking airflow.

gains heat. In a ductless system, air passes over the evaporator or indoor coil and immediately enters the conditioned space. This improves the efficiency of ductless systems when compared with ducted systems.

The thermistor monitors the air temperature and the coil temperature and continuously reports these temperatures to the control board. The control board is programmed to adjust the compressor motor speed based on this information to maintain the proper temperature in the space.

The components that connect the outdoor unit to the indoor unit include refrigerant lines (line set), electrical conductors, and a condensate drain. Electrical power is supplied to the indoor units through the high-voltage lines, so a separate electrical connection is not needed. The design of a ductless system includes minimum and maximum lengths for the line set.

Wall-mount unit

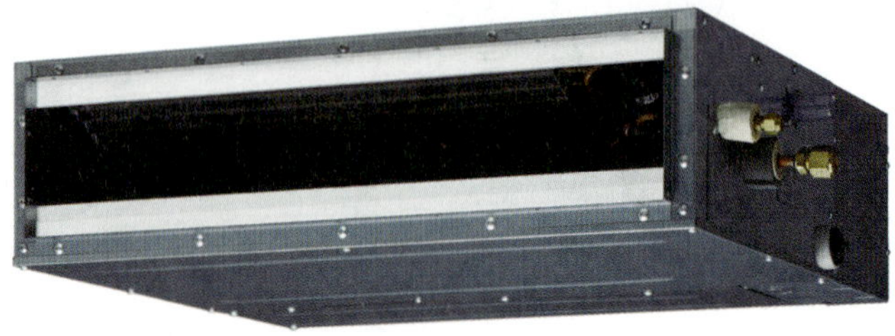

In-line duct unit

Ceiling cassette

Floor unit

Fujitsu General America, Inc.

Figure 30-19. Several types of indoor units are used in ductless systems.

Ductless split systems range from simple to complex:

- A single-zone mini-split system consists of a single outdoor unit and a single indoor unit.
- A single-zone, multi-split system consists of a single outdoor unit and multiple indoor units that are controlled individually. One outdoor unit can connect to as many as eight indoor units.
- In a ***variable refrigerant flow (VRF) system***, one or more outdoor units supply a large number of indoor units. VRF systems may be heat pump systems or heat recovery systems. In heat recovery systems, some units can be set to heat mode while other units are set to cool mode. In heat pump systems, all units must be set to the same mode (heat or cool). Small VRF systems are used in residential and light commercial applications, and large VRF systems are used in large commercial applications.

PRO TIP

Manufacturer Training for Ductless Systems

Ductless systems must be installed, maintained, and serviced as specified by the system's manufacturer. These procedures vary from one manufacturer to another. Most manufacturers offer training for service technicians, and some require their systems to be installed and serviced by approved technicians. Always refer to the manufacturer's information when working on ductless systems.

30.6 Troubleshooting Heat Pumps

As described in previous chapters, systematic troubleshooting skills must be developed. The same five important steps in systematic troubleshooting discussed with gas and electric furnaces applies to heat pump systems as well:

- Know what the equipment *should* do (sequence of operation).
- Know what the equipment *is* doing.
- Know what the equipment is *not* doing.
- Eliminate areas that *are* functioning properly.
- Concentrate on areas that are *not* functioning properly.

The customer interview usually helps narrow the problem to a specific area. Heat pump issues typically fall into six categories:

- No heat or no cooling
- Not enough heat
- Too much heat
- Noise
- Odor
- Excessive operating cost

The following are some of the symptoms and possible causes of heat pump issues.

No Heat

Possible Causes:

- Wrong thermostat settings
- Broken or loose thermostat wires
- Defective thermostat
- No power supply (check fuses, circuit breakers)
- Low voltage at transformer secondary (less than 22 V)
- Low voltage at transformer primary (more than 10% drop)
- Defective transformer (check fuse)

No Cooling

Possible Causes:

- Low-voltage transformer problem
- Faulty thermostat or subbase
- Wrong thermostat settings
- Defective low-ambient thermostat
- Defective compressor contactor
- Defective compressor
- Low-pressure safety switch open
- High-pressure safety switch open
- Defective compressor start components (capacitors and relay)
- Defective reversing valve or coil

Not Enough Heat

Possible Causes:

- Wrong thermostat settings
- Vibration at thermostat
- Thermostat influenced by warm air
- Defective thermostat
- Wrong setting on outdoor thermostat
- Wrong location of outdoor thermostat
- Defective outdoor thermostat
- Dirty or clogged air filter
- Blower motor problem
- Defective high-limit control
- Defective sequencers
- Defective heating elements

Too Much Heat

Possible Causes:

- Thermostat in cold draft
- Defective thermostat
- Defective sequencer
- Control circuit "shorted"
- Compressor contactor welded shut

Noise

Possible Causes:

- Loose blower bearings
- Blower bearing dry and squeaking
- Blower wheel off-center
- Loose cabinet panels
- Loose relay mountings
- Noisy contactor (if voltage is correct, replace it)
- Humming transformer (replace it)
- Loose ductwork

Odor

Possible Causes:

- Dirty air filters
- Stagnant water or sludge in humidifier (if equipped)
- Control transformer burning out
- Overheated wiring or overloaded circuit
- Relay shorted or burned
- Dirty heating elements

Excessive Cost of Operation

Possible Causes:

- Thermostat problems
- Outdoor thermostat problem
- Defective heating relay
- Dirty air filter
- Insufficient home insulation
- Excessive air leakage from home (fireplace, vents)
- Defective blower motor with high amperage
- Dirty blower wheel
- Wrong humidity
- Unit cycling on low-pressure safety switch
- Unit cycling on high-pressure safety switch
- Defective compressor valve reeds
- Reversing valve stuck or leaking
- Defective defrost cycle

While many of the possible causes for heat pump issues are electrical, it is important to remember that an incorrect refrigerant charge may cause issues such as no heat, no cooling, low heat, or too much heat. The best way to check the charge in a system is to recover the charge, weigh it in, and add any additional refrigerant or remove any excess refrigerant per manufacturer's instructions.

Chapter Review

Summary

30.1 Air-source heat pumps use an indoor coil and an outdoor coil that offer heating and cooling capabilities based on refrigerant flow. A reversing valve is used to control direction of refrigerant flow. During the cooling cycle, the heat pump functions like an ordinary air conditioner. The indoor coil is an evaporator, and the outdoor coil is a condenser. However, when the unit is switched to the heating cycle, the indoor coil becomes the condenser, and the outdoor coil becomes the evaporator.

30.1 The measurement of heat output (Btu) divided by its power input (Btu/watt) is called the *coefficient of performance (COP)*. Energy efficiency ratio (EER) is stated as the ratio of energy produced (output) to energy used (input). The balance point of a heat pump is the point at which the heating ability of the device equals the heat loss of the building. Airflow through each coil of a heat pump must be the same. All these factors must be considered in the design and installation of a heat pump system.

30.2 A typical heat pump is composed of a four-way reversing valve, scroll compressor (or compressor with accumulator and crankcase heater), TEV or metering orifice, and a special filter-drier.

30.3 Thermostats used in heat pumps are similar to digital thermostats used in other systems but have additional wiring and settings that need to be set during installation.

30.3 Heat pumps require a defrost cycle to remove frost from the outdoor coil. One method involves heating the outdoor coil. Other methods include using pressure differential or temperature switches, time-initiated/temperature-terminated defrost systems, and outdoor defrost thermostats.

30.4 Basic components included in a heat pump schematic include the compressor, outdoor fan, transformer, thermostat, auxiliary heat relay, reversing valve, compressor contactor, and indoor blower relay. Always check each unit's wiring schematic and pictorial representation for wiring methods, additional components, and sequences of operation.

30.5 A ductless split system is a heat pump or air-conditioning system in which the evaporator and indoor blower can be located in the space to be conditioned. These systems eliminate the need for ductwork. Variations of these systems include: single-zone, mini-split; single-zone, multi-split; and variable refrigerant flow (VRF) systems.

30.6 When troubleshooting heat pumps, a technician should know what the equipment *should* do (sequence of operation), what the equipment *is* and is *not* doing, eliminate areas that *are* functioning properly, and concentrate on areas that are *not* functioning properly. Heat pump issues typically fall into six categories: no heat or no cooling, not enough heat, too much heat, noise, odor, and excessive operating cost.

Know and Understand

1. What component of a heat pump controls the direction of refrigerant flow? (30.1)
   A. Outdoor coil
   B. Indoor coil
   C. Reversing valve
   D. Air handler
2. What is the EER of a heat pump that produces 70,000 Btu with an input of 4400 W at an outdoor temperature of 65°F? (30.1)
   A. 0.24
   B. 15.9
   C. 1033.5
   D. 15,909.1
3. What term describes when the heating ability of a heat pump equals the heat loss of the building? (30.1)
   A. Coefficient of performance
   B. Balance point
   C. Energy efficiency ratio
   D. Pressure regulation threshold

4. *True or False?* During the heating cycle, faster airflow permits better heat pickup in the outdoor coil. (30.1)
5. *True or False?* In heat pump systems, refrigerant flow through the compressor is always the same. (30.2)
6. A reversing valve's pilot valve automatically controls _____ at each end of the piston. (30.2)
   A. defrost
   B. low-side pressure
   C. high-side pressure
   D. pressure (high or low)
7. What type of compressor is best for heat pumps as it can handle pumping liquid refrigerant? (30.2)
   A. Reciprocating compressor
   B. Scroll compressor
   C. Screw compressor
   D. Rotary compressor
8. How many bypass lines and check valves should be installed in a heat pump system with 2 TEVs (one installed at each coil)? (30.2)
   A. 1 bypass line, 2 check valves
   B. 2 bypass lines, 0 check valves
   C. 2 bypass lines, 2 check valves
   D. 3 bypass lines, 2 check valves
9. Which of the following describes proper operation of a heat pump with a metering orifice? (30.2)
   A. Refrigerant flow is metered in one direction with a fixed pressure drop and allowed to flow freely in the opposite direction.
   B. Refrigerant flows through the orifice during use and through bypass lines and check valves when not in use.
   C. Refrigerant flows freely in both directions.
   D. In cooling mode, the metering device meters refrigerant so that the outdoor coil acts as the evaporator and the indoor coil acts as the condenser.
10. A heat pump thermostat may include an emergency heat mode to energize the _____. (30.3)
   A. defrost cycle
   B. auxiliary heat
   C. outdoor fan
   D. indoor fan
11. What device prevents compressor operation when the outdoor temperature drops between 10°F to 15°F (–12°C to –9°C)? (30.3)
   A. Outdoor thermostat
   B. Low-ambient thermostat
   C. Indoor thermostat
   D. Defrost timer
12. For a defrost cycle, a heat pump automatically switches to _____ mode. (30.3)
   A. cooling
   B. heating
   C. auxiliary heat
   D. off
13. What device terminates the defrost cycle? (30.3)
   A. Reversing valve
   B. Indoor thermostat
   C. Defrost timer
   D. Outdoor defrost thermostat
14. *True or False?* Heat pump schematics are standardized and generally do *not* vary from manufacturer to manufacturer. (30.4)
15. Which of the following does *not* describe a function performed by the outdoor fan and defrost relay upon startup of the defrost cycle? (30.4)
   A. A NC contact opens and turns off the outdoor fan.
   B. A NO contact closes, energizing the reversing valve coil.
   C. A NC contact opens, de-energizing the relay coil.
   D. A second NO contact closes, completing a 24 V circuit to the auxiliary heat relay.
16. Variable refrigerant flow (VRF) systems differ from other ductless split systems in that they _____. (30.5)
   A. have multiple thermostats
   B. can have one outdoor unit supply multiple indoor units
   C. may be used in heat recovery systems that allow some units to be set to heat mode while others are set to cool mode
   D. use variable-speed compressors to increase efficiency

17. A service technician is troubleshooting a heat pump system that operates properly but results in high utility costs for the customer. The technician observes that the outdoor coil has some ice accumulating on the unit. Where might the issue likely be found? (30.6)
    A. In the air filter
    B. In the defrost system
    C. In the indoor fan
    D. In the power supply or control board

Apply and Analyze

1. Explain the functions of the indoor and outdoor coils of a heat pump during the cooling cycle. (30.1)
2. Explain the functions of the indoor and outdoor coils of a heat pump during the heating cycle. (30.1)
3. Why do heat pumps include an accumulator in the suction line? (30.2)
4. Which TEV is functioning during the heating cycle of a heat pump? (30.2)
5. Explain how an outdoor thermostat is wired to prevent auxiliary heat from operating during a normal heating or cooling cycle. (30.3)
6. How is the defrost cycle typically terminated? (30.4)
7. Explain how ductless split systems supply conditioned air without ductwork. (30.5)
8. A service technician is troubleshooting a heat pump system that is not operating its auxiliary heat in low ambient conditions. The technician opens the outdoor thermostat and observes that the thermostat's NC contacts remain closed even with outdoor temperature below the auxiliary heat setpoint. What may be the likely cause of the issue? (30.6)

Critical Thinking

1. Imagine you are a service technician taking a service call over the phone. The customer explains they have noticed a large vapor cloud coming out of the outdoor unit of their newly installed heat pump system. They report it only occurs roughly every 90 minutes. What should you tell the customer? (30.3)
2. Imagine you are a service technician troubleshooting a ductless split system. The customer reports that one of their indoor wall-mounted units is not running during the cooling cycle. They also report that before the wall-mounted unit failed that the air blowing from the unit was not very cold. You suspect a refrigerant leak. How should you go about confirming a refrigerant leak and repairing the leak? (30.5)

C_shortt/Shutterstock.com

31 Ductwork

LEARNING OUTCOMES

After studying this chapter, you will be able to:

31.1 Identify the different classifications of furnaces.

31.2 Recognize the four different configurations of air distribution systems.

31.3 Describe various distribution system components.

31.4 Identify the design considerations of an air distribution system and how these considerations can impact system efficiency.

31.5 Lay out and begin installation on duct systems, including starting collars and takeoffs.

31.6 Join, cut, and install round metal pipe.

31.7 Work with rectangular ductwork to install an extended plenum system.

31.8 Recognize support options for various types of ductwork.

31.9 Identify when to use insulated flexible duct and how to make proper connections.

31.10 Install and join wall stacks.

31.11 Install registers, diffusers, and grilles.

31.12 Explain why and how to use mastic and fiberglass membrane.

TECHNICAL TERMS

A-coil
boot
branch line
cubic feet per minute (cfm)
diffuser
downflow furnace
drive cleat
extended plenum
fiberglass reinforcing membrane
forced-air system
horizontal furnace
insulated flexible duct
mastic
multiposition furnace
perimeter system
plenum
radial system
reducing extended plenum
register
register boot
return air
scrim
S-hook
starting collar
supply air
takeoff
trunk line
U-channel
upflow furnace
vapor barrier

Introduction

The information contained in this chapter provides the knowledge needed to work with heating and air-conditioning ductwork. While the illustrations in this chapter show typical situations, keep in mind that adaptation may be necessary to conform with state and local codes. Persons working with air ducts commonly use special hand tools such as the ones shown in **Figure 31-1**. Other basic hand tools are needed such as a trouble light, battery-operated drill, chisel, gloves, lineman's pliers, and long-nose pliers.

31.1 Furnace Classifications

Ductwork begins at the furnace or air handler. Before the ductwork and distribution system is designed and installed, the classification of the furnace must be determined. All forced-air furnaces are classified according to the direction of the airflow for the supply air. These classifications are upflow, downflow, horizontal, and multiposition.

Upflow Furnace

In an ***upflow furnace***, the return air enters from the bottom and circulates upward through the unit to exit at its top. Upflow furnaces are normally located in a basement or closet, **Figure 31-2**. The supply air duct system is attached to the furnace ***plenum*** (sometimes called the *bonnet*), a sheet metal chamber anchored to the top of the furnace with sheet metal screws. The air-conditioning evaporator coil is mounted on top of the furnace, but inside the plenum. The supply air duct is often insulated to help maintain the temperature of the treated air. Return air enters the bottom of the furnace. The blower serves for both heating and air-conditioning modes, but airflow direction remains the same.

Downflow Furnace

A ***downflow furnace*** is usually used in a building with a crawl space under the first floor, or in a mobile home, **Figure 31-3**. The air-conditioning evaporator coil is mounted on the bottom of the furnace, inside the plenum. Supply air travels from the plenum through ductwork to the living spaces. Return air enters the top of the furnace. The blower serves for both heating and air conditioning.

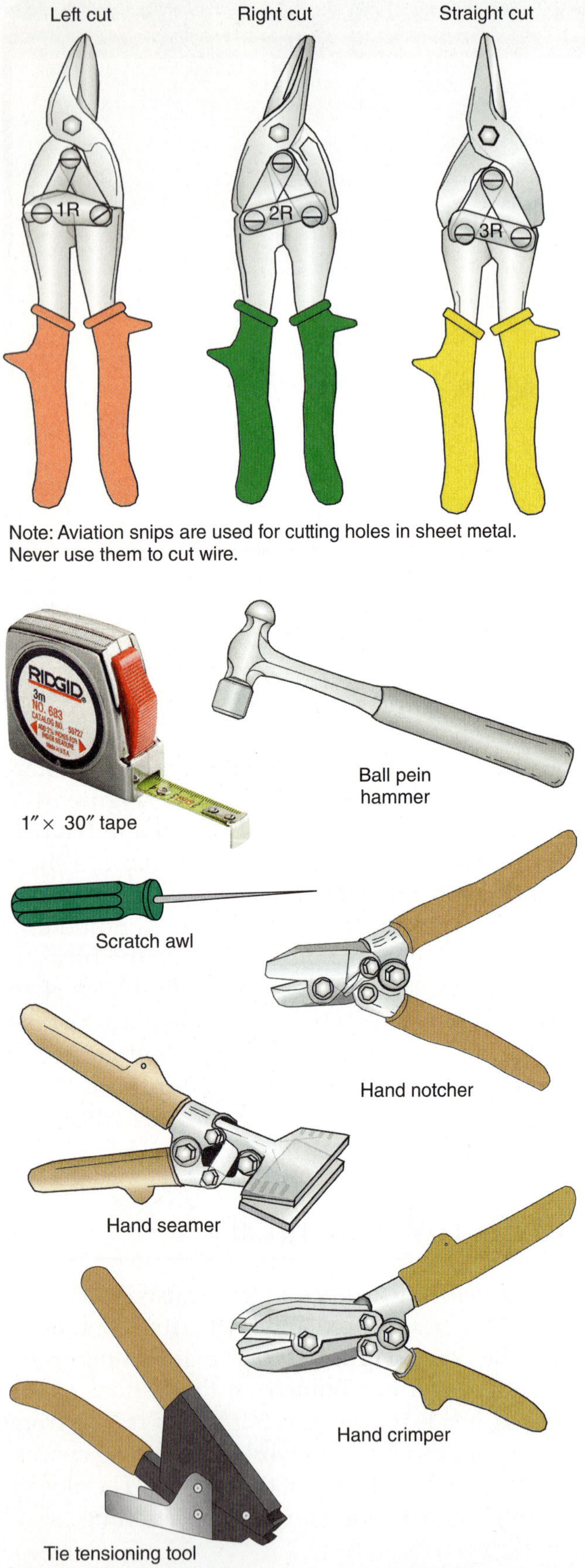

Goodheart-Willcox Publisher

Figure 31-1. Hand tools often are named according to the job they perform.

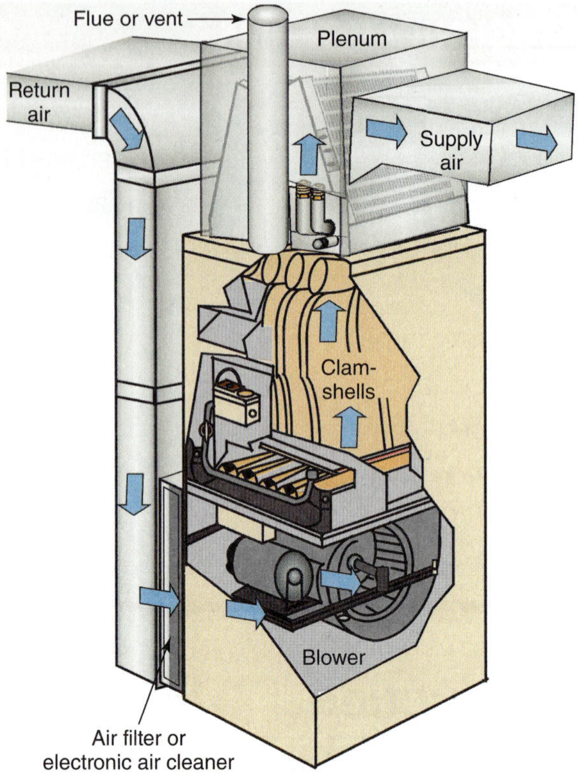

Goodheart-Willcox Publisher

Figure 31-2. An upflow furnace has a top-mounted plenum to distribute supply air.

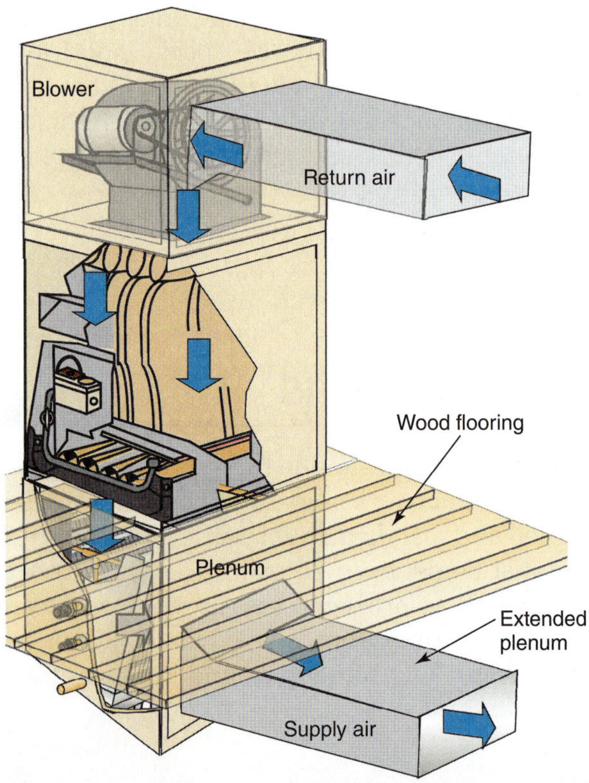

Goodheart-Willcox Publisher

Figure 31-3. In a downflow furnace, air enters above and exits below. The plenum is located below the furnace.

Horizontal Furnace

In a ***horizontal furnace***, return air enters through the one end, and the supply air exits at the other end. These are designed for use in a crawl space or attic, **Figure 31-4**. This means the ductwork is located in the crawl space or in the attic, along with the furnace. These horizontal furnaces are often electric, but gas or a heat pump can be used.

Multiposition Furnace

A ***multiposition furnace*** is designed to be used in either upflow, downflow, or horizontal applications just by making a few modifications to the furnace according to the installation instructions. The main benefit of a multiposition furnace is that it can reduce inventory cost because a single furnace can replace three different types of furnaces, **Figure 31-5**.

Air-Conditioning Evaporators

The most popular type of air-conditioning evaporator for residential systems is the A-coil. An ***A-coil*** is an evaporator that consists of two

Image courtesy of Johnstone Supply

Figure 31-5. A multiposition furnace can be configured to be used in an upflow, downflow, or horizontal application.

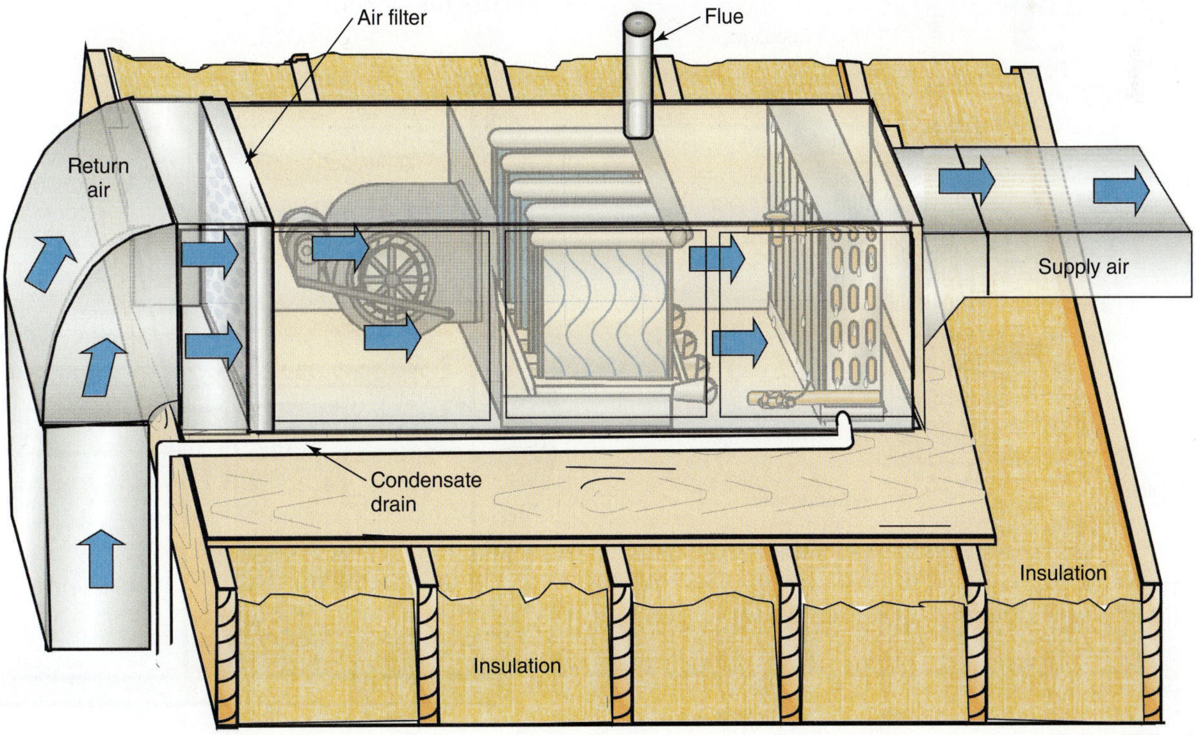

Goodheart-Willcox Publisher

Figure 31-4. A horizontal furnace is usually located in an attic or a crawl space.

evaporator coils arranged to form the letter *A*. Two other types of evaporators primarily used for commercial and industrial systems are the *slant-coil* and *upright-coil*. Commercial and many industrial air-conditioning systems operate the same as residential types. The components are simply bigger (and easier to repair), **Figure 31-6**.

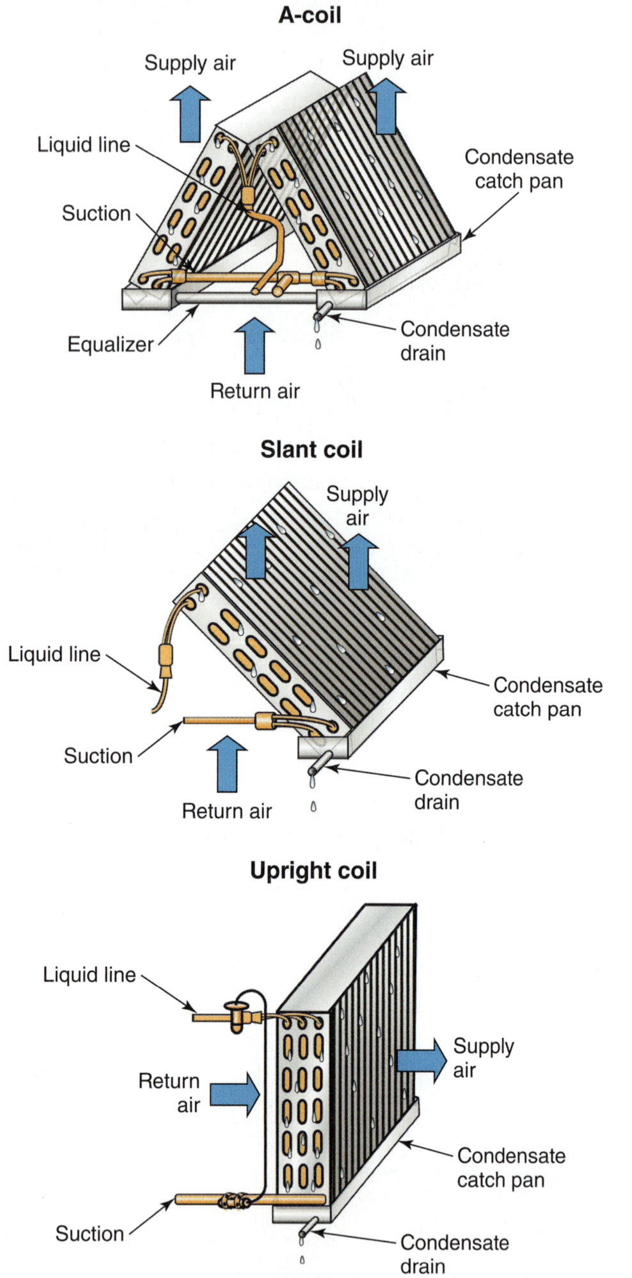

Goodheart-Willcox Publisher

Figure 31-6. The three types of air-conditioning evaporators. In residential installations, the A-coil type is most commonly used.

31.2 Distribution Systems

Before the development of forced-air systems, gravity furnaces were used. These furnaces had to be located centrally, with short runs of ductwork from the furnace to the air outlets. All supply air outlets were located in or near the inside walls of rooms. Return air inlets were located in or near the baseboards of outer walls. Some houses still have gravity furnaces.

Gravity furnaces have become rare, and many heating and cooling systems are now ***forced-air systems***. These central heating and cooling systems use a furnace that is centrally located in the building. Basic components of a forced-air system include a furnace unit, an evaporator coil, an outdoor condensing unit for air conditioning, ductwork, registers, diffusers, and grilles, **Figure 31-7**.

The ductwork consists of separate supply air and return air sections. The ***supply air*** section distributes conditioned air to living spaces, while the ***return air*** section brings air from the living spaces back to the furnace to be heated or cooled again. There are four types of duct systems commonly used for supply air distribution:

- Radial system
- Extended plenum system
- Reducing extended plenum system
- Perimeter system

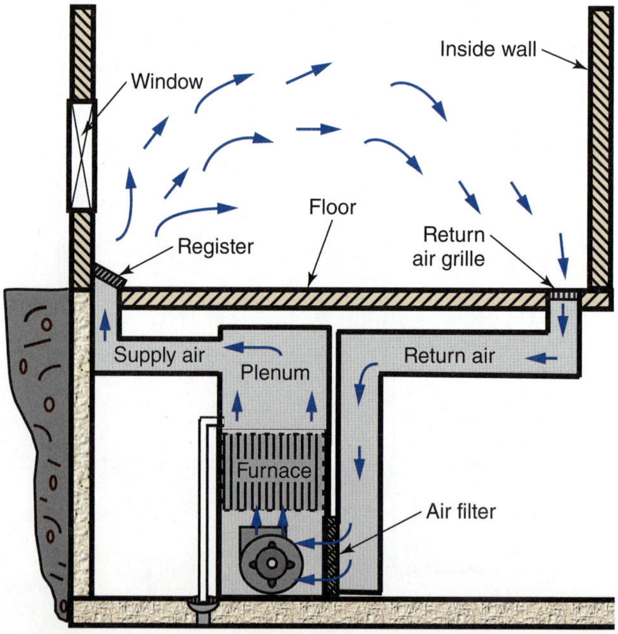

Goodheart-Willcox Publisher

Figure 31-7. This diagram shows the many components included in a forced-air system.

Radial System

A ***radial system*** is one that uses round pipe, made of either sheet metal or flexible duct, running directly from the plenum to the ***register boot*** (component that is used to connect the pipe to the outlet registers). There is a separate run of round pipe from the furnace plenum to each register boot. These piping runs radiate from the furnace plenum, **Figure 31-8**. This system is practical where the piping runs are located in an attic or crawl space. The radial system is most economical and easiest to install, but is not practical if the furnace cannot be centrally located. Any building that is odd-shaped or long cannot use the radial system.

Extended Plenum System

An ***extended plenum*** distribution system combines rectangular duct with round pipes, **Figure 31-9**. The furnace plenum is extended with rectangular duct, and all round pipes branch out at right angles from the extended plenum. This permits short lengths of round pipe to reach the registers. For a neat appearance, the round pipes are normally installed in the spaces between the

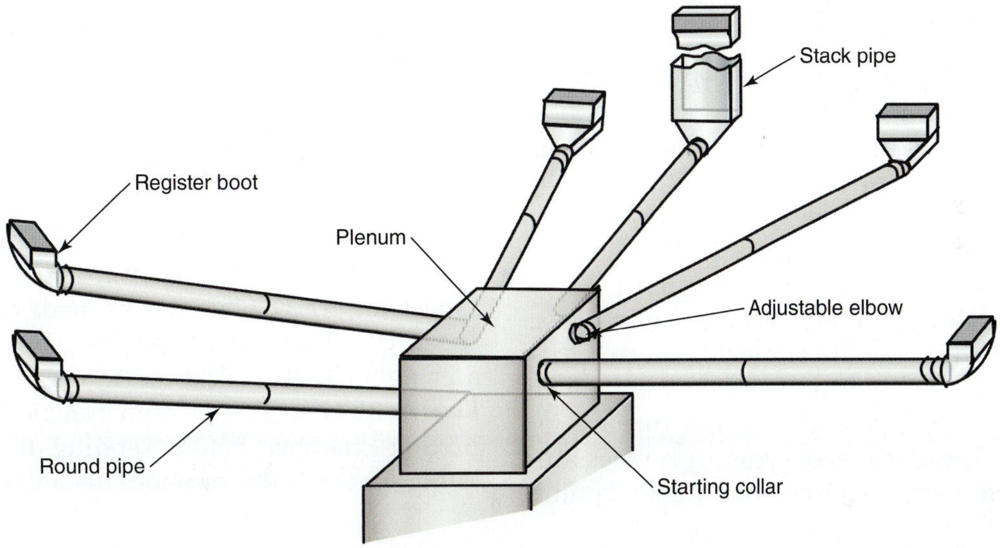

Goodheart-Willcox Publisher

Figure 31-8. The radial piping system uses round pipes running straight to the register boots. In this example, the stack pipe will carry treated air to an upper floor for distribution.

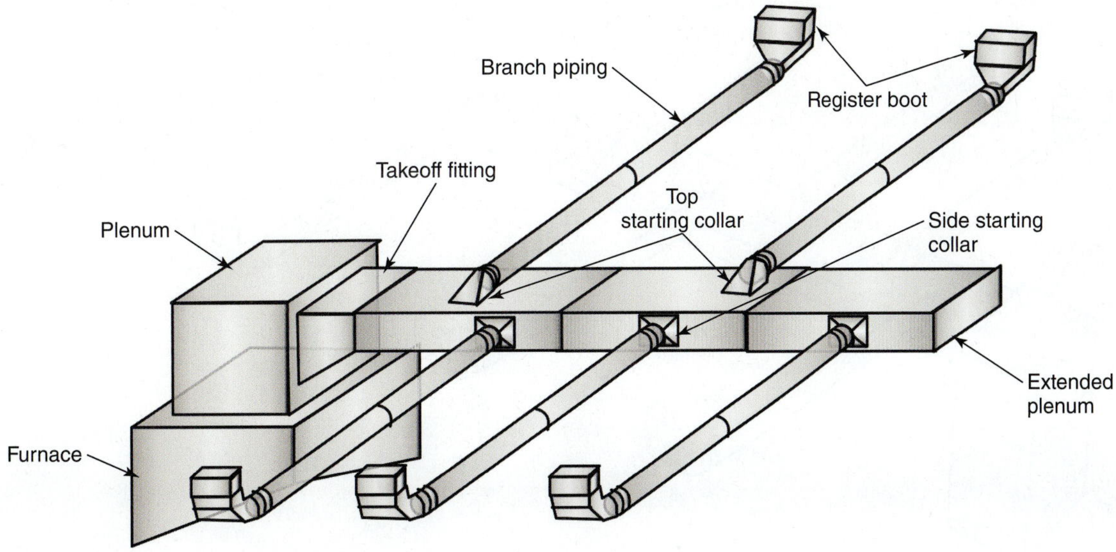

Goodheart-Willcox Publisher

Figure 31-9. Round branch piping extends outward from an extended plenum formed with rectangular duct.

flooring joists or in attics. It is necessary to cut a hole and add a register boot at the end of each run.

Reducing Extended Plenum System

The ***reducing extended plenum*** system is similar to the extended plenum system. The difference is that the ductwork reduces in size at least once to have even airflow at the end of long plenum runs, **Figure 31-10**. This reduction helps to keep duct air pressure constant throughout its length.

Perimeter System

A ***perimeter system*** uses feeder ducts to pressurize a loop of forced air around the outer perimeter of a house. The perimeter loop contains numerous supply air outlets. These outlets may be in a baseboard along outside walls or in floors or ceilings near outer walls, **Figure 31-11**. The supply air "blankets" the outer wall against drafts in heating mode and shields against radiant heat in cooling mode. This air circulates inward to a return air grille located in an inside wall, where it is drawn off and returned to the furnace for retreatment. The perimeter system provides sufficient and uniform air circulation for maximum comfort. Perimeter systems are often used in houses that are built on concrete slabs.

31.3 Distribution System Components

The supply air distribution system includes the plenum, the main ***trunk line***, and all the branches serving different living spaces. Each branch terminates in a ***register***, which is a device that directs the airflow and usually controls its volume. Another system of air ducts serves to return air from the living spaces for heating or cooling. This system is called "return air," and includes another main trunk line and all its branches.

Plenum (Bonnet)

The first fitting to be installed in any air distribution system is the *plenum*, sometimes called a *bonnet*, **Figure 31-12**. It may be made of galvanized sheet metal or fiberglass duct board, forming a chamber that is attached to the furnace outlet, **Figure 31-12A**. The plenum functions as an air-mixing chamber before sending the air to the living spaces. It also provides the necessary surface

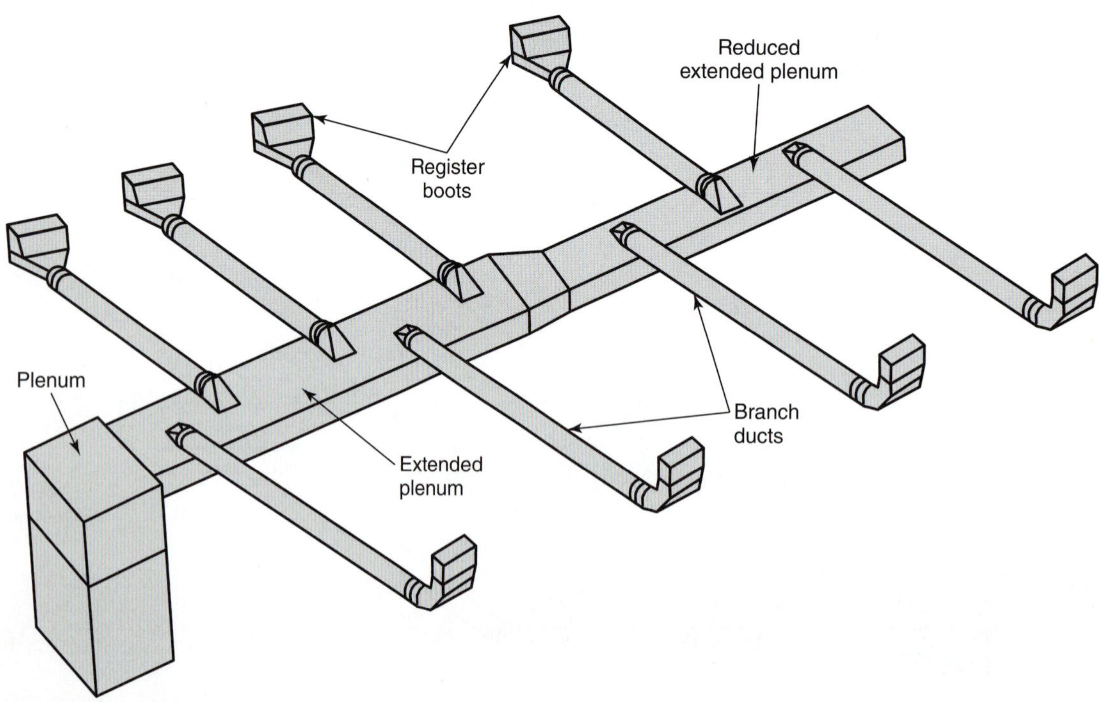

Goodheart-Willcox Publisher

Figure 31-10. A reducing extended plenum system reduces the main duct at least once to keep proper airflow and pressure.

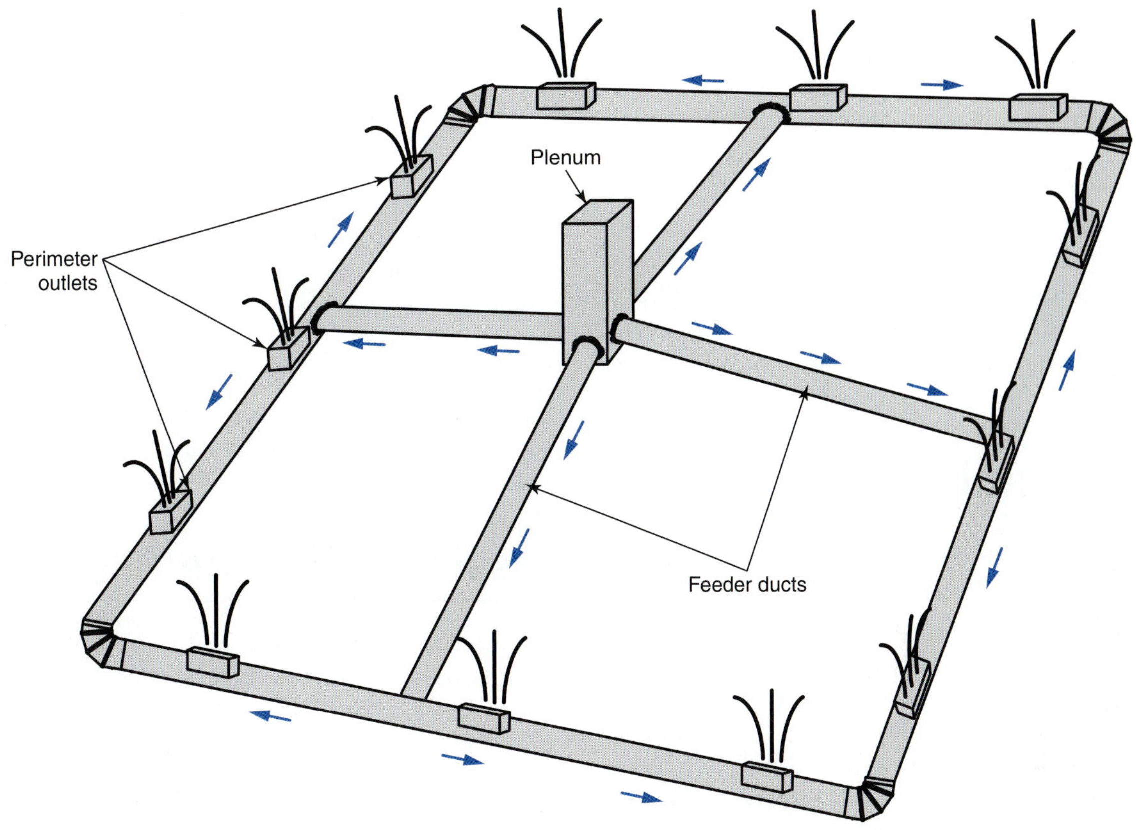

Figure 31-11. Perimeter systems are used in houses built on concrete slabs.

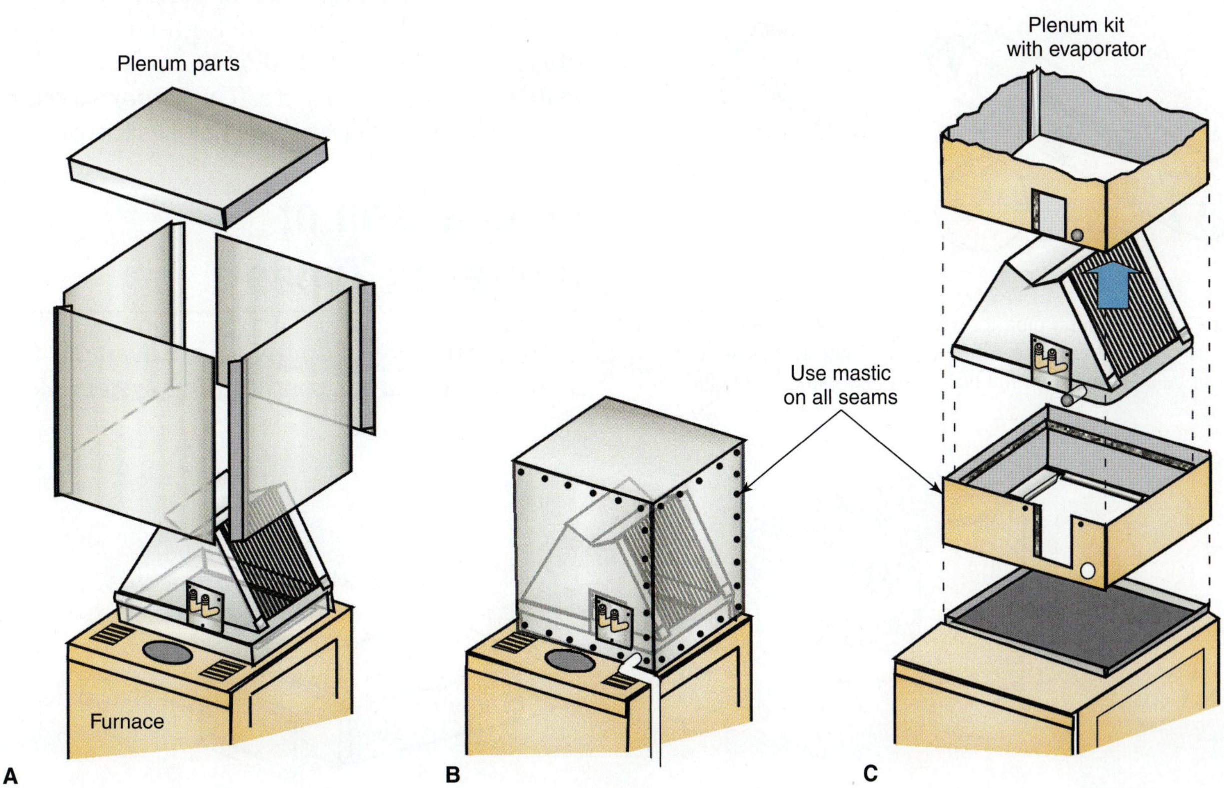

Figure 31-12. A plenum can be fabricated in the field (A, B) or it may be assembled from a factory kit (C).

for attaching takeoffs and starting collars. An air-conditioner's evaporator may be located inside the plenum and the supply air duct attaches to the plenum, **Figure 31-12B**. Gas furnace manufacturers often offer a plenum kit along with the necessary evaporator. For adding air conditioning to a previously installed furnace, some additional ductwork may be required. After all parts are assembled and screwed together, the seams must be secured and sealed with mastic (a nontoxic adhesive material used for permanently sealing joints and seams), **Figure 31-12C**.

Round Metal Pipe

Round pipe can be used as a main trunk line or a branch line in both supply air and return air systems. Round metal pipe has the most efficient characteristics and is commonly used. Standard round pipe comes in 30-, 28-, or 26-gauge, 3″–14″ in diameter, and in 2′, 3′, 5′, and 10′ lengths. All round pipe is factory crimped on one end for easy insertion into another length of pipe, **Figure 31-13**.

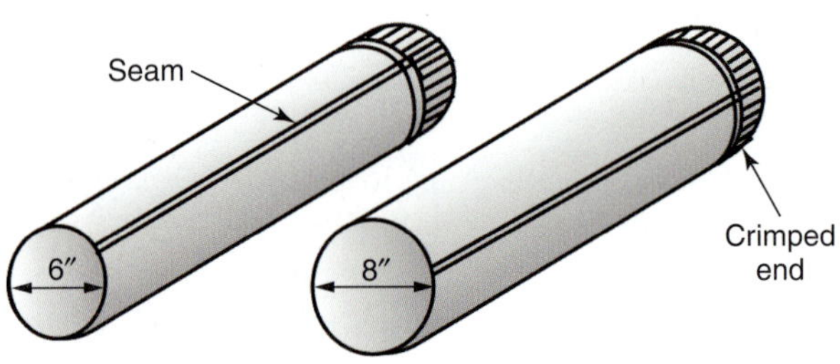

Goodheart-Willcox Publisher

Figure 31-13. Round pipe for branch lines can easily be snapped together at the seam. The crimped area at one end allows the pipe to be slipped inside the next section of pipe, forming a joint.

Rectangular Duct

Rectangular duct is 28-, 26-, or 24-gauge, galvanized sheet metal, and is mainly used as a trunk line, or "extended plenum." It is used as a space-saver and can be easily routed in overhead space between floor joists, or in an attic. ***Branch lines*** consisting of round pipe come off the extended plenum to transport air to a register. A branch line begins with a starting collar or takeoff installed on the extended plenum (main trunk line). The end of the extended plenum must be blocked with an end cap that is anchored with sheet metal screws and sealed with mastic, **Figure 31-14**.

Insulated Flexible Duct

As shown in **Figure 31-15**, ***insulated flexible duct*** consists of several layers, each with a specific function. The *core* is usually two plies of polyester film used to encapsulate the galvanized steel wire helix. The purpose of the *wire helix* is to help flexible duct keep its shape. The *insulation* is a fiberglass blanket, varying in thickness, depending on the R-value (measure of a material's resistance to heat). Many states have adopted energy codes for flexible duct that require certain R-values for code approval. ***Scrim*** is a high-tensile-strength, fiberglass mesh that provides protection for the insulation and inner core. The ***vapor barrier*** is a plastic jacket or sleeve that encloses the product.

Comparison of Ductwork Types

Several types of ductwork are available for use in a heating and air-conditioning system. Because

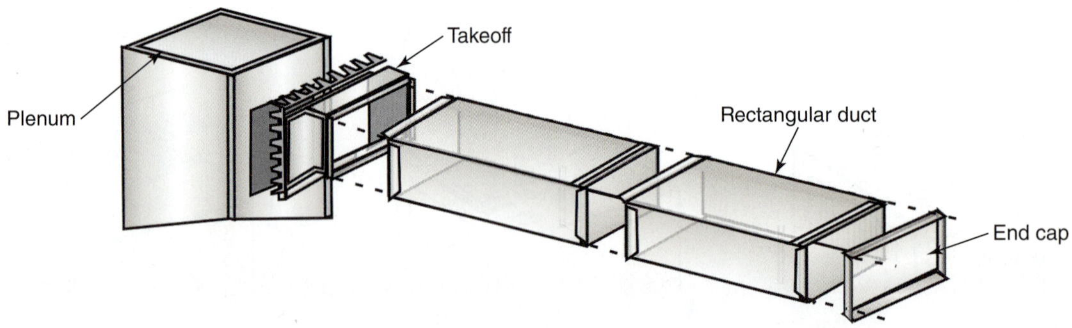

Goodheart-Willcox Publisher

Figure 31-14. An extended plenum, formed from rectangular duct, has a takeoff fitting at one end and an end cap at the other end.

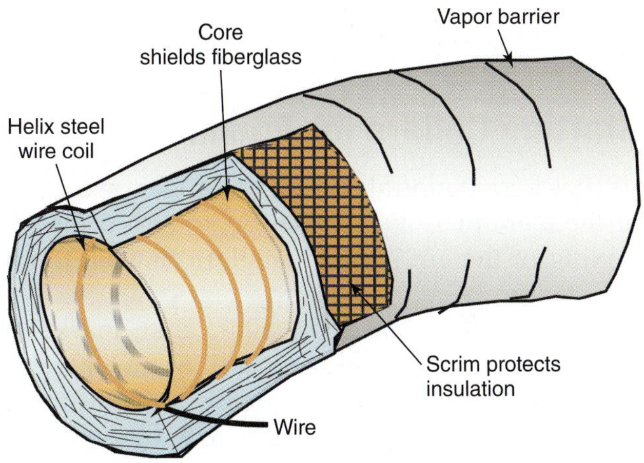

Goodheart-Willcox Publisher

Figure 31-15. This cutaway view shows the construction of a typical insulated flexible duct.

they were all designed for a particular application, they have different strengths and weaknesses. **Figure 31-16** lists some pros and cons of the major duct types.

31.4 System Efficiency

An efficient residential heating and cooling system will move air at the proper velocity and quantity to each room. It will also direct the air to enter each room at the proper location. Air velocity is an important consideration. Air that is moving too fast creates uncomfortable drafts and makes more noise as it comes through the system. Air moving too slowly will not have enough speed to be thrown across the room or up the wall. A proper heat load calculation and duct design needs to be performed. Various methods are available to perform heat load calculations. Do *not* use rules of thumb. No two houses are the same.

Sizing the Extended Plenum

An extended plenum is sized according to the ***cubic feet per minute (cfm)*** of air that is required through each round pipe it serves. Rectangular duct used for extended plenums is constructed from 28-, 26-, or 24-gauge, galvanized sheet metal, 8″ deep, 60″ long, and comes in standard widths of 10″, 14″, 18″, 24″, and 28″. One of these standard sizes should provide the needed capacity, but any length of duct can be cut to a shorter length.

Major Duct Types

Round Metal Pipe

Pros	Cons
1. Excellent airflow characteristics	1. If required, insulation must be installed
2. Longer-lasting than flexible duct	N/A

Insulated Flexible Duct

Pros	Cons
1. Pre-insulated	1. May not be as long-lasting as metal duct
2. Ease of installation	2. Some reduction in airflow
3. Reduced labor costs	3. Must be fully extended (cannot make sharp bends)
4. Eliminates some fittings	N/A

Rectangular Duct

Pros	Cons
1. Spacesaver; can use space between floor studs	1. If required, insulation must be installed
2. Longer-lasting than flexible duct	N/A

Goodheart-Willcox Publisher

Figure 31-16. Advantages and disadvantages of three types of air ducts.

Provide Adequate Airflow

A gas furnace located in a small utility room or closet may not receive enough air for proper combustion. Check with local and national mechanical and gas codes for proper combustion air duct installation.

31.5 How to Lay Out and Install a Duct System

The heating and cooling units should be matched to the heat load calculation performed on the building. In addition, they should be able to distribute a set quantity of air measured in cubic feet per minute (cfm) to the living area through the duct system. The cfm for air-conditioning systems is calculated at 400 cfm per tonnage of air conditioning. Therefore, a living space with a 3-ton air conditioner, for example, needs to have ductwork that can handle 1200 cfm of air (400 cfm × 3 tons).

Two or three main trunk lines can come off the plenum to the living areas to achieve a total capacity of 1200 cfm, **Figure 31-17**.

Examples:

Two 12″ round duct
main trunks at 620 cfm each = 1200 cfm

Three 10″ round duct
main trunks at 400 cfm each = 1200 cfm

Note that these are only loosely estimated examples. For exact calculations, cfm performance data charts provided in installation manuals should be referenced. Duct sizes must also be matched to account for system static pressure and fan speeds.

With the main trunk lines installed, begin branching from the mains with smaller lines running to the various rooms. Always keep a constant rate of airflow (cfm) to the various rooms. A branch line must properly reduce in size to maintain a constant airflow from the trunk line.

Installing Starting Collars and Takeoffs

Starting collars and takeoffs are installed on a plenum to begin a "run." A ***starting collar*** is used with round duct, while a ***takeoff*** is used with rectangular duct.

Dimensions	Approx. Cfm	Sq in
4″ Round	30	12.57″
5″ Round	60	19.64″
6″ Round	100	28.27″
2¼″ × 10″	60	23.00″
2¼″ × 12″	70	27.00″
3¼″ × 10″	100	33.00″
3¼″ × 12″	120	39.00″
7″ Round	150	38.48″
3¼″ × 14″	140	46.00″
8″ Round	200	50.27″
8″ × 8″	260	64.00″
9″ Round	290	63.62″
10″ Round	400	78.54″
12″ × 8″	440	96.00″
12″ Round	620	113.09″
16″ × 8″	660	128.00″
14″ Round	930	153.93″
16″ Round	1300	201.06″

Goodheart-Willcox Publisher

Figure 31-17. Approximate cfm of airflow (at a velocity of 700 ft/min) for different-size ducts.

Starting Collars for Round Duct

It is best to mount the starting collar on the side of the plenum, 6″ below the top, to ensure proper mixing of the air. Never install a collar on top of the plenum.

To install a starting collar, first mark the proper size opening with a pencil or metal scribe, **Figure 31-18A**. Next, pierce the metal surface with a sharp tool and use snips to cut the opening. Then, insert the collar into the opening and fold the tabs down inside the plenum. Finally, secure the collar with three sheet metal screws and seal it with mastic, **Figure 31-18B**.

Takeoffs for Rectangular Duct

Manufactured takeoffs are available in many sizes, such as 8″ × 8″, 12″ × 8″, 14″ × 8″, or 16″ × 8″, and mount easily on any side of the plenum. It is best to mount the takeoff 6″ below the top of

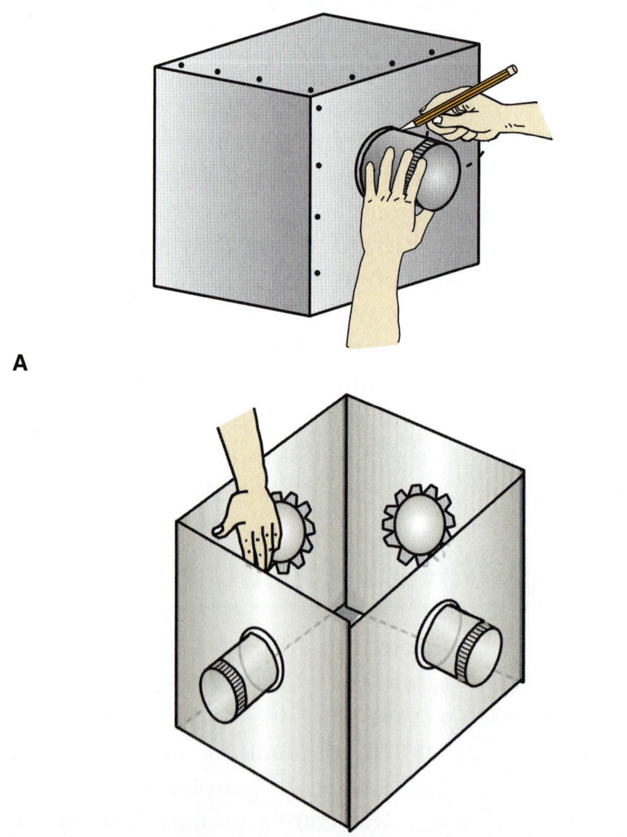

Goodheart-Willcox Publisher

Figure 31-18. Installing a starting collar for round ductwork. A—Mark and cut the opening for the starting collar on the plenum. B—Insert the collar and bend tabs over. Secure the collar with three metal screws and seal with mastic.

the plenum to ensure proper mixing of the air, **Figure 31-19**. When a furnace is located at one end of the house, the extended plenum may run to the opposite end. When the furnace is located at the center of the house, two takeoffs are used to provide an extended plenum on each side of the furnace.

To install the takeoff, first mark off the proper size opening. Next, pierce the metal surface using a drill or sharp tool and cut the opening with snips. Then, insert the takeoff into the opening, and fold the tabs down inside. Finally, secure the takeoff with three sheet metal screws and seal it with mastic.

31.6 Working with Round Metal Pipes

Standard round pipe is made of 30-, 28-, or 26-gauge, galvanized metal, **Figure 31-20**. Round pipe is usually shipped nested, with the seams open, **Figure 31-20A**. Fittings are usually shipped preformed and ready to install. Each pipe length has an easily connected snap-lock seam. Round the pipe until the two seam edges meet. Insert the tongue on one seam edge into the groove on the other seam edge. Start at one end of the pipe and work down the seam toward the other end until the entire seam is snapped shut, **Figure 31-20B**.

Joining Round Pipe Lengths

Each pipe length has a raw end and a crimped end. Lengths are joined by sliding the raw end of one length over the crimped end of another length. Line up seams for the best appearance and push lengths together for a tight joint. Fittings are made similarly. The crimped ends always point in the direction of airflow. Secure each joint with three drive screws evenly spaced around the joint and seal with mastic, **Figure 31-21**.

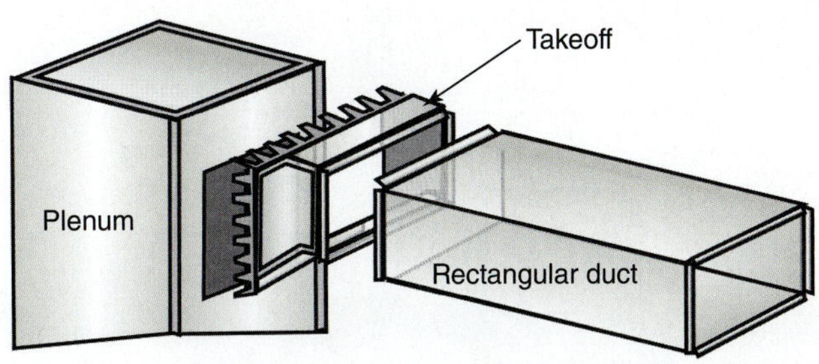

Goodheart-Willcox Publisher

Figure 31-19. The top of the takeoff should be located 6″ below the top of the plenum.

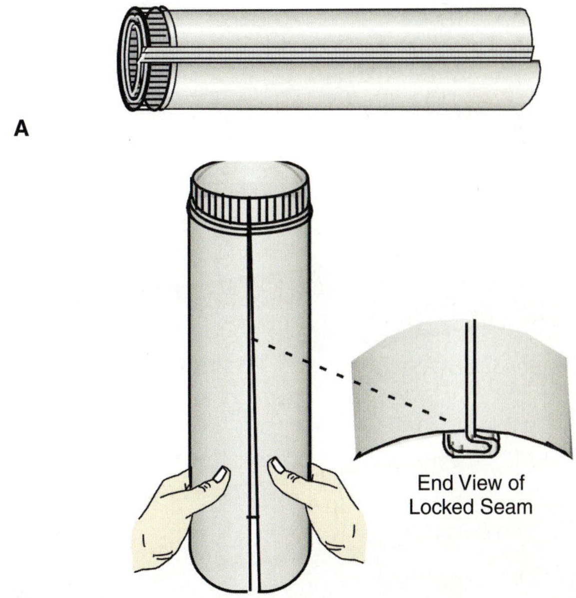

Goodheart-Willcox Publisher

Figure 31-20. Round metal pipe. A—Nested pipe lengths with open seams. B—Self-locking seam snaps together when compressed or squeezed with normal hand pressure.

Cutting Round Pipe

The last length of duct in a run usually must be cut, **Figure 31-22**. Do *not* snap the seam together until after the cut is made. Hold the last piece of unformed pipe in place and mark where it is to be cut. The cutting mark should be on the raw (not crimped) end. Be sure to allow enough extra length for the cut end to slip over the crimped end of the preceding piece, up to the bead. Draw a straight cutting line across the pipe, then make the cut with tin snips. When making the cut, the tin snips will squeeze the snap-lock seam edge closed. Before forming the pipe, the closed end of the seam will need to be pried open again, using a screwdriver.

If a piece of pipe that is already formed needs to be cut, use a hacksaw. Make a continuous line around the pipe to follow while cutting. Support the round pipe between two wood blocks or other flat pieces, then saw with smooth, even strokes, beginning at the seam. Because of the hacksaw frame, the saw will only be able to saw partway through the pipe. Rotate the pipe and start again with the saw on top. Continue rotating and sawing in this manner until the cut is complete.

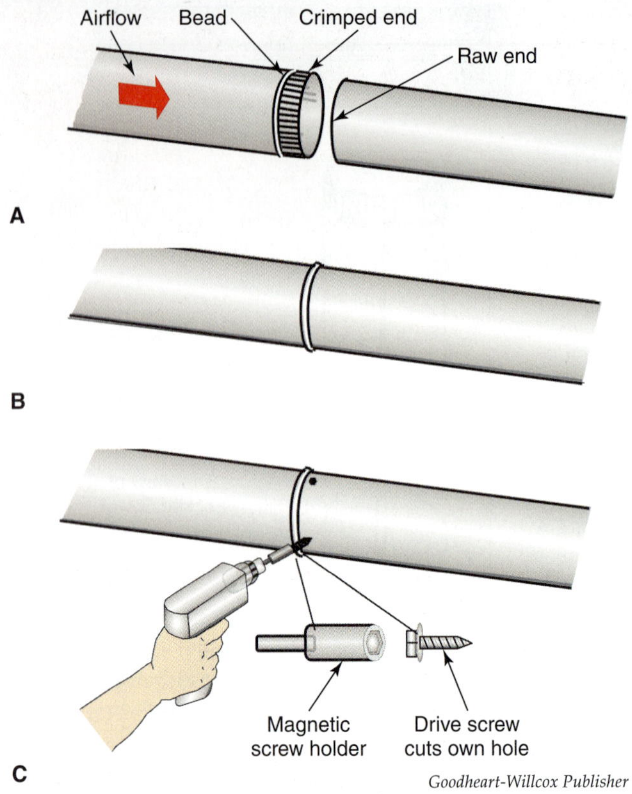

Figure 31-21. Joining pipe lengths. Making connections with round pipe should be step-by-step. A—Each length has a crimped end and a raw end. B—The crimped end slides into the raw end of the next pipe until seated against the bead. C—Three drive screws should be used to secure each joint. Temporary wire supports are often used to hold pipe steady while connections are being made.

Adjustable Elbows

Adjustable elbows can be conformed to the desired angle. A 90° adjustable elbow can be twisted to form any angle from 0°–90°, **Figure 31-23.** A 45° adjustable elbow is also available. Insert the crimped end to move with the direction of airflow. Install the elbow and secure the joint with sheet metal screws, then seal it with fiberglass tape and mastic. Do *not* use ordinary duct tape.

Sealing Collars

Fiberglass reinforcing membrane will wrinkle when it is wrapped around a joint between round and rectangular ducts. To make the membrane material lie flat, use a utility knife to make a series of slices about every 2″ along its edge. Wrap the uncut part along the round duct. When the material is wrapped around the round duct, the tabs of the cut side will fan out and lie flat against the rectangular duct, **Figure 31-24.**

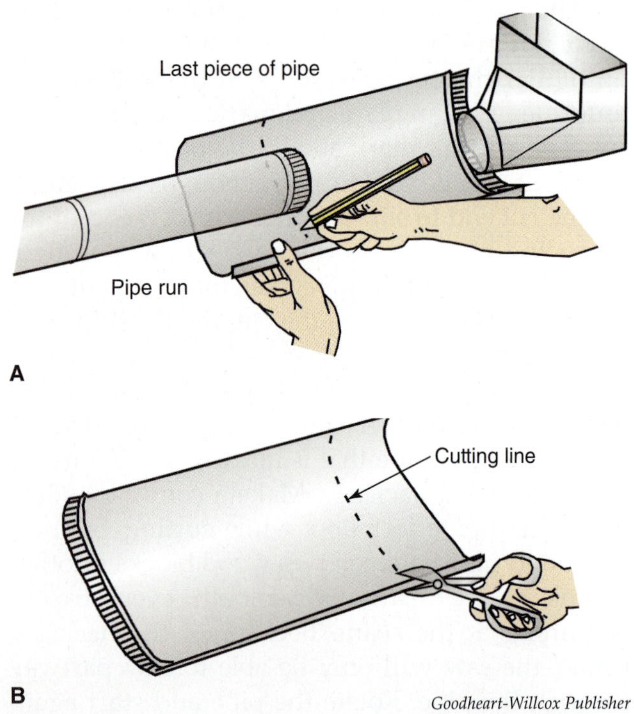

Figure 31-22. In a branch run, the last piece must usually be cut to size. A—Mark the cut with a pencil or scratch awl on the raw end. B—Make the cut with tin snips or aviator snips.

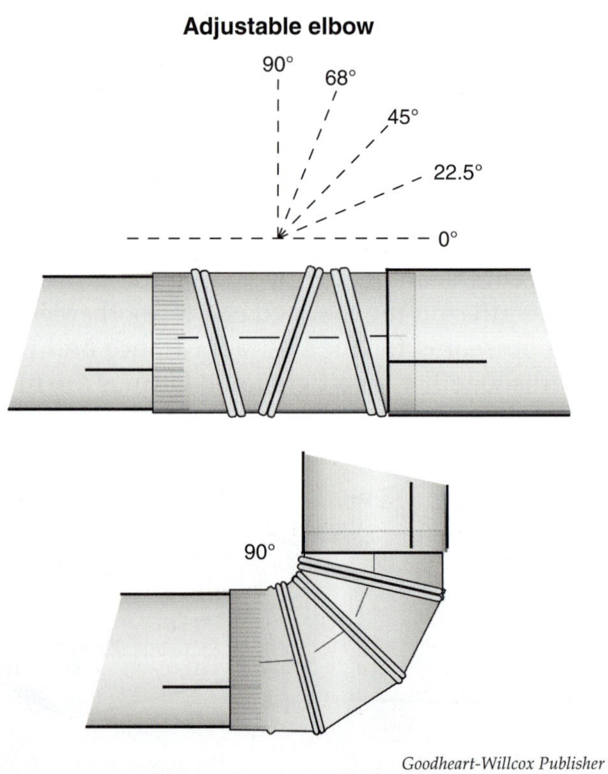

Figure 31-23. Adjustable elbows are convenient for making all types of turns. They are available in all round duct sizes.

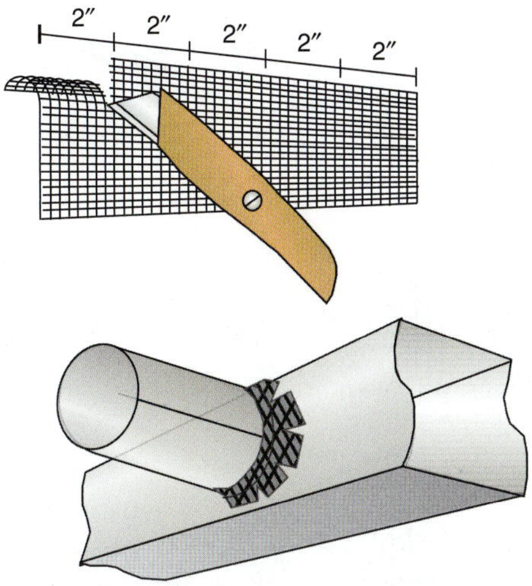

Goodheart-Willcox Publisher

Figure 31-24. After securing the collar with sheet metal screws, seal the collar with membrane and mastic. This prevents leaks due to vibration. Cutting slits in the membrane will help it lie flat.

31.7 Extended Plenum System

Begin with the takeoff fitting in the plenum, and work outward. When a branch pipe comes off the top of the extended plenum, install the starting collar before assembling the rectangular duct. Side starting collars or takeoffs can be installed in the extended plenum after the run is completed, **Figure 31-25**.

Support the duct at intervals with hangers or straps. Be sure to keep ducts at least 1″ from joists and other combustible materials. Close the end with an end cap. After completing the extended plenum with starting collars or takeoff fittings attached, install branch runs to connect with outlets. Insulate all supply ducts and pipes that are in unheated areas.

Working with Rectangular Duct

Rectangular duct is made of 28-, 26-, or 24-gauge galvanized sheet metal in 4′ lengths. Each duct length is made up of two half sections, as shown in **Figure 31-26**. To form a length, place the two half sections on the floor with the "S-hook" ends up. Pick up one piece and engage its edge channels with the channels on the other piece, then slide the two together.

After assembling a section of duct, hammer each channel snug for a tight fit. This can be done as shown in **Figure 31-27**. Hang the assembled duct over a sturdy board suspended between two sawhorses so the seam lies flat on the board edge. Use a ball pein or sheet metal hammer to tighten the seam.

Rectangular duct is joined with S-hooks and drive cleats, **Figure 31-28**. One end of each section has formed ***S-hooks*** on the long edges and U-channels for the ***drive cleats*** on the short edges. The other end has raw long edges, with ***U-channels*** on the short edges. Join pieces by

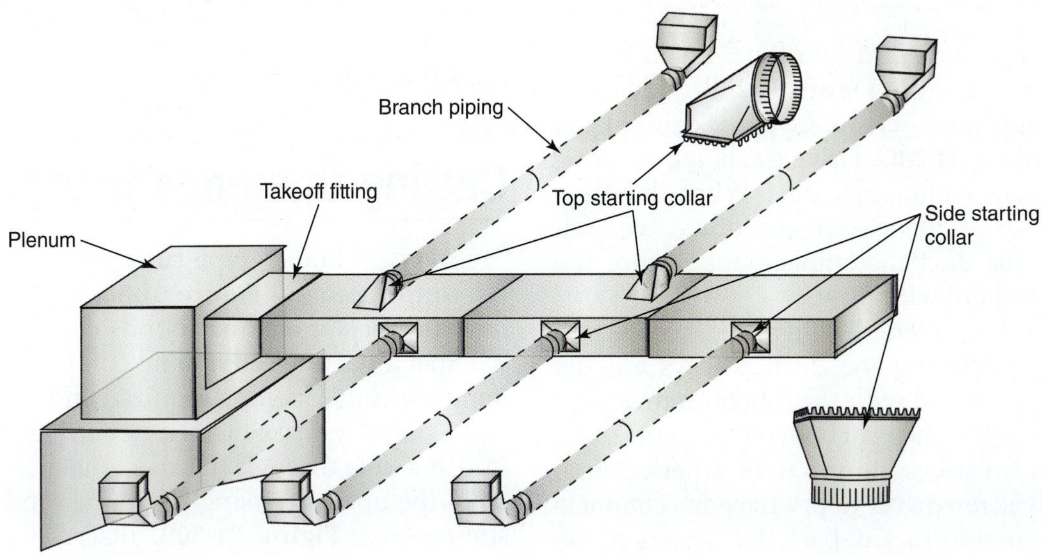

Goodheart-Willcox Publisher

Figure 31-25. An extended plenum system uses side and top takeoff fittings to connect branch lines. A centrally located furnace may have two extended plenums, one on each side.

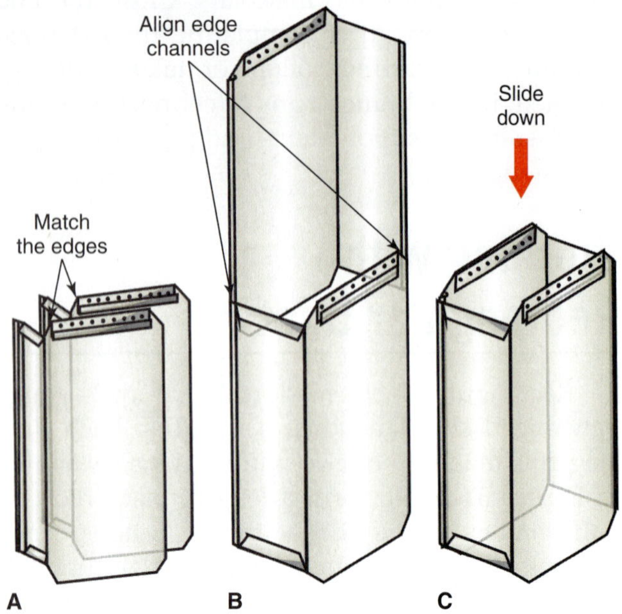

Goodheart-Willcox Publisher

Figure 31-26. How to assemble rectangular duct. A—Match the ends of the two half sections. B—Start the edge channels. C—Slide the two sections together.

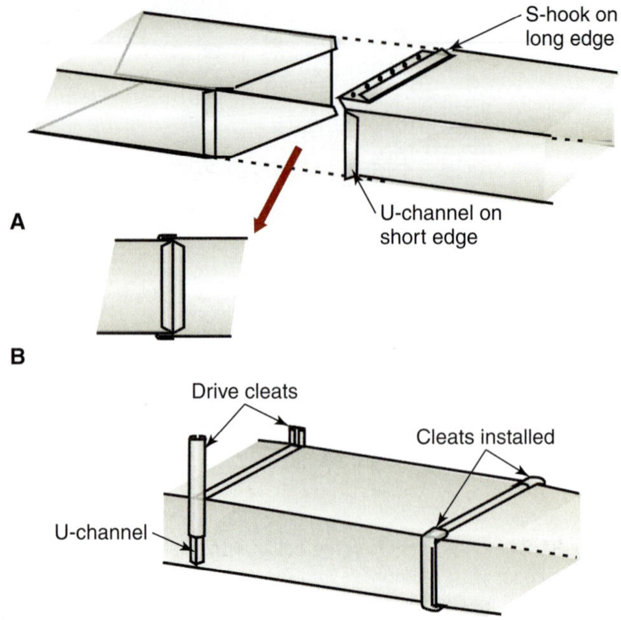

Goodheart-Willcox Publisher

Figure 31-28. Joining rectangular duct using S-hooks and drive cleats.

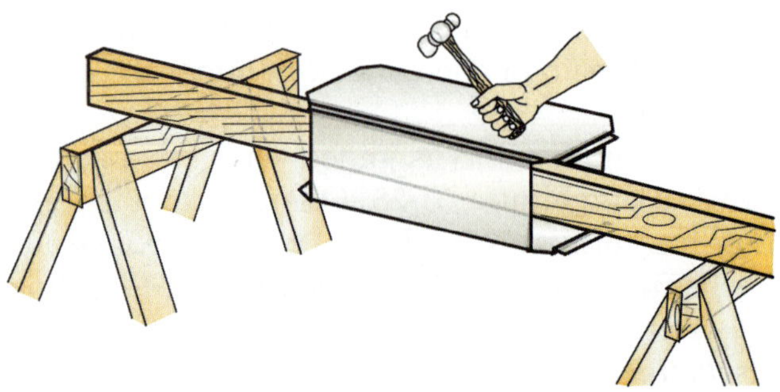

Goodheart-Willcox Publisher

Figure 31-27. Use a ball pein hammer on each seam for a tight fit.

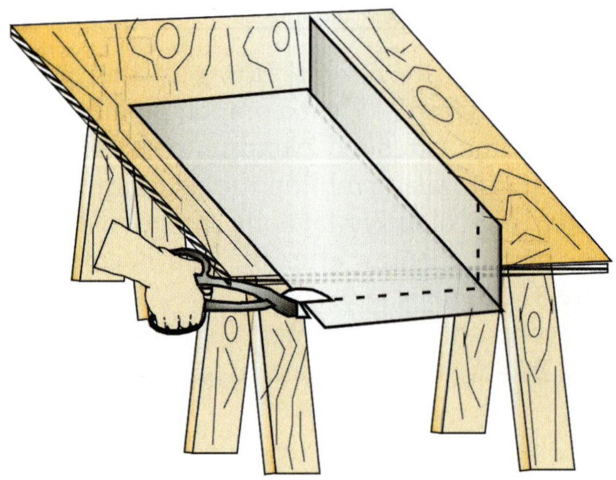

Goodheart-Willcox Publisher

Figure 31-29. Measure carefully, then use snips to cut duct pieces to size.

inserting a raw long end into the S-hooks of the next piece, as shown in **Figure 31-28B**. Slide drive cleats through the U-channels to lock the pieces together, **Figure 31-28C**. Drive cleats are 10″ long and used on 8″ rectangular duct. When the drive cleats are in place, bend each end of a drive cleat down over the duct to complete and secure the joint. Seal the joint with mastic.

Where a shorter length is required, cut the two pieces before forming the duct. Always cut the raw end, not the end with the S-hook connection. Measure carefully, then cut both pieces to the same length with tin snips, **Figure 31-29**. After cutting, use a sturdy screwdriver to pry the edge channels to their original form. Cut back the corners of the new end and use a hand seamer to form a new U-channel on the short edge of each piece.

Cutting Openings in Plenum

To begin branch pipe runs, openings must be cut in the plenum, **Figure 31-30**. These openings may be rectangular or round, depending upon whether a starting collar or takeoff is used. Make a template of the opening required and draw its outline accurately on the metal to be cut, **Figure 31-30A**. Punch a hole through the sheet metal within limits of the opening, using a hammer and chisel or similar tool, **Figure 31-30B**. Insert aviator snips and cut to the outline. Then cut around the outline until the opening is complete, **Figure 31-30C**.

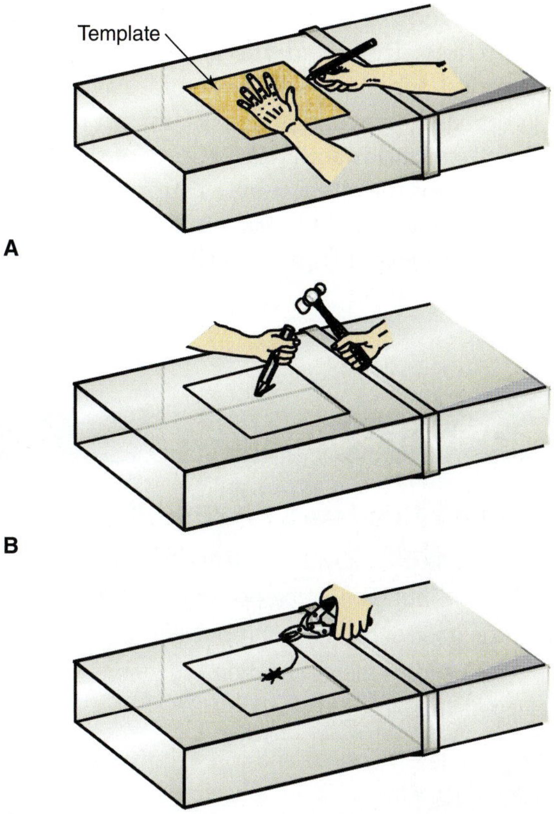

Goodheart-Willcox Publisher

Figure 31-30. Installing a top-mounted takeoff. A—Draw the outline for the opening. B—Punch a hole to use when starting the cutting snips. C—Cut around the outline.

When a branch pipe will be run from the top of an extended plenum, the opening must be cut and the takeoff or starting collar installed before the plenum is fastened in place. Openings on the sides of the plenum can be made at any time.

Rectangular 90° Elbows

Rectangular elbows are not adjustable but are available in 90° "long-way" or 90° "short-way" forms, **Figure 31-31**. They are also available in 45° configurations. A long-way elbow changes the direction of the long dimension, and the short-way elbow changes the direction of the short dimension. To form any smaller angle needed, 45° and 90° rectangular elbows can be cut.

31.8 Duct Support

All duct should be run in a straight line and supported at regular intervals (typically, 2′–5′) to prevent any sags. The best support material is

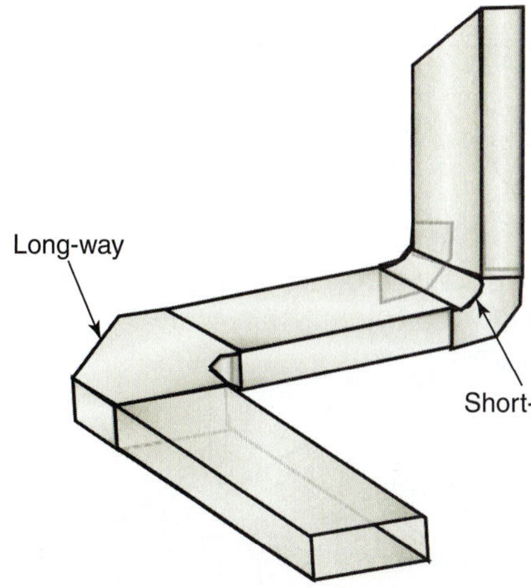

Goodheart-Willcox Publisher

Figure 31-31. Rectangular elbows are made in two varieties: long-way and short-way. They are not adjustable.

woven polypropylene strap that comes in various widths. It is sometimes called "webbing strap" and is wide enough so it will not bite into flex duct. Perforated metal strap is a second choice, with steel wire being the least desirable support option. Steel wire is still often used to support round pipe for branch runs. The wire should be wrapped once around the pipe before anchoring to a joist, **Figure 31-32**.

Keep pipes at least 1″ from joists and other combustible material. To prevent heat loss, metal duct is often wrapped with insulation, **Figure 31-33**. If insulating material is applied after the duct is already in place, the insulation will have to be slit

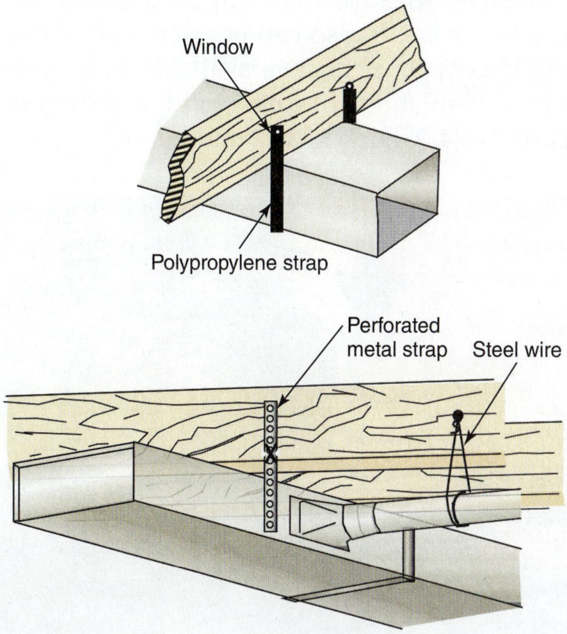

Goodheart-Willcox Publisher

Figure 31-32. Three methods of duct support include woven polypropylene strapping, perforated metal strap, and steel wire.

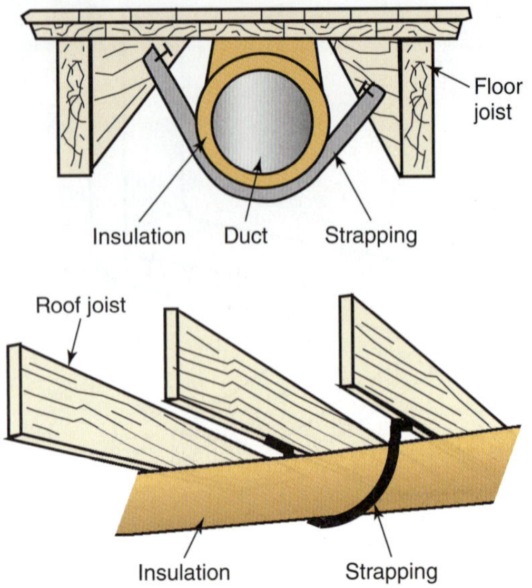

Goodheart-Willcox Publisher

Figure 31-33. Duct is often located in a basement and hidden between floor joists. When located in an attic, it is hung with strapping attached to roof joists. Insulation is often wrapped around a duct to prevent heat loss.

or stripped to get around the supporting strap or wire. This opening must be resealed to maintain insulating effectiveness.

31.9 Working with Insulated Flexible Duct

Insulated flexible duct can be used for main trunk line runs. It also can be used as branching duct. This type of duct is available in a variety of sizes, insulation thickness, and specifications, **Figure 31-34**. Check local building codes for the exact R-value required in a given area. To prevent reduced airflow, insulated flexible duct must be fully extended when installed. Avoid any abrupt turns. When a sharp turn is necessary, cut the flex duct at the turn and install a 90° elbow of the same diameter. Do not use steel wire to support flexible duct. Instead, use a wide fabric strapping material and anchor it to wood framing with staples.

Goodheart-Willcox Publisher

Figure 31-34. Different sizes and insulation thicknesses are available in insulated flexible duct.

PROCEDURE

Connecting Insulated Flexible Duct

Making connections with insulated flexible duct is not difficult, but it must be performed properly to obtain correct airflow and prevent leaks. **Figure 31-35** shows the connection procedure step-by-step.

1. Cut completely around and through the layers of the duct with knife or scissors. Use heavy-duty snips or side cutters to cut the wire helix, **Figure 31-35A**.

A

Goodheart-Willcox Publisher

Figure 31-35A. Connecting lengths of insulated flexible duct. Cut through the layers of duct.

2. Peel back the jacket and insulation from the core, **Figure 31-35B**.

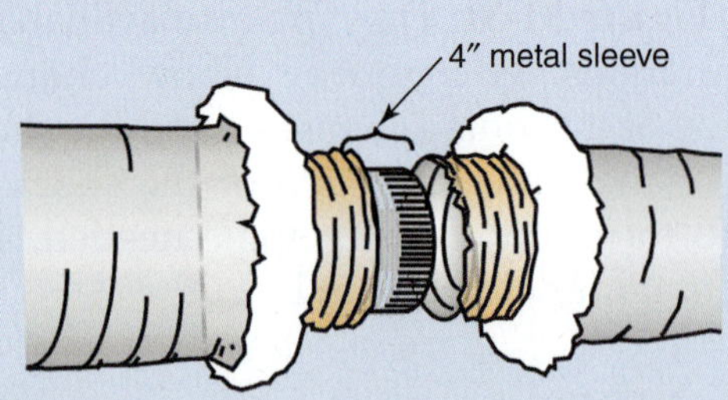

B

Goodheart-Willcox Publisher

Figure 31-35B. Peel back the jacket and insulation from core to allow inserting a standard 4″ metal sleeve.

3. Insert a standard metal sleeve, the same size as the duct, between the two ends.

(Continued)

PROCEDURE

Connecting Insulated Flexible Duct *(continued)*

4. Butt the two cores together over the sleeve.
5. Tape the core ends together with at least two wraps of duct tape.
6. Secure each core end onto the sleeve with two approved clamps, Figure 31-35C.

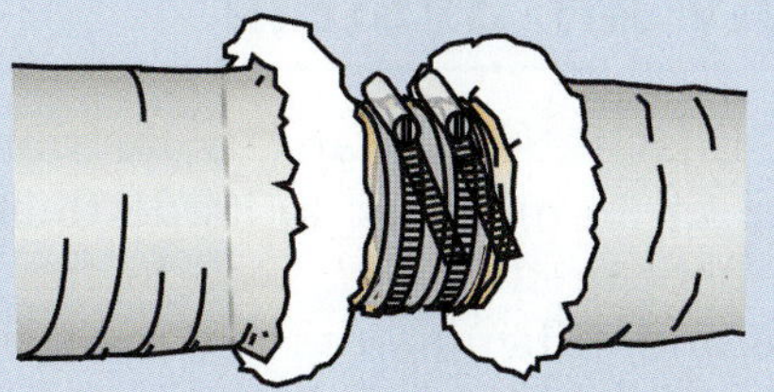

C

Goodheart-Willcox Publisher

Figure 31-35C. After butting the ends together over the sleeve, tape the cores together with at least two wraps of duct tape. Secure with two approved clamps.

7. Pull the jacket and insulation back over each core.
8. Tape the jackets together with two wraps of tape, Figure 31-35D.

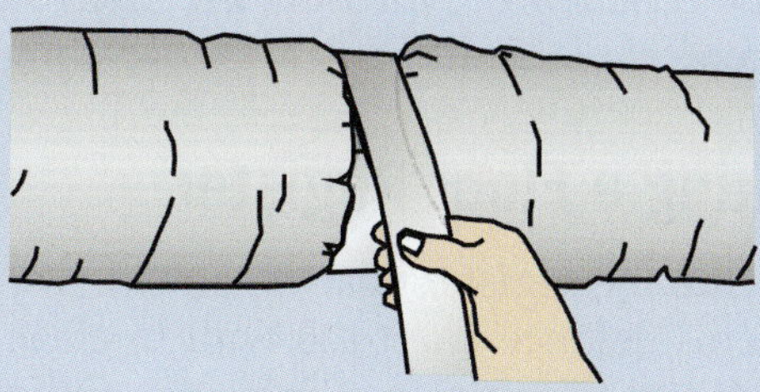

D

Goodheart-Willcox Publisher

Figure 31-35D. Pull jacket and insulation back over cores and tape the jackets together with two wraps of tape. Seal with mastic.

9. Seal the seam with mastic.

31.10 Wall Stack

Wall stacks transport air vertically, from floor to floor, Figure 31-36. In residential applications, stacks carry air to high-wall or second-floor registers from a basement or crawl space. They can also be used to carry air to baseboard or low-wall registers from attic piping.

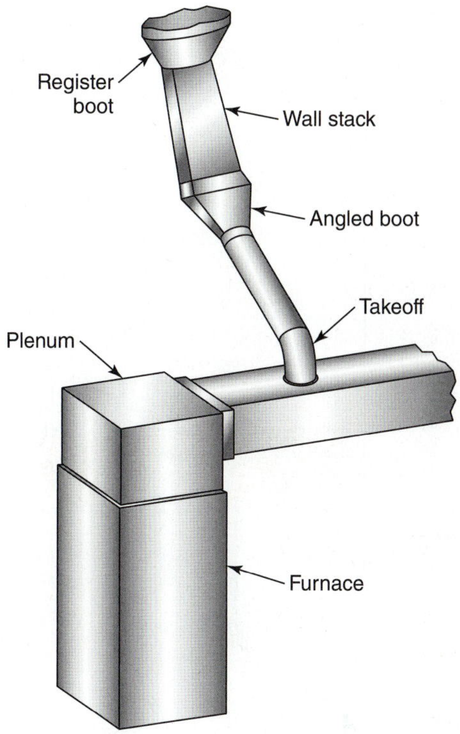

Goodheart-Willcox Publisher

Figure 31-36. Wall stacks carry air to high-wall or second floor registers. They can be used as part of the supply or return air systems.

Installing Wall Stack

When installing wall stacks, work from the basement, if possible. As each section is attached, push the stack up into the stud space and attach the next section, Figure 31-37. Continue until the top of the stack reaches the register boot or fitting to which it will connect. Reduce the length of the last section as required. When a wall stack must be lowered into place from above, enlarge openings as necessary and lower the stack into the stud space as it is assembled section by section.

Joining Wall Stack

To join two sections of wall stack duct together, S-cleats can be used to secure the long sides of the duct. Sheet metal screws should be applied on the shorter sides of the ducting and can be used on the longer dimension, if needed. The seams are then sealed with mastic. The same care should be taken when connecting with the stack heads or boots, Figure 31-38.

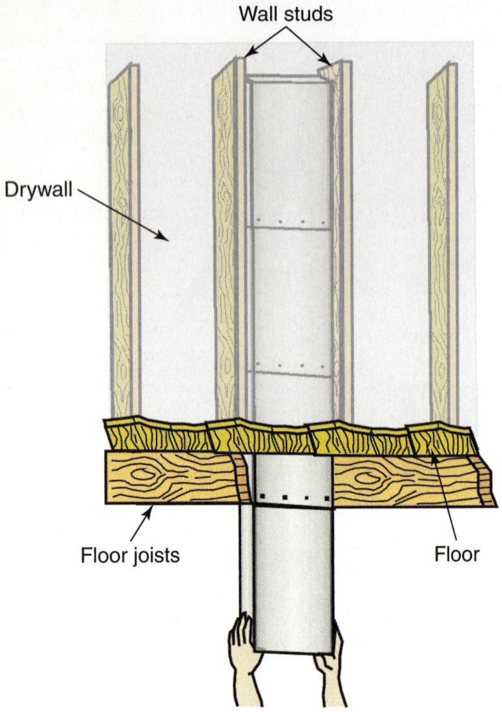

Goodheart-Willcox Publisher

Figure 31-37. In residential installations, wall stack is generally pushed up into the wall space from the basement. When cutting openings for a stack, use a flashlight and mirror to check for obstructions between studs.

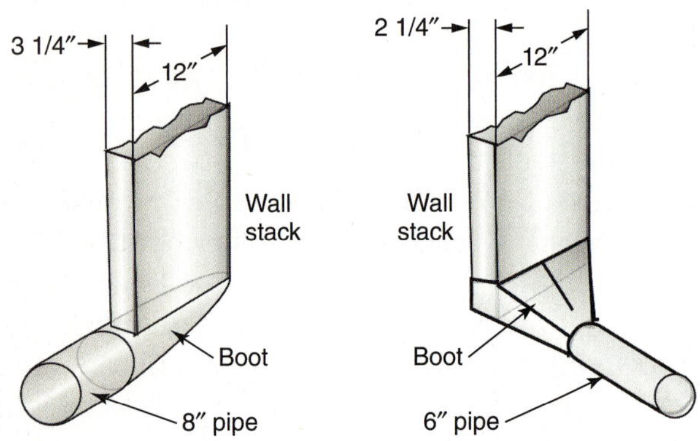

Goodheart-Willcox Publisher

Figure 31-38. Wall stack duct is designed to fit between the studs inside a wall. Different sizes include 2 1/4″ or 3 1/4″ deep by 10″, 12″, or 14″ wide. Standard length of a stack section is 24″.

31.11 Installing Supply and Return Air Systems

The final fittings to install in an air distribution are those used to move air to the destination and return it to the furnace or air handler for conditioning. Installation of these systems depends on the space and design of the building.

Supply Air System

The final fitting in an air distribution duct system used to move air to its destination is the register boot or the ***boot***. Due to the numerous locations and difficult areas that registers must fit, boots are available in a variety of shapes and sizes, **Figure 31-39**. Supply air is released into a room through registers or diffusers. Registers have dampers or vanes for controlling the amount of air, while diffusers do *not*. However, ***diffusers*** are designed to spread or diffuse air more widespread and uniformly than registers, **Figure 31-40**. For most effective air movement, registers are located in the baseboard on an outside wall or in the floor near the baseboard. Diffusers are often located in more central places like in a room's ceiling.

Locating a Floor Register

Locate the register near an outside wall, but at sufficient distance from the wall to clear draperies or other obstructions. Draw an outline of the opening required, then drill a hole in the center of the planned opening. Measure from the hole to the sides of the opening, then go into the basement or crawl space and find the hole position from below. Measure the same distance from the hole to be sure it will clear floor joists or other obstructions. If not, the opening must be shifted, **Figure 31-41**.

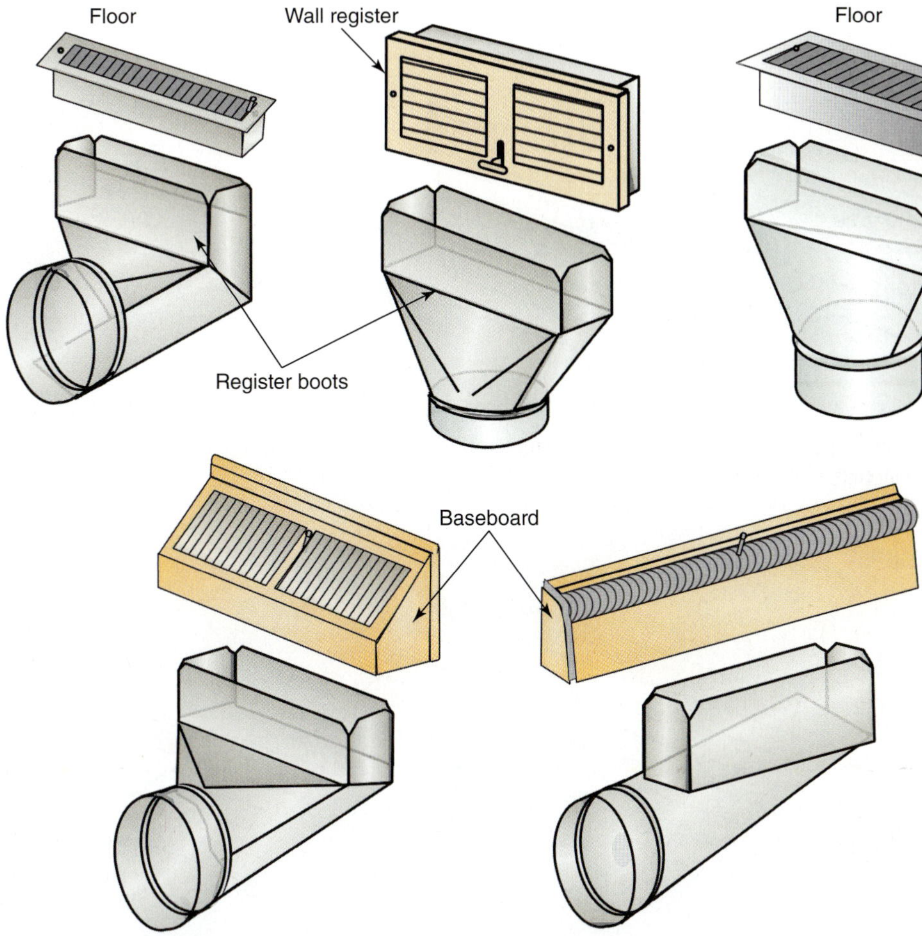

Goodheart-Willcox Publisher

Figure 31-39. Placement of supply registers is important to ensure a pleasant room temperature. A variety of shapes and sizes are available to meet varying requirements.

Goodheart-Willcox Publisher

Figure 31-40. Diffusers dispense air in multiple directions for uniform distribution.

Installing a Floor Register

A typical one-piece register fits into a 12″ × 4″ rectangular opening in a register boot. After locating the proper position, cut a hole in the floor just large enough for the boot to fit snugly. The metal edges of the boot should be level with the floor. Secure the boot in position by nailing it to the floorboards. Insert the register into the boot and anchor it with screws at each end, **Figure 31-42**.

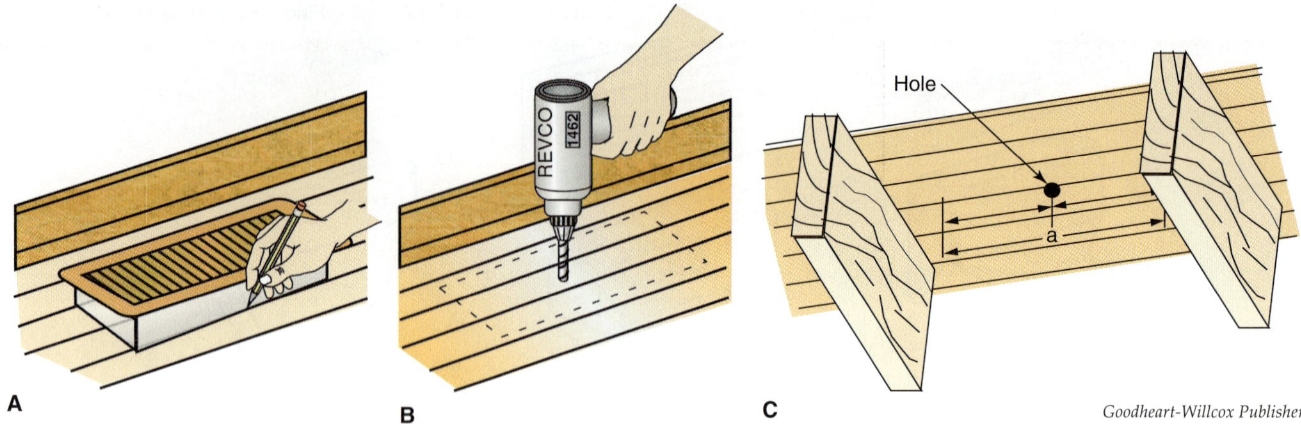

Goodheart-Willcox Publisher

Figure 31-41. Preparing to cut a floor register opening. A—Measure and trace the opening for the register. B—Drill a centered hole through floor. C—Check clearances from below.

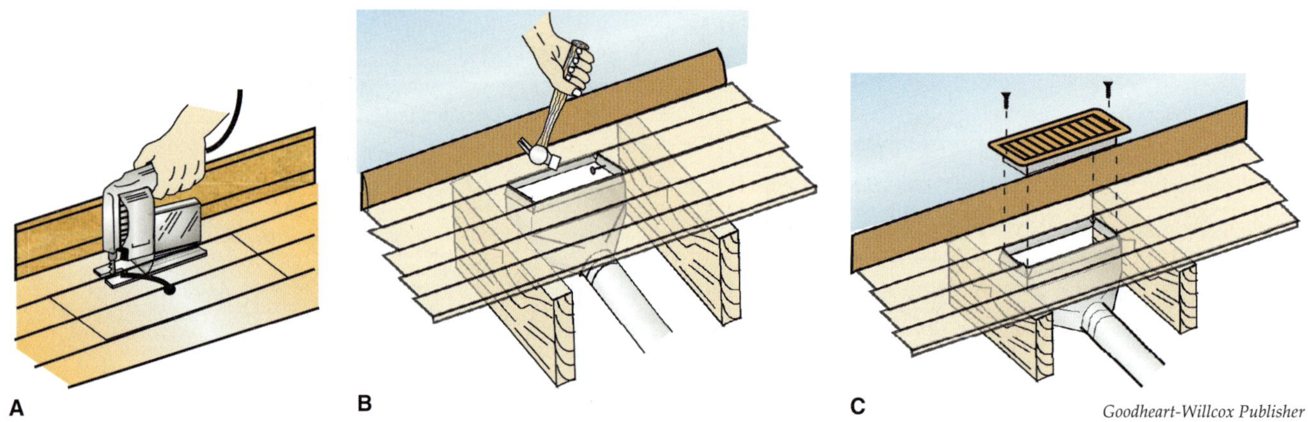

Goodheart-Willcox Publisher

Figure 31-42. Installing a floor register. A—Cut along the outline with a jigsaw, circular, or reciprocating saw. B—Secure the register boot to floor with nails. C—Install the floor register into the boot opening and secure it with screws.

Return Air System

In houses where air can easily circulate to one location, a central return air grille is adequate. In two-story homes or long ranch designs, a more efficient, yet more expensive system routes air through separate piping from individual rooms back to the furnace for retreatment. The system filter or electronic air cleaner is normally installed where the return air duct attaches to the furnace, **Figure 31-43**.

Return Air Grilles

Return air grilles are nonadjustable and are *not* equipped with dampers. They should permit unrestricted flow of return air back to the furnace. Return air grilles are available in various sizes. A return air grill is normally positioned on an inside wall, in a baseboard on an inside wall, or in the floor next to an inside wall, **Figure 31-44**.

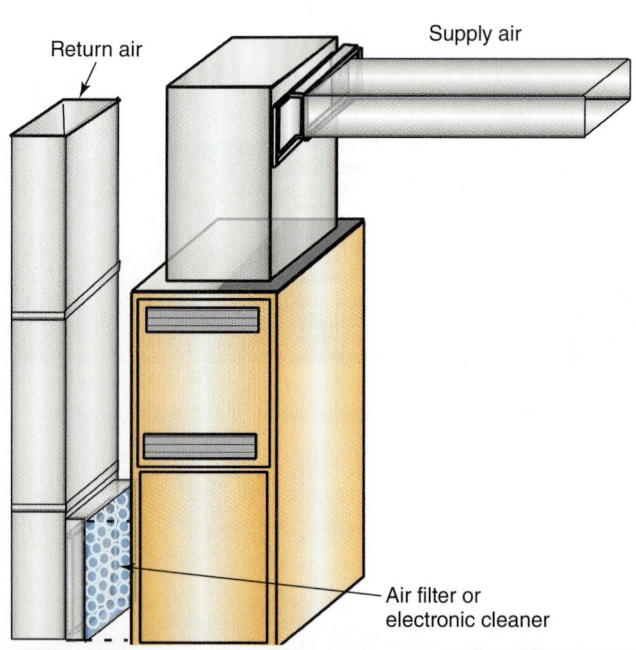

Goodheart-Willcox Publisher

Figure 31-43. For proper circulation and air purification, return air systems must included an air filter

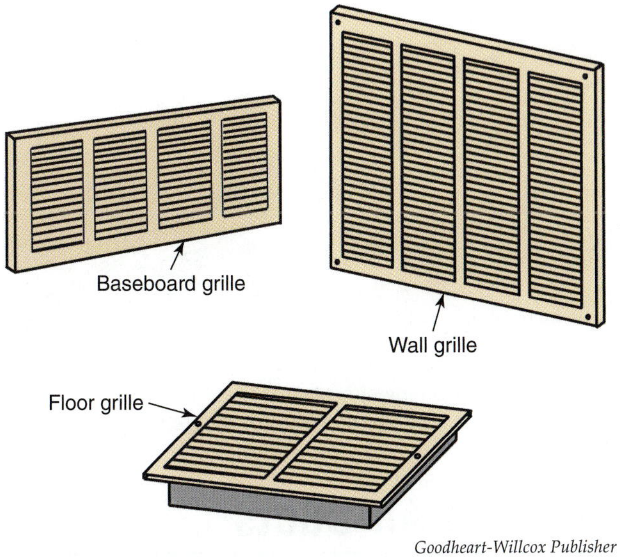

Figure 31-44. Return air grilles may be located in the floor, ceiling, wall, or baseboard.

When installing wall-type grilles, the floor plate inside the partition must be cut for passage of return air to the duct or enclosed joist space below. It is easier if the floor plate can be cut from below the floor. Using the stud space as a return-air duct requires sealing off the stud space just above the grille opening with a 2″ × 4″ wood block as shown in **Figure 31-45**.

Installing floor type or baseboard type grilles is much the same as installing supply air registers in those surfaces, **Figure 31-46**. Floor type and baseboard grilles require larger holes than do registers.

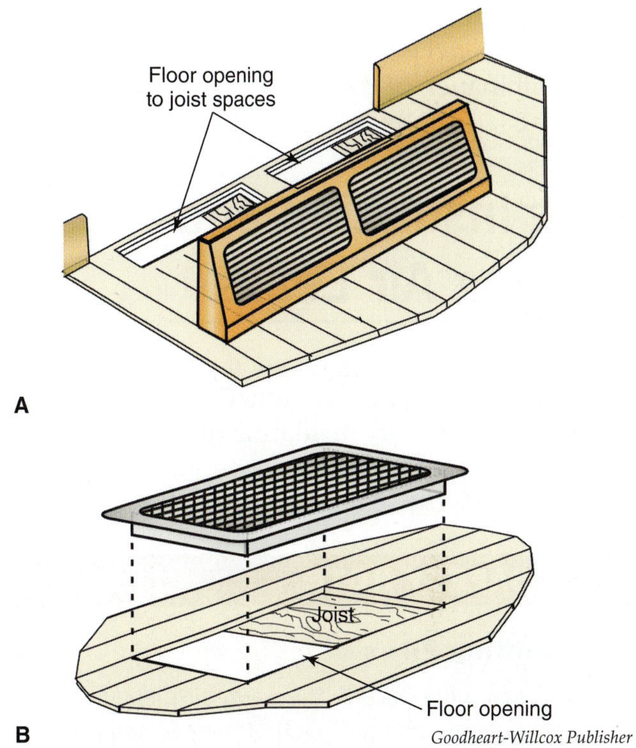

Figure 31-46. A—If a large baseboard grille is used, double stud and joist spaces should be prepared. B—Floor type grilles are commonly used when the furnace is located in the basement.

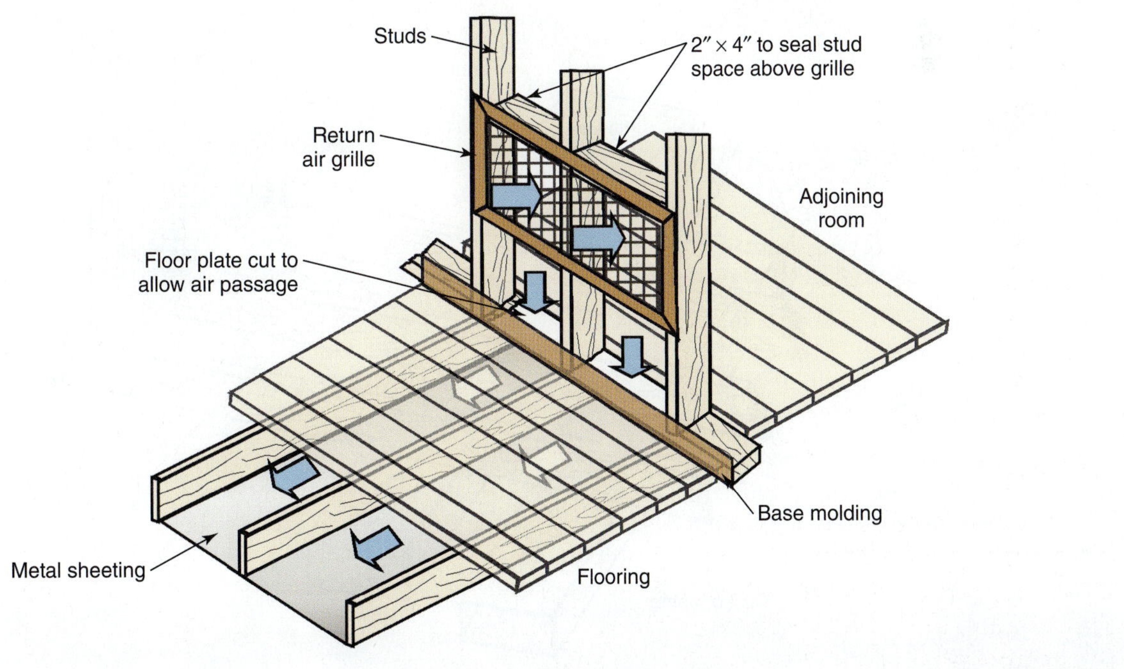

Figure 31-45. With wall-type grille openings, air is returned to the furnace through stud spaces, sheeted joist spaces, or ducts.

Basement Return-Air System

With the return-air grilles installed and the return takeoff fittings attached to the return-air plenum, it is only necessary to complete the runs from the grilles to the furnace. This may be done through sheeted joist and stud spaces, direct ducts, or a combination of the two, **Figure 31-47**.

31.12 Insulation and Sealing Air Leaks

Insulate all supply ducts and pipes installed in an attic, crawl space, or other unheated areas. Carefully wrap the insulating material around each pipe and secure with proper glue or by applying tape along the seam, **Figure 31-48**.

SAFETY NOTE

Preventing Fires

To avoid a fire hazard, the warm-air plenum and branch ducts must be separated from wood joists, floors, and other structural elements by at least 1″ of clear space or insulation.

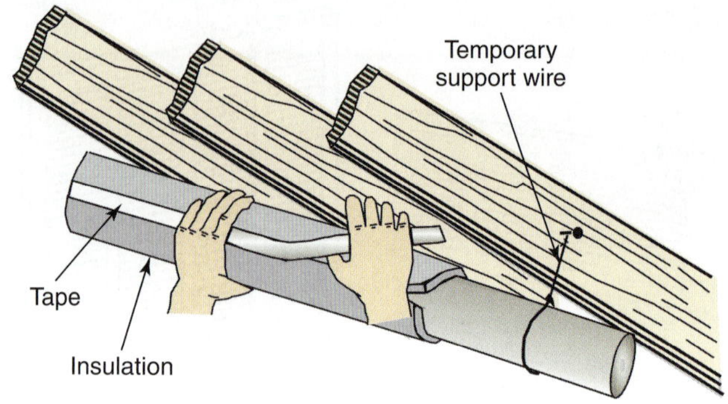

Goodheart-Willcox Publisher

Figure 31-48. Use tape or sealant to close insulation on round ducts. Remove temporary support wires and replace them with strapping.

Leaks in Air Ducts

Over time, fabric-based duct tape grows brittle and the glue dries out. This permits sections of duct to come loose and leak air. Air leaks from poorly sealed ducts can be one of the biggest energy and money losers in a home. Many areas no longer allow the use of fabric-based duct tape. Duct tapes and mastics used for sealing duct joints, seams, and connections must be compliant with the local residential building code. Check local codes for the correct tapes and mastics to use.

Ducts in homes with high energy-efficiency ratings are usually sealed with mastic and fiberglass reinforcing membrane, **Figure 31-49**. ***Mastic*** is a

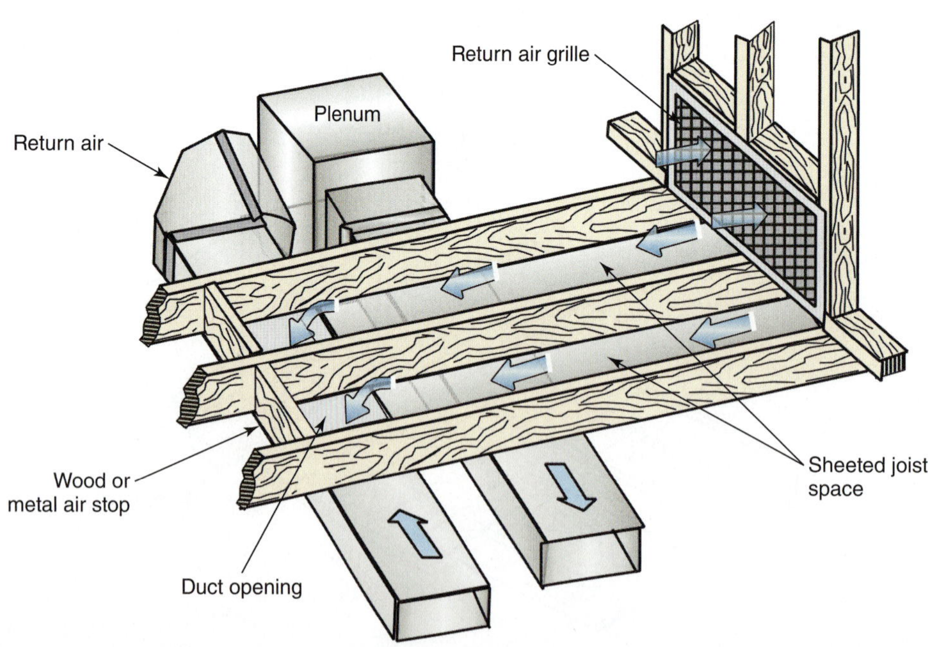

Goodheart-Willcox Publisher

Figure 31-47. For return air passages, sheeting (or panning) is easier to install than ducts.

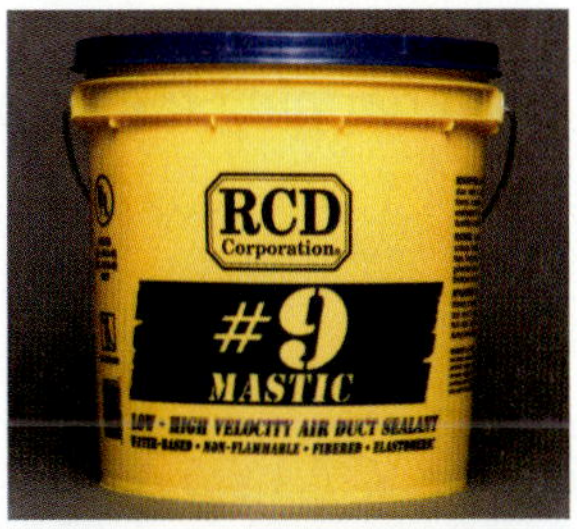

A

B

Courtesy of RCD Corporation

Figure 31-49. A—Mastic is available in 1-, 2-, and 5-gallon recyclable plastic pails. B—Fiberglass reinforcing membrane is designed for use with mastic. A roll is 3″ × 50′ or 3″ × 150′.

nontoxic adhesive-material used for permanently sealing joints and seams for all types of air duct materials. Most mastic is water-based, nonflammable, fiber-reinforced, fast drying, and waterproof. Easy application and cleanup with soap and water are attractive features of the material. Mastic can be applied by brush, trowel, caulking gun, or with a gloved hand. It typically dries to touch in 1 or 2 hours, but unusually wet conditions may require 4–5 hours drying time. Insulation can be installed over mastic that is still wet, but care is needed to prevent excessive movement and loss of seal.

Fiberglass Reinforcing Membrane

Fiberglass reinforcing membrane is used with mastic for sealing air leaks in insulation, air ducts, plenums, and fittings, **Figure 31-49B**. This inorganic woven fiberglass membrane will not rot, is lightweight, and conforms to irregular surfaces. A typical roll is 3″ wide × 50′ long or 3″ wide × 150′ long. This product is designed to be used with mastic as a reinforcement when there is a gap of 1/4″ or more. If the membrane has a sticky side, apply it over the clean metal surface before applying the mastic. If the membrane does not have a sticky side, apply a thin layer of mastic and place the membrane in this layer. Add another layer of mastic until the membrane is completely covered.

How to Use Mastic and Fiberglass Reinforcing Membrane

Before applying mastic or fiberglass reinforcing membrane, the duct surface must be clean and dry. Use a cloth to wipe dust from the surface of the duct, **Figure 31-50**. If an oily film or grease is present, wipe the duct clean with a damp cloth. If necessary, use a grease-cutting cleaner on the cloth, then wipe thoroughly dry.

Sealing Joints with Less Than 1/4″ Gaps

Load the brush with mastic and coat the entire joint with a continuous strip of the adhesive. Use the end of the brush to work mastic into the joint. Spread mastic at least 1″ on each side of the joint, **Figure 31-51**. The mastic should be thick enough to hide the metal surface of the duct (about 1/16″ thick).

Sealing Joints with Gaps Greater Than 1/4″

If the gap in the duct connection is wider than 1/4″, use fiberglass reinforcing membrane

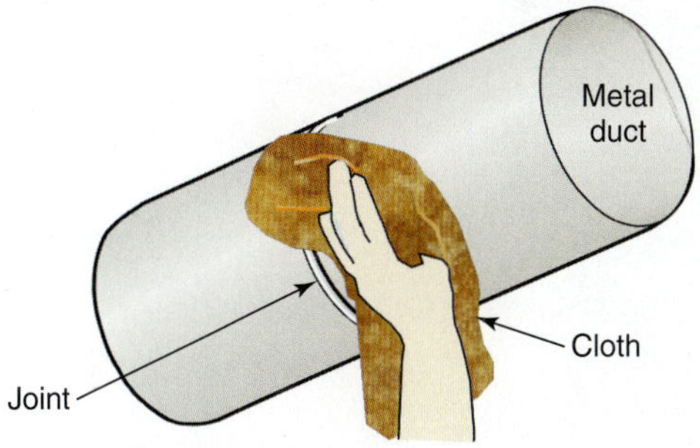

Goodheart-Willcox Publisher

Figure 31-50. Mastic will not stick to a dirty surface. Wipe dust off the surface and clean any greasy residue.

in addition to mastic. Cut enough membrane to cover the joint, and press the membrane over the joint, **Figure 31-52**. If necessary, first brush a thin layer of mastic on the joint to hold the membrane in place. Finally, cover the membrane with mastic in a layer about 1/16″ thick, overlapping the membrane on each side.

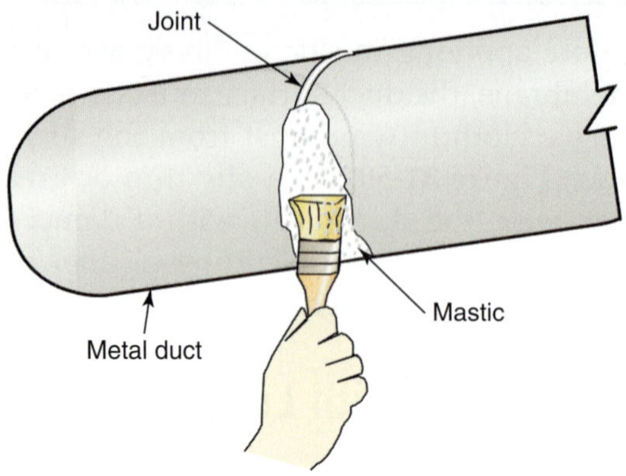

Goodheart-Willcox Publisher

Figure 31-51. Joints with a gap of less than 1/4″ can be sealed with mastic only.

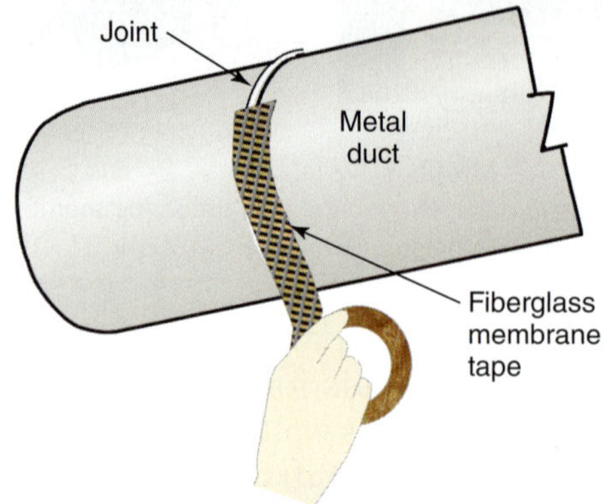

Goodheart-Willcox Publisher

Figure 31-52. Gaps of more than 1/4″ require use of reinforcing membrane.

Chapter Review

Summary

31.1 Forced-air furnaces can be classified as either upflow, downflow, horizontal, or multiposition.

31.2 Air distribution systems may be radial, extended plenum, reducing extended plenum, or perimeter types. Radial systems use round pipe that runs directly from the plenum to the register boot. Extended plenum systems combine rectangular duct with round pipes. Reducing extended plenum systems are similar to extended plenum systems except the ductwork is reduced in size at least once to keep air pressure constant throughout the length of the duct. Perimeter systems use feeder ducts to pressurize a loop of forced air around the outer perimeter of a house.

31.3 The supply air distribution system includes the plenum, the main trunk line, and all the branches serving different living spaces. Each branch terminates in a register. Another system of air ducts serves to return air from the living spaces for heating or cooling, which includes a main trunk line and all its branches. Round metal pipe, rectangular duct, or insulated flexible duct are the materials used for ductwork. Each material has its own strengths and weaknesses and is selected depending on the needs of a particular application.

31.4 Proper heat load calculations, plenum sizing, and ductwork location are important considerations when trying to optimize system efficiency.

31.5 First, main trunk lines are installed, then branch lines are installed. Branch lines must properly reduce in size to maintain constant airflow from the trunk line. Starting collars and takeoffs are installed on a plenum to begin a "run." A starting collar is used with round duct, while a takeoff is used with rectangular duct.

31.6 Round metal pipe has a raw end and a crimped end. Lengths are joined by sliding the raw end of one length over the crimped end of another length and then secured with screws and mastic. Unformed pipe can be cut using tin snips or already formed pipe can be cut using a hacksaw. Adjustable elbows can be used to conform pipe to a desired angle, and collars can be sealed using fiberglass reinforcing membrane.

31.7 Installation of an extended plenum system begins with installing a takeoff fitting on the plenum and then working outward on branch lines. Rectangular duct is joined with S-hooks and drive cleats. Openings are cut into the plenum for branch lines and rectangular elbows can be used to achieve changes in duct direction.

31.8 All duct should be run in a straight line and supported at regular intervals (typically, 2′–5′) to prevent any sags. Woven polypropylene strap, perforated metal strap, and steel wire are three options (in order from most to least desirable) for supporting ductwork.

31.9 Insulated flexible duct may be used for main trunk lines or as branching lines. This type of duct should be fully extended when installed, and can be cut to make connections of multiple lengths of flexible duct.

31.10 Wall stacks transport air vertically, from floor to floor. When installing wall stacks, work from the basement, if possible. Wall stack sections can be joined using S-cleats.

31.11 The supply air system consists of boots and registers or diffusers to bring supply air into a room. Registers are located in the baseboard on an outside wall or in the floor near the baseboard. Diffusers are often located in more central places like a room's ceiling. The return air system consists of a return air grille (or several) that routes air back to the furnace for retreatment.

31.12 All supply ducts and pipe should be insulated with insulating material and secured with glue or tape. Air duct leaks can be sealed with mastic and fiberglass reinforcing membrane, depending on the size of the gap. Mastic is used on gaps less than 1/4″, and fiberglass reinforcing membrane is used in addition to mastic on gaps greater than 1/4″.

Know and Understand

1. In a(n) _____ furnace, return air enters the top of the furnace and supply air travels from the bottom of the furnace to the living spaces. (31.1)
   A. upflow
   B. downflow
   C. horizontal
   D. multiposition
2. *True or False?* Horizontal furnaces can only be used in electric heating systems. (31.1)
3. In a residential installation, the _____ type of evaporator is most often used. (31.1)
   A. A-coil
   B. slant coil
   C. interlaced coil
   D. upright coil
4. A(n) _____ system combines rectangular duct with short lengths of round pipes. (31.2)
   A. radial
   B. extended plenum
   C. reducing extended plenum
   D. perimeter
5. A(n) _____ system uses feeder ducts to pressurize a loop of forced air to serve as supply air "blankets" along the outer wall of a home. (31.2)
   A. radial
   B. extended plenum
   C. reducing extended plenum
   D. perimeter
6. The first fitting to be installed in any air distribution system is the _____. (31.3)
   A. main trunk line
   B. register boot
   C. plenum
   D. return air duct
7. Round metal pipe can be used as a _____ in supply air and return air systems. (31.3)
   A. takeoff fitting only
   B. main trunk line
   C. branch line
   D. main trunk line or branch line
8. The end of the extended plenum must be blocked with a _____. (31.3)
   A. register boot
   B. vapor barrier
   C. sealed end cap
   D. piece of scrim
9. *True or False?* An extended plenum is sized according to the cubic feet per minute (cfm) of air that is required through the main trunk line. (31.4)
10. Starting collars are used with _____ duct, and takeoffs are used with _____ duct. (31.5)
    A. rectangular, round
    B. insulated flexible, round
    C. rectangular, insulated flexible
    D. round, rectangular
11. It is best to mount a takeoff _____. (31.5)
    A. on top of the plenum
    B. 6″ below the top of the plenum
    C. 6″ up from the bottom of the plenum
    D. at the bottom of the plenum
12. Which of the following statements describes a proper method of working with round metal pipe? (31.6)
    A. Lengths are joined by sliding the crimped end of one length over the raw end of another length.
    B. The raw end should always point in the direction of airflow.
    C. Already formed pipe should be cut with a hacksaw.
    D. A cut should be marked on the crimped end of the pipe.
13. Round duct should be sealed with _____. (31.6)
    A. duct tape
    B. fiberglass reinforcing membrane
    C. glue
    D. elbows
14. Short edges of rectangular ducts are joined with _____. (31.7)
    A. drive cleats
    B. S-hooks
    C. drive screws
    D. mastic and fiberglass reinforcing membrane
15. *True or False?* When cutting rectangular duct, always cut the raw end, not the end with the S-hook connection. (31.7)
16. Which of the following materials is the most preferred support for ductwork? (31.8)
    A. Perforated metal strapping
    B. Steel wire
    C. Woven polypropylene strapping
    D. Wooden brackets

17. Insulated flexible duct should be supported by _____. (31.9)
    A. steel wire
    B. perforated metal strap
    C. drive screws driven into wood framing
    D. fabric strapping anchored to wood framing
18. To join two sections of wall stack duct together, _____ can be used to secure the long sides of the duct. (31.10)
    A. drive cleats
    B. U-channels
    C. S-cleats
    D. duct tape
19. *True or False?* Diffusers have dampers or vanes for controlling the amount of supply air entering a room, while registers do *not*. (31.11)
20. Which of the following accurately describes return air grilles? (31.11)
    A. Equipped with adjustable vanes
    B. Equipped with dampers
    C. Available in one standard size
    D. Installed in or on an inside wall, or in the floor next to an inside wall
21. Which of the following statements describes the proper procedure for making a long-term seal of an air leak in ductwork? (31.12)
    A. Use fabric-based duct tape.
    B. Use mastic for sealing gaps more than 1/4″.
    C. Use fiberglass reinforcing membrane for sealing gaps less than 1/4″.
    D. Use both mastic and fiberglass reinforcing membrane for gaps more than 1/4″.

Apply and Analyze

1. Explain the main benefit of using a multiposition furnace. (31.1)
2. What's the difference between an extended plenum system and a reducing extended plenum system? (31.2)
3. List two functions of the plenum (bonnet). (31.3)
4. Explain how to install a starting collar on a plenum. (31.5)
5. Explain how to install a takeoff on a plenum. (31.5)
6. Describe how to cut the last length of round pipe duct that is unformed. (31.6)
7. Explain how to cut an opening in an extended plenum system for branch pipe runs. (31.7)
8. Why must insulated flexible duct be fully extended when installed? (31.9)
9. Explain how to properly locate the placement of a floor register. (31.11)
10. Explain how to properly seal a duct air leak greater than 1/4″ wide. (31.12)

Critical Thinking

1. Elaborate on why you think gravity furnaces have largely been replaced by forced-air systems. (31.2)
2. Using the table in **Figure 31-17**, calculate the amount of air (in cfm) leaving the plenum for a radial duct system with three 4″ round ducts and five 5″ round ducts. (31.5)

Glossary

A

ABS. A type of plastic used to make pipe and fittings that are used only for drain, waste, and venting. *ABS* stands for *acrylonitrile-butadiene-styrene*. (7)

absolute pressure. Any pressure above a perfect vacuum, expressed in terms of *pounds per square inch absolute (psia)*. (12)

absolute zero. A temperature condition in which all heat is removed, and molecular motion stops completely. This occurs at –460°F (–273°C). (11, 12)

absorb. To soak up. (14)

Acceptable Exposure Limit (AEL). An airborne-exposure limit with a time-weighted average (TWA) to which workers may be exposed repeatedly during an 8- or 12-hour day, or 40-hour week, without adverse effects. (15)

access opening. A small stub on a hermetic compressor that allows access to the normally sealed refrigerant circuit. (20)

accumulator. In a refrigeration system, a small reservoir installed in the suction line and used as a safety device to prevent liquid floodback by trapping any liquid refrigerant that did not evaporate during passage around the evaporator. (13, 14)

acetone. A liquid solvent used in acetylene cylinders that is absorbed by a porous substance within. (8)

acetylene. A fuel gas that produces the highest temperature for soldering, brazing, cutting, and welding. (8)

acid. A corrosive chemical compound. Acid may form within refrigeration systems in the presence of oxygen and moisture. Acid reacts with metal parts and contributes to the formation of sludge. (17)

A-coil. An evaporator consisting of two coils arranged together to form the letter *A*. This is the most popular type of air conditioning evaporator for residential systems. (31)

across-the-line starter. A contactor that has built-in overload protectors, used to operate and protect three-phase motors. Also known as a *line starter* or a *magnetic starter*. (25)

ACR tubing. Copper tubing used for air conditioning and refrigeration work. It is measured by outside diameter (OD), is available in soft copper or hard-drawn copper, and has three classifications according to wall thickness: Types K, L, and M. (6)

active recovery. A recovery method that uses a recovery unit to draw refrigerant out of a system either not operating or with refrigerant already removed. This method must be used on systems that hold more than 15 pounds of refrigerant. (16)

adapter. A special fitting used to connect a pair of pipes or tubes that are made of different materials. (7)

adjustable head pressure control valve. A control valve that works as a condenser pressure regulator, throttling liquid refrigerant leaving the condenser. This causes liquid refrigerant to accumulate in the condenser and raises its pressure, maintaining a set minimum pressure in the condenser. These are commonly used to keep head pressure from falling too low during low ambient temperature. (19)

adsorb. To collect substances on a surface in a condensed layer. (14)

air filter. In a furnace or air handler, a cleaning device used to capture dust, lint, dirt, and other impurities from the circulating airstream. (28)

air-source heat pump. A heat pump that relies on ambient air as a heat sink and as a heat source. (30)

air space. The area being cooled. (17)

alkylbenzene (AB) lubricants. Synthetic compressor oils manufactured from propylene and benzene. AB lubricants are used with CFCs, HCFCs, and blends including CFCs and HCFCs. (21)

alloy. A compound of two or more metals. This term can also be used to describe solder or brazing filler metal material, which often consists of an alloy. (8)

alternating current (ac). Electric current that reverses direction many times each second. (22)

ambient temperature. The temperature of the air or other fluid surrounding an object, often a motor or condenser. (12)

American National Acme Thread. A form of thread with a 29° angle that is used for feed screws, jacks, and vises. (5)

American National Standard Thread. A group of three main series of threads, consisting of National Fine (NF), National Coarse (NC), and National Pipe (NP). (5)

ammeter. An instrument used to measure electron flow (electrical current). These may be in either *in-line* or *clamp-on* style. (22)

Note: The number in parentheses following each definition indicates the chapter in which the term can be found.

ampacity. The amount of current a conductor can safely carry without becoming overheated. (22)

amperage interrupting capacity (AIC). The point beyond which a circuit protection device can no longer interrupt current flow, which may result in the device violently arcing and exploding before starting a fire. Also called *overcurrent interrupting limit.* (23)

ampere. The unit of measure for electrical current flow that designates a current value with the letter *A*. In math equations, amperage may be designated with the letter *I*, standing for *current intensity*. (22)

annealed. A term used to describe metal (such as copper tubing) that is softened by heating it to a bright cherry red color and permitting it to cool. This allows it to be bent easily. (6)

arbor. The spindle or axle used to attach a cutting tool, such as a saw blade to a motor. (4)

area of a circle. The measurement of the surface of a circle that is calculated as the pi (π) value of 3.1416 multiplied by the circle's radius squared ($A = \pi r^2$). (10)

armature. The movable arm of a relay. (25)

aspirator hole. A small hole drilled in the side of the accumulator dip tube. It permits small quantities of oil to enter the outlet tube and be drawn back to the compressor. (14)

atmospheric air. A mixture of various gases and water vapor in the earth's atmosphere. (27)

atmospheric pressure. The pressure exerted by air upon objects on the earth's surface. Atmospheric pressure is greatest at sea level and is generally 14.7 pounds per square inch (101.3 kilopascals). (12)

atom. The smallest physical particle of any substance that can exist by itself and retain its chemical properties. The three types of atomic particles that compose an atom are the neutron, proton (+), and electron (–). (11, 22)

automatic expansion valve (AEV). A pressure-type refrigerant control that is installed in the liquid line at the evaporator inlet. It maintains a constant pressure in the evaporator during system operation. (18)

automatic gas valve. A valve used to start and stop the flow of gas to the burners in a gas furnace. (28)

auxiliary heat. In a heat pump, another form of heat that augments the heat pump's heat production and may also be used in the defrost cycle. It may also be called *supplemental heat* and is usually provided by staging electric heating elements. At or below a heat pump's balance point, auxiliary heat is turned on. (30)

axial. A term used to describe propeller-like fan blades that move air in a direction parallel to the rotating shaft. (24)

azeotrope. A mixture of two pure refrigerants that behaves as a single component refrigerant with a single boiling point at a particular pressure and temperature. Azeotropes do *not* separate in either the gaseous or liquid state. Also called *azeotropic refrigerant* or *azeotropic refrigerant mixture (ARM).* (15)

azeotropic refrigerant. A mixture of two refrigerants that combine to form a third, unique refrigerant having its own individual characteristics. These refrigerants exhibit a single boiling temperature different from either of the refrigerants of which it is composed. (15)

B

backfire. A condition that causes the brazing torch flame to extinguish with a loud cracking sound. (9)

backseated. A service valve position in which the valve stem is turned all the way back (turned left or counterclockwise) so that the service gauge port is blocked off but flow between the valve inlet and outlet is unobstructed. The opposite position is frontseated. (14)

balanced port valve. A type of TEV that equalizes opening and closing forces, compensating for a wide range of head pressure changes and ensuring more uniform operation of the thermostatic expansion valve. (18)

balance point. In a traditional refrigeration system, the point at which the high-side pressure of a system rises to a saturation point at a rate equal to that of gas entering the condenser. In a heat pump, the point at which the heating ability of the heat pump equals the heat loss of the building. At or below a heat pump's balance point, auxiliary heat is turned on. (17, 30)

balling. An undesirable condition in which balls or globules of solder or brazing filler metal form due to base metal temperature being too low, having dirty joint surfaces, or the torch flame being directed onto the filler metal and melting a drop from the end. (8)

basic refrigeration cycle. The process in which a circulating refrigerant absorbs unwanted heat at one location and carries that heat to another location where it can be discarded. (13)

bearing. Lubricated support for a rotating shaft. (24)

bimetallic. A term used to describe a disc designed to make and break a set of contacts. The disc is made by fusing together two thin circular discs of different metals (usually steel and copper) that expand at different rates when heated. (24)

bleedover. A process that involves allowing a small amount of high-side pressure into the low-pressure side of a capillary tube system during the off cycle. Bleeding over reduces the high-side pressure and starting load on the compressor. (18)

bleed resistor. A protective resistor that permits electrons trapped in one side of a capacitor (the charged side) to slowly migrate to the empty side (the uncharged side) until the two sides are equal. (24)

blind hole. A pilot hole that does not penetrate completely through a component. (5)

blind rivet. A permanent metal fastener used to join two pieces of sheet metal with a strong mechanical connection that will not loosen with vibration. Also called a *pin rivet*. (5)

blower relay. In a heating and cooling system, a control relay that energizes a blower motor during the cooling cycle. Many systems use a two-speed blower motor that the blower relay runs in high speed for cooling and low speed for heat. In gas furnaces, blower relays are generally used in older furnaces that do *not* have an integrated control module. Also called *fan relays*. (28)

blower section. The furnace cabinet area containing the blower motor and the squirrel cage-type centrifugal fan that circulates air through a furnace for conditioning (heating or cooling). (28)

boiling point. Temperature and pressure at which water and other liquids change state and become a gas. (11, 12)

bolt. A fastener that is used with a nut and normally used for fastening heavy metal parts together. (5)

bolt extractor. A threaded tool used to remove a broken bolt or screw. (5)

boot. Another name for register boot. See *register boot*. (31)

Boyle's Law. A gas law that explains the relationship between pressure and volume. It states: "The pressure of a gas varies inversely (opposite) as the volume, provided the temperature remains constant." (12)

branch line. A secondary run of ducting that extends from the trunk line and runs to a register boot. (31)

brazing. The process of joining two pieces of base metal with the use of a filler alloy at a temperature higher than 840°F (448°C). The base metals are heated by a torch, and the molten alloy is drawn by capillary action into the very narrow space between base metals. (8, 9)

brazing alloy. A metal alloy that melts and flows between the two joining pieces of metal during a brazing process. Filler metals are available in a variety of forms, including foil, tape, sheet, wire, flux-cored or solid bar, preform, or paste. (9)

British thermal unit (Btu). The basic unit used to measure quantity of heat. One Btu is the amount of heat required to raise the temperature of 1 lb of water by 1°F. (11)

Brown and Sharpe Worm Thread. A form of thread with a 29° angle that is used to mesh with worm gears and transmit motion. (5)

building engineer. A technician that works in-house at a particular commercial building to maintain its HVAC equipment. Locations include apartments, offices, hospitals, and factories. (1)

busbar. A metal bar that serves as a common connector for electrical circuits. (22)

bushing. A sleeve around a shaft that acts as a bearing. (24)

bypass line. Any length of tubing running in parallel with some component to allow refrigerant to bypass that component. For example, in a heat pump using two TEVs, a bypass line allows refrigerant to bypass the expansion valve when its coil is operating as a condenser and does not need a pressure drop. (30)

C

cadmium-bearing alloy. A brazing alloy containing cadmium, which is a toxic metal when molten. A cadmium-bearing alloy emits highly poisonous cadmium oxide fumes that can cause illness or death. (9)

callback. A return visit necessary to correct a problem that was not resolved on the initial service call. (2)

capacitor. An electrical device that can store an electrical charge. Capacitors may be used to help start or run electric motors by changing the phase of the electrical signal, which affects the magnetic field produced by the start winding. (24)

capacitor-start, capacitor-run (CSCR) motor. A single phase motor that runs as a CSIR during start-up and as a PSC for the rest of normal operation. It is the most powerful of the single-phase induction motors, providing both high starting torque and high running torque. The CSCR motor is used when the motor must start under a heavy load and carry a heavy load during the run cycle. (24)

capacitor-start, induction-run (CSIR) motor. A split-phase motor with a special capacitor that shifts the power phase in the start winding. CSIR motors are commonly used in situations where the motor must start under a heavy load, but once started, can operate with just the run winding. (24)

capacity control. In mechanical refrigeration, a controlling method for reducing the heat displacement ability of the system. In some cases, capacity control is intended to prevent a compressor from cycling on and off rapidly (short cycling) during reduced load conditions. In other cases, capacity control may be used to control evaporator pressure and its corresponding temperature. There are various methods for system capacity control and the capacity control of specific components in a system. (19, 27)

capillary action. A phenomenon in which a liquid moves against gravity to fill a small space. An example of capillary action is the molten alloy in soldering or brazing automatically filling the gap between two pieces of base metal. (8)

capillary tube. A length of copper tubing with a tiny, accurately sized inner diameter, used as a refrigerant metering device in small refrigeration systems. (18)

cap screw. A setscrew that has a head. (5)

capture. To temporarily trap or store refrigerant in a designated area within a refrigeration system while a repair is made to a different part of the system. (16)

carburizing flame. An oxyacetylene flame that results from supplying excess acetylene. Also called a *reducing* flame. (9)

cartridge fuse. A larger-capacity fuse that fits into special holders. (23)

Celsius. The temperature scale developed by the Swedish astronomer Anders Celsius, used today by most countries of the world. The scale measures the freezing point of water as 0° and the boiling point as 100°. (10)

Celsius scale. See *Celsius*. (10)

centigrade. Literally, *hundred steps*. Original name for the Celsius scale, which has exactly 100 divisions between the freezing and boiling points of water. (12)

centrifugal (squirrel cage) blade. A type of fan blade with a particular blade arrangement ("squirrel cage") that draws air into its center and flings it outward. (24)

centrifugal switch. A device that opens or closes a switch when a motor exceeds or drops below a designated speed. It actuates based on centrifugal force. (24)

chain-type pipe wrench. A pipe wrench with an adjustable chain used to grip pipe. (7)

chamfer. A beveled edge used in forming flared copper tubing. (6)

change of state. A phenomenon that occurs when the temperature and speed of the moving molecules of a substance reach a certain level. At this precise temperature, the molecules will rearrange themselves into a different pattern, changing a solid, liquid, or gas from one physical state to another. (11)

charging. Adding refrigerant to a system. (17)

Charles's Law. A gas law that explains the relationship between pressure and volume. It states: "At a constant pressure, the volume of a confined gas varies directly as the absolute temperature. At a constant volume, the absolute pressure varies directly as the absolute temperature." (12)

cheater. An improper object, such as a pipe extension, used to increase leverage of a wrench. (4)

check valve. A valve that only allows fluid flow in one direction. These are commonly used to prevent fluid flow in the wrong direction. (19)

chlorofluorocarbon (CFC). A refrigerant that is composed of chlorine, fluorine, and a hydrocarbon (such as methane, ethane, or propane). They are classified as *CFCs* due to their composition (chlorine-fluorine-carbon). (15)

circuit. An electrical pathway through which electrons travel. (22)

circuit protection device. An electrical safety device, such as a fuse or a circuit breaker, used to protect against overcurrent. (23)

Clean Air Act (CAA). A federal law passed in 1990 that contains severe restrictions and penalties for venting refrigerants to the atmosphere. (16)

clearance pocket. Space formed at the top of a reciprocating compressor cylinder to prevent the piston from striking the valve reed. (20, 21)

close nipple. A short section of pipe threaded along its entire length. Also called an *all-thread nipple*. (7)

coefficient of performance (COP). A measurement of heat pump efficiency obtained by dividing heat output (in Btu) by power input (in Btu/watt). (30)

cold. A lack of heat. (11)

cold anticipator. A small resistance heater, mounted inside a thermostat, that causes the thermostat switch to close just before room temperature reaches the cut-in level. Unlike a heat anticipator, a cold anticipator is wired in parallel with the thermostat contacts. (28)

combination fan/limit control. A control device that regulates operation of a furnace's blower motor and provides protection against overheating of the furnace. It is composed of two temperature-operated switches: the fan switch (NO) and the limit switch (NC). (28)

combustion. A process of burning. In fuel-fired heating systems, such as an oil or gas furnace, controlled combustion is used to produce usable heat for comfort climate control. Combustion can produce undesirable byproducts and requires an adequate supply of air. (28)

commercial refrigeration technician. A technician whose primary function is to service and repair commercial refrigeration systems. A technician may specialize in supermarkets, restaurants, ice machines, or another industry focus. (1)

common. An electrical term used to describe the wire that connects one end of the run winding and one end of the start winding of a single-phase motor. It can also describe the neutral (grounded, not ground or grounding) wire in an electrical circuit. (24)

compliant. A term meaning *yielding*. It describes compressors (used in heat pumps and other applications) that can pump liquid without damage. It eliminates the need for installing an accumulator and crankcase heater to prevent slugging/liquid floodback. (20)

compound gauge. A special pressure gauge that displays two scales and makes it possible to read pressures both above and below atmospheric pressure. (12)

compression fitting. A pipe or tubing fitting used for heating applications that involve only low pressure. (6)

compression ratio. A measurement obtained by dividing the suction pressure (in absolute pressure, psia) into the condensing (head) pressure (in absolute pressure, psia). It refers to the relationship between the low-side pressure and the high-side pressure of a system. Also called *pumping ratio*. (21)

compressor. A device that "squeezes" low-temperature, low-pressure refrigerant vapor into a small volume, which will result in a high-temperature, high-pressure gas. Considered the "heart" of a refrigeration system. (13)

condensate pan. A container located under an evaporator to catch defrost water. (27)

condensation. Changing from a gas to a liquid. Heat is released in the process. (13)

condense. Change to a liquid state. (27)

condenser. A heat exchanger designed to remove heat from the superheated refrigerant vapor, causing the vapor to condense (change state) back to a liquid, and then subcool. (13)

condensing unit. A collection of the system equipment necessary to reject a refrigerant's absorbed heat and reclaim the refrigerant vapor by condensing it back into a liquid. This consists of a compressor, condenser, hot gas discharge line, condenser fan, electrical panel box, and some accessory components. (14)

conductance. Ease with which current flows. (22)

conduction. The movement of heat through a substance from one end to the other. (11)

conductor. An electrical wire or pathway that carries electrical energy from place to place. (22)

connecting rod. A part that attaches the piston to the crankshaft in a reciprocating compressor. (20)

constructive criticism. The act of giving feedback with the goal of helping the other person learn or improve something. (2)

contactor. An electromagnetic device used to control multiple heavy-duty contacts that are opened or closed at the same time. A contactor is like a relay with high-current capacity contacts. (25)

contacts. Electrical switch components that are used to open or close to energize or de-energize a circuit. (22)

contaminants. Sludge, acid, corrosion, and other foreign substances in a refrigeration system that can cause various system problems. (17)

continuity. The existence of a complete path for electron movement from one point to another. (22)

control circuit. An electric circuit that is used to control a power circuit operating at high voltage and higher current than the control circuit. In a relay's control circuit, a low-voltage (24 V) may be used to open or close a line-voltage (120 V) circuit. (25)

controls technician. A technician whose primary function is to install, program, and service the automated controls used in the HVAC systems of commercial buildings. (1)

convection. The movement of heat by means of a carrier or flowing fluid, such as air or water. (11)

corrosion. Rusting and related chemical deterioration of metal. (7)

coulomb. The measure of the number of electrons flowing past a given point in one second. One coulomb of electricity equals 6.25×10^{18} electrons flowing past a given point in one second. (22)

counter-electromotive force (c-emf). A reactive voltage generated by current through a magnetic coil that causes an increased resistance, decreasing current flow. Also called *counter emf* or *back-electromotive force (back emf)*. (22)

counter sales representative. Individual working behind the counter of a supply house. Supply house representatives need to be knowledgeable about different parts and manufacturers. (1)

coupling. A fitting used to connect two lengths of hard copper tubing or two lengths of piping. (6)

CPVC. A type of plastic used to make pipe and fittings that are used for hot and cold waterlines or drains, provided local building codes permit. *CPVC* stands for *chlorinated polyvinyl chloride*. (7)

cracked. A valve position in which a valve is opened slightly by turning the stem about one or two turns off the backseated position. (14)

crankcase heater. Electric heater used to keep the crankcase oil warm to help prevent condensing of refrigerant vapor in the crankcase in cold ambient temperature. (21)

crankcase pressure regulator (CPR). A control valve used to protect the compressor from excessive suction pressure, most often at startup or after a defrost cycle. When necessary, it holds back refrigerant to maintain a limited pressure in the compressor's crankcase. (19)

crankshaft. A device used to change the rotary (circular) motion of the electric motor into a reciprocating (up-and-down) motion in a reciprocating compressor. (20)

crest. The top edge of two adjoining screw threads. (5)

crimp. To squeeze. In electrical work, crimping involves forming a mechanical connection of a wire to a terminal connector. (23)

crimp-on pipe fitting. A pipe fitting that is crimped into place. These types of fittings are used on potable water, hydronic hot water heating systems, and condensate drain lines. (7)

cross-charged. A term that refers to a sensing bulb that is charged with a refrigerant different from the one used in the refrigeration system. (18)

crossover igniter. A slotted length of metal used to ignite several main burners from a single pilot flame. (28)

cubic feet per minute (cfm). A measure of the volumetric flow. In HVACR, cfm is a measure of the air volume that a fan can move. (31)

cubic measure. The measure of volume, found by multiplying length, width, and height (L × W × H). (10)

current. The flow of electrons in a conductor. (22)

customer interview. The process of asking a customer questions about the problems they are experiencing with their HVAC system. (2)

customer relations. The way in which a business interacts with the public, particularly customers. In HVACR, good customer relations depends on having workers with good people skills. (2)

cut-in (CI). The setting value that causes a control's contacts to close, turning on the heating or refrigeration system. (13, 26)

cut-out (CO). The setting value that causes a control's contacts to open, turning off the heating or refrigeration system. (26)

cycling on the overload. A situation in which an overload condition causes the overload to trip, shutting down the motor. After each trip, it requires more and more time for the overload device to cool and allow a restart. (24, 25)

D

Dalton's Law. A gas law of partial pressures. It states, "In a mixture of gases, each gas behaves as if it occupies that space alone." To obtain the total pressure of a confined mixture of gases, the pressure for each gas in a mixture must be added together. (12)

data plate. A metal plate mounted on a compressor that contains coded information revealing construction details of the unit. (21)

decimal number. A whole number, plus any parts (or fractions) of the whole. A decimal point separates the whole number from the parts. (10)

decimal point. The dot used to separate the whole number from the parts in a decimal number. (10)

deep vacuum. A method of system evacuation in which the system is drawn down to a vacuum of 500 microns in one long pull. (17)

defrost cycle. A sequence of operations used to remove frozen moisture from an evaporator and restore it to full efficiency. (27)

defrost termination thermostat. A control used to disconnect the defrost heater when evaporator temperature reaches about 50°F (10°C). This device prevents the heating of an already defrosted evaporator and stops the system from adding an unnecessary heat load that would reduce a system's efficiency. (27)

defrost timeclock. A control device used to regulate evaporator defrosting by using a timer and two or more switches to automatically turn off certain system components (the compressor and/or evaporator fan) and turn on other components (the electric defrost heater or hot gas solenoid valve). (27)

dehumidifier. A refrigeration system designed specifically to remove moisture from air without affecting air temperature. (27)

dehydrate. Dry out or remove moisture. (6)

delta. A type of three-phase transformer with windings similar in shape to a triangle. (23)

denominator. The bottom number of a fraction, indicating the number of parts required to make a whole. (10)

density. A measurement of how closely the molecules of a substance are packed. (21)

desiccant. A drying agent, commonly used in refrigeration drier or filter-drier. (14)

desuperheating TEV. A thermostatic expansion valve used as a control valve that injects enough high-side liquid refrigerant into the bypassed discharge hot gas to cool it down to the recommended suction temperature. A desuperheating TEV works with a hot gas bypass valve as a form of system capacity control. Also called a *liquid injection valve*, based on its function. (19)

dew point. In referring to air, this is the point at which the air becomes saturated (holds all the moisture it is capable of holding) for a given temperature. For zeotropic blended refrigerants, dew point is the saturated vapor pressure for zeotropic blends, as listed on a P/T card. The dew point indicates the pressures at which the first drop of liquid condenses from a vapor. (27)

diameter. The length of a line across the widest part of the perimeter of a circle. (10)

diaphragm. Flexible disk or bellows that responds readily to pressure changes. (19)

die. A tool used to cut threads around the outside of a piece of metal. (5)

dielectric. An insulating material used in such devices as capacitors. (24)

difference of opinion. The event of two people having different ideas about a topic or issue. (2)

differential. The difference between he cut-in (CI) and cut-out (CO) points on a thermostat. (26)

differential valve. See *receiver pressure regulator*. (19)

diffuser. In air distribution duct systems, a supply air outlet that spreads or diffuses air in a widespread and uniform manner. Diffusers are completely stationary, while supply air registers have movable vanes or dampers. In a centrifugal compressor, a small passage through which compressed vapor is discharged from the impeller to the volute. (31)

digit. A number from 0 to 9. (10)

digital multimeter (DMM). An instrument used to measure various electrical variables such as voltage, resistance, amperage, capacitance. Some models may be able to measure temperature as well. (4)

dimensions. Measurements of length, width, and depth. (10)

dip tube. A tube that extends to about one-half inch from the bottom of the liquid receiver to ensure that only liquid refrigerant enters the liquid line receiver outlet. (14)

direct current (dc). Electric current that flows in one direction. (22)

direct drive. An open-type compressor drive arrangement in which the rotor is coupled directly to the compressor's crankshaft. (20)

direct spark ignition system. A type of gas furnace ignition system that uses a spark igniter to light the burners. Direct spark ignition systems do *not* use a pilot flame at all. (28)

discharge bypass valve. A modulating valve that controls hot gas bypass to automatically maintain a desired minimum evaporator pressure. This provides a false load to help control a system operating under varying or low load conditions. Another name for *hot gas bypass valve*. (19)

discharge line stub. One of the stubs on a hermetic compressor. It feeds the compressor's high-pressure exhaust to the hot gas discharge line leading to the condenser. (20)

discharge port. A hole through the valve plate of a reciprocating compressor where hot gas is discharged. (21)

discharge service valve (DSV). A service valve connected to a compressor's outlet and the hot gas discharge line. Since it is located on the high-pressure side of the system, it allows technicians to take high-side pressure readings and to control the flow of vapor leaving the compressor. (14)

disconnects. A heavy-duty, manually operated set of contacts used for disconnecting large loads from a power supply. (22)

disc-type compressor valve. A valve design that uses discs to virtually eliminate the clearance pocket at the top of the piston stroke, which minimizes the re-expansion of gas. This design increases compressor capacity up to 25% by improving volumetric efficiency. (20)

disposable cylinder. A refrigerant container that is discarded after use. This type of container may not be refilled. (17)

dissipated. Lost, such as the loss of some superheat as refrigerant travels through the hot gas discharge line into the condenser. (13)

double-pole breaker. A two-switch circuit breaker used to disconnect both hot wires on 230 V, single-phase branch circuits. (23)

double-pole, double-throw (DPDT) switch. A switch that has two poles and two contacts for each pole. This switch makes it possible to control variety of loads from one location. (22)

double-pole, single-throw (DPST) switch. A switch that has two poles and a contact for each pole. This switch makes it possible to open or close two circuits at the same time. (22)

downflow furnace. A type of furnace in which the return air enters the top and circulates downward and through the unit. (31)

downhill joint. A type of brazing connection in which the fitting is below the point where the alloy is applied. (9)

draft. Air movement. This term is also used in furnaces to describe the movement of flue gases. (28)

dressed. A term used to describe the head of a chisel or other tool that has been ground smooth to eliminate jagged metal. (4)

drive cleat. A fastening device used to hold together sections of rectangular duct. (31)

dual element. A type of fuse designed to permit an overload of short duration, but to blow instantly if a short circuit (high overcurrent) occurs. (23)

dual-pressure motor control. A motor control combining the low-pressure and high-pressure safety controls into a single unit. Each side of the control operates independently. (26)

dual-voltage motor. A motor manufactured to operate on either of two supplied voltages. These commonly include single-phase and three-phase motors. (24)

ductless split system. An air-conditioning or heat pump system in which the evaporator and indoor blower are located in the space to be conditioned. (30)

E

elbow. In tubing and piping, a coupling that bends to change direction (usually in 45° or 90° angles). In ductwork, a fabricated component used to make turns in duct runs. See *rectangular elbow* and *round adjustable elbow*. (6)

electrical load. An energy-conversion device that converts electrical energy to another form of energy. Examples of loads are light bulbs, heater elements, and motors. In context, sometimes simply called a *load*. (22)

electrical power. The rate at which electrons are harnessed to perform a useful function. (22)

electrical symbols. Standardized simple drawings used for ease of communication in schematic diagrams. (23)

electric defrost. An automatic method of defrosting using electric resistance heaters to melt accumulated frost and ice. (27)

electric furnace. A furnace that uses forced convection to distribute heated air that has been warmed by electric heating elements. (29)

electric heating element. An electric resistance wire that generates heat when voltage is applied and current flows through it. These are used in electric furnaces, electric duct heaters, some heat pumps, and other devices with different applications, such as heating conditioned space air, defrosting coils, and warming compressor crankcase for safety. (29)

electricity. A form of energy in which electrons move from one atom to another atom. (22)

electric shock. A discharge of electrical current passing through the body. (3)

electrodes (capacitor). In a capacitor, the two conductive layers of aluminum foil separated by an insulating layer of specially treated paper. (24)

electromagnet. A magnet that can be turned on and off. These are commonly used in many electrical devices and controls. (25)

electromagnetism. The phenomenon of generating a magnetic field around a conductor (wire) when current is flowing through the wire. (25)

electromotive force (emf). The stimulus that causes free electrons to reel out of orbit and travel to an adjoining atom. Sometimes called *voltage*, *potential difference*, or *electrical pressure*. (22)

electron. An atomic particle orbiting the nucleus of an atom, carrying a negative charge (–). (22)

electronic expansion valve (EEV). A highly efficient metering device operated by electronic controls. (18)

electronic leak detector. A refrigerant leak detection device that consists of a probe and a case containing the electronics. This instrument is used to detect the presence of halide refrigerant gas. (17)

electronic scale. A weighing instrument with displays that can be set to show the amount of refrigerant being withdrawn from a cylinder and charged into a system. (17)

elongation. A term used to describe the stretch factor of a filler metal. (8)

emergency heat switch. A switch on a heat pump's thermostat used to manually energize the auxiliary heat. (30)

emergency heat. When a heat pump stops normal operation due to low ambient temperature or heat pump malfunction and the system warms a building with only auxiliary heat functioning. (30)

energy auditor. A technician whose primary function is to inspect buildings to find where improvement can be made to make the building more energy efficient. (1)

energy efficiency ratio (EER). A numeral rating of the energy efficiency of appliances based on the ratio of energy produced (output) to energy used (input). (30)

enthalpy. The heat content of a vapor or liquid in Btu/lb., as given on a saturation chart for refrigerants. (15)

EPA certification. The process of establishing that a technician has the required knowledge of refrigerant recovery techniques and EPA regulations prohibiting the venting of refrigerants to the atmosphere. EPA certification is acquired by passing an EPA-approved test and is mandated by the Clean Air Act. (16)

epoxy. A compound that chemically hardens. Commonly used as a filler, leak stopper, and an adhesive. (8)

erratic. A term used to describe a valve that does not maintain a particular setting for any length of time. (26)

evaporation. The process of changing from a liquid to a gas. Heat is absorbed in the process. (13)

evaporator. In a refrigeration or air conditioning system, the heat exchanging device located inside the area where heat is to be removed. Low-pressure liquid refrigerant here absorbs heat to evaporate and often increase temperature (superheat). (13)

evaporator fan delay thermostat. A control device used to delay the restarting of an evaporator fan after a defrost cycle until evaporator temperature drops to a set value. (27)

evaporator pressure drop. A pressure drop that develops in the evaporator due to the length of its tubing, the return bends in the tubing, and the vacuum (suction) action of the compressor. (18)

evaporator pressure regulator (EPR). A special control valve used to prevent evaporator pressure from falling below a set limit. When multiple evaporators share a single suction line, an EPR is often used to maintain an evaporator at a different temperature than the others. For this reason, an EPR may also be called a *two temperature valve*. (19)

evaporator unit. The equipment of a refrigerant system that is necessary to absorb heat and vaporize the liquid refrigerant. It consists of the evaporator, refrigerant metering device, evaporator fan, and some accessory components (14)

exponent. A numeral that is placed above and to the right of a base number as an abbreviation to indicate a multiplication process. Also called a *power*. For example, *2* is the exponent of *5* here: 5^2. (10)

extended plenum. An air duct system that uses runs of rectangular duct to extend the plenum and runs of round duct to reach individual supply air registers. (31)

externally equalized. A type of TEV that transfers evaporator pressure from the evaporator outlet to the underside of the valve diaphragm. (18)

F

factor. The base number used with an exponent. (10)

Fahrenheit. The temperature scale developed by German scientist Gabriel Fahrenheit, in which the freezing point of water is 32° and the boiling point is 212°. (10)

Fahrenheit scale. See *Fahrenheit*. (12)

false load. An artificial loading of the low side to control a system operating under varying or low-load conditions. One method of producing a false load is to bypass hot gas refrigerant from the compressor's discharge line using a discharge (hot gas) bypass valve. This heat load is *false*, because it is not being drawn from the conditioned space but rather injected into the low side from the high side. (19)

fan cycle control. A pressure-type control used to control a condenser fan or fans. When head pressure drops below acceptable limits, the fans are turned off. This allows condenser heat to accumulate and raise head pressure back up. (26)

fan switch. A control used to operate the blower motor in some gas furnaces. When temperature in the heat exchanger reaches a set cut-in point, a fan switch's normally open contacts close to turn on the furnace blower. (28)

feeder service. Wiring directly from the main disconnect or electrical load center to the electric furnace. This includes two hot wires and one ground wire for 240 V, single phase power. (29)

ferrous. A term used to describe iron-containing metals. Ferrous metals will readily accept magnetism and are commonly used in electromagnetic devices. (8, 25)

fiberglass reinforcing membrane. An inorganic woven fiberglass membrane that is lightweight and conforms to irregular surfaces. It is commonly used with mastic for sealing air leaks in insulation, air ducts, plenums, and fittings. (31)

filter-drier. A device used to absorb moisture from the system and catch any foreign particles circulating with the refrigerant. (14)

First Law of Thermodynamics. Energy cannot be created or destroyed, but energy can be converted from one form to another. This law applies to heat and other forms of energy, such as electrical, mechanical, light, chemical, and atomic. (11)

fixed bore (piston) metering device. A refrigerant metering device that operates on the principle of restriction much like a capillary tube. In heat pumps, a fixed bore contains a piston that allows free flow of refrigerant in one direction and restricted flow of refrigerant in the other direction. Heat pumps using fixed bores as refrigerant metering devices have one installed on the indoor coil and another on the outdoor coil. (18)

flame rectification. A process that uses a flame rod to confirm that the pilot flame is lit in an intermittent pilot ignition system. (28)

flare bonnet. A copper insert used to convert an ordinary flare nut into a cap nut. (6)

flare cap nut. A female nut that is used to seal off a male-threaded fitting. (6)

flare elbow. A fitting used to connect two flare nuts of the same size, while providing an accurate bend of either 45° or 90°. (6)

flare fitting. A mechanical fitting with threads and a 45° angle beveled contact area for the flared end of a tube. A flare nut screws into the threads and holds the flared tubing against the flare fitting. (6)

flare nut. A piece of hardware that holds flared tubing against a flare fitting. It secures the connection when its internal threads are screwed into the external threads of the flare fitting. Flare nuts are sized by the hole through which the tubing is inserted. (6)

flare plug. A fitting used to seal a flare nut or similar female-threaded opening. (6)

flare tee. A flare fitting with three connections that makes it possible to connect a branch onto an existing line of copper tubing. (6)

flare union. A fitting used to connect two flare nuts of the same size. (6)

flaring. A process of expanding or spreading copper tubing to give the end of the tube a funnel shape with a 45° angle. (6)

flaring block. A special tool used to hold tubing for enlarging in the flaring and swaging processes. (6)

flashback. A condition in which the brazing flame burns back inside the torch tip. This condition is revealed by a shrill hissing or squealing sound. (9)

flash gas. The vapor that results from the sudden pressure drop (from high to low) as liquid refrigerant passes through the TEV orifice. (18)

floodback. Liquid refrigerant in the suction line, resulting from too much liquid refrigerant being metered into the evaporator. (18)

flooded start. A condition in which a compressor starts and must pump *liquid* rather than *vapor*. This could result from several different causes, such as refrigerant migration from cold ambient temperature or a refrigerant metering device that leaks or is poorly seated during the off cycle. (18)

flow point. See *liquidus*. (8)

flutes. 1) The body of a twist drill bit composed of two spiral grooves. 2) The threaded grooves of a tap. (4, 5)

flux. A multipurpose chemical used to treat the clean surface of the base metal when soldering or brazing. (8)

flywheel. A pulley (sometimes weighted) that is attached to the external crankshaft of a motor. (20)

forced-air system. A heating and/or cooling system that uses a furnace that is centrally located in the building and ductwork extending throughout the building to distribute blown (forced) conditioned air. (31)

forced convection. A term describing heating and cooling systems that use a fan or blower to circulate air. Also used to describe evaporators and condensers equipped with a fan. (11, 13)

forced-feed system. A lubrication method that uses an oil pump to deliver oil, under pressure, to all bearing surfaces and other critical parts in a compressor. (21)

formula. A mathematical expression used as a standard to determine a value. (10)

fraction. In mathematics, part of a whole, often written with one number (numerator) above another number (denominator). (10)

fractional horsepower. A term describing a motor rated at less than one horsepower. (24)

fractionation. The separation of parts in a zeotropic blend. Zeotropic refrigerants fractionate when the most volatile component boils into a vapor more quickly than the less-volatile components. (15)

freeze-up. A condition in which an evaporator becomes clogged with frost, preventing airflow. (27)

frequency. The number of complete cycles of alternating current that occur in one second. Measured in units of *hertz (Hz)*. (22)

frontseated. A service valve position in which the valve stem is turned all the way forward (turned right or clockwise) so that it is blocking either the inlet or outlet (depending on the way the valve is installed in the system). The other valve opening is open to the service gauge port. The opposite position is backseated. (14)

full-floating. A term used to describe a piston pin that is free to rotate within the connecting rod and the piston. (20)

full load amperage (FLA). The operating ("running") amperage of a motor after initial startup. (23, 24)

fuse. A device designed to detect excessive electrical current and melt a fusible element to open the circuit before danger arises. (22)

fusible link. In an electric furnace, a safety device that is factory-installed and connected in series in one of the "hot" legs of a heating element. If a limit switch fails and the temperature of the heating element becomes abnormally high, the fusible link will melt, opening the line voltage circuit. (29)

G

galvanized. A term used to describe items that have been zinc-coated. Commonly used on steel pipe. (7)

gas furnace. A furnace that uses forced convection to distribute heated air that has been warmed by the combustion of a gas. Also called a *gas-fired furnace.* (28)

gas manifold. In a gas furnace, a pipe that distributes gas to the burners through an orifice (outlet hole) in each burner. (28)

gauge pressure. Pressure expressed in *pounds per square inch gauge (psig).* Readings in psig measure only pressures above atmospheric pressure, as most pressure gauges are calibrated to read 0 psi at atmospheric pressure. (12)

grounded. Connected to the earth. This term also describes the neutral wire in an electrical circuit. *Grounded* can also be used to describe a device or component (such as a motor) that is *shorted to ground* (has developed an internal electrical short to the device's frame) or a *ground fault.* (22)

ground fault. An internal electrical problem that permits electron flow to the frame and makes a device's frame electrically live. See also *grounded* and *shorted to ground.* (22)

ground fault circuit interrupter (GFCI). A safety device designed to protect a circuit from overcurrent and to protect people from potentially hazardous ground faults arising from the use of defective appliances or portable tools. (23)

ground-source heat pump. A heat pump that relies on the earth or a body of water as a heat sink and as a heat source. Also called *geothermal heat pumps.* (30)

H

halide refrigerants. Refrigerants that combine a halogen with a hydrocarbon compound, such as acetylene, methane, or ethane. (15)

halide torch. A device used for detecting halogenated refrigerant leaks. (17)

halogens. The chemical elements fluorine, chlorine, iodine, and bromine. (15)

hand shut-off valve. A manually operated valve that makes it possible to quickly isolate sections of a system for servicing. Used on commercial and industrial systems. (19)

hard-drawn copper tubing. A hard and rigid form of copper tubing that cannot be easily bent. Hard-drawn copper tubing is available in Type L and Type K thicknesses. (6)

hard solder. A term referring to silver-bearing brazing alloys that have melting temperatures ranging from 1100°F to 1500°F (593°C to 816°C). (9)

hard start kit. A relay and a capacitor in a single package, used to resolve hard-start problems by giving the motor more torque with another capacitor. (24)

hazard. A potentially harmful situation or substance. (3)

hazardous waste. Waste material considered to require proper handling or disposal under law. (16)

head pressure. The pressure on the high side of the system. (17)

head pressure control valve. A modulating valve used to solve low head pressure problems caused by low ambient temperatures. These may be adjustable or nonadjustable. (19)

head-pressure safety control. A control device that will shut off a refrigeration system if head pressure exceeds proper limits. (26)

heat anticipator. A small resistance heater, mounted inside a mechanical thermostat, that causes the thermostat switch to open just before room temperature reaches the desired level (cut-out). Unlike a cold anticipator, a heat anticipator is wired in series with the thermostat contacts. (28)

heat exchanger. Any device that transfers heat from one medium to another. In heating, it is a section of a furnace where heat is transferred from the products of combustion or electric heating element to the circulating air. In a refrigeration system, it is a component that transfers heat from the warm liquid line to the cool suction line, sometimes called a *suction line-liquid line heat exchanger.* (14)

heating, ventilation, air conditioning, and refrigeration (HVACR). A broad industry that uses the control of temperature, humidity, and other factors to condition buildings and spaces for comfort, product preservation, and numerous other applications. (1)

heating-cooling thermostat. A thermostat that provides control of both heating and cooling systems from one device. (29)

heat load. Air to be cooled. (17)

heat pump. A refrigeration system capable of reversing refrigerant flow so that it can function to produce cooling or heating at its two refrigerant coils. (30)

heat sink. Any heavy metal device that draws heat away, such as a valve or compressor that draws heat away from the heated area in soldering or brazing. (8)

heat stress. A bodily response to excess heat that can result in heat stroke, exhaustion, cramps, or rashes. (3)

height. Linear measure of one dimension (from base to top) of an object. (10)

hermetic compressor. A compressor that contains an electric motor that is connected directly to the compressor within a sealed shell. (20)

hex-head. A type of bolt or other fastener with a six-sided head for use with a wrench. (4)

high-pressure safety control. A pressure-sensitive switch that stops the compressor if head pressure exceeds its setpoint. (29)

high-pressure side. The part of a refrigeration system that is under high pressure during operation. The high side begins at the compressor outlet and ends at the refrigerant metering device. (13)

high-temperature applications. A term that refers to refrigeration systems for air conditioning (comfort cooling), which have a low compression ratio and are designed to maintain temperatures in the 70s. (15)

horizontal furnace. A type of furnace in which the system air passes through horizontally. These are normally installed in an attic or crawl space. (31)

horizontal joint. A type of brazing connection in which the tubing and the fitting are on the same level and are heated equally. (9)

hot-gas bypass valve. A modulating valve that controls hot-gas bypass to automatically maintain a desired minimum evaporator pressure. This provides a false load to help control a system operating under varying or low load conditions. Also called *discharge bypass valve.* (19)

hot surface igniter. Used in gas furnaces, an ignition device that is made of a special high-resistance material that becomes extremely hot when voltage is applied. (28)

hot surface ignition system. A type of gas furnace ignition system in which the main gas valve is opened and immediately lit by an igniter. Hot surface ignition systems do *not* use a pilot flame at all. (28)

hot-gas defrost. A defrost method in which hot gas is taken directly from the compressor's discharge line and delivered directly to the evaporator inlet through a bypass line controlled by a solenoid valve. (27)

hot-gas discharge line. Copper tubing connecting the compressor to the condenser. It is the part of the system under the highest pressure and at the highest temperature. (13)

humidistat. A moisture-sensitive switch used to control system operation by responding to humidity (27)

humidity. Moisture in the air. (27)

hunt and surge. A condition in which a TEV fluctuates from fully open to fully closed. This may result when a TEV is oversized. (18)

hydraulic swage. A type of swaging tool used to expand tubing using the appropriate-sized expander head to match the tubing size. This swaging tool is operated with one hand and applies hydraulic pressure to expand the tubing. (6)

hydrocarbon (HC). A refrigerant that contains only hydrocarbons. These refrigerants have zero ozone depletion potential (ODP) and very low global warming potential (GWP) but are highly flammable. (15)

hydrochlorofluorocarbon (HCFC). A refrigerant that contains hydrogen atoms that cause the compound to be less stable in the atmosphere (and thus, less damaging) than CFCs. (15)

hydrofluorocarbon (HFC). A refrigerant that contains no ozone-depleting chlorine atoms. (15)

hydrofluoro-olefin (HFO). A refrigerant that contains olefins (hydrocarbons with at least one double bond) and have zero ozone depletion potential (ODP) and low global warming potential (GWP). (15)

hydrostatic expansion. The expanding of a liquid when heat is added. (14, 17)

I

Ice Melting Equivalent (IME). The amount of heat absorbed in melting one ton (2000 lb) of ice at 32°F (0°C) in exactly 24 hours. (11)

improper fraction. A fraction in which the numerator is the same value or larger than the denominator. (10)

incomplete combustion. Partial burning of fuel that produces substances dangerous to human health. (28)

indoor coil. On a heat pump, the refrigerant-circulating heat exchanger coil that is located within the building's air handler/furnace. An indoor coil can function as a condenser (for the heating cycle) or an evaporator (for the cooling cycle). (30)

induced magnetism. Magnetism that is generated by an electric current. An example occurs in an electric motor. The moving magnetic field of the stator induces voltage in the rotor. This voltage causes an electric current, which generates a magnetic field (induced magnetism) in the rotor. (24)

induced voltage. Voltage that is produced in a conductor by a magnetic field. (23)

inert. Chemically inactive. (9)

in. Hg. The abbreviation used to indicate inches of mercury in vacuum measurements. *Hg* is the chemical symbol for mercury, which is derived from the Latin term *hydrargyrum* meaning *liquid silver*. (12)

initiate. To begin an action or sequence of events (such as a defrost cycle). (27)

inorganic refrigerants. Refrigerants that are considered "expendable," where the vapor is not recovered for reuse. The liquid is vaporized and released to the atmosphere. (15)

inside diameter (ID). The distance across the inside of a pipe or piece of tubing. Abbreviated *ID*. (6)

installer. A technician whose primary function is the installation of new systems and components in the HVACR industry. (1)

insulated flexible duct. Flexible ducting formed from plastic sheeting with a spiral wire stiffener to maintain its round cross-section. An insulation layer is applied to minimize heat transfer through the plastic sheet material. This type of duct material may be used for main trunk lines or branch ducts. It is available in a variety of sizes, insulation thickness, and specifications. (31)

insulation. A nonconductive covering for electrical wires, intended to prevent shocks, energy loss, and short circuits. (23)

insulator. A substance that is a poor heat conductor. Examples of insulators are cork, wood, fiberglass, mineral wool, and polyurethane foams. Electrical insulators are materials that have very few free electrons and exhibit high resistance to electron flow. (11, 22)

integral horsepower. A term referring to motors rated at one or more horsepower. (24)

integrated control module. In gas furnaces, a printed circuit board with various control devices, including built-in timers that turn on the blower after the main burner has been lit. These devices are used in newer gas furnaces and are replacing combination fan/limit switches. Integrated control modules control blower operation based on time, while combination fan/limit operate a blower based on temperature. (28)

intensity. Strength, such as that of the magnetic field measured by a clamp-on ammeter. (22)

intermittent pilot ignition system. A type of gas furnace ignition system that lights a pilot flame just before lighting the main burners. It conserves fuel by only having a pilot flame on intermittently, when it is needed to light the main burners. (28)

internally equalized. A type of TEV in which a special passage built into the valve body transfers evaporator pressure to the underside of the valve diaphragm. (18)

International Thread. A standardized metric thread used in most parts of the world. (5)

interstage pressure. The pressure of gas when leaving the first stage of a two-stage compressor. (21)

J

jaw-type pipe wrench. A pipe wrench with serrated jaws to grip pipe. (7)

jumper wire. A short wire installed to complete a connection from terminal to terminal. Often used to bypass a device for a given purpose, such as in troubleshooting a system. (25)

K

Kelvin scale. The temperature scale using Celsius divisions, with absolute zero at 0K. The freezing point of water is 273K, and the boiling point of water is 373K. Note that temperature values expressed in Kelvin do *not* use the degree symbol (°) as used with temperature values in other scales. (12)

kilojoule (kJ). In refrigeration work, the unit used to measure quantities of heat. (11)

L

ladder diagram. An electrical diagram that shows the logic of system operation. The power supply lines are drawn vertically parallel to form the side rails of a ladder, and electrical loads are drawn horizontally between the power source lines, forming rungs. (1, 23)

laminated. Assembled in layers. (24)

latent heat. Heat energy that causes a change of state, but no temperature change. (11)

latent heat of condensation. The amount of heat removed that changes a gas (or vapor) into a liquid but does not change the substance's temperature. (11)

latent heat of freezing. The amount of heat removed that changes a liquid to a solid but does not change the substance's temperature. Also called *latent heat of fusion*. (11)

latent heat of melting. The amount of heat added to change a solid to a liquid but does not change the substance's temperature. (11)

latent heat of sublimation. The amount of heat added to change a substance directly from a solid to a vapor (without passing through the liquid phase) but does not change the substance's temperature. (11)

latent heat of vaporization. The amount of heat added to change a liquid to a vapor but does not change the substance's temperature. Also called *latent heat of evaporation*. (11)

latent heat value. A value used in selecting a refrigerant, equal to the number of Btus absorbed for each pound of liquid evaporated. (15)

lead end. The end of a motor where the electrical connections are located. Opposite of the shaft end. (24)

lever-type tubing bender. A tool calibrated to allow accurate short-radius bends of up to 180° of copper tubing using a lever action. (6)

limit switch. A safety device that actuates when a set limit is reached. In a gas furnace, a limit switch cuts off the power supply if the furnace becomes overheated (based on its temperature limit). A gas furnace limit switch may be housed in a single unit with a fan switch (combination fan/limit switch), or it may be found on an integrated control module. In an electric furnace, a limit switch uses a normally closed, bimetallic disc enclosed in a metal housing that is mounted on the frame near the heating elements. It de-energizes heating elements if the temperature exceeds the switch's setting. (28, 29)

line voltage. The power supply. Line voltage terminals are designated by the letter *L* and a number, such as L_1, L_2, and L_3. (22)

linear measure. The distance from one point to another in a dimension. (10)

liquid charging. Charging a system with refrigerant in a liquid state. (17)

liquid injection valve. See *desuperheating TEV*. (19)

liquid line. The copper tubing connecting the condenser outlet (or liquid receiver) to the refrigerant metering device. (13)

liquid receiver. Installed in larger systems, this is an accessory that serves as a storage tank for excess liquid refrigerant. (14)

liquid receiver service valve (LRSV). A service valve located on a liquid receiver, most commonly found between the liquid line and the liquid receiver outlet that allows the technician to take pressure readings and to control refrigerant flow. See *king valve* and *queen valve*. (14)

liquidus. The point where a filler metal melts and flows as a liquid in soldering and brazing. (8)

load. See *electrical load*, *evaporator load*, *false load*, and *heat load*, *low evaporator load*, *single-phase load*, and *three-phase load*. (19)

load center. Distribution panel that supplies electrical power to several branch circuits. (23)

load rating. A circuit protection device's current carrying-capacity during normal operating conditions. (23)

locked rotor amperage (LRA). High current flow that occurs when an electric motor starts. (24)

lockout. The process of locking the main electrical switch in the open position to safely perform equipment maintenance. (22)

lockout/tagout (LOTO). A safety procedure that uses a lock and tag to isolate or shutoff sources of electricity and communicate that someone is working on the equipment. (3)

logging. The collection of refrigeration lubricant (oil) remaining in the evaporator due to low refrigerant velocity. If too much lubricant remains in the evaporator, it can result in a lack of lubrication in the compressor. (26)

low-ambient thermostat. A heat pump control thermostat located outdoors that opens its NC contacts to stop compressor operation when outdoor temperature is so low that the system would run more efficiently on auxiliary heat than as a heat pump. (30)

lowest terms. The reduced terms of a fraction when the numerator and the denominator are changed to the smallest possible numbers without changing the value of the fraction. (10)

low evaporator load. A condition in which the heat load cannot reach the evaporator surface. (18)

low-loss fitting. A hose fitting required by the Environmental Protection Agency in the use of recovery units. Low-loss fittings prevent loss of refrigerant to the atmosphere when connecting or disconnecting hoses from the recovery unit to a refrigeration system. (16)

low-pressure motor control. A device used on commercial systems to control power supply to the contactor coil. The control responds to low-side (suction) pressure at the compressor. (26)

low-pressure safety control. A pressure-sensitive switch that stops the compressor if suction pressure falls below its setpoint. (29)

low-pressure side. The part of a refrigeration system that is under low pressure during operation. The low side begins at the outlet of the refrigerant metering device and ends inside the compressor. (13)

low-side test pressure. The upper pressure limit that may be exerted on the low side of a refrigeration system during leak testing. (17)

low-temperature applications. A term that refers to refrigeration systems for use as frozen food cases, walk-in freezers, and ice cream storage. (15)

lubrication. The process of providing oil or other friction-reducing material to critical points of mechanical devices to lessen wear and prevent overheating. (21)

M

machine screw. A small bolt with a screw-type head. (5)

magnetic field. Lines of magnetic flux generated by a magnet or an electric current. (23)

magnetism. The ability of a natural material or a field of force generated by an electric current to attract certain metals. (25)

major diameter. The widest measurement from the outside edges of the screw threads. (5)

malleable fitting. A pipe fitting that is made of cast iron that has been annealed (softened by heating.) (7)

mandrel. A nail-like pin designed to break off a blind rivet when the rivet head is formed to fasten two pieces of sheet metal together. (5)

manifold gauge. A pressure-checking instrument that has both compound and high-pressure gauges, control valves, and hose connectors from the service valves. (4, 12)

manometer. An instrument used to measure air pressure in inches of water column (in. WC). (4)

markup. See *profit margin*. (10)

masonry anchor. An expanding device that accepts a fastener for mounting objects on a masonry surface. (5)

mastic. A nontoxic adhesive-material used for permanently sealing joints and seams for all types of air duct materials. (31)

mechanical skill. The ability to work well with one's hands. (1)

medium-temperature applications. A term that refers to refrigeration systems for use as domestic refrigerators, dehumidifiers, commercial walk-in coolers, watercoolers, dairy cases, meat cases, and produce cases. (15)

melting point. See *solidus*. (8)

melting temperature. The temperature at which an alloy starts to melt (solidus). (8)

mercury barometer. A measuring device for atmospheric pressure. It consists of a hollow glass tube that is sealed on one end and open at the other. The glass tube is filled with mercury, turned upside down, and the open end placed in a dish half-filled with mercury. (12)

metering orifice. A refrigerant metering device that operates on the principle of restriction much like a capillary tube. In heat pumps, a metering orifice contains a piston that allows free flow of refrigerant in one direction and restricted flow of refrigerant in the other direction. Heat pumps using metering orifices as refrigerant metering devices have one installed on the indoor coil and another on the outdoor coil. (30)

microfarad. A measuring unit for the amount of energy a capacitor can store. Microfarad ratings are used for both start and run capacitors. This unit may be represented by *mfd* or *μfd*. (24)

migrate. Travel from one place to another. (18)

mineral oil (MO). A refrigeration lubricant made by refining crude oil. MO is used in CFC and HCFC systems, since it is *not* compatible with HFC refrigerants. (21)

minor diameter. The smallest measurement from the inside edges of the screw threads. (5)

miscibility. An attribute describing a substance's capability to mix with another substance. (21)

mixed decimal number. A decimal number with digits before and after the decimal point. (10)

mixed number. A figure expressed as a whole number and a fraction used together. (10)

modulates. Adjusts to changing conditions. (18)

moisture indicator. A disc or other element that is highly sensitive to moisture, gradually changing color to reflect the moisture content in the refrigerant. These are often found in a sight glass along the liquid line. (14)

molecule. A group of atoms bonded together, representing the smallest unit in a chemical compound that can take part in a chemical reaction. (11)

Molly bolt. An expanding metal fastener used to anchor relatively light objects to hollow walls. (5)

motor overload protector. A device that protects the motor windings from damage caused by overheating due to overloaded conditions or poor ventilation. Often simply called *overload*. (24)

motor vehicle air conditioning (MVAC). A category of work involving repairs, leak testing, and "topping off" air conditioning systems low on refrigerant, as well as other vehicle repairs that require dismantling the air conditioner. This work is regulated by Section 609 of the Clean Air Act and requires all persons who perform such services to be certified. (16)

mullion heaters. Resistance heaters that prevent condensation of moisture around refrigerator doors during periods of high humidity. (27)

multiflame tip. A brazing tip that produces several small flames, which tend to wrap around the tubing being heated. (9)

multimeter. An electrical test meter that performs multiple functions, such as the measurement of voltage, resistance, current, and other variables depending on circumstances. (22)

multiplexed. The use of several components connected in parallel in a refrigeration system. This setup is commonly seen with several evaporators connected to a single suction line and a single compressor or group of compressors. Evaporator pressure regulator valves make it possible to operate each evaporator at a different temperature. (19)

multiplexed compressors. Several compressors connected in parallel and operating independently or simultaneously as controlled based on system capacity. (19)

multiposition furnace. A type of furnace that is designed to be used in upflow, downflow, or horizontal applications just by making a few modifications to the furnace according to the installation instructions. An alternative to *multiposition* is *multipoise*. (31)

multi-speed motor. A motor that can be operated at different speeds, such as an air conditioner fan with high, medium, and low settings. (24)

mushroomed. A term used to describe the head of a chisel or other tool that is flattened and spread out from being struck. (4)

N

naphthenic. A term describing mineral oils used in refrigeration systems. Naphthenic oils are not compatible with most of the new HFC refrigerants developed to lessen ozone depletion. (21)

National Electrical Code (NEC). A standard for electrical wiring practices and materials. (22)

natural convection. A term describing heating and cooling systems that use the natural movement of air in which lighter warm air rises and heavier cold air falls. (11)

NC. See *normally closed*. (19)

negative criticism. The act of unfairly criticizing or issuing a harsh judgment. (2)

net oil pressure. The pressure added to oil by the pump (not including crankcase pressure). (21, 26)

neutral flame. An oxyacetylene flame that results from supplying equal amounts of oxygen and acetylene. (9)

neutron. An atomic particle in an atom's nucleus that has no charge. (22)

nipple. A short length of pipe that is threaded on each end and used to connect fittings that are close together. (7)

NO. See *normally open*. (19)

nominal size. A pipe measurement term based on the insider diameter or bore of the pipe. Used to measure steel and plastic pipe. (7)

nonadjustable head pressure control valve. A three-way modulating valve that controls head pressure and is controlled by discharge pressure and low ambient temperature. These are commonly used to keep head pressure from falling too low during low ambient temperature. (19)

noncondensable. A foreign substance in a refrigeration system that remains a vapor and cannot be condensed. This substance can be responsible for high head pressure and other system problems. (14)

normally closed (NC). A term describing a switch, contact, or valve that is in a closed state until energized or actuated. Abbreviated as *NC*. (19, 22)

normally open (NO). A term describing a switch, contact, or valve that is in an open state until energized or actuated. Abbreviated as *NO*. (19, 22)

nucleus. The center of an atom. (22)

number. A figure or word that denotes quantity. (10)

numerator. The top number of a fraction, indicating the number of parts available. (10)

nut. An internally threaded piece of metal hardware, usually square or hexagonal in shape, that is used in conjunction with a bolt for a mechanical connection. (5)

O

Occupational Safety and Health Administration (OSHA). A government organization—created as part of the Occupational Safety and Health Act of 1970—responsible for worker safety and health. (3)

off cycle. The time when an HVACR system's compressor is *not* operating. (13)

off-cycle defrost. A defrost method in which the evaporator fan runs continuously while the rest of the system is off, so circulating air melts accumulated frost from the evaporator. (27)

offset. A short-distance change in the direction of tubing, commonly made by using street elbow fittings. (6)

ohm. The unit of electrical resistance. It is signified with an omega symbol (Ω). (22)

ohmmeter. A device used to accurately check resistance. (22)

Ohm's Law. A summary of the exact relationships among voltage, amperage, and resistance. It states that a circuit's amperage (I) times its resistance (R) equals voltage (E). Mathematically, it appears as $E = I \times R$. Ohm's Law is used to predict and control the activity of electricity. (22)

oil passages. Lubrication holes drilled in the crankshaft, connecting rods, and other components of a pump or compressor. (21)

oil pressure safety control. A control that stops the compressor if oil pressure falls below safe limits. (21, 26)

oil pump (cooling system). In a refrigeration system, a device operated by a reciprocating compressor's crankshaft that delivers oil through special oil passages drilled in the crankshaft, connecting rods, and other components to lubricate the compressor. (21)

oil pump discharge pressure. The pressure at an oil pump's discharge. This value can be measured on an oil pump's Schrader valve. Oil pump discharge pressure is composed of crankcase pressure and net oil pressure combined together. (21)

on cycle. The time when an HVACR system's compressor is operating. (13)

open capacitor. A defective capacitor in which an electrical connection is broken, usually when one of the terminals becomes separated from its coil. (24)

open circuit. A lack of continuity in an electrical circuit or a circuit with a broken path. Electrons cannot flow in an open circuit. An open circuit may result from an open electrical contact (switch, disconnect, circuit breaker), a blown fuse, a disconnected/broken wire, or any number of other causes. (22)

open-type compressor. A compressor that is driven by an external source of power, usually an electric motor. The motor may be directly connected to the compressor, or it may use a V-belt to transmit the turning power. (20)

open winding. A condition resulting from a broken wire in the motor winding that is not touching the motor shell. This can be detected by measuring infinity (OL) with ohmmeter across a motor's winding terminals. (24)

outdoor coil. On a heat pump, the refrigerant circulating heat exchanger coil that is located outdoors. An outdoor coil can function as a condenser (for the cooling cycle) or an evaporator (for the heating cycle). (30)

outdoor defrost thermostat. A heat pump control thermostat used to control the defrost cycle. Its contacts close at a certain low outdoor coil temperature to allow defrost to begin and then open at a certain high outdoor coil temperature to terminate the defrost cycle. (30)

outdoor fan and defrost relay. A heat pump component that initiates the defrost cycle by opening one set of NC contacts to turn off the outdoor fan and closes two sets of NO contacts to turn on auxiliary heat and energize the reversing valve for cooling cycle refrigerant flow. (30)

outdoor thermostat. In a two-stage heating system, a temperature-sensitive control device mounted outdoors that is used in conjunction with a regular (single-stage) indoor thermostat to initiate the second stage of heating. In a heat pump, a thermostat that prevents auxiliary heat from turning on unless outdoor temperature reaches a low setpoint. (29)

outside diameter (OD). The distance across the outside of a pipe or piece of tubing. Abbreviated *OD*. (6)

overcharge. An excess of refrigerant in a refrigeration system. (14)

overcurrent. A situation in which there is current in excess of the normal flow for a given circuit. Too much current flows through a wire and damages the insulation. Overcurrent can be an overload (more current than normal) or a short circuit (very high current). (23)

overload. Electrical current in excess of the normal flow for a given circuit. Also, the term *overload* may be used to refer to the electrical safety device designed to protect a compressor's motor from an overload. See *motor overload protector*. (18, 23)

oversized. A term used to describe the design of refrigeration systems. It refers to refrigeration systems that have more capacity than the minimum requirement needed to permit the system to lower the temperature to a certain point and then shut off. (26)

oversizing. Specifying a transformer with a secondary that is able to carry more than the minimum required amperage for an application, so it can accommodate the load(s) connected to it. (23)

oxidizing flame. An oxyacetylene flame that results from supplying excess oxygen. (9)

oxyacetylene torch. A torch burning a mixture of oxygen and acetylene as a heat source for brazing, welding, and cutting. (9)

oxygen regulator. An adjustable device used to maintain the cylinder pressure and the working pressure that flows to the torch. (9)

ozone depletion potential (ODP). The potential all chlorofluorocarbon refrigerants have to destroy the protective ozone layer in the stratosphere. An ODP number is assigned to each refrigerant indicating the risk to the ozone layer caused by the chlorine in refrigerants. (15)

P

parallel. Two electrical loads along separate, multiple paths that are connected to the same input and output. Rather than being connected in series (one after another), parallel connections follow the same route along separate but equal pathways, like lanes on a highway. Components in parallel have the same voltage applied, though each component's current will depend on its resistance. (22)

partial vacuum. Any pressure less than atmospheric but higher than a perfect vacuum. (12)

passive recovery. A system-dependent method of recovering refrigerant from a mechanical refrigeration system in which the compressor pumps the refrigerant into a recovery cylinder. This method must be used on systems with less than 15 pounds of refrigerant. (16)

people skills. The ability to work well with people. (2)

perfect vacuum. A theoretical state in which all atmospheric pressure is removed. (12)

perimeter system. A type of forced-air distribution system that uses feeder ducts to pressurize a loop of forced air around the outer perimeter of a house. The perimeter loop has supply outlets in rooms along the outer walls of the building. (31)

permanent split-capacitor (PSC) motor. A split-phase motor with a capacitor in series with its start winding to increase its running torque. (24)

permeation. The process of gas seeping through the walls of refrigeration hoses. (17)

personal protective equipment (PPE). Equipment worn to minimize exposure to hazards that can cause injuries or illnesses. (3)

phase. A conducting loop in a generator. In a three-phase arrangement, each phase is 120° out of phase with each other. (22)

physical states. The three forms in which most substances can exist — solid, liquid, or gas. (11)

pickup voltage. The point at which back-emf (counter-emf) is sufficiently high to allow energizing of the relay coil. (25)

pictorial diagram. A system diagram that shows the physical appearance, component locations, wiring connections, and internal arrangement of a piece of equipment. (1, 25)

pilot assembly. A gas furnace component consisting of a pilot burner, a spark electrode, and a sensing probe. Pilot assemblies are used in intermittent ignition systems. (28)

pilot flame. A small gas flame used to ignite the flame on the main burners of some gas furnaces. (28)

pilot valve. A small valve whose action determines the action of a larger valve or device. For example, a heat pump's reversing valve uses a solenoid-operated pilot valve that directs high pressure refrigerant to one end of the reversing valve body and low pressure refrigerant to the other end. This action determines the position of the sliding valve's piston within the reversing valve, which determines the flow of refrigerant through the reversing valve. (30)

pipe. A term used to describe conduits with thick-walled materials, such as steel or plastic. (6)

pipe compound. Paste material normally brushed onto the threads of pipe and fittings before assembly to ensure a strong, leakproof seal. Sometimes referred to as *pipe dope* by plumbers. (7)

pipe cutter. A tool that clamps around the outside of a pipe and uses a wheel blade to rotate and cut around the pipe. Depending on the wheel blade, these may be used on steel pipe or plastic pipe. (7)

pipe die. A special tool used to cut threads on the outside of steel pipe. (7)

pipe fitting. A component used to join sections of pipe. (7)

pipe reamer. A tool used to remove burrs from the inside of cut ends of pipe. (7)

pipe thread. Thread that is tapered 1/16″ for every inch of length. (7)

pipe vise. A tool used to securely hold steel pipe in position for cutting and threading. (7)

piston. The device that moves up and down inside a reciprocating compressor's cylinder to compress refrigerant gas. (20)

piston rings. Sealing rings that fill the gap between a piston and the compressor cylinder wall. Piston rings are typically used only on compressors rated at 10 horsepower or higher. (20)

place value. In mathematics, the special position occupied by a digit located to the left or right of a decimal point. (10)

plastic range. The range of temperatures between the solidus and liquidus points of a solder. (8)

plenum. The central collecting chamber for conditioned air leaving a furnace or air-conditioning unit. Trunk or branch line ducts are connected to the plenum. Also called a *bonnet*. (31)

pointer flutter. Pressure pulsations that cause a pressure gauge's pointer to swing above and below the actual pressure reading. (12)

points. Teeth around the opening of a box-end or socket wrench that grip the edges of the nut or bolt head. (4)

pole. An electrical switch component used to open or close contacts. (22)

poles. One or more pairs of stationary electromagnets, positioned at opposite sides of a circle inside a motor. The two poles in each pair have opposite magnetic polarity. (24)

polyol ester oils (POEs). A family of synthetic lubricants that are compatible with all refrigerants (CFCs, HCFCs, and HFCs) and are miscible with mineral oil (MO) and alkylbenzene (AB) lubricants. (21)

polyvinyl chloride. See *PVC.* (7)

positive temperature coefficient (PTC) thermistor. A thermistor that reacts rapidly to an increasing temperature change by greatly increasing its resistance to current flow. (25)

potential difference. The imbalance of electrons that exists when the electrical charge (positive versus negative) between two points is not in balance. Sometimes called *voltage, electromotive force (emf),* or *electrical pressure.* (22)

potential relay. A relay that operates on the back-emf (counter-emf or voltage or potential) that is generated by the start winding. Also called a *voltage relay.* (25)

power factor. A measure of transformer capacity. This can be computed by multiplying VA × 1.73 and converting the answer to kVA. (23)

pressure control. A device that functions as a switch, turning a refrigeration system on and off as a result of changing low-side refrigerant pressure to maintain temperatures within the desired range. (17)

pressure drop. A decrease of pressure in a tubing or piping system. These restrict flow and result from a variety of causes. (13, 18)

pressure-operated switch. A switch that responds to rising or falling system pressures. A pressure-operated switch is often used as the primary control of a system. (26)

pressure regulator. A device used to reduce cylinder pressure to working pressure. (8)

primary air. Combustion air that is drawn into a furnace's gas burner by the jet of gas entering the burner. Primary air is mixed thoroughly with gas inside the burner before delivery to the ports. (28)

primary control. In air-conditioning and refrigeration systems, a term referring to the control that is used to start and stop the condensing unit when certain temperatures or pressures are reached. (26)

primary winding. A transformer winding through which the incoming current flows. (23)

process stub. A short copper tube in a refrigeration system to which a Schrader valve can be brazed. (14)

professionalism. The act of being respectful, courteous, and carrying oneself in a businesslike manner. (2)

proper fraction. A fraction in which the numerator is smaller than the denominator. (10)

proton. An atomic particle in an atom's nucleus that has a positive charge (+). (22)

psi. Pounds per square inch. A measure of pressure. (6)

psia. Pounds per square inch absolute, used for pressure measurements. Readings expressed in psia are 14.7 psi higher than readings expressed in psig (pounds per square inch gauge). See *absolute pressure.* (12)

psig. Pressure expressed in pounds per square inch gauge. See *gauge pressure.* (12)

psychrometer. An instrument used to measure air-related variables such as air temperature, relative humidity, and dew point. (4)

P/T card. A special pocket-size pressure-temperature chart showing the pressure and temperature relationship for several refrigerants. A small portable saturation chart for various refrigerants. Also referred to as a *saturation chart.* (15)

PTC relay. A relay that uses a positive temperature coefficient (PTC) thermistor as a sensor and switching device. PTC relays are commonly used in hard start kits. (25)

P-trap. A special tube or pipe fitting named for the shape into which it is bent. P-traps are commonly used as drain fittings in plumbing. (6)

pull a vacuum. To lower the pressure in a refrigeration system below atmospheric pressure. This is done using a vacuum pump before a system is charged with refrigerant. (12)

pumpdown. To *pumpdown the system* is the process of removing all refrigerant from the low-pressure side and pumping it over to the high-pressure side of the system to capture the majority of the refrigerant charge in the liquid receiver. (14, 16)

pumpdown cycle. A system operation in which all refrigerant is removed from the low-pressure side of the system following each run cycle to prevent liquid migration to the compressor during the off cycle, avoid high suction pressure at startup, prevent liquid slugging, and providing better oil return flow to the compressor crankcase. (19, 26)

purging. The necessary release of small amounts of refrigerant from the hose ends after connection to a system. This process removes atmospheric moisture and air from the service hoses, which prevents contamination of the system. (16)

push-on pipe fitting. A pipe fitting that is attached merely by pushing into place. These types of fittings are used on potable water, hydronic hot water heating systems, and condensate drain lines. (7)

PVC. A type of plastic used to make pipe and fittings that are used for cold water supply or drain lines. *PVC* stands for *polyvinyl chloride.* (7)

Q

quotient. A result derived from the division of one number by another. (10)

R

radial system. An air distribution ducting system that uses round pipe runs that extend from the supply plenum to each register. (31)

radial. In a circular pattern. (21)

radiation. The transfer of heat by waves similar to light waves or radio waves. (11)

radius. The length of a line from the center of a circle to its perimeter (half the diameter). (10)

Rankine scale. A temperature scale using Fahrenheit divisions, with absolute zero at 0°R. The freezing point of water is 492°R, and the boiling point of water is 672°R. (12)

ratchet tubing bender. A tubing bender that bends copper tubing using a ratcheting action. (6)

reactance. A form of resistance found in situations involving devices that produce a voltage (emf) or store voltage, either of which acting in direct opposition to the supply voltage. (22)

reamed. A description of tubing or pipe that has been scraped with a tool to remove burrs from a cut surface. (6)

recalibration. Periodic adjustment of a gauge or meter to ensure accuracy. (12)

receiver pressure regulator. A differential valve that bypasses hot gas refrigerant from the compressor discharge line into the liquid receiver to maintain a set minimum pressure in the liquid receiver. This valve is used in conjunction with an adjustable head pressure control valve to deal with low head pressure during low ambient temperature. Sometimes simply called *differential valve*. (19)

receptacle. An electrical outlet that allows easy connection and disconnection of electrical equipment. Sometimes called *duplexes*. (22)

reciprocating. A term used to describe a motion in one direction and then in the opposite direction, such as back and forth or up and down. Many HVACR compressors operate using a reciprocating motion to draw in and then compress refrigerant. (13)

reciprocating compressor. A compressor that uses an electric motor to turn a crankshaft that operates reciprocating pistons within cylinders to draw in and compress refrigerant. These operate much like an automobile engine. Reciprocating compressors are one of the most widely used types of compressors in HVACR. (20)

reclaim. To restore a recovered refrigerant to a level equal to new product specifications as determined by chemical analysis. A refrigerant must be reclaimed if it is to change ownership. (16)

recover. To remove refrigerant in any condition from a system and store it in an approved cylinder or external container. (16)

recovery cylinder. A refillable, heavy-duty, certified pressure vessel used in the recovery of refrigerant. Recovery cylinders must meet Department of Transportation (DOT) specifications. (16)

recycle. To clean recovered refrigerant for reuse by removing moisture and contaminants. Recycling is accomplished by repeatedly passing the recovered refrigerant through one or more filter-driers. (16)

red iron oxide. Rust that forms when oxygen from the air and moisture combine. (17)

reducing. The process of changing a fraction to its lowest terms by lowering the numerator and the denominator to their smallest possible numbers without changing the value of the fraction. (10)

reducing coupling. A fitting that has openings of different sizes, allowing a larger-diameter piece of tubing or piping to be connected to tubing or piping with a smaller diameter. (6)

reducing extended plenum. An air duct system similar to the extended plenum system (using runs of rectangular duct to extend the plenum and runs of round duct to reach individual supply air registers); however, part of the rectangular extended plenum is reduced to a smaller size, which helps to keep duct air pressure constant throughout its length for even airflow at the end of long plenum runs. (31)

reference. An individual who will provide important information about you to a prospective employer. A reference can be a teacher, school official, previous employer, or any other adult outside your family who knows you well. (1)

refrigerant. A substance that readily changes from a liquid to a gas, then is condensed back to a liquid for recirculation. In refrigeration systems, refrigerants are the vehicles that transport heat from one location to another. (15)

refrigerant distributor. On larger systems, a device that directs liquid refrigerant through the different circuits of an evaporator. The refrigerant distributor receives refrigerant directly from the thermostatic expansion valve and divides it equally among all evaporator circuits. (18)

refrigerant metering device. A device that meters the flow of liquid refrigerant into the evaporator. It divides the high side from the low side. (13, 18)

refrigeration. The process of removing unwanted heat and carrying it away from the space. (11)

register. A supply air outlet that has an adjustable damper or vanes. (31)

register boot. A device used to make the transition from round pipe or flexible duct to a supply air register. Also called a *boot* or *register box*. (31)

relative humidity. The amount of moisture in the air compared to the amount of moisture the air would hold if the air was saturated (filled to its capacity of moisture). The ratio of actual water vapor content to the maximum possible water vapor content is expressed as a percentage. (27)

relay. An electrically operated switch. A relay can be used to automatically disconnect the start winding and/or the start capacitor of a motor or control a variety of devices. Relays often use low voltage control wiring to switch higher voltage power wiring. (25)

remainder. In mathematics, the number left over in addition to the quotient when the division of one number by another yields a mixed number. (10)

resistance. In electricity, anything opposing (working against) the movement of electrons. (22)

resistive. A type of resistance that remains constant (does not change) whether current is applied or not. (22)

restrict. To limit or hold back. To restrict refrigerant flow decreases refrigerant flow. (6)

résumé. A document that briefly outlines your education, work experience, and other qualifications for work. (1)

return air. Air from conditioned space that is brought back to the furnace to be heated or cooled again. Sometimes referred to as *air on*. (31)

reverse-cycle defrost. A defrost method that reverses the evaporator fan's direction of rotation, so that warm store air is circulated in the refrigerated case and used to defrost the evaporators. (27)

reversing valve. A four-way valve that is used to control the direction of refrigerant flow in a heat pump. It usually consists of a solenoid-operated pilot valve and a larger valve body containing a piston-actuated sliding valve that connects to the four refrigerant tubes. A solenoid-operated pilot valve directs high pressure refrigerant to one end of the reversing valve body and low pressure refrigerant to the other end. This action determines the position of the sliding valve's piston within the reversing valve, which determines the flow of refrigerant through the reversing valve to and from the four refrigerant tubes. (30)

revolutions per minute (rpm). A unit used to measure the speed of motors and other rotating devices. (24)

riser. A vertical section of the tubing or piping. (26)

root. The bottom of the "V" formed by adjoining screw threads. (5)

rotary compressor. A compressor that uses a rotor that interacts with rotating or stationary blades to compress gas. These are classified as either *rotating vane* or *stationary blade*. (20)

rotation. Circular movement around a center point (for example, the turning of a motor shaft). (24)

rotor. Rotating part of a motor. (24)

rough measurement. A distance estimate that is simpler to make, but less accurate, than a measurement made with a ruler or tape measure. (10)

R&R system. A system that houses recovery and recycle units for refrigerants and performs both operations. (16)

Rule of 10. The principle of temperature difference maintained in the evaporator of a refrigeration system. While a refrigeration system is running, the refrigerant in the evaporator is about 10°F (5.5°C) colder than the evaporator tubing, and the evaporator tubing is about 10°F (5.5°C) colder than the air. (26)

run capacitor. A capacitor used on single-phase motors to create better running torque. (24)

run winding. The stator coil windings in a motor. (24)

S

safety data sheet (SDS). A document provided by chemical manufacturers, distributors, and importers that communicates safe handling information on hazardous chemicals. The information is broken down into 16 sections. (3)

safety ground. The equipment-grounding conductor that connects to *earth ground*. This wire is often green, green with yellow stripes, or bare copper. (22)

saturated. Full. Regarding air and humidity, this term may describe a substance that is holding as much moisture as possible. Regarding refrigerant or another substance holding as much heat as possible before it changes physical states (by vaporizing or condensing). (27)

saturated condition. A term that refers to the boiling/condensing point of a substance, which is dictated by a specific combination of temperature and pressure. (11)

saturation. A condition in which a refrigerant can change its state and exist as a liquid, a mixture of liquid and vapor, or a vapor, at the same pressure and temperature. (15)

saturation point. The boiling (or condensing) point of a substance. A substance at its saturation point is at a pressure-temperature condition such that it is ready to change its state. (11)

schematic. An orderly method of presenting electrical circuits and components. A schematic diagram shows the types of components involved and how they are connected in the circuit. (22)

schematic diagram. See *schematic*. (23)

Schrader valve. A pin-action, stem-type access valve installed in a system for service access. The valve is only opened when its pin is depressed. (14)

screw. A threaded fastener used to hold parts securely while providing a means for the parts to be readily removed. (5)

screw thread. A helical ridge of uniform section formed inside of a hole (such as a nut) or on the outside of a fastener (such as a screw or bolt). (5)

scrim. The part of an insulated flexible duct that is made of a high-tensile-strength, fiberglass mesh that provides protection for the insulation and inner core of the duct. (31)

scroll compressor. A type of compressor that involves the orbiting of one scroll (spiral) that is intertwined with another scroll to form a series of crescent-shaped pockets that reduce in size and increase in pressure. During compression, one scroll remains stationary while the other scroll orbits around the space between it and the fixed scroll. (20)

secondary air. Combustion air that is drawn into a gas furnace's gas burner chamber after ignition of the flame to help support complete combustion. (28)

secondary control. A safety device used to shut off a system or turn on a safety mechanism if certain conditions exist. (26)

secondary winding. A transformer winding in which an electrical current is induced by the generated magnetic field in the core. (23)

Second Law of Thermodynamics. Heat always travels from hot to cold. (11)

semihermetic compressor. A compressor that contains an electric motor and compressor that can be accessed by opening the casing. It is a field service ready version of a hermetic compressor, combining the benefits of the hermetic compressors. (20)

sensible heat. Heat that causes a change in temperature, but not a change of state. (11)

sensing bulb. A device filled with a volatile fluid that is used to sense heat. A thermostatic expansion valve's sensing bulb is placed on the suction line at the evaporator outlet and controls the operation of the TEV. (18)

sensing element. A component of a thermostat that consists of a bellows (movable part), a capillary tube (connecting link), and a sensing bulb that holds a refrigerant. The location of the sensing bulb determines what temperature is being controlled. (26)

sequencer. In an electric furnace, a control device used to stage (turn on in sequence) electric heating elements. It uses a temperature-responsive bimetallic element for the normally open switch and includes a small heater to activate (close) its switches after a set amount of time. (29)

series. An electrical term for components connected one after the other in a circuit. The same current level flows through each component, though each component's voltage drop depends on its resistance. (22)

series connection. An arrangement in which flow is through one section after another (moving in a series). In refrigeration, an example of a series connection is in certain evaporators in which the refrigerant must travel through one shelf of tubing before going to the next. (13)

series-parallel. Electrical circuit in which some components are connected in series, while others are connected in parallel. (22)

serrated. Having a toothed or ridged edge that keeps material from slipping while being cut. (4)

service technician. A technician whose primary function is the regular repair and maintenance of various HVACR systems. (1)

setpoint. The cut-in or cut-out point of a system control device. (17, 26)

setscrew. A headless fastener used to anchor pulleys and fan blades to motor shafts. (5)

shaded-pole motor. A specially designed, low-cost single-phase motor that is often used to operate small fans. (24)

shading coil. A single closed loop of wire on each pole of a shaded pole motor. The loop produces a tiny magnetic field that is "out of step" with the main poles, aiding in starting the motor. (24)

shaft end. The end of a motor where the rotor shaft exits. Opposite of the lead end. (24)

shank. 1) The shaft of a tool, such as a screwdriver. 2) The unthreaded portion of a tap. (4, 5)

sheet metal fabricator. A technician whose primary function is to shape and manipulate sheet metal into a variety of useful products using hand tools and various machines. (1)

S-hook. The long edges of one end of a section of rectangular sheet metal ductwork. S-hooks are used for joining two sections of rectangular sheet metal ductwork. The plain long edged end of another sheet metal piece is pushed into the S-hooked long edged end of another duct section for joining. (31)

short circuit. A condition involving an unintentional electrical path with very low (or no) resistance. This usually allows an unrestricted current to flow (that is often above the circuit's design amperage). (22)

short cycle. A term describing a condition in which a refrigeration system turns on and off rapidly. (19, 24)

shorted capacitor. A defective capacitor in which the aluminum coils touch, resulting in a short circuit. (24)

shorted to ground. A term describing a device or component (such as a motor) that has developed an internal electrical short to the device's frame (or ground). See *grounded* and *ground fault*. (24)

shorted winding. A motor condition in which two wires on the same winding short out or touch each other to shorten the path of the current flow through that winding. (24)

shoulder nipple. A short nipple that includes a very short section of unthreaded pipe. (9)

sight glass. A small window placed in the liquid line on commercial systems to provide a view of the flowing liquid. Many also include a moisture indicator element. (14)

single-phase load. An electrical load that operates on one phase of electrical power that may be provided either by two hot wires (and a ground) or by a hot wire and a neutral wire. (22)

single-phasing. A condition in which a three-phase motor loses one of its phases but continues to run, only now as a single-phase motor. Single-phasing can damage three-phase motors. (24)

single-pole breaker. A single-switch circuit breaker used to disconnect the hot wire on 120V, single-phase branch circuits. (23)

single-pole, double-throw (SPDT) switch. A switch that has one pole and operates two contacts (one normally open, the other normally closed). (22)

single-pole, single-throw (SPST) switch. A switch with one pole that closes or opens one contact. (22)

slip. The percentage of relative motion of a motor, indicating the difference between synchronous speed and actual motor speed under full load. (24)

sludge. A dark, gummy mass caused by a chemical breakdown of oil that combines with other materials, such as carbon, metals, oxides, or salts. (17)

slugging. A condition in which the compressor must pump liquid, rather than vapor. Liquid slugging can be the pumping of liquid refrigerant or refrigeration lubricant. Slugging can cause severe damage in reciprocating compressors. (21)

socket. An individual, cylinder-shaped box-end tool, used with a handle to perform the same turning tasks as performed by other wrenches. (4)

soft copper tubing. A bendable form of copper tubing that has been annealed. Soft copper tubing is available in three thicknesses: Type K, L, and M. (6)

solder. An alloy used in soldering, usually consisting of tin and another low-melting-point metal, such as lead, that acts as a filler metal between the base metals being joined. (8)

soldering. The process of joining two pieces of base metal using a filler metal alloy and applied heat at a temperature less than 840°F (448°C). The base metals are heated by a torch and the molten filler metal alloy is drawn by capillary action into the very narrow space between base metals. (8)

solenoid valve. A valve that uses electric current to generate electromagnetism that opens or closes the valve. Solenoid valves are widely used in HVACR and other industries for automatic flow control of refrigerant, water, or other gases and liquids. (19, 25)

solidus. The melting point of a filler metal alloy in soldering or brazing. (8)

soluble. Easily dissolved. (17)

solvent cement. A chemical solution used to soften and join plastic pipe and fittings. A different solvent cement is required for each plastic type and should be used only with that material. (7)

solvent-welding. A chemical process used to join plastic pipe. Commonly referred to as *cementing*. (7)

spark igniter. A gas furnace ignition device that produces an electric arc across its electrodes to automatically light a flame. (28)

specific heat. The amount of heat required to raise the temperature of one pound of a substance one degree Fahrenheit. (11)

spinner swage. A type of swaging tool designed to be inserted into a drill and uses high speed to create friction that heats up the copper tubing, which then allows the tubing to be expanded by the tool. (6)

splash system. A lubrication method in which the crank throw dips into the oil and splashes it around inside a reciprocating compressor. (21)

split-phase motor. A single-phase motor that has a start winding installed in the stator to produce a magnetic field out of step with the main winding. (24)

spring bender. A spring designed to be slipped over copper tubing to completely cover the area of the bend and prevent kinking. (6)

square measure. A measurement of two dimensions, length and width. (10)

square unit. Means used to express the surface area of a shape, such as square inches or square centimeters. (10)

squirrel cage rotor. The most common rotor type used in induction motors. Its body consists of a laminated soft iron cylinder enclosing and attached to the motor shaft. (24)

stable. Unchanging; not subject to breaking down. (21)

stage. To cycle on or off in sequence. (26)

standard conditions. Uniform standards for evaluating refrigerants. Standard conditions have been established as 5°F (15°C) evaporating temperature and 86°F (30°C) condensing temperature. (15)

standing pilot. A continuously lit pilot flame in a gas furnace. (28)

standing pilot ignition system. A type of gas furnace ignition system that has a continuously lit pilot flame (standing pilot) that is always ready to light the main burners when the thermostat calls for heat. (28)

standing pressure test. A method of leak detection that involves pressurizing the system and testing the following day to determine if pressure loss has occurred. (17)

start capacitor. A capacitor used on single-phase motors to create better starting torque. (24)

starting collar. A device used to start a round duct branch off the plenum, flat surface, or a rectangular run of ductwork. A hole is cut into the plenum or rectangular duct and the starting collar inserted. Then, tabs are folded inside to secure the collar. (31)

start relay. A relay that is designed and used to start and stop a single-phase motor. A variety of start relays are available, including current relays, potential relays, and PTC relays. (25)

start winding. An additional stator winding that establishes another magnetic field in the stator that is "out-of-step" with the main (run) winding. (24)

starve. A term that describes the depriving an evaporator of adequate refrigerant. This situation can occur by a variety of causes. (18)

stator. Assembly of stationary electromagnets in a motor. (24)

step-down transformer. A transformer with fewer coils in the secondary winding than in the primary winding, designed to decrease (step down) voltage. (23)

step-up transformer. A transformer with more coils in the secondary winding than in the primary winding, designed to increase (step up) voltage. (23)

stinger leg. A higher-voltage wire from a three-phase, delta transformer. Also called a *wild leg*. (23)

stranded. Electrical term used to describe a group of wires twisted together. (22)

strap-type pipe wrench. A pipe wrench that uses a strap tightened around the pipe to avoid tooth marks. (7)

street elbow. An angled fitting with a female end and a male end, frequently used to make offsets in tubing or piping. (6)

struck tool. A tool that is struck with a hammer, such as a punch or cold chisels. (4)

stubs. Short copper-coated steel pipes welded to openings in the housing of a hermetic compressor for attaching tubing or valves. (20)

subcooled. A liquid at a temperature below its saturation point. (11, 13, 15)

sublimate. A substance changing physical states directly from a solid to a gas (without passing through the liquid phase) by the removal of heat. (11)

suction line. Copper tubing that connects the evaporator outlet to the compressor intake. (13)

suction line pressure drop. The pressure drop between an evaporator outlet and a compressor inlet. This should be limited to about 2 psig (14 kPa) if proper line sizing and good piping practice were followed when installing the system. (18)

suction line stub. The short copper-coated steel pipe welded to an opening in the housing of a hermetic compressor for attaching the suction line tubing. (20)

suction service valve (SSV). A service valve located on the low-pressure side of the system between the suction line and compressor. It allows a technician to take low-side pressure readings and to control refrigerant flow to the compressor. (14)

superheat. A term used to describe a gas with a temperature above its saturation point. (13, 15)

superheated vapor. A gas that has been heated to a temperature that is above its boiling point. (11)

supply air. Conditioned air from a furnace or air handler that is blown into a conditioned space. Sometimes referred to as *air off*. (31)

surface area. A square measure found by multiplying length times width. (10)

swaging. Enlarging the diameter of one end of a length of soft copper tubing so the end of another length with the same diameter can be slipped into it. (6)

swing space. Room needed to turn a nut or bolt with a wrench. (4)

switch. A device that stops and starts the flow of electrons to an electrical load. (22)

synchronous speed. The speed of a stator's rotating magnetic field in a motor, which is based on the electrical supply's frequency and the number of poles in the stator winding (represented in the rpm formula). A motor's actual speed is slightly lower than this due to rotor magnetic field lag and the motor's load. (24)

T

tagout. The practice of placing tags on all operator switches to inform others that repairs are in process. (22)

takeoff. A device used to start a rectangular duct branch off the plenum, flat surface, or a rectangular run of ductwork. Takeoffs are used much in the same way as a starting collar. They can be used to route air off any side of the plenum or used as a transition off a rectangular pipe run. Different configurations are used to provide the lowest resistance and the largest selection to fit any application, including top takeoff and side takeoff. (31)

tandem compressors. Two individual compressors with an interconnected housing that replaces the individual end bells. (21)

tap. A tool used to cut threads inside a hole. (5)

tapered. Sloped or steadily reduced in diameter. (7)

taps. On a multispeed single-phase motor, wires or terminals that make it possible to select the number of poles being used. (24)

tare weight. The weight of a container before being filled. (8)

technical training instructor. A teacher that equips students with the knowledge and basic set of skills in a particular technical area. (1)

tee. A fitting used to make a 90° branch while continuing the main run. These fittings are used in tubing, piping, and ductwork and may have a variety of ends for different connections. (6)

telescoping. The process of fitting different sizes of tubing inside one another. (6)

temperature. A measurement that describes the intensity (heat level) of a substance (how hot or cold a substance is). (12)

temperature difference (td). A condition in which a difference of temperature exists between substances. A temperature difference must exist between ambient air and the refrigerant in order to have heat transfer from hot to cold, such as in evaporators and condensers. (13, 17)

temperature glide. The fluctuation of temperature at a given pressure in a zeotropic blend, caused by fractionation between the evaporator inlet and outlet or fractionation between the condenser inlet and outlet. (15)

temperature range. The specific span of temperatures within which controls are designed to operate. (26)

terminal. An electrical connection point on such devices as transformers and motors. (23)

terminal connector. A small metal tip that crimps onto wire ends to make connections simpler and faster. (23)

thermistor. A temperature-sensitive resistor that varies its resistances with temperature. Thermistors are used as control devices and are classified as either PTC (positive temperature coefficient) or NTC (negative temperature coefficient). (25)

thermocouple. A device that converts heat energy to electrical energy. In a gas furnace, it is used as safety device that prevents the main gas valve from opening when it does *not* sense the presence of a pilot flame. When the thermocouple senses a pilot flame, it generates and sends an electrical signal to the gas valve allowing the furnace to ignite the main burners. (28)

thermodynamics. The science that deals with the mechanical action of heat. (11)

thermometer. A device used to measure temperature (heat level). Bulb thermometers usually consist of a glass tube calibrated in degrees. Newer thermometers are often electronic using various sensors and digital displays. (12)

thermostat. A temperature-activated switch that opens and closes its contacts according to temperature changes. An HVAC system thermostat combines a temperature-activated switch with various other controls that govern the operation of the numerous components of a heating and cooling system. (26, 28)

thermostatic expansion valve (TEV). A refrigerant metering device that maintains a constant level of superheat at the evaporator outlet by metering the flow of liquid refrigerant. Also known as a *superheat valve*. (18)

thermostat overshoot. Continued furnace operation after the thermostat is satisfied. This results in conditioned space temperature above the cut-out point. A heat anticipator is used to reduce or compensate for thermostat overshoot. See *heat anticipator*. (28)

thread angle. The V-angle of the screw threads. The sides of the threads normally form an angle of 60°. (5)

threaded rod. A metal rod with its entire length threaded. It can be cut to any length with a hacksaw. (5)

threads per inch (TPI). A measurement determined by counting the number of crests (or roots) per inch of threaded section. (5)

three-phase load. Electrical devices designed for connection to all three hot wires of a three-phase system. (22)

three-phase power generation. Electric power production using a generator that rotates three different conducting loops at the same time. Since most of these use three phases, it is commonly called *three-phrase* or *3-phase*. It is also called *polyphase generation*. (22)

three-pole breaker. A three-switch circuit breaker used to disconnect all three hot wires on a three-phase circuit. (23)

throw. Movement of the pole in an electrical switch. (22)

tinning. A term used to describe the spreading action of solder. (8)

toggle bolt. A threaded fastener with spring-loaded wings that is used to fasten heavier objects to a hollow wall. The wings spread out and bear against the inside of the wall. (5)

ton of refrigeration effect. Heat removal rate equal to 288,000 Btu in 24 hours (or 12,000 Btu per hour or 200 Btu per minute). Air-conditioning and refrigeration systems are often rated by tons of cooling, such as a one-ton system. (11)

topping off. The process of adding enough refrigerant to bring a system up to full capacity. (15)

torque. Turning force, usually measured in foot pounds (ft lb) or inch pounds (in lb). (4, 24)

transformer. An electrical device that makes it possible to increase (step up) or decrease (step down) voltage. (23)

triple evacuation. A method of system evacuation that involves pulling three vacuums and injecting nitrogen into the system after the first and second vacuums. The reason for breaking the vacuum with nitrogen is that the inert gas will help to dry out the system allowing the vacuum pump to perform better. (17)

trunk line. The main run of supply ducting that extends from a plenum. (31)

tubing. A term used to describe conduits with thin-walled materials, such as copper. (6)

tubing cutter. A tool used to make an accurate 90° cut on either hard or soft copper tubing. (6)

tubing expander. A type of swaging tool that uses a lever-actuated expanding mechanism to swage the end of soft copper tubing. (6)

two-stage compressor. A compressor that is divided internally into low (or first) and high (or second) stages. A two-stage compressor has a higher compression ratio than a regular compressor and is often used in ultralow-temperature applications. (21)

two-stage heating system. A heating system that has two levels of heat production. In an electric furnace, two different sets of electric heating elements can be energized: one set for low heat and both sets for higher heat. (29)

two-stage heating thermostat. A thermostat that controls a heating system that has two stages of heat production. Two thermostat switches monitor temperature and determine which mode of operation should be used. (29)

two-temperature system. A system in which two evaporators, operating at different temperatures, are connected to a single suction line. (19)

Type K. A classification of copper tubing that is based on wall thickness. Type K has the thickest walls of copper tubing. It is used in commercial refrigeration systems and in applications where abuse or corrosion might occur. (6)

Type L. A classification of copper tubing that is based on wall thickness. Type L tubing has walls of medium thickness. It is the most common ACR tubing and used in residential and commercial applications. (6)

Type M. A classification of copper tubing that is based on wall thickness. Type M has the thinnest walls of copper tubing. It is used in some water-carrying applications but not for refrigerant-carrying applications. (6)

U

U-channel. The short edges on each end of a section of rectangular sheet metal ductwork. When joining sections of rectangular sheet metal ductwork, the U-channels are held together by drive cleats that grip the length of the U-channels. (31)

ultralow-temperature applications. A term that refers to special refrigeration systems designed for use in research laboratories and industrial applications requiring extremely low temperatures. (15)

ultrasonic leak detector. A leak detection instrument that hears the "hiss" of a refrigerant leak and responds with audible and visual indications. (17)

undercharged. A term describing an insufficient amount of refrigerant in a refrigeration system. (14, 18)

undersized. A term used to describe wiring or a transformer secondary coil that is too small to carry the expected current. (23)

Unified Thread. The present standard thread form. Interchangeable with American National threads. (5)

union. A tubing or piping fitting that provides a leakproof joint that can be disconnected at the mechanical connection. (6)

universal joint. A type of swivel for a socket wrench handle that permits reaching around objects. (4)

unloader. A piston-operated valve assembly that uses a solenoid to control pressure that will hold open or closed suction valves on one or more cylinders as a means of compressor capacity. (21)

upflow furnace. A type of furnace in which the return air enters the unit from the bottom and circulates up through the unit. (31)

uphill joint. A type of brazing connection in which the fitting is above the point where the alloy is applied. (9)

V

Vac. Abbreviation for *volts, alternating cycle* and used as a unit of rating in capacitors. (24)

vacuum pump. A device used to remove all gas and moisture from a refrigeration system. (17)

valence electron. Refers to any electron orbiting farthest from an atom's nucleus. Also known as a *free electron.* Valence electrons are more lightly held by the nucleus and can move from one atom to another. (22)

valve plate. A specially milled steel plate located between the head and the piston cylinders of a compressor. It holds the valve reeds and serves as a division point between the low-side pressure and high-side pressure. (20)

valve reed. A thin piece of steel located on top of the compressor cylinder and under the valve plate. The valve reed flexes to open and close as gas is compressed inside the cylinder. The *suction valve reed* admits refrigerant into the cylinder, and the *discharge valve reed* releases compressed refrigerant from the cylinder. (20)

vapor barrier. In building construction, a thin film applied to a building's exterior to prevent moisture from migrating through building materials. Also, a plastic jacket or sleeve that encloses an insulated flexible duct. (31)

vapor charging. Charging a system by adding refrigerant vapor to the low-pressure side. (17)

variable. A factor that changes. (12)

variable refrigerant flow (VRF) system. A type of ductless split system where one or more outdoor units supply a large number of indoor units. VRF systems may be heat pump systems or heat recovery systems. (30)

venting. Releasing refrigerants into the atmosphere. The venting of ozone-depleting refrigerants is prohibited by the Clean Air Act. (16)

vertical riser. A vertical portion of the suction line, usually used with an oil trap, in a commercial refrigeration system. (13)

vibration eliminator. A device used at the compressor to prevent transmission of noise and vibration to refrigerant piping. (21)

viscosity. The thickness of oil at a given temperature. This relates to the ability of an oil or lubricant to flow. Concerning viscosity, it is desirable for an oil to maintain good lubrication despite temperature changes. (21)

volt. A measuring unit for potential difference or electromotive force (emf). (22)

voltage. The potential difference or the strength of electromotive force (emf) that causes electrons to move. Sometimes called *potential difference, electromotive force (emf),* or *electrical pressure.* (22)

voltage drop. The loss of electromotive force (voltage) between a power source and the electrical load. (22)

voltmeter. An instrument used to measure potential difference (voltage) between two specific points. (22)

volume. The space within a container or space defined by the dimensions of length, width, and height. Always expressed in cubic terms. (10)

volumetric efficiency. A comparison (expressed as percentage) of the amount of gas actually pumped by a reciprocating compressor to the amount of gas it should pump, according to piston displacement and length of stroke. (21)

W

washer. A disc of metal or other material with a hole through the center, used to extend the gripping area of a fastener or to prevent a nut from loosening. (5)

water vapor. Water in a gaseous form. (27)

watt. A measuring unit for electrical power, calculated by multiplying amperage times voltage. (22)

watt-hour meter. An electrical meter that measures how many thousands (kilo) of watts are used per hour (kWh) of the electrical energy used by a household or business. (22)

wattmeter. A special meter designed to measure the power (wattage) used by an appliance or house. (22)

whole number. A positive number that contains no fractional parts. (10)

wire. A single strand or filament of drawn metal used as a conductor of electricity. (23)

wire nut. A solderless screw-on connector used to make terminal connections. (23)

work-hardening. A process in which soft copper tubing hardens and becomes brittle as a result of vibration, oxidation, and bending. (6)

working pressure. A test pressure at which fuel gases are used. Working pressure is reduced from the far higher cylinder pressure by a pressure regulator. Sometimes referred to as *flow pressure*. (8)

wrought fitting. A fitting joined by soldering or brazing to either hard-drawn or soft copper tubing. Wrought fittings are usually made of copper but sometimes of brass. (6)

wye. A type of three-phase transformer with windings similar in shape to the letter *Y*. Sometimes referred to as *star* configuration. (23)

Z

zeotrope. A refrigerant blend formed by combining two or more single-component refrigerants. Zeotropes do *not* have a specific pressure for each degree of temperature change and thus exhibit temperature glide. Zeotropes are designed to replace ozone-depleting CFC and HCFC refrigerants. Also called *zeotropic refrigerant* or *zeotropic blend*. (15)

zerk fitting. A fitting used to inject grease to lubricate a bearing. (24)

zero potential. No voltage. (22)

Index

D

E

I

J

K

L

P

Q

R

S

T

U

V

W

Y

Z